International Conference PhysicA.SPb/2017

Journal of Physics: Conference Series Volume 1038

St. Petersburg, Russia
24 - 26 October 2017

Editors:

Nikita S. Averkiev
Sergey A. Poniaev
Grigorii S. Sokolovskii

ISBN: 978-1-5108-6491-7
ISSN: 1742-6588

Printed from e-media with permission by:

Curran Associates, Inc.
57 Morehouse Lane
Red Hook, NY 12571

CURRAN ASSOCIATES INC.
proceedings.com

Some format issues inherent in the e-media version may also appear in this print version.

Printed by Curran Associates, Inc. (2018)

For permission requests, please contact the Institute of Physics
at the address below.

Institute of Physics
Dirac House, Temple Back
Bristol BS1 6BE UK

Phone: 44 1 17 929 7481
Fax: 44 1 17 920 0979

techtracking@iop.org

Additional copies of this publication are available from:

Curran Associates, Inc.
57 Morehouse Lane
Red Hook, NY 12571 USA
Phone: 845-758-0400
Fax: 845-758-2633
Email: curran@proceedings.com
Web: www.proceedings.com

Table of contents

Volume 1038

International Conference PhysicA.SPb/2017

24–26 October 2017, Saint-Petersburg, Russian Federation

Accepted papers received: 22 May 2018
Published online: 18 June 2018

Preface

Papers

Astronomy and Astrophysics

Biophysics

Devices and materials of the THz and microwave ranges

Mathematical physics and numerical methods

Nano-structured and thin film materials

Nuclear and elementary particle physics

Optics and Spectroscopy

Physics and technology of energy conversion

Physics of ferroics

Physics of quantum structures

IOP Conf. Series: Journal of Physics: Conf. Series **1038** (2018) 011001 doi:10.1088/1742-6596/1038/1/011001

International Conference PhysicA.SPb/2017

The International Conference PhysicA.SPb was held 24-26 October 2017 in Saint-Petersburg, Russia. The Conference continues the tradition of Saint-Petersburg Seminars on Physics and Astronomy originating from mid-90s. Since then PhysicA.SPb maintains both scientific and educational quality of contributions delivered to the young audience. This is the main feature of the Conference that makes it possible to combine the whole spectrum of modern Physics and Astronomy within one event.

PhysicA.SPb/2017 has brought together over 300 young scientists and their colleague professors from many universities and research institutes across whole Russia as well as from Australia, Belarus, Germany, Finland, Ireland, Netherlands, United Kingdom, and France. Oral and poster presentations were combined into well-defined sections among which one should name Astronomy and Astrophysics, Optics and spectroscopy, Physics of ferroics, Nanostructured and thin-film materials, Mathematical physics and numerical methods, Devices and materials for the THz and microwave ranges, Biophysics, Optoelectronic devices, Surface phenomena, Physics and technology of energy conversion, Plasma physics, hydrodynamics and aerodynamics, Nuclear and elementary particle physics, Impurities and defects in solids, and Physics of quantum structures.

This issue of the *Journal of Physics: Conference Series* presents the extended contributions from participants of PhysicA.SPb/2017 that were peer-reviewed by expert referees through processes administered by the Presiders of the Organising and Programme Committees to the best professional and scientific standards. This was made possible by the efforts of the Sectional Editors of this Issue: Prof. Petr Arseev (Lebedev Physical Institute), Prof. Alexander Ivanchik (Ioffe Institute), Prof. Polina Ryabochkina (Ogarev Mordova State University), Prof. Yuri Kusraev (Ioffe Institute), Dr. Sergey Nekrasov (Ioffe Institute), Dr. Nikolay Bert (Ioffe Institute), Dr. Nikita Gordeev (Ioffe Institute), Dr. Alexey Popov (Ioffe Institute), Dr. Prokhor Alekseev (Ioffe Institute), Dr. Mikhail Dunaevskii (Ioffe Institute), Dr. Alexandra Kalashnikova (Ioffe Institute) and Prof. Dmitry Khokhlov (Moscow State University).

The Editors: **Nikita S. Averkiev, Sergey A. Poniaev and Grigorii S. Sokolovskii**

IOP Conf. Series: Journal of Physics: Conf. Series **1038** (2018) 011002 doi:10.1088/1742-6596/1038/1/011002

Peer review statement

All papers published in this volume of *Journal of Physics: Conference Series* have been peer reviewed through processes administered by the proceedings Editors. Reviews were conducted by expert referees to the professional and scientific standards expected of a proceedings journal published by IOP Publishing.

Pulsars with bow shocks: model constraints of the pulsar wind Lorentz factor

A E Petrov[1], A M Bykov[1,2], S M Osipov[1], A M Krassilchtchikov[1] and K P Levenfish[1]

[1]Ioffe Institute, 26 Politekhnicheskaya st., St. Petersburg 194021, Russia
[2]Peter the Great St. Petersburg Polytechnic University, 29 Politekhnicheskaya st., St. Petersburg 195251, Russia

E-mail: a.e.petrov@mail.ioffe.ru

Abstract. Pulsar wind nebulae with bow shocks (BSPWNe) can efficiently accelerate electrons and positrons of the pulsar wind (PW). The particles injected into such nebulae at the PW termination shock can gain energy due to the Fermi acceleration in the colliding shock flow between the two shocks. Monte-Carlo modeling of the PW particle transport through the nebula reveals a significant deformation of their spectrum between the shocks in comparison with the injected spectrum. The maximal energies achieved by the hard component of the spectrum ($f(E) \propto E^{-p}$ with $p < 2$) depend on the flow velocities. Comparison of the obtained spectra of the accelerated PW particles in BSPWNe and their synchrotron emission maps with the observational data obtained in the optical, far-UV and X-ray bands allows one to constrain global parameters of particular BSPWNe such as Lorentz-factors of their PWs.

1. Introduction

Multiwavelength observations of pulsar wind nebulae (PWNe) have revealed a vast variety of their morphologies, which can strongly depend on the considered energy range. Spatially resolved emission spectra provide an evidence of evolution of the particle energy distributions in the nebulae.

An example of such evolution is the nebula around PSR J0437-4715 (hereafter J0437). The bow shock (BS) of the pulsar moving at a supersonic velocity $u_{\rm psr} \approx 104$ km s^{-1} through the interstellar medium (ISM) was first detected in H_α [1] and later also seen in the far ultraviolet (FUV) range (1250 – 2000 Å, 6.2 – 9.9 eV) [2]. The observed FUV bow shock luminosity is $L_{\rm FUV} \sim 5\times10^{28}$ erg s^{-1} – an order of magnitude higher than the H_α one. However, in the 0.5-7 keV X-rays the BS is absent, and only a faint PWN with luminosity $L_X \sim 3\times10^{28}$ erg s^{-1} is observed [2].

The energy dependence of the J0437 morphology could be understood within a model of the synchrotron emission of the relativistic PW particles accelerated in the colliding shock flows (CSFs) between the termination shock (TS) of the PW and the BS. In [3] a Monte-Carlo model of propagation of PW particles through the BSPWN was discussed. The simulation shows that a hard component of particle spectra with $f(E) \propto E^{-p}$, $p < 2$ is formed between the two shocks.

The efficiency of Fermi acceleration in the CSFs depends on the flow velocities. As discussed in [3], trans-relativistic flows are the most efficient sites for the diffusive shock acceleration. Proper velocities of the pulsars $u_{\rm psr}$ usually do not exceed (1-2)$\times10^3$ km s^{-1} [4], so the relativistic velocity of the shocked PW u_{pw} should be the key parameter of the acceleration process [3]. A simple estimation given in [3] shows that in the absence of severe radiative energy losses, the maximal energy gained by the particles accelerated in CSFs scales $\propto \sqrt{u_{pw}}$.

Simulation of BSPWN emission for a range of probable PW velocities allows one to impose constraints on u_{pw}, which is important for studies of non-trivial energy conversion processes in the PW. Here we apply the model developed in [3] to reproduce the observed PWN morphology at various energies and constrain the PW velocity in the J0437 BSPWN.

2. Model Overview

Here we present a brief overview of the Monte-Carlo (MC) model of particle transport in a BSPWN discussed in details in [3].

The MC model takes into account the main geometrical features of a BSPWN and allows us to consider particle escape from the acceleration site, that is essential for modeling of particle acceleration in astrophysical environments. Concentration on the high energy particles emitting synchrotron radiation in the FUV and X-ray bands allows a significant model simplification. Transport of those particles with large mean free paths (mfps) is almost not affected by the complicated structure of flows near the contact discontinuity (CD) between the CSFs. Along with dissipation of magnetic inhomogeneities by CD and deceleration of the flows this grounds the spatial structure of the simulation region (see the sketch in Fig. 1).

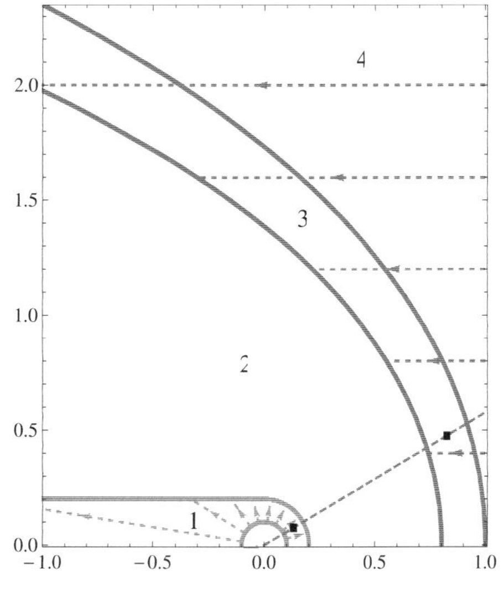

Figure 1. The model geometry: half of the axial section. Region 1 (between the orange curves) – the shocked PW, 2 — region near the contact discontinuity, 3 (between the red curves) — the postshock flow of the ISM matter, 4 — the unperturbed ISM. Black squares mark locations chosen for demonstration of particle distribution function in Fig. 2. The angle between the gray dashed line and the symmetry axis is 30°.

Particles are injected at the termination shock (TS) of the PW (sphere of radius r_{ts}) and propagate through the axisymmetric system of regions with fixed values of mean magnetic induction, distribution of flow velocities and diffusion parameters. The system has an axial symmetry with axis that coincides the pulsar proper motion velocity vector and is bounded by a cylindrical free escape boundary (FEB) surface. Particle propagation is treated in the pulsar rest frame with consideration that particle scatterings are isotropic in the local plasma rest frame. In Region 2 near the CD the particle advection is neglected. It is separated by a surface given by (2) from Region 1 – the PWN, where the radially directed flow velocity is given by (1), u_0 is a free parameter of the model (here r is the distance from the pulsar, θ is the angle between its proper velocity and a considered direction; a_{in} – a model parameter).

$$u_{pw} = u_0 \left(\frac{r_{ts}}{r}\right)^2 \tag{1}$$

$$r(\theta) = \begin{cases} a_{in}, \text{ if } \theta \leq \pi/2 \\ a_{in}/\sin\theta, \text{ if } \theta > \pi/2. \end{cases} \tag{2}$$

$$r(\theta) = R \csc\theta \sqrt{3\,(1 - \theta\,\cot\theta)}\ . \tag{3}$$

The borders of Region 3 (the area near the bow shock) are described by equation (3), giving the shape of a BS in the thin shell approximation [5]. The apexes of the inner and outer border positions are free parameters of the model. The flow velocity is uniform in both Regions 3 and 4 (the unperturbed ISM) and equal to $u_{\rm psr}$.

The growth of cosmic ray (CR) driven instabilities caused by the proton acceleration at the BS can yield a significant growth of magnetic field fluctuations near the BS. It can occur at the wavelengths up to the scale $L_{cor} \gg R_{max}$, where R_{max} is the gyroradius of protons of maximum achieved energy [6]. Hence, the mfps of PW particles propagating through the BS region are likely defined by a Bohm-like diffusion coefficient up to the Lorentz-factor γ_1 defined as $r_g(\gamma_1) = L_{cor}$, where $r_g(\gamma)$ is the gyroradius of a particle with the Lorentz-factor γ in the local mean magnetic field. In the model γ_1 is considered as a free parameter, and the mfps in Regions 2 and 3 are $\sim r_g$ for $\gamma < \gamma_1$ and correspond to a short-scale scattering mode $\propto \gamma^2$ otherwise (see [3] for more details). In the PWN we use mfp $\propto r_g$, while in the ISM an estimation obtained from the galactic CR propagation models is employed (see, e.g. [7]).

The Monte-Carlo modeling allows one to obtain the particle distribution function at the grid of bins in the phase space (see [3]). The PWNe emission is affected by the Doppler boosting, which is taken into account. The particle acceleration is related to anisotropy of particle distribution (e.g., currents), which may be strong enough in the case of relativistic flows. Binning of all particle momentum components allows one to obtain the distribution function averaged over an arbitrary small solid angle element along a given direction. That makes possible calculation of particle emission with an account of an anisotropy of their distribution. In practice, the accuracy of the obtained particle angular distribution is limited by available computational resources.

Due to the axial symmetry of the source its appearance for the observer depends on its orientation. The angle Θ between the pulsar proper motion velocity and the direction from the observer to the pulsar is a free parameter of the model. We use $\Theta = 90°$.

3. Simulated particle and emission spectra

In the left panel of Figure 2 the simulated particle spectra are shown. The particle energy distributions at the locations indicated by black squares in Figure 1 are presented. Distribution functions $f(E)$ are normalized so that the mean concentration $n = \int f(E)\,dE$ ($E = mc^2\gamma$ is the particle energy, m – the electron mass) and multiplied by E^2. Red color corresponds to the PW velocity at the TS $u_0 = 0.2\,c$, green — to $0.4\,c$, blue — to $0.6\,c$. The black curve shows the spectrum close to the one injected at the TS. The latter was set as a power law $f(E) \propto E^{-2.3}$ for $\gamma_{min} = 4 \times 10^5 \leq \gamma \leq \gamma_{max} = 2 \times 10^8$.

As expected, a hard component $f(E) \propto E^{-p}$ with $p < 2$ is formed in some range of energies $E_{\rm min}(u_0) < E < E_{\rm max}(u_0)$ in both discussed regions (in the case of Region 3 the spectra at low energies are also modified by the flow modulation). One can see that $E_{\rm max}$ grows with increasing u_0.

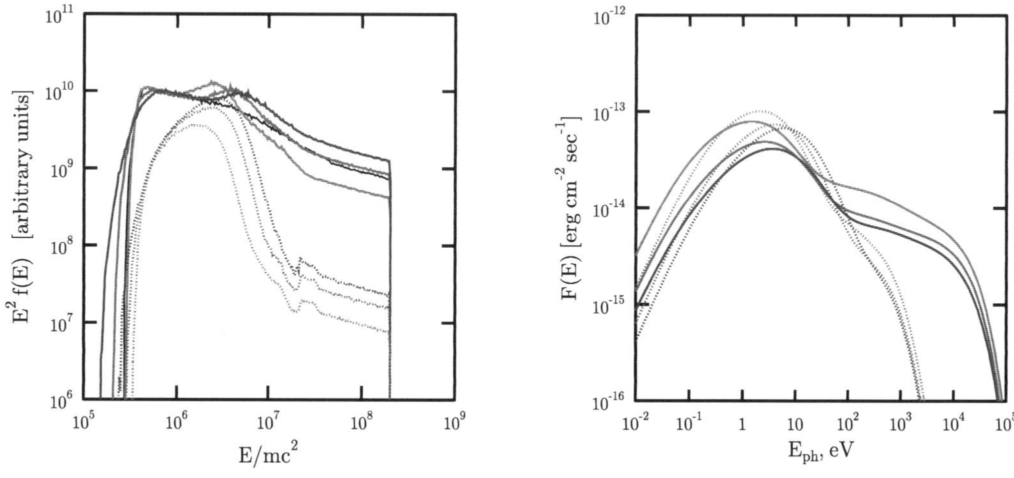

Figure 2. Left: Particle energy distribution in the head parts of the PWN (solid lines) and at the bow shock (dotted lines). Right: Synchrotron spectra of the PWN (solid lines) and of the bow shock region (dotted lines). The red curves correspond to u_0 =0.2 c, green – to 0.4 c and blue – to 0.6 c. The black curve in the left image shows the spectrum close to the one injected at the TS.

In the right panel of Figure 2 the modelled spectra of synchrotron emission (in units $\sim$ erg cm^{-2} sec^{-1}) from the PWN (solid lines) and from the BS region (dotted lines) are shown. The colors correspond to the same values of u_0 as in the left panel, and the fluxes for each setup are normalized in such a way that the observed FUV flux from the 32 arcsec^2 region at the bow shock of J0437 used in [2] is reproduced. Like in the case of particle energy distributions, the range of relatively hard spectrum achieves larger photon energies with the growth of u_0. It also should be emphasized, that in agreement with the observed structure of J0437, at soft X-ray energies (about a few keV) the simulated flux from the BS is much smaller than one from the PWN. In the following section we demonstrate it one more time with simulated images of J0437.

4. Synchrotron images

In Figure 3 the simulated synchrotron images of the object similar to J0437 are shown. Images show the intensity maps of the source, integrated over the spectral range – $(6.2 - 9.9)$ eV in the FUV and $(0.5-7)$ keV in the X-rays. The intensity is given in units of (photons cm^{-2} sec^{-1} sr^{-1}) and is normalized in the same way as the integrated fluxes discussed above.

One can see that in all cases the observed morphological features of J0437 are well represented: the bow shock is seen in the FUV images and is absent in the X-ray images. The increasing efficiency of particle acceleration with the growth of u_0 leads to a relative dimming of the PWN images.

To quantitatively compare the model with the observational data we have calculated the luminosities of the PWN in the X-ray band and of the bow shock region in the FUV band and in the 32Å band around the Hα line (6550 – 6582 Å). The ratios of the luminosities are summarized in Table 1.

In the case of $u_0 = 0.2\,c$ the ratio $L_{\rm FUV}/L_{\rm X}$ is significantly smaller than the observed value that is about 1.7. This is likely due to a modest particle acceleration in the case of a relatively small pulsar wind velocity. However, in all cases the simulated ratios $L_{\rm FUV}/L_{\rm H\alpha}$ are

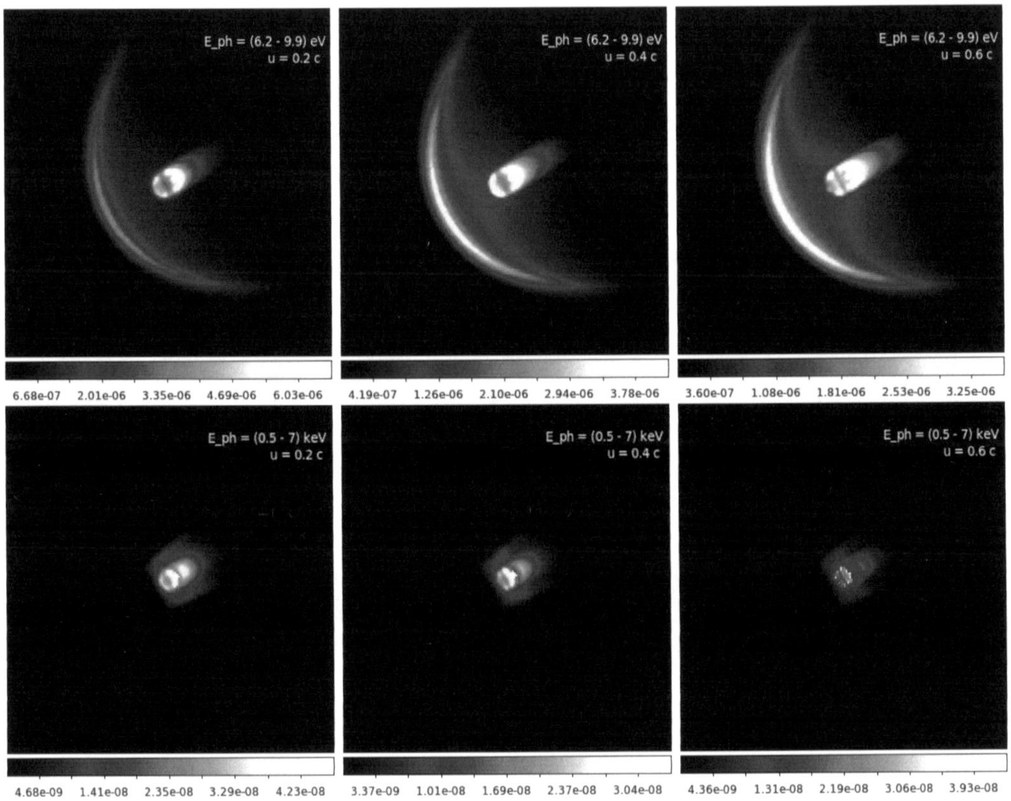

Figure 3. Simulated synchrotron images of a source similar to J0437-4715 in the FUV (6.2 – 9.9 eV; top row) and X-ray (0.5 – 7 keV, bottom row) bands. Left column: $u_0 = 0.2\,c$, middle — $0.4\,c$, right — $0.6\,c$. The intensities for each value of u_0 are normalized in such a way that the flux from the bow shock reproduces the observed one.

Table 1. The ratios of the modelled PWN and BS luminosities.

	$u_0 = 0.2\,c$	$u_0 = 0.4\,c$	$u_0 = 0.6\,c$
$L_{\rm FUV}/L_{\rm X}$	1.2	1.9	2.2
$L_{\rm FUV}/L_{\rm H\alpha}$	82	126	158

overestimated compared to the observed $L_{\rm FUV}/L_{\rm H\alpha} \sim 10$. This does not mean that the model contradicts the observations as the main contribution to the observed Hα emission is likely provided by the ISM gas heated at the bow shock, which we do not consider.

5. Conclusions

The ability of the Monte-Carlo model of relativictic PW particle transport through a BSPWN developed in [3] to constrain global parameters of pulsar wind has been demonstrated. The model allows one to simulate propagation of a particle population injected into a PWN at the TS through the zone of colliding flows between the TS and the bow shock where particles can be efficiently accelerated. The acceleration causes reshaping of particle energy distribution – a spectral component harder than injected spectrum is formed. The maximal energy achieved by this component depends on the PW velocity. The model allows to obtain the particle momentum distribution and calculate the spatial distribution of synchrotron emission radiated by the accelerated particles with an account for Doppler boosting and anisotropy of the distribution. In particular, an object similar to the BSPWN of J0437 has been simulated. The model synchrotron images in the FUV (6.2 – 9.9 eV) and X-ray (0.5 – 7 keV) ranges reproduce the observed properties of the J0437 morphology for a whole range of PW velocities: the bow shock observed in the FUV is almost absent in the X-rays. However, the simulated ratios of the bow shock FUV luminosty to the PWN X-ray luminosity show, that PW velocities $u_0 \leq 0.2\,c$ appear too low to provide sufficiently strong particle acceleration and thus can be excluded for the modelled source.

Acknowledgments
A.E.P. and A.M.B. acknowledge support from RAS Presidium program No. 28. Most of the numerical modeling was performed using computational resources of Peter the Great St. Petersburg Polytechnic University Supercomputing Center (http://scc.spbstu.ru).

References
[1] Brownsberger S and Romani R W 2014 *ApJ* **784** 154 (*Preprint* 1402.5465)
[2] Rangelov B, Pavlov G G, Kargaltsev O, Durant M, Bykov A M and Krassilchtchikov A 2016 *ApJ* **831** 129 (*Preprint* 1605.07616)
[3] Bykov A M, Amato E, Petrov A E, Krassilchtchikov A M and Levenfish K P 2017 *Space Sci. Rev.* **207** 235–290 (*Preprint* 1705.00950)
[4] Hobbs G, Lorimer D R, Lyne A G and Kramer M 2005 *MNRAS* **360** 974–992 (*Preprint* astro-ph/0504584)
[5] Wilkin F P 1996 *ApJ* **459** L31
[6] Bykov A M, Brandenburg A, Malkov M A and Osipov S M 2013 *Space Sci. Rev.* **178** 201–232 (*Preprint* 1304.7081)
[7] Strong A W, Moskalenko I V and Ptuskin V S 2007 *Annual Review of Nuclear and Particle Science* **57** 285–327 (*Preprint* arXiv:astro-ph/0701517)

The optical depth of gamma radiation due to interaction with the thermal bremsstrahlung of hot gas in galaxy clusters.

A N Popov[1], **D P Barsukov**[1] **and A V Ivanchik**[1]
[1] Ioffe Institute, Saint Petersburg, Russia
E-mail: bars.astro@mail.ioffe.ru

Abstract. The interaction of high energy gamma quantum to thermal bremsstrahlung photons of hot intracluster gas with producing electron-positron pair is considered. It is supposed that galaxy cluster is relaxed and electron number density and temperature profiles are spherical symmetric. The dependence of optical depth on gamma quantum energy and distance to cluster center is considered. It is shown that the optical depth due to considered interaction is about $10^{-8} - 10^{-5}$ at gamma quantum energy 100 MeV - 1 TeV.

1. Introduction

The high energy gamma quanta from cosmological sources like blazar and active galactic nuclei are subjected to interaction with cosmological photon background [1]. At gamma quantum energy range $E \sim 100$ GeV $-$ 100 TeV the interaction with optical and infrared photons of Extragalactic Background Light (EBL) is dominated [2, 3, 4]. At energy $E \sim 100$ TeV-10^7 TeV the interaction with Cosmic Microwave Background (CMB) photons is dominated [5, 6]. At very high energy $E > 10^7$ TeV the interaction with radio photons of Cosmic Radio Background (CRB) becomes important [5, 6]. In this paper we consider the interaction of gamma quantum to thermal bremsstrahlung photons of hot intracluster gas with producing electron-positron pair.

2. Model

The cross section of interaction of gamma quantum to thermal photon with producing electron-positron pair is [7]:

$$\sigma = \frac{\pi}{2} r_e^2 (1 - v^2) \left((3 - v^4) \ln \left(\frac{1+v}{1-v} \right) - 2\, v\, (2 - v^2) \right) h(s) \tag{1}$$

where $r_e = \frac{e^2}{mc^2}$ is classical electron radius, m is mass of electron, $h(s)$ is Heaviside function ($h(s) = 1$ at $s > 0$ and $h(s) = 0$ at $s < 0$),

$$v = \sqrt{1 - 1/s} \quad \text{and} \quad s = \frac{E\epsilon}{m^2c^4}(1 - \cos\Psi), \tag{2}$$

E is energy of gamma quantum, ϵ is energy of thermal photon, radiated by intracluster gas, Ψ is angle between its impulses, see figure 1. For simplicity, we assume that gamma quantum path

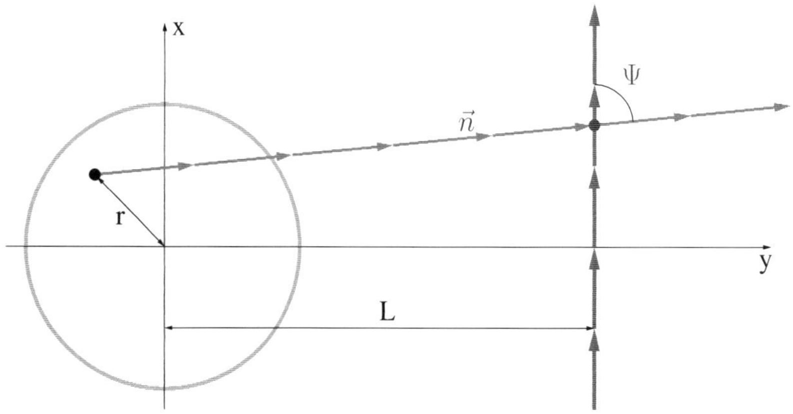

Figure 1. A sketch to illustrate the considered geometry. The intracluster gas is shown by yellow circle, gamma quantum path is shown by green line, thermal photon path is shown by red line.

is straight line: $\vec{x}(s) = L\vec{e}_y + \vec{e}_x s$ at distance L to cluster center, see figure 1. Optical depth τ is calculated by formula

$$\tau = \int_{-\infty}^{+\infty} ds \int_{-\infty}^{+\infty} d^3p \, (\sigma(\Psi) \cdot (1 - \cos\Psi) \cdot f(\vec{x}(s), \vec{p})\,) \tag{3}$$

where $f(\vec{x}, \vec{p})$ is distribution function of thermal photons at point $\vec{x}$, $\vec{p}$ is photon impulse, $\epsilon = pc$.

We consider only relaxed clusters with spherically symmetric gas distribution, i.e. electron number density n and temperature T depend only on distance r to cluster center. Also we assume that intracluster gas is optically thin and hence [8]

$$f(\vec{x}, \vec{p}) = \frac{2\pi}{\bar{h}} \frac{c^2}{\epsilon^3} \int_0^{+\infty} \varepsilon_\nu \left(\vec{x} - \lambda\vec{n}, \vec{p}\right) d\lambda \tag{4}$$

where $\vec{n} = \vec{p}/p$, $\varepsilon_\nu(\vec{x}, \vec{p})$ is volume emissivity. We take into account only thermal bremsstrahlung from intracluster gas and, for simplicity, neglect helium and metals contribution. Hence ε_ν may be written as [8]

$$\varepsilon_\nu(\vec{x}, \vec{p}) = \frac{8}{3}\sqrt{\frac{2\pi}{3}}\, mc^2\, r_e^3 \sqrt{\frac{mc^2}{T(r)}}\, n^2(r)\, g(\epsilon, T(r))\, \exp\left(-\frac{\epsilon}{T(r)}\right) \tag{5}$$

where $g(\epsilon, T)$ is Gaunt factor. For simplicity, we assume that it is equal to [9]

$$g(\epsilon, T) = \frac{\sqrt{3}}{\pi} K_0\left(\frac{\epsilon}{2T}\right) \exp\left(\frac{\epsilon}{2T}\right) \tag{6}$$

where $K_0(x)$ is Macdonald function.

3. Results

At first let us to consider galaxy cluster Abell 2204 [10]. The electron number density and temperature profile used in calculation are shown in table 1 [10]. We assume that angular distance to cluster is $D = 541.6$ Mpc ($z = 0.1523$), so $r_{200} = 11.2'$ radius of cluster corresponds

Table 1. Electron number density $n(r)$ and temperature $T(r)$ profile of Abell 2204, used in calculation. Spatial distance $(r/1\ \mathrm{Mpc}) = (1.76\ \mathrm{Mpc}/11.2')\,(r/1')$. Data are taken from [10].

Bin		1	2	3	4	5	6	7	8
Radius	(arcmin)	0-0.5	0.5-1.5	1.5-2.5	2.5-3.6	3.6-4.9	4.9-6.7	6.7-9.2	9.2-12.8
$n(r)$	$(10^{-3}\mathrm{cm}^{-3})$	24.4	5.30	1.79	0.86	0.47	0.27	0.11	0.06
$T(r)$	(keV)	2.7	7.2	10.4	12.9	9.0	5.8	4.0	5.7

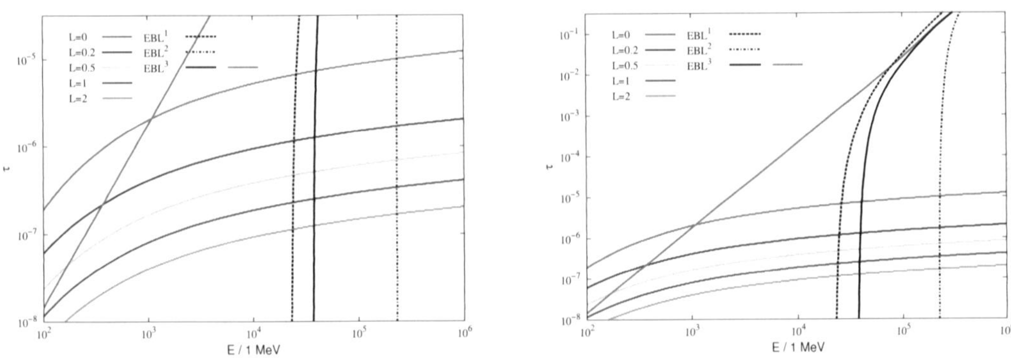

Figure 2. The dependence of optical depth τ on gamma quantum energy E in case of Abell 2204 is shown. The optical depth τ due to interaction with EBL photons is shown by black lines. For simplicity, we assume that EBL spectrum does not depend on z. Dashed line (EBL^1) corresponds to EBL spectrum taken from [12], dot-dashed line (EBL^2) corresponds to EBL spectrum taken from [13], black solid line (EBL^3) corresponds to EBL spectrum taken from [2], gray solid line corresponds to its extrapolation by power law to large energies. The left and right graphs differ in the scale only.

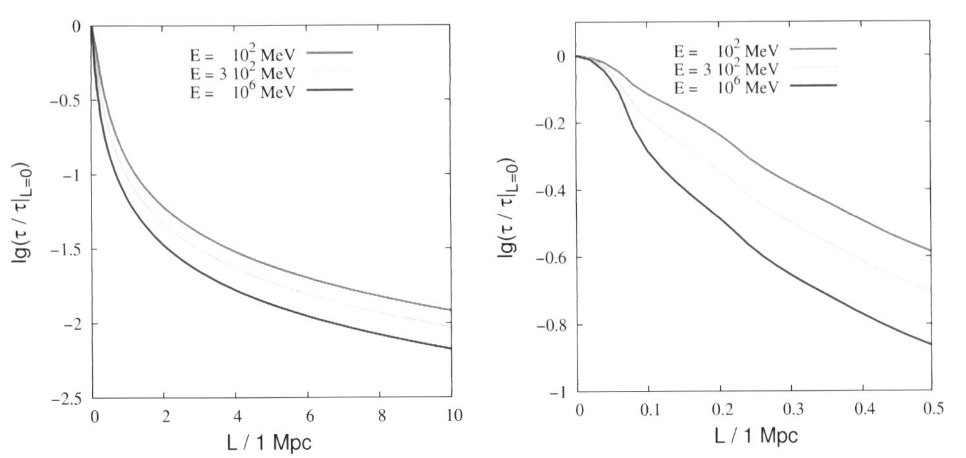

Figure 3. The dependence of normalized optical depth $\tau/\left.\tau\right|_{L=0}$ on distance L to cluster center in case of Abell 2204 is shown. The left and right graphs differ in the scale only.

to 1.76 Mpc [10]. The dependence of optical depth τ on gamma quantum energy E is shown in figure 2. The dependence of normalized optical depth $\tau/\left.\tau\right|_{L=0}$ on distance L to cluster center

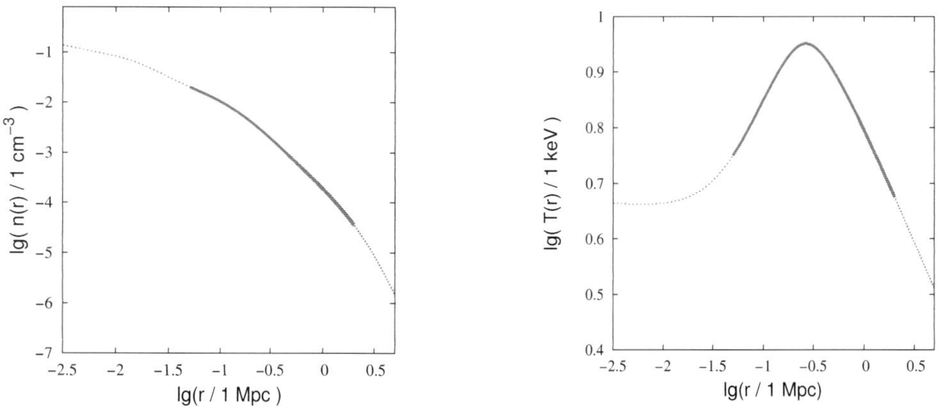

Figure 4. The electron number density $n(r)$ and temperature $T(r)$ profiles of galaxy cluster Abell 478 taken from [11] are shown in left and right panels correspondingly. Observed profiles are shown by solid lines, its extrapolations are shown by dashed line.

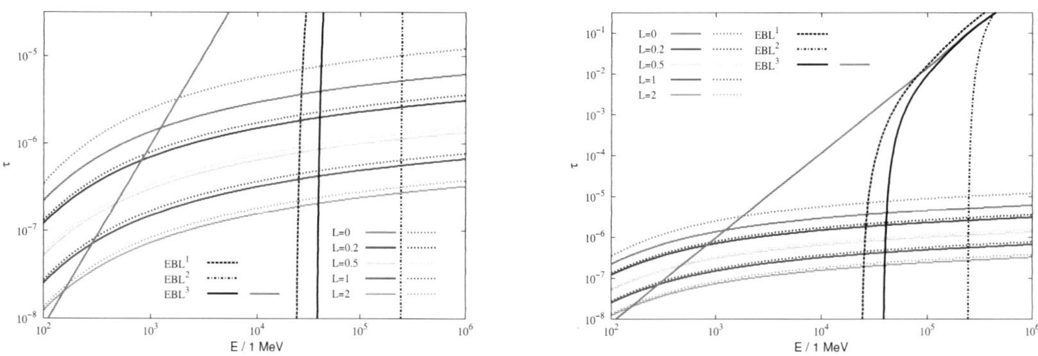

Figure 5. The same as in figure 2, but the dependence of optical depth τ on gamma quantum energy E in case of Abell 478 is shown. The optical depth calculated by observed profiles $n(r)$ and $T(r)$ is shown by solid line, the depth calculated by extrapolated profiles is shown by dashed line.

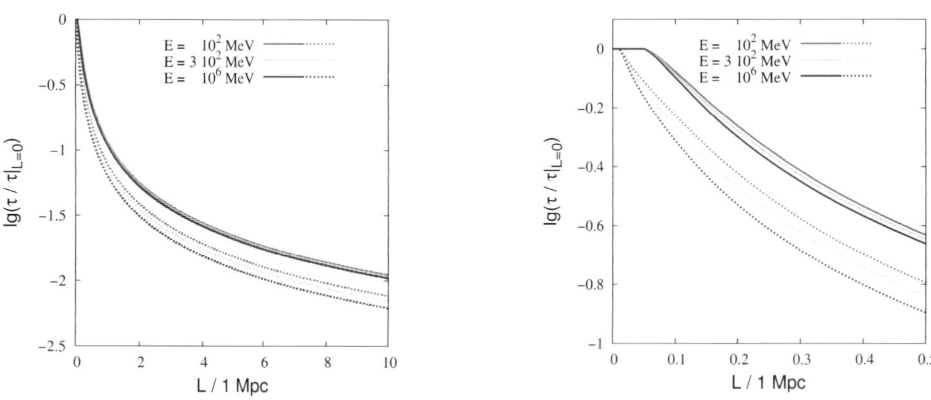

Figure 6. The same as in figure 5, but dependence of normalized optical depth $\tau / \tau|_{L=0}$ on distance L to cluster center is shown

is shown in figure 3, where $\tau|_{L=0}$ is optical depth in case of gamma quantum passed through cluster center.

Now we consider galaxy cluster Abell 478 with redshift $z = 0.0881$ [11]. The electron number density and temperature profiles taken from [11] are shown in figure 4. The dependence of optical depth τ on gamma quantum energy E is shown in figure 5. The dependence of normalized optical depth $\tau/\ \tau|_{L=0}$ on distance to cluster center is shown in figure 6. It is easy to see that normalized optical depth profile depends on gamma quantum energy E only slightly. Plateau at $L = 0$ in figure 6 is because of we exclude contribution of gas in cluster center.

In this paper we consider only gamma quantum energy range $E \sim 10^2$ MeV $-$ 1 TeV. In this range the input of interaction with CMB and CRB photons to optical depth is negligible small [6]. We receive that optical depth due to gamma quantum interaction with thermal bremsstrahlung photons of hot intracluster gas is negligible compared to input of interaction with EBL-photons at gamma quantum energy $E > 10^2$ GeV but it may give at least a comparable input to optical depth at energy range $E \sim 10^2$ MeV $-$ 10 GeV.

Acknowledgments
We sincerely thank S.V. Bobashev for help, comments and useful discussions.

References
[1] Ruffini R, Vereshchagin G V and Xue S S 2016 *Astrophysics and Space Science* **361** id 82
[2] Franceschini A, Rodighiero G and Vaccari M (2008) *A & A* **487** 837-52
[3] Sinha A, Sahayanathan S, Misra R, Godambe S and Acharya B S 2014 *ApJ* **795** 91-8
[4] Dwek E and Krennrich F 2013 *Astroparticle Physics* **43** 112-33
[5] De Angelis A, Galanti G and Roncadelli M 2013 *MNRAS* **432** 3245-49
[6] Gould R J and Schreder G P 1967 *Phys. Rev.* **155** 1408-11
[7] Gould R J and Schreder G P 1967 *Phys. Rev.* **155** 1404-7
[8] Lang K R 1980 *Astrophysical Formulae. A Compendium for the Physicist and Astrphysicist* (Berlin: Springer-Verlag)
[9] Zheleznyakov V V 1997 *Radiation in astrophysical plasmas* (Moscow: Yanus-K)
[10] Basu K et al 2010 *A & A* **519** id A29
[11] Vikhlinin A, Kravtsov A, Forman W, Jones C, Markevitch M, Murray S S and Van Speybroeck L 2006 *ApJ* **640** 691-709
[12] Finke J D, Razzaque S and Dermer C D 2010 *ApJ* **712** 238-49
[13] Elbaz D, Cesarsky C J, Chanial P, Aussel H, Franceschini A, Fadda D and Chary R R 2002 *A & A* **384** 848-65

Modeling of the physical selenocentric surface using modern satellite observations and harmonic analysis methods

A O Andreev[1], N Y Demina[1,2], Y A Nefedyev[1,2], S A Demin[2], A A Zagidullin[1]
[1] Kazan Federal University, Engelhardt Astronomical observatory, Zelenodolski region, st. Observatory, 422526 Russia
[2] Kazan Federal University, Institute of Physics, Kazan, 420008 Russia

E-mail: alexey-andreev93@mail.ru

Abstract. On the basis of satellite observations taken during “Apollo”, “Clementine”, “Kaguya”, “LRO”, “GRAIL”, and “SMART-1” space missions a model of lunar physical surface of 18th order of expansion into a series of harmonic coefficients was constructed. In order to expand the data on relief into spherical functions a step-by-step regression was applied. All the constructed models included only significant elements. The regression analysis of the models of lunar relief for various expansion orders was carried out; as a result, it was determined that increase in order of expansion did not play a prominent role from a certain stage. The influence of model’s overdetermined structure on values of the parameters determined was also investigated.

1. Introduction

The modern experimental studies of external and internal characteristics of the Moon are related to the plans of its exploration in the near future. During the last two decades, the Moon has been the object of comprehensive investigation, as evidenced a number of space experiments such as “Lunar Laser Ranging” (LLR) (1969–2012), “Lunar Prospector” (1998–1999) [1], and “Clementine” [2] space missions. A few space programs aimed at thorough study of the Earth’s natural satellite were launched at the beginning of 21st century. It is necessary to note such space projects as “SMART–1” (European satellite, 2003–2006), “Kaguya” (Japanese satellite, 2007–2009) [3], “Chang–1” (Chinese spacecraft, 2007–2009), “Chang–2” (Chinese spacecraft, 2010) [4], “Chandrayaan–1” (Indian satellite, 2008–2009), “Chandrayaan–2” (Indian satellite, 2013) [1], “LRO–LCROSS” (U.S. satellites, 2009–2012) [1], “GRAIL” (The Gravity Recovery and Interior Laboratory, 2011, program on study gravitational field and internal structure of the Moon as well as modelling its heat history) [5]. The program included construction of selenographic mapping system, study of the Moon’s internal structure, investigation of the physical libration of the Moon subtle effects in rotational motion, use of intersatellite tracking in order to analyze lunar gravitational field.

The use of physics of complex systems methods and regression analysis allow studying and simulating multi-parametric celestial objects, which is of particular importance in terms of solution of planetary astrophysics problems. A number of modern space experiments are aimed at studying the Moon [6–10]. This fully applies to the Russian missions “Luna–25, –26, –27, –28” (2017–2020+) which imply obtaining a large amount of information on internal structure, inertial coordinate system, gravitational field, and dynamical figure of the Moon. It should be noted that the modern digital models of the lunar physical surface do not have the precise coordinate and time support. This occurs due to the fact that during the processing of topographic satellite observations of selenocentric data the reference surface relies on a satellite orbit’s coordinate system rather than the dynamic selenocentric coordinate system [11]. Nevertheless, based on the amount of satellite observational data obtained during space studies, expanding of astrophysical parameters in a series of spherical functions, and multi-parametric harmonic analysis it is possible to construct the dynamic selenocentric digital model of the lunar physical surface. In case of the Earth, the solution may be derived using the large amount of both space and ground-based observations. However, for other celestial bodies such capabilities are significantly limited, since there are no observations taken directly from their surfaces.The greatest amount of space information is collected for the Moon. And there is a great advantage in the presence of observations taken on the Earth, which allows space selenographic observations binding to the inertial coordinate system.

2. The problem of the Moon macrofigure model construction

Currently, the entire data on lunar topography falls into two categories. On the one hand, the data obtained by the lunar sphere laser scanning from satellite describes the lunar relief well, but does not provide the accurate coordinates of the reference objects on the Moon. The other type of data provides the accurate coordinates of the reference objects by their direct binding to the stars, but does not describe the lunar relief with the sufficient accuracy. Besides, all these systems have different reference systems and orientation of coordinate axes. On the other hand, the data obtained during the space missions refer to quasi-dynamic coordinate system in which the reference center is the lunar center of mass, but the coordinate axes do not coincide with the Moon's axes of inertia. The majority of modern selenodetic catalogues also refer to quasi-dynamic coordinate system, since neither their reference centers coincide with the Moon's center of mass, nor the coordinate axes coincide with the lunar axes of inertia. Currently, there is no dynamic selenocentric coordinate system obtained on the basis of space observations covering sufficient part of the lunar surface. Besides, despite the high accuracy of determining the lunar physical relief by space missions, the reference surface of this relief presents a vague figure. Thus, it cannot be said that satellite topographic maps are models of full value with their certain reference surface of topographic data. To achieve the objectives of the present work the method of harmonic analysis based on expanding topographic data in spherical functions was used.

3. Selenocentric macrofigure construction method

In order to investigate reliability of topographic data of the reference surface one may:

1) Directly compare macroreliefs of certain lunar topographic maps;

2) compare coordinates of the Moon's center of mass position in relation to the lunar center of figure obtained from various sources of topographic data.

When conducting these investigations, it is implied that all the sources of hypsometric data involved in calculation share equally accurate. In case of selenographic catalogues, there is high accuracy of the presented coordinates of reference objects on the lunar surface and there is low accuracy at relief description, but in case of satellite altimetry, there is sufficient accuracy at physical relief description and low accuracy of observational data binding to the selenocentric coordinate system. As for the problem of determining the position of the lunar center of mass in relation to the lunar center of figure, this is of great importance for both studying the Moon's origin, its structure, evolution, and precise solving the near-Moon navigational problems. This work was performed in [11].

In the present paper, the data from "Apollo", "LRO", "Kaguya", "Clementine", "SMART–1", and "GRAIL" space missions was used. The lunar megarelief model was constructed by expanding variations of the lunar surface points vectors in series of spherical functions. When expanding selenographic observational data in spherical functions the step-by-step regression method was used. The software developed for this purpose included only basic harmonic elements.

In order to achieve effective expansion order the model of the lunar physical surface was analyzed using the regression method. It was found that from the certain stage further increase in order did not play a significant role in the expansion. The problem of determining the most practical order of the harmonic expansion was also considered, and it was established that the 18th order of expansion fully satisfied the assigned task. As a model describing relief behavior the expansion of altitude function $R(\lambda_i, \varphi_i)$ in spherical harmonics as regression model was used:

$$R(\lambda_i, \varphi_i) = F(\bar{C}_{nm}, \bar{S}_{nm}, \bar{P}_{nm}) + \varepsilon, \quad (1)$$

where

λ, φ – latitude, longitude – known coordinates of lunar objects;

$\bar{C}_{nm}, \bar{S}_{nm}$ – normalized harmonic amplitudes;

$\overline{P}_{nm}$ – normalized associated Legendre functions;
ε – random regression error.

Solution of the over determined system (1) for various sources of hypsometric information was obtained within regression modelling approach whichstipulated use of a number of quality statistics including external measures, meeting basic conditions of Ordinary least square diagnostics, adaptation in case of their violation in addition to usual stages (model (1) postulation and amplitudes $\bar{C}_{nm}, \bar{S}_{nm}$ estimation). As computational schemes of Ordinary least square the Gauss-Jordan and Householder schemes were used. Direct use of model (1) for certain areas of the sphere (hemisphere etc.) might be complicated due to multicollinearity of the expansion coefficients; this is why in the present work the method of estimating model (1) amplitudes by preliminary expansion of a segment up to the full sphere which allows completeeliminating the effect of multicollinearity was used. Then, noise harmonics were removed by step-by-step regression.

4. Analysis of the Moon macrofigure construction results

The modern methods of constructing the Moon macrofigure models were analyzed. As a result, it was concluded that despite the high accuracy of presenting the physical relief obtained from space missions data, construction of lunar macrofigure models remained an unsolved problem due to the difficulties with determining coordinate grid's dimension and inaccuracies of the lunar physical surface coordinates counting. The main method used for the study of the lunar macrorelief is numerical and analytical one, which involves expansion of selenographic data in harmonic series of spherical functions and application of regression analysis, described above. The model of lunar macrofigure was constructed on the basis of observations taken during "Apollo", "LRO", "Kaguya", "Clementine", "SMART–1", and "GRAIL" space missions.The new data on global macrorelief of the lunar surface was obtained by multi-parametric harmonic analysis of radius-vectors (absolute heights) of 282 215 selenocentric points (Global Lunar Model –"GLM"). As a result, on the basis of this data a model of 18th order of expansion in spherical functions was developed. The section of global model at selenographic longitude of $\lambda = 20^0$ is shown in figure 1 (line 1). For comparison, the sections for the macrofigure obtained by missions "Clementine" (line 2) and "Kaguya" (line 3) at selenographic longitude of $\lambda = 20^0$ are presented as well.

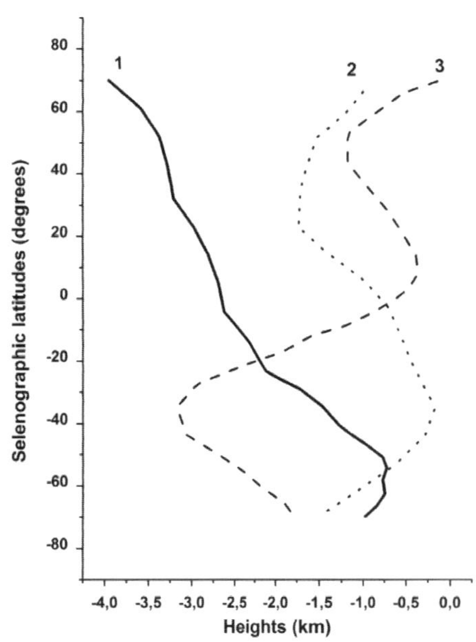

Figure 1. Comparison of lunar relief sections based on "GLM" (line 1), "Clementine" (line 2), and "Kaguya" (line 3) at selenographic longitude $\lambda = 20^0$. *X*–axis corresponds to height (km), *Y*–axis corresponds to latitude (degrees).

The results of analysis of section at longitude $\lambda = 20^0$ (figure 1):

1) The mean level of heights in the southern hemisphere is higher than in the northern one (for about 0.5–1.0 km);

2) correlation coefficients for hypsometric curves "GLM"–"Clementine", "GLM"–"Kaguya", are 0.65 and –0.81 respectively;

3) standard deviation σ is determined by formula

$$\sigma = \sqrt{\frac{1}{n}\sum_{i=1}^{n}(x_i - y_i)^2},$$

where n is the number of measurements of 2 curves deviations, for "GLM"–"Clementine" deviation is 0.9 km, for "GLM"–"Kaguya" deviation is 1.7 km.

In figure 2 the 3D–model of the near side of the Moon of the mean radius of 1 736.34 ± 1.20 km with the asperity of 0.9–1.5 km in the central part is presented. North of the parallel of 10^0 radius-vectors are being gradually reduced and at latitudinal zone of $30^0 - 70^0$ become, on average, 0.9 km less than lunar mean radius.

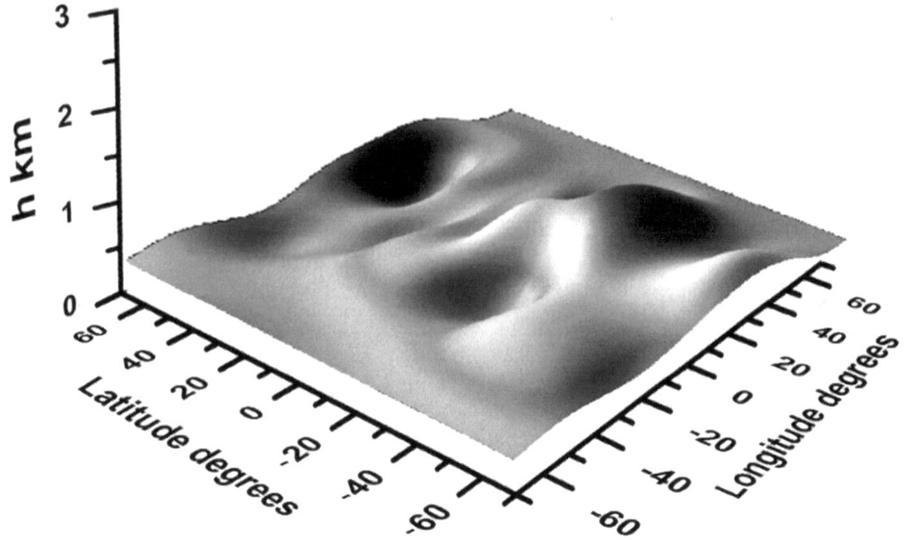

Figure 2. Selenocentric model of the near side of the Moon.

5. Summary and conclusions

In this paper, the model of the Moon's physical surface was analyzed by regression method in order to determine the effective order of expansion. It was established that increase in the order of expansion from the certain stage did not play a significant role. Thus, the aim of this work was to find the most practical order of harmonic expansion of the lunar physical surface. It was found that the 18th order of expansion fully satisfied the assigned task. The optimal structure for the constructed model was analyzed as well as the impact of the obtained structure's over determination on the parameters calculated. As a result, the model of the lunar physical surface for the 18th order of expansion was developed.

The main conclusions are as follows:

1) Analysis of dynamic and geometrical figures of the Moon showed that the most important problems were construction of selenocentric dynamic reference networks, development of lunar topocentric and gravimetric models, establishment of mutual position of lunar center of mass and center of figure, setting selenographic coordinates reference system for navigational orientation, and reference surface determination for lunar surface mapping [12].

2) The modern topographic models built on space observations have uncertain reference systems, and thus, inaccurate coordinates of the objects presented in them. Without reference catalogue of lunar objects best covering the area studied, the problem of space navigational networks thickening and widening is impracticable [13].

3) On the basis of the conducted analytical data analysis one may conclude the modern multiple processing of various sets of space data is necessary, since the processing methods and approaches, according to which the reduction of global selenodetic reference networkis usually performed [14], keep improving.

4) The methods and data obtained in the present study are going to allow developing lunar coordinate systems that will serve as a basis for selenographic data reduction and provide navigational support for the Moon's exploration and other space projects [15–20].

Acknowledgments

The work is performed according to the Russian Government Program of Competitive Growth of Kazan Federal University. This work was supported by the Russian Foundation for Basic Research, project nos. 16-32-60071 mol_a_dk (N.D.), 18-02-01098 a, 18-32-00895 mol_a.

References

[1] Kirk R L, Archinal B A, Gaddis L R and Rosiek M R 2007 Cartogaphy for Lunar exploration: current status and planned missions *Proceedings of the 23rd International Cartographic Conference* (International Cartographic Association) pp 1–12

[2] Smith D E, Zuber M T, Neumann G A and Lemoine F G 1997 *J. Geophys. Res.-Planets* **102** 1591

[3] Noda H, Araki H, Goossens S, et al. 2008 *Geophys. Res. Lett.* **35** L24203

[4] Li C L, Ren X, Liu J J, et al. 2010 *Science China Earth Sciences* **53** 1582

[5] Sharifi M A and Seif M R 2017 *Journal of the Earth and Space Physics* **43** 489

[6] Robinson M S, Brylow S M, Tschimmel M, et al. 2010 *Space science reviews* **150** 81

[7] Ping J S, Huang Q, Yan J G, et al. 2009 *Science in China Series G: Physics, Mechanics and Astronomy* **52** 1105

[8] Zuber M T, Smith D E, Lemoine F G and Neumann G A 1994 *Science* **266** 1839

[9] Spudis P D, Bussey D B J, Baloga S M, et al. 2010 *Geophys. Res. Lett.* **37**

[10] Araki H, Tazawa S, Noda H, et al. 2009 *Science* **323** 897

[11] Nefedyev Y A, Valeev S G, Mikeev R R, Andreev A O and Varaksina N Y 2012 *Advances in Space Research* **50** 1564

[12] Varaksina N Y, Nefedyev Y A, Churkin K O, Zabbarova R R and Demin S A 2015 *Journal of Physics: Conference Series* **661** 012014

[13] Rizvanov N G, Nefed'ev Yu A and Kibardina M I 2007 *Solar System Research* **41** 140

[14] Petrova N, Zagidullin A, Nefedyev Y, Kosulin V and Andreev A 2017 *Advances in Space Research* **60** 2303

[15] Andreev A O, Demina N Y, Demin S A, Nefedyev Y A and Churkin K O 2016 *Nonlinear Phenomena in Complex Systems* **19** 271

[16] Nefedjev Yu A and Rizvanov N G 2002 *Astronomische Nachrichten* **323** 135

[17] Nefedyev Y, Andreev A, Demina N, Demin S and Andreeva Z 2017 The method of moonquakes selenophysical parameters analysis *International Multidisciplinary Scientific GeoConference Surveying Geology and Mining Ecology Management, SGEM* vol 17 pp 961–966
[18] Nefedjev Yu A and Rizvanov N G 2005 *Astronomy and Astrophysics* **444** 625
[19] Demina N, Nefedyev Y, Churkin K, Demin S and Andreev A 2017 Analysis of 430322 lunar occultations *International Multidisciplinary Scientific GeoConference Surveying Geology and Mining Ecology Management, SGEM* vol 17 pp 885–890
[20] Nefedyev Y A, Bezmenov V M, Demin S A, Andreev A O and Demina N Y 2016 *Nonlinear Phenomena in Complex Systems* **19** 102

IOP Conf. Series: Journal of Physics: Conf. Series **1038** (2018) 012004 doi:10.1088/1742-6596/1038/1/012004

Analysis of orbital theories for the construction of the numerical theory of the lunar physical librations

A.A.Zagidullin[1], N.K.Petrova[1,2], V.S.Usanin[1], A.O. Andreev[1], Y. A.Nefedyev[1]
[1] Kazan Federal University, Institute of Physics, Kazan, 420008 Russia
[2] Kazan State Power Engineering University, Kazan, 420034 Russia

E-mail: arhtur.zagidullin@ya.ru

Abstract: When using the numerical approach to construct the theory of physical libration of the Moon, it is necessary to solve the problem of comparing the numerical and analytical theories of lunar orbital motion. During these studies, the impact on the parameters of physical libration from the Moon's center of mass motion is determined. The paper contains the results of determining the effects caused by distinction in the position of lunar center of mass whose location is obtained according to two various lunar orbital theories. The analytical theory by Gutzwiller and Schmidt [1] constructed within the main problem of the Moon's motion and DE432 numerical theory taking into account a large number of factors, that extend beyond the main issue and that would be complicated or even impossible to consider when obtaining the analytical solution, are compared. Some reductions had been carried out in order to bring both theories to a unified reference system. Then the positions of the Moon's center of mass obtained in those theories were compared at an interval of 800 years. As a result, it was established that the amplitude of residual differences at the interval did not exceed 80 arc seconds in longitude and 10 arc seconds in latitude. The main reason for the obtained differences is neglecting planetary perturbations in the analytical theory. Other effects to distinguish the numerical theory from the analytical one are: motion of plane of the ecliptic, the Earth's and Moon's flattening, tidal effects etc. Those effects also cause the difference between the theories but slightly.

1. Introduction

Currently, the problem of studying the dynamics and physics of the Moon is relevant in connection with the plans on its exploration in the near future. On the one hand, study of physical libration of the Moon (PLM) is necessary to analyze and interpret the data on long-term lunar laser ranging and for the planned experimental observation of the Moon's rotation using the equipment placed on its surface. On the other hand, simulation of libration for various parameters of the Moon's interior provides a sense of the structure and physical characteristics of the lunar body [3]. At Kazan Federal University there is some long-term experience in constructing the theory of PLM. However, at the present time, when investigating the Moon's rotation, the numerical approach providing a large number of various effects requires consideration. This method provides higher accuracy at comparing theory and observations. That is why at certain stage of constructing the theory of PLM, there is a need to transit from analytical description of the lunar center of mass motion (theory by Gutzwiller and Schmidt) to the numerical theory (dynamic ephemeris DE432 (JPL NASA)).

At the transition from theory [1] to theory [2] of PLM, the correct method of revealing the impact on physical libration from the divergence in position of the center of mass for these 2 types of theories was developed.

2. Problem on investigation of orbital theories statement

Theory of PLM is being constructed within the main problem of PLM whose essence lies in the fact that physical body of the Moon is considered rigid, orbital motion does not depend on the rotational one, and the point Sun and Earth are taken as perturbing sources. At the first stage of constructing the numerical theory of PLM, the analytical theory by Gutzwiller and Schmidt [1] was used to obtain the solution for similar analytical model of PLM [4]. This solution was carried out and satisfactory agreement with the analytical one was achieved [5]. The next stage was related to transit to numerical analytical ephemeris DE432 [2].

It should be noted that the analytical theory by Gutzwiller and Schmidt is constructed within the main problem, solved with Hill-Brown method, and adapted by W, Eckert for computer calculations. The main problem in the Moon's motion considers 3 point bodies: the Moon, Earth, and Sun. In this theory, the ecliptic coordinate system is chosen as a reference system. Because due to precession from planets the ecliptic is exposed to secular motion, at significant time intervals the ecliptic coordinate system is not inertial. Authors of PLM theory [1] estimated maximum error of its each term to be 2 cm in radial direction and 20 cm in other directions. This is the internal accuracy of the theory. Comparison with observations of PLM theory [1] was not carried out.

Numerical dynamic ephemerides of DE series are constructed within post-Newtonian formalism. These models consider accelerations caused by tidal deformation of the Earth as well as shapes of the Earth and Moon. The equations in them are given for all planets of the Solar system, this is why indirect impacts from other planets and impacts from the main asteroid belt and Kuiper belt are taken into account in each ephemeris. Quasi-inertial ICRF system is chosen as a reference system. According to the analysis conducted by the authors of this series of ephemerides, the center of mass position is determined to within 1 meter.

At the transition to numerical theory of orbital motion, divergence from the results [5] in longitude at the very first stage of integration exceeded 200 arc seconds and at interval of a year reached 20000 arc seconds. There was also a marked tendency of residual differences with angular coefficient of 4.6 seconds. The reason for those enormous divergences could be both incorrect transition to dynamical ephemeris and significant impact of the divergences in these two types of theories, since the dynamical ephemeris included a large number of effects extending beyond the main problem.

This is why in the present paper search for the correct transition from ecliptic coordinate system of theory [1] to the one of theory [2] became a central problem.

3. Method and orbital theories of physical libration of the Moon results analysis

The numerical ephemeris constructed in inertial coordinate system reduction to the rotating ecliptic coordinate system, in which theory by Gutzwiller and Schmidt is constructed, was carried out on the basis of complex precession matrix. After all the reduction adjustments had been introduced, the following results were obtained: at time interval of about 800 years the amplitude of residual differences in longitude was between -40 and +80 arc seconds and in latitude did not exceed 13 arc seconds. As a result, one may conclude that such an enormous difference in parameters in the considered theories of PLM is explained by neglecting planetary perturbations in analytical theory of PLM.

In the present work, algorithm of calculating the Earth's orbit elements in relation to the Moon's center of mass using dynamic ephemeris DE432 [5] is developed as well.

The algorithm of comparing orbital theories may be divided into 2 stages: the main problem effects consideration and the effects extending beyond the main problem consideration. Model of "the main problem" in lunar motion (see Fig. 1) is described by simplifying restrictions as follows:

- The lunar body is under the influence of gravitational pull of 2 point masses: the Earth and Sun;
- Heterogeneity of the lunar body is not considered; the lunar body is taken as rigid;
- The Moon's orbital motion does not affect the rotational one.

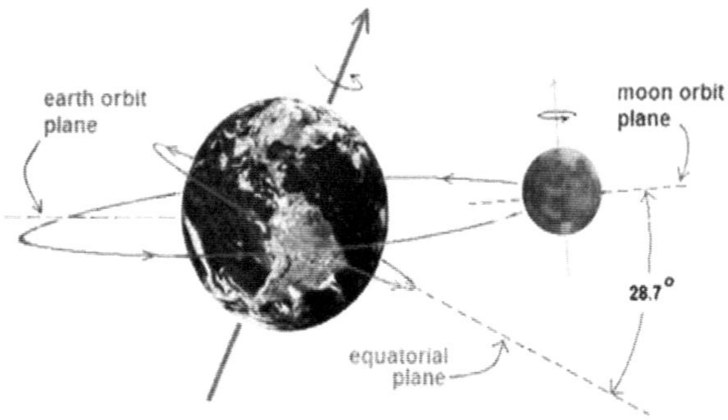

Fig. 1 The Earth – Moon dynamic system

Within the main problem reductions are carried out as follows:

- Consideration of non-linear (secular) terms of Delaunay arguments expansion, which was not performed in the analytical theory of PLM. Over the period of 800 years their impact become noticeable causing divergence in compared longitudes of 832 seconds. This is explained by slow but existing secular acceleration of the Moon;
- Transition from equatorial coordinate system to the ecliptic one used in the analytical theory of PLM;
- Transition from inertial coordinate system to the rotating one used in the numerical theory of PLM;
- Consideration of total precession in the analytical theory:
  - Transition to non-rotating coordinate system;
  - Consideration of total precession from planets;
  - Transition to rotating coordinate system.

After that, reductions to consider factors extending beyond the main problem were performed:

- Impact on the Moon's motion from the Earth's flattening;
- Resonance impact on the Moon from Venus.

Results of comparing residual differences at 2 types of reduction implementation are presented in Fig. 2 – 4.

In Fig. 2 results of comparing solutions for the Moon's latitude (upper line) and longitude (lower line) before and after conducting reductions for consideration of secular acceleration but without taking into account precession from planets (left and right diagrams correspondingly) are given. Obviously, there is no improvements for latitude, while for longitude there is a decrease in differential amplitude and even at the end of integration interval the difference increases up to 180 seconds compared with 300 seconds in non-reduced case.

Consideration of joint impact from secular acceleration and precession from planets (Fig. 3) leads to significant decrease in amplitudes of residual differences both in latitude (± 10 arc seconds) and longitude (between -30 to +70 arc seconds). Similar and even slightly better effect is achieved when the Earth's flattening is taken into account: ± 6 arc seconds in latitude and the same (between -30 to +70 arc seconds) in longitude.

IOP Conf. Series: Journal of Physics: Conf. Series **1038** (2018) 012004 doi:10.1088/1742-6596/1038/1/012004

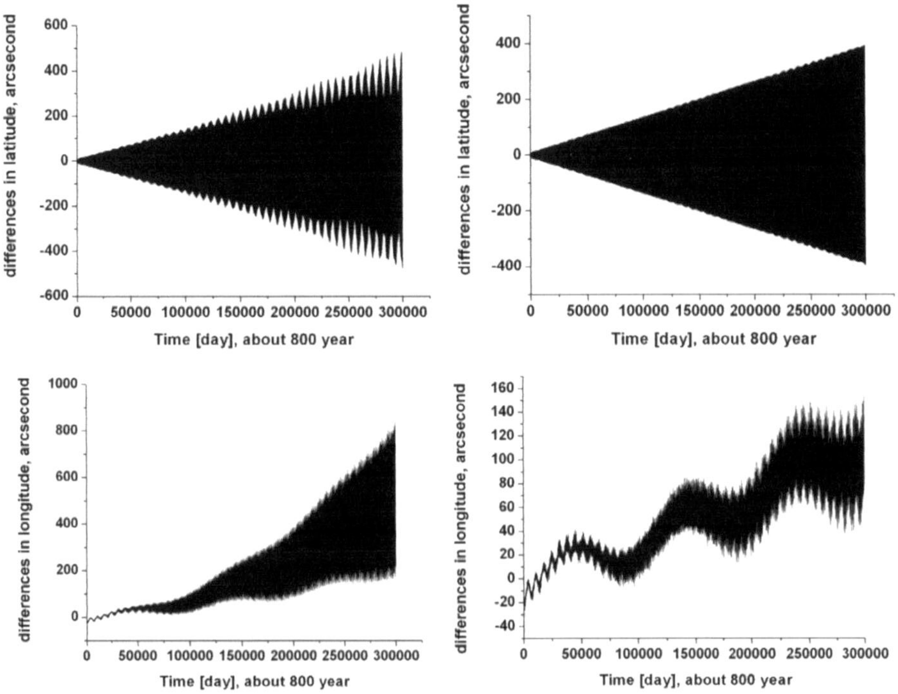

Fig. 2 Left line presents residual differences when comparing numerical and analytical theories of MPL in both latitude and longitude before reductions. Right line presents the same differences after considering all reductions.

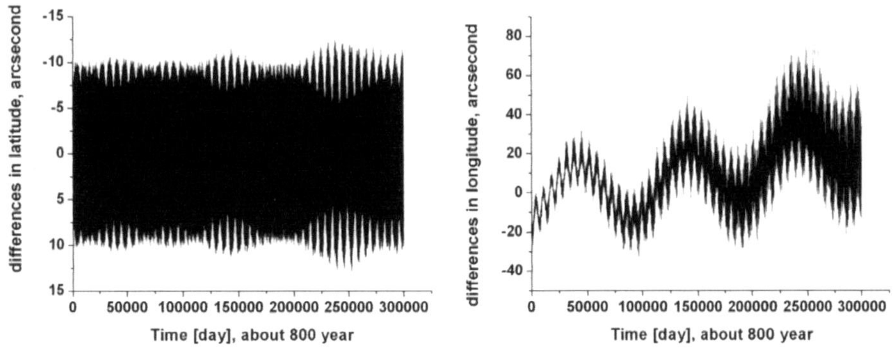

Fig. 3 Consideration of secular acceleration and precession from planets. To the left there is difference in latitude, to the right there is difference in longitude.

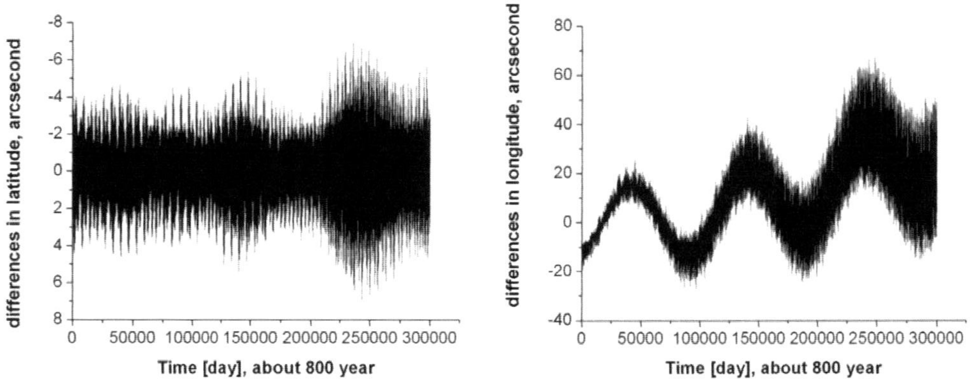

Fig. 4 Consideration of Earth's flattening. To the left there is difference in latitude, to the right there is difference in longitude.

4. Summary and conclusions

This paper provides a short description of analytical theory by Gutzwiller and Schmidt for lunar orbital motion used for MPL analytical theory construction and numerical ephemeris DE432 used for MPL numerical theory construction. Construction method, considered perturbation factors, and accuracy of MPL orbital theories are analyzed. The emphasis is placed on developing an algorithm of bringing both considered theories to the single coordinate system. As a result, the system of MPL orbital theories has been brought to the mean ecliptic J2000, in which the analytical theory is constructed, and to the coordinate system of DE432 theory. During reductions, the effects of precession from planets making the planes of ecliptic to diverge for about 400 arc seconds at integration interval of 800 years are taken into account. The Moon's center of mass position at the same time interval has been determined for both theories as well; their divergences have been estimated. Based on comparative analysis, it has been concluded that at time intervals of about 800 years non-linear terms of higher order of Delaunay classic arguments expansion must not be neglected. Otherwise, there arises a false secular acceleration in the motion of the lunar center of mass: about 830 seconds for 800 years. Taking into account all the considered facts has led to decrease in amplitudes of residual differences down to 80 arc seconds in longitude and 16 arc seconds in latitude. All the results of the MPL theories comparison are presented in the figures. Based on the diagrams analysis, the main reason for the divergences of various theories is neglecting planetary perturbations in the analytical approach. Other effects, such as the Earth's flattening and resonance lunar orbit's perturbations from Venus, do have influence but significantly lower one compared with planetary perturbations.

Acknowledgments

The work is performed according to the Russian Government Program of Competitive Growth of Kazan Federal University. This work was supported by the Russian Foundation for Basic Research, project nos. 16-02-00496 a, 18-02-01098 a, 18-32-00895 mol_a.

References

[1] Gutzwiller M.C, and Schmidt D S 1986 *Astronomical Papers prepared for the use of the American Ephemeris and Nautical Almanac* (U.S. Government Printing Office, Washington, D.C.) Vol. XXIII Part 1 pp 1–272
[2] Folkner W M, Williams J G et al. 2014 *IPN Progress Report* pp 1-18.
[3] Petrova N, Zagidullin A, Nefedyev Y, Kosulin V, Andreev 2017 *Advances in Space Research* Vol. 60 Issue 10 pp 2303-2313.
[4] Petrova N 1996 *Earth, Moon and Planets* Vol. 73 No 1 pp 71-99

[5] Zagidullin A, Petrova N, Nefed'ev Yu 2016 *Table of content ISBN 978-5-00015-013-9 Abstracts Moscow IKI N. 7MS3-PS-33*

Heliospheric modulation potential reconstructed by means of the radiocarbon data from the beginning of 11th century AD till the middle of the 19th century AD

A I Kuleshova[1], V A Dergachev[2], I V Koudriavtsev[2,1], Yu A Nagovitsyn[1] and M G Ogurtsov[2,1]

[1]Central Astronomical Observatory at Pulkovo of RAS, 65 Pulkovskoe shausse, St Petersburg 196140, Russia
[2]Ioffe Institute, 26 Politekhnicheskaya, St Petersburg 194021, Russia

E-mail: velendia@yandex.ru (Kuleshova A I); igor.koudriavtsev@mail.ioffe.ru (Koudriavtsev I V)

Abstract. This paper presents the results of reconstruction of the heliospheric modulation potential based on the content of ^{14}C cosmogenic isotope in natural archives from the beginning of 11-th to mid 19th century. This time period includes a number of grand minima of solar activity such as Oort, Wolf, Spoerer, Maunder, Dalton and climatic Little Ice Age. Medieval maximum of solar activity has happened during this time interval as well during this period there were variations both in the atmospheric carbon dioxide concentration and in the global temperatures, which were taken into account in the reconstruction of the ^{14}C production rate in the terrestrial atmosphere. As a result, values of the modulation potential were reconstructed using different temperature proxies.

1. Introduction

Data on relative content of cosmogenic isotopes ^{14}C and ^{10}Be obtained by measuring isotope content in tree rings (^{14}C) and ice cores (^{10}Be) allow to study variations of the intensity of galactic cosmic rays (GCR) over pre-instrumental epoch – the period, for which there are no direct measurements of the GCR flux. The work [1] presents the results of such a reconstruction for the past 8000 years. It shown that the intensity of the GCR during the Spoerer and Maunder minima of solar activity (SA) was significantly higher than during the Dalton minimum. A similar result for the ^{14}C isotope production rate in the atmosphere of the Earth has been given in the paper [2]. However, the issue about the impact of climate change on the results of such reconstructions remains open. The necessity to consider the influence of climate change on the ^{14}C reconstruction has been noted in the paper [3] because the ^{14}C transfer rate between natural reservoirs is a temperature dependent and this leads to the natural redistribution of carbon dioxide (CO_2) between various reservoirs.

In present work we consider the reconstruction of the heliospheric modulation potential, which describes the variation of GCR intensity during the time interval from the beginning of 11-th century till the mid-19th century. During this epoch the extrema of SA were established, such as Oort (≈1050AD), Wolf (≈1300AD), Spoerer (1400-1510AD), Maunder (1645-1700AD), Dalton (≈1810AD) minima and Medieval maximum, as well as the Little Ice Age (LIA) – the period, during which there were changes both in the concentration of CO_2 in the atmosphere and in the global temperature. In [4] the reconstruction of the ocean surface temperature (SST) near Antarctica during the last 12,000 years is presented. As was shown in [4] the reconstruction of the SST during the last 12 millennia demonstrated clearly that decrease of the temperature in the middle of the 2-nd millennium BC could reach up to 2^0C and decline coincided with the Little Ice Age. The authors of the work [5] also showed that decrease of the global SST took place during second millenium. Such changes of the oceanic temperature should be taken into account in calculations.

Unfortunately, the accuracy and temporal resolution of the data do not allow us to use them for detailed calculations. However, taking into account the fact that SST variation coincided with the global climate change, we may assume that variations of SST are proportional to the variations of global surface air temperature, which reconstructions have greater accuracy and good temporal resolution. In this paper we used the various certain temperature reconstructions.

2. Reconstruction of the heliospheric modulation potential based on radiocarbon data

Methods of calculating the heliospheric modulation potential $\varphi(t)$ is described in detail in the works [2,6], where take into account the influrence of the geomagnetic field [7], which counteract to the penetration of GCR particles in the Earth's atmosphere. The corresponding geomagnetic dipole moment was submitted in paper [2]. Calculation of the potential $\varphi(t)$ reduces to the solution of the following equation (see,e.g.[6]):

$$Q = \sum_{i=1}^{2} Y_i(E)J_i(E,\varphi)(1-f(E))dE \, , \tag{1}$$

where Q is the ^{14}C isotope production rate in the Earth's atmosphere; Y_i is yield function of ^{14}C isotope production in the atmosphere [8]; J_i is the spectrum of protons and α-particles of the primary cosmic rays; f describes the influence of the geomagnetic field on the penetration of GCR particles in the Earth's atmosphere, E is the kinetic energy of the GCR particles (GeV/nucleon). The specific form of these functions is given in [6].

Reconstruction of the production rate of the cosmogenic isotope ^{14}C (Q) currently is performed by means of one of the two models. In the first case reconstruction of Q(t) is performed using 3-D model Bern3D-LPJ [9], which includes the dynamic model of vegetation, oceanic sediments and tree-dimension model of the ocean.

The second way of Q(t) calculation is use of the five-reservoir radiocarbon model (see, e.g. [10]), which takes into account the radiocarbon exchange between the atmosphere, biosphere, humus, mixed and the deep layers of the ocean and this model we used in [11,12]. Also when we reconstruct ^{14}C production rate we take into account climatic effects – the change of CO_2 concentration in the Earth's atmosphere and global temperature changes. We will use reconstruction of CO_2 concentration in the Earth's atmosphere according to [13] (fig. 1a), which describes change of this concentration in the second millennium AD in more detail. This is principally important during the Little Ice Age, when abundance of CO_2 in terrestrial atmosphere dropped. According to this reconstruction sharp decrease of CO_2 during the Little Ice Age started in 1540 and reaches minimum value in 1620. During the same time (1530-1605) decrease of the relative abundance of the ^{14}C isotope in the atmosphere (Δ^{14}C) took place [14] (fig. 1b).

It is necessary note that the "classic" five reservoir model needs some modernization in order to take into account influence of changes of the global temperature on the rate of radiocarbon transition between the natural reservoirs. This modernization was started in the work [15]. In this work the rate of ^{14}C transfer from the mixed layer of the ocean and the atmosphere (λ_{mOa}) was simulated by expression $\lambda_{mOa}=(1+k\Delta T)\lambda^0_{mOa}$, where ΔT is anomaly of the global temperature; k is the temperature coefficient. It is shown that the value of the temperature coefficient k should be ~ $0.1K^{-1}$ [15].

In our calculations we considered temperature changes in of the surface layer of the ocean as synchronous with the changes of the surface temperature in the vicinity of the Earth. We will use different temperature reconstructions for the global temperature [16-21]. All these series were averaged over 30 years in order to remove short-term weather variations.

Thus, when we used these temperature reconstructions and the temperature coefficient $k=0.1K^{-1}$, then the change of the rate of transition of the isotope ^{14}C from the surface layer to the atmosphere (it is described by the value $k\Delta T$) can reach 3% - 8%. We have noted above that decrease of the temperature of the surface layer of the ocean in the middle of the second millennium AD could reach $1\text{-}2^0C$, while the partial pressure of the carbon dioxide in water increases by ~ 4% when temperature of water increases by 1 ^0C [22]. That is why variations of the rate of the radiocarbon transition (3% - 8%) from upper layer of the ocean to the atmosphere look like fairly substantiated.

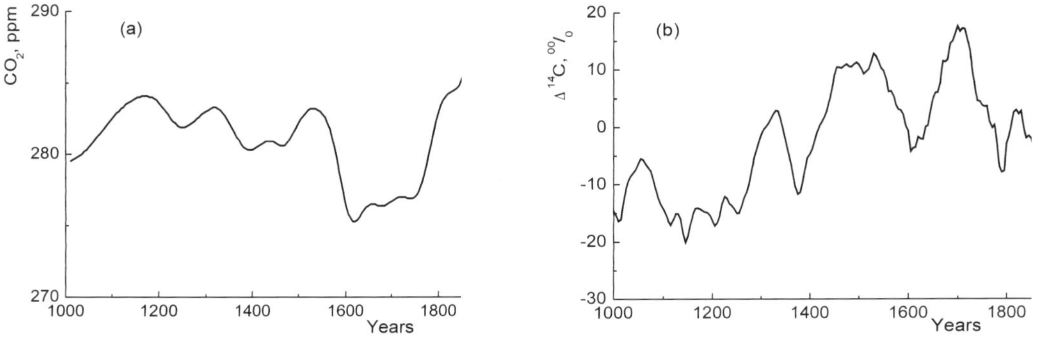

Figure 1. (a) - Reconstruction of the CO_2 concentration in the Earth's atmosphere according to [13]; (b) – $\Delta^{14}C$ according to [14].

Procedure of calculation of Q(t) is similar to that which was used in calculations performed for the time interval after the end of 14th century AD [12]. The calculation of φ(t) is performed using the equation (1). In present work we consider 1010 AD as the starting point. This is determined by the data on the variations of the CO_2 concentration in terrestrial atmosphere [13] used in calculations. The results of reconstruction of φ(t) are shown in Figures 2.

One should note, that the initial conditions for equations, describing the five-reservoir exchange system, were chosen proportional to initial conditions (see [10,13]), so that the values of *Q*(t) during the Dalton minimum coincide with the values obtained by [2].

As it is seen from this Figure, minimum values of the reconstructed modulation potential φ(t) during the Maunder and Dalton minima are closed when the reconstruction [17,19,21] are used. When the reconstructions [16,18,20] are used the values of φ(t) during the Maunder minimum are higher.

When the reconstruction of Crowley and Lowery [19] is used the value of the reconstructed modulation potential during the solar minima of Oort, Wolf, Spoerer, Maunder and Dalton can drop down to 0.35, 0.3, 0.15, 0.27 and 0.23 GV correspondingly. During the Medieval maximum of SA the reconstructed values of φ(t) are quite high (0.4-0.6 GV) when the temperature proxy of Crowley and Lowery [19] is used. In should be noted that the value of the reconstructed modulation potential during Oort minimum of SA was close to 0.2-0.4 GV.

3. Conclusion

Deep minima of the Sun's activity are clearly observed in all the reconstructions of the solar heliospheric potential obtained in present work. All the reconstructions (besides the reconstruction, obtained using the proxy [21]) show that the mean level of solar activity during the Maunder minimum was higher than during the minimum of Dalton. It should be noted, however, that divergences between different reconstructions are appreciable. Thus, more precise information about the temporal behaviour of the temperature of the ocean is necessary for more reliable reconstruction of the solar activity in the past.

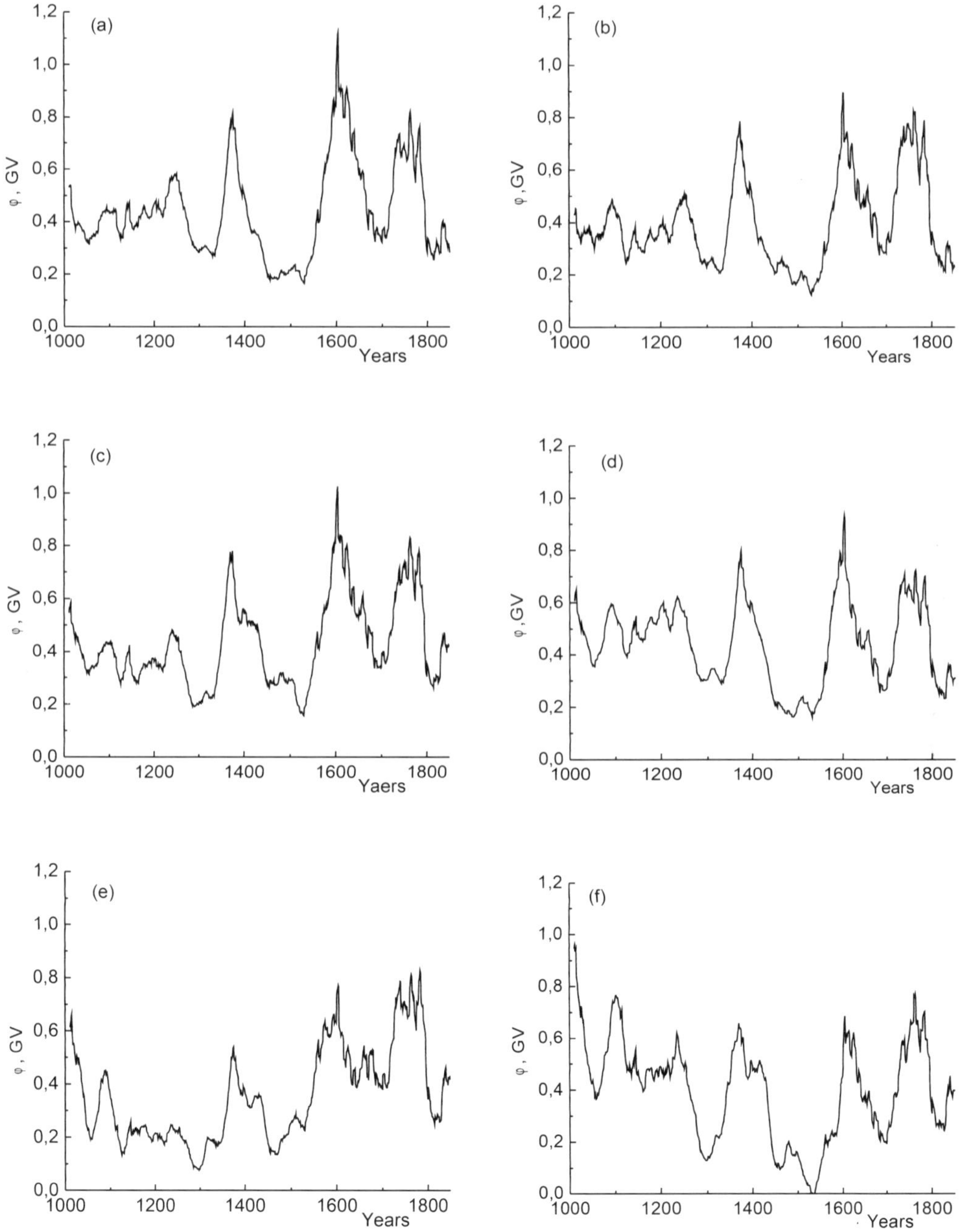

Figure 2. Reconstructions of heliospheric modulation potential obtained with using the temperature reconstructions after Mann et al. [16] (a), Jones et al. [17] (b), Briffa [18] (c), Crowley and Lowery [19] (d), Esper et al. [21] (e) and Moberg et al. [21] (f)

Acknowledgments

The work of Kuleshova A.I., Kudriavtsev I.V., Nagovitsyn Y.A., Ogurtsov M.G. was partly supported by RFBR (grant No 16-02-00090), and the program of the Presidium of RAS.

References

[1] Kocharov G E , Vasil'ev V A , Dergachev V A and Ostryakov V M 1983 *Soviet Astronomy Letters* **9** 110-112
[2] Usoskin I G et al 2014. *Astron. & Astrophys.* **562** L10
[3] Dergachev V A and Ostryakov V M 1978 *Proc. of the 6th All-Union Meeting on Problem "Astrophysical Phenomena and Radiocarbon"* (Tbilisi) (Tbilisi: TGU press) 177-182 (in Russian)
[4] Shevenell A E et al 2011 *Nature* **470** 250-254
[5] McGregor H V et al 2015 *Nature Geoscience* **8** 671-678
[6] Kovaltsov G A, Mishev A and Usoskin I G 2012 *Earth Planet. Sci. Lett.* **337** 114-120
[7] Licht A, Hulot G, Gallet Y and Thйbault E 2013 *Physics of the Earth and Planetary Interiors* **224** 38–67
[8] Poluianov S V, Kovaltsov G A, Mishev A L and Usoskin I G 2016 *J. Geophys. Res. Atmos.* **121** 8125-8136
[9] Roth R and Joos F 2013 *Climate of the Past* **9** 1879–1909
[10] Dorman L I 1978 *Proc. of the 6th All-Union Meeting on Problem "Astrophysical Phenomena and Radiocarbon"* (Tbilisi) (Tbilisi: TGU press) 49-96 (in Russian)
[11] Kuleshova A I et al 2015. *Geomagnetism and Aeronomy* **55** 1071-1075
[12] Kudryavtsev I V, et al 2016 *Geomagnetism and Aeronomy* **56** 998–1005.
[13] Etheridge D M et al 1998 http://cdiac.ornl.gov/ftp/trends/co2/lawdome.smoothed.yr75
[14] Reimer P J et al 2009 *Radiocarbon* **51** 1111–1150
[15] Koudriavtsev I V et al 2014 *Geohronometria* **41** 216 -222
[16] Mann M E, Bradley R S and Hughes M K 1999 *Geophysical Res. Let.* **26** 759–762
[17] Jones P D, Briffa K R, Barnett T P and Tett S F B 1998 *The Holocene* **8** 455-471
[18] Briffa K R 2000 *Quaternary Science Rev.* **19** 87-105
[19] Crowley T J and Lowery T S 2000 *Ambio* **29** 51–54
[20] Esper J, Cook E R and Schweingruber F H 2002 *Science* **102** 23413-23419
[21] Moberg et al. 2005 *Nature* **433** 613–617
[22] Takahashi T et al 1993 *Global Biogeochemical Cycles* **7** 843–878.

IOP Conf. Series: Journal of Physics: Conf. Series **1038** (2018) 012006 doi:10.1088/1742-6596/1038/1/012006

Distance Estimate of Tycho's SNR

A V Kozlova [1,2], **S I Blinnikov** [3,4]

[1] Ioffe Institute, 194021 St. Petersburg, Russia
[2] Sternberg Astronomical Institute, Moscow M. V. Lomonosov State University, 119234 Moscow, Russia
[3] Institute for Theoretical and Experimental Physics (ITEP), 117218 Moscow, Russia
[4] All-Russia Research Institute of Automatics (VNIIA), 127005 Moscow, Russia

E-mail: ann_kozlova@mail.ioffe.ru

Abstract. We present an approach to finding distances to young supernova remnants. Our method is based on hydrodynamical simulations of Tycho's SNR using the SUPREMNA code. For the explosion models, we use the classical W7 deflagration model and the delayed detonation model that was previously shown to provide good fits to the X-ray emission of Tycho's SNR. Combining our hydrodynamical simulations with the Chandra high-resolution images of the remnant obtained in 2015, we derive a distance estimate to Tycho's SNR of 2.8 ± 0.4 kpc.

1. Introduction

Supernova remnants (SNRs) play a significant role in the enrichment of the interstellar medium (ISM) with heavy elements, they are a source of energy injection and likely the sources of Galactic cosmic rays. Studying these objects is essential for our understanding a wide range of sections in physics and astronomy, for example, such important field as supernova explosion mechanisms, cosmological parameters, and distances to host galaxies (in case of Type Ia SN). Reliable distance determination to SNRs is a key in constraining their physical parameters like the size, the age and the explosion energy, which helps us to study their environment and ISM.

Tycho's supernova remnant is a member of a small subclass of Galactic SNRs known as the "historical supernovae". It is the remnant of a standard Type Ia [1] (i.e., a thermonuclear explosion of a white dwarf star in a close binary) SN observed by Tycho Brahe in 1572. The remnant has been widely studied across the electromagnetic spectrum (radio: [2], optical: [3], X-rays: [4], γ-rays: [5]). However, the distance remains uncertain with a spread of published values mostly between 2 and 4 kpc [6].

The rest of this paper is organized as follows. In Section 2, we report on details of modeling the X-ray emission from Tycho's SNR. We present the results and compare them with *Chandra* observations in Section 3. Finally, in Section 4, we conclude with a few remarks.

2. Simulations

Our goal is to model the X-ray emission from the Tycho's SNR using the grid of synthetic spectra computed with SUPREMNA [7]. It is a one dimensional spherically symmetric hydrodynamical code, which is aimed at the macroscopic scale study of SNRs. The code uses an implicit Lagrangean formulation and takes into account the most important physical processes, such as time-dependent ionization, the influence of radiative losses, the account of electron thermal

IOP Conf. Series: Journal of Physics: Conf. Series **1038** (2018) 012006 doi:10.1088/1742-6596/1038/1/012006

conduction and nonthermal particles [8]. We attempt to reproduce the observed X-ray emission, calculate the SNR dynamics, and derive distance estimate by comparing shock velocity with the observed expansion rate.

We base our ejecta models on two Type Ia SN explosion models: a classical deflagration W7 model [9], which results from a thermonuclear, subsonic flame propagation through the white dwarf, and a modern delayed detonation "deldet" model [10, 11], which assumes that the flame starts as a deflagration and turns into a detonation later on, as they appear to give the good overall agreement with the general characteristics of SN and well reproduce the X-ray emission of Tycho's SNR. For W7 explosion we use 300 zones: 99 for the ejecta and the rest for the ISM; for "deldet" model the hydrodynamical mesh consists of 200 zones, 100 of which is for the ejecta. We expand the original explosion models up to the age of 15–35 years and then start the simulations of an SN expansion into uniform ISM. During the calculation, we record the history of velocity, the temperature, the density and ion composition changes for each mesh zone. The resulting ionization stages for 15 most abundant elements (for which the calculations of kinetics were performed in the code) at the age of Tycho's remnant are used to evaluate an X-ray spectrum.

The synthetic spectra are mainly characterized by the SN explosion model, the age of the SNR, the density of the ISM, and non-Coulomb energy exchange between electrons and ions (q). Of the four parameters, the age is known to be 443 yr for the Tycho's SNR (at the time of *Chandra* observation used). Thus, only two remaining parameters can be varied for each of explosion models. We performed our hydrodynamical calculations for different values of the ISM density in a $(0.5 - 5) \times 10^{-24}$ g/cm^3 range. The parameter q specifies the fraction of artificial viscosity Q, added to the pressure of ions ($(1 - q)Q$ added to electrons). If only the collisional energy exchange is taken into account, then $q = 1$ and we get the usual system of equations with the heating of only ions at the front. The detailed description of this approach is presented in [7, 8]. We have sampled this parameter space with three points (0.5, 0.8, 1) to see if there is an appreciable non-Coulomb energy exchange due to the presence of magnetic field and plasma instabilities. Calculated spectra were converted to XSPEC table models with this interpolation parameters and fitted to *Chandra* X-ray observation. Then with the best-fitting values for interpolation parameters a new hydrodynamical model was calculated.

3. Results and Comparison with observation

Chandra archival data over a 15-year baseline from 2000 to 2015 are available for Tycho's SNR. We obtained data from three observations, with the longest exposure times: Obs. ID 3837 (146 ks, taken on 29 April 2003), Obs. ID 10095 (173 ks, taken on 23 April 2009), and Obs. ID 15998 (147 ks, taken on 22 April 2015). To analyze datasets we used the CIAO 4.8 and CALDB 4.7.2 [12]. We reprocessed *Chandra* data with `chandra_repro` and extracted source and background spectra using `specextract`. Since our code is one dimensional, we focus on the western part of Tycho, avoiding irregular ejecta clumps in the southeast and the interactions with a denser ISM toward the northeast [13, 14]. The background-substracted spectra of the same region, which we choose to study, from three different datasets are shown in Figure 1. Differences in the energy spectra are almost absent at $E \geq 1.5$ keV, so below we only consider the analysis of latest available observation (Obs. ID 15998).

We used XSPEC 12.9.0 [15] for spectral analysis, and performed model fitting in the 1.5-7.0 keV band. We choose this band in order to avoid the uncertain absorption features from interstellar matter, and, at the same time, to include in our analysis a Si line at ~ 1.86 keV. Results of the spectra simulations for different explosion models are shown in Figure 2.

Taking into account the simplicity of the models, limitations of one-dimensional calculations, and the uncertainties in the atomic data, we do not expect a valid fit from the statistical point of view, but nevertheless, we report the value of reduced χ^2 as an indicator of goodness of fit.

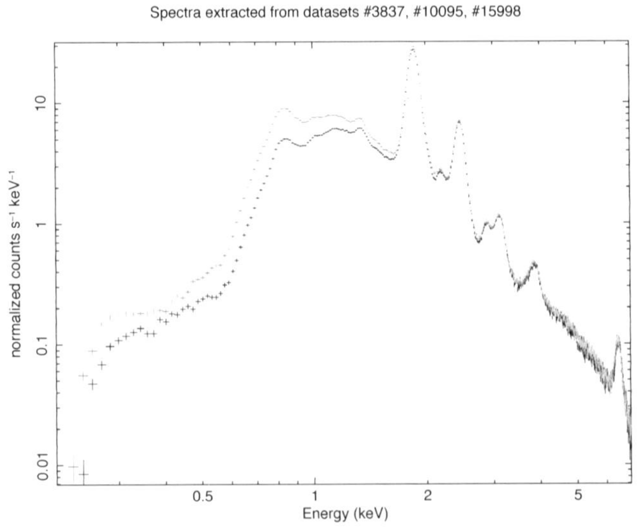

Figure 1. The background-substracted spectra of the western part of Tycho, extracted from Obs. ID 3837 (green), Obs. ID 10095 (red), and Obs. ID 15998 (black).

We find that the W7 model does not have enough Fe to reproduce the Fe Kα flux in the Tycho. This model gives a good fit to X-ray spectra of Tycho ($\chi^2 = 56.6$), only when it requires the high ISM density of $\sim 10^{-23}$ g/cm^3, which is not in agreement with the observed properties of the remnant [16]. The model based on "deldet" explosion matches well with Tycho X-ray spectra ($\chi^2 = 63.7$) and reproduces the observed parameters. Thus, we choose the "deldet" model to be the preferred one. Note that we add a power-law component to account for X-ray synchrotron radiation contribution to the spectrum because there is no self-consistent theory for it to implement in the code [17].

Using the predicted by SUPREMNA shock velocity v_s and the proper motions μ on the western side of the remnant derived by Williams [18], the distance to the remnant was evaluated:

$$d(\text{parsecs}) = \frac{v_s(\text{km/s})}{4.74\mu(\text{arcsec/yr})}$$

Table 1. Estimates of the distance to Tycho's SNR.

Model name	Distance (kpc)
"deldet"	2.8 ± 0.4
W7, $\rho_{\text{ISM}} \sim 10^{-24}$g/cm^3	2.6 ± 0.4
W7, $\rho_{\text{ISM}} \sim 10^{-23}$g/cm^3	1.8 ± 0.3

The calculated distances are summarized in Table 1.

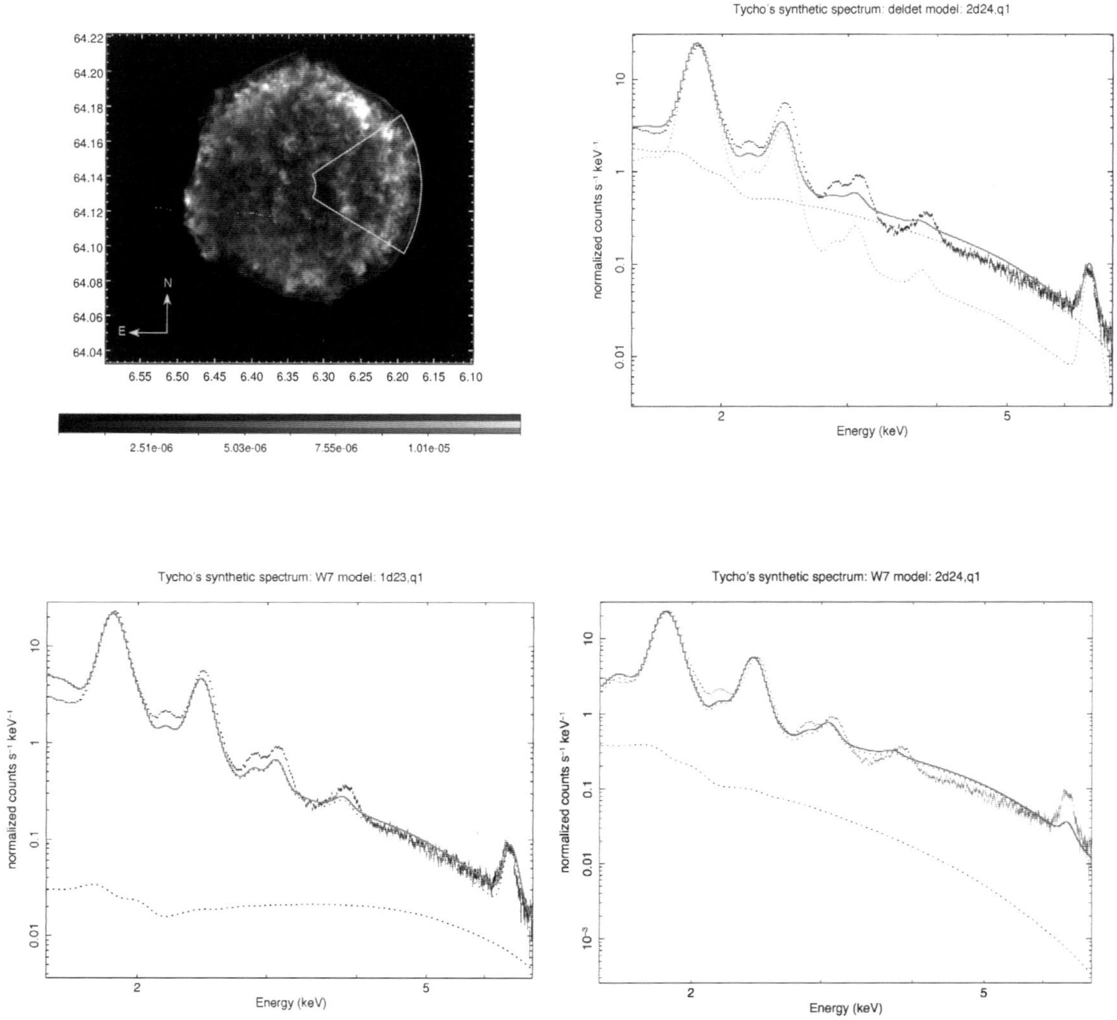

Figure 2. *Top row, left:* A three-color *Chandra* image of the Tycho's SNR obtained from the observation ID 10095, with the source spectrum extraction region indicated in green. Red corresponds to low-energy ($0.5 - 1.2$ keV), green to middle-energy ($1.2 - 2.0$ keV), and blue to high-energy ($2.0 - 7.0$ keV). Comparison between the ejecta models and the X-ray spectrum of region. "deldet" model, ISM density $\rho = 2 \times 10^{-24}$ g/cm^3 (*Top row, right*). W7 model, $\rho = 2 \times 10^{-23}$ g/cm^3 (*bottom row, left*) and W7 model $\rho = 2 \times 10^{-24}$ g/cm^3 (*bottom row, right*). The power-law component is displayed (dotted line).

4. Conclusion

We have modeled the evolution, performed a hydrodynamical calculation of X-ray spectrum of Tycho's SNR and compared the model X-ray emission with archived *Chandra* observation. We have found the delayed detonation model, expanding into a uniform ISM of $\rho = 2 \times 10^{-24}$ g/cm^3 to be the preferred one for Tycho's SNR. This is in agreement with results of [19]. Predicted by SUPREMNA shock velocities are slightly higher than those obtained previously by Fang et al [20] in a 3D hydrodynamical simulation. This difference is mainly due to the density configuration used in simulations (uniform medium of constant density vs a latitude-dependent wind-blown cavity). Based on "deldet" model the distance to the SNR is estimated to be 2.8 ± 0.4 kpc. This estimate is in agreement with other shock-wave methods [6].

The errors can be improved with the application of more detailed models to different regions of the remnant. Our method of distance determination is more robust than other shock-wave methods which rely on line emission of knots which have irregular structure.

In the future, we plan to modify SUPREMNA code for self-consistent treatment of cosmic rays.

Acknowledgments

The work by A. K. (modeling X-ray emission of Tycho's SNR and deriving distance estimates) is supported by the RSF grant 16-12-10519. We thank Ping Zhou for useful references and discussion.

References

[1] Krause O, Tanaka M, Usuda T, Hattori T, Goto M, Birkmann S and Nomoto K 2008 *Nature* **456** 617–619
[2] Dickel J R, Murray S S, Morris J and Wells D C 1982 *ApJ* **257** 145–150
[3] Ghavamian P, Raymond J, Hartigan P and Blair W P 2000 *ApJ* **535** 266–274
[4] Sato T and Hughes J P 2017 *ApJ* **840** 112
[5] Archambault S *et al.* 2017 *ApJ* **836** 23
[6] Hayato A *et al.* 2010 *ApJ* **725** 894–903
[7] Sorokina E I, Blinnikov S I, Kosenko D I and Lundqvist P 2004 *Astronomy Letters* **30** 737–750
[8] Kosenko D, Blinnikov S I and Vink J 2011 *A&A* **532** A114
[9] Nomoto K, Thielemann F K and Yokoi K 1984 *ApJ* **286** 644–658
[10] Röpke F *et al.* 2012 *ApJ* **750** L19
[11] Seitenzahl I R, Ciaraldi-Schoolmann F, Röpke F K, Fink M, Hillebrandt W, Kromer M, Pakmor R, Ruiter A J, Sim S A and Taubenberger S 2013 *MNRAS* **429** 1156–1172
[12] Fruscione A *et al.* 2006 CIAO: Chandra's data analysis system *Society of Photo-Optical Instrumentation Engineers (SPIE) Conference Series* (*Proc. SPIE* vol 6270) p 62701V
[13] Decourchelle A, Sauvageot J L, Audard M, Aschenbach B, Sembay S, Rothenflug R, Ballet J, Stadlbauer T and West R G 2001 *A&A* **365** L218–L224
[14] Reynoso E M, Velázquez P F, Dubner G M and Goss W M 1999 *AJ* **117** 1827–1833
[15] Arnaud K A 1996 XSPEC: The First Ten Years *Astronomical Data Analysis Software and Systems V* (*ASP Conf. Ser.* vol 101) ed Jacoby G H and Barnes J p 17
[16] Katsuda S, Petre R, Hughes J P, Hwang U, Yamaguchi H, Hayato A, Mori K and Tsunemi H 2010 *ApJ* **709** 1387–1395
[17] Cassam-Chenaï G, Hughes J P, Ballet J and Decouchelle A 2007 *ApJ* **665** 315–340
[18] Williams B J, Chomiuk L, Hewitt J W, Blondin J M, Borkowski K J, Ghavamian P, Petre R and Reynolds S P 2016 *ApJ* **823** L32
[19] Badenes C, Borkowski K J, Hughes J P, Hwang U and Bravo E 2006 *ApJ* **645** 1373–1391
[20] Fang J, Yu H and Zhang L 2018 *MNRAS* **474** 2544–2550

Definitions of energy for the description of gravity as the splitting theory

D. A. Grad*, **S. A. Paston**†, **A. A. Sheykin**‡
Saint-Petersburg State University, Saint-Petersburg, Russia
E-mail: *d.grad@spbu.ru, †s.paston@spbu.ru, ‡a.sheykin@spbu.ru

Abstract. We study the definitions of energy, naturally arising in the splitting theory, which is the field theoretic formulation of the Regge-Teitelboim gravity. The latter regards our spacetime as a surface embedded in a flat bulk. The splitting theory describes embedded spacetime in the language of the some field theory in a flat bulk. We consider the Noether energy-momentum tensor (EMT) and the metric EMT defined by the variation with respect to the metric of a flat bulk. We discuss a localizability of energy. Then using these EMTs we calculate the full energy of an isolated massive body. We compare the results with the standard general relativity results obtained from the Einstein energy-momentum pseudotensor (pEMT) and from the Møller pEMT. Finally, we propose the several ways of correction of the definitions of the energy in the splitting theory.

1. Introduction

One of the oldest problems of general relativity is the problem of gravitational energy definition. It is well-known that the Noether procedure applied to the Lagrangian of general relativity gives the Noether current, which is energy-momentum pseudotensor rather than energy-momentum tensor. So, it can be eliminated at any given point of spacetime and the gravitational energy turns out to be nonlocalizable, what seems to be physically unsatisfaying situation. For the comprehensive review of the energy problem in general relativity see [1, 2, 3].

The splitting theory [4] provides the coordinate-independent field theoretic description of gravity. Thus, in this paper we want to discuss the various definitions of energy in the splitting theory, trying to solve the known problems of gravitational energy.

2. The splitting theory

The splitting theory [4] describes gravity as a field theory in a flat bulk of higher dimensions. This theory considers $N-4$ real fields $z^A(y)$, $A, B, \dots = 1, \dots, N-4$ in N-dimensional Minkowski space with the Lorentzian coordinates y^a, $a, b, \dots = 0, \dots, N-1$ in a such a way that:

(i) Each field configuraion corresponds to the some splitting of a bulk into a system of 4D surfaces $\mathfrak{S} : z^A(y) = \text{const}$.

(ii) Surfaces do not interact and do not intersect.

(iii) Any surface can be viewed as our spacetime.

According to the Friedman theorem [5], there exists local and isometrical embedding of d-dimensional pseudoriemannian manifold with p timelike and $d-p$ spacelike directions in N-dimensional pseudoeuclidean space with at least p timelike and at least $d-p$ spacelike directions, where

$$N \geq \frac{d(d+1)}{2}. \tag{1}$$

It is worth noting that the minimal number of dimensions coincides with number of independent components of the metric of d-dimensional manifold. In our case $d = 4$ (with one timelike and three spacelike directions), so $N = 10$ is enough, and here and elsewhere we take $N = 10$ with one timelike and other spacelike directions, so $A, B, \dots = 1, \dots, 6$ and $a, b, \dots = 0, \dots, 9$.

The original action of the splitting theory can be written as the sum of the Einstein-Hilbert actions of each surface over all of the surfaces:

$$S = \int dz\, S_{\mathfrak{S}}(z) = \int dz d^4x \sqrt{-g}\left(-\frac{1}{2\varkappa}R\right), \tag{2}$$

where we introduce the internal coordinates $\{x^\mu\}, \mu = 0, 1, 2, 3$ on the surfaces. One can inverse the set of functions $\{x^\mu(y), z^A(y)\}$ and obtain the embedding functions $y^a(x, z)$ of $4D$-surface $\mathfrak{S}$ in a bulk at fixed z. The metric of a such a surface $\mathfrak{S}$ in (2) is induced

$$g_{\mu\nu} = \eta_{ab}\partial_\mu y^a \partial_\nu y^b, \quad \partial_\mu = \frac{\partial}{\partial x^\mu}. \tag{3}$$

Changing the variables $\{x^\mu, z^A\} \to \{y^a\}$ in (2), we get (see details in [4])

$$S = -\frac{1}{2\varkappa}\int dy \sqrt{|w|}R, \tag{4}$$

where $w = \det w^{AB}$, $w^{AB} = \eta^{ab}\partial_a z^A \partial_b z^B$, $\partial_a = \frac{\partial}{\partial y^a}$. Here $\sqrt{|w|}$ arises from the Jacobian of the change of the variables. The other important quantities, including the scalar curvature R, in the splitting theory are also constructed from the $\partial_a z^A(y)$, η_{ab}, w^{AB} and $w_{AB} = (w^{AB})^{-1}$:

(i) The projector onto the surface

$$\Pi_{ab} = \eta_{ab} - w_{AB}\partial_a z^A \partial_b z^B. \tag{5}$$

(ii) The second fundamental form of the surface

$$b^{a,}{}_{cd} = \Pi^b_c \Pi^e_d \partial_a \Pi^a_b. \tag{6}$$

(iii) The Riemann tensor, the Ricci tensor, the scalar curvature

$$R_{abcd} = b^{e,}{}_{ac} b_{e,bd} - b^{e,}{}_{ad} b_{e,bc},\ R_{ac} = \eta^{bd} R_{abcd},\ R = \eta^{ac} R_{ac}, \tag{7}$$

and the Einstein tensor

$$G_{ac} = R_{ac} - \frac{1}{2}\Pi_{ac}R. \tag{8}$$

If we add the matter, the original action of the splitting theory will be

$$S = -\frac{1}{2\varkappa}\int dy \sqrt{|w|}R + \int dy \sqrt{|w|}L_{\mathrm{m}}, \tag{9}$$

where $L_{\rm m}$ is the contribution of matter. It consists of matter fields defined in a bulk in a such a way that their excitations propagate only along the surfaces and hence there is no interaction between the matter fields on the different surfaces. It was shown in [6] that the original action (9) is the most natural choice of the action in the splitting theory. This action gives the Regge-Teitelboim equations [7, 8, 9] for each surface in the field theoretic form:

$$\left(G^{cd} - \varkappa T_{\rm m}^{cd}\right) b^{a,}{}_{cd} = 0, \tag{10}$$

where $T_{\rm m}^{cd}$ is the EMT of matter.

Finally, it should be noted that the splliting theory has the symmetry under "the renumeration of the surfaces"

$$\tilde{z}^A(y) = f^A(z^B(y)), \tag{11}$$

where $f^A(z)$ are arbitrary functions, because $\tilde{z}(y)$ and $z(y)$ correspond to the same splitting of a bulk. This symmetry shows that physics depends only on how we perform the splitting of a bulk into a system of the surfaces, but not on the actual values of fields $z^A(y)$ on each surface. With respect to this renumeration $\sqrt{|w|}$ transforms multiplicatively:

$$\sqrt{|w|} \to J\sqrt{|w|}, \quad J = \det \frac{\partial f^A}{\partial z^B}. \tag{12}$$

Thus, the action (9) is not invariant under the renumeration of the surfaces, but the equations of motion (10) are, nevertheless, possess such an invariance. For the detailed construction of the splitting theory see [4].

3. The Noether EMT

The Lagrangian corresponding to the action (9) is

$$\mathfrak{L} = -\frac{\sqrt{|w|}}{2\varkappa} R + \sqrt{|w|} L_{\rm m} \tag{13}$$

and it contains second derivatives of fields $z^A(y)$ as it can be seen from (5), (6), (7). In this case the Noether procedure gives the following EMT

$$\tau^b{}_a = \frac{\partial \mathfrak{L}}{\partial \partial_b z^A}\partial_a z^A + \frac{\partial \mathfrak{L}}{\partial \partial_c \partial_b z^A}\partial_c \partial_a z^A - \partial_c \frac{\partial \mathfrak{L}}{\partial \partial_c \partial_b z^A}\partial_a z^A - \mathfrak{L}\delta^b_a. \tag{14}$$

The calculation gives [10] the EMT as the sum of gravitational and matter parts:

$$\tau^{ba} = -\frac{\sqrt{|w|}}{\varkappa} G^{ba} + \sqrt{|w|} T_{\rm m}^{ba} - \partial_c A^{cba}. \tag{15}$$

The quantity A^{cba} depends on matter fields, which usually decrease rapidly enough in the spatial directions, so the last term doesn't give a contribution to the conserved energy

$$E = \int d^9 y\, \tau^{00} = \int d^9 y \sqrt{|w|} \left(T_{\rm m}^{00} - \frac{1}{\varkappa} G^{00}\right), \tag{16}$$

where $d^9 y = dy^1 \dots dy^9$.

Since there is no interaction between matter on the different surfaces in the splitting theory (see after (9)) and hence no energy exchange between the surfaces, we can pick out a contribution of a single surface to the full energy. To do that, we must rewrite energy expression (17) in the curvilinear coordinates $\{x^\mu, z^A\}$ in a bulk and assume that $x^0 = y^0$ (choice of time on the surface). Thus, we obtain for the Noether energy

$$E = \int dz d^3x \sqrt{-g}\left(T_{\rm m}^{00} - \frac{1}{\varkappa}G^{00}\right). \tag{17}$$

Hence one can choose the Noether energy of a single surface $\mathfrak{S}$:

$$E_{\mathfrak{S}} = \int d^3x \sqrt{-g}\left(T_{\rm m}^{00} - \frac{1}{\varkappa}G^{00}\right). \tag{18}$$

Now let us discuss the localizability of the obtained energy density τ^{00}. It is 4D-localizable in sense that it is explicitly diffeomorphism-invariant as the action (9) of the splitting theory does not contain the internal coordinates x^μ on the surfaces $\mathfrak{S}$. But in the splitting theory we have additional specific symmetry (11) of "the renumeration of the surfaces". The Noether energy density τ^{00} contains $\sqrt{|w|}$, which transforms multiplicatively with respect to this symmetry (see (12)). Thus, τ^{00} transforms in the same way, so it can be localized on a single surface $\mathfrak{S}$, because J remains constant along the surface. However, τ^{00} is nonzero only for the solutions of the splitting theory, which are the solutions of the Regge-Teitelboim equations (10), but are not the solutions of the Einstein equations.

4. The metric EMT

As the splitting theory has the form of a some field theory in a flat spacetime, it is possible to define the metric EMT, namely to vary the action with respect to ambient space metric. For the action (9) this procedure gives [10] the EMT as the sum of gravitational and matter parts:

$$\tau^{ab} = -\frac{\sqrt{|w|}}{\varkappa}G^{ab} + \overline{\tau}^{ab} + \sqrt{|w|}T_{\rm m}^{ab}, \tag{19}$$

where

$$\overline{\tau}^{ab} = \frac{1}{\varkappa}\partial_c\partial_d\left(\sqrt{|w|}(\Pi^{ab}\Pi^{dc} - \Pi^{ac}\Pi^{db})\right). \tag{20}$$

The full energy will be

$$E = \int d^9y\, \tau^{00} = \int d^9y \sqrt{|w|}\left(T_{\rm m}^{00} - \frac{1}{\varkappa}G^{00}\right) + \int d^9y\, \overline{\tau}^{00}. \tag{21}$$

Let us pick out a contribution of a single surface as it was done in the previous section. Rewriting (22) in the curvilinear coordinates $\{x^\mu, z^A\}$ in a bulk and assume that $x^0 = y^0$, we obtain (see details in [10])

$$E = \int dz d^3x \sqrt{-g}\left(T_{\rm m}^{00} - \frac{1}{\varkappa}G^{00}\right) + \frac{1}{\varkappa}\int dz d^3x\, \partial_i\left(\sqrt{-g}\frac{\partial x^i}{\partial y^c}\frac{\partial x^0}{\partial y^b}\psi^{cb0}\right), \tag{22}$$

where

$$\psi^{cba} = \Pi_d^e\partial_e\left(\sqrt{|w|}(\Pi^{ab}\Pi^{dc} - \Pi^{ac}\Pi^{ab})\right). \tag{23}$$

According to (22) one can choose a contribution of a single surface $\mathfrak{S}$ for the metric energy:

$$E = \int d^3x \sqrt{-g}\left(T_{\rm m}^{00} - \frac{1}{\varkappa}G^{00}\right) + \frac{1}{\varkappa}\int d^2\sigma_i \sqrt{-g}\frac{\partial x^i}{\partial y^c}\frac{\partial x^0}{\partial y^b}\psi^{cb0}. \tag{24}$$

IOP Conf. Series: Journal of Physics: Conf. Series **1038** (2018) 012007 doi:10.1088/1742-6596/1038/1/012007

We have applied Gauss's theorem in the second term and now the integration is perfomed over the infinitely remote spatial 2D surface laying in $\mathfrak{S}$.

Now let us discuss the localizability of the metric energy density. As the Noether energy density, the metric τ^{00} is 4D-localizable, but it contains $\partial_a\sqrt{|w|}$, which transforms inhomogenously under "the renumeration of the surfaces" (11) :

$$\partial_a\sqrt{|w|} \to J\left(\partial_a\sqrt{|w|} + \sqrt{|w|}\partial_a z^C \frac{\partial^2 f^B}{\partial z^C \partial z^A}\frac{\partial z^A}{\partial f^B}\right). \tag{25}$$

Thus, the metric energy density can't be localized even on a single surface, but not because of 4D coordinate transformations as it takes place for the energy density in general relativity.

5. The full energy of an isolated massive body

Let us calculate the full energy of field configuration, which corresponds to the solution of the Einstein equations with the spherical massive matter with mass M. The Noether energy (18) vanishes for any Einsteinian solution, so it is zero in this case. It can be nonzero only for the "extra" solutions, which are the solutions of the Regge-Teitelboim equations (10), but are not the solutions of the Einstein equations.

The metric energy (24) gives a nontrivial answer (for details see [10])

$$E_{\mathfrak{S}} = 2M. \tag{26}$$

This suprising result of the splitting theory can be related to the choice of action (9), which is analogous to the Einstein-Hilbert one in general relativity. Let us remind that the Einstein-Hilbert action in general relativity leads to the pEMT, which gives $M/2$ for the energy of an isolated massive body (this pEMT is exactly the one half of the original Møller pEMT [11]). At the same time, there exists another, more physically, the Einstein pEMT [12], which gives M for the energy of an isolated mass. The latter pEMT can be derived by the Noether procedure from the well-known noncovariant action, which contains only first order derivatives of the metric and differs from the Einstein-Hilbert action by a certain surface term. Thus, it will be a good idea to find more successful (probably localizable) definitions of energy in the splitting theory by moving from the action (9) to the another action, which differs from the original one by a surface term. Herewith, the metric EMT doesn't change, but the Noether EMT gains an additional contribution. Two variants of such a surface term can be proposed.

First option. The most reasonable choice is the Gibbons-Hawking-York term [13, 14], which can be added at the stage of constructing of the action of the splitting theory (cf. with (2)):

$$S = \int dz\, S_{\mathfrak{S}}(z) = \int dz \int\limits_{\mathfrak{S}} d^4x\, \sqrt{-g}\left(-\frac{1}{2\varkappa}R\right) + \frac{1}{\varkappa}\int dz \int\limits_{\partial\mathfrak{S}} d^3\hat{x}\, \sqrt{|\hat{g}|}K, \tag{27}$$

where $\hat{g}_{ik}$ is the metric on $\partial\mathfrak{S}$ and K is the trace of the extrinsic curvature of $\partial\mathfrak{S}$.

Second option. Another possibility of choice of the additional surface term is moving to the form of the action with only first order time derivatives of $z^A(y)$ [15]. To do this, one should, when building the action (2), assume that $x^0 = y^0$ and should use the ADM action [16]:

$$S = \int dz\, S_{\mathfrak{S}}(z) = -\frac{1}{2\varkappa}\int dz \int d^4x\, \sqrt{-g}\left(\hat{R} + ((K^i_i)^2 - K_{ik}K^{ik})\right), \tag{28}$$

where $\hat{R}$ is the scalar curvature of the 3D surface $x^0 = \text{const}$, and K_{ik} is the extrinsic curvature of this surface. The resulting action can be written in terms of fields $z^A(y)$ and it will contain

IOP Conf. Series: Journal of Physics: Conf. Series **1038** (2018) 012007 doi:10.1088/1742-6596/1038/1/012007

only first order time derivatives of $z^a(y)$. But the price of this is the loss of the Lorentz invariance in a bulk (preffered time direction appears) for the obtained action.

Probably, one of these ways will lead to the localizable energy for the description of gravity as the splitting theory and also will give the physically reasonable value of the full energy of an isolated massive body, which is equal to M.

Acknowledgments
The authors are grateful to R. V. Ilin for the useful discussions.

References
[1] Pauli W 1958 *Theory of Relativity* (New York: Pergamon, reprinted Dover 1981)
[2] Petrov A N 2008 *Classical and Quantum Gravity Research* ed Christiansen M N and Rasmussen T K (New York: Nova Science) chap 2, pp 79–160 (*Preprint* arXiv:0705.0019)
[3] Faddeev L D 1982 *Soviet Physics Uspekhi* **25** 130
[4] Paston S A 2011 *Theoretical and Mathematical Physics* **169**(2) 1611–1619 (*Preprint* arXiv:1111.1104)
[5] Friedman A 1961 *J. Math. Mech.* **10** 625–649
[6] Sheykin A A and Paston S A 2016 *Physics of Atomic Nuclei* **79**(11-12) 1494–1496 (*Preprint* arXiv:1704.06883)
[7] Regge T and Teitelboim C 1977 *Proceedings of the First Marcel Grossmann Meeting, Trieste, Italy, 1975* ed Ruffini R (North Holland, Amsterdam) pp 77–88 (*Preprint* arXiv:1612.05256)
[8] Deser S, Pirani F A E and Robinson D C 1976 *Phys. Rev. D* **14** 3301–3303
[9] Paston S A and Franke V A 2007 *Theoretical and Mathematical Physics* **153**(2) 1582–1596 (*Preprint* arXiv:0711.0576)
[10] Grad D A, Ilin R V, Paston S A and Sheykin A A 2017 *International Journal of Modern Physics D* **26**(14) 1750188 (*Preprint* arXiv:1707.01074)
[11] Moller C 1958 *Annals of Physics* **4**(4) 347 – 371
[12] Einstein A 1916 *Sitzungsber. preuss. Akad. Wiss.* **2** 1111
[13] Gibbons G W and Hawking S W 1977 *Phys. Rev. D* **15**(10) 2752–2756
[14] York J W 1972 *Phys. Rev. Lett.* **28**(16) 1082–1085
[15] Sheykin A A and Paston S A 2016 *International Journal of Modern Physics: Conference Series* **41** 1660128 (*Preprint* arXiv:1511.09268)
[16] Arnowitt R, Deser S and Misner C 1962 *Gravitation: an introduction to current research* ed Witten L (New York: Wiley) chap 7, pp 227–265 (*Preprint* arXiv:gr-qc/0405109)

Isotopic terrestrial imprints of solar superflares

Frolov D.A.[1], Ostryakov V.M.[1], Pavlov A.K.[2], Struminsky A.B.[3], Vasilyev G.I.[2]

[1]Peter the Great St.Petersburg Polytechnic University, 195251 Polytekhnicheskaya 29, St.Petersburg, Russia
[2]Ioffe Institute, 194021 Polytekhnicheskaya 26, St.Petersburg, Russia
[3]Space Research Institute, 117997, 84/32 Profsoyuznaya Str., Moscow, Russia

E-mail: anatoli.pavlov@mail.ioffe.ru

Abstract. The extreme cosmic event that occurred in AD 775 was detected using ^{14}C measurements in tree rings and ^{10}Be, ^{36}Cl abundances in polar ice cores. Perhaps it is the most powerful solar proton event in the past several thousands of years. Simulation of isotope production with the spectra of solar flares observed in the modern era (23.02.56, 04.08.72 etc.) showed that to produce the measured amount of radionuclides, the particle fluence in the AD 775 event has to be by tens - hundreds times greater than the modern powerful solar events. The results of calculations of long-lived cosmogenic radionuclides (^{14}C, ^{10}Be, ^{36}Cl) production in the Earth's atmosphere are presented.

1. Introduction

Production of cosmogenic radionuclides in the Earth's atmosphere is caused by nuclear reactions of cosmic rays (CRs). The concentration of long-lived ^{14}C, ^{10}Be and ^{36}Cl isotopes in natural archives has reflected the history of atmospheric irradiation. Their records demonstrate temporal variability which can be partly attributed to the known astrophysical phenomena [1]. For example, correlations of the nuclide concentration records with long-term CR intensity variations due to the changes in solar activity and geomagnetic field were comprehensively studied [2]. However, another class of extraterrestrial sources – the impulsive sporadic events such as powerful solar proton events (SPEs) and gamma-ray bursts (GRBs), could momentarily increase the production rate by several orders of magnitude. Therefore, they could produce a sharp rise of the atmospheric concentration of the nuclides. Such an enigmatic sharp increase of the ^{14}C relative concentration ($\delta^{14}C$) in tree rings dated AD 775 was discovered by Miyake et al. [3]. It was later confirmed by other groups in different tree ring collections [4, 5, 6] and in corals [7]. Data analysis from both hemispheres confirms duration of the initial event which was probably shorter than one year [6]. These results stimulated intensive discussion on a possible origin of the unexpected phenomenon. Several working hypothesis even such exotic as a comet impact on the Sun and Earth [7] have been proposed since. At present, only two of them have survived as realistic. They are a solar superflare with very hard energy spectrum of protons [4] and GRB of Galactic origin [8, 9]. For SPE hypothesis the isotope production mainly depends on the particle superflare spectra and (in a minor degree) on the geomagnetic field strength at that period.

In this paper we model production of ^{14}C, ^{10}Be and ^{36}Cl by SPEs with different spectra. We demonstrate also that these spectra result in the significant difference of the isotopic imprints (we mean both global production rate, Q, and the local one in the polar troposphere, Q_T). Additionally, we study the influence of the geomagnetic field variations because of significant screening effect for low-energy particles penetrating the Earth's atmosphere.

2. Method

We used the standard GEANT4 toolkit which is based on the Monte Carlo method for simulations of atmospheric interactions of high-energy charged particles. The GEANT4 includes several intranuclear cascade models (BIC and BERT) and cross section data in a wide energy range, which are continuously updated. As was shown in our previous paper[10] the BERT model results agree better with previous simulations of the ^{10}Be atmospheric production by GCRs, and the BIC model better fits actual estimates of ^{14}C global production rate. Production rates were calculated for 0-130 km altitudes with 1 km step with particle flux penetrating the Earth's atmosphere being isotropic. These yield functions were further convoluted with the SPE spectra taking into account geomagnetic rigidity cutoff.

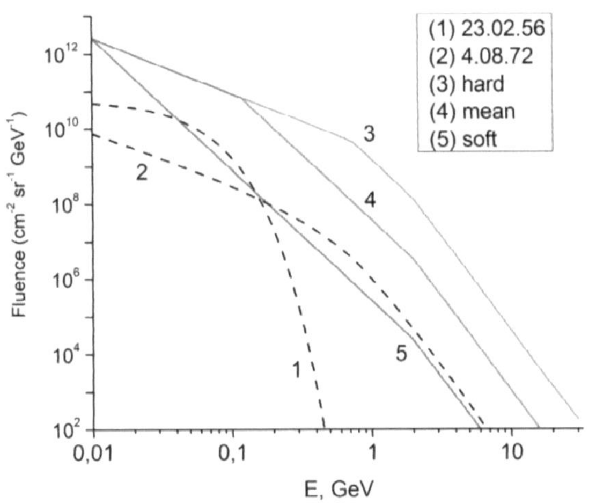

Fig. 1. Spectra of SPEs used for calculations.

3. Isotopic imprints of powerful SPEs

Under the term "isotopic imprint" of a certain event we shall understand the resulted peculiarity of isotopic composition fixed in terrestrial natural archives. To model possible imprints, caused by SPEs, it is necessary to assign the flux and energy spectra of protons. Direct measurements of solar accelerated particles started in the 50-s of the last century. Early observations were carried out with neutron monitors located on the ground and thus sensitive to the high-energy part of the spectra of penetrating atmosphere particles. Later on the low-energy part of solar particles was studied by the instruments deployed in outer space.

Atmospheric production rate of a nuclide is a function of the incident particle energy. It is known that for individual SPEs of a given total energy, injected into the atmosphere, the solar proton flux and energy spectra may differ significantly. Therefore, we present calculations of the nuclide production for a number of observed powerful flares with the hardest (23.Feb.1956) and the softest (04.Aug.1972) energy spectra (see, e.g., [11], [12], [13]) and also for three theoretical ultimate spectra (Fig. 1).

According to our simulations the required SPE fluence for the considered flares has to be higher by factors (α) presented in Table 1 in order to fit ^{14}C spike. Ratio of the global productions Q(^{14}C)/Q(^{10}Be) for the considered events is almost constant, and Q(^{10}Be)/Q(^{36}Cl) ratio rises by 1-14 times with the growth of the characteristic rigidity (or mean particle energy) of the event (Table 1).

Table 1. Modeled global production Q. Here α is the factor by which SPE fluence should be multiplied to account for the AD 775 ^{14}C excess signal.

SPE spectrum	1956	1972	mean	soft	hard	[14]
α	89	334	1.4	176	0.087	
^{10}Be, 10^6at/cm^2	4.9	3.5	4.8	4.4	4.8	0.9
^{36}Cl, 10^6 at/cm^2	0.46	3.4	0.79	2.4	0.34	0.3
$^{14}C/^{10}Be$	41	57	42	46	42	233
$^{10}Be/^{36}Cl$	11	1.0	6.0	1.8	14	2.9
^{10}Be excess	18	13	18	16	18	3.4
^{36}Cl excess	10	76	18	53	7.6	6.3
^{10}Be/GCR	5.9	4.2	5.8	5.3	5.8	
^{36}Cl/GCR	10	75	17	52	7.5	

As was shown in [10] the tropospheric ^{10}Be production by GCRs provides the observed deposition rate at central stations of Greenland and Antarctica (at the mean tropopause height 9-10 km). Also, it provides deposition variations during the 11-yrs solar cycle without significant contribution of the stratospheric ^{10}Be. One can see from Fig. 2 that its production in polar troposphere at mean tropopause height 9-10 km can account for the additional instant deposition of ^{10}Be for the AD 775 event assuming its spectrum close to that of the 1956 flare (or even harder). For the events similar to the 1972 flare the explanation is possible only under assumption of main contribution of the stratospheric ^{10}Be. Analogous inference for ^{36}Cl is impossible because even maximal tropopause height (12 km) does not result in the observed deposition rate for AD 775 at any particle spectra. The deposition rate pulse of ^{36}Cl in Central Greenland could be accounted for only at prevailing contribution of the stratospheric ^{36}Cl produced by superflare. Similar result was obtained for the GCR modeling [10].

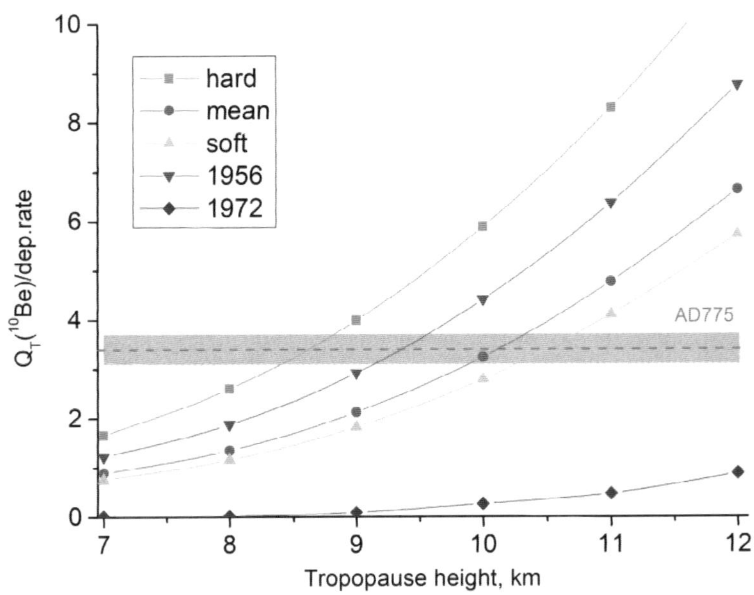

Fig. 2. Tropospheric production rate $Q_T(^{10}Be)$ (in experimental deposition rate units, see [14]) for different tropopause heights. Grey bar is the AD 775 ^{10}Be excess with errors based on data from [14].

4. Conclusions

The main results of the present paper are as follows:

1. To account for the AD 775 event by solar superflare the particle fluence has to be by 89 times larger than that of the 23.Feb.1956 event (hard spectrum) and 334 for the 04.Aug.1972 event (soft spectrum).

2. The $Q(^{14}C)/Q(^{10}Be)$ production ratio for SPE turns out to be 40-60 weakly dependent on the incident particle spectra and geomagnetic field if $0<M(t)/M_0<1.3$. At the same time the $Q(^{10}Be)/Q(^{36}Cl)$ ratio lies within the range 1-14 mainly depending on the SEP energy distribution.

3. The global production of ^{36}Cl for all simulated event spectra exceeds its deposition rate by 1.2-12 times for the AD 775 event.

4. The tropopause height and its variations after the event (AD 775) may significantly affect the amplitude of the signal in ^{10}Be. To explain the observed deposition rate for the AD 775 event the troposphere height must be in the range of 8.5-10.5 km.

Acknowledgments

A.K. Pavlov, D.A. Frolov and A.B. Struminsky were partially supported by the Russian Foundation for Basic Research (grant 16-02-00328).

References

[1] Lal D., Peters B. 1967 *Handbuch der Physik* **46/2** 551
[2] Beer J., McCracken K., Steiger R. 2012 *Cosmogenic radionuclides: theory and applications in the terrestrial and space environments*. Springer Science & Business Media
[3] Miyake F., Nagay K., Masuda K., Nakamura T. 2012 *Nature* **486** 240
[4] Usoskin I.G., Kromer B., Ludlow F. et al. 2013 *Astron. and Astrophys.* **552** L3
[5] Jull A.J.T., Panyushkina I.P., Lange T.E. et al. 2014 *Geophys. Res. Lett.* 41 doi:10.1002/2014GL059874
[6] Guttler D., Adolphi F., Beer J. et al. 2015 *Earth and Planet. Science Lett.* **411** 290
[7] Yi Liu, Zhao-feng Zhang, Zi-cheng Peng et al. 2014 *Scientific Reports* **4** 3728 doi:10.1038/srep03728
[8] Hambaryan V.V., Neuhäuser R. 2013 *MNRAS* **430** 32
[9] Pavlov A.K., Blinov A.V., Vasilyev G.I. et al. 2013 *Astronomy Lett.* **39** 571
[10] Pavlov A.K., Blinov A.V., Frolov D.A. et al. 2017 *J. of Atmosph. and Solar-Terrestr. Phys.* **164** 308
[11] Rojdev K., Atwell B., Tylka A.J., Dietrich W.F., Badavi F. 2011 *AIAA International Conference on Environmental Systems* AIAA5253
[12] Tylka A.J., Dietrich W.F. 2008 *37 COSPAR Scientific Assembly* D23-0003-08
[13] Usoskin I.G., Kovaltsov G.A., Mironova I.A., Tylka A.J., Dietrich W.F. 2011 *Atmos. Chem. Phys.* **11** 1979
[14] Mekhaldi F., Muscheler R., Adolphi F. et al. *Nat. Commun.* 2015 **6** 8611

IOP Conf. Series: Journal of Physics: Conf. Series **1038** (2018) 012009 doi:10.1088/1742-6596/1038/1/012009

Quasistationary fluid motions in magnetized neutron stars

D D Ofengeim, M E Gusakov, and E M Kantor
Ioffe Institute, 26 Politekhnicheskaya Street, St. Petersburg 194021, Russia
E-mail: ddofengeim@gmail.com

Abstract. In the recent work by Gusakov, Kantor & Ofengeim [Phys. Rev. D 96, 103012 (2017)] a new method to calculate the quasistationary evolution of magnetic field in the core of a neutron star was proposed. Here we further develop it, focusing on a simple case of neutron-proton-electron (npe) core composition with purely poloidal magnetic field B. We find that the meridional flow of the npe-fluid can be unexpectedly large. We estimate the typical timescale τ of the field evolution due to dragging of the magnetic field lines by this flow and show that τ can be as small as $\sim (100 - 1000)$ yr for magnetars with $B \sim 10^{16}$ G.

1. Introduction

One of the most intriguing problems of neutron star (NS) physics is understanding of interrelation between various classes of neutron stars with often very different surface magnetic fields. What are the factors that make their fields so distinct? A lot of effort has been devoted to answer this question. Clearly, the magnetic field at the surface should be related in some way to the magnetic field in the internal layers of NSs. Up until now the main body of research has been concentrated on the magnetic field evolution in the crust (e.g., [1]). The evolution in the core has received less attention partly because such an analysis is substantially more complicated, while the physics involved is not well understood, and partly because it is often believed (but see, e.g., [2]) that the typical timescales of the magnetic field evolution in the core are much larger than those in the crust (see, e.g., recent papers [2], [3], and [4] for different approaches to the problem). This work continues our studies of the magnetic field evolution in the NS core. Recently we formulated a self-consistent approach to this problem [5]. In the present work we further develop it and apply the ideas from [5] to the simplified case of normal (nonsuperfluid and nonsuperconducting) NS core, composed of only three particle species (neutrons n, protons p, and electrons e). The magnetic field in the core is taken to be axisymmetric and purely poloidal. Note that purely poloidal configuration of the magnetic field in the whole star is known to be unstable (e.g. [6]; [4] and refs. therein). Stable models seem to have mixed poloidal-toroidal structure (in principle, the toroidal component can be localized in the stellar crust). The formalism presented below can be applied to magnetic fields with toroidal component in the core, but such an analysis is beyond the scope of this short paper.

2. Evolution equations

Let $n_a(r)$ and $\mu_a(r)$ be the unperturbed (i.e., in the absence of the magnetic field) number densities and chemical potentials of particle species $a = n, p, e$, respectively. Since we assume that an unperturbed star is non-rotating, $n_a(r)$, $\mu_a(r)$ and other unperturbed quantities depend on the radial coordinate only (hereafter we use the spherical coordinates r, θ, φ with the origin at the stellar centre). These dependencies follow from the solution to the Tolman-Oppenheimer-Volkoff equations. For a degenerate matter of NS core we have $\sum_a \mu_a n_a = \varepsilon + P$, where $\varepsilon = \rho c^2$ and P are the energy density and pressure, respectively.

An applied magnetic field drives the system out of diffusive and beta-equilibrium, which disturbs chemical potentials, $\delta\mu_a \sim (B^2/P)\mu_a$, and induces motions of each particle species a with the velocities $\boldsymbol{u}_a$. For realistic internal fields, $B \lesssim 10^{16}$ G [7], we have $\delta\mu_a \ll \mu_a$. It is convenient to define a total flow velocity, $\boldsymbol{U} = \sum_a \mathscr{X}_a \boldsymbol{u}_a$, where $\mathscr{X}_a = \mu_a n_a/(\varepsilon + P)$, and the diffusion velocities $\boldsymbol{w}_a = \boldsymbol{u}_a - \boldsymbol{U}$. For an npe NS core $\mathscr{X}_n = n_n/n_b$, where $n_b = n_n + n_p$ is the total baryon number density, $\mathscr{X}_e = \hbar c(3\pi^2 n_e)^{1/3} n_e/(\varepsilon + P)$, and $\mathscr{X}_p = 1 - \mathscr{X}_n - \mathscr{X}_e$. For future convenience we also express the electric field $\boldsymbol{E}$ in the form: $\boldsymbol{E} = \boldsymbol{E}_{\rm com} - \boldsymbol{U} \times \boldsymbol{B}/c$, where $\boldsymbol{E}_{\rm com}$ is the electric field in the frame comoving with the fluid ($\boldsymbol{U} = 0$).

To simplify analysis of the magnetic field evolution in the NS core, we, following the papers [4, 5, 8], neglect the effects of General relativity, although the equation of state is assumed to be relativistic. We also assume that NS interiors are thermally relaxed (the evolution is sufficiently slow) so that the (redshifted) core temperature is uniform, $\widetilde{T} = T\sqrt{g_{00}}$, where g_{00} is the time component of the metric tensor, and T is the local temperature. Then, in the quasistationary approximation, the evolution equations can be written as [5]

$$e_a n_a \left(\boldsymbol{E}_{\rm com} + \frac{\boldsymbol{w}_a}{c} \times \boldsymbol{B}\right) = n_a \boldsymbol{\nabla}\delta\mu_a + \sum_b J_{ab}\left(\boldsymbol{w}_a - \boldsymbol{w}_b\right), \tag{1a}$$

$$\operatorname{div} n_n \left(\boldsymbol{U} + \boldsymbol{w}_n\right) = -\operatorname{div} n_p \left(\boldsymbol{U} + \boldsymbol{w}_p\right) = -\operatorname{div} n_e \left(\boldsymbol{U} + \boldsymbol{w}_e\right) = \lambda\Delta\mu, \tag{1b}$$

$$e n_e \left(\boldsymbol{w}_p - \boldsymbol{w}_e\right) = \frac{c}{4\pi} \operatorname{rot} \boldsymbol{B}, \tag{1c}$$

$$\frac{\partial \boldsymbol{B}}{\partial t} = -c \operatorname{rot} \boldsymbol{E}_{\rm com} + \operatorname{rot}(\boldsymbol{U} \times \boldsymbol{B}). \tag{1d}$$

Here $J_{ab}(\boldsymbol{w}_a - \boldsymbol{w}_b)$ is the friction force between the species a and b, and $\lambda\Delta\mu$ (with $\Delta\mu \equiv \delta\mu_p + \delta\mu_e - \delta\mu_n$) is the rate of non-equilibrium Urca processes, which is valid under assumption $\Delta\mu \ll kT$ (k is the Boltzmann constant)[1]. The coefficients J_{ab} and λ are taken from [9] and [10, 11], respectively. Generally, they depend on n_a and T and, therefore, on the radial coordinate r. Since $J_{np}, J_{ep} \gg J_{ne}$, the latter quantity is neglected below. For simplicity we neglect perturbations of gravitational force in equation (1a) (Cowling approximation). However, it could be easily accounted for if necessary [5]. Below we shall demonstrate how to find $\boldsymbol{U}$ and $\boldsymbol{E}_{\rm com}$ from equations (1a) — (1c) for a given $\boldsymbol{B}$. Knowledge of these quantities opens a possibility to self-consistently study the evolution of the magnetic field with time using the Faraday equation (1d).

The calculation of $\boldsymbol{U}$ and $\boldsymbol{E}_{\rm com}$ proceeds in 5 steps. The *first step* consists in obtaining $\Delta\mu$ and $\delta\mu_n$ from equation (1a), summed over the index a,

$$n_b \boldsymbol{\nabla}\delta\mu_n + n_e \boldsymbol{\nabla}\Delta\mu = \frac{1}{4\pi} \operatorname{rot} \boldsymbol{B} \times \boldsymbol{B} = \boldsymbol{f}_L, \tag{2}$$

where $\boldsymbol{f}_L$ is the Lorentz force density acting on the fluid. In an axisymmetric and poloidal magnetic field we are interested in here one has $f_{L\varphi} = 0$ and the solution to (2) can be written as [5]

$$\begin{pmatrix} \hat{\mathbf{P}}_l \Delta\mu \\ \hat{\mathbf{P}}_l \delta\mu_n \end{pmatrix} = \begin{pmatrix} n_e & n_b \\ n_e' & n_b' \end{pmatrix}^{-1} \begin{pmatrix} r\hat{\mathbf{P}}_l \int_0^\theta f_{L\theta} \mathrm{d}\theta \\ \partial_r \left(r\hat{\mathbf{P}}_l \int_0^\theta f_{L\theta} \mathrm{d}\theta\right) - \hat{\mathbf{P}}_l f_{Lr} \end{pmatrix}, \qquad l \geqslant 1, \tag{3}$$

[1] As shown in [5], the case $\Delta\mu \gtrsim kT$ can be realized only for superstrong magnetic fields $> 10^{16}$ G, which are not considered here.

where the operator $\hat{\mathbf{P}}_l(\cdot)$ extracts the l'th Legendre component of the function $(\cdot)$, i.e. $\hat{\mathbf{P}}_l(\cdot) = (2l+1)/2\int_0^\pi(\cdot)P_l(\cos\theta)\sin\theta\mathrm{d}\theta$, where P_l is the l'th Legendre polynomial. Note that the components with $l \geqslant 1$ are independent of the neutron star temperature $\widetilde{T}$. Note also that the components $\hat{\mathbf{P}}_0\Delta\mu$ and $\hat{\mathbf{P}}_0\delta\mu_n$ are still left undetermined.

The *second step* consists in expressing $\boldsymbol{w}_a$ and $\boldsymbol{E}_{\rm com}$ from equations (1a) for neutrons and electrons, equation (1c), and the condition $\sum_a \mathscr{X}_a\boldsymbol{w}_a = 0$, following from the definitions of $\boldsymbol{w}_a$ and $\boldsymbol{U}$. For the diffusion velocities the result is

$$\boldsymbol{w}_n = -(1-\mathscr{X}_n)\frac{n_n}{J_{np}}\boldsymbol{\nabla}\delta\mu_n + \frac{c}{4\pi e n_e}\mathscr{X}_e\,\mathrm{rot}\,\boldsymbol{B}, \tag{4a}$$

$$\boldsymbol{w}_p = \mathscr{X}_n\frac{n_n}{J_{np}}\boldsymbol{\nabla}\delta\mu_n + \frac{c}{4\pi e n_e}\mathscr{X}_e\,\mathrm{rot}\,\boldsymbol{B}, \tag{4b}$$

$$\boldsymbol{w}_e = \mathscr{X}_n\frac{n_n}{J_{np}}\boldsymbol{\nabla}\delta\mu_n - \frac{c}{4\pi e n_e}(1-\mathscr{X}_e)\,\mathrm{rot}\,\boldsymbol{B}, \tag{4c}$$

while for the electric field

$$\boldsymbol{E}_{\rm com} = -\frac{\boldsymbol{\nabla}\delta\mu_e}{e} + \frac{cJ_{ep}}{4\pi e^2 n_e^2}\mathrm{rot}\,\boldsymbol{B} - \frac{\mathscr{X}_n n_n}{cJ_{np}}\boldsymbol{\nabla}\delta\mu_n\times\boldsymbol{B} + \frac{1-\mathscr{X}_e}{4\pi e n_e}\mathrm{rot}\,\boldsymbol{B}\times\boldsymbol{B}, \tag{5}$$

where $\delta\mu_e$ is a function of $\delta\mu_n$ and $\Delta\mu$, and is specified by choosing equation of state.

In the *third step* we derive the poloidal components of $\boldsymbol{U}$ from the continuity equations (1b) for, e.g., neutrons and protons. For poloidal $\boldsymbol{B}$ the terms with $\mathrm{rot}\,\boldsymbol{B}$ in equations (4) have azimuthal components only and due to axisymmetry of the system they do not contribute in (1b). This leads to the relation $\mathrm{div}\,\boldsymbol{U} = -(n_b'/n_b)U_r$, where the prime $'$ means d/dr. Then, employing the method from Sec. III C 1 of [5], we find

$$U_r = \frac{n_b}{n_b' n_e - n_e' n_b}\left[\mathrm{div}\left(\mathscr{X}_n\frac{n_e n_n}{J_{np}}\boldsymbol{\nabla}\delta\mu_n\right) + \lambda\Delta\mu\right], \tag{6a}$$

$$U_\theta = -\frac{1}{r n_b\sin\theta}\frac{\partial}{\partial r}\left(r^2 n_b\int_0^\theta U_r(r,\theta')\sin\theta'\mathrm{d}\theta'\right) + \frac{\xi(r)}{\sin\theta}, \tag{6b}$$

where $\xi(r)$ is an arbitrary function.

The *fourth step* consists in accounting for the fact that $U_\theta(\theta=0)=0$ and $U_\theta(\theta=\pi)=0$ for any r. The first condition is satisfied with $\xi=0$; the second one is equivalent to $\hat{\mathbf{P}}_0 U_r = 0$, or

$$\frac{1}{r^2}\frac{\mathrm{d}}{\mathrm{d}r}\left(r^2\mathscr{X}_n\frac{n_e n_n}{J_{np}}\frac{\mathrm{d}}{\mathrm{d}r}\hat{\mathbf{P}}_0\delta\mu_n\right) + \lambda\hat{\mathbf{P}}_0\Delta\mu = 0. \tag{7}$$

Using this equation together with the 0-th Legendre component of equation (2),

$$n_b(\hat{\mathbf{P}}_0\delta\mu_n)' + n_e(\hat{\mathbf{P}}_0\Delta\mu)' = \hat{\mathbf{P}}_0 f_{Lr}, \tag{8}$$

one can calculate $\hat{\mathbf{P}}_0\delta\mu_n$ and $\hat{\mathbf{P}}_0\Delta\mu$, and hence fully determine the quantities $\delta\mu_n$ and $\Delta\mu$ (see step 1). However, in order to solve equations (7) and (8) one has to specify three boundary conditions.

One of these conditions, following from the requirement of finiteness of the first term in (7), is $(\hat{\mathbf{P}}_0\delta\mu_n)'\big|_{r=0} = 0$. This condition is equivalent to $(\hat{\mathbf{P}}_0\Delta\mu)'\big|_{r=0} = 0$, since $\hat{\mathbf{P}}_0 f_{Lr}\big|_{r=0} = 0$ due to the symmetry of the system. The other two conditions should be the values of $\hat{\mathbf{P}}_0\Delta\mu$ and $\hat{\mathbf{P}}_0\delta\mu_n$ at some point in the star. Possible options for these conditions are discussed in the end of Appendix D in [5].

However, in the two limiting (yet important) cases equations (7) and (8) can be easily solved. The first is the case of low temperatures, when the second term in (7) can be neglected. Then $\hat{\mathbf{P}}_0\delta\mu_n \approx C_1$ and $\hat{\mathbf{P}}_0\Delta\mu \approx C_2 + \int_0^r \hat{\mathbf{P}}_0 f_{Lr}/n_e \mathrm{d}r$, where C_1 and C_2 are some constants. The opposite case [large temperatures; the first term in (7) can be neglected] results in $\hat{\mathbf{P}}_0\Delta\mu \approx C_1 + \int_0^r \hat{\mathbf{P}}_0 f_{Lr}/n_b \mathrm{d}r$ and $\hat{\mathbf{P}}_0\Delta\mu \approx 0$. As shown in [5], at intermediate temperatures, when both terms in equation (7) are equally important, dissipation of magnetic field due to ambipolar diffusion (driven by np friction) and due to non-equilibrium Urca processes are close to one another. According to figures 1 and 2 of [5] this occurs at the redshifted temperature $\widetilde{T} \sim (0.4 - 0.9) \times 10^8$ K, if the direct Urca process is allowed and at $\widetilde{T} \sim (5 - 10) \times 10^8$ K if it is forbidden, and substituted by the next most powerful beta-process – the modified Urca one. Since λ and J_{np} rapidly vary with $\widetilde{T}$, the range of temperatures, where both limiting solutions are inaccurate, is thin. For instance, when the direct Urca is forbidden, the low-temperature limit is already valid at $\widetilde{T} \lesssim 3 \times 10^8$ K. Similarly, the high-temperature limiting solution can be safely used at $\widetilde{T} \gtrsim 2 \times 10^9$ K.

The *fifth step* consists in solving the problem of U_φ that is easy for poloidal magnetic field. Below, let us distinguish the poloidal and toroidal parts of vectors with the superscripts (p) and (t), respectively. Then $\boldsymbol{U}^{(t)} = U_\varphi \boldsymbol{e}_\varphi$, and we already know $\boldsymbol{U}^{(p)}$, $\boldsymbol{E}^{(p)}_{\rm com}$, and $\boldsymbol{E}^{(t)}_{\rm com}$ from the steps 2 — 4. In case of $\boldsymbol{B}^{(t)} = 0$ equation (1d) splits into two equations,

$$\frac{\partial \boldsymbol{B}^{(p)}}{\partial t} = \mathrm{rot}\left(\boldsymbol{U}^{(p)} \times \boldsymbol{B}^{(p)} - c\boldsymbol{E}^{(t)}_{\rm com}\right), \qquad \frac{\partial \boldsymbol{B}^{(t)}}{\partial t} = \mathrm{rot}\left(\boldsymbol{U}^{(t)} \times \boldsymbol{B}^{(p)} - c\boldsymbol{E}^{(p)}_{\rm com}\right). \tag{9}$$

From the second of these equations we see that if, initially, $\boldsymbol{B}^{(t)} = 0$, it will remain so if $\boldsymbol{U}^{(t)} = -c(\boldsymbol{E}^{(p)}_{\rm com} + \boldsymbol{\nabla}\chi) \times \boldsymbol{B}^{(p)}/(\boldsymbol{B}^{(p)})^2$, where $\chi(r,\theta)$ is an arbitrary scalar function. Note that, as follows from (9), the evolution of the magnetic field $\boldsymbol{B} = \boldsymbol{B}^{(p)}$ in this case is independent of the toroidal velocity of the flow. Thus, applying steps 1 — 5 to a given poloidal field $\boldsymbol{B}(r,\theta)$, we can find $\partial_t \boldsymbol{B}$ and hence follow the evolution of the magnetic field with time.

However, numerical calculations (e.g., [6]; see also [4] and references therein) indicate that pure poloidal field is unstable. In the more general case of mixed poloidal-toroidal axisymmetric configuration of the magnetic field, we can derive an equation for U_ϕ from the condition $\partial_t f_{L\varphi} = 0$ [5]. Solving it, we can completely determine $\boldsymbol{U}$, so that $\partial_t \boldsymbol{B}$ could again be found from (1d).

3. Numerical example

To illustrate the results of the previous section, we consider a neutron star with the HHJ equation of state [12] and mass $M = 1.4\ \mathrm{M}_\odot$, that gives the circumferential radius $R = 12.2$ km. All the non-perturbed quantities like $n_a(r)$, $\varepsilon(r)$, etc. are obtained from the solution of the Tolman-Oppenheimer-Volkoff equations. The direct Urca process is forbidden in the whole NS volume.

We use a simple analytic model for the poloidal field $\boldsymbol{B}$ taken from [4] (their model A):

$$\boldsymbol{B} = \frac{\boldsymbol{\nabla}\Psi \times \boldsymbol{e}_\varphi}{r\sin\theta}, \qquad \Psi = B_{\rm max} R^2 \left(\frac{r^2}{2R^2} - \frac{3r^4}{5R^4} + \frac{3r^6}{14R^6}\right)\sin^2\theta. \tag{10}$$

Here Ψ is a magnetic flux function, $B_{\rm max}$ is the maximum value of the magnetic field, which is achieved in the NS centre. It is proportional to the magnetic field at the pole, $B_p = (8/35)B_{\rm max}$. It is the simplest model of a dipole poloidal field which is matched with a vacuum dipole magnetic field outside the star. Its configuration, as well as the direction of the Lorentz force density are shown in figure 1a.

Figures 1b,c present $\Delta\mu$ and $\delta\mu_n$ in units of $k\widetilde{T}$, derived from equation (3) adopting the low-temperature approximation to the solution of equations (7) and (8). This is a good approximation for a chosen $\widetilde{T} = 2 \times 10^8$ K. To plot the figures, we assume $B_p = 10^{14}$ G

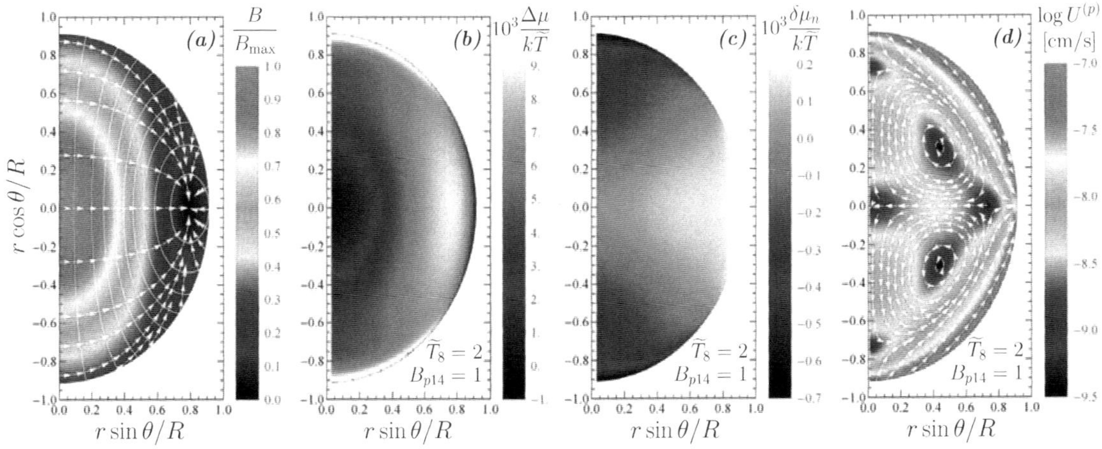

Figure 1. Properties of the NS model considered in the text. All figures show a meridional cross-section of the NS core; on each plane the left vertical axis is the symmetry axis. *(a)* The magnetic field model (equation 10). Different colors denote strength of the magnetic field in units of $B_{\rm max}$. Thin white lines are the field lines, arrows show direction of $\boldsymbol{f}_L$. *(b)* and *(c)* The ratios $\Delta\mu/k\widetilde{T}$ and $\delta\mu_n/k\widetilde{T}$ for $\widetilde{T}=2\times10^8$ K and $B_p=10^{14}$ G calculated in the low-temperature approximation (it is assumed that the direct Urca process is forbidden). *(d)* Poloidal velocity of the flow $\boldsymbol{U}^{(p)}$ for the same conditions. Arrows show direction of $\boldsymbol{U}^{(p)}$, its magnitude is shown by colors in $\log_{10}$ scale.

(corresponding to $B_{\rm max}\approx 4.4\times10^{14}$ G), which is a typical magnetar field [13]. Inspection of the figures reveals that, typically, $\Delta\mu\sim(n_b/n_e)\delta\mu_n$.

The ratio $\Delta\mu/kT$ was also calculated in [4] for a very similar NS model. The results of [4] are presented in the left panel of their figure 2 and differ substantially from those shown in figure 1b. The reason for the discrepancy is some approximations made in [4], which, as we argue in Appendix B of [5], are incorrect.

Figure 1d is the most interesting and it displays the poloidal flow velocity $\boldsymbol{U}^{(p)}$ derived from equations (6). The same friction coefficient J_{np} as in [4] and [5] is employed. While one could expect such a geometry of the flow for a dipole field $\boldsymbol{B}$ with the equatorial symmetry (e.g., a similar configuration was proposed in [14]), the typical velocities appear to be surprisingly large, and they grow rapidly near the crust-core interface. The reason for this behavior is easy to understand. According to (6), U_r and U_θ contain a number of terms depending on the derivatives $n_a'(r)$, $n_a''(r)$, $n_a'''(r)$, and $n_a''''(r)$. Near the crust-core interface, one has $|n_a^{(k+1)}(r)|\gtrsim(10-15)|n_a^{(k)}(r)/R|$ for $k=0...3$, where a superscript (k) denotes k'th derivative with respect to r. As a result, for a chosen fiducial values of B_p and $\widetilde{T}$ one has $U^{(p)}\gtrsim(10-100)w_a$, because the diffusion velocities depend on the first and second derivatives $n_a'(r)$ and $n_a''(r)$ only. An actual numerical value of the ratio $U^{(p)}/w_a$ depends on B_p and $\widetilde{T}$, but for $B_p\gtrsim10^{13}$ G the flow velocity $U^{(p)}$ is always much larger than w_a. The second interesting observation following from the discussion above is that $U^{(p)}$ is not only very large near the crust-core interface but this property is, to a large extent, independent of an equation of state employed (derivatives of $n_a(r)$ grow in the vicinity of the crust for any equation of state), as well as on the configuration of the magnetic field (at least, for sufficiently large $\boldsymbol{B}$). Implication of these results is discussed in the next section.

4. Discussion and conclusions

The self-consistent method of calculation of $\partial_t\boldsymbol{B}$ for a given $\boldsymbol{B}$ in the quasistationarity approximation, was formulated in [5]. In this work it is further developed and applied to study non-equilibrium processes induced in npe NS core by applied poloidal magnetic field. One of our main results is the analytic expression (6) for the meridional flow velocity $\boldsymbol{U}^{(p)}$. It is driven by frictional entrainment of neutrons by protons that interact with the magnetic field. The net flow of npe-fluid appears to be substantially larger than the diffusion flows of individual particle species. This result is of conceptual importance since it contradicts a standard point of view that the matter of neutron star cores, which is mainly composed of neutral particles, is (roughly) at rest in magnetized neutron stars (e.g., [4, 8]).

Comparing the terms $-c\,\mathrm{rot}\,\boldsymbol{E}_{\rm com}$ and $\mathrm{rot}\,[\boldsymbol{U}\times\boldsymbol{B}]$ in the Faraday equation (1d), and using the expression (5) for $\boldsymbol{E}_{\rm com}$, one can demonstrate that $c\,|\mathrm{rot}\,\boldsymbol{E}_{\rm com}| \ll |\mathrm{rot}\,[\boldsymbol{U}\times\boldsymbol{B}]|$ for $B \gtrsim 10^{13}$ G, provided that $w_a \ll U$. This means that a sufficiently strong magnetic field is frozen in the npe-fluid to a good approximation.

Let us consider several scaling relations for the quantities derived above. From equation (2) we obtain $\Delta\mu$, $\delta\mu_n \propto B^2$, where B is some typical value of the magnetic field in the core (e.g., $B_{\rm max}$). The n-p friction coefficient $J_{np} \propto \widetilde{T}^2$ and $\lambda \propto \widetilde{T}^6$ for modified Urca process (see [5], section VI). Then we have $U \propto B^2/\widetilde{T}^2$ in the low-temperature limit and $U \propto B^2\widetilde{T}^6$ in the high-temperature limit. Defining now the characteristic timescale of the field evolution as $\tau \sim B/\partial_t B \sim R/U$ and taking into account that the velocity U near the crust-core interface can be as large as $\sim 10^{-7}$ cm/s for $B \sim B_{\rm max} \approx 4.4\times 10^{14}$ G and $\widetilde{T} \sim 2\times 10^8$ K, we get the following estimate, valid in the low-temperature case

$$\tau \gtrsim (1-10)\,\mathrm{Myr}\left(\frac{\widetilde{T}_8}{B_{14}}\right)^2, \tag{11}$$

where $B_{14} = B/(10^{14}\,\mathrm{G})$ and $\widetilde{T}_8 = \widetilde{T}/(10^8\,\mathrm{K})$. Note that this is a timescale of nondissipative transformation of the magnetic field, in contrast to the timescale $\tau_{\rm B}$, defined in [5], which is purely dissipative. As one can see from this estimate, the typical evolution timescale for magnetars with the highest interior field $B \sim 10^{16}$ G [7] is of the order of $\sim 100 - 1000$ yr.

In the end let us mention that the approach to quasistationary evolution of the magnetic field developed here, can be extended in a number of ways. In particular, it can be generalized to include various additional particle species (e.g., muons and hyperons), baryon pairing, and non-axisymmetric magnetic field [5].

Acknowledgments

This work is supported by Russian Science Foundation (grant No. 14-12-00316). The authors are grateful to P S Shternin for criticism and to D G Yakovlev for encouragement.

References

[1] Viganò D, Rea N, Pons J A, Perna R, Aguilera D N and Miralles J A 2013 *MNRAS* **434** 123–141
[2] Bransgrove A, Levin Y and Beloborodov A 2018 *MNRAS* **473** 2771–2790
[3] Castillo F, Reisenegger A and Valdivia J A 2017 *MNRAS* **471** 507–522
[4] Passamonti A, Akgün T, Pons J A and Miralles J A 2017 *MNRAS* **465** 3416–3428
[5] Gusakov M E, Kantor E M and Ofengeim D D 2017 *Phys. Rev. D* **96**(10) 103012
[6] Braithwaite J and Spruit H C 2006 *A&A* **450** 1097–1106
[7] Beloborodov A M and Li X 2016 *ApJ* **833** 261
[8] Goldreich P and Reisenegger A 1992 *ApJ* **395** 250–258
[9] Yakovlev D G and Shalybkov D A 1990 *Soviet Astronomy Letters* **16** 86
[10] Reisenegger A 1995 *ApJ* **442** 749–757
[11] Yakovlev D G, Kaminker A D, Gnedin O Y and Haensel P 2001 *Phys. Rep.* **354** 1–155
[12] Heiselberg H and Hjorth-Jensen M 1999 *ApJL* **525** L45–L48
[13] Mereghetti S, Pons J A and Melatos A 2015 *Sp. Sci. Rev.* **191** 315–338
[14] Urpin V A and Ray A 1994 *MNRAS* **267** 1000

IOP Conf. Series: Journal of Physics: Conf. Series **1038** (2018) 012010 doi:10.1088/1742-6596/1038/1/012010

Deep optical observations of the gamma-ray pulsar J2055+2539 with the GTC

D M Beronya[1], Yu A Shibanov[1], D A Zyuzin[1] and S V Zharikov[2]

[1] Ioffe Institute, Politekhnicheskaya 26, St. Petersburg, 194021, Russia

[2] Instituto de Astronomía, Universidad Nacional Autónoma de México, Apdo. Postal 877, Ensenada, Baja California, 22800, México

E-mail: daria.beronya@gmail.com

Abstract. The report gives a broad outline of deep, down to a 28 magnitude limit, optical observations of the γ-ray pulsar J2055+2539 (total exposure time $\approx$ 7.9 ks), which were obtained with the Gran Telescopio Canarias in the g'-band. To derive the most precise pulsar position we utilized recent *Chandra* observations, where the pulsar X-ray counterpart is detected with high significance. Using two reference stars detected both in the optical and X-rays we improved the absolute astrometry accuracy for the *Chandra* image of the PSR J2055+2539 field. The resulting *Chandra* pulsar position error circle of about $0\farcs2$ is compatible with the 1σ *Fermi* position of the pulsar ($0\farcs5$), obtained from timing analysis.

No optical counterpart was detected within the refined pulsar error circle. However, we managed to correct a shallow upper limit on the optical flux of the pulsar, which had been derived from preliminary observations. The new value $g' = 26.6$. It is brighter than the limit stated above due to the presence of a nearby star, $0\farcs8$ from the pulsar, with $g' = 20.3$. The inspection of the pulsar multiwavelength spectrum suggests a break between the optical and X-rays.

1. Introduction

PSR J2055+2539 (hereafter, J2055) was discovered in a blind frequency search using data from the *Fermi* Large Area Telescope (LAT) in 2009 [1]. The pulsar is located high above the Galactic plane ($b = -12\fdg52$), that implies a low interstellar absorption and thereby facilitates the observations in other spectral wavelengths. Based on the timing analysis of the *Fermi* LAT data it was determined that the pulsar has a spin period P of 0.32 s and a spin period derivative $\dot{P}$ of 4 $\times$ 10^{-11}. Thus, the pulsar seems relatively old with a characteristic age $\tau = 0.5P\dot{P}^{-1}$ of 1.24 Myr. J2055 is one of the least energetic *Fermi* LAT pulsars with a spin-down energy loss rate $\dot{E} = 5 \times 10^{33}$ erg s^{-1}. However, it makes the list of the 100 brightest γ-ray sources with the observed γ-ray energy flux of about 5.5×10^{-11} erg cm^{-2} s^{-1} [2]. The pulsar distance of 0.58 kpc can be estimated from γ-ray pseudo-distance relation [1]. That all makes this source appealing in terms of multiwavelength search.

Shortly after the discovery the follow-up observations were carried out using the *Arecibo* radio telescope at 327 MHz, however no pulsed emission have been detected [3]. Short-exposure data from XMM-*Newton*, obtained in 2009 and 2010, enabled to identify the pulsar with the X-ray source XMMU J205549.4+253959 [1]. Additional XMM-*Newton* observations, performed in 2013 for a total time of 135.2 ks, revealed a bright cometlike tail extending to the south of the pulsar (figure 1) [3]. In 2015, the high spatial resolution observations of J2055 were performed with the ACIS instrument on board the *Chandra* observatory (figure 2).

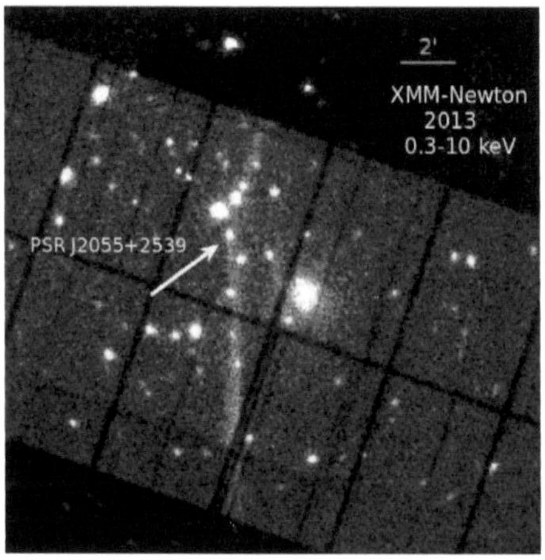

Figure 1. The X-ray image of the PSR J2055+2539 field resulting from merging all XMM-*Newton* observations [3].

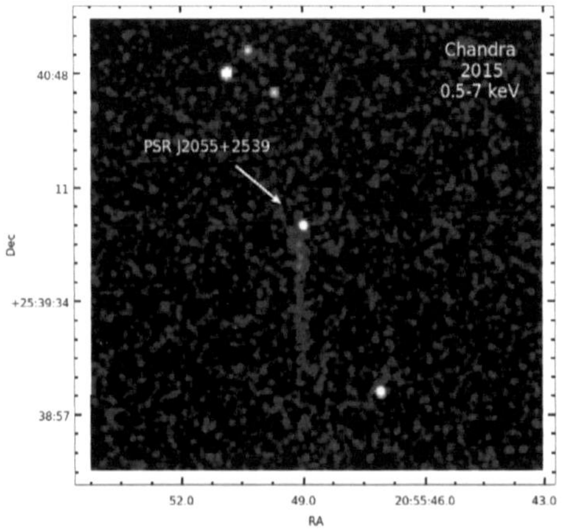

Figure 2. 2$'$.5 × 2$'$.4 image fragment of the PSR J2055+2539 field, derived from the archival data of 96.8 ks *Chandra* observations (ObsID 16957).

The presence of the cometlike tail gave a reason to suspect that the pulsar velocity is high enough to produce a detectable H_α bow shock nebula around the pulsar. In view of that possibility optical observations are particularly valuable. In 2013 we conducted the preliminary optical search for J2055 and its nebula with the *BTA* using broadband V, R_c and narrowband H_α filters. Neither the pulsar, nor the bow shock was detected [4]. The derived optical flux upper limits were close to a magnitude of 23 ($V = 22.8$, $R_c = 23.2$).

2. Observations of the PSR J2055+2539 field with the GTC

We undertook deep optical observations with the Gran Telescopio Canarias (*GTC*) for a more detailed inspection of the J2055 vicinity. The data were obtained in three *GTC* observations, executed on 16, 19 and 24 August, 2014. The pulsar field was imaged in the Sloan g$'$ band ($\lambda_C = 481.5$ nm, FWHM = 153 nm) with the Optical System for Imaging and low Resolution Integrated Spectroscopy (OSIRIS). Nine exposures of 291 s each and one short frame of 10 s were taken on every night of the observations. In total, about 8 ks of time were obtained for the target, with an average seeing of 0$''$.7.

Standard processing steps such as bias subtraction, flatfielding, image aligning and stacking were done using `IRAF` packages. The resulting image is depicted in figure 3. To perform the astrometric calibration of the resulting image we used 12 reference stars included in the USNO-B1.0 catalogue and then carried out the fit by means of `ccmap/ccsetwcs` algorithms. For the 7 most reliable stars left after iterative removal of outliers, we obtained 1σ rms errors of 0$''$.08 and 0$''$.09 in RA and Dec, respectively. Taking into account the nominal USNO accuracy of 0$''$.2, we found the resulting error of the astrometric calibration of about 0$''$.23. Combining this value with the XMM-*Newton* or *Fermi* LAT position uncertainties, we estimated the 3σ pulsar position errors in the *GTC* image of 5$''$.4 for the former instrument and 1$''$.65 for the latter one. Both error circles are shown in figure 3.

High spatial resolution of the *Chandra* image enabled us to strongly constrain J2055 position. The nominal accuracy of the *Chandra* astrometry is 0$''$.6. In order to correct the *Chandra* absolute astrometry we chose two optical sources J205550.31+254055.5 and J205550.85+254048.2, contained both in the SDSS7 catalogue (the first one is also known as USNO-B1.0 1156-0496126). These objects were identified with two X-ray sources (RA 20:55:50.3, Dec +25:40:55 and RA 20:55:50.8, Dec +25:40:48, respectively) in the *Chandra* image. We used a couple of mutually aligned *GTC* images of 10 s and 600 s to obtain more accurate reference stars coordinates. The referencing between the *Chandra* and *GTC* images was performed with the `celldetect/wcs_match` tools from the `CIAO` package. Considering the distance of the X-ray counterparts from the ACIS aimpoint of $1'$, we neglected any distortion effects. The total astrometry errors and the improved WCS coordinates of the pulsar are given in the third column of table 1. The resulting 3σ *Chandra* error ellipse is shown in figure 3.

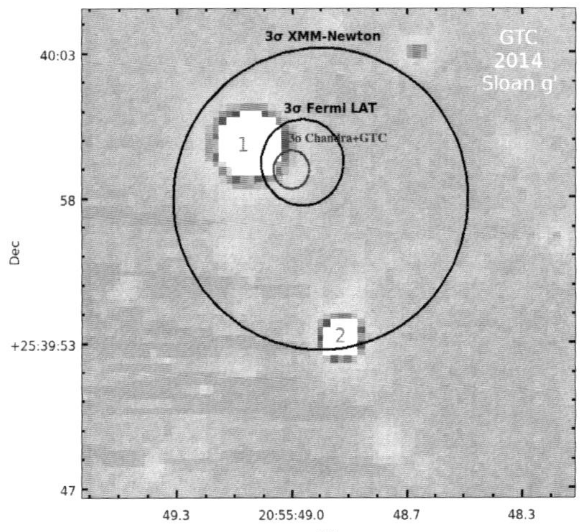

Figure 3. $0'.3 \times 0'.3$ image fragment of the J2055 field, observed with the *GTC*. The black circles indicate the 3σ pulsar position uncertainties computed from the XMM-*Newton* observations and *Fermi* timing analysis. The blue ellipse marks the J2055 3σ position error, derived from *Chandra* absolute astrometry correction.

Table 1. XMM-*Newton*, *Fermi* and *Chandra* (after astrometric correction using the *GTC* images) positions of J2055. 1σ errors are taken in brackets.

	XMM-*Newton*	*Fermi* LAT	*Chandra*+*GTC*
RA_{J2000}	$20^{\mathrm{h}}55^{\mathrm{m}}48^{\mathrm{s}}.89(1''.8)$	$20^{\mathrm{h}}55^{\mathrm{m}}48^{\mathrm{s}}.94(0''.5)$	$20^{\mathrm{h}}55^{\mathrm{m}}48^{\mathrm{s}}.97(0''.22)$
Dec_{J2000}	$+25^{\circ}39'57''.8(1''.8)$	$+25^{\circ}39'59''.1(0''.5)$	$+25^{\circ}39'58''.85(0''.23)$

We found out that the corrected 3σ *Chandra* error ellipse region is contaminated by a nearby bright star $0''.8$ from the pulsar position (object 1 in figure 3). No reliable counterpart for J2055 was revealed in that ellipse. To provide a detection limit measurements we performed the photometric calibration using the standard star from the SA 109 field [5], observed at the third night of the *GTC* observations. The photometric zeropoint is 28.70 ± 0.05. Then we derived the 3σ detection limit [6] in the g'-band of 26.6 magnitude. We also conducted the aperture photometry for two bright stars within the XMM-*Newton* 3σ error circle, marked 1 and 2 in figure 3. For these objects the magnitudes of 20.3 and 22.3 respectively are consistent with SDSS7 measurements.

3. Multiwavelength spectrum of PSR J2055+2539

The XMM-*Newton* spectral data of J2055 are fitted well by an absorbed power law model, resulting in the absorbing column density N_H of $(2.18 \pm 0.26) \times 10^{21}$ cm^{-2} and the photon index $\Gamma = 2.36 \pm 0.14$ [3]. To include the derived upper limit on the J2055 optical brightness in multiwavelength spectral analysis we carried out a magnitude to flux conversion, correcting it for the effects of interstellar absorption [7], using relation between N_H and the optical extinction A_V from [8], and then taking into account $AB_{off} = 0$ for the SDSS photometric system [9]. We obtained the conservative A_V of 1.5. Exponential cut-off power law is an appropriate model for the *Fermi* data [2]. Figure 4 illustrates the unabsorbed multiwavelength spectrum of J2055. The dereddened *GTC* upper limit of 0.47 μJy confirms the suggestion of the spectral break between the optical and X-rays, which is typical for rotation powered pulsars [4].

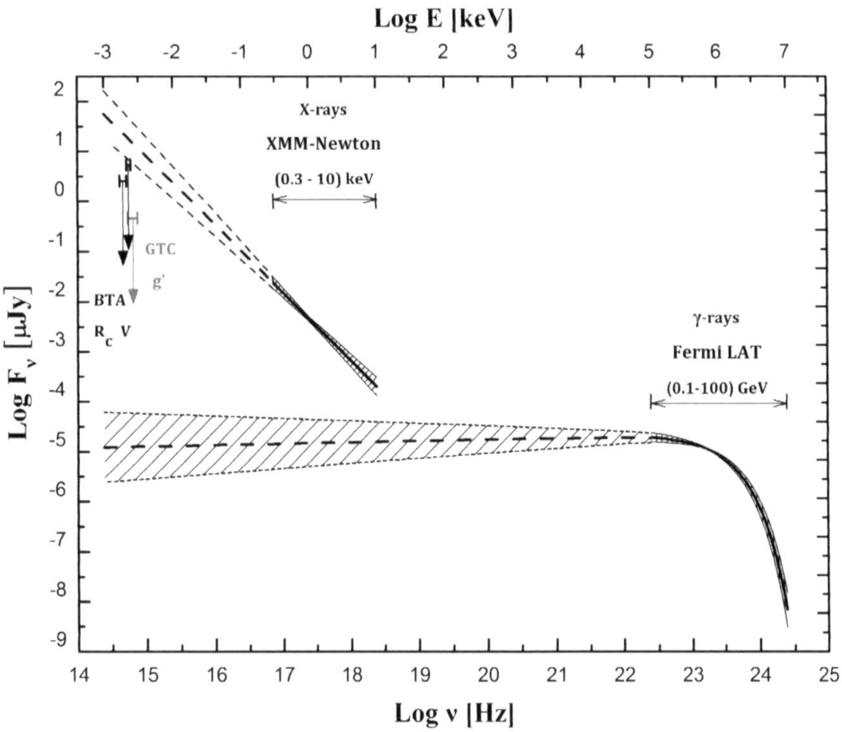

Figure 4. Unabsorbed spectrum of PSR J2055+2539. The instrument detection ranges are marked by horizontal bars. Solid lines indicate the best spectral fits of the XMM-*Newton* and *Fermi* LAT data, extrapolated toward the optical range with dashed lines. Shaded regions correspond to 1σ uncertainties of fitting parameters. Horizontal bars with arrows mark the 3σ detection limits according to the *GTC* (the red one) and *BTA* data (the black ones).

4. Conclusions

We conducted the deep, down to $g'_{lim} = 28$, optical observations of PSR J2055+2539 with the *GTC*. We performed the correction of the *Chandra* absolute astromerty using two reference stars, detected both in the optical and X-rays. This improvement resulted in the most precise pulsar position with the 1σ uncertainty of roughly $0''.2$. However, within the corresponding 3σ error ellipse there is no reliable counterpart of J2055 in the *GTC* image. The conservative detection limit on the pulsar emission $g' = 26.6$. It is brighter than the limit stated above due to the presence of the nearby relatively bright star with $g' = 20.3$, located $0''.8$ from the pulsar. The multiwavelength spectrum including this upper limit confirms the spectral break between the optical and X-rays, that was suggested in [4]. Further optical searches for the pulsar require a high spatial resolution which can be provided using either ground-based telescopes equipped with adaptive optics or the *HST*.

Acknowledgments
DMB and DAZ were partially supported by the RF Presidential Programme MK-2566.2017.2. The work is based on observations made with the Gran Telescopio Canarias (*GTC*), located at the Spanish Observatorio del Roque de los Muchachos of the Instituto de Astrofísica de Canarias, on the island of La Palma.

References
[1] Saz Parkinson et al. 2010 Eight γ-ray pulsars discovered in blind frequency searches of Fermi LAT data *ApJ* **725** 571
[2] Acero et al. 2015 Fermi Large Area Telescope Third Source Catalog *ApJS* **218** 23
[3] Marelli et al. 2016 The tale of the two tails of the oldish PSR J2055+2539 *ApJ* **819** 40
[4] Beronya et al. 2015 Search for the optical counterparts of the γ-ray pulsars J2055+2539, J2043+2740, J1957+5033 *JPCS* **661** 1 012001
[5] Landolt A U 1992 UBVRI photometric standard stars in the magnitude range 11.5-16.0 around the celestial equator *AJ* **104** 340
[6] Zharikov S and Mignani R P 2013 On the PSR B1133+16 optical counterpart *MNRAS* **435** 2227
[7] Cardelli et al. 1989 The Relationship Between Infrared, Optical And Ultraviolet Extinction *ApJ* **345** 245
[8] Tian et al. 2013 The Galactic NH - AV Relation and its Application to Historical Galactic SNRs *NCS* **1** 291 (*Preprint* 1306.6506)
[9] Frei Z and Gunn J E 1994 Generating Colors and k Corrections From Existing Catalog Data *ApJ* **108** 4

IOP Conf. Series: Journal of Physics: Conf. Series **1038** (2018) 012011 doi:10.1088/1742-6596/1038/1/012011

The influence of baryon-photon ratio on 21 and 92 cm brightness temperature

D N Kosenko[1] and A V Ivanchik[1]
[1]Ioffe Institute, Politekhnicheskaya 26, St.-Petersburg, 194021, Russia
E-mail: kosenkodn@yandex.ru

Abstract.
An epoch before first stars and galaxies formation, so-called cosmic Dark Ages, still remains unobserved. One of the possibilities for observation phenomena took place during Dark Ages is studying hyperfine transition of such elements as hydrogen and deuterium. In this article there were examined physical processes in the Universe after recombination and evolution of 21 and 92 cm brightness temperature and studied influence of baryon-photon ratio on this radiation.

1. Introduction
Cosmic Dark Ages is an epoch after recombination (380 000 years after Big Bang [1]) lasted to first stars and galaxies formation. Imaging Universe during this period is difficult because, firstly there are no bright galaxies or quasars. Secondly barionic matter in the Universe during this epoch consists of almost cold neutral hydrogen and helium atoms, therefore there are almost no recombination lines, compton scattering of CMB (cosmic microwave background) photons and so on.

Nevertheless the possibility of experimental study of Dark Ages is given by hyperfine transitions in hydrogen and deuterium (H and D respectively) (21.1 cm and 91.6 cm lines). During the observations radio telescope measures antenna temperature, which is associated with brightness temperature of the source (T_b), which depends on intensity of the source. In this paper the where built the evolution of hydrogen and deuterium brightness temperature.

In this work there where used following parameters describing the standard ΛCDM cosmology: $\Omega_m = 0.31$, $\Omega_\Lambda = 0.69$, $\Omega_b = 0.049$ (matter, dark energy and baryon densities divided by the critical density respectively), $H_0 = 67$ km s^{-1}Mpc^{-1} - the Hubble constant [1].

2. The spin temperature
Effective spin temperature T_s defines distribution of atoms with different spin states:

$$\frac{n_+}{n_-} = \frac{g_+}{g_-} exp\left(-\frac{T_\star}{T_s}\right) \tag{1}$$

Here $T_\star = h\nu_{+-}/k_B$, $h\nu_{+-}$ is the difference between higher and lower energy levels ($T_\star^H = 0.0682$ K, $T_\star^D = 0.0157$ K), $\mathbf{F} = \mathbf{S} + \mathbf{I}$ is the total spin angular momentum, $\mathbf{S}$ is spin of the electron, $\mathbf{I}$ is nuclear spin, so F_{+-} are total angular momentum quantum numbers, $g_+ = 2F_+ + 1$, $g_- = 2F_- + 1$ are the statistical degeneracy factors and $g_+^H/g_-^H = 3$, $g_+^D/g_-^D = 2$ for H and D respectively.

Spin temperature is determined by following factors: absorption of 21 cm CMB photons, absorption and emission of $Ly\alpha$ photons (Wouthuysen-Field effect) and atomic spin-change collisions, so from [2] (x is H or D)

$$T_s^x = \frac{(1+\chi^x)T_k T_{CMB}}{T_k + \chi^x T_{CMB}} \tag{2}$$

where T_k is kinetic gas temperature, $\chi^x = \chi_c^x + \chi_\alpha^x$ is sum of parameters for spin-change collisions and scattering of $Ly\alpha$ photons:

$$\chi_c^x = \frac{C_{+-}^x T_\star^x}{A_{+-}^x T_{CMB}}; \qquad \chi_\alpha^x = \frac{P_{+-}^x T_\star^x}{A_{+-}^x T_{CMB}}$$

$P_{+-} \propto P_\alpha$, P_α is $Ly\alpha$ scattering rate, C_{+-}^x is the collision rate, A_{+-}^x - Einstein coefficient. At $z \gg 10$, before first galaxies formation, P_α is small so it can be neglected [3].

For neutral hydrogen: $n_B/n_H = 1.3302$ [4] (there were H, D, ^{3}He and ^{4}He taken into consideration), $n_B = \eta_B n_\gamma (1+z)^3$($\eta_B$ is baryon-photon ratio), $A_{+-}^H = 2.85 \times 10^{-15}$ s^{-1}. Collisional coefficient:

$$C_{+-}^H = k_{10}^{HH}(T_k)n_H + k_{10}^{eH}(T_k)n_e + k_{10}^{pH}(T_k)n_p$$

where k_{10}^{HH} is the scattering rate between two hydrogen atoms [5], [6], k_{10}^{eH} is the scattering rate between hydrogen atoms and electrons [7], k_{10}^{pH} is the scattering rate between hydrogen atoms and protons [8] (here collisions with ^{4}He were neglected). For deuterium: $n_D = 2.61 \times 10^{-5} n_H$ [1] (here $n_D = n_{DI} + n_{DII}, n_H = n_{H_I} + n_{HII}$), $A_{+-}^D = 4.7 \times 10^{-17}$ s^{-1}. Collisional coefficient:

$$C_{+-}^D = \sqrt{\frac{8k_B T_k}{\mu_{DH}\pi}}\sigma_{+-}^{DH} n_H$$

$\mu_{DH} = \dfrac{m_D m_H}{m_D + m_H} \approx \dfrac{2}{3} m_H$ is reduced mass, σ_{+-}^{DH} is spin-change cross-section [9].

At fig.1 there were plotted H and D spin temperatures (T_s^H and T_s^D respectively), gas kinetic temperature (T_k) and CMB temperature (T_{CMB}). Before $z \sim 300$: $T_k = T_{\text{CMB}}$; due to collisions (because $\chi_c \gg 1$) $T_s = T_k$. At $30 \lesssim z \lesssim 300$ (for hydrogen) $5 \lesssim z \lesssim 300$ (for deuterium) gas was cooling adiabatically due to Universe expansion: $T_k \propto (1+z)^2$. So $T_k < T_{\text{CMB}}$ ($T_{\text{CMB}} \propto (1+z)$) and $T_s < T_{\text{CMB}}$ (due to collisions). At $z \lesssim 30$ (for

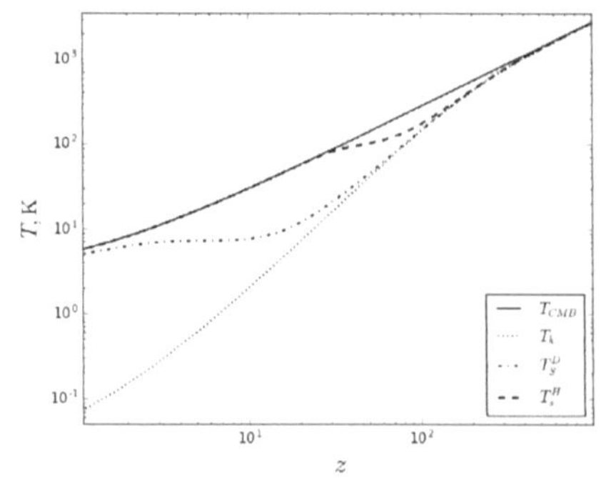

Figure 1. H and D spin temperatures (T_s^H, T_s^D), gas kinetic temperature (T_k) and CMB temperature as functions of z

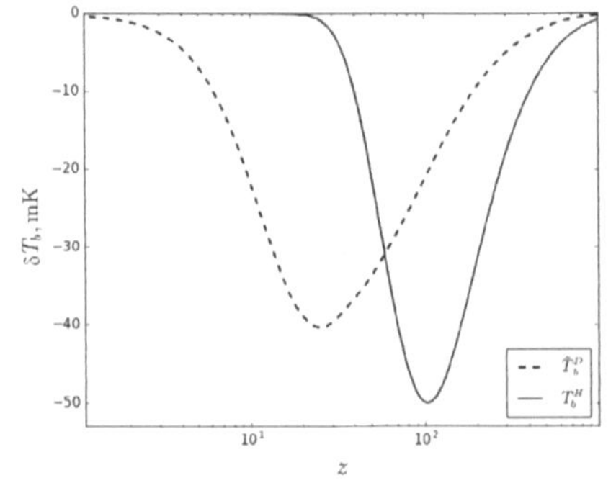

Figure 2. The H and D brightness temperature $\left(T_b^D = \tilde{T}_b^D \frac{n_D}{n_H}\right)$

hydrogen) $z \lesssim 5$ (for deuterium) because of Universe expansion gas density decreases so collisions become ineffective and T_s drives to T_{CMB} (T_s^D is coupled to T_{CMB} at lower z because lifetime of the excited level is longer ($A_{+-}^H / A_{+-}^D = 61.35$) and $\overline{\sigma}_{+-}^{DH} \gg \overline{\sigma}_{+-}^{HH}$ at low temperatures [8]).

3. Gas kinetic temperature

It is necessary to calculate gas kinetic temperature T_k to obtain T_s (following equation (2)). Gas kinetic temperature is defined by the following equation [10]:

$$\frac{dT_k}{dz} = \frac{2T_k}{1+z} - \frac{x_e}{1+f_{He}+x_e}\frac{8\sigma_T u_{CMB}}{3m_e c}\frac{T_{\text{CMB}}-T_k}{(1+z)H(z)} \tag{3}$$

where $f_{He} = 0.08$ is helium fraction, $\sigma_T = 6.65 \times 10^{-25}\text{cm}^2$ is Thomson cross-section, u_{CMB} is the energy density of the CMB, x_e is free electron fraction.

Here the first term describes adiabatic cooling due to Universe expansion, the second term describes heating due to Compton scattering of CMB photons from the residual free electrons.

4. Ionization of H, D, ^{3}He, ^{4}He

The second term in the equation (3) depends on free electron fraction x_e; it is necessary to calculate ionization fraction to find x_e.

To calculate ionization fraction of H, D, ^{3}He and ^{4}He, there were solved system of balance equations:

$$\frac{d}{dt}\left[\frac{n_{HI}}{n_H}\right] = \Gamma_{HI} n_e \frac{n_{HI}}{n_H} + \beta_{HI} n_e \frac{n_{HI}}{n_H} - \alpha_{HII} n_e \frac{n_{HII}}{n_H} \tag{4}$$

$$\frac{d}{dt}\left[\frac{n_{HeII}}{n_{He}}\right] = \Gamma_{HeI} n_e \frac{n_{HeI}}{n_{He}} + \beta_{HeI} n_e \frac{n_{HeI}}{n_{He}} - \beta_{HeI} n_e \frac{n_{HeII}}{n_{He}} - \\ - (\alpha_{HeII} + \xi_{HeII}) n_e \frac{HeII}{n_{He}} + \alpha_{HeIII} n_e \frac{n_{HeIII}}{n_{He}} \quad (5)$$

$$\frac{d}{dt}\left[\frac{n_{HeIII}}{n_{He}}\right] = \Gamma_{HeII} n_e \frac{n_{HeII}}{n_{He}} + \beta_{HeII} n_e \frac{n_{HeII}}{n_{He}} - \alpha_{HeIII} n_e \frac{n_{HeIII}}{n_{He}} \quad (6)$$

$$n_e = n_{HII} + n_{HeII} + 2n_{HeIII} \quad (7)$$

Here n_e is free electron fraction, n_{HI}, n_{HII} are number densities of neutral and ionized hydrogen, $n_H = n_{HI} + n_{HII}$ is the total number density of hydrogen; $n_{HeI}, n_{HeII}, n_{HeIII}$ are number densities of neutral, singly-ionized and doubly-ionized helium, $n_{He} = n_{HeI} + n_{HeII} + n_{HeIII}$ is total number density of helium.

Also, Γ_x is photo-ionization rate, α_x is recombination rate, ξ_x is dielectronic recombination rate, β_x is collisional ionization rate; they where taken from [11]. These functions depend on gas kinetic temperature, so it is necessary to solve them with equation (3).

5. The brightness temperature

The brightness temperature is the temperature that black body would have to emit observed intensity. To find it let us consider the equation of radiative transfer for intensity I_ν in the absence of scattering along the path:

$$\frac{dI_\nu}{ds} = -\alpha_\nu I_\nu + j_\nu \quad (8)$$

where $ds = cdt$, j_ν is an emission coefficient, α_ν is an absorption coefficient.

$$\frac{dI_\nu}{d\tau_\nu} = -I_\nu + S_\nu$$

here $\tau_\nu = \int \alpha_\nu(s)ds$ is an optical depth, $S_\nu = j_\nu/\alpha_\nu$.

To simplify calculations there were used Rayleigh-Jeans limit, appropriates for frequencies ν that much smaller then the peak frequency of CMB spectrum: $I_\nu = 2k_B T\nu^2/c^2$. So after rewriting (8) we can get equation of radiative transfer for light from radio source with brightness temperature T_b through the cloud with spin temperature $T_s = \frac{c^2}{2k_B\nu^2}\frac{j_\nu}{\alpha_\nu}$. So the observed brightness temperature at frequency ν:

$$T_b^{obs} = T_b e^{-\tau_\nu} + T_s(1 - e^{-\tau_\nu}) \quad (9)$$

Taking into consideration that optical depth is small at relevant redshifts ($10 \lesssim z \lesssim 1000$) ($x$ is H or D):

IOP Conf. Series: Journal of Physics: Conf. Series **1038** (2018) 012011 doi:10.1088/1742-6596/1038/1/012011

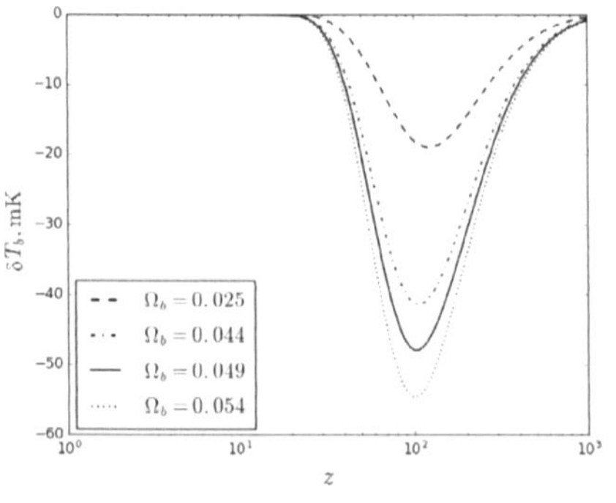

Figure 3. 21 cm brightness temperature with different values of baryon density Ω_b (the modern value of $\Omega_b = 0.049$ [1] - solid curve)

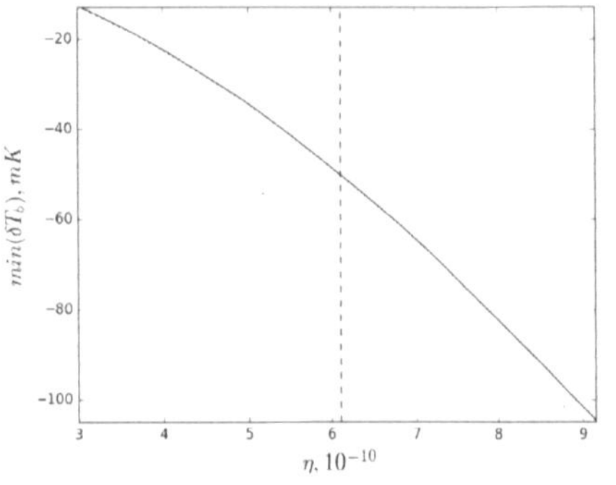

Figure 4. Dependence of 21 cm brightness temperature on different values of $\eta = n_B/n_\gamma$; dashed line corresponds to the modern value $\eta = 6.1 \times 10^{-10}$

$$\delta T_b^x = \frac{T_s^x - T_{\rm CMB}}{1+z}(1 - e^{-\tau^x}) \approx \frac{T_s^x - T_{\rm CMB}}{1+z}\tau^x \tag{10}$$

Here we used CMB as a background radio source, so the brightness temperature of the source T_b is equal to $T_{\rm CMB}$. From [9]:

$$\tau^x = \frac{g_+^x c \lambda^2 h A_{+-}^x n_x}{8(g_+^x + g_-^x)\pi k_B T_s^x H(z)} \tag{11}$$

where $H(z) = \sqrt{\frac{8\pi}{3} G\rho_c (\Omega_m(1+z)^3 + \Omega_{rad}(1+z)^4 + \Omega_\Lambda + \Omega_{curv}(1+z)^2)}$, $G = 6.67 \times 10^{-8}\ s^2 cm^{-1} g^{-1}$, ρ_c is critical density. In figure 2 there were plotted brightness temperatures δT_b^H and $\tilde{T}_b^D = \frac{n_H}{n_D}\delta T_b^D$. $\delta T_b = 0$ when $T_s = T_{CMB}$ (i.e. firstly when $T_k = T_{CMB}$ ($z \gtrsim 300$) and secondly when collisions became inefficient ($z \lesssim 30$ for H and $z \lesssim 5$ for D)), $\delta T_b < 0$ during $30 \lesssim z \lesssim 300$ for H and $5 \lesssim z \lesssim 300$ for D because of $T_k < T_s < T_{CMB}$. Minimum of δT_b^D is at the lower z in contrast with δT_b^H, it happens because T_s^D is coupled with T_{CMB} later then T_s^H.

6. Dependence of 21 cm brightness on baryon-photon ratio

On the fig.3 there were plotted dependencies of 21 cm brightness on z with different values of baryon density Ω_b; solid curve corresponds to the modern value of $\Omega_b = 0.049$. On the fig.4 there was plotted dependence of 21 cm brightness on baryon-photon ratio $\eta = n_B/n_\gamma$; dashed line corresponds to the modern value of $\eta = 6.1 \times 10^{-10}$. From [12]:

$$\eta = 273.4 \times 10^{-10}\Omega_b h^2$$

$h = 0.67$ is dimensionless Hubble constant.

Without loss of generality let us consider that η increases, so n_H will increase too ($n_H \propto \eta$) and x_e will decrease therefore gas kinetic temperature T_k will decrease. Coefficient describes collisions $\chi_c \propto C_{+-}$ will increase; it drives spin temperature T_s to T_k, thus T_s will decline as T_k. So $\tau \propto n_H/T_s$ will decrease and $\delta T_b = \dfrac{T_s - T_{\rm CMB}}{1+z}\tau$ will decrease too.

7. Conclusions

In this paper there were considered physical processes during cosmic Dark Ages and there were calculated H and D brightness temperature, gas kinetic temperature and CMB temperature. At fig.1 there were plotted dependence of H and D spin temperature (T_s^H and T_s^D respectively), gas kinetic temperature (T_k) and CMB temperature ($T_{\rm CMB}$) on redshift z.

In this work there were examined dependencies of 21 cm brightness temperature on baryon density (fig.3) and dependence on baryon-photon ratio $\eta = n_B/n_\gamma$(fig.4). Ω_B and η are the key parameters in modern cosmology and hence we can see a strong dependence on these parameters.

These measures potentially could become a new instrument for independent measurement of barion-photon ratio during Dark Ages.

References

[1] Planck colloboration 2016 *Astronomy& Astrophysics* **594** A13
[2] Field G B 1958 *Proc.IRE.* **46** 240-250
[3] Furlanetto S R, Oh S P and Briggs F H 2006 *Physical Reports* **433** 181-301
[4] Steigman G 2006 *Journal of Cosmology and Astroparticle Physics* 10 016
[5] Allison A C and Dalgarno A 1969 *Astrophys.J.* **158** 423
[6] Zygelman B 2005 *Astrophys.J.* **622** 1356-1362
[7] Furlanetto S R and Furlanetto M R 2007 *Mon. Not.R.Astron. Soc.* **374** 547-555
[8] Furlanetto S R and Furlanetto M R 2007 *Mon. Not.R.Astron. Soc.* **379** 130-134
[9] Sigurdson K and Furlanetto S R 2006 *Phys.Rev.Lett.* **97** 091301
[10] Seager S, Sasselov D D and Scott D 2000 *The Astrophysical Jornal Supplement Series* **128** 407-430
[11] Mo H et al 2010 *Galaxy formation and Evolution* (Cambridge, UK.: Cambridge University Press)
[12] Ivanchik A V, Balashev S A, Varshalovich D A and Klimenko V V 2015 *Astronomy Reports* **59** 100-117

Primordial deuterium abundance at $z_{\rm abs} = 2.504$ towards Q1009+2956

E O Zavarygin[1], J K Webb[1,2], S Riemer-Sørensen[3,4] and V Dumont[5,1]

[1] School of Physics, University of New South Wales, Sydney, NSW 2052, Australia
[2] DAMTP, Centre for Mathematical Sciences, Wilberforce Road, Cambridge, CB3 0WA, UK
[3] Institute of Theoretical Astrophysics, The University of Oslo, Oslo NO-0316, Norway
[4] ARC Centre of Excellence for All-sky Astrophysics (CAASTRO), NSW 2016, Australia
[5] Department of Physics, University of California, Berkeley, California 94720-7300, USA

E-mail: `e.zavarygin@gmail.com`

Abstract. The $z_{\rm abs} = 2.504$ Lyman Limit absorption system (LLS) towards Q1009+2956 has previously been used to estimate the primordial deuterium abundance. Since the initial measurement by Burles & Tytler [1], this quasar has been observed extensively with the Keck telescope, providing a substantial increase in signal-to-noise (from 60 to 147 at continuum level of Ly-α at $z_{\rm abs} = 2.504$). We explore a set of different models for the absorption system and find that the deuterium feature is contaminated by Ly-α absorption from a low column density H I cloud. This significantly limits precision to which one can estimate the D/H ratio in this LLS. Our final result for this system D/H$= 2.48^{+0.41}_{-0.35} \times 10^{-5}$ has the same relative uncertainty of 17% as the previous estimate by [1] despite the far higher signal-to-noise of our dataset. A weighted mean of 13 D/H measurements available in the literature (including our result) gives the primordial deuterium abundance of $({\rm D/H})_{\rm p} = (2.545 \pm 0.025) \times 10^{-5}$ and the corresponding baryon density of the Universe of $\Omega_{\rm b}h^2 = 0.02174 \pm 0.00025$ marginally inconsistent with the 2015 Planck CMB data. More measurements of the deuterium abundance in quasar spectra are needed to check this tension.

1. Introduction

Most of the elements in the Universe were produced during its late evolution in stars, supernovae and cosmic rays, whilst light elements such as hydrogen, helium, lithium and their isotopes are primordial, produced during the Big Bang nucleosynthesis epoch (BBN), the first few minutes after the Big Bang (e. g. [2]). The BBN yields of the light elements abundances depend on physical conditions at that time, allowing us to explore physics in the early Universe.

For the standard BBN scenario, the relative abundance of the light species is unambiguously determined by a single parameter: the baryon-to-photon ratio η. Given its sensitivity to and monotonic dependence on η, the primordial abundance of deuterium (D/H ratio) is generally accepted as the best "baryometer" amongst all the primordial elements.

The most reliable measurements of the primordial deuterium abundance come from absorption in quasar spectra by metal-poor gas clouds on the lines of sight (e. g. [1, 3]). Atoms of neutral hydrogen (H I) and deuterium (D I) in gas clouds intersecting the line of sight absorb the flux from a quasar, due to Lyman series transitions. The values of the ionisation potential of H I and D I are almost identical ($\Delta E = 0.0037\,{\rm eV}$ [4]). The corresponding ionization correction should be less than 0.1% [5] making D/H = D I/H I a reasonable approximation.

Here we present the main results of a new analysis of the previously analysed metal-poor LLS at $z_{\rm abs} = 2.504$ towards the quasar Q1009+2956. Since the previous analysis by [1], this quasar has been observed extensively with the Keck telescope, and the data quality available is now substantially improved. The full version of the analysis is available at [6].

2. Data

The dataset used in this work consists of 22 individual exposures observed with the High Resolution Echelle Spectrometer (HIRES [7]) mounted on the Keck telescope. The initial data reduction was done using the MAKEE software designed by Tom Barlow. Echelle orders were combined and continuum-fitted using UVES_popler software [8]. Then, using new custom-made code, the one-dimensional normalised spectra for each of the 22 exposures were checked (and corrected) for relative velocity offsets. The 22 datasets were then co-added into four distinct one-dimensional spectra (with signal-to-noise of 57, 47, 78 and 27 near Lyman-α of the LLS), determined by the instrumental settings used during the observations: the spectral resolution (resolving power $R = 37000$ and 49000) and binning in the dispersion direction (unbinned and 2×binned). In the modelling process, these four spectra were fitted simultaneously as described below.

3. Analysis

3.1. Fitting

The absorption features were fitted with Voigt profiles using VPFIT 10.2 [9]. The following assumptions were made. Redshifts of H I and D I are tied in each component. The ratio of D/H is assumed to be equal in all components which is a good approximation at low metallicities [10, 11], as is the case here ([Si/H] = −2.5 [1]). All column densities are allowed to vary freely in the fitting procedure and can reduce to zero if a component at some particular velocity is not required by the data itself. A more stringent constraint on D/H can be obtained by solving for *summed* H I and D I column densities for the main components of the LLS. A velocity shift free-parameter is included in the model taking into account any possible residual velocity off-sets between the spectra. A zero level is fitted for transitions with saturated lines in order to correct for any possible imperfect night-sky subtraction.

3.2. Transitions used

The entire H I Lyman series of the $z_{\rm abs} = 2.504$ LLS are covered by our dataset. Out of them, we select those which are not severely contaminated by absorption lines from other redshifts: Lyman α, β, γ, 6, 13, 14, 21-24. Due to the relatively low H I, column density and because of line blending, the deuterium absorption appears most prominently in D I Lyman α, as Fig. 1 and 2 illustrate. In models using metal species, the following transitions were also used: C II 1334, C III 977, C IV 1548 and 1550, Si IV 1393 and 1402. We do not see absorption by other species commonly detected in higher H I column density absorption systems such as O I, Fe II, Mg II.

3.3. Blending of the Lyman α line

Fitting the selected Lyman series transitions reveals that there is residual absorption in both wings of H I Lyman α (see Fig. 1). This absorption appears to be fairly symmetric with respect to the center of the LLS. We adopt two blending models: multiple relatively narrow ($\sim 10-40$ km/s) blends on either side of the H I Lyman α and a single broad (large b parameter, ~ 105 km/s) line sitting on top of H I Lyman α.

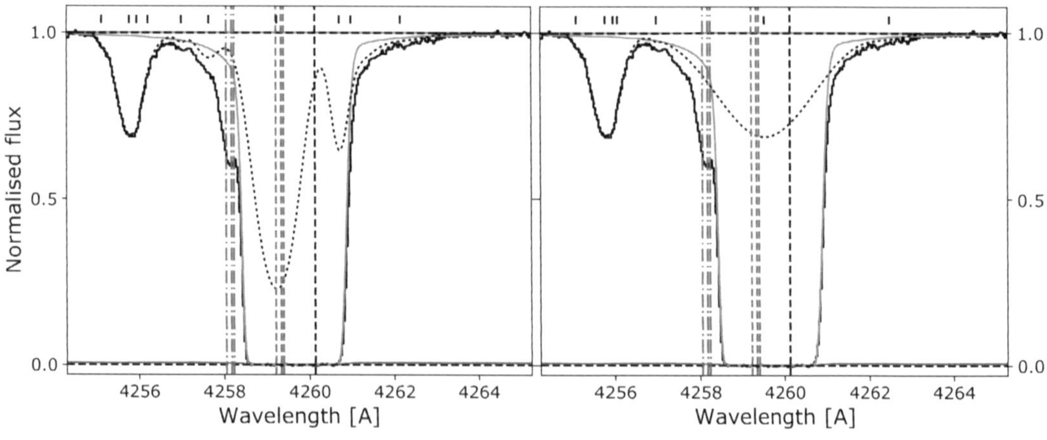

Figure 1. The Ly α transition of the $z_{\rm abs} = 2.504$ LLS. The spectrum illustrated (black histogram) was obtained by a variance-weighted co-addition of the data from all 22 exposures. It is for illustration only - see Section 2. The continuous green line is formed from the three H I components (vertical red dashed lines) in models 6a (left panel) and 6b (right panel) that give rise to the D I absorption feature (vertical blue dash-dotted lines), plus one moderate column density H I component (vertical black dashed line) required to account for the additional absorption redwards of the three main components. The 1σ error array is shown by the purple line near $y = 0$. The grey (very narrow) shaded area near $y = 1$ corresponds to the $\pm 1\sigma$ uncertainty in the continuum. The absorption due to additional low-column density blends is illustrated by the black dotted line. The positions of the blends are indicated with vertical ticks. A single broad ($b \sim 105$ km/s) blend in model 6b (right panel) replaces four relatively narrow blends in model 6a (left panel).

3.4. Different models

We explore a set of models to test the sensitivity of the inferred D/H to the underlying kinematic structure of the LSS. For each blending model, discussed above, we consider six kinematic models, varying the number of absorbing components and the transitions used:

1) one-component model, fit to Lyman series only;
2) two-component model, fit to Lyman series only;
3) three-component model, fit to Lyman series only;
4) two-component model, C II used;
5) three-component model, C II used;
6) three-component model, C II, C III, C IV and Si IV used.

We use subscripts "a" and "b" to indicated models using multiple relatively narrow blends and a single broad line to model wings of H I Lyman α: models 1a–6a and 1b–6b, respectively. To avoid degeneracy, when fitting Lyman series only (models 1–3), we tie the main H I and D I components thermally. Thermal broadening is supported by fitting to metal transitions only and to all the selected transitions making this assumption reasonable [1, 6]. When metal transitions are used (models 4–6), we solve for both temperature and $b_{\rm turb}$ parameter. When all the metals are fitted simultaneously, three velocity components are required.

3.5. Fitting results

Fig. 2 illustrates the overall profile for model 6a over-plotted with the spectrum co-added from the whole dataset. Fig. 3 illustrates the final D/H values for all 12 models considered. The

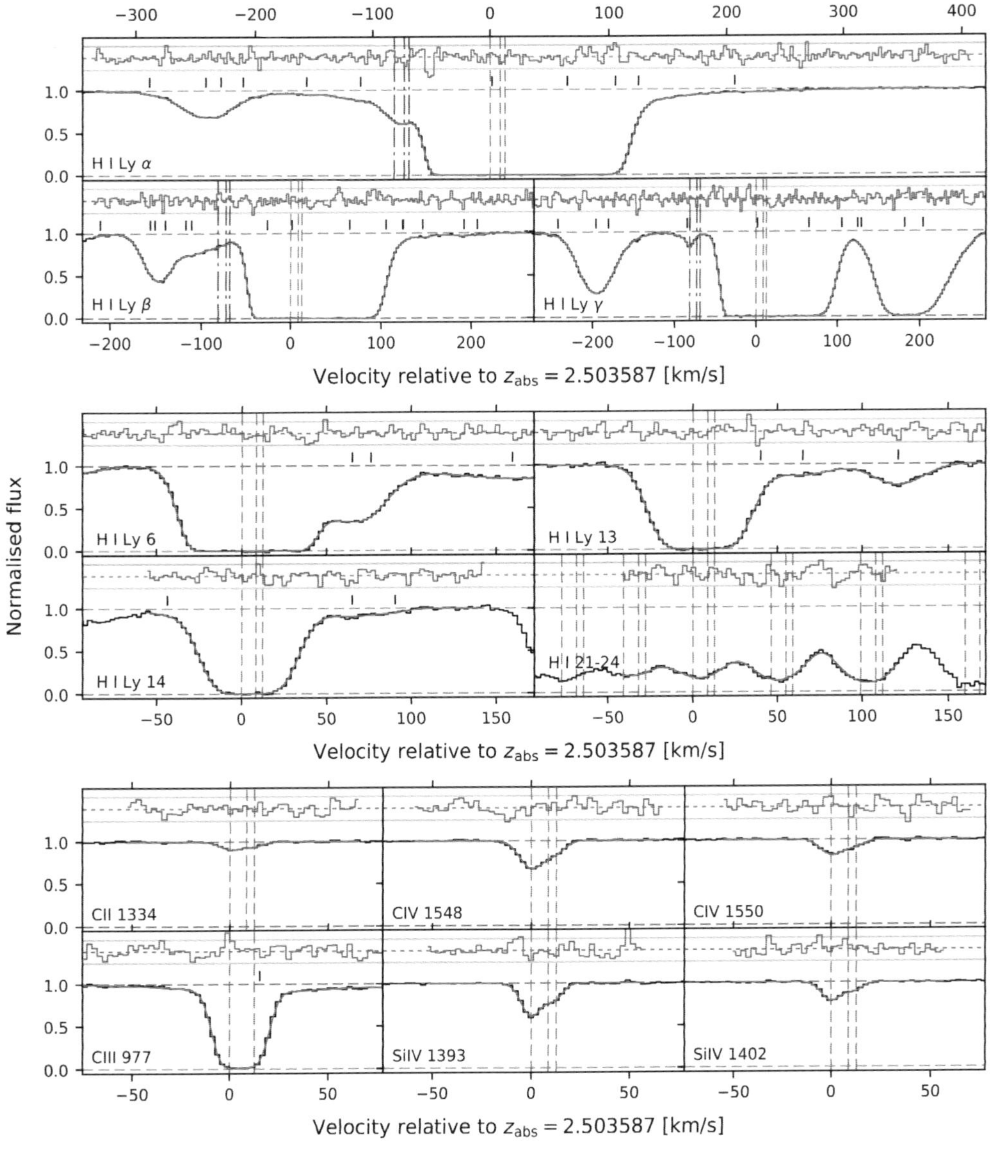

Figure 2. The overall model profile corresponding to model 6a (red solid line) for the transitions used in the analysis (indicated in the bottom left corner of each panel) over-plotted with the co-added spectrum combined from the whole dataset of 22 exposures (black histogram). This spectrum is only used for illustration, fitting was done to the four co-added spectra described in Section 2. The normalised residuals are illustrated above the spectra, with horizontal blue lines indicating the $\pm 2\sigma$ range. The zero point for the x-axis, in velocity units, corresponds to the bluemost (i.e. lowest redshift) component in model 6a. The red dashed vertical lines indicate the position of the three main H I components. The blue dash-dotted vertical lines in the top three panels indicate the corresponding D I lines. The shorter vertical lines above each fit show the position of the additional blends included in the model.

original estimate by [1] and the CMB prediction by [12] are also illustrated.

Using multiple narrow blends to model the residual absorption in the wings of H I Lyman α

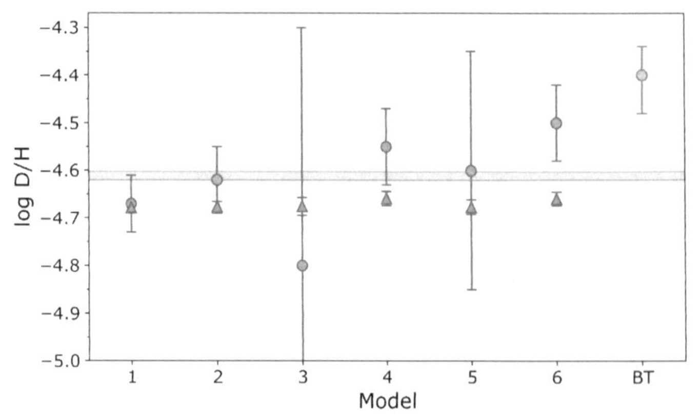

Figure 3. D/H values with their 1σ confidence intervals (y-axis) for all considered models (indicated on the x-axis; numeration is given in Section 3.4). The blue points and the orange triangles correspond to models 1a–6a and 1b–6b, respectively. The green point at far right is the estimate by [1]. The CMB prediction by [12] is shown with the horizontal band.

significantly degrades the precision since the position of one blend coincides closely to the strongest D I line, Lyman α. However, we see that D/H values for models 1a–6a are in good agreement with each other. Model 3a is a 3-component fit to the hydrogen series only, that is, no metals have been used. Model 5a is also a 3-component fit, using the hydrogen series plus only one metal transition, C II, which is rather weak (Fig. 2). These two models give a conspicuously larger D/H uncertainty compared to the other models (Fig. 3), primarily because the deuterium column density is poorly constrained. Using additional metal transitions for three-component models (3a, 5a, and 6a) reduces the degeneracy and provides a better constraint on D/H. Models using a broad line to model the wings of H I Lyman α also give consistent D/H values among the b models. However, unlike models 1a–6a, a broad blend removes the severe parameter degeneracy, resulting in far smaller D/H uncertainties in models 1b–6b. Whilst the broad line based models disagree with the CMB prediction by [12], we seek to derive D/H *independently* from other constraints, including the CMB prediction.

3.6. Systematic error and final D/H

Our two sets of D/H models agree with each other within the errors although the broad-blend models (the "b" models) values fall systematically low compared to the narrow-blend models, "a". For our final constraint, we initially consider models 6a and 6b, since they use more transitions, and thus more information, to constrain the LLS velocity structure. The D/H values for these models are $\log \mathrm{D/H} = -4.50 \pm 0.08$ and -4.660 ± 0.014, respectively. We make a crude estimate of the systematic error associated with choice of blending model to be half the difference between these two D/H values, i.e. $\sigma_{\mathrm{sys}}(\log \mathrm{D/H}) = 0.08$. This value is really a rough estimate of the total error rather than just a systematic term, so is conservative. Since the σ_{sys} error is random in nature, we add it in quadrature to the statistical-only D/H uncertainties of models 6a and 6b. Our final D/H estimate for this LLS is a weighted average of the two values:

$$\log(\mathrm{D/H}) = -4.606 \pm 0.066 \qquad \text{or} \qquad \mathrm{D/H} = 2.48^{+0.41}_{-0.35} \times 10^{-5}$$

and is in excellent agreement with the CMB prediction by [12] of $\mathrm{D/H} = (2.45 \pm 0.05) \times 10^{-5}$.

Note, the precision of our result is 17% which is the same as that of the previous D/H estimate by [1], despite the far higher S/N of our dataset. However, the D/H we derive is lower compared to [1].

4. The primordial deuterium abundance and $\Omega_b h^2$

We combine the measurement described in this paper with 14 other measurements of the D/H ratio using quasar spectra from the literature – 12 listed in Table 4 of [13], excluding a previous measurement of the LLS reported here, plus two additional measurements from [14, 15].

To accommodate the possibility that scatter amongst these 15 measurements exceeds the nominal uncertainties, we apply the method of Least Trimmed Squares (LTS) to identify and discard outliers. An LTS trimming fraction of 0.85 is adopted, i. e. we discard 15% of measurements (which in our case is obviously two measurements). The discarded measurements turn out to be those from [14, 16]. The weighted mean of the remaining 13 points provides a high precision estimate of the primordial deuterium abundance and is

$$(\mathrm{D/H})_{\mathrm{p}} = (2.545 \pm 0.025) \times 10^{-5}. \tag{1}$$

This value is in good agreement with the CMB prediction by [12]. For the 13 D/H measurements in our final (i.e. after LTS) sample, $\chi^2/\mu = 0.97$ (where $\mu = 12$ is the number of degrees of freedom), so the scatter amongst these 13 points is consistent with the published errors. If we exclude the measurement presented in this paper and re-calculate the weighted mean D/H from the remaining 12 measurements, the value is only marginally different: $(\mathrm{D/H})_{\mathrm{p}} = 2.546 \times 10^{-5}$, with the same uncertainty as Eq. 1.

We use our estimate of the weighted mean primordial deuterium abundance (Eq. 1) and the fitting formula for the BBN calculations (eq. 6 from [17]) to infer the baryon density of the Universe, the statistical uncertainty, and the uncertainty from nuclear data parameters:

$$\Omega_{\mathrm{b}} h^2 = 0.02174 \pm 0.00013_{\mathrm{QSO}} \pm 0.00021_{\mathrm{nucl}}.$$

This compares to the Planck 2015 TT+lowP+lensing value [18] (see their table 4) of $\Omega_{\mathrm{b}} h^2 = 0.02226 \pm 0.00023$, i.e. there is a marginal tension at the 1.6σ level with our result. It is interesting to see that the overall uncertainty is now dominated by imprecision from nuclear data rather than the spectral quality of the high redshift quasar data. The latter will need experimental improvement if we are to check on this marginal 1.6σ discrepancy.

Further details about this analysis can be found in [6]. We are refining the fit at this stage so the final result for D/H in this LLS may alter slightly.

Acknowledgments

The conference travel was funded via the Postgraduate Research Student Support (PRSS) Scheme of UNSW. EOZ is supported by an Australian Government Research Training Program (RTP) Scholarship. This work is based on observations collected with the Keck Observatory Archive (KOA), which is operated by the W. M. Keck Observatory and the NASA Exoplanet Science Institute (NExScI), under contract with the National Aeronautics and Space Administration. Part of the analysis and figures were done using publicly available python packages: matplotlib [19], numpy [20], astropy [21]. This research made use of NASA's Astrophysics Data System. EOZ is also grateful to GitHub for the Student Developer Pack.

References

[1] Burles S and Tytler D 1998 *ApJ* **507** 732–744 (*Preprint* astro-ph/9712109)
[2] Weinberg S 2008 *Cosmology* (Oxford University Press)
[3] Burles S and Tytler D 1998 *ApJ* **499** 699–712 (*Preprint* astro-ph/9712108)
[4] Morton D C 2003 *ApJS* **149** 205–238
[5] Cooke R and Pettini M 2016 *MNRAS* **455** 1512–1521 (*Preprint* 1510.03867)
[6] Zavarygin E O, Webb J K, Dumont V and Riemer-Sørensen S 2017 *ArXiv e-prints* (*Preprint* 1706.09512)
[7] Vogt S S *et al.* 1994 *Instrumentation in Astronomy VIII* (*Society of Photo-Optical Instrumentation Engineers (SPIE) Conference Series* vol 2198) ed Crawford D L and Craine E R p 362

[8] Murphy M T 2016 UVES_popler: POst-PipeLine Echelle Reduction software URL https://doi.org/10.5281/zenodo.56158
[9] Carswell R F and Webb J K 2014 VPFIT: Voigt profile fitting program Astrophysics Source Code Library (*Preprint* 1408.015)
[10] Fields B D, Olive K A, Silk J, Cassé M and Vangioni-Flam E 2001 *ApJ* **563** 653–659 (*Preprint* astro-ph/0107389)
[11] Sembach K R *et al.* 2004 *ApJS* **150** 387–415 (*Preprint* astro-ph/0311177)
[12] Coc A, Petitjean P, Uzan J P, Vangioni E, Descouvemont P, Iliadis C and Longland R 2015 *Phys. Rev. D* **92** 123526 (*Preprint* 1511.03843)
[13] Riemer-Sørensen S, Kotuš S, Webb J K, Ali K, Dumont V, Murphy M T and Carswell R F 2017 *MNRAS* **468** 3239–3250 (*Preprint* 1703.06656)
[14] Srianand R, Gupta N, Petitjean P, Noterdaeme P and Ledoux C 2010 *MNRAS* **405** 1888–1900 (*Preprint* 1002.4620)
[15] Cooke R, Pettini M and Steidel C C 2017 *ArXiv e-prints* (*Preprint* 1710.11129)
[16] Pettini M and Bowen D V 2001 *ApJ* **560** 41–48 (*Preprint* astro-ph/0104474)
[17] Balashev S A, Zavarygin E O, Ivanchik A V, Telikova K N and Varshalovich D A 2016 *MNRAS* **458** 2188–2198 (*Preprint* 1511.01797)
[18] Planck Collaboration 2016 *A&A* **594** A13 (*Preprint* 1502.01589)
[19] Hunter J D 2007 *Computing in Science and Engineering* **9** 90–95
[20] Van Der Walt S, Colbert S C and Varoquaux G 2011 *CSE* **13** 22–30 (*Preprint* 1102.1523)
[21] Astropy Collaboration 2013 *A&A* **558** A33 (*Preprint* 1307.6212)

IOP Conf. Series: Journal of Physics: Conf. Series **1038** (2018) 012013 doi:10.1088/1742-6596/1038/1/012013

On Dynamics and Spectra of Fine Spatial Structures in the Vela Pulsar Wind Nebula

G.A. Ponomaryov[1], **K.P. Levenfish**[1], **A.M. Krassilchtchikov**[1], **Yu.A. Kropotina**[1], **A.E. Petrov**[1]

[1] Ioffe Institute - Politechnicheskaya 26, 194021, St. Petersburg, Russia

Abstract. Evolution of X-ray surface brightness and spectra of fine spatial structures in the Vela pulsar wind nebula (PWN) are studied. Variability of morphology and the spectrum of a small-scale structure at the base of the southern jet of the nebula – a bright transverse *bar* is investigated for the first time. The size of the bar located within $4''$–$5''$ from the Vela pulsar is $\sim 4.5'' \times 1.5''$. It is shown that brightness, shapes, and positions of the *bar* and the southern jet base vary on the timescale of 1–2 weeks, the *bar* can disappear and emerge again within 10 days. The power-law photon index of the jet base is much harder than that of the bar, 1.20 ± 0.12 vs 1.65 ± 0.15. The latter index is close to 1.6, a typical value for synchrotron emission of electrons accelerated at an ultrarelativistic shock. We speculate that bright blobs seen on both sides of the southern jet base may have the same nature as the bright blobs in the Crab PWN, which are believed to mark the position of the termination shock of the pulsar wind.

1. Introduction

Pulsar wind nebulae (PWNe) are bubbles of relativistic magnetized plasma blown out by pulsars, rotating magnetized neutron stars. The nebulae are fed by the $e^{\pm}$-winds of the pulsars and by their rotational energy. They appear as extended areas of bright emission (mostly, of synchrotron origin) around pulsars and form through the interaction of the pulsar wind (PW) with the surrounding medium (a supernova remnant, or the interstellar medium). PWNe show power-law spectra from radio-waves to gamma-rays and are thought to be powerful particle accelerators.

The distance of the Vela PWN is about 290 pc. It is energized by the relatively young (11 kyr) pulsar PSR B0833-45. X-ray observations of the nebula with the Chandra X-ray Observatory have revealed several bright and variable structures the most prominent being the two arcs and two extended jets with bright bases [9, 10]. Observations with Chandra allow one to resolve extended X-ray features of arcsecond scale.

2. Observations and data analysis

To study the dynamics of small scale X-ray structures of the Vela PWN we have analyzed the data of 11 observations performed with Chandra/ACIS in 2009-2010, each of 40 kiloseconds long (http://cxc.harvard.edu/cda; obs. ID 10132–10139, 12073–12075). We used standard routines from CIAO v.4.9 and XSpec v.12.9.1 to reprocess and analyse the data.

X-ray maps of the PWN in the 0.5–8 keV band were made for each observation. Their summed image is shown in Fig. 1. Following previous studies [10, 8] we chose for analysis several emitting structures rendered in Fig. 2: the wings of the outer (1a, 1b) and inner (2a, 2b) arcs, the central bright spot on the outer arc (3), the bases of northern (4) and southern (5) jets and

IOP Conf. Series: Journal of Physics: Conf. Series **1038** (2018) 012013 doi:10.1088/1742-6596/1038/1/012013

a thin transverse bar (6) at the beginning of the southern jet's base. The position of the PSR is marked on the maps by a cross. The binning and smoothing parameters applied to produce the images are given in the Appendix (table 1).

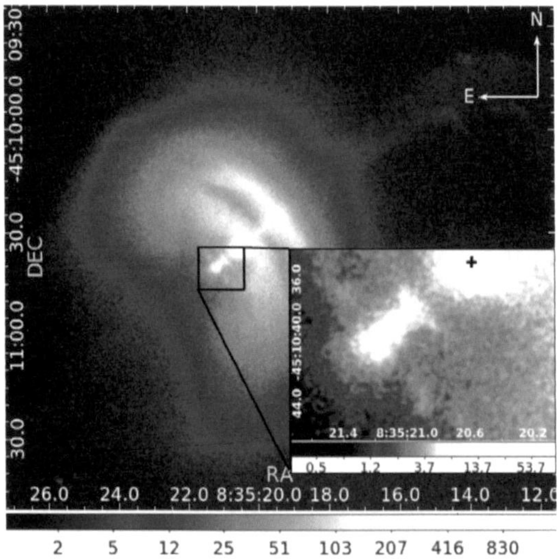

Figure 1. X-ray summed map of the Vela PWN in 0.5-8.0 keV range.

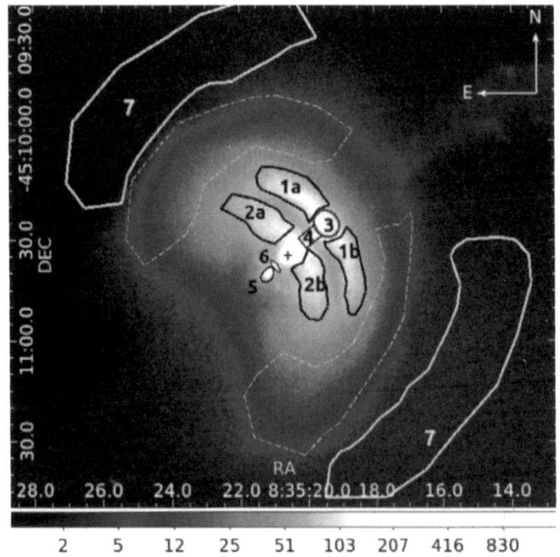

Figure 2. Structures of the Vela PWN selected for the analysis.

2.1. Morphology
The position of the PW termination shock (TS) in the Vela nebula is unknown. In the well studied Crab PWN the TS appears on the X-ray map as a bright ring (an ellipse in the sky plane projection), whose distant (SE) part consists of several blobs marked in Fig. 3 by letters A – D (notice the inverted color map). Similar blobs emerge on the X-ray maps of the Vela PWN on both sides of the southern jet's base. The structures are highly variable (alike Crab's blobs) and therefore look blurred on the summed image in Fig. 4, in contrast with individual images, like that shown in Fig. 5 for observation 10132. If the nature of these blobs is the same as in Crab, they might as well reveal the position of the TS.

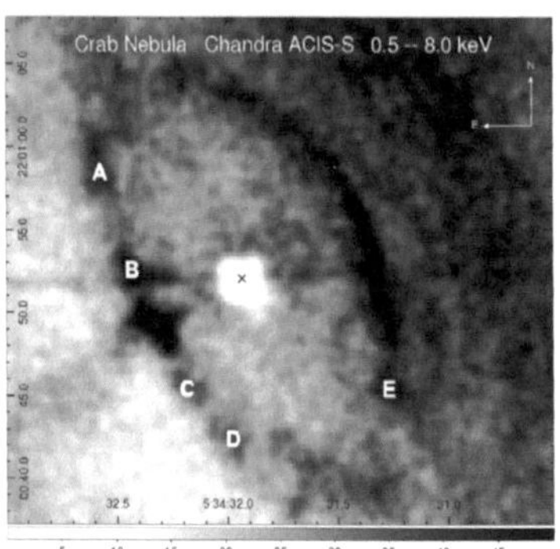

Figure 3. X-ray image of the Crab PWN[7]. The dark ring is thought to represent the pulsar wind TS with apparent bright blobs some of which are marked by capital letters.

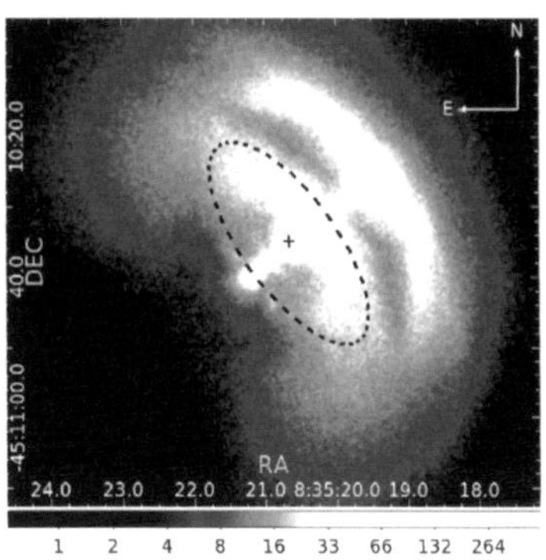

Figure 4. A possible geometry of the TS in the Vela PWN shown by the dashed ellipse. The SE edge of the ellipse outlines the bright blobs visible in the image of Fig. 5

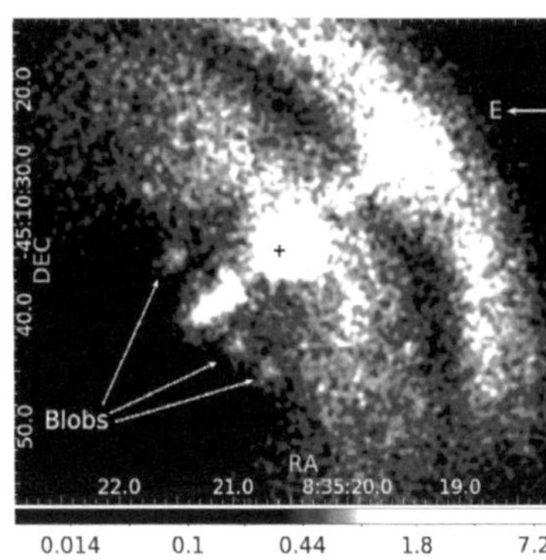

Figure 5. An X-ray image of the Vela PWN (obs. ID 10132). The bright blobs in the vicinity of the base of the SE jet are indicated by arrows.

Much longer exposures are needed to test this suggestion, due to the faintest spatial extent of the blobs and their fast variability. Such a hypothesis seems non-controversial at least geometrically: in Fig. 4 we inscribe into the Vela's X-ray map a TS ring whose distant (SE) edge touches the blobs, the nearest (NW) edge does not go beyond the diffuse emission of the inner arc, with the pulsar at the center and with the minor axis matching the PWN's symmetry axis.

2.2. The spectral and brightness variability of the prominent PWN structures

Using 11 observations of the Vela PWN we studied the spectral and surface brightness variability of its prominent structures in the 0.5–8 keV range. The emission of the PWN is mostly of synchrotron origin, so we have fitted the spectra with a power-law model corrected for the absorption of the X-rays in the interstellar medium [11]:

$$I(E) = A\left(\frac{E}{1\ keV}\right)^{-\Gamma} \exp\left(-\sigma(E)\,N_{\rm H}\right). \quad (1)$$

The hydrogen column density $N_{\rm H}$ was fixed at $3.2{\cdot}10^{20}$ cm^{-2} as determined from the observation of the Vela pulsar with the Chandra LETG spectrometer [9]. The normalization A and the photon index Γ were determined as best-fit parameters. The area "7" in Fig. 2 was taken as the background for the spectral analysis (see comments in the Appendix).

Because of relatively large statistical errors the variations and correlations of photon indices with surface brightness cannot be traced with confidence, see Fig. 6, 7. One can discuss only the relative hardness of the spectra. If the X-rays emitting $e^{\pm}$ in PWNe would be accelerated at the wind TS only, then one would expect the spectra to have typical $\Gamma \sim 1.5$–1.6. However, if $e^{\pm}$ are pre-accelerated at the TS and are further accelerated in converging flows (see comments in the Appendix), they will acquire even harder spectra $\Gamma \sim 1$ [1].

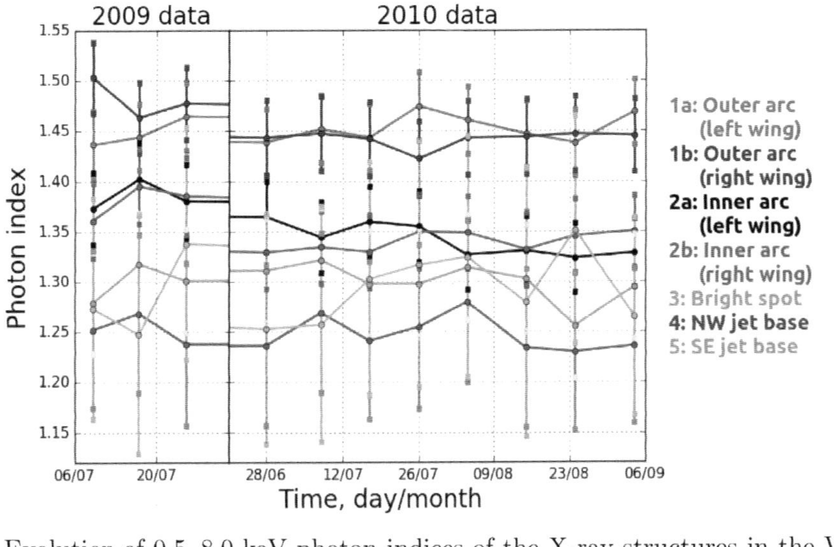

Figure 6. Evolution of 0.5–8.0 keV photon indices of the X-ray structures in the Vela PWN. The black vertical line indicates the 11 months gap between observations of 2009 and 2010. The error bars show 3σ errors.

Indeed, the photon indices we obtained ($\sim 1.2 - 1.5$) are substantially harder, in good agreement with the indices derived by [10, 5, 6]. The softest emission with $\Gamma \sim 1.5$ comes from the *outer arc* and its index is indeed close to that predicted for the emission of the electrons accelerated by Fermi mechanism at the TS of the relativistic pulsar wind. The emission of the

IOP Conf. Series: Journal of Physics: Conf. Series **1038** (2018) 012013 doi:10.1088/1742-6596/1038/1/012013

other nebular structures is harder: $\Gamma \sim 1.3$–1.4 for the *inner arc* and ~ 1.2–1.3 for the *jets' bases* and the *bright spot* on the *outer arc*.

The energy fluxes (erg·cm^{-2}·s^{-1}) were determined from spectral fits and divided by the areas of the studied regions. No signicant evolution of surface brightness of individual structures is observed, except the PWN *arcs* which swapped their relative brightness in the 11 months gap between the observations of 2009 and 2010; the right wing of the *inner arc* also got brighter by $\sim 15\%$.

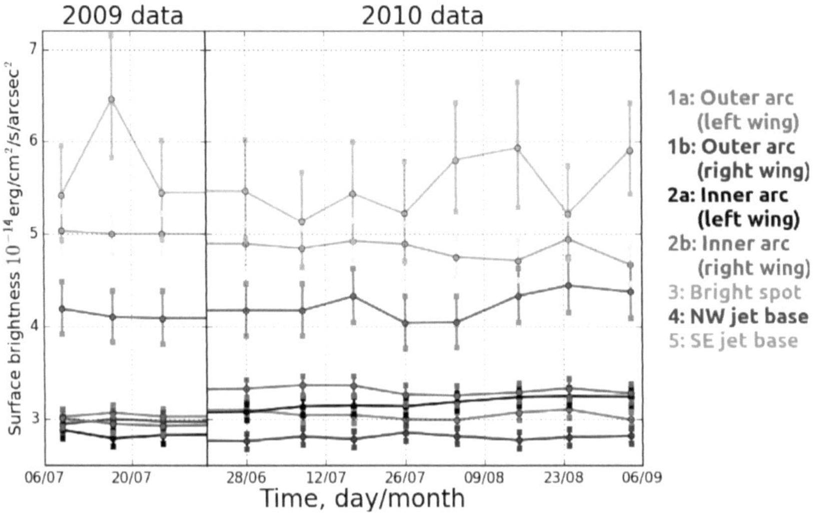

Figure 7. Evolution of unabsorbed 0.5–8.0 keV surface brightness (in units: 10^{-14} erg·cm^{-2}·s^{-1}·arcsec^{-2}) of bright structures of the Vela PWN. The black vertical line separates observations of 2009 and 2010. The error bars show 3σ errors.

2.3. The Dynamics and Spectra of the Fine PWN Structures

For the first time we studied the dynamics and spectra of the immediate vicinity of the Vela pulsar. In the inset of Fig. 1 the region with the *base* of the southern jet and the *bar* is enlarged. Both structures are unstable and change their shape, brightness, and orientation in ~ 10 days. The *bar* can disappear and form again, as shown in Fig. 8. For example, in the image no. 10137 it is apparent, while in the image no. 10136 taken 9 days earlier it is almost indistinguishable. The *base* of southern jet seems to "wiggle": its inclination to the symmetry axis is slightly different in 2009 and 2010 observations. In some of 2010 observations the *base* seems curved a little.

The *bar* spectrum is obtained by fitting the combined data of 10 observations (obs. ID 10132–12074; 12075 was not used because the *bar* is absent on it). The angular size of the *bar* is $\sim 4.5'' \times 1.5''$, so it spans about 3×9 pixels of the ACIS detector. Therefore, statistics of the individual observations, i.e., the number of X-ray photons collected in this small area was subject to substantial uncertainties. From ~ 600 to 1000 counts were collected in the *bar* during the individual exposures. To compare, we obtained a combined spectrum for the southern jet *base*.

As mentioned above, the choice of background regions for the spectral analysis is not obvious because of the highly structured morphology of the nebula and the large mean free path of the X-ray emitting particles. We tried two different backgrounds shown in Fig. 9. With background I we got $\Gamma = 1.64^{+0.14}_{-0.13}$ (the errors here and below at the 3σ level) for the *bar* and $\Gamma = 1.14 \pm 0.06$ for the *jet base*. With background II we obtained $\Gamma = 1.60^{+0.08}_{-0.07}$ and $\Gamma = 1.25 \pm 0.04$, respectively. The fit qualities are given in the Appendix (Table 2). The estimates consistent with both

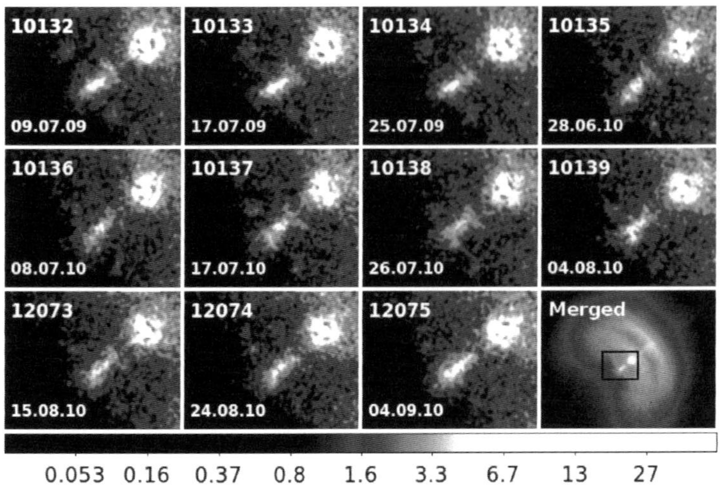

Figure 8. The X-ray maps of the vicinity ($22.8'' \times 16.5''$ area) of the Vela pulsar. The evolution of the southern jet *base* and the transverse *bar* is shown. First three images were made in 2009, the other eight – in 2010. In the lower right image the studied area is shown with a black box. The intensity scale is given for the images of individual observations, not for the merged one.

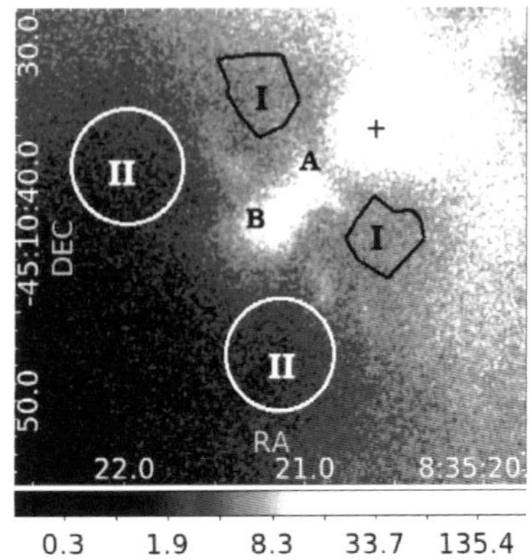

Figure 9. The summed X-ray image of the vicinity of the Vela pulsar: the *bar* (A) and the *base* of the southern jet (B) are shown. Black and white contours show the background areas used for spectral analysis of these structures.

background choices are: $\Gamma = 1.65 \pm 0.15$ for the *bar* and $\Gamma = 1.20 \pm 0.12$ for the *base* of the southern jet.

Notice that in Fig. 6 the photon indices of the individual observations of the southern jet's base appear softer (~ 1.25–1.35, though with larger uncertainties), since they were derived using yet another background (the region "7" in Fig. 2). In [10] the photon index of the southern jet base (termed there the "inner counter-jet") was estimated as $\Gamma = 1.20$ (with the *rms* deviation of 0.10) with the background consisting of two circles of 10.6 arcsec^2 each in the immediate vicinities northeast and southwest of the jet base. Another estimate $\Gamma = 1.36 \pm 0.04$ was recently given in [6], though the authors did not specify the background choice. Nevertheless, the spectrum of the *bar* is much softer than of the *jet base* and other PWN structures and its photon index resembles that of the emission of electrons accelerated at an ultrarelativistic collisionless shock.

3. Conclusions

Evolution of X-ray surface brightness and spectra of fine spatial structures in the Vela pulsar wind nebula (PWN) have been studied. The spectrum and dynamics of the transverse $4.5'' \times 1.5''$ *bar* within $4''$–$5''$ from the Vela pulsar, at the base of the southern jet of the nebula, was investigated. The *bar* can disappear and form again in ~ 10 days. Its power law photon index is much softer than that of the adjacent jet's *base* – 1.65 ± 0.15 vs 1.20 ± 0.12, respectively – and is close to 1.6, a typical value for synchrotron emission of electrons accelerated at an ultrarelativistic shock. Brightness, shape, and positions of the *bar* and the southern jet *base* vary on a timescale of 10 days. The *base* seems to "wiggle" – it curves and changes its inclination to the PWN's symmetry axis. We speculate that bright *blobs* seen on both sides of the southern jet *base* may have the same nature as the bright *blobs* on the terminaton shock ring of the Crab PWN, and thus might reveal the position of the termination shock of the Vela pulsar wind.

Acknowledgements

G.P. and K.L. appreciate A.M. Bykov for enlighting discussions, and Yu.A. Shibanov for consultations on Chandra data processing. We thank the referee for the comments to the paper. A.K. and Yu.K. acknowledge support from RAS Presidium program No. 28.

References

[1] Bykov A M, Amato E, Petrov A E, Krassilchtchikov A M,. Levenish K P 2017 *Space Sci. Rev.* **207** 235
[2] Bykov A M, Gladilin P E, Osipov S M 2013 *MNRAS* **429** 2755
[3] Chevalier R A, Reynolds S P 2011 *ApJ Lett* **740** 26
[4] Dodson R, Lewis D, McConnell D, Deshpande A A 2003 *MNRAS* **343** 116
[5] Kargaltsev O, Pavlov G G, Sanwal D, Garmire G P 2002 *in: Neutron Stars in Supernova Remnants, ASP Conf. Series* **271** 181
[6] Kargaltsev O, Klingler N, Chastain S, Pavlov G G 2017 *arXiv*:1711.02656 *(to appear in AIP Conf. Series)*
[7] Krassilchtchikov A M, Bykov A M, Castelletti G M, Dubner G M , Kargaltsev O Yu and Pavlov G G. 2017 *IOP Conf. Series: Journal of Physics: Conf. Series* **798** 012003
[8] Levenfish K P, Bykov A M, Durant M, Kargaltsev O Y, Kropotina Y A, Pavlov G G, Krassilchtchikov A M, & Uvarov Y A 2013 *Mem. S.A.It.* **84** 588
[9] Pavlov G G, Zavlin V E, Sanwal D, Burwitz V, Garmire G P 2001 *ApJ* **552** L129
[10] Pavlov G G, Teter M A, Kargaltsev O, Sanwal D 2003 *ApJ* **591** 1157
[11] Wilms J, Allen A, McCray R 2000 *ApJ* **542** 914

Appendix

The choice of backgrounds in highly structured maps of extended X-ray emission is not obvious. It is especially non-trivial for the Vela PWN as it is rich in fine structures, which change their spatial extent and position, as well as brightness and spectral hardness, from observation to observation. Further complication comes from the probable presence of a weak bow shock embracing the whole nebula from the North [1, 4, 3]. The bow shock is formed where a mildly relativistic downstream of the TS of the pulsar wind encounters a weakly supersonic flow (of the Mach number $\sim$1.3), which hits the PWN from the North. The flow probably results from the passage of the asymmetric reverse shock of the supernova over the PWN. As such, the converging flows are shown to be highly effective in accelerating particles, which emit very hard X-ray spectra of $\Gamma \sim 1$ [2, 1]. Owing to their large mean free path, the X-ray emitting particles accelerated in the converging flows in Vela PWN may be partially responsible for the spectral hardness of its fine structures. For the same reason, the X-ray emission of the individual structures may overlap and appear as the diffuse emission embracing the whole nebula (which may account for $\sim 70\%$ of the total emission of the PWN [9]). Since we are interested in the intrinsic spectra of the fine structures, two background regions "7" right outside the bright PWN were considered to subtract the contributions of the diffuse area and the bow shock emission, and to account for gradual weakening of the latter (as one moves away from the bow shock). However, if the spectra are fitted with the background shown by the dashed contours in Fig. 2), no significant difference is found in spectra and brightness (deviations were less than few %).

Table 1. Parameters used for the analysis of the X-ray maps of the Vela PWN in DS9. All Vela maps are made with the colormap "b". Gaussian function was used to smooth the images.

fig.	bin	contrast	bias	radius(px)	fig.	bin	contrast	bias	radius (px)
1	0.5	1.26	0.42	2	5	0.15	5.47	0.55	6
1 (inset)	0.1	2.4	0.43	5	8	0.25	3.43	0.56	3
2	0.5	1.26	0.42	2	9	0.15	2.7	0.43	3
4	0.25	2.6	0.45	3					

Table 2. Photon indices of the *bar* and *SE jet base* (3 sigma errors)

	bar		*SE jet base*	
background	Γ	C-stat/(1020 dof)	Γ	C-stat/(1020 dof)
I	$1.64^{+0.14}_{-0.13}$	1.03	1.14 ± 0.06	1.08
II	$1.60^{+0.08}_{-0.07}$	0.95	1.25 ± 0.04	1.01

Evolution of anisotropic distributions of weakly charged heavy ions downstream collisionless quasiperpendicular shocks

J A Kropotina[1,2], **A M Bykov**[1,2], **A M Krassilchtchikov**[1] **and K P Levenfish**[1]

[1]Ioffe Institute, 194021, St. Petersburg, Russia
[2]Peter the Great St. Petersburg Polytechnic University, 195251, St. Petersburg

E-mail: `juliett.k@gmail.com`

Abstract. Hybrid simulations of quasiperpendicular collisionless shocks (QRCS) in multicomponent plasmas have revealed that weakly charged ions can keep anisotropic nonequilibrium distributions at relatively long distances downstream the shock front. However, analytic considerations suggest that such distributions lead to growth of the Alfven ion cyclotron (AIC) instability, which in turn governs isotropisation. Hence, we have conducted targeted hybrid simulations of QRCS with increased accuracy, which appeared to agree with the analytical predictions and suggest that the nonequilibrium distributions eventually relax due to the AIC instability. On another hand, downstream electromagnetic instabilities generated by anisotropic particle distributions may affect the energy balance, therefore influencing macroscopic shock parameters, such as the compression ratio. Simulations of QRCS with parameters close to those of the heliospheric termination shock measured by the Voyager mission, showed that an admixture of just 0.7% (by mass) of oxygen decreases the compression ratio by 3-4%. Thus, one might conclude that along with the effect of the pickup ions, the multi-species composition of the interplanetary plasma can add to the observed decrease of the shock compression.

1. Introduction

Shock waves provide a universal mechanism of conversion of kinetic energy of supersonic flows into the thermal energy of plasma, as well as into the energy of amplified magnetic fields and of suprathermal particles. In the strong enough collisionless shocks (typically, of a Mach number above a few depending on the shock obliquity) the electron resistivity cannot provide energy dissipation fast enough to create a quasi-steady shock transition on a microscopic scale [1–3]. The structure of the shock transition is governed by the instabilities of the ion flows and such shocks are called *supercritical*(see, e.g., [4]; [5]; [6]). At the microscopic scale the front of a supercritical shock is a transition region occupied by magnetic field fluctuations of an amplitude $\delta B/B \gtrsim 1$ and characteristic frequencies of about the ion gyro-frequency. Generation of the fluctuations is due to instabilities in the interpenetrating multi-flow ion movements. The width of the viscous transition region of a collisionless shock typically reaches a few hundred ion inertial lengths defined as $l_i = c/\omega_{\rm pi} \approx 2.3 \times 10^7\, n^{-0.5}$ cm. Here $\omega_{\rm pi}$ is the ion plasma frequency and n is the ionized ambient gas number density measured in $\rm cm^{-3}$. The width of the collisionless shock transition region is smaller by many orders of magnitude than the Coulomb mean free path in the interstellar or intergalactic medium [7], so particle relaxation is due to collective electromagnetic field fluctuations. In such an enviroment nonequilibrium states can arise and be kept for distances much larger than the characteristic shock transition lengths. Particularly, quasi-stationary anisotropic velocity distributions of a small admixture of weakly charged heavy ions downstream QRCS (i.e., the shocks with initial magnetic field directed nearly transversely to the shock normal) were found by [8, 9] via hybrid plasma simulations.

Here we present a detailed modeling of AIC instability caused by such ion distributions. Due to the high nonlinearity of a collisionless shock and the strong coupling between electromagnetic fluctuations and the motions of charged particles, pure analytic studies of such processes are possible only in some special cases and still provide only approximate results. Hence we combine an analytic approach with numerical hybrid simulations of QRCS (see [8, 9] for more detailed discussions of this approach). In section 2 the hybrid code and initial parameters of a simulation are briefly described. In section 3 the resulting nonequilibrium distributions are shown and both analytical and numerical treatments of AIC instability are discussed, as well as their possible application of to the physics of the heliospheric termination shock.

2. Hybrid Simulations of Shocked Plasma Flows

A 3d second-order accurate particle-mesh hybrid code with exact magnetic field divergence conservation was used for simulations of QRCS evolution. The main feature of a hybrid code is that it combines hydrodynamic and particle-in-cell approaches, treating ions as particles and electrons as a neitralizing massless fluid ([10]). This allows one to self-consistently model processes governed by ions at time scales up to several thousand inverted ion gyrofrequencies (Ω_i^{-1}. A standard system of equations for a hybrid code is given in [11]. These equations are solved within a cyclic leapfrog approach with particle coordinates and velocities given at the odd timesteps and fields – at the even ones. A "Reconstruct-Solve-Average" algorithm including finite-volume discretisation, piesewise-parabolic reconstruction and linearized Roe-type Riemann solver (see, e.g., [12]) was implemented in the code to achieve the exact magnetic field divergence conservation. Triangular-shaped cloud representation of the ions and Boris particle pusher (see, e.g., [13]) were employed to move particles and calculate induced currents with second-order accuracy in time and space. The shock was initialized via the standard piston method, where a superalfvenic flow reflects from a conducting wall and forms the front moving oppositely to the initial flow.

3. Evolution of Heavy Ion Distributions

3.1. Nonequilibrium phase spaces and distributions

According to [8,9], nonequilibrium heavy ion distributions appear for various QRCS parameters, provided that the ion charge-to-mass ratio and relative density is low enough. For an illustrative example shown in Fig. 3.1 plasma parameters typical for a supernova reverse shock were chosen: the upstream Alfven Mach number (in the downstream frame) $M_A = 9.0$, the inclination angle (between the direction of the initial magnetic field and the shock normal) $\theta = 90^o$ (so the shock was exactly perpendicular), and the ratio of thermal to magnetic pressure $\beta = 1.0$. The reflecting wall was at x =0, the initial flow moved to the left, and the initial number of model particles per cell (ppc) was equal to 200. Plasma consisted of 99% O IV and 1% Si II (by mass). Figure 3.1 shows the resulting phase spaces of the main element (left panels) and heavy ion admixture (middle panels) at $t = 100\Omega_i^{-1}$, where $\Omega_i = q_{main}B/m_{main}c$ is the main ions sort (O IV) gyrofrequency. The corresponding spatially averaged velocity distributions are given at the right panels. It is clear that while the main element phase spaces and distributions are isotropic and homogenious, the Si II ones show peculiar quasi-periodic structure, associated with heavy ions gyromotion downstream the shock. The averaged Si II velocity distributions are substantially anisotropic with preferred magnetic field direction (which is along z in this simulation).

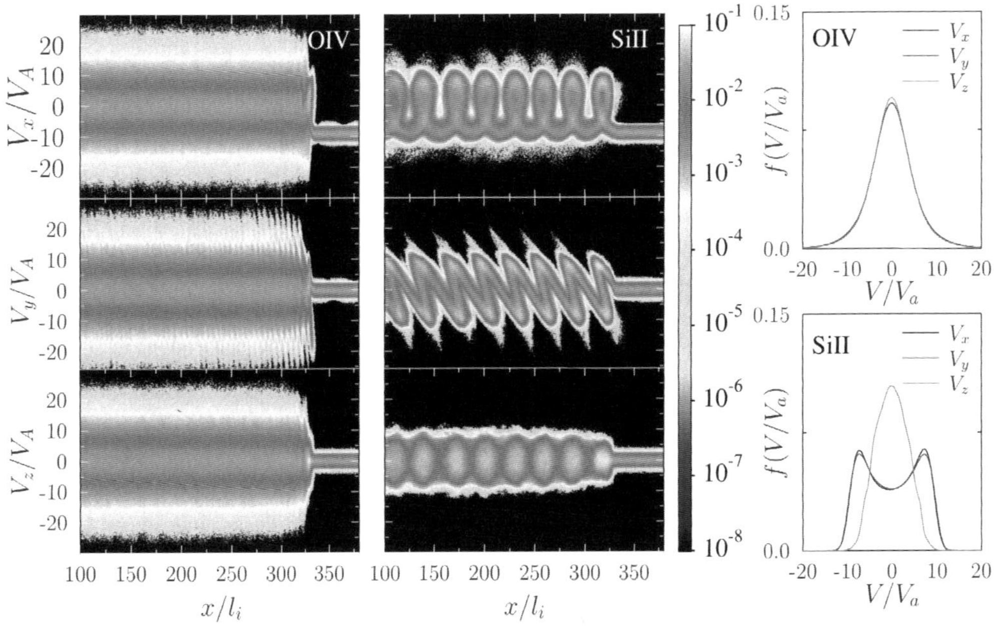

Figure 1. The O IV and Si II $x - V$ phase spaces and velocities distributions in shock with $M_a = 9.0$, $\theta = 90^o$ and $\beta = 1.0$ at $t = 100\Omega_i^{-1}$. All lengths are given in the main element l_i, while velocities are in the far upstream $V_a = B_0/\sqrt(4\pi\rho)$.

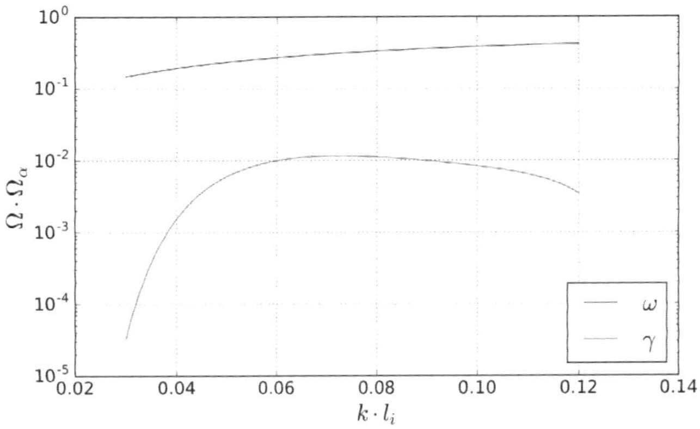

Figure 2. Solution of the dispersion equation for medium with macroscopic parameters and SiII distribution anisotropy taken from the simulated downstream region of the same shock as in figure 3.1. Positive γ indicate instability growth.

3.2. Alfven ion cyclotron instability

In paper [14] it was shown (with application to thermonuclear fusion) that temperature anisotropy of hot minority ions population with isotropic equilibrium background can drive the AIC instability, which leads to the hot minority distribution relaxation. In the kinetic approach the distribution details appeared to be unimportant, so just the distribution moments are included in the following dispersion equation:

$$\Omega^2 - c^2k^2 + \omega_{pe}^2 \frac{\Omega}{kV_{T\|e}} Z(\xi_e^\pm) + \omega_{pi}^2 \frac{\Omega}{kV_{T\|i}} Z(\xi_i^\pm) + \omega_{p\alpha}^2 \frac{\Omega}{kV_{T\|\alpha}} Z(\xi_\alpha^\pm) + \omega_{p\alpha}^2 (\frac{T_{\perp\alpha}}{T_{\|\alpha}} - 1)[1 + \xi_\alpha^\pm Z(\xi_\alpha^\pm)] = 0 \quad (1)$$

where $Z(\xi) = \pi^{-0.5} \int_{-\infty}^{\infty} dt \cdot exp(-t^2)/(t-\xi)$, $\xi_j^\pm = \Omega \pm \Omega_j/kV_{T\|j}$, $\Omega_j = q_jB/m_jc$, $\omega_{pj}^2 = 4\pi n_j q_j^2/m_j$ are jth species gyro- and plasma frequencies and $V_{T\|j}$, $T \parallel j$, $T_\perp j$ - their thermal velocities and temperatures with respect to the averaged magnetic field direction, i index corresponds to the main sort, while α - to the admixture. The solution of this equation (real and imaginary parts of Ω) is plotted versus k in Figure 3.1 for parameters and SiII anisotropy from model downstream region. The wave vector is given in inversed l_i and frequencies in Ω_α. It can be seen that there persists positive increment area, which means the instablity growth, though the maximum increment for such low SiII density is only $10^{-2}\Omega_\alpha$, which gives the characteristic growth time about $\tau \sim 500\Omega_i^{-1}$. In order to numerically confirm these analytical results we made the following test run of the hybrid code: one period of heavy ions phase space was placed in a periodic box with uniform averaged downstream magnetic field and equilibrium downstream main sort population. In order to avoid numerical effects and make visible the expected tiny AIC instability effect we increased number of ppc up to 2000. The resulting phase spaces and velocity distributions evolution for ppc=2000 are shown in Figure 3.2, while the space-averaged magnetic field variance evolution for cases of ppc=200 and ppc=2000 is given in Figure 3.2. It is clear that for the case ppc=200 numerical noise is greater than the effect we consider, while for ppc=2000 there can be seen even two instabilities. The first one grows up very rapidly and is substantially damped up to $t = 200\Omega_i^{-1}$. It gives rise to magnetic field fluctuations propagating along x axis, so it cannot be associated with AIC instability, which propagates along the averaged magnetic field direction (z in our case). Probably it is governed by heavy ions spatial inhomogenity, which was not included in analytic studies and is valuably reduced with time, as can be seen from Figure 3.2. The second, slowly growing, type of instability gives rise to magnetic field fluctuations, transverse to the magnetic field direction and propagating along it with nearly alfven speed. Its visible growth is roughly consistent with characteristic time $\tau \sim 500\Omega_i^{-1}$, obtained in kinetic theory approach. Figure 3.2 also indicates slow velocity distribution evolution towards isotropy. So it can be concluded that anisotropic heavy ion distributions drive AIC instability, which eventually leads to their relaxation on scales much less that Coulomb ones, but much greater than characteristic shock transition lengths.

3.3. Termination shock simulations

The effect of slow heavy ions temperature relaxation downstream QRCS leads to increased magnetic field fluctuations level in the vicinity of shock transition. This effect can therefore change the total energy balance and affect the Rankine-Hugonio relations. As Supernova remnant observations are indirect and cannot reveal such a local effect, we tested this hypothesis on the shock with parameters close to the Termination shock ones, measured by Voyager2 ([15]): $M_a = 4.0$ in downstream frame, $\theta = 82^o$, $\beta = 0.01$. Actual Solar wind composition includes about 85% of H II, 14.2% of He III and about 0.7% of heavier ions (by mass). So three test runs were done to estimate the heavy ions impact on the shock compression ratio: the first – for the pure Hydrogen shock, the second – for the shock consisting of H II and He III with Solar abundancies and the third one with admixture of 0.7% of O VII. For all cases the averaged downstream density was measured. The resulting values for the proton and proton-helium shocks coincided within the uncertainty, while addition of only 0.7% of O VII appeared to decrease the downstream density by 3-4%. The O VII distributions appeared to be anisotropic

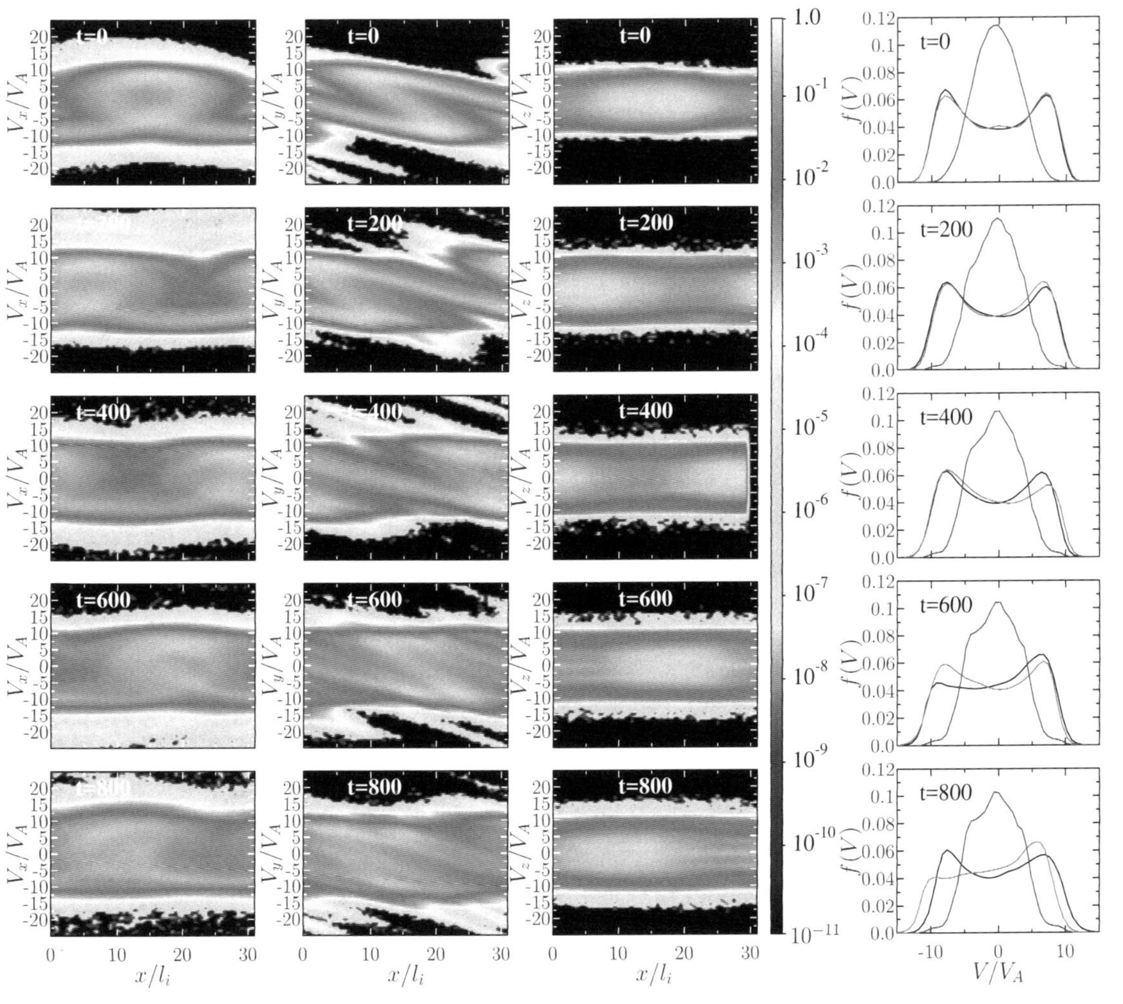

Figure 3. The O IV and Si II $x - V$ phase spaces and velocity distributions evolution in the test run(see text)

in the latter case, and the corresponding level of magnetic field fluctuations was higher than without Oxygen. This may lead to the conclusion that pickup ions are not the only reason of the observed low Termination shock compression ratio ([15]), but plasma composition can also affect this parameter. However, to check this statement, the full model including both pickup ions and heavy ions is needed, which is out of the scope of this paper.

4. Conclusions

The two approaches such as solution of plasma kinetic dispersion equation and hybrid modeling with increased number of particles per cell (ppc = 2000) follow to the same conclusion that anisotropic heavy ions distribution, arising downstream QRCS, lead to emerging of alfven ion cyclotron instability, which might grow rather slowly in case of small heavy ions relative density. Anyway, this instability has two consequences:

(i) Relaxation of anisotropic ion distributions to isotropic maxwellian ones on scales much less than the Coulomb length for typical astrophysical objects

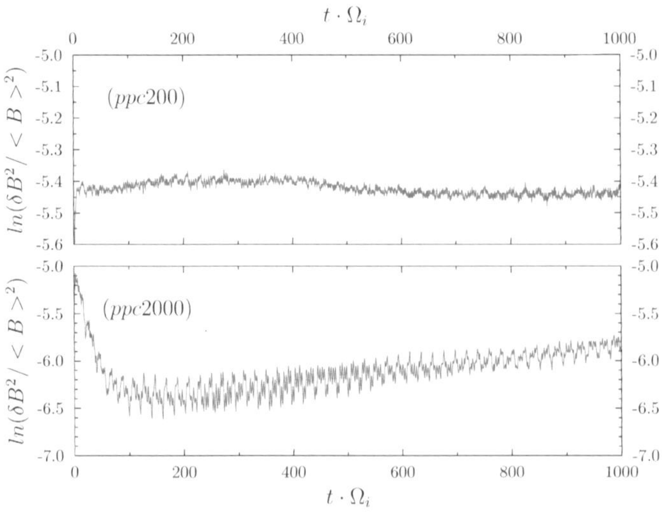

Figure 4. Relative magnetic field variance in the same run as in Figure 3.2

(ii) Increasing level of magnetic field fluctuations in the vicinity of transition, which can lead to changing of macroscopic shock parameters. The latter effect was illustrated on the example of the Termination shock.

5. Acknowledgements

A.M. Bykov and J.A. Kropotina are grateful to Peter the Great St. Petersburg Polytechnic University for the access to the computation resources of the Tornado cluster and also to the Moscow Joint Supercomputer Center for the access to the computation resources of MVS-10P, which together made possible the resource-intensive simulations with increased accuracy. A.M. Bykov and J.A. Kropotina were supported by the RSF grant 16-12-10225.

References

[1] Sagdeev R Z 1966 *Reviews of Plasma Physics, New York: Consultant Bureau* vol 4 ed Leontovich M A pp 23–46
[2] Tidman D A and Krall N A 1971 *Shock waves in collisionless plasmas*
[3] Treumann R A 2009 *Astronomy and Astrophysics Reviews* **17** 409–535
[4] Kennel C F, Edmiston J P and Hada T 1985 *Washington DC American Geophysical Union Geophysical Monograph Series* **34** 1–36
[5] Lembege B and others/ 2004 *Space Science Reviews* **110** 161–226
[6] Burgess D *et al.* 2005 *Space Science Reviews* **118** 205–222
[7] Bykov A M, Paerels F B S and Petrosian V 2008 *Space Science Reviews* **134** 141–153 (*Preprint* 0801.1008)
[8] Kropotina and others Y A 2015 *Technical Physics* **60**(2) 231–239
[9] Kropotina and others Y A 2016 *Technical Physics* **61**(4) 517–524
[10] Winske D and Omidi N 1996 *JGR* **101** 17287–17304
[11] Matthews A P 1994 *J. Comput. Phys.* **112**(1) 102–116 ISSN 0021-9991
[12] R J LeVeque D Mihalas E D E M 2006 *Computational Methods for Astrophysical Fluid Flow* (Springer Science and Business Media)
[13] Birdsall C K and Langdon A B 1991 *Plasma Physics via Computer Simulation*
[14] Lashmore-Davies C N, Dendy R O and Kam K F 1993 *Plasma Physics and Controlled Fusion* **35** 1529–1540
[15] Richardson J D, Kasper J C, Wang C, Belcher J W and Lazarus A J 2008 *Nature* **454** 63–66

Equation of state of the intergalactic medium in the early Universe

K N Telikova, S A Balashev and P S Shternin

Ioffe Institute, 26 Politeknicheskaya st., St. Petersburg, 194021, Russia

E-mail: ks.telikova@mail.ru

Abstract. Spectroscopy of the Lyα forest in quasar spectra proved to be a useful tool for probing the intergalactic gas. We developed the automatic program for Voigt profile fitting of Lyα forest lines. We run this code on 9 high resolution ($\sim$ 50000) quasars spectra with a high signal-to-noise ratio ($\sim 50 - 100$) from the Keck telescope archive and obtained the sample of single well-fitted Lyα lines. Fitting the joint 2d distribution of column density and Doppler parameter from this sample by physically reasonable model we estimate a power law index, γ, of the equation of state of the intergalactic medium in the redshift range $z \sim 2 - 3$. We found that our measurement is in an agreement with measurements by other groups obtained with Voigt profile fitting technique.

1. Introduction

It is well known that the intergalactic medium (IGM) contains a dominant fraction of baryons in the Universe [1]. After the recombination epoch at a redshift $z \sim 1100$ the intergalactic medium (IGM) had remained completely neutral until first stars, galaxies and quasars began to produce the ionizing background radiation. The IGM at this epoch mostly consisted from hydrogen and helium and consequently the IGM reionization included two major stages. The first stage is the reionization of H I which was completed at $z > 6$ [2] and dominated by starlight escaped from the galaxies [3]. At this epoch the helium was already singly ionized. The second stage is the reionization of He II which occurs at later epochs, $z < 3$ [4], and is driven by quasars radiation [5]. Despite the considerable progress, our understanding of the reionization process is incomplete. For instance, we still poorly know the properties of the ionizing ultraviolet background, precise redshifts of the reionization events and their impact on the thermal state of IGM. It is assumed [6] that the equation of state – the temperature-density, $T - \rho$, relation – after the H I reionization epoch is set by the competing processes of photo-heating and adiabatic cooling and obeys a power-law form

$$T = T_0 \left(\frac{\rho}{\bar{\rho}}\right)^{\gamma-1} \equiv T_0 \Delta^{\gamma-1}, \tag{1}$$

density where T_0 is the temperature at the mean density $\bar{\rho}$, Δ is the overdensity and according to theoretical predictions. Numerical calculations suggest the index γ is about 1.6 well after the reionization [6, 7].

The small residual neutral fraction of hydrogen remained after the reionization (at $z < 6$) leaves traces in the spectra of background quasars observed as series of the absorption features

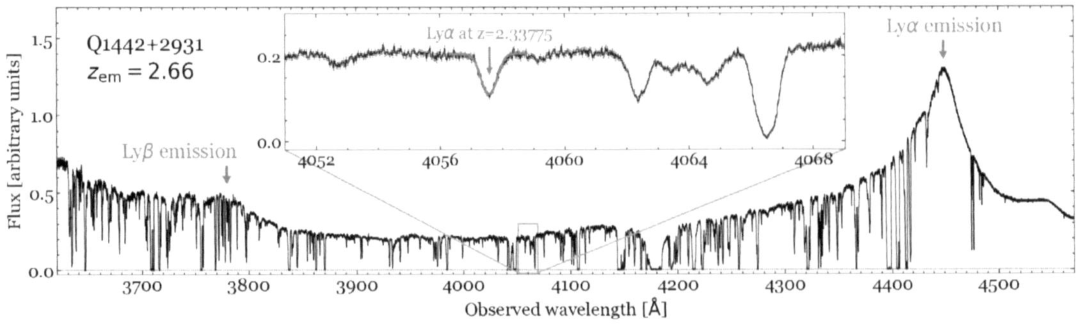

Figure 1. The high resolution spectrum of the Q1442+2931 obtained with Keck telescope. The Lyα forest is the region bluewards the Lyα emission line of quasar. It consists of numerous Ly series H I absorption lines associated with IGM. *Inlay*: zoom of a small part of the spectrum; the red curve shows the fit to the H I Lyα absorption line at $z \approx 2.33775$.

called the Lyα forest (figure 1). An analysis of these lines allows to probe the thermal state of the IGM and its evolution with redshift. Various methods were suggested based on the statistical properties of the Lyα forest. These include analysis of the flux probability distribution function [8, 9], the power-spectrum of the transmitted flux [10, 11], the average local curvature [12, 13] and wavelet decomposition of the Lyα forest [14, 15]. In this paper we employ another widely-used technique which is based on a decomposition of the Lyα forest into individual components and their subsequent Voigt profile analysis [16–22].

2. Data and analysis

Table 1 collects basic parameters of 9 selected high-resolution quasar spectra with high signal-to-noise ratio (SNR) we used for the analysis. Most of them was taken from KODIAQ database [23] and one spectrum, Q1442+2931, was reduced by hand from the Keck telescope archival data using MAKEE package[1]. An unabsorbed continuum level for the Q1442+2931 spectrum was estimated using B-spline interpolation, spectra from KODIAQ database are already normalized, but in some cases we adjusted continuum manually.

We developed an automatic procedure for analysis of the H I absorption lines in the Lyα forest. In each quasar spectrum we used the spectral region between Lyα and Lyβ emission lines with offset of 12000 km/s bluewards from Lyα emission lines. This allowed us to avoid confusion with Lyβ forest lines and to minimize proximity effects for quasar. Each individual H I absorber in the Lyα forest was described by the Voigt profile, with the column density, N, Doppler parameter, b, and redshift, z. Instead of fitting each Lyα forest line in specified redshift range with multicomponent profile, we searched for the solitary lines that can be fitted with only one component (see example in the inlay of figure 1). We constructed the dense grid of parameters (z, N, b), see table 2, and used χ^2 criterion to compare model profiles calculated on that grid with observed ones. The Lyα forest lines in the probed redshift range, $z \sim 2-3$, frequently blend with each other. Therefore, we used a sort of outlier rejection routine to include in analysis also the lines with weak blends in one of the wings. In addition we checked that the higher-order Lyman-series absorption lines are in agreement with the estimated parameters from corresponding Lyα line. Finally, we excluded lines with parameters lying on the grid edges to get rid of the boundary effects on the line sample density estimation in the parameters space.

[1] `http://www.astro.caltech.edu/~tb/makee/`

Table 1. Quasar spectra used for the analysis.

Quasar	$z^{\dagger}_{\rm em}$	$z^{\ddagger}_{\rm range}$	SNR	Resolution
J010311+131671	2.72	2.06 – 2.68	107	36000
J014516–094517	2.72	2.03 – 2.70	83	36000
J020950–000506	2.83	2.15 – 2.79	82	48000
J082619+314848	3.09	2.44 – 2.58	44	48000
J101155+294141	2.65	2.11 – 2.60	63	36000
Q1442+2931	2.66	2.08 – 2.62	117	48000
J155152+191104	2.84	2.09 – 2.80	96	36000
J162548+264658	2.52	2.11 – 2.50	43	36000
J170100+641209	2.73	2.06 – 2.71	70	36000

† quasar redshift
‡ redshift range used for analysis of the Lyα forest.

Table 2. Parameters of the grid.

Parameter	Range	Step
Redshift, z	Individual	8.12×10^{-6}
Column density, $\log N$ [cm^{-2}]	13 – 15	0.02
Doppler parameter, b [km s^{-1}]	12 – 27	0.48

The choice of steps in the grid was based on the Fisher matrix estimate.

In this way we obtained the representative sample of "solitary" Lyα forest lines in the specified range of (N, b) parameters.

Since our selection procedure is not the line profile fitting, it does not return the line parameter uncertainties σ^N and σ^b. These were obtained analytically via the Fisher matrix for the χ^2 likelihood function with account for resolution and SNR of each spectra. We checked that this gives reasonably close estimates of the expected uncertainties would the real fit be performed. The resulting sample in (N, b) parameter space is shown in figure 2. The mean redshift of the Lyα lines in our sample is $\bar{z} = 2.35$.

The obtained (N, b) distribution shows a prominent lower envelope. Its physical origin is as follows [16]. The parameter b characterizes the width of the velocity distribution of hydrogen atoms, which contains thermal and turbulent motions inside each absorber. Since thermal broadening is unavoidable, for each N there exists a minimal value of b which corresponds to a pure thermal broadening

$$b \to b_{\rm th} \equiv \sqrt{2k_{\rm B}T/m}, \tag{2}$$

where $k_{\rm B}$ is the Boltzmann constant and m is the hydrogen atomic mass.

The position of the low boundary of the $N - b$ distribution can be connected to the temperature-density relation of the IGM, given by eq (1). In assumption of uniform background radiation, the H I column density N is related to the overdensity Δ as [24, 25][2]

$$N = 1.3 \times 10^{14}\, \Delta^{3/2} \frac{T^{-0.22}}{\Gamma_{-12}} \left(\frac{1+z}{3.4}\right)^{9/2} \ {\rm cm}^{-2}, \tag{3}$$

[2] The value of the prefactor in eq (3) and the power exponent of T there depend on the adopted value of case-A hydrogen recombination coefficient. Here we take the same prescription as in [25].

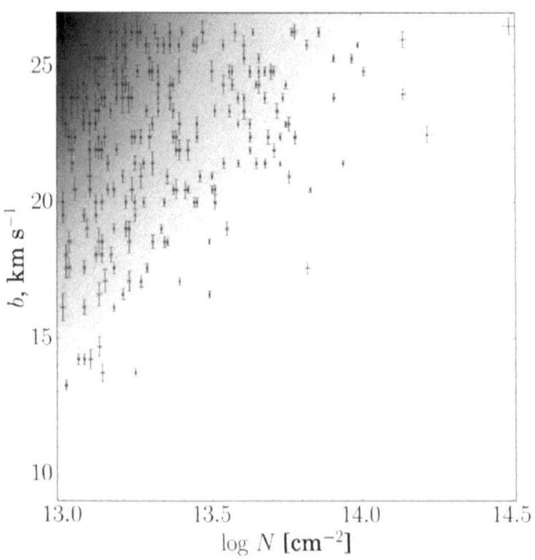

Figure 2. Error crosses show the obtained absorption lines sample. The gradient-filled area shows the best-fit probability density function of the (N, b) sample.

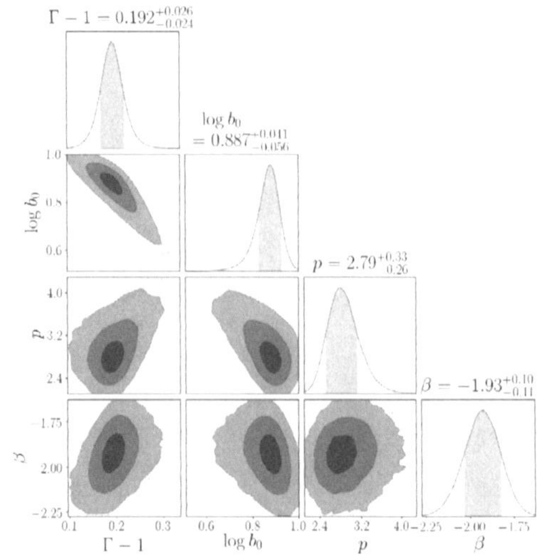

Figure 3. Marginal posterior distributions of the parameters Γ, $\log b_0$ [km s^{-1}], p and β. Dark and light blue areas shows 1σ and 2σ levels for 1D distributions respectively.

where Γ_{-12} is the hydrogen photoionization rate in units of 10^{-12} s^{-1}. Eqs (1), (2) and (3) allow to predict the $b_{\rm th} - N$ dependence

$$b_{\rm th} = b_0 \left(\frac{N}{10^{12}\ {\rm cm}^{-2}}\right)^{\Gamma-1}, \quad \Gamma - 1 = \frac{\gamma - 1}{3 - 0.44(\gamma - 1)}, \tag{4}$$

where the normalization constant b_0 depends on T_0, Γ_{-12} and z. Once b_0 and Γ are found, the parameters $\gamma(z)$ and $T_0(\Gamma_{-12}, z)$ can be inferred.

Techniques that employ decomposition of Lyα forest lines are usually based on direct determination of the lower boundary of $N - b$ distribution. This is not a straightforward task. Various iterative techniques are proposed to eliminate the turbulent contribution to width of the lines [17, 19, 20]. Most popular statistics is based on the so-called 2σ rejection algorithm [16], which in some cases is calibrated with the hydrodynamical simulations.

Here we propose to fit the whole (N, b) distribution. We use microturbulence assumption that thermal and turbulent motions are uncorrelated and can be characterized by the Gaussian distributions with widths $b_{\rm th}$ and $b_{\rm turb}$, respectively. The convolution of these distributions is also Gaussian with the characteristic width $b = \sqrt{b_{\rm th}^2 + b_{\rm turb}^2}$. We assume that turbulent motions are independent on the thermal state of the gas and its density. Besides, we suggest that distributions over N and $b_{\rm turb}$ have power-law shapes with indexes β and p, respectively. Notice that for N distribution this is a standard assumption and index β is well measured, see, e.g., [18]. In this way the 2D probability density distribution has the form

$$f(N, b) = \int f_N(N) f_{\rm turb}(b_{\rm turb}) \delta\left(b - \sqrt{b_{\rm th}^2 + b_{\rm turb}^2}\right) {\rm d}b_{\rm turb}, \tag{5}$$

where f_N and $f_{\rm turb}$ are the probability density distributions over N and $b_{\rm turb}$, respectively. Thus the function f depends on the parameter vector $\boldsymbol{\Theta}$=(Γ, $\log b_0$, p, β) in addition to N and b.

To estimate the posterior distributions of the parameters $\mathbf{\Theta}$ we used the Bayesian framework. Flat priors were chosen for all parameters. Our $(N,\, b)$ sample probes only a limited part of the full distribution $f(N, b)$, namely the region specified by the searching procedure box (see table 2). The estimated uncertainties σ_i^N and σ_i^b in the sample are heteroscedastic, therefore the probability of the data point to fall in this box is different for each line, since this probability depends on the observed (N, b) values and not on the unknown true values $(\widetilde{N}, \widetilde{b})$. As a result, the correctly normalized likelihood function is [26]

$$\mathcal{L}\,(N_i,\, b_i|\,\mathbf{\Theta}) = \prod_i \frac{\int f(\widetilde{N},\widetilde{b}) f_i(N_i, b_i|\,\widetilde{N},\widetilde{b})\,\mathrm{d}\widetilde{N}\,\mathrm{d}\,\widetilde{b}}{\int f(\widetilde{N},\widetilde{b}) I(N,b) f_i(N,b|\,\widetilde{N},\widetilde{b})\,\mathrm{d}N\,\mathrm{d}b\,\mathrm{d}\widetilde{N}\,\mathrm{d}\widetilde{b}}, \tag{6}$$

where $f_i(N, b|\,\widetilde{N},\widetilde{b})$ is the bivariate normal distribution of parameters N, b with the mean $\widetilde{N}$, $\widetilde{b}$ and the dispersion σ_i^N, σ_i^b. $I(N,\, b)$ is the selection function equal to one if a pair $(N,\, b)$ falls in our analysis box and zero otherwise. We performed the maximum likelihood estimate of the distribution parameters $\mathbf{\Theta}$ employing the Markov Chain Monte Carlo (MCMC) simulations with the affine sampler emcee [27].

3. Results and discussion

The marginal posterior distributions of the fit parameters are shown in figure 3. According to eq (4), the estimated power-law index in the temperature-density relation (1) of the IGM at the average redshift $z = 2.35$ is $\gamma - 1 = 0.53 \pm 0.07$ at the 68 per cent confidence level. Using eq (3) we find the temperature at the mean density

$$\log T_0\,[K] = (3.98 \pm 0.05) - (0.35 \pm 0.04) \log \Gamma_{-12} \tag{7}$$

which, however, depends on the photoionization rate Γ_{-12}. From the analysis of the mean opacity of the Lyα forest lines it was constrained that $(T_0[K])^{-0.72}/\Gamma_{-12} = (1.88 \pm 0.08) \times 10^{-3}$ at $z = 2.4$ [28]. Their calculations also depend on the equation of state of the IGM, however, their adopted value for $\gamma - 1 = 0.6$ is rather close to our 0.53 (see their figure 3), so we do not expect a significant bias. Adopting the result of [28] we can constrain both $T_0 = 1.25^{+0.05}_{-0.19} \times 10^4$ K and $\Gamma_{-12} = 0.60^{+0.08}_{-0.03}$ with a strong anticorrelation between these parameters.

The sample is much smaller than some samples derived from the multicomponent fit (see e.g. [19, 20]), but it allows us to avoid a lack of knowledge of "true" multicomponent models and correlations of the parameters of such compound models. This can be comprehended if one compares our sample with sample, obtained on the same quasar spectra [19]: at each column density region the Rudie et al detect much more lines with low b values than we did. But finally they rejected most of the lines with low b values by some arbitrary routine.

In previous sections we suggest that broadening of absorption lines determined by thermal and turbulent motions. Turbulent broadening is driven by peculiar velocities of the absorbed intergalactic clouds and in our approach it is independent on column density N. But in general real turbulent mechanisms may depend on N. For example we did not take into account expected dependence of b on N due to pressure smoothing scale and differential Hubble flow along to physical extent of intergalactic clouds [29–31].

4. Summary

We developed the automatic procedure for searching and fitting Lyα forest absorption lines and applied it to 9 high-resolution quasar spectra with high SNR in the redshift range $z \sim 2 - 3$. Analysis of the full distribution of the absorption line parameters allowed us to probe the equation of state of the IGM at $\bar{z} = 2.35$. The power-law index of the temperature–overdensity relation $\gamma - 1 = 0.53 \pm 0.07$ is in a broad agreement with the results of other authors. We also obtained the relation (7) between the temperature T_0 at the mean density $\bar{\rho}$ and the hydrogen photoionization rate Γ_{-12}. Combining this result with measurements of [28] we were able to separately constrain T_0 and Γ_{-12}. The proposed automatic program will allow us to analyze in the future much more quasar spectra available in Keck and VLT archives. Such increase of statistic is strongly important to precisely estimate the redshift evolution of the main parameters described the IGM equation of state – γ, T_0 and Γ_{-12}.

References

[1] Meiksin A A 2009 *Reviews of Modern Physics* **81** 1405–69
[2] McGreer I D, Mesinger A and D'Odorico V 2015 *MNRAS* **447** 499–505
[3] Robertson B E, Ellis R S, Furlanetto S R and Dunlop J S 2015 *Astroph. J. Lett.* **802** L19
[4] Worseck G, Prochaska J X, McQuinn M, Dall'Aglio A, Fechner C, Hennawi J F, Reimers D, Richter P and Wisotzki L 2011 *Astroph. J. Lett.* **733** L24
[5] Madau P and Haardt F 2015 *Astroph. J. Lett.* **813** L8
[6] Hui L and Gnedin N Y 1997 *MNRAS* **292** 27–42
[7] Puchwein E, Haardt F, Haehnelt M G and Madau P 2018 *ArXiv e-prints* (*Preprint* 1801.04931)
[8] Viel M, Bolton J S and Haehnelt M G 2009 *MNRAS* **399** L39–3
[9] Lee K G *et al.* 2015 *Astroph. J.* **799** 196
[10] Zaldarriaga M, Hui L and Tegmark M 2001 *Astroph. J.* **557** 519–26
[11] Rorai A *et al.* 2017 *MNRAS* **466** 2690–709
[12] Becker G D, Bolton J S, Haehnelt M G and Sargent W L W 2011 *MNRAS* **410** 1096–112
[13] Boera E, Murphy M T, Becker G D and Bolton J S 2014 *MNRAS* **441** 1916–33
[14] Lidz A, Faucher-Giguère C A, Dall'Aglio A, McQuinn M, Fechner C, Zaldarriaga M, Hernquist L and Dutta S 2010 *Astroph. J.* **718** 199–230
[15] Garzilli A, Bolton J S, Kim T S, Leach S and Viel M 2012 *MNRAS* **424** 1723–36
[16] Schaye J, Theuns T, Leonard A and Efstathiou G 1999 *MNRAS* **310** 57–70
[17] Schaye J, Theuns T, Rauch M, Efstathiou G and Sargent W L W 2000 *MNRAS* **318** 817–26
[18] Ricotti M, Gnedin N Y and Shull J M 2000 **534** 41–56
[19] Rudie G C, Steidel C C and Pettini M 2012 *Astroph. J. Lett.* **757** L30
[20] Hiss H, Walther M, Hennawi J F, Oñorbe J, O'Meara J M and Rorai A 2017 *ArXiv e-prints* (*Preprint* 1710.00700)
[21] Gaikwad P, Srianand R, Choudhury T R and Khaire V 2017 *MNRAS* **467** 3172–87
[22] Rorai A, Carswell R F, Haehnelt M G, Becker G D, Bolton J S and Murphy M T 2018 *MNRAS* **474** 2871–83
[23] O'Meara J M, Lehner N, Howk J C, Prochaska J X, Fox A J, Peeples M S, Tumlinson J and O'Shea B W 2017 *Astron. J.* **154** 114
[24] Schaye J 2001 *Astroph. J.* **559** 507–15
[25] Bolton J S, Becker G D, Haehnelt M G and Viel M 2014 *MNRAS* **438** 2499–507
[26] Kelly B C 2007 *Astroph. J.* **665** 1489–506
[27] Foreman-Mackey D, Hogg D W, Lang D and Goodman J 2013 *Pub. Astron. Soc. Pacific* **125** 306
[28] Faucher-Giguère C A, Lidz A, Hernquist L and Zaldarriaga M 2008 *Astroph. J.* **688** 85-107
[29] Gnedin N Y and Hui L 1998 *MNRAS* **296** 44–55
[30] Kulkarni G, Hennawi J F, Oñorbe J, Rorai A and Springel V 2015 *Astroph. J.* **812** 30
[31] Garzilli A, Theuns T and Schaye J 2015 *MNRAS* **450** 1465–76

The spatial structure of the Galaxy subsystems as it looks from an analysis of the system of galactic planetary nebulae

Lomara Maksimova[1,2] **and Alexander Kholtygin**[1]
[1]Sobolev Astronomical Institute of Saint-Petersburg State University, St. Petersburg, 198504 Russia
[2]Pulkovo Observatory, Pulkovskoe shosse 65-1, St. Petersburg, 196140 Russia

E-mail: lomara.maksimova@gmail.com

Abstract. We recalculated the distance to 554 galactic Planetary Nebulae (PNe) using the relation between the ionized mass of the nebula μ and the optical thickness parameter τ obtained by Akimkin et al. [1]. We classified these nebulae according to the Peimbert [9] types modified by Quireza et al. [11]. The obtained by us distances to galactic PNe were used to obtain the scale heights h for different subsystem of the Milky Way. Next values if h were estimated: $h = 208 \pm 10$ pc for thin disk objects, $h = 600 \pm 54$ pc for thick disk ones, and $h = 1378 \pm 180$ pc for halo PNe.

1. Introduction

Planetary Nebulae (PNe) are the descendants of the stars with initial masses on main sequence in a range from 0.8 to $8M\odot$. PNe are located in different Galaxy subsystems and they have a wide variety of spatial and kinematic properties, chemical composition, and morphology.

Studying of the spatial distribution of galactic PNe is often used test the different models of the Galactic evolution [3]. In the present time there is no agreement in the understanding of the structure of Galactic disk. Bovy et al. [4] claim that there is no clear separation between the thin and thick disks and they have the common scale height. On the other hand Girardi et al. [5] conclude that there are different scale heights both for thin and thick disks. An analyzing of the vertical distribution of PNe can help us to clarify its structure.

Unfortunately the distances to PNe which can help us to solve this problem are badly known. More or less exact distances to selected PNe which are found from the trigonometric parallaxes of their central stars are known only for nearly 40 objects. For the great majority of galactic PNe one has to use the distances based mainly on the Shklovsky/Daub statistical scale [6] which can be very unreliable. The individual statistical distances to PNe obtained by various methods are often strongly different.

In the present work, we used a new calibration of the existing distance scales for Galactic disk PNe obtained by Akimkin et al. [1]. They used the spatially-kinematic modeling of the galactic PNe ensemble proposed by Nikiforov and Bobrova [2]. Then we reclassified PNe according to the Peimbert types using the modified classification by Quireza et al. [11] based on Bayesian posterior probability estimations. Finally we recalculated the distance to 554 galactic PNe and obtained the estimations for scale heights for thin disk, thick disks, and halo.

2. Distances to PNe

Different statistical distance scales are based on the calibration of the relation between ionized mass $\mu = (2.266 \times 10^{-21} D^5 \theta^3 F)^{1/2}$ and the optical thickness parameter $\tau = \log(4\theta/F)$ of the PNe. Here D is the distance to the PN in parsecs, θ is the nebular radius in arcsec, and F is the nebular flux at 5 GHz in janskys.

Akimkin et al. (1) recalibrated the known distance scales to PNe basing on the kinematic model of the galactic PNe by Nikiforov and Bobrova [2] and the high-precision determination of the distance to the center of the Galaxy. Using the corrected distances to disk PNe the obtained the refined relation between μ and $= \tau$:

$$\log \mu = \begin{cases} 0.08\tau - 1.04, & \tau \leq 2.89, \\ 0.69\tau - 2.80, & \tau > 2.89. \end{cases} \quad (1)$$

This relation together with the observational data on the angular sizes of the nebulae and their radio fluxes we used to obtain the precise distances to 554 PN which are in the catalogue by Stanghellini et al. [7] (SSV).

Received distances were compared in Fig. 1 with those taken from the catalogue by Stanghellini et al. [7] (SSV). These authors used the distances to the Magellanic Clouds PNe for the calibration. However, the exactness of the distances to the Magellanic Clouds themselves is not very high [8]. Moreover the properties of the PNe in the Magellanic Clouds can strongly differ from those for galactic PNe.

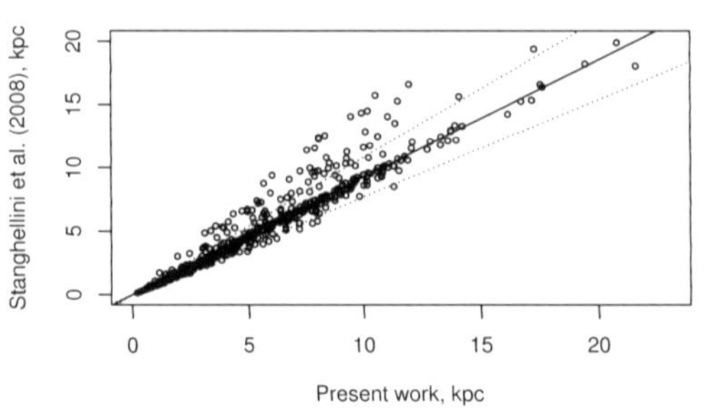

Figure 1. A comparison of distances D to PNe (in kpc) derived in the present paper and determined by Stanghellini et al. [7]. The solid line shows the mean ratio of the scales and the dotted one marks one standard deviation σ.

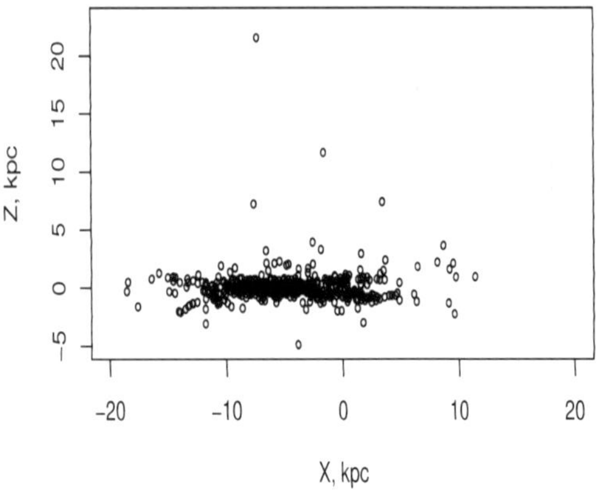

Figure 2. Distribution of PNe in the (X,Z) plane.

The mean ratio of the SSV distances to those obtained by us is 0.927 ± 0.007. It means that our scale of the galacic PNe distances appeared to be a bit longer than that given by SSV.

In Fig. 2 we plot the projection of the galactic PNe with obtained by us distances into the galactic plane (X,Z), where axis Z is directed from the center of the Galaxy to the north pope and the axis X is directed from the galactic center to the Sun.

We compile the catalog of the calculated distances and other parameters of all 554 galactic PNe. A small part of the catalogue is given in table 1. In the fist column we present the name of

PNe, parameters τ, θ and F are given in columns 2-4. The obtained by us distances are printed in the 5th column. In the last 2 columns the distances z of PNe to the galactic plane and the distances to the center of the Galaxy R_g are given.

Table 1. First 6 line in our catalog of the galactic PNe distances

Name	τ	θ $''$	F	D [kpc]	z [kpc]	R_g [kpc]
NGC 40	3.46	18.2	0.460	1.365	0.232	8.651
NGC 246	5.31	112.0	0.248	0.595	−0.574	7.976
NGC 650-1	5.24	69.2	0.110	0.930	−0.171	8.526
NGC 1360	5.82	192.0	0.222	0.457	−0.370	8.107
NGC 1501	4.08	25.9	0.224	1.335	0.151	9.013
NGC 1514	4.59	50.2	0.262	0.903	−0.238	8.746

3. Results and discussion

3.1. Classification of PNe

PNe in the Galaxy have a wide variety of spatial and kinematic properties, chemical composition, and morphology. So it is important to understand what the galactic subsystems belongs any PNe in our catalogue. Peimbert [9] divided all galactic PNe into four types, from the type I with the most massive progenitors to type IV which includes the PNe in halo. Later was defined type V which includes the bulge objects [10]. Each of these types correspond to the stellar population of the Galaxy belonging to different subsystems. PNe of types I and IIa are objects of thin disk. Types IIb and III includes the objects of thick disk and finally all halo PNE belong to type IV.

Due to some uncertainties in Peimbert's classification it was extended and modified by Quiresa et al. [11]. Using modified criteria by Quiresa et al. [11] we determine that in our catalogue 36 objects belong to type I and 193 PNe belongs to type IIa. Type IIb in the catalogue is presented by 58 objects, whereas 119 PNe are the type III objects. 58 PNe appeared to be the halo objects (type IV), other 90 PNe are bulge objects. Our classification was based on distances z of the PNe to the galactic plane and their radial velocities. We also used the statistical package R[1].

3.2. Scale heights

The observed distribution of PNe along the z coordinate axis can be described by an exponential density law:

$$N(z) = N_0 \exp\left(-\frac{|z|}{h}\right), \tag{2}$$

where $N(z)$ is the number of PNe in the layer $z - \Delta Z/2, z + \Delta Z/2$, N_0 is the normalization coefficient, and h is the scale height for the considered class of objects. The value of ΔZ were variable. The scale height is different for different types of object and generally have to increases with the age of the object.

Using this formula we obtained the scale heights for thin disk, thick disk and halo. The dependences of values $\log N(z)$ on z are shown in figures 3, 4, 5, respectively. At large galactic heights ($z > 5$ kpc), the number of nebulae is very small. It explains the large deviation of the $N(Z)$ value from the exponential law at $z > 10$ kpc in figure 5.

[1] https://www.r-project.org/

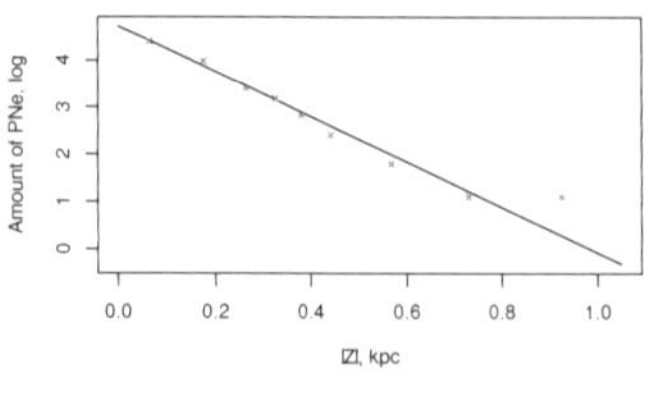

Figure 3. $N(z)$ dependence for thin disk objects.

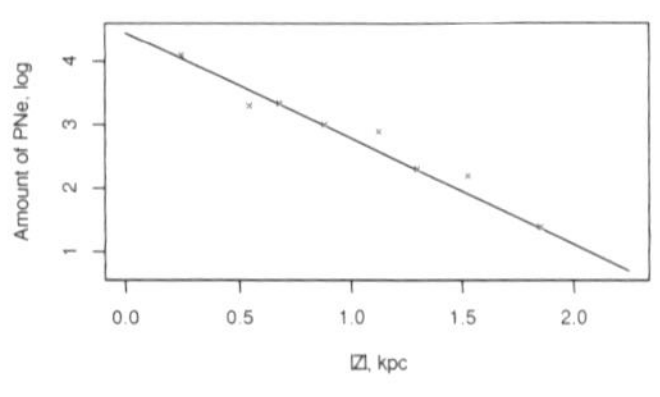

Figure 4. $N(z)$ dependence for thick disk PNe.

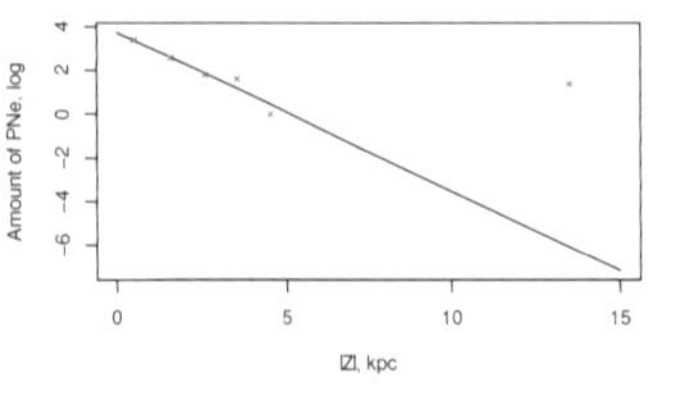

Figure 5. $N(z)$ dependence for halo nebulae.

Table 2. Estimations of scale height, pc. Asterisk marks the results obtained in this work.

Thin disk	Thick disk	Halo
208 ± 10*	600 ± 54*	1378 ± 180*
197 ± 10 [12]	640 ± 32 [13]	1350 ± 0 [14]

Results of our determination of the scale heights for different galactic subsystems and a comparison with the same values obtained by other authors are presented in table 2. We can see good agreement of our scale heights h (1st column) with those obtained by Bobylev & Baikova [12] for thin disk PNe which used PNe within distance of 2 kpc from the the Sun only.

Du et al. [13] analyzed a sample of over 1400 main sequence stars. An agreement of their value of h with ours (column 2) let us to conclude that the z distribution of the PNe progenitors weakly changes during evolution of this stars. Milanova& Kholtygin [14] also investigated the galactic PNe distribution for different Peimbert types. Their scale heights for halo objects is in good agreement with our determination.

References

[1] Akimkin V V, Nikiforov I I and Kholtygin A F 2012 *Astron. Astroph. Trans.* **27** 365-368
[2] Nikiforov I I and Bobrova A Yu 1999 *Kinem. Phys. Cel. Bodies* Appl. **2** 29-33
[3] Marigo P 2001 *A&A* **370** 194-217
[4] Bovy J, Rix H-W, Schlafly E F, Nidever D L, Holtzman J A, Shetrone M and Beers T C 2016 *ApJ* **823** 30
[5] Girardi L, Groenewegen M A T, Hatziminaoglou E and Costa L 2015 *A&A* **436** 895915
[6] Daub C T 1982 *ApJ* **260** 612-624
[7] Stanghellini L and Shaw R A 2008 *ApJ* **689** 194-202
[8] Nikiforov I I 2004 *in Order and Chaos in Stellar and Planetary Systems* ASP Conf. Ser. **316** 199-208
[9] Peimbert M 1978 *in Planetary Nebulae, Observation and Theory* ed Y Terzian (Dordrecht: Reidel), IAU Symp. **76** 215
[10] Maciel W J 1989 *in Planetary Nebulae* ed S Torres-Peimbert (Dordrecht), IAU Symp. **131** 73
[11] Quireza C, Rocha-Pinto H J and Maciel W J 2007 *A&A* **475** 217-231
[12] Bobylev V V and Bajkova A T 2017 *Astr. Lett.* **43** 304-315
[13] Du C, Zhou X, Ma J, Chen A B-C, Yang Y, Li J, Wu H, Jiang Z and Chen J 2003 *A&A* **407** 541549
[14] Milanova Yu V and Kholtygin A F 2009 *Astr. Lett.* **35** 518-533

Modeling of X-ray images of Tycho's supernova remnant

M E Kalyashova[1,2], A M Bykov[2,1] and S M Osipov[2]

[1] Peter The Great St. Petersburg Polytechnic University, 29 Politekhnicheskaya st.,St. Petersburg 195251, Russia

[2] Ioffe Institute, 26 Politekhnicheskaya st., St. Petersburg 194021, Russia

E-mail: `filter-happiness@yandex.ru`

Abstract. High-resolution *Chandra* X-ray observations of Tycho's supernova remnant (SNR 1572) have revealed several series of bright, nearly parallel stripes in some regions of the SNR clearly seen in 4-6 keV band images. The observed radiation of Tycho's SNR is most likely the synchrotron radiation of electrons with energies well above TeV. In this paper we present the modeling of synchrotron X-ray images of Tycho's SNR in order to reveal the physical mechanism of the observed coherent structures formation. It is shown that the mirror instability, which evolves in plasma near SNR shock as a result of anisotropic distribution function of accelerated particles, can be the reason of the bright stripes.

1. Introduction

Tycho's supernova remnant is one of the most famous objects in space, which has been observed with different instruments in different energy ranges. Eriksen et al.[1] observed Tycho in 2009 with the Chandra X-ray Observatory in 4-6 keV energy range. The observations have revealed several groups of nearly-regularly spaced stripes. The exact location, size and brightness of stripes are given in [1]. Tycho's image with the sets of stripes is shown in Figure 1.

The problem is to understand what is the origin of the observed coherent stripes and what can we learn from the observations. Since the stripes are prominent mostly in 4-6 keV X-ray range, they are likely produced by synchrotron radiation of ultrarelativistic electrons. Earlier, Bykov et al.[2] performed modeling of the stripes, caused by synchrotron radiation of shock-accelerated electrons in the magnetic field, amplified by Bell instability. Laming et al.[3] suggested that the stripes are connected with the Alfven wave propagating in the oblique SNR shock.

Local maxima and minima in the synchrotron emission can be provided with the modulation of the magnetic field or electron distribution function, or both. We are taking into account a specific mirror-type instability, which arises when there is an anisotropy of parallel and perpendicular pressure in plasma. The mirror instability doesn't provide strong enough amplification of the magnetic field to make visible stripes, but it affects also the distribution function of electrons, background plasma density and the magnetic field, amplified by Bell instability, as will be shown later. We suggest that the observed stripes are the result of all these effects combined. Furthermore, unlike Bell instability, the mirror instability is long-wavelength, so it can provide a good correspondence with the observational data.

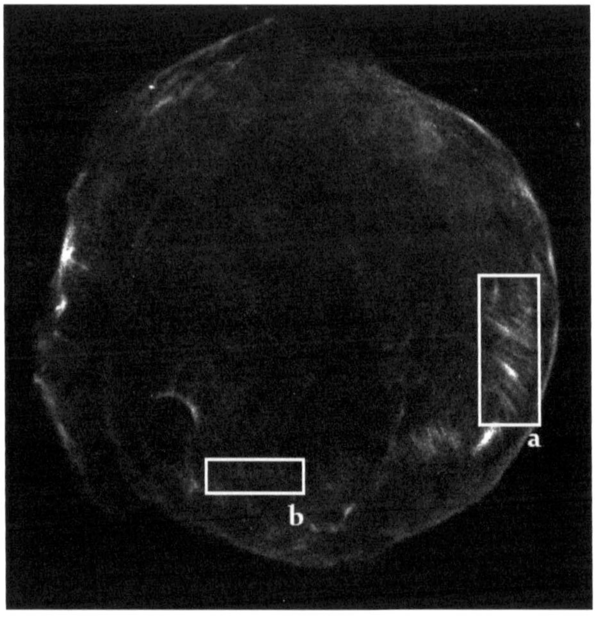

Figure 1. *Chandra* image of Tycho's SNR in 4.0-6.0 keV. a) The main ensemble of stripes b) The fainter series of stripes.

2. Mirror instability

The mirror instability is observed in the Earth's magnetosphere. The mirror mode propagates perpendicular to the magnetic field and grows, when the perpendicular pressure is larger than the parallel pressure. The physical mechanism of the mirror instability in thermal plasma is described in [4]. The instability region was recorded by spacecraft as oscillations of the magnetic field and plasma density, as illustrated in Figure 2. The mirror instability can also occur in SNRs.

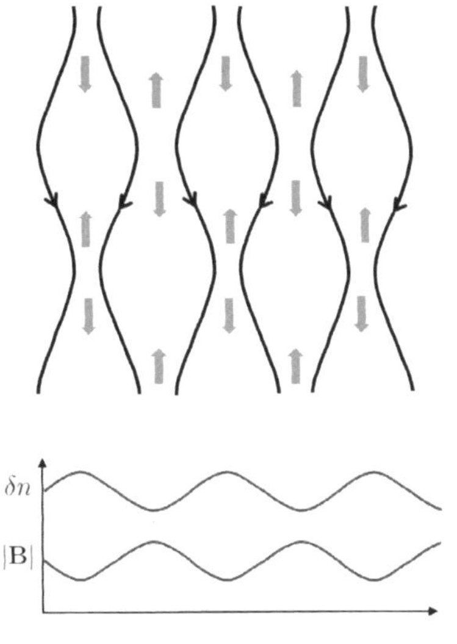

Figure 2. The sketch of the mirror-unstable region. Grey arrows point at particle motions in the magnetic mirrors. The insert at the bottom illustrates the magnetic field and plasma density in the mirror-unstable region. [4]

In our model we consider the mirror instability in SNR driven by particles accelerated at the shock wave[5]. The mirror instability occurs, when the pressure of the energetic particles is anisotropic, or, in other words, when the quadrupole anisotropy of the energetic particles distribution function is non-zero.

The distribution function of accelerated particles with the quadrupole anisotropy has the form:

$$f_0^{ap}(p,\mu) = \frac{n_{ap}N(p)}{4\pi}\left[1 + \frac{\chi}{2}(3\mu^2 - 1)\right], \tag{1}$$

where χ is the anisotropy parameter, n_{ap} is the concentration of accelerated particles, p is their momentum, $\int_0^\infty N(p)p^2dp = 1$ and $\mu = \cos\theta$, where θ is an angle between particle velocity v and z-axis (which is parallel to the background magnetic field: $\mathbf{e}_z = \mathbf{B_0}/B_0$). Calculations (see [5]) show that near the shock wave $\chi \simeq -0.05$.

The dispersion relation for the mirror instability can be derived from linearized fluid mechanics equations (for background plasma) and kinetic equation (for energetic particles), where anisotropic distribution function (1) should be used. All the parameters of plasma are defined as $\xi = \xi_0 + \delta\xi, \delta\xi \propto \exp(i\mathbf{kr} - i\omega t)$. The wavevector is $\mathbf{k} = \mathbf{e}_\perp k_\perp + \mathbf{e}_\| k_\|$, where the parallel direction is $\mathbf{e}_z = \mathbf{e}_\|$, and $\mathbf{e}_x = \mathbf{e}_\perp$ is the transverse direction. We examine the long-wavelength regime, i.e. $kr_g \ll 1$, where $r_g = cp/eB_0$ is the gyroradius of accelerated particles, $B_0 = 3\mu G$ is the background (interstellar) magnetic field, c is the speed of light, e is the particle charge. The dispersion relation has the form

$$\omega_{mir}^2 = \left(v_a^2 + a_0^2 + 2\frac{\delta P}{\rho_0}\right)k^2 \tag{2}$$

Here, v_a is the Alfven velocity, $a_0 = \sqrt{\dfrac{\gamma_g p_0}{\rho_0}}$, where γ_g is the adiabatic index, p_0 and ρ_0 are unperturbed pressure and density of background plasma. δP is the difference between parallel and perpendicular pressure of energetic particles.

$$\delta P = P_\| - P_\perp = \frac{3\chi}{5}P_0, \quad \text{where} \quad P_0 = \frac{n_{ap}}{3}\int_0^\infty vp^3N(p)\,dp\,. \tag{3}$$

Equation (2) shows that the magnetic field modes ($\delta\mathbf{B} \propto \exp(i\mathbf{kr} - i\omega t)$) are growing, when $\delta P \propto \chi < 0$. The complete derivation of the formula (2) can be found in [5].

The mirror instability affects also the background plasma density and the distribution function of accelerated electrons which transforms into $f = f_0 + \delta f$, where

$$\delta f \simeq \frac{\delta B_z}{B_0}\frac{n_{cr}N(p)}{4\pi}\frac{3\chi}{2}\sin^2\theta \tag{4}$$

The background plasma density modulation can be found as $\delta\rho/\rho_0 \simeq \delta B_z/B_0$.

Aside from the mode of the magnetic field, amplified by the mirror instability, there is the short-wavelength random magnetic field in the remnant, amplified mostly by Bell instability. Magnetic field is frozen into plasma, which leads to the magnetic field fluctuations as a result of plasma density modulation: $< B_{rand}^2 >^{1/2} \propto \rho = \rho_0 + \delta\rho$. This effect provides the stripes contrast.

3. Modeling

We model electron distribution function near the front of a collisionless shock wave using Monte Carlo (MC) simulations (the method is described by Bykov et.al. in [5]). Simulations show that the best amplified mode as a result of the mirror instability is the mode with $k_0 r_{g0} = 2 \cdot 10^{-7}$, where $r_{g0} = mcu_{sh}/eB_0$, $u_{sh} = 5 \cdot 10^8 cm/s$ is a shock velocity, m is the particle mass. The growth rate of this mode gives $\delta B_z/B_0 \simeq 3$ and fluctuations of other parameters depend on this

IOP Conf. Series: Journal of Physics: Conf. Series **1038** (2018) 012017 doi:10.1088/1742-6596/1038/1/012017

rate. The total magnetic field in our code is the sum of the background field, mode, amplified by the mirror instability and random magnetic field, also modulated in the region where there are fluctuations of the background plasma density.

$$B^2 = B_0^2 + B_{mirr}^2 + < B_{rand}^2 > \tag{5}$$

The rate of the total magnetic field is defined by MC simulations.

To get the intensity of synchrotron radiation of electrons we use the well-known formula from [6]. We compute the intensity of synchrotron radiation in a small part of the remnant, where the main ensemble of stripes was discovered. We have a simulation box with the size of 100x300x1200 pixels. To imitate the observed picture we integrate the intensity along the line of sight.

4. Results

The result of the modeling is shown in Figure 3.

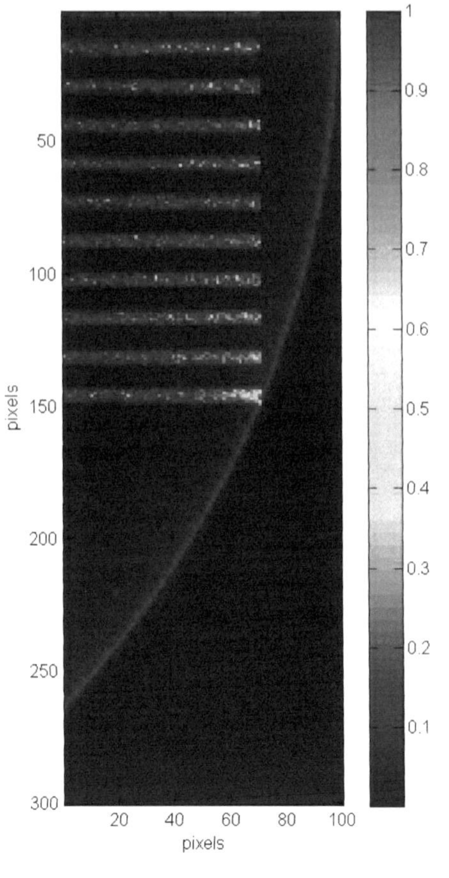

Figure 3. Simulated image of the intensity of synchrotron emission of Tycho's SNR in 4.0-6.0 keV in relative units. 1 pixel corresponds to 0.01 pc.

The wavevector of the mirror mode $\mathbf{k} = \mathbf{e}_\perp k_\perp + \mathbf{e}_\parallel k_\parallel$, taking into account that $k_\perp \gg k_\parallel$, is directed upwards in the picture plane (see Figure 3), i.e. is perpendicular to the line of sight. Hence, the magnetic field and the distribution function of electrons change in that direction. This geometry gives us maxima and minima of synchrotron radiation.

We can compare our result with the observations. The two most important parameters of stripes are distance between stripes and brightness contrast of stripes (i.e. the ratio between intensities of stripe and non-stripe regions). The comparison is given in Table 1. As one can see, the parameters that we got with our model correspond well with the observations.

Table 1. Stripes parameters comparison

Point of comparison	Observations	Modeling
Distance between stripes	$4.6 \cdot 10^{17}$ cm	$5.2 \cdot 10^{17}$ cm
Brightness contrast ($I_{stripe}/I_{non-stripe}$)	4–10	5–15

5. Conclusions

The results, presented in this paper, are in a reasonably good correspondence with the observations by the Chandra X-ray Observatory. Hence, our model gives a satisfactory explanation for appearing of a highly-ordered structure in the radiation of Tycho's supernova remnant. That means that the mirror instability can evolve in supernova remnants' plasma and should be considered as a possible mechanism of amplification and modulation of the magnetic field and distribution function of accelerated particles in supernova remnants.

Acknowledgments

S.M.O. and A.M.B. acknowledge support from RAS Presidium program No. 28.

The scientific results reported in this article are based in part on observations made by the Chandra X-ray Observatory and published previously in cited articles.

The results of the work were obtained using computational resourses of Peter the Great Saint-Petersburg Polytechnic University Supercomputing Center (http://www.spbstu.ru)

References

[1] Eriksen K A, Hughes J P, Badenes C, Fesen R, Ghavamian P, Moffett D, Plucinsky P P, Rakowski C E, Reynoso E M and Slane P 2011 *ApJ* **728** L28
[2] Bykov A M, Ellison D C, Osipov S M, Pavlov G G and Uvarov Y A 2011 *ApJ* **735** L40
[3] Laming J M 2015 *ApJ* **805** 102
[4] Treumann R A and Baumjohann W 1997 *Advanced Space Plasma Physics* (London:Imperial College Press)
[5] Bykov A M, Ellison D C and Osipov S M 2017 *Phys. Rev.* E **95** 033207
[6] Ginzburg V L and Syrovatskii S I 1965 *ARA&A* **3** 297

IOP Conf. Series: Journal of Physics: Conf. Series **1038** (2018) 012018 doi:10.1088/1742-6596/1038/1/012018

Graviatom of superheavy dark matter as a source of gravitational radiation

M A Misyura[1], **A A Grib**[1]

[1] Department of Theoretical Physics, Herzen University, St Petersburg 191186, Russia

E-mail: max.misyura94@gmail.com

Abstract. The nature of dark matter in modern physics is an important problem. Therefore, in this paper we consider a quantum system consisting of two identical particles. These particles are candidates for the role of dark matter. We will call them graviatoms and they could be formed in the early Universe. In this paper we have found the energy spectrum of the gravitatom from particles of dark matter, the Bohr radius, the intensity of gravitational radiation, the frequency of radiation, and the lifetime of a state. This lifetime is great for superheavy particles. Estimates of these values for different values of the masses of particles of dark matter are given.

1. Introduction

Dark matter is a hypothetical type of matter in astronomy and cosmology and theoretical physics. This matter has never been directly observed, because it does not emit or interact with observable electromagnetic radiation, such as light, and is thus invisible to the entire electromagnetic spectrum. There are several theoretical models that provide a set of possible candidates for the role of dark matter [1],[2],[3]. Also in the article [4], a hypothesis was proposed that particles of dark matter are neutral, scalar, superheavy particles with the mass of the order of Grand Unification 10^{15} GeV. Similar ideas were considered later in [5]. The main argument in favor of this hypothesis is that was shown earlier in the work on the creation of particles from vacuum by the gravitational field of the early Universe in Friedmann cosmology [6] that the number of particles with such the mass is close to the observed number of Eddington-Dirac particles of protons in the Universe $N = 10^{80}$. Therefore, the decay of such particles in the early Universe at high energies leads to an explanation of this number. However, as shown in [7], at lower energies, the superheavy particles become stable and these particles form a modern dark matter. In the early universe, pairs of particles of dark matter were born, between which, in addition to the Grand Unification, there was also a gravitational interaction. Therefore, it is of interest to consider so-called the graviatom of the positronium type, which is a bound state of pairs of particles of dark matter, taking into account only the gravitational interaction. The definition of the graviatom was proposed M. L. Fil'chenkov in [8]. Such graviatoms were formed in the early universe at the birth of pairs and can be preserved until the present era.

2. Graviatom

The quantum system consisting of two identical particles is described by a wave function that is a solution of the stationary Schr̈dinger equation

$$\hat{H}\psi = E\psi \tag{1}$$

Since the system is connected only by gravitational interaction, the Hamiltonian operator has the form in the frame of reference associated with the laboratory

$$\hat{H} = -\frac{\bar{h}}{2m_1}\triangle_{\overline{r_1}} - \frac{\bar{h}}{2m_2}\triangle_{\overline{r_2}} - \frac{Gm^2}{|\overline{r_1} - \overline{r_2}|} \tag{2}$$

We pass to the mass center reference system, and in this approximation we neglect the kinetic energy operator of the mass center because we are only interested in the internal states of the system

$$\hat{H} = -\frac{\bar{h}}{2\mu}\triangle_{\overline{r}} - \frac{Gm^2}{r},\ \mu = \frac{m}{2},\ r = |r_1 - r_2| \tag{3}$$

The solution of the stationary Schr̈dinger equation is the wave function which is normalized to unity

$$\int\limits_V \psi^*(r,\theta,\phi)\psi(r,\theta,\phi)dV = 1, \tag{4}$$

$$\psi(r,\theta,\phi) = R_{nl}(r)Y_{lm}(\theta,\phi), \tag{5}$$

- where $R_{nl}(r)$ $Y_{lm}(\theta,\phi)$ are the radial and angular wave functions that are identical to the corresponding functions of the hydrogen atom. The energy spectrum of such a system, which depends on the principal quantum number n, is given by the formula

$$E_n = -\frac{m^2\alpha_g^2}{4n^2},\ \alpha_g = \frac{GMm}{\bar{h}c} \tag{6}$$

The constant α_g is the gravitational equivalent of the fine structure constant, G is the gravitational constant, and c is the speed of light. The Bohr radius of such a system will be equal to

$$a_b = \frac{\bar{h}}{Gm^3} \tag{7}$$

For the mass of particles equal to $m_1 = m_2 = 10^{15}$ GeV, the eigenvalues of the total energy operator of the system

$$E_n = -\frac{1,123\cdot 10^7}{n^2}\ eV,\ a_b = 5,889\cdot 10^{-23}\ m, \tag{8}$$

Due to quadrupole transitions, this graviatom radiates a graviton with frequency

$$\omega_{mn} = \frac{E_m - E_n}{\hbar} \tag{9}$$

To obtain the magnitudes of the intensity of gravitational radiation, the formula obtained for quadrupole transitions of the hydrogen atom in [9]was used

$$I_{mn} = \frac{8Gm_p\bar{h}\omega_{mn}^3 f_{mn}}{c^3}, \tag{10}$$

Due to the fact that the masses of particles are the same, intensity formula will be changed slightly

$$I_{mn} = \frac{16Gm\hbar\omega_{mn}^3 f_{mn}}{c^3}, \tag{11}$$

- where f_{mn} is the oscillator strength, which depends on the square of the matrix element of the quadrupole moment operator D_{mn}

$$f_{mn} = \frac{m\omega_{mn}^3 D_{mn}}{9\hbar c^2}, \tag{12}$$

Next, we introduce one more quantity characterizing the lifetime of the τ_{mn} level with energies E_n for the corresponding $n \to m$ transition [10]

$$\tau_{mn} = \frac{\hbar\omega_{mn}}{I_{mn}}, \tag{13}$$

For the quadrupole transition $3d \to 1s$, the square of the matrix element of the quadrupole moment operator D_{mn} will have the form [11]

$$D_{mn} = \left(\int_0^{2\pi} \int_0^{\pi} \int_0^{\infty} \psi_m^* r^2 \left(\frac{3}{2}\cos^2(\theta) - \frac{1}{2} \right) \psi_n r^2 \sin(\theta) dr d\theta d\varphi \right)^2 \tag{14}$$

As a result of the calculations, the values were obtained: α_g, Bohr radius a_b, emission frequency ω, transition energy E, gravitational radiation intensity I and the transition lifetime as a function of the particle mass. The values are presented in the table 1.

Table 1. Characteristics of graviatom.

$3d1s$	1 GeV	200 GeV	10^9 GeV	10^{15} GeV
α_g	6.7×10^{-39}	2.68×10^{-34}	6.7×10^{-21}	6.7×10^{-9}
a,m	5.89×10^{22}	7.36×10^{15}	5.89×10^{-5}	5.89×10^{-23}
ω,s^{-1}	1.52×10^{-53}	4.85×10^{-42}	1.52×10^{-8}	1.51×10^{22}
E,eV	9.98×10^{-69}	3.2×10^{-57}	9.98×10^{-24}	9.98×10^{6}
I,W	1.37×10^{-332}	1.43×10^{-286}	1.37×10^{-152}	1.37×10^{-32}
τ,s	1.17×10^{245}	3.57×10^{210}	1.17×10^{110}	1.17×10^{20}

3. Conclusion

Depending on the mass, the following changes in magnitudes are observed, the Bohr radius from cosmological values passes into the microcosm for superheavy particles. The energy of a given transition of such an atom varies from very small values to the binding energy of light nuclei for particles with a mass of the order of Grand Unification, and the frequency is the frequency of the gamma radiation of an electromagnetic wave. The gravitational intensity is small for all mass values. And the lifetime has values greater than the lifetime of the observable Universe. The last two quantities have a strong power-law dependence on the mass.

4. Acknowledgments

Authors gratefully acknowledges partial supports from the Russian Foundation for Basic Research (grant # 18-02-00461).

References

[1] Bergström L 2012 *Annalen der Physik* **524** 479
[2] Garrett K and Duda G 2011 *Advances in Astronomy* **2011** 22
[3] Feng J L 2010 *Annual Rev. of Astron. and Astrophys.* **48** 495
[4] GRIB A and DOROFEEV V Y 1994 *Inter. J. of Modern Phys.* D **03** 731
[5] Dubrovich V, Fargion D and Khlopov M 2004 *Astroparticle Phys.* **22** 183
[6] Grib A A and Pavlov Y V 2002 *Inter. J. of Modern Phys.* A **17** 4435
[7] Grib A A and Pavlov Y V 2002 *Inter. J. of Modern Phys.* D **11** 433
[8] Laptev Y P and Fil'chenkov M L 2006 *Astron. & Astrophys. Transactions* **25** 33
[9] Fil'chenkov M L and Laptev Y P 2016 *Kvantovaya gravitaciya: Ot mikromira k megamiru* (Lenand)
[10] Blohincev D I 1976 *Osnovy kvantovoj mekhaniki* (Nauka)
[11] Bethe A H and Salpeter E E 1957 *Quantum mechanics of one-and two-electron atoms* (Springer, Berlin, Heidelberg)

Use of multiparametric analysis of meteor showers for their parental bodies' genetic parameters determination

N Y Demina[1,2], A O Andreev[2], Y A Nefedyev[1,2], S A Demin[1], L A Nefediev[1]
[1] Kazan Federal University, Institute of Physics, Kazan, 420008 Russia
[2] Kazan Federal University, Engelhardt Astronomical observatory, Zelenodolski region, st. Observatory, 422526 Russia

E-mail: vnu_357@mail.ru

Abstract. In this paper, the possibility of multiparametric analysis application for studying genetic connections between meteor showers and their parental bodies is considered. The modern methods of determining such connections take into account some parameters of parental bodies, such as size, orbital elements, and chemical composition. Obviously, all of these parameters refer to different physical substances. In this regard, simultaneous estimation of the impact of those parameters may be carried out with a given probability using regression analysis. The aim of this work is to develop a method of robust estimation of genetic parameters and use it for asteroid and meteor genetic connections analysis.

1. Introduction

Currently, there are an increasing number of researches concerned with the study of structural parameters of small bodies of the Solar system and genetic connections between them. Such studies were carried out for the Taurid meteor shower [1]. At the present time, it is established that small planets with short orbital periods may be the parental bodies of the meteor showers. Such small bodies come from 2 areas of the Solar system: the asteroid belt which is between the orbits of Mars and Jupiter, and Kuiper belt located beyond Neptune's orbit. IAU has not yet introduced the exact distinction between terms "small planet", "asteroid", and "comet". All these objects are meteoroids. The amount of unbound water inside a small planet depends on where it is located in the Solar system. Small bodies in the inner Solar system have lost water and this is why are capable of constructing stony structures after whose decay meteorites are formed. Starting from Jupiter's orbit and outwards to the edge of the Solar system, there are volatile ices in small bodies. Such small celestial bodies are comets. Thus, the difference between "comet" and "asteroid" is quite minorand can only be seen at the asteroid belt. Some simple asteroids may contain volatile ices and turn into comets at certain conditions. Near Earth Objects (NEO) include comets and asteroids whose orbits' perihelion distance $q \leq 1.3$ AU, which can make them approach the Earth's orbit. This occurs when a small body's orbit is in resonance with Jupiter, in this case perihelion distance reduces and at the same time aphelion of the meteoroid remains at the asteroid belt. For comets, such process takes place due to the resonance with Neptune. Such comets refer to "Centaurs" type. In the sequel, orbits of comets evolve down to $q \leq 1.3$ AU. As they lose momentum at close collision, their aphelion approaches to orbit of planet while perihelion moves away. As it shown here, evolution processes of small bodies of the Solar system occur continuously, and as a result, under the force of solar gravity and due to the change of temperature gradient they turn into meteor showers. That is why the problem of determining parental bodies of meteor showers is aquite grounded one.

2. Modern analysis of meteoroids disintegration

The first fifteen asteroid-like NEO were found in 1898 (433 Eros), 1911 (719 Albert), 1918 (887 Alinda), 1929 (1627 Ivar), 1924 (1036 Ganymed), 1932 (1862 Apollo, the first Earth-crossing asteroid) and (1221 Amor), 1936 (2101 Adonis), 1937 (69230 Hermes), 1947 (2201 Oljato), 1948 (1685 Toro) and (1863 Antinous), 1949 (1566 Icarus), 1950 (29075), and 1951 (1620 Geographos). When Whipple (1938) managed to determine first meteoroid orbits in his multistationary photographic project, he stated that some of short orbital periods are close to the ones of asteroids of Apollo, Adonis (which he called Anteros), and Hermes types. They had the similar small inclination of orbit and minor axis of orbit of just 1.91 AU, which was less than that of Jupiter Family Comets. The problem of the proposed meteor showers of asteroid origin is related to long-term periods of exposure to cosmic rays. They are the periods during which a meteor was far from a larger parental body that generated neutrons when was exposed to cosmic rays. Those periods vary between 1 to 30 million years. During these periods meteor showers lose cohesion with asteroid, fly away and thus may be separated from parental asteroid. Asteroid meteor shower is only possible if small planet degenerates after collision before its fragments fall on the Earth. It was discovered that the most probable source of meteor showers formation wascomets disintegration. Observations of Lirid, Leonid, Andromedid, Draconid, Orionid etc. showed there were connections with the comets [2–5]. A few comets that could be parental bodies for Taurids were found. The modern studies have shown that comet disintegration processes are not gradual and long-term ones but discrete [6]. Through processes of products disintegration, the dynamical parameters of asteroids and comets may be investigated. Knowing how representative parental bodies and meteor showers are, one may obtain important results [7]. At the same time, comet disintegration can be considered as a fractal process. Only Near Earth Asteroids (NEA) can be considered as parental bodies of meteor showers [8]. Orbital velocity of asteroids should be less than 40 km/s [9].

3. Genetic identification of meteoroid stream

Figure 1 shows statistical distribution of meteors for genetic predisposition. Here, the first column contains 80 NEA whose parental bodies are not identified. 6 meteor showers are considered to be of cometary origin, 5 of comet and asteroid origin, and 21 of asteroid origin. It is generally believed that comet disintegration process is the most probable source of meteor shower formation. All NEAs were studied with multiparametric regression analysis with taking into account spectral composition, eccentricity of orbit, diameter, and mass of the asteroids. Regression model for determining genetic connections between NEAs and meteor showers may be written as follows:

$$G = d_0 + d_1A_1 + d_2A_2 + d_3A_3 + d_4A_4 \,. \tag{1}$$

Parameters d_i are partial correlation coefficients. $(d_i)^2$ is interpreted as dispersion portion G. In case of correlating predictors, there is a problem of uncertainty in estimates that become dependent on the order of introducing the predictors in the model. In such cases, application of correlation and step-by-step regression analyses is required.

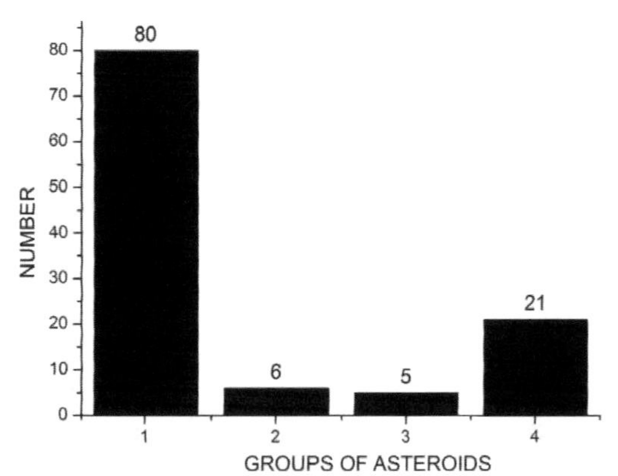

Figure 1. Statistical distribution of meteoroid stream. X – groups of asteroids, Y – number of asteroids.

The method is as follows. The probability of asteroid distribution W (p) for a certain parameter p is defined as follows. Let us consider N observations of various asteroids in a group. Each asteroid is characterized by p_i (i=1...N) parameter. Let p change between p_{min} and p_{max}. Let us introduce a step of the parameter p change as dp. Within each interval (p_i, $p_i + dp$) let us isolate a subgroup of n_j asteroids falling into this interval of the parameter. The number of those groups is

$$k = \text{Int}\,[(p_{max} - p_{min})/dp],\ \sum_{j=1}^{k} n_j = N.$$

Then $W(p_j) = n_j p_j / \sum_i^N p_i$. Besides, the value n_j/N gives the distribution of the number of asteroids in subgroups depending on the parameter p change. As a result, the following distributions were obtained (figure 2).

Speaking of non-linear models of regression analysis, it is important to notice whether we consider non-linearity on independent parameters (from formal point of view, easily converged to linear regression) or non-linearity on the parameters estimated (causing strong additional difficulties). In case of non-linearity of the first type, it is important to mark appearance of coefficients of A_1, A_2, A_3 kinds demonstrating the presence of multicollinearity. The complete multicollinearity leads to the uncertainty of the values of parameters, while the partial multicollinearity leads to estimates instability. The instability manifests as increase in statistical uncertainty – estimates dispersion. This means that specific results of estimation may significantly differ for different selections despite the selections are homogeneous. The solution of model (1) showed that the most probable parental bodies of the meteor showers observed were asteroids from Apollo group. The solution was based on multiparametric regression analysis of eccentricity of orbits, size, mass, and chemical and mineralogical composition of NEAs. Distribution for both eccentricity and diameter is given in figure 2 and confirms the hypothesis on Apollo group's priority during the process of meteor showers formation.

4. Summary and conclusions

In the present paper, it is obtained that meteoroids disintegrating in the Earth's atmosphere may be directly connected with their parental bodies if they belong to certain meteor showers. The physical and chemical properties of meteoroids coming from different parental bodies may be compared by light curve, meteors' height, and their spectrum. Thus, analysis of meteorite data may identify differences in physical structure and chemical composition of various small bodies in the Solar system (comets and asteroids). The methods and data obtained in the present paper are going to allow studying dynamics and kinematics of small bodies of the Solar system [10–13].

Acknowledgments

The work is performed according to the Russian Government Program of Competitive Growth of Kazan Federal University. This work was supported by the Russian Foundation for Basic Research, project nos. 16-32-60071 mol_a_dk (N.D.), 18-02-01098 a, 18-32-00895 mol_a.

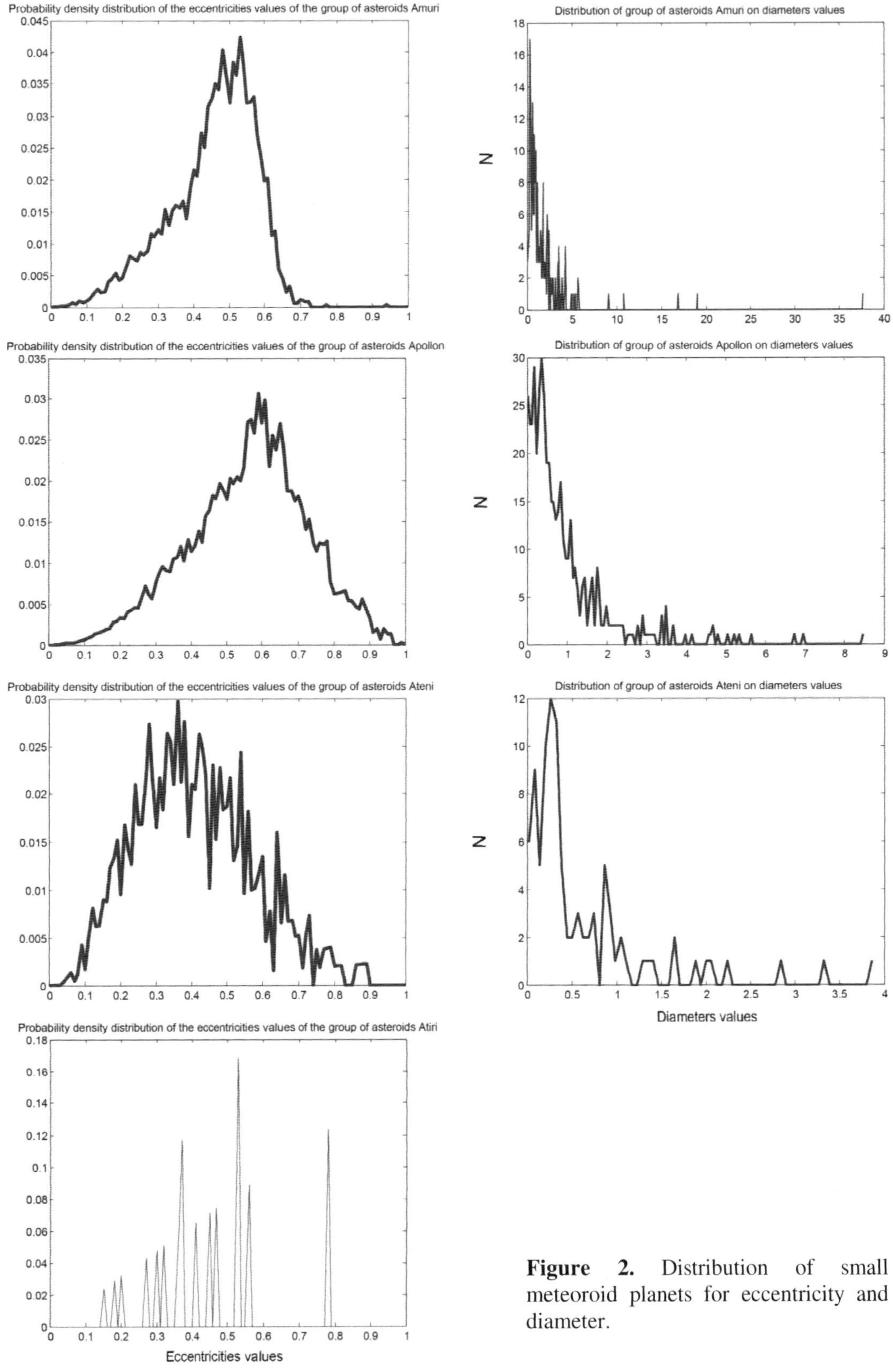

Figure 2. Distribution of small meteoroid planets for eccentricity and diameter.

References

[1] Porubčan V, Kornoš L and Williams I P 2006 *Contributions of the Astronomical Observatory Skalnaté Pleso* **36** 103
[2] Sokolova M, Nefedyev Y, Sergienko M, Demina N and Andreev A 2016 *Advances in Space Research* **58** 541
[3] Sokolova M, Nefedyev Y and Varaksina N 2014 *Advances in Space Research* **54** 2415
[4] Sokolova M, Kondratyeva E and Nefedyev Y 2013 *Advances in Space Research* **52** 1217
[5] Usanin V, Nefedyev Y and Andreev A 2016 *Advances in Space Research* **58** 2400
[6] Jenniskens P 2008 *Earth, Moon and Planets* **102** 505
[7] Trigo-Rodríguez J M and Llorca J 2006 *Monthly Notices of the Royal Astronomical Society* **372** 655
[8] Jenniskens P 2006 *Meteor showers and their parent comets* (Cambridge, U.K.: Cambridge University Press) p 803
[9] Drolshagen G, Dikarev V, Landgraf M, Krag H and Kuiper W 2008 *Earth, Moon and Planets* **102** 191
[10] Usanin V S, Nefedyev Y A and Andreev A O 2017 *Advances in Space Research* **60** 1101
[11] Nefedjev Y A and Rizvanov N G 2005 *Astronomy and Astrophysics* **444** 625
[12] Andreev A, Nefedyev Y, Demin S, Demina N and Andreeva Z 2017 Scientific and educational center of space researches and technologies in Kazan federal university *International Multidisciplinary Scientific GeoConference Surveying Geology and Mining Ecology Management, SGEM* vol 17 pp 175–182
[13] Andreeva Z, Andreev A, Nefedyev Y, Demin S and Demina N 2017 Teaching courses on space geodesy, astronomy, and natural science conceptions at Kazan federal university *International Multidisciplinary Scientific GeoConference Surveying Geology and Mining Ecology Management, SGEM* vol 17 pp 189–194

The fractal analysis of the topography and gravitational field of Venus

S A Demin[1], A O Andreev[2], N.Y. Demina[1,2], Y.A. Nefedyev[1,2]
[1] Kazan Federal University, Institute of Physics, Kazan, 420008 Russia
[2] Kazan Federal University, Engelhardt Astronomical observatory, Zelenodolski region, st. Observatory, 422526 Russia

E-mail: serge_demin@mail.ru

Abstract. The purpose of this paper is to present the results of structural analysis of gravitational and topographic parameters of Venus using the data from space missions, including "Magellan". The model gravitational potentials are presented as analytical functions of coordinates. The model is constructed on the basis of variations of Venus' artificial satellites orbits. As a result, fractal correlations of Venus' geoid anomalies and gravitational potential in both longitude and latitude as well as the mean value of fractal dimensions are calculated. The mean fractal dimension of Venus topographic model in latitude is $\overline{D} = 1.061$, in longitude is $\overline{D} = 1.037$; the mean fractal dimension of Venus gravitational potential model in latitude is $\overline{D} = 0.96$, in longitude is $\overline{D} = 1.053$.

1. Introduction

Currently, the basic methods of analyzing processes in complex systems are statistical ones [1–5]. They have restrictions of probabilistic and statistical approaches, since not every process and notevery result (form and structure of system) is of probabilistic nature. In contrast to this, fractal analysis allows studying structure of complex objects taking into account their qualitative specificity. Determination of fractal dimension allows studying both the structure itself and the connection between the structure and processes of its formation. The problem of developing methods of determining fractal structures of complex objects is relevant in this regard. Thus, anomalies of Venus' gravitational field and variations of Venus' physical surface are multi-parametric systems whose analysis should be conducted using the methods of physics of complex systems, and fractal analysis is one of them [6]. In this paper, fractal structures of Venus were studied on the basis of NASA "Magellan" space mission data [7]. Scientific purposes of "Magellan" space mission were studying relief, chemical processes as well as modelling the internal structure of Venus. The equipment of the artificial satellite allowed scanning almost entire surface of Venus using theradar system with synthetic aperture of *S*-range (12 cm) and microwave radiometer. These devices also allowed investigating topography on the basis of measurements taken by the special radar – altimeter [7]. Inaccuracies in some data obtained by "Magellan" were taken into account using the information which had been collected during "Venera 15", "Venera 16", and "Pioneer Venus 1" space missions. The data on gravitational field of Venus obtained during the space missions was used as well. In astrophysics, gravitational potentials presented as analytical functions of coordinates are used. They are determined by analyzing celestial bodies' artificial satellites orbits evolution and gravitational effects that may be measured. Such models allow studying gravitational field at some distance from celestial body and judging mass distribution in its interior. It is also necessary to consider the fact that choice of basic reference surface on Venus is specified by certain value of potential or point on the surface through which the geoid passes. To construct models of gravitational potential anomalies we used the data from [8].

2. The method of Venusian structures fractal analysis

The models of Venus' physical surface and gravitational field anomalies are of complex structure, and to study them the methods of multi-parametric analysis are required. Such models are fractal structures. Thus, the best method of analyzing such systems is fractal geometry. Significant differences between fractal dimension for the Venus' surface and gravitational field anomalies models and their real physical values show that there is some complex distribution of model structure in space. In this paper, while constructing a model of fractal object the Weierstrass-Mandelbrot fractal function was used. On the one hand, complex physical systems cannot be described as a single fractal and represent multifractals consisting of a set of interconnected fractals of their own dimensions. On the other hand, in a fractal model, each part of it repeats the entire model on structure and does not change at various scales, that is recursive. Use of multifractal analysis allows studying a system as a spectrum of fractal dimensions. Such method provides high accuracy when describing complex fractal structures by investigating local areas. In the present paper, for determining and analyzing fractal dimensions the Minkowski mathematical algorithm, which is a simplified option of Hausdorff-Besicovitch dimension and provides high reliability and accuracy, was used.

Let M be a limited set in the plane and let us define $N(\sigma)$ as a minimal number of an area being investigated covers whose combination covers M [9]. In general case, the set M has a dimension of D=dim M, $0 \le D \le 2$, if at $\sigma \to 0$ the number of covers $N(\sigma)$ increases as R/σ^D, where R is a positive constant called D-measure of the set M. For R in the general case we may write:

$$R = \lim_{\sigma \to 0} \sigma^D N(\sigma). \tag{1}$$

Expression (1) is not convenient when calculating, since it contains the value of D-measure. The fractal dimension D may be determined according to the following expression [9]:

$$D = \dim M = \lim_{\sigma \to 0} \frac{\ln N(\sigma)}{\ln \frac{1}{\sigma}}, \tag{2}$$

as

$$\lim_{\sigma \to 0} \frac{\ln N(\sigma)}{\ln \frac{1}{\sigma}} = \lim_{\sigma \to 0} \frac{\ln \frac{R}{\sigma^D}}{\ln \frac{1}{\sigma}} = \lim_{\sigma \to 0} \frac{D \ln \frac{1}{\sigma} + \ln R}{\ln \frac{1}{\sigma}} = D. \tag{3}$$

To analyze Venus' fractal structure the data obtained during "Magellan" space mission [7] was used. The model of gravitational potentials presented as analytical functions of coordinates was used. The model included Venus' artificial satellite orbit variations and other measurements of its gravitational field. The constructed gravimetric model of Venus showed the difference between real and normal gravity force variations. Using this model, one may study mass distribution in Venus interior and investigate gravitational field near the surface of the planet. The basic reference surface is specified by the point on its surface coinciding with the certain value of potential. Depending on gravimetric potential, the reference surface may be both under and above the physical surface of Venus similarly to terrestrial potential [8]. The obtained topographic and gravimetric models were investigated using the Minkowski algorithm. The study of the obtained fractal dimensions allowed conducting comparative analysis of various geophysical parameters of Venus.

3. Harmonic models of topography and gravitational field of Venus

Using harmonic analysis of expansion in spherical functions of the topographical data received from boards of «Magellan» a topographical model of Venus was constructed [5, 10]:

$$h(\varphi, \lambda) = \sum_{n=0}^{N} \sum_{m=0}^{n} (\bar{C}_{nm} \cos m\lambda + \bar{S}_{nm} \sin m\lambda) \cdot \bar{P}_{nm}(\cos\varphi) + \varepsilon, \tag{4}$$

where $h(\varphi, \lambda)$ – function of altitude;
φ, λ – latitude, longitude (known parameters);
$\bar{C}_{nm}$, $\bar{S}_{nm}$ – normalized harmonic amplitudes;
$\bar{P}_{nm}$ – normalized associated Legendre functions;

ε – random regression error.
This equation was also used for the analysis of positional observations taken at Engelhardt Astronomical Observatory [11].

On the basis of gravimetric measurements a harmonic model of Venus' gravitational potential was constructed [12]:

$$U(r,\varphi,\lambda) = \frac{GM}{r}\left[1 + \sum_{n=2}^{\infty}\left(\frac{R_0}{r}\right)\sum_{m=0}^{n}(\bar{C}_{nm}\cos m\lambda + \bar{S}_{nm}\sin m\lambda)\cdot\bar{P}_{nm}(\sin\varphi)\right] + \varepsilon, \quad (5)$$

where G – gravitational constant;
GM – the product of the Gravitational constant and the Mass of Venus (324 858.601 km^3/sec^2) [13];
r, φ, λ – spherical coordinate of reference point;
R_0 – equatorial radius of planet.

4. Study of Venusian fractal structures and the results of analysis

The topographical model of Venus was cut by planes on meridians with longitudes: 0, 20, 40, 60, 80, 100, 120, 140, 160, 180, 200, 220, 240, 260, 280, 300, 320, 340 degrees in order to create macroprofiles of the planet's surface. As an example, at longitude of 180 degrees this may be seen in figure 1. The system of units on both X and Y axis correspond to the measured data and are not critical, since the profiles are considered as self-similar structures.

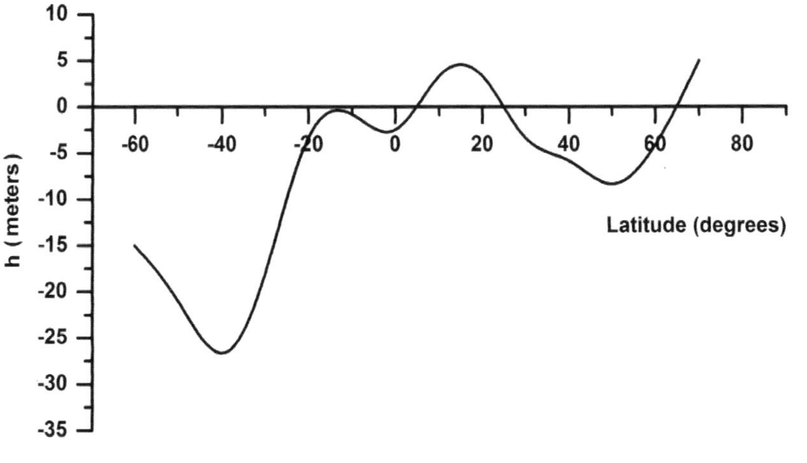

Figure 1. Macroprofile of Venus topography at longitude of 180 degrees. X-axis corresponds to latitude (φ) in degrees, Y-axis corresponds to altitude in meters (h).

Similarly to this, the model of gravity anomalies was cut by planes on meridians with longitudes: 0, 20, 40, 60, 80, 100, 120, 140, 160, 180, 200, 220, 240, 260, 280, 300, 320, 340 degrees in order to create macroprofiles of the planet's gravity. As an example, at longitude of 300 degrees this may be seen in figure 2.

IOP Conf. Series: Journal of Physics: Conf. Series **1038** (2018) 012020 doi:10.1088/1742-6596/1038/1/012020

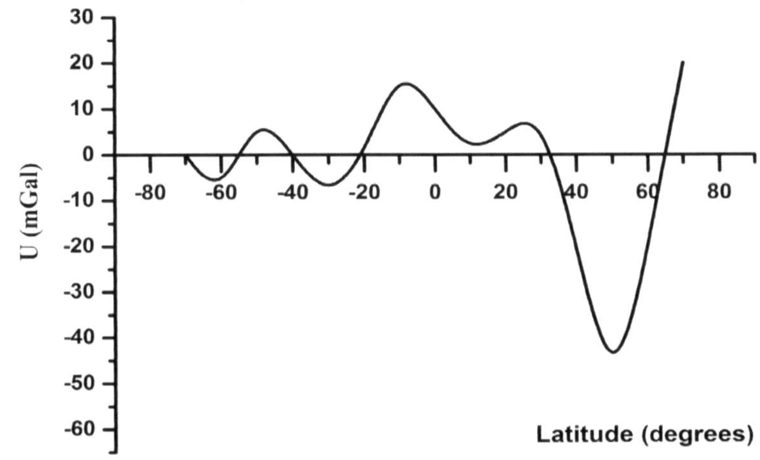

Figure 2. Macroprofile of Venus gravity anomalies on the longitude of 300 degrees. *X*-axis corresponds to latitude in degrees (φ), *Y*-axis corresponds to units of acceleration mGal (0.001 cm/sec^2).

For each cut, fractal dimensions were calculated according to equation (2), where N – number of covers in the area investigated. The division into covers starts with value of 24, for number of covers to be integer. The obtained values for the topographical model and the model of gravitational field anomalies are presented in figure 3.

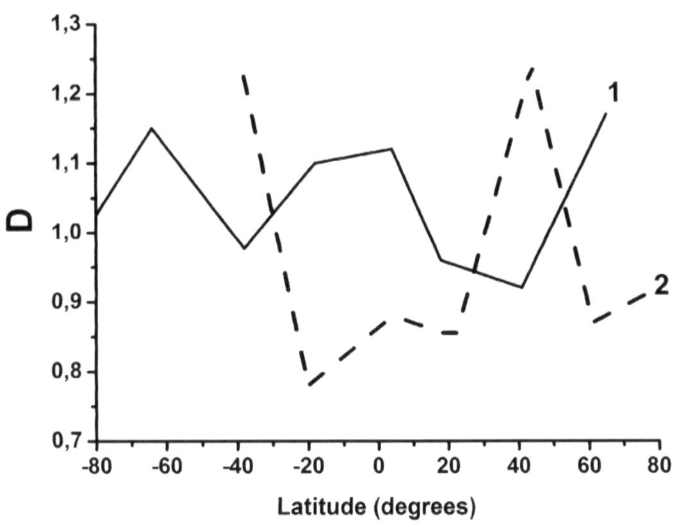

Figure 3. Fractal dependence of topographical model and gravity anomalies in Venus latitude. *X*-axis corresponds to latitudes in degrees (φ), Y-axis corresponds to fractal dimension values *D*. Line 1 corresponds to fractal dimensions of Venus topographical model, line 2 corresponds to fractal dimensions of gravity anomalies on Venus.

Analysis of figure 3 shows that for the topographical data the variations of fractal dimensions vary between 0.90 and 1.20. Considering the fact that the more significantly topological dimensions of a structure differ from fractal ones, the more complex formation of the system is, the similarity of topographical formations on Venus is sufficiently defined. The variations of gravity anomalies fractal dimensions change between 0.76 and 1.23. Thus, the gravitational field of Venus may be described as a topological dissimilar system with a complex structure in latitude measurements.

The next stage is determination of fractal dimensions in longitude. To conduct this, cuts of Venus models by Venusian meridians in certain Venusian longitudes within 0–350 degrees with a step of 25 degrees were made. As a result, the curve in figure 4 was plotted.

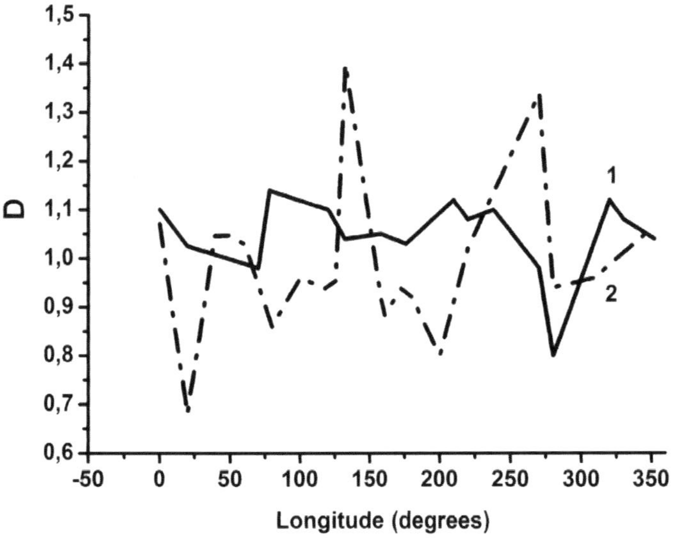

Figure 4. Dependence of fractal dimension of topographical model and gravity anomalies on longitude of Venus. *X*-axis corresponds to longitude in degrees (λ), *Y*-axis corresponds to values of fractal dimension *D*. Line 1 corresponds to fractal dimensions of Venus topographical model; line 2 corresponds to fractal dimensions of gravity anomalies on Venus.

It should only be noted that topographical fractal dimensions behave more evenly while gravitational fractal dimensions show more anomalies, especially in figure. Analysis of figure 4 shows that for the topographical data the variations of fractal dimensions vary between 0.82 and 1.16. The variations of gravitation anomalies fractal dimensions change between 0.68 and 1.42. The latter indicates that gravitational field in Venusian longitude is of heterogeneous structure.

5. Summary and conclusions

In this work, the fractal dependencies for Venusian topographic model and anomalies of gravitational potential in both longitude and latitude are determined. The values of mean fractal dimensions are obtained: mean fractal dimensions of Venusian topographic model in latitude $\overline{D} = 1.061$, in longitude $\overline{D} = 1.037$; mean fractal dimensions of Venusian gravitational potential anomalies in latitude $\overline{D} = 0.96$, in longitude $\overline{D} = 1.053$.

As seen from the analysis of fractal dimensions for Venusian topographic model, the physical surface of Venus is close to sphere. Venusian gravitational field has a more heterogeneous and complex structure.

It should be noted that the method of fractal analysis allows introducing independent estimates of Venusian macrostructure, which leads to the new approaches to interpreting physical processes taking place on Venus. The further use of fractal comparative analysis when processing the data of space observations allows obtaining the important results that will help to solve many problems of space investigations.

Acknowledgments
The work is performed according to the Russian Government Program of Competitive Growth of KazanFederal University.This work was supported by the Russian Foundation for Basic Research, project nos. 16-32-60071 mol_a_dk (N.D.), 18-02-01098 a,18-02-00895 mol_a.

References
[1] Sokolova M, Nefedyev Y, Sergienko M, Demina N and Andreev A 2016 *Advances in Space Research* **58** 541
[2] Demin S A, Panischev O Y and Nefedyev Y A 2014 *Kinematics and Physics of Celestial Bodies* **30** 63
[3] Demin S A, Panischev O Y and Nefedyev Y A 2014 *Nonlinear Phenomena in Complex Systems* **17** 177
[4] Demin S A, Panischev O Y and Nefedyev Y A 2015 *Journal of Physics: Conference Series* **661** 012003
[5] Andreev A O, Demina N Y, Demin S A, Nefedyev Y A and Churkin K O 2016 *Nonlinear Phenomena in Complex Systems* **19** 271
[6] Turcotte D L 1987 *Journal of Geophysical Research* **92** 597
[7] Hensley S, Mitchell K, Nunes D, Shaffer S, Deen R, Parcheta C and Rusert M 2016 *Proceedings of EUSAR 2016: 11th European Conference on Synthetic Aperture Radar* (*Hamburg, Germany, June 6-9, 2016*) (Berlin: VDE Verlag GmbH) 7559360
[8] Ferrari A 1979 *Revs. of Geophys. and Space Phys.* **17** 1663
[9] Zhikov V V 1996 *Soros Educational Journal* **12** 109
[10] Nefedyev Y, Andreev A, Demin S, Demina N and Andreeva Z 2017 Fractal analysis of the earth topographic models using multi-parametric harmonic analysis *International Multidisciplinary Scientific GeoConference Surveying Geology and Mining Ecology Management, SGEM* vol 17 pp 913–918
[11] Rizvanov N G, Nefed'ev Y A and Kibardina M I 2007 *Solar System Research* **41** 140
[12] Busarev V V, Shevchenko V V and Surdin V G 2007 The physical conditions near the Moon and planets of the Solar System *Model of the Cosmos* ed M I Panasyuk (Moscow: KDU) pp 794–861
[13] Konopliv A S and William L S 1996 Venus Gravity Handbook *JPL Publication* **96** 1–66

IOP Conf. Series: Journal of Physics: Conf. Series **1038** (2018) 012021 doi:10.1088/1742-6596/1038/1/012021

Astrometric and photometric baseline observations of the asteroid 2014 JO25

S N Petrova[1], A V Devyatkin[1], D L Gorshanov[1] and V N L'vov[1]

[1]The Central Astronomical Observatory of the Russian Academy of Sciences, Pulkovskoe chaussee 65, Saint-Petersburg, 196140, Russia

E-mail: stalkered@yandex.ru

Abstract. Ground-based baseline observations of the potentially hazardous asteroid 2014 JO25 were made. The data were processed and the accuracy of the orbital elements of the asteroid was estimated on the basis of the obtained observations and the data from the Minor Planet Center. Also, the estimation of *BVRI* color indices of the asteroid was made.

1. Introduction
In April 2017, in the frame of the observational campaign organized by the Central (Pulkovo) Astronomical Observatory of the Russian Academy of Sciences (CAO RAS), ground-based baseline observations of the near-Earth object 2014 JO25 were made during its flyby. The aim of the investigation was to construct a precise orbit of the object in question using short-time simultaneous observation of two or more observatories. Also, if possible, the estimation of the axial rotation period of the asteroid as well as the attempt to determine its taxonomical class were planned.

2. Baseline astronomic observations
Ground-based baseline astronomic observations of a celestial body are observations performed simultaneously at two or more observation sites with well-known coordinates located at a certain distance (preferably thousands of kilometers) from each other. This method of observations provides an opportunity to measure by triangulation the distances to the object and to construct a three-dimensional model of its motion [1].

The aim of this investigation was to construct a precise orbit of the asteroid 2014 JO25 based on the data obtained during short-time (1-3 hours) quasi-simultaneous baseline observations.

Six observatories took part in this observational campaign: Pulkovo Observatory of RAS, Mountain Astronomical Station of Pulkovo Observatory (near Kislovodsk), Ussuriisk Astrophysics Observatory of RAS, Zvenigorod Observatory of Institute of Astronomy of RAS, Gissar and Sanglokh observatories of Institute of Astrophysics of Tajikistan. The largest base estimated 6820 km. Parameters of the telescopes and the coordinates of the observation sites are provided in the Table 1.

Table 1. Observatories and telescopes that participated in the campaign of observations of 2014 JO25 in April, 2017.

Observatory MPC code	Coordinates (lat., long.)	Telescope	Aperture (mm) Focus (mm)	Scale (″/pix)
Ussuriysk C15	+43°42′ 132°10′	GAS-250	250 748	3.3
Kislovodsk C20	+43°45′ 42°40′	MTM-500M	500 4100	1.2
Pulkovo 084	+59°46′ 30°20′	ZA-320M	320 3200	1.8
Zvenigorod 102	+55°42′ 36°46′	Zeiss-600	600 7200	0.5
Gissar 190	+38°29′ 68°41′	AZT-8	700 2820	1.8
Sangloh 192	+38°16′ 69°13′	Zeiss-1000	1000 13000	0.3

3. Asteroid 2014 JO25

Asteroid 2014 JO25 was discovered in 2014 by A. D. Grauer at the Mt. Lemmon Survey. It belongs to Apollo group of the near-Earth objects and its minimal orbit intersection distance is estimated as 0.0118 AU. 2014 JO25 is classified as a "Potentially Hazardous Asteroid" by the Minor Planet Center [2].

Object's closest approach to Earth (within 4.6 lunar distances or 0.0118 AU) occurred on April 19, 2017. During its flyby, radar observations were made at the Goldstone and Arecibo observatories. The results of these observations showed that the asteroid is in fact a contact-binary object and has a peanut-like shape. The axial rotation period was approximately estimated as 4.5 hours and the largest dimension of the asteroid is at least 650 m [3].

4. Astrometric study

EPOS [4] and APEX-II [5] software systems were used to process the astrometric data and UCAC4 was used as the guide star catalog.

After processing of the data obtained by this campaign, the orbit of 2014 JO25 was constructed and its accuracy was compared to the accuracy of the orbit that can be constructed using the data of not simultaneous observations made by different observatories from 2014 to 2017 (collected on the Minor Planet Center (MPC) web-site). The Table 2 shows comparison of the MPC orbit and the orbit constructed on the base of 1.5-hour simultaneous observations of Ussuriisk and Kislovodsk observatories.

5. Photometric study

APEX-II software package [5] was used to process the photometric data and 2MASS was used in the process of the photometrical reduction.

Observations in *BVRI* bands of the international photometric system were made to determine color-indices of the asteroid. Based on the data, an attempt to determine the taxonomical class of the asteroid was made. The color-indices of 2014 JO25 asteroid obtained from our observations have best coincidence with spectra of Q, V or G class (by Tholen classification [6]). The figure 1 illustrates comparison of spectra of 2014 JO25 asteroid and Q class [7].

By this stage of the investigation, the axial rotation period of the asteroid 2014 JO25 has been obtained by means of radar observations in the Goldstone and Arecibo observatories and it was estimated to be approximately 4.5 hours.

The lightcurve obtained from the longest observation series is shown on the figure 2. The Scargle method [8] was used to determine the rotational period from this data. The result shows that the previously estimated axial rotation period is right and is approximately 4.5±0.2 hours. But further investigation is needed.

Table 2. Comparison of 2014 JO25 asteroid orbital elements for the epoch JD2458000.5 of MPC orbit and orbit constructed using the short-time base-line observations.

Elements	MPC	Base-line	Δ
M, (°)	58.788880	59.143965	0.355085
ω, (°)	49.571260	49.811066	0.239806
Ω, (°)	30.652780	30.639300	–0.013480
i, (°)	25.269930	25.518579	0.248649
e	0.88543290	0.88397329	–0.00145961
a, (AU)	2.068265602	2.060720772	–0.00754483
q, (AU)	0.236955192	0.239098660	0.002143467

Where: *M* — the mean anomaly; ω — the argument of perihelion; Ω — the ascending node; *i* — the inclination; *e* — the eccentricity; *a* — the semi-major axis; *q* — the perihelion distance; Δ — the differences.

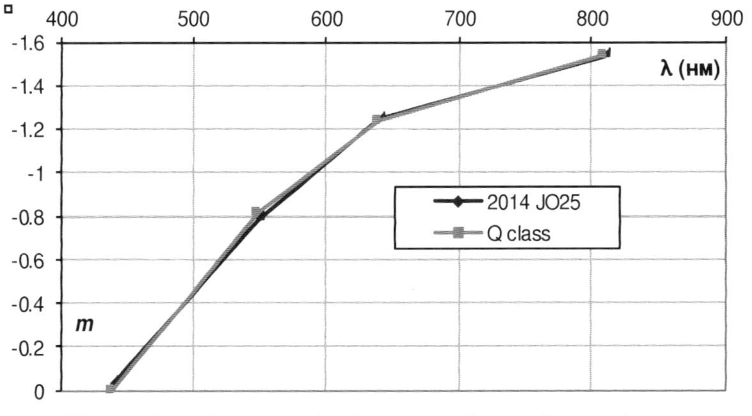

Figure 1. Spectrum of 2014 JO25 determined using *BVRI* observations and spectrum Q-class asteroids.

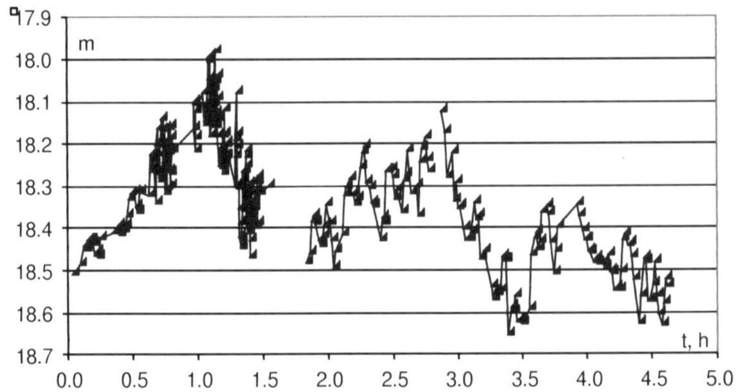

Figure 2. Lightcurve obtained by processing the observations made in Zvenigorod on April 19, 2017.

6. Conclusion

In this study, ground-based quasi-simultaneous baseline observations of 2014 JO25 asteroid were made in April, 2017. Six observatories participated in the campaign. The results of the campaign showed that using short-time (1-2 hours) simultaneous observations from at least two observatories it's possible to construct the orbit of the celestial body in question with the high precision. The accuracy of this constructed orbit is close to the orbit published on the MPC web-site that was constructed with more than 1.5 thousand observations obtained by many observatories from 2014 to 2017.

The taxonomical class (Tholen classification) of 2014 JO25 was presumed to be Q, V or G.

References

[1] Devyatkin A V, Gorshanov D L, L'vov V N, Tsekmeister S D, Chubey M S 2016 Astrometric Baseline Observations of Solar System Bodies. *Kinematics and Physics of Celestial Bodies* **32** no 5 pp 241–244

[2] https://echo.jpl.nasa.gov/asteroids/2014JO25/2014JO25_planning.html

[3] http://www.naic.edu/~pradar/press/2014JO25.php

[4] L'vov V N *et al* 2012 The use of the Epos software package for research of the Solar System objects *Solar System Research* **46** issue 2 pp 177-179

[5] Gorshanov D L , Kouprianov V V *et al* 2010 Apex I and Apex II Software Packages for the Reduction of Astronomical CCD Observations *Solar System Research* **44** no 1 pp 68–80

[6] Tholen D J 1984 Asteroid Taxonomy from Cluster Analysis of Photometry Ph.D. thesis University of Arizona Tuscon

[7] Dandy C L, Fitzsimmons A, Collander-Brown S J 2003 Opticalcolorsof 56 near-Earth objects: trends with size and orbit *Icarus* **163** pp 363–373

[8] Scargle J D 1982 Studies in Astronomical Time Series Analysis. II. Statistical Aspects of Spectral Analysis of Unevenly Spaced Data *Ap.J.* **263** pp 835-853

Electron and ion acceleration by relativistic shocks: particle-in-cell simulations

V I Romansky[1,2], **A M Bykov**[1,3] **and S M Osipov**[1]

[1] Ioffe Institute, 26 Politekhnicheskaya st., St. Petersburg 194021, Russia
[2] Sternberg Astronomical Institute, Moscow State University Universitetsky pr., 13, Moscow 119234, Russia
[3] Peter the Great St. Petersburg Polytechnic University, 29 Politekhnicheskaya st., St. Petersburg 195251, Russia

E-mail: romanskyvadim@gmail.com

Abstract. The problem of particle acceleration by collisionless astrophysical shocks is of fundamental importance for the cosmic ray origin problem and for the high energy astrophysics as a whole. Fast magnetohydrodynamical shocks produced by relativistic jets are considered as very likely sites of cosmic ray acceleration. While cosmic ray acceleration by both non-relativistic and ultra relativistic shocks was studied in some details, the shocks of moderate Lorentz factors have attracted much less attention so far, while they are important for modeling gamma-ray burst afterglows and a number of other interesting objects. In this paper we present simulations of relativistic shock waves in electron-proton plasmas obtained with an implicit particle-in-cell code for the initial study the particle injection efficiencies.

1. Introduction

One of the possible sources of high-energy cosmic rays (CRs) are shock waves in astrophysical collisionless plasmas which are believed to accelerate particles via the diffusive shock acceleration (DSA) mechanism, see, e. g. [1], [2]. The particle acceleration in the vicinity of shock front is a sophisticated and nonlinear problem, because the CRs strongly influence on the structure of the shock. Their pressure is comparable with the pressure of background plasma and significantly modifies the shock front (see, e.g., [3]). Also the instabilities in the anisotropic plasma can result to a strong magnetic field amplification. The spectra of synchrotron emission observed from supernova remnants (SNRs) indicate that the magnetic field in their shells is about 100 times stronger than the interstellar field (see, e.g., [4],[5]). One of the most powerful methods to explore the processes in collisionless plasma on microscopic scales is particle-in-cell (PIC) simulation.

2. Shock wave simulation

We studied the evolution of moderately relativistic shocks and spectra of accelerated particles. We chose relativistic shocks with a relatively low Lorentz factor ($\gamma = 1.5$) because the maximum energy of the produced CRs cosmic rays increases with the shock Lorentz factor, while the efficiency of acceleration decreases, because it becomes more difficult for particle to cross fast moving front [9]. So we assumed that an intermediate case of moderately relativistic shocks would provide the most efficient acceleration.

To perform the simulations we developed the implicit PIC code [6], based on the scheme suggested by Lapenta et al. [7] and improved for the relativistic case by Noguchi et al.[8]. The code is fully three-dimensional and parallelized with MPI technology. The electric field divergence is corrected after every several time steps and the shortwave harmonics Fourier filtering is applied to suppress the numerical Cherenkov instability and improve energy conservation.

In the model setup the homogeneous plasma flows into the simulation box through its right boundary and collides with the reflecting superconducting wall at the left boundary thus launching a shock wave. This is a common way to initialize a shock (e. g. [11], [12]), because an initialization using the Rankine-Hugoniot conditions does not take into account the microscopic distribution, and a shock, created in such a way, would probably fall apart into several discontinuities.

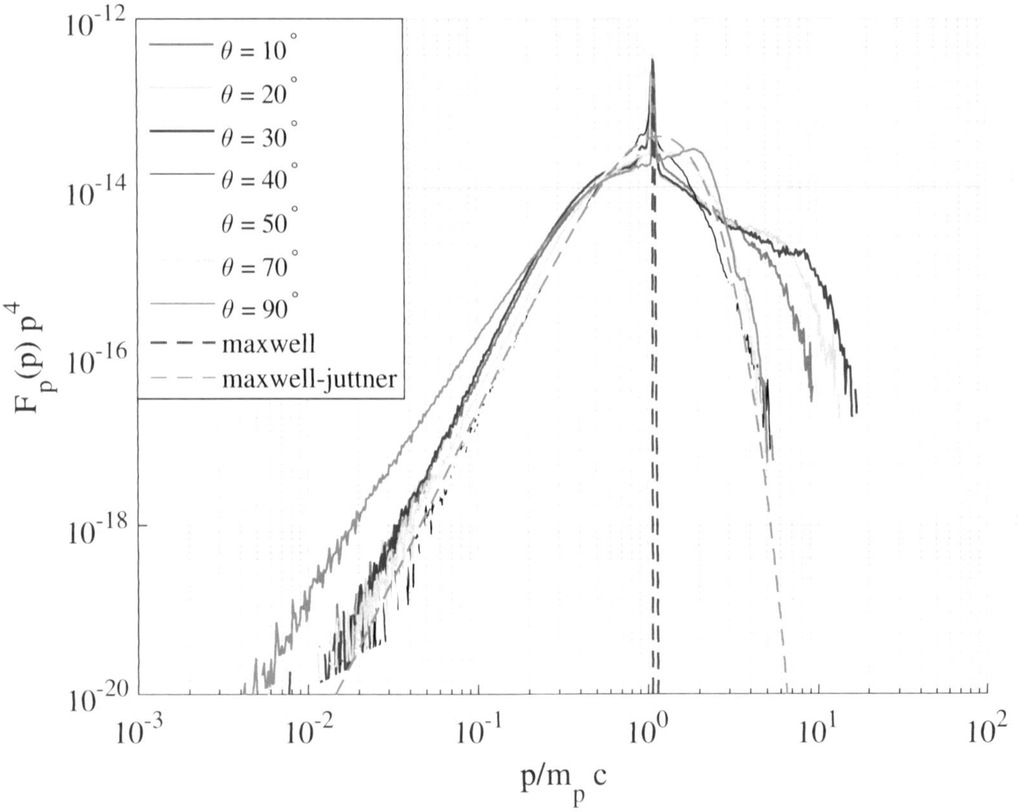

Figure 1. Distribution of protons in a relativistic shock wave with Lorentz factor 1.5 at various inclination angle θ.

The simulations are one-dimensional and have the following parameters of the initial flow: Lorentz factor $\gamma = 1.5$, the number densities $n_e = 10^{-4}\text{cm}^{-3}$, $n_p = 10^{-4}\text{cm}^{-3}$, the temperature, $T_e = T_p = 5 \cdot 10^8$K, the magnetic field $B = 10^{-4}$G, the full size of the box $L = 2 \cdot 10^{12}$cm, the number of cells $N = 2 \cdot 10^4$. The electron mass m_e is increase by 100 times to enlarge electron scales of the plasma, it is common approach in PIC simulations see e.g. [11] to reduce required computational resources. The full time of simulation is $T = 5000{\omega_p}^{-1}$, where ω_p is electron plasma frequency.These values allow one to define the dimensionless parameter magnetization

$\sigma = B^2/4\pi\gamma(n_p m_p + n_e m_e)c^2 = 0.003$. Inclination angle θ is the angle between the flow velocity and the magnetic field. We present results for the particle spectrum in several simulations at various θ. Stable shocks formed in all simulations and the total energy of the system conserved to better than 3%, which indicates that the code provides a correct model of the shock waves.

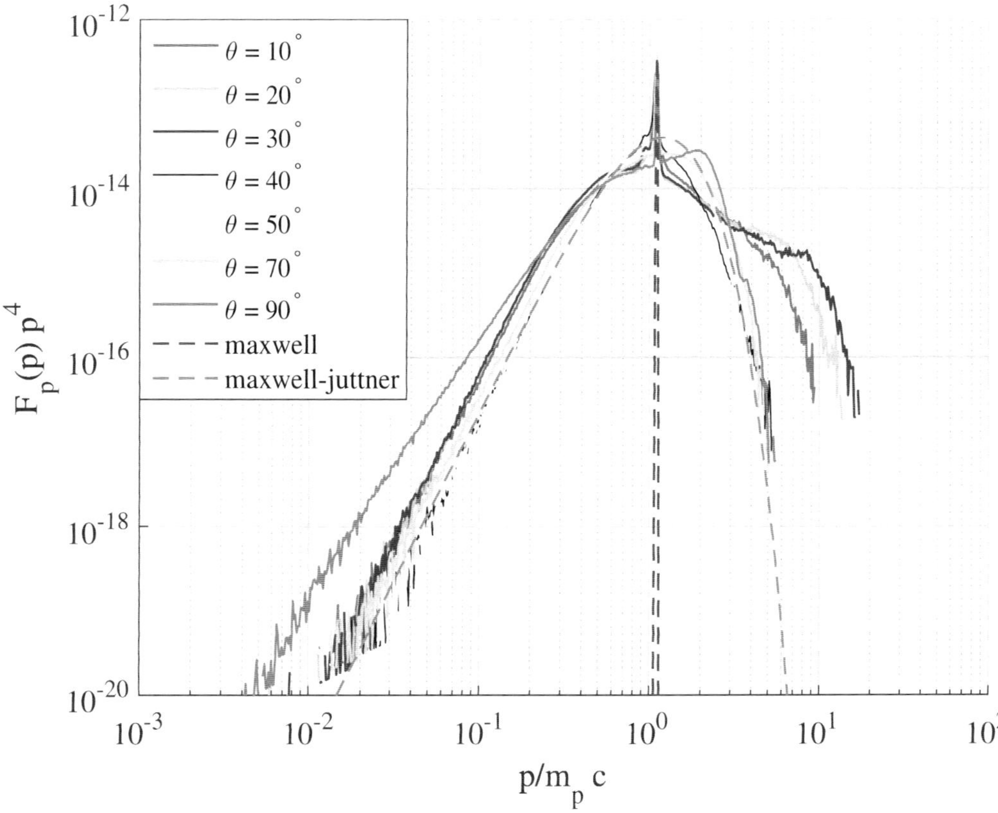

Figure 2. Distribution of electrons in a relativistic shock wave with Lorentz factor 1.5 at various inclination angle θ.

One can see in Figures 1 and 2 the particle spectra for various inclination angles θ. Each spectrum consists of three parts: the narrow peak of cold but moving fast initial upstream flow, the wide peak of hot downstream and the non-thermal accelerated component. The initial upstream flow spectrum is compared with the Maxwell distribution, shifted by the initial momentum, while the hot downstream spectrum is compared with the Maxwell-Jüttner distribution. The small-scale oscillations of spectra are produced by numerical errors and should not be taken into account. Figures show that the spectrum of accelerated particles strongly depends on he inclination angle θ. If θ is less than the critical value, defined by the equation $c \cdot \cos(\theta_{crit}) = v_{shock}$, where all values are measured in the upstream rest frame, particles can escape from the front and cross it several times to gain more energy (see, e.g., [10]), if θ is greater, shock becomes superluminal - particles need to move along the field lines faster than light to escape from the front, and acceleration is suppressed. It is difficult to evaluate the critical angle *a priori*, because the compression relation of the shock which will be created in the simulation is unknown. However the estimate for the case of a strong wave (the compression relation equals 4) provides $\theta_{crit} = 45^\circ$ in the downstream frame, which is consistent with the results of our simulations.

Also, one can see that for the angles less then the critical one, concentrations of accelerated particles are higher and extend to higher energies for larger θ. The spectrum of non-thermal component is about 10 times higher for the protons than for the electrons at the same energy. It is consistent with other PIC simulations of the shock waves [11].

These results show some new interesting effects for angles close to the critical value (black and yellow solid lines)- the electrons are still accelerating while the protons are not. It can be interpreted as electrons accelerate in the short-wavelength turbulence of the shock precursor. Due to the small gyroradii of the electrons in initial flow, the short-wavelength turbulence scatter them more efficiently than protons. However, further research is necessary to study this phenomenon in detail.

3. Conclusions

In this paper we presented results of the PIC simulations of particle acceleration by moderately relativistic shock waves in space plasmas. Simulations show, that injection of particles into the acceleration process strongly depends on the angle between the flow bulk velocity and the magnetic field. The injection efficiency as a function of the inclination angle has been studied. Injection of protons is more efficient than that of electrons for the inclination angles below the critical angle where the transition to the superluminal shock regime occurs. On another hand the electron acceleration is more efficient in shocks with the inclination angles close to the critical angle.

Acknowledgments

V I Romansky acknowledges support from RSF grant 16-12-10519. The numerical results presented here were obtained using computational resources of Peter the Great Saint-Petersburg Polytechnic University Supercomputing Center (http://www.scc.spbstu.ru).

References

[1] Bell A R 1978 *MNRAS* **182** 147
[2] Blandford R D and Ostriker J P 1978 *ApJ* **221** L29
[3] Bykov A M, Ellison D C, Osipov S M and Vladimirov A E 2014 *ApJ* **789** Issue 2, 137
[4] Berezhko E G, Ksenofontov L T and Völk H J 2003 *A&A* **412** L11
[5] Uchiyama Y, Aharonian F A, Tanaka T, Takahashi T and Maeda Y 2007 *Nature* **449** 576
[6] Romansky V I, Bykov A M, Osipov S M and Gladilin P E 2017 *Journal of Physics: CS* **929** id 012014
[7] Lapenta G, Brackbill J U and Ricci P 2006 *Phys. Plasmas* **13** 055904
[8] Noguchi K, Tronci C, Zuccaro G and Lapenta G 2007 *Phys. Plasmas* **14** 042308
[9] Ellison D C, Warren D C and Bykov A M 2013 *ApJ* **776** Issue 1, 46
[10] Pelletier G, Bykov A M, Ellison D C and Lemoine M 2017 *Space Science Reviews* **207** Issue 1-4, pp. 319-360
[11] Sironi L and Spitkovsky A 2011 *ApJ* **741** Issue 1, 39
[12] Iwamoto M, Amano T, Hoshino M and Matsumoto Y 2017 *ApJ* **840** Issue 1 article id. 52, 14 pp.

IOP Conf. Series: Journal of Physics: Conf. Series **1038** (2018) 012023 doi:10.1088/1742-6596/1038/1/012023

Investigation of native cells in liquid using the high aspect ratio nanowhisker probes by means of atomic force microscopy

M V Zhukov[1,2], F E Komissarenko[1,4], V I Chubinskiy-Nadezhdin[3], M M Khalisov[1,5]

[1] ITMO University, 197101 Kronverkskiy pr., 49, Saint-Petersburg, Russian Federation
[2] Institute for Analytical Instrumentation RAS, 198095 Ivan Chernykh st., 31-33, Saint-Petersburg, Russian Federation
[3] Institute of Cytology RAS, 194064 Tikhoretsky avenue 4, Saint-Petersburg, Russian Federation
[4] St. Petersburg Academic University RAS, 194021 Khlopina st., 8/3, Saint-Petersburg, Russian Federation
[5] Institute of Physiology by I.P. Pavlov RAS, Makarova embankment, 6, Saint-Petersburg, Russian Federation

E-mail: cloudjyk@yandex.ru

Abstract. Specialized Pt/C single nanowhisker probes for atomic force microscopy have been used to study the native cellular bio-objects (fibroblasts) in PBS liquid. Optimal parameters of nanowhisker fabrication were found such as accelerating voltage 5 kV, exposure time 5-10 sec, aperture of electron microscope 20 μm. The preliminary results of our study showed an improvement of image contrast and adhesion forces when using nanowhisker probes compared to the standard Si probes.

1. Introduction

Atomic force microscopy (AFM) is one of the few methods that allow to study native (living) cell. Unlike common optical methods for investigation of cellular structures, AFM allows to generate three-dimensional topography images of topology with nanoscale accuracy as well as to obtain the additional information about the specimen (roughness, adhesion force, stiffness, etc.). However, there is a problem of further enhancement of the method resolution to visualize the structure of complex bio-objects, in particular, the fine structure of cell membranes [1, 2].

One of the main elements affecting spatial resolution of the resulting images is the AFM probe (nanoneedle or tip) which interacts with the sample. To increase the AFM spatial resolution for detailed diagnostics of bio-objects the modification of the shape and composition of standard probe are carried out. One of the promising directions in this area is the formation of rod like nanostructures at the apex of the standard probes also known as nanowhiskers [3, 4]. However, there are still few works on controllable fabrication of single oriented nanowhiskers as well as their application in a liquid medium.

Therefore, the purpose of this work was to fabricate and apply specialized probes with single oriented nanowhiskers to study the fine structure of cellular native bio-objects in liquid by means of atomic force microscopy.

2. Experimental setup

The Pt/C nanowhiskers (NW) were fabricate at the top of the standard probes by electron beam induced deposition of precursor gases in a vacuum chamber of electron microscope [5]. The fabrication and visualization of the NW were performed by scanning electron microscope (SEM) CrossBeam Neon 40 (Carl Zeiss, Germany).

The investigation of cells was carried out by scanning probe microscope BioScope Catalyst (Bruker, Germany), combined with an inverted optical microscope Axio Observer D1m (Carl Zeiss, Germany). The investigation of native cells was carried out in PBS (phosphate-buffered saline) liquid using the PeakForce QNM mode [6]. TCS SP5 (Leica, Germany) confocal microscope was utilized to study the location of the cells by using a TRITC fluorescent dye and a 100x/1.4 N.A lens.

3. Results and Discussions

Nanowhisker was fabricated at the top of the probe by decomposition and precipitation of the gas precursor $C_9H_{16}Pt$ components under an electron beam in vacuum [6]. The optimal length of NW was about 300-400 nm with average diameter of about 40-60 nm. NW were fabricated with an angle to the tip axis ~ 22-25 degrees to reduce double-tap contact with the surface (compensation of the tip holder angle in microscope).

The optimal growth parameters of NW were follows: accelerating voltage 5 kV, exposure time 5-10 sec, aperture of electron microscope 20 μm (Fig. 1, b). The aperture of 20 μm made possible to achieve a whisker thickness of about 40-60 nm while the exposure value affects the height of the whiskers.

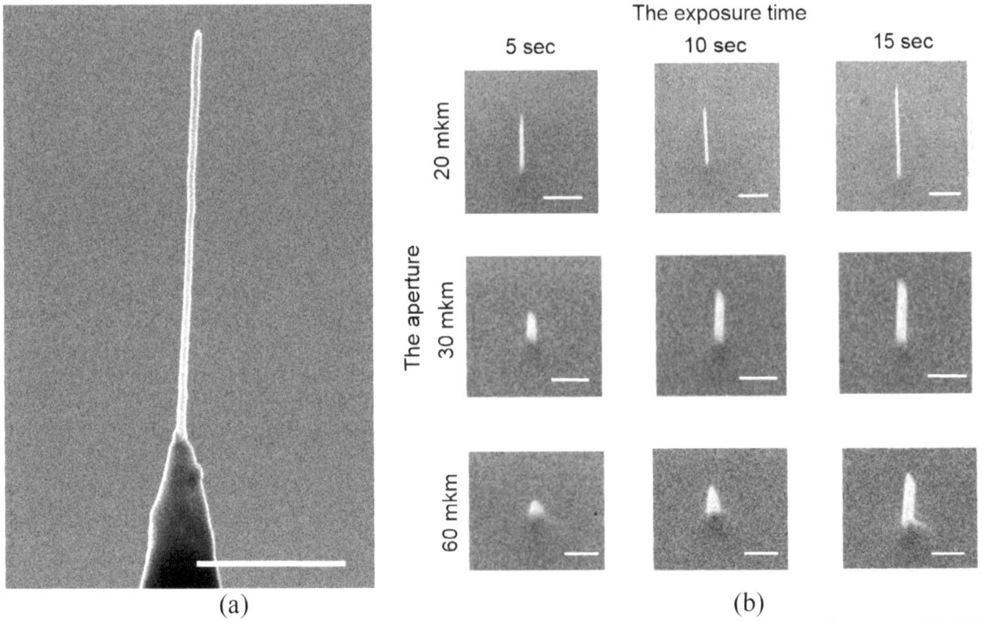

Figure 1. SEM image of NW formed at the standard probe apex (a), and SEM images of NW after variation of fabrication parameters of NW - exposure time and aperture (b). The scale bar is 500 nm.

The minimum value of accelerating voltage (5 kV) was chosen to achieve the optimal thickness of NW which allows to increase the penetrating ability (image contrast) and the mechanical stability of NW during scanning process.

To stain actin structures the mouse fibroblasts were plated on coverslips 1-2 days before the experiments. Cells were fixed with 3.7% paraformaldehyde solution (10 minutes, room temperature), permeabilized 0.1%. Fluorescence of the dye was excited by a laser with a wavelength of 543 nm, the emission was recorded in the range of 580-670 nm (Fig. 2, a). Fibroblast cells were studied close to native state by means of AFM in PBS liquid (Fig. 2, b).

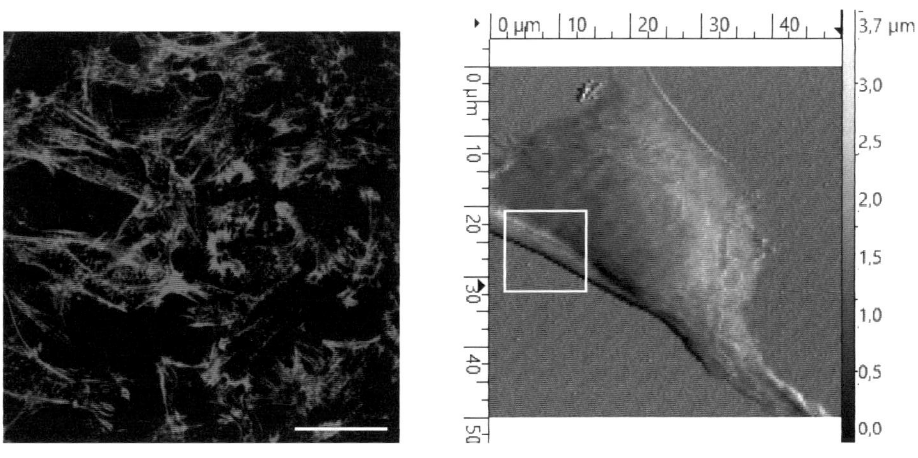

(a) (b)

Figure 2. Confocal image of actin cytoskeleton in transformed fibroblasts of SV40 mouse (a) and AFM image of single fibroblast close to native state (b). Staining of F-actin with fluorescent dye TRITC-phalloidin. The scale bar is 30 µm. The contrast of AFM image was increased by peak force error mode, the white frame shows the comparison area of the probes.

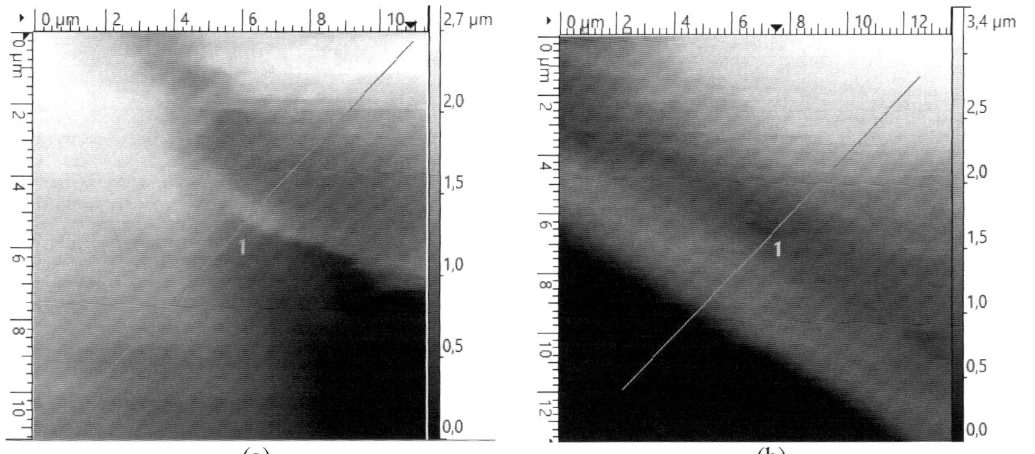

(a) (b)

Figure 3. AFM topography images of fibroblast visualized by the standard (a) and the NW (b) probes in PBS buffer.

An improvement of the penetrating ability using NW probes was found in the study of fibroblast membranes (Fig. 3, 4). Thus, the NW probes visualized the same section with a height difference of 3.4 µm while the standard probes showed height difference of about 2.7 µm in the same region. As seen in the cross-sections, the height of the projections is greater when used NW probes (small protrusion ~ 1,3 µm, large protrusion up to ~ 3 µm) compared to standard probes (Fig. 4, a, b).

The maximum value of adhesion force was about 2.4 nN for the probe with NW and of about 1.1 nN for the standard Si probe. It should be noted that the cell can change during the study therefore it is necessary to measure the flat area. Measurement of flat surface area revealed extension of adhesion force using the NW probes (~ 0,9 nN) in 2 times in comparison to the standard probes (~ 0,45 nN) (Fig. 5, 6).

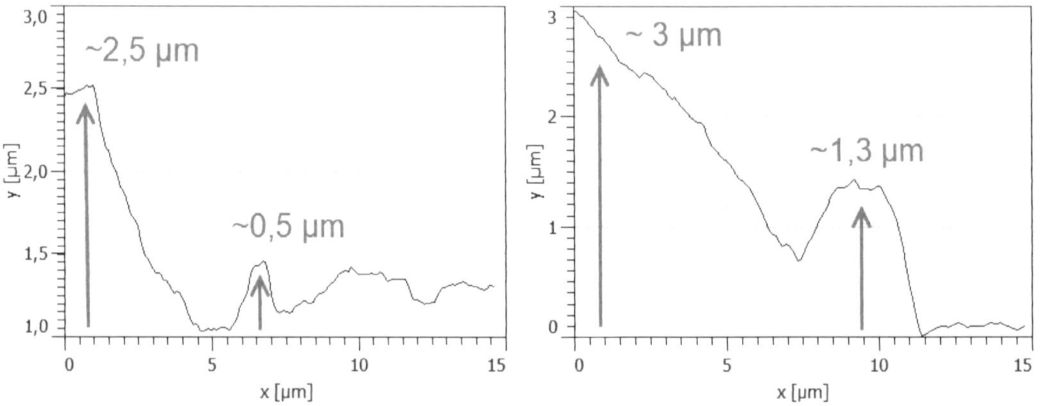

(a) (b)

Figure 4. Diagonal cross-sections of the native fibroblast surface from the AFM image visualized by the standard (a) and the NW (b) probes in PBS buffer.

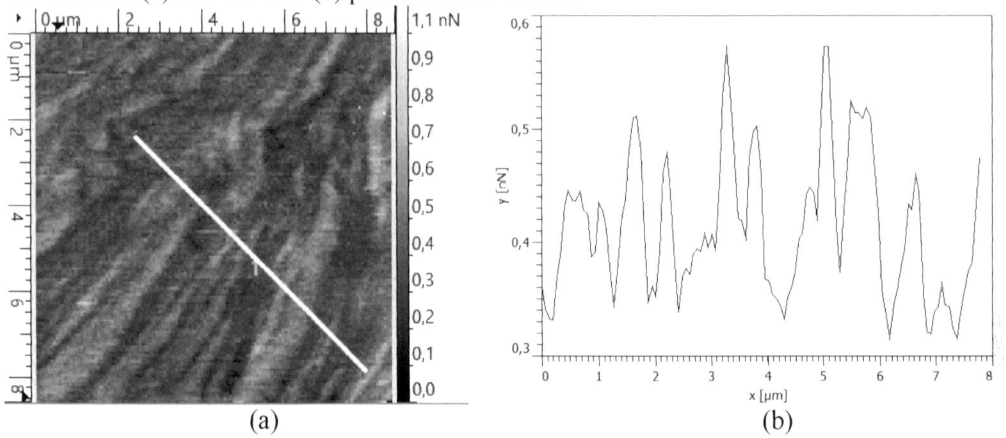

(a) (b)

Figure 5. AFM images of adhesion forces distribution (a) and cross-section area (b) measured on the fibroblast (central area) by standard probe in PBS buffer.

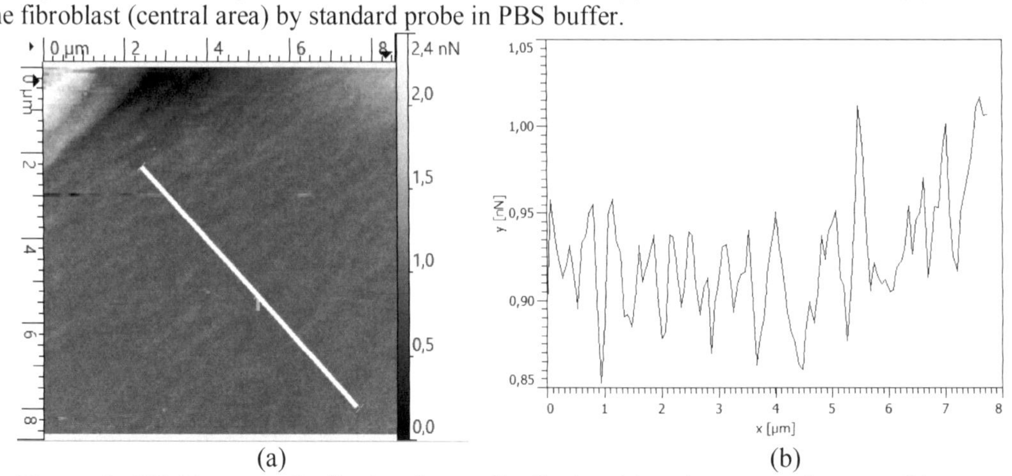

(a) (b)

Figure 6. AFM images of adhesion forces distribution (a) and cross-section area (b) measured on the fibroblast (central area) by NW probe in PBS buffer.

Thus, NW probes are shown to improve the accuracy of heights visualization and it corresponds to the previous data of adhesion forces to dry erythrocyte membranes [7] that reveals a better penetration ability and hydrophilic features of NW probes. Application of the specialized NW probes can produce better results in image contrast in comparison to the standard probes while studying the native cells in liquid medium by means of atomic force microscopy. It is difficult to analyze lateral resolution due to modification of cells by tips and relocation of their surface, therefore, it requires additional research.

4. Conclusions

Optimal parameters of nanowhisker fabrication were found to study the fine structure of cellular native bio-objects. It was found that application of probes with nanowhiskers in PBS liquid improve image contrast and adhesion forces in the study of native fibroblast cells. The results obtained in this study can be used for a more detailed and accurate study of native objects in liquid with complex structure by means of atomic force microscopy in such areas as pharmacology and medicine.

Acknowledgments

The work carried out with the support of Russian Foundation for Basic Research (16-32-00806), the leading universities of the Russian Federation (grant 074-U01).

References

[1] Yuri L Lyubchenko 2011 *Micron* **42** 196–206
[2] Kailas L, Ratcliffe E C, Hayhurst E J, Walker M G, Foster S J, Hobbs J K 2009 *Ultramicroscopy* **109** 775780
[3] Zhukov M V, Kukhtevich I V, Levichev V V, Mukhin I S and Golubok A O 2014 *J. Phys.: Conf. Ser.* **541** 012042
[4] Beard J D and Gordeev S N 2011 *Nanotechnology* **22** 175303
[5] Levichev V V, Zhukov M V, Mukhin I S, Denisyuk A I, Golubok A O 2013 *Technical Physics* **58** 1043–1047
[6] Bede Pittenger and Andrea Slade 2013 *Microscopy today* **21** 12–17
[7] Zhukov M V, Mukhin I S, Levichev V V and Golubok A O 2015 *Technical Physics Letters* **41** 149–152

Statistical memory effects in human stride dynamics

O.Yu. Panischev, S.N. Panischeva, S.A. Demin, R.R. Latypov
Institute of Physics, Kazan Federal University, 420008 Kazan, Kremlevskaya Street, 18 Russia
E-mail: opanischev@gmail.com

Abstract. In paper the Memory functions formalism (MFF) is used to analyze the dynamics of walking stride interval of healthy young adults, healthy old adults and Parkinson's disease. Using the non-Markovian parameter, we determine the degree of manifesting the statistical memory effects in dynamics of the walking stride interval for the considered volunteers. The stride dynamics of the first two volunteer groups demonstrates the low memory effects reflected in high values of non-Markovian parameter.The Parkinsonian patients demonstrate the non-Markovian behavior of stride dynamics with strong memory effects. The method can be used to study the signal dynamics of the different nature.

1. Introduction

At present, the studying of the statistical memory effects in evolution and dynamics of natural objects is interest for theoretical and computational physics [1, 2]. At the same time, the efforts of physicists working in this field are aimed to searching the statistical and numerical indicators to optimize such measures for describing the discrete time evolution of complex systems of different nature. The solving these problems becomes especially urgent when identifying and quantitatively describing the anomalous functioning of living systems from discrete time series. The time series analysis is an important tool to study the nature of complex systems in physical research. In this paper, the Memory functions formalism (MFF) [3, 4] is used to analyze the dynamics of walking stride interval of healthy young adults, healthy old adults and Parkinson's disease.

2. Memory function formalism

Within the framework of MFF [3, 4], we consider studied time series as the sequence $\{x_j\}$ of a random value X :

$$X = \{x(T), x(T+\tau), x(T+2\tau), \ldots, x(T+(N-1)\tau)\} = \{x_0, x_1, x_2, \ldots, x_{N-1}\},$$

$$x_j = x(T+j\tau), \qquad \langle X\rangle = \frac{1}{N}\sum_{j=0}^{N-1} x_j, \qquad \delta x_j = x_j - \langle X\rangle, \qquad \sigma^2 = \frac{1}{N}\sum_{j=0}^{N-1}\delta x_j^2.$$

Here T is the initial time point, τ is the time interval of signal discretisation, $\langle X\rangle$ is the mean value of X, δx_j is fluctuation. To describe the probabilistic relations in the sequence of X we use the normalized time-dependent correlation function (TCF):

$$a(t = m\tau) = M_0(m\tau) = \frac{1}{(N-m)\sigma^2}\sum_{j=0}^{N-m-1}\delta x_j \delta x_{j+m}, \qquad (1)$$

which is rewritten further as a scalar product of vectors of the initial state

$$\mathbf{A}_k^0 = \mathbf{A}_k^0(0) = \{\delta x_0, \delta x_1, \dots, \delta x_{k-1}\}$$

and of system state in time moment t

$$\mathbf{A}_{m+k}^m = \mathbf{A}_{m+k}^m(t) = \{\delta x_m, \delta x_{m+1}, \dots, \delta x_{m+k-1}\}$$

as follows:

$$a(t) = \frac{\langle \mathbf{A}_k^0(0)\mathbf{A}_{m+k}^m(t)\rangle}{\langle |\mathbf{A}_k^0(0)|^2\rangle}. \quad (2)$$

Using the the technique of projection operators [4] we introduce relations for the kinetic and relaxation parameters (by solving the finite-difference Liouvilles equation for the state vectors):

$$\lambda_n = \frac{\langle \mathbf{W}_{n-1}\hat{L}\mathbf{W}_{n-1}\rangle}{\langle |\mathbf{W}_{n-1}|^2\rangle}, \qquad \Lambda_n = i\frac{\langle |\mathbf{W}_n|^2\rangle}{\langle |\mathbf{W}_{n-1}|^2\rangle},$$

where the $\hat{L}$ is Liouville's quasioperator, the dynamic orthogonal variables:

$$\mathbf{W}_0 = \mathbf{A}_k^0(0), \qquad \mathbf{W}_1 = (i\hat{L} - \lambda_1)\mathbf{W}_0, \qquad \mathbf{W}_2 = (i\hat{L} - \lambda_2)\mathbf{W}_1 - \Lambda_1\mathbf{W}_0 - \dots, \quad (3)$$

the memory functions of the n order:

$$M_n(t = m\tau) = \frac{\langle \mathbf{W}_n\{1 + i\tau\hat{L}_{22}\}^m\mathbf{W}_n\rangle}{\langle |\mathbf{W}_n|^2\rangle},$$

and their power spectra:

$$\mu_n(\nu) = \left|\tau\sum_{j=0}^{N-1} M_n(j\tau)\cos 2\pi\nu j\tau\right|^2. \quad (4)$$

The dynamical orthogonal variables for $n > 1$ are used A relaxation time of the initial TCF and memory functions of the n order are determined as follows:

$$\tau_a = \Delta t\sum_{j=0}^{N-1} a(t_j),\ \tau_{M_1} = \Delta t\sum_{j=0}^{N-1} M_1(t_j), \dots,\ \tau_{M_n} = \Delta t\sum_{j=0}^{N-1} M_n(t_j). \quad (5)$$

The set of dimensionless values will determine the statistical spectrum of non-Markovian parameter, the informational measure of memory:

$$\{\varepsilon_i\} = \{\varepsilon_1, \varepsilon_2, \dots, \varepsilon_{n-1}\}, \quad \varepsilon_1 = \frac{\tau_a}{\tau_{M_1}},\ \varepsilon_2 = \frac{\tau_{M_1}}{\tau_{M_2}}, \dots,\ \varepsilon_n = \frac{\tau_{M_{n-1}}}{\tau_{M_n}}. \quad (6)$$

In this paper we use the first point of the frequency-dependent case of non-Markovian parameter:

$$\varepsilon_1 = \left\{\frac{\mu_0(\nu)}{\mu_1(\nu)}\right\}^{\frac{1}{2}}. \quad (7)$$

it is important to note that in this paper we use the values of the first order. See [4, 5] for details about the MFF.

3. Experimental data

Walking stride interval time series [6] included are from 15 subjects: 5 healthy young adults (23 – 29 years old), 5 healthy old adults (71 – 77 years old), and 5 older adults (60 – 77 years old) with Parkinson's disease. Subjects walked continuously on level ground around an obstacle-free path. The stride interval was measured using ultra-thin, force sensitive resistors placed inside the shoe. The analog force signal was sampled at 300 Hz with a 12 bit A/D converter, using an ambulatory, ankle-worn microcomputer that also recorded the data. Subsequently, the time between foot-strikes was automatically computed. The method for determining the stride interval is a modification of a previously validated method that has been shown to agree with force-platform measures, a "gold" standard.

4. Discussion

Using the MFF toolkit, particularly the non-Markovian parameter, it was possible to determine the degree of manifesting the statistical memory effects in dynamics of the walking stride interval for the considered volunteers. Figure 1 shows the frequency spectra of the non-Markovian parameter for typical representatives of each group of subjects. The key to our analysis is the value of the non-Markovian parameter at zero frequency. If $\varepsilon \gg 1$, then the Markovian components dominate in dynamics of stochastic processes. The short range (low) statistical memory effects is manifested. The decrease of this information measure characterizes the expansion of memory lifetime. $\varepsilon \sim 1$ means that the processes are characterized by long range (high) statistical memory. When $\varepsilon > 1$ we can consider the studied processes to be quasi-Markovian with moderate (intermediate on time existence) statistical memory. Thus, the introduced quantitative criterion allows us to parameterize the memory effects intensity and the velocity of loss of relaxation processes. The stride dynamics of the first two volunteer groups demonstrates the low memory effects reflected in high values of non-Markovian parameter. At that the age-related alterations in stride dynamics are manifested by increasing the memory effects. The stride dynamics for the second group corresponds by quasi-Markovian case. The Parkinsonian patients demonstrate the non-Markovian behavior of stride dynamics with strong memory effects.

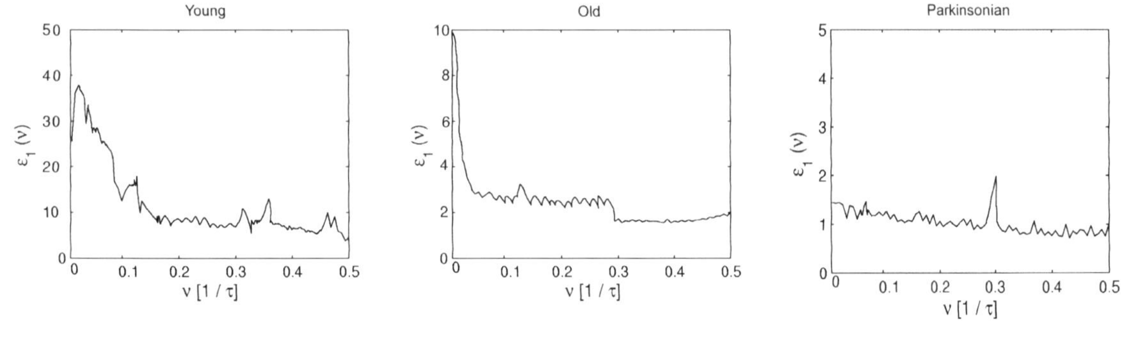

Figure 1. Non-Markovian parameter for considered groups of volunteers

5. Conclusion

In this paper, we have considered age- and pathological-related alterations in human locomotor activity. We have found the degree of statistical memory effects in stride dynamics at ageing and pathology. The strong memory effects has been found to reflect the pathological abnormalities in gait dynamics. Memory function, formalizm allows to discover the alterations in studied dynamics that are connected to different factors. The method can be used to study the signal dynamics of the different nature.

6. Acknowledgments

The work is performed according to the Russian Government Program of Competitive Growth of Kazan Federal University.

References

[1] Hausdorff J M, Cudkowicz M.E. et al. 1998 *Mov. Disord.* 13 428
[2] Peng C K, Havlin S., Stanley H E et al. 1995 *Chaos* 6 82
[3] Demin S A, Yulmetyev R M, Panischev O Yu and Hänggi P 2008 *Phys. A* **387** 2100
[4] Demin S A, Panischev O Yu and Demina N Yu 2016 *J. Phys. Conf. Ser.* **741** 012073
[5] Yulmetyev R, Hänggi P and Gafarov F 2000 *Phys. Rev. E* **62** 6178
[6] https://physionet.org/physiobank/database/gaitdb/

IOP Conf. Series: Journal of Physics: Conf. Series **1038** (2018) 012025 doi:10.1088/1742-6596/1038/1/012025

Collective effects in human EEGs at cognitive activity

S.N. Panischeva, O.Yu. Panischev, S.A. Demin, R.R. Latypov

Institute of Physics, Kazan Federal University, 420008 Kazan, Kremlevskaya Street, 18 Russia

E-mail: opanischev@gmail.com

Abstract. In this paper we use the Memory function formalism to study the collective phenomena in EEG signals from subjects with musical abilities ("musicians") and from subjects with a low level of these abilities ("non-musicians") at solving the three cognitive tasks. We have revealed the spectral features in the EEG collective dynamics of these groups, as well as sensor pairs, which have the a different interaction of signals.

1. Introduction

The most important factor determining the evolution of complex systems is a presence of collective effects arising from the interaction or redistribution of certain relationships between parts of a composite object. In many cases, it is impossible to carry out the objective analysis of functioning of these systems, without taking into account group phenomena. There are different approaches to studying the collective phenomena of complex systems: frequency and phase synchronization [1, 2], stochastic synchronization [3], the "generalized synchronization" relationship [4], detrended cross correlation analysis [5]. All of them, one way or another, are connected with the research of unique features of composite objects: certain quantitative and qualitative relationships between the elements of the system, the consistency of the dynamics of components under external influences, specific phenomena of synchronization. One of the most amazing and complex objects in the functioning of which the effects of synchronization and synchronization of a large number of signals are most clearly manifested is the human brain.

In this paper we use the Memory function formalism (MFF) [6, 7] to study the collective phenomena in EEG signals from subjects with musical abilities ("musicians") and from subjects with a low level of these abilities ("non-musicians") at solving the three cognitive tasks.

2. Memory function formalism

Within the framework of MFF [6, 7], we consider studied time series as the sequences $\{x_j\}$ and $\{y_j\}$ of random values X and Y :

$$X = \{x(T), x(T+\tau), x(T+2\tau), \dots, x(T+(N-1)\tau)\} = \{x_0, x_1, x_2, \dots, x_{N-1}\},$$
$$Y = \{x(T), y(T+\tau), y(T+2\tau), \dots, y(T+(N-1)\tau)\} = \{y_0, y_1, y_2, \dots, y_{N-1}\}.$$

$$z_j = z(T+j\tau), \qquad \langle Z\rangle = \frac{1}{N}\sum_{j=0}^{N-1} z_j, \qquad \delta z_j = x_j - \langle Z\rangle, \qquad \sigma_z^2 = \frac{1}{N}\sum_{j=0}^{N-1}\delta z_j^2.$$

Here T is the initial time point, τ is the time interval of signal discretisation, $\langle X\rangle$ is the mean value of X, δx_j is fluctuation, $Z \equiv X, Y$. To describe the probabilistic relations between the sequences of X and Y we use the normalized time-dependent cross correlation function (CCF):

$$c(t = m\tau) = M_0^{XY}(m\tau) = \frac{1}{(N-m)\sigma_x\sigma_y} \sum_{j=0}^{N-m-1} \delta x_j \delta y_{j+m}, \tag{1}$$

which is rewritten further as a scalar product of vectors of the initial state

$$\mathbf{A}_k^0 = \mathbf{A}_k^0(0) = \{\delta x_0, \delta x_1, \ldots, \delta x_{k-1}\}, \ \mathbf{B}_k^0 = \mathbf{B}_k^0(0) = \{\delta y_0, \delta y_1, \ldots, \delta y_{k-1}\}$$

and of system state in time moment t

$$\mathbf{A}_{m+k}^m = \mathbf{A}_{m+k}^m(t) = \{\delta x_m, \delta x_{m+1}, \ldots, \delta x_{m+k-1}\}$$
$$\mathbf{B}_{m+k}^m = \mathbf{B}_{m+k}^m(t) = \{\delta y_m, \delta y_{m+1}, \ldots, \delta y_{m+k-1}\}$$

as follows:

$$c(t) = \frac{\langle \mathbf{A}_k^0(0)\mathbf{B}_{m+k}^m(t)\rangle}{\langle \mathbf{A}_k^0(0)\mathbf{B}_k^0(0)\rangle}. \tag{2}$$

Using the the technique of projection operators we introduce relations for the kinetic and relaxation parameters:

$$\lambda_n^{XY} = \frac{\langle \mathbf{W}_{n-1}^X \hat{L} \mathbf{W}_{n-1}^Y\rangle}{\langle \mathbf{W}_{n-1}^X \mathbf{W}_{n-1}^Y\rangle}, \qquad \Lambda_n^{XY} = i\frac{\langle \mathbf{W}_n^X \mathbf{W}_n^Y\rangle}{\langle \mathbf{W}_{n-1}^X \mathbf{W}_{n-1}^Y\rangle},$$

the dynamic orthogonal variables:

$$\begin{aligned} \mathbf{W}_0^X &= \mathbf{A}_k^0(0), \quad \mathbf{W}_1^X = (i\hat{L} - \lambda_1^{XY})\mathbf{W}_0^X, \quad \mathbf{W}_2^X = (i\hat{L} - \lambda_2^{XY})\mathbf{W}_1^X - \Lambda_1^{XY}\mathbf{W}_0^X - \ldots, \\ \mathbf{W}_0^Y &= \mathbf{B}_k^0(0), \quad \mathbf{W}_1^Y = (i\hat{L} - \lambda_1^{XY})\mathbf{W}_0^Y, \quad \mathbf{W}_2^Y = (i\hat{L} - \lambda_2^{XY})\mathbf{W}_1^Y - \Lambda_1^{XY}\mathbf{W}_0^Y - \ldots, \end{aligned} \tag{3}$$

the memory functions of the n order:

$$M_n^{XY}(t = m\tau) = \frac{\langle \mathbf{W}_n^X \{1 + i\tau\hat{L}_{22}\}^m \mathbf{W}_n^Y\rangle}{\langle \mathbf{W}_n^X \mathbf{W}_n^Y\rangle},$$

and their power spectra:

$$\mu_n^{XY}(\nu) = \left| \tau \sum_{j=0}^{N-1} M_n^{XY}(j\tau) \cos 2\pi\nu j\tau \right|^2. \tag{4}$$

3. Experimental data

Eight subjects [8] were considered and grouped in two different sets, musicians (number of subjects = four with mean age 28.5 years, each with at least 5 years of musical training) and non-musicians (number of subjects = four with mean age 28 years with no musical training). All subjects were instructed to listen attentively for several minutes via earphones to music (two sonatas for violin and piano by Brahms and Beethoven, respectively, and French Suite No. 5 by Bach). A piece of computer music by Martin and a text of neutral content (a short story, "Versuendigung gegen die Nachwelt" by H. Weigel, read by Christiane Hoerbiger) were also presented. In addition, the subjects were asked to perform a spatial imagination task, which involves mental rotation of figures. The EEGs were recorded for 90 s with a sampling rate of 128 Hz from 19 gold-cup electrodes equally distributed over the scalp according to the so-called 10-20 system as displayed in Figure 1 with respect to the averaged signals from both ear-lobes. Moreover, periods of EEG at rest were recorded before, during and after each task. Their durations were the same as those of the musical pieces. Eyes were closed during listening to music and text. See [8] for detail.

4. Discussion

Analyzing the types of power spectra (Figure 2 of the initial CCF demonstrates the dominance of low-frequency dynamics during the performance of mental rotation in both groups of subjects, which indicates the absence of differences in the interaction of the cerebral cortex areas in the performance of this task. However, when listening to classical music and listening to the text, differences in the EEG are found among musicians and non-musicians for certain pairs of electrodes. These pairs are electrodes numbered 1 and 19, 1 and 12, 8 and 17 for text; 2 and 18, 7 and 19, 7 and 13 for listening to music. For the remaining pairs of electrodes, there are no significant differences in the interaction of parts of the cerebral cortex.

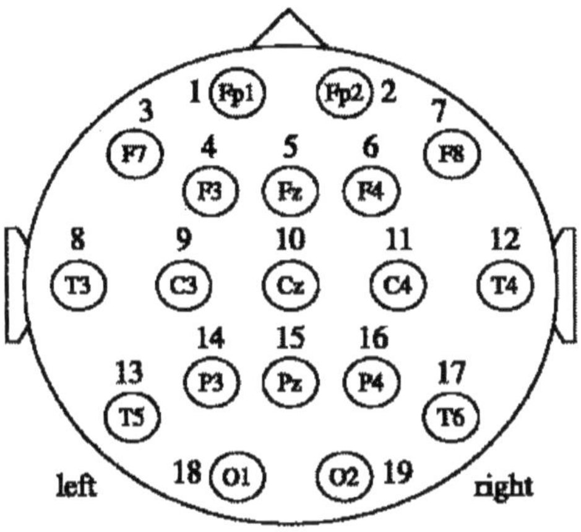

Figure 1. Position of the 19 electrodes and their designations according to the international 10-20 electrode placements systems

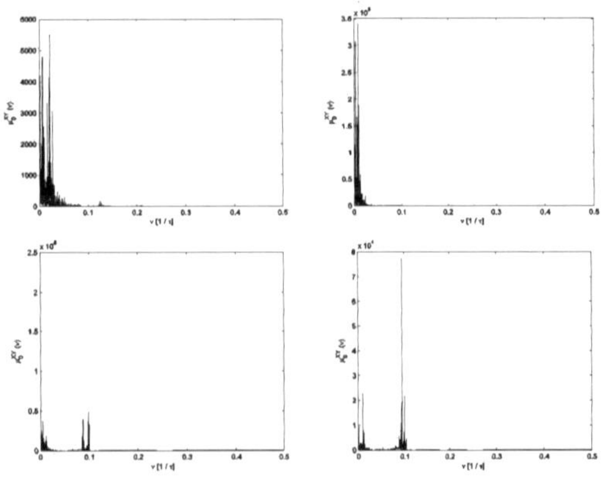

Figure 2. Types of power spectra of the initial CCF: low-frequency peak, very low frequency peak, high frequency peak, two peaks

5. Conclusion

We have studied the EEG signals of two groups of subjects: people having ("musicians") and not having ("non-musical") musical abilities when solving several cognitive tasks. We have revealed the spectral features in the EEG collective dynamics of these groups, as well as sensor pairs, which have the a different interaction of signals. The presented results will be of interest for biophysics, cognitive psychology, neurophysiology and other fields related to the study of amazing features of the bioelectric activity of the human brain.

6. Acknowledgments

The work is performed according to the Russian Government Program of Competitive Growth of Kazan Federal University.

References

[1] Rosenblum M, Pikovsky A and Kurths J 1996 *Phys. Rev. Lett.* **76** 1804
[2] Quiroga R Q, Kraskov A, Kreuz T and Grassberger P 2002 *Phys. Rev. E* **65** 041903
[3] Afraimovich V S, Verichev N N and Rabinovich M I *Izv. Vyssh. Uchebn. Zaved. Radiofiz* 1986 **29** 1050
[4] Rulkov N F, Sushchik M M, Tsimring L S and Abarbanel H D I 1995 *Phys. Rev. E* **51** 980
[5] Podobnik B and Stanley H E 2008 *Phys. Rev. Lett.* **100** 084102
[6] Panischev O Yu, Demin S A and Bhattacharya J 2010 *Phys. A* **389** 4958
[7] Demin S A, Panischev O Yu and Demina N Yu 2016 *J Phys. Conf. Ser.* **741** 012073
[8] Bhattacharya J and Petsche H 2005 *Sig. Proc.* **85** 2161

IOP Conf. Series: Journal of Physics: Conf. Series **1038** (2018) 012026 doi:10.1088/1742-6596/1038/1/012026

Study of DNA interaction with cobalt disulfophthalocyanine

R Tikhomirov[1], V Demidov[2] and N Kasyanenko[1]

[1] Faculty of physics SPbSU, Peterhof, Saint-Petersburg, 198504, Ulyanovskaya, 13, Russia

[2] Pro-Bright Neva, 192289, Saint-Petersburg, Sofiyskaya ul., 93, St. Petersburg, Russia

E-mails: [1] romka_tikhomirov@mail.ru, nkasyanenko@mail.ru,

[2] vndemidov@mail.ru

Abstract. DNA interaction with disulfophthalocyanine of cobalt (CodSPC) was studied by UV-VIS spectrophotometry and low gradient viscometry. The thermal denaturation of DNA in complexes with CodSPC was compared with free DNA denaturation. The small increase in melting temperature indicates that the secondary structure of DNA in complexes is stable. As a results of experimental research the external binding of CodSPC to DNA was determined. The only one type of binding is accompanied by stacking of heterocycles at high concentrations of CodSPC. An increase in viscosity of DNA solutions with the growth of CodSPC concentration can be explained by intermolecular DNA contacts via CodSPC.

Most of the antitumor drugs are aimed at binding to the DNA molecule in the cell. This helps to prevent cell division or leads cancer cells to apoptosis. Different types of interaction between drugs and double-stranded DNA can be distinguished as major or minor groove binding, external binding with phosphates and interacalation. DNA interaction with phthalocyanine, porhyrine and their compounds with metals has a significant interest because of their spectral features, which have medical application, such as photodynamic therapy [1]. Metal phtalocyanines can be used also as pigments and catalysts. The absorption in infrared area of spectrum, high quantum yield of fluorescence and an ability to cleave DNA promote their application in medical and biological practice. An aggregation of phthalocyanines in water solutions is a well-known fact. For this reason the usage of phthalocyanines in chemotherapy is limited. Nevertheless, an addition of sulfonic, carboxylic or amino groups to the peripheral sites can reduce aggregation between molecules [2, 3]. The type of the coordination metal also influences on aggregation. Although study of cationic complexes of phthalocyanines are in the focus of investigators. The anionic complexes are much less studied [3]. In this paper the possibility of negative DNA interaction with anionic CodSPC was regarded. We used UV-VIS spectrophotometry, low gradient viscomery and thermal denaturation of DNA (DNA melting). Also binding constant was calculated with Wolf-Shimer equation [4].

1. Experimental section

1.1. Materials

High molecular calf thymus DNA (Sigma Aldrich, CAS 73049-39-5; D1501-1G Type I "Highly Polymerized", fibrous preparation) with molecular mass $9*10^6$ that was determined from the intrinsic viscosity of DNA in 0,15 M NaCl was used. Water solution of NaCl (chemically pure) in concentration 0.005 M was used as solvent for the preparation of DNA solution. DNA concentrations was determined by spectral method. For that the difference in absorption at 270 and 290 nm for DNA denaturated in 6% $HClO_4$ (in 15 min at 100° C) was measured. This procedure allows one to independently determine the molar extinction coefficient for a given DNA sample by an absorbance at 260 nm for native molecule (without thermal denaturation). The sodium salt of anionic complex of cobalt disulfophthalocyanine was used. The structure of the compound is shown in Fig.1. Sulfonic groups make CodSPC water soluble. The Tris buffer was used in all our experiments. Consequently in this work all solutions of CodSPC (with or without DNA) contain Tris buffer.

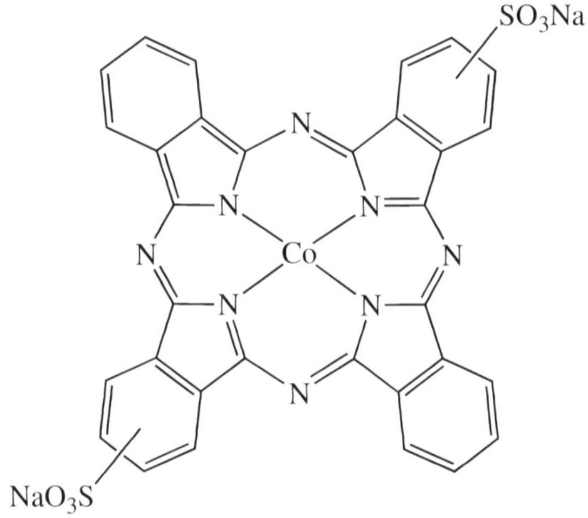

Fig.1 Structure of cobalt disulfophthalocyanine (CodSPC).

1.2 Methods.

1.2.1 UV-VIS absorption spectrophotometry

All experiments were conducted in 0,005 M NaCl and Tris buffer solutions. Spectrophotometer SF-56 (Russia) with operating range from 190 nm – 1100 nm was used. Quartz cuvettes were used with optical pans 1 cm.

Spectrophotometric titration allows to determine binding constant according to Wolfe-Shimer equation:

$$\frac{[DNA]}{(\varepsilon_a-\varepsilon_f)} = \frac{[DNA]}{(\varepsilon_b-\varepsilon_f)} + \frac{1}{K_b(\varepsilon_b-\varepsilon_f)} \quad , \qquad (1)$$

where [DNA] is the DNA concentration in moles of base pairs, ε_a corresponds to the $\frac{A}{[compound]}$, where A is the absorption of the compound at the selected wavelength within the analyzed band, [*compound*] is the concentration of compound, ε_b and ε_f are molar extinction coefficient of free compound and compound fully bonded to DNA respectively. The plot of $\frac{[DNA]}{(\varepsilon_a-\varepsilon_f)}$ value on [DNA] called Wolfe-Shimer plot can give the binding constant value [4].

1.2.2 Low-gradient viscometry

IOP Conf. Series: Journal of Physics: Conf. Series **1038** (2018) 012026 doi:10.1088/1742-6596/1038/1/012026

In order to determine the specific viscosity η_r-1 of DNA solutions at constant concentration of DNA the low-gradient rotation Zimm-Crothers type viscometer was used. All solutions were prepared with constant DNA concentration C_{DNA}=0.005% (determined from spectral data). All systems contain NaCl salt (0,005M) and Tris buffer in small concentrations.

1.2.3 Thermal denaturation study (DNA melting)
Two systems were prepared in 0.005 M NaCl with Tris buffer. Specord 200 Plus spectrophotometer was used.

2. Results and discussion

2.1 Absorption spectroscopy
The absorption spectrum of CodSPC has bands in UV and visible areas (Fig. 2A).Shortwave UV band in intersects with DNA absorption spectrum. It is difficult to use for the analysis of binding. Near UV band has maximum at 345 nm (so-called Soret or B band). This band associates with the nature of complexing metal. Visible red split band with main maximum at 693 nm and unresolved maximum around 595 nm (Q-bands) associate with electronic structure of heterocycles. It was shown that the addition of DNA into CodSPC solution causes change in spectrum of compound. Indeed, all bands undergo a change: one can see hypsochromic shift with hyperchromism in Soret and minor Q bands and hypochromism with blue shift in main Q band (Fig. 2). In this way, spectral data show that DNA interacts with CodSPC in a solution. Spectrophotometric titration demonstrates the presence of isosbestic points in Soret and Q bands, which indicates only one type of CodSPC binding to DNA.
Titration is the most convenient way to determine the binding constants [5]. Aggregation of phthalocyanines in solutions without DNA was not recognized. But according to the Fig.2 C the hyperchromism in Q band can indicate some association of chromophores. We could not see this effect for Soret band. It is known that B-band is common for all phthalocyanines and related to the π-π electron transitions [6]. One type of binding, according to isosbestic point [7] associates with decrease in absorption of phtalocyanines. Although there is no significant changes in the DNA absorption (we calculate this value believing that absorption of CodSPC in this region does not change).
According to the literature such spectral changes support the idea about external binding of CodSPC to DNA [8]. To estimate the binding constant, the Wolfe-Shimmer equation was applied (Fig.3). According to the data obtained, the value of K_b is quite small for the intercalation and huge for an electrostatic binding [2, 5]. However, the calculated binding constant is correct for an external interaction. The values of hypochromism and hypsochromism as well as K_b are presented in the Table 1.

Table 1 Hypochromic (%) effect, hypsochromic ($\Delta\lambda$) shift for C_{DNA}= 0,007% binding constant and shift in DNA melting temperature

Complex	Hypochromic effect (%)	Hypsochromic Effect (nm)	K_b, M^{-1}	ΔT_m
$Co(C_8H_4)_2(C_8H_3SO_3Na)_2N_8$	4.38	1.6	$3,6*10^{-4}$	3.5

IOP Conf. Series: Journal of Physics: Conf. Series **1038** (2018) 012026 doi:10.1088/1742-6596/1038/1/012026

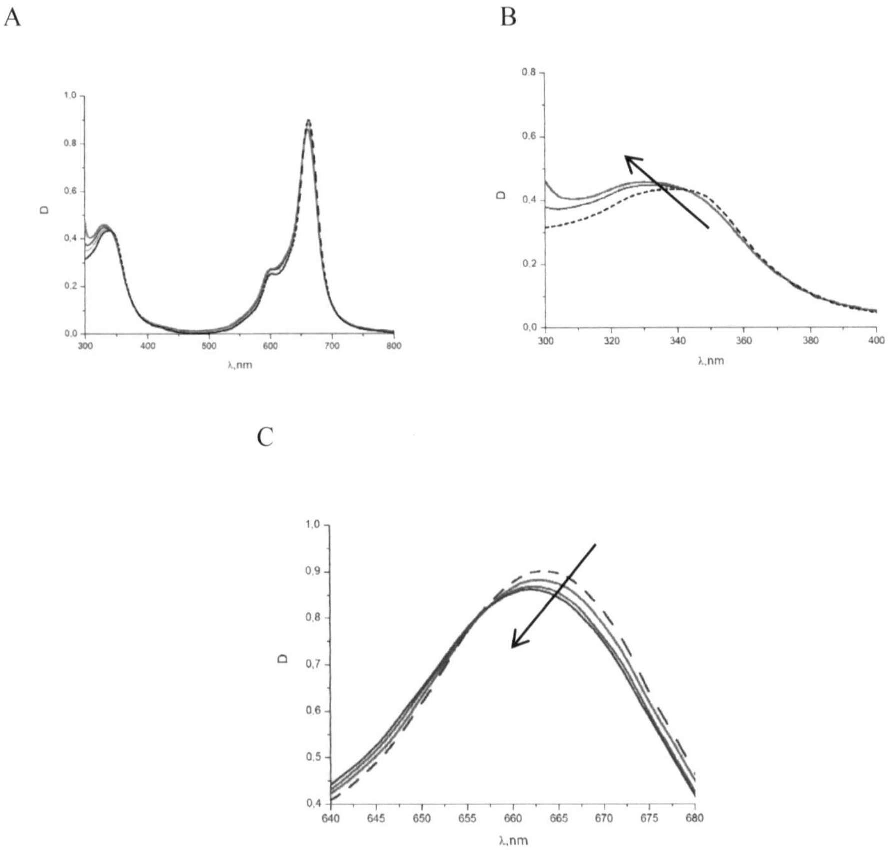

Fig.2 CodSPC absorption spectrum with (solid lines) and without DNA (segment line) in 0,005 M NaCl solution with Tris buffer (a), the results of titration in B-(b) and Q-(c) bands. [CodSPC] = 2,5*10^{-5}M a) and two bands of compound absorption (b and c). DNA concentration was changed from 0 to 0.015%

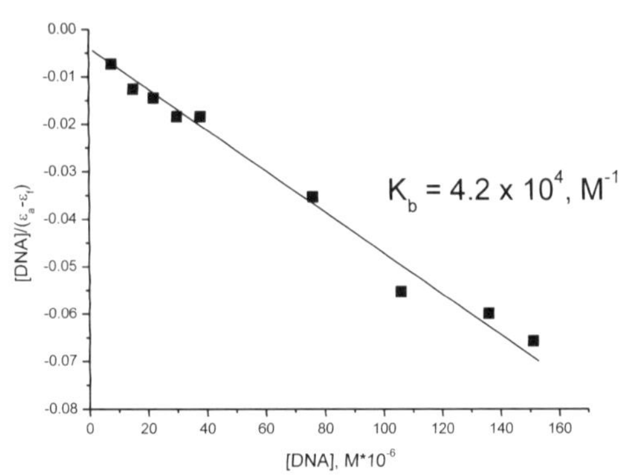

Fig.3 Wolfe-Shimmer plot in order to determine the binding constant (see in text)

2.2 DNA melting study

Another method which can give an information about stability of DNA secondary structure during binding is DNA thermal denaturation studies. According to the experimental data, the intercalation is accompanied by a large change in DNA melting temperature, while groove binding or an electrostatic interaction cause only small shift in DNA melting temperature respectively [9]. In our experiment free DNA in 0,005 M NaCl (C = 0,005%) and complex of DNA with CodSPC (C_{CoPc} = 5.0*10^{-5}M) were heated from 25 °C to 98 °C. The dependences of absorption at 260 nm (maximum of DNA absorption band) on T are shown in Fig. 4. Analysis of the melting curves provides us information about melting temperature (middle point of transition). The small increase in the melting temperature of DNA was jserved in complexes (see Table 1). In Fig. 4 the curve with open circles indicates free DNA melting, and the curve with the black one shows melting of DNA in complexes The data do not contradict with an idea about external binding of CodSPC to DNA.

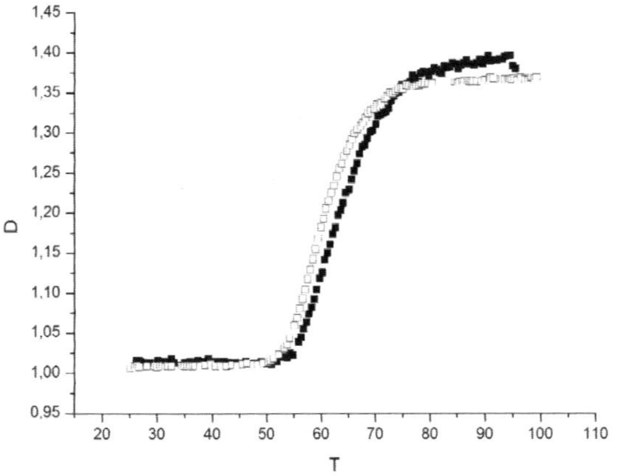

Fig.4 Melting curves of free DNA (pen circles) and DNA in complexes with CodSPC (black circles)

2.3 Low-gradient viscometry

The binding of small molecules to DNA can effect on DNA volume in a solution and on DNA rigidity [10]. To investigate the influence of CodSPC on DNA tertiary structure we use low gradient viscosity. The measurements were carried out at room temperature (21° C). DNA concentration in all systems was constant, C_{DNA}=0.005%. The compound CodSPC was added into DNA solution by the mixing of DNA and CodSPC solutions in same solvents. Different concentrations of CodSPC cause increase in solution viscosity. The dependence of reduced viscosity of DNA solutions on CodSPC concentration is presented in Fig. 5. The data show an increase (28%) in the reduced viscosity of DNA solutions with CodSPC respect to DNA solution without compound.

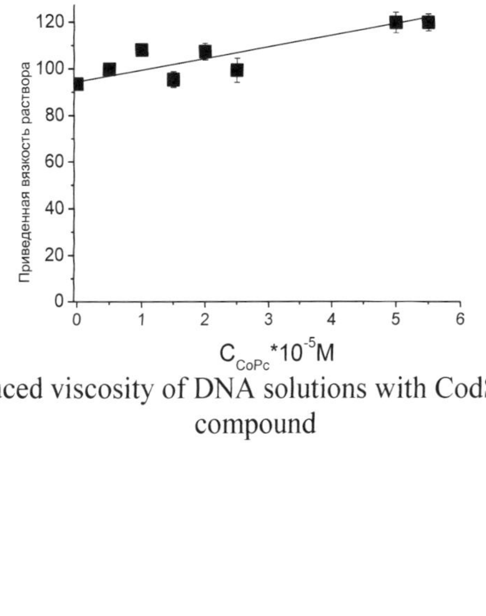

Fig.5 Dependence of reduced viscosity of DNA solutions with CodSPC on concentration of the compound

The intercalation of heterocycles of CodSPC can induce the observed increase in viscosity, but spectral data and melting experiment indicate the external binding. We can assume that the only one type of binding - the external binding of CodSPC to DNA, is accompanied by stacking of heterocycles outside of DNA helix. An increase in viscosity of DNA solutions with the growth of CodSPC concentration can be explained by intermolecular DNA contacts via CodSPC. For better understanding further investigation is necessary.

3. Conclusions

Our study show that CodSPC in experimental condition interacts with DNA. Only one type of binding is realized. The binding is accompanied by changes in CodSPC spectrum and by increase in viscosity of DNA solutions. The binding does not cause destabilization of DNA secondary structure. For better understanding of processes of binding further experiments should be carried out. The support of this work was partially provided by Saint-Petersburg State University Research Park Center of Diagnostics of Functional Materials for Medicine, Pharmacology and Nanoelectronics.

References

[1] Bhattacharya S, Mandal G and Ganguly T 2010 detailed spectroscopic investigations to reveal the nature of interaction of anionic porphyrin with calf thymus DNA *J. Photochem. Photobiol.* B **101** 89-96

[2] Hassani L, Hakimian F, Safai E and Fazeli Z 2013 antibacterial effect of cationic porphyrazines and anionic phthalocyanine and their interaction with plasmid DNA *J. Mol. Struct.* **1052** 221-7

[3] Rasouli N and Sohrabi N 2016 a study on the aggregation and calf thymus DNA binding characteristics of anionic cobalt(II) tetrasulfonated phthalocyanine *Phys. Chem. Res.* **4** 83-94

[4] Wolfe A, Shimer G.H and Meehan T 1987 polycyclic aromatic hydrocarbons physically intercalate into duplex regions of denatured DNA *Biochem.* **26** 6392-6

[5] Barut B, Demirbas Ü, Özel A and Kantekinb H 2017 novel water soluble morpholine substituted Zn(II) phthalocyanine: synthesis, characterization, DNA/BSA binding, DNA photocleavage and topoisomerase I inhibition *Int. J. Biol. Macromol.* **105** 499-508

[6] Gurinovich G.P, Sevchenko A.N and Soloviev K.N 1963 porhyrins spectroscopy *Phys. Usp.* **79** 181-2

[7] Scatchard G 1949 *Ann NY Acad. Sci.* **51** 660

[8] Gaier A.J, Ghimire S, Fix S.E and McMillan D.R 2014 internal versus external binding of cationic porphyrines to single-stranded DNA *Inorg.Chem.* **53** 5467-73

[9] Lopez Zeballos N.C, Gauna G.A, Garcia Vior M.C. Awruch J and Dicelio L.E 2014 interaction of cationic phthalocyanine with DNA, importance of the structure of the substituents *J. Photochem. Photobiol.* B **136** 29-33

[10] Tsvetkov V.N, Eskin V.E and Frenkel C 1964 macromolecules structure in the solution *Sci.* 162-70, 189-94

DNA Damage Induced by Gamma-Radiation Revealed from UV Absorption Spectroscopy

SA Tankovskaia[1], OM Kotb [1,2], OA Dommes[3], SV Paston[1]

[1]Department of Molecular Biophysics and Polymer Physics,
St. Petersburg State University, Petergof, Ulyanovskaya st. 3, St. Petersburg, Russia
[2]Department of Physics, Faculty of Science, Zagazig University, Sharkia Gov Zagazig, 44519 Egypt
[3]Institute of macromolecular compounds, 199004 Saint-Petersburg, Bolshoy pr. 31, Russia

E-mail: tasva-ara1@yandex.ru

Abstract. DNA damages are the main cause of radiation-induced cell death, mutations and carcinogenesis. The study of influence of environmental conditions on the radiation damage of DNA structure is important for the wide range of medical problems. The present work is devoted to the study of alterations in DNA structure caused by γ-irradiation at the variation of counterions concentration and dose rate.The DNA solutions of the ionic strengths 0.005M and 0.15M NaCl were exposed to γ-radiation with the doses of 0-100 Gy, dose rates2.3Gy/s and 0.3 Gy/s. The lesions in DNA structure (base damages, partial denaturation) were studied by the methods of UV absorption spectroscopy and DNA melting.Radiation-induced base damages decrease at the rise of NaCl concentration in the solution and at the lowering of the dose rate. The stability of DNA secondary structure increase at the rise of ionic strength of irradiated solution. The influence of radiation dose rate on the DNA helicity reveals definitely only at the dose of 100 Gy. The relative fall of the DNA melting temperatureafter the irradiation slightly depends on the dose rate and NaCl concentration.

1. Introduction

Ionizing radiation damages all the molecules in a living cell but justinjuries in the DNA are most fatal because of the unique role of this molecule. Unrepaired damages in DNA structure cause cell death, mutations, malignant degeneration [1-4]. Radiation-induced DNA lesions are extensively studied, the most frequent types of them are single- and double strand breaks (SSB and DSB), inter- and intrastrand cross-links, destruction, modification and release of nucleobases, local breakage of hydrogen bonds (partial denaturation) [1-4].To reveal radiation-induced structural damages in DNA the model aqueous DNA solutions are widely applied in experiments [1,2]. It is possible because the secondary structure of the macromolecule in a cell and in an aqueous solution is the same and represents the B-form [5]. Besides the mechanisms of ionizing radiation action on DNA molecule in a cell and in a solution are identical: mainly it is so-called indirect action, i.e. DNA injury over its interaction with the products of water radiolysis [1,2]. The present work is devoted to the study of alterations in DNA structure caused by γ-irradiation of its aqueous solutions.

DNA conformation, its ionic surrounding, water content, temperature, presence of some low molecular compounds (radioprotectors or radiosensitizers) influence the radiation effect on the DNA, but the molecular mechanism of many of these processes are not yet clarified [1, 6-11]. Counterions concentration in the DNA solution plays an important role in the formation of the tertiary structure of the macromolecule and stability of its secondary structure, positive ions are integrated into the hydration layer of DNA [5, 12, 13]. Nevertheless there are only a few works dealing with the influence of ionic strength of solution on the DNA radiosensitivity [14-16]. We compared the effect of γ-radiation on DNA molecule in aqueous solutions of the ionic strengths μ_1=0.005MNaCl and μ_2=0.15MNaCl at the different radiation doses and dose rates.

The UV absorption spectroscopy can provide essential information about the DNA structure. Hyperchromic effect indicates the degree of DNA helicity [17]. The method of A.S. Spirin [18] allows to find the concentration of chromophores (nucleobases) in the DNA solution and thus estimate the amount of nucleobases destructed under the action of radiation. DNA melting process is very sensitive to any defects in the primary and secondary structure of DNA [19].In the present work these methods were applied to reveal DNA damages induced by the γ-irradiation of DNA solutions.

2. Experimental section

2.1. Materials

DNA (Sigma) from calf thymus with a molecular mass of $(11.2\pm0.6)\cdot10^6$ Da was used. All aqueous solutions were prepared using double distilled deionized water. NaCl and $HClO_4$ were reagent-grade.

2.2. Radiation exposure

The DNA solutions were exposed to ^{60}Co γ-radiation under the aerobic conditions at 20°C on "Issledovatel" apparatus at B. P. Konstantinov Institute of Nuclear Physics (St.- Petersburg). The dose rates were I_1=2.3Gy/s and I_2=0.3 Gy/s. The DNA concentration in the irradiated solutions was C_{DNA}= $(45–50)\cdot 10^{-3}$ g/l.

2.3. UV absorbance spectroscopy and Spirin' method

The hydrolysis of DNA solutions was carried out by adding 3ml of 6% $HClO_4$ to 1ml of DNA solution, subsequent exposure in a boiling water bath during 20 min and quick cooling to 0°C. The concentration of DNA in the solution was determined from the difference in UV absorption of hydrolyzed solutions at wavelengths 270 nm and 290 nm (D_{270} and D_{290}) measured at 20°C (Spirin' method) [18]:

$$C = \frac{10.1(D_{270} - D_{290})}{0.19}\frac{V_2}{V_1}$$

where V_1 is the volume of the DNA solution and V_2 is the volume of the solution after hydrolysis. The hyperchromicity of DNA was calculated by the formula:

$$\delta = \frac{D_{260}^{hyd} - D_{260}}{D_{260}},$$

where D_{260} is the optical density of the studied DNA solution in the maximum of the DNA absorption spectrum and D_{260}^{hyd} is the optical density of the studied DNA solution of the same DNA concentration after the hydrolysis. The hyperchromicity of the native DNA is about 0.4 [17]. The values of experimental errors for C and δ were estimated from 3 experiments with unirradiated DNA solution.

UV absorbance spectra were measured on SF-56 spectrophotometer (Russia) in rectangular quartz cells of 1 cm optical path. Absorbance of corresponding solvent was subtracted from the DNA solution spectrum.

2.4. DNA melting

The melting curves of DNA samples were obtained by measuring the dependence of the optical density of DNA solution at λ=260 nm on the temperature ($D_{260}(T)$). The experiment was carried out on Specord 210 Plus (Analytik Jena, Germany) with the Peltier equipment in a 0.5° step mode, the velocity of heating was 1° per minute. Data were collected with the help of the software WinASPECT (Analytik Jena,Germany) supplied with the instrument, further data processing were carried out via OriginPro. The melting curves $D_{260}(T)$ were smoothed by the Savitzky-Golay method. Normalized melting curves f(T) for DNA solutions were obtained using the following equation [20]:

$$f(T) = \frac{D_{260}(T) - D_{260}^{min}}{D_{260}^{max} - D_{260}^{min}},$$

where D_{260}^{min} and D_{260}^{max} are the minimal and maximal values of the optical density at 260 nm, corresponding to the native and denaturated states of DNA in the solutions. The melting temperature of DNA is determined as the point of maximum of first derivative of the melting curve $df(T)/dT$. The value of experimental error for T_m was estimated from 3 experiments with unirradiated DNA solution.

3. Results and discussions

In all DNA sites where the lesions of the primary structure occur, the secondary structure of the macromolecule disturbs. One can estimate the level of DNA helicity measuring the hyperchromic effect – i.e. the relative growth of DNA UV absorbance resulted from the denaturation of the macromolecule. In this study the denaturation of DNA was carried out by heating of the solution up to 97°C and also by hydrolysis. An example of DNA absorption spectra before and after the denaturation is shown in the Figure 1.It can be seen that hydrolysis causes the more complete DNA denaturation, so for the determination of the hyperchromicity of DNA we used the value D_{260}^{hyd} corresponding to the denaturated state of DNA.

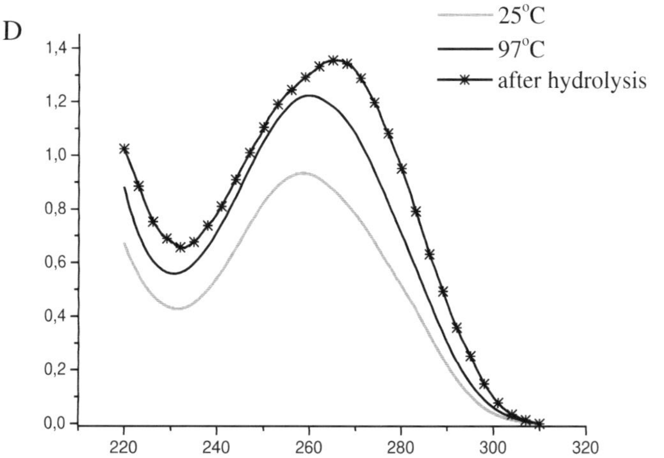

Figure 1. UVabsorbance spectra of DNA solutions measured at different conditions after γ-irradiation with the dose of 30 Gy and dose rate I_1=2.3 Gy/s;μ=0.005MNaCl.

The obtained values for the hyperchromicity of DNA in the solutions γ-irradiated at the different conditions are given in the Table 1. One can see that at the larger ionic strength the degree of DNA helicity bigger both in the control solutions of native DNA and also after the radiation exposure. The stabilizing role of monovalent cations on the native DNA secondary structure is well-estebilshed [5, 12]. It is interesting to note, that Na^+ ions also prevent the radiation-induced breakage of DNA secondary structure. One should mention that the increase of NaCl concentration in the DNA solution leads to the decrease of relative changes of the size of the coil of double-stranded DNA caused by γ-irradiation [15, 16].

An influence of radiation dose rate on the DNA helicity is not so obvious (Table 1). At the doses of 30 Gy and 50 Gy the values of DNA hyperchromicity in the solutions irradiated with the dose rates differing almost 10 times are close. And only at the dose of 100 Gy the degree of DNA helicity at I_1=2.3 Gy/s is sufficiently smaller then at I_2=0.3 Gy/s for the solutions of both ionic strengths studied. Apparently at the large concentration of lesions the speed of their accumulation becomes influential.

Investigation of radiation-induced base damages reveals the same tendencies: at the larger ionic strength DNA nucleobases are more steady, and lowering of the dose rate leads to the decrease of base damages at all doses (Table 2). The origin of Na^+ ions influence on nucleobases radioresistance is not clear for the present. We can propose that the observed effect causes by an alteration of the DNA hydration layers at the variation of counterions concentration [5, 12]. More certain conclusions require further research.

Table 1. Hyperchromicity of DNA after the γ-radiation exposure.

(Dose±2), Gy	$\delta \pm 0.02$			
	μ=0.005 M NaCl		μ=0.15 M NaCl	
	I_1=2.3 Gy/s	I_2=0.3 Gy/s	I_1=2.3 Gy/s	I_2=0.3 Gy/s
0	0.46		0.49	
30	0.40	0.42	0.46	0.45
50	0.37	0.36	0.45	0.42
100	0.24	0.34	0.37	0.41

Table 2. Relative amount of radiation-induced nucleobases degradation. C_0 is the nucleobases concentration before irradiation and C is the nucleobases concentration after the irradiation.

(Dose±2), Gy	$\frac{C_0-C}{C_0} \pm 0.005$			
	μ=0.005 M NaCl		μ=0.15 M NaCl	
	I_1=2.3 Gy/s	I_2=0.3 Gy/s	I_1=2.3 Gy/s	I_2=0.3 Gy/s
30	0.050	0.045	0.026	0.004
50	0.076	0.053	0.042	0.014
100	0.109	0.094	0.070	0.023

One of the widespread methods of study of DNA structural changes is DNA melting [19]. The helix-coil transition of DNA is very sensitive to the state of secondary structure of the macromolecule, to its interaction with biologically active agents, and also to defects of the primary structure of DNA [21-24]. The data obtained by the method of spectrophotometric melting of DNA in γ-irradiated solutions are shown in Table 3. An example of DNA melting curves and their first derivatives in the control solution (before irradiation) and after the expose to γ-radiation with the dose of 30 Gy at the different dose rates are presented in the Figure 2.One can see that at this dose the melting temperatures T_m of irradiated DNA are equal with the accuracy of the experimental error (Table 3).As it was expected T_m in the solutions of μ=0.15 M are larger than T_m in the solutions of μ=0.005 M, which indicates the stabilization of DNA secondary structure by counterions. But the radiation effect on T_m does not repeat the pattern observed in our previous results. At μ=0.15 M the relative decrease of T_m after the irradiation is slightly bigger than at μ=0.005 M. Considering the possible DNA lesions in irradiated solutions we can divide them on two groups: 1) the lesions which decrease T_m (SSB, DSB, destruction, modification and release of nucleobases) and 2) the lesions which increase T_m (inter- or intramolecular crosslinks) [19, 23]. We suppose that we observe the combined influence of these two groups of lesions on the DNA melting temperature. Further investigations can help to segregate these radiation effects and estimate them separately.

Table 3. Melting temperatures T_m of DNA in solutions exposed with the γ-radiation at the dose rate I_1=2.3 Gy/s.

μ, M NaCl	(Dose±2), Gy	(T_m±0.5), °C	(ΔT_m±1), °C
0.005	0	57.3	
	30	50.6, 50.3*	7
	50	47.0	10
0.15	0	84.3	
	30	77.5	8
	50	73.5	12

*Dose rate I_2=0.3 Gy/s

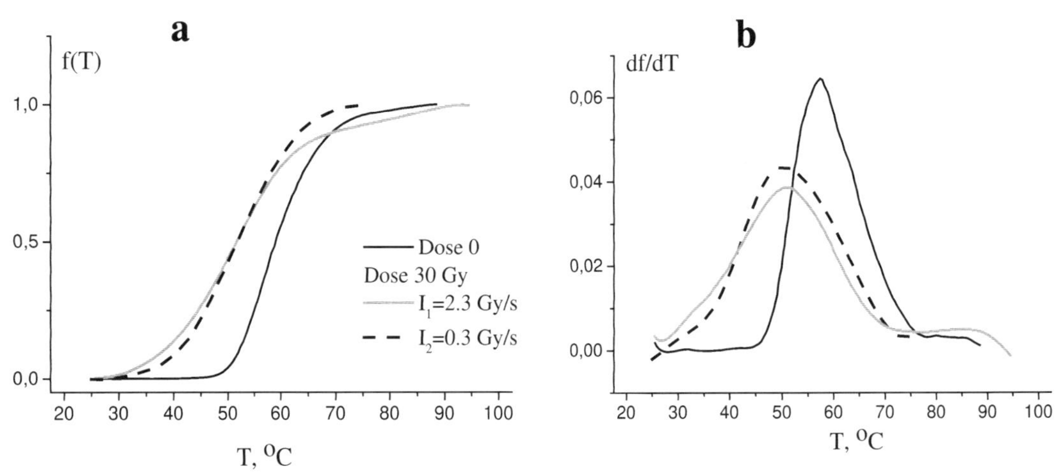

Figure 2. The melting curves (a) and their first derivatives (b) of the native DNA and DNA exposed to γ-radiation with the dose of 30 Gy at the different dose rates; μ=0.005MNaCl.

4. Conclusions

Radiation-induced base damages decrease at the rise of NaCl concentration in the solution (from 0.005M to 0.15M) and at the lowering of the dose rate (from 2.3 Gy/s to 0.3 Gy/s). The stability of DNA secondary structure increases at the rise of ionic strength of irradiated solution. The melting temperature of DNA T_m in irradiated solutions diminishes monotonously with the radiation dose growth. The relative fall of T_m after the irradiation slightly depends on the dose rate and NaCl concentration. It can be explained by the superposition of effects from different DNA lesions which decrease and increase the DNA melting temperature.

5. Acknowledgements

A part of this work was performed using the equipment of the Centre for Optical and Laser Materials Research (COLMR) and Center for Thermogravimetric and Calorimetric Research in Research park of St. Petersburg State University.

This work was supported by the RFBR, project no. 15-08-06876.

References

[1] Kudryashov Yu B 2008 *Radiation Biophysics (Ionizing Radiations)* New York: Nova Science Publishers, Inc.
[2] Coggle J E 1983 *Biological Effects of Radiation* Taylor & Francis Ltd, London
[3] Cadet J, Douki T, Gasparutto D, Ravanat J-L 2005 *Radiat. Phys. Chem.* **72** 293
[4] Han W, Yu K N 2010 *Advances in Genetics Research* 4 Ed.: Urbano K V, New York: Nova Science Publishers, Inc.
[5] Saenger W 1984 *Principles of Nucleic Acid Structure* Springer Advanced Texts in Chemistry. N.Y. Berlin HeidelbergTokyo: Springer-Verlag
[6] Milano M T, Bernhard W A 1999 *Rad. Res.***151** 39.
[7] Ito T, Baker S C, Stickley C D, Peak J G, Peak M J 1993 *Int. J. of Radiat. Biol.***63** 289.
[8] Wanek J, Ruhli F J 2016 *Radiat. Environ. Biophys.* **55** 229.
[9] Paston S V, Zamotin V V 2009 *J. of Struct. Chem.* **50** 970.
[10] Paston S V, Tarasov A E 2011 *J. of Struct. Chem.* **52** 1209.
[11] Ershov D S, Paston S V, Kartsova L A, Alekseeva A V, Ganzha O V, Kasyanenko N A. 2011 *Structural Chemistry* **22** 475.
[12] Kas'yanenko N A 2006 *J. of Struct. Chem.***47** 163.
[13] Feig M, Pettitt B M 1999 *Biophysical Journal* **77** 1769.
[14] Uyesugi D F, Trumbore C N 1983 *Int. J. Radiat. Biol* .**44** 627.
[15] Frisman E V, Zarubina O P 1993 *Biophys. Chem.* **46** 37.
[16] Paston S V, Zyryanova I M, Zaichikova Yu V, Zamotin V V 2007 *J. of Struct. Chem.* **48** 734.
[17] Cantor C R, Schimmel P R 1980 *Biophysical Chemistry. Part 2* San Francisco: W. H. Freeman and Company.
[18] SpirinAS 1958 *Biokchimiya (USSR)***23** 656.
[19] Lazurkin Yu S,Frank-KamenetskiiMD, TrifonovEN1970*Biopolymers***9** 1253.
[20] Chikhirzhina E V, Starkova J T, Polyanichko A M 2014 *Journal of Spectroscopy* Article ID 387138, http://dx.doi.org/10.1155/2014/387138
[21] Eichhorn G L, Shin Y A 1968 *J. Am. Chem. Soc.* **90** 7323.
[22] Chikhirzhina E V, Starkova T Yu, Kostyleva E I, Chikhirzhina G I, Vorobiev V I, Polyanichko A M 2011 *Cell and Tissue Biology* **5** 536.
[23] Lando D Y, Fridman A S, Kabak A G, Akhrem A A 1997 *Journal of Biomolecular Structure and Dynamics* **15** 141.
[24] Duguid J G, Bloomfield V A 1995 *Biophysical Journal* **69** 2642.

One-channel EEG monitor for tracking the depth of narcosis

V A Simon, V A Gerasimov, D K Kostrin, L M Selivanov and A A Uhov

Department of electronic instruments and devices, Saint Petersburg Electrotechnical University "LETI", 5 Professor Popov Street, Saint Petersburg, 197376, Russia

vsev.simon@gmail.com

Abstract. Complications in designing of one-lead electroencephalographic (EEG) monitor for tracking the deepness of narcosis are discoursed. Connection between the level of narcosis and statistical attributes of the bioelectric signal is disclosed. Electric circuit of the developed EEG monitor is discussed. Obtained recordings of the EEG and ECG signals are shown.

1. Introduction

It is important for success of the invasive medical intervention to make the human body impervious to any form of external influence such as surgical incision or suturing. Narcosis has analgesic effect which prevents execution of spontaneous protecting mechanisms of the body and occurrence of the pain shock. For this reason local, regional and general narcosis became ubiquitous in clinical practice.

Local narcosis [1] is applied in particular for the analgesia of small organs or body tissue areas: in dentistry, odontology, cosmetic surgery, and also in military medicine (camp conditions). Regional narcosis is broadly similar to the local technique and used for the anesthesia of large organs: arms, legs, spinal column; also during the cesarean section.

In case of complex life-critical surgical intervention the general narcosis is implemented. General narcosis leads to myorelaxation, deceleration of the neural activity, absence of pain and loss of memory (amnesia) during operation. During the surgical operation with general narcosis it is necessary to control the arterial pressure, oxygen saturation of blood, body temperature, chemical composition of the inhaled and exhaled gases (capnogram monitoring in particular), and also to record an electrocardiogram (ECG) and electroencephalogram (EEG) [2].

2. Problem statement

General narcosis requires the monitoring of the depth of unconscious state, in which is immersed the nervous system to promptly define anesthesia awareness during operation and also to prevent critically deep coma [3] [4]. However the deepness of narcosis can be determined only by means of the specialized clinical accessories.

This task can be resolved with the usage of the electroencephalographic signal (EES) that reflects the state of the brain neurons. When the depth of narcosis is increased, the band of the EES moves lower through the axis of frequencies due to intensification of the delta rhythm. The delta waves features magnitude in the order of hundreds of microvolts and typical band from 1 to 4 Hz. Computer aided identification of the narcosis stages requires estimation of the entropy of the EES in frequency and temporal domains and normalization of the calculated values to a unified measure [5–7].

One of the measures that became de-facto standard in clinical monitoring is the Bispectral (BIS) Index, established by Aspect Medical (now part of Covidien) [8]. The algorithm which used to compute BIS index discovers the correspondence between various harmonics in the EES [9]. Specifically, BIS index reflects the level of synchronization between the sine waves in the EES, along with the conventional Fourier spectra of the signal.

Another commonly used measure is the Entropy index – a statistical attribute of the signal evaluating the state of chaos in the EES [10]. In accordance with the Entropy's approach, the EES is not a simple combination of the harmonic waves but a nonlinear, stochastic system. The Entropy index exploits several statistical methods: Shannon [11], spectral [12] and approximate [13] entropy. A mass-produced Entropy monitoring module was developed by Datex-Ohmeda (GE Healthcare) [14].

In [15] was justified that both the BIS index and the Entropy are adequate methods to define current depth of sedation and to discern the different levels of narcosis.

3. Proposed solution

A professional multilead EEG monitor is not an indispensable instrument for recording the EES with the goal of acquiring data for tracking the level of narcosis. It is quite satisfactory to employ the EEG monitor with just one bipolar lead recording the electric potential difference of two electrodes installed on the forehead. The third electrode is applied as a reference point and connected to the earlobe.

Electric circuit of the designed one-lead EEG monitor (figure 1) includes several stages: high common-mode rejection instrumentation amplifier INA333 with several pull-up resistors ($R_{\text{pull-up}}$) and the gain resistor R_G, common-mode voltage buffer [16–18], 0.2 Hz low-cut filter, mains hum rejection filter [19], 150 Hz roof filter, and a successive approximation voltage-to-code transducer the serial data output [20]. The power supply for this circuit includes two accumulators (each has typical output voltage of 3.7 V). First accumulator is connected to the EEG monitor and discharged, thus provides long-play energy source for stable functioning of the device. Meanwhile, the second accumulator gathers the electric charge from the mains power line through the charging unit composed from a step-down transformer and diode rectifier (charging voltage of 4.2 V). When the output voltage of the first accumulator is reduced to 3.3 V, the EEG monitor is switched to the second full charged accumulator; herewith the first accumulator is switched to the charging unit. This charging technique provides reliable noise-free source of electric power and ensures the galvanic isolation of the EEG monitor from the industrial power line [21].

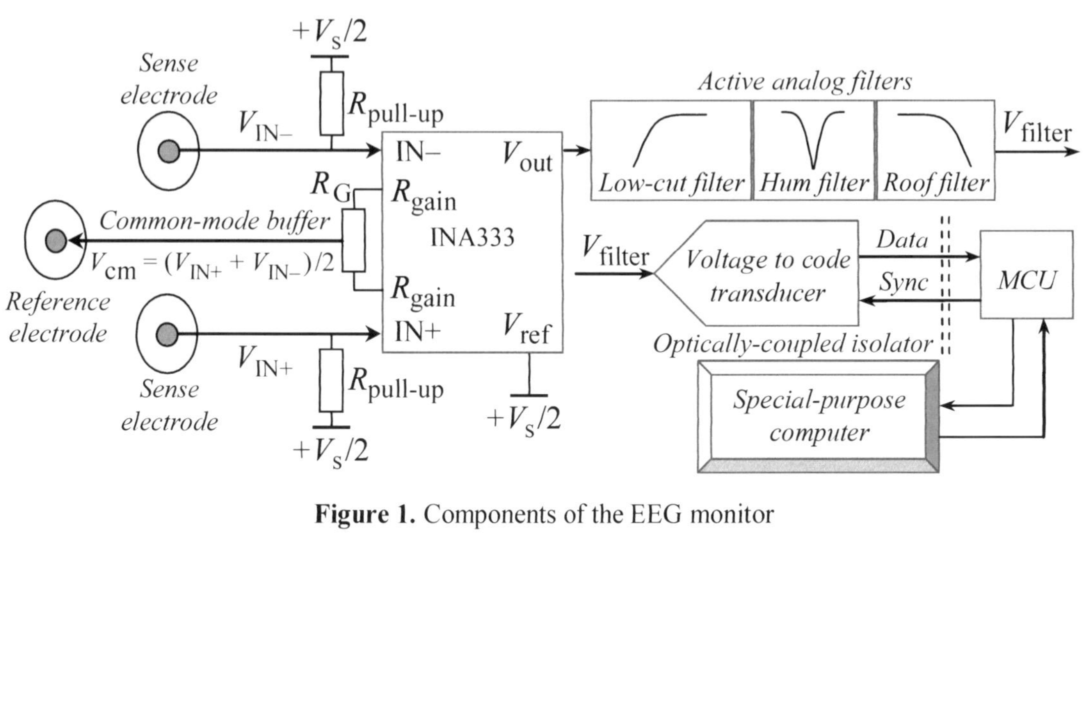

Figure 1. Components of the EEG monitor

The EEG monitor is connected via optically-coupled isolator to a microcontroller (MCU) with the core based on the ARM Cortex M3 architecture. The EES transformed by the voltage-to-code transducer from the analog field to digital domain is received by the MCU and reconfigured into data stream suitable for the RS-232 standard. Then MCU transmits EEG data to a special-purpose computer with the Qseven Conga-QMX6 (ARM Cortex A9 core) controller board running Linux-like operating system. The program running within Linux environment computes the entropy index of the EES and displays how this index is changed over time [6] [7]. Computed entropy index as a function of time, the EES recording itself and the Fourier spectra of the signal are contributed to making the right estimation of the required concentration of anesthetic agent during operation.

4. Discussion of results

With the aim to verify the functionality of the designed EEG monitor the acquisition of the EES and also the ECG signal (ECS) was conducted; the recordings of the received signals are shown below. Sample frequency during the experiment was 2 kHz: it is more than enough for applying of the moving average filter (32 points for the ECS and 16 points for the EES). The all waves of the PQRST-complex of the ECG [22] are clearly visible in figure 2. The EEG registered on the test subject with open eyes (figure 3) contains waves with higher frequencies than the EEG recorded in case of closed eyes (figure 4). Low disturbance superposing the acquired EES includes the mains hum 100 Hz and 150 Hz waves accompanied by small amount of thermal and fluctuation noise. The software adaptive algorithm could be exploited to reduce the level of noise over the signal.

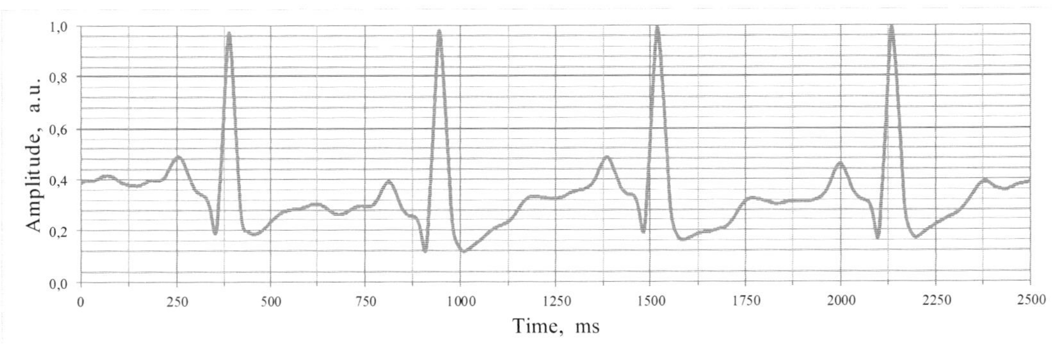

Figure 2. Fragment of the ECS record

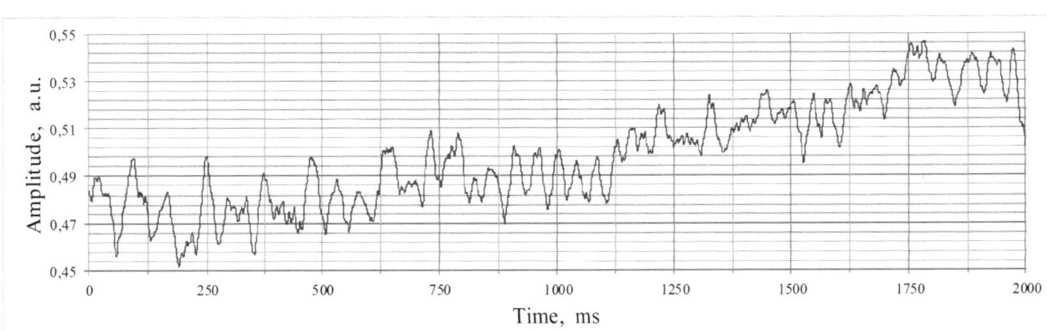

Figure 3. Fragment of the EES record (eyes open)

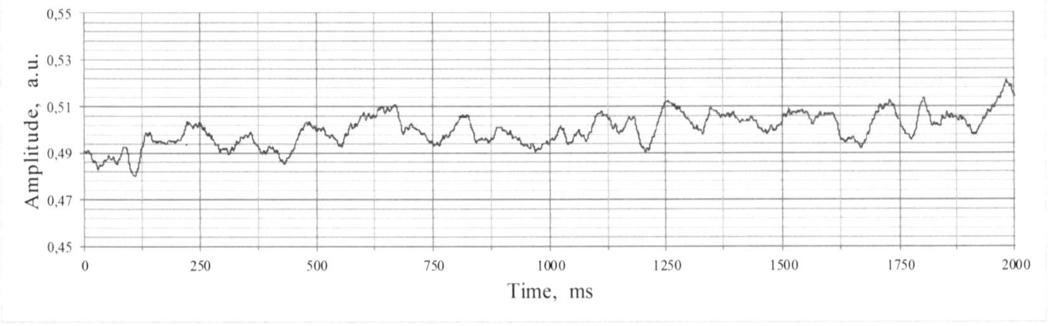

Figure 4. Fragment of the EES record (eyes closed)

5. Conclusion

Designed EEG monitoring module allows continuous and reliable recording of the bioelectric activity of the human body such as EES and ECS. It is assumed to integrate the developed EEG monitor into the narcosis monitoring module «Sapus» which is produced by the research and manufacturing facility «Red Guardsman» located in St. Petersburg, Russian Federation.

6. Acknowledgements

Authors acknowledge financial support by the comprehensive project for the high-technology manufacturing with the partnership of the Russian institutions for higher education on the theme «Development and Design of the Manufacturing Facility of the Electronic Atomizer for Anesthetic Agents and the Narcosis Monitoring System» in accordance with the 218th Resolution of the Russian Federation, contract №02.G25.31.0192 from 27th April 2016. Authors also acknowledge the good collaboration with the engineers and managers of the science-production company «Red Guardsman».

References

[1] Andreae M H and Andreae D A 2012 *Cochrane Database of Systematic Reviews* **10** CD007105
[2] *Recommendations for standards of monitoring during anaesthesia and recovery 2015* URL: https://www.aagbi.org/sites/default/files/Standards of monitoring 20150812.pdf
[3] Musizza B and Ribaric S 2010 *Sensors* **10** 10896–935
[4] Fiehn A 2002 *Annals of Cardiac Anaesthesia* **5** 149–55
[5] Nemirko A P, Manilo L A, Kalinichenko A N, and Volkova S S 2010 *Biotechnosphere* **3** 3–10
[6] Manilo L A and Volkova S S 2013 *Pattern Recognition And Image Analysis (Advances In Mathematical Theory And Applications)* **1** 92–7
[7] Kalinichenko A N, Manilo L A, and Nemirko A P 2015 *Pattern Recognition And Image Analysis (Advances In Mathematical Theory And Applications)* **25** 632–41
[8] Technology overwiev: Bispectral Index (White Paper) 1997 Aspect Medical Systems, Inc.
[9] Sigl J C and Chamoun N G 1994 *J. of Clinical Monitoring* **10** 392–404
[10] Viertio-Oja H, Sarkela M, Talja P, Tolvansen-Laakso H, and Yli-Hankala A 2000 *Anesthesiology* **93** A1369
[11] Bruhn J, Lehmann L E, Roepcke H, Bouillon T W, and Hoeft A 2001 *Anesthesiology* **95** 30–5
[12] Ellerkmann R K, Liermann V M, Alves T M, Wenningmann I, Kreuer S, Wilhelm W, Röpcke H, Hoeft A, and Bruhn J 2004 *Anesthesiology* **101** 1275–82
[13] Bruhn J, Röpcke H, and Hoeft A 2000 *Anesthesiology* **92** 715–26
[14] Viertiö-Oja H, Maja V, Särkelä M, Talja P, Tenkanen N, Tolvanen-Laakso H, Paloheimo M, Vakkuri A, Yli-Hankala A, and Meriläinen P 2004 *Acta Anaesthesiol. Scand.* **48** 154–61
[15] Schmidt G N, Bischoff P, Standl T, Hellstern A, Teuber O, and Schulte Esch J 2004 *Anesthesiology* **101** 1283–90

IOP Conf. Series: Journal of Physics: Conf. Series **1038** (2018) 012028 doi:10.1088/1742-6596/1038/1/012028

[16] Gerasimov V A, Afanasjev P V, Bokhov O S, Kostrin D K, Luchinin V V, Selivanov L M, and Uhov A A 2015 *Biotechnosphere* **3** 2–6

[17] Uhov A A, Gerasimov V A, Selivanov L M, Kostrin D K, and Simon V A 2016 *Proc. IEEE NW Russia Young Researchers in Electrical and Electronic Engineering Conf. (Saint Petersburg)* pp 797–9

[18] Simon V A 2016 *Transactions of XII Russian-German Conf. on Biomedical Engineering (Suzdal)* pp 144–7

[19] Simon V A, Gerasimov V A, Kostrin D K, Selivanov L M, and Uhov A A 2017 *Proc. of Saint Petersburg Electrotechnical University (Izvestia SPbETU "LETI")* **5** 3–9

[20] Simon V A, Gerasimov V A, Kostrin D K, Selivanov L M, and Uhov A A 2017 *J. Phys.: Conf. Ser.* **881** 012021

[21] Uhov A A, Gerasimov V A, Selivanov L M, Dudnikov S Yu, and Simon V A 2017 *Proc. IEEE Conf. of Russian Young Researchers in Electrical and Electronic Engineering (Saint Petersburg)* pp 1619–21

[22] Gacek A and Pedrycz W 2012 *ECG Signal Processing, Classification and Interpretation: A Comprehensive Framework of Computational Intelligence* (London: Springer-Verlag)

IOP Conf. Series: Journal of Physics: Conf. Series **1038** (2018) 012029 doi:10.1088/1742-6596/1038/1/012029

Super-resolution microscopy of *Mollicutes* cells

V S Polinovskaya[1,2, a], A D Vedyaykin[1,2], I E Vishnyakov[2], V.A. Ivanov[2] and M A Khodorkovskii[1]

[1] Peter the Great St.Petersburg Polytechnic University, Saint-Petersburg, 195251, Russia

[2] Institute of Cytology, Russian Academy of Sciences, Tikhoretsky av. 4, Saint Petersburg 194064, Russia

E-mail: vasilisapolinovskaya@gmail.com

Abstract. Super-resolution microscopy is especially useful for visualization of structures in small cells which sizes are comparable to the diffraction limit, for instance, in *Mollicutes* species. In this work we obtained super-resolution images of FtsZ and IbpA protein structures in fixed *Acholeplasma laidlawii* & *Mycoplasma gallisepticum* cells using immunofluorescence staining.

1. Introduction

Mycoplasmas (*Mollicutes* class) are the smallest known organisms that can live on artificial media. Mycoplasmas are able to cause acute and chronic diseases of human, animals and plants [1]. All species of *Mollicutes* lack peptidoglycan layer and show various cell morphology [2]. They also demonstrate reduced genomes which lost a considerable portion of their ancestors' non-essential genes, thus mycoplasmas are considered to be close to the concept of the "minimal cell" [3]. FtsZ protein is conserved in most mycoplasmas species. FtsZ is a well-known prokaryotic tubulin homologue being a key protein involved in division process in a majority of bacteria. In most bacteria it assembles into a ring structure, the so-called Z-ring, at the site of cytokinesis and recruits downstream proteins to form a large protein machine — the divisome, which major role is considered to be the synthesis of cell-wall [4]. The role of FtsZ in mycoplasmas is not yet clear, in particular due to the lack of cellwall. Also it is believed that Mollicutes have several other than FtsZ cytoskeleton-like proteins, that may help to maintain the shape of the cell. The study of structures formed by cytoskeleton-like proteins in mycoplasmas is complicated by small size of their cells. That's why super-resolution microscopy techniques seem to be promising solution to study their internal organization. Single-molecule localization microscopy (SMLM) was previously successfully used to obtain images of FtsZ structures in *Escherichia coli* cells using immunofluorescence staining [4-6]. In current work SMLM was utilized to visualize FtsZ protein of two species – *Acholeplasma laidlawii* & *Mycoplasma gallisepticum* and small heat-shock protein IbpA of *A. laidlawii* with a resolution well below the diffraction limit.

[a] To whom any correspondence should be addressed.

2. Materials and methods

2.1. Bacterial strains and sample preparation

The following strains of microorganisms were used in the work: *Acholeplasma laidlawii* PG8 (Collection of cell cultures of the Institute of Cytology RAS), *Mycoplasma gallisepticum* S6 (kindly provided by Gleb Fisunov, Federal Research and Clinical Centre of Physical-Chemical Medicine Federal Medical Biological Agency, Moscow, Russia). Cells were grown at 37°C in custom-made medium, based on 25 mM Tris-HCl, pH=7.4, and supplemented with ampicillin (100 µg/ml), peptone (2% w/v), NaCl (0.5 % w/v), KCl (0.13 % w/v), horse serum (10% v/v), yeast dialysate (5% v/v), glucose (1%) and phenol red indicator. After change in medium color from red to yellow, cells were harvested by centrifugation, washed in PBS and fixed by addition of formaldehyde (2%) and glutaraldehyde (0.1%) for 10 minutes at room temperature followed by 50 minutes on ice. After that the cells were washed in PBS three times and immobilized on poly-L-lysine coated coverslips. Then cells were permeabilized using 0.1% Triton X-100 in PBS for 5 minutes.

Next cells were blocked using 2% w/v BSA in PBS (PBS-BSA) and after that with addition of non-rabbit antibodies in order to avoid nonspecific binding of target antibodies, both steps for 0.5 hour. Then samples were incubated overnight with custom-made rabbit anti-FtsZ or anti-IbpA polyclonal antibodies at the final dilution from 1:20 to 1:100 in PBS-BSA. Cells were washed 5 times with 0.01% Tween-20 in PBS, incubated with goat anti-rabbit secondary antibodies (Alexa 647 conjugated F(ab')2 fragments of goat anti-rabbit antibodies (LifeTechnologies)), diluted 1:100 in PBS-BSA for 1 hour at room temperature and washed another 5 times as described above. To prevent sample drift during data acquisition, custom-made active stabilization system was used [7]. To track sample position, polystyrene microspheres (2.1 µm, Spherotech) were immobilized on the coverslip surface.

The antibodies used in the experiments were preliminarily obtained by immunizing rabbits with appropriate recombinant FtsZ and IbpA proteins to obtain an immune response. After collection, the blood serum was affinity purified using FtsZ proteins cross-linked with the CNBr-Activated Sepharose 4B (GE Healthcare Life Sciences) resin according to the producer protocol.

2.2. Experimental setup and data processing

All images were obtained using custom set-up based on AxioImager.Z1 (Carl Zeiss) motorized microscope which was described previously [8]. Images were acquired by EM-CCD camera (Andor iXon 897) at 100× total magnification. One CCD pixel corresponded to 108 nm in the focal plane. LF635/LP-B-000 (Semrock) filter set was used to visualize Alexa 647 fluorescence (excitation filter was removed to enable dye activation using a 405 nm laser.

Raw images were acquired using MicroManager [9] in PBS-Tris buffer with pH 7.5, containing 10% w/v glucose, 10 mM 2-mercaptoethylamine combined with 50 mM 2-mercaptoethanol, 2 mM cyclooctatetraene and an oxygen scavenging system (2.5 mM protocatechuic acid and 50 nM protocatechuate 3,4-dioxygenase) [10]. Alexa 647 fluorescence was excited using 635 nm diode laser with a power density of approximately 1 kW/cm2 at the focal plane and 2000-10000 frames containing individual Alexa 647 molecule images were acquired.

Data processing was performed using ImageJ, SMLM reconstruction was carried out using ThunderSTORM plugin for ImageJ [11].

3. Results and discussion

Using SMLM in combination with immunofluorescence staining we were able to obtain fine super-resolution images of IbpA protein in *A. laidlawii*. After 1.5 hour incubation upon conditions of cold shock IbpA protein in *A.laidlawii* cells seems to form some well-ordered structures such as filaments, arcs or rings (see Figure 1). This behavior may be due to either binding of IbpA to some structures in cell or to tendency of IbpA to multimerize. Further work is needed to explain this observation.

The same approach was used to visualize FtsZ protein in two mycoplasmas – *A.laidlawii* & *M.gallisepticum* (see Figure 2). It is clear that SMLM technique allows to visualize structures in more details that are unreachable in the case of conventional fluorescence microscopy. Preliminary data does not yet allow us to draw specific conclusions based on the structures formed by FtsZ and IbpA proteins. We can only conclude that in some cells these proteins form structures resembling cytoskeleton elements. To interpret the patterns of protein distributions in cells obtained by the LM method, it is necessary to accumulate more experimental data, as well as data verification using other methods (for example, electron microscopy).

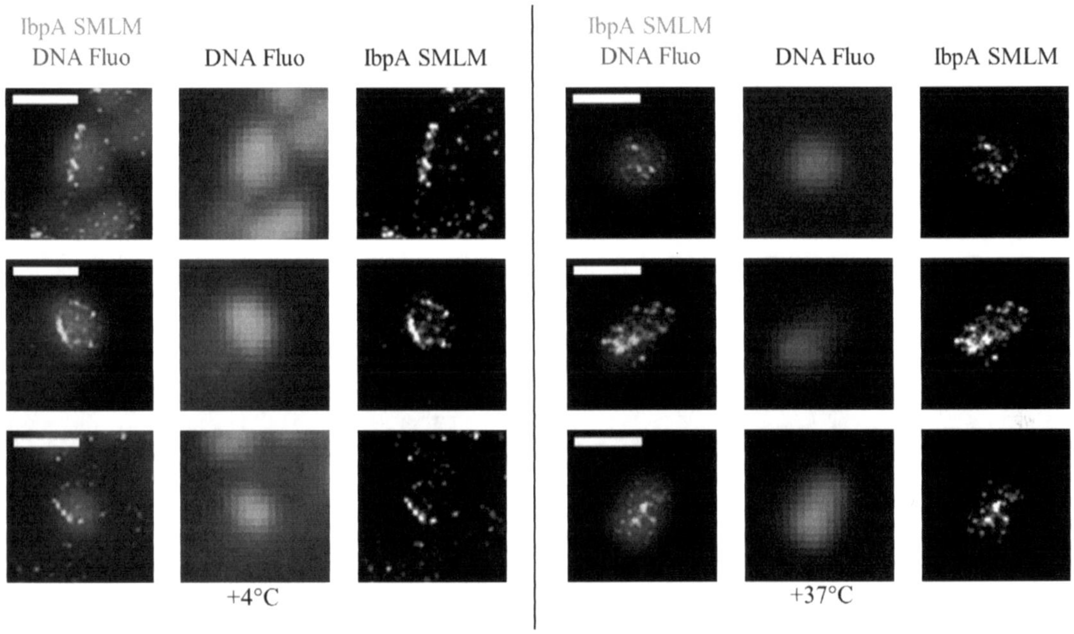

Figure 1. SMLM visualization of IbpA protein in *Acholeplasma laidlawii* cells. Left panel – cells before fixation were incubated for 1.5 hours at +4°C, right – at +37°C. DNA Fluo – conventional fluorescence image of DNA stained with YOYO-1 dye; IbpA SMLM – super-resolution image of IbpA protein. Scale bar corresponds to 1 μm.

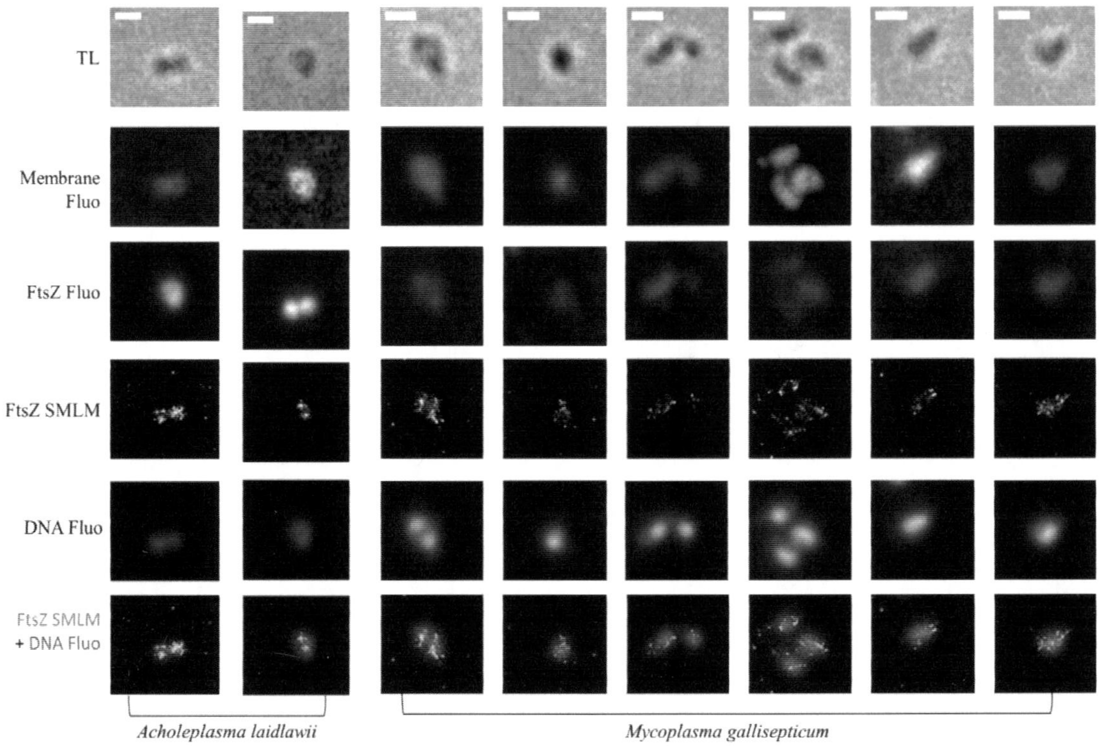

Figure 2. SMLM visualization of FtsZ protein in *Acholeplasma laidlawii* (left) and *Mycoplasma gallisepticum* (right) cells. TL – transmitted light image, Membrane Fluo – conventional fluorescence image of membrane, FtsZ Fluo – conventional fluorescence image of FtsZ, DNA Fluo – conventional fluorescence image of DNA stained with DAPI dye; FtsZ SMLM – super-resolution image of FtsZ protein. Scale bar corresponds to 1 μm.

4. Conclusion

Using the method of single-molecule localization microscopy, we visualized FtsZ and IbpA proteins in two mycoplasma species. The data obtained show that the method of localization microscopy makes it possible to visualize intracellular structures in mycoplasmas that could not be resolved using conventional fluorescence microscopy. Further study of these proteins using SMLM and other techniques may help to better understand their function in corresponding species.

Acknowledgements

The work was supported by Russian Science Foundation, research project No. 17-74-20065 (visualization of IbpA protein in *Acholeplasma laidlawii*) and Russian Foundation for Basic Research, research project No. 15-04-07472 (visualization of FtsZ protein in *Acholeplasma laidlawii* & *Mycoplasma gallisepticum*). The work was carried out using scientific equipment of the Center of Shared Usage «The analytical center of nano- and biotechnologies of SPbPU»

References

[1]. G.F. Browning, C. Citti, *Mollicutes: Molecular Biology and Pathogenesis*. 2014: Caister Academic Press.

[2]. M. Miyata, H. Ogaki. Cytoskeleton of mollicutes. // J Mol Microbiol Biotechnol, 2006. 11(3-5): p. 256-64.

[3]. C.A. Hutchison, 3rd, R.Y. Chuang, V.N. Noskov, N. Assad-Garcia, T.J. Deerinck, M.H. Ellisman, J. Gill, K. Kannan, B.J. Karas, L. Ma, J.F. Pelletier, Z.Q. Qi, R.A. Richter, E.A. Strychalski, L. Sun, Y. Suzuki, B. Tsvetanova, K.S. Wise, H.O. Smith, J.I. Glass, C. Merryman, D.G. Gibson, J.C. Venter. Design and synthesis of a minimal bacterial genome. // Science, 2016. 351(6280): p. aad6253.

[4]. A.D. Vedyaykin, A.V. Sabantsev, M.A. Khodorkovskii, A.R. Kayumov, I.E. Vishnyakov. Recombinant FtsZ Proteins from Mollicutes Interact with Escherichia coli Division Machinery. // BioNanoScience, 2016. 6(4): p. 443-446.

[5]. A.D. Vedyaykin, A.V. Sabantsev, I.E. Vishnyakov, N.E. Morozova, V.S. Polinovskaya, M.A. Khodorkovskii. 3D super-resolution microscopy of bacterial division machinery. // 3rd International School and Conference on Optoelectronics, Photonics, Engineering and Nanostructures (Saint Petersburg Open 2016), 2016. 741.

[6]. A.D. Vedyaykin, I.E. Vishnyakov, V.S. Polinovskaya, M.A. Khodorkovskii, A.V. Sabantsev. New insights into FtsZ rearrangements during the cell division of Escherichia coli from single-molecule localization microscopy of fixed cells. // Microbiologyopen, 2016. 5(3): p. 378-386.

[7]. A.D. Vedyaykin, I.E. Vishnyakov, V.S. Polinovskaya, I.T. Artamonova, M.A. Khodorkovskii, A.V. Sabantsev. [NEW INFORMATION ABOUT THE STRUCTURES FORMED BY FtsZ PROTEIN IN ESCHERICHIA COLI CELLS DURING DIVISION PROCESS OBTAINED BY SINGLE-MOLECULE LOCALIZATION MICROSCOPY]. // Tsitologiia, 2015. 57(11): p. 823-30.

[8]. A.D. Vedyaykin, A.V. Sabantsev, I.E. Vishnyakov, S.N. Borchsenius, Y.V. Fedorova, A.S. Melnikov, P.Y. Serdobintsev, M.A. Khodorkovskii. Localization microscopy study of FtsZ structures in E. coli cells during SOS-response. // Journal of Physics: Conference Series, 2014. 541(1).

[9]. A. Edelstein, N. Amodaj, K. Hoover, R. Vale, N. Stuurman. Computer control of microscopes using microManager. // Curr Protoc Mol Biol, 2010. Chapter 14: p. Unit14 20.

[10]. N. Olivier, D. Keller, P. Gonczy, S. Manley. Resolution doubling in 3D-STORM imaging through improved buffers. // PLoS One, 2013. 8(7): p. e69004.

[11]. M. Ovesny, P. Krizek, J. Borkovec, Z. Svindrych, G.M. Hagen. ThunderSTORM: a comprehensive ImageJ plug-in for PALM and STORM data analysis and super-resolution imaging. // Bioinformatics, 2014. 30(16): p. 2389-90.

Photoluminescence spectroscopy features in the study of green leaves drying process

V.B. Fadeenko [1], V.Yu. Rud' [1,2], Yu.V.Rud' [3], A.P. Glinushkin [2], V.Ch. Shpunt [3], William Hogland [4]
[1]Peter the Great Saint Petersburg Polytechnic University, Saint Petersburg 195251, Russia
[2]Department of Ecology, All-Russian Research Institute of Phytopathology, 143050, Moscow Region, Odintsovo district, B.Vyazyomy, Russia
[3]Department of Solid State Physics, Ioffe Physico-Technical Institute, Russian Academy of Sciences, St Petersburg 194021, Russia
[4] Faculty of Health and Life Sciences, Linnaeus University SE-391 82 Kalmar, Sweden

Abstract. The presented work demonstrates new results of studying the photoluminescence kinetics of green leaves of Brassica rapa L. that were separated from the parent plant and in fact is the logical development of our studies. We found that the time dependence of its intensity includes 2 stages characterized by the fact that in the first one there is an increase in intensity, reaching a maximum and then decrease, but with long drying times in conditions of constant room temperature, it does not fall below its characteristic value for a living plant.

1. Introduction

The importance of studying organization and the functioning of living systems is very relevant for solving practical and theoretical problems in the development of biological and agricultural sciences in connection with the threats of overpopulation of the planet Earth. The results of such studies make it possible to increase the efficiency of plant breeding methods leading to the development of sustainable agriculture. In this connection, there is a growing need to use scientific and experimental methods that are well-proven in other scientific issues in research. Thanks to the application of such methods, it was found that when green leaves are excited with optical radiation with a quantum energy of 1.96 eV, the intense red photoluminescence arises consisting of two overlapping bands [1]. This led to the expansion of the application of methods for studying various characteristics of optical properties of plants in which green plant objects would be studied as ordinary solid-state semiconductor samples. In this case, the analysis of the results is carried out proceeding from the representations established for solid objects and to obtain new information about plant objects. That is why the use of well-developed in application to solid-state physics methods for the investigation of photoluminescence [1] for the study of living green plants seems to us entirely innovative and perspectival popular [2, 3]. The presented work demonstrates the results of studying the photoluminescence kinetics of green leaves of Brassica rapa L. that were separated from the parent plant and in fact is the logical development of our studies [1-3]. This allows us to establish new aspects of the decay of biological processes in green leaves at room temperature.

2. The method

The study of photoluminescence was carried out on green leaves of various types of plants. As a source for photoluminescence excitation used ILA 120-1 Carl Zeiss argon laser with excitation energies hw_{exc} = 2.1; 2.50; 2.54; 2.6; 2.71 eV. For each of the above energies, the power density was 50-100 mW/cm^2. The photoluminescence radiation then fed to the MDR-3 monochromator with a grating 600 lines/mm and detected by a photomultiplier.

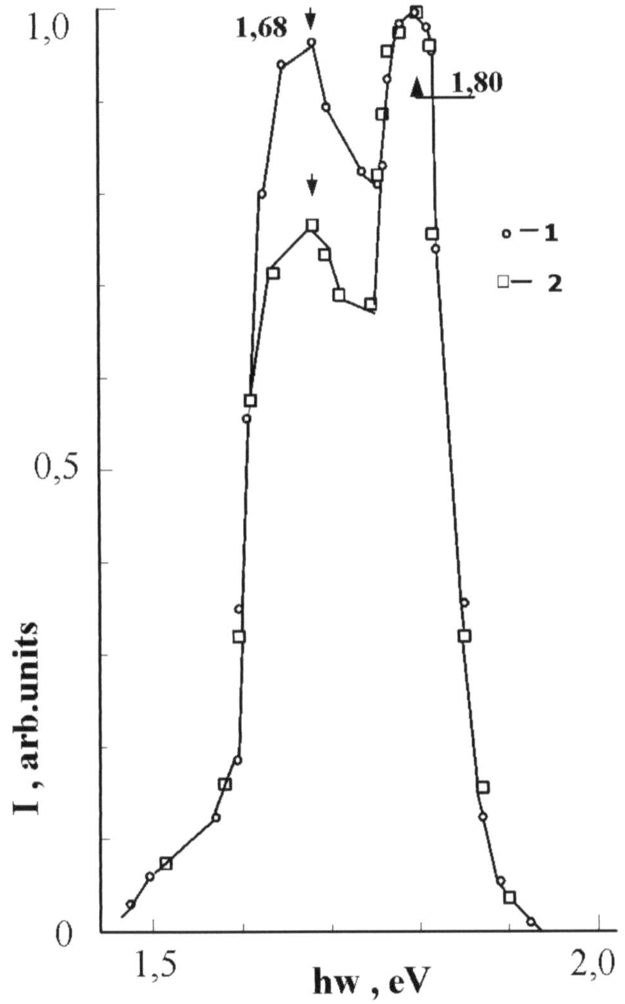

Figure. 1. The spectral dependences of photoluminescence of leave Brassica rapa L. at T=300K ($hw_{exc.}$, eV : for curves 1- 2.705, for curve 2- 2.410). Spectral resolution setup is 1 meV.

The photoluminescence red visually observed at all of these energies, photoluminescence excitation spectra for green leaves in all cases represent two closely spaced bands (Fig. 1). In this regard, it is important to note that from fluorescence studies it is known that regardless of the wavelength of the incident light, the chlorophyll fluoresces in the red part of the spectrum. This indicates that red luminescence can be associated with chlorophyll associated with the life cycle of the plant world. Thus both energy bands for different types of plants did not differ from each other. It is seen that the values of half-widths of the component and the ratio of their intensities I_2/I_1 is close in leaves of different plants. When you change the flux density of the exciting radiation the intensity of the photoluminescence field detection of peaks is described by a linear dependence of I~L . It is important to note that the important thing for photoluminescence analysis in physics, the ratio of the photoluminescence peak of the solid and half-width at half-maximum values were also close to all types of plants [1].

This photoluminescence consists of two overlapping bands. This work logically continues previous studies on the properties of the plant world with the help of widely recommended methods of semiconductor spectroscopy. In this work, we present the results of studying regularities in the transition of plant from a living to a dead condition. To do this, the photoluminescence spectrum of the living leaf in a single system with the plant is first studied, and then the spectral dependence of the same leaf, but already separated from the living plant, is studied. This allows us to study the processes of decay of living biological processes in a leaf separated from a living plant.

3. Results and discussions

These studies were conducted on a wide range of plants. As an example, the data of spectral dependences of the intensity of stationary photoluminescence of Brassica rapa L. leaves are presented and analysed. It should be noted at once that all the studied patterns of photoluminescence are practically identical for different plants. Photoluminescence in experiments was excited by the emission of helium neon or argon lasers [2]. We studied the spectral dependences of photoluminescence with a spectral resolution of not lower than 1 meV at room temperature, immediately after separation of the green leaf from the plant, depending on the time of their location in the separated state.

We find when the Brassica rapa L leaf is still in the system of a single whole with the plant is accepted for the origin of the sample $t = 0$. The further period of time up to 30 days already refers to the condition of the leaf separated from the green plant. The main regularity of these studies revealed that the spectral contour of red photoluminescence is practically insensitive to the decay process of leaves. Indeed, it was found that the spectral position of both maxima in the spectral dependence of the photoluminescence $hw_1 = 1.6$ eV and $hw_2 = 1.80$ eV and their half width at half height ($\delta_1 = 50$-55 meV and $\delta_2 = 20$-25 meV), respectively, remain constant all the time (up to 30 days) of isothermal leaves decay at room temperature. Hence, we can conclude that for green leaves in the green plant and for the decayed leaves, separated from the plant, the spectral contour of red photoluminescence remains constant. It is also important to note the fact that those radiative processes that lead to the appearance of red photoluminescence in the green leaf remain for the decay condition after separation from the living plant.

We calculated that the ratio of photoluminescence intensities I_1 / I_2 in the region of maxima I_1 and I_2 with the time of leaves decay remains practically permanent. It is also important to note that the short-wavelength maximum hw_2 with increasing decay time becomes less marked in the spectrum with respect to the dominant long-wavelength maximum hw_1. The time dependences of the intensities of red photoluminescence at a fixed leaf temperature have a similar appearance, which can be explained by the preservation in time of their correlation.

Based on the isothermal time dependences of the intensities, we can draw an important conclusion that the transition to the decay condition in terms of photoluminescence patterns is a complex process in which at least 2 different stages can be distinguished. The first stage lasts about 23 hours and is accompanied by the ignition of photoluminescence. We also got the result that this stage is characterized by the fact that the intensity of stationary photoluminescence increases almost twofold. In the process of drying the leaf separated from the green plant at a constant temperature within 1.5-12 hours, it was found that the complex kinetics [3, 4, 5] of establishing stationary values of I_1 and I_2 after the inclusion of exciting radiation is manifested. This process is characterized by the fact that after the excitation radiation was turned on, the photoluminescence intensity I_1 and I_2 decreased in such a way that the stationary value of I_∞ was established in a time of 30-400 s. It is important to note that in this case, the time for achieving stationary intensities with increasing leaf decay time greater than 1.5 hours initially increased, then passed through a maximum in the vicinity of 6-8 hours exposure. After that, there was a sharp drop, so sharp that in the region of 12-15 hours the stationary value of the intensity of the photoluminescence was established almost instantaneously, which indicates a practically complete absence of any kinetics. We found that the process of intensity decrease takes place in accordance with the exponential law [6, 7]. It is important to note that at times of drying of 6 hours, the relaxation time is 30 s, and at a drying time of 8 hours, the relaxation time is already 100 s. At a drying time of 9 h, the relaxation time decreases again, and so that for times of the order of 12 h the kinetics after the inclusion of the photoluminescence is practically not detected at all.

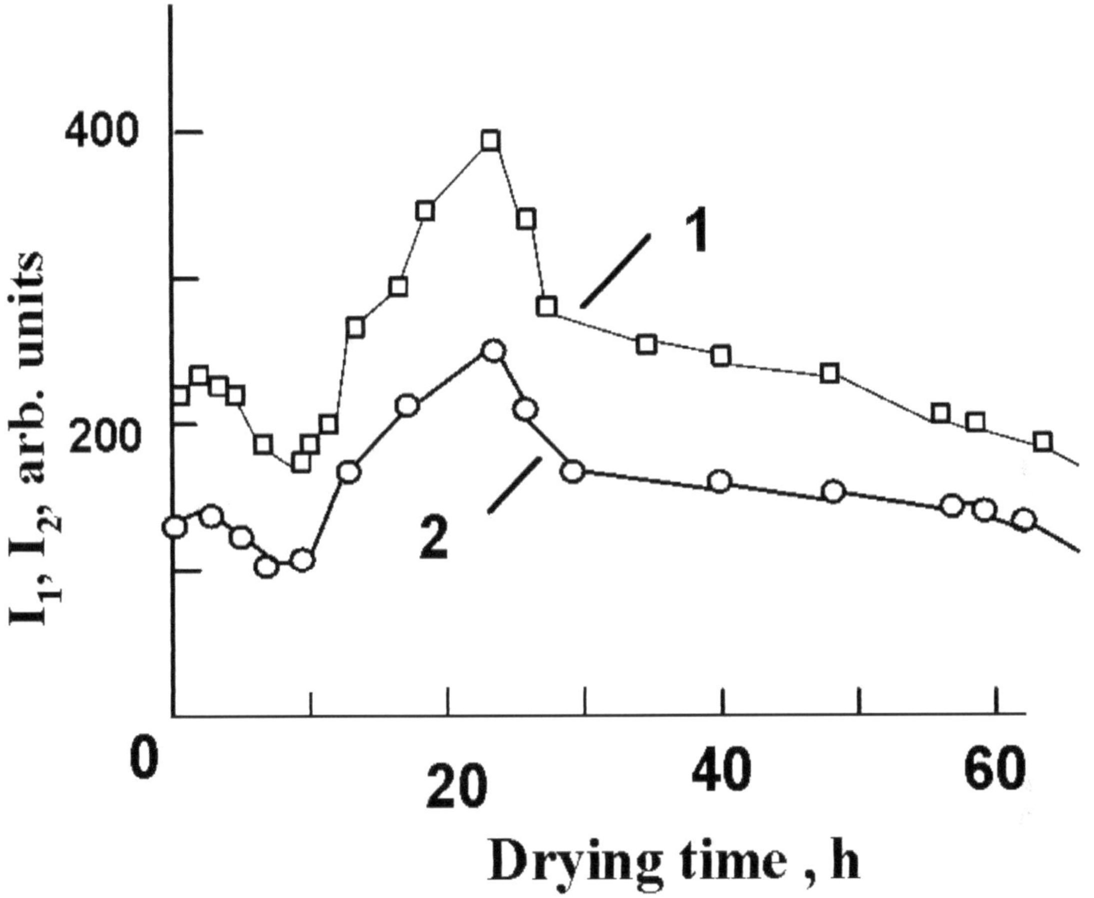

Figure 2. Dependences I_1= f(t) and I_2=f(t), where I – intensity of photoluminescence, t- drying time for Brassica rapa L. leaf .

It was found that in the region of the manifestation of the kinetics in the photoluminescence decrease, a section of the brightly expressed minimum, the "saddle" (Fig. 2), is found on the curves for I_1 and I_2, which can be related to a drop in the radiation intensity with a characteristic time of relaxation. The revealed feature can be associated with the manifestation of the amplification of the non- radiative recombination channel in connection with the decay of the leaf (the dying of life in it).

4. Conclusion

Thus, we found that in leaves separated from the green plant during photoexcitation bright red photoluminescence continues to be present, the time dependence of its intensity includes 2 stages characterized by the fact that in the first one there is an increase in intensity, reaching a maximum and then decrease, but with long drying times in conditions of constant room temperature, it does not fall

IOP Conf. Series: Journal of Physics: Conf. Series **1038** (2018) 012030 doi:10.1088/1742-6596/1038/1/012030

below its characteristic value for a living plant. The founded features can cause a sharp splash of research in the study of the possibilities of practical use of luminescent properties of plant objects.

References

[1] Kesamanly F P, Rud' V Yu, Rud' Yu V 1999 *Semiconductors* **5** 483

[2] Kudryashova I S, Rud' V Yu, Shpunt V Ch, Rud'Yu V, Glinushkin A P, 2016 *Journal of Physics: Conference Series* **741**-012106

[3] Shpunt V Ch, Rud' Yu V, Tsendin K D 1993 *Journal Technical Physics* **19** 13 41

[4] Pelant I 2012 Oxford; New York: Oxford University Press 542

[5] Valenta J., Ostatnicky T., Pelant I., Elliman R.G., Linnros J., Honerlage B. 2004 *J. Appl. Phys.* 96 9 5221

[6] Pryakhin Yu Yu, Kartabaeva B B, Sidorova O P , Glinushkin A P 2015 *Russian Agricultural Science* Review 6 6-2 29

[7] Glinushkin A P, Rud' Yu V, Shpunt V Ch, Rud' V Yu, Kudryashova I S, Mozhaiko A A 2015 *Biotics* **6** 7 29

IOP Conf. Series: Journal of Physics: Conf. Series **1038** (2018) 012031 doi:10.1088/1742-6596/1038/1/012031

Inhibition of Taq polymerase activity by singlet oxygen generation at photodynamic therapy

V.V. Klimenko[1,2], N.E. Kaydanov[1], A.K. Emelyanov[1,3], N.A. Verlov[1,3], S.V. Shmakov[1], N.A. Knyazev[1,2,4], A.A. Bogdanov[2,3,5]

[1] St Petersburg Academic University, St Petersburg, Russian Federation
[2] St. Petersburg clinical scientific and practical center for special types of medical care (oncology-oriented), St. Petersburg, Russian Federation
[3] Petersburg Nuclear Physics Institute NRC Kurchatov Institute, Gatchina, Russian Federation
[4] Institute of Cytology, St Petersburg, Russian Federation
[5] ITMO University, St. Petersburg, Russian Federation

E-mail: klimenko.vova@gmail.com

Abstract. The goal of this study was to assess the ability of singlet oxygen generated at PDT to inhibit the activity of the model enzyme - Taq polymerase. Experimentally were determined the laser irradiation dose and the singlet oxygen concentration (over 300 μM) that results to complete inhibition of Taq polymerase activity. The results show that enzymatic proteins have an order of magnitude higher susceptibility than DNA molecules to damage by singlet oxygen.

1. Introduction

Photodynamic therapy (PDT) is a clinically approved method for the treatment of cancer which can have cytotoxic activity against tumor cells [1-3]. The damage mechanism of tumor cells at PDT strongly depend on the photosensitizer (PS) localization (lipid membrane, intracellular organelles or nucleus) [4]. The chlorine and porphyrin PS drugs used in clinical practice (Radachlorin, NPe6, Photofrin, BPD and mTHPC), mainly accumulate in the cell cytoplasm (not in the nucleus), binding to protein components and localizing the damaged area. PS can accumulate in endoplasmic reticulum, Golgi apparatus, mitochondria, lysosomes, and plasma membrane. Upon the laser irradiation with specific wavelength PS can generate an excited state of the molecular oxygen - singlet oxygen witch induce further oxidation of biological objects. In the same time, it is possible to damage DNA molecules near the surface of the nuclear membrane, which can lead to mutation formation. In this study was investigated of the damage of enzymatic protein (Taq DNA polymerase) and single strand DNA (ssDNA) molecules under photodynamic treatment with Radachlorin PS. It was observed that the inhibition of Taq polymerase enzymatic activity was correlated with the laser irradiation dose and the concentration of singlet oxygen generated at PDT. The obtained results demonstrated that enzymatic proteins have an order of magnitude higher sensitivity to damage by singlet oxygen than DNA molecules.

2. Materials and methods

2.1. DNA damage by singlet oxygen

ssDNA was taken as a model system to investigate DNA oxidation at photodynamic therapy. cDNA was obtain from K562 cells using the GeneJET RNA purification Kit (Thermo Scientific, USA) and Maxima First Strand cDNA Synthesis Kit with dsDNase (Thermo Scientific, USA) according to the manufacturer's instructions. Concentration of cDNA was measured on NanoDrop 2000c (Thermo Scientific, USA). 20µl water solution of cDNA 9.6 ng/µl and 3.8 µM Radachlorin photosensitizer (Rada-Pharma Co, Ltd., Moscow, Russia) in individual tubes was prepared. This solution was irradiated with 662 nm laser «LAHTA – MILON» (MILON Laser, St. Petersburg, Russia) at power density I = 50 mW/cm^2 with energy doses: 2.5 J/cm^2, 5 J/cm^2, 7.5 J/cm^2, 10 J/cm^2, 15 J/cm^2. cDNA damage after photodynamic treatment was analyzed using polymerase chain reaction - diagnosis method (PCR). The PCR was performed in a 40 µL volume using 20 µl water solution of cDNA (9.6 ng/µl) after irradiation, 0.25 µM forward (5'-TGGTTCATCATCATTCAACGGTGG-3') and reverse (5'-GTTCCCGTAGGTCATGAACTCAG-3') primers [5], Taq polymerase (1.25 units per reaction), dNTP (200 µM) and 10x Mg^{2+} Tris-HCl buffer («Sileks», Russia). Used primers were specific to the *BCR-ABL* gene fragment of cDNA. The PCR protocol contained initially denaturated for 1 minute at 95°C followed by 35 cycles of 95°C for 15 seconds, 60°C for 15 seconds, and 72°C for 15 seconds, with a final extension at 72°C for 5 minutes using a Thermocycler T100 (Bio-Rad, USA). The PCR products (374 b.p.) were separated by electrophoresis in 8% polyacrylamide gel. Results were analyzed with Gel Doc XR+ (Bio-Rad, USA). The results were reported as the mean values with standard deviation from three experiments.

2.2. Inhibition of Taq polymerase at PDT

Taq polymerase («Evrogen», Russia) was added to 20µl Taq Buffer in 9 individual tubes. Taq polymerase concentration in experiments was equal 11.9 µg/ml (0.136 µM). The molar concentration of protein was measured on NanoDrop 2000c spectrophotometer. Then water solution of Radachlorin (0.5-2.0 µM) was added to each samples. This solution was mixed and irradiated with 662 nm laser at power density I = 20 mW/cm^2 with energy doses: 0.5, 1.25, 2.5, 3.75, 5, 7.5 and 10 J/cm^2. Then the 20µprepped solution of "control cDNA", dNTP and forward and revers primers for target gene BCR-ABL [5], was added to treated samples. The PCR was performed according to the previous protocol using a Thermocycler T100. The PCR products for BCR-ABL gene (374 bp) were separated by electrophoresis in 8% polyacrylamide gel. The intensity of bands was used as a quantitative characteristic of the polymerase activity. The intensity value of bands was measured in the control and in irradiated samples. Taq polymerase replication activity was presented in percent as band intensity in experimental samples with irradiation relative to bend intensity in control sample with PS and without irradiation.

2.3. Detection of singlet oxygen generation at PDT

Chemical trap method was used to register singlet oxygen generated in water under photodynamic interaction [6]. Chemical trap 1,3-diphenylisobenzofuran (DPBF) was prepared in 1.5 ml water (or PDS) with 0.2M SDS in 1 cm quartz cuvette. The concentration [DPBF] = 42 µM and Radachlorin concentration [PS] = 3.5 µM (extinction coefficient ε_{655} = 28 300 M^{-1}cm^{-1} in PBS). This solution was irradiated with 662 nm laser «LAHTA – MILON» at power I = 10 mW/cm^2 during t = 100 sec. Absorption spectra were measured each 10 second on NanoDrop 2000c. The decrease of absorbance on 414 nm wavelength corresponds to the interaction the chemical traps DPBF with singlet oxygen. Absorption change with time corresponds to the singlet oxygen generation rate, taking into account the efficiency of singlet oxygen quenching by DPBF in solution. The efficiency of singlet oxygen quenching by DPBF depended on singlet oxygen lifetime, DPBF concentration and quenching rate constant.

3. Results

3.1. Inhibition of Taq polymerase activity at PDT

Fig.1 shows the DNA polyacrylamide gel which presented Taq polymerase activity after photodynamic interaction. The Taq polymerase samples were incubated with three concentration of Radachlorin (0.5 μM, 1 μM, 2 μM) and then irradiated under laser at 662 nm. For a low PS concentration equal 0.5 μM, there was no complete inhibition of the polymerase activity. For concentrations 1.0 and 2.0 μM were observed complete inhibition of Taq polymerase activity at irradiation doses 2.5 and 7.5 J/cm^2, respectively. On Fig.2 presented the relative values of Taq polymerase replication activity on the irradiation doses for 0.5, 1.0 and 2.0 μM Radachlorin. At a low PS concentration and small light dose an increase the Taq polymerase activity (PCR product) was observed. It can be suggested that this effect may be associated with the appearance of an unstable oxidized enzyme but more active for PCR product formation.

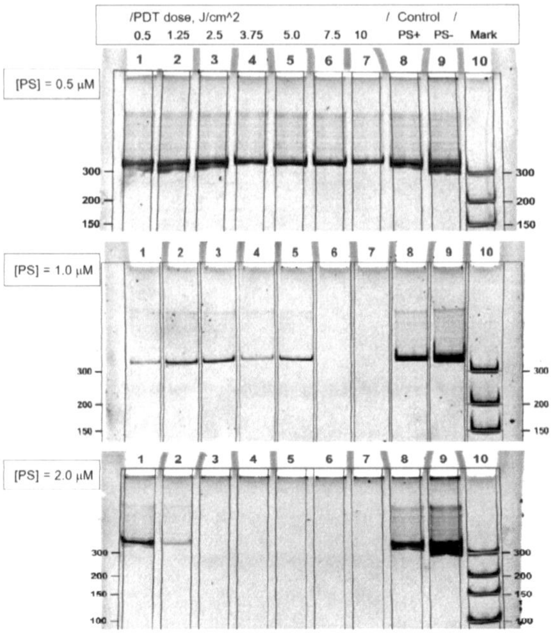

Figure 1. DNA polyacrylamide gel electrophoresis after photodynamic interaction on the Taq polymerase solution. Lanes 1-7 — PDT with irradiation doses 0.5, 1.25, 2.5, 3.75, 5.0, 7.5, 10 J/cm^2; Lane 8 — control PS+; Lane 9 — control PS-; Lane 10 — 300-bp DNA ladder.

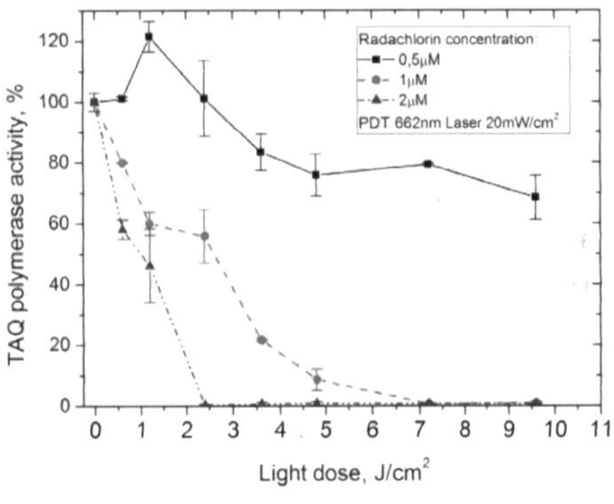

Figure 2. Dose dependent inhibition Taq polymerase replication activity (relative PCR product) after photodynamic treatment with 0.5, 1 and 2.0 μM of Radachlorin PS and irradiation doses 0.5-10 J/cm^2.

3.2. Direct ssDNA damage by singlet oxygen at PDT

The experimental results of ssDNA damage at PDT were obtained in our previous study [5]. Fig. 3 shows picture of DNA polyacrylamide gel electrophoresis after photodynamic treatment and intensity of BCR-ABL DNA fragment. Fig. 4 shows dependence of target PCR product (374 bp) from the irradiation dose. The second fragment of DNA (about 50 b.p.) was primer (forward and reverse) dimers. This is a nonspecific product of PCR reaction and it was increased at decreased amount of target DNA. The PCR product (DNA replication activity) was presented as relative intensity I/I0, where I is intensity of the analyzed band, I0 is intensity of the control band.

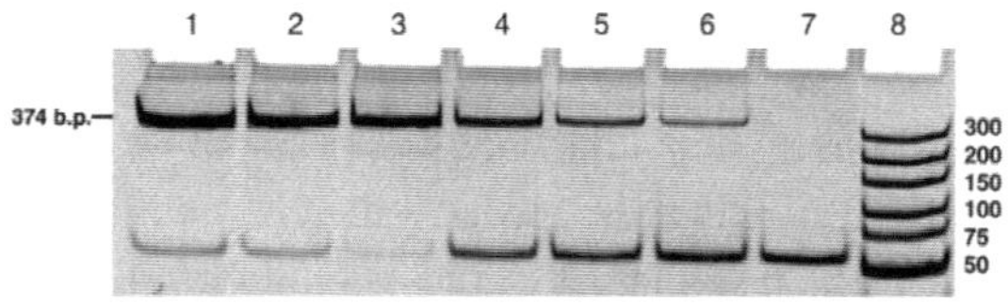

Figure 3. DNA polyacrylamide gel electrophoresis after photodynamic treatment (target gene BCR-ABL). Lane 1 — control PS-; Lane 2 — control PS+; Lanes 3-7 — PDT with irradiation doses 2.5, 5, 7.5, 10, 15 J/cm^2; Lane 8 — 300-bp DNA ladder.

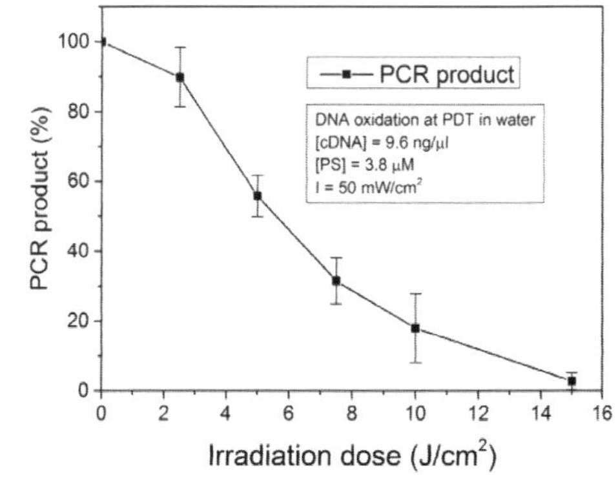

Figure 4. Irradiation dose dependent PCR product (DNA replication activity) after photodynamic treatment with 3.8 μM of Radachlorin PS and irradiation doses 2.5-15 J/cm^2.

This data demonstrated that DNA replication activity depends from irradiation doses at PDT. The analysis of DNA damages can be applied to investigate the efficiency of laser irradiation modes at photodynamic therapy.

3.3. DPBF detection of singlet oxygen generation

Fig. 5 shows photodegradation of DPBF by laser irradiation with Radachlorin in water 0.2M SDS. Absorbance decrease on $\lambda = 414$ nm was measured. Fig. 6 shows the calculation concentration dependences of oxidized DPBF and calculated concentration of $^1O_2$ from irradiation time. Using the development of the methods for the direct study of kinetics of the reaction of singlet oxygen [7, 6], equation (1) was obtained that describes the interaction of singlet oxygen with a reactive substrate DPBF:

$$\frac{1}{\varepsilon}\frac{dA}{dt} = \frac{d\left[{}^1O_2\right]}{dt}\frac{k_{ox}\left[DPBF\right]}{\tau_\Delta^{-1} + k_{ox}\left[DPBF\right]} \tag{1}$$

where A is absorption of DPBF at 414 nm without value of PS absorption, ε is molar absorption coefficient DPBF equal to 2.35×10^4 M^{-1} cm^{-1} (414 nm), τ_Δ is life time of $^1O_2$ in water equal to 4.0 μs, k_{ox} is a constant of oxidation of DPBF equal to 2.8×10^9 M^{-1} s^{-1} in water with SDS [6], [DPBF] is concentration of active DPBF traps. The same experimental data was received in ethanol [5].

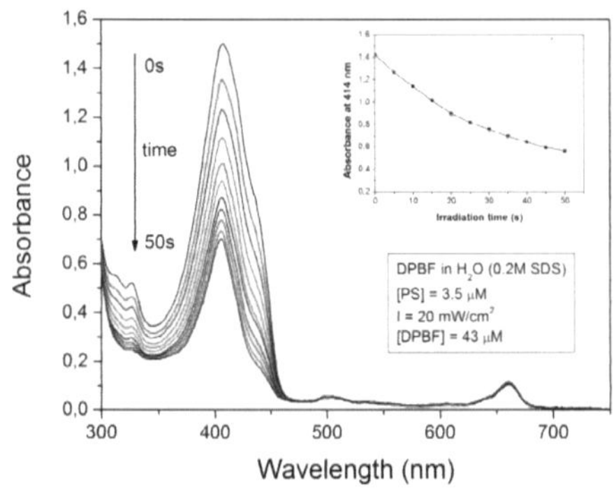

Figure 5. Photodegradation of DPBF by laser irradiation at 662 nm (20 mW/cm^2) in water with 3.5 μM Radachlorin .

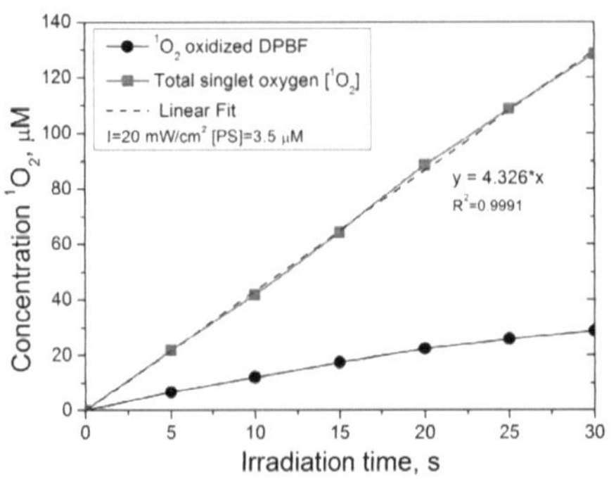

Figure 6. Time dependences of concentration of DPBF oxidized by singlet oxygen (black line) and total singlet oxygen concentration (red line).

Using the equations of macroscopic singlet oxygen model [8, 9] the equation (2) was obtained for calculated the singlet oxygen generation rate. ξ is the parameter of singlet oxygen generation efficiency including in itself the PS extinction coefficient at certain wavelength and singlet oxygen quantum yield of PS. This parameter was calculated using linear approximations of data from Fig. 6. This parameter is the proportionality constant between the concentration of PS, the irradiation intensity, oxygen concentration and the rate of singlet oxygen generation. ξ has value of singlet oxygen generation rate per I = 1 mW/cm^2 irradiation and 1 μM of photosensitizer per second. β is the ratio of the triplet photosensitizer decay rate to the rate of the triplet photosensitizer quenching by $^3O_2$ (8.6 μM for chlorine PS).

$$\frac{d[^1O_2]}{dt} = \xi \cdot [PS] \cdot I \cdot \left(\frac{[^3O_2]}{[^3O_2] + \beta} \right) \quad (2)$$

Table 1. Photochemical parameters: singlet oxygen generation rates of Radachlorin PS and life time of $^1O_2$ molecule in ethanol and water upon laser irradiation of at 662 nm.

Solvent	ξ, mW^{-1}cm^{2}s^{-1}	τ_Δ, μs	Ref.
Ethanol	0.067 ± 0.007	13.5	[5]
Water + 0.2M SDS	0.043 ± 0.003	4	[5, 10]
PBS + 0.2M SDS	0.062 ± 0.003	3.5	[10, this study]

This value can be used to recalculate generation of singlet oxygen in solutions with other concentrations of PS and irradiation intensity. This constant was further used in our study to calculate cumulative singlet oxygen concentration at photodynamic therapy.

3.4. Taq polymerase damage mediated by singlet oxygen

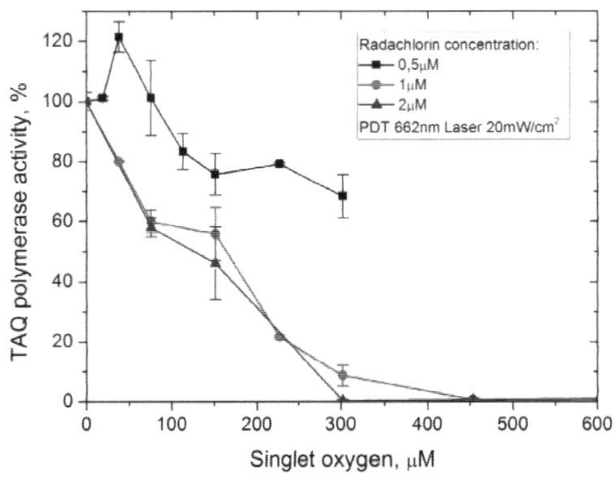

Figure 7. Singlet oxygen dose-dependent inhibition Taq polymerase activity after photodynamic treatment with 0.5, 1.0 and 2.0 μM of Radachlorin and irradiation doses 0.5-10 J/cm^2.

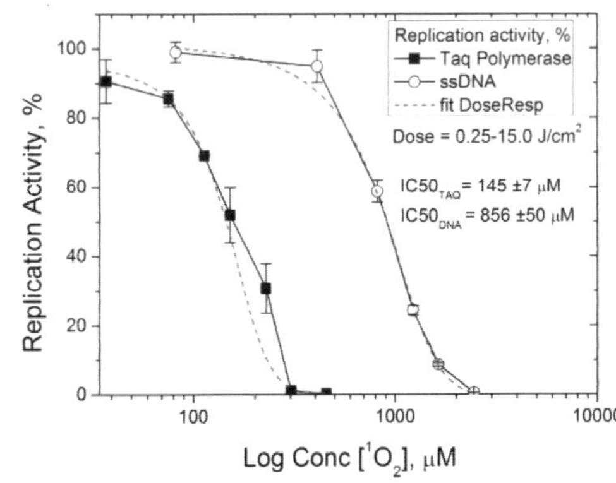

Figure 8. Singlet oxygen dose-dependent inhibition Taq polymerase activity and DNA replication activity after photodynamic treatment with Radachlorin PS and irradiation dose 0.25-15 J/cm^2.

The mathematical modeling of singlet oxygen generation at PDT allowed calculating singlet oxygen concentration and PS photobleaching [8-10]. This modeling were applied to calculated singlet oxygen doses which induced the inhibition of Taq polymerase activity and the damaged of ssDNA (oxidation ssDNA). Fig.7 demonstrated the dependency of Taq polymerase activity on singlet oxygen concentration after photodynamic treatment. At a low dose of the PS concentration equal 0.5 μM, there was no complete inhibition of the polymerase activity. Perhaps this effect is due to PS binding to the enzyme, and at a low PS concentration, there are free molecules of the enzyme that are not interacted with active singlet oxygen. Using the dose-response fitting were determined the IC50 values of singlet oxygen concentration which inhibited 50% Taq polymerase activity and reduced the ssDNA replication activity. The data of replication activity from singlet oxygen concentration was presented on Fig.8. The IC50 values of [$^1O_2$] were 145±7 μM for Taq polymerase and 856±50 μM for ssDNA.

4. Discussion

Experimentally were determined the laser irradiation dose and the singlet oxygen concentration (over 300 μM) that results in complete inhibition of enzymatic activity of Taq polymerase. Also were determined the irradiation doses at PDT, which reduce the replication activity of single-stranded DNA molecules [5]. The results show that enzymatic proteins have an order of magnitude higher susceptibility to damage by singlet oxygen than DNA molecules and direct damage the DNA in the nucleus of tumor cells is unlikely. Obtained data confirm that the protein have high rate constants for reaction with singlet oxygen. Using the singlet oxygen quenching rate constants by proteins [11] can be determined the concentration of molecule damages which induces complete inhibition of polymerase activity. The understanding dose dependences of molecular damages at PDT are an important step in assessing the photodynamic effect on tumor cells [4, 9].

Acknowledgments

The reported study was funded by RFBR according to the research project № 16-04-01858 a.

IOP Conf. Series: Journal of Physics: Conf. Series **1038** (2018) 012031 doi:10.1088/1742-6596/1038/1/012031

References

[1] Allison R R, Sibata C H, 2010 *Photodiagnosis Photodyn.* **7** 61–75.
[2] Dolmans D, Fukumura D, Jain R K, 2003 *Nature Reviews Cancer* **3** 380–387.
[3] Wilson B, Patterson M, 2008 *Phys. Med. Biol.* **53** 61–109.
[4] Castano A P, Demidova T N, Hamblin M R, 2005 *Photodiag Photodyn Ther* **2** (1) 1-23.
[5] Klimenko V V, Kaydanov N E, Emelyanov A K, Bogdanov A A, 2017 *J. Phys. Conf. Ser.* **917** 042013
[6] Krasnovsky A A Jr., Kozlova A S, Roumbala Ya V, 2012 *Photochem. Photobiol. Sci.* **11** 988-997
[7] Wilkinson F, Helman P, Ross A, 1995 *J Phys Chem Ref Data* **24** 663-1021.
[8] Wang K K, Finlay J C, Busch T M, Hahn S M, Zhu T C, 2010 *J Biophoton* **3** 304–318.
[9] Klimenko V V, Knyazev N A, Moiseenko F V, Rusanov A A, Bogdanov A A, Dubina M V 2016 *Photodiagnosis Photodyn. Ther* **13** 101–107.
[10] Klimenko V V, Shmakov S V, Kaydanov N E, Knyazev N A, Kazakov N V, Rusanov A A, Bogdanov A A, Dubina M V, 2017 *Proc. SPIE* **10417** 1041703.
[11] Gimenez R E et al. 2016 *Free Radic. Biol. Med.* **94** 99-109.

Features of magnetic field stabilization in caesium atomic clock for satellite navigation system

A A Petrov[1], N M Grebenikova[1], N A Lukashev[1], V V Davydov[1, 2, 3], N V Ivanova[2], N S Rodygina[2], A V Moroz[1]

[1]Higher School of applied physics and space technologies, Peter the Great Saint - Petersburg Polytechnic University, Saint - Petersburg 195251, Russia
[2]The Bonch-Bruevich Saint - Petersburg State University of Telecommunications, Saint - Petersburg 193232, Russia
[3]Department of Ecology, All-Russian Research Institute of Phytopathology, 143050, Moscow Region, Odintsovo district, B.Vyazyomy, Russia

alexandrpetrov.spb@yandex.ru

Abstract. In this paper one of several directions of caesium atomic clock modernisation is presented. This paper deals with theoretical model which takes into account features of the development of magnetic field stabilization system for caesium atomic clock and describes the affect of this stabilization system on frequency stability of caesium atomic clock. New magnetic field stabilization system allows eliminating one of the most important perturbing factors affecting on long-term frequency stability. Experimental research of the caesium atomic clock's metrological characteristics with magnetic field stabilization system showed improvement long-term frequency stability on 10%.

1. Introduction

The accurate measurement of time and frequency is vital to the success of many fields of science and technology. For example atomic physics (atom-photon interactions, atomic collisions, and atomic interactions with static and dynamic electromagnetic fields), geodesy, radio-astronomy (very long baseline interferometry), pulsar astronomy, various communication equipment and metrological services rely on high frequency stability and uniform timescales [1-5]. The same is valid for the operation of satellite navigation systems.

Global navigation satellite constellations such as the European Galileo, Russian GLONASS, and the USA Global Positioning Systems use atomic frequency standards for precision time-keeping and stable frequency generation.

One of the central problems in the satellite system is the problem of mutual synchronization of the satellite time scales up to nanoseconds and less [2, 3, 5]. The error of the navigation signals emitted by the different satellites at 10 ns causes an additional error in determining the location of the consumer to 10-15 meters.

The problem solution of high-precision synchronization of the on-board time scales required the implementation of highly stable on-board caesium and rubidium frequency standards on satellites, as well as the creation of ground-based devices for comparing time scales.

The concept of development of space navigation system and development of the metrological services need to modernize the currently used quantum frequency standards or developing new ones. Development and commissioning of new atomic clock is very long and costly process, which in the most cases hasn't enough funds and time. Therefore, in most cases, for specific tasks related to the operating conditions of frequency standards modernization is occur [6, 7].

The process of quantum frequency standards modernization includes various directions: changing the weight and dimensions, reducing energy consumption, improvement metrological characteristics. And for frequency standards characterized by the fact that modernization may not be for the whole construction and may be for individual units or blocks. In present work one of the several directions of improvement metrological characteristic is considered.

2. Principles of caesium atomic clock operation

The work of a caesium atomic clock is based on the principle of adjustment a highly stable voltage-controlled quartz crystal oscillator (VCXO) to quantum frequency transition of atoms of caesium-133 [1, 7, 8]. Figure 1 shows a block diagram of a caesium atomic clock.

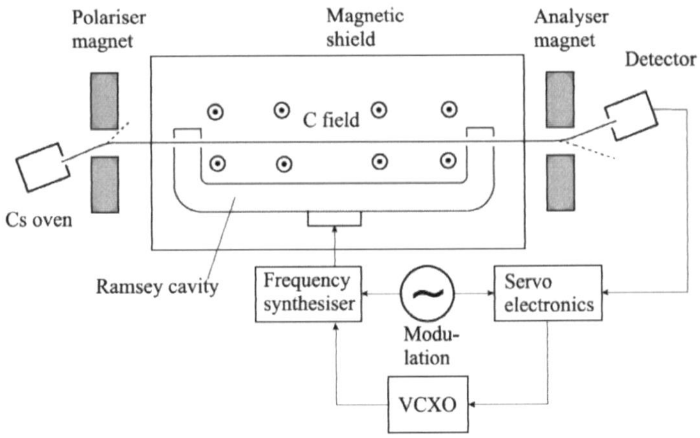

Figure 1. Block diagram of a caesium atomic clock.

The output signal frequency 5MHz of the VCXO is supplied to the frequency synthesiser. Frequency synthesiser consists of frequency converter, mixer signals and multiplier signals. In the frequency converter input signal frequency 5MHz is converted to the signal frequency 12,631772 MHz and supplied to the input of mixer signals. In the multiplier signals input signal frequency 5 MHz is multiplied to the frequency 270 MHz and then to frequency 9180 MHz. This signal frequency 9180 MHz is also supplied to the input of mixer signals. As a result, the output signal of the frequency synthesiser is the signal of ultrahigh frequency 9192,631772 MHz. This signal is supplied in the Ramsey cavity.

In caesium atomic clock with the help of magnet polarizer 2 the atoms are prepared such that they are either in the $|F=4, m_F=0>$ or in $|F=3, m_F=0>$ state. Afterwards the atoms interact with an electromagnetic field that induces transitions into the former unoccupied state.

A magnetic field is used to separate energetically the otherwise degenerate magnetic sub-levels in order to allow the excitation of the clock transition $|F = 3, m_F = 0> \rightarrow |F = 4, m_F = 0>$ isolated from the other transitions. By convention such a field is referred to as the C field as it is applied between the fields of the polariser and the analyser.

The magnitude of the C field is chosen as a compromise between two conflicting requirements. First, it has to be large enough to separate the otherwise overlapping resonances. Second, the C field shifts the resonance frequency quadratically which has to be corrected. However, in a larger field the

IOP Conf. Series: Journal of Physics: Conf. Series **1038** (2018) 012032 doi:10.1088/1742-6596/1038/1/012032

frequency of the clock is influenced to a larger extent by fluctuations of the magnetic field. In the scheme of a commercial Cs clock the C field is often generated by a coil with windings in the paper plane wound around the Ramsey resonator and hence, points perpendicularly to that plane. Owing to the dependence of the frequency of the clock transition from the magnetic field, efficient magnetic shielding has to be provided in order to attenuate the ambient magnetic field and the magnitude of the associated fluctuations.

The atoms in the former unoccupied state are detected and allow one to determine the frequency of the interrogating field where the transition probability has a maximum. The observed transition frequency is corrected for all known frequency offsets that would shift the transition frequency from the unperturbed transition and is used to produce a standard frequency or pulse per second every 9192631772 cycles [1, 7, 8].

Scanning the frequency ν of the atomic resonance leads to a detector current like the one shown on the figure 2. The signal shows the Ramsey resonance structure on a broader, so-called, Rabi pedestal.

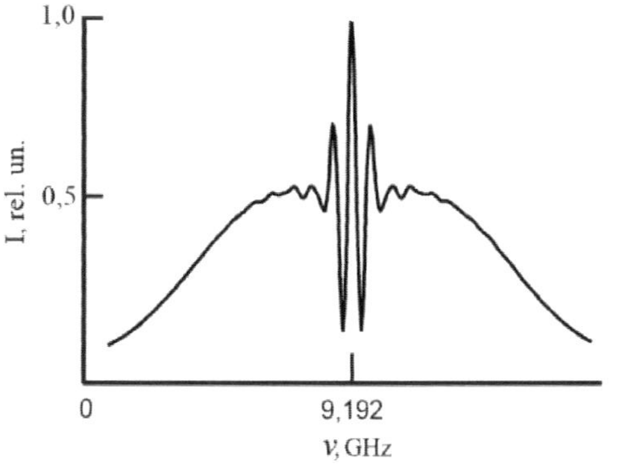

Figure 2. Ramsey resonance structure on the Rabi pedestal.

The central feature with the maximum at the transition frequency ν_0 is used to stabilize the frequency of the crystal oscillator to the atomic transition frequency. To this end, the frequency from the synthesizer is modulated across the central peak. The signal from the detector is phase-sensitively detected in the servo electronics, integrated and this servo signal is used for stabilising the frequency of the VCXO. From this suitable output frequencies are derived, such as 5MHz or a 1 PPS signal.

3. Features of C field stabilization

The stable isotope Cs-133 has a two hyperfine states F = 4 and F = 3 which are split in the magnetic field into 16 components. In accordance with the selection rules seven transitions between the components of hyperfine sublevels are possible [9, 10]. These are presented in figure 3.

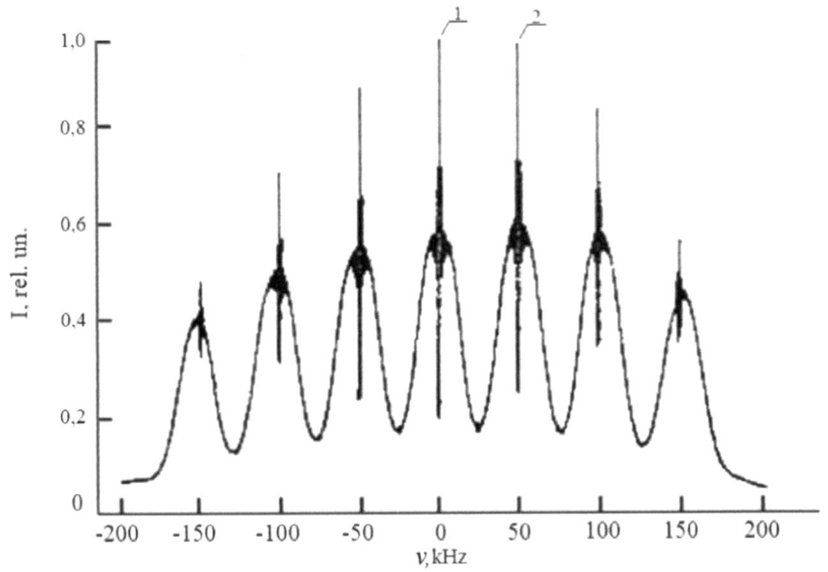

Figure 3. Microwave resonances in the Ramsey cavity.

The central resonance |F =3, m_F=0> → |F =4, m_F=0> (marked on the figure 3 as a «1») due to the Zeeman effect expose a quadratic frequency shift. With the help of formula (1) we can calculate a frequency shift.

$$\Delta f_{B^2} \approx 4.2745 * 10^{-2}\ \text{Hz} * \left(\frac{6*10^{-6}\text{T}}{\text{uT}}\right)^2 = 1{,}5388\ \text{Hz}. \quad (1)$$

For a typical value of C field equals 6 uT the frequency shift equals 1,5388 Hz. This frequency shift corresponds to a relative frequency shift of $3 \cdot 10^{-10}$.

How we can see from formula (1) the accuracy of the output signal of quantum frequency standard is dependent on the shift of the central resonance. It should be noted that not only the central resonance is exposed the frequency shift, but also all six transitions (3, m_F) ↔ (4, m_F), which $\Delta m_F = 0$. To express these changes as a function of magnetic induction B and atomic constants use the equation Bright Rabi:

$$E(F, m_F) = -\frac{hv}{2(2I+1)} - g_I \mu_B B m_F + \varepsilon \frac{hv}{2}\left(1 + \frac{4m_F}{2I+1}x + x^2\right)^{\frac{1}{2}}, \quad (2)$$

where E(F, m_F) is the change energy of atoms in the ground state; I is the quantum number of nuclear spin; g_I is the factor Lande for electron; μ_B is the Bohr magneton; m_F is the magnetic quantum number; $x = \frac{(g_j+g_I)\mu_B B}{hv}$.

The relation (2) can be used for calculation the frequency shift of any transition between two hyperfine sublevels, depending on the magnetic field. Revealing this expression, we find that the first member is proportional to the magnetic field B. For caesium beam primary frequency standards we must consider the quadratic member of this expression.

In theory, the frequency shifts can be taken into account in the calculation of the functional dependence on magnetic field values and the atomic constants using equation Bright-Rabi. But in practice, any changes of the magnetic field shift the resonance frequency. And values of these frequency shifts cannot be accounted for in advance.

Thanks to the development of a new frequency synthesiser design [7, 8 10, 11] the range of generated output frequencies has been expanded. It allowed detuning output synthesiser's frequency to

IOP Conf. Series: Journal of Physics: Conf. Series **1038** (2018) 012032 doi:10.1088/1742-6596/1038/1/012032

the neighboring resonance frequency of spectral line that makes it possible to adjust the C-field in quantum frequency standard.

Now in caesium atomic clock the magnetic field is maintained by the active stabilization system. For this purpose the neighboring transition |F =3, m_F=1> ↔ |F =4, m_F=1> (marked on the figure 3 as a «2») is used. The method of C-field adjustment is similar to the method of frequency adjustment to the main maximum. For this purpose, the average value of the sampling frequency ν_i is changed from the value of the ν_{Cs} to $\nu_{Cs} + \Delta\nu$, where ν_{Cs} is the frequency of the main transition of the caesium atom, $\Delta\nu$ - difference between the transitions for a preset value of the magnetic field. Then the value of magnetic field is adjusted in a such way that the frequency of transition |F =3, m_F=1> ↔ |F =4, m_F=1> match the preset frequency. This adjustment set up automatically several times per minute. The value of the applied field is automatically maintained at a predetermined level. In this case effects associated with any changes in the magnetic field (for example, long-term drift of the current source, temperature dependence, effect of external magnetic field, etc.) are excluded.

This feature gave us the opportunities to close the ring-locked loop frequency oscillator on the neighboring resonance transition.

Alternately closing the ring-locked loop at the central and the neighboring transition we adjust the frequency of the VCXO to the frequency of the central atomic transition, and support the constant value of the magnetic field inside the Ramsey cavity.

4. Experimental research of caesium atomic clock output signal

Experimental research of caesium atomic clock output signal consisted of measuring the output frequency and calculation the Allan variance, which allows to evaluate the frequency instability. But before that, the correct functioning of the magnetic field stabilization system was verified. The figure 4 shows the dependence of the coil current change, which determines the value of the magnetic field from the temperature change.

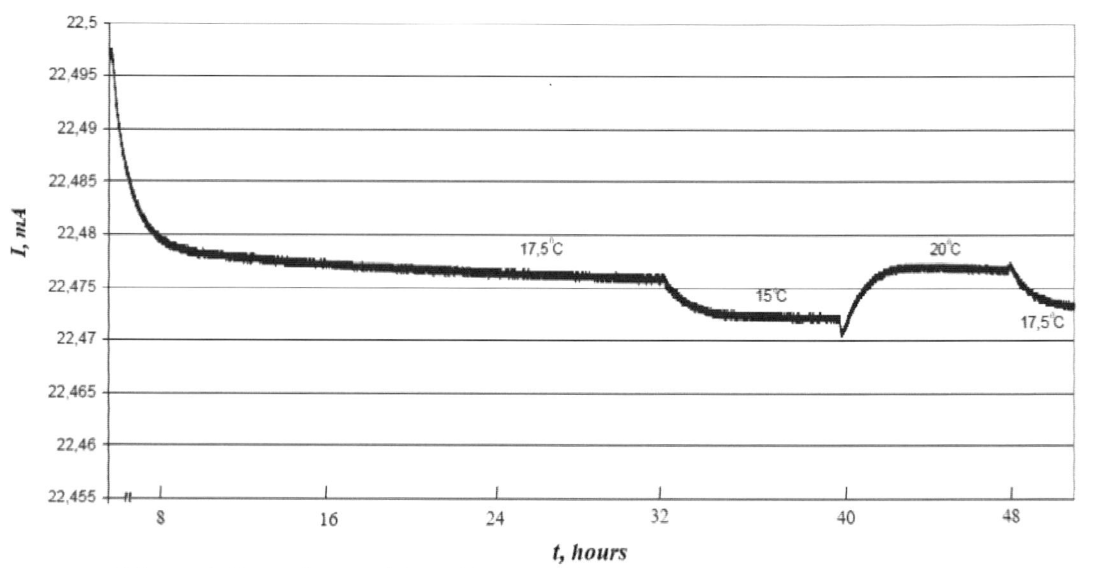

Figure 4. Dependence of the coil current change from the temperature change.

From this figure 4 we can see that in different external temperatures the operation modes of atomic clock are different. As a result of temperature changes, the temperature-dependent parameters of all electronic components are change. The output frequency of the frequency synthesizer, which corresponds to the frequency transition of caesium atom-133 is shifted. Magnetic field stabilization system begins to change the coil current so as to compensate shift of the central resonance frequency of caesium atom -133.

Thanks to the magnetic field stabilization, the current value stabilizes at a constant level after external temperature changes. Also in this way frequency shifts associated with any magnetic field changes are excluded.

Figure 5 shows Allan variance (σ_y) via time (t) for caesium atomic clock with magnetic field stabilization system (2) and previous design (1).

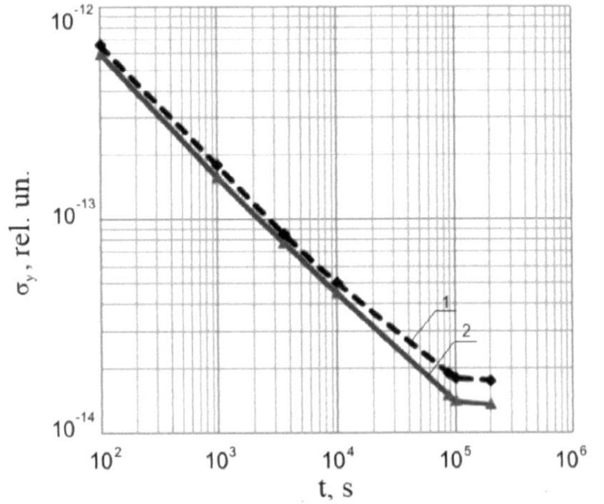

Figure 5. Allan variance via time for caesium atomic clock with magnetic field stabilization system (2) and previous design (1).

From these results it is clear that the use of magnetic field stabilization system make it possible to obtain a better long-term frequency stability of quantum frequency standard output signal. It is mainly achieved through by compensating the frequency shift of central resonance of caesium-133.

5. Conclusion

Frequency shifts associated with magnetic field changes and features of magnetic field stabilization in caesium atomic clock are considered. Magnetic field stabilization system for caesium atomic clock is developed. New magnetic field stabilization system allows eliminating one of the most important perturbing factors affecting on long-term frequency stability. Experimental research of correct functioning of the magnetic field stabilization system was successfully verified. Experimental research of the caesium atomic clock's metrological characteristics with magnetic field stabilization system showed improvement long-term frequency stability on 10%.

6. References

[1] Riechle Fr 2004 *Frequency Standards. Basics and Applications* (WILEY-VCH Verlag GmbH Co.KGaA, Weinheim)

[2] C. Audoin and B. Guinot 2002 *Measurement of Time, Time Frequency and the Atomic Clock* (Cambridge University Press, Cambridge, UK).

[3] McCllelland T, White Y and Wilson D 2011 Proc 43rd Annual PTTI Systems and Applications Meeting (California) vol 1 (USA, Hyatt Regency Long Beach) p 325-340

[4] Davydov R V, Saveliev I V, Lenets V A, Tarasenko M Yu, Yalunina T R, Davydov V V, Rud' V Yu 2017 *Lecture Notes in Computer Science (including subseries Lecture Notes in Artificial Intelligence and Lecture Notes in Bioinformatics)* **10531 LNCS** 177-183.
[5] Pakhomov A A 2007 *Journal of Communications Technology and Electronics* **52** 1114 – 1118
[6] Petrov A A, Davydov V V, Myazin N S, Kaganovskiy V E 2017 *Lecture Notes in Computer Science (including subseries Lecture Notes in Artificial Intelligence and Lecture Notes in Bioinformatics)* **10531 LNCS** 561-568.
[7] Petrov A A, Davydov V V 2015 *Lecture Notes in Computer Science (including subseries Lecture Notes in Artificial Intelligence and Lecture Notes in Bioinformatics)* **9247** 739-744
[8] Petrov A A, Davydov V V 2017 *Journal of Communications Technology and Electronics* **62** 289 – 293
[9] Semenov V V, Nikiforov N F, Ermak S V, Davydov V V 1991. *Soviet Journal of Communications Technology and Electronics* **36(4)** 59-63
[10] Petrov A A, Vologdin V A, Davydov V V, Zalyotov D V 2015 *Journal of Physics: Conference Series* **643(1)** 012087
[11] Petrov A A, Davydov V V 2016 *Journal of Physics: Conference Series* **769(1)** 012065

A Low Phase Noise Tunable Microwave Spin Wave Optoelectronic Oscillator

A B Ustinov[1], A A Nikitin[1], V V Lebedev[2], A A Serebrennikov[2], A V Shamray[2], A V Kondrashov[1], B A Kalinikos[1]

[1]St. Petersburg Electrotechnical University, Department of Physical Electronics and Technology, St. Petersburg, 197376 Russia
[2]Ioffe Physical-Technical Institute of the Russian Academy of Science, St.Petersburg, 194021, Russia

E-mail: ustinov_rus@yahoo.com

Abstract. Experimental investigation of the optoelectronic oscillator constructed with a spin-wave bandpass filter was carried out. A phase noise of -140 dBc/Hz at the frequency offset of 10 kHz was obtained in the frequency range 3-12 GHz.

During the past decade optoelectronic oscillators attract an increased research interest as a source of low noise microwave signals [1-4]. Usually the microwave photonic oscillators have a ring circuitry consisting of the microwave and optical paths. Eigen-mode spectrum of the ring oscillators consists of a series of spectral lines. In other words, it is a frequency comb. In order to obtain self-generation of continuous wave signal with low phase noise a high-Q resonators are typically utilized as a filter for selection of the single mode from the frequency comb. Tunable microwave filters could be used to control the generated signal frequency. In order to obtain a wide range of frequency tuning the ferromagnetic microwave filters made of yttrium iron garnet (YIG) spheres were utilized usually [2,5,6]. It is noteworthy that a principle of operation of the YIG sphere filter is based on the ferromagnetic resonance phenomenon. Recently we proposed to use a narrow-band microwave delay line based on an epitaxial YIG film as a frequency selective tunable element, i.e. as a spin-wave bandpass filter [7].

Purpose of the present work is experimental study of the phase noise of the optoelectronic oscillator constructed with a spin-wave bandpass filter. A block-diagram of the oscillator is shown in Fig. 1. The main components of the oscillator are a laser diode, an electro-optic modulator (EOM) of Mach-Zehnder type, an optical fiber, a photodetector, a DC-block, microwave amplifiers, an isolator, a variable attenuator, a spin-wave bandpass filter (SWBPF), and a directional coupler. Coaxial cables connect components of the microwave path. Optical radiation wavelength of the laser was 1.55 μm. Its power was 100 mW. Bandwidth of the EOM was 12 GHz. Half-wave voltage of the EOM was 3.3 V. Upper operating frequency of the photodetector was 25 GHz. The amplifiers bandwidth was 1-18 GHz.

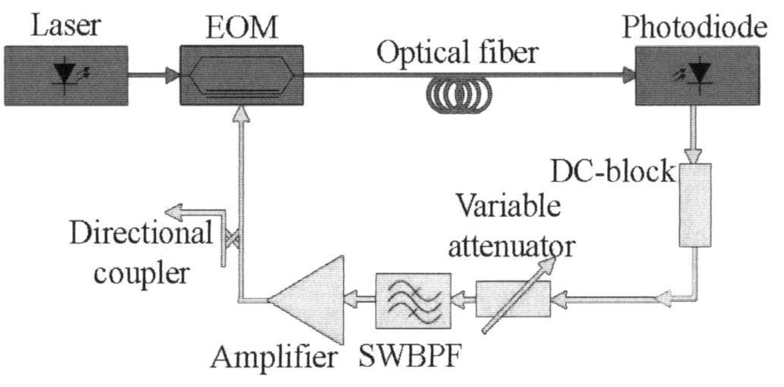

Figure 1. Block-diagram of the optoelectronic oscillator

Several SWBPFs were made in the form of YIG-film delay lines. The 2 mm wide YIG films of different thickness and saturation magnetization were utilized in the filters. The short circuited microstrip transducers were used for excitation and detection of the spin waves. The distance between transducers was 3 mm. The film was magnetized by a spatially uniform bias magnetic field directed in the film plane and perpendicular to the spin wave propagation. Such a geometry provides the conditions for excitation of surface magnetostatic spin waves [8]. The magnetic field was varied in order to provide frequency tuning of the amplitude-frequency characteristic of the filter as is shown in Fig. 2.

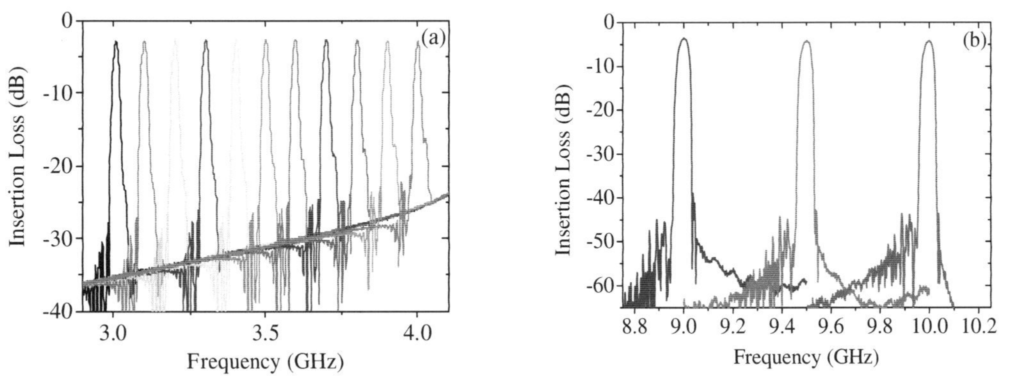

Figure 2. Amplitude-frequency characteristics of the spin-wave filters fabricated with a substituted YIG film having $4\pi M_s = 850$ G (a) and with a pure YIG film having $4\pi M_s = 1750$ G (b). Different characteristics corresponds to different bias magnetic fields.

Principle of operation of the oscillator is as follows. The laser emits a continuous wave optical radiation, which is used as a carrier signal. The electro-optical modulator modulates the optical radiation by a microwave signal coming from the MW path. The modulated optical signal propagates through the optical fiber. The photodetector demodulates the optical signal and transmits its envelope to the MW path. The DC-block removes the direct current level from the output photocurrent. The resulting alternative current signal is applied to the input microstrip transducer and excites the spin wave. The spin wave propagates in accordance with its dispersion. The bandpass is defined by YIG film geometry and bias magnetic field. After detection at the output transducer microwave signal is amplified by the microwave amplifiers and supplied to the driving port of the electro-optic modulator, closing the ring circuit. A pair of the microwave amplifiers compensates total propagation losses of the microwave signal in the ring. The variable attenuator is used for control of these losses. A small portion of the generated microwave signal is extracted from the ring by the directional coupler for monitoring. The monitoring was carried out with Rohde&Schwarz FSWP26 phase noise analyzer.

The measurements of the phase noise spectra were carried out for different lengths of optical fiber up to 2.2 km. We used a commercial SMF optical fiber having a core diameter of 8 μm. A preliminary characterization of the optical path showed losses for microwave signal from driving input of the optical modulator to the photodetector output of approximately 15 dB in frequency range up to 12 GHz. Test measurements of the microwave amplifier demonstrated an output power at 1 dB suppression of 24 dBm and the gain of 26 dB. These values were almost constant in the nominal frequency range of 1-12 GHz. The spin-wave filters provided in this range insertion losses from 2.5 to 5 dB.

Figure 3 shows by solid line the measured phase noise spectrum. The results demonstrate that the minimum values of the phase noise at the frequency offset of 10 kHz were around -140 dBc/Hz for frequencies of 3977 MHz (Fig. 3(a)) and for 10044 MHz (Fig. 3(b)). These values were achieved for the longest optical fiber of 2.2 km. The phase noise characteristic demonstrates the comb of spectral lines at $\Delta f = 1/\tau$ corresponding to the neighboring modes of the oscillator.

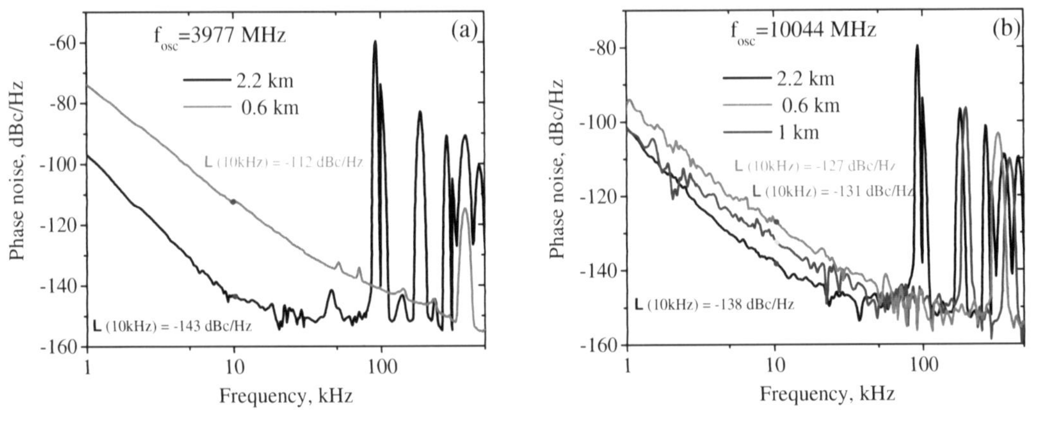

Figure 3. Phase noise spectra of the optoelectronic oscillator with different length of optical delay lines as indicated.

In conclusion, a low phase noise of -140 dBc/Hz at the frequency offset of 10 kHz was obtained in broad frequency range for microwave optoelectronic oscillator fabricated with spin-wave bandpass filter based on YIG film. The obvious improvement in phase noise values of the oscillator is achieved in comparison with [7]. Our results confirm also that the phase noise remains low in broad frequency range. The oscillator could find a wide variety of application in radar and telecommunication systems.

The work on the development of the experimental prototype oscillator was supported by Russian Science Foundation (project № 16-12-10440). The work on experimental investigation of the oscillator was supported by Ministry of Education and Science of Russian Federation (Project "Goszadanie").

References

[1] Salik E, Yu N, Maleki L 2007 *IEEE Photonics Technology Letters* **19** 444-446

[2] Belkin M E, Loparev A V, Semenova Y et al 2011 *Microwave and Optical Technology Letters* **53** 2474-2477
[3] Liang W, Eliyahu D, Ilchenko V S, Savchenkov A A, Matsko A B, Seidel D, Maleki L 2015 *Nature Communications* **6** 7957
[4] Mikitchuk K, Chizh A, Malyshev S 2016 *IEEE J. of Quant. Electronics* **52** 1-8
[5] Eliyahu D and Maleki L 2003 *Proc. of the 2003 IEEE Int. Frequency Control Symp. and PDA Exhibition Jointly with the 17th European Frequency and Time Forum* 405-410
[6] Zhu D, Pan S, Ben D 2012 *IEEE Photonics Technology Letters* **24** 194-196
[7] Ustinov A B, Nikitin A A, Kalinikos B A 2015 *IEEE Magnetics Letters* **6** 3500704.
[8] Stancil D D, Prabhakar A 2009 *Spin Waves: Theory and Applications* (Springer)

IOP Conf. Series: Journal of Physics: Conf. Series **1038** (2018) 012034 doi:10.1088/1742-6596/1038/1/012034

The control of electrophysical properties of GaAs pHEMT heterostructures

G Yakovlev[1], M Mironova[2], V Zubkov[1] and A Dudin[2]
[1]Department of micro- and nanoelectronics, St. Petersburg State Electrotechnical University "LETI", Prof. Popov str. 5, St. Petersburg, 197376, Russia
[2]JSC "Svetlana-Rost", Engels str. 27, St. Petersburg, 194156, Russia

E-mail: gy@etu.ru

Abstract. AlGaAs/InGaAs/GaAs pHEMT heterostructures were investigated by means of electrochemical capacitance-voltage (ECV) and photoluminescence (PL) techniques coupled with a numerical simulation. A set of GaAs pHEMT structures was fabricated using various doping techniques: modulated and δ-doping. The results of the PL spectra simulation and ECV free charge carrier distributions for investigated structures are presented. The analysis and comparison of QW filling in case of different doping mechanism of donor layer was done. Based on obtained results, we suggest optimized heterostructures in order to obtain higher electron localization and occupation of the quantum states.

1. Introduction

HEMT transistors are the fastest active elements of microelectronics and nanoelectronics. The new record results associated with the improved technology and the use of new materials, such as GaN and InP, are published regularly [1]. The range of application areas for such devices is constantly expanding beyond broadband communication and special purpose devices [2]. Semiconductor heterostructures of modern HEMT devices are a set of epitaxial layers of heterogeneously doped materials of various composition. The improvement of such devices with the purpose of increasing operating frequencies and currents steadily leads to high requirements for reproduction accuracy of the sequence, composition, and doping level of the layers. Therefore, the right choice and optimization of diagnostic techniques for such structures along with numerical simulation is an important and relevant task.

In present work we have investigated a set of pHEMT samples of various design based on GaAs, using several complementary methods: Hall measurements, electrochemical capacitance-voltage profiling [3] and photoluminescence measurements at various temperatures. Basing on self-consistent solution of Schrodinger and Poisson equations we calculated PL spectra [4] for investigated heterostructures and compared them with the experimental data.

The main goal of our investigation was to explain any possible deviations of the experiment data from the theoretical one and to study the features of free charge carriers spatial distribution in GaAs pHEMT heterostructures.

2. Samples and experimental technique

In this work we investigated three types of pHEMT heterostructures grown by the MBE (figure 4). Samples №1 and №2 had different type of donor layer (emitter): modulated (bulk) and δ- one. The asymmetrical $Al_{0.22}Ga_{0.78}As/In_{0.22}Ga_{0.78}As/GaAs$ quantum well (QW) of these samples had 14 nm width and located at depth of 39 nm and 31.5 nm, respectively. In this case, sample №1 had a sufficiently thick donor layer. Sample №3 has two-side doping of 12 nm QW, which was located at a depth of 54.5 nm. Each donor layer had thin undoped $Al_{0.22}Ga_{0.78}As$ spacer layers on both sides.

GaAs cap layer
$Al_{0.7}Ga_{0.3}As$ stop layer
GaAs smoothing layer
$Al_{0.22}Ga_{0.78}As$ donor layer
$In_{0.22}Ga_{0.78}As$ QW
GaAs buffer
GaAs S/I substrate

Sample №1

GaAs cap layer
$Al_{0.7}Ga_{0.3}As$ stop layer
GaAs smoothing layer
δ-$Al_{0.22}Ga_{0.78}As$ donor layer
$In_{0.22}Ga_{0.78}As$ QW
GaAs buffer
GaAs S/I substrate

Sample №2

GaAs cap layer
$Al_{0.9}Ga_{0.1}As$ stop layer
GaAs smoothing layer
$Al_{0.22}Ga_{0.78}As$ donor layer
$In_{0.22}Ga_{0.78}As$ QW
$Al_{0.22}Ga_{0.78}As$ donor layer
GaAs buffer
GaAs S/I substrate

Sample №3

Figure 1. The sequence of layers in investigated samples.

The ECV measurements were performed at room temperature using ECVPro (Nanometrics). A 0.1M Tiron solution was chosen as an electrolyte [5]. The area of the electrolytic rectifying contact was 0.1 cm^2, etching current was maintained at level of 0.5 mA/cm^2. During ECV profiling the samples were etched gradually with 1 nm step. The ac test signal frequency was 300 Hz. In some of experiments, the Agilent E4980A LCR-meter was used [6].

Photoluminescence spectra were measured at 77 K and 300 K using conventional synchronous detection with a lock-in selective amplifier, monochromator and photodetector. A solid-state single-mode laser module with a continuous output power of 40 mW at wavelength of 532 nm was used as an excitation light source. The laser beam was modulated and focused on a sample surface into a 100 microns diameter spot. Obtained signal was amplified by a selective nanovoltmeter and detected with a processing unit.

3. The experimental and simulation results and discussion

In this part of the paper both the experimental ECV results of measured free charge carriers depth distributions and PL spectra simulation results are presented.

3.1. ECV results

The apparent free charge carrier depth distribution profile of sample №1 with modulated doping of QW was measured at several measurement points (figure 2). As could be seen from the figure, the obtained ECV data shows high reproducibility. There are two concentration peaks on the measured profile. The first peak located at a depth of 30 nm corresponds to the highly doped donor region, the second at a depth of 44 nm - response from the QW. The free charge carriers concentration of the donor layer is $1.5 \cdot 10^{18}$ cm^{-3}, the concentration in the QW is $2 \cdot 10^{18}$ cm^{-3}, which corresponds to a concentration of a 2DEG of $1.6 \cdot 10^{12}$ cm^{-2}. The results are well correlated with Hall measurements, the peaks position corresponds to the regions of charge carrier localization according to structure's specification. A slight mismatch (less than 5 nm) with respect to the position of quantum well associated peak is observed, it is explained by QW skew [7]. It is important to note that the peak corresponding to the donor layer in such structures could be registered only by the ECV technique. According to the literature, conventional CV gives only one (attributed to QW) peak at the concentration profile [8]. It is shown in figure 3. The presence of two peaks is explained by the smaller Schottky barrier in case of electrolytic rectifying contact [9].

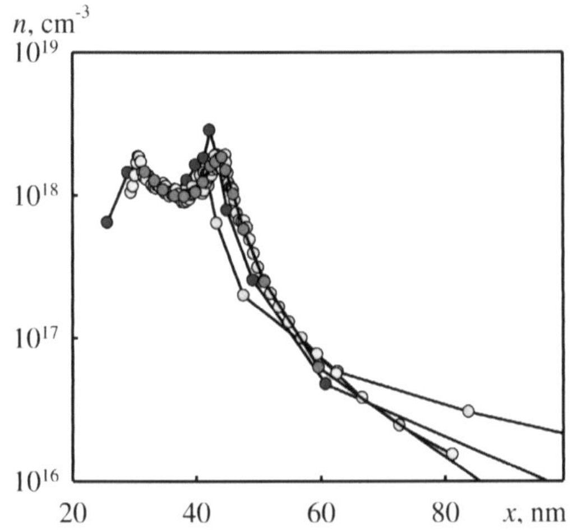

Figure 2. Apparent free charge carrier profile of sample №1 (modulated-doped GaAs pHEMT) measured in 4 different measurement points.

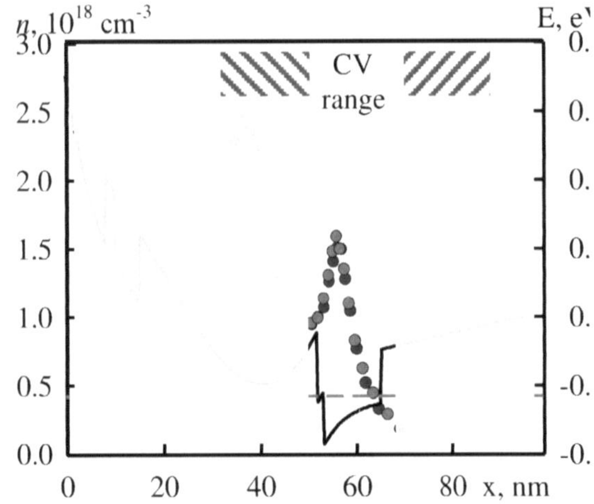

Figure 3. Available working range of conventional CV technique for GaAs pHEMT structures.

Figure 4 shows the depth distribution profile of free charge carriers in sample №2 with δ-doping, measured at several points. As above, there are two peaks in the concentration profile corresponding, in this case, to the δ-layer (29 nm) and the quantum well (37 nm). The peak concentration of free charge carriers in the δ-layer and the QW are $3 \cdot 10^{18}$ cm^{-3} and $2 \cdot 10^{18}$ cm^{-3} ($1.6 \cdot 10^{12}$ cm^{-2}), respectively. The observed concentration value after the QW response monotonically decreases according to the Debye law [10] to a semi-insulating substrate with a concentration of ~ 10^{10} cm^{-3}.

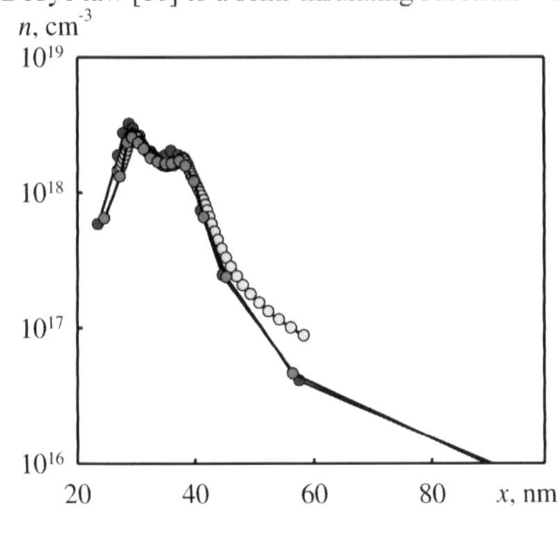

Figure 4. Apparent free charge carrier profile of sample №2 (δ-doped GaAs pHEMT) measured in 3 different measurement points.

Comparing the observed profiles of free charge carriers for samples №1 and №2, it could be seen that there is no obvious advantage in using a δ-doped emitter in terms of 2DEG concentration for the considered layer configuration and doping levels. Hall measurements of these samples also show the same concentration of 2DEG in a channel for both doping techniques. However, a smaller thickness of upper layers in the case of δ-layer structure design allows to obtain a greater value of the steep subthreshold slope characteristic, and therefore better controllability for the final device at the same values of operating current and power. This also helps to reduce the cutoff bias.

We investigated and analyzed the regularity of changing of the concentration peak associated with the quantum well of sample №3. The change in the amplitude is caused by a change of a QW filling degree due to the top layers etching. It is shown on figure 5. This feature is very useful on practice, since it could be used for controlled thinning of the HEMT gate region in order to saturate the QW with charge carriers, increase the channel conductivity and device operating current. For investigated HEMT structure design, the optimal value of thinning is 16 nm corresponding to the boundary of the smoothing and barrier (donor) layers. The estimated by ECV 2DEG density in QW is $2 \cdot 10^{12}$ cm^{-2}, according to Hall measurements this value was $2.14 \cdot 10^{12}$ cm^{-2}. Note that ECV etching is carried out only in a small region of 3 mm diameter, thereby simulating conditions of a real thinning technology of gate region.

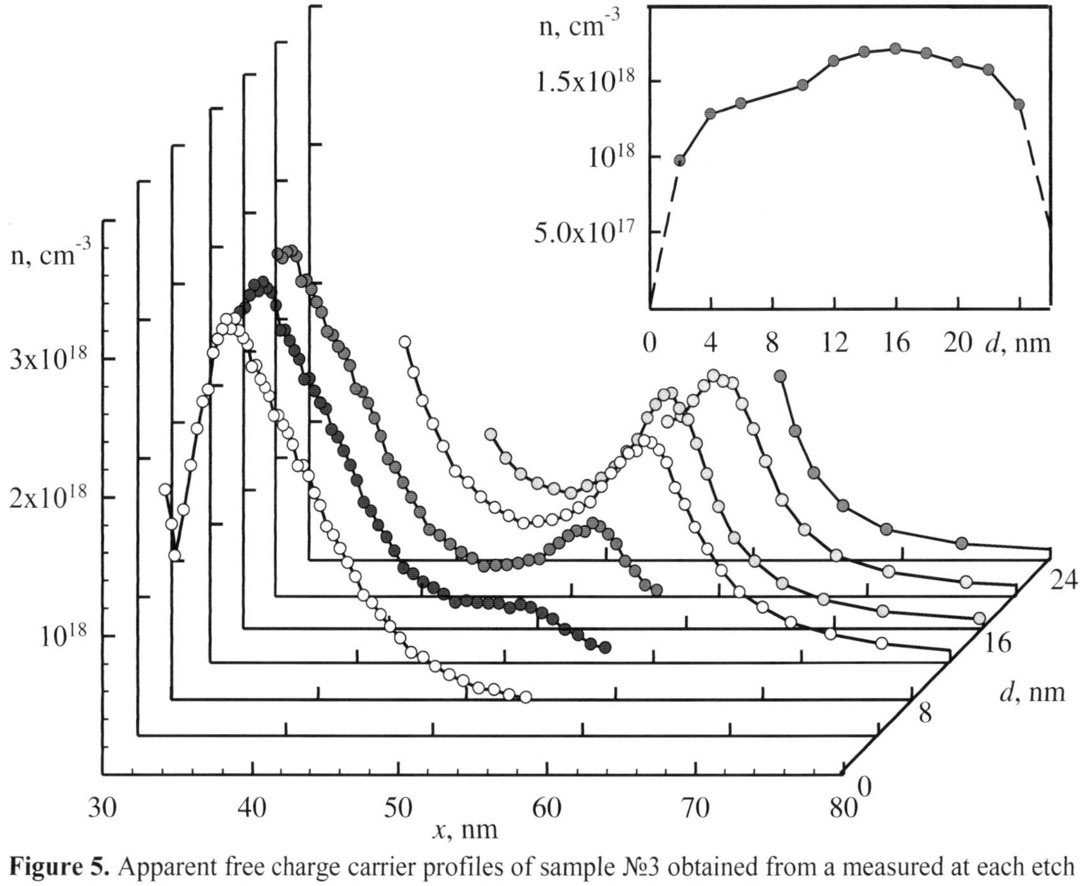

Figure 5. Apparent free charge carrier profiles of sample №3 obtained from a measured at each etch step CV plot. Inserted graph: dependence of QW filling from the etching depth.

3.2. PL simulation results

GaAs has direct band gaps corresponding to the near-infrared region of the spectrum. This leads to highly informative optical measurements (primarily, photoluminescence spectra) of such pHEMT heterostructures, which can be effectively used for structure characterization along with electrical methods. PL is non-destructive and rapid technique of semiconductor structures analysis, widely used to determine grown layers composition, for evaluation heterostructure quality and many other parameters [11, 12].

For detailed analysis of the PL spectra of investigated structures we performed a numerical self-consistent solution of Schrodinger and Poisson equations. Stresses due to the structure pseudomorphic growth were taken into account using the deformation potentials theory and data introduced in [13].

Basing on the suggested theoretical model we calculated PL spectra for the heterostructures and compared them with the experimental data

We assumed parabolic dispersion laws both in conduction and valence bands determined by electron and heavy hole effective masses:

$$E_n(\mathbf{k}) = E_n + \frac{\hbar^2(k_x^2 + k_y^2)}{2m_e^*}, \quad (1)$$

$$E_m(\mathbf{k}) = E_m - \frac{\hbar^2(k_x^2 + k_y^2)}{2m_h^*} \quad (2)$$

where E - energy, $\mathbf{k}$ – wave vector, $\hbar$ – Plank constant, m – effective mass.

In our theoretical model we have such expression for PL intensity:

$$I(\hbar\omega) \sim \sum_{m,n,\mathbf{k},\mathbf{k}'} \left|M_{mn}(\mathbf{k},\mathbf{k}')\right|^2 \times \delta\left(\hbar\omega - \left(E_n(\mathbf{k}') - E_m(\mathbf{k})\right)\right) \times \left(1 - f\left(E_m(\mathbf{k}), F_P\right)\right) \times f\left(E_n(\mathbf{k}'), F_N\right) \quad (3)$$

where I – PL intensity, M – transition matrix element, f – probability density function, F – quasi Fermi level.

Matrix element is:

$$M_{mn}(\mathbf{k},\mathbf{k}') = \frac{1}{2}\left\langle m,\mathbf{k}\middle|\hat{H}_{e\omega}\middle|n,\mathbf{k}'\right\rangle, \quad (4)$$

$$\left|M_{mn}(\mathbf{k},\mathbf{k}')\right|^2 = \left|M_{mn}\right|^2 \delta_{\mathbf{kk}'} \quad (5)$$

where $\hat{H}$ – Hamiltonian operator, δ – Kronecker delta-function.

Probability density function is:

$$f\left(E_m(\mathbf{k}), F_P\right) = \frac{1}{1 + \exp\left(\frac{E_m(\mathbf{k}) - F_P}{kT}\right)} \quad (6)$$

Taking into account (1), (2) and (4) - (5), the expression (3) for calculating the PL intensity may also be formatted as:

$$I(\hbar\omega) \sim \sum_{m,n} \left|M_{mn}\right|^2 \Theta(\hbar\omega - (E_n - E_m)) \times \quad (7)$$

$$\times \frac{1}{1 + \exp\left(-\frac{E_m - \frac{m_e}{m_e + m_h}(\hbar\omega - E_n + E_m) - F_P}{kT}\right)} \times \frac{1}{1 + \exp\left(\frac{E_n - \frac{m_h}{m_e + m_h}(\hbar\omega - E_n + E_m) - F_N}{kT}\right)}$$

where Θ - Heaviside step function:

$$\Theta(\hbar\omega - (E_n - E_m)) = \frac{1}{1 + \exp\left(\frac{\hbar\omega - (E_n - E_m)}{\Gamma}\right)} \quad (8)$$

In our model we used following adjustable parameters: Γ – spreading parameter, F_P – holes quasi Fermi level, ΔE_V – valence band offset. Conduction band offset was $\Delta E_C = 0.85x - 0.3x^2$.

The experimental and simulated PL spectra of structures similar to sample №1 is shown on figure 6. We observe two PL peaks with different intensity in the 1000 nm region. These peaks are formed due to recombination of electrons from the first and the second energy subbands in conduction band and holes in valence bands. It could be seen that the simulation results are in a good agreement with the experimental data. The difference in PL spectra between these structures is explained by the different growth temperature of the QW layer. With a rise of growth temperature, increases segregation and diffusion of indium atoms, which leads to enlarge of the interface spreading. The reduction in QW width due to the indium segregation is accompanied by the long-wavelength peak predominance, which can be explained by a weak filling of the second energy subband in the relatively narrow QW. At the same time, the peak separation is less dependent on the growth parameters and stays about 40 nm.

Figure 7 shows experimental and simulated PL spectra (at various temperatures) of pHEMT structure with higher doping level of donor layer. With decrease of the temperature the spectrum thermal spreading also decreases, so we could observe two peaks in PL spectra. If we will compare PL spectra, shown in figure 6 and figure 7, we could conclude that the form of PL spectra is in a strong dependence of donor doping level.

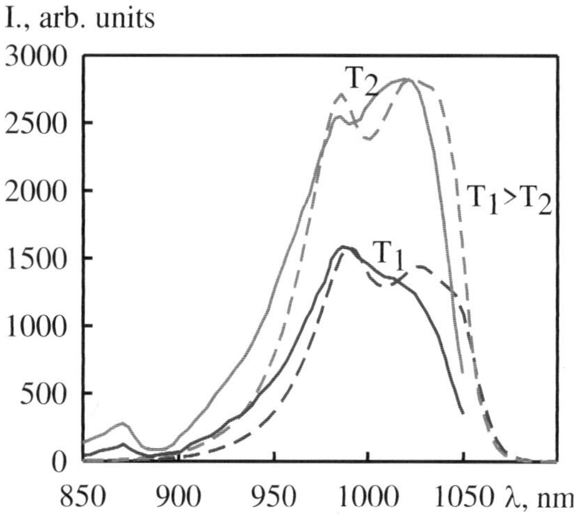

Figure 6. The experimental (300K) and simulated PL spectra of GaAs pHEMT structures with a different QW growth temperature.

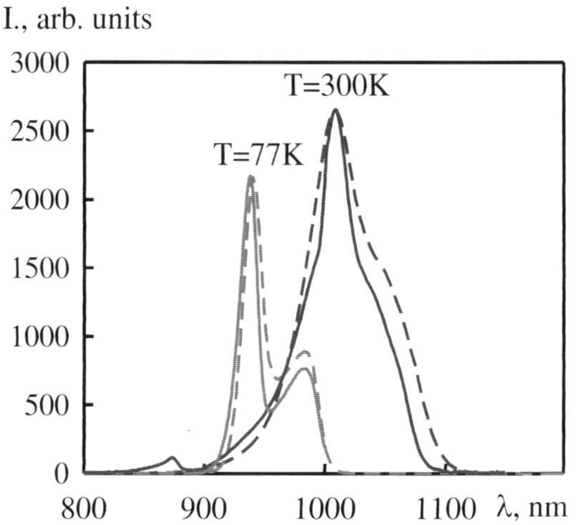

Figure 7. The experimental and simulated PL spectra (at various temperatures) of GaAs pHEMT structures with a highly-doped donor layer.

4. Conclusion

In this work a set of GaAs pHEMT heterostructures was investigated by means of ECV and PL techniques coupled with a numerical simulation. The apparent free charge carriers depth distribution profiles are obtained, the position of the donor layer and 2DEG concentration in the QW are estimated. The obtained results show high repeatability and well correlated with Hall measurements. The comparison of the apparent free charge carriers profiles of GaAs pHEMT structures with different types of donor layer doping (modulated and δ-) showed identical QW filling. Despite the apparent absence of δ-doping advantages, less thickness of top layers, allows to obtain a greater value of the steep subthreshold slope characteristic, and therefore ability of better control of the final device with the same operating current and power. This also helps reduce the cutoff bias. It was shown that by using the ECV technique one could optimize the etched thickness of a cap layers to obtain the best concentration of free charge carriers in QW of a pHEMT device and to enlarge it's operating current. The comparison of experimental and simulated PL spectra established the dependencies of PL peaks shape, position and intensity on the structure properties, primarily donor doping level. These dependencies may ideally point out the deviation of grown structure from the standard one.

Acknowledgments

The research was supported by Saint Petersburg Electrotechnical University "LETI" (project M.5.1.2).

IOP Conf. Series: Journal of Physics: Conf. Series **1038** (2018) 012034 doi:10.1088/1742-6596/1038/1/012034

References

[1] Aadit M, Kirtania S, Afrin F, Alam K and Khosru Q 2017 *Different Types of Field-Effect Transistors – Theory and Applications* (Rijeka: InTech) ed M M Pejovic and M M Pejovic chapter 3 pp 45-64

[2] Xiao-Guang H, De-Gang Z and De-Sheng J 2015 *Chin. Phys. B* **24** 067301

[3] Dorokhin M, Zaitsev S, Rykov A, Zdoroveishchev A, Malysheva E, Danilov Yu, Zubkov V, Frolov D, Yakovlev G and Kudrin A 2017 *Technical Physics* **62** 1545

[4] Dudin A, Kogan I, Yakovlev G, Mironova M, Schukov I, Frolov D, Zubkov V and Glinskii G 2017 *Proc. of 2017 11th Int. Workshop on the Electromagnetic Compatibility of Integrated Circuits (EMCCompo)* (St. Petersburg: IEEE) p 108

[5] Shestakova L, Yakovlev G and Zubkov V 2017 *J. Phys. Conf. Ser.* **816** 012022

[6] Frolov D and Zubkov V 2016 *Semicond. Sci. Technol.* **31** 125013

[7] Vasilevskii I, Galiev G, Klimov E, Mokerov V, Shirokov S, Imamov R and Subbotin I 2008 *Semiconductors* **42** 1084

[8] Tarasova E, Obolenskaya E, Hananova A, Obolensky S, Zemliakov V, Egorkin V, Nezhenzev A, Saharov A, Zazul'nokov A, Lundin V, Zavarin E and Medvedev G 2016 *Semiconductors* **50** 1599

[9] Dudin A, Mironova M, Yakovlev G, Frolov D, Kogan I, Schukov I, Zubkov V and Glinskii G 2017 *Applied Physics* **3** 78

[10] Yakovlev G, Zubkov V, Frolov D, Zubkova A, Sterlyadkin O and Sorokin S 2016 *Semiconductors* **50** 320

[11] Khabibullin R, Vasil'evskii I, Galiev G, Klimov E, Ponomarev D, Gladkov V, Kulbachinskii V, Klochkov A and Uzeeva N 2011 *Semiconductors* **45** 657

[12] Galiev G, Vasil'evskii I, Klimov E, Klochkov A, Lavruhin D, Pushkarev S and Maltsev P 2015 *Semiconductors* **49** 234

[13] Zubkov V, Melnik M, Solomonov A, Tsvelev E, Bugge F, Weyers M and Tränkle G 2004 *Phys. Rev. B* **70** 075312

IOP Conf. Series: Journal of Physics: Conf. Series **1038** (2018) 012035 doi:10.1088/1742-6596/1038/1/012035

Features of use direct and external modulation in fiber optical simulators of a false target for testing radar station

Tarasenko M.Yu.[1], Lenets V.A.[1], Malanin K.Yu.[2], Akulich N.V.[2], Davydov V.V.[1]

[1] Peter the Great Saint-Petersburg Polytechnic University, Polytechnical st. 29, 195251, St. Petersburg, Russia

[2] AO "Zaslon", Koli Tomchaka st. 9 H, 196084, St. Petersburg, Russia

margaritai.tarasenko@gmail.com

Abstract. The purpose of this paper was to develop and research the fiber optical simulators of false target with direct and external modulation for testing radar station for the X-range. The modes of testing radar station by simulator depending on external conditions were studied. Temporal and spectral responses have been received. The optimum modes of simulators depending on external conditions have been defined.

1. Introduction

One of the most perspective directions of radio photonics is transmission of analog signals of very high frequency on communication fiber optical systems (FOS). Numerous experiments, and also operating experience showed that analog FOS can effectively be used in different paths of the multielement active phased antenna grids (APAG), switching devices of radar stations, etc. [1]. Use of analog fiber optical systems allows solving the most difficult problems connected to support of reliable operation of the send-receive antenna systems placed on moving platforms of the upper masts of the ships, different airplanes, etc. [2, 3]. But first developers need to set up and check entire antenna system for working capacity, both in the conditions of the laboratory or a polygon, and on the place of its forthcoming dislocation. And in the last two cases the flight purposes for different reasons can be absent.

The solution of this problem is the use of false target simulators, but as the most part of signals is transmitted over the fiber-optic communication line, one should use the reliable fiber optical simulator. Fiber optical simulator has exceptional feature in comparison with other models of simulators working at other physical principles. Delay factor of an analog signal essentially doesn't depend on its frequency. It allows creating false targets even when using broadband sounding pulses. Frequency restrictions on a very high frequency of a signal are only connected to technical features of electrooptical modulators and photoreceiving modules [4]. Remaining advantages of fiber optical simulators: immunity to electromagnetic interferences, low losses (about 0,2 dB/km for external modulation and about 0.35 dB/km for direct modulation in the single-mode fiber), complete galvanic isolation, mechanical flexibility, and also small mass-dimensional characteristics compared with other simulators are well studied.

The most difficult solvable task of fiber optical simulators is determination of the optimum modes of their operation when testing the radar station. It is related to different features used in direct or external modulation in the transmitting modulex [5, 6].

We examined fiber optical simulators of false target with direct and external modulation. Prototypes for this purpose were made and the following characteristics were received: oscillograms of the reflected pulses in case of different duty cycles used in radar station; amplitude-frequency characteristics, spectrum of two types of modulators and response characteristics. Research has been done for two types of FOS: the nature of losses in frequency range from 2 to 18 GHz, response and spectral characteristics. The research looked into very high frequency signals of different forms which are used in radar stations. Research was conducted at different temperatures in the conditions of light modulation phase displacement that corresponds to real operating conditions of FOS on a polygon, etc.

The response characteristic of simulators at different temperatures formed out to be the most interesting as the purpose was to reveal an optimum operation mode depending on external conditions of radar station. Those experimental installations are shown in fig. 1, where the developed samples of simulators are placed into a heat chamber.

2. Experimental methods

The fiber optical simulator of false target is based on the optical fiber coil with receiving and transmitting module. The main difference between two types of simulators is in the transmitting module: the laser with wavelength of 1310 nanometers is used for direct modulation and modulation is performed by a pumping current, for external modulation – a 1550 nm wavelength is used and the external modulator [7]. The difference between two types of simulators [8-10] can be seen in fig. 1 (a, b), where the experimental installations of FOS with two types of modulation are provided.

Based on the features of the tested radar station, FOS's frequency range is 2 to 18 GHz, provided the required signal power and system performance.

a)

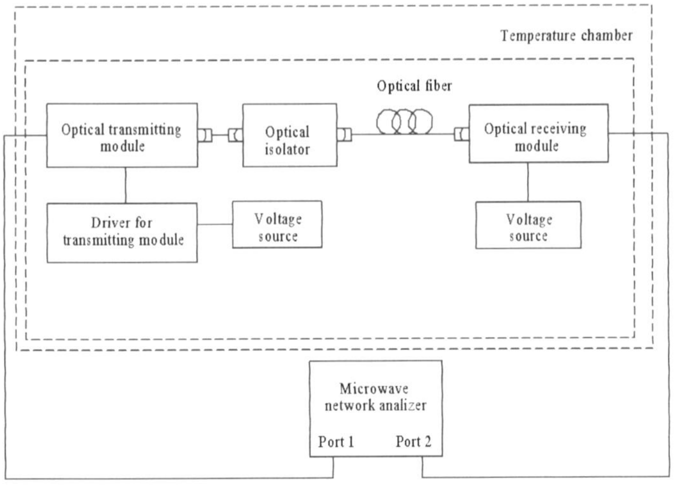

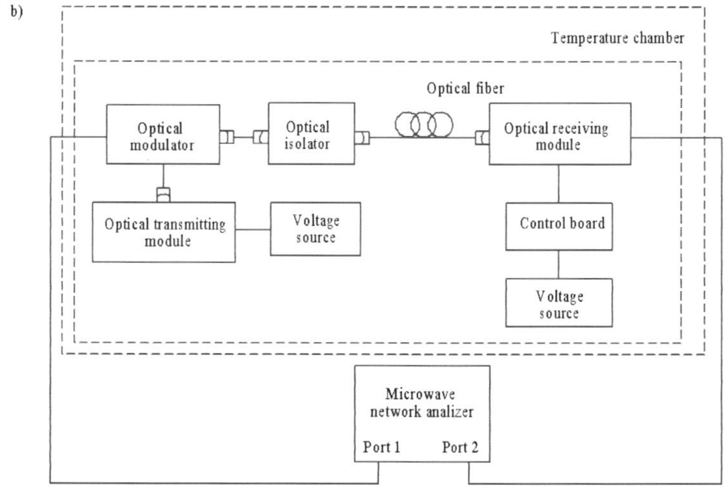

Fig. 1. Installation for measurement of parameters FODL with internal (a) and external (b) modulation.

The received results of measurements FOS's parameters have shown that this scheme of experimental installations (fig. 1 a, b) is the most rational from side of carrying out researches.

3. Experimental results and discussion

The conducted researches showed that operation of simulators of false targets both types significantly doesn't change in case of different forms of signals and different duty cycles. So, time and spectral characteristics of both simulators aren't perceptible.

Frequency response of FOS with direct and external modulation at three temperatures are shown in fig. 2 (a, b).

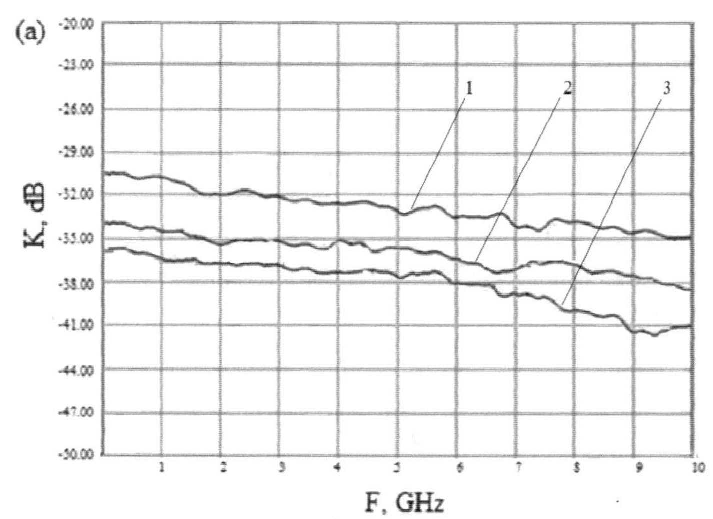

(b)

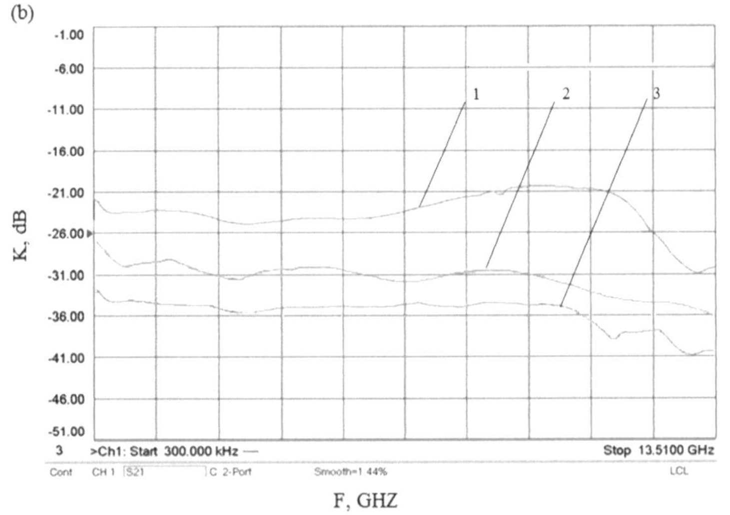

Fig. 2. Frequency response of FOS with direct (a) and external (b) modulation. Lines 1, 2, 3 corresponds to temperatures of 253 K, 293 K and 313 K.

The analysis of received frequency response for various ways of modulation shows that with increase in temperature when using external modulation the bandwidth is reduced by transfer analog by the signal microwave compared with direct modulation. It once again proves that it is necessary to use the system of thermostabilization allowing to provide a working point of the transfer characteristic at the segment with the maximum slope for the external modulator.

The response characteristic of simulators is of the greatest interest. Dependences of power output from input for three temperatures (253 K, 293 K and 313 K) are shown in fig. 3. In view of features of a measuring equipment and assembled simulators it was succeeded to take characteristics in range from -20 dBm to 10 for direct modulation and from -20 dBm to 18 dBm for external modulation.

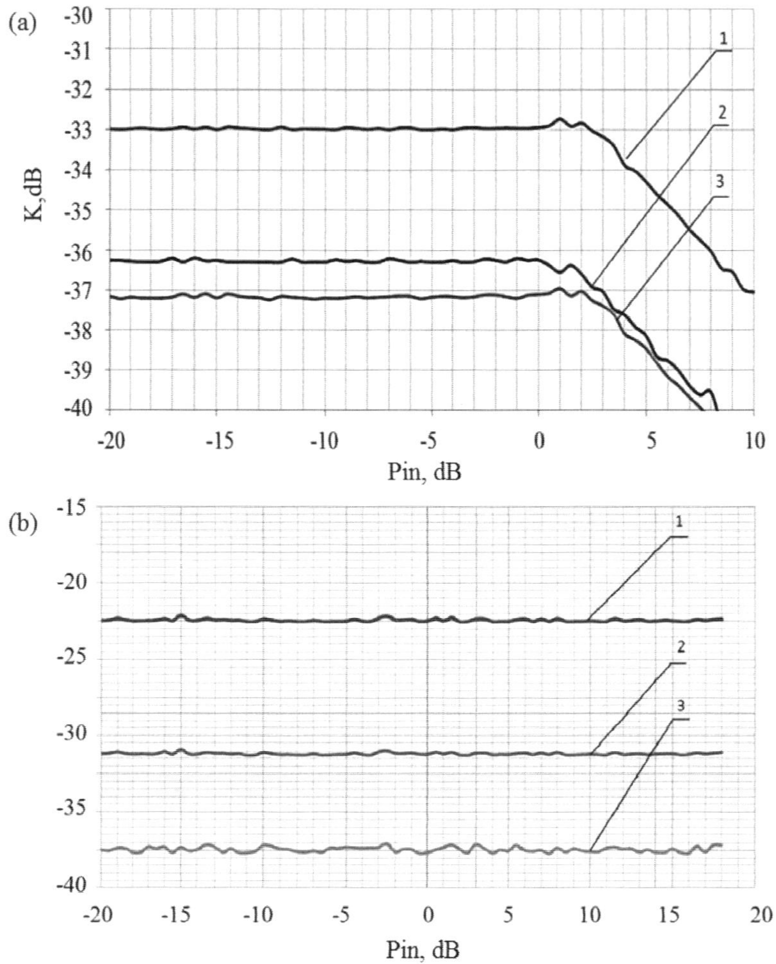

Fig. 3. Dynamic response of FOS with direct (a) and external (b) modulation at a frequency of 9 GHz. Lines 1, 2, 3 corresponds to temperatures of 253 K, 293 K and 313 K.

The analysis of the obtained experimental results allow to draw the following conclusions. The mode of external modulation is more sensitive to change of temperature that leads to increase in dynamic transmission losses of a very high frequency signal (violation of the mode of thermostabilizing on 283 K can lead to errors). In case of transmission rate more than 5 Gbit\s the mode of direct modulation enters distortions to the transmitted very high frequency signal, and also losses sharply increase, the linear range in a response characteristic is almost absent. This fact can be explained finite lifetime of carriers of charges and photons in the active layer of the laser diode. It was set that observed distortions of the form of pulses of a very high frequency of signals in case of direct modulation are connected to chromatic dispersion of the spurious frequency shift keying arising in case of distribution of radiation on an optical fiber because of existence.

4. Conclusions

The defined features are necessary to consider when testing radar station with signals of very high frequency and small on-off time ratio. In case of small delay periods of optical signal τ optical compensators of chromatic dispersion are necessary. In case of big τ compensating of dispersion needs to be carried out by electronic methods.

The received results allowed us to develop an algorithm of choice of an optimum operation mode of FOS taking into account features of the modes of modulation and to justify need for extension of its operation capabilities to use in its construction the transferring modules with two types of modulation.

Nowadays the range of applied frequencies tends to extend up to tens of gigahertz in the modern APAG. Therefore it is necessary to use more wide-band simulators. Our research showed that it is possible. Application of the corresponding receiving and transferring modules (it is more preferable with external modulation) is necessary for operation at frequencies up to 40 GHz that will allow to apply the universal fiber optical systems to APAG successfully.

References

[1] Davydov V V, Sharova N V, Fedorova E V, Gilshteyn E P, Malanin K Y, Fedotov I V, Vologdin V A, Karseev A Yu 2015 *Lecture Notes in Computer Science* **9247** 712
[2] Velichko M A, Naniy O E, Susyan A A 2005 *Light wave Russian Edition* **4** 21
[3] Ermolaev A N, Krishpents G P, Davydov V V, Vysoczkiy M G 2016 *Journal of Physics: Conference Series*. **741(1)** 012171
[4] Davydov V V, Ermak S V, Karseev A Yu, Nepomnyashchay E K, Petrov A A, Velichko E N 2014 *Lecture Notes in Computer Science* **8638** 694
[5] Friman R L 2012 *Fiber–Optic Systems for Communications* (Hoboken, New Jersey: Wiley) p 496
[6] Davydov V V, Dudkin V I, Karseev A Yu 2014 *Optical Memory & Neural Networks (Information Optics)* **23(3)** 170
[7] Marpaung D, Roeloffzen C, Heideman R, Leinse A, Sales S, Capmany J 2013 *Integrated Microwave photonics, Laser & Photonics Reviews* **7(4)** 506
[8] Davydov R V, Saveliev I V, Lenets V A, Tarasenko M Yu, Yalunina T R, Davydov V V, Rud' V Yu 2017 *Lecture Notes in Computer Science (including subseries Lecture Notes in Artificial Intelligence and Lecture Notes in Bioinformatics)* **10531 LNCS** 177
[9] Davydov V V, Dudkin V I, Velichko E N, Karseev A Yu 2015 *Journal of Optical Technology (A Translation of Opticheskii Zhurnal)* **82(3)** 132
[10] Davydov V V, Dudkin V I, Karseev A Yu 2014 *Optical Memory & Neural Networks (Information Optics)* **23(4)** 259

Investigation of the temperature effect on the dynamic parameters of ultrafast silicon carbide current switches

A A Smirnov, S A Shevchenko, B V Ivanov, V A Ilyin, A V Afanasyev

Saint-Petersburg Electrotechnical University "LETI", Saint-Petersburg 197376, Russia
E-mail: smirnov.eltech@gmail.com, sashevchenko.rte@gmail.com, BVIvanov@etu.ru, ilyincmid@gmail.com, a_afanasyev@mail.ru

Abstract. The results of a theoretical and experimental study of temperature effect on a switching process of a 4H-SiC drift step recovery diode are presented. The effect of the injected charge losses lowering at high temperatures is demonstrated.

1. Introduction

Drift step recovery diodes (DSRDs) are the most popular high power superfast semiconductor switches that may be implemented in nano- and subnanosecond pulse generators amongst others. The DSRDs usage is very common due to their reliability, high switching velocities, relatively simple manufacturing process and the possibility of combining the diodes into high voltage stacks [1]. The main principle underlying a typical DSRD operation cycle is based on the superfast recovery effect that occurs in a *p-n* junction after injection of non-equilibrium charge carries into the base region of the structure with a short forward current pulse and their further removal with reverse current. It is widely known that silicon-based DSRDs are capable of switching voltages of kilovolt orders at 1 V/ps switching velocity.

Yet in [2, 3] it was shown experimentally that implementation of silicon carbide (4H-SiC) in DSRDs fabrication leads to a major increase of switching velocities up to 3-4 V/ps. Apart from gaining high switching velocities 4H-SiC DSRDs are able to operate at high repetition frequencies up to dozens of megahertz, which results in heavy heating of the diode structure. One of the main advantages of 4H-SiC is its high thermal conductivity, which determines the upper temperature limit at which the diode's operation is stable, thus 4H-SiC DSRDs have huge perspectives of their implementation in high voltage pulse generators with subnanosecond rise times working at high pulse repetition frequencies.

However, at the moment both the maximum temperatures the diodes can operate at, and the temperature affection on the parameters of the voltage pulses shaped have not been studied. In addition, despite 4H-SiC provide the major switching velocity gain, one of the general issues the 4H-SiC DSRDs suffer is a vast loss of injected charge [4, 5] which may constrain the maximum switching velocity and decrease the energy efficiency of the device. Also, being a distinctive effect for 4H-SiC, incomplete dopants ionization may lead to severe losses of charge at room temperature. All the conditions listed above determine the relevance of our study of the temperature effect on the dynamic parameters of the superfast 4H-SiC current switches and, in particular, on the injected charge losses.

2. Instrumentation

The diodes under examination have p^+-p-n^+ structure with the thickness of the epilayers, respectively: 2, 18 and 350 um. Aluminum and nitrogen were used as dopants with concentrations in epilayers respectively: $2\cdot10^{19}$, $5\cdot10^{15}$ and $5\cdot10^{18}$ cm^{-3}.

The theoretical study was carried out by simulation in Synopsys Sentaurus TCAD software with the following effects accounted for: bandgap narrowing at high doping levels, Shockley-Read-Hall and Auger recombination, impact ionization at high fields, incomplete dopants ionization and several mobility models including carriers scattering on ionized impurities and carrier velocity saturation at high fields.

The 4H-SiC DSRDs' operation has been studied in a proprietary stand, which consisted of an ultrashort pulse generator with inductive energy accumulation with the diodes implemented as opening switches, a set of microwave attenuators (Aeroflex-Weinschel and Barth Electronics) and a stroboscopic oscilloscope Tektronix DAS8300 with a 20 GHz band.

The heating of the diode structure was carried out by setting a high pulse repetition rate whilst the cooling was executed by using non-flammable circuit freeze spray. The temperature of the diode was observed by the means of non-contact measurements of injection electroluminescence spectra technique during the diode's operation [6]. A mobile optical spectrometer [7] was used in order to define the temperature of the structure by spectral maximum wavelength detection.

3. Results and discussion

The diode's operation had been studied by its recovery process simulation in a circuit containing a voltage source, the diode under examination and a 50Ω load connected in series. For 70 nanoseconds the diode is biased forward with a linearly rising voltage pulse up to 1800 V, which is accompanied by injection of non-equilibrium carriers into the base region of the structure. After that, the voltage at the source changes its polarity and remains constant while the voltage drop at the diode is low as the injected carriers extraction takes place. By the end of non-equilibrium carriers removal the extraction of majority carriers from base with saturated velocity occurs simultaneously with a fast voltage rise at the diode, which corresponds to its switching. The results of simulation of the diode's recovery processes at different temperatures are presented in Figure 1.

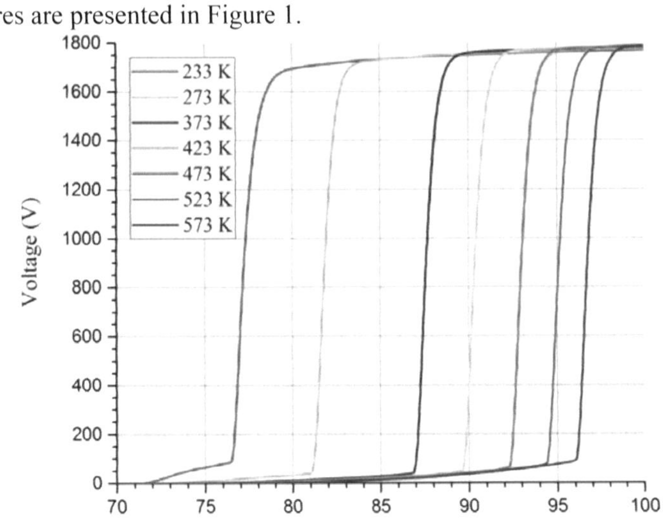

Figure 1. The DSRD under examination switching processes at various temperatures

If we define the loss of charge as the relation Q_-/Q_+ between the extracted non-equilibrium charge Q_- and injected charge Q_+, which are defined by the areas under current transient response at different stages of the diode's recovery process [5]. One can clearly observe that the temperature rise leads to better charge accumulation in the structure. The percentage of ionized impurities in p^+- and n^+-emitters increases with temperature from 3% at 233 K to 27% at 573 K, which outcomes in a significant growth of p^+-p and p- n^+ junctions injection ratios. Hence the higher the temperature is, the more charge may be injected into the base region of the structure and the longer it takes to extract it, which is approved by the shapes of non-equilibrium electron-hole plasma distributions at different temperatures at the end of the injection stage (Figure 2). The charge losses estimation is presented in Table 1.

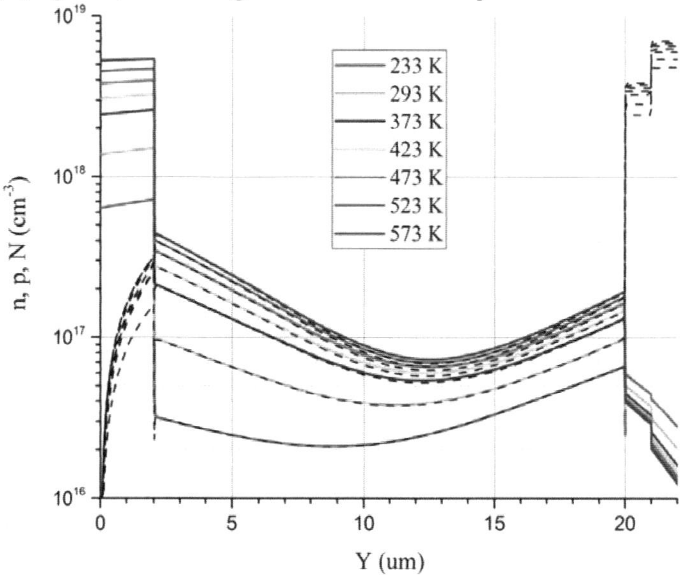

Figure 2. Distributions of holes (coloured lines) and electrons (dashed lines) at the end of the forward current pulse at various temperatures

Table 1. Injected charge losses estimation at various operating temperatures

T, K	233	293	373	423	473	523	573
T_D, ns	6,48	11	16,85	19,6	22,3	24,4	26
Q_-, nC	233	396	606	705	802	879	936
Q_-/Q_+, %	19	31	48	56	64	70	74

The diode's performance had been observed in the ultrashort pulse generator circuit (Figure 3). The operation principle of the circuit is described in detail in [8]. A power MOSFET was implemented as the S1 switch whereas the DSRD D1 under examination was used as the opening switch. It is also worth noting that during the experiment the MOSFET was set to a heat sink in order to exclude its overheating, so the S1 voltage pulse shape was stable.

During the diode's operation in pulse generators it is essential to implement minimal forward current pulses durations of in order to diminish the charge losses. In case of the following study the forward current pulse duration was defined by the characteristics of the circuit and set up to 70 nanoseconds. The values of capacitances and inductances in the circuit were chosen to guarantee non-equilibrium carriers' extraction to end at the same moment the diode current reaches its maximum value, the latter process being accompanied by superfast switching of the diode.

In the Figure 4 the comparison between the results of theoretical and experimental study is given. The time dependence U_{S1} represents the voltage pulse shaped at the MOSFET S1. The results of the simulations and the experimentally obtained voltage pulses are shown in Figure 4 as well. The difference between the amplitude of the voltage pulse obtained in the experiment (1400 V) and the one the examined DSRD may provide (1800 V) is determined by the parameters of the attenuators array implemented as the 50Ω load: the first attenuator in the array was the high power Aeroflex-Weinschel (model 49-40-34) with 40 dB attenuation and 5 kW maximum input peak power. With the voltage pulse amplitude 1400 V the peak power the 50Ω load is to dissipate is 39 kW, which exceeds the maximum allowable power by nearly 8 times, so in order to prevent damaging the attenuator the voltage pulse amplitude was not raised higher than 1400 V.

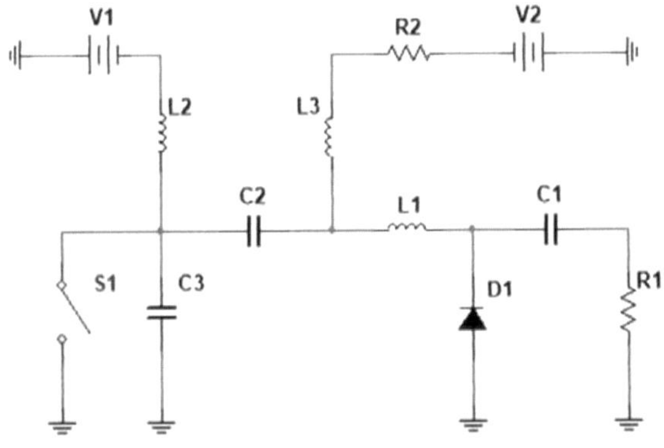

Figure 3. The ultrashort pulse generator circuit the diode was examined in

As one can observe from Figure 4, the rise of temperature affects the diode's operation in the same way it was demonstrated in simulation (Figure 1): in case of high temperatures the higher it is, the better the injected charge is stored, which results in the prolonged period of carriers' extraction from the base region; in case of low temperature the charge accumulation is the worst. Therefore, the moment the superfast switching should occur is offset in case of any other temperature than 293 K concerning the optimum when the diode current reaches its peak value. The amplitude of the generated voltage pulse decreases from 1400 V at room temperature to 650 V at 573 K and the mean switching velocity lowers from 2 to 0.7 V/ps.

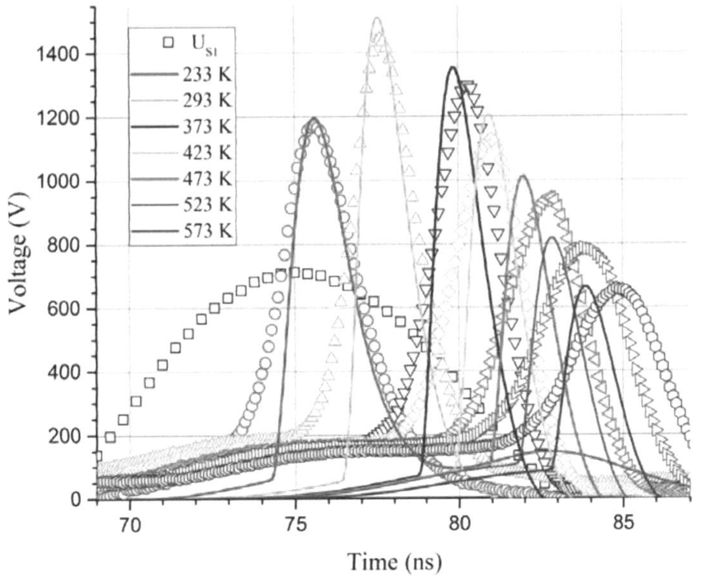

Figure 4. The results of simulations (solid lines) and experimentally observed voltage pulses (markers) at different temperatures

IOP Conf. Series: Journal of Physics: Conf. Series **1038** (2018) 012036 doi:10.1088/1742-6596/1038/1/012036

It is also worth noting that during the experiment all the parameters of the circuit were constant, but it is possible to achieve the optimal moment of time for the diode to switch at various temperatures by adjusting the circuit parameters. Thus, for every duration of the charge extraction process it is essential to tune the circuit.

4. Conclusion

The analysis given in the following paper establishes the fact that the vast loss of injected charge usually observed in 4H-SiC drift step recovery diodes strongly depend on the percentage of ionized doping. The results of the simulation demonstrated that at high temperatures the injection of non-equilibrium carriers occurs more effectively, which outcomes in better charge accumulation in the diode structure. The results of the experiments approve both the adequacy of the model and the theoretical statements concerning temperature affection on charge losses in 4H-SiC DSRDs.

References

[1] Grekhov I V, Mesyats G A, Nanosecond semiconductor diodes for pulsed power switching, Physics-Uspekhi (Advances in Physical Sciences), vol. 48, no. 7, pp. 703-712, 2005

[2] Afanasyev A V, Ivanov B V, Ilyin V A, Kardo-Sysoev A F, Kuznetsova M A, Luchinin V V. Superfast drift step recovery diodes (DSRDs) and vacuum field emission diodes based on 4H-SiC, Materials Science Forum, Vols. 740-742, pp. 1010-1013, 2013

[3] Afanasyev A V, Ivanov B V Smirnov A A, et. al. High Voltage Subnanosecond Silicon Carbide Opening Switch, Proc. 57th International Scientific Conference on Power and Electrical Engineering (RTUCON), pp. 51-52, 2016

[4] Kyuregyan A S. Theory of Drift Step Recovery Diodes. Solid State Electronics, 2004, Vol. 49, №6, pp. 720–727

[5] Ivanov B V, Smirnov A A, Shevchenko S A. A Study of Charge Losses in 4H-SiC Drift Step Recovery Diodes (DSRD), Proc. of EIConRusNW, pp. 51-52, 2016

[6] Afanasyev A V, Ivanov B V, Ilyin V A, Smirnov A A, Kardo-Sysoev A F, Temperature dependence of silicon carbide drift step recovery diodes injection electroluminescence, Journal of Physics: Conference Series, Vol. 741, Number 1 (012175), 2016

[7] Kostrin D K, Ukhov A A. Hardware-software complex for spectrometric investigation of parameters of light-emitting diodes. Biotechnosphera. 2013. Vol. 3. pp 21-25

[8] Afanasyev A V, Demin Y A, Ivanov B V, Ilyin V A, Luchinin V V, Sergushichev K A, Smirnov A A, Kardo-Sysoev A F. Megawatt generator of nanosecond pulses based on silicon carbide drift step recovery diode. Izvestiya Vysshyh uchebnih zavedeniy Rossii. Radioelectronica . 2015. Vol. 3. PP. 21-24

IOP Conf. Series: Journal of Physics: Conf. Series **1038** (2018) 012037 doi:10.1088/1742-6596/1038/1/012037

New method for testing of antenna phased array in X frequency range.

V A Lenets[1, 2], M Yu Tarasenko[1, 2], V V Davydov[1, 4], N S Rodygina[3], A V Moroz[1]

[1]Higher School of applied physics and space technologies, Peter the Great Saint Petersburg Polytechnic University, Saint Petersburg 195251, Russia
[2] NTC JSC "Zaslon", Saint Petersburg 196084, Russia
[3]The Bonch-Bruevich Saint - Petersburg State University of Telecommunications, Saint Petersburg 193232, Russia
[4]Department of Ecology, All-Russian Research Institute of Phytopathology, 143050, Moscow Region, Odintsovo district, B.Vyazyomy, Russia

lenets03@mail.ru

Abstract. The results of the research of the developed fiber-optic transmission systems for analog high frequency signal are represented. On its basis, a new method to identify various structural defects in the active phased antenna arrays is elaborated.

1. Introduction

Now there is a constant improvement of exploited fiber-optic transmission systems (FOTS) of information both trunk and local use and development of new FOTS [1-5]. But in recent years, analog FOTS for transmission of very high frequency signals (radiophotonics) became an important component of development of local area telecommunication networks [2, 6-8]. Analog FOTS are greatly applicable in radar-tracking systems of different function [9-12]. Because in locations of radar-tracking systems the large number of various equipment and power stations are concentrated. In addition, use of FOTS can reduce the weight and size of the radar systems themselves, in case of their use on aircraft, etc.

Nowadays, various modifications of FOTS have been developed using direct and external modulation of laser radiation by an analog microwave signal [1, 2, 8-11]. Their designs and characteristics differ depending on the tasks to be solved with their use in radiolocation under various working conditions [1, 2, 8-11, 13-15]. The most difficult task is the tuning of active phased antenna arrays (APAA) at the stage of designing radar stations.

One of the problem is the distortion of the signal in the transmission from the receiver to the high frequency analyzer at different APAA modes of operation during testing in an anechoic chamber (AC). These distortions are connected with the position of coaxial lines for transmitting signals from nearby high-voltage cables entering the APAA, and also location of the amplifier (without it the signal fades) in the area of direct radiation pattern of high power electromagnetic radiation (since the distance between the transmitter and receiver is smaller than 3 m) [9, 10, 13-16]. In addition, the placement of any signal transforming microwave devices (e.g. network analyzer) in the AC violates its anechoic properties, which makes use of the camera ineffective.

One solution to this difficult problem is to develop special construction of FOTS, where the analog signal from the receiving device placed in AC, transmits the received electromagnetic radiation of APAA in the area of instrumentation location through the region with a complex electromagnetic environment. Previously developed models of FOTS, as the conducted research has shown, cannot be used for these purposes, since their design does not consider the specific features of APAA testing in AC.

2. Fiber-optic transmission system and measurement technique

Since the distance between the active elements of APAA and the AC receiving device is not more than 3 m, it is necessary to exclude the direct effect of the pattern of the antenna radiation on the case of the optical transmission module. For this purpose, we developed a new design of a measuring probe in the form of a horn. A small design of the transmitting laser module was chosen. For this module, a small power driver has been designed and manufactured. That made it possible to place the transmitting laser module with the power driver behind the receiving horn, so that direct electromagnetic radiation from the antenna pattern did not fall on their cases. Such a construction of the FOTS had never been used before.

In Fig. 1 a block diagram of the stand for testing APAA in AC is represented. In developed by us the fiber-optic communication line the compact high-frequency laser module 1 was used ("Dilaz" company) – it is the transmitter with direct modulation with wavelength of 1310 nm. The receiver is an optical module 2 ("Dilaz" company). The choice of a transmitter with direct modulation is caused by a small length of optical line (350 meters long) and its parameters more stable to temperature changes [1, 2, 6, 9, 10]. As the APAA testing with use of FOTS developed by us must be held in case of ambient temperatures from 233 to 323 K (real operating conditions of radar station) [4, 9, 10, 12]. A part of temperature conditions is provided when testing APAA in AC. Remaining temperature conditions are checked when testing APAA in the conditions of a polygon in the presence of opportunities.

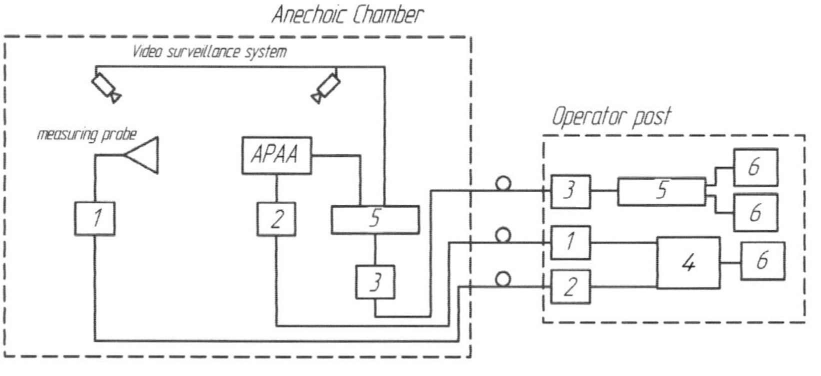

Figure 1. Block diagram of the test stand for testing APAA in AC: 1- laser module; 2 –receiving module, 3 SFP module, 4 vector network analyzer, 5 – router, 6 – personal computer

The data obtained from measuring probe is coming over FOTS to vector network analyzer 4 with the subsequent processing on the personal computer 6.

Since the signals from the microwave generators are also transmitted to APAA via cables placed near the lines with high voltage for a variety of its devices (in AC there is only one entrance - it's also the output for all switching elements), so for this purpose it is also desirable to use FOTS. This will improve noise immunity in addition to significantly reduce the size of the connections, and unload the platform on which an APAA is placed. Additionally, higher accessibility of the equipment in case of maintenance and testing of APAA will be provided. The load on the mechanical elements providing movement of a platform when scanning the APAA pattern will be less.

3. Results and their discussion

As an example, in fig. 2 amplitude-frequency characteristic (AFC) in case of transmission of analog signals of very high frequency via FOTS developed by us for different temperature conditions of its operation are provided.

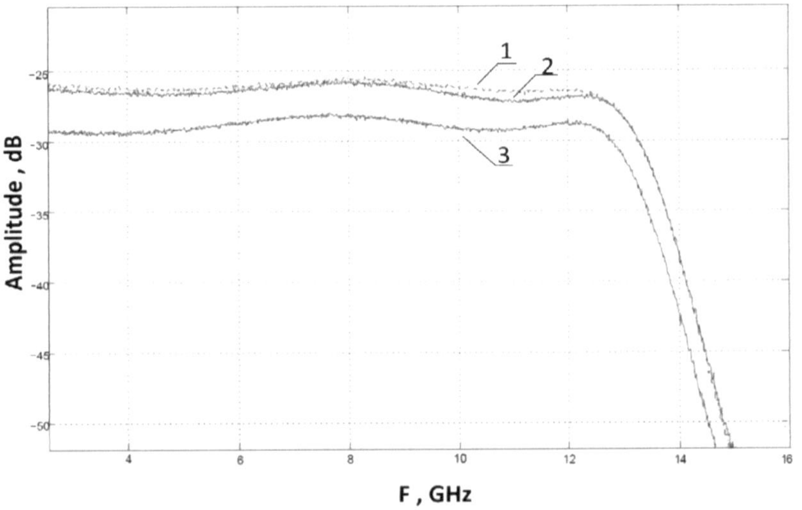

Figure 2. The amplitude - frequency characteristic. Graphs 1, 2 and 3 corresponds to T in K: 288.1; 298.2; 328.3.

The analysis of the received experimental result shows that losses of power during transmission of a very high frequency of a signal on FOTS developed by us in the frequency range of operation of APAA in the range of temperatures of its operation – aren't essential. It allows using FOTS developed by us for testing of APAA, as in the conditions of AC, and also at a polygon and the place of operation.

As an example, in fig. 3 the dynamic characteristics for FOTS a developed by us at a temperature of 298.4 K are presented. This temperature regime is the main one for testing APAA in AC.

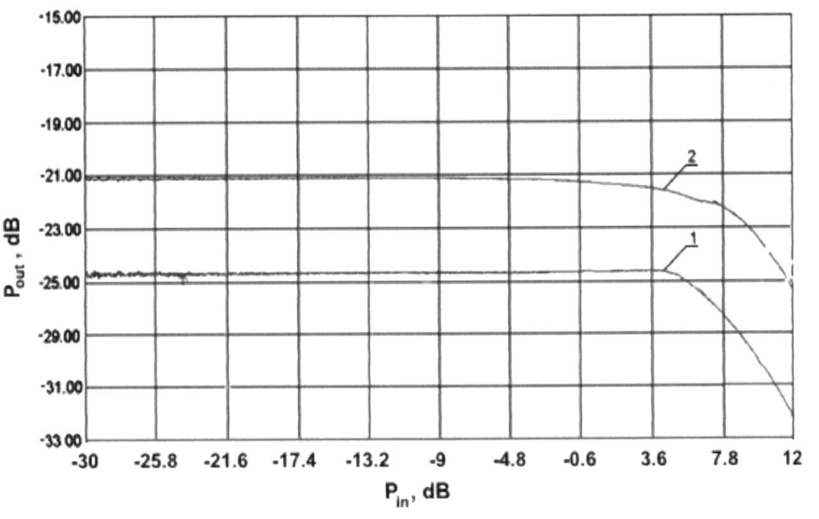

Figure 3. The dynamic characteristics of FOTS. Graphs 1 and 2 corresponds to analog signal frequency in GHz: 9.1; 13.1.

The obtained result shows, that the developed FOTS could be used for radar station testing on frequency range of its exploitation [3, 9, 10, 12]. The output power is changed by less than 1 dB while changing the input power of analog microwave signal in wide range.

As the experiments showed, the use of the developed FOTS allowed us to research besides range of radiation of all APAA in different operation modes, also the ranges of radiation of its single active elements. These

IOP Conf. Series: Journal of Physics: Conf. Series **1038** (2018) 012037 doi:10.1088/1742-6596/1038/1/012037

studies weren't possible earlier because of the existence of different noises, in particular the noise in the transferring path.

As an example, fig. 4 shows the emission spectra of a single active element APAA, transmitted to the control sector from the registration device of high frequency signal over coaxial cable, and developed by authors FOTS.

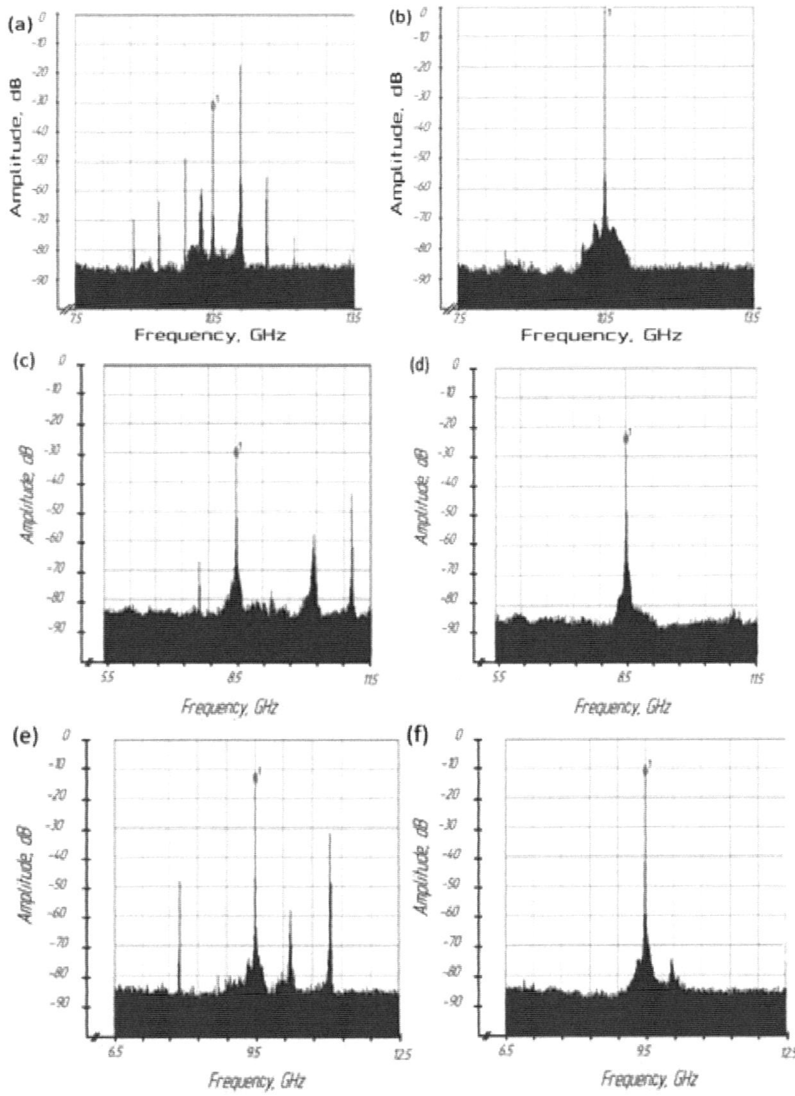

Figure 4 (a, b, c, d, e, f). The emission spectra of a single active element on 10.5, 9.5, 8.5 GHz frequency transmitted: **(a, c, e)** through a coaxial cable, **(b, d, f)** through a FOTS.

Analysis of the results, shows that the use of the developed FOTS allowed us to completely eliminate transmission noise from unpredictable RF disturbances and different induced signals, taking account of which and their further compensation (for example, subtraction) requires numerous additional measurements.

The use of multiple transformations to compensate interference causes additional distortions, both in the spectral line of a single active element and in its pattern. While using FOTS, changes in the registered pattern of the APAA single elements are connected only with the change of parameter of the very high frequency signal radiated by the antenna or defects of montage, and in the emitter itself. At the presence of the defect of the installation construction or in the emitter itself, the change of the pattern form takes place (for example reduction of amplitude at a maximum, etc.). Comparing the constructed patterns of different emitters allows determining these defects.

4. Conclusions

The carried out research showed that the use of developed FOTS allows us to solve the problems considered earlier arising during testing APAA in AC with use the coaxial cable for transmission of a very high frequency signal. Besides that, the use of FOTS allowed the development of a new technique of determination defects both in assembling of APAA, and in the construction of an individual emitter. It should be noted that our new technique with the use of FOTS for the first time made it expedient to test each emitter of APAA as a part of all operating construction of the APAA, in contrast to the previously used techniques.

The received results confirmed feasibility of the FOTS use for testing APAA and validity of a new developed technique to determine its various defects.

In addition, APAA works in range from 20 to 42 GHz. At these frequencies the testing of APAA with coaxial cable in the anechoic chamber is impossible to carry out. Using this range makes the use of a coaxial cable ineffective for testing an APAA in an anechoic chamber. In addition, the transmission of microwave signals over the optical fiber on sea ships from various antennas (from the mast deck) to the hull of the ship (to the processing device) is becoming of extremely need over distances of 100 meters or more. At the moment, various investigations have been directed at its solutions. One of the directions of our further research will be the search for an optimal solution to this problem.

5. References

[1] Reinhold N 2016 *Essentials of Modern Optical Fiber Communication* (Springer)
[2] Silver M 2015 *Optical Fiber Communication Systems* (Clanrye International)
[3] Petrov A A, Davydov V V 2017 *Journal of Communications Technology and Electronics* **62** 289 – 293
[4] Petrov A A, Davydov V V 2015 *Lecture Notes in Computer Science (including subseries Lecture Notes in Artificial Intelligence and Lecture Notes in Bioinformatics)* **9247** 739-744
[5] O'Mahony M J 2006 *IEEE OSA Journal of Light wave Technology* **24** 4684 – 4696
[6] Agrawal G P 2014 *Light wave technology: telecommunication systems* (NJ. Wiley-Inter science)
[7] Marpaung D, Roeloffzen C, Heideman R, Leinse A, Sales S., Capmany J 2013 *Integrated Microwave photonics (Laser & Photonics Reviews)* **4** 506-511
[8] Davydov V V, Dudkin V I, Karseev A Yu 2014 *Optical Memory & Neural Networks (Information Optics)* **23** 259 – 264
[9] Davydov V V, Sharova N V, Fedorova E V, Vologdin V A, Karseev A Yu 2015 *Lecture Notes in Computer Science (including subseries Lecture Notes in Artificial Intelligence and Lecture Notes in Bioinformatics)* **9247** 712-721
[10] Davydov V V, Ermak S V, Karseev A Yu, Nepomnyashchay E K, Petrov A A, Velichko E N 2014 *Lecture Notes in Computer Science (including subseries Lecture Notes in Artificial Intelligence and Lecture Notes in Bioinformatics)* **8638 LNCS** 694-702
[11] Friman R K 2012 *Fiber-optic communication systems* (NJ. Wiley-Inter science)
[12] Ermolaev A N, Krishpents G P, Davydov V V, Vysoczkiy M G 2016 *Journal of Physics: Conference Series* **741(1)** 012071
[13] Davydov V V, Dudkin V I, Karseev A Yu 2013 *Optical Memory & Neural Networks (Information Optics)* **22** 112 – 117
[14] Davydov V V, Velichko E N, Dudkin V I, Karseev A Yu 2015 *Journal of Optical Technology* **82** 132-135
[15] Tarasenko M Yu, Davydov V V, Akulich N V, Lenets V A, Yalunina T R 2017 *Lecture Notes in Computer Science (including subseries Lecture Notes in Artificial Intelligence and Lecture Notes in Bioinformatics)* **10531 LNCS** 227-232

IOP Conf. Series: Journal of Physics: Conf. Series **1038** (2018) 012037 doi:10.1088/1742-6596/1038/1/012037

[16] Davydov V V, Dudkin V I, Karseev A Yu 2014 *Optical Memory & Neural Networks (Information Optics)* **23** 170 – 176

An inexpensive MP3-Player based 45-kHz band noise generator for engineering and scientific applications

D D Stupin[1], **S V Koniakhin**[1], **and N A Verlov**[1,2]

[1]St. Petersburg Academic University, Khlopina 8/3, 194021 St. Petersburg, Russia
[2]Petersburg Nuclear Physics Institute NRC "Kurchatov Institute", Gatchina, St. Petersburg 188300, Russia

E-mail: `Stu87@ya.ru`, `Stupin@spbau.ru`

Abstract. In this study we propose a home-made setup based on MP3 Player for producing high quality noise-like shape voltage with 45-kHz frequency band. The proposed generator's output signal is investigated both theoretically and experimentally.

1. Introduction

Noise-like-shape voltage is widely used as the test signals [1, 2], for «black box» investigation [3], for random quantities simulation, and for device diagnostics [4, 5, 6]. However the price of the commercially available noise generators is very high. Moreover, the most number of generator schemes produce very wide-frequency band output, which complicates its digitizing (Kotelnikov-Nyquist theorem), so the problem of the band limitation is also actual. Here we propose the simple low-cost noise-shape voltage generator with uniform spectrum limited in the 45-kHz band. This frequency range is important, *e.g.*, for impedance spectroscopy and ultrasound applications.

2. Working principle

To obtain the output in the 45-kHz band we have used the commercially available MP3 Player (Quba, China, 25-kHz band, dimensions $5\times5\times5$ cm^3) and AD835 analog four-quadrant multiplier (Analog Devices, USA, dimensions $5\times4\times1$ mm^3) for providing the heterodyne transformation [7, 8]. To be specific, we have prepared the audio-file, one stereo channel (*e.g.*, left channel) of which contains 25-kHz band noise-like data, another channel contains single-sine data with frequency 20 kHz. The outputs of the both channels were then multiplied by AD835, which results in generating noise-like voltage with 45-kHz band (figure 1). The audio-file was created using Adobe Audition v 1.5 software.

From the mathematical point of view, for the modulated $V(t)$ we have

$$V(t)\sin(2\pi f_B t) = V(t)\frac{e^{2i\pi f_B t} - e^{-2i\pi f_B t}}{2i} \doteqdot \frac{V(f+f_B) - V(f-f_B)}{2i}, \tag{1}$$

where f is frequency, $f_B = 20$ kHz is single-sine output frequency, t is time, $\doteqdot$ means the Fourier transformation, $V(f)$ is the Fourier image of $V(t)$. Multiplying by the exponent the in time domain corresponds to the shift in the frequency domain. So, the resulting output signal spectrum will be sum of the left- and right- shifted by 20 kHz $V(t)$ spectra, which leads to

IOP Conf. Series: Journal of Physics: Conf. Series **1038** (2018) 012038 doi:10.1088/1742-6596/1038/1/012038

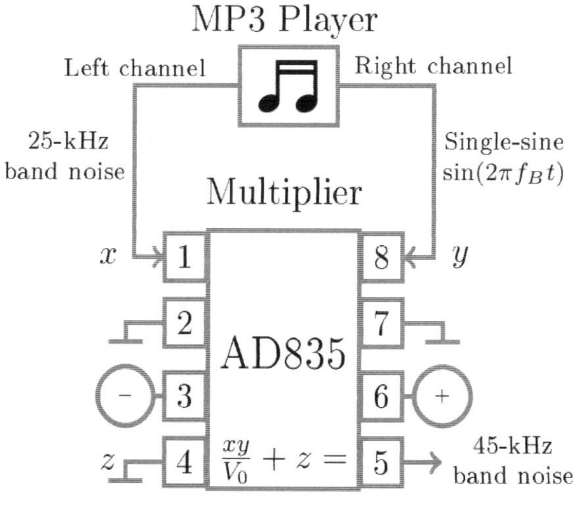

Figure 1. The proposed noise generator scheme. The MP3 Player produces 25-kHz band output from one channel (x), and sine-shape output with 20 kHz frequency on another (y). Further, both MP3 Player's outputs are multiplied by the AD835 scheme, which results in generating 45-kHz band noise-shape output. The input $z = 0$, because it is connected to the ground. The $V_0 = 1.05$ V is scaling voltage value. The pins 3 and 6 are connected to negative and positive voltage supply, respectively. Here we want to warn, that AD835 polarity reversal disturb the scheme. The pins 2 and 7 are connected to ground. Pin 5 produces the multiplier output.

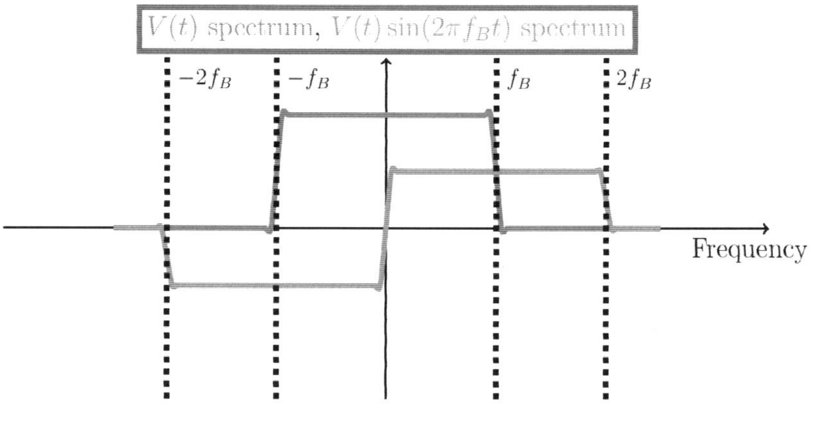

Figure 2. The scheme of the heterodyne transformation (Eq. 1) of the spectrum (except for sign). Blue line is the real part of the original spectrum, red line is the imaginary part of the transformed spectrum. Here original spectrum is located in [-f_B;f_B] kHz band and f_B is also the heterodyne frequency. It should be noticed, that both spectra are located in negative and positive frequency domain, because $V(t)$ and $V(t)\sin(2\pi f_B t)$ are real functions.

the desired extension of the frequency band up to 45 kHz. With out loss of generality figure 2 illustrate spectrum transformation for the case when modulation frequency f_B is equal to $V(t)$ frequency band.

3. Experimental approbation and discussion

For the measurements of the generator's and MP3 Player's outputs we have used the L-Card E20-10 analog-to-digital converter (ADC) (L-Card, Russia). The sample rate was 500 kHz, data collecting time was 500 ms. The signal's spectra were calculated from the time-domain data by fft routine in the MatLab package. The output signal frequency response measurements of the proposed generator and MP3 Player in the range up to 250 kHz are given in figures 3(a-c). For comparison on figure 3(d) presented spectrum of the output's noise-like voltage produces by commercially available arbitrary shape voltage generator AKIP-3413-3 (AKIP, Russia).

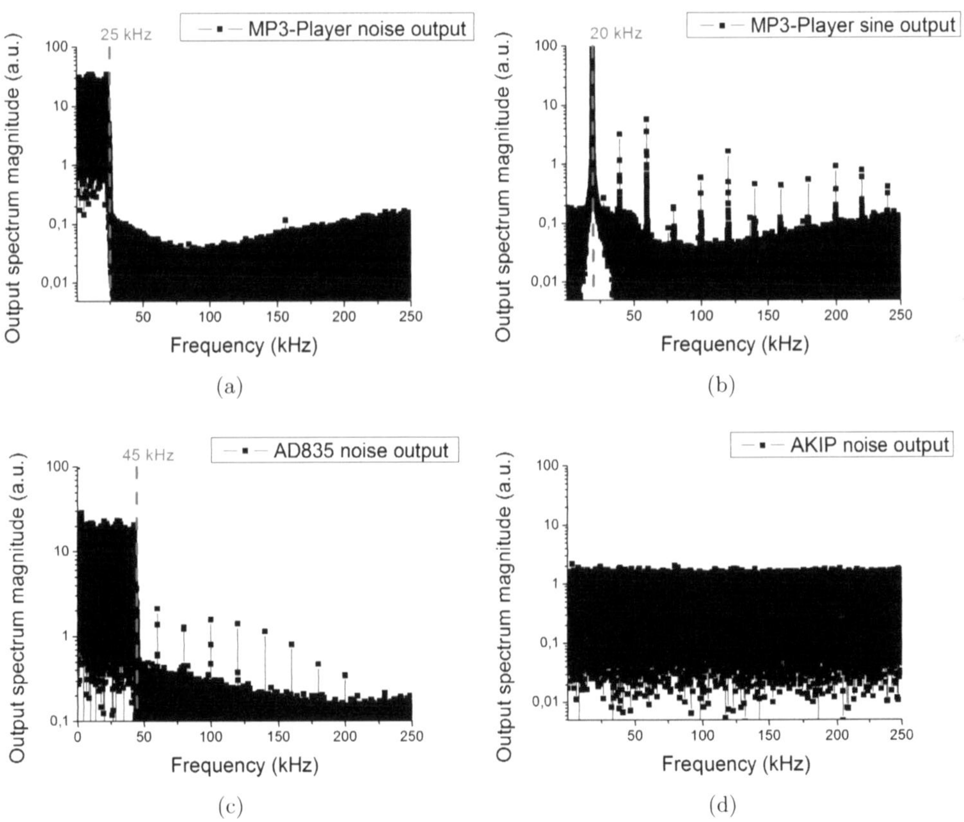

Figure 3. The obtained experimental spectra from various sources. (a) Spectrum of the MP3-Player noise-like data output; (b) spectrum of the produced by MP3-player sine-like 20-kHz output; (c) spectrum of the assembled noise generator output; (d) commercially available noise generator AKIP-3413-3 output's spectrum. The harmonics amplitudes of the proposed generator output spectrum in the range up to 45 kHz are two orders greater that the higher harmonics amplitudes, contrary to AKIP's output. The appearance of the negligible low amplitude heterodyne obertones in AD835's output, probably, is connected with AD835's inter-channel interference.

One can see that the proposed generator's output's spectrum high-amplitude harmonics are mainly located in the range up to 45 kHz and all of them has the steadily none-vanishing amplitude. In the other spectrum part (>45 kHz) the harmonics have the two orders lower amplitude, contrary to AKIP's output spectrum, which uniform in all measurement range. It should be noticed, that the generator band limitation allows to not use the over-sampling technique and the high-cost input low-pas filters at the ADC inputs, which simplifies overall

IOP Conf. Series: Journal of Physics: Conf. Series **1038** (2018) 012038 doi:10.1088/1742-6596/1038/1/012038

circuit and makes digitizing experimental results more affordable [9].

4. Conclusion
The developed noise generator produces fine and stable output signal with approximately uniform spectrum in the 45 kHz range. It can be successfully used for such engineering and scientific problems as producing exiting voltage for electrical impedance spectroscopy, ultra-sonic waves generation for acoustic and music research, as external noise source for modeling high-noise conditions. Additionally, our device is portable and simple to bring into use, which makes it very perspective as a prototype for creating the portable electrical impedance spectroscopy bio-sensors based on available in market and widespread electronic components.

Acknowledgments
The authors acknowledges Andrew L. Chernev, Ivan N. Terterov, and Michael V. Dubina for fruitful discussions. This study was unded by Ministry of Education and Science of Russian Federation 16.9790.2017/BCh.

References
[1] Rivnay J, Leleux P, Hama A, Ramuz M, Huerta M, Malliaras G G and Owens R M 2015 *Scientific reports* **5**
[2] Popkirov G and Schindler R 1992 *Review of scientific instruments* **63** 5366–5372
[3] Stupin D D, Koniakhin S V, Verlov N A and Dubina M V 2017 *Phys. Rev. Applied* **7**(5) 054024
[4] Gupta M S 1975 *Proceedings of the IEEE* **63** 996–1010
[5] Denda W 1993 *Noise as source of information (in Russian)* (Mir)
[6] Denda W 1989 *Rauschen als Information (in German)* (Hüthig)
[7] Lubyako L, Skalyga N and Kuftin A 2009 *Technical Physics* **54** 1332–1338
[8] Titerich N 1968 *Noise generators and noise characteristics measurement (in Russian)* , ch. 3-6 ed (Energia)
[9] Alan V Oppenheim Ronald W Schafer J R B 1999 *Discrete-Time Signal Processing* , ch. 4.9.1, 2nd ed (Prentice Hall)

IOP Conf. Series: Journal of Physics: Conf. Series **1038** (2018) 012039 doi:10.1088/1742-6596/1038/1/012039

The transformation of Hermite-Gauss beams with embedded optical vortex by lens system

E O Monin [1], **A V Ustinov** [1,2]
[1] Samara National Research University, 34, Moskovskoye shosse, Samara, 443086, Russia
[2] Image Processing Systems Institute of RAS – Branch of the FSRC "Crystallography and Photonics" RAS, 151, Molodogvardeyskaya street, Samara, 443001, Russia

Abstract. In this paper the transformation of Hermite-Gauss beams with embedded vortex phase by lens system is investigated theoretically and numerically. A particular attention is devoted to the formation of vortex phase singularities in focal area. It is shown, that under appointed relations between HG mode indices and the number of embedded optical vortex the vortex phase singularity in a focal plane centre may disappear.

1. Introduction

Vortex beams are laser beams with the vortex phase singularity which is characterized by zero intensity in a point of phase uncertainty [1-7]. Well known vortex beams are Laguerre-Gauss modes [8-12] and Bessel modes [13-18]. Hypergeometric modes [19-22] and Zernike functions [23-25] contain angular harmonic components also. All said distributions are separable in the polar coordinates system – they are representable as a product of the function depending on radius only, and angular harmonic component $\exp(im\varphi)$, m is optical vortex order.

Beams which are separable in Cartesian coordinates system, such as Hermite-Gauss modes (HG) [26-28], Airy beams [29-34] are not vortex ones, but their certain superpositions are used for transformation to vortex beams [35, 36]. However, Airy beams with embedded vortex phase are examined lately [37-39]. Besides, in paper [40] vortex HG beams depending on complex argument were examined. In paper [41] the optical element matched with HG modes was used as a low-frequency grating for a multiplication of first-order phase singularity.

In this paper we investigate HG beams with embedded optical vortex of an arbitrary order. A particular attention is devoted to a picture in the focal plane and to the formation of vortex phase singularities depending on relation of HG mode indices and order of embedded optical vortex. We obtain an analytical condition of the disappearance of vortex phase singularity in the focal plane centre. Also we have examined a propagation of such beam through paraxial lens system on base of usage of fractional Fourier transformation.

2. Theoretical foundations.

Let us consider the Hermite-Gauss beams with embedded vortex phase (embedded vortex Hermite-Gaussian, EVHG):

$$\Psi_{nms}(x,y)=\exp\left(-\frac{x^2+y^2}{2\sigma^2}\right)H_n\left(\frac{x}{\sigma}\right)H_m\left(\frac{y}{\sigma}\right)(x+iy)^s, \tag{1}$$

where $H_n(x)$ is Hermite polynomial.

Hermite polynomials have a particular property, namely all terms have equal power parity. However, now we shall not take this in account, and, for commonness, we shall examine a case of the arbitrary polynomial of two variables. An individual term looks like $P_{pqs}(x,y)=A(r)x^p y^q(x+iy)^s$, where $A(r)$ is an arbitrary axial-symmetric function depending on radius only. For a convenience of following analysis $P_{pqs}(x,y)$ is written in polar coordinates:

$$P_{pqs}(r,\varphi) = A(r)\left(r\cos\varphi\right)^{p}\left(r\sin\varphi\right)^{q} r^{s}\exp\left(is\varphi\right). \tag{2}$$

Let us consider the Fourier transformation of function (2):

$$F_{pqs}(\rho,\theta) = \frac{2\pi}{\lambda f}\int_{0}^{2\pi}\int_{0}^{\infty} P_{pqs}(r,\varphi)\exp\left[-\frac{i2\pi}{\lambda f}\rho r\cos(\varphi-\theta)\right] r\,\mathrm{d}r\,\mathrm{d}\varphi, \tag{3}$$

where λ is the wavelength of an illuminating light, f is focal length of a parabolic lens implementing the Fourier transformation.

It is well known that vortex beams retain the vortex phase singularity and the zero intensity on optical axis under propagation in homogeneous medium and under passing through parabolic lens [1-7]. However in anisotropic media [42, 43] or at an astigmatic transformation [35, 36, 44] such beams lose the axial symmetry of intensity distribution, in particular, non-zero value may appear on the axis.

Since the beam (1) has not axial symmetry, then under its propagation or passing through lens it may undergo changes similar to anisotropic or astigmatic transformation.

Let us clarify whether zero value will be in the focal plane centre (on axis). For this purpose we may use results from Ref. [45]. Zero amplitude on axis will be under conditions:

$$\begin{array}{l} (s > p+q)\ or\ (p+q+s\ \ is\ odd),\quad s>0 \\ p\ or\ q\ is\ odd,\quad s=0 \end{array} \tag{4}$$

These conditions are *sufficient*, but the condition $(p+q+s\ \ is\ odd)$ may not be *necessary* (at its violation amplitude on axis may be zero).

3. The numerical simulation

In this section we present simulation results for EVHG beams prescribed by expression (1).

3.1 Field distribution in the focal plane.

Table 1 contains distribution in the focal plane obtained on base of expression (3) for different values *n, m* and *s*.

Table 1. Distributions in the focal plane for different values *n, m* and *s* (amplitudes are shown).

	n=0, m=1	n=1, m=1	n=1, m=2	n=2, m=2	n=3, m=3
s=0	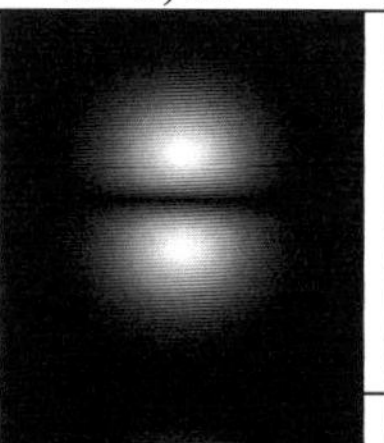	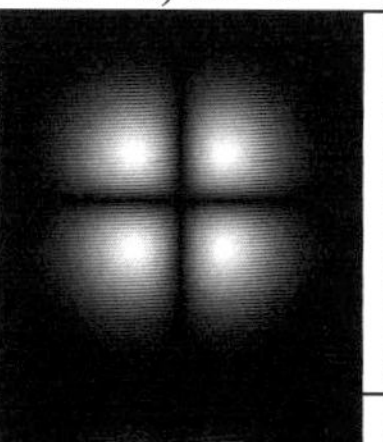	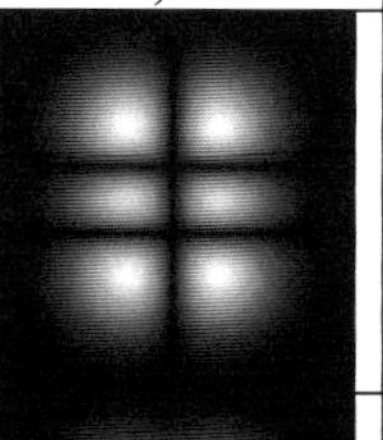	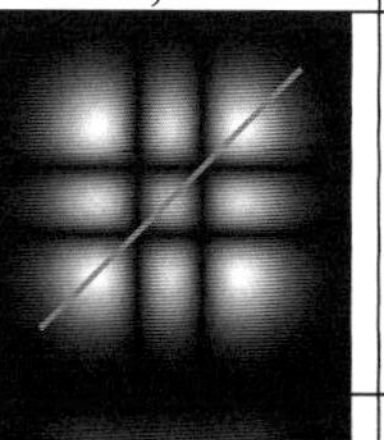	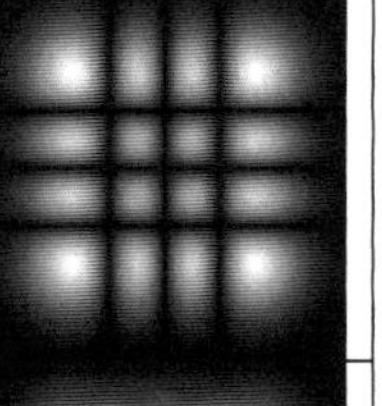
s=1	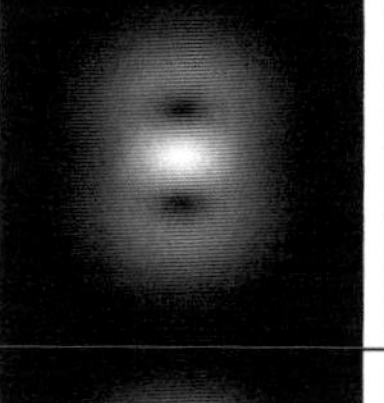	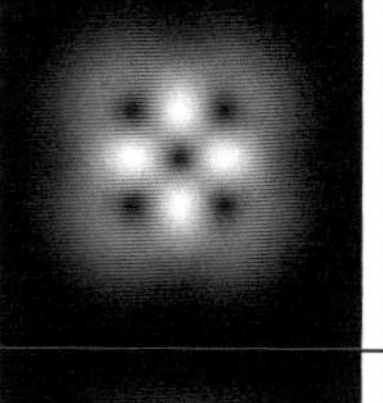	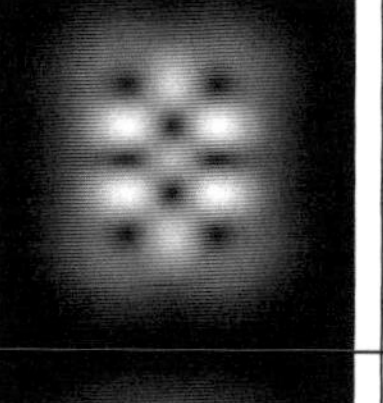	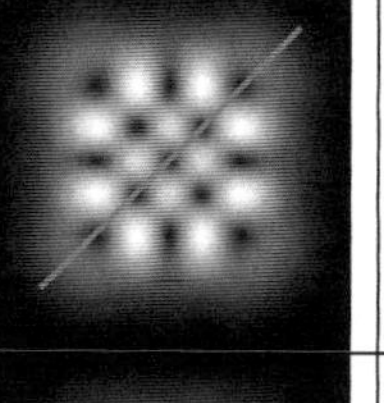	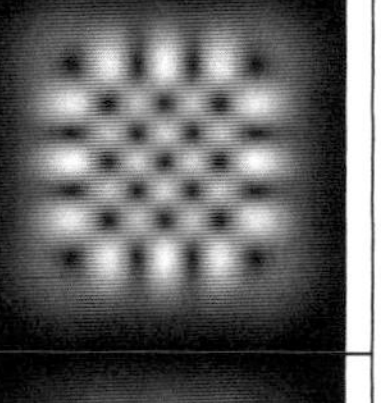
s=2	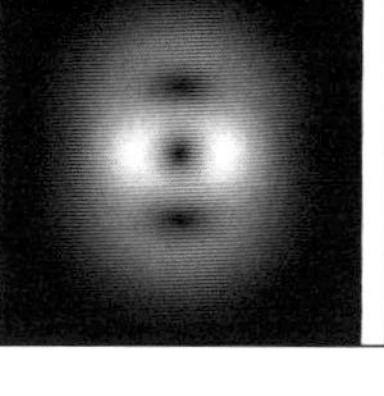	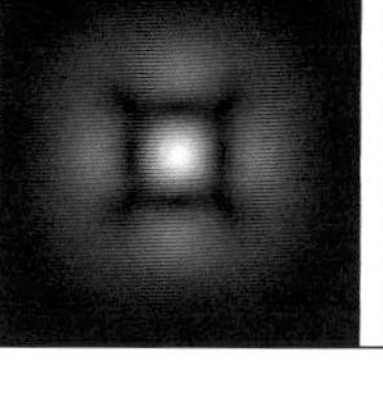	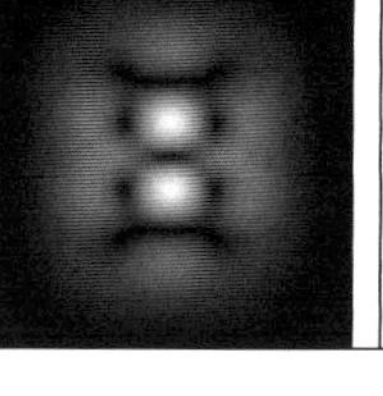	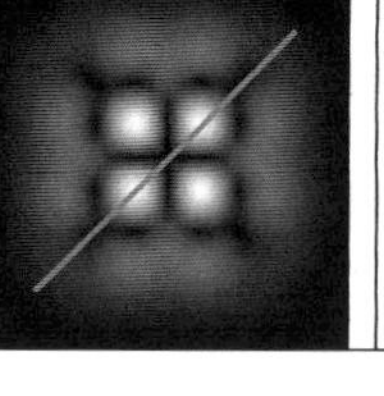	

The first row of Table 1 contains distributions for Fourier-images of conventional HG modes (s=0). In this case distributions are similar to initial functions and they have a binary phase. According to formula (4) value in centre is zero if at least one index is odd.

The second row of Table 1 contains distributions of spatially-spectral picture for HG beams with embedded optical vortex of order one (s=1). In this case we have set of separated optical vortex of order one instead of singular lines (linear phase jumps) Number of vortices is defined by formula:

$$N_{s=1} = (n+1)(m+1) + nm \,. \tag{5}$$

For determining what value will be in centre (zero or nonzero) one may use formula (4).

The third row of Table 1 contains focal pictures for HG beams with embedded optical vortex of order two (s=2). As we see, in this case the distinction between sufficient and necessary condition takes place at n=2, m=2, s=2.

Thus, theoretical discussions listed above are conformed to calculation results completely.

An interesting result is obtained at embedding of the optical vortex into HG modes with one index equal zero (columns 1 and 2). In this case the number of zero intensity points which are easily found out by sight, is connected with the vortex order unambiguously:

$$N_s = m + s \,. \tag{6}$$

Formula (6) is true for vortices orders s<m+2. Consequently, Hermite-Gauss modes with one zero and other large enough index may be used for determination of the optical vortex order.

3.2 The beam propagation through lens system.

In conclusion let us consider the propagation of beam (1) through paraxial lens system. This process may be described with usage of fractional Fourier transformation [46, 47]. In polar coordinates its expression looks like [48]:

$$E(r,\varphi,z) = -\frac{ik}{2\pi f \sin(\alpha z)} \exp\{ikz\} \exp\left\{\frac{ikr^2}{2f\tan(\alpha z)}\right\} \times \\ \times \int_0^{2\pi}\int_0^{\infty} E_0(\rho,\theta) \exp\left\{\frac{ik\rho^2}{2f\tan(\alpha z)}\right\} \exp\left\{-\frac{ik}{f\sin(\alpha z)}\rho r\cos(\theta-\varphi)\right\} \rho\, d\rho\, d\theta, \tag{7}$$

where $\alpha = \pi/(2f)$; $k = 2\pi/\lambda$ is the wavenumber, λ is the wavelength of an illuminating light.

Table 2. Longitudinal distributions (diagonal sections as shown by red line in Table 1) for different values of n, m and s (amplitudes are shown).

	n=1, m=1	**n=2, m=2**	**n=3, m=3**
s=0	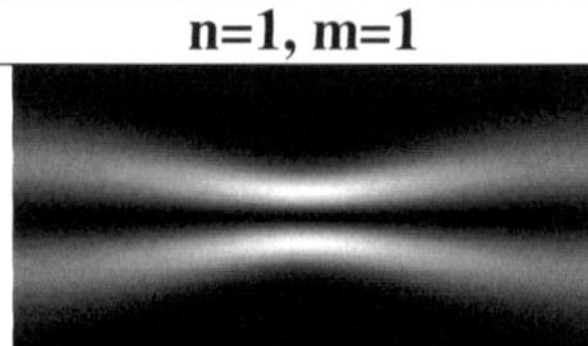	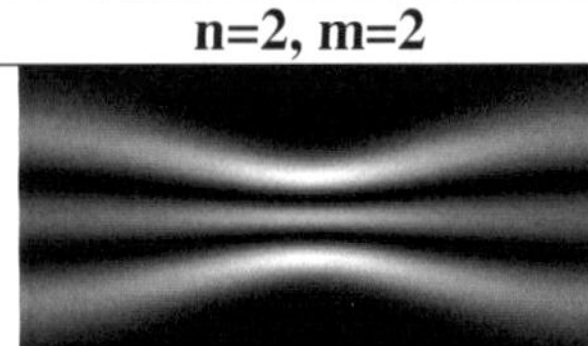	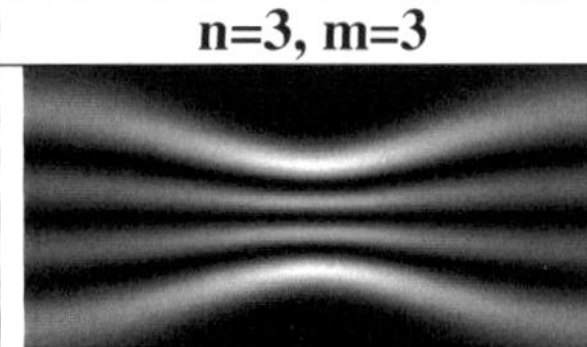
s=1	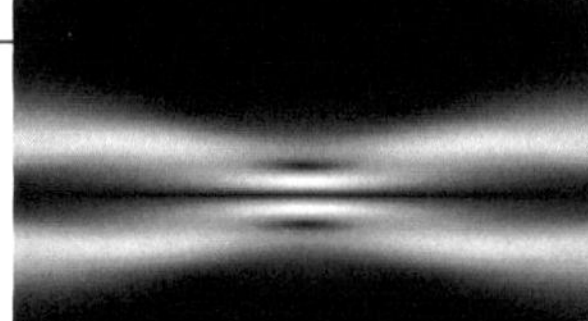	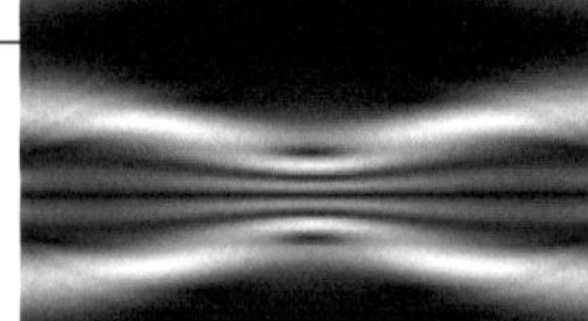	
s=2	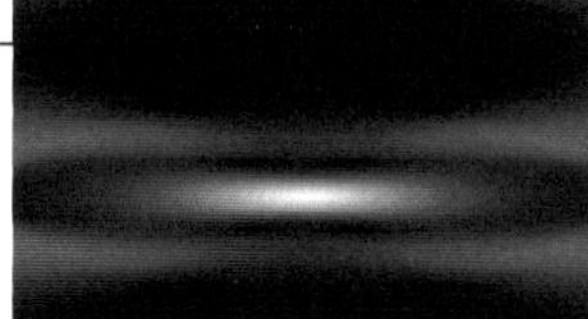	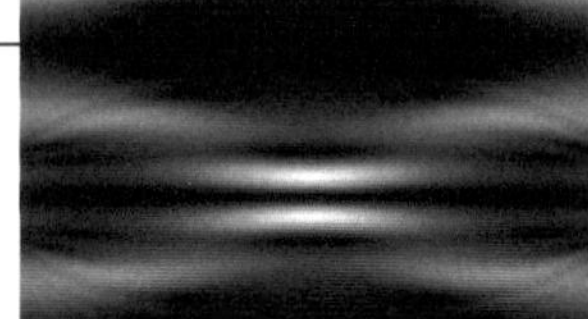	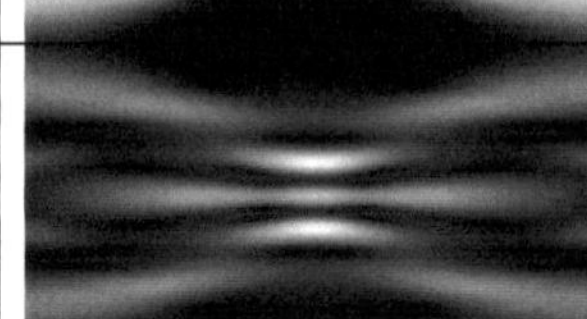

The operator (7) allows to obtain a beam shape (taking into account a scale) in any paraxial domain – both in Fresnel diffraction domain and in far domain. Table 2 shows passing of field (1) from input plane (z=0) through focal plane (z=f) to output plane (z=2f).
In output plane (z=2f) in common case the distribution which is centrally symmetrical to initial one is being formed, but the amplitude of beam (1) is symmetrical initially, therefore the distribution *coincides* with initial (z=0). This explains a symmetry of longitudinal pictures in Table 2 relative to the plane z=f, so as examination of changes from z=0 to z=f is quite enough.
An analysis of images in Table 2 allows us to arrive at the following conclusions. At vortex absence the distribution is similar to initial one along the whole length from z=0 to z=f. At vortex presence there is passing between the domain with distribution similar to initial one, and the domain with distribution similar to one in focal plane. The fact itself is obvious, but the longitudinal distribution shows that domains with other structure are absent.

4. Conclusion

In this paper we have examined HG modes with embedded vortex phase. As against axially symmetrical distributions the vortex phase (and zero intensity value on optical axis) for such beams is not remained under propagation or under passing through parabolic lens.
Theoretical and numerical analysis of Fourier transformation for HG modes with embedded vortex phase has been implemented. We have verified theoretical conditions for zero intensity value in the centre of focal plane by the numerical simulation.
With aid of the fractional Fourier transformation the modeling of HG beams passing through paraxial lens system is implemented. Results show that even at the vortex presence, there is a *continuous* transformation from initial distribution to focal one.

5. Acknowledgements.

This work was financially supported by the Russian Foundation for Basic Research (grants 16-07-00825, 16-47-630546).

6. References

[1] Nye J F and Berry M V 1974 *Dislocations in wave trains.* (Proc. R. Soc. London vol 336) pp 165-190
[2] Coullet P, Gil G and Rocca F 1989 *Optical vortices, Opt. Commun.* **73** pp 403–408
[3] Bazhenov V, Vasnetsov M V and Soskin M S 1990 *Laser-beam with screw dislocations in the wavefront, JETP Lett.* **52** pp 429– 431
[4] Khonina S N, Kotlyar V V, Soifer V A , Jefimovs K and Turunen J 2004 *Generation and selection of laser beams represented by a superposition of two angular harmonics, Journal of Modern Optics* **51** pp 761–773
[5] Soskin M S 2001 *Progress in Optics* (Amsterdam, North Holland: Elsevier Science vol 42) pp 219-276
[6] Molina-Terriza G, Torres J P and Torner L 2007 *Twisted photons, Nat. Phys.* **3** pp 305–310
[7] Andrews D L 2008 *Structured Light and Its Applications: An Introduction to Phase-Structured Beams and Nanoscale Optical Forces, Elsevier Inc.* p 373
[8] Allen L, Beijersbergen M W, Spreeuw R J C and Woerdman J P 1992 *Orbital angular momentum of light and the transformation of Laguerre-Gaussian laser modes, Phys. Rev.* A **45** pp 8185–8189
[9] Kotlyar V V, Soifer V A and Khonina S N 1997 *Rotation of Gauss-Laguerre multimodal light beams in free space, Technical Physics Letters* **23** pp 657-658
[10] Kennedy S A, Szabo M J, Teslow H, Porterfield J Z and Abraham E R I 2002 *Creation of Laguerre-Gaussian laser modes using diffractive optics. Physical Review* A **66** 043801

[11] Matsumoto N, Ando T, Inoue T, Ohtake Y, Fukuchi N and Hara T 2008 *Generation of high-quality higher-order Laguerre-Gaussian beams using liquid-crystal-on-silicon spatial light modulators J. Opt. Soc. Am.* **25** pp 1642-1651
[12] Vallone G 2015 *On the properties of circular beams: normalization, Laguerre-Gauss expansion, and free-space divergence, Opt. Lett.* **40** p 1717
[13] Arlt J and Dholakia K 2000 *Generation of high-order Bessel beams by use of an axicon, Opt. Commun.* **177** pp 297–301
[14] Khonina S N, Kotlyar V V, Skidanov R V, Soifer V A, Jefimovs K, Simonen J and Turunen J 2004 *Rotation of microparticles with Bessel beams generated by diffractive elements, Journal of Modern Optics* **51** pp 2167–2184
[15] McGloin D and Dholakia K 2005 *Bessel beams: diffraction in a new light, Contemp. Phys.* **46** pp 15–28
[16] Kotlyar V V, Kovalev A A, Khonina S N, Skidanov R V, Soifer V A, Elfstrom H, Tossavainen N and Turunen J 2006 *Diffraction of conic and Gaussian beams by a spiral phase plate, Appl. Opt.* **45** pp 2656-2665
[17] Duocastella M and Arnold C B 2012 *Bessel and annular beams for materials processing, Laser Photonics Rev.* **6** pp 607–621
[18] Vaity P. and Rusch L. 2015 *Perfect vortex beam: Fourier transformation of a Bessel beam, Optics Letters* **40** pp 597-600
[19] Kotlyar V V, Skidanov R V, Khonina S N and Soifer V A 2007 *Hypergeometric modes, Optics Letters* **32** pp 742-744
[20] Karimi E, Zito G, Piccirillo B, Marrucci L and Santamato E 2007 *Hypergeometric-Gaussian modes. Opt. Lett.* **32** pp 3053-3055
[21] Kotlyar V V, Kovalev A A, Skidanov R V, Khonina S N and Turunen J 2008 *Generating hypergeometric laser beams with a diffractive optical element, Applied Optics* **47** pp 6124- 6133
[22] Skidanov R V, Khonina S N and Morozov A A 2013 *Optical rotation of microparticles in hypergeometric beams formed by diffraction optical elements with multilevel microrelief, Journal of Optical Technology* **80** pp 585–589
[23] Khonina S N, Kotlyar V V, Soifer V A, Wang Y, Zhao D 1998 *Decomposition of a coherent light field using a phase Zernike filter* Proceedings of SPIE 3573, pp 550-553
[24] Carpenter J, Thomsen B and Wilkinson T 2012 *Mode division multiplexing of modes with the same azimuthal index, IEEE Photon. Technol. Lett.* **24** pp 1969–1972
[25] Porfirev A P and Khonina S N 2016 *Experimental investigation of multi-order diffractive optical elements matched with two types of Zernike functions* Proc. SPIE **9807** 98070E-9p
[26] Khonina S N, Kotlyar V V, Soifer V A, Lautanen J, Honkanen M and Turunen J 2000 *Generation of Gauss-Hermite modes using binary DOEs, Proceedings of SPIE Int. Soc. Opt. Eng.* **4016** pp 234-239
[27] Bastiaans M J and Alieva T 2005 *Generating function for Hermite–Gaussian modes propagating through first-order optical systems, Journal of Physics A: Mathematical and General* **38** L73
[28] Ren Y-X, Fang Z-X, Gong L, Huang K, Chen Y and Lu R-D 2015 *Digital generation and control of Hermite–Gaussian modes with an amplitude digital micromirror device, J. Opt.* **17** p 125
[29] Grossman J G, Casperson L W, Stafsudd O M and Sutter Jr. L V 1984 *Propagation of Airy-Hermite-Gaussian waveguide modes in free space, Appl. Opt.* **23** pp 48–52
[30] Siviloglou G A and Christodoulides D N 2007 *Accelerating finite energy Airy beams, Opt. Lett.* **32** pp 979–981
[31] Banders M A and Gutierrez-Vega J C 2007 *Airy-Gauss beams and their transformation by paraxial optical systems, Opt. Express* **15** pp 16719–16728
[32] Khonina S N and Volotovsky S G 2008 *Bounded 1D Airy beams: laser fan, Computer Optics* **32** pp 168–174

[33] Khonina S N 2011 *Specular and vortical Airy beams, Opt. Commun.* **284** pp 4263–4271
[34] Khonina S N and Ustinov A V 2017 *Fractional Airy beams, Journal of the Optical Society of America A* **34** pp 1991-1999
[35] Abramochkin E G and Volostnikov V G 1991 *Beam transformation and nontransformed beams, Opt. Commun.* **83** pp 123–135
[36] Beijersbergen M W, Allen L, Van der Veen H E and Woerdman J P 1993 *Astigmatic laser mode converters and transfer of orbital angular momentum, Opt. Commun.* **96** pp 123 –132
[37] Mazilu M, Baumgartl J, Cizmar T and Dholakia K 2009 *Accelerating vortices in Airy beams.* Proc. SPIE, p 7430
[38] Dai H T, Liu Y J, Luo D and Sun X W 2010 *Propagation dynamics of an optical vortex imposed on an Airy beam. Opt Lett* **35** p 4075
[39] Rui-Pin C, Khian-Hooi C, Sailing H 2013 *Dynamic control of collapse in a vortex Airy beam, Scientific Reports* **3** p p1406-1415
[40] Kotlyar V V, Kovalev A A and Porfirev A P 2015 *Vortex Hermite–Gaussian laser beams, Optics Letters* **40** pp 701-704
[41] Porfirev A P and Khonina S N 2017 *Simple method for efficient reconfigurable optical vortex beam splitting, Optics Express* **25** pp 18722-18735
[42] Hacyan S and Jáuregui R 2009 *Evolution of optical phase and polarization vortices in birefringent media, Journal of Optics A: Pure and Applied Optics* **11** 085204
[43] Khonina S N, Paranin V D, Ustinov A V and Krasnov A P 2016 *Astigmatic transformation of Bessel beams in a uniaxial crystal, Optica Applicata* **46** pp 5-18
[44] Khonina S N, Kotlyar V V, Soifer V A, Jefimovs K, Paakkonen P and Turunen J 2004 *Astigmatic Bessel laser beams, Journal of Modern Optics* **51** pp 677–686
[45] Ustinov A V 2017 *An analysis of the spatial spectrum of Hermite-Gauss beams with embedded optical vortex, Computer Optics* **41** pp 617–625
[46] Alieva T, Bastiaans M J and Calvo M L 2005 *Fractional transforms in optical information processing, J. Appl. Signal Processing* **10** pp 1-22
[47] Kirilenko M S, Zubtsov R O and Khonina S N 2015 *Calculation of eigenfunctions of a bounded fractional Fourier transform, Computer Optics* **39** pp 332-338
[48] Mossoulina O A, Kirilenko M S and Khonina S N 2016 *Simulation of vortex laser beams propagation in parabolic index media based on fractional Fourier transform, Journal of Physics* **741** pp 12142-12148

IOP Conf. Series: Journal of Physics: Conf. Series **1038** (2018) 012040 doi:10.1088/1742-6596/1038/1/012040

Stability analysis of the lattice Boltzmann schemes with body force action

Sergey A Mikheev, Gerasim V Krivovichev

Faculty of Applied Mathematics and Control Processes Saint Petersburg State University, 7/9 Universitetskaya nab., Saint Petersburg, 199034, Russian Federation

E-mail: helps2@rambler.ru, g.krivovichev@spbu.ru

Abstract. Stability analysis of lattice Boltzmann equations for the case of body force action is realized. The analysis is performed by von Neumann method and is based on the linearized equations. The stability is estimated by the values of the areas of the stability domains in parametric space. As the results of the analysis, the recommendation on the use of the models in practical simulations may be performed. The proposed approach to the analysis may be extended to the investigation of the other cases of body force actions, especially in the cases of multiphase and multicomponent flows.

1. Introduction

Lattice Boltzmann method (LBM) in the last several decades is considered as a powerful tool for computational modelling of complex flows with multiple phases and components [1]. These types of flows may be characterized by the existence of the body force, which model interphase or intercomponent interactions. Other types of flows with the body force action are realized in the cases of flows with vector fields of external body forces, such as gravitational or electromagnetic forces. So for the practical simulations it is necessary to correctly represent the model of the body force action into computational schemes used in the LBM.

The first models of body force action were proposed in the early years of the development of the LBM. In [2, 3] models based on the perturbation of the equilibrium flow regime are proposed for the simulation of multiphase and multicomponent flows. In [4, 5] the models for the representation of the body force term in the discrete Boltzmann equations are proposed. Ladd et al [6] and Guo et al [7] proposed the models for terms with body forces, which are based on the expansions on the generalized Hermite polynomials. The models are based on the theory of the discretization of kinetic equations in the velocity space of kinetic equations [8]. In [9] the so-called exact difference method for incorporating of body force term is proposed. The model is based on the representation of body force term by a difference of the values of equilibrium distribution functions on perturbed and unperturbed flows. Comprehensive reviews of the models used for bode force action modeling are presented at [1, 10].

Despite the widely development of body force action models used in the LBM, investigation of their stability properties and comparison of this property are still not realized. So the presented paper is dedicated to the stability analysis of LBE's with body force action, proposed by different authors.

The structure of the paper is the following. In Section 2 discrete Boltzmann equations and LBE's are discussed. In Section 3 the models of body force action widely used in the LBm are presented. In Section 4 the problem of the stability investigation is stated and some numerical results of the analysis are discussed. Some concluding remarks are made in Section 5.

2. Lattice Boltzmann equations

Let us consider the Boltzmann kinetic equation with the relaxation collision term (Bhatnaghar — Gross — Krook equation):

$$\frac{\partial f}{\partial t} + \mathbf{V}\nabla f + \mathbf{a}\nabla_{\mathbf{V}} f = -\frac{f - f^{(eq)}}{\lambda}, \tag{1}$$

where $f = f(t, \mathbf{r}, \mathbf{V})$ is a particle distribution function, t is a time, $\mathbf{r}$ is a vector of space variables, $\mathbf{V}$ is a vector of particle velocity, $\mathbf{a} = \mathbf{F}/m$, $\mathbf{F}$ is a body force, m is a particle mass, $\lambda > 0$ is a relaxation time, $f^{(eq)} = f^{(eq)}(\rho, \mathbf{U})$ is a Maxwellian distribution function, ρ is a gas density, $\mathbf{U}$ is a gas velocity. These macrovariables may be calculated as a moments of the distribution function [8]:

$$\rho(t, \mathbf{r}) = \int f(t, \mathbf{r}, \mathbf{V})d\mathbf{V}, \quad \rho(t, \mathbf{r})\mathbf{U}(t, \mathbf{r}) = \int \mathbf{V} f(t, \mathbf{r}, \mathbf{V})d\mathbf{V}.$$

After the discretization of (1) on the velocities the following grid in the velocity space is considered: $\mathbf{V}_i = V\mathbf{e}_i$, $i = \overline{1, n}$, where $V = l/\delta t$ is a typical particle velocity, l is a mean free path, δt is a mean free time, $\mathbf{e}_i$ are the dimensionless vectors. After the computation of f at nodes $\mathbf{V}_i$ the following system on $f_i(t, \mathbf{r}) = f(t, \mathbf{r}, \mathbf{V}_i)$ is obtained:

$$\frac{\partial f_i}{\partial t} + \mathbf{V}_i\nabla f_i + (\mathbf{a}\nabla_{\mathbf{V}} f)_i = -\frac{f_i - f_i^{(eq)}}{\lambda}, \quad i = \overline{1, n}. \tag{2}$$

In cases of fluids and dense gases, number n is not so high for the exact approximation of $(\mathbf{a}\nabla_{\mathbf{V}} f)_i$, so for the approximation of this terms the model representations are used. In this paper the so-called D2Q9 lattice is considered. This lattice is defined by the following set of $\mathbf{e}_i$: $\mathbf{e}_1 = (0, 0)$, $\mathbf{e}_2 = (1, 0)$, $\mathbf{e}_3 = (0, 1)$, $\mathbf{e}_4 = (-1, 0)$, $\mathbf{e}_5 = (0, -1)$, $\mathbf{e}_6 = (1, 1)$, $\mathbf{e}_7 = (-1, 1)$, $\mathbf{e}_8 = (-1, -1)$, $\mathbf{e}_9 = (1, -1)$.

After the approximation of (2) on time and space grids, the following LBE's are obtained:

$$f_i(t_j + \delta t, \mathbf{r}_{kl} + \mathbf{V}_i\delta t) = f_i(t_j, \mathbf{r}_{kl}) + \Delta f_i - \frac{1}{\tau}\left(f_i(t_j, \mathbf{r}_{kl}) - f_i^{(eq)}(\rho, \mathbf{U})\right), \tag{3}$$

where t_j is a node of the time grid, $\mathbf{r}_{kl}$ is a node of the grid on space variables, $\tau = \lambda/\delta t$ is a dimensionless relaxation time and terms Δf_i model the influence of the body force.

In the case of the absence of body forces, (3) is rewritten as:

$$f_i(t_j + \delta t, \mathbf{r}_{kl} + \mathbf{V}_i\delta t) = f_i(t_j, \mathbf{r}_{kl}) - \frac{1}{\tau}\left(f_i(t_j, \mathbf{r}_{kl}) - f_i^{(eq)}(\rho, \mathbf{U})\right). \tag{4}$$

3. Body force action models

At the time interval $[t_j, t_j + \delta t]$ the body force changes the momentum at the node $\mathbf{r}_{kl}$ on the value $\Delta(\rho\mathbf{U}) = \mathbf{F}\delta t$. If the case of weakly-compressible flows is considered, so density at the node $\mathbf{r}_{kl}$ may be approximated by a constant, so the velocity change in this node is equal to $\Delta\mathbf{U} = \mathbf{F}\delta t/\rho$.

In this paper the following widely-used models of body force action are considered:

1) Model of Shan and Chen. This model is proposed in [2, 3] and is based on the following assumption: the body force action at one time step leads to the equilibrium state of the system

of particles with velocities $\mathbf{V}_i$. So at this approach $\delta t = \lambda$ and the velocity value after the body force action is computed as:

$$\mathbf{U}^{(eq)} = \mathbf{U} + \frac{\mathbf{F}\lambda}{\rho}.$$

To the next time moment system is evolved from the equilibrium state, which corresponds to the velocity $\mathbf{U}^{(eq)}$. So, according to this approach the body force terms Δf_i are not introduced and the model is based on the discrete equations (4) in the case, when in right part of (4) $\mathbf{U} = \mathbf{U}^{(eq)}$. The actual velocity is calculated by the correction based on the following expression [3, 10]:

$$\mathbf{U}^* = \mathbf{U} + \frac{\mathbf{F}\delta t}{2\rho}. \tag{5}$$

2) Model of L.S. Luo. The model may be realized in the case of the complete kinetic equation (1) and is based on the following assumption:

$$\nabla_{\mathbf{V}} f \approx \nabla_{\mathbf{V}} f^{(eq)}, \tag{6}$$

which is based on the asymptotic representation of f, where the main part of f is defined by its equilibrium part $f^{(eq)}$. According to (6), the following representation of Δf_i is obtained:

$$\Delta f_i = \rho W_i \left(\frac{(\mathbf{V}_i - \mathbf{U})}{\theta} + \frac{(\mathbf{V}_i \cdot \mathbf{U})}{\theta^2} \mathbf{V}_i \right) \Delta \mathbf{U},$$

where $\theta = kT/m$, k is a Boltzmann constant, W_i are the parameters of the equilibrium distribution function and T is a temperature. From this model the actual velocity is obtained as: $\mathbf{U}^* = \mathbf{U}$.

3) Model of Guo et al. The model, proposed in [7], is based on the representation of body force term by expansion on Hermite polynomials. According to this approach, after the discretization Δf_i are represented by:

$$\Delta f_i = \left(1 - \frac{1}{2\tau}\right) \rho W_i \left(\frac{(\mathbf{V}_i - \mathbf{U}^*)}{\theta} + \frac{(\mathbf{V}_i \cdot \mathbf{U}^*)}{\theta^2} \mathbf{V}_i \right) \Delta \mathbf{U}.$$

4) Model of Kupershtokh et al. In [9] the following representation of $\nabla_{\mathbf{V}} f^{(eq)}$ is obtained: $\nabla_{\mathbf{V}} f^{(eq)} = -\nabla_{\mathbf{U}} f^{(eq)}$. After the changing of velocity on one time interval of length δt on $\Delta \mathbf{U}$, the following expression of the body force term is obtained:

$$\mathbf{a} \nabla_{\mathbf{V}} f \delta t = -(f_i^{(eq)}(\rho, \mathbf{U} + \Delta \mathbf{U}) - f_i^{(eq)}(\rho, \mathbf{U})),$$

and the following approximation is presented in [9]:

$$\Delta f_i = \rho W_i \left(\frac{(\mathbf{V}_i - \mathbf{U})}{\theta} + \frac{(\mathbf{V}_i \cdot \mathbf{U})}{\theta^2} \mathbf{V}_i \right) \Delta \mathbf{U} + \rho W_i \left(\frac{(\mathbf{V}_i \cdot \Delta \mathbf{U})^2}{2\theta^2} - \frac{(\Delta \mathbf{U})^2}{2\theta} \right).$$

The actual velocity is computed by using of (5).

4. Stability analysis

4.1. Statement of the problem

Let us consider dimensionless forms of all the proposed schemes. The stability is investigated in the cases of spatially homogeneous stationary flows in unbounded domain, corresponds to following values of macrocharacteristics: $\rho = 1$, $U_x = U$, $U_y = 0$, $U \in [0, 1]$. The case of the constant body force $\mathbf{F} = (F_x, F_y)$ is considered. Components of $\mathbf{F}$ are represented as: $F_x = F\cos(\varphi)$, $F_y = F\sin(\varphi)$, where $\varphi \in [0, 2\pi)$ is the angle between vectors $\mathbf{F}$ and $\mathbf{U}$. The case of $F \in [0, 1]$ is considered. So for the stability investigation stability domains in the parameter space (U, F, φ) are considered.

Solutions of the LBE's are presented in following form:

$$f_i(t_j, \mathbf{r}_{kl}) = \overline{f}_i + \delta f_i(t_j, \mathbf{r}_{kl}), \tag{7}$$

where δf_i are the perturbations of the unperturbed solutions $\overline{f}_i$, which are calculated as: $\overline{f}_i = f_i^{(eq)}(\rho, \mathbf{U}) = const$, where $\mathbf{U}$ corresponds to the stationary regime discussed above. After the substitution of (7) into LBE's and its linearization the system for δf_i may be obtained. The solution of this system may be presented in following form [11]:

$$\delta f_i(t_j, \mathbf{r}_{kl}) = F_i(t_j)\exp(\mathbf{i}\boldsymbol{\Theta}\mathbf{r}_{kl}), \tag{8}$$

where $\mathbf{i}^2 = -1$, $\boldsymbol{\Theta} = (\theta_x, \theta_y)$, $\theta_\alpha \in [-\pi, \pi]$, $\alpha = x, y$.

After the substitution of (8) into linearized LBE's, the following system of linear difference equations for F_i is obtained:

$$F_i(t_{j+1}) = \sum_{s=1}^{9} G_{is}F_s(t_j), \tag{9}$$

where the components G_{is} for the case of eqs. (4) are presented as:

$$G_{is} = \begin{cases} \left(\left(1 - \frac{1}{\tau}\right) + \frac{1}{\tau}\frac{\partial f_i^{(eq)}}{\partial f_s}(\overline{\mathbf{f}}) \right)\exp(-\mathbf{i}\boldsymbol{\Theta}\mathbf{e}_i), & i = s, \\ \left(\frac{1}{\tau}\frac{\partial f_i^{(eq)}}{\partial f_s}(\overline{\mathbf{f}}) \right)\exp(-\mathbf{i}\boldsymbol{\Theta}\mathbf{e}_i), & i \neq s, \end{cases}$$

and for the case of eqs. (3) the following expressions are written:

$$G_{is} = \begin{cases} \left(\left(1 - \frac{1}{\tau}\right) + \frac{1}{\tau}\frac{\partial f_i^{(eq)}}{\partial f_s}(\overline{\mathbf{f}}) + \frac{\partial(\Delta f_i)}{\partial f_s}(\overline{\mathbf{f}}) \right)\exp(-\mathbf{i}\boldsymbol{\Theta}\mathbf{e}_i), & i = s, \\ \left(\frac{1}{\tau}\frac{\partial f_i^{(eq)}}{\partial f_s}(\overline{\mathbf{f}}) + \frac{\partial(\Delta f_i)}{\partial f_s}(\overline{\mathbf{f}}) \right)\exp(-\mathbf{i}\boldsymbol{\Theta}\mathbf{e}_i), & i \neq s. \end{cases}$$

So the problem of the stability of unperturbed solutions of systems (3) or (4) is reduced to the problem of the stability of the null solution of a system (9). This solution is stable, when the following inequality on the absolute values of the eigenvalues of matrix $\{G_{is}\}$ is realized [11]:

$$|\lambda(\tau, U, F, \varphi, \theta_x, \theta_y)| \leq 1. \tag{10}$$

For the finding of the eigenvalues of $\{G_{is}\}$ the QR-algorithm realized in EISPACK package is used.

4.2. Results of computations

At the first stage we try to compare all of the proposed models at the case of $\tau = 1$ (because the model of Shan and Chen is valid only in this case). For the construction of the stability domains the following uniform grid $M \times N$ nodes is constructed: $\Omega = \{(F_p, \varphi_r) | F_p = ph_F, \varphi_r = rh_\varphi, h_F = (F_M - F_1)/(M-1), h_p = (\varphi_N - \varphi_1)/(N-1), p = \overline{1, M}, r = \overline{1, N}\}$. For every node of Ω the eigenvalues of matrix $\{G_{is}\}$ are obtained for the grids, constructed in the domain, where parameter U and vector $\boldsymbol{\Theta}$ are changed. If inequality (10) in node (F, φ) is realized for all the

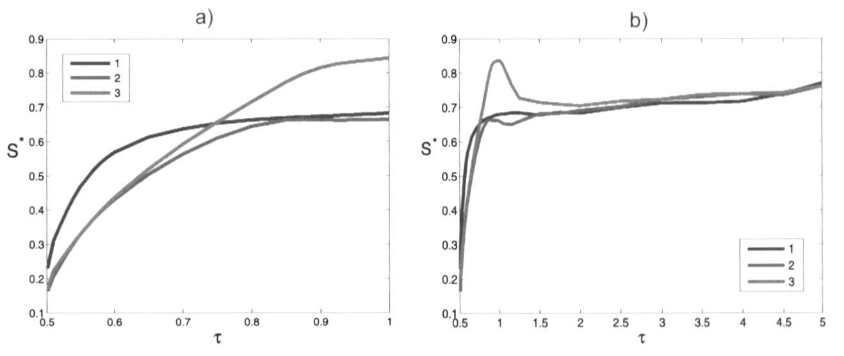

Figure 1. Plots of the maximal values $S^*(\tau)$ for considered models in the cases of $\tau \in [1/2, 1]$ (a) and $\tau \in [1/2, 5]$: 1 — model proposed by Luo; 2 — model proposed by Guo et al; 3 — model proposed by Kupershtokh et al

nodes in the domains, where U and $\mathbf{\Theta}$ are changed, so this node is included to the stability domain. The area of the stability domain $S(F, \varphi)$ is calculated by rectangular quadrature rule.

All calculations were performed at the grids with 100×100 nodes. At table 1 the maximal values of S and its arguments $(F^*, \varphi^*) = \arg\max S(F, \varphi)$ are presented.

Table 1. The results of the numerical stability analysis: the maximal values of the areas of the stability domains $S^* = \max\limits_{(F,\varphi)} S(F, \varphi)$ and its arguments for the case of $\tau = 1$.

Model	S^*	F^*	φ^*
Shan and Chen model	0,8480	1	3,1733
Luo model	0,6735	0,3434	3,1733
Guo et al model	0,6621	0,6465	3,1733
Kupershtokh et al model	0,8480	1	3,1733

As it can be seen, the largest values of S are achieved for the case of $\varphi \approx \pi$, that corresponds to the case of the opposite directions of vectors $\mathbf{F}$ and $\mathbf{U}$. The values of S^* and F^* are the same at the cases of Shan and Chen model and Kupershtokh et al model. The smallest value of S takes place in the case of Guo et al model.

At the second stage the models, which may be realized in the case, when τ may be varied, are investigated. For these cases we compute the values of $S^*(\tau) = \max\limits_{(F,\varphi)} S(F, \varphi, \tau)$ and $F^*(\tau)$ corresponds to this maximal value. We consider the cases of $\tau \in [1/2, 1]$, which is widely used in practical calculations and case of $\tau \in [1/2, 5]$ for investigation of the role of τ on the stability on LBE's with the body force action. At fig. 1 the plots of $S^*(\tau)$ are presented in both cases, corresponding plots of $F^*(\tau)$ are presented in fig. 2. As it can be seen, at the values of τ close to 1/2, the maximal area of the stability domain is realized for Luo's model, but in this case the values of $F^*(\tau)$ are smaller, when in the case of Kupershtokh et al model. When $\tau \in [1/2, 1]$, maximal values of $F^*(\tau)$ takes place in the case of Guo et al model, but in a bounded interval. For a wide range of the τ values the maximal values of S^* and F^* are realized in the case of Kupershtokh et al model. In the cases of large values of τ all the models have the same behavior. So due to the realized analysis at values of $\tau \approx 1/2$ the models proposed in [5, 9] may be recommended. In the cases of the highest values of τ model proposed in [9] may be recommended to practical simulations due to its stability properties.

The typical examples of the practical problems, where τ is a very close to 1/2, are the problems of solid–liquid phase transitions, such as melting or freezing [12, 13]. In these problems models

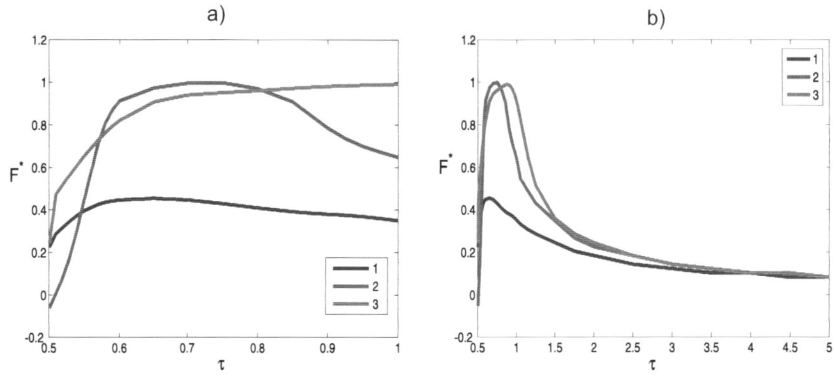

Figure 2. Plots of the maximal values $F^*(\tau)$ for considered models in the cases of $\tau \in [1/2, 1]$ (a) and $\tau \in [1/2, 5]$: 1 — model proposed by Luo; 2 — model proposed by Guo et al; 3 — model proposed by Kupershtokh et al

proposed in [5, 9] may be used. An example of the model with a wide range of τ (from 1/2 to 2) is the problem of interaction of multicomponent fluids [14, 15, 16]. For example, in [14] the case of $\tau = 1$ is considered for the simulation. So, for this kind of problem, the models, proposed in [2, 3] may be used. For the general case of multicomponent systems, the model proposed in [9] may be used for simulation due to the better stability properties in a wide range of τ.

5. Conclusion

In the presented paper stability analysis of LBE's for the case of body force action is realized. The analysis is performed by von Neumann method and is based on the linearisation of LBE's. The stability is estimated by the values of the areas of the stability domains in parametric space. as the results of the analysis, the recommendation on the use of the models in practical simulations may be performed. The proposed approach to the analysis may be extended to the investigation of the other cases of body force actions, especially for the cases of multiphase and multicomponent flows.

References

[1] Li Q, Luo K H, Kang Q J, He Y L, Chen Q and Liu Q 2016 *Progr. in Energy and Combus. Sc.* **52** 62
[2] Shan X and Chen H 1993 *Phys. Rev. E* **47** 1815
[3] Shan X and Chen H 1994 *Phys. Rev. E* **49** 2941
[4] He X, Chen X and Doolen G D 1998 *Phys. Rev. E* **57** R13
[5] Luo L S 1998 *Phys. Rev. Lett.* **81** 1618
[6] Ladd A J C and Verberg R 2001 *J. Stat. Phys.* **104** 1191
[7] Guo Z, Zheng C and Shi B 2002 *Phys. Rev. E* **65** 046308
[8] Shan X, Yuan X-F and Chan H 2006 *J. Fluid. Mech.* **550** 413
[9] Kupershtokh A L, Medvedev D A and Karpov D I 2009 *Comp. Math. Appl.* **58** 965
[10] Huang H, Krafczyk M and Lu X 2011 *Phys. Rev. E* **84** 046710
[11] Richtmyer R D and Morton K W 1994 *Difference Methods for Initial-Value Problems* (New York: John Wiley and Sons)
[12] Li D, Ren Q, Tong Z X and He Y L 2017 *Int. J. Heat and Mass Transf.* **112** 795
[13] Huo Y and Rao Z 2017 *Int. J. Heat and Mass Transf.* **122** 292
[14] Joshi A S, Peracchio A A, Grew K N and Chiu W K S 2007 *J. Phys. D* **40** 2961
[15] Xu H and Dang Z 2017 *Numer. Heat Transf. B* **72** 250
[16] Succi S 2001 *The Lattice Boltzmann Equation for Fluid Dynamics and Beyond* (Oxford: Clarendon Press)

Parallel realization of the computational algorithm based on the implicit lattice Boltzmann equations

Elizaveta A Prokhorova, Gerasim V Krivovichev

Faculty of Applied Mathematics and Control Processes Saint Petersburg State University, 7/9 Universitetskaya nab., Saint Petersburg, 199034, Russian Federation

E-mail: proxliza@mail.ru, g.krivovichev@spbu.ru

Abstract. The approach to parallel realization of the computational algorithm based on the implicit lattice Boltzmann equations is proposed and discussed. Program realization is performed in C++ language using the OpenMP technology. The program is tested on the lid-driven cavity flow problem. The obtained speedup for the cases of the computations with various processes numbers demonstrate the perspectives for realization of the algorithm on systems based on graphics processing units and systems with hybrid architecture.

1. Introduction

Lattice Boltzmann method (LBM) in the last decades is considered as an efficient tool for computational modeling of fluid and gas flows defined by a wide range of the Knudsen number [1, 2]. Despite the traditional approach applied in computational fluid dynamics, LBM is based on the solution of the problems for *kinetic* equations. Widely used difference schemes are obtained by discretization along the projections of its characteristics on the physical space. These schemes are called *lattice Boltzmann equations* (LBE's). The great advantage of the algorithms, based on LBE's, is based on its high parallelism and possibilities to realization on the modern multiprocessor systems on various architectures.

In the last years, algorithms of LBM are successfully parallelized and its program realizations are implemented in various programming languages and multiprocessor systems. In [3] LBM code is written in C language and is parallelized using OpenMP libraries. Domain decomposition method of data parallelism is adopted. In [4] five different parallel libraries, such as matlabpool, pMatlab, GPU-Matlab, OpenMP and OpenMP+OpenMPI are used to parallel realization of the LBM code. The results of the computations demonstrate that parallel simulations using Matlab subroutines perform almost equally. Parallel simulations using C language with OpenMP libraries were ten times faster. Most of the performed realizations deal with the computations using Graphics Processing Units (GPU's). In [5] GPU compatible version of the so-called bundle data layout is proposed to achieve the performance of the LBM algorithm. Szoke et al in [6] provides a detailed comparison between the Unified Paralell C (UPC) library and the Computer Unified Device Architecture (CUDA) development kit through the parallelization of the algorithm of two-dimensional LBM. In [7] algorithm of the entropic LBM is realized on the hybrid system, which consists of multicore Central Processing Units (CPU's) with GPU's. The resulting computational performance is compared with the results obtained after computation

IOP Conf. Series: Journal of Physics: Conf. Series **1038** (2018) 012041 doi:10.1088/1742-6596/1038/1/012041

on a single CPU core and a single GPU device. Wang et al in [8] proposed an approach to the acceleration of the volumetric LBM for the problems of computational haemodynamics.

In the papers, considered above, the algorithms are based on the *explicit* LBE's. This type of LBE's is mathematically formulated as an explicit difference equations, which are *conditionally stable*. It restricts the values of the relaxation parameter of the gas, so the range of the considered physical processes is restricted. This problem may be solved by an application of the *implicit* LBE's.

Presented paper is devoted to the parallel realization of the computational algorithm based on implicit LBE's proposed by Krivovichev in [9]. These equations may be considered as an extension of the LBE's proposed in [10].

The structure of the paper is follows. In Section 2 implicit LBE's are presented and its stability condition is discussed. In Section 3 the parallel algorithm is presented. Program realization and performance of the algorithm are discussed in Section 4. Some concluding remarks are made in Section 5.

2. Implicit lattice Boltzmann equations

Implicit LBE's are obtained as discrete analogues of the following system of kinetic Bhatnagar — Gross — Krook equations for the modeling of the dynamics of the ensemble of particles with velocities $\mathbf{V}_i$:

$$\frac{\partial f_i}{\partial t} + \mathbf{V}_i \nabla f_i = -\frac{1}{\lambda}(f_i - f_i^{(eq)}), \tag{1}$$

where $f_i = f_i(t, \mathbf{r})$ are the distribution functions of particles with velocities $\mathbf{V}_i = V\mathbf{e}_i$, where $V = l/\delta t$, l is a mean free path, δt is a mean free time, $\mathbf{e}_i$ are the dimensionless vectors, t is a time, $\mathbf{r}$ is a vector of space variables, λ is a relaxation time, $f_i^{(eq)}$ are the equilibrium distribution functions. In this paper two-dimensional case is considered and the following set of vectors $\mathbf{e}_i$ (named D2Q9 lattice) is used: $\mathbf{e}_1 = (0, 0)$, $\mathbf{e}_2 = (1, 0)$, $\mathbf{e}_3 = (0, 1)$, $\mathbf{e}_4 = (-1, 0)$, $\mathbf{e}_5 = (0, -1)$, $\mathbf{e}_6 = (1, 1)$, $\mathbf{e}_7 = (-1, 1)$, $\mathbf{e}_8 = (-1, -1)$, $\mathbf{e}_9 = (1, -1)$.

System (1) may be rewritten in integral form on the time interval $[t, t + \delta t]$ [11]:

$$f_i(t + \delta t, \mathbf{r} + \mathbf{V}_i\delta t) - f_i(t, \mathbf{r}) = -\frac{1}{\lambda}\int_0^{\delta t}(f_i(t + \xi, \mathbf{r} + \mathbf{V}_i\xi) - f_i^{(eq)}(t + \xi, \mathbf{r} + \mathbf{V}_i\xi))d\xi. \tag{2}$$

In [9] system (2) is rewritten in equivalent form with parametric coefficients:

$$f_i(t + \delta t, \mathbf{r} + \mathbf{V}_i\delta t) - f_i(t, \mathbf{r}) = -\frac{1-\sigma}{\lambda}\int_0^{\delta t}(f_i(t + \xi, \mathbf{r} + \mathbf{V}_i\xi) - f_i^{(eq)}(t + \xi, \mathbf{r} + \mathbf{V}_i\xi))d\xi-$$

$$-\frac{\sigma}{\lambda}\int_0^{\delta t}(f_i(t + \xi, \mathbf{r} + \mathbf{V}_i\xi) - f_i^{(eq)}(t + \xi, \mathbf{r} + \mathbf{V}_i\xi))d\xi, \tag{3}$$

where $\sigma \in [0, 1]$ is a dimensionless parameter.

After the computation of integral in (3) by simple quadrature rules and discretization of (3) on time and space grids, the following system of implicit difference equations is obtained

$$f_i(t_j + \delta t, \mathbf{r}_{kl} + \mathbf{V}_i\delta t) - f_i(t_j, \mathbf{r}_{kl}) = A(\sigma, \tau)(f_i(t_j, \mathbf{r}_{kl}) - f_i^{(eq)}(\mathbf{f}(t_j, \mathbf{r}_{kl})))+$$

$$+B(\sigma, \tau)(f_i(t_j + \delta t, \mathbf{r}_{kl} + \mathbf{V}_i\delta t) - f_i^{(eq)}(\mathbf{f}(t_j + \delta t, \mathbf{r}_{kl} + \mathbf{V}_i\delta t))), \tag{4}$$

where coefficients A and B are defined by quadrature rule used for the computation of integrals in (3), t_j is a node of time grid constructed with step δt, $\mathbf{r}_{kl}$ is a node of space grid constructed with step l. In [9] six systems of equations (4) are obtained.

The expression for the apparent viscosity of LBE's (4) may be obtained by application of the Chapman — Enskog method to the first differential approximation of (3) [12] and is written as follows:

$$\nu = \left(1 + \frac{A-B}{2}\right)\frac{\tau}{3}\frac{l^2}{\delta t}. \tag{5}$$

The stability conditions of LBE's (4) may be obtained from the condition of non-negativity of ν: $\nu \geq 0$. For example, for the LBE's defined by the coefficients $A = -(1-\sigma)/\tau$ and $B = -\sigma/\tau$, this condition is written as following inequality: $\sigma \geq 1/2 - \tau$. In the presented paper, we consider this system of implicit LBE's for the performing of all calculations.

3. Parallelization of the algorithm

The main computational problem solved during the calculations based on LBE's (4) is a problem of the solution of the systems of nonlinear algebraic equations for the computation of $f_i(t_j + \delta t, \mathbf{r}_{kl} + \mathbf{V}_i\delta t)$. These systems are formed by (4) and are written in every node of two-dimensional grids. For the convenience, let us rewrite (4) in following form:

$$f_i(t_j + \delta t, \mathbf{r}_{kl}) - f_i(t_j, \mathbf{r}_{kl} - +\mathbf{V}_i\delta t) = A(\sigma,\tau)(f_i(t_j, \mathbf{r}_{kl} + \mathbf{V}_i\delta t) - f_i^{(eq)}(\mathbf{f}(t_j, \mathbf{r}_{kl} + \mathbf{V}_i\delta t)) +$$

$$+B(\sigma,\tau)(f_i(t_j + \delta t, \mathbf{r}_{kl}) - f_i^{(eq)}(\mathbf{f}(t_j + \delta t, \mathbf{r}_{kl}))), \tag{6}$$

So at one time step $[t_j, t_j+\delta t]$ system (6) must be solved in every internal node $\mathbf{r}_{kl}$ of the space grid for the determination of the values $f_i(t_j + \delta t, \mathbf{r}_{kl})$. If the case of D2Q9 lattice is considered ($i = \overline{1,9}$) and space grid of $N_x \times N_y$ nodes is used, at every time step $(N_x - 2) \times (N_y - 2)$ systems of nine nonlinear algebraic equations must be solved. It must be noted, that at moment $t = t_j + \delta t$ these systems are *independent on each other*, so the algorithm is highly parallelizable on data.

The following approach to the parallelization of the algorithm may be proposed: at every time step the marches on nodes of fixed space coordinate (x or y) may be considered *independently* and every march may be realized by a specific computational process. In the presented paper march on coordinate x is realized (fig. 1). During the computations these processes may be performed independently in cycle on k from 2 to $N_x - 1$ and results must be synchronized during the march on x at every time step by means of the technology of the programming realization.

In every grid node system (6) is solved by the Newton — Raphson method. Let us rewrite (6) in node $\mathbf{r}_{kl}$ in compact form. Let us introduce new variables $z_i = f_i(t_j + \delta t, \mathbf{r}_{kl})$, $i = \overline{1,n}$, so (6) may be rewritten as:

$$\mathbf{F}(\mathbf{z}) = \mathbf{0},$$

where $F_i = (1-B)z_i + Bf_i^{(eq)}(\mathbf{z}) - \varphi_i$, $\varphi_i = (1+A)f_i(t_j, \mathbf{r}_{kl} - \mathbf{V}_i\delta t) + Af_i^{(eq)}(\mathbf{f}(t_j, \mathbf{r}_{kl} - \mathbf{V}_i\delta t))$.

The Newton — Raphson method is based on the following iterative procedure:

$$\mathbf{z}^{s+1} = \mathbf{z}^s - [\nabla\mathbf{F}(\mathbf{z}^s)]^{-1}\mathbf{F}(\mathbf{z}^s), \tag{7}$$

where $\nabla\mathbf{F}$ is a Jacobi matrix defined by following components:

$$\frac{\partial F_i}{\partial z_j}(\mathbf{z}^s) = \begin{cases} 1 - B + B\frac{\partial f_i^{(eq)}}{\partial z_j}(\mathbf{z}^s), & i = j, \\ B\frac{\partial f_i^{(eq)}}{\partial z_j}(\mathbf{z}^s), & i \neq j. \end{cases}$$

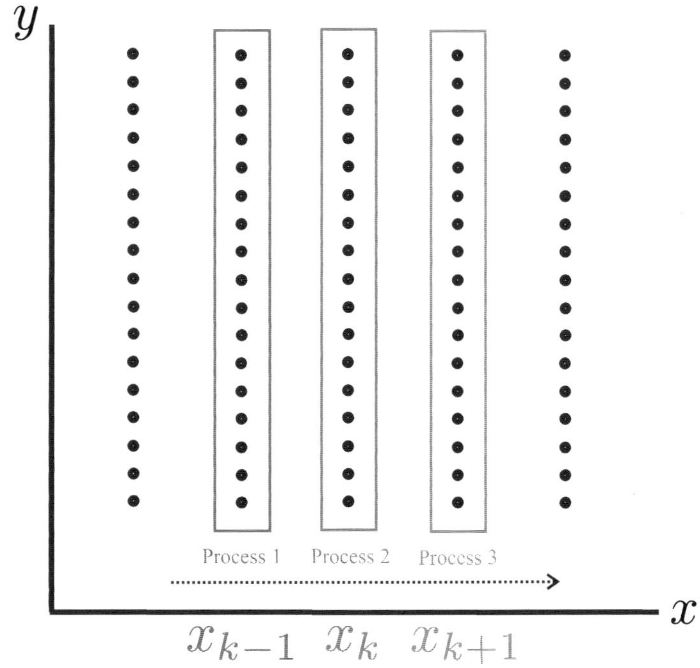

Figure 1. Schematic representation of the approach to parallel realization of computational algorithm

For the avoiding of the procedure of the inversion of the Jacobi matrix, the following equivalent of the procedure (7) is used:

$$\mathbf{z}^{s+1} = \mathbf{z}^s + \delta\mathbf{z}^s,$$

where $\delta\mathbf{z}^s = \mathbf{z}^{s+1} - \mathbf{z}^s$ may be obtained after the solution of the following system of linear algebraic equations:

$$[\nabla\mathbf{F}(\mathbf{z}^s)]\delta\mathbf{z}^s = -\mathbf{F}(\mathbf{z}^s). \tag{8}$$

In our work system (8) is solved by Gaussian method.

Initial member of the iterative sequence $\mathbf{z}^0$ is obtained by apllication of the *explicit* LBE's [11]:

$$f_i(t_j + \delta t, \mathbf{r}_{kl}) - f_i(t_j, \mathbf{r}_{kl} - \mathbf{V}_i\delta t) = -\frac{1}{\tau}(f_i(t_j, \mathbf{r}_{kl} - \mathbf{V}_i\delta t) - f_i^{(eq)}(\mathbf{f}(t_j, \mathbf{r}_{kl} - \mathbf{V}_i\delta t))),$$

so the procedure "predictor-corrector" is realized during the computations.

4. Program realization

Due to the proposed algorithm, the parallelization is realized by processes, which may be realized by *threads*. So the program realization may be effectively performed by using the procedures of OpenMP technology. Using the OpenMP directives the parallel performance may be realized by introduction of parallel sections in the code. It may be realized at the following way: the section of code that is meant to run in parallel regimes is marked accordingly, with a compiler directive that will cause the threads to form before the section is executed.

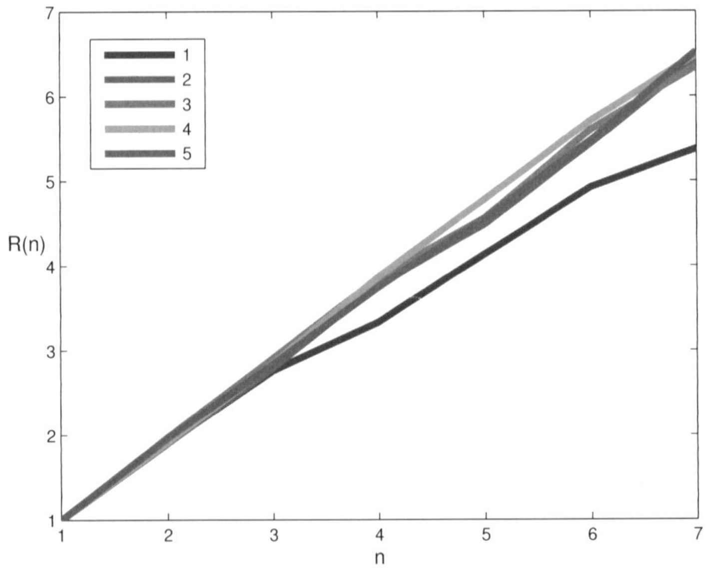

Figure 2. Plots of speedup for the cases of different spatial grids: 1 — 50 × 50 nodes; 2 — 100 × 100 nodes; 3 — 150 × 150 nodes; 4 — 200 × 200 nodes; 5 — 250 × 250 nodes

The realization is performed using C++ language. The march procedure on the coordinate x is realized by directive `pragma omp for` for the cycle on k.

The program realization is tested on the numerical solution of the problem of lid-driven cavity flow, stated in [14]. The performance of the parallel realization of the algorithm is estimated by computations on space grids with 50 × 50, 100 × 100, 150 × 150, 200 × 200 and 250 × 250 nodes. After the computations on every grid the *speedup* is used for the estimation of the algorithm:

$$R(n) = \frac{T_1}{T_n},$$

where n is a number of threads, T_1 is a time of computation for the case of one thread.

All computations are realized on a computational cluster of the Faculty of Applied Mathematics and Processes of Control of Saint-Petersburg State University. At fig. 2 the plots of speedup $R(n)$ for the case of every space grid are presented. As it can be seen, the speedup dependency on n is close to linear, so the proposed algorithm is effective for the parallel realization and perspective for realization on system with multicore GPU's. For this purpose authors wish to realize the proposed parallel algorithm for computations using CUDA technology. After this step and after testing of the obtained software, the authors try to make their code free to download according to GNU license agreement.

5. Conclusion

In the presented paper the approach to parallel realization of the computational algorithm based on implicit LBE's is proposed and discussed. Program realization is performed in C++ language using the OpenMP technology. The program is tested on the lid-driven cavity flow problem. The obtained speedup for the cases on the computations with various processes numbers demonstrate the perspectives for realizations on GPUs and systems with hybrid architectures.

Acknowledgements
The reported study was funded by Russian Foundation of Basic Research according to research project No. 16-31-00021 mol_a.

References
[1] Levesley J, Gorban A N and Packwood 2011 *Approximation algorithms for complex systems* ed E H Georgoulis, A Iske and J Levesley (Berlin: Springer-Verlag) p 127
[2] Bo Y, Wang P, Guo Z and Wang L-P 2017 *Comp. and Fluids* **155** 9
[3] Basha M, Sidik M A C and Beriache M 2017 *Int. J. of Heat and Mass Transf.* **115** 158
[4] Basha M and Sidik M A C 2018 *Int. J. of Heat and Mass Transf.* **116** 715
[5] Robertsen F, Westerholm J and Mattila K 2017 *Int. J. of High Perf. Comput. Appl.* **31** 246
[6] Szoke M, Jozsa T I, Koleszar A, Moulitsas I and Konoszy L 2017 *ACM Trans. on Math. Soft.* **44** 8
[7] Ye Y, Li K, Wang Y and Deng T 2015 *Comp. and Fluids* **110** 114
[8] Wang Z, Zhao Y, Sawchuk A P, Dalsing M C and Yu H 2015 *Comp. and Fluids* **115** 192
[9] Krivovichev G V 2014 *Appl. Math. Sci.* **8** 5003
[10] Sankaranarayanan B, Shan X, Kevrekidis I G and Sundaresan S 2002 *J. Fluid Mech.* **452** 61
[11] He X and Luo L-S 1997 *Phys. Rev. E* **55(6)** R6333
[12] Sofonea V 2003 *J. Comp. Phys.* **184**, 422
[13] Wolf-Gladrow D A 2000 *Lattice-Gas Cellular Automata and Lattice Boltzmann Models* (Berlin: Springer)
[14] Ghia U, Ghia K N and Shin C T 1982 *J. Comp. Phys.* **48** 387

IOP Conf. Series: Journal of Physics: Conf. Series **1038** (2018) 012042 doi:10.1088/1742-6596/1038/1/012042

Estimation of hydrogen permeability parameters by the results of a 'cascade' penetration experiment

Yury V Zaika[1,2] **and Natalia I Rodchenkova**[1]

[1] Institute of Applied Mathematical Research, Karelian Research Centre, Russian Academy of Sciences, Pushkinskaya St. 11, 185910 Petrozavodsk, Russia
[2] Petrozavodsk State University, Lenina St. 33, 185910 Petrozavodsk, Russia

E-mail: zaika@krc.karelia.ru, nirodchenkova@krc.karelia.ru

Abstract. High-purity hydrogen is required for clean energy and a variety of chemical technology processes. Different alloys, which may be well-suited for use in gas-separation plants, were investigated by measuring specific hydrogen permeability. One had to estimate the parameters of diffusion and sorption to numerically model the different scenarios and experimental conditions of the material usage (including extreme ones), and identify the limiting factors. This paper presents a nonlinear model of fast hydrogen permeability in accordance with the specifics of the experiment and the parametric identification algorithm.

1. Introduction

Studies on the interaction of hydrogen isotopes with structural materials are mainly necessitated by problems in the energy industry (including hydrogen one) [1] and metal protection from hydrogen corrosion. Different alloys, which may be well-suited for use in gas-separation plants, were investigated by measuring specific hydrogen permeability. One had to estimate the parameters of diffusion and sorption to numerically model the different scenarios and experimental conditions of the material usage (including extreme ones), and identify the limiting factors. The limiting factors are diffusion processes as well as physical and chemical phenomena at the surface. The transfer parameters depend on the technological features of material batch production. It is therefore unreasonable to target at 'tabular data'. Instead, effective algorithms of solving the inverse problem of parametric identification are necessary. The model matches the method of hydrogen permeability and takes into account only the main factors for the membrane filtering problem. This research is based on the article [2] and on the experimental data [3].

2. Mathematical model

2.1. Distributed transfer model

The sample of a structural material acts as a vacuum chamber barrier. The pressure of molecular hydrogen is built up in steps at the inlet side. The declining pressure in the input chamber and increasing pressure in the output chamber are measured.

Consider hydrogen transfer through the sample (ℓ is the plate thickness and S is its area). The sample degassing is performed in advance. The temperature T is constant throughout the experiment. The membrane is thin and the material has a high hydrogen permeability coefficient so we restrict ourselves to the standard diffusion equation:

$$\partial_t c(t,x) = D(T)\partial_x^2 c(t,x),\ t>0, \quad c(0,x)=0,\ x\in[0,\ell]. \tag{1}$$

Here, $c(t,x)$ is the concentration of diffusing dissolved (atomic) hydrogen. The diffusion coefficient D depends on the sample temperature T in an Arrhenius way: $D = D_0\exp\{-E_D/[RT(t)]\}$. Nonlinear boundary conditions are derived from the flux balance:

$$-\frac{dQ_{\text{in,out}}}{dt} = \left[2s(T)\mu(T)p_{0,\ell}(t) - b(T)c_{0,\ell}^2(t)\right]S = \mp SD\frac{\partial c}{\partial x}\Big|_{x=0,\ell}. \tag{2}$$

Here, $Q_{\text{in}}(t)$, $Q_{\text{out}}(t)$ are the amounts of hydrogen atoms in the input chamber of volume V_{in} and output chamber of volume V_{out}, $c_0(t)\equiv c(t,0)$, $c_\ell(t)\equiv c(t,\ell)$. The identity sign is frequently used in the sense of equality by definition. The negative sign on the right side is used for the indices in, 0. Within the considered operating temperature range ($T\in[500,800]$K) the gaseous hydrogen is in molecular form, but for consistency (considering that atomic hydrogen diffuses through the metal) we use atoms as the unit. The incident particle flux density J_p is related to the pressure p by the Hertz–Knudsen formula: $J_p = p/\sqrt{2\pi mkT}$ (k is the Boltzmann constant, m is the mass of a hydrogen molecule). In the context of the experiment it is convenient to select the following measurement units $[\ell]$ = cm, $[p]$ = Torr. Then we numerically obtain the dependence $J_p = \mu p$, $\mu(T)\approx 2.474\cdot 10^{22}/\sqrt{T}$ $\big([\mu] = 1_{\text{H}_2}/(\text{Torr cm}^2\,\text{s}),\ [T]=\text{K}\big)$. The processes of physical adsorption, chemisorption, dissociation of molecules into atoms, and dissolution take place on the surface. Only a small part of 'incident' H atoms will, however, be absorbed into the membrane volume. This is taken into account by the factor $2s$. One can write s (as a parameter of the model) instead of $2s$, but it is more convenient to interpret the probability factor s as the fraction of absorbed H atoms within the $2s$ notation. Thus, $2s\mu p$ is the resulting flux of atoms through the surface into the bulk without differentiation into more elementary stages.

Hereinafter, $J_{0,\ell} = bc_{0,\ell}^2$ are the desorption flux densities, b is the desorption coefficient. If a constant saturation pressure of molecular hydrogen p_{s} = const is maintained at a constant temperature T on both sides of the membrane, the equilibrium concentration $\bar{c}$ of the dissolved atomic diffusionally mobile hydrogen is finally established. By equating the derivatives in the model (2) to zero, we get $\bar{c}\propto\sqrt{p_{\text{s}}}:\ \bar{c}=\Gamma\sqrt{p_{\text{s}}},\ \Gamma\equiv\sqrt{2s\mu/b}$.

Let us define the experimental conditions more precisely. The volumes $V_{\text{in,out}}$ comprise several liters, the thickness of the membrane is $\ell <$ mm, the area S is about 1 cm^2, the inflow pressure $p_0(0)$ is within several hundreds Torr. It now remains to find the magnitudes of Q_{in}, Q_{out}. Within the time of transfer through the membrane the gas is in the thermodynamical quasi-equilibrium, wherefore we use the formula $N = pV/(kT)$. Here, N is the number of gas particles occupying the volume V at the temperature $T = T_V$ and the pressure p (in the SI system $[p]$ = Pa, $[V]$ = m^3, $[k]$ = J/K). Taking into account Torr = 133.322 Pa and Pa = J/m^3 (formally), we get the expressions for the corresponding pressures and volumes in the conditions (2) $Q = 2N = \alpha pV/T$, $\alpha\approx 1.931\cdot 10^{19}$. Here, p, V, T are the values of the variables in the selected units (Torr, cm^3, K).

The membrane is situated in the tube between the inlet and the outlet chambers. The membrane temperature should be taken for the kinetic constant $\mu(T)$. The gas inside the volumes $V_{\text{in,out}}$ may get heated up. During the preliminary experiment it is recommended to fill the chambers with a practically impermeable metal membrane between them with ambient gas. Then heat the tube and record the pressure rises. Within the framework of the ideal gas approximation this procedure enables estimation of the increments of the gas temperature inside the chambers. The corresponding gas temperatures are the ones to be used in the formula given for Q (and the subsequent ones, only excluding the value μ). The need of such a refinement arises from the characteristics of this particular experimental unit. Such an adjustment of the values of T should not cause difficulties in further calculations. Besides, this procedure has relatively little effect on the final calculation of the model pressures taking into account the measurement errors and relatively large volumes V.

2.2. *Fast hydrogen permeability model*

A quasi-stationary state is quickly established when the membrane is thin and the material has a sufficiently high hydrogen permeability coefficient: the traps are saturated and the diffusant concentration distribution is practically linear with respect to the thickness. Since near-to-surface concentrations cannot be measured, the Richardson approximation is usually used to analyze the penetrating flux: $J(t) = -D\partial_x c = D\ell^{-1}\big[c_0(t) - c_\ell(t)\big] \approx J_R(t) = D\Gamma\ell^{-1}\big[\sqrt{p_0(t)} - \sqrt{p_\ell(t)}\big]$. Let us formulate the problem of modelling the concentrations $c_{0,\ell}$ using the pressures $p_{0,\ell}$ (the problem is also of interest *per se*) without the quasi-equilibrium simplification $c(t) = \Gamma\sqrt{p(t)}$.

The quasi-stationary state is achieved within a time t_0, which is short compared to the total experiment time (the traps become saturated and $\partial_x c = -[c_0(t) - c_\ell(t)]/\ell$). So, the original model (1)–(2) can be simplified (taking into account the formula $Q = \alpha pV/T$):

$$\dot{p}_{0,\ell}(t) = \mp\beta_{0,\ell}\big[c_0(t) - c_\ell(t)\big], \quad \beta_{0,\ell} \equiv SD\left[\alpha V_{\text{in,out}}\ell\right]^{-1}T, \tag{3}$$

$$2s\mu p_{0,\ell}(t) - bc_{0,\ell}^2(t) = \pm D\ell^{-1}\big[c_0(t) - c_\ell(t)\big], \; t \geq t_0 > 0. \tag{4}$$

The dimensionless normalized variables are convenient for numerical simulation

$$X_{0,\ell}(t) = 1 + 2\ell c_{0,\ell}(t)bD^{-1}, \quad a_{0,\ell}(t) = 4\ell^2\Gamma^2 p_{0,\ell}(t)b^2D^{-2} - 1. \tag{5}$$

In addition, the system of equations (4) is compactly written in the form $a_0 + 2X_\ell = X_0^2$, $a_\ell + 2X_0 = X_\ell^2$. For the variable $X \equiv X_\ell$ we obtain the incomplete quartic equation $[X^2 - a_\ell]^2 = 4[2X + a_0]$. The explicit expression is somewhat cumbersome and we will anyway have to numerically integrate the first equation of (3) in the form $\dot{p}_0 = f(p_0)$. Therefore, we shall aim to derive equations for $X_{0,\ell}$, since information about the dynamics of $c_{0,\ell}$ is also important.

Differentiate the equations (4) with respect to time and substitute the pressure derivatives from (3). For the variables $X_{0,\ell}$ we get the system:

$$\dot{X}_0(t) = -2sM_0\big[X_0 - X_\ell\big] \cdot \frac{X_\ell - V_{\text{in}}V_{\text{out}}^{-1}}{X_0X_\ell - 1}, \quad M_0 \equiv \frac{\mu ST}{\alpha V_{\text{in}}}, \tag{6}$$

$$\dot{X}_\ell(t) = 2sM_\ell\big[X_0 - X_\ell\big] \cdot \frac{X_0 - V_{\text{in}}^{-1}V_{\text{out}}}{X_0X_\ell - 1}, \quad M_\ell \equiv \frac{\mu ST}{\alpha V_{\text{out}}}, \quad t \geq t_0. \tag{7}$$

Let us formulate step-by-step the numerical algorithm of modelling the pressures $p_{0,\ell}(t)$ ($t \geq t_0$) for the current values of D, b, s coefficients.

1. We fix $t = t_0$: omit fast transient processes. For the variable $X \equiv X_\ell$ we choose the root of the biquadratic polynomial $[X^2 - a_\ell(t_0)]^2 - 4[2X + a_0(t_0)]$. From physical considerations it follows that $c_\ell(t_0) > \bar{c}_\ell(t_0)$ and thus $X > 1 + \sqrt{1 + a_\ell} = 1 + 2\ell\Gamma\sqrt{p_\ell(t_0)}\,bD^{-1}$.
2. The system of equations $a_0 + 2X_\ell = X_0^2$, $a_\ell + 2X_0 = X_\ell^2$ ($t = t_0$) yields the value of $X_0(t_0)$.
3. We numerically integrate the ODE system (6), (7)($X_{0,\ell}(t_0)$). The change in (5) defines the concentrations $c_{0,\ell}(t)$, which are used to calculate the pressures $p_{0,\ell}(t)$, $t \geq t_0$ from (4).

3. Numerical modelling of hydrogen permeability

Experimental practices employ various modifications of the penetration method and thermal desorption spectrometry (TDS). A successive use of various methods often causes impurities to appear on the sample surface. Experimental data are the input for the inverse problems of parametric identification, which are sensitive to the level of error. It is therefore advisable to aggregate experiments to improve the accuracy and informative value of the measurements. We suggest the following set-up of the 'cascade' experiment.

A sufficiently high pressure of gaseous hydrogen gas was built up in steps at the inlet side. The penetrating flux was determined by mass-spectrometry in the vacuum maintained at the outlet side. This is a penetration method. Its advantage is a reliable determination of the diffusion coefficient by the Daines - Berrer method (based on lag time). It allows distinguishing between the bulk and the surface processes in the model, keeping in mind that surface parameters are significantly more difficult to estimate. When the steady state level of the penetrating flux is registered, we increase the inlet pressure and wait until a new steady state value is established. Using (at least) three pressure jumps at the inlet side we record the steady state flux values at the outlet side, thus determining 'the degree of rectilinearity' of the isotherm. Then the pumping for vacuum building is stopped and the experiment proceeds as the 'communicating vessels' method (this stage is described above). When pressure values become nearly equal (the sample is almost uniformly saturated with hydrogen) it is possible to turn off the heating, create the vacuum at both sides of the membrane and begin slowly reheating the sample (TDS-experiment).

3.1. Numerical modelling of the penetration experiment

The proposed model is adapted to the experimental conditions and the data range for alloys based on V group metals with high hydrogen permeability, in particular, data for vanadium alloys which are presented in [2–4]. We fix $T = 673\,\mathrm{K}$, $\ell = 0.05\,\mathrm{cm}$, $\mu(T) = 2.474 \times 10^{22}/\sqrt{T}\ 1_{\mathrm{H}_2}/(\mathrm{Torr\,cm^2\,s})$, $D = 2 \times 10^{-5}\,\mathrm{cm^2/s}$, $\Gamma = 2 \times 10^{20}\,1_{\mathrm{H}}/(\mathrm{cm^3}\sqrt{\mathrm{Torr}})$, $\Phi = D\Gamma = 4 \times 10^{15}\,1_{\mathrm{H}}/(\mathrm{cm\ s}\sqrt{\mathrm{Torr}})$. We set the value $s = 1.2 \times 10^{-4}$ and calculate the corresponding desorption coefficient $b = 2\mu s/\Gamma^2 = 5.7 \times 10^{-24}\,\mathrm{cm^4/s}$. The input pressures $\bar{p}_{1,2,3} = \{30, 50, 70\}$ Torr were built up in steps at the inlet, and maintained to achieve steady state fluxes at the outlet. The concentration at the membrane outlet side is near zero and at the inlet side a stationary concentration is quickly established (but it is lower than the equilibrium one): $\tilde{c} < \bar{c}$. Within the model we determine $\bar{c}_i$ and $\tilde{c}_i$ by the formulas

$$\bar{c}_i = \sqrt{2\mu s\bar{p}_i b^{-1}},\ 2\mu s\bar{p} - b\tilde{c}^2 = \bar{J} = D\tilde{c}\ell^{-1} \Rightarrow \tilde{c}_i = -a + \sqrt{a^2 + \Gamma^2\bar{p}_i},\ a \equiv D[2b\ell]^{-1}.$$

Stage I. The boundary value problem of the penetration method is:

$$c_t = Dc_{xx},\ 0 < x < \ell,\ t > 0,\quad c(t,0) = \tilde{c}_1,\ c(t,\ell) = 0,\ t \geq 0,\quad c(0,x) = 0,\ x \in [0,\ell].$$

The penetrating flux is $J_1(t) = -Dc_x|_\ell = D\tilde{c}_1\ell^{-1}\big[1 + 2\sum_{n=1}^{\infty}(-1)^n \exp\{-n^2\pi^2Dt/\ell^2\}\big]$.

From the computational point of view it is convenient to introduce the dimensionless time $t' = Dt/\ell^2$ which is oriented at the characteristic diffusion time ℓ^2/D. As small $t \to 0$ a singularity appears if we directly use the partial sum. So we provide another expression for $f(t') \ll 1$ (using the properties of the Jacobi theta function (θ_3)):

$$f(t') = 1 + 2\sum_{n=1}^{\infty}(-1)^n \exp\{-n^2\pi^2t'\} = \frac{2}{\sqrt{\pi t'}}\sum_{m=2n-1}\exp\left\{\frac{-m^2}{4t'}\right\}\ (n \in \mathbb{N}).$$

Stage II ($t_* \to t_0 = 0$ is the new t time zero, $t \geq 0$):

$$c_t = Dc_{xx},\ 0 < x < \ell,\ t > 0,\quad c(t,0) = \tilde{c}_2,\ c(t,\ell) = 0,\quad c(0,x) = \tilde{c}_1(\ell - x)\ell^{-1},\ x \in [0,\ell].$$

The penetrating flux is $J_2 = D\ell^{-1}\big[\tilde{c}_1 + (\tilde{c}_2 - \tilde{c}_1)\big[1+2\sum_{n=1}^{\infty}(-1)^n \exp\{-n^2\pi^2Dt/\ell^2\}\big]\big]$.

Stage II ends with $c(t_*, x) = \tilde{c}_2(\ell - x)\ell^{-1}$.

Stage III. The formulas are similar to the cyclic interchange $\tilde{c}_1 \to \tilde{c}_2$, $\tilde{c}_2 \to \tilde{c}_3$.

The result of connecting the stages into a single 'experimental' curve of the penetrating flux $J(t)$ (conventionally, $t = (t_1, t_1^* + t_2, t_2^* + t_3)$, $J = (J_1, J_2, J_3)$) is shown in Fig. 1.

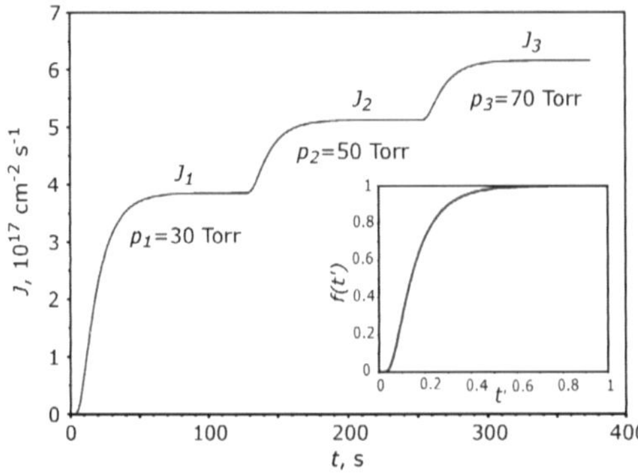

Figure 1. Establishment of steady state fluxes.

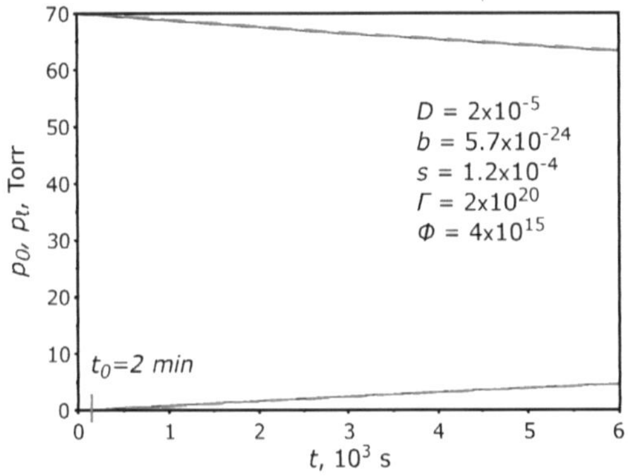

Figure 2. Dynamics of pressures $p_{0,\ell}(t)$.

3.2. The hydrogen penetration experiment without vacuum pumping

When the steady state permeability value is established during the penetration experiment, continuous pumping at the outlet and maintenance of constant pressure at the inlet are stopped. The aggregated experiment moves to the stage of 'communicating vessels': inlet pressure declines and outlet pressure grows ($p_{0,\ell}(t)$ are measured). We are so far talking about the direct problem of modelling hydrogen pressures inside the volumes $V_{\rm in,out}$.

The additional input data are $S = 0.5\,\mathrm{cm}^2$, $V_{\rm in} = 1500\,\mathrm{cm}^3$, $V_{\rm out} = 2200\,\mathrm{cm}^3$, $p_0(0) = \bar{p}_0 = \bar{p}_3$, $\bar{c} = \Gamma\sqrt{\bar{p}_0}$. We use here the data for $V_{85}Ni_{15}$ alloy [3].

If we use the ODE system (6), (7) instead of the 'full' model (1)–(2)$\big(c(0,x) = \tilde{c}_3(\ell - x)/\ell\big)$, standard software packages will suffice, for example, Scilab (we substitute the values of the hydrogen temperature inside the chambers $V_{\rm in,out}$ into the expressions for $M_{0,\ell}$). To this end one should skip the initial time t_0 within several minutes until a quasi-stationary (not quasi-equilibrium) mode is established. Fig. 2 shows that there is no significant additional error during the numerical simulation. We wait until the sample is uniformly saturated (pressure values become nearly equal) before moving to the TDS experiment. In this article we confine ourselves to aggregating only two types of experiments and referencing [5] (and [6] for powder).

4. Parametric identification procedure

4.1. Determination of the lag time

The flux curve $J_1(t)$ asymptotically moves to the stationary value $\bar{J}_1 = D\tilde{c}_1/\ell$. Hence, $Q(t) = \int_0^t J(\tau)\,d\tau \approx \bar{J}_1[t - \ell^2/(6D)]$ $(t \geq t_* = t_1^*)$. The intersection of the asymptote with the t axis gives so-called lag time $\tau_0 = \ell^2/(6D)$ which allows estimating the diffusion coefficient (the Daines - Berrer method). Analytically, $\tau_0 \approx t_* - \int_0^{t_*} J_1(\tau)\bar{J}_1^{-1}d\tau$, where $J_1 \approx \bar{J}_1$, $t \geq t_*$. Note that a relative magnitude which does not require absolute values of the penetrating flux in any measurement units ($\bar{J}_1 = \sup J_1(t)$) is under the integral sign. In addition, the value τ_0 does not depend on $\tilde{c}_1$. It is usually assumed that the locally equilibrium concentration $\bar{c}_1 = \Gamma\sqrt{\bar{p}_1}$ is quickly established at the inlet, so one can additionally estimate the solubility $\Gamma = \sqrt{2s\mu/b}$ and the permeability $\Phi = D\Gamma$ using the corresponding value $\bar{J}_1 = D\bar{c}_1/\ell$. This assumption is not used in this article. We assume that according to the experimental conditions the stationary inlet concentration $c_0(t) \approx \tilde{c}_1 < \bar{c}_1$ $(t \geqslant \varepsilon,\ \varepsilon \ll 1)$ is considered to be quickly established given that $c_\ell(t) \approx 0$. The value of $\tilde{c}_1$ as such is yet to be clarified. Thus, only the estimate of the diffusion coefficient D is considered reliable at this stage.

With a new zero-time reference ($t_* \to t_0 = 0$), integrating the expression $J_2(t)$ we get $\int_0^t [J_2(\tau) - \bar{J}_1]\,d\tau \approx [\bar{J}_2 - \bar{J}_1]\cdot[t - \ell^2(6D)^{-1}]$, where $\bar{J}_i = D\tilde{c}_i/\ell$, $t \geq t_* = t_2^*$. Formally, we obtain the same expressions for the lag time and the estimate of D if we change both the zero time and the flux baseline $\bar{J}_1$ value overstatement. There is no additional information here (about the target values of the surface parameters b and s), but the triple penetration experiment allows to refine D. For the considered values of the model parameters we have $\tau_{01} \approx \tau_{02} \approx \tau_{03} \approx 20.8\,\mathrm{s}$.

4.2. Isotherm. Initial estimates of b, s.

In experimental practices analyze the isotherm, i.e. the curve of the steady state penetrating flux $\bar{J}$ dependence on the inlet pressure $\bar{p}$ when vacuum pumping is performed at the outlet side. If one targets at the Sieverts' law and the (quasi)equilibrium concentration $\bar{c} = \Gamma\sqrt{\bar{p}}$ at the inlet side (hence $\bar{J} = D\bar{c}/\ell$), then it is natural to plot the dependence $\bar{J} = \bar{J}(\sqrt{\bar{p}})$.

Let us analyze the steady state flux balance equation: $2s\mu\bar{p} - b\tilde{c}^2 = \bar{J} = D\tilde{c}\ell^{-1} \Rightarrow$

$$\tilde{c} = a\big[-1 + \sqrt{1 + a^{-2}\Gamma^2\bar{p}}\,\big],\ a \equiv D(2\ell b)^{-1}, \quad \bar{J} = D\tilde{c}\ell^{-1} = D\ell^{-1}a\big[-1 + \sqrt{1 + a^{-2}\Gamma^2\bar{p}}\,\big].$$

Asymptotic analysis shows that the dependence $\bar{J}(\sqrt{\bar{p}})$ has a parabolic shape ($\bar{J} \propto \bar{p}$) at low $\bar{p}$:

$$a^{-2}\Gamma^2\bar{p} \equiv x^2 \ (x \propto \sqrt{\bar{p}}) \Rightarrow \tilde{c} = a\big[-1 + \sqrt{1 + x^2}\,\big] = a[0.5\,x^2 + \ldots]$$

and a straight line form at high inlet pressures $\bar{p}$: $a^{-2}\Gamma^2\bar{p} \gg 1 \Rightarrow \bar{J} = -D^2(2\ell^2 b)^{-1} + D\Gamma\ell^{-1}\sqrt{\bar{p}}$.

Using the straight-line segment of the isotherm we find $D\Gamma/\ell$ (the slope of the straight line) and knowing the estimate of D we determine the initial approximation of the solubility coefficient Γ. From the intersection of the straight line with the ordinate axis we find b. Knowing the values of $\Gamma = \sqrt{2s\mu/b}$ and b compute s and $\Phi = D\Gamma$.

4.3. The final stage of parametric identification algorithm

The penetration method with vacuum pumping is characterized by a significant measurement error, and data on the penetrating flux are required (and this, in turn, requires a more accurate determination of the vacuum system characteristics). The model of the dissolved hydrogen concentration jump at the inlet side is not very precise either. We are brought to a conclusion that the 'communicating vessels' stage, where molecular hydrogen pressures are measured over a long time, is characterized by a much higher accuracy of measurements.

Thus, the first stage of the aggregated experiment is perceived as preliminary estimation of the coefficients D, b, s. It is essential that the solution of the inverse problem is unique, since the results obtained for thin laboratory membranes are extrapolated (recalculated) to the dimensions of real-life structures. The results are 'fine-tuned' by means of local variation of the preliminary values of D, b, s in the fast hydrogen permeability model (the ODE system).

Acknowledgments

This work was supported by the Russian Foundation for Basic Research, Project 15-01-00744.

References

[1] Yukhimchuk A A (Ed) 2009 *Hydrogen isotopes. Fundamental and applied studies* (Sarov: Russian Federal Nuclear Center) p 697 [in Russian]

[2] Kozhakhmetov S, Sidorov N, Piven V, Sipatov I, Gabis I and Arinov B 2015 Alloys based on Group 5 metals for hydrogen purification membranes *J. Alloys and Compounds* **645** S36–40

[3] Dolan M D 2010 Non-Pd BCC alloy membranes for industrial hydrogen separation *J. Membrane Science* **362** 12–28

[4] Voyt A, Sidorov N, Sipatov I, Dobrotvorskii M, Piven V and Gabis I 2017 Hydrogen solubility in $V_{85}Ni_{15}$ alloy *Int. J. Hydrogen Energy* **42** 3058–63

[5] Zaika Yu V and Kostikova E K 2017 Computer simulation of hydrogen thermal desorption by ODE-approximation *Int. J. Hydrogen Energy* **42** 405–15

[6] Zaika Yu V and Rodchenkova N I 2009 Boundary-value problem with moving bounds and dynamic boundary conditions: diffusion peak of TDS-spectrum of dehydriding *Appl. Math. Modelling* **33** 3776–91

IOP Conf. Series: Journal of Physics: Conf. Series **1038** (2018) 012043 doi:10.1088/1742-6596/1038/1/012043

Estimation of localization of point sources from a printed circuit board in the near field

I V Skvortsov[1], V V Bochkarev[1] and R R Latypov[1]

[1] Kazan Federal University, Institute of Physics, Kazan, 420008, Russia
E-mail: skvorcov_ilya@mail.ru

Abstract.
This article describes algorithms for localizing sources of electromagnetic radiation from a printed circuit board in the near field. The sources of radiation are represented in the form of a set of the simplest oscillators - Hertz dipoles. Comparison of the two methods of localization of radiation sources is carried out using the Tikhonov regularization and the LASSO regression. The model examples show that the application of LASSO regression gives better results.

1. Introduction

The main problem of electromagnetic compatibility is the ability of technical means to function in conditions of unintended disturbances. The most important characteristics of such disturbances are location of their source and radiation power. The most well-known method for detecting radiation sources is to scan the amplitudes of the component fields at each point of space [1, 2]. This method is commonly used for measurements in the far field, but it can also be used in the near field to improve accuracy of measurement and reduce the requirements for the experimental setup [2]. To ensure electromagnetic compatibility of electronic devices, it is important to be able to determine the location of the radiation sources from the printed circuit boards. There are situations where, according to the known value of the field at each measuring point and the distance to the plane of the object under study, it is necessary to determine the regions of the most active radiation. As a rule, to solve such an inverse problem, a model is used in which the sources of radiation from the printed circuit board are replaced by a set of Hertz dipoles [3-5]. Equations for electric and magnetic Hertz dipoles are well known in the near and far fields [6]. Hertz dipoles are located in the grid nodes.

The inverse problem of determining the location of radiation sources by measuring amplitudes of the electromagnetic field is incorrectly posed. As a rule, Tikhonov regularization is used in such cases [7, 8]. However, recently, LASSO regression (least absolute shrinkage and selection operator) is often used in various fields of physics [9-12]. The main advantage of this approach is that the method allows obtaining low-dimensional solutions (with a small number of non-zero coefficients). In this paper, we propose to use LASSO regression to solve the problem of finding the most active regions of radiation from a printed circuit board.

2. Model of equivalent dipoles

Usually, an equivalent model of a printed circuit board is a combination of electric or magnetic dipoles located in the plane of an object. To build a dipole model of a printed circuit board, the results of modelling the tangential components of the electric or magnetic fields in the measurement plane are used [7, 8]. In our model, we use only magnetic dipoles.

Let us consider an elementary dipole with a dipole moment $\vec{p}$, which is at the point (x_0, y_0, z_0) in the plane of the object. The magnetic field generated by an elementary dipole at the observation point (x, y, z) and located at a distance r, is determined by the expression [6]:

$$\vec{H}(\vec{r}) = j\omega\left(jk + \frac{1}{r}\right)\cdot\left(\vec{p}\times\vec{l}_r\right)\cdot G(\vec{r}), \tag{1}$$

Here $G(\vec{r}) = \frac{e^{-jkr}}{4\pi r}$ is the Green's function in free space, $\vec{p} = p_x\cdot\vec{l}_x + p_y\cdot\vec{l}_y + p_z\cdot\vec{l}_z$ is the magnetic dipole moment, $\vec{l}_r = \frac{\vec{r}}{r}$ is the unit radial vector.

If the dipole is located in the XOY plane, the expression has the form:

$$(\vec{p} \times \vec{l}_r) = p_y \cdot \vec{l}_x \cdot \frac{z-z_0}{r} - p_x \cdot \vec{l}_y \cdot \frac{z-z_0}{r} + p_x \cdot \vec{l}_z \cdot \frac{y-y_0}{r} - p_y \cdot \vec{l}_z \cdot \frac{x-x_0}{r} \quad (2)$$

The expressions for the components of the magnetic field have the form:

$$H_x = j\omega \frac{e^{-jkr}}{4\pi r^2}\left(jk + \frac{1}{r}\right)(z - z_0) \cdot p_y = A_x(r) \cdot p_y, \quad (3)$$

$$H_y = -j\omega \frac{e^{-jkr}}{4\pi r^2}\left(jk + \frac{1}{r}\right)(z - z_0) \cdot p_x = A_y(r) \cdot p_x, \quad (4)$$

$$H_z = j\omega \frac{e^{-jkr}}{4\pi r^2}\left(jk + \frac{1}{r}\right)(y - y_0) \cdot p_x - j\omega \frac{e^{-jkr}}{4\pi r^2}\left(jk + \frac{1}{r}\right)(x - x_0) \cdot p_y = A_{zy}(r) \cdot p_y - A_{zx}(r) \cdot p_x \quad (5)$$

Equations (3,4) can also be written briefly as

$$\begin{bmatrix} 0 & A_x \\ A_y & 0 \end{bmatrix} \begin{bmatrix} p_x \\ p_y \end{bmatrix} = \begin{bmatrix} H_x \\ H_y \end{bmatrix} \quad (6)$$

The components of the magnetic field and the transition matrix A are assumed to be known, but the dipole moments p are unknown. The system of equations (6) is expressed in matrix form as $A \cdot p = H$, and its formal solution can be written as $p = A^{-1} \cdot H$ (where A^{-1} is the inverse of A). In our case, the matrix A is normally ill-conditioned, and therefore a small change in the right-hand side in (6) will result in a large change in the solution. Thus, the inverse problem of determining the location of radiation sources by measuring amplitudes of the electromagnetic field is incorrectly posed. To solve incorrectly posed problems, regularization procedure is usually used.

3. Ridge regression (Tikhonov regularization) and LASSO regression

Regularization is used to find a stable solution of ill-posed problems by taking into account some a priori information. For example, using Tikhonov regularization, a solution with a minimal Euclidean norm is chosen from all possible solutions that ensure a small disparity of equations (6). To find such a solution, it is necessary to minimize the functional $\Omega(y, \lambda)$:

$$\Omega(p, \lambda) = \|A \cdot p - H\|_2^2 + \lambda \|p\|_2^2, \quad (7)$$

where λ is the regularization parameter. This optimization problem has an explicit solution [13]:

$$p_\lambda = \left(A^H A + \lambda^2 I\right)^{-1} A^H \cdot H \quad (8)$$

Among other possible methods of regularization, LASSO regression is of particular interest for this task. In this case, a solution with a minimum L1-norm is chosen from possible solutions. Such problem is equivalent to minimizing the functional

$$\Omega(p, \lambda) = \|A \cdot p - H\|_1^2 + \lambda \|p\|_1 \quad (9)$$

This regularization method makes allows obtaining solutions of low dimensionality (as a rule, most of the coefficients in the solution vector are equal to zero).

There is no explicit solution to the problem of minimizing the functional (9), but there is an effective numerical method [14]. Thus, minimizing the LASSO Ω (p, λ) functional and choosing the parameter λ, we obtain vectors p for two tangential components of the electromagnetic radiation at each point.

4. The application of regularization for the estimation of radiation sources

In real conditions, the sources of electromagnetic radiation on the circuit board are unevenly distributed over the entire plane of the object. Sources are usually located in certain areas. Electromagnetic radiation in this case is often regarded as a superposition of radiation from point uncorrelated sources [7, 8]. A set of thermal noise elements and components of the device can serve as an example of point uncorrelated sources. The number of sources can usually be considered to be much less than the number of grid nodes.

To test the algorithms under consideration, let us define a model with two point sources located close to each other with coordinates $A_1(50; 50)$ and $A_2(50; 60)$ and a radiation power of 6 and 8 conventional units. The grid size is 10,000 × 10,000 nodes. The distance from the measurement plane to the object under study is 5 times larger than the distance between the grid nodes. The wavelength is set 300 times higher than the grid spacing. The distribution of the magnetic field generated by the two sources is shown in Fig. 1.

Figure 1. The module of the magnetic field strength from two closely spaced sources

The region with the highest radiation intensity is in the centre of Fig. 1. Axes x and y are oriented in the plane of measurement.

To solve the inverse problem and determine the position of the sources, we used Tikhonov regularization and LASSO regression. The estimates of the power distribution of radiation sources obtained in two ways are shown in Fig. 2. Fig.3 shows the vertical sections of the distributions from Fig.2. It can be seen that the LASSO regression estimate gives a better source localization than Tikhonov's regularization.

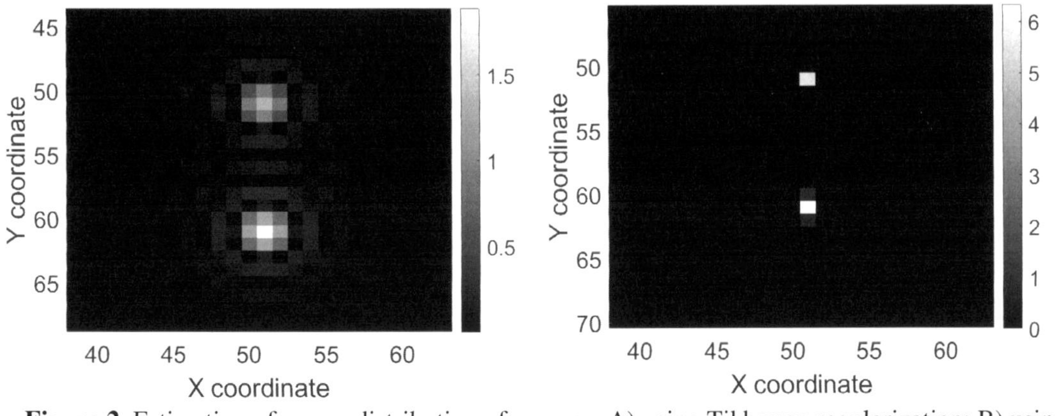

Figure 2. Estimation of power distribution of sources: A) using Tikhonov regularization; B) using LASSO regression

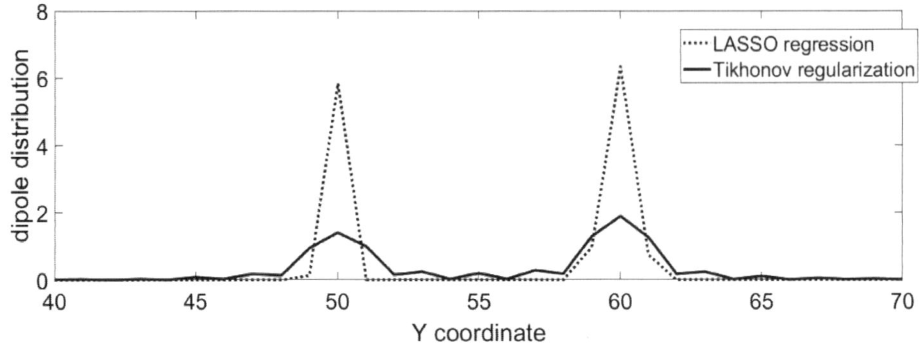

Figure 3. Estimation of the power distribution of sources along the x = 51 coordinate using Tikhonov regularization and LASSO regression

Let us estimate quantitatively the quality of reconstruction of the sources location. As can be seen from Fig. 2A and Fig.3, the point source of radiation after the reconstruction spreads into a bell-shaped distribution. The displacement of the center of gravity of the "bell" and its half-width are good quantitative characteristics of distortions. The position of the center of gravity can be found by formulas

$$x_0 = \frac{\sum_{i,j} x_i |p_{ij}|^2}{\sum_{i,j} |p_{ij}|^2}, \quad y_0 = \frac{\sum_{i,j} y_i |p_{ij}|^2}{\sum_{i,j} |p_{ij}|^2} \tag{10}$$

where the summation is performed in the immediate vicinity of the "bell". Having determined the position of the centre of gravity, we calculate the half-width of the "bell" Δ in accordance with the following expressions:

$$\Delta = \sqrt{\Delta_x^2 + \Delta_y^2}, \; \Delta_x^2 = \frac{\sum_{i,j} (x_i - x_0)^2 |p_{ij}|^2}{\sum_{i,j} |p_{ij}|^2}, \; \Delta_y^2 = \frac{\sum_{i,j} (y_i - y_0)^2 |p_{ij}|^2}{\sum_{i,j} |p_{ij}|^2} \tag{11}$$

Also, the quality characteristic of the localization can be the fraction of the radiation power attributable to the maximum point in the estimated distribution. The larger this fraction, the better the source localization and the greater the conformity of the estimate to the initially assigned distribution.

Table 1 shows the values of numerical parameters characterizing the quality of source localization using Tikhonov regularization and LASSO regression. The calculations were carried out at a signal-to-noise ratio of 500, or 54 dB (normal white noise was added to the model values of the field strength). As can be seen from the table, the use of LASSO regression yields significantly better results.

Table 1. The quality of the estimation of the distribution of radiation sources using the Tikhonov regulation and LASSO regression. The two columns indicate the parameters for the first and second sources.

	Displacement of the center of gravity		Range of values, Δ		Fraction of the radiation power attributable to the maximum point, %	
	1st	2nd	1st	2nd	1st	2nd
Tikhonov regularization	(51;50.5)	(51;60.6)	7.4	8.1	20.1	19.8
LASSO regression	(51;51)	(51;61)	0.13	0.12	85.5	87.2

5. Conclusion

The article considers two algorithms of estimating the localization of electromagnetic radiation sources from a printed circuit board in the near field. To restore the power distribution of sources from measurements of the field strength, Tikhonov regularization and LASSO regression were used. Comparison on model examples showed that LASSO regression gives a more accurate estimation of source localization than Tikhonov's regularization. Also, estimates of source power distributions using LASSO regression are more resistant to noise.

Acknowledgments

The work is performed according to the Russian Government Program of Competitive Growth of Kazan Federal University

References

[1] Gregson S McCormick J Parini C 2007 Principles of Planar Near-Field Antenna Measurements. (London:The Institution of Engineering and Technology)
[2] Weng H Beetner D G DuBroff R E Shi J 2007 IEEE Transactions on Electromagnetic Compatibility **49(3)** 805–815
[3] Sarkar T K Taaghol A 1999 IEEE Transactions on Antennas and Propagation **47(3)** 566-573
[4] Tong X Thomas D Nothofer A Sewell P Christopoulos C 2010 IEEE Trans. Electromagn. Compat **52(2)** 462–470
[5] Nishina B Qiang C 2016 IEEE Transactions on Antennas and Propagation **64(4)** 1334 - 1341
[6] Sophocles J Orfanidis 2002 Electromagnetic Waves and Antennas (Piscataway:Rutgers University)
[7] Tong X 2010 Simplified equivalent modelling of electromagnetic emissions from printed circuit boards (Nottingham: University of Nottingham)
[8] Gorbunova A Baev A Konovalyuk M Kuznetsov Y 2015 IEEE International Symposium on Electromagnetic Compatibility (EMC) 450 – 455
[9] Do H N Choi J Lim C Maiti T 2015 IEEE Conf. of Advanced Intelligent Mechatronics 984-89
[10] Wu Y Chen Y Shi Y Lu C Song D 2016 IEEE Conf. on Digital Signal Processing 190-93
[11] Jones L K 2009 IEEE Transactions of Information theory **55(12)** 5700-27
[12] Zhang B Geng J Lai L 2015 IEEE Transactions of Signal Processing **63(9)** 2209-24
[13] Tikhonov A N 1963 **151(4)** Soviet Mathematics 1035–1038
[14] Tibshirani R 1996 Journal of the Royal Statistical Society. Series B **58 (1)** 267–88

IOP Conf. Series: Journal of Physics: Conf. Series **1038** (2018) 012044 doi:10.1088/1742-6596/1038/1/012044

Formation of an active part of inertial mass based piezoelectric nanogenerator

A A Semenova, N A Lashkova, A I Maximov, V A Moshnikov

Educational-scientific laboratory of Nanomaterials, Micro- and nanoelectronics department, St Petersburg Electrotechnical University «LETI», Saint-Petersburg 197376, Russia

E-mail: anastassiyan@mail.ru

Abstract. Zinc oxide piezoelectric nanorstructured coatings were formed on tantalum and aluminum substrates by the chemical solution methods. The methods mentioned above are suitable for formation of active parts of inertial mass based piezoelectric nanogenerator (PENG). This PENGs design using leads to an enhanced output due to the increased effective active parts area. Obtained nanostructured ZnO coatings were investigated by scanning electron microscopy and atomic force microscopy.

1. Introduction

In accordance with the increasing energy demands a large number of alternative energy converters are being developed. Recently, converters of mechanical impact pulses energy into electrical current – piezoelectric nanogenerators (PENG) – are widely disseminated [1, 2].

One of the most prospective materials for a PENGs active parts formation is zinc oxide ZnO – wide bandgap (E_g = 3.36 eV at T = 300 K) semiconductor compound of A^2B^6 group, which crystallizes in wurzite modification under normal conditions [3]. Zinc oxide has piezoelectric properties due to the lack of the inversion center and to the partially ionic nature of the relationship between the cations Zn^{2+} and anions O^{2-}.

In most papers [4, 5] for the ZnO nanorods (NR) based PENG is used a construction in which the nanorods are vertically oriented to a stiff substrate and upper electrode. During the deformation of $1D$-nanostructures in the 6th order axis direction the electrical signal is formed. The main disadvantage of this design is the applications constraint due to the required horizontal devices position and a uniform mechanical loads distribution on upper electrode surface.

Flexible substrates application narrows the range of methods suitable for the active part formation due to low temperature resistance. However, with the using of PENGs in a flexible manner the possibility of uneven vibration impact appears, as well as the implementation of the nanogenerator with a partially suspended fixed design [6, 7].

Among a wide variety of methods for a ZnO based PENGs active part formation chemical bath deposition was used [8] because of the low synthesis temperatures and the possibility of nanostructures formation on both sides of a substrate in a single technological cycle.

2. Piezoelectric nanogenerator design

The structure of the piezoelectric generator created in this work is shown in figure 1. PENG is performed on a flexible dielectric substrate (marker 1) with lowers electrodes coated sides. The necessity of their separation is caused by opposite polarities of the voltage generated by the upper and lower (marker 2) active parts of the nanogenerator.

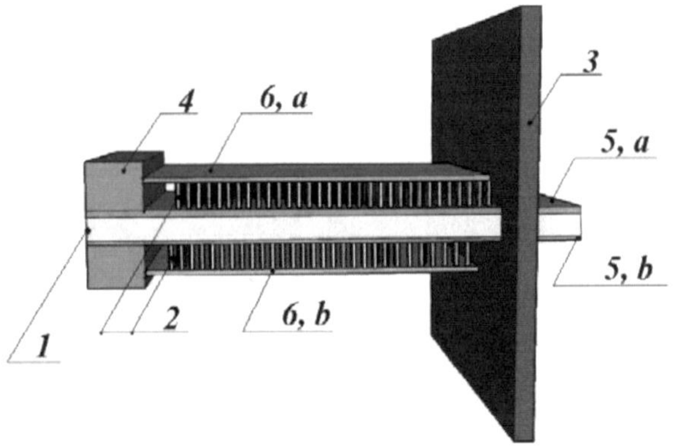

Figure 1. The ZnO nanorods based piezoelectric nanogenerator scheme.

One of the parts is rigidly fixed in the wall, located in the vertical plane (marker 3). Vibration impact is transmitted through the wall on the fixed part of the PENG. Increasing the nanogenerator sensitivity to a mechanical impact is achieved by the inertial mass application (marker 4) made of a dielectric polymeric material.

Lower electrodes terminals are located behind the wall (marker 5, *a* and *b*). The opposite contact to the active parts is carried out using flat electrodes (marker 6, *a* and *b*) fixed in the inertial mass.

The equivalent circuit diagram of the considered device is presented in figure 2. Terminals 1 and 2 (figure 2) correspond to the markers 6, *a* and 5, *a* (figure 1), respectively (contacts to the "top" active part of the PENG). Terminals 3 and 4 (figure 2) correspond to the markers 5, *b* and 6, *b* (figure 1).

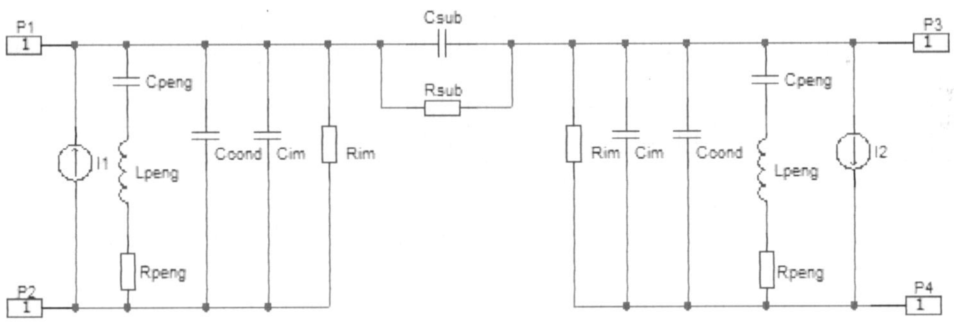

Fig. 2. The equivalent electrical circuit of ZnO nanorods based piezoelectric nanogenerator.

The main difference between the construction mentioned above and a one active part based PENG consists of increased parasitic capacitances C_{im}, corresponded to the inertial mass, and C_{sub}, corresponded to the substrate capacitance. The capacitance reduction is achieved by applying materials with a low dielectric permeability ε at high frequencies and decreasing of electrodes unused area (5, *a* and *b* in figure 1).

During substrate selection penetration of the material is also needed to be taken into account (due to the high voltages (up to 200 V) supplied to the generator).

3. Experimental

The PENGs active part formation was carried out in two stages: seed layers deposition via liquid phase ionic layering (SILAR) and nanorods synthesis by chemical solution deposition. Seed layers application is necessary for higher nanorods orientation achievement, as well as improving their adhesion to a substrate.

In the paper tantalum and aluminum substrates were used as the bottom electrode.

3.1. Deposition of ZnO seed layers

SILAR method (liquid phase ionic layering, «Successive Ionic Layer Adsorption and Reaction») [9] consists of the sequential immersion of a substrate into appropriate solutions and proceeds of the following stages:

- creation of the initial solution containing synthesized substances ions;
- ions adsorption on a substrate during solution immersion;
- rinsing the substrate for excess ions removing from the surface;
- reaction in the interaction of adsorbed ions with an aqueous agent;
- drying process for a film securing on a substrate surface.

Zinc nitrate hexahydrate $Zn(NO_3)_2{\bullet}6H_2O$ with the 0.1 M concentration and 4.4% aqueous solution of ammonia were used for the solution creation with zinc-ammonium complexes $[Zn(NH_3)_4]^{2}$:

$$Zn(NO_3)_2 + NH_4OH \rightarrow [Zn(NH_3)_4]^{2+} + nH_2O + 2NO_3^- \qquad (1)$$

Within washing in aqueous solution at low temperature (20°C) adsorbed complexes $[Zn(NH_3)_4]^{2+}$ partially react with zinc hydroxide formation:

$$[Zn(NH_3)_4]^{2+} + 4H_2O \rightarrow Zn^{2+} + 4NH^{4+} + 4OH^- \qquad (2)$$

$$Zn^{2+} + 2OH^- \rightarrow Zn(OH)_2 \qquad (3)$$

Increased (more than 90^0) temperature is used for zinc oxide synthesis:

$$Zn(OH)_2 + O_2 \rightarrow ZnO + H_2O + O_2 \qquad (4)$$

Occurring on the substrate surface during the synthesis overall process reaction can be written:

$$[Zn(NH3)_n]^{2+} + nH_2O \rightarrow ZnO_{solid} + nNH^{4+} + (n\text{-}2)OH^- \qquad (5)$$

Air atmosphere drying process at $T = 350^0$C within 1 min is the final stage of an each cycle.

3.2. ZnO nanorods formation

ZnO nanorods were produced on the obtained thin films via chemical bath deposition method. $Zn(NO_3)_2$ at 25 mM was perfomed as a Zn^{2+} precursor. Hexamethylenetetramine was used (HMTA) $C_6H_{12}N_4$ for the buffer solution creation and consequently providing the possibility of carrying out synthesis at the low (less than 100°C) temperature due to its slow two-step OH^- groups release.

Appeared by hydrolysis of zinc precursor cations Zn^{2+} react with OH^- anions with various hydroxides formation: $ZnOH^+$, $Zn(OH)_2$, $[Zn(OH)_3]^-$ and $[Zn(OH)_4]^{2-}$. Reactions take place in the solution are described in details at [10]. The charged reaction products may be adsorbed on different charged crystal faces of growing structures or nucleation layer crystallites with the further *hydroxide-oxide* transformation. $Zn(OH)_2$ don't participate in 1*D*-nanostructures growth, because as a result of its flow the uncharged zinc hydroxide is formed.

Dielectric insulation of the nanorods, as well as control of the morphology of the obtained structures was achieved by addition to stock solution *L*-Cysteine ($C_3H_7NO_2S$) used as the zwitterionic surface-active substance (surfactant) with concentration 1 surfactant molecule per 10 Zn^{2+} cations. $C_3H_6NO_2^+$ adsorption mainly on ZnO nanorods lateral surfaces leads to an increase in 1*D*-nanostructures growth rate in the *c*-axis direction. Also surfactant hydrophobic parts eliminate adsorbed water layer appearance on the ZnO crystal surfaces, thereby preventing electrostatic leakage on the lateral faces of the nanorods.

Investigation of obtained nanostructures was carried out using atomic force microscopy (AFM Therma NTEGRA NT-MDT) and scanning electron microscopy (SEM Seron Technology AIS 2300C).

4. Results and discussion

IOP Conf. Series: Journal of Physics: Conf. Series **1038** (2018) 012044 doi:10.1088/1742-6596/1038/1/012044

Investigation of zinc oxide seed layers via AFM revealed a large roughness of the crystallites (figure 3), meanwhile a large contribution to the structures morphology is made by the relief of the substrate. Deposition of crystallites occurs mainly between the rolling stripes.

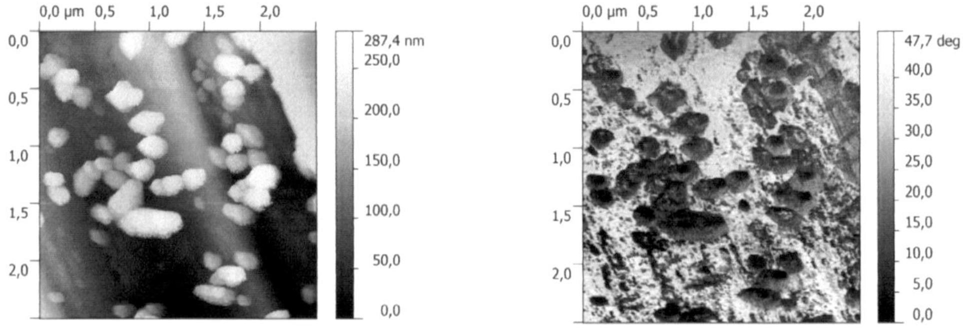

(a) **(b)**

Figure 3. AFM-data of ZnO thin film (1 layer): *a* - topography, *b* –phase contrast.

A further number of cycles increasing leads to the formation of uniformly distributed ZnO crystallites.

On the obtained seed layers were formed zinc oxide 1*D*-nanostructures (figure 4). Usage of tantalum substrates (figure 4, *a*), in comparison with aluminum, does not lead to the porous structures formation. This effect is explained due to the high inertia of Ta and its oxides formed via heating SILAR layers.

Obtained on Ta substrates ZnO 1*D*-nanostructures have a hexagonal cut and the average aspect ratio of nanorods *X* (the ratio of the diameter to the length) is 0.12. Thickness increasing of the seed layer leads to a probable mechanical stresses rising resulting to delamination of zinc oxide nanostructures.

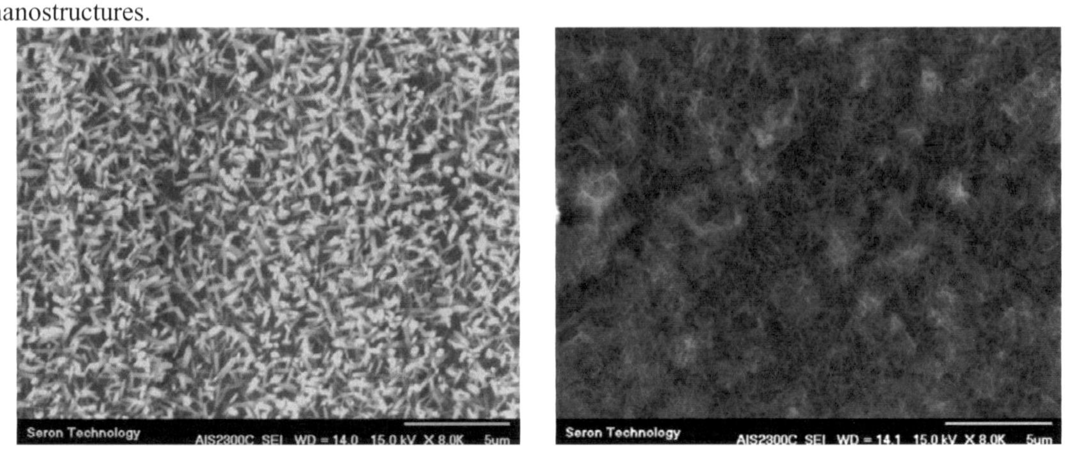

(a) **(b)**

Figure 4. SEM-data of ZnO nanostructures on substrates (3 seed layers): *a* - Ta, *b* – Al.

During the process of chemical deposition the etching of aluminum oxide takes place (figure 4, *b*). Thus, the etching of the Al_2O_3 layer occurs via the reaction with the formation of ammonium tetrahydroaluminate:

$$Al_2O_3 + 2NH_4OH + 3H_2O \leftrightarrow 2NH_4[Al(OH)_4] \quad (6)$$

$$NH_4[Al(OH)_4] \leftrightarrow Al(OH)_3\downarrow + NH_4OH \quad (7)$$

passing by the reaction (7) in aluminium hydroxide in an aqueous solution during heating with time.

On substrate areas near aluminum plates in the cavities of the porous layer the formation of tetrahydroaluminate ammonium takes place according to reaction (6). This process is parallel to the surface passivation with the further Al_2O_3 layer formation at the interface solution-aluminum in accordance with the processes:

$$2Al + 6H_2O \leftrightarrow 2Al(OH)_3\downarrow + 3H_2\uparrow \quad (8)$$

$$2Al(OH)_3 \leftrightarrow Al_2O_3 + 3H_2O \quad (9)$$

It must also be noted that aluminum substrates application leads to the ZnO nanostructures Al^{3+} cations doping, further carrier concentration increasing and reduction of piezoelectric signal time formation. By contrast, due to the high inertia Ta substrates doesn't significantly impact on the zinc oxide crystallite nanostructures properties.

Conclusion

The ZnO based PENG construction mentioned in the paper can improve output characteristics due to the two-layer structure of an active part. Such zinc oxide nanostructured coatings can be obtained within a single technological cycle with the methods of a liquid phase ionic layering and a chemical bath deposition.

In comparison with aluminum subastrates, tantalum wafers application allows obtaining zinc oxide NR uniform arrays all over the substrates surface and thus reducing the parasitic capacitance C_{sub}.

References

[1] Crossley S, Kar-Narayan S 2015 *Nanotechnology* **26** 344001
[2] Minary-Jolandan M, Minary-Jolandan R A, Espinosa H D 2011 *MRS Communications* **1** 45
[3] Özgür Ü, AlivovYa I, Liu C et al. 2005 *J. Appl. Phys.* **98** 1
[4] Xi Y, Song J, Xu Sh et al. 2009 *Journal of Materials Chemistry* **19** 9260
[5] Han X, Du W, Yu R et al. 2015 *Advanced Materials* **27** 7963
[6] Hu Y, Wang Z L 2015 *Nano Energy* **14** 3
[7] Francesca L Boughey, Timothy Davies, Anuja Datta 2016 *Nanotechnology* **27** 28LT02
[8] Zang C, Avdeeva A V, Muradova A G et al. 2014 *Uspehi v himii i v himicheskoi tehnologii* **28** (6) 73
[9] Matyushkin L B, Reshetnikova A A, Andronov A O et al. 2017 Semiconductors **51** (5) 615
[10] Bobkov A A, Lashkova N A, Maximov A I et al. 2017 *Semiconductors* **51** (1) 61

The study of metal-oxide sol-gel nanocomposites using scanning probe microscopy and X-ray photoelectron spectroscopy

A S Lenshin [1], E V, Maraeva [2], S S Nalimova [2], A N Beltyukov [3]

[1]. Voronezh State University, Universitetskaya pl. 1, Voronezh, 394006 Russia

[2].St. Petersburg State Electrotechnical University, st. Prof. Popova 5, St. Petersburg, 197376 Russia

[3]. Physico-Technical Institute of Ural Department of RAN, Kirova st. 132, 426000, Izhevsk, Russia

lenshinas@mail.ru

Abstract. With the use of atomic-force microscopy (AFM) technique and X-ray photoelectron spectroscopy (XPS) morphology and composition of the surface in SnO_2-SiO_2 nanocomposites were investigated. The samples were obtained by sol-gel method for their use as sensor agents. It was found that the obtained nanocomposites differ from each other as by the distribution of cluster sizes at the surface layer as by the pores size in a dependence on the composition and preparation technique. The study by X-ray photoelectron microscopy demonstrated that in the composites obtained with the use of the above-named technique tin was present mainly in the form of dioxide with a little admixture of non-oxidized metallic tin. This fact can have a certain impact on the functional characteristics of the sensor structures.

1. Introduction

Currently a great attention of researchers and engineers is focused on the materials for multi-sensor systems, multi-functional optoelectronics as well as those semiconducting devices based on the use of these materials [1,2]. Creation of gas-sensitive composite materials with a high sensitivity and selectivity including those ones on the basis of tin matched with silicon technology is also of an important interest for non-invasive medical express-diagnostics [2-4]. Additionally, deposition of thin metal-containing layers can improve current flow in the system. In this case nanostructures and nanocomposites on the basis of silicon and metal oxides such as sol-gel SnO_2-SiO_2 composites, the structures of metal oxide/por-Si, can demonstrate some new and technically attractive adsorption and optical properties [5-7].

Composition and properties of metal-containing layers are varied in a wide range depending on the deposition technique, thickness and morphology of the obtained film [6,7]. In the work we investigated morphology and composition of SnO_2-SiO_2 nanocomposites formed by sol-gel technique under different technological parameters applying X-ray photoelectron spectroscopy and atomic-force microscopy for the analysis.

Sol-gel technique was based on hydrolysis and condensation of the precursor molecules. In a dependence of the used precursors it was possible to obtain nanomaterials in a wide range of compositions as well as the size and shape of the particles [8].

2. Experimental technique

Nanocomposites of SnO_2-SiO_2 with different composition were obtained by sol-gel technology from the alcohol solutions of the components. In order to prepare solutions for the reaction iso-propyl alcohol, distilled water, tin chloride (II) bi-hydrous and tetraethoxysilane (TEOS) were applied. The prepared solutions were deposited onto glass and quartz substrates. pH value of the original solutions-sols was of 2-3. The temperature of anneal was of 600 °C. AFM-images of the nanocomposite samples were obtained with microscope INTEGRA (NT-MDT, Russia) with the use of silicon probes of NSG brand having radius of curvature equal 10 nm in semi-contact mode. Image processing with the use of the program Gwiddion allowed estimation of the spread for the particles over their dimensions.

Investigations of the samples by XPS technique were performed with the laboratory spectrometer produced by SPECS company. XPS spectra were excited with the use of Mg K_α irradiation. Analysis of the composition in the studied composites applying this method was conducted in the surface layers of ~ 1-2 nm in thickness after cleaning of the sample surface by argon ions with the energy of 4 keV (for 3 s.); ion current density was of 30 $\mu A/cm^2$.

Processing of the experimental data was performed with the use of the program suite Origin 9.0. A technique for determination of the background as well as its subtraction according to the algorithms proposed by Shirley [9], were also realized applying the same program. While determining binding energies of the core levels for the elements comprising nanocomposites C1s line of the natural hydrocarbon contaminations on the surface of any sample that was not subjected to a special cleaning procedure was taken as a reference one. It was reduced to the energy of E_b[C1s]=285 eV. Identification of the elements core levels according to XPS data and their chemical state was made with the use of X-ray photoelectron spectra database at National Institute of Standards (USA) [10].

3. Results and discussion

Figure 1a represents a typical morphology of the surface in nanocomposite sample with a composition of 15%SnO_2-85%SiO_2. Figure 1b provides AFM-image of the nanocomposite sample with 85%SnO_2-15%SiO_2 composition. Figure 1c illustrates morphology of the sample surface of 85%SnO_2-15%SiO_2 composition obtained under the speed of centrifuge rotation 1000 rpm, while in all other cases the speed of rotation was of 5000 rpm. A decrease in the speed of rotation resulted in the formation of porous structure.

The obtained data mean that nanocomposite samples with a tin dioxide content of 85 mol. % are characterized by porous structure. Pore sizes for these samples depends on the speed of centrifuge rotation. Decrease of the speed resulted in the formation of porous structure with a large pore size. Therefore, a speed of a centrifuge rotation among some other technological conditions of sol- gel synthesis proves to be an important parameter determining the size of pores.

Processing of the images in the program suite known as Gwiddion allowed an estimation of a spread of the particles by their sizes (fig. 2) It should be noted that these distributions are characteristic ones just for the selected scanned area of 5 x 5 μm in size.

IOP Conf. Series: Journal of Physics: Conf. Series **1038** (2018) 012045 doi:10.1088/1742-6596/1038/1/012045

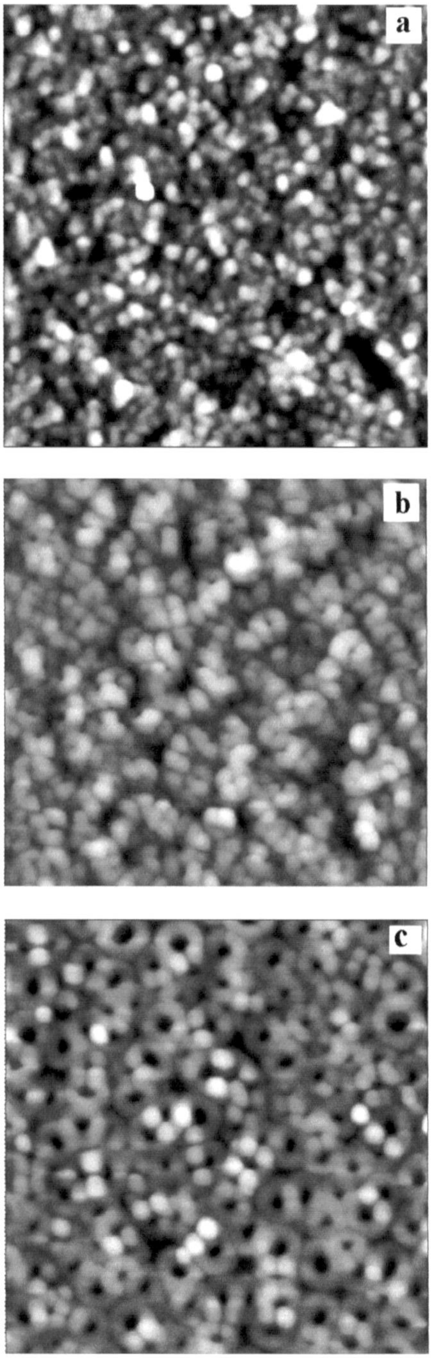

Figure 1. a) AFM-image of the layer of 15%SnO_2-85%SiO_2. b) 85%SnO_2-15%SiO_2, obtained under the speed of centrifuge rotation equal to 5000 rpm, c) AFM-image of 85%SnO_2-15%SiO_2 layer obtained under 1000 rpm. The scan size area is 5x5 μm, height range 0-20nm.

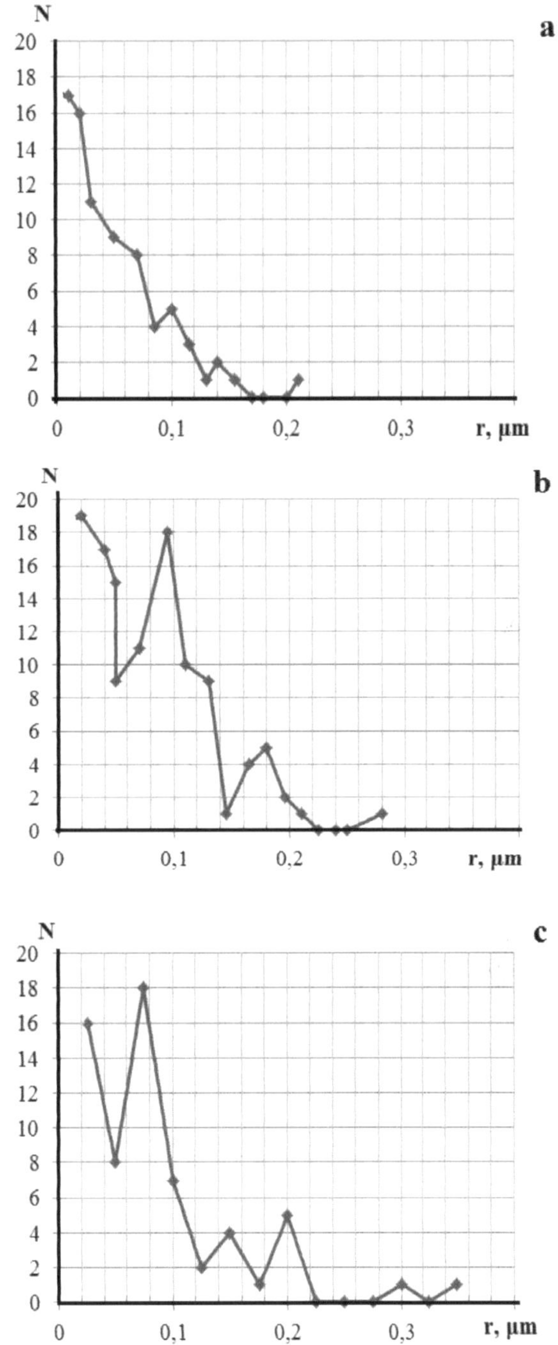

Figure 2. Distribution of the particles by their sizes in the samples obtained by sol-gel technique: a) 15%SnO_2-85%SiO_2, b) 85%SnO_2-15%SiO_2 (5000 rpm), c) 85%SnO_2-15%SiO_2 (1000 rpm).

It was found that the distribution of the particles by their sizes mainly depends on the sample composition. The number of particles in the layer of 15%SnO_2-85%SiO_2 composition is monotonously reduced with the increase of their size. For the samples of 85%SnO_2-15%SiO_2 composition minimum can be observed in the range of ~ 50 nm. The changes in the particles distributions after different treatments can be explained by the fact that with an increase of tin content in the sol the independent polymer chains of tin dioxide are being formed, that leads to a phase separation in the material. It was discussed in details previously [11].

XPS of Sn 3d5/2 and O1S spectra with their decomposition into the spectral components are presented in Fig 3 for sol-gel nanocomposites of SnO_2-SiO_2 (85%:15%); centrifugal separation was performed at 3000 rpm and 4000 rpm, respectively.

Analysis of the shape and position for XPS Sn 3d5/2 spectra and their decomposition into spectral components demonstrated that Sn in the surface layer of nanocomposites is present mainly in the form of oxide SnO_x, where x is close 2 (spectral component with E_b = 486,5-487 eV). At the same time relatively small amount of non-oxidized metal tin is also observed (spectral component with Eb=485 eV) [10,12], that can nevertheless play a considerable role, for example, in the adsorption processes on the surface of nanocomposites and result in the change of the functional characteristics of devices using these materials. Due to the relatively small concentration of silicon in the composite XPS Si 2p spectra of the samples were rather low-intensive. Therefore, these spectra were not presented in our work.

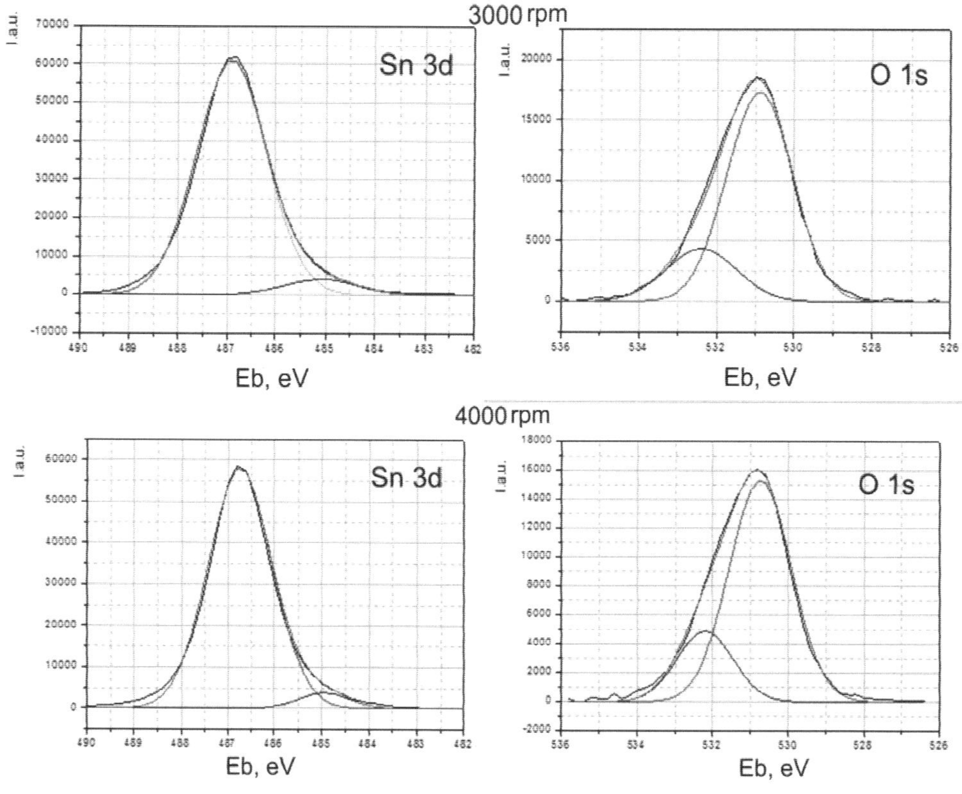

Figure 3. XPS of Sn 3d5/2 and O1S spectra for sol-gel nanocomposites of SnO_2-SiO_2 (85%:15%) with a decomposition into the components; centrifugal separation was performed at 3000 rpm and 4000 rpm.

Analysis of O1s spectra in nanocomposites confirmed the presence of the presumable amount of tin dioxide in the surface layer of the foil (spectral component with Eb = 530.5 eV). The obtained XPS spectra also include two components. The first one more intensive peak is related with oxygen bound with tin. The second one, less intensive component is due to the oxygen in silicon oxide SiO_x and oxygen adsorbed on the surface with Eb=532,5 eV [12,13].

4. Summary

Morphology and composition of the surface in SnO2-SiO2 nanocomposites were investigated. It was found that the distribution of the particles by their sizes mainly depends on the sample composition. The number of particles is monotonously reduced with the increase of their size. It was demonstrated that in the composites obtained with the use of the above-named technique tin was present mainly in the form of dioxide with a little admixture of non-oxidized metallic tin.

Acknowledgements:
The work concerned with the study of XPS spectra was performed under support of the grant provided by the President of RF (MK-4865.2016.2). Some part of the work related to the synthesis of the samples by sol-gel technique and analysis of AFM images was performed in the frameworks of the design part of the state task № 16.897.2017/PCh.

References

[1] Zharova Y A, Fedulova G V, Astrova E V, Baldycheva A V, Tolmachev V A and Perova T S 2011 Fabrication technology of heterojunctions in the lattice of a 2D photonic crystal based on macroporous silicon *Semiconductors* 45 1103
[2] Kupriyanov L Y 1996 *Semiconductor sensors in physico-chemical studies* (Amsterdam; Boston: Elsevier)
[3] Lenshin A S, Kashkarov V M, Spivak Y M and Moshnikov V A 2012 Investigations of nanoreactors on the basis of p-type porous silicon: Electron structure and phase composition *Mater. Chem. Phys.* 135 293–7
[4] Moshnikov V A, Gracheva I, Lenshin A S, Spivak Y M, Anchkov M G, Kuznetsov V V and Olchowik J M 2012 Porous silicon with embedded metal oxides for gas sensing applications *J. Non-Cryst. Solids* 358 590–5
[5] Len'shin A S, Kashkarov V M, Goloshchapov D L, Seredin P V, Polumestnaya K, Maraeva E V, Soldatenko S, Yurakov Y and Domashevskaya E P 2012 Composition and reactivity of porous silicon nanopowders *Inorg. Mater.* 48 965–70
[6] Lenshin A S, Seredin P V, Minakov D A, Kashkarov V M, Agapov B L, Domashevskaya E P, Kononova I E, Moshnikov V A, Terebova N S and Shabanova I N 2014 Specific features of the sol-gel formation and optical properties of 3d metal/porous silicon composites *Semiconductors* 48 551–5
[7] Sockaja N V, Dolgih O V, Kashkarov V M, Len'shin A S, Kotljarova E A and Makarov S V 2010 Surface modification of composites with metal nanoparticles *Inorg. Mater.* 46 1192–11
[8] Maksimov A I, Moshnikov V A, Tairov Ju M and Shilova O A 2008 Fundamentals of sol-gel nanocomposite technology (Saint-Petersburg: SPb.: OOO «Tehnomedia» / Izd-vo «Jelmor»,)
[9] Shirley D A 1972 High-Resolution X-Ray Photoemission Spectrum of the Valence Bands of Gold *Phys. Rev. B* 5 4709–14
[10] NIST X-ray Photoelectron Spectroscopy Database
[11] Levitskiy V S, Lenshin A S, Maximov A I, Maraeva E V, Moshnikov V A 2012 Formation peculiarities of metal oxide porous structures in SiO_2 - SnO_2 and SiO_2 - CoO sol-gel systems *Sorption and Chromatographic Processes* 12 (5) 725-733
[12] Lenshin A S, Kashkarov V M, Domashevskaya E P, Seredin P V, Ryabtsev S V, Bel'tyukov A N and Gil'mutdinov F Z 2017 Composition of nanocomposites based on thin layers of tin on porous silicon formed by magnetron sputtering *Phys. B Condens. Matter* 504 1–8
[13] Lenshin A S, Kashkarov V M, Domashevskaya E P, Bel'tyukov A N and Gil'mutdinov F Z 2015 Investigations of the composition of macro-, micro- and nanoporous silicon surface by ultrasoft X-ray spectroscopy and X-ray photoelectron spectroscopy *Appl. Surf. Sci.* 359 550–9

Influence of endometallofullerene on the conductive characteristics of polyphenylenoxide

A A Kononov, R A Castro

Physics Department, The Herzen State Pedagogical University of Russia, Moika River Embankment 48, St. Petersburg 191186, Russia

E-mail: rakot1991@mail.ru

Abstract. The paper investigates endometallofullerene impact on conductive characteristics of poly(phenyleneoxide) within a wide range of frequencies and temperatures. It is revealed that the introduction of iron into the polymer matrix leads to an increase in the specific conductivity of the material and changes the parametres of its temperature dependence. The revealed regularities can be accounted for electro-donor properties of iron embedded in fullerene molecules.

1. Introduction

Polymers are considered as ideal matrices for the development of new materials that possess desired properties and can be used to produce membranes of broad application, namely, when concentrating and separating suspensions, purifying products from impurities, regenerating valuable components to be reused, obtaining desalinated and purified water, etc.

The object of the present research is polyphenylenoxide (PPO) modified by endometallofullerene C_{60}-Fe that is introduced into the polymer matrix. To date, endohedral carbon clusters (endometallofullerenes) that contain metal atoms inside a fullerene molecule are fullerenes derivatives that are becoming an intensive and promising field of research.

Endometallofullerenes (EMF) represent a completely new type of carbon clusters that are distinctively different from hollow fullerenes. The metal atom embedded inside a fullerene molecule changes its electronic properties significantly. However, research into the properties of these derivatives is still limited due to challenges in synthesis, low output of the material and poor solubility as well as extremely high production costs. Nevertheless, due to its magnetic properties, given it is introduced inside the fullerene molecule iron is extremely promising for enhancing polymer conductivity and development of new materials with specific properties: superconductors, organic ferromagnetics, laser and ferroelectric materials, pharmaceuticals.

2. Experimental

The paper presents the results of studies of dielectric and conductive characteristics of polyphenylenoxide (PPO) composites with 1% content of fullerene C_{60} samples and those of composite PPO + 1% EMF Fe-C_{60}.

The composites were obtained by mixing solutions of PPO in chloroform (concentration 2 wt.%) and fullerene C_{60}/endofullerene C_{60}-Fe in toluene (concentration 0.14 wt.%) in the amounts that provide the desired filler content in the composite. The solution obtained in this manner was left for 3-

4 days for interaction between the polymer and the fullerene molecules to take place. Then, the solution of the composite was processed with ultrasound for 40 min. and filtered through the Shot's filter to remove dust impurities. The films were obtained by the method of toluene evaporation from the mixture the polymer and the filler on the surface of cellophane at a temperature of 40°C. [1]

The measurement of the dielectric spectra was conducted on the spectrometer "Concept 81" produced by the company NOVOCONTROL TechnologiesGmbH& Co with the automatic ALPHA-ANB high performance frequency analyzer. The samples were thin layers with a thickness of 60-125 microns and a diameter of 15 mm. The measurement of the parameters of the PCM samples was performed in the temperature range from 273 K to 523 K and within the range of frequencies between $5*10^{-2}$ and $1*10^{6}$Hz. The voltage applied to the sample equalled 1.0 V.

3. Results and discussions

Frequency dependence of the complex conductivity σ'(f) in the range $10^{-1} - 10^{6}$ obtained for the films of the systems at temperatures of T=273K...523K indicates growth of σ'(f) with an increase in frequency. The transition from the frequency-independent zone to the frequency-dependent one was identified, which indicates the beginning of conductivity relaxation.

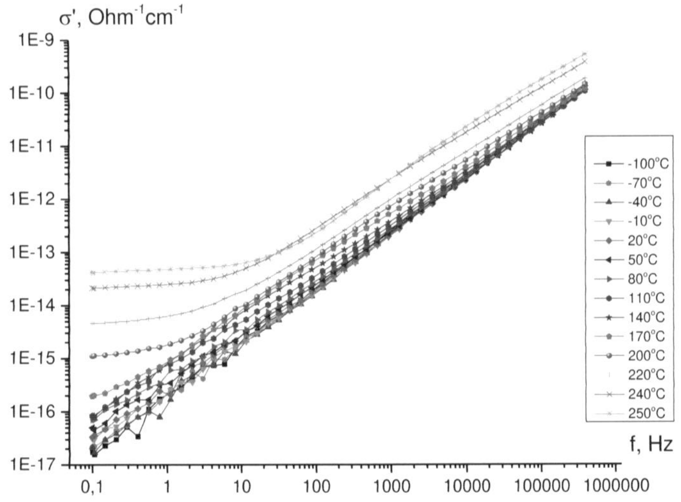

Figure 1. The dependence of sample PPO+1%C_{60} conductivity on frequency in the external field at different temperatures.

The comparison of the spectra of PPO+C_{60} (Figure 1) and PPO+C_{60} –Fe (Figure 2) samples showed an increase in conductivity that is ten-fold times much within the whole frequency interval, which can be a cause for clearly observed electron-donor properties of iron embedded in the fullerene matrix. Thus, EMF not only display obvious electron-acceptor properties, like hollow fullerene, but also demonstrate properties of electron-donors.

According to the current experimentally-based conception, electrical conductivity in polymeric dielectrics is mainly of ionic character. It is known that alternating current conductivity for amorphous semiconductors is expressed as [2]:

$$\sigma' = \sigma_{ac}(\omega) = \sigma_t(\omega) - \sigma_{dc}(\omega) = A\omega^s, \tag{1}$$

where ω is angular frequency (ω = $2\pi f$ – test frequency), $\sigma_t(\omega)$ – measured total conductivity, $\sigma_{dc}(\omega)$ –conductivity measured at constant current, s is an exponent and A is a constant that is dependent on temperature.

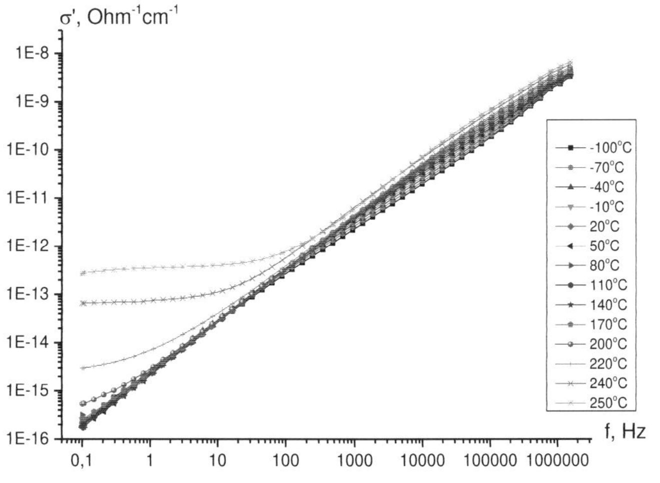

Figure 2. The dependence of conductivity of PPO+1%C_{60} –Fe sample on frequency in the external field at different temperatures.

Values of the exponent s were calculated for the samples under investigation by nonlinear approximation of the experimental curves (Figure 3).

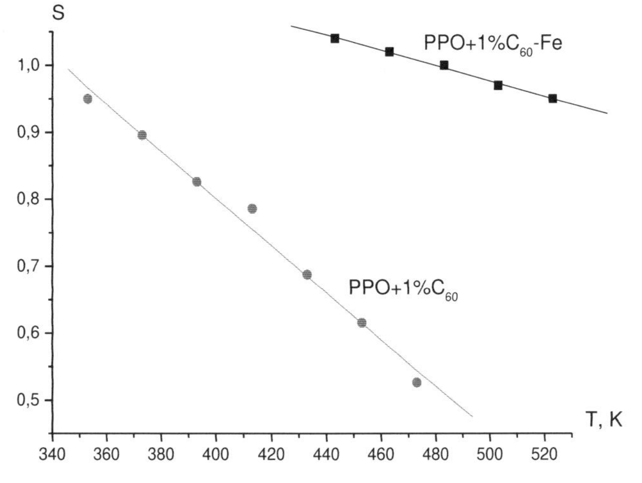

Figure 3. The dependence of the exponent *s* on temperature for the studied systems.

It was found out that s decreases with an increase in temperature in the range between 1 and 0.5, which can be regarded as experimental confirmation that conductivity is provided by hopping mechanism. It is assumed that in the model of hopping mechanism with accidental energy value [3] the elementary act of polarization is the overbarrier transition of one carrier from one state into another, and the activation energy of the center is caused by Coulomb's interaction.

The temperature dependence of the specific conductivity σ' on various frequencies has also been investigated for the studied samples.

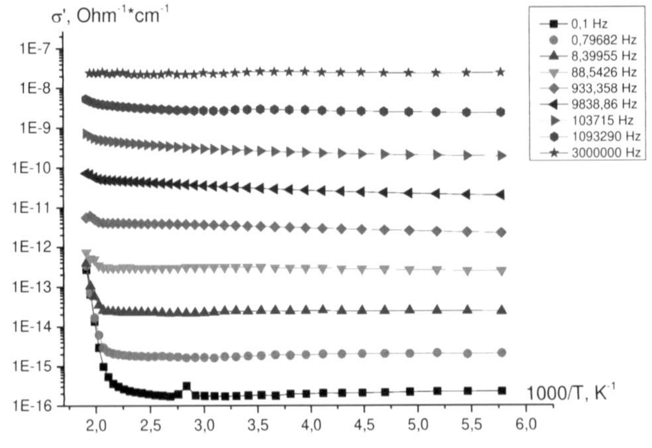

Figure 4. The dependence of the conductivity of PPO+1%C_{60} –Fe sample on temperature for different frequencies in the external field.

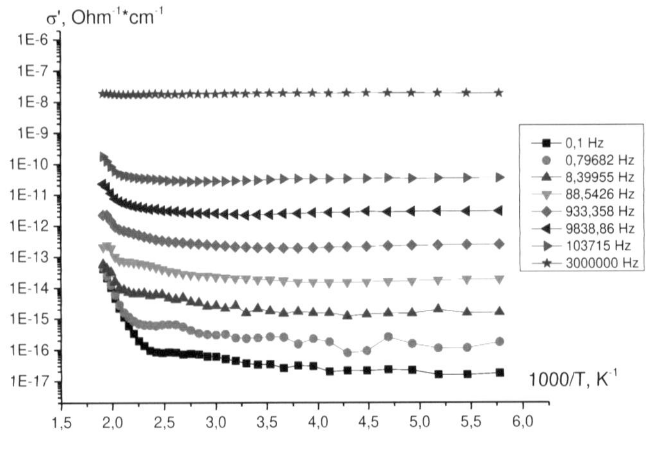

Figure 5. The dependence of the conductivity of PPO+1%C_{60} sample on temperature for different frequencies in the external field.

Figure 4 shows the graph of σ' dependence on 1000/T of the sample PPO+1%C_{60}–Fe at temperatures T=173K...523K. As it can be seen from the figure, this dependence has a complex character with the zone of exponential dependence, according to Arrhenius's law. In the zone of high temperature, nonlinear increase in σ' along with the growth of temperature was revealed.

Multiple regions of the exponential dependence and the changing nature of temperature dependence for PPO+1%C_{60}-Fe samples in Figure 5 can be caused by either the presence of impurities and defects in the sample structure or by transition process from mixed conductivity to ionic conductivity.

IOP Conf. Series: Journal of Physics: Conf. Series **1038** (2018) 012046 doi:10.1088/1742-6596/1038/1/012046

Apparently, at low temperatures, the source of free charges that are non-bound chemically with macromolecules is ionic impurities with low molecular weight, the mobility of which is limited by high viscosity of the medium. At high temperatures, conductivity becomes intrinsic.

4. Conclusions

The comparison of the spectra of PPO+C60 and PPO+C60 –Fe samples showed an increase in conductivity that is ten-fold times much within the whole frequency interval. It is revealed that the introduction of iron into the polymer matrix leads to an increase in the specific conductivity of the material and changes the parametres of its temperature dependence. It was found out that *s* decreases with an increase in temperature in the range between 1 and 0.5, which can be regarded as experimental confirmation that conductivity is provided by hopping mechanism. The revealed regularities can be accounted for electro-donor properties of iron embedded in fullerene molecules.

References

[1] Polotskaya G A et al 2017 *Russian Journal of Applied Chemistry* 90 9 1549–1557
[2] Mott N F, Davis E A 1979 *Calendon Press Oxford* 589
[3] Yakuphanoglu F et al 2003 *Physica B: Condensed Matter* 334 3 443-450

Gas sensing properties of nanocomposites with ZnO nanowires

A A Bobkov, V I Gorshanov

Micro- and nanoelectronics department, Saint Petersburg State Electrotechnical University, 5 Prof. Popova Street, Saint Petersburg, Russia

Abstract. In this work gas sensitivity of zinc oxide nanostructures to reducing gases (acetone and isopropyl alcohol) was investigated; impedance spectroscopy method was used for resistance measurements in a wide range of frequencies. Special attention is payed to the idea that increasing surface area of the samples leads to a corresponding increase in sensitivity. In order to check it, gas sensitivity of ZnO – SiO_2 nanocomposites was measured first; after that the samples were used to form ZnO nanowires on their surface, and the comparison of gas sensitivity of samples with and without nanowires is presented.

1. Introduction

Zinc oxide is a II-VI compound semiconductor with a direct bandgap of 3.4 eV at room temperature, which has found various applications in microelectronics and optoelectronics [1]. It has also been widely investigated as a gas sensing material; although a few metal oxides change their resistance after being subjected to gas, ZnO is one of the prominent ones [2].

When zinc oxide sample is in the atmosphere, oxygen species get adsorbed by the surface of the semiconductor. In such case oxygen acts as a trap for electrons, thus the surface conductivity gets lower than the bulk conductivity. However, when the sample is exposed to a reducing or oxidizing gas, the oxygen species start interacting with the gas species; after the interaction the concentration of oxygen on the semiconductor surface changes, which leads to changes in surface conductivity. In general, when a n-type metal oxide semiconductor (like ZnO or SnO_2) is exposed to a reducing gas, its surface conductivity increases; when it is exposed to an oxidizing gas, its conductivity decreases. The situation is opposite with p-type metal oxides [3].

As long as all the interactions responsible for gas sensing properties happen on the surface of semiconductor, using nanostructured materials which have bigger surface area has often [4,5] been considered an option for increasing gas sensitivity of metal oxides. This work compares gas sensing properties of ZnO – SiO_2 nanocomposite samples before and after ZnO nanowires were grown on the nanocomposite surface.

2. Experimental methods

2.1. Samples synthesis

ZnO – SiO_2 nanocomposite films were obtained using sol-gel synthesis; the method had been shown in [5] and [6]. Inorganic salt $Zn(NO_3)_2{\bullet}6H_2O$ was used as the ZnO precursor; in order to improve the adhesion TEOS was used. The resulting sol was kept at room temperature for several days. After that the solution was spin-coated on silicon substrates (3000 RPM during 15 seconds); the spin-coating was followed by 30 minutes annealing of the samples carried out at a temperature of 500 °C. Monocrystalline silicon plates, which had been previously chemically purified, were used as substrates.

The acquired layers were used to form ZnO nanowires via hydrothermal method, which had been shown in [7] and [8]. As the growth solution, an aqueous solution of zinc acetate, hexamethylenetetramine (HMTA) and cetyltrimethylammonium bromide (CTAB) was used. The samples were put into the solution and kept in it for 90 minutes at a temperature of 85°C. After the end of the process the samples were rinsed with distilled water and air dried. The resulting nanowires are shown in figure 1.

Figure 1. SEM-image of nanowires obtained by hydrothermal synthesis on ZnO – SiO_2 nanocomposite film.

2.2. Measurements method

In order to measure gas sensitivity of the samples impedance spectroscopy method was used. The impedance spectroscopy implies measuring real and imaginary resistance of a sample in a wide range of frequencies (during this work the measurements were conducted between 100 Hz and 500 kHz); after that an equivalent electric circuit is often matched to the Im(Re) graph [9].

In this work, in order to compare sensitivity properties of different samples, we used a formula that is to describe the relative change of complex resistance of a sample in the air and in the gas:

$$S(f) = \frac{\Delta Z}{Z_a} = \frac{Z_a - Z_g}{Z_a} = \frac{\sqrt{Re^2(Z_a) + Im^2(Z_a)} - \sqrt{Re^2(Z_g) + Im^2(Z_g)}}{\sqrt{Re^2(Z_a) + Im^2(Z_a)}},$$

where Z_a is the complex resistance of a sample in the air; Z_g is the complex resistance of a sample in the gas. In order to conduct the measurements, contact pads were formed using electrically conductive adhesive containing silver.

During the measurements samples' temperature was 200 °C; the gas concentration was 3000 ppm.

After the measurements the highest sensitivity of samples and corresponding frequencies were determined.

3. Results and discussion

The results of measuring gas sensitivity of the samples are shown in table 1; the sensitivity was calculated according to the previously given formula.

Table 1. Comparison of gas sensitivity of the samples with and without nanowires

Gas	Parameter	Without nanowires	With formed nanowires
Acetone	Maximum sensitivity, relative units	0.15	0.219
	Frequency of max. sensitivity, kHz	0.4	1.4 – 2.1
Isopropyl alcohol	Maximum sensitivity, relative units	0.045	0.188
	Frequency of max. sensitivity, kHz	0.4	1.3 – 2.5

As can be seen from the table, after the nanowires were grown on nanocomposite samples, the sensitivity improved noticeably. Sensitivity to acetone increased 1.46 times, while sensitivity to isopropanol increased more than 4.17 times.

This result corroborates the fact that the processes responsible for gas sensitivity properties of metal oxides all take place on the surface of materials and increase in surface area which happens with usage of nanostructured materials also increases gas sensitivity.

The study was performed by a grant from the Russian Science Foundation (project 17-79-20239).

References

[1] Klingshirn C, Hauschild R, Priller H, Decker M, Zeller J and Kalt H 2005 *Superlattices and Microstructures* **38** 4–6 209–22
[2] Korotcenkov G 2007 *Materials Science and Engineering* B **139** 1 1–23
[3] Fine G F, Cavanagh L M, Afonja A and Binions R 2010 *Sensors* **10** 6 5469–502
[4] Hjiri M, Mir L E and Leonardi S G 2014 *Chemosensors* **2** 2 121–30
[5] Bobkov A A 2014 *Young Scientist* 7 115–18 (in Russian)
[6] Bobkov A A and Karpov S S 2012 *Young Scientist* 9 21–5 (in Russian)
[7] Bobkov A A et al. 2015 *Physics and Technology of Semiconductors* **49** 10 (in Russian)
[8] Lashkova N A, Maksimov A I, Ryabko A A, Bobkov A A, Moshnikov V A and Terukov E I 2016 *Physics and Technology of Semiconductors* **50** 9 (in Russian)
[9] Barsoukov E and Macdonald J R 2005 *Impedance Spectroscopy: Theory, Experiment, Applications* (Hoboken: John Wiley & Sons)

IOP Conf. Series: Journal of Physics: Conf. Series **1038** (2018) 012048 doi:10.1088/1742-6596/1038/1/012048

Investigation of the photoluminescence of porous silicon layers obtained under various technological conditions

A.I. Pastukhov, A.O. Belorus, Yu. M. Spivak and V.A. Moshnikov

St. Petersburg State Electrotechnical University «LETI», Saint Petersburg, Russian Federation

drony102@mail.ru

Abstract.The article considers the information about optical properties of porous silicon layers. Samples were obtained using electrochemical and photoelectrochemical anodic etching methods. The monocrystalline n-type (phosphorus doping) silicon was used as the initial material for preparation. It is shown, that the photoluminescence intensity and wavelength, surface morphology of porous silicon samples are hinge on many technological conditions. The results can be used to estimate an advisability of porous silicon particles application as biomarkers; LED and other optical devices creation.

1. Introduction

Porous silicon (PSi) is an essential material in terms of creating different devices, due to its various properties, which depend on many technological factors [1-4]. It is known, that there is a possibility to control material characteristics such as dielectric permeability, intensity and wavelength of luminescence, optical transmission and other. Thereby, one can take advantage of PSi as the material for photonics, optoelectronics, electronics, energetics and biomedical application. Thus, there is an opportunity to improve the efficiency of PSi layers and nanoparticles application in many areas by varying and combining technological conditions of preparation.

Silicon is a non-direct-gap semiconductor and its application as the material for LED devices is impractical because of inefficient luminescence. However, porous silicon layers, obtained by certain technological conditions, have an effective photoluminescence (in some cases – electroluminescence) in a visible, near ultraviolet and infrared range [5].

The investigation deals with the preparation conditions of photoluminescent PSi layers for biomedical purpose (targeted drug delivery, theranostics, etc). Special attention in the experiment is given to the hydrophilic/hydrophobic nature of porous layers. In previous works [6, 7] it is shown, that silicon nanoparticles must have hydrophilic properties of the surface for efficient drug encapsulation.

It is important to investigate the photoluminescence mechanism and spectrum of PSi layers for biomedical application. It has been revealed, that incident wavelength (in electrochemical etching process) results in change of surface parameters, morphology and skeleton distribution.

Thus, the study presents the investigation of preparation conditions (technological conditions) influence on obtaining biocompatible photoluminescent PSi layers, using electrochemical and photoelectrochemical anodic etching methods.

2. Experimental

Porous silicon samples were prepared using electrochemical and photoelectrochemical anodization (electrolyte – aqueous solution of hydrofluoric acid with addition of isopropyl alcohol). The monocrystalline n-type (phosphorus doping) silicon wafers were used as the initial material. The preparation conditions are represented in table 1 and 2.

Table 1. Technological conditions of electrochemical etching

Silicon parameters	Resistivity, Ω·cm	Anodization time, min	Anodization current density, mA/cm^2
n-Si <111>	4.5	20	20
			40
			60
			80

Table 2. Technological conditions of photoelectrochemical etching

Silicon parameters	Resistivity, Ω·cm	Anodization time, min	Incident wavelength, nm	Anodization current density, mA/cm^2
n-Si <111>	4,5	20	650	20
				40
				60
				80

It is shown, that the variable technological parameters of anodic etching are the current density and incident wavelength (presence or absence). Anodization time is constant – 20 minutes; incident power – 250 mW, wavelength – 650 nm.

The radiation source (laser) was used to generate more electron-hole pairs at the sample surface/electrolyte boundary, which directly stimulates the anodic dissolution process [8,9].

The spectra of the samples were studied by a spectrofluorimeter; the excitation wavelength of the photoluminescence of porous silicon layers is 405 nm.

3. Results and discussion

The experimental dependence of radiation intensity as a function of the photoluminescence wavelength for electrochemical (a) and photoelectrochemical (b) anodic etching is represented in Figure 1. As can be seen, the samples, obtained using photoelectrochemical etching technique, have more intensity photoluminescence in the shortwave region in comparison with the samples, obtained by electrochemical anodic etching. This phenomenon is connected with the intensification of various processes (dissolution, oxidization, etc) under the influence of light, which results in porosity, morphology and other parameters. The presence of additional peaks in the short-wavelength region, probably, is associated with various recombination mechanisms. The absence of a shift in the luminescence peak at a wavelength of 714 nm indicates that recombination occurs predominantly on structural defects. The samples, obtained by photoelectrochemical etching, have a peak at a wavelength of 644 nm, which we assume to be associated with a number of factors: the change in the dimensions of silicon nanocrystals, the contribution of recombination to the thermolysis of charge carriers.

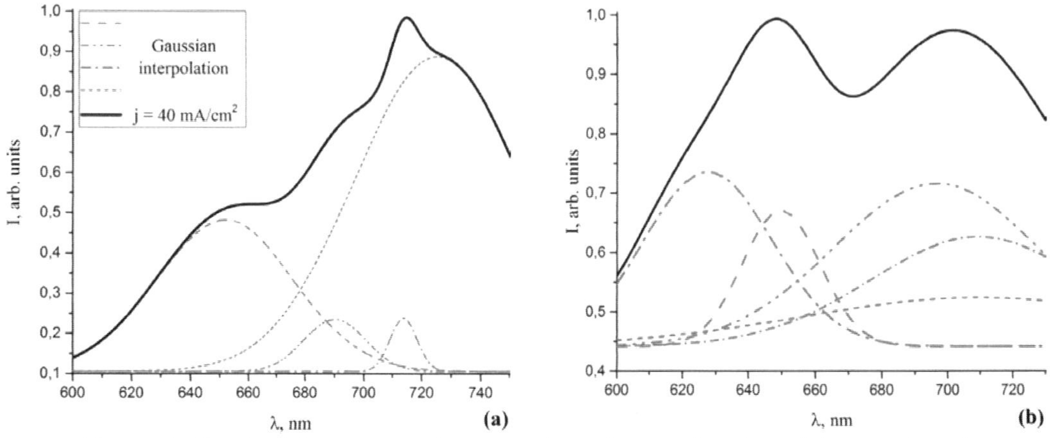

Figure 1. Graph of the photoluminescence intensity as a function of the wavelength (a) – electrochemical etching, (b) – photoelectrochemical etching

The experimental dependence of the photoluminescence intensity of porous silicon layers, obtained by photoelectrochemical anodic etching, as a function of the exposure time by a source of radiation (wavelength – 405 nm) is shown in Figure 2.

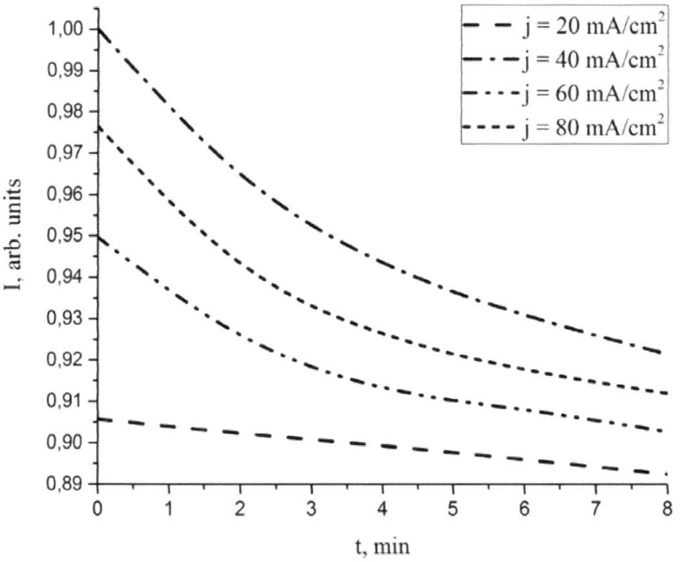

Figure 2. Graph of the photoluminescence intensity (of a single peak at ~700 nm) as a function of the exposure time

As can be seen from the figure, the luminescence intensity of the samples decreases with increasing exposure time. This circumstance can be explained if one assumes that under radiation, the oxidation (on the air) of the sample occurs more rapidly, which directly affects the photoluminescence efficiency of the layers of porous silicon.

4. Summary

The samples, which were obtained by electrochemical and photoelectrochemical anodic etching at various current densities (20…80 mA/cm^2) from initial monocrystalline silicon <111>, were studied. It is shown, that radiation, which is used in the process of photoelectrochemical etching, has a significant contribution to the characteristics of the samples. It is assumed that the photoluminescence spectral band shift of the samples to the short-wavelength region is associated with a decrease in the size of silicon nanocrystals and electron thermolysis (the recombination mechanism). In both cases (electrochemical and photoelectrochemical etching), the peak at 714 nm is stable, which is connected with recombination on the defects. It should also be noted that the samples obtained by photoelectrochemical etching are characterized by more intense photoluminescence.

Thus, one can control porous silicon parameters (the radiation wavelength, the intensity of photoluminescence) by changing technological conditions of preparation (using illumination λ = 650 nm, which influences on the anodic side of silicon). These results are essential in terms of creating various light-emitting devices and methods, where PSi is used as luminophore (biomarkers, for example).

5. Acknowledgments
The work was carried out as part of the project event 5.1.2 "Increase of the publication activity of scientific and scientific-pedagogical workers" to increase competitiveness among the world's leading research and educational centers program.

6. References
[1] Canham L T 1990 Silicon Quantum Wire Array Fabrication by Electrochemical and Chemical Dissolution of Wafers *Appl Phys* lett (№) 57(10) pp 1046–1048
[2] Canham L T 1997 Properties of Porous Silicon *EMIS datareview series* (IEE Press London) lett (№) 18
[3] Spivak Yu M, Moshnikov V A, Pshchelko N S, Kadi Y S, Nalimova S S 2014 Morphological Feature Analysis of the Prospective Combined Gas Sensitive Sensor Elements *Smart Nanocomposites* vol 5 (1) p 5
[4] Moshnikov V A, Gracheva I E, Lenshin A S, Spivak Yu M, Anchkov M G, Kuznetsov V V, Olchowik J M 2012 Porous Silicon with Embedded Metal Oxides for Gas Sensing Applications *Journal of Non-Crystalline solids* vol 358 pp 590-595
[5] Korotcenkov G, Cho B K 2010 Silicon Porosification: State of the Art *Critical reviews in solid state and materials sciences* lett (№) 35:3 pp 153–260
[6] Pastukhov A I, Belorus A O, Bukina Ya V, Moshnikov V A, Spivak Yu M 2017 Influence Of Technology Conditions On The Surface Energy Of Porous Silicon Using The Method Of Contact Angle *2017 IEEE Conference of Russian Young Researchers in Electrical and Electronic Engineering (EICONRUS)* pp 1183-1185
[7] Spivak Yu M, Belorus A O, Somov P A, Bespalova K A, Moshnikov V A, Tulenin S S 2015 Porous Silicon Nanoparticles for Targeted Drug Delivery: Structure and Morphology *Journal of Physics: Conference Series* vol 543 (1)
[8] Asano T, Higa K, Aoki S, Tonouchi M, Miyasato T 1992 Effects of Light Exposure During Anodization on Photoluminescence of Porous Si *Jpn J Appl Phys* lett (№) 31 (Part 2) p 373
[9] Föll H, Carstensen J, Foca E, Leisner M 2007 Understanding and Controlling Pore Etching in Semiconductors *ECS Transactions* pp 309-322

Mathematical model of segmented capillary electromigration performances depending on stationary phase physicochemical properties

A Yu Shmykov[1], V A Kornienko[2], A N Krasovskii[1], S V Mjakin[2], N A Bubis[1], L Sh Boridko[1], L M Kuznetzov[3], A A Fedorov[1], V E Kurochkin[1]

[1] Institute for Analytical Instrumentation, RAS, St. Petersburg, Russia
[2] St. Petersburg State Institute of Technology (Technical University), Russia
[3] St. Petersburg State University of Economics, Russia
E-mail: korn.verka@gmail.com, shmykov.alexey@gmail.com

Abstract. Segmented capillaries comprising two segments with different surface charges of polymer layers are prepared. One segment with a neutral surface charge is responsible for separation and another one with negatively charged surface provides the electro-osmotic flow adjustment. The average electro-osmotic flow rate is found to correlate with the length of negatively charged segment. A mathematical model is suggested to describe electromigration performances of segmented capillaries depending on the segments length ratio also taking into account the stationary phase/coating with definite surface properties (ξ–potential) or electroosmotic flow rate.

1. Introduction.

The improvement of separation techniques in electromigration analytical procedures relates to obtaining stationary phases [1, 7] on the internal surface of fused silica capillaries, dynamic coatings [2], pseudo-stationary phases [3-5], additives [6] yielding complexes, ion pairs, solvation, micelles, etc. All the considered approach have certain advantages and drawbacks described in numerous reviews and monographs [1,2,6,7].

So, a disadvantage of stationary phases inside capillaries is the complexity of obtaining a uniform stationary phase in the whole bulk or along the entire inner surface of the capillary [7]. In the case of dynamic coatings, the reproducibility of their uniform deposition upon the internal surface of the capillary is limited by the capillary conditioning procedure and deteriorates with each consecutive deposition or rinsing [2]. If a dynamic coating is permanently present in a buffer solution, e.g. at DNA sequencing, the reproducibility of separation results is methodically limited by 100 experiments [1]. The development of pseudo-stationary phases in capillaries generally relate to self-organization of magnetic particles slurries into an integral system upon magnetic field application. However, the main problem in this case determined by non-steady behavior of magnetic nanoparticles as a stationary phase due to instability of suspending nanoparticles is still not addressed [3-5].

The addition of special additives into buffer solutions allows the optimization of separation selectivity and efficiency; however this approach is limited by the instability of ξ–potential at the inner surface of the capillary. This drawback intrinsic to dynamic coatings [2,6] is determined by the complexity of adjustment of ionization and adsorption activity adjustment for surface silanol groups in fused silica capillaries [1,2,7]. Various kinds of silanol groups have different ionization contants and adsorption activity those results in the instability of electric performances of the capillary surface [7, 8]. Even a small adsorption of admixtures or separated substances significantly deteriorates the separation quality at capillary electrophoresis. For this reason the development of approaches to shielding silanol groups on the surface of capillaries is crucial for the further improvement of electromigration analytical procedures [1,2,7].

In our earlier studies the surface functional composition and surface structure of atactic polystyrene coatings on fused silica plates [9,10] and electromigration performances of capillaries with polystyrene coatings [11] were characterized in order to obtain uniform polymer layers (stationary phases) in capillaries with adjustable hydrophilic-hydrophobic properties. Certain features and trends of polystyrene coating surface energy changes with the polymer concentration variation in the solution were revealed [9], correlations between electromigration performances of polystyrene-coated fused silica capillary columns and components of polystyrene coating surface energy were established [11] and optimal concentration and temperature-kinetic conditions for obtaining polystyrene coatings with adjustable electokinetic performances were defined [9,10,11]. However, the developed methods for the synthesis of stationary phases, like most of the existing approaches to covalent stationary phases in capillaries, afford either electro-osmotic flow (EOF) stabilization within a wide range of background electrolyte pH values or enhance the separation efficiency [2,7,12,13]. In order to address both of these goals, in [12,14] a novel promising approach was suggested on the basis of the development of segmented capillaries (SC) comprising two segments with different polymer layer surface charges, e.g. neutral-anionic or neutral-cationic pairs. The increased separation efficiency in SC is determined by the involvement of a segment with a neutral (non-charged) surface providing the retention and separation selectivity in the absence of irreversible adsorption in couple with another segment bearing a negative (or positive) surface charge affording the adjustment and, if necessary, increase of EOF. In order to study the effect of the ratio between the lengths of the considered segments upon SC electromigration performances, single polymer (PS or PSSA) modified and segmented (involving a neutral polystyrene (PS) and negatively charged polystyrenesulfonic acid (PSSA)) segments onto the internal surface of fused silica capillary followed by a comparative analysis of EOF marker mobility.

2. Method

2.1. Polymer coating synthesis

PS and PSSA polymer coatings on the capillary internal surface were synthesized by the consecutive filling of fused silica capillaries with the corresponding reaction mixtures followed by chemical and thermal initiation of polymerization reactions according to the procedures earlier descried in [9-11]. The internal surface of SC was modified using a similar procedure taking into account the requirement to provide the target PSSA:PS ratios 2:1±5% and 1:2±5% with the capillary length up to the detection window l_0 = 45 cm (Fig. 1.).

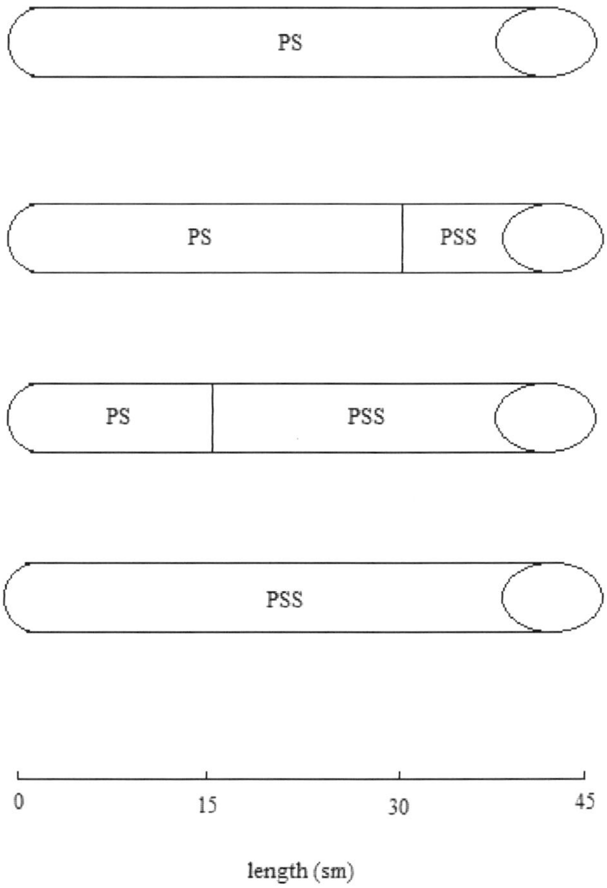

Figure 1. Modified and segmented capillaries with PS and PSSA coatings.

2.2. Characterization of electromigration performances

EOF rate was measured using a capillary electrophoresis installation Nanophor-01 (Institute for Analytical Instrumentation, Russian Academy of Sciences) using a standard procedure [1] for the characterization of surface electric properties of fused silica capillary columns and ξ-potential on the sorbent surface. The electro-osmotic mobility μ_{eo} was determined according to the equation:

$$\mu_{eo} = \frac{l_0}{E \cdot t} \quad (1)$$

where l_0 is the capillary length to the detector, t is EOF marker migration time (s) and E is the electric field strength[1].

Electromigration performances of the prepared modified and segmented capillaries were comparatively studied by measuring EOF.

2.3. Model separation of cations

Preliminary studies were performed in respect of the efficiency of modified capillaries with negatively charged polymer coatings in the separation of inorganic cations (K^+; Na^+; Mg^{2+}; Ca^{2+}) in comparison with non-modified fused silica capillary. The selection of inorganic cations mixtures for this research was determined by the problems relating to separation efficiency using standard complexating reagents affecting the resulting peak symmetry [6].

3. Results

3.1. Effect of segment lengths on EOF

Elecrophoregrams for EOF marker DMSO in segmented capillaries are shown in Fig. 2.

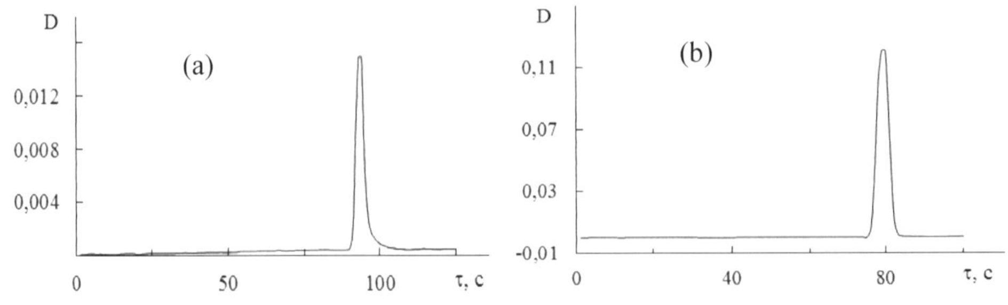

Figure 2. EOF electrophoregrams in segmented capillaries with PS:PSSA segments ratio 1:2 (a) and 2:1 (b). Effective length 45 cm, total length 50 cm, internal diameter 75 μm, mobile phase — Bicine (0.03 M, pH 9.0), working voltage 15 kV, EOF marker – dimethylsulfoxide, spectrophotometric characterization at λ = 214±2 nm.

The obtained EOF marker profiles in both single polymer coated and segmented SC demonstrate the peak symmetry reproducible retention time.

The data on EOF marker mobility measured in modified and segmented capillaries suggest that EOF rate linearly grows with PSSA (negative): PS (neutral) segments length ratio.

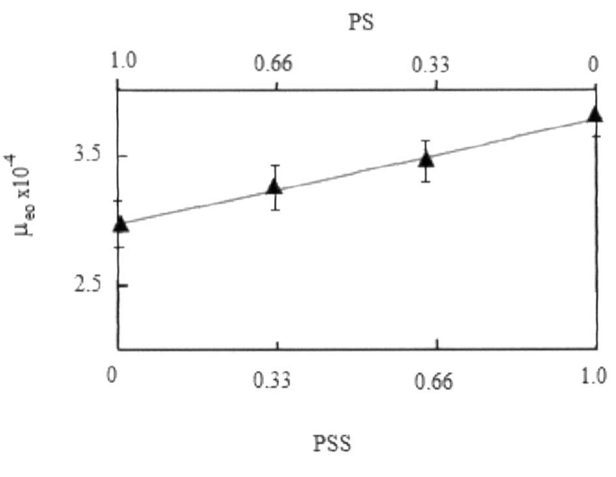

Figure 3. EOF rate change depending on PS/PSS-filled capillary segments length ratio. Effective length 45 cm, overall length 50 cm, internal diameter 75 μm, mobile phase 0.03 M Bicin pH 9.0, working voltage 15 kV, EOF marker – dimethylsulfoxide, spectrophotometric detection at λ = 214±2 nm.

The analysis of linear correlation in Fig. 2 allowed us to develop a mathematical model to describe and predict EOF rate μ_{eo} in segmented capillaries depending on the segments lengths and stationary phase with known surface properties including ξ-potential, EOF rate, etc.

$$\mu_{eo} = \frac{\mu_{eo1} l_1}{l_0} + \frac{\mu_{eo2} l_2}{l_0} \qquad (2)$$

where μ_{eo1} and μ_{eo2} are EOF rates in the columns 1 and 2 respectively, l_1, l_2 – segment length; l_0 – overall column length.
According to the suggested model, EOF rate μ_{eo} in SC is equal to the average value between the intrinsic rates μ_{eo1} and μ_{eo2} for each of the modified capillaries. The obtained data are in agreement with the results reported in [12] regarding monolithic and printed columns for capillary electrochromatography.

3.2. Model separation of cations
The efficiency of inorganic anions mixture separation was comparatively studied for a non-modified capillary with anionic coatings and non-modified fused silica capillaries (i.e. capillary elecrophoresis CE technique) (Fig. 4.).

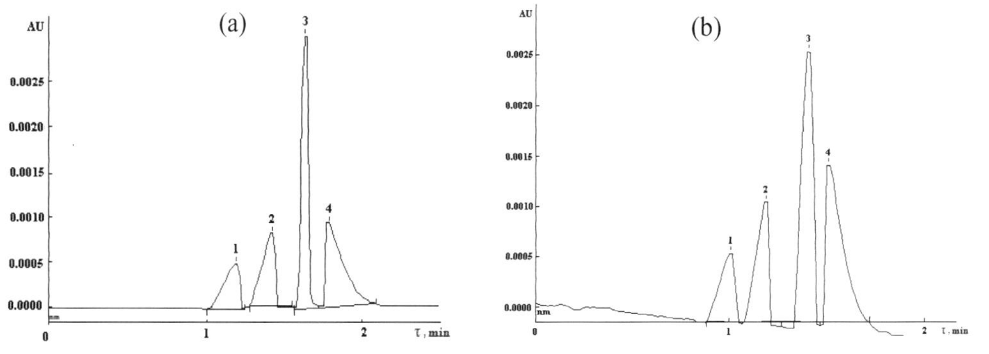

Figure 4. Electrophoregrams of cations mixture separation using a non-modified capillary (a) and hollow anionic capillary column (b). Separation conditions: Nanophor 01 installation, internal diameter of capillaries 50 μm; sample injection within 5 s at 0.05 bar pressure; voltage – 25 kV; background electrolyte – 40 mM benzimidazole, 20 mM tartaric acid, 10 mM 18-crown-6; UV-detection at 267 nm. The observed peaks relate to K^+ (1); Na^+ (2); Mg^{2+} (3); Ca^{2+} (4) 0.05 M solutions.

The obtained data indicate that the application of modified capillary with anionic coating provides a decrease in migration time and dispersion of K^+, Na^+, Mg^{2+} and Ca^{2+} peaks as well as an enhanced resolution in comparison with non-modified fused silica capillaries. More detailed data on the comparison of efficiency and selectivity of cations and transport proteins in non-modified, single polymer coated and segmented capillaries will be published in the nearest time.

4. Conclusion
Generally, the obtained results indicate that the suggested approach based on the development of capillary columns involving segments bearing different surface charges responsible for various functions (retention/separation and electroosmotic flow stabilization) is promising to provide reproducible steady performances for enhanced electrophoretic separation of various target mixtures.

5. Acknowledgements.
Support of this work by the Russian Foundation for Basic Research (grant no. 16-03-00625) is gratefully acknowledged.

References
[1] Belen'kii B G 2009 *High-Performance Capillary Electrophoresis* (St. Petersburg: Nauka) 2009 chapter 1 pp 3-33.
[2] Horvath J and Dolník V 2001 *Electrophoresis* **22** 644-655.
[3] Nilsson C et al. 2006 *Anal. Chem.* **78** 6088–6095.

IOP Conf. Series: Journal of Physics: Conf. Series **1038** (2018) 012049 doi:10.1088/1742-6596/1038/1/012049

[4] Neiman, B et al. 2002 *Anal. Chem.* **74** 3484–3491.
[5] Kuo I T et al. 2005 *Electrophoresis* **26** 2643–2651.
[6] Timerbaev A R 1997 J. *Chromatogr.* **792** 495-518.
[7] Yan Li et al. 2004 *Electrophoresis* **25** 2242-2256.
[8] Ailer R 1982 *The chemistry of silica* (M.: Mir) 706.
[9] Krasovskii A N 2009 *Nauchnoe Priborostroenie* **19** 51–58.
[10] Kurochkin V E et al. 2012 *Nauchnoe Priborostroenie* **22** 30–33.
[11] Shmykov A Yu et al. 2016 *Russian Journal of Applied Chemistry* **89** 1978–1984.
[12] Rarenga S and El Rassi Z 2011 *Electrophoresis* **32** 1033-1043.
[13] Zhang M and El Rassi Z J. 2006 *Proteome Res.* **5** 2001–2008.
[14] Yang C and El Rassi Z 1999 *Electrophoresis* **20** 18–23.

Synthesis of ternary chalcogenide colloidal nanocrystals in aqueous medium

D S Mazing, I S Chernaguzov, A I Shulga, O A Korepanov, O A Aleksandrova, V A Moshnikov

Saint Petersburg Electrotechnical University "LETI", ul. Prof. Popova, 5, 197376 Russia

E-mail: dmazing@yandex.ru.

Abstract. Colloidal nanocrystals of $AgInS_2$, Zn-Ag-In-S were synthesized directly in an aqueous medium by injection method. Two ligands, L-Glutathione and sodium citrate, were used for preparation of precursor solutions to achieve a balance of the reactivity of the two types of cations. Obtained samples are characterized by asymmetric photoluminescence band, tunable depending on the conditions of the synthesis.

1. Introduction

Semiconductor quantum dots (QD) play important role in founding new functional materials with better optical and electrical properties. The majority of QDs for visible and near-infrared range is based on binary compounds. These QDs contain toxic heavy metals such as Cd, Pb or Hg. Using heavy metals can limit the application of QD technology in household appliances and especially in biomedicine. Nanocrystals of ternary chalcogenide compounds I-III-VI ($CuInS_2$, $AgInS_2$) and its quaternary modifications may be a substantially less toxic alternative.

They are characterized by a wide tuning range of absorption and emission bands, a large Stokes shift and long lifetime of the excited state. The most significant drawback of this type of nanocrystals is the large full-width at half maximum of the emission band (usually> 80 nm). Synthesis of ternary chalcogenide nanocrystals in a nonpolar organic medium has recently seen significant progress [1,2]. However, in view of biomedicine applications direct production of the nanoparticles in aqueous solutions is of great interest [3,4].

2. Materials and methods

Silver nitrate ($AgNO_3$), indium(III) nitrate hydrate ($In(NO_3)_3{\bullet}4.5H_2O$), L-Glutathione, sodium citrate, sodium sulfide hydrate ($Na_2S{\bullet}9H_2O$), zinc nitrate hydrate ($Zn(NO_3)_2{\bullet}6H_2O$) were used as purchased without additional purification.

Absorption spectra were acquired on a PE-5400UV UV–vis spectrophotometer (LLC “Ekohim”). Photoluminescence measurements were carried out on spectrofluorimeter based on monochromator MDR-206 (LOMO) using laser excitation $\lambda = 445$ nm.

3. Experiment and results

Adjusting the reactivity of various cationic precursors is critical to avoid phase separation in the synthesis of multicomponent $CuInS_2$ and $AgInS_2$ nanocrystals. Hydrophobic 1-dodecanethiol have been extensively used to balance the reactivity of copper or silver and indium precursors in organic media. However, the difference of reactivity between $Ag^+(Cu^+)$ and In^{3+} ions is expected to be even more significant in aqueous solution. According to the hard and soft acids and bases (HSAB) theory, $Ag^+(Cu^+)$ is a soft acid, whereas In^{3+} is hard. Water-soluble thiols, such as L-glutathione and 3-mercaptopropionic acid, are soft bases (due to the presence of a sulfhydryl group) and their complex with In^{3+} hard acid have a low stability. As a result, phase separation may occur if water-soluble thiols are used as the sole complexing agent. To reduce the reactivity of the In^{3+} ion, a hard ligand which contains polycarboxylate groups, such as sodium citrate may be introduced [4]. Therefore, in this work both L-glutathione and sodium citrate are used as stabilizing agents for the synthesis of water-soluble colloidal nanocrystals of $AgInS_2$ and Zn-Ag-In-S.

The synthesis protocol was based on the method described in [4]. Colloidal nanocrystals of $AgInS_2$ were prepared with various ratios of cation precursors [Ag]:[In]. For [Ag]:[In] = 1:4 feeding ratio 0.005 mmol of $AgNO_3$ and 0.02 mmol of $In(NO_3)_3{\cdot}4.5H_2O$ were dissolved in 5 ml of H_2O and loaded into 10 ml flask. Thereafter 0.01 mmol of L-glutathione and 0.08 mmol of sodium citrate (200 μl of an aqueous solution) were added to this solution. The anionic precursor solution was obtained by dissolving 0.04 mmol of $Na_2S{\cdot}9H_2O$ in 500 ul distilled water. The sulfur precursor solution was injected into the stock solution at room temperature, then heated up to 95 ° C for 40 minutes. For ZnS shell deposition initial solution of the nanocrystal cores was cooled to room temperature and 0.02 mmol of $Zn(NO_3)_2{\cdot}6H_2O$ in 200 ul of distilled water was injected into the flask followed by dropwise addition of 0.02 mmol $Na_2S{\cdot}9H_2O$. The mixture was heated up to 95°C for 40 minutes. The same synthesis protocol was used for other ratios of cation precursors.

Optical properties of colloidal nanocrystals of ternary compounds have strong composition dependence. Synthesis of $AgInS_2$ was done at various molar ratios of metal cations. These aqueous samples were investigated using absorption and photoluminescence spectroscopies. Figure 1 shows the PL spectra of samples with different feeding ratio of cation precursors.

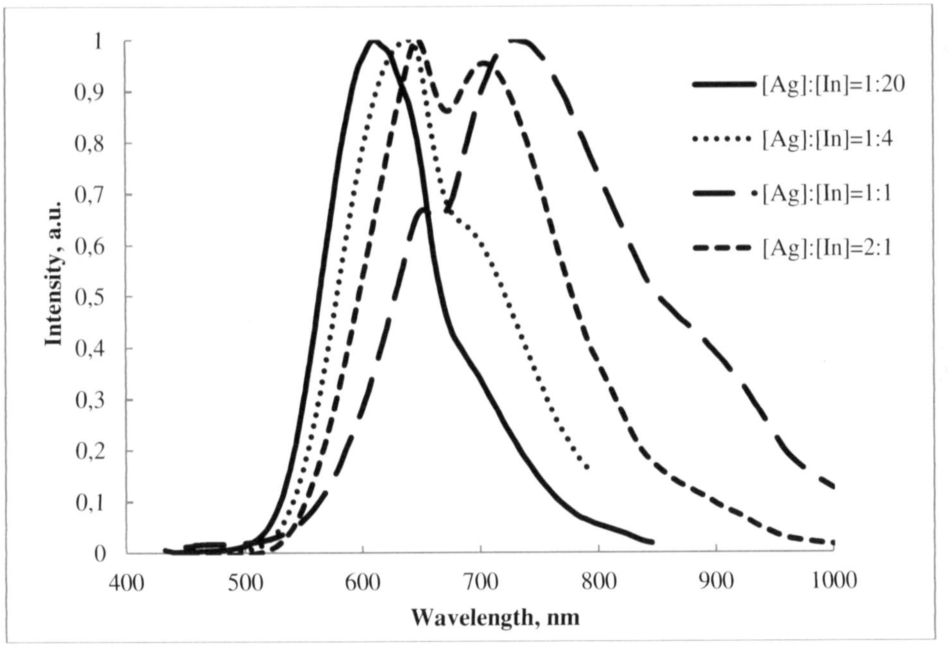

Figure 1. Photoluminescence spectra of colloidal nanocrystals of $AgInS_2$ prepared with different Ag^+ to In^{3+} molar ratios.

Emission band of synthesized nanocrystals consists of at least two subbands with low energy band tending to become more intensive relative to higher energy band with increasing Ag to In ratio. Surface states and internal traps have a huge impact on the observed broad and asymmetric PL spectrum. This issue was considered in detail in work [5] on $AgInS_2$ colloidal nanocrystals obtained in a nonpolar medium. The authors of this work investigated the changes in the emission spectra as a function of the growth time (size) of nanocrystals using time-resolved photoluminescence spectroscopy. They have shown that the two main emission bands making up the spectrum can be associated with radiative recombination through low-lying surface traps (the lifetime of the excited state is ~ 50 ns) for relatively high-energy segment of the spectrum and deep defect levels for the lower energy region (the lifetime of the excited state is ~ 220 ns). The intensity of the bands was redistributed in favor of emission through deep centers during the growth of nanocrystals [5].

The ratio of cations in the reaction volume was varied in our experiment, and the observed change in the shape of the PL spectrum in terms of the conclusions made in [5] is counterintuitive, taking into account that a decrease in the mole fraction of Ag^+ should lead to an increase in the concentration of silver vacancies and substitutional impurities of In_{Ag}^{2+} type which form deep acceptor levels. On the other hand there is possibility that concentration variation of cationic species affects not only resulting composition but also mean size of crystallites. As smaller QDs have higher surface area to volume ratio this may account for the dominance of shorter wavelength PL band at lower [Ag]:[In] precursor content signifying higher input of surface defects in resulting PL. To clarify the mechanisms of PL and the evolution of spectra in our hydrophilic samples, it makes sense to make measurements by the method of time-resolved photoluminescence spectroscopy, which will allow us to estimate the lifetime for different sections of the PL band and to draw a conclusion about emission nature.

In this experiment, the Zn-In-S cores were first obtained, after which various amounts of silver dopant were introduced. For this 0.02 mmol of $Zn(NO_3)_2 \cdot 6H_2O$ and 0.02 mmol of $In(NO_3)_3 \cdot 4.5H_2O$ in 5 ml of H_2O and loading into 10 ml flask. Thereafter 0.04 mmol of L-glutathione and 0.08 mmol of sodium citrate (200 μl of an aqueous solution) were consequentially added to this solution. The anionic precursor solution was obtained by dissolving 0.04 mmol of $Na_2S \bullet 9H_2O$ in 500 ul of distilled water. The sulfur precursor solution was injected into the stock solution at room temperature, then heated up to 95°C for 40 minutes. $AgNO_3$ precursor was dropwise added to the resulting solution of the cores so that the percentage mole ratio of Ag to In was 2.5%. The solution was heated up to 95 ° C for 4 hours. The same synthesis protocol was used for 5% and 10% ratios of cation precursors. Figure 2 shows the absorption and PL spectra of samples obtained with different amount of silver precursor.

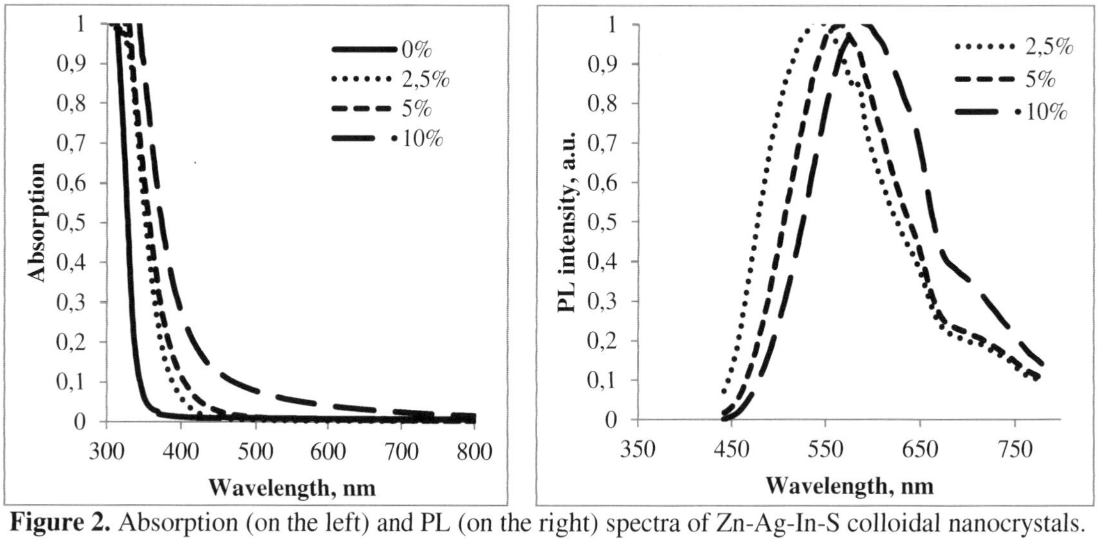

Figure 2. Absorption (on the left) and PL (on the right) spectra of Zn-Ag-In-S colloidal nanocrystals.

As can be seen from the photoluminescence spectra with increasing content of Ag^+, the absorption edge and the emission maximum shift to the long-wavelength region. It is suggested that Ag^+ has a higher rate of cation exchange with Zn^{2+} than with In^{3+} cations, which may result from bond of sulfur with indium being stronger than with zinc. The highest occupied molecular orbital (HOMO) of Zn-In-S consists of a hybrid of S 3p and In 5s5p orbitals, while the levels of the lowest unoccupied molecular orbital (LUMO) are formed from S 3p hybridized with Zn 3d. After the addition of Ag^+ cations of Zn^{2+} are partially replaced and the HOMO levels rise due to the strong interaction of the wave function of Ag 4d and S 3p. As a result band gap is reduced [6].

IOP Conf. Series: Journal of Physics: Conf. Series **1038** (2018) 012050 doi:10.1088/1742-6596/1038/1/012050

4. Conclusion

Colloidal nanocrystals of $AgInS_2$ and Zn-Ag-In-S were directly synthesized in aqueous medium employing L-glutathione and sodium citrate as ligands. Photoluminescence of synthesized nanocrystals consists of two or more subbands and is of defect nature. It exhibits strong dependence on the ratio of cationic precursors so that with increasing silver content emission band shifts to longer wavelength region. Additional investigations by means of time resolved photoluminescence spectroscopy may help to elucidate exact mechanism of emission and it dependence on the cation ratio.

Acknowledgments

The study was financially supported by the Russian Science Foundation (project № 17-79-20239).

References

[1] Aldakov D, Lefrançois A and Reiss P 2013 *J. Mater. Chem. C.* **24** 3756-3776

[2] Mazing D S, Shulga A I, Matyushkin L B, Aleksandrova O A and Moshnikov V A 2017 *Optics and Spectroscopy* **122(1)** 110-113

[3] Zhang B, Wang Y, Yang C, Hu S, Gao Y, Zhang Y, Wang Y, Demir H V, Liu L and Yong K T 2015 *Phys. Chem. Chem. Phys.* **17(38)** 25133-25141

[4] Chen Y, Li S, Huang L and Pan D 2013 *Inorg. Chem.* **52(14)** 7819-7821

[5] Mao B, Chuang C, Wang J, and Burda C 2011 *J. Phys. Chem. C.* **115** 8945–8954

[6] Song J, Ma C, Zhang W, Yang S, Wang S, Lv L, Zhu L, Xia R and Xu X 2016 *J. Mater. Chem. B.* **4(48)** 7909-7918

IOP Conf. Series: Journal of Physics: Conf. Series **1038** (2018) 012051 doi:10.1088/1742-6596/1038/1/012051

Thermal effect in magnetic capillary columns

A Y Shmykov, A LBulyanitsa and N A Esikova

Institute for Analytical Instrumentation, RAS, St. Petersburg, Russia

E-mail: shmykov.alexey@gmail.com, antbulyan@yandex.ru, elpis-san@yandex.ru

Abstract. The linearized Poisson-Boltzmann equation is used to estimate the radial distribution of potentials in a coaxial capillary and confirmed by experimental data. ξ-potentials are estimated for a capillary column with a sorbent containing magnetic nanoparticles. The thermal power in the column is calculated on the basis of the strength of an electric field with a certain electric conductivity. The column heating is calculated taking into account a thermal effect of the electric field for two estimates of the thermal diffusivity coefficient in the radial cross-section of a column comprising two layers of compounds with different physical properties.

Introduction

Presently the application of magnetic nanopartices (MNP) in the capillary electrochromatography (CEC) is focused on pseudo-stationary phases (PSP-CEC) [1], open-tubular (OTCEC) [2-7] and monolithic capillary columns [8-10]. In OT CEC and monolithic columns MNP are either used to modify the sorbent surface or incorporated into its polymeric matrix. This approach provides the adjustment of polarity, charge, ionization degree, adsorption activity, hydrophilic-hydrophobic properties and other performances of sorbents in capillary columns depending on the reaction mixture composition, polymerization conditions and physicochemical properties of MNP. Unfortunately, studies reported so far on this issue do not consider the effect of the magnetic field generated by MNPs in/on the polymer matrix upon the electrochromatorgaphic separation since MNPs are either not magnetizedor taken into account only indirectly on the basis of empirical data, hence the corresponding conclusions are without the consideration of magnetic field effects. The interpretation of magnetic field effect upon the analytical performances of the columns is complicated primarily by an insufficient reproducibility of the sorbent structure and MNPs distribution uniformity at the sorbent synthesis in capillary columns. Presently no available publications present data on the application of capillary columns with MNPs in capillary electrochromatography providing the estimation of MNP magnetic field effects upon the analyte separation, transport, etc. Nevertheless, MNP incorporation into sorbents and magnetic field generated by MNPs in the sorbent polymer matrix determines a new fundamental factor in the electrokinetic separation method. MNP-generated magnetic field in the sorbent polymer matrix affects such basic parameters of electromigration separation as the analyte and electro-osmotic flow profiles, dielectric permittivity, spatial distribution of the potential, ξ-potential, etc. The presence of MNPs in the sorbent at simultaneous effects of electric and magnetic fields can be also supposed to affect the separation process due to the temperature change in the capillary column.

This study is aimed at the development of mathematical modeling and assessment of MNP effects in the polymer sorbent layer upon the thermal diffusivity in an electromagnetic field.

The main objectives of this research are the study of the radial heat transfer and estimation of MNP presence effect upon this process schematically is illustrated in Figure 1.

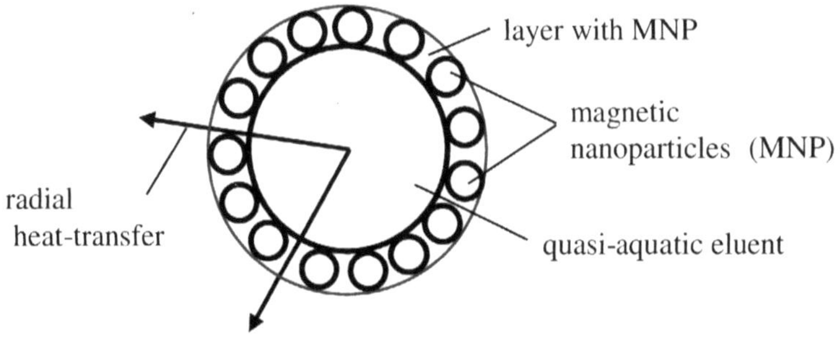

Figure 1. Scheme of heat-transfer in the capillary

The model description

Electric field acts as a heat source in capillary column. A possible heat care is calculated taking into account the thermal diffusivity coefficient at certain (ambient) temperature at the outer boundary of the capillary. The thermal power of the capillary column is determined according to the strength of longitudinal electric field (E) with a certain electric conductivity [11].

The mathematical model based on a linearized Poisson-Boltzmann equations (1) and (2). It has an analytical solution in the form of a linear combination of modified Bessel functions, market as I_0 and K_0. This approach allowed us to estimate a radial distribution of potentials $\phi(x)$ (3) and electro-osmotic flow (EOF) rate (4) based on the following defining parameters: R – outer radius of the coaxial capillary, c_0 – concentration of charged particles in the buffer (defining the solution pH value), T – temperature, ε - dielectric permittivity of the medium, as well as Faraday (F), Boltzmann (k_b), universal electric (ε_0) constants and electron charge (e). Argument x is the R-normalized radial coordinate, Δ_r– radial component of Laplace operator, ξ - value of the corresponding ξ-potential.

$$\Delta_r\phi = \alpha^2\phi \qquad (1)$$

with conventional boundary conditions: first - ϕ-function is limited for x tending to zero; second – at x=1 ϕ=1 (i.e. the potential at boundary is equal to ξ-potential).

$$\alpha^2 = \frac{2Fec_0R^2}{\varepsilon\varepsilon_0 k_b T} \qquad (2)$$

$$\phi(x) = \frac{(K_0(\alpha)-K_0(\alpha\rho))I_0(\alpha x)+(I_0(\alpha\rho)-I_0(\alpha))K_0(\alpha x)}{I_0(\alpha\rho)K_0(\alpha)-I_0(\alpha)K_0(\alpha\rho)}\xi \qquad (3)$$

$$V(x) = E\frac{\varepsilon\varepsilon_0}{\eta}(\xi-\phi(x)) \qquad (4)$$

The heat balance calculation (particularly the estimation of the effect of capillary heating promoted by the electric field) requires the determination of thermal diffusivity coefficient for the material comprising areas with different physical properties (polymer matrix and MNPs). The studied performances were averaged similar to the Lichtenecker procedure (power relationship with k=1) for the characterization of dielectric properties of composites [12].

The equation (5) affords averaging of the thermal diffusivity coefficient (α) taking into account the contributions of water-like buffer, polymer coating and MNPs in the capillary, while the equation (6) gives average thermal conductivity coefficient (k), medium density (ρ) and specific heat capacity (C), followed by the calculation of the thermal diffusivity coefficient according to the determined average values.

$$\alpha^{(1)} = \alpha_1(1-\delta) + \alpha_2\delta \quad (5)$$

$$\alpha^{(2)} = \frac{k_1(1-\delta)+k_2\delta}{(\rho_1(1-\delta)+\rho_2\delta)(C_1(1-\delta)+C_2\delta)} \quad (6)$$

where δ is a relative thickness of the near-wall layer; indices "1" and "2"correspond to the water-like buffer and near-wall layer with nanoparticles, respectively.

Results

The validity of linearized Poisson-Boltzmann equation application for the estimation of potentials distribution along the channel (capillary column) cross-section and EOF determination was tested using the equations (1) - (4) and results of mathematical processing for earlier obtained [13] experimental data on the estimation of EOF mobility value in capillary columns with different charges of the sorbent surface and various pH values of the buffer.

The equations (1) - (4) are applicable to the considered problem in the case EOF mobility changes for buffers with pH 4, 7 and 9 (7) is only determined by the difference in ξ-potentials (charges on the sorbent surface). The distribution of potentials along the column cross-section remained constant at a given pH value. Consequently, the ratio (7) should be complied

$$\frac{\mu(7)-\mu(4)}{\mu(9)-\mu(4)} = const \quad (7)$$

where μ(7), μ(9), μ(4) are EOF marker mobility in the buffers with pH 4, 7 and 9 [13].

The adequacy of the mathematical model confirmed by performance of the condition (7) for two types of columns with the error is not more than 2%.

The expressions (3) and (4) in combination with μ(7) value [1] were used to estimate ξ-potential on the sorbent surface. Upon x value, based averaging of ϕ(x) according to (3) the average ratio ϕ(x)/ ξ is 0.4994. Given the dynamic viscosity coefficient of the water-like buffer is equal to $0.87\text{x}10^{-3}$ m^2/s [14] and temperature is 26°C, the estimatedξ-potential is 115 mV.

A priori MNP role in the electroosmotic effect can be desceibed as follows. Due to their presence in the wall layer on the capillary surface, relative permittivity (ε) significantly decreases in urn resulting in the inreasre of the potential gradient and consequently in EOP rate gradient. The transferred substance is further concentrated near the channel axis promoting the increase in average convective rate and decrease of the peak width relating to the sample components. Actually magnetic forces at the used rates and magnetic field strengths are negligible even in comparison with the diffusion of large molecules.

The transversal electric field strength was approximately estimated as ξ-potential divided by the half of the coaxial gap (65 μm). The overestimated transversal electric field strength of 18 V/cm is about 11 times less compared with the axial longitudinal strength 220 V/cm. Thus the radial component can be neglected and the thermal power can be assessed only according to the axial component of vector E.

The results of thermal diffusivity coefficient estimation according to models (5) and (6) at different relative (R-normalized) thickness of near-wall layer with MNPs (comprising 25%vol. of the polymer matrix) are shown in the Figure 2.

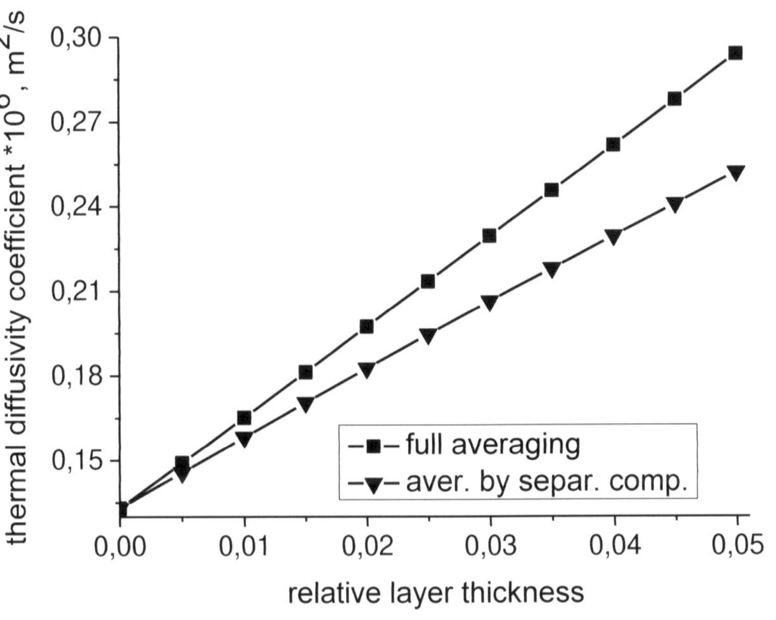

Figure 2. Dynamics of thermal diffusivity coefficient as a function of MNP-containing near-wall layer thickness: ■- full averaging of the of thermal diffusivity $\alpha^{(1)}$; ▼ - averaging by separate components $\alpha^{(2)}$

The figure indicates that thermal diffusivity coefficient grows with the polymer sorbent matrix thickness (with linear increase for $\alpha^{(1)}$ and non-linear growth of $\alpha^{(2)}$). Upon the addition of MNPs to the polymer matrix in OTCEC or modifying coating on the internal surface of the capillary in the migration analysis procedures the near-wall sorbent layer with 25 vol.% MNP content results in almost double increase in the thermal diffusivity coefficient thus providing a more intensive cooling of the capillary. The changes in electric field thermal power can be only determined by the transversal (radial) field component which contribution is only about 1%.

Conclusion

The performed calculations indicated that the presence of MNPs in the sorbent provides an increase of the transversal electric field strength consequently resulting in the enhancement of its thermal power in the capillary column. The increase of MNP in the near-wall layer of the sorbent further strengthens this effect, however the transversal electric field contribution into the thermal effect is only about 1% of the total thermal power of the electromagnetic field. Heating by longitudinal (axial) electric field in the capillary column does not exceed several hundredths of a degree per second. Nevertheless, during the sample separation within 100 s in the absence of external cooling, the eluent and analyte heating achieves up to 5-6°C and can linearly grow with the analysis time that is critical in respect of biological samples separation. Their identification can be complicated due to significant changes in viscosity and diffusion coefficient upon temperature changes even by several degrees resulting in the change of release time for the corresponding component of the sample.

Acknowledgements

The studywas carried out and financially supported within the framework of the State contract (No. 007-00229-18-00).

References

[1] Nilsson C and Nilsson S 2006 *Electrophoresis* **27** 76–83

[2] Zhu Y et al 2012 *Electrophoresis* **33** 340–347
[3] Liang R et al 2012 *J. Chromatogr. A* **1266** 95–102
[4] Zhu Y et al 2013 *Talanta* **104** 173–179
[5] Wang X et al 2014 *J. Chromatogr. A* **1362** 301–308
[6] Wang W et al 2015 *J. Chromatogr. A.* **1411** 92–100
[7] Qin S et al 2011 *Chin. J. Chromatogr.* **29** 942–946
[8] Carrasco-Correa E et al 2015 *J. Chromatogr. A.* **1385** 77–84
[9] Krenkova J and Foret F 2013 *Anal. Bioanal. Chem.* **405** 2175–2183
[10] Krenkova J and Foret F 2011 *J. Sep. Sci.* **34** 2106–2112
[11] Kukhling K H 1982 (*Handbook of Physics* Moscow: Mir) p 475
[12] Edvabnik V G 2015 *Sovremennyye problem nauki i obrazovaniya.* **1-2**
[13] Shmykov A Yu et al 2016 *Russian Journal of Applied Chemistry* **89** 1978–1984
[14] Xiangchun X et al 2004 *Int. J. of Heat and Mass transfer* **47** 3145-3157

IOP Conf. Series: Journal of Physics: Conf. Series **1038** (2018) 012052 doi:10.1088/1742-6596/1038/1/012052

The analyses of the parameters of microporous structure in metal-oxide nanomaterials by comparative sorption methods

E V Maraeva, V A Moshnikov and P V Groshev
Department of Micro-and Nanoelectronics, Saint-Petersburg Electrotechnical University «LETI», ul. Prof. Popova, 5, Saint Petersburg 197376, Russia

Abstract. The article is concerned with the investigation of nanomaterials based on SiO_2 and SnO_2 obtained with the use of sol-gel technology. The specifics of applying the comparative adsorption methods for studying such porous structure parameters of nanomaterials as micropores volume are discussed.

1. Introduction

Recently the adsorption methods of analyses are among the most popular methods for studying the porous structure parameters in nanomaterials [1-3]. As recommended by IUPAC, the porous systems are classified by predominant pore size into microporous (pore diameter up to 2 nm), mesoporous (2 to 50 nm), and macroporous (over 50 nm). Recently the micropores are additionally classified into ultramicropores and supermicropores with a boundary between them of about 1 nm. At the same time only ultramicropores (<1 nanometers) are considered as "the true microporous" where the special mechanism of volume filling with a phase change of an adsorbate from a status of the adsorbed gas in the condensed phase without the intermediate education mono - or a polymolecular layer operates on walls of pores. Supermicropores are intermediate between the "true" micropores and mesopores, here the formation of the condensed layer on a surface of pore walls can precede a phase change.

In metal oxide sensors [4, 5], the macropores can facilitate the feed of gas or liquid materials to nanoreactors, as well as carry away the reaction products. The micropores appear to be a key player in adsorption processes of semiconducting gas sensors. In mesh structures, there is a microporous system that blocks and unblocks the conductivity. That is why the measurement of micropores volume in metal-oxide nanomaterials is an actual problem.

2. Experiment

In this study the sol-gel technology [6, 7] was used to synthesize the porous nanomaterials based on silicon dioxide and stannic oxide. Tetraethoxysilane served as a source (precursor) of silicon dioxide, and inorganic water-soluble tin salt $SnO_2 \cdot 2H_2O$ was used as a source of stannic oxide. The propyl and butyl alcohols served as solvent. To obtain powders, 10 % aqua ammonia (NH_4OH) was added to the solutions-sols. Adding ammonia is accompanied by changes in pH of the solution ranged from 2 to 8, and an immediate condensation of the hydrolysis products takes place resulting in a much faster gelation process. After the gelation, evaporation of the solvent at the indoor temperature was performed, and then the powders received were heat-treated for 30 minutes at the temperature of 600 °C.

To control properties of the synthesized nanomaterial surface, the nitrogen thermal desorption and atomic force microscopy were used. Specific surface area and pore distribution measurements were made using Sorbi MS (CJSC «META», Novosibirsk, Russia). Before the measurements, all samples were pretreated in the SorbiPrep sample preparation station. The operating principle of the station is based on the degassing dispersed and porous materials by heating in a stream of an inert gas (helium). The station allows you to set the temperature and time of heating in the range of 50–400° C, 0–99 minutes.

3. Results and discussion

For the measurement of external surface area and the volume of micropores it is necessary to use the comparative methods of analyses. In general terms the sorption equation looks like:

$$a(h) = a_\mu(h) + S{\cdot}\alpha(h) + a_{kk}(h) \quad (1)$$

The first member of sum describes sorption in the micropores, the second one describes sorption in the mesopores and the last one shows the capillary condensation contribution. The specific surface of macropores is very small (from 0.5 to 2 m^2/g), so the adsorption on the surface of the pores of this type can essentially be neglected.

For the nitrogen when relative pressures are less than 40 % (when capillary condensation doesn't take place), the equation looks like follow:

$$a(h) = V_\mu + S{\cdot}\alpha(h), \quad (2)$$

where V_μ is a value of the micropores volume, S is the specific surface area of mesopores, $\alpha(h)$ – unit-area adsorption. This is a linear equation, and it can be presented as a straight curve. The value of V_μ can be found as cut-off of approximate line in the axis of ordinate, the value of S – from the slope of the curve. For the pattern without micropores this curve passes through the origin.

In the figure 1 examples of evolution of the data obtained by the instrument Sorbi, in coordinate a-α are given.

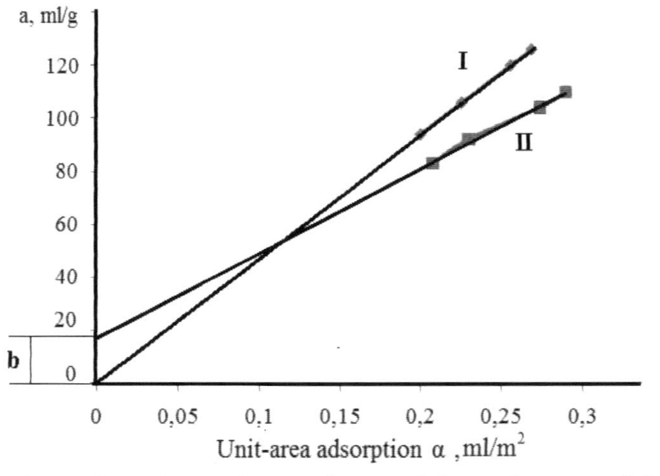

Figure 1. Typical a - α dependences for samples containing and not containing micropores

The line I in the figure 1 corresponds to the standard sample of Al_2O_3 without micropores, the line II corresponds to the sample with microporous system. The volume of micropores in the samples like II is calculated on a ratio:

$$V_\mu = k{\cdot}b, \quad (3)$$

where b is the value which is split off by the line II on ordinate axis, k is a constant depending on the ratio of density of an adsorbate in gas and liquid phases (for nitrogen value of b = 0.00015).

For investigation a set of metal-oxide patterns preheated in various conditions before the measurements was chosen. For all the patterns the thermal desorption measurements were made using Sorbi MS. For example figure 2 presents the adsorption isotherms for a set of samples with composition $0.85SiO_2$-$0.15SnO_2$ and $0.5SiO_2$-$0.5SnO_2$ preheated in SorbiPrep in various conditions.

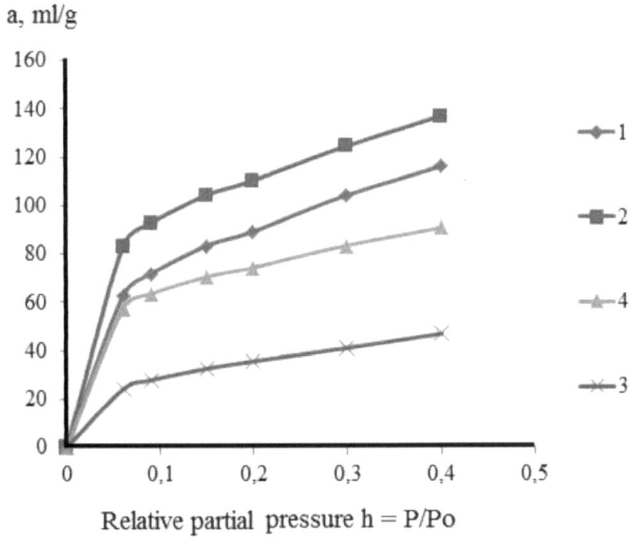

Figure 2. Adsorption isotherms for a set of samples: 1 – $0.85SiO_2$-$0.15SnO_2$ without pretreating, 2 – $0.85SiO_2$-$0.15SnO_2$ with pretreating at 150°C, 40 min, 3 – $0.50SiO_2$-$0.50SnO_2$ without pretreating, 4 – $0.50SiO_2$-$0.50SnO_2$ with pretreating at 150°C, 40 min

Figures 3, 4 illustrate a - α dependences based on the analyses of adsorption isotherms shown in the figure 2.

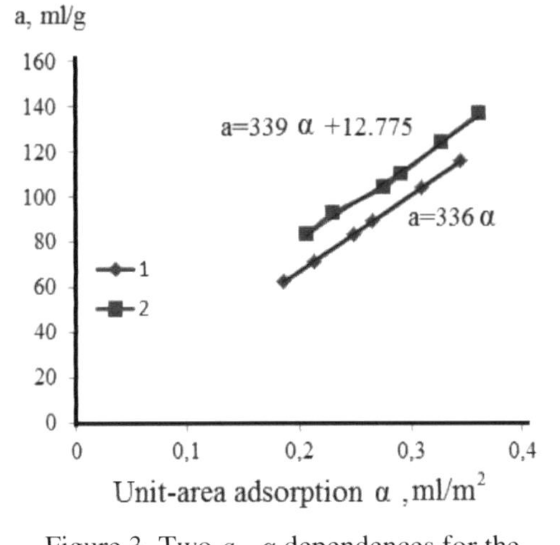

Figure 3. Two a - α dependences for the samples with composition $0.85SiO_2$-$0.15SnO_2$

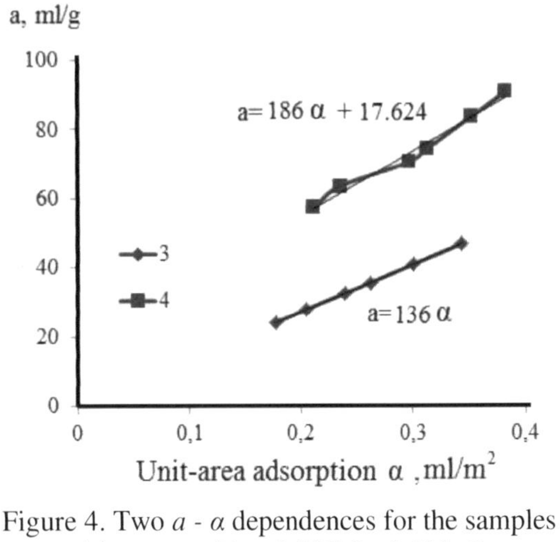

Figure 4. Two a - α dependences for the samples with composition $0.50SiO_2$-$0.50SnO_2$

From the figures 3, 4 it can be seen that the samples number 2 and 4 which were not preheated in SorbiPrep station are characterized by microporous system with volume 0.011 cm^3/g for $0.85SiO_2$-$0.15SnO_2$ sample and 0.009 cm^3/g for $0.5SiO_2$-$0.5SnO_2$ sample. Most likely, this is driven by H_2O molecules removal from the porous structure during the thermal annealing. The parameters of porous structure, including the specific surface area (S) and external surface area (STSA) are shown in the table 1.

Table 1. The parameters of porous structure in the samples based on tin and silicon dioxides

Sample	Composition	Solvent	SorbyPrep conditions	S, m^2/g	STSA, m^2/g	V, cm^3/g
1	$0.85SiO_2$-$0.15SnO_2$	butanol	-	336	336	0
2	$0.85SiO_2$-$0.15SnO_2$	butanol	150 °C 40 min	403	379	0.011
3	$0.50SiO_2$-$0.50SnO_2$	butanol	-	136	136	0
4	$0.50SiO_2$-$0.50SnO_2$	butanol	150 °C 40 min	270	237	0.009
5	$0.15SiO_2$-$0.85SnO_2$	butanol	150 °C 40 min	120	120	0
6	$0.15SiO_2$-$0.85SnO_2$	propanol	150 °C 40 min	30	30	0
7	$0.10SiO_2$-$0.90SnO_2$	propanol	150 °C 40 min	27	27	0

From the table 1 it can be seen that heat treatment of samples 1-4 at T = 150 °C and t = 40 min in SorbiPrep station leads to the microporous system release. Such preheating conditions were chosen for all other samples as optimal. It was noticed that the presence of a system of micropores is observed only in the samples of composition with molar share of silicon dioxide more than 0.5. The maximal specific surface area is 403 m^2/g for $0.85SiO_2$-$0.15SnO_2$ sample and 225 m^2/g for $0.5SiO_2$-$0.5SnO_2$ sample. However, for gas-sensitivity it is necessary to obtain the layers with high content of solid-state phase [4]. These samples were also investigated and the results are presented in table 1.

4. Conclusions

In result, it was found that the most important parameters that influence on microporous system appearing are mole ratio of the components and time-temperature conditions of preheating the patterns before the measurements. In some cases it was noticed that microporous system can be found in the patterns after longstanding preheating. The study of metal oxide nanomaterials by comparative sorption methods allows us to confirm the assumption of the existence of a multilevel system of pores in the gas-sensitive layers, and to develop the technological regimes for obtaining of the layers for gas sensors of new generation with the most advanced characteristics, where the surface development and the presence of multilevel porous system plays a decisive role.

Acknowledgments

This study was supported by the Ministry of Education and Science of the Russian Federation within the framework of the project № 16.897.2017/PCh.

References

[1] Schneider P, Hudec P, Solcova O. 2008 *Microporous and Mesoporous Materials* **115 (3)** 491
[2] Passe-Coutrin N, Altenor S, Cossement D, Jean-Marius C, Gaspard S. 2008 *Microporous and Mesoporous Materials* **111** 517
[3] Belorus A O, Maraeva E V, Spivak Yu M and Moshnikov V A 2015 *Journal of Physics: Conference Series* **586** 012017
[4] Abrashova E V, Gracheva I E, Moshnikov V A 2013 *Journal of Physics: Conference Series* **461** 012019
[5] Vasilyev V A, Averin I A, Pecherskaya R.V., Pronin I A, Karmanov A A RF Patent for Invention No. 2506659 (2014)

[6] Maximov A I, Moshnikov V A, Tairov Yu M and Shilova O A 2008 *Basics of sol-gel technology of nanocomposites, in Russian* (St. Petersburg: Elmor)
[7] Maraeva E V, Istomina M S, Moshnikov V A, Nalimova S S, Maximov A I, Alexeev N I, Semenov K N 2016 *Journal of Physics: Conference Series* **690** 012031

IOP Conf. Series: Journal of Physics: Conf. Series **1038** (2018) 012053 doi:10.1088/1742-6596/1038/1/012053

Epitaxial GaN nanotripods: morphology and crystal structure

G.A. Sapunov[1], A.D. Bolshakov[1], V.V. Fedorov[1], A.M. Mozharov[1], D.A. Kirilenko[3], A.A. Sitnikova[3], I.S. Mukhin[1,2]

[1]St.Petersburg Academic University, 194021, St. Petersburg, Russia
[2]ITMO University, 197101, St. Petersburg, Russia
[3]Ioffe Physical-Technical Institute of the Russian Academy of Sciences, 194021 St. Petersburg, Russia

Abstract. Abstract. The goal of this work is to study the morphology and crytal structure of novel GaN epitaxial nanostructures (Y-shape nanotripods) grown on silicon (111) by plasma assisted molecular beam epitaxy (PA-MBE). Prior to the nanowire synthesis epitaxial GaN nanoparticles were formed on the Si(111) via nitridation of Ga nanodroplets on Si substrate surface (droplet epitaxy technique). Effect of the seeding layer on futher nanostructure growth, GaN/Si heterointerface formation, crystal structure and morphology is studied by use of different microscopic techniques.

1. Introduction

Growth of group-III nitride nanostructures, such as GaN nanowires (NWs), attracts a lot of attention due to their importance both in fundamental research and in development of future optoelectronic nano-scaled devices [1]. Growth of GaN NWs on Si attracts a lot of attention due to several facts. Small contact area with the substrate, commonly observed for the epitaxial nanostructures, can provide efficient lateral strain relaxation and defect-free NW growth, as it has been demonstrated for GaN NWs [2]. Another feature of the GaN NWs formation on Si is the self-induced technique. This technique doesn't involve the substrate surface patterning and catalyst particles use which makes it attractive both from technological and further device application points of view.

Majority of today's GaN on Si growth technologies inevitably leads to formation of thin silicon nitride (Si_3N_4) layer on the substrate surface. In our study we attempted Ga-droplets deposition on the silicon surface. In this report, we will demonstrate that use of the GaN seeding layer prepared by the droplet epitaxy [3,4] can lead to formation of oriented arrays of tripod-shaped nanoparticles (tripods) along with NWs. Previously it was reported, that GaN nanotripods and hexapods can be formed during HVPE on c-plane Al_2O_3 and PA-MBE [5] on pillar patterned GaN templates [6]. In the papers above, the seeding layers were not used. As far as the authors know, formation of tripod-shaped GaN nanoparticles in PA-MBE on Si deposition process is observed for the first time in our study. It worth to mention, that hexapod shaped nanoparticles were not observed in our work.

2. Experimental

GaN nanostructures were grown by plasma-assisted molecular beam epitaxy (PAMBE) in Veeco GEN-III MBE machine with use of Riber valved RF-plasma nitrogen source (13.56 MHz). Silicon

(111) p-type wafers with a 4° miscut oriented towards (110), cleaned with a Shiraki method were used as substrates. After thermal annealing (10 min @ 950° C) atomically clean Si surface was obtained as confirmed by observation of a 7x7 RHEED reconstruction pattern. Due to the 4° miscut high temperature annealing leads to coalescence of silicon surface steps, i.e. step bunching [7], and ordered step-terrace arrays appear on substrate with average terrace width of ~80 nm.

To monitor the atomic or molecular beam equivalent pressure (BEP) from nitrogen source or effusion cells conventional ion gauge was used. Total nitrogen BEP was in the range of 2..3x10^{-7} Torr. To maintain nitrogen-rich (N-rich) conditions Ga source BEP was kept sufficiently small - 1..2x10^{-8} Torr. Ga source BEP was kept constant during deposition of the seeding layer and growth process.

In order to obtain direct nucleation of GaN on Si, deposition of the Ga droplet seeding layer was performed prior to ignition of the RF-plasma. In the following stage droplet layer was annealed under the nitrogen activated nitrogen flux up to growth temperature (790-810 °C). After the nitridation further deposition of GaN was carried out.

Ga droplets were formed by exposing the reconstructed Si (111)-(7x7) substrate surface at 200 °C to a Ga molecular beam. Equivalent Ga thickness was estimated by monitoring the Ga-induced surface reconstruction with RHEED. It is known that deposition of 0.3 ML of Ga leads to formation of $\sqrt{3}\times\sqrt{3}$ -R30° Ga-reconstructed surface [8] and deposition of 0.6 ML leads to formation of 6.3x6.3-Ga induced superstructural phase [9] (1 ML here means $6.81 \cdot 10^{14}$ atoms/cm^2, the surface density of the truncated (111) plane of bulk Si) [10]. Seeding layers with equivalent thickness in the range of 0.6-2.5 ML were deposited while monitoring the resulting surface reconstruction in-situ with RHEED. Diffuse semi-circular halos of liquid Ga was observed for the exposition with equivalent thickness more than 1.5 ML. Relatively low substrate temperature and small Ga exposure were chosen to obtain dense arrays of nano-droplets [11].

Structure and morphology of the synthesized nanoheterostructures were studied with scanning electron microscopy (Zeiss SUPRA 25-30-63) and high-resolution transmission electron microscopy (HRTEM) using a JEOL 2100F microscope (200kV). The samples for HRTEM studies were prepared by standard methods involving ion sputtering at the last stage.

3. GaN tripod-shaped nanostructure

It is well known that deposition of GaN on Si (111) substrate in N-rich conditions usually lead to formation of vertical GaN NWs [2]. Our growth series revealed that use of nano-droplet seeding layer can result in formation of well oriented tripod-shaped GaN nanoparticles grown along with NWs on Si (111) substrate. SEM images of the obtained structures are presented in Fig. 1. According to the literature [5,6] tripod and hexapod formation were observed under the certain conditions, and transition between GaN nanocolumn axial growth and tripod formation can be observed [6] while in our research both NWs and tripods growth can occur simultaneously and surface densities for both NWs and nanotripods strongly depend on equivalent thickness of the seeding layer and growth parameters. As can be seen from top view SEM images (Fig. 1) tripod-shaped nanoparticles have planar structure and are formed with three equally sized elongated branches lying at an angle of 120° to each other. Single and double branched V-shaped GaN nanoparticles can be found along with the tripod-shaped ones. According to [5] GaN tripod growth is related with the formation of nanoparticle in the zinc blende (ZB) phase and further nucleation and growth of the wurtzite (WZ) nanorods on the {111} facets of the ZB core. We assume that in our case formation of small ZB nanoparticles occurs during the seeding layer growth in droplet epitaxy process.

According to analysis of SEM images tripod-shaped nanoparticles demonstrate several preferential in-plane orientations which correspond to alignment of the nanoparticle branches along with the $<1\bar{1}0>$ and $<11\bar{2}>$ directions of Si. The approximate particle orientation is determined by comparing the directions of branches and wafer primary flat (or cleavage direction). Crystal rotational twinning corresponding to 180° rotations around normal to the Si substrate and mosaic spread of in-plane orientation in the range of ±5° are observed. In-situ registered RHEED pattern shows six-fold symmetry corresponding to the twinned wurtzite structure.

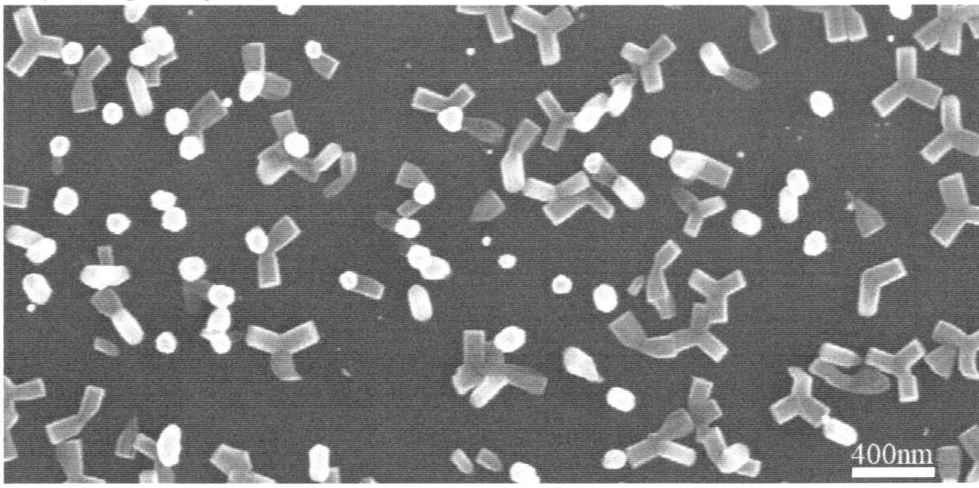

Figure 1. Top SEM images of tripod-shaped GaN on Si substrate

More detailed analysis of crystal structure and morphology of the tripod-shaped nanoparticles was performed with HRTEM. As can be seen from the cross-section TEM-images (Fig. 2 a) direct Si-GaN heterointerface cannot be obtained: a thin (~1.5nm) amorphous layer is observed at the heterointerface. Formation of the SiN_x layer under the nanostructures takes place most probably during the seeding layer formation: redistribution of the Ga species from the droplets and subsequent exposure of the silicon surface to the nitrogen flux. Another process leading to the substrate surface nitridation is volume diffusion of nitrogen atoms from the GaN into the Si surface layer [15].

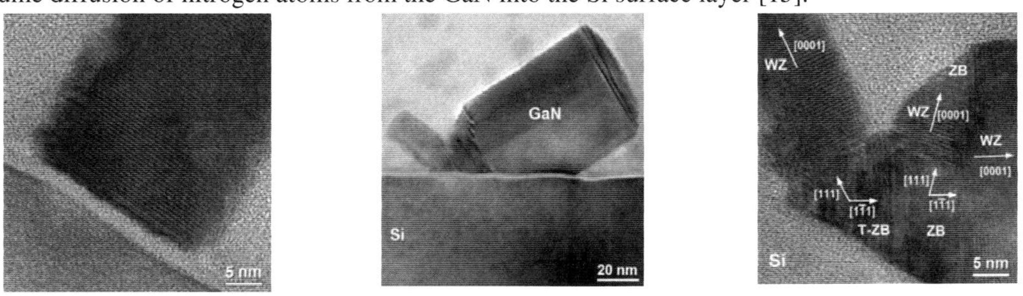

(a) (b) (c)

Figure 2. TEM images of the synthesized nanostructures: a) vertical NW, b) inclined NW, c) HRTEM of the bottom part of inclined NW

According to the RHEED and HRTEM nanoparticles have a hexagonal wurtzite GaN crystal structure. GaN c-planes (0001) can be seen on close-up image in Fig. 2 (c). Linear contrast features at the HRTEM image (Fig. 2 (b)) are related to planar defects perpendicular to the (0001) direction, in particular, stacking faults in the hexagonal lattice caused by disorder in stacking sequence and monolayer-thick inclusions of zinc-blende GaN polytype.

In Fig. 2 one can find TEM (b) and close-up HRTEM (c) images of inclined nanocolumns which are forming the branches of nanotripod. According to the HRTEM image (Fig. 2 c) zinc-blende phase can be found in the central part of the tripod-shaped island.

We assume, that zinc-blende GaN nanoparticle can be formed during seeding layer formation by droplet layer epitaxy. It is known [18-19], that low temperature and excess Ga can lead to the nucleation of zinc-blende GaN islands. After formation lateral {111} facets of the zinc-blende nanoislands play role of nucleation sites and formation of inclined nanocolumns and nanotripods islands occurs.

4. Conclusions

Tripod-shaped nanoparticles (nanotripods) were grown on silicon (111) PA-MBE with use of seeding layer prepared by droplet epitaxy. Nanotripods demonstrate several preferential in-plane orientations with twinning corresponding to 90° and 180° rotations around normal to the Si substrate. The direct Si-GaN heterointerface without SiN_x layer under the nanostructures cannot be obtained even with use of Ga-droplet seeding layer. GaN nanoparticles forming the seeding layer crystallizes in zinc-blende phase and acts as a seed for further growth of inclined NWs or tripod-shaped nanoparticles which are crystallized in hexagonal wurtzite GaN crystal structure.

Acknowledgements

This work was supported by the Russian Foundation for Basic Research (grants 16-32-00560, 15-02-06839), Council for grants of the Russian Federation President (grant MK-3632.2017.2), grants 16.2593.2017/4.6, 16.2593.2017/8.9 of the Ministry of Education and Science of the Russian Federation and Leading universities of the Russian Federation grant 074U01.

References

[1] M. Mandl et al. 2013 *Phys. status solidi - Rapid Res. Lett.* vol. **7** no. 10 pp. 800–814
[2] V. Consonni 2013 *Phys. status solidi - Rapid Res. Lett* vol. **7** no. 10 pp. 699–712
[3] R. K. Debnath et al. 2009 *J. Cryst. Growth* vol. **311** no. 13 pp. 3389–3394
[4] I.-S. Yu, C.-P. Chang, C.-P. Yang, C.-T. Lin, Y.-R. Ma, and C.-C. Chen 2014 *Nanoscale Res. Lett.* vol. **9** no. 1 p. 682
[5] S. Lee, Y. Sohn, C. Kim, D. R. Lee, and H. H. Lee 2010 *J. Appl. Crystallogr.* vol. **43** no. 6 pp. 1300–1304
[6] P. Wang et al. 2017 *Adv. Funct. Mater.* 1604854
[7] H. Hibino and T. Ogino 1994 *Phys. Rev. Lett.* vol. **72** no. 5 pp. 657–660
[8] A. Kawazu and H. Sakama 1988 *Phys. Rev. B* vol. **37** no. 5 pp. 2704–2706
[9] J. Čechal, M. Kolíbal, P. Kostelník, and T. Šikola 2007 *J. Phys. Condens. Matter* vol. **19** no. 1 p. 16011
[10] P. Kumar, J. Kuyyalil, and S. M. Shivaprasad 2010 *Appl. Phys. Lett.* vol. **97** no. 22 p. 221913
[11] J. Ristić et al. 2008 *J. Cryst. Growth* vol. **310** no. 18 pp. 4035–4045
[12] K. Yamane, T. Kobayashi, Y. Furukawa, H. Okada, H. Yonezu, and A. Wakahara 2009 *J. Cryst. Growth* vol. **311** no. 3 pp. 794–797
[13] A. Dadgar 2015 *Phys. status solidi* vol. **252** no. 5 pp. 1063–1068
[14] Ž. Gačević, D. Gómez Sánchez, and E. Calleja 2015 *Nano Lett.* vol. **15** no. 2 pp. 1117–1121
[15] T. A. Rawdanowicz and J. Narayan 2004 *Appl. Phys. Lett.* vol. **85** no. 1 pp. 133–135
[16] T. Wang 2016 *Semicond. Sci. Technol.* vol. **31** no. 9 p. 93003
[17] B. M. Shi, M. H. Xie, H. S. Wu, N. Wang, and S. Y. Tong 2006 *Appl. Phys. Lett.* vol. **89** no. 15 p. 151921
[18] L. T. Romano, B. S. Krusor, R. Singh, and T. D. Moustakas 1997 *J. Electron. Mater.* vol. **26** no. 3 pp. 285–289
[19] F. Schuster et al. 2015 *Nano Lett.* vol. **15** no. 3 pp. 1773–1779

IOP Conf. Series: Journal of Physics: Conf. Series **1038** (2018) 012054 doi:10.1088/1742-6596/1038/1/012054

Formation of fractal microstructures in conductive layers of indium-tin oxides and zinc oxide

A A Bobkov[1], V F Borodzulya[2], A V Solomonov[1] , I I Mikhailov[1], V A Moshnikov[1], S A Tarasov[1], I A Lamkin[1], T D Lebedeva

[1] Saint Petersburg Electrotechnical University «LETI» (ETU)

[2] Peter the Great St.Petersburg Polytechnic University

iimihalov@gmail.com

Abstract. In this paper, we have studied the fractal microstructures that form within conducting layers of oxide compositions and the effects that happen during electrical breakdown. We have determined that application of additional polymer films on top of the layers of oxide compositions allows one to visualize the breakdown processes taking place in the structure. The enlarged "polymer photography" that results in this makes it possible to assess the quality of contact layers without using high-resolution optical equipment.

1. Introduction

Despite positive outlook of the development of flexible electronics and photonics, there are still problems that slow down the progress in this area. In particular, there are difficulties in creating contact areas that can withstand multiple bends. Contacts with a fractal structure can solve this problem. However, the creation of such contacts using conventional methods of microelectronic engineering is a complex and expensive task. Therefore, it was logical to search for a way of creation of this fractal structure during a physical process. The effects that appeared during the flow of high-density currents through conventional film materials made it possible to study the possibility of creation of fractal film micro- and nanostructures. The film materials used for these experiments are normally used in transparent conductive electrode coating and are made on the bases of oxide compositions.

2. Samples and techniques

Structures based on indium-tin oxide and zinc oxide served as test samples (Figure 1). The thickness of the substrate was 0.5 - 3 mm. The thickness of the oxide compositions films varied in the range of 1-10 μm. Some of the samples were coated with organic polymeric layers based on polycarbonate (PC) and polymethyl methacrylate (PMMA) with a thickness of 100-200 nm.

The setup used to study the effects of high density currents (Figure 2) had a nanoelectrode based on the eutectic composition Ga / In. This electrode allowed creation of high local field intensities and maintained diameter of contact area between the electrode and the film close to ~ 60 μm [1].

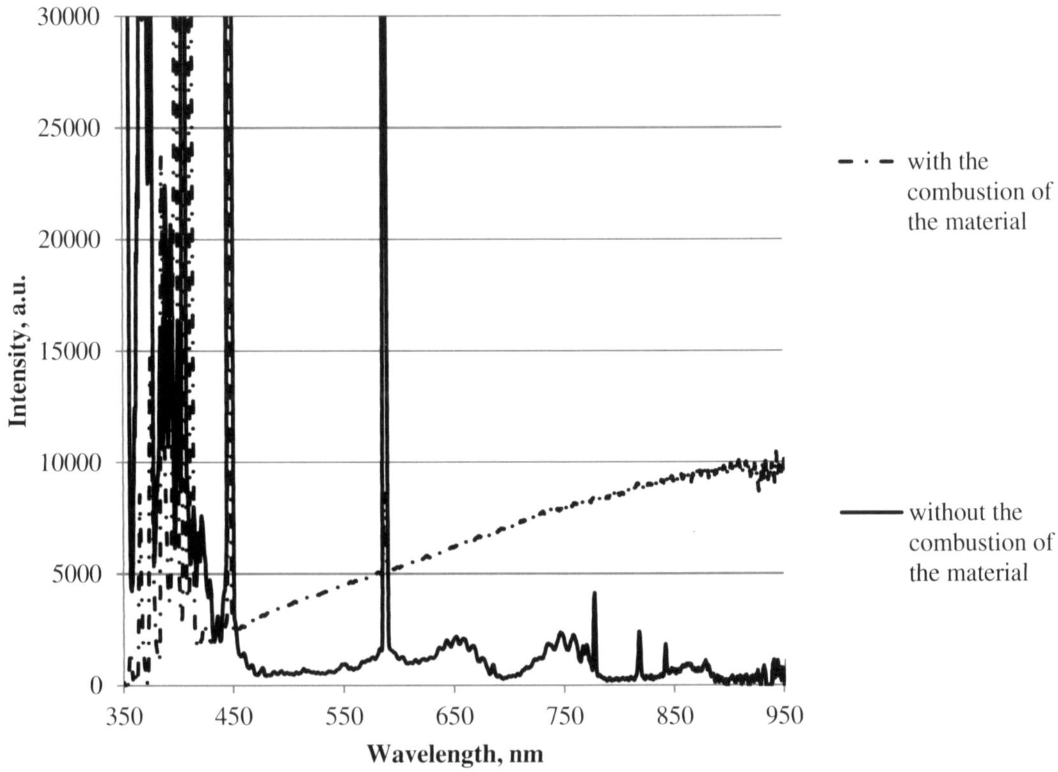

Figure 4. Radiation spectrums that appear from the breakdown of the ITO layer with and without combustion of the material.

The most obvious were destructions of thin ITO layers when a thin polymeric layer that increased the contrast of the image were applied to the ITO surface. In addition, in this case the electrical contact between the ITO and the filamentary electrode occured only after electrical breakdown and destruction of the polymer dielectric film in the region of contact of the threadlike electrode. The polymer film broke through at a voltage of ~ 100 V. Hence, the voltage after the breakdown was applied to the ITO. That consequently allowed a significant current flow through the ITO. Its flow through the ITO layer was localized in a narrow (about 10 μm) track. Its heating led to destruction and removal of the polymer from the track and eventually to the destruction of the ITO in the track region.

Thus, the application of additional polymer films over transparent oxide layers allows visualization of the breakdown processes occurring in the structure. The resulting "polymer photography" makes it possible to assess the quality of the oxide layer.

4. Conclusion

The focus of this research was to study processes that occur during the formation of fractal microstructures in conductive oxide compositions formed on glass substrates. A setup was developed for this research. It allowed usage of high-intensity local field intensity and a nanoelectrode based on a Ga / In eutectic composition to provide a micrometer-sized contact spot diameter. The created fractal micro- and nanostructures were studied using the methods of optical spectrometry, metallographic computer analysis, and atomic force microscopy. This research has shown that in thin ITO layers (thickness of 100 300 nm) the destructions have the form of individual short tracks with lengths of 200-400 μm. In ITO with thickness of more than 300 nm the destructions have the shape of a multi-turn helix with a diameter of ~ 1 mm. The specific features of current flow processes in the structures studied were largely determined by the fact that the liquid gallium-indium eutectic was used as the lead electrode. This liquid contact performed self-oscillatory motions along a spiral trajectory when an electric current passed through it. Unlike conventional electrodes, the liquid-contact electrode did not burn out, and the contact point moved from the region of the evaporated material to the neighboring region practically along a circular path because of the tendency of the elongated electrode to return to the most advantageous vertical position. The step size of the turn was determined by the amplitude of the voltage pulse applied to the structure. In most structures, the spiral trajectory of breakdown was supplemented by numerous branches, which in turn have additional processes at the next level of the fractal hierarchy. Analysis of the emission spectrum that occur during the formation of spiral fractal structures shows that a line spectral characteristic with dominant peaks appears, whose position is determined by the parameters of the oxide layer.

The work was carried out by the Ministry of Education of the Russian Federation in the field of scientific activity, project number 16.1750.2017 / 4.6.

References

[1] Formation of thin-film fractal micro- and nanostructures by the electric discharge method / Tarasov S. A, Borodzyulya VF et al. // Science and education: technology of success. International scientific conference. Publishing house SPbGETU «LETI» 2016, C. 84–92

[2] Tadtaev P. O., Moshnikov V. A., Tarasov S. A., Solomonov A. V., Borodzyulya V. F., Sudar N. Thin-film Fractal Nanostructures Formed by Electrical Breakdown // Proceedings of the 2017 IEEE Russia Section Young Researchers in Electrical and Electronic Engineering Conference, ElConRus 2017, p. 1434-1437

[3] Gracheva IE, Maksimov AI, Moshnikov VA Analysis of the features of the structure of fractal nanocomposites based on tin dioxide by atomic force microscopy and X-ray phase analysis // Surface. X-ray, synchrotron and neutron studies. 2009. № 10. C. 16-23.

IOP Conf. Series: Journal of Physics: Conf. Series **1038** (2018) 012055 doi:10.1088/1742-6596/1038/1/012055

Study of the electric field strength in planar multigraphene/SiC field emission nanostructures with different arrangement of the electrode planes

I L Jityaev[1] and A M Svetlichnyi[1]
[1]Southern Federal University, Institute of Nanotechnologies, Electronics, and Equipment Engineering, Taganrog, 347922, Russia

E-mail: izhityaev@sfedu.ru

Abstract. The paper describes simulation of planar field emission nanostructures on the basis of graphene on a semiinsulating silicon carbide. The effect of arrangement of the electrode planes on the field strength and field gain is shown. The model is considered with a nanoscale interelectrode distance and a potential difference of 10 V.

1. Introduction

Developments in the field emission electronics suggest a transition to terahertz electronics. Graphene films are promising for these purposes. The unique geometric and electrical properties of graphene contribute to characteristic improvement of the devices on its basis [1-4]. The properties of graphene films depend on the preparation method [5-7]. The most technological method is the thermal destruction of SiC in vacuum [7-9]. Graphene films suitable for electronic application can be fabricated with this method. A different number of graphene layers (multigraphene) can be obtained by controlling the technological parameters of SiC annealing [10, 11]. Silicon carbide substrate has high thermal conductivity, mechanical strength, resistance to aggressive environment [12]. An increase in the number of layers in the multigraphene film enhances the durability of the field emission device.

One of the directions of emission electronics is field emission nanoelectronics. Graphene is a good material for the elements of field emission nanoelectronics [13-20]. The nanometer thickness of the multigraphene makes it possible to develop planar-edge emitters with a high field gain [21-23]. A promising method for formation of nanometer interelectrode gaps and ultrasharp emitters is the method of focused ion beams [24, 25]. In this case, the minimum interelectrode distance is determined by the ion beam width. The transition to nanometer interelectrode distances helps to reduce the threshold voltages and reduce the possibility of emitter destruction due to the ion bombardment.

The design of nanoscale field emission structures should be carried out taking into account the influence of design parameters of the emission cell on the distribution of the electric field strength in the interelectrode gap. One of the important parameters of planar-edge emission structures is the relative arrangement of the electrode planes. The change in the arrangement of the electrode planes of the planar-edge emission structures leads to a change in the anode area which receives the emitted electrons and, as a result, changes in the current density. A high local current density can contribute to a process of critical overheating, leading to the destruction of individual sections [26-28].

The combination of the use of original design and technological solutions and the advantages of multigraphene and silicon carbide allow the development of stable energy-efficient devices. The object of studying is the simulation of the electric field strength in nanoscale field emission structures with planar-edge cathode and determination of the influence of different electrode planes arrangement on field gain.

2. Design

Field emission occurs when high electric field strength is created at the cathode surface. Emitters with a high curvature of the surface are used to obtain such fields. Planar-edge emission structures with a tip form of the emitters and a nanometer-wide interelectrode distance are considered in this work. This form of emitters based on graphene on SiC helps to reduce the threshold voltages down to few volts [19]. Geometric parameters of the model were chosen with taking into account the dimensions of the experimental emitter samples based on the multigraphene films on semiinsulating silicon carbide and are shown in Table 1. General view of the investigated field emission structure is shown in Figure 1.

Table 1. Simulation parameters.

Parameter	Symbol	Value
Emitter length	l	400 nm
Rounding-off radius of the cathode top	r	5 nm
Thickness of the multigraphene film	w	3 nm
Inclination angle of lateral face of the emitter relative to the central axis	α	10°
Initial interelectrode distance	R	5 nm

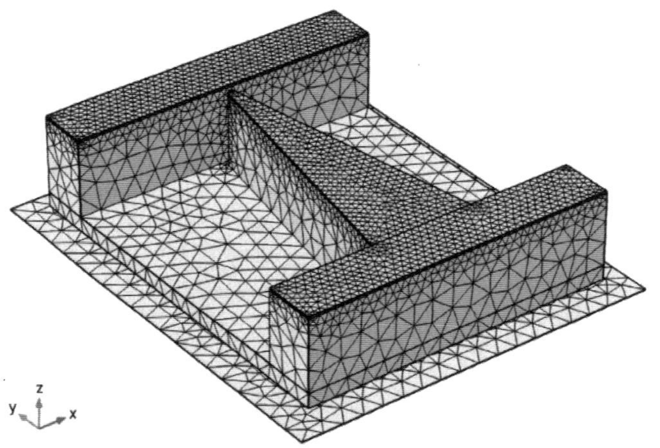

Figure 1. Model of planar field emission structure with tip-shaped emitter based on multigraphene films on a semiinsulating silicon carbide.

3. Simulation and results

Three-dimensional simulation was applied for the determination of the electric field strength E in nanosized interelectrode gap. Laplace equation for the potential was solved numerically by the finite element method with Dirichlet boundary conditions: for the cathode Uk = const; for the anode Ua = const; for the boundary of the model Us = 0. The height difference p between the anode plane relative to the cathode was varied in the range: -50 to +50 nm. The change of the plane arrangement of the electrodes led to an increase in the interelectrode distance. Figures 2, 3 show the distribution of the electric field strength in the interelectrode gap of the emission structure with different arrangement of the electrode planes.

IOP Conf. Series: Journal of Physics: Conf. Series **1038** (2018) 012055 doi:10.1088/1742-6596/1038/1/012055

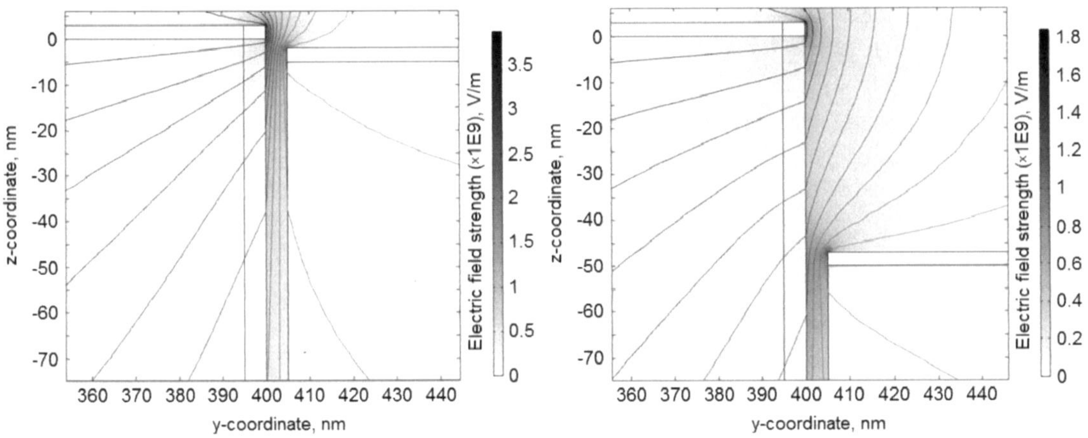

a) b)

Figure 2. The distribution of the electric field strength when the cathode plane is higher than the anode plane: (a) p = -5 nm; (b) p = -50 nm

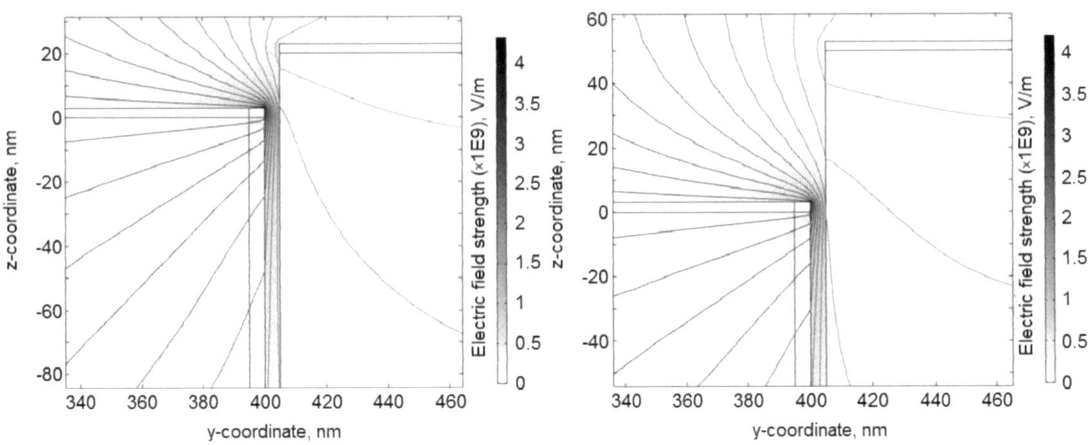

a) b)

Figure 3. The distribution of the electric field strength when the cathode plane is lower than the anode plane: (a) p = +20 nm; (b) p = +50 nm

Dependences of the electric field strength on the arrangement of the electrode planes are obtained on the basis of the simulation results (Figure 4). The field strength is within the range 10^9 V/m. An increase in the difference of the plane arrangement leads to a decrease in the electric field at the emitter surface. It was found that a decrease in the anode plane with respect to the cathode at 50 nm leads to a decrease in the electric field strength by ~ 60%. The decrease in the cathode plane with respect to the anode at 50 nm leads to a decrease in the electric field strength by less than ~ 10%.

The field gain β is calculated with taking into account the obtained values of the electric field strength and the corresponding dependence on the arrangement of the electrode planes was plotted (Figure 5). The field gain is $\beta=E/E_0$, where E_0 is the electric field strength calculated as for a plane-parallel capacitor at identical interelectrode distances. The obtained dependences show that the electric field gain increases with increasing the height difference p between the anode plane relative to the cathode. Thus, the efficiency of the design of the tip emitters decreases when nanoscale interelectrode distances $R \to r$. If the anode plane is 50 nm lower with respect to the cathode, the electric field gain will multiplicate in

~ 5 times. If the anode plane is 50 nm higher with respect to the cathode, the electric field gain will multiplicate in ~ 10 times.

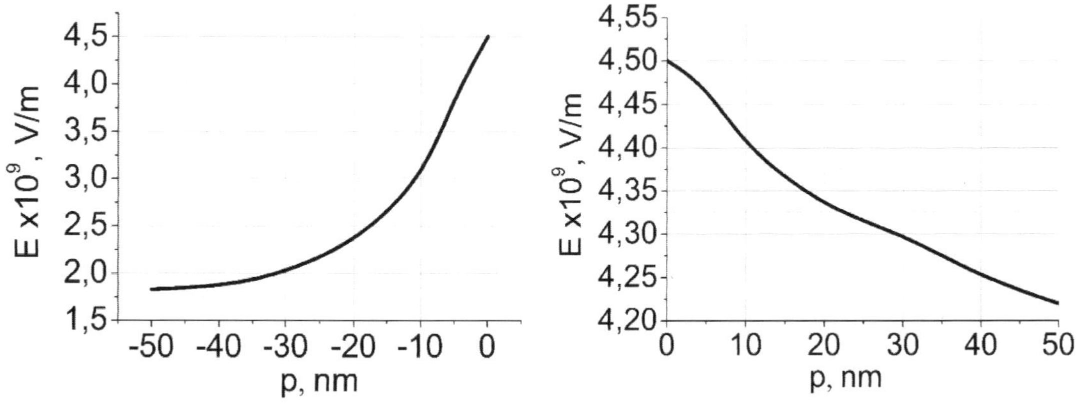

a) b)

Figure 4. The dependences of the electric field strength on the arrangement of the electrode planes when: (a) the cathode plane is higher than the anode plane; (b) the cathode plane is lower than the anode plane

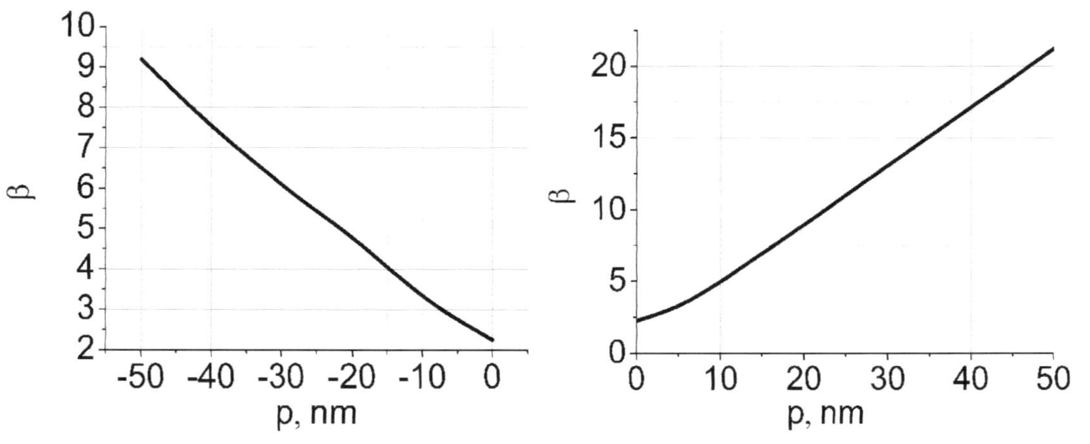

a) b)

Figure 5. The dependences of the field gain on the arrangement of the electrode planes when: (a) the cathode plane is higher than the anode plane; (b) the cathode plane is lower than the anode plane

4. Conclusions

The obtained simulation results allowed to estimate the influence of the design parameters of the emission cell on the electric field distribution in the nanosized interelectrode gap. The effect of the arrangement of the electrode planes on the electric field strength and field gain was shown. The obtained values of the electric field strength correspond to the values of nanoscale emission structures. It is necessary to take into account the presented changes in the electric field strength and field gain in designing nanoscale emission structures, but find a compromise between obtaining maximum electric field strength and equipment durability.

Acknowledgments
This work was funded by Internal grant of the Southern Federal University No. VnGr-07/2017-26.

References
[1] Novoselov K S, Geim A K, Morozov S V, Jiang D, Zhang Y, Dubonos S V, Grigorieva I V and Firsov A A 2004 *Science* **306** 666
[2] Lin Y-M, Dimitrakopoulos C, Jenkins K A, Farmer D B, Chiu H-Y, Grill A and Avouris Ph 2010 *Science* **327** 662
[3] Hong S K, Kim J E, Kim S O, Choi S Y and Cho B J 2010 *IEEE Electron device letters* **31** 1005
[4] Huang M, Pascal T A, Kim H, Goddard W A and Greer J R 2011 *Nano Letters* **11** 1241
[5] Suk J W, Lee W H, Lee J, Chou H, Piner R D, Hao Y, Akinwande D and Ruoff R S 2013 *Nano Letters* **13** 1462
[6] Park S J and Ruoff R S 2009 *Nature Nanotechnology* **4** 217
[7] Hass J, de Heer W A and Conrad E H 2008 *Journal of Physics: Condensed Matter* **20** 323202
[8] Nagase M, Hibino H, Kageshima H and Yamaguchi H 2008 *Journal of Physics: Conference Series* **100** 052006
[9] Konakova R V, Kolomys O F, Okhrimenko O B, Strelchuk V V, Volkov E Yu, Grigoriev M N, Svetlichnyi A M and Spiridonov O B 2013 *Semiconductors* **47** 812
[10] Lebedev A A, Kotousova I S, Lavrent'ev A V, Lebedev S P, Dement'ev P A, Petrov V N, Smirnov A N and Titkov A N 2010 *Physics of the Solid State* **52** 855
[11] Lebedev A A, Agrinskaya N V, Lebedev S P, Mynbaeva M G, Petrov V N, Smirnov A N, Strel'chuk A M, Titkov A N and Shamshur D V 2011 *Semiconductors* **45** 623
[12] Lezhenin F F, Gnesin G G 1967 *Soviet Powder Metallurgy and Metal Ceramics* **6** 114
[13] Fursey G N, Petrick V I and Novikov D V 2009 *Technical Physics* **54** 1048
[14] Yafyasov A, Bogevolnov V, Fursey G, Pavlov B, Polyakov M and Ibragimov A 2011 *Ultramicroscopy* **111** 409
[15] Konakova R V, Okhrimenko O B, Svetlichnyi A M, Ageev O A, Volkov E Yu, Kolomiytsev A S, Jityaev I L and Spiridonov O B 2015 *Semiconductors* **49** 1278
[16] Volkov E, Jityaev I and Kolomiitsev A 2015 *IOP Conf. Series: Materials Science and Engineering* **90** 012031
[17] Konakova R V, Okhrimenko O B, Kolomys A F, Strel'chuk V V, Svetlichnyi A M, Ageev O A, Volkov E Yu, Kolomiitsev A S, Zhityaev I L and Spiridonov O B 2016 *Journal of Superhard Materials* **38** 235
[18] Jityaev I L, Svetlichnyi A M, Kolomiytsev A S, Volkov E Yu, Polyakova V V and Ageev O A 2017 *IOP Conference Series: Materials Science and Engineering* **256** 012021
[19] Jityaev I L, Ageev O A, Svetlichnyi A M, Kolomiytsev A S and Spiridonov O B 2016 *Journal of Physics: Conference Series* **741** 012011
[20] Jityaev I L, Ageev O A, Svetlichnyi A M and Spiridonov O B 2016 *AIP Conference Proceedings* **1772** 040010
[21] Lee S W, Lee S S and Yang E-H 2009 *Nanoscale Research Letters* **4** 1218
[22] Yamaguchi H, Murakami K, Eda X G, Fujita T, Guan P, Wang W, Gong Ch, Boisse J, Miller St, Acik M, Cho K, Chabal Y J, Chen M, Wakaya F, Takai M and Chhowalla M 2011 *ACS Nano* **5** 4945
[23] Jityaev I L, Ageev O A, Svetlichnyi A M, Kolomiytsev A S and Spiridonov O B 2016 *Journal of Physics: Conference Series* **741** 012011
[24] Prével B, Benoit J-M, Bardotti L, Mélinon P, Ouerghi A, Lucot D, Bourhis E and Gierak J 2011 *Applied physics letters* **99** 083116
[25] Avilov V I, Ageev O A, Jityaev I L, Kolomiytsev A S and Smirnov V A 2016 *International Conference on Micro- and Nano-Electronics 2016* **10224** 102240T
[26] Bonard J-M and Klinke Ch 2003 *Physical review B* **67** 115406
[27] Dyuzhev N A, Makhiboroda M A, Kretov V I, Churilin M N and Rudnev V Yu 2012 *Russian Microelectronics* **41** 387
[28] Svetlichnyi A M and Jityaev I L 2017 *JETP Letters* **106** 613

Study of Strained Superlattices Grown by MOCVD Method

I V Fedorov[1,2], R V Levin[2,3], L A Sokura[2] and L V Danilov[2]

[1] Saint-Petersburg National Research University of Information Technologies, Mechanics and Optics, Kronverkskiy pr., 49, lit. A, St. Petersburg 197101, Russia

[2] Ioffe Institute, 26 Polytekhnicheskaya, St. Petersburg 194021, Russia

[3] Submicron Heterostructures for Microelectronics Research and Engineering Center of the RAS, 26 Polytekhnicheskaya, St. Petersburg 194021, Russia

E-mail: kingwash@yandex.ru , lev13@yandex.ru

Abstract. This study is dedicated to the development of the technology of InAs/GaSb superlattice by MOCVD. There was obtained an InAs/GaSb superlattice consisting of 10 pairs of alternating layers of InAs and GaSb grown at the temperature of 500°C. The obtained structures were studied by transmission electron microscopy (TEM) and photoluminescence (PL). Also, a light-emitting structure with quantum well was obtained to substantiate the developed technology.

Introduction

Human vision is limited by the visible spectrum. Expansion of the boundaries of human vision is possible by creation of devices that operate in short and mid-wavelength IR. For example, night vision goggles enable humans to see in the darkness, and thermal viewer can be used for malfunctions detection in a heating system or in thermal intelligence for military purposes.

Materials that can be used in production of short and mid-wavelength IR devices include strained InAs/GaSb superlattices. The advantages of such a material system as InAs/GaSb are the possibility to reduce tunneling currents, the swift Auger recombination and, as a consequence, the possibility to increase the operating temperature of the device for the given structure [1-2].

Experiment

The main method of InAs/GaSb superlattices growing is the method of molecular beam epitaxy [3]. In this study we showed the possibility of InAs/GaSb superlattices growing by metalorganic chemical vapor deposition (MOCVD), developed the technology of InAs/GaSb superlattice growing by MOCVD and resolved the following objectives: finding the best conditions to grow InAs/GaSb superlattice and examination of the structural and optical-energy characteristics of the structures obtained as the result.

All the structures were grown with AIXTRON AIX-200. Sources of the growth elements were the following compounds: TMIn, TEGa, TMS and AsH_3.

The structures were grown on n-GaSb substrate (001) at the temperature of 500°C. The pressure in the reactor was 76 Torr. The substrate rotated at the rate of 100 revolutions per minute. The carrier gas was purified H_2 with the dew point not worse than −100°C, and the total flow through the reactor was 5.5 litres per minute. The grown structures were studied by transmission electron microscopy (TEM) and photoluminescence (PL).

The Results of the Experiment

The effect of pauses between growth of the individual layers on the sharpness of heterointerface was studied. (Pause is the time when all the flows of TMIn, TEGa, TMSb, AsH_3 are interrupted, except the flow of H_2.) For this study, a structure including 6 pairs of alternating InAs and GaSb layers was grown under the growth temperature of 500°C. The flows of compounds were interrupted after each layer of InAs or GaSb was grown, while the flow of H_2 was still continued for a time period (t). This time period was defined as the result of the following ratio: $t = V_p / G$, where V_p is the volume of the reactor, cm^3, and G is hydrogen flow rate, cm^3/sec. The pause time between 2 pairs of the InAs and GaSb layers, when the flowing of the compounds of TEGa and Arsine was stopped, was 1 minute. The pause time for layer switching was: 10 seconds for the first 2 pairs of the InAs and GaSb layers, 30 seconds for the second 2 pairs and 50 seconds for the third 2 pairs.

The results of this study showed that all the InAs/GaSb borders were of equal quality, regardless of the exposure time or the direction of switching (GaSb→InAs or InAs→GaSb). The image of the sample is shown in Figure 1. For the InAs/GaSb superlattices growing, an average pause between alternating layers was chosen, which was 30 seconds.

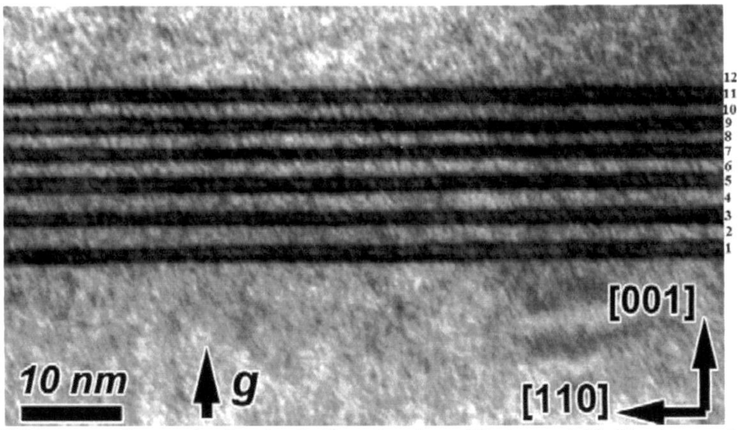

Figure 1. TEM image of superlattice including 6 pairs of InAs and GaSb alternating layers. Pause between alternating layers 1 to 4 was 10 seconds, between alternating layers 5 to 8 – 30 seconds, and between alternating layers 9 to 12 – 50 seconds.

The InAs/GaSb superlattice consisting of 10 pairs of alternating InAs and GaSb layers was grown at the temperature of 500°C as the result of finding the best growing conditions. The obtained layers of InAs and GaSb have clearly defined borders and thicknesses on the electron micrograph (Figure 2). The growth rate of InAs was 0.2 nm per sec. The growth rate of GaSb was 0.03 nm per sec. The thickness of the superlattice`s layers varied in the range of 1.4 nm to 1.9 nm for InAs and 2.5 nm to 3.0 nm for GaSb.

IOP Conf. Series: Journal of Physics: Conf. Series **1038** (2018) 012056 doi:10.1088/1742-6596/1038/1/012056

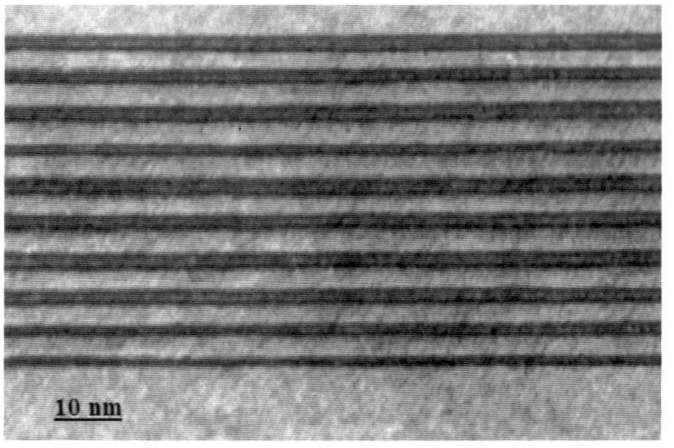

Figure 2. TEM image of a superlattice consisting of 10 pairs of alternating InAs and GaSb layers. Pause time between alternating layers was 30 seconds.

The PL spectra of the superlattice of 10 pairs of alternating InAs and GaSb layers was measured at the temperature T = 77 K. The pumping source used for the PL measurements was diode laser with the wavelength of ~ 1μm. In the range of 0.45–0.9 eV the superlattice had the following peaks: the maximum = 0.801 eV, (this peak is related to interband transition in the GaSb layers) and the maximum = 0.725 eV (this peak is related to the GaSb substrate). In the range of 0.27–0.45 eV the superlattice had a peak with the maximum = 0.315 eV and full width at half maximum = 78 meV, which is most probably related to the radiative recombination in the InAs/GaSb superlattice layers. Figure 3 shows the PL spectrum for the superlattice of 10 pairs of layers in the energy range of 0.27–0.45 eV.

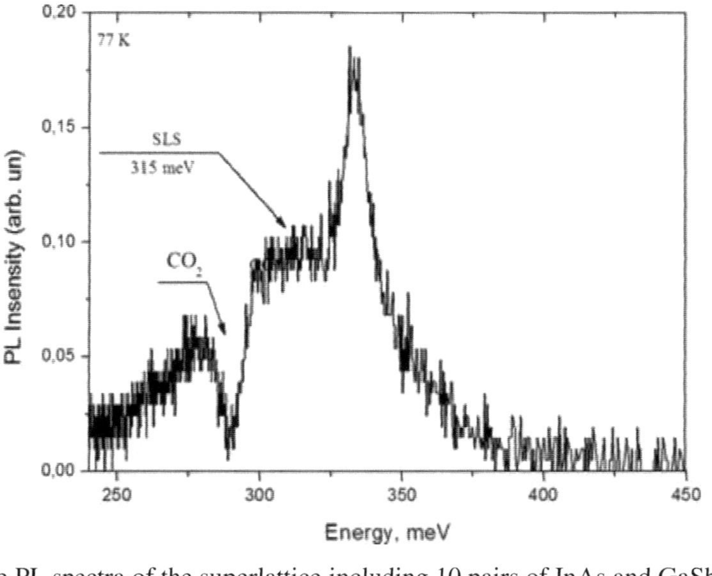

Figure 3. The PL spectra of the superlattice including 10 pairs of InAs and GaSb alternating layers.

A light-emitting structure with quantum well was grown. This structure (Figure 4) consisted of a substrate of n-GaSb ($n = 5–7•10^{17} cm^{-3}$), a buffer layer of n-GaSb:Te ($d = 0.5–0.9$ μm, $n = 5•10^{17} cm^{-3}$), an undoped InAs quantum well ($d = 5$ nm) and a capping layer p-GaSb:Si ($d = 1$ μm, $p = 3–4•10^{17} cm^{-3}$).

IOP Conf. Series: Journal of Physics: Conf. Series **1038** (2018) 012056 doi:10.1088/1742-6596/1038/1/012056

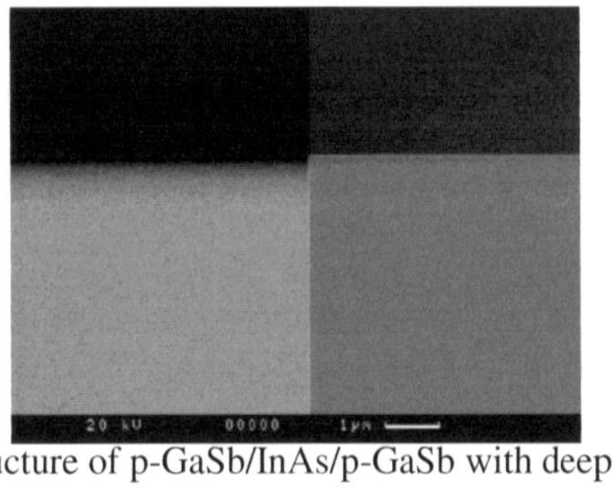

Figure 4. The structure of p-GaSb/InAs/p-GaSb with deep quantum well.

The spectra of electroluminescence (EL) was measured in the spectral range of 2–5 μm depending on the feed current. The samples were placed under a pulse current of a rectangular shape and the frequency of 512 Hz. The amplitude of the current varied in the range ±(0–200mA). Radiation spectra were measured at 77–300K using a PC controller on the basis of a DK-480 monochromator. For radiation recording, a cooled InSb photodiode (Judson Co) was used. As can be seen from the electroluminescence spectra, the resulting structure has a high emission intensity (Figure 5).

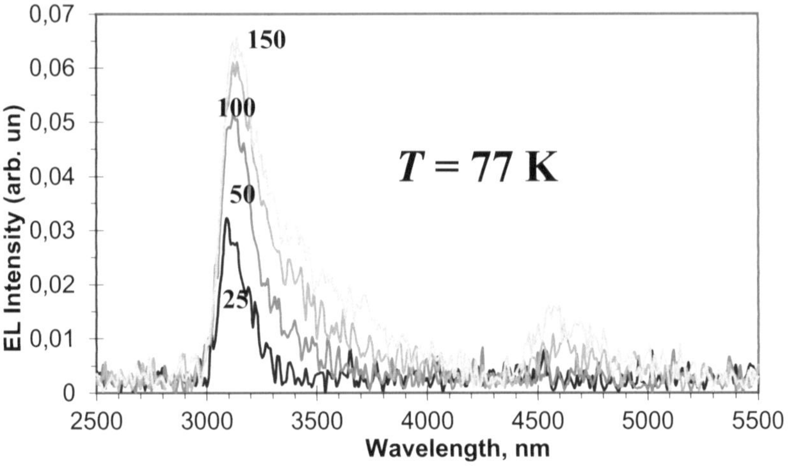

Figure 5. EL spectra of the n-GaSb/n-InAs/p-GaSb structure at different currents in the range of 25–150 mA and T = 77K.

Sublinear dependence of the radiation intensity on the driving current at 77 and 300 K was studied. We suppose that the peak of EL with maximum = 0.41 eV (77 K) and maximum = 0.37 eV (300 K) corresponds to the radiative recombination of InAs layer.

The authors thank B. A. Matveev and M. A. Remennoy for their assistance in the research. The work was executed with partial support of the RFBR grant № 16-08-01130-a.

References

[1] Grein C H, Young P M and Ehrenreich H 1992 *Appl. Phys. Lett.* **61**, 2905
[2] Smith D L and Mailhiot C 1987 *J. Appl. Phys.* **62** 2545
[3] Razeghi M and Nguyen B M 2014 Advances in mid-infrared detection and imaging: a key issues review. *Rep. Prog. Phys.* **77**, 082401

Investigation of photoinduced nucleation and growth of silver nanoparticles

M Churo and L Matyushkin

Department of micro- and nanoelectronics, St. Petersburg Electrotechnical University LETI, Prof. Popov str. 5, St. Petersburg 197376, Russia

E-mail: mashachuro@yandex.ru, leva.matyushkin@gmail.com

Abstract. The influence of electromagnetic radiation with maxima wavelengths of 455, 520 and 630 nm on silver nanoparticles growth was investigated. The corresponding possibility of the plasmon resonance spectrum reconstruction in the visible range is shown. Investigations of the radiation frequency influence on the shape and size of the particles both at the stages of growth and at post-synthetic processing have been carried out. The possibility of transformation of spherical particles with a diameter of 20 nm into triangular shaped objects with a triangle edge length of 60–90 nm is shown. To study the optical characteristics of silver particles transmission spectroscopy was used as the primary diagnostic technique. The results of photometric measurements are in agreement with the data of scanning electron microscopy, which shows the transformation of the particles shape as a result of electromagnetic radiation.

1. Introduction

Nanoparticles of noble metals are of broad practical interest due to their unique optical properties [1, 2, 3, 4]. A lot of the applications are due to the plasmon resonance effect, which is the collective oscillation of the electron density relative to the crystal lattice at the resonant excitation frequency. Being excited in a nanoparticle, the oscillation is spatially limited and therefore called localized plasmon resonance. In this paper, the investigation of the plasmon resonance effect is carried out for the example of silver nanoparticles characterized by the highest intensity of plasmon resonance and high extinction coefficient [5]. Since silver in a nanostructured form efficiently interacts with external radiation of the visible range, these properties can be used to synthesize the nanoparticles themselves efficiently.

The study of electromagnetic radiation effect on the nucleation and growth of metal particles is of particular interest. With the close arrangement of ions and charged clusters relative to the surface of nanoparticles, electron density oscillations that occur under the action of external radiation can lead to the reduction of small charged particles and their adsorption on the surface of the metal nanoparticle. As a result, the new electromagnetic field distribution is obtained leading to a gradual increase in the number of particles. The particles shape can be controlled using different radiation frequencies and corresponding various geometry of electron density oscillations, which in turn leads to a rearrangement of the plasmon resonance spectrum of the ensemble of metal nanoparticles.

2. Experiment

The synthesis procedure used is based on the well-known citrate method of silver particles synthesis, but the thermal influence is replaced by photoinduced influence [6]. As sources of radiation, we used an RGB-LED strip with 14.4 W total power. The radiation from the group of light-emitting diodes corresponding to the blue light had a wavelength of the maximum of the spectrum $\lambda_{ex} = 455$ nm, green — $\lambda_{ex} = 520$ nm and red — $\lambda_{ex} = 630$ nm. Optical density spectra of colloidal solutions formed as a result of exposure to radiation were measured with PE-5400UV spectrophotometer (Ecohim LTD, Russia).

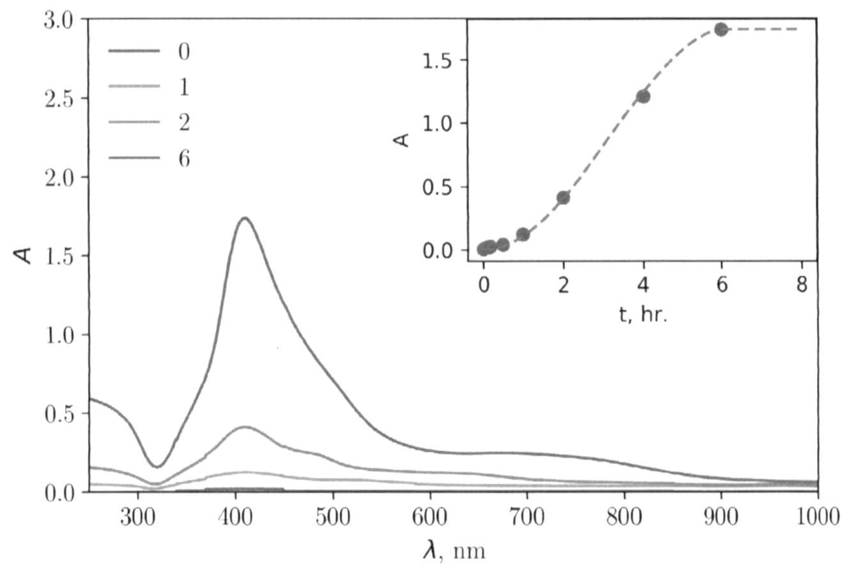

Figure 1. Evolution of the optical density spectrum of silver nanoparticles under exposure to radiation with $\lambda_{ex} = 455$ nm (numbers in the legend are hours of exposure). The inset shows the dependence of optical density of main maximum on the exposure time.

The mixed solution of silver nitrate and sodium citrate was irradiated with radiation at wavelength $\lambda_{ex} = 455$ nm (Fig. 1); in the legend time is given in hours. Inset in Fig. 1 represents the dependence of the optical density maxima on the time of exposure. The 430 nm peak (Fig. 1) of plasmon resonance is associated with spherical nanoparticles of small size. The appearance of the longwave tail can be a result of the formation of larger nanoparticles or non-spherical particles [7]. When the solution was exposed to radiation with $\lambda_{ex} = 370$ nm, nanoparticles also form. However, when the solution was exposed to a radiation with $\lambda_{ex} = 520$ nm and 630 nm, nanoparticle nucleation was not observed. Thus, we can assume that there is a certain threshold value of the photon energy at which the nucleation of silver nanoparticles begins to occur.

The size and shape of nanoparticles can be influenced by radiation not only during their synthesis but also after the synthesis, since free, unreacted ions and silver clusters remain in the solution. For a long time, under the influence of electromagnetic radiation, nanoparticles grow in solution, while silver ions remain in it. The wavelength of external radiation determines the shape and size of the nanoparticles obtained. Synthesis of initial particles of 20 nm in diameter monodisperse silver nanoparticles was produced by citrate-ascorbic acid procedure [8].

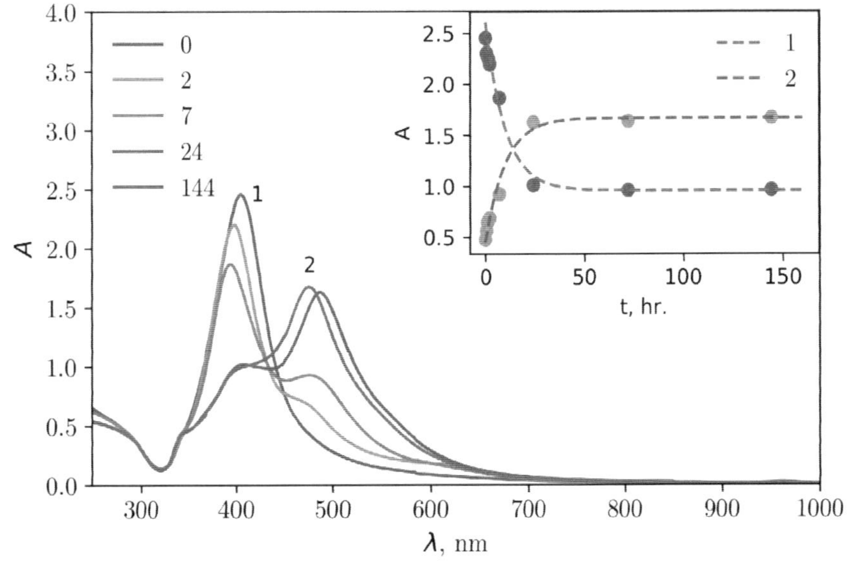

Figure 2. Evolution of the optical density spectrum of silver particles with long-term effects of radiation with $\lambda_{ex} = 455$ nm. The inset shows the dependence of optical density of maxima on the exposure time.

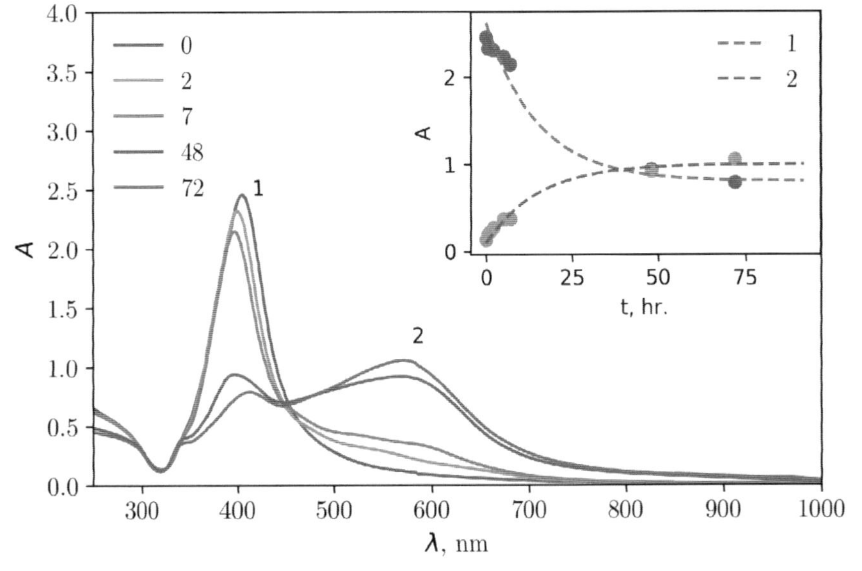

Figure 3. Evolution of the optical density spectrum of silver particles with long-term effects of radiation with $\lambda_{ex} = 520$ nm (numbers in the legend are hours of exposure). The inset shows the dependence of optical density of maxima on the exposure time.

When using electromagnetic radiation with a wavelength of 455 nm, the photoinduced synthesis lasted for six days until all the silver ions reacted in the solution. On the optical density spectrum (Fig. 2) during the first day, the first peak (1 in Fig. 2) shifts toward higher frequencies, which may be due to the etching of nanoparticles surface by chlorine ions. At the same time, an additional peak (2 in Fig. 2) of plasmon resonance appears at a lower frequency, probably due to the appearance of larger spherical nanoparticles or with a change in the shape of the particles. And the first peak almost completely disappears. The dependence of the optical density on the time of exposure to radiation has a similar character for both peaks. The tuning of the plasmon resonance spectrum mainly occurs during the first 20 hours and completely stops after three days.

When silver nanoparticles solution was exposed to radiation with $\lambda_{\mathrm{ex}} = 520$ nm, the reconstruction of the plasmon resonance spectrum is observed. The result is in Fig. 3. As with exposure to $\lambda_{\mathrm{ex}} = 455$ nm radiation, the second peak (2 in Fig. 3) appears, but it corresponds to the longer wavelength. The dependence of the optical density on time also has an exponential character.

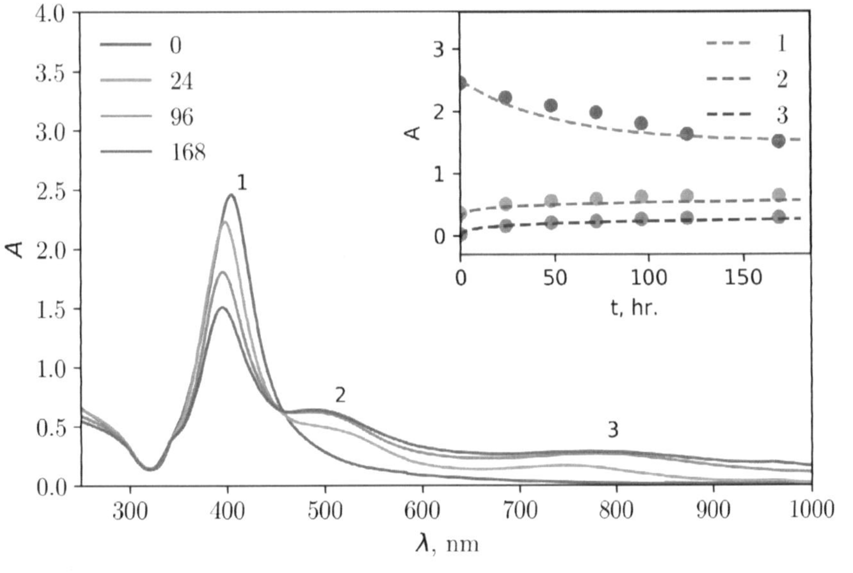

Figure 4. Evolution of the optical density spectrum of silver particles with long-term effects of radiation with $\lambda_{\mathrm{ex}} = 630$ nm (numbers in the legend are hours of exposure). The inset shows the dependence of optical density of maxima on the exposure time.

The appearance of two additional longwave weak broad peaks (2, 3 in Fig. 4) is noticeable on the optical density spectrum under long-term exposure to radiation from $\lambda_{\mathrm{ex}} = 630$ nm as a result of photoinduced synthesis. This effect is due to the appearance of several plasmonic modes, i.e. with the change in the shape of the particles.

3. Results and discussion

From the optical density spectra it can be said that photoinduced synthesis in the absence of nuclei leads to the formation of mainly spherical particles, while in post-synthetics, the formation of nonspherical nanoparticles is observed. Let us study in more detail the effect of post-synthesis on the particle size and shape.

It can be concluded that the closer the wavelength to the maximum of the main plasmon resonance, the more clearly the formation of additional optical density maxima shifted to the wavelength of the excitation. On the other hand, the greater the difference in the wavelengths of the plasmon resonance and the exciting radiation, the more extended the extinction spectrum can be obtained.

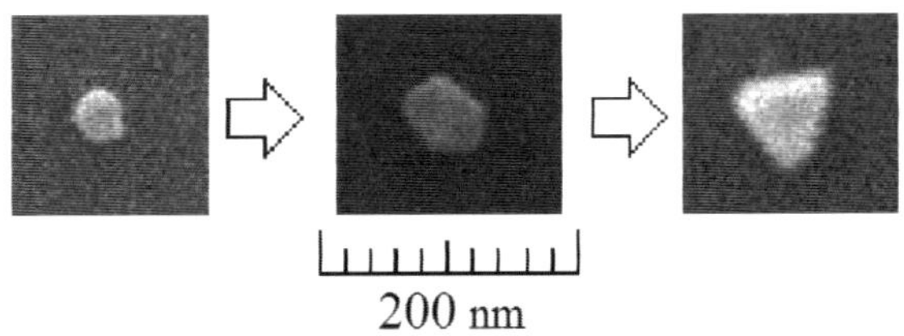

Figure 5. Changing the shape of a particle under the influence of radiation (scanning electron microscopy of individual particles)

In the image of particles obtained with scanning electron microscopy (image is obtained in secondary electrons), it was determined that the spherical nanoparticles have a diameter of 25–80 nm. The formation of larger spherical particles is due to the fact that in addition to the effect of radiation, the solution also heats up because of the absorption of electromagnetic radiation. There are also non-spherical particles in solution. On the basis of the obtained results, the non-spherical particles have the form of a triangular shape with an average rib length of 70 nm. Figure 5 shows the transformation of a spherical nanoparticle into a triangular shape as a result of exposure.

4. Conclusions

As a result of work silver nanoparticles of various sizes and shapes were synthesized under the influence of electromagnetic radiation of different spectral range. The effect of radiation on the processes of growth and nucleation of silver particles was investigated. To nucleate particles, it is necessary that the radiation energy is above a threshold value that exceeds 2.4 eV.

In the post-synthetic effect of radiation, a change in the optical density spectrum is observed: the appearance of an additional peak at a frequency of a lower frequency of electromagnetic radiation. The plasmon resonance spectrum of the obtained particles varies in the visible range.

With the help of scanning electron microscopy, images of silver particles of various shapes and sizes formed during post-synthetic exposure were obtained. As a result of radiation, triangular shaped objects with a rib length of 70–90 nm are formed.

References

[1] Lohse S E and Murphy C J 2012 *Journal of the American Chemical Society* **134** 15607–15620
[2] Cobley C M, Skrabalak S E, Campbell D J and Xia Y 2009 *Plasmonics* **4** 171–179
[3] Linic S, Aslam U, Boerigter C and Morabito M 2015 *Nature materials* **14** 567–576
[4] Nakayama K, Tanabe K and Atwater H A 2008 *Applied Physics Letters* **93** 121904
[5] Evanoff D D and Chumanov G 2005 *A European Journal of Chemical Physics and Physical Chemistry* **6** 1221–1231
[6] Jin R, Cao Y C, Hao E, Métraux G S, Schatz G C and Mirkin C A 2003 *Nature* **425** 487–490
[7] Garcia M A 2011 *Journal of Physics D: Applied Physics* **44** 283001
[8] Li H, Xia H, Wang D and Tao X 2013 *Langmuir* **29** 5074–5079

IOP Conf. Series: Journal of Physics: Conf. Series **1038** (2018) 012058 doi:10.1088/1742-6596/1038/1/012058

X-Ray analysis of compacted and sintered UHMWPE reactor powders

M Dermeneva[1,2], E Ivan'kova[3], V Marikhin[1], L Myasnikova[1] and M Yagovkina[1], E Radovanova[1]

[1] Ioffe Institute, 194021 Polytekhnicheskaya st. 26, Saint-Petersburg, Russian Federation

[2] ITMO University, 197101 Kronverkskiy pr. 49, Saint-Petersburg, Russian Federation

[3] Institute of Macromolecular Compounds of Russian Academy of Sciences, 199004 Bolshoy pr. 31, Saint-Petersburg, Russian Federation

E-mail: mdermeneva@inbox.ru

Abstract. X-ray diffraction analysis of solid-state compaction/sintering-induced changes in crystalline structure of lab-scale (M332) and commercial UHMWPE (X600) reactor powders has been carried out. The profile fitting programs TOPAS-5 and Peakfit 4.1 have been used to deconvolute the overlapping peaks. The linear FWHM of (110) and (002) X-ray peaks have been taken for calculation of the transverse (D_{110}) and longitudinal (D_{002}) crystallite sizes with the help of Sherrer method. The growth of crystallite sizes in cold compacted powders with sintering has been detected in both directions. Besides, the sintering has resulted in the generation of the planar a-texture, more pronounced in the lab-scale UHMWPE samples. As known, orientation hardening of the sintered reactor powders is an alternative route to UHMWPE high performance fibers distinct from the common gel-technology. It has been found that the highly oriented thin film threads produced from highly textured sintered material demonstrated the extremely high tensile strength up to 3.0 GPa, close to that of gel-fibers (3.5GPa) [1] while X600 less textured powders has shown low compactability and drawability.

1. Introduction

UHMWPE is a unique material due to its exceptional physical and mechanical properties, such as low friction, highly abrasive resistance, chemical inertness and the record specific strength, in particular. It determines the wide range of its applications, from bullet-proof jackets to artificial joints, implants, membranes, fishing nets and sportive goods. The gel-technology for production of UHMWPE high performance fibres, much needed in both engineering and science, was developed [2-5]. However, processing of low concentration UHMWPE solution via gel-technology is expensive and environmentally harmful because of recuperation of the large solvent volume.

Then, in current times, the alternative solvent-free route to high performance fibres adopted from powder metallurgy is intensively developed. The so called "dry" technique involves the cold compaction of the "never processed" UHMWPE reactor powders and sintering followed by orientation drawing, the sintering temperature being lower than that of melting (T_m). This technique was materialized on lab-scale in a number of the scientific groups [6-8]. The small-scale production of high-performance UHMWPE film-threads under trademark Tensylon was started not so long ago. However the tensile strength of these films threads is slightly lower (2,5GPa) than that of the most strong gel-fibers Dyneema and Spectra produced by leading firms DSM (the Netherlands) and Honeywell (USA), correspondingly (3-3,6 GPa) [1].

As known, not all the reactor powders can be successfully processed to high performance fibers via solvent-free method. It is widely believed that the main factor controlling their drawability is the density of entanglements which catastrophically increases provided that sintering is carried out at $T> T_m$. But the structural changes inevitably occur even on sintering at $T< T_m$. which also can influence the effectiveness of the orientation hardening. In the present work the changes in crystalline structure of two different UHMWPE reactor powders upon sintering are investigated with the help of X-ray technique and the influence on their drawability and mechanical characteristics of the end-products is studied.

2. Experimental method and materials

The objects of investigations were UHMWPE compacted and sintered films produced from two different UHMWPE reactor powders M332 and X600. The synthesis of M332 reactor powder ($\overline{M}_w$ $=2,5\cdot 10^6$) was performed at the Institute of Physical-Organic Chemistry of the National Academy of Sciences of Republic Belorussia, Minsk, in a slurry process in n-heptane at 60^0C on colloidal size particles of catalyst system (a high-performance titanium chloride catalyst modified with an organic magnesium compound). Almost nothing is known about commercial X600 reactor powder (China) except its $\overline{M}_w$ ($\overline{M}_w$ $=6,0\cdot 10^6$) because of confidentiality.

According to the principle of powder metallurgy, reactor powders were compacted and then sintered. Powder compaction was performed by compression in hydrostatic conditions in a Carver hydraulic press (model 3912) at room temperature at a pressure of 95 MPa for 15 minutes. The tablets of 1 mm thick and 20 mm in diameter were produced from about 210 mg of the powder.

The sintering of compacted tablets was performed by compression of the compacted tablets between two aluminium foils at a temperature of 130 ° C under a pressure of 95 MPa for 30 minutes. During sintering the pressure was falling down to 46 MPa because increasing the area of the sintering sample. The obtained films of near 150 μm hick were cut into narrow strips of 1mm wide and subjected to multi-stage high temperature zone drawing, the temperature in the process increasing stepwise from 110°C to 140°C. The film threads drawn to different draw ratios were tested on calibrated Instron 1120 equipment.

Characterization

Wide angle X-ray diffraction (WAXS) data from compacted and sintered samples were collected in Bragg-Brentano mode on a D2 Phaser, Bruker (Cu $_{Ka}$ radiation, λ=1.54 A) operating at 30 kV and 10 mA using an PSD detector. Signal accumulation time for each step was 4 sec in an angle range from 10 to 30 degrees (the region of 110 and 200 orthorhombic and three monoclinic peaks, if any), 40 sec for a range from 30 to 60 degrees (the range of 220 orthorhombic peak) and 60 sec from 60 to 80 degrees (the range of 002 orthorhombic peak). The analysis of X-ray diffraction profiles was performed by the programs TOPAS-5 and Peakfit 4.1. In the first case the fitting was performed when the X-ray peaks are approximated by a complicate combination of 8 different functions. In the latter, Pearson VII was used. The broadening of the reflections of higher orders was analyzed using Sherrer method [9], which takes into account a dependence of X-ray peak shape on crystallite sizes and crystalline distortions. The fitted 110 and 220 diffractions were used to calculate the transvers crystallite sizes (D_{110}) and the crystallite distortions (g-factor)) by Sherrer formula: D_{hkl}= 0.89 λ / (FWHM· cosθ), where 0, 89 is the form-factor, λ is the X-ray ($Cu_{K\alpha}$) wavelength (1.54 Å), the FWHM

IOP Conf. Series: Journal of Physics: Conf. Series **1038** (2018) 012058 doi:10.1088/1742-6596/1038/1/012058

(Full Width Half Maximum) is the linear half-width of correspondent X-ray peak, and θ is the Bragg angle. The 002 diffraction was used to calculate the longitudinal crystallite sizes (D_{002}).

3. Results

The typical X-ray scattering from compacted reactor powders are shown in Figure 1.

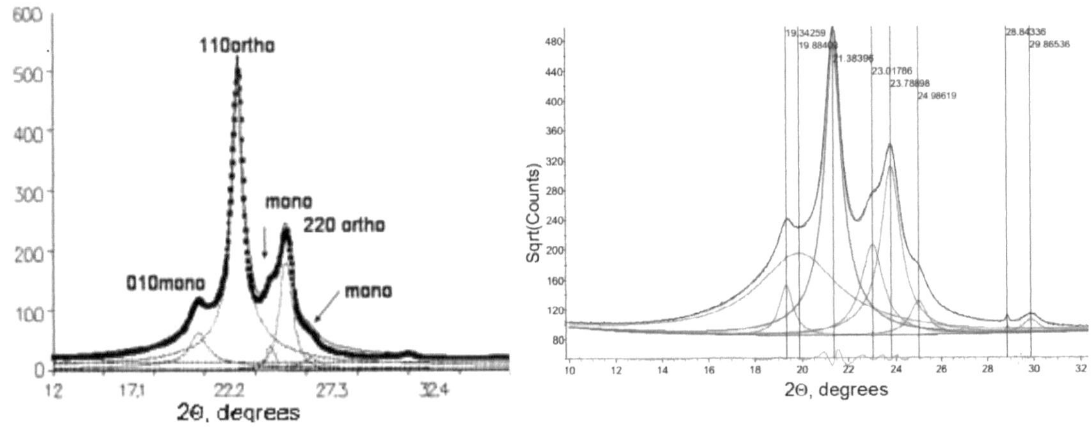

Figure 1. Deconvolution of WAXS M332 compacted powder by Peakfit (at left) and by TOPAS (at right)

It is well seen that besides from the usual 110 and 200 PE orthorhombic peaks the three peaks from monoclinic crystalline phase are easily recognized. Since monoclinic phase is only stable under stress one should conclude that compaction results in developing stress in the compacted powder. The appearance of monoclinic phase in the UHMWPE powders when compacted was earlier observed in our experiments [10] and also by other authors. The estimates of monoclinic phase content differ a lot [11-1]. The mess in the results is caused by both methodological causes as well as meta-stability of monoclinic phase. Besides, there is an uncertainty in deconvolution of the overlapped peaks in the range of 19.6° because both monoclinic 010 peak and amorphous halo contribute to this region. Put then aside the monoclinic phase to concern with orthorhombic crystallites, all the more so it disappeared after sintering in both M332 and X600 films (see Figure 2).

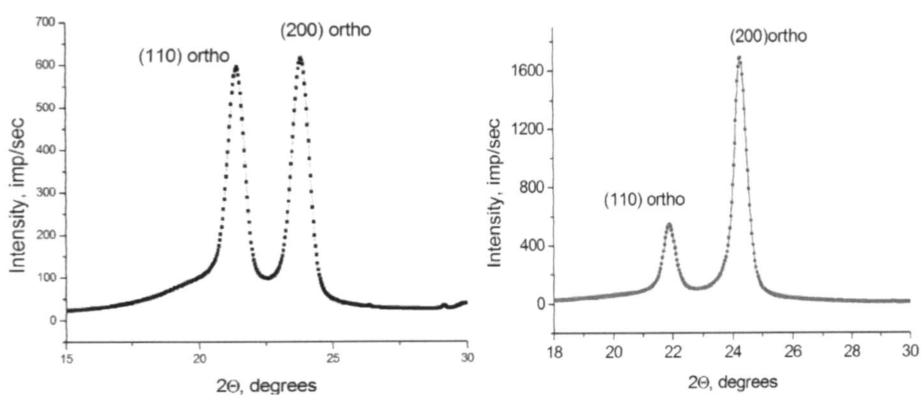

Figure 2. WAXS of sintered X600 film (at the left) and M332 (at the right).

First what immediately springs to eyes is that abnormal relatonship between intensity of 110 and 200 ortorhombic reflexes (I_{200}/I_{110}). It is 1,13 for X600 sintered film, and 1,53 for M332 film instead of table value 0.7. This evidences the generation of plane "a"-texture in the compacted samples upon sintering. The role of this structural transformation will be discussed later on.
Along with the observed phenomena sintering the compacted sample is accompanied by growth of transverse and longitudinal crystallite sizes.

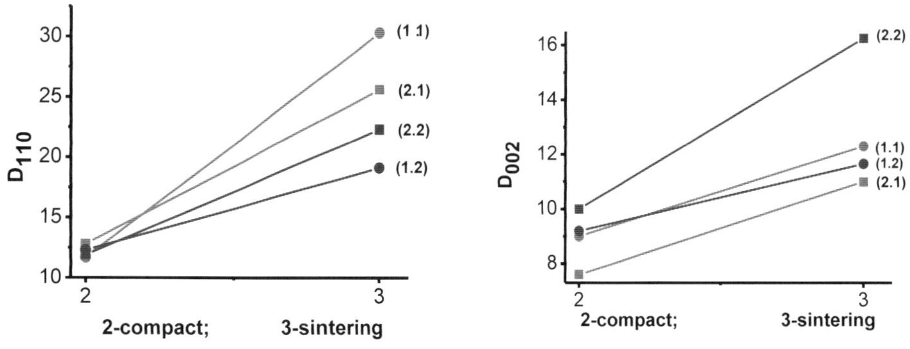

Figure 3. Changes in transverse (D_{110}) (at the left) and longitudinal (D_{002}) crystallite sizes (at the right) on sintering calculated for M332 and X600 on the base of TOPAS and PeakFit fitting (M332: 2.1 and 1.1.; X600: 2.2 and 1.2, correspondingly).

The larger crystallite sizes D_{110} obtained for M332 with PeakFit than those with TOPAS are obvious because in the first case the broadening of higher order (220) were taken into account while TOPAS only gave the estimate on the broadening of the first order 110 peak. For all other cases when the crystallite sizes were calculated for the first order reflection, TOPAS data were larger than those obtained with the help of Peakfit. Anyway, despite the significant scatter in the crystallite sizes estimated one can emphasize a certain tendency in the presented data: the sintering actually results in growth of crystallite sizes. The size D_{110} in M332 samples grow faster than in X600.
Thus with the help of X-ray study a difference in crystalline structure behavior was manifested.
The difference in drawability and ultimate mechanical characteristics of the samples investigated were also discovered. The results of the mechanical testing of the film threads of various draw ratios produced from the sintered films obtained from UHMWPE powders are given in Figure 4.

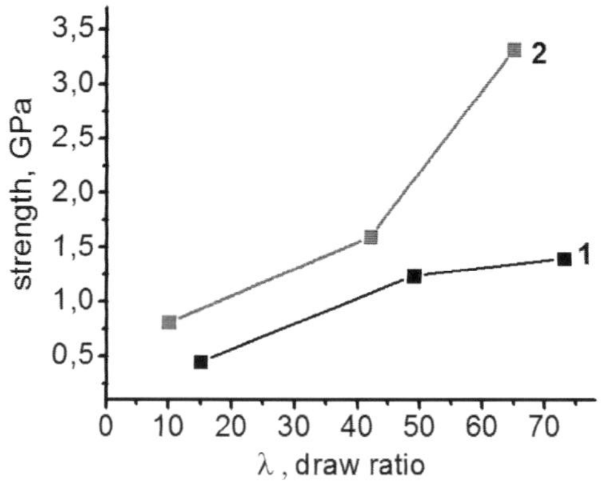

Figure 4. Dependence of tensile strength of oriented film threads from X600 (1) and M332 (2) on draw ratio.

It is seen that both sintered films could be drawn to very high orientation elongation however, they have different rate hardening. Despite X600 film thread could be drawn even more (DR=73) than M332 film threads (DR=63) the former has lower tensile strength while the tensile strength of the latter reach very high values (3.3 GPa) comparable to the best gel-fibers.

4. Discussion

What is the reason of so different structure- mechanical response of UHMWPE on one and the same processing? First, we believe that the best results obtained for M332 ultimately oriented threads is due to distinct a-texture convenient for re-arrangement of fold crystal into the highly oriented fibrils.
At the same time, to receive an additional information on the powder investigated we studied the initial powder with the help of scanning electron microscopy. The morphology of X600 and M332 reactor powders appeared to be catastrophically different (see Figure 5).

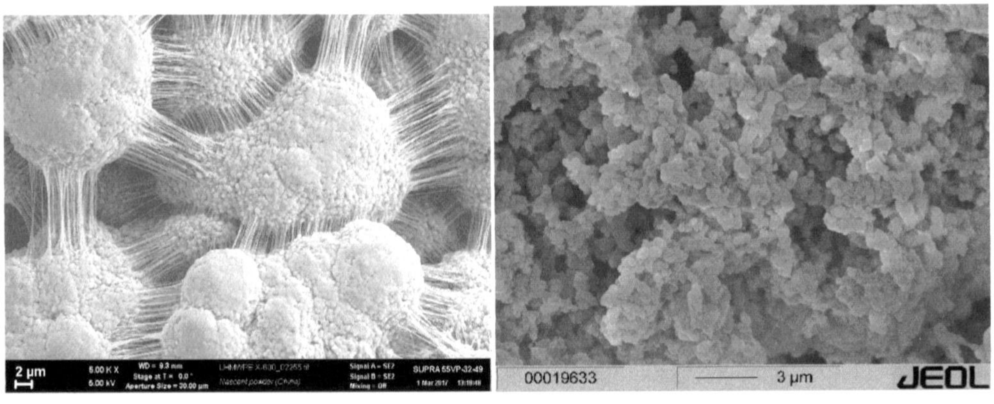

Figure 5 SEM of X600 UHMWPE reactor powder (at the left) and that of M332(at the right)

The morfology of X600 is two fold. It consists of a lot of fibrillar entities and small nodulus of obviously lamellar structure. The existence of fibrillar bandles evidences high rate of synthesis because these fibrils could be formed during desintegration of catalists under the tension of fast growing polymer mass. As distinct from X600, M332 reactor powder has characteristic homogeneous morphology that comprises lamellar nodulus of 0.5-1μm size. One can propose that the fibrillar bandles prevent to fully re-arrangement of initial morphology to a-texture more convenient to effective orintation hardening.

5. Conclusions

It is shown that the "dry" route to high performance UHMWPE fibers can be successfully realized, and highly oriented super strong film threads can be obtained, provided that the initial reactor powder can be re-formed by compaction/sintering into the "a"-textured film convenient for effective orientation hardening. X-ray analysis that follows after changes in crystalline structure upon sintering gives a valuable information on the structural transformations.

6. Acknowledgements

The characterizations of compacted and sintered samples produced from the UHMWPE reactor powders were performed using equipment owned by the Joint Research Center "Material science and characterization in advanced technology" (Ioffe Institute, St.-Petersburg, Russian Federation).

7.References

[1] Karacan I 2006 *J. Appl. Polym. Sci.* **101** p 1317
[2] Smith P and Lemstra P J 1979 *Die Macromolekulare Chemie* **180** p 2983

IOP Conf. Series: Journal of Physics: Conf. Series **1038** (2018) 012058 doi:10.1088/1742-6596/1038/1/012058

[3] Smith P, Lemstra P J 1980 *J. Mater. Sci.* **15** p 505

[4] Fang X, Wyatt T, Hong Y, Donggang and Yao 2016 *Pol. Eng. and Sci.* **56** (6) p 697

[5] Jian' T, Shyu W D, Lin W T, Chen K N and Jian T 2003 Polym. Eng. and Sci. **43** (11) p 1765

[6] Rotzinger B P, Chanzy H D, Smith P 1989 *Polymer* **30** p 1814

[7] Smith P, Lemstra P J, Booij H C 1981 *J. Polym. Sci. Phys. Edn.* **19** p 877

[8] Deplancke T, Lame O, Rousset F, Seguela R and Vigier G 2015 Macromolecules **48** (15) p 5328

[9] Balta-Calleja F J, Vonk CG 1989 *X-ray Scattering of Synthetic Polymers* (Amsterdam: Elsevier Science Publishers) p 241

[10] Myasnikova L et al 2011 *Reactor powder morphology* ed P J Lemstra and L P Myasnikova(New York: Nova Science Publishers) chapter 5 p 93

[11] Aulov V A, Shcherbina M A, Chvalun S N, Makarov S V, Kuchkina I O, Pantyukhin A A, Bakeev N F, Pavlov Yu S 2004 *Polym. Sci. Series* A **46** (6) p 620

[12] Joo Y L, Han O H, Lee H K, Song J K 2000 *Polymer* **41** (4) p 1355

IOP Conf. Series: Journal of Physics: Conf. Series **1038** (2018) 012059 doi:10.1088/1742-6596/1038/1/012059

3D model of short-range order of one-hour milled cellulose

M V Smirnov[1] **and S V Loginova**[1]

[1]Petrozavodsk state university, Petrozavodsk, Republic of Karelia, Russia

E-mail: max-17000@yandex.ru

Abstract. The main purpose of this work was to investigate the structure of short-range order of microcrystalline cellulose (MCC) milled for one hour. Usually mechanical milling produces amorphous segments of crystalline cellulose. Investigation of short-range order was conducted by Debay's method to get X-ray diffraction (XRD) pattern when the coordinates of atoms in a cluster are known. The samples were studied by XRD. Crystallinity degree was measured using Segal's method. Cellulose Iα and Iβ were used as the initial structures to form clusters. The profile factor (R_p) was used, as the evaluation factor. The results showed that the cluster based on cellulose Iβ insufficiently characterized the short-range order of crystalline-amorphous cellulose (R_p>18%). Consequently, we examined a unit cell consisting of one cellobiose fragment. The final modeling cluster had the size of 35Å×22Å×29Å. The cluster consisted of 3 layers orientated randomly in relation to one another. The result of comparison of theoretical and experimental XRD patterns revealed that R_p was 11.4%. Therefore, the structure of short-range order of one-hour ground cellulose can be characterized by disordered cellulose chains with the length of 21 Å.

1. Introduction
Cellulose is a linear polymer with crystalline and amorphous regions alternating along the axis of microfibril [1]. Cellulose chains are located randomly in amorphous regions so interaction during reaction takes place more easily than in crystalline regions. Most of reactions takes place on plane of crystallites and in amorphous domains without affecting the intracrystalline regions [2].
The long-range order of crystalline in a mircofibril results from intermolecular interaction such as van der Waals's force and hydrogen bonds. A physical or chemical influence may destroy intermolecular forces in crystallites of cellulose. For example, the higher the time of mechanical milling, the higher increase of amorphous degree in cellulose fibers. XRD pattern of crystalline amorphous cellulose has amorphous galo and peaks belonging to crystalline regions. Amorphous galo defines the presence of the short-range order in the structure of cellulose. The method of randomly disordered layers (Debay's method) gives information about the structure of short-range order [3]. Therefore, the aim of this paper is to use Debay's method to investigate the influence of mechanical milling on the structure of the short-range order for one-hour milled MCC.

2. Experimental details
5 ml of MCC was introduced into an agate bowl and the milling was performed in a planetary ball mill (Pulverisette 7 premium line, Fritsch, Germany) using agate balls (5 mm diameter) during one hour. A rotation speed of 750 rpm was used.

XRD measurements of the MCC and one-hour milled cellulose were carried out using a DRON-4.0 automated diffractometer with MoK_{α} radiation. Scattering intensities were measured over an angular range of 2^0 to 70^0 with a step size of 0.1^0 and an angular range of 70^0 to 145^0 with a step size of 0.2^0 in the reflection and transmission geometry.
Debay's method used to obtain the distribution curves of scattering intensity from randomly disordered clusters. Debay's formula is given as in [3]:

$$I(s) = \frac{1}{N_{\mathrm{f}}}\left[\sum_{i=1}^{N} f_i^2 + 2\sum_{i}^{N-1}\sum_{j=i+1}^{N}\frac{1}{2}\left(f_i f_j^* + f_i^* f_j\right)\frac{\sin(sr_{ij})}{sr_{ij}}\exp\left(-0.5\sigma_{ij}^2 s^2\right)\right] \quad (1)$$

where I(s) – scattering intensity; N_f – the number of formula units per cluster; N – the number of atoms in the cluster system; f_i^2 – the atomic scattering factor for i atom; s – diffraction vector; r_{ij} - the distance between i and j atoms, σ_{ij} – dispersion of r_{ij}. The method is described in more detail in [3].
Crystalline degree is measured by Segal's method [4]. Segal's method is empirical and evaluates the ratio of the difference between the maxima intensity of reflection with (200) index (I_{200}) and the intensity of amorphous galo (I_{am}) to I_{200}. Crystalline degree is calculated as follows:

$$CD(\%) = \frac{I_{200} - I_{\mathrm{am}}}{I_{200}} * 100 \quad (2)$$

The fitting quality of the experimental data have been assessed by the profile factor defined as follows:

$$R_p(\%) = \sum_j \left|I_j - I_j^*\right| \Big/ \sum_j I_j * 100 \quad (3)$$

where I_j is the experimental point and I_j^* is the calculated point and the summation includes all points of data.

3. Results and Discussion

The XRD pattern for the sample MCC and one-hour milled cellulose in the reflection geometry is shown in figure 1. The XRD pattern of MCC has been analyzed using Rietveld method. The obtained results show that MCC had the structure of cellulose Iβ with antiparallel orientation of cellulose chains in monoclinic symmetry. The refined cell parameters were a=7.84±0.02 Å, b=8.13±0.03 Å, c=10.34±0.02 Å, α=β=90^0, γ=96.2±0.2^0. Crystalline degree of MCC was 80±5%.

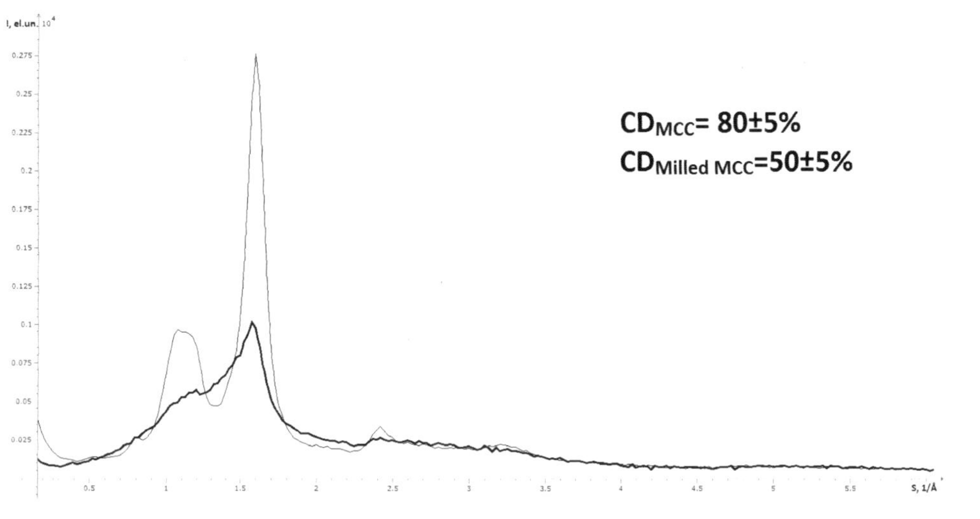

Figure 1. XRD pattern of MCC (—) and one-hour milled cellulose (▬).

IOP Conf. Series: Journal of Physics: Conf. Series **1038** (2018) 012059 doi:10.1088/1742-6596/1038/1/012059

Figure 1 shows that MCC scattering intensity of peaks decreases sharply after one hour milling. One-hour milling led to 30% reduction of the crystalline index measured according to equation (2). Hence, mechanical milling resulted in the increase of the amorphous regions number in MCC, but one hour of milling was not enough to transform MCC into amorphous cellulose, as there are some weak peaks on the XRD pattern. Scattering from amorphous part of cellulose is maximal in case of using the transmission geometry [1]. Therefore XRD pattern of one-hour milled cellulose obtained in the transmission geometry was used for comparison with XRD pattern of models obtained by Debay's method.

Next step was to construct clusters, get the distribution curve of scattering intensity of the sample (calculated by equation (1)) and do compare experimental and theoretical XRD patterns. A cluster was constructed by translating the unit cell of crystalline phase in three directions (a,b,c – the periods of unit cell). Cellulose Iβ, in which the unit cell consists of two cellulose chains rotated from each other by 180^0, and cellulose Iα with one cellulose chain per unit cell was taken as the structure of different clusters. The values of the profile factor (calculated by equation (3)) for different sizes of clusters are shown in table 1.

Table 1. The profile factors for different sizes of clusters based on cellulose Iα and Iβ.

	A size of cluster	1a×1b×1c	2a×2b×2c	3a×3b×3c	3a×2b×2c
Cellulose Iβ	R_p, %	19.7	24.0	37.1	29.3
Cellulose Iα	R_p,%	25.7	12.9	17.9	12.7

As can be seen from table 1, cellulose Iβ had the highest values of profile factor. It is due to the fact that there are two cellulose chains in unit cell that influence reflections of the diffraction pattern. The bigger the size of a cluster, the higher the intensity peaks. The cluster based on cellulose Iα allows us to reach significantly better results. So for the cluster of the size of 3a×2b×2c, the profile factor was minimal and equal to 12.7%. Consequently, the structure of short-range order crystalline-amorphous cellulose was characterized within the framework of the cellulose Iα model.

In order to improve fitting quality of experimental and theoretical XRD patterns, the initial cluster was divided into some layers. At the next step each layer was shifted in three directions (XYZ) and was rotated in plane of YZ taking into account the dispersion as to others layers. Operations of disorder are shown in figure 2b.

The initial cluster with the dimensions of 3a×2b×2c was divided into 3 layers with the size of 1a×2b×2c. Primary distance between layers along X was the period *a* of MCC unit cell. The rotation angle of second layer relative to the other ones was 180^0 to correspond the cellulose Iβ model of MCC.

As the consequence of modulating disorder parameters, it was found that the distances along X between the first and the second, and the third and the second layers were 7.34 Å and 7.80 Å, respectively. Rotation angle equaled -5^0 first layer as to the second one and 195^0 second layer as to the third layer. Shift of the first layer relative to the second and third layers along the direction of Y,Z was -0.5 Å and -1 Å, respectively. The comparison between experimental and theoretical XRD patterns is shown in figure 2a. The profile factor was R_p=11.4%.

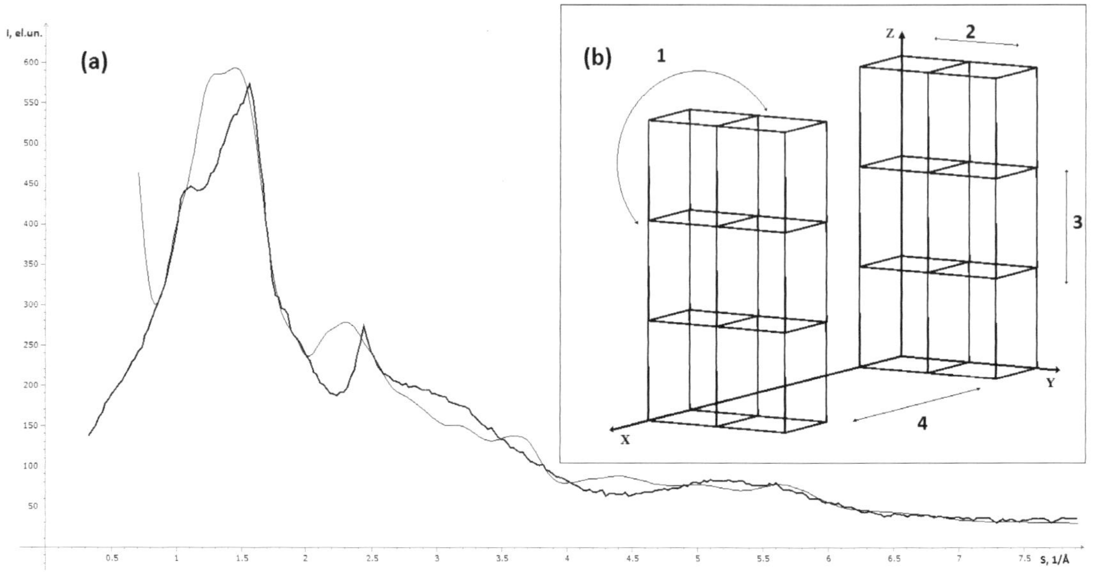

Figure 2. (a) — the comparison between the distribution curves of scattering intensity for the sample (**—**) and theoretical XRD pattern for the disorder of cluster (—); (b) — the ways of disorder of layers from each other: 1 – the rotation operation; 2, 3, 4 – the shift operation along Y, Z, X.

Figure 2a shows that experimental XRD pattern of the one-hour milled cellulose can be described within the framework of a randomly disordered cluster model. The final cluster with the dimensions of 35 Å × 22 Å × 29 Å characterized the structure of short-range order. There were three layers consisting of disordered cellulose chains with the length of 21 Å in the cluster. The number of formula units in the cluster was 24. The cluster included 504 atoms.

4. Conclusion

Mechanical milling of MCC during one hour resulted in the cellulose amorphization as the amorphous domains prevailed on XRD patterns and crystalline degree reduced to 30%. The structure of short-range order of one-hour ground cellulose can be described by the cluster with the dimensions of 35 Å × 22 Å × 29 Å in which the three layers were shifted relative to each other on 7.34 Å and 7.80 Å, consequently. The rotate angles of the first layer as to the second one and the second layer as to the third layer were -5^0 and 195^0. Thus, the structure of one-hour milled cellulose can be characterized by disordered cellulose chains with antiparallel orientation.

References

[1] Reginaldo A Festucci-Buselli, Wagner C Otoni and Chandrashekhar P Joshi 2007 *Braz. J. Plant. Physiol.* **19** 1–13
[2] Ciolacu D, Ciolacu F and Popa I. V. 2011 *Cellulose Chem. Technol.* **45** 13–21
[3] Guinier A 1956 *Théorie et technique de la Radiocristallographie* (Paris: Dunod) p 736
[4] Terinte N, Ibbett R and Schuster K C 2011 *Lenzinger Berichte* **89** 118–31

IOP Conf. Series: Journal of Physics: Conf. Series **1038** (2018) 012060 doi:10.1088/1742-6596/1038/1/012060

Electrical properties of copper iodide nanoparticles embedded into porous alumina matrix

N O Alexeeva, S E Gango, N I Puchkov, V G Solovyev, A V Cvetkov
Pskov State University, Lenin square, 2, 180000 Pskov, Russia

E-mail: muxanin@mail.ru

Abstract. $CuI/\text{por}Al_2O_3$ nanocomposite was prepared by synthesis of copper iodide (CuI) nanoparticles within porous alumina ($\text{por}Al_2O_3$) matrix. Electrical conductivity and Seebeck coefficient of bulk copper iodide and those of $CuI/\text{por}Al_2O_3$ samples were measured at different temperatures. Optical characteristics of the samples were obtained by spectral ellipsometry. Type and the temperature dependence of electrical conductivity of the materials under study are discussed.

1. Introduction

Porous anodic aluminium oxide films formed by the electrochemical oxidation of aluminium have been widely studied to protect and decorate the surface of aluminium, to make inorganic membranes etc. During last two decades, porous alumina ($\text{por}Al_2O_3$) with relatively regular pores, narrow distributions of pore sizes and interpore spacings prepared by two-step anodization [1] has become a popular dielectric matrix for the design of various nanocomposite materials.

Recently, we studied electrical properties of ferroelectric sodium nitrite nanoparticles, as well as those of semiconductor iodine species embedded into porous alumina [2, 3]. In this paper, electrical properties of wide-gap semiconductor copper iodide (CuI) [4, 5] within porous alumina fabricated under different anodizing parameters were studied.

2. Experimental procedure

We used sample preparation process which was divided into three stages. First of all, fabricating alumina templates using two-step anodization at a constant voltage of 40 V in 0.3 M oxalic acid at $8\,^0C$ or at a constant voltage of 160 V in 0.1 M phosphoric acid at $3\,^0C$ was carried out [2]. At the second stage, host matrices – inorganic $\text{por}Al_2O_3$ membranes – were made by dissolving of aluminium layer in $CuCl_2$ solution and by removal of barrier oxide layer in phosphoric acid at $60\,^0C$ [6]. Finally, copper iodide was synthesized in the channels of $\text{por}Al_2O_3$ membranes by diffusion of two reagents – $CuSO_4$ and KI – in opposite directions (see experimental setup in figure 1) according to the following reaction equation:

$$2CuSO_4 + 4KI \rightarrow 2CuI + I_2 + 2K_2SO_4 . \qquad (1)$$

One could estimate the volume fraction (about 15%) of the guest substance in the $CuI/\text{por}Al_2O_3$ composite from the dielectric permittivity values of the materials under study obtained by the spectroscopic ellipsometer "Ellips-1891" (Novosibirsk) [7].

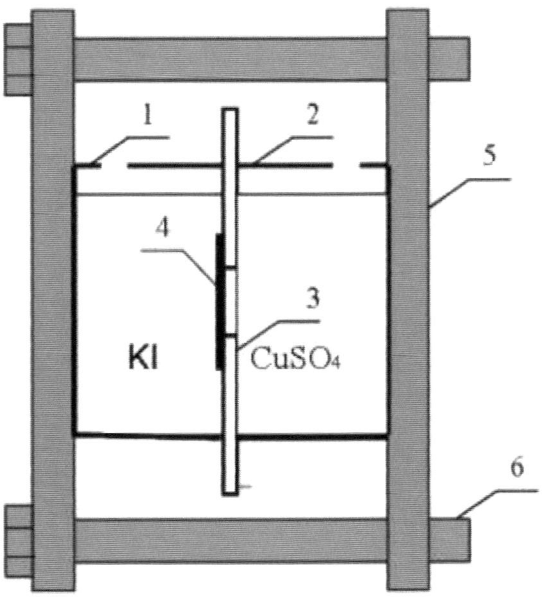

Figure 1. Experimental setup for synthesis of CuI nanoparticles in porAl_2O_3 membranes: 1, 2 – vessels for reagents; 3 – glass plate; 4 – sample; 5 – metal wall; 6 – bolt.

Alternating-current (AC) conductivity of the samples under study was measured at the frequency of 1 kHz with graphite electrodes. Thermal electromotive force of thin samples was measured by pulse method proposed in [8]. Photoluminescence was excited at room temperature by pulse nitrogen laser AIL-3 (wavelength value $\lambda = 337\,\text{nm}$).

3. Results and discussion

Arrhenius plots demonstrating temperature dependences of the AC (1 kHz) electrical conductivity for bulk CuI guest substance and CuI/porAl_2O_3 nanocomposite are shown in figures 2 and 3. One can observe rather low value and strong temperature dependence of the electrical conductivity during the samples' heating. On the other hand, when the samples' temperature decreases from rather high values, electrical conductivity becomes larger and being characterised by lower values of activation energy (0.1 – 0.3 eV). This behaviour is typical for the high-temperature superionic $\alpha-$ phase of CuI (this phase can coexist with low-temperature $\gamma-$ and $\beta-$ phases during rapid crystallization of copper iodide [4, 5]).

Our measurements of bulk polycrystalline copper iodide Seebeck coefficient give the typical value for p-type semiconductor: $S_{CuI} \approx +0.12\frac{mV}{K}$, specific thermal electromotive force value for the CuI/porAl_2O_3 nanocomposite being about two times smaller. The most striking feature is the negative Seebeck coefficient of the CuI/porAl_2O_3 nanocomposite based on the porAl_2O_3 matrix prepared by anodization in the oxalic acid (when the initial alumina matrix is fabricated in phosphoric acid, the mechanism of charge transport remains unchanged: $S_{CuI/\text{por}Al_2O_3} > 0$). We suppose that these effects are due to the high concentration of donors (e.g., anion vacancies [5]) appearing in the nanocomposite as the result of the reaction (1), because its duration for the samples prepared in oxalic acid was much longer, than that for those prepared in phosphoric acid.

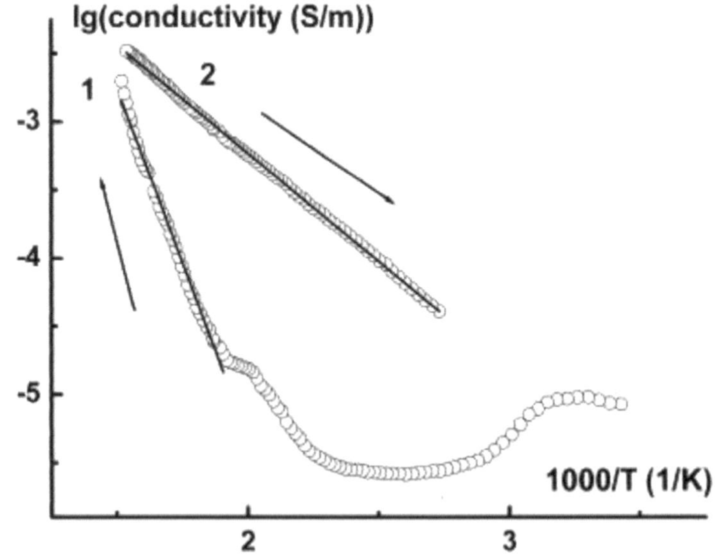

Figure 2. Temperature dependence of the alternating-current (1 kHz) electrical conductivity for polycrystalline *CuI*. Inclined arrows indicate heating (curve 1) and cooling (curve 2) of the sample.

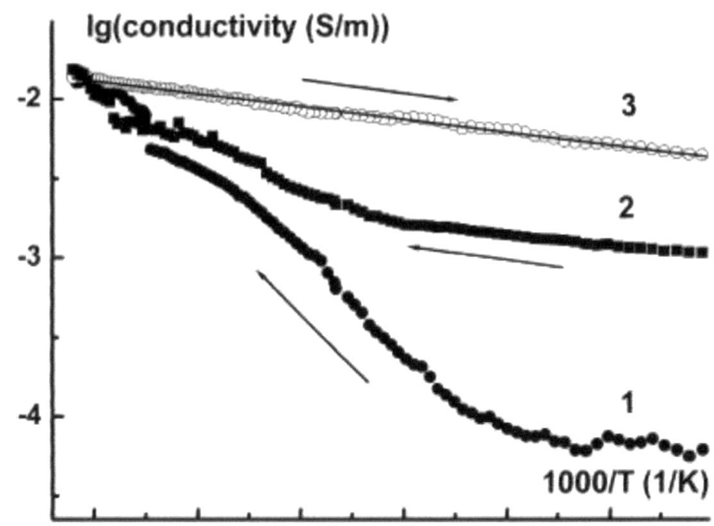

Figure 3. Temperature dependence of the alternating-current (1 kHz) electrical conductivity for $CuI/porAl_2O_3$ nanocomposite. Inclined arrows indicate first heating (curve 1), second heating (curve 2) and cooling (curve 3) of the sample.

It should be noted that iodine vacancies may cause red photoluminescence in copper iodide [5]; similar optical phenomenon under UV light excitation was observed in our *CuI* samples, indeed.

4. Acknowledgments

This work was supported by the Ministry of Education and Science of Russian Federation (Project No. 576, Task No. 2014/700).

References

[1] Masuda H, Fukuda K 1995 *Science* **268** 1466
[2] Alexeeva N O 2013 *Journal of Physics: Conference Series* **461** 012023
[3] Alexeeva N, Cema G, Podorozhkin D, Solovyev V, Trifonov S, Veisman V 2015 *Journal of Self-Assembly and Molecular Electronics* **2** 27
[4] Gruzintsev A N, Zagorodnev W N 2012 *Semiconductors* **46** 35
[5] Gruzintsev A N, Zagorodnev W N 2012 *Semiconductors* **46** 149
[6] Lillo M, Losic D 2009 *Journal of Membrane Science* **327** 11
[7] Shvets V A, Spesivtsev E V, Rykhlitskii S V, Mikhailov N N 2009 *Nanotechnologies in Russia* **4** 201
[8] Gango S E, Markov V N, Solov'ev V G 1998 *Instruments and Experimental Techniques* **41** 858

Kinetic limitations of stress relaxation and generation in GaN/AlN and AlGaN: Si/AlN heterostructures grown on c-sapphire by plasma-assisted molecular beam epitaxy

O A Koshelev, D V Nechaev, S I Troshkov, V V Ratnikov, V N Jmerik, and S V Ivanov
Ioffe Institute, 26 Polytekhnicheskaya, St. Petersburg 194021, Russia

Abstract. The paper describes experimental study of stress relaxation and generation in (1-2)-µm-thick GaN and AlGaN layers grown on AlN/c-Al_2O_3 buffer layers by low temperature (<720 °C) metal-rich plasma-assisted molecular beam epitaxy (PA MBE). The atomically smooth undoped GaN layers demonstrate only gradual relaxation of the compressive stress, which is probably related to thermodynamically driven inclination of threading dislocations (TDs).The slower stress relaxation at the lower growth temperature is explained by kinetic limitation of this process. The switch of compressive to tensile stress in the less-strained undoped $Al_{0.7}Ga_{0.3}N$ layers, attributed mostly to the same effect of TD inclination, occurs in the low-temperature PA MBE conditions at much larger thickness (~0.6 with) as compared to MOVPE ones. Introduction of high Si doping ($n \sim 10^{19}cm^{-3}$) reduces noticeably the initial compressive stress in the AlGaN film due to substitution of small Si atoms in the group-III sublattice. At larger thickness, Si atoms seem to effect the TD propagation and suppress generation of tensile stress related to TD inclination, which makes possible to grow ~1-µm-thick $Al_{0.7}Ga_{0.3}N$: Si films without cracking.

Introduction

(Al,Ga)N-based binary and ternary compounds are basic materials for novel devices of ultraviolet (UV) photonics and high-frequency/ power electronics [1, 2]. Due to low commercial competitiveness of native substrates for these compounds, the device heterostructures are used to grow on various heterosubstrates (*c*-Al_2O_3, 6H-SiC and Si (111) *etc.*). An AlN/*c*-sapphire heterostructure with moderate values of crystallographic lattice mismatch $(a_{Al2O3}-a_{AlN})/a_{AlN}$= -11.7% and thermal mismatch $\Delta\alpha_{th}= \alpha_{AlN} - \alpha_{Al2O3}= -4.12\cdot10^{-6}K^{-1}$ [3] is the most suitable template for mass-production of devices in which the substrate transparency (for UV-optoelectronics), large diameter and low cost are the key factors when choosing a substrate.

In general, significant stresses generating at all stages of growth of the heterostructures are relaxed in different ways, including the generation of various bulk extended defects, the formation of developed surface morphology, *etc*. Among the former, the misfit dislocations at the AlN/*c*-Al_2O_3 interface are the most important, since threading dislocations (TDs) arising from them can reach the upper (active) regions of the device heterostructures and deteriorate their optical and electrical properties. To reduce the concentration of TDs the AlN buffer layers are commonly grown for a thickness of a few microns, which is sufficient for effective interaction of TDs with each other as well as with filtering super lattices or interlayers [4, 5].As a result, the TD density can be reduced to a level of 10^8-10^9 cm^{-2}.On the other hand, the increasing thickness of AlN leads, as a rule, to transition from the initial compressive mismatch stress to tensile one [6], which, in general, can result in cracking of the film. To overcome this issue, the growth of compressively strained GaN or AlGaN layers on the

AlN buffer layers is used to compensate of the harmful tensile stress. However, unfortunately, these layers also demonstrate a fairly fast transition to an incremental tensile stress [7, 8].

Plasma-assisted molecular beam epitaxy (PA MBE) is one of the main technologies currently being developed for production of various III-Nitrides-based devices. Among its unique features is the possibility of epitaxial growth of (Al,Ga)N compounds at extremely low substrate temperatures (<<1000 °C), which causes specific kinetics and thermodynamics of epitaxial growth of the alloys. In principle, the growth temperature should play an important role in the relaxation and generation of stresses in the heterostructures, but this influence for (Al,Ga)N/*c*-Al_2O_3 heterostructures has not been sufficiently studied. Thus, only one paper by Okumura et al. [9] has reported a critical thickness of 900-nm necessary to introduce dislocations into an AlN buffer layer grown on a 6H-SiC substrate, which is more than two orders of magnitude greater than the thickness estimated in accordance with a Matthews–Blakeslee model. Even less is known about influence of Si atoms on the stresses in n-AlGaN: Si layers grown by PA MBE, although creation of an additional tensile stress during MOVPE growth of AlGaN: Si epilayers has been studied by several groups and a few explanations of this phenomenon have been proposed [9, 10].

This paper describes the behavior of stresses in the GaN and AlGaN layers grown by PA MBE on AlN/c-Al_2O_3 buffer layers. The former is studied as a function of the substrate temperature (T_S) whereas the influence of high Si doping on the stress variation is evaluated for the latter.

Experimental

All samples were grown on *c*-sapphire substrates by PA MBE setup Compact 21T. The peculiarities of growth of AlN nucleation and buffer layers by migration enhanced epitaxy and metal modulated epitaxy (MME), respectively, were described in our previous papers [11, 12]. The nitrogen flux of 0.47 ML/s, defining the growth rate at the employed metal-rich conditions, was kept constant during growth of all structures. The bottom buffer AlN layers in all structures were grown with the same thickness of about 1μm at T_S=780 °C and slightly metal-rich conditions with a flux ratio of Al/N~ 1.1. The 1-μm-thick GaN layers in two GaN/AlN heterostructures were grown by MME at different temperatures of 690 °C (GaN-LT-sample) and 720 °C (GaN-HT-sample) by using the Ga/N flux ratios of 2.2 and 2.9, respectively. The two samples with undoped and Si-doped $Al_{0.7}Ga_{0.3}N$ layers with thicknesses of 1 and 1.8 μm, respectively, were grown on the AlN buffer layers by a standard PA MBE at T_S=700 °C using continuous fluxes with a III/N ratio of about 2.

Laser reflectometry and reflection high energy electron diffraction (RHEED) were used for *in situ* control of growth rate and surface morphology. The stress in the heterostructures was evaluated *in situ* by a home-made multi beam optical stress sensor (MOSS), as described in [13]. The incremental stresses $\sigma(h_l)$ were calculated by using the Stoney differential equation

$$\sigma(h_l) = \frac{M_s h_s^2}{6}\frac{\partial k}{\partial h_l},$$

Where M_S is a biaxial elastic modulus of substrate (602 GPa for *c*-Al_2O_3 [14]), h_l and h_s are the thicknesses of substrate and epitaxial layer, respectively, while $\frac{\partial k}{\partial h_l}$ denotes the change of substrate curvature with respect to the layer thickness change. The surface morphology of the samples was studied by optical (OM) and scanning electron (SEM) microscopies. Measurements of the width of *x*-ray diffraction symmetric and skew-symmetric peaks were used to estimate the screw and edge TD densities in AlN layers. The doping concentration was measured by C-V technique.

Results and discussion

The inset in Fig.1 shows typical SEM image and RHEED pattern observed for GaN-LT and GaN-HT layers. They indicate the atomically smooth surface morphology for both of the layers due to appropriate choice of Ga-rich conditions for their MME growth. Also, these layers demonstrate almost the same initial compressive stress of about 10 GPa corresponding to the mismatch stress for GaN/AlN heterostructures $\sigma = M_{GaN}\,\varepsilon = 10.8$ GPa, where the biaxial elastic modulus $M_{GaN} = 450$ GPa and the crystallographic mismatch $\varepsilon = -2.4\%$ [14]. However, Fig. 1a demonstrates the different behavior of the substrate curvature (stress × thickness) vs thickness dependences during growth of these layers.

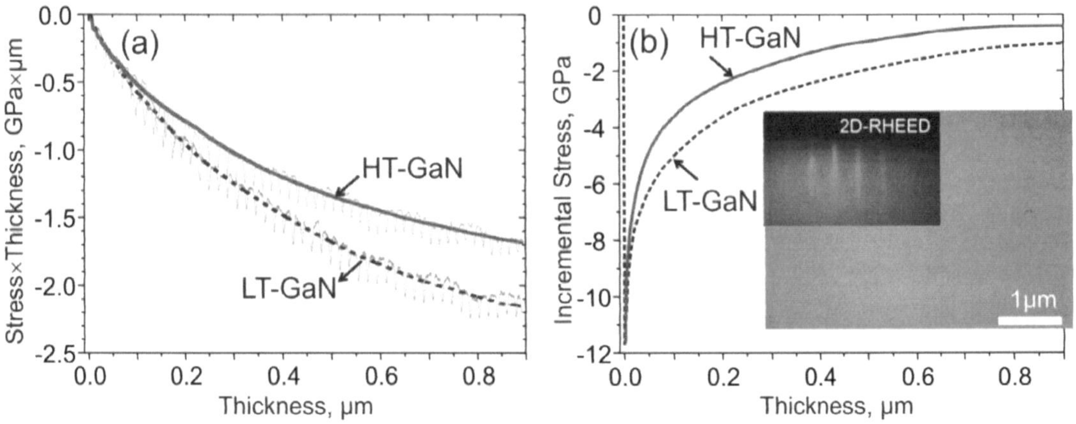

Figure1. (a) Curvatures (stress × thickness) and (b) incremental stresses (first derivative of the plots in (a)) vs thickness for LT-GaN (dashed) and HT-GaN (solid) layers grown on relaxed AlN/*c*-Al_2O_3. The thin grey curves in (a) are the experimental data of curvature change during MME growth. Insertion in (b) shows typical SEM image and streaky RHEED pattern observed for both GaN layers grown by MME.

The slower stress relaxation in LT-GaN compared to HT-GaN, possessing the same surface morphology, can be related apparently with kinetic limitations of this process. Moreover, taking into account XRD data for AlN buffer layers, revealing an approximately the same rather high initial TD density ~ 10^9-10^{10} cm^{-2} in the both samples, one can assume that the inclination of TDs with respect to the (0001) growth direction is a major mechanism of the stress relaxation in the compressively strained GaN layers [15]. However, the thermodynamic model by Romanov and Speck [15], considering the energy balance approach in the layers with various TD configurations, can only predict that the inclination of TDs results in decrease in dislocation and system energy, which can be easily realized in case of high temperature epitaxial technology like MOVPE. Slowing down this process with reduction of T_S in PA MBE indicates an existence of the kinetic barrier for the TD inclination, which efficiently prevents the stress relaxation in the GaN overlayers. The using of Ga-rich MME process providing an atomically-smooth featureless surface morphology of the PA MBE films plays a role of additional factor minimizing the density of surface defects which may facilitate the TD bending by local changing the surface crystalline orientation or nucleate new TDs in the lattice-mismatched GaN.

Figure 2 shows the changes of the compressive stresses in undoped and highly Si-doped $Al_{0.7}Ga_{0.3}N$ layers during growth on AlN buffer layers. The corresponding initial stress values of 3.7 GPa and 2.3 GPa are significantly lower as compared to the GaN/AlN heterosystem due to the smaller lattice mismatch (~ 0.8%). In addition, high Si doping obviously reduces the compressive stress from the very beginning of growth presumably owing to small covalent radii of Si atoms incorporated into group-III sublattice. The RHEED patterns exhibit only slight degradation of the initial streaky pattern which acquires additional spotty features at the AlGaN growth. These facts indicate that the relaxation proceeds through the weak generation of misfit dislocations at the AlGaN/AlN interface and/ or some 2D-3D transition of surface morphology of the AlGaN layers facilitating at low temperature the inclination of TDs inherited from the AlN buffer [15]. Importantly, that these mechanisms explain the gradual relaxation of the compressive stresses, but they cannot be used for clarifying the reasons for transition from compressive to tensile stresses observed at the thicknesses of 0.6 and ~0.9 μm for undoped and Si-doped layers, respectively (Fig. 2a).

A few mechanisms responsible for emergence of so-called intrinsic tensile strain in nominally compressively strained AlGaN/AlN heterostructures have been proposed. Firstly, it can be caused by a gradual coalescence of islands in the imperfect grained layers in accordance with a Nix-Klemens model [16], though this mechanism seems to prevail at the interfaces with higher lattice mismatch (e.g. AlN/c-Al_2O_3) and is not dominant in the considered case. Also, the relaxation of compressive strain in the AlGaN layer through the inclination of pre-existing TDs was suggested to result in generation of equibiaxial misfit dislocations compensating efficiently the initial compressive strain [17, 18]. It should be emphasized that, in accordance with [15], the angle of inclination remains unchanged during the layer growth and it can even lead to generation of significant tensile stress after some critical thickness, leading finally to plastic relaxation of the layer through cracking.

In addition, the stress developing in the doped AlGaN layers is further complicated by influence of Si atoms, as seen in Fig. 2a. There are also several mechanisms of the introduction of additional tensile strain in Si-doped AlGaN layers. The most obvious explanation of this phenomenon is less dimension of Si atoms than Ga ones that leads to hydrostatic pressure in the doped films. As a result, the doped film has the smaller lattice parameters as compared to the undoped layer, which is displayed in the smaller initial stress in the AlGaN: Si layer, mentioned above. Also, Forghani et al. [10] proposed an alternative mechanism of inducing the tensile strain in the doped AlGaN films related to a selective incorporation of Si atoms around edge TDs resulting in transformation the strain dipole field into predominantly tensile strained dipoles.

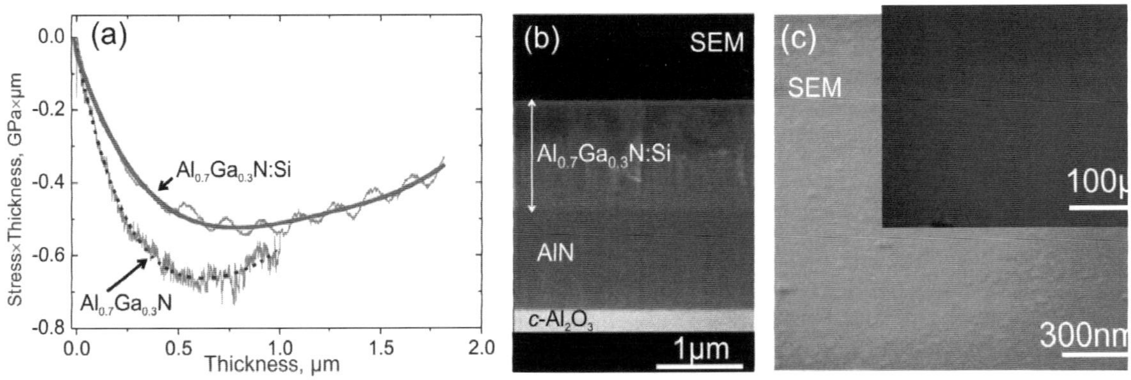

Figure2. (a) Substrate curvature (stress × thickness) versus thickness for $Al_{0.7}Ga_{0.3}N$: Si and $Al_{0.7}Ga_{0.3}N$ layers grown on AlN/-Al_2O_3 buffer layers. Cross-sectional (b) and plan-view (c) SEM images of the $Al_{0.7}Ga_{0.3}N$: Si layer. The inset in (c) shows its surface image obtained by OM.

However, the processes of relaxation of compressive stress and generation of tensile one in the undoped and highly Si-doped $Al_{0.7}Ga_{0.3}N$/AlN heterostructures grown by PA MBE (Fig. 2a) appears to differ strongly from similar structures grown by MOVPE [10, 19, 20]. The undoped PA MBE film exhibits the compressive incremental stress (negative slope in curves in Fig. 2a) up to the thickness of ~0.6 μm, when the curvature slope changes sign to the positive one. In contrast, the $Al_xGa_{1-x}N$/AlN heterostructures grown by high temperature MOVPE at T_S= 1100 °C [19] demonstrate the compressive stress only for much higher lattice mismatch (x= 0.2-0.38), whereas the higher Al-content layers show negligible curvature change (at x= 0.8) or even a positive transient slope (at x= 0.86) from the very beginning [19]. This can be explained by the low-temperature growth conditions inherent to

PA MBE, which slowdown the TD inclination process considered as the most probable for generation of tensile stress in the heterostructures with small lattice mismatch.

The effect of high Si doping in PA MBE, well pronounced at the initial stage of the AlGaN: Si growth, also shows different behavior as compared to MOVPE reports [19, 20]. Firstly, it does not reverse the stress sign immediately as it does in low-Al-content AlGaN: Si layers grown by high-temperature MOVPE [19], which is probably related to much lower fraction of Ga atoms exchanged by Si ones. Second, it slows down the stress relaxation process shifting the tensile-stress onset thickness by approximately 30-40 % (to 0.9-1.0μm) as compared with the undoped AlGaN layer case. The latter is assumed to be caused by interaction of the high-density Si atoms with TDs, which was thoroughly studied in Ref. [10]. However, in contrast to high-temperature MOVPE case, in the low-temperature PA MBE with the suppressed surface and bulk Si diffusion, the Si atoms seem to affect the TD propagation process, e.g. by blocking or changing the inclination angle of TDs, rather than creating the tensile strained dipoles along the edge TDs. As a result, the compressively stressed Si-doped high-Al-content AlGaN films with an electron concentration up to 10^{19} cm^{-3}can be grown by PA MBE on AlN buffers up to much larger thickness without film cracking as compared to the undoped layers.

Conclusions

We have demonstrated the strong influence of kinetic limitations on different processes of compressive stress relaxation and generation of intrinsic tensile stress in the undoped GaN and $Al_{0.7}Ga_{0.3}N$ (undoped and highly doped with Si) layers grown on AlN/c-Al_2O_3 buffer layers by low temperature (T_S<720 °C) PA MBE. This allowed us to grow 1-μm-thick GaN and Si-doped $Al_{0.7}Ga_{0.3}N$ ($n=10^{19}cm^{-3}$) layers on the AlN buffers, preserving the average compressive stress. These heterostructures serve as constituent elements of various UV-optoelectronic and electronic devices where the heterostructure cracking and substrates bowing are the critical issues.

Acknowledgement

This work has been supported by Russian Science Foundation (Grant #14-22-00107).

References

[1] Shatalov M, Sun W, Jain R, Lunev A, Hu X, Dobrinsky A, Bilenko Y, Yang J, A Garrett G A, Rodak L E, Wraback M, Shur M, Gaska R 2014 Semicond. Sci. Technol. **29** 084007.

[2] Johnson W, Piner E L 2012 Springer Series in Materials Science GaN HEMT Technology in GaN and ZnO-based Materials and Devices vol 156 Pearton S (Springer-Verlag Berlin Heidelberg).

[3] Ferreyra R A, Zhu C, Teke A, Morkoç H 2017 Springer Handbook of Electronic and Photonic Materials Kasap S, Capper P (Springer International Publishing AG).

[4] Bai J, Dudley M, Sun W H, Wang H M, Khan M A 2006 Appl. Phys. Lett. **88** 051903.

[5] Jmerik V N, Lutsenko E V, Ivanov S V 2013 Phys. Status Solidi A **210** 439.

[6] Radhavan S, Redwing J M, 2004 J. Appl. Phys. **96** 2995.

[7] Romanov A E, Beltz G E, Cantu P, Wu F, Keller S, DenBaars S P, Speck J S 2006 Appl. Phys. Lett. **89** 161922.

[8] Brunner F, Mogilatenko A,Kueller V,Knauer A,Weyers M 2013 J. Cryst. Growth **376** 54.

[9] Okumura H, Kimoto T, Suda J 2012 Appl. Phys. Express **5** 105502.

[10] Forghani K, Schade L, Schwarz U T, Lipski F, Klein O, Kaiser U, Scholz F 2012 J. Appl. Phys. **112** 093102.

[11] Nechaev D V, Aseev P A, Jmerik V N, Brunkov P N, Kuznetsova Y V, Sitnikova A A, Ratnikov V V and Ivanov S V 2013 J. Crys. Growth **378**, 319.

[12] Jmerik V N, Mizerov A M, Nechaev D V, Aseev P A, Sitnikova A A, Troshkov S I, Kop'ev P S and Ivanov S V 2012 J. Crys. Growth **354** 188.

[13] Zolotukhin D S, Nechaev D V, Ivanov S V, Zhmerik V N 2017 Technical Physics Letters. **43** 262.
[14] Hearne S, Chason E, Han J, Floro J A, Figiel J, Hunter J, Amano H, Tsong I S T 1999 Appl. Phys. Lett.**74** 356.
[15] Romanov A E, Speck J S 2003 Appl. Phys. Lett. **83** 2569.
[16] Nix W D, Clemens B M 1999 J. Mater. Res. **14** 3467.
[17] Romanov A E, Beltz G E, Cantu P, Wu F, Keller S, DenBaars S P, Speck J S 2006 Appl. Phys. Lett. **89** 161922.
[18] Follstaedt D M, Lee S R, Allerman A A, Floro J A 2009 J. Appl. Phys. **105** 083507.
[19] Brunner F, Mogilatenko A, Kueller V, Knauer A, Weyers M 2012 J. Cryst. Growth **376** 54.
[20] Cicek E, McClintock R, Vashaei Z, Zhang Y, Gautier S, Cho C Y, Razeghi M 2013 Appl. Phys. Lett. **102** 051102.

The influence of activation and growth time on the geometry and structural perfection of multi-walled carbon nanotubes

O I Il'in[1], M V Il'ina[1], N N Rudyk[2], A A Fedotov[2], D I Levshov[3], O A Ageev[4]

[1] Southern Federal University, Department of Nanotechnologies and Microsystems, Taganrog, 347922, Russia
[2] Southern Federal University, Laboratory of Nanobiotechnology Problems and New Materials, Taganrog, 347922, Russia
[3] Southern Federal University, Faculty of Physics, Rostov-on-Don, 344090, Russia
[4] Southern Federal University, Research and Education Center "Nanotechnologies", Taganrog, 347922, Russia

oiilin@sfedu.ru

Abstract. We studied the influence of the activation time on the phase composition and geometric parameters of catalytic centers (CC). We found, that the treatment of substrate in an ammonia atmosphere allows to restore oxidized nickel. The activation time and the presence of ammonia plasma do not affect the phase composition of catalyst, but has a significant effect on the geometric dimensions of catalytic centers. Also we studied the influence of growth time on the geometry and structure of vertically aligned carbon nanotubes (VACNT). When the growth time ranged from 5 to 15 min we observed the formation of new walls, which resulted in an increase of the outer diameter. Within the range of 15 to 30 minutes, however, there was a decrease in the diameter of carbon nanotubes, which is associated with the disturbance of structural perfection and disorientation of carbon nanotubes caused by wall undercutting in the plasma.

1. Introduction

The unique properties of VACNT provide extended opportunities for their use as functional elements of emission nanoelectronics devices [1], memory cells [2, 3] and interconnections [4]. The key factors for the wide use of CNTs in these devices are high aspect ratio [5] and the metallic nature of their conductivity [6]. Thus there is definitely a clear need for research in process conditions to grow VACNT with controlled geometry, structure and properties.

It is well known that plasma enhanced chemical vapor deposition (PECVD) is considered to be the most promising method of CNT formation [7, 8], which allows to grow aligned CNTs on substrates with catalytic centers [3]. The processes of deposition of the catalyst film and the growth of CNTs are usually carried out on different equipment. This leads to the oxidation of the catalyst film upon contact with the atmosphere. Hence, the necessary step in the CNTs production is the activation, i.e. the restoration of the oxidized catalyst film with a hydrogen or ammonia atmosphere in PECVD process. But the influence of the activation parameters on the structure and geometrical dimensions of CC has not been studied sufficiently and requires further research.

We should note that a large number of interrelated parameters of PECVD (such as temperature, pressure, flow and type of process gas, catalyst film material, etc.) have a significant impact on VACNT growth processes [9-13]. One of the most important parameters determining the aspect ratio of VACNT is growth time [14, 15].

However, the influence of this parameter does not appear to have been sufficiently studied, so our aim was to study the influence of activation and growth time of VACNT obtained by PECVD on their geometry and structural perfection.

2. Experiments and Methods

We had grown VACNT by vertex mechanism in the PECVD module of a cluster nanotechnology complex NANOFAB (NT-MDT, Russia). RCA-cleaned Si (100) wafers were used as substrates. To eliminate the interaction between catalyst centers material and the substrate, a buffer sublayer of the metal film was created. Ni/Cr films have been used as a combination of metals for the catalyst/sublayer, because they provide high homogeneity of the parameters of carbon nanotubes [6-8].

20 nm thick Cr film was used as a buffer sublayer which we deposited from a chromium target (99,95% purity, Kurt J. Lesker Co.) in an atmosphere of Ar with the substrate heated to 100°C. A catalyst layer film of 10 nm thick Ni was deposited from a nickel target (99,995% purity, Kurt J. Lesker Co.) in an atmosphere of Ar with the substrate heated up to 220°C. We deposited Cr and Ni films using a magnetron sputtering set AUTO 500 (BOC Edwards, UK) at a chamber pressure of 83.1 Pa.

We activated and grew VACNTs in the flows of NH_3 (210 sccm), at a chamber pressure of 4.5 Torr and a temperature of 750 ° C. We used C_2H_2 (70 sccm) as the carbon-containing gas. We carried out the activation of the experimental samples with DC plasma initiation (45 W) and without it. We changed the activation time in the range from 1 to 5 min. We varied the growth time t for our samples from 5 to 30 minutes.

We investigate the phase composition of the structure of the nickel catalytic layer by X-ray absorption spectroscopy (XANES, EXAFS) at a specialized source of synchrotron radiation (SIC "Kurchatov Institute", Moscow, Russia). The obtained X-ray absorption spectra were compared with reference samples.

We studied the VACNT samples using a scanning electron microscope (SEM) Nova NanoLab 600 (FEI, Netherlands). The diameter and height of the VACNT array obtained were evaluated by means of statistical processing of SEM images. The structural analysis of the VACNT arrays was conducted by a Raman spectrometer Renishaw InVia Reflex (Renishaw plc, UK) in the backscattering geometry at 514 nm excitation wavelength.

3. Results and Discussion

The analysis of the X-ray absorption spectra of the catalyst film prior to the activation process showed that the nickel was predominantly in an oxidized state with a volume of pure Ni ~ 30%. Processing the substrate in an ammonia atmosphere for $1 \div 5$ minutes, allowed us to restore the oxidized nickel with a metal content of 90-95%. We observed that the activation time and the ammonia plasma initiation did not affect the phase composition of the catalyst.

However, the analysis of the catalyst film by AFM and SEM methods showed a significant influence of the activation time on the geometric dimensions of the formed catalytic centers.

It was found that with increasing activation time, the diameter (Fig. 1a) and height (Fig. 1b) of the catalytic centers diminished, which was associated with the increase of the temperature exposure time (750 °C) during activation.

Thus, during the annealing phase, the difference between the temperature expansion coefficients of the film and the substrate promotes the appearance of a mechanical stress on their contact. As a result, the fragmentation of the metal film with the formation of catalytic centers takes a place. Presumably, as temperature exposure time increases, the simultaneous processes of sublimation and surface diffusion take place, which leads to diameter and height decrease, as well as an increase of CC homogeneity. The presence of plasma in the activation process does not influence the CC dimensions (Fig. 1).

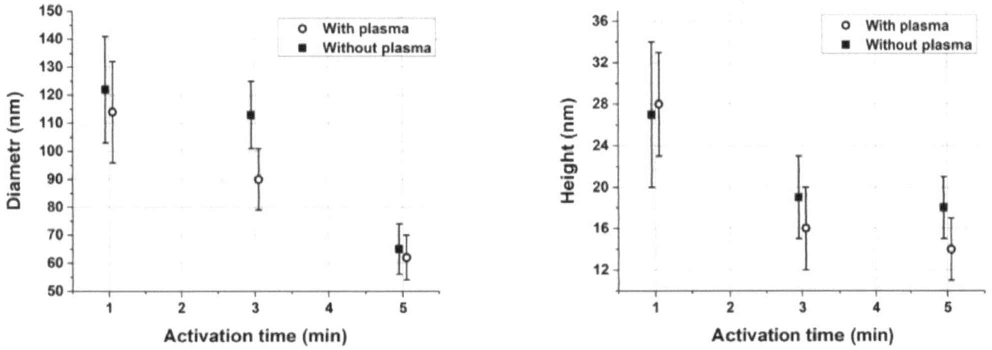

Figure 1. The dependence of diameter (a) and height (b) of catalytic centers on activation time.

In order to avoid the formation of nanofibers on a large size CC, the further growth of the experimental samples of the VACNT was conducted with 5 min activation time.

Having analyzed the SEM images of our samples (Fig. 2), we were able to determine the dependence of VACNT diameter and height on growth time.

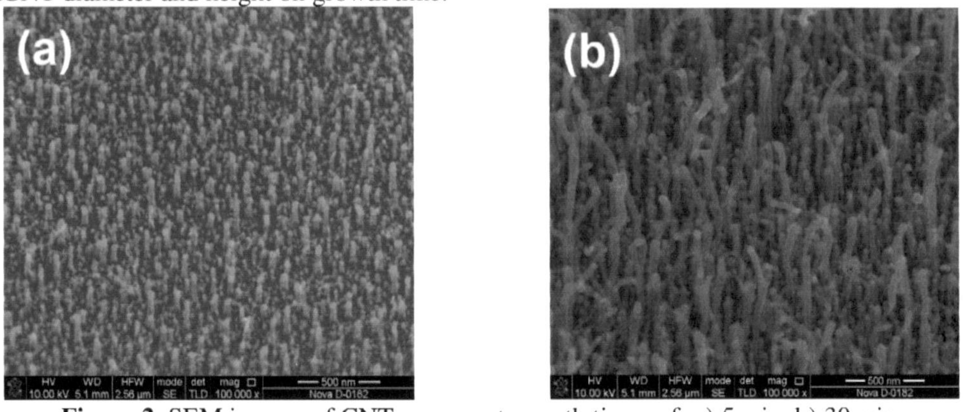

Figure 2. SEM images of CNTs grown at growth times of: a) 5 min, b) 30 min.

With an increase in growth time *t* from 5 to 15 min CNT diameter increased from 35 to 58 nm (Fig. 3a), which may be associated with the formation of new walls of multilayer VACNTs. However, when we increased growth time *t* from 15 to 30 min, the CNT diameter decreased to 54 nm (Fig. 3a), which may be due to the undercutting of nanotubes caused by ammonia plasma (Fig. 4).

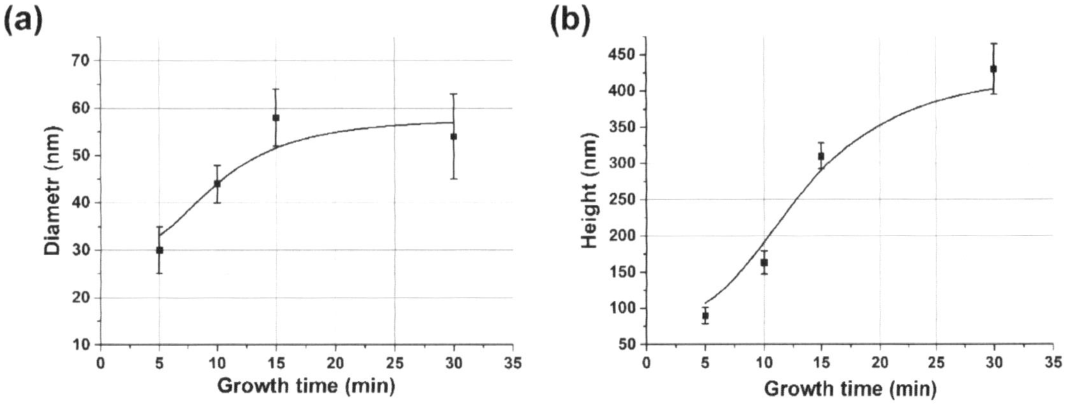

Figure 3. The dependence of CNT diameter (a) and height (b) on growth time.

We observed an increase in VACNT height from 90 to 310 nm occurring within the whole range of growth times (Fig. 3b). However, we noticed an interesting fact that at t = 30 min carbon nanotubes began to deviate from the normal position and started getting disoriented (Fig. 2b). This may also be associated with the undercut of the nanotubes at their base, which causes their bending.

In Figure 5 we compare the Raman spectra of the investigated samples measured at 514 nm excitation wavelength. We clearly observe the presence of D- and G-bands. The high intensity of D-band implies the presence of defects in the samples. [16, 17].

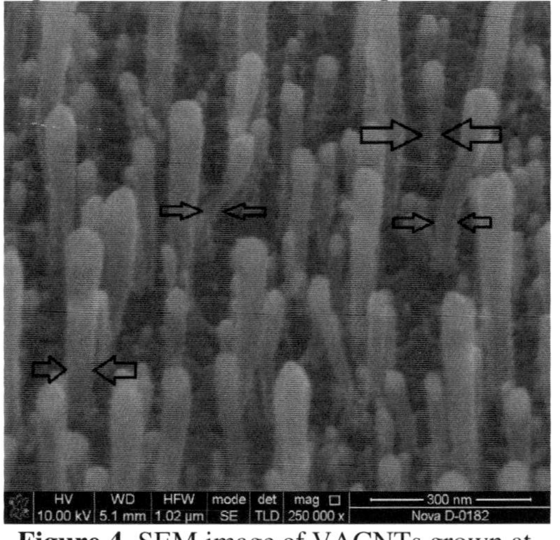

Figure 4. SEM image of VACNTs grown at t = 15 min.

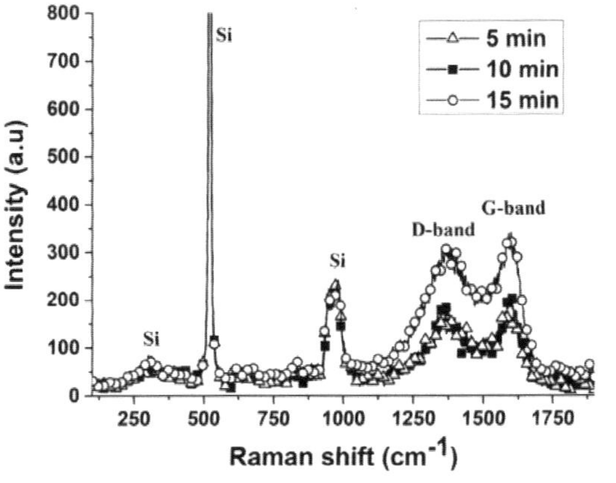

Figure 5. Raman spectra of the investigated samples measured at 514 nm.

Having analyzed the intensity ratio I_D/I_G of the D and G bands for different samples, we were able to estimate the impact of growth time on the defectiveness of VACNT. I_D/I_G ratios were 0.95, 0.86 and 0.91 for t = 5, 10 and 15 minutes, respectively. We found that the CNTs grown for 5 minutes contained the greatest number of defects (I_D/I_G = 0.95). This fact can be accounted for by the influence of catalytic center material which occupies most of the volume of the whole nanotube. The presence of catalytic centers at the tops of nanotubes might cause the asymmetric distortion of the nanotubes' layers and, therefore, led to an increase in D-mode intensity. Having increased growth time to 10 minutes, we observed an increase in the aspect ratio and the volume of the nanotube, which reduced the I_D/I_G ratio to 0.86. Further increase in growth time led to a subsequent increase in the value of I_D/I_G, which occurred due to the breakdown of carbon bonds as a result of the etching of nanotube walls in an ammonia plasma. SEM images of VACNTs grown at $t \geq$ 15 min (Fig. 5) confirmed the effect of plasma undercutting of CNT base.

Thus we found that the growth time has a significant impact on the geometry and defectiveness of VACNT, but it is by no means the only key parameter for the nanotube growth. Based on the results obtained, we can assume that in order to grow less defective VACNTs characterized by high aspect ratios, we need to reduce the ammonia plasma power and increase the pressure in the chamber to speed up the carbon gas transfer to catalytic centers.

We should note, however, that the structural imperfection of VACNTs caused by undercutting can be used to create emission structures based on graphenated multi-wall carbon nanotubes [18].
Besides, previous studies have shown [19] that multilayer VACNTs grown under similar conditions have a resistivity of $8.32 \pm 3.18 \cdot 10^{-4}$ Ohm·m, which indicates the metallic nature of their conductivity.

Acknowledgments
This work was funded by the Russian Foundation for Basic Research according to the research project No. 16-29-14023 ofi_m and Internal grant of the Southern Federal University No. VnGr-07/2017-26.

The results were obtained using the equipment of the Research and Education Center and Center of Common Using "Nanotechnologies" of the Southern Federal University.

References

[1] Li D, Cheng Y, Zhang H, Wang Y, Sun J and Dong M 2016 *Vacuum* **123** 69–75
[2] Li L and Wen D 2016 *Org. Electron.* **34** 12–7
[3] Ageev O A, Blinov Y F, Il'in O I, Konoplev B G, Rubashkina M V., Smirnov V A and Fedotov A A 2015 *Phys. Solid State* **57** 825–31
[4] Vollebregt S and Ishihara R 2016 *Carbon N. Y.* **96** 332–8
[5] Lekawa-Raus A, Patmore J, Kurzepa L, Bulmer J and Koziol K 2014 *Adv. Funct. Mater.* **24** 3661–82
[6] Bandaru P R 2006 *J. Nanosci. Nanotechnol.* **7** 1–29
[7] De Volder M F L, Tawfick S H, Baughman R H and Hart A J 2013 *Science (80-.).* **339**
[8] Ando Y, Zhao X, Sugai T and Kumar M 2004 *Mater. Today* **7** 22–9
[9] Saeidi M 2014 *J. Cryst. Growth* **404** 34–8
[10] Löffler R, Häffner M, Visanescu G, Weigand H, Wang X, Zhang D, Fleischer M, Meixner A J, Fortágh J and Kern D P 2011 *Carbon N. Y.* **49** 4197–203
[11] Kim C, Lee H-R and Kim H T 2016 *Mater. Chem. Phys.* **183** 315–9
[12] Shah K A and Tali B A 2016 *Mater. Sci. Semicond. Process.* **41** 67–82
[13] Cassell A M, Ye Q, Cruden B A, Li J, Sarrazin P C, Ng H T, Han J and Meyyappan M 2004 *Nanotechnology* **15** 9–15
[14] Gohier A, Minea T M, Point S, Mevellec J-Y, Jimenez J, Djouadi M A and Granier A 2009 *Diam. Relat. Mater.* **18** 61–5
[15] Niu Z and Fang Y 2006 *J. Cryst. Growth* **297** 228–33
[16] Saito R, Hofmann M, Dresselhaus G, Jorio A and Dresselhaus M S 2011 *Adv. Phys.* **60** 413–550
[17] Chen Y, Li C, Tu J and An L 2013 *Carbon N. Y.* **55** 375
[18] Gautier L-A, Le Borgne V and El Khakani M A 2016 *Carbon N. Y.* **98** 259–66
[19] Ageev O A, Il'in O I, Rubashkina M V., Smirnov V A, Fedotov A A and Tsukanova O G 2015 *Tech. Phys.* **60** 1044–50

IOP Conf. Series: Journal of Physics: Conf. Series **1038** (2018) 012063 doi:10.1088/1742-6596/1038/1/012063

MBE growth of thin AlGaAs nanowires with a complex structure on strongly mismatched SiC/Si(111) substrate

R R Reznik[1,2,4,5,7], I V Shtrom[1,4], I P Soshnikov[1,3,4], S A Kukushkin[6], D A Zeze[5,7] and G E Cirlin[1-5]

[1]St-Petersburg Academic University – Nanotechnology Research and Education Centre RAS, Khlopina 8/3, 194021, St-Petersburg, Russia
[2]Peter the Great St.Petersburg Polytechnic University, Polytechnicheskaya 29, 195251, St-Petersburg, Russia
[3]Ioffe Physical Technical Institute RAS, Politekhnicheskaya 26, 194021, St-Petersburg, Russia
[4]Institute for Analytical Instrumentation RAS, Rizhsky 26, 190103, St-Petersburg, Russia
[5]ITMO University, Kronverkskiy pr. 49, 197101 St-Petersburg, Russia
[6]Institute of Problems of Mechanical Engineering Russian Academy of Science, Bolshoj 6, 199178 St-Petersburg, Russia
[7]Department of Engineering, Durham University, Durham DH1 3LE, United Kingdom

E-mail: moment92@mail.ru

Abstract. The possibility of AlGaAs nanowires MBE growth on a silicon substrate with a nanometer silicon carbide buffer layer was demonstrated for the first time. Under the same experimental conditions (including the same composition), the diameter of the AlGaAs/SiC/Si nanowires are smaller than nanowires diameter grown on a silicon substrate. Based on the photoluminescence analysis suggests that when AlGaAs/SiC/Si NWs are grown, a physical complex structure appears due to the self-organized formation of regions with different molar fractions of aluminum in the solid solution.

1. Introduction

The relevance of investigations nanowires (NWs) is explained by the necessity to solve and important problem, i.e. to create new non-planar semiconductor nanomaterials and nanosystems with controlled properties on silicon (Si). We have previously demonstrated the ability to grow high-quality GaN and InN NWs on a silicon substrate covered with a thin silicon carbide (SiC) buffer layer [1-3]. These and A^3B^5 materials are particularly needed (and are already used) for the development of novel devices for microelectronics, optoelectronics, analytical biomedicine, emission cathodes, probes for scanning tunnelling microscopy, high efficiency solar energy converters, etc. [4]. Because of their large diameter which usually exceeds or is comparable to the de Broglie wavelength of a bulk material, NWs are not true one-dimensional, but quasi-one-dimensional nanomaterials. Hence, it is essential to reduce their diameter to fully exploit them as quantum materials. As shown in the literature [5,6], the critical diameter of the catalyst droplet, under which the NWs formation can be initiated, depends on the ratio of lattice constants between the NWs and substrate materials and decreases with an increasing ratio. Thus, it is expected that further mismatch increase of the lattice constant between the substrate and NWs materials will lead to a decrease of the NWs diameter.

In this work, in order to reduce the AlGaAs NWs diameter, hybrid Si(111) substrates with a nanometer SiC buffer layer were used for growth by molecular beam epitaxy (MBE). For these hybrid substrates, the lattice constant mismatch with these A^3B^5 compounds is significantly greater than that of a conventional silicon substrate, e.g. 44% for AlGaAs and 4% for Si and AlGaAs).

2. Experimental procedures

MBE growth was carried out using Riber Compact 21 MBE growth system equipped with an additional combined vacuum chamber for the deposition of Au (metallization chamber). SiC/Si(111) hybrid substrates were used for MBE growth. Growth was performed in two stages: after the cleaning of SiC/Si wafer surface at the temperature of 950°C, the sample temperature was lowered to 550°C before a ~0.1–0.2 nm thick gold (Au) film deposition onto the substrate surface in the metallization chamber. The temperature was maintained for a minute to allow Au droplets to be form. The sample was subsequently cooled to room temperature and then transferred to the growth chamber without breaking of ultrahigh-vacuum conditions. The substrate temperature in the growth chamber was increased to growth temperature - 510°C before Al, Ga, and As shutters were opened for AlGaAs NWs growth under As-stabilized growth conditions. The AlGaAs growth rate was maintained constant at 1 monolayer per second (ML/s) while Ga and Al flux were chosen in accordance with the required molar fraction x=0.3 in the $Al_xGa_{1-x}As$ solid solution. Total growth time was 20 min.

The surface morphology was studied by scanning electron microscopy (SEM). The photoluminescence (PL) was excited by a continuous-wave neodymium laser (wavelength 532 nm) at an excitation power density of about 10 W/cm^2. A Hamamatsu R928 photomultiplier served as a photodetector. The PL signal was excited and detected in the normal configuration. Measurements were made at a temperature of 10 K with nanowire-containing samples placed in a closed cycle helium cryostat.

3. Results

Figure 1 shows the SEM image of AlGaAs NWs array grown on hybrid SiC/Si(111) substrate. It seen that 20 minutes of growth is enough to synthesize 4 μm long AlGaAs NWs. The study of AlGaAs NWs array morphological properties has shown that the NWs formed in various directions, with average surface density - 6×10^7 cm^{-2}. Furthermore, the AlGaAs NWs have a conical shape, with a diameter of 80 nm at the bottom and 15 nm at the top of NWs due to the difference between the migration coefficients of Ga and Al adatoms, which is leading to an uneven lateral growth [7]. Nevertheless, the average diameter of AlGaAs NWs grown on the SiC/Si(111) substrate is substantially smaller than the diameter of the NWs grown on Si (100–120 nm [8]).

Figure 1. SEM image of AlGaAs NWs on hybrid SiC/Si(111) substrate.

A PL study of AlGaAs NWs arrays on the hybrid SiC/Si(111) substrate showed interesting results. The low-temperature PL spectra of these NWs are contrasted against that of AlGaAs NWs grown on Si(111) with the same growth conditions are shown in Figure 2. In line with previous reports [8], AlGaAs NWs with a sufficient Al content exhibit a core–shell structure when grown on a Si(111) substrate, where the content of Al being almost nominal in the shell and substantially lower in the core [7, 8]. It was found that for the AlGaAs grown of SiC/Si(111), the spectrum displays two main peaks at 1.6-eV and 1.81 eV. The peak at 1.6-eV corresponds to an aluminum content of 0.1 consistent published data [9]. In turn, 1.81 eV, which, to our knowledge, has not been reported to date could correspond to AlGaAs with another fraction of aluminum. The PL band at 1.6 eV is attributed to the emission from the rod whereas the higher energy band at 1.81 eV is associated to an emission from the shell. This band corresponds to an emission from AlGaAs with an Al content of about 0.22, which is in good correlation with our previous publication [8]. However, a fluctuation of the Al content can be observed in the shell of the AlGaAs NW. A possible reason why spatially inhomogeneous shells of the AlGaAs NW are formed, with its various parts having different compositions in the solid solution, was described in [10–12]. It was shown in these studies that, in the growth of AlGaAs, NW shells can be formed in a self-organized way as aluminum-enriched regions, alternating layers with different compositions and thicknesses of several angstroms, and AlGaAs quantum dots with an aluminum content lower than the average composition in the NWs. As a result of the quantum-confinement effect, the PL peak from these quantum dots is shifted to the region of 1.8–1.95 eV, which is in agreement with the energy of the additional PL peak observed in our AlGaAs NWs on SiC/Si substrate. In all of these cases, a broad (~120 meV) band is observed, probably constituted by a large number of discrete lines associated with an emission from single levels of quantum dot (0D) objects that are inhomogeneous in their composition and size. However, we could not explicitly discern separate discrete levels because of the insufficient resolution of the optical system used to support optical analysis. To investigate the hypothesis propose, i.e. the PL originates from 0D objects of the AlGaAs shell, we are currently studying the nanostructural properties of our AlGaAs/SiC/Si NWs by transmission electron microscopy (TEM).

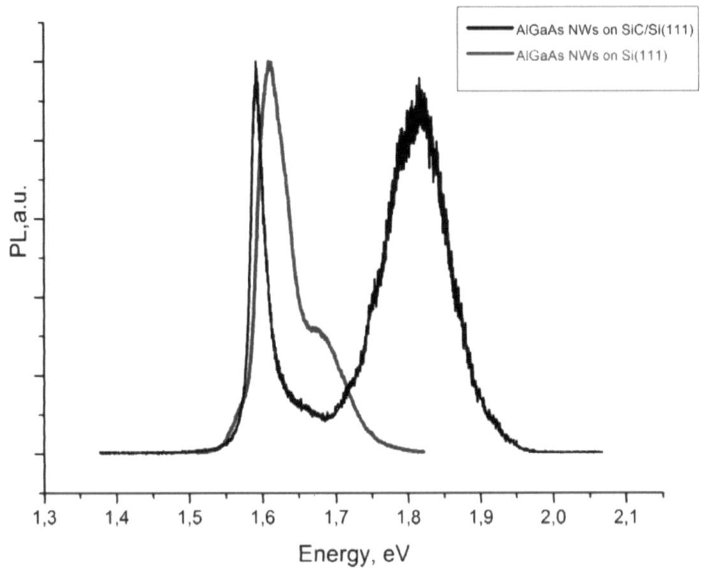

Figure 2. PL spectra measured at T = 10 K of AlGaAs NWs on hybrid SiC/Si(111) and Si(111) substrates.

4. Conclusions

In summary, the possibility of AlGaAs NWs growth by MBE on a silicon substrate with nanometer silicon carbide buffer layer was demonstrated for the first time. The morphology studies confirmed that an increase in the lattice mismatch between the substrate and the nanowires lead to decrease in the diameter of the NWs. Furthermore, these optical studies results allow us to suggest that, during AlGaAs NWs growth, a physical complex structure appears due to the self-organized formation of regions with different molar fractions of aluminum in the solid solution.

Acknowledgments

The nanowire samples were grown under the support of Russian Science Foundation (Project No 14-12-00393). Measurements of the NWs structural properties have received support from the European Union's Horizon 2020 research and innovation programme under the Marie Skłodowska-Curie grant agreement No 722176. The study of NWs optical properties were financially supported by the Ministry of Education and Science of the Russian Federation under State assignment no. 16.24834.2017/4.6).

References

[1] Reznik R R, Kotlyar K P, Ilkiv I V, Soshnikov I P, Kukushkin S A, Osipov A V, Nikitina E V and Cirlin G E MBE growth and optical properties of GaN nanowires on SiC/Si(111) hybrid substrate *AIP Conference Proceedings* **1748** 040003 (2016)

[2] Reznik R R, Kotlyar K P, Ilkiv I V, Soshnikov I P, Kukushkin S A, Osipov A V, Nikitina E V and Cirlin G E The use of SiC/Si(111) hybrid substrate for MBE growth of GaN nanowires *Journal of Physics: Conference Series* **741** 012027 (2016)

[3] Reznik R R, Kotlyar K P, Ilkiv I V, Soshnikov I P, Kukushkin S A, Osipov A V, Nikitina E V and Cirlin G E Growth and Optical Properties of Filamentary GaN Nanocrystals Grown on a

Hybrid SiC/Si(111) Substrate by Molecular Beam Epitaxy *Physics of the Solid State* **58** no10 1952–1955 (2016)

[4] Dubrovskii V G, Cirlin G E and Ustinov V M Semiconductor nanowhiskers: Synthesis, properties and applications *Semiconductors* **43** no 12 1539-1584 (2009)

[5] Chuang L C, Moewe M, Chase C, Kobayashi N P and Chang-Hasnain C Critical diameter for III-V nanowires grown on lattice-mismatched substrates *Appl. Phys. Lett.* **90** no 043115 (2007)

[6] Cirlin G E, Dubrovskii V G, Soshnikov I P, Sibirev N V, Samsonenko Yu B, Bouravleuv A D, Harmand J C and Glas F Critical diameters and temperature domains for MBE growth of III–V nanowires on lattice mismatched substrates *Phys. Status Solidi RRL* **3** no 4 112-114 (2009)

[7] Cirlin G E, Shtrom I V, Reznik R R, Samsonenko Yu B, Khrebtov A I, Bouravleuv A D and Soshnikov I P Hybrid AlGaAs/GaAs/AlGaAs nanowires with a quantum dot grown by molecular beam epitaxy on silicon *Semiconductors* **50** no 11 1421-1424 (2016)

[8] Dubrovskii V G, Shtrom I V, Reznik R R, Samsonenko Yu B, Khrebtov A I, Soshnikov I P, Rouvimov S, Akopian N, Kasama T and Cirlin G E Origin of Spontaneous Core–Shell AlGaAs Nanowires Grown by Molecular Beam Epitaxy *Cryst. Growth and Des.* **16** no 12 7251- 7255 (2016)

[9] Pavesi L and Guui M Photoluminescence of AlxGa1−xAs alloys *J. Appl. Phys.* **75** no 10 4779-4842 (1994)

[10] Heiss M, Fontana Y, Gustafsson A, Wüst G, Magen C, O'Regan D D, Luo J W, Ketterer B, Conesa-Boj, A. V. Kuhlmann, J. Houel, E. Russo-Averchi, J. R. Morante, M. Cantoni, N. Marzari, J. Arbiol S, Zunger A, Warburton R J and Fontcuberta i Morral A Self-assembled quantum dots in a nanowire system for quantum photonics *Nat. Mater.* **12** no 5 439-444 (2013)

[11] Rudolph D, Funk S, Döblinger M, Morkötter S, Hertenberger S, Schweickert L, Becker J, Matich S, Bichler M, Spirkoska D, Zardo I, Finley J J, Abstreiter G and Koblmüller G Spontaneous Alloy Composition Ordering in GaAs-AlGaAs Core–Shell Nanowires *Nano Lett.* **13** no4 1522-1527 (2013)

[12] Jeon N, Loitsch B, Morkoetter S, Abstreiter G, Finley J, Krenner H J, Koblmueller G and Lauhon L J Alloy Fluctuations Act as Quantum Dot-like Emitters in GaAs-AlGaAs Core–Shell Nanowires *ACS Nano* **9** no 8 8335-8343 (2015)

IOP Conf. Series: Journal of Physics: Conf. Series **1038** (2018) 012064 doi:10.1088/1742-6596/1038/1/012064

The characterisation of nanostructured porous silicon/silver layers via Raman spectroscopy

R S Smerdov[1], Yu M Spivak[2], V S Levitsky[3], V A Moshnikov[2]
[1]Department of General and Technical Physics, Saint Petersburg Mining University, Saint Petersburg 199106, Russia
[2]Department of Micro- and Nanoelectronics, Saint Petersburg Electrotechnical University "LETI", Saint Petersburg 197376, Russia
[3]R&D Center for Thin-Film Technologies in Energetics under Ioffe Institute

Abstract. The investigation of porous silicon/silver nanostructured composite layers (PS/Ag) via Raman spectroscopy technique has been carried out. Interpretation of PS/Ag Raman spectra has been proposed. Data from films acquired by varying anodization time showed a significant shift of a Raman peak primarily located at the frequency expected for monocrystalline silicon towards lower energy. In addition to the single peak a number of low-frequency bands emerged in the Raman spectra of silver-functionalised porous silicon. These effects have been interpreted as due to the influence of functionalizing silver nanoparticles and synthesis parameters on the structure and properties of composite material.

1. Introduction

Porous silicon (PS) and PS-based composite nanostructures are of great interest due to unique properties of such materials including photoluminescence, electroluminescence and the dependence of refraction index on porosity. It is well established that porous silicon is a promising material for a number of applications associated with micro- , nano- and optoelectronics, targeted drug delivery, and gas sensing [1] by virtue of its high specific surface area and extensive capability for surface functionalization. Composite porous silicon – silver materials can also act as a promising substrate for identification and structural characterization of various biological and chemical structures due to the presence of pronounced surface enhanced Raman scattering effect (SERS) [3] caused by collective oscillations in silver nanoparticle electron density. It should be noted that silver-functionalized porous silicon layers can be used in order to implement optical filters for fiber-optic communication systems [2], since the contemporary thin-film filters are expensive due to the necessity of ultra-high vacuum maintaining during the synthesis process.

2. Experimental details

PS layers were obtained with anodization of monocrystalline (111) n-type silicon wafers with resistivity of 4.5 Ω·cm. Pre-treatment of the wafers was performed with volatile liquids (acetone) and ultrasonication for 15 minutes. The anodization process was conducted in an electrochemical cell filled with HF-Isopropyl alcohol-Distilled water solution. Anodization time varied in the range of 0.5 to 40 min. The deposition of silver was carried out from an aqueous solution of silver nitrate in a three-electrode electrochemical cell. The silver chloride electrode was used as the reference electrode. Porous silicon acted as a working electrode. The deposition of silver on the surface of porous silicon samples was performed by applying a potential of 700 mV for one minute.

3. Results and discussion

Raman spectroscopy technique has been extensively used in order to characterize both crystalline and amorphous silicon and silicon based nanostructured materials [4]. Raman shift is equal to 521 cm^{-1} and 480 cm^{-1} (transverse optical TO – mode) for crystalline (c-Si) and amorphous (a-Si) samples respectively [3]. Thus, if the structure of a particular sample resembles c-Si and a-Si then the corresponding Raman spectra should be close to those of crystalline and amorphous silicon.

Representative Raman spectra of composite PS/Ag layers depending on anodization time are shown in figure 1(a). Initial porous silicon Raman spectrum is depicted as a reference. It should be noted that the Raman shift of the original sample is 521 cm^{-1}which is indicative of its monocrystalline structure. It is evident than an increase in anodization time (from 0.5 to 40 min) leads to a shift in position of a Raman peak initially located at 521 cm^{-1} towards lower energies. In this context according to the confinement model [5] it is assumed that this effect is related to the vibrational confinement caused by the formation of silicon nanocrystallites in the composite material structure due to the deposition of silver nanoparticles. The respective sizes of nanocrystalline silicon clusters (L) corresponding to a certain Raman shift could be estimated by using equation (1)

$$\Delta\omega = -A\left(\frac{a}{L}\right)^{\gamma}, \tag{1}$$

where $\Delta\omega$ is the frequency down shift of the sample referring to the c-Si line, $a = 5.431$ Å is the lattice constant of Si, $A = 47.41$ cm^{-1} and $\gamma = 1.44$ are the parameters describing vibrational confinement for spherical Si nanocrystallites. Calculated sizes of Si clusters are summarized in table 1.

Table 1. The results of calculations of Si nanocrystallite sizes.

Silicon matrix anodization time, min	Frequency downshift, cm^{-1}	Value of nanocrystal size calculated according to (1), **Å**
0.5	1.4	62.7
1	1.4	62.7
10	2.3	44.4
20	2.3	44.4
40	3.3	34.6

According to equation (1) an increase in anodization time accompanied by the increase in Raman peak frequency downshift (with respect to the c-Si line) leads to a decrease in silicon nanocrystal sizes. It should be noted that such modification of porous silicon crystalline structure characterized with a downshift of the 521 cm^{-1} Raman peak is presumably related to an effect of stretching as a result of the deposition of silver nanoparticles inside the pores in the silicon surface.

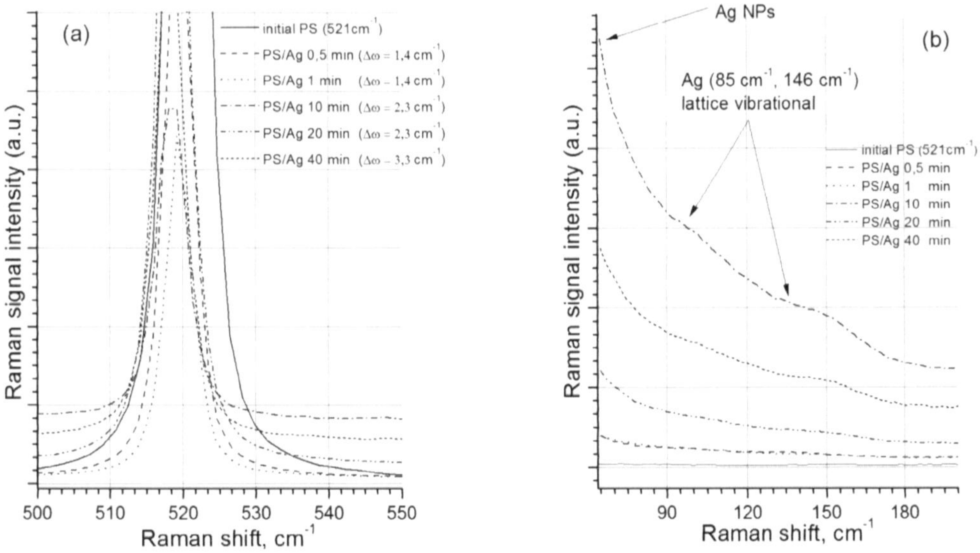

Figure 1(a, b). (a) PS/Ag Raman spectra (500-550 cm^{-1}); **(b)** PS/Ag Raman spectra (60-200 cm^{-1}).

Figure 1(b) shows Raman spectra of PS/Ag layers in the range of 60 to 200 cm^{-1}. According to [6] the Raman bands at 85 cm^{-1} and 146 cm^{-1} have been attributed to silver lattice vibrations. Duval et al. [7] states that Raman spectra of both metal nanoparticles and composite materials based on those contain a low-frequency peak the nature of which is related to the mechanical oscillations of nanoparticles under the influence of incident optical radiation. In order to estimate the dimensions of silver nanoparticles deposited on the PS surface we used Lamb theory [7] for the vibration modes of spherical particles. This particular model accounts for the existence of spheroidal and torsional vibration modes. According to Duval only the lowest-energy modes appear to be active and the frequencies of the lowest-energy spheroidal and torsional modes are virtually equal. This fact is related to the presence of strong coupling between surface vibrational and electronic modes whose energy is close to that of incident photons. Herein by using Lamb theory it is possible to derive an equation for the frequency of the lowest-energy spheroidal mode as (2)

$$\omega_2 = 0.85\frac{v_t}{dc}, \tag{2}$$

where v_t is the transverse sound velocity ($v_t = 1660$ m/s, a value, averaged on the different directions in the crystal [7]), d is the diameter of the particle, c is the speed of light in vacuum, the ω_2 is the observed peak position. Equation (2) leads to the diameter values of Ag nanoparticles deposited on PS surface of approximately 0.7 nm.

It is worth noting that anodization time of 10 minutes leads to the formation of an optimal morphology of silicon surface thus providing a significant increase in Raman signal intensity (see figure 1(a,b)) probably related to the surface enhanced Raman scattering (SERS) effect [3].

4. Conclusion

The porous silicon matrix structure prior to deposition of functionalizing silver particles is similar to that of monocrystalline silicon. The deposition of silver nanoparticles results in the deformation of silicon lattice, accompanied by the appearance of bands being uncharacteristic for the unmodified porous silicon in the Raman spectra of investigated samples. Also the process of silver deposition on the PS matrix surface results in the formation of silicon nanocrystals, accompanied by a shift of the peak corresponding to monocrystalline silicon (521 cm^{-1}) towards lower energy. The calculation of the characteristic dimensions of Si nanocrystals is conducted in accordance with the confinement model [5], depending on anodization time of a silicon matrix. The nature of low-energy Raman bands has been associated with the formation of silver nanoparticles (0.7 nm in diameter) inside the pores on the surface of a silicon matrix. Thus the increase in anodization time of PS leads to a rise in the thickness of the porous layer as well as in the pore diameter.

5. Acknowledgments

The technological part of this study was carried out at the Department of Micro- and Nanoelectronics, Saint Petersburg Electrotechnical University "LETI".

References

[1] Moshnikov V.A., Gracheva I.E., Lenshin A.S., Spivak Y.M., Anchkova M.G., Kuznetsov V.V., Olchowik J.M. 2012 *J. of Non-Crystalline Solids* **358** 590
[2] Smerdov R.S., Loboda V.V., Spivak Yu.M., Moshnikov V.A. 2016 *St. Petersburg State Polytechnical University Journal Computer Science. Telecommunications and Control Systems* **3(247)** 12
[3] Mock J. J., Barbic M., Smith D. R., Schultz D. A., Schultz S. 2002 *J. Chem. Phys.* **116** 6755
[4] Feng Z.C., Tsu R. 1994 *Porous Silicon* (World Scientific Publishing)
[5] Zi J., Büscher H., Falter C., Ludwig W., Zhang K. et al. 1996 *Appl. Phys. Lett.* **69** 200
[6] Martina I., Wiesinger R., et al. 2012 *e-PS* **9** 1
[7] Duval E., Boukenter A., and Champagnon B. 1986 *Phys. Rev. Lett.* **56** 2052

IOP Conf. Series: Journal of Physics: Conf. Series **1038** (2018) 012065 doi:10.1088/1742-6596/1038/1/012065

Choice of technological conditions for synthesis of sensing materials based on polyacrylonitrile on flexible substrates

T.V. Semenistaya, A.V. Ivanenko

Southern Federal University, Engineering-technological Academy, Institute of Nanotechnologies, Electronics, and Equipment Engineering, Str. Shevchenko, 2, Taganrog, Russia, 347922

semenistayatv@sfedu.ru

Abstract. Cd-containing PAN films with modifying additive $CdCl_2$ have been fabricated by pyrolysis under the influence of incoherent infrared radiation under low vacuum conditions using different temperature and time of heat treatment. The influence of modifying additive concentration and of technological conditions on the films electrical and gas sensitive properties have been investigated. AFM-investigation of the films surface morphology has been fulfilled. The technological conditions for the formation process of the sensitive layer material for hydrogen sulphide sensor have been revealed experimentally.

1. Introduction

Currently, portable electronic devices such as mobile phones, computers, digital cameras, various sensors, are becoming more and more multifunctional and tend to reduce size, weight, the ability to flex and twist to meet the demands of today's market. The development of technologies related to the creation of portable power sources and energy storage is still at a rather low level. Technologies for creating gas sensors on flexible media are actively developing all over the world.

Flexible portable sensors that detect dangerous gases (hydrogen, nitrogen dioxide, hydrogen sulphide), as well as high levels of ultraviolet radiation are in demand. These electronic devices can be attached to the body in the form of patches or sewed into clothing and special equipment [1]. The use of such sensors will make it possible to more effectively monitor toxic and explosive gases in transport, industrial plants and thermal power plants, inaccessible places (mines, caves), thereby reducing risks for workers and the public [2-5].

Polyacrylonitrile (PAN) is a promising material for creating such flexible sensor devices. The technology of fabrication gas sensitive PAN films that does not require high-tech equipment and the ability to exhibit gas-sensing properties at room temperatures opens the prospect of creating flexible sensors. The gas-sensitive metal-containing PAN thin films have been fabricated [6] that are sensors for detecting to CO, NO_2, Cl_2, NH_3, H_2S [7-10].

2. Experiment

To form cadmium-containing PAN films pyrolysis method under the influence of IR radiation under low vacuum conditions was used. Electrically conductive films were obtained from film-forming solutions of PAN with a modifying additive content of cadmium chloride ($CdCl_2$) from 0.0% to 1.0 % in dimethylformamide. The prepared film-forming solution was applied to the substrates by

IOP Conf. Series: Journal of Physics: Conf. Series **1038** (2018) 012065 doi:10.1088/1742-6596/1038/1/012065

centrifugation. Then, the samples were dried in oven at T = 160 ° C for 30 minutes. The temperature-time IR-annealing regimes were selected experimentally, because the intensity and duration of exposure to IR radiation makes it possible to control the properties of the film material, changing the molecular structure of the polymer [6]. Halogen lamps KG-220 with high radiation in the area of 0.8 μm – 1.2 μm were used as the source of radiation. Uniform heating of the samples was supplied with up and down position of the halogen lamps along the longitudinal perimeter outside the quartz reactor. The sample in the graphite cassette was fixed on special quartz holders inside the reactor. The thermoelectric couple chromium-aluminium inside the graphite cassette was used to control the intensity of IR-radiation. The intensity of the radiation in the first stage of IR annealing corresponded to temperatures of 250 °C, 300 °C for 20 min, and the intensity of the radiation in the second stage of IR annealing varied 350 °C, 400 °C, 450 °C for 5 min. Electrically conductive properties of PAN are manifested as a result of thermal treatment of the polymer. Further, the films were cooled down for 40-60 minutes.

Thickness measurement of the films was carried out using MII-4 interferometer.

The film surface morphology was observed by atomic force microscopy (AFM) using scanning probe microscope Solver P47 Pro (NT-MDT). The 5 × 5 μm^2 area AFM-images of the films are processed with Image Analysis program.

Resistance measurement was carried out using E6-13A Terraommetr. Sensitivity is a change of measured resistance per analyte concentration unit. The gas sensitivity coefficient (S) was defined as the ratio $S = (R_o - R_g)/R_o$, when $R_o > R_g$, where R_o is the films resistance in air, R_g is the films resistance in the atmosphere of analyte.

3. Results and discussion

Influence of cadmium compounds concentration in the organic matrix of the film is obvious: the film thickness increases with the concentration of the modifying additive (Fig. 1).

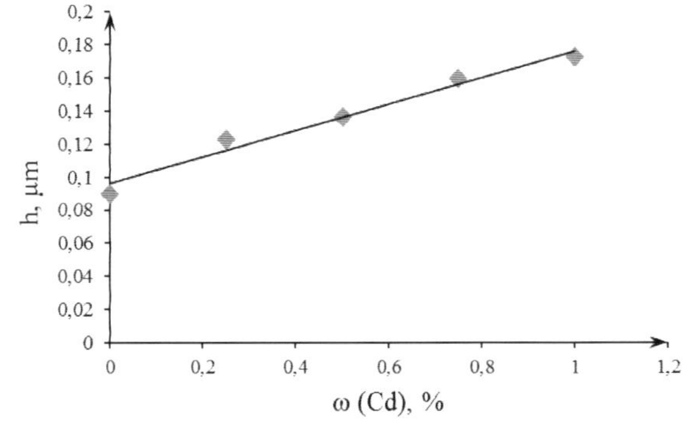

Figure 1. Thickness of the Cd-containing PAN films vs. weight concentration of modifying additive

As observed from Fig. 2 the Cd-containing PAN films have a dense homogeneous structure and developed surface morphology.

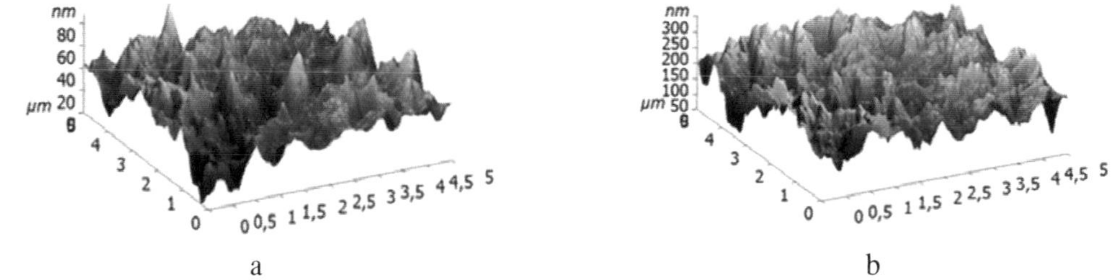

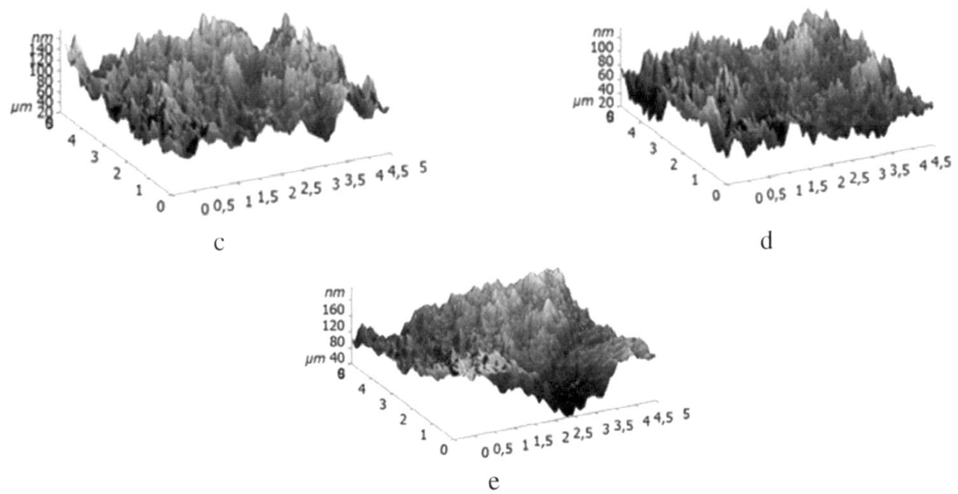

Figure 2. AFM-images of surface of Cd-containing PAN films, weight concentration of the modifying additive (ω, %): a) 0.0, b) 0.25, c) 0.5, d) 0.75, e) 1.0

For the Cd-containing PAN film (ω(Cd) = 0.25%) the highest peak-to-peak value of 300.5 nm and the highest average roughness value of 31,9 nm were obtained, corresponding to such film surface morphology, which creates the necessary energy background for the adsorption of the detected gas.

The measured values of resistance of the obtained samples are in the range 10^8-10^{10} Ω. Fig. 3 depicts there is reduction in resistance of cadmium-containing samples with concentration of the modifying additive.

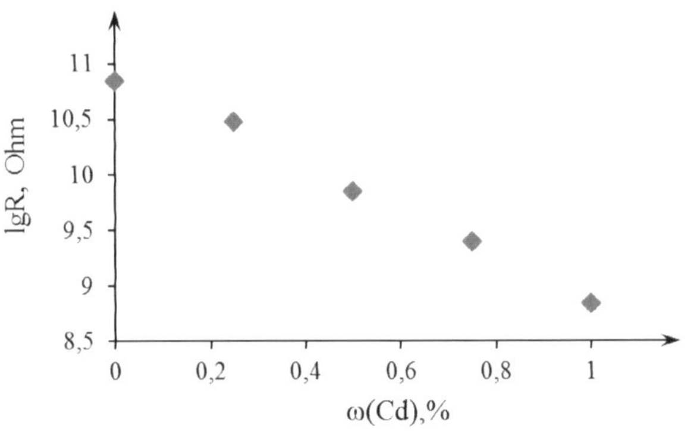

Figure 3. Dependence of the Cd-containing PAN films resistance on weight concentration of modifying additive

The presence of transition metal compounds affects the electrical conductivity of the samples. The films become more electrically conductive with an increase in the number of charge carriers.

Temperature dependences of the resistance of cadmium-containing PAN films in the range from 18 °C to 100 °C were studied. A typical relation between the resistance and temperature of the fabricated films is shown in Fig. 4. The value of the determination coefficient R^2 is close to 1, which means that the approximation curve best describes the experimental dependence of the Cd-containing PAN film resistance on temperature. The experimental curve gives the approximating function y = f (x), where x corresponds to the heating temperature of the film.

To study the dependence of the gas sensitivity coefficient of the films on the concentration of the detected gas, the films with the best characteristics were selected. Investigating the sensor properties of the films, we tested the samples with different contents of the modifying additive (Fig. 5). As it can be seen, samples with a cadmium concentration of 0.25% turned out to be the best. This can be explained by the fact that the ratio of the amount of the modifying additive and the organic matrix makes it possible to form the structure of the composite material and the morphology of the film surface which meets the conditions for the best sorption of the detected gas. To determine the range of concentrations of the gas to be detected in which the film material is sensitive, the measurements shown in Fig. 6 were done.

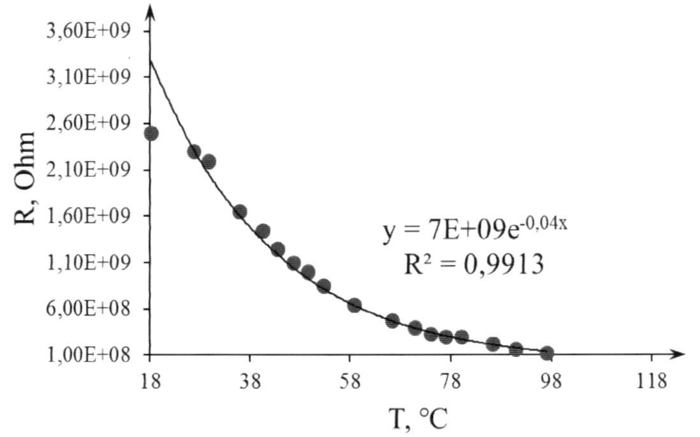

Figure 4. Cd-containing PAN films resistance dependence on temperature (weight concentration of the modifying additive ω(Cd) = 0.5%)

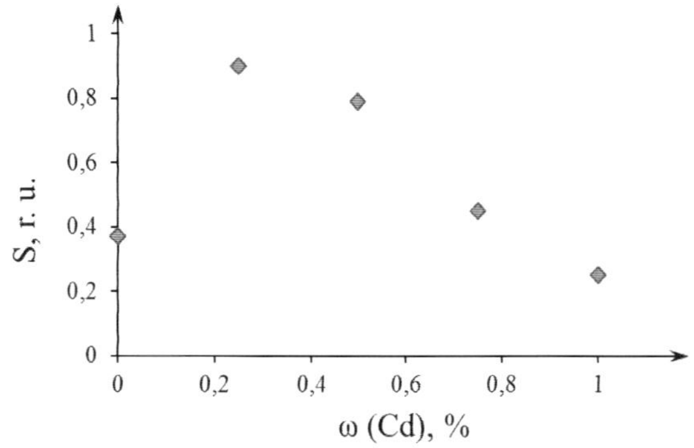

Figure 5. Sensitivity of the Cd-containing PAN films vs. weight concentration of modifying additive

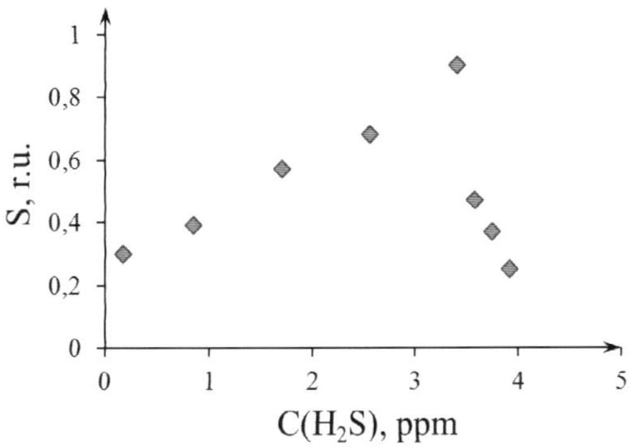

Figure 6. Sensitivity of the Cd-containing PAN films vs. concentration of hydrogen sulfide

Temperature dependences of gas sensitivity coefficient of the obtained samples showed that the maximum values of the gas sensitivity coefficient are reached at room temperature.

Polyacrylonitrile is the prospective material as gas-sensing layer of sensors as it offers low cost and relative simple structure. The design of environmentally-sensitive products involves selecting appropriate materials, technology and manufacturing equipment. Advantage of PAN-based material is the possibility of obtaining films with functional properties and there is no need to use complex technological equipment.

4. Conclusion

Thin cadmium-containing PAN films (h = 0.09 μm – 0.17 μm) with a resistance of $7\cdot10^8$ to $3\cdot10^{10}$ Ω were obtained by pyrolysis method. It has been found that the temperature dependence of the resistance of the films has an exponential character.

It is known that doping gas sensitive materials with cadmium compounds allows achieve sensor sensitivity to H_2S [11], formaldehyde (CH_2O) [12], ethanol (C_2H_5OH) [13], trimethylamine $(CH_3)_3N$ [14], liquefied petroleum gas (LPG), [15]. H_2S sensor based on polyaniline–CdS nanocomposites that demonstrated maximum response for 100 ppm at room temperature (300 K) has been fabricated [16]. It has been created [17] the 3.0 wt% Cd-doped SnO_2 based sensor that showed selectivity toward H_2S at 275 °C, giving a response of about 31–10 ppm.

In our work it has been found that the cadmium-containing PAN films are sensitive toward H_2S at room temperature, which is not worse than the sensors presented in [11-17]. The best gas sensitivity at concentrations of the detected gas of 0.17 -3.9 ppm showed a sample of 0.25% Cd. Potential applications of the Cd-containing PAN films include environmental monitoring of gas emissions as non-heated semiconductor sensor.

Acknowledgments
This work was performed with the financial support of the Ministry of Education and Science of the Russian Federation, agreement No. 14.575.21.0126 (identification number of the project RFMEFI57517X0126). The equipment of the Research and Educational Center of "Nanotechnologies" and of the Research and Educational Center "Microsystem Technology and Multisensory Monitoring Systems" of Southern Federal University was used for this study.

References

[1] Honda W, Harada S, Arie T, Akita S and Takei K 2014 Adv Funct Mater 24 3299–3304
[2] Kenry Y J C and Lim C T 2016 Microsys Nanoeng 2 16043
[3] Kahn N, Lavie O, Paz M, Segev Y and Haick H 2015 Nano Lett 15 7023–7028
[4] Gutruf P, Walia S, Sriram S and Bhaskaran M 2015 Adv Electron Mater 1 1–6
[5] Wang C 2016 Flex Electron Mater Dev 1–44
[6] Semenistaya T V 2016 Springer Proc Phys 175 61–77
[7] Semenistaya T V, Plugotarenko N K and Petrov V V 2015 Appl Mech Mater 727–728 145–149
[8] Yalovega G and Semenistaya T 2016 Solid State Phenom 257 175–178
[9] Semenistaya T V, Petrov V V, Bednaya T A and Zaruba O A 2015 Mater Today Proc 2 77–84
[10] Semenistaya T V, Petrov V V and Lu P 2013 Adv Mater Res 804 135–140
[11] Merdrignac-Conanec O, Bernicot Y, Guyader J 2000 Sens. Actuators, B 63 86-90
[12] Zeng W, Liu T, Wang Z, Tsukimoto S, Saito M, Ikuhara Y 2009 Sensors 9 9029-9038
[13] Sivalingam D, Gopalakrishnan J B, Rayappan J B B 2012 Mater. Lett. 77 117-120
[14] Chua X, Zhoua S, Donga Y, Suna W, Ge X 2011 Mater. Chem. Phys. 131 27-31
[15] Bulakhe R N, Lokhande C D 2014 Sens. Actuators, B 200 245-250
[16] Raut B T, Godse P R, Pawar S G, Chougule M A, Bandgar D K, Patil V B 2012 Meas 45 94-100
[17] Sun P, Zhou X, Wang Ch, Wang B, Xua X, Lu G 2013 Sens. Actuators, B 190 32–39

Magnetostriction in $Fe_{80-x}Co_xP_{14}B_6$ amorphous ribbons evaluated by Becker-Kersten method

V S Severikov[1], A M Grishin[1,2,3], V S Ignakhin[1]

[1] Petrozavodsk State University, 185910 Petrozavodsk, Karelian Republic, Russia
[2] KTH Royal Institute of Technology, SE-164 40 Stockholm-Kista,Sweden
[3] INMATECH Intelligent Materials Technology, SE-127 51 Skärholmen, Sweden

E-mail: severva3@gmail.com

Abstract. Becker-Kersten method, which involves observing hysteresis *M-H* loops under mechanical stress, was applied to measure magnetostriction properties in amorphous rapid quenched ribbons $Fe_{80-x}Co_xP_{14}B_6$. It is shown that magnetostriction constant increases with the growth of cobalt atomic content from $(1.75 \pm 0.13)\times10^{-6}$ for $x = 25$ to $(1.60 \pm 0.05)\times10^{-5}$ for $x = 40$.

1. Introduction

Due to excellent magnetosoft characteristics $Fe_{80-x}Co_xP_{14}B_6$ metallic glasses are attractive for various commercial applications such as electricity transformers, magnetic field concentrators for sensor applications, magnetic screens etc. [1-3]. Strongly pronounced magnetoelastic effect in $Fe_{80-x}Co_xP_{14}B_6$ amorphous ribbons enables design and fabrication of multicomponent strain gauges [4]. Magnetic properties, thermal stability and crystallization kinetics of $Fe_{80-x}Co_xP_{14}B_6$ metallic glasses were thoroughly investigated [1-3, 5, 6]. For example, the maximum relative differential permeability in as quenched ribbons is about 110000, the saturation induction B_s = 1.45 T, quasistatic coercive field as low as 4 A/m, Curie temperature above 700 K, low hysteresis loss of about 0.26 W/kg in the saturation mode at frequency of 100 Hz, and significantly higher thermal stability due to increased by 60 K crystallization temperatures compared to FeNi-glasses [1-3].

Nevertheless, magnetostriction phenomenon in $Fe_{80-x}Co_xP_{14}B_6$ metallic glasses has not been studied in details and magnetostriction constants have not been measured. The aim of this article is to report experimental results on magnetostriction constants determined by Becker-Kersten method in amorphous rapidly quenched $Fe_{80-x}Co_xP_{14}B_6$ ribbons with various Fe-to-Co ratio compositions.

2. Experimental

The method of Becker and Kersten [7] was demonstrated to be useful to perform reasonably accurate measurements and reliably quantify very small magnetostriction effect specific for amorphous magnetosoft metallic glasses [8].

At first, we consider an ideal elastic and magnetically isotropic ferromagnetic amorphous thin ribbon with no internal stress. As the typical dimensions of the ribbons used in present experiments were 300 mm long, 1 mm wide and 30 μm thick, when a tensile stress $\sigma > 0$ and magnetic field H applied parallel to the ribbon axis the uniaxial magnetoelastic anisotropy energy E_{me} is induced along the ribbon length [9]:

$$E_{me} = \tfrac{3}{2}\lambda\,\sigma\,\sin^2\Theta - M_s H\cos\Theta \ . \qquad (1)$$

Here λ is the magnetostriction constant, M_s is the saturation magnetization, and θ is the angle between the ribbon axis and magnetization. Minimizing U_{me} with respect to θ we obtain

$$\cos\Theta = M_s H / 3\lambda\sigma \ ,$$

and equilibrium magnetization M along magnetic field H and ribbon length appeared to be equal

$$M = M_s \cos\Theta = {M_s}^2 H / 3\lambda\sigma \ . \qquad (2)$$

The observed magnetization M increases with the growth of magnetic field H and reaches the maximum value M_s when cosθ = 1 at magnetic field

$$H = H_k(\sigma) = 3\lambda\sigma / M_s \qquad (3)$$

called field of magnetic anisotropy.

In practice, as quenched ribbons possess also some cast-induced uniaxial anisotropy independent upon σ. It has a perpendicular to the ribbon length character that is confirmed by some finite strength of magnetic field required to apply at "easy" direction to saturate the unloaded ribbon. Presumably finite cooling rate and insufficient thermal contact between quenched ribbon and rotating cooper wheel can be the cause of inherited transversal stresses. As a result, the observed magnetization M can reach the maximum available value M_s when magnetic field H overcomes the effective field of magnetic anisotropy $\widetilde{H}_k = H_{k0} - H_k(\sigma)$. Here H_{k0} notifies specific field of uniaxial magnetic anisotropy in unloaded ribbons. Fig. 1 depicts schematically idealized M-H dependences in two cases: negative and positive magnetostriction.

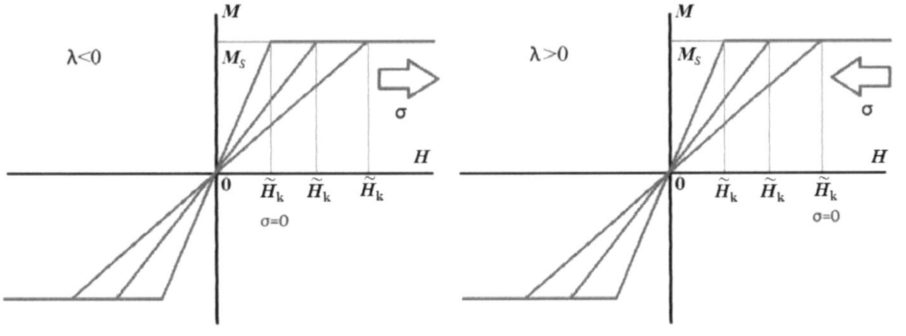

Figure 1. Illustration how to determine the effective anisotropy field $\widetilde{H}_k$ for samples with negative and positive magnetostriction at applied tensile load. The arrows show the direction of curve transformation with increasing tensile stress σ.

Becker and Kersten method is based on the registration of magnetic hysteresis loops M vs. H under various mechanical stresses σ and consequent determination the dependence $\widetilde{H}_k$ vs. σ from the series of loops. To utilize Becker-Kersten method straitghforwardly we should replace a real hysteresis loop with the "averaged" curve by collapsing the ascending and descending hysteretic M-H branches to each other. Also, the kink on the magnetization curve corresponding to the anisotropy field $\widetilde{H}_k$ is often not so sharp as it is shown in Fig. 1 since hysteretic curve is smooth and rounded. For this reason it is conventionally accepted to define $\widetilde{H}_k(\sigma)$ at the level M = 0.9 M_s [8-9]. From Eq. (3) it follows that magnetostriction constant λ can be calculated from the slope (derivative) of $\widetilde{H}_k$ vs. σ dependence as follows:

$$\lambda = -\tfrac{1}{3} M_s \frac{d\widetilde{H}_k}{d\sigma} \ . \qquad (4)$$

The experimental setup is shown in Fig. 2. The amorphous ribbon was passed through a coil of the magnetometer 1, was twisted and tensioned by means of two stepping motors 3.1 and 3.2. Stepper motor 3.1 sets the ribbon tension by moving the slider with a worm gear. Stepper motor 3.2, mounted on the movable carriage, creates a torsion strain. Controller 4 governs both stepper motors. Linear mechanical tension of ribbon is controlled by a beam load cell 2 at the rated load of 3 kgf.

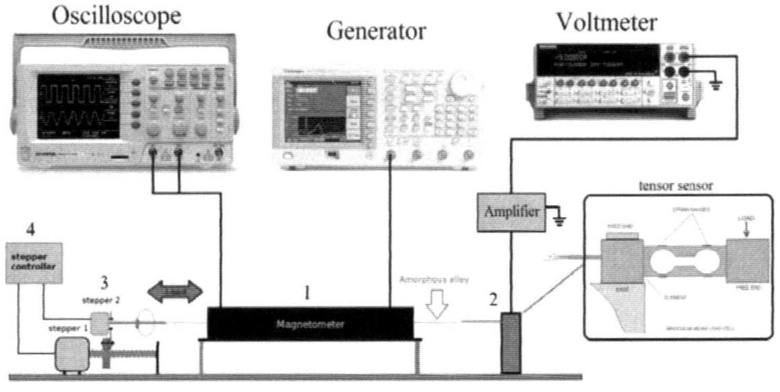

Figure 2. The experimental setup.

3. Results and discussion

In Fig. 3 the "averaged" magnetization curves for amorphous ribbons $Fe_{40}Ni_{40}P_{14}B_6$ and $Fe_{40}Co_{40}P_{14}B_6$ are presented. The commercial composition $Fe_{40}Ni_{40}P_{14}B_6$ (Metglas2826) ribbon was used as a reference sample to compare the calculated magnetostriction constant value with a literature data (e.g. measured by SAMR method [10]). As clearly seen from the Fig. 3, both samples demonstrate positive magnetostriction ($\widetilde{H}_k$ decreases with a tensile load growth).

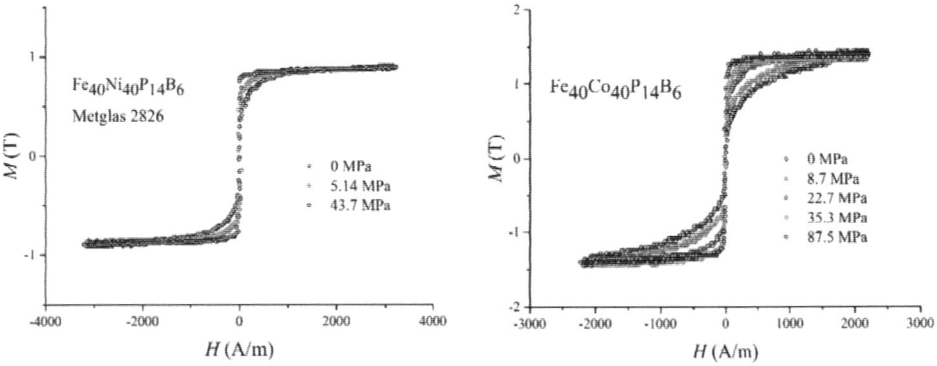

Figure 3. Series of hysteresis loops under various mechanical stresses in as quenched $Fe_{40}Ni_{40}P_{14}B_6$ and $Fe_{40}Co_{40}P_{14}B_6$ amorphous ribbons.

The dependences $\widetilde{H}_k(\sigma)$ for $Fe_{40}Ni_{40}P_{14}B_6$ and $Fe_{40}Co_{40}P_{14}B_6$ amorphous ribbons are shown in Fig. 4. For both compositions the anisotropy field $\widetilde{H}_k$ drops with increasing mechanical stress that is typical for materials with a positive magnetostriction constant. At low applied loads, both FeNi- and FeCo-glass follow linear $\widetilde{H}_k$ vs. σ dependence. At $\sigma = 0$ it starts from the cast-induced field H_{k0} equals to 785 A/m and 1215 A/m in $Fe_{40}Ni_{40}P_{14}B_6$ (Metglas2826) and $Fe_{40}Co_{40}P_{14}B_6$ metallic glasses, correspondingly. Magnetostriction constant λ was calculated using Eq. (4) from the slope of initial region of $\widetilde{H}_k$ vs. σ curve. These results are summarized in Table I. Taken as an etalon amorphous as-quenched $Fe_{40}Ni_{40}P_{14}B_6$ (Metglas2826) ribbon shows $\lambda = (1.60 \pm 0.15)\times10^{-5}$ that is satisfyingly close to the tabulated value [10]. Also, in Table I we present the calculated values of anisotropy energy $K = \frac{3}{2}\lambda\sigma_S$ where σ_S equals, respectively, to 16 MPa and 36 MPa for FeNi- and FeCo-glass. Tensile stress σ_S values define the onset of saturation $\widetilde{H}_k(\sigma)$ regime.

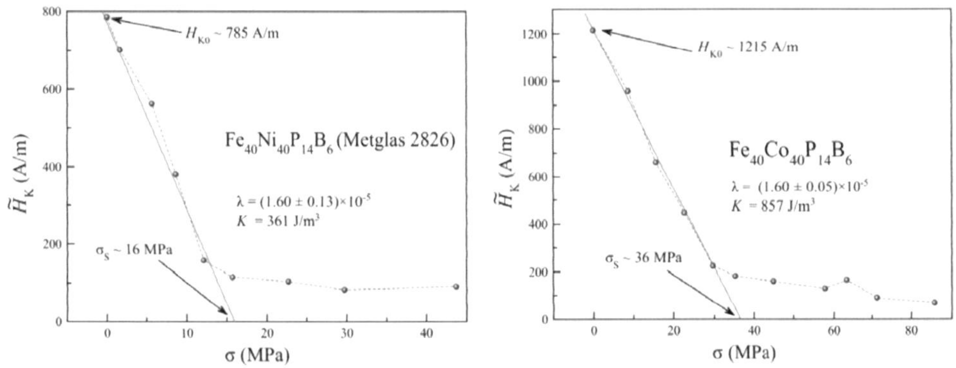

Figure 4. Dependence of effective field of magnetic anisotropy $\widetilde{H}_k$ on the mechanical stress σ applied to as-quenched amorphous ribbons.

Table I. Magnetostriction constant λ and magnetoelastic anisotropy energy $K = \frac{3}{2}\lambda\sigma_S$ in $Fe_{80-x}Co_xP_{14}B_6$ and $Fe_{40}Ni_{40}P_{14}B_6$ amorphous ribbons measured by Becker-Kersten method.

Composition	λ	K (J/m^3)
$Fe_{40}Ni_{40}P_{14}B_6$	$(1.6 \pm 0.13)\times10^{-5}$	360 ± 30
$Fe_{40}Co_{40}P_{14}B_6$	$(1.6 \pm 0.05)\times10^{-5}$	857 ± 30
$Fe_{52}Co_{28}P_{14}B_6$	$(1.0 \pm 0.1)\times10^{-5}$	483 ± 50
$Fe_{55}Co_{25}P_{14}B_6$	$(1.75 \pm 0.13)\times10^{-6}$	863 ± 65

4. Conclusions

The Becker-Kersten method which involves observing the hysteresis loops under mechanical stress is applied to measure magnetostriction in amorphous $Fe_{80-x}Co_xP_{14}B_6$ ribbons. We recorded and analyzed a series of hysteresis loops for several compositions under various tensile stresses. The value of magnetostriction constant for $Fe_{80-x}Co_xP_{14}B_6$ amorphous ribbons is increasing with the growth of cobalt atomic content: $\lambda = (1.75 \pm 0.13)\times10^{-6}$ in $Fe_{55}Co_{25}P_{14}B_6$, $\lambda = (1.0 \pm 0.1)\times10^{-5}$ in $Fe_{52}Co_{28}P_{14}B_6$, and $\lambda = (1.60 \pm 0.05)\times10^{-5}$ in $Fe_{40}Co_{40}P_{14}B_6$.

Acknowledgements
This work was carried out within the Russian Federal Targeted Program "R&D for priority areas of the development of S&T complex of Russia 2014-2020", contract 14.583.21.0067, unique identifier RFMEFI58317X0067.

References

[1] Hollmark M., Tkatch V.I., Grishin A.M., Khartsev S.I., Processing and properties of soft magnetic $Fe_{40}Co_{40}P_{14}B_6$ amorphous alloy, *IEEE transactions on magnetic* **37** (4), 2278-2280 (2001).
[2] Tkatch V.I., Grishin A.M., Khartsev S.I., Delayed nucleation in $Fe_{40}Co_{40}P_{14}B_6$ metallic glass, *Materials Science and Engineering* **A337** (1), 187-193 (2002).

[3] Prahova D.A., Grishin A.M., Ignahin V.S., Lugovskaya L.A., Osaulenko R.N., Melt-spun Fe-Co-P-B metglasses: structure, crystallization kinetics, magnetic properties, *Journal of Physics: Conference Series* **769**, 012034 (2016).
[4] Severikov V.S., Grishin A.M., Ignahin V.S. Study of inverse magnetostrictive effect in metallic glasses $Fe_{80-x}Co_xP_{14}B_6$, *Journal of Physics: Conference Series* **929**, 012049 (2017).
[5] Kostyrya S.A., Idzikowski B., Tkatch V.I., Popov V.V., Rassolov S.G., Thermal stability of amorphous structure and magnetic properties $Fe_{80-x}Co_xP_{14}B_6$ (x = 20–40) ribbons. *Physica Status Solidi* (b) **243** (1), 339-342 (2006).
[6] Popov V.V., Tkatch V.I., Rassolov S.G., Aronin, A.S., Effect of replacement of Ni by Co on thermal stability of $Fe_{40}Co_{40}P_{14}B_6$ metallic glass, *Journal of Non-Crystalline Solids* **356** (25), 1344-1348 (2010).
[7] Becker R., Kersten M., Die magnetisierung von nickeldraht unter starkem zug, *Zeitschrift für Physik*, **64** (9-10), 660-681(1930).
[8] O'Dell T.H., Magnetostriction measurements on amorphous ribbons by the Becker-Kersten method, *Physica status solidi* (a) **68** (1), 221-226 (1981).
[9] Hayakawa M., Aso K., A simple apparatus for measurement of small magnetostriction of amorphous alloys, *Journal of Physics* E*: Scientific Instruments*, **17** (12), 1220-1223 (1984).
[10] Narita K., Yamasaki J., Fukunaga H., Measurement of saturation magnetostriction of a thin amorphous ribbon by means of small-angle magnetization rotation, *IEEE Transactions on Magnetics* **16** (2), 435-439 (1980).

IOP Conf. Series: Journal of Physics: Conf. Series **1038** (2018) 012067 doi:10.1088/1742-6596/1038/1/012067

Experimental research of π^0 meson production in U+U collision at 192 GeV

A. Ya. Berdnikov, Ya. A. Berdnikov, D. O. Kotov, P. V. Radzevich, S. V. Zharko
Peter the Great St.Petersburg Polytechnic University (SPbPU), Russia, 195251, St.Petersburg, Polytechnicheskaya, 29
E-mail: radzevichp@gmail.com

Abstract. To study properties of quark-gluon plasma(QGP) is one of the main tasks of modern High Energy Physics. One of the signatures of QGP formation is jet quenching, which was observed as a suppression of particle yields (compared to the yield of the same particles in proton-proton collisions) at high transverse momentum region in central collisions of ultra-relativistic heavy nuclei. Jet quenching is the effect of the final state of heavy nuclei collision which is usually explained through parton energy loss in hot and dense medium. Experimentally the jet quenching is studied with nuclear modification factor. This paper presents invariant transverse momentum spectra and nuclear modification factors for π^0 mesons in different classes of centrality in U+U collisions at $\sqrt{s_{NN}}$ =192 GeV. Spectra and factors are measured in a wide p_T range up to 18 GeV/*c* for different classes of centrality. In central U+U collisions suppression of neutral π-mesons is the same as the suppression of neutral π-mesons in Au+Au collisions at the similar collision energy and similar numbers of participants.

1. Introduction

High transverse momentum (p_T) hadrons are produced from the fragmentation of hard scattered quarks and gluons (partons) in the collisions of ultra-relativistic protons (p+p). This phenomenon is well described with the framework of perturbative quantum chromodynamics (pQCD) [1]. Over against, production of high-p_T hadrons in the collisions of ultra-relativistic heavy nuclei (A+A) is modified by the hot and dense medium – quark-gluon plasma (QGP) and cannot be described with pQCD formalism. High-p_T partons traverse through the produced medium and lose their energy, which results in production of hadrons with reduced energy. This effect is called jet-quenching [2,3] and is manifested by suppressed production of hadron yields relative to yields measured in elementary p+p collisions.

The measurement of π^0 meson production spectra allows to study jet-quenching effect with a good precision, because π^0 can be identified up to the highest p_T with relatively small uncertainties. In 2012, PHENIX experiment collected data from U+U collisions at $\sqrt{s_{NN}}$ =192 GeV. This collision system provides the largest energy density available at RHIC, which gives a unique possibility to obtain additional restrictions on the parameters of various theoretical models and measure properties of QGP with a higher accuracy.

2. Data Analysis

This paper reports on the measurement of neutral pion production in U+U collisions at $\sqrt{s_{NN}}$=192 GeV at RHIC. The whole analysis uses 9.4×10^8 events taken by the PHENIX experiment [4]. Collision centrality is determined from the correlation between the number of charged particles detected in the Beam-Beam Counters (BBC, $|\eta|<0.35$). A Glauber model Monte-Carlo simulation of the BBC responses is used to estimate the associated average number of participating nucleons ($<N_{part}>$) and binary nucleon-nucleon collisions ($<N_{coll}>$) for each centrality bins. There are two sets of data due to different deformation parameterization of U ion in the Wood-Saxon distribution [5,6]. Neutral pions are measured in the $\pi^0 \rightarrow \gamma\gamma$ decay channel with the photons reconstructed in the Electromagnetic

Calorimeter (EMCal), which is located in the two central arms of PHENIX ($|\eta| \leq 0.35$). The EMCal [7] consists of two subsystems: six sectors of lead-scintillator sandwich calorimeter (PbSc) and two sectors of lead-glass Cherenkov calorimeter (PbGl). Different types of the EMCal have different energy resolution, segmentation and linearity, and the use of both types of them provides excellent cross-check of the final results.

A detailed description of the analysis, including extraction of the raw π^0 yield, correction for acceptance, assignment of detector energy resolution and dead areas, estimation of π^0 reconstruction efficiency have been described in another paper [8].

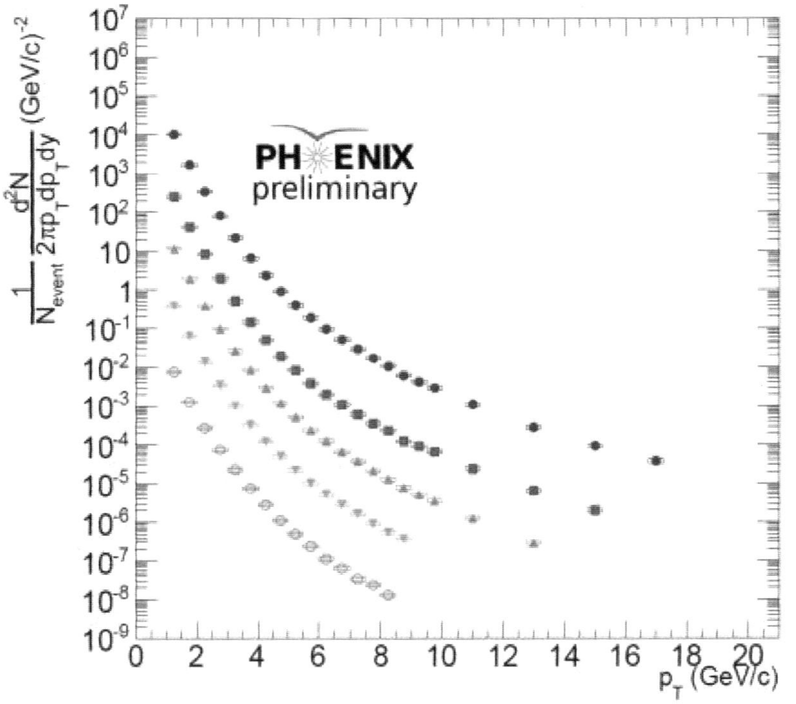

Figure 1. Invariant transverse momentum spectra measured for π^0 mesons in U+U collisions at 192 GeV in minimum bias(●), 0-20%(■), 20-40%(▲), 40-60%(▼), 60-80%(○) centrality intervals. Statistical uncertainties are smaller than the marker size. Open boxes correspond to systematic uncertainties.

3. Results

Figure 1 shows the π^0 invariant p_T-spectra for four different centrality classes and minimum bias collisions. Invariant transverse momentum spectra of π^0-mesons are measured in a wide p_T range (up to 18 GeV/*c* in minimum bias collisions). The main sources of systematic uncertainties are yield extraction, efficiency corrections, and energy scale, none of which exhibit a significant centrality dependence. The PbSc and PbGl detectors have quite different systematic uncertinties with all but one of them (off-vertex π^0) uncorrelated. Obtained results show that main systematic uncertainty comes from possible mismatch in absolute energy scales (5% at low and intermediate p_T to 10% at high p_T) and reproduction of cluster merging effect between data and simulation (influence only at the highest p_T, ~15%).

To quantify the comparison of spectra in heavy ion and *p+p* collisions, the nuclear modification factor is used. Nuclear modification factors of π^0 mesons is calculated with formula:

$$R_{AA}(p_T) = \frac{1}{\langle N_{coll} \rangle} \frac{dN_{AA}(p_T)}{dN_{pp}(p_T)}, \quad (1)$$

where $dN_{AA}(p_T)$ $(dN_{pp}(p_T))$ – particle yield, measured in $A+A$ $(p+p)$ collisions, $\langle N_{coll}\rangle$ – average number of binary inelastic nucleon-nucleon collisions.If $R_{AA}<1$ ($R_{AA}>1$), particle production is suppressed (enhanced) in comparison to elementary nucleon-nucleon collisions. Contrariwise, if $R_{AA}=1$, there are no collective effects in $A+A$ collisions.

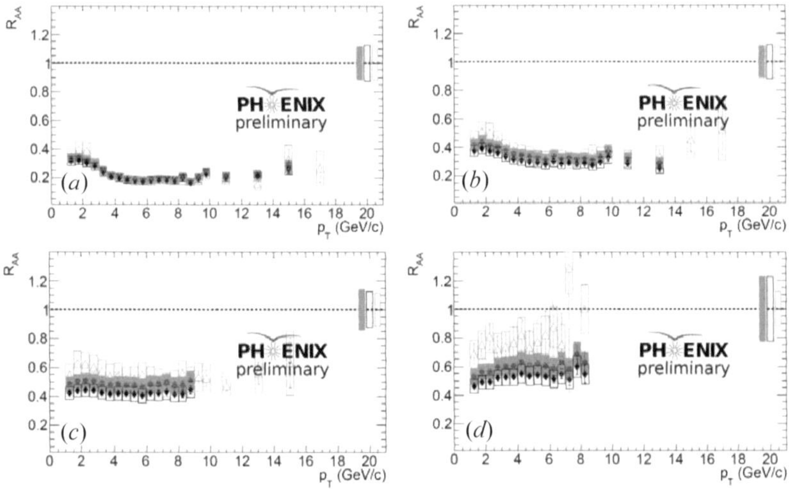

Figure 2. Comparison of R_{AA} values in U+U Set 1(●), U+U Set 2 (♦) ($\sqrt{s_{NN}}$ = 192 GeV) and Au+Au (Δ) ($\sqrt{s_{NN}}$ = 200 GeV) [9] collisions at similar N_{coll} for different centrality bins in 0-20% (*a*), 20-40% (*b*), 40-60% (*c*), 60-80% (*d*). Error bars and open boxes show statistical and systematic uncertainties, respectively. Boxes at unity show scaling uncertainty.

Figure 2 shows π^0 nuclear modification factors measured for different sets of the collision parameters [10,11] in minimum bias U+U collisions and different U+U centrality intervals as a function of p_T. R_{AA} are compared to the previous results obtained in Au+Au collisions at 200 GeV [9] at similar values of N_{coll}. Presented values of R_{AA} are consistent between two collision systems within the uncertainties of the measurements.

Figure 3 shows the integrated nuclear modification factor in U+U and Au+Au[9] for π^0 as a function of numbers of nucleons participating in the nuclei interactions (N_{part}). The suppression increases monotonically with N_{part} without any sign of saturation within uncertainties. In peripheral collisions (N_{part} < 80) π^0 mesons seem to be slightly more supressed but still agree within uncertainties.

4. Summary

PHENIX has measured neutral pion production in U+U collisions at $\sqrt{s_{NN}}$=192 GeV at mid rapidity in the transverse momentum range of $1<p_T<18$ GeV/*c*, using RHIC Run-12 data. In central collisions the yield is suppressed by a factor of ~5 at $p_T>5$GeV/*c* compared to the binary scaled *p*+*p* reference. The suppression prevails with little or no change up to 18 GeV/c. In central U+U collisions π^0 production is suppressed in the same way as in Au+Au at similar number of binary collisions, which suggests that suppression level depends on the size of the interaction region and not on its shape. In peripheral U+U collisions low p_T π^0 mesons seem to be slightly more suppressed but still agree within uncertainties.

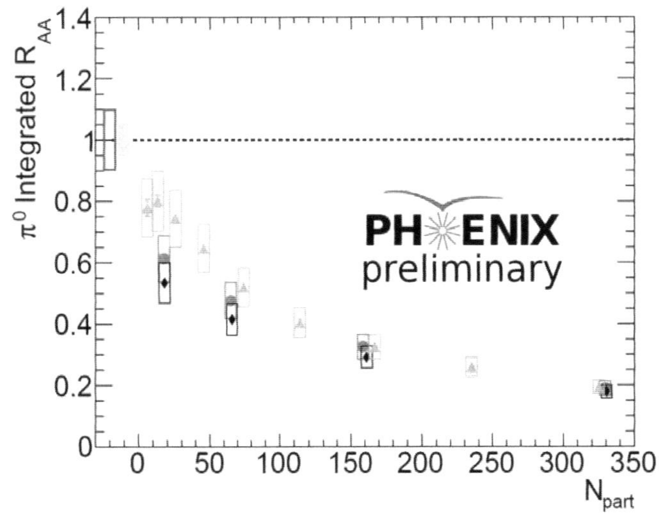

Figure 3. Integrated nuclear modification factor in U+U SET 1 (●), SET 2 (♦) and Au+Au (▲) collisions at p_T>5 GeV/c for π^0 as a function of numbers of participants. Error bars and open boxes show statistical and systematic uncertainties, respectively. Boxes at unity show scaling uncertainty.

Acknowledgments

We acknowledge support from Russian Ministry of Education and Science, state assignment 3.1498.2017/4.6

References

[1] D. de Florian and W. Vogelsang 2005 Phys. Rev. D71, 114004
[2] M. Gyulassy and M. Plumer1990 Phys. Lett. B243, 432
[3] X.-N. Wang and M. Gyulassy 1992 , Phys. Rev. Lett. 68, 1480
[4] K. Adcox et al. 2003 Nucl. Instrum. Meth. A499, 469
[5] A. Adare et al. 2008 Phys..Lett. B679, 440
[6] A. Adare et al. 2015 Phys..Lett. B749, 215
[7] L. Aphecetche et al. 2003 Nucl. Instrum. Meth. A499, 521
[8] S. S. Adler et al. 2007 Phys. Rev. C76, 034904
[9] A. Adare et al. 2008 Phys.Rev.Lett. 101, 232301
[10] A. Adare et al. 2009 Phys. Lett. B 679, 440
[11] A. Adare et al. 2015 Phys. Lett. B 749, 215

IOP Conf. Series: Journal of Physics: Conf. Series **1038** (2018) 012068 doi:10.1088/1742-6596/1038/1/012068

Nuclear modification factors of light mesons in Cu+Au collisions

A. Berdnikov, Ya. Berdnikov, D. Kotov, P. Radzevich and S. Zharko

Peter the Great Saint Petersburg Polytechnic University, 29 Polytechnicheskaya st., Russia

E-mail: zharkosergey94@gmail.com

Abstract. Strongly interacting quark-gluon plasma (sQGP) is a state of nuclear matter at extremely high values of temperature with quarks and gluons as degrees of freedom. Jet-quenching is one of the evidences for sQGP formation in relativistic heavy ion collisions (A+A) and is manifested by suppression of particle yields at high transverse momentum relative to yields of the same particles measured at elementary proton-proton collisions at the same collision energy. Systematic experimental research of jet-quenching is provided by the measurement of nuclear modification factors of light mesons (π^0, η, K_s^0, ω etc.) in different A+A systems. Asymmetric system of Cu+Au collisions at $\sqrt{s_{NN}} = 200$ GeV is characterised by similar energy density but different collision geometry when compared to symmetric systems (Au+Au or Cu+Cu). This paper presents measurements of π^0 and η mesons nuclear modification factors in Cu+Au collisions. Results are compared with ones measured in Au+Au and Cu+Cu collisions at the same collision energy.

1. Introduction

Strongly interacting quark-gluon plasma (sQGP) presents a liquid-like state of quasi-free quarks and gluons (partons) at extremely high values of temperature ($T \approx 170$ MeV, $\sim 10^{12}$ K). First theoretical predictions of quark-gluon plasma were carried out by E. Shuryak [1], and first statements of the sQGP formation were made by experiments [2–5] in central collisions of relativistic heavy ions (A+A) at the Relativistic Heavy Ion Collider, RHIC (BNL, USA) [6].

Jet-quenching [7, 8] is the one of the effects, which indicates the sQGP formation. It is manifested by suppression of yields of jets and leading hadrons in A+A relative to binary scaled yields measured in elementary proton-proton (p+p) collisions. This effect suggests that high energy partons produced in hard scattering processes lose their energy in the quark-gluon medium, which results in the production of hadrons with the reduced transverse momentum (p_T) values.

Nuclear modification factor (R_{AA}) is the main observable of jet-quenching and other collective effects in heavy ion collisions:

$$R_{AA}(p_T) = \frac{1}{N_{coll}} \frac{dN_{AA}(p_T)}{dN_{pp}(p_T)}, \tag{1}$$

where dN_{AA} (dN_{pp}) – particle invariant differential yields measured in A+A (p+p) collisions, $\langle N_{coll} \rangle$ – average number of binary inelastic nucleon-nucleon collisions. Values of R_{AA} are calculated in different intervals of the collision centrality, which represents the magnitude of the nuclei overlap. Centrality of 0–20% corresponds to central collisions with a large nuclear overlap region, and centrality of 60–90% corresponds to the most peripheral collisions, where only few nucleons participate in the nuclei interactions.

Systematic study of jet-quenching includes measurement of p_T-spectra and R_{AA} of different particles in different systems of colliding nuclei. The PHENIX spectrometer provides excellent capabilities for measurement of light neutral mesons (π^0, η, K_s^0, ω etc.). Pions are copiously produced in the nuclei interactions thus π^0 yields can be measured in a wide p_T-range with relatively small uncertainties. η mesons have a four-times larger mass and a hidden strangeness thus the measurement and comparison of R_{AA} of π^0 and η provide an opportunity to study the dependence of jet-quenching on particles mass and quark content of the fragmented particles. Study of the particle yield ratios (η/π^0, K_s^0/π^0 etc.) can help to obtain more detailed information about fragmentation mechanisms of particles with different properties.

The data for Cu+Au collisions at $\sqrt{s_{NN}}$ = 200 GeV has been delivered by RHIC in 2012. It is the first asymmetric collision system of heavy relativistic ions characterised by similar energy density but different collision geometry when compared to symmetric systems (Au+Au or Cu+Cu). Measurement of light neutral mesons R_{AA} in Cu+Au collisions provides an opportunity to discriminate between parameters of various theoretical models of parton energy loss in the sQGP medium.

2. Data Analysis

All results presented in this paper are carried out with the PHENIX spectrometer central arms, each covers $\pi/2$ of azimuth and ±0.35 interval of pseudo-rapidity (η). Subsistems of the PHENIX spectrometer are described in detail elsewhere [9]. In the analysis π^0 and η mesons are reconstructed in their main decay channel ($\pi^0 \to \gamma\gamma$, $\eta \to \gamma\gamma$) with the electromagnetic calorimeter (EMCal) [10].

The EMCal of the PHENIX consists of the two technologically-different subsystems. Lead-scintillator (PbSc) and lead-glass (PbGl) subsystems have different energy resolution ($\delta E/E = 2.1\% \oplus 8.1\%/\sqrt{E}$ for PbSc and $0.8\% \oplus 5.9\%/\sqrt{E}$ for PbGl), granularity ($\delta\eta \times \delta\phi \approx 0.01 \times 0.01$ for PbSc and 0.008×0.008 for PbGl), linearity and respect to hadronic background, thus simultaneous use of both types provides an important cross-check of the results in the framework of the one experiment.

The centrality of analysed events is determined by two beam-beam counters (BBC) [11] located along the beam axis in the $3.0 < |\eta| < 3.9$ interval of pseudo-rapidity. A set of geometric parameters including N_{coll}, number of participating nucleons (N_{part}), impact parameter (b) for each centrality interval is calculated using the response of BBC along with the Glauber-model based Monte-Carlo simulation [12].

Invariant mass distributions for $\gamma\gamma$ pairs were accumulated in different intervals of the meson p_T and event centrality. Considered γ clusters are required to satisfy several analysis cuts: γ energy is equal to 0.4 GeV to reduce charged hadrons contamination. To reduce the combinatorial background an asymmetry cut was applied for each $\gamma\gamma$ pair: $\alpha < 0.8$, where $\alpha = |E_{\gamma 1} - E_{\gamma 2}|/(E_{\gamma 1} + E_{\gamma 2})$, $E_{\gamma 1}$ and $E_{\gamma 2}$ are energies of γ clusters in the pair.

Peaks of reconstructed π^0 and η seat on a large combinatorial background. The shape of the uncorrelated part of the background is estimated with so-called event-mixing technique, when $\gamma\gamma$ pairs are formed from two different events with similar values of the collision centrality. Estimated shape is normalized to foreground integral in the mass range 0.3–0.4 GeV/c^2 and then subtracted. Resulting distributions are fitted to a sum of the Gaussian and second order polynomial to estimate meson yield and remaining background, respectively.

Derived yields are corrected for the detector effects and limited acceptance with the reconstruction efficiency, which is obtained from GEANT 3 based Monte-Carlo simulation. The influence of high occupancy effects in the detector is accounted with embedding simulated π^0 and

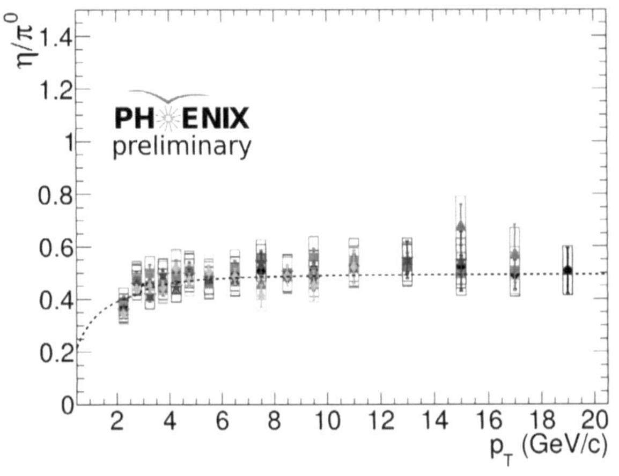

Figure 1. Ratios of η and π^0 yields measured in 0–93% (●), 0–20% (■), 20–40% (▲), 40–60% (★) and 60–90% (◆) centrality intervals of Cu+Au collisions. Dashed curve shows m_T-scaling function. Error bars and open boxes show statistical and systematic uncertainties, respectively

η mesons into real Cu+Au events. Besides that all analysis cuts are applied in the simulation in the same way as in the analysis of real data. Full procedure of π^0 and η reconstruction is carried out separately for PbSc and PbGl subsystems. Final results are obtained by averaging PbSc and PbGl with weights defined by uncertainties uncorrelated between PbSc and PbGl subsystems.

To estimate systematic uncertainties we varied detector performance in simulation and analysis cuts. Among them there are uncertainties from meson raw yield extraction approach, possible mismatch in absolute energy scale, resolution and acceptance of the EMCal between real data and simulations, imperfect reproduction of photon reconstruction algorithms and conversion in detector subsystems in simulation. The main uncertainty at intermediate p_T is the mismatch in absolute energy scale: the mismatch in ~1% of the meson masses in data and simulation results in ~8% uncertainty in yield measurement.

Yields of π^0 are suffered from so-called cluster merging effect at highest p_T ($p_T >$ 12–15 GeV/c). Two γs from the decay of high-p_T π^0 have rather small opening angle and cannot be reconstructed as two individual cluster in the EMCal due to its finit granularity. Thus such π^0s couldn't be reconstructed. The mismatch in the reproduction of this effect in the simulation is

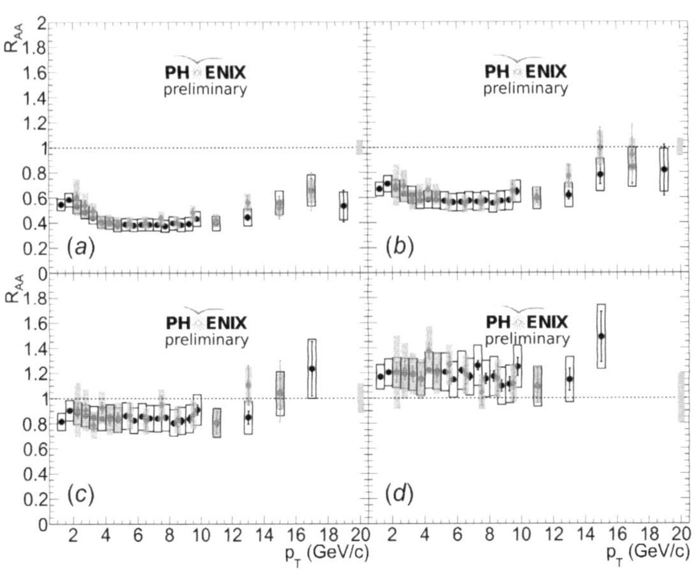

Figure 2. Nuclear modification factors of π^0 (●) and η (◆) mesons as a function of p_T measured in 0–20% (a), 20–40% (b), 40–60% (c) and 60–90% (d) centrality intervals of Cu+Au collisions at $\sqrt{s_{NN}} = 200$ GeV. Error bars and open boxes show statistical and systematic uncertainties, respectively. Boxes at unity show scaling uncertainty.

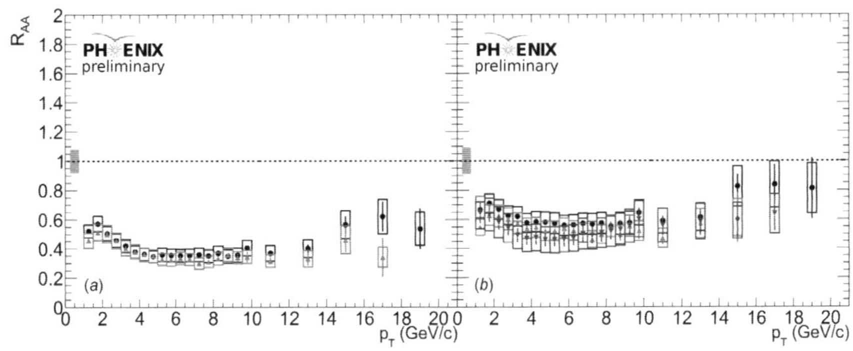

Figure 3. Comparison of π^0 meson nuclear modification factors measured as a function of p_T in Cu+Au (●), Au+Au (▲) [16] and Cu+Cu (▼) [17] collisions at 200 GeV. Panel (a) is for 0-10% Cu+Au with $N_{part} = 177$ and 20-30% Au+Au with $N_{part} = 167$. Panel (b) is for 20-40% Cu+Au with $N_{part} = 80.4$, 40-50% Au+Au with $N_{part} = 74.4$ and 0-10% Cu+Cu with $N_{part} = 98.2$. Error bars and open boxes show statistical and systematic uncertainties, respectively. Boxes at unity show scaling uncertainty.

followed by ~15% uncertainty. For η mesons this effect can be observed only at $p_T > 60$ GeV/c, which far exceeds the limit of η yield measurements.

3. Results

Figure 1 presents η/π^0 ratios measured as a function of meson p_T in different centrality intervals of Cu+Au collisions. Values of the ratios are the same for all centrality intervals within uncertainties and consistent with m_T-scaling curve [13], normalized to 0.5 at highest p_T. Obtained results are consistent with previous measurements in lepton-lepton, hadron-hadron, hadron-nucleus and nucleus-nucleus collisions in a wide interval of collision energies [14], which suggests that η meson fragmentation processes are independent from presence of the sQGP medium.

Comparison of π^0 and η R_{AA} measured as a function of meson p_T in different centrality intervals of Cu+Au collisions is presented in the Figure 2. Values of R_{AA} are the same within uncertainties in all centrality intervals. Meson yields are suppressed twice in central Cu+Au collisions. A hint of enhancement of π^0 and η yields is observed in peripheral Cu+Au collisions. A similar pattern of R_{AA} is observed for reconstructed jets in the same collision system [15].

Figure 3 shows comparison of π^0 R_{AA} measured in Cu+Au, Au+Au [16] and Cu+Cu [17] collisions at $\sqrt{s_{NN}} = 200$ GeV and similar values of N_{part}. For all collision systems π^0 R_{AA} values are the same within uncertainties in the whole p_T range, which suggests that π^0 suppression level depends on the size of the nuclear overlap region but not on its shape. R_{AA} of η mesons are also the same within uncertainties in Cu+Au and Au+Au [14, 18] collisions as presented in the Figure 4.

4. Summary

PHENIX has measured π^0 and η nuclear modification factors in Cu+Au collisions at $\sqrt{s_{NN}} = 200$ GeV as a function of meson p_T and event centrality. Production of π^0 and η mesons shows similar suppression in all centrality intervals, which suggests that suppression occurs on the parton level before fragmentation processes, i.e. the suppression is the result of the energy loss of partons in the QGP medium. This effect could not be reproduced with the Perturbative QCD calculations yet, but with different semi-phenomenological models, for example [7,8]. Production of π^0 and η is suppressed twice in central Cu+Au collisions, a hint of enhancement of π^0 and η

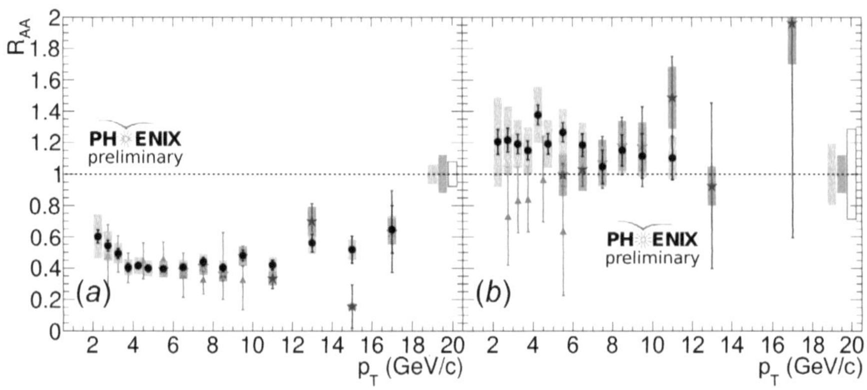

Figure 4. Comparison of η meson nuclear modification factors measured as a function of p_T in Cu+Au (●), Au+Au (published in 2007, ▲) [14] and Au+Au (published in 2010, ★) [18] collisions at 200 GeV. Panel (*a*) is for 0-20% Cu+Au with N_{part} = 154, 20-60% Au+Au (published in 2007) with N_{part} = 100 and 20-40% Au+Au (published in 2010) with N_{part} = 140. Panel (*b*) is for 60-90% Cu+Au with N_{part} = 8.85, 60-92% Au+Au (published in 2007 and 2010) with N_{part} = 14.5 and 0-10% Cu+Cu with N_{part} = 14.5. Error bars and open boxes show statistical and systematic uncertainties, respectively. Boxes at unity show scaling uncertainty.

yields is observed in peripheral Cu+Au collisions. A similar suppression pattern is observed for reconstructed jets in the same collision systems. R_{AA} values of π^0 and η are the same within uncertainties in different collision systems at the same collision energy and similar values of N_{part}, which suggests that suppression level mostly depends on the nuclei overlap size but not on its geometry. Measured values of η/π^0 ratios are consistent with previously measured in different systems, which suggests that η meson fragmentation processes are independent from presence of the quark-gluon medium.

Acknowledgments

We acknowledge support from Russian Ministry of Education and Science, state assignment 3.1498.2017/4.6.

References

[1] Shuryak E V 1980 *Phys. Rept.* **61** 71
[2] Arsene I et al 2005 *Nucl. Phys. A* **757** 1-27
[3] Back B et al 2005 *Nucl. Phys. A* **757** 28-101
[4] Adams J et al 2005 *Nucl. Phys. A* **757** 102-83
[5] Adcox K et al 2005 *Nucl. Phys. A* **757** 184-283
[6] Baym G 2002 *Nucl. Phys. A* **698** 23
[7] Arnold P B et al 2002 *JHEP* **206** 30
[8] Lokhtin I P, Snigirev A M 2006 *Eur. Phys. J. C.* **45** 211
[9] Adcox K et al 2003 *Nucl. Inst. Meth. A* **499** 469-07
[10] Aphecetche L et al 2003 *Nucl. Inst. Meth. A* **499** 521-36
[11] Allen M et al 2003 *Nucl. Inst. Meth. A* **499** 549-59
[12] Loizides C 2016 *Phys. Rev. C* **94** 024914
[13] Khandai P K et al 2011 *Phys. Rev. C* **84** 054904
[14] Adler S S et al 2007 *Phys. Rev. C* **75** 024909
[15] Timilsina A 2016 *Nucl. Phys. A* **956** 537
[16] Adare A et al 2008 *Phys. Rev. Lett.* **101** 232301
[17] Adare A et al 2008 *Phys. Rev. Lett.* **101** 162301
[18] Adare A et al 2010 *Phys. Rev. C* **82** 011902(R)

Optimization of the methods for measuring color characteristics of light-emitting diodes in laboratory conditions

A A Pavlova, A N Ramazanov, V A Simon, D K Kostrin and A A Uhov

Department of electronic instruments and devices, Saint Petersburg Electrotechnical University "LETI", 197376, Saint Petersburg, Russia

E-mail: ainapavlova-leti@yandex.ru

Abstract. In this work the features of measuring color characteristics of light-emitting diodes are described. Methods allowing increasing the accuracy of measurements in the laboratory conditions are shown. The results of determination of the color characteristics of white light-emitting diodes using the described technique are analyzed.

1. Introduction

Light-emitting diodes (LEDs) due to their reliability, low energy consumption and long service life are currently almost universally applied to replace previously used sources of optical radiation. Application areas of LEDs are not only devices providing light of one color, such as traffic lights, but also various lighting systems used in residential and industrial premises and also for illumination of sidewalks and roads [1, 2].

Extension of the application areas of LEDs increases the need for the accurate measurement of their parameters. Characteristics of LEDs from different manufacturers vary considerably; also spread of parameters even within the same batch is possible. That is why measurement of light and color characteristics, as well as the possibility of visual examination of emission spectra of the investigated LED is very important. During the evaluation of LEDs the photometric, radiometric, colorimetric, goniometric and operational characteristics should be analyzed [3–5].

The aim of this work is to develop and study simplified express methods for measuring some key color parameters of light-emitting diodes using compact optical spectrometer, which allow achieving results with an acceptable accuracy.

2. Characteristics of light-emitting diodes and methods of their measurement

Let us consider the colorimetric characteristics of LEDs, defining the perception of the color of light by a human eye. Human color perception is very complex, because it depends not only on the physical properties of light, but also from the surrounding objects, properties of the emitter and the psychological state of the observer [6, 7]. Colorimetric parameters of LEDs are usually expressed as chromaticity coordinates and dominant wavelength, for white LEDs the value of correlated color temperature is also used.

The measurement of LED parameters should be conducted in strictly regulated by the International Commission on Illumination (CIE) conditions [8]. However, in laboratory conditions and also for the tuned technological process, it is possible to use a simplified method of measurement. The most popular instruments for the study of LEDs are spectrometers constructed using CCD (charge-coupled

IOP Conf. Series: Journal of Physics: Conf. Series **1038** (2018) 012069 doi:10.1088/1742-6596/1038/1/012069

device) photodetectors. To analyze the parameters of LEDs an optical resolution of the device of 1,0...1,5 nm is sufficient [4].

It should be noted that prior to achieving the emission spectrum of the investigated sample it is necessary to perform an operation of obtaining the reference spectrum from the source of A type (incandescent lamp with tungsten spiral). For this task the program Aspect2010 developed in Saint Petersburg Electrotechnical University "LETI" to control optical spectrometers ISM3600 [9, 10] can be used. For the implementation of the spectral correction the reference spectrum of the A type source with the subtraction of a dark spectrum should be obtained using an adjustable device. Next, in the program a prepared ideal spectrum of a reference source (i.e. spectrum in a form in which it should be observed using the spectrometer if it did not have its non-linear spectral dependence) is used. Later, using these two spectra (real and ideal), the program calculates a correction function, by which all obtained spectra are automatically multiplied. Figure 1 shows fragments of the emission spectra of white LED with and without the correction of the spectrometer spectral dependence. It can be seen that this procedure leads to a significant change in the spectrum and, consequently, to a change in the results of calculation of the chromaticity coordinates and other parameters of a radiation source.

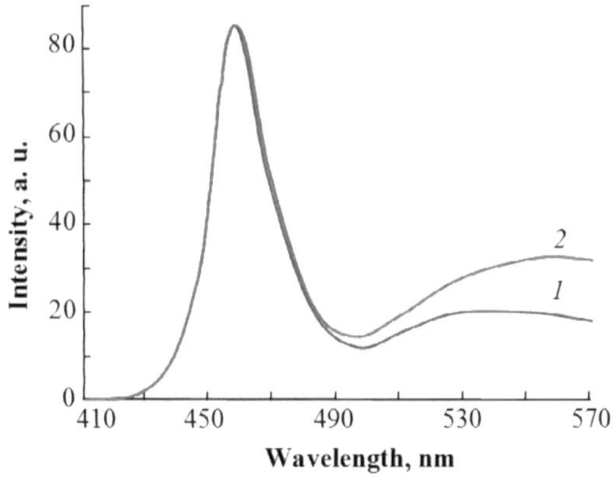

Figure 1. Emission spectra of a white LED with (*1*) and without (*2*) spectral dependence correction.

It is also necessary to take into account the dark signal of the CCD photodetector and to take care of the noise filtering in the region of wavelengths 380...780 nm, in which the integration when calculating the color of light takes place [11]. CCD photodetector has some noise level consisting of several components. The first component is a "geometric" noise, emerging due to the topology of a sensor; it can be programmatically deleted using linear interpolation of odd and even pixels. The next component is an intrinsic noise. For its reduction it is possible to make averaging of the multiple spectra taken.

Also the processing of the received data of the radiation spectral distribution using a spline digital filter is quite effective. During the measurement of LED parameters the influence of the background signal close to the zero level leading to "whitening" of the perceived color can be further reduced. In this case, when performing color measurements signal with a value less than selected will be zeroed. To provide that the ambient light has no effect on the results of the LED color characteristics determination, LED and the optical input of the spectrometer should be placed under an opaque cover [12].

LED parameters are largely affected by the temperature – both ambient and achieved during the self-heating of a semiconductor crystal. The shift of the LED emission spectrum with increasing temperature occurs in the direction of longer wavelengths. The color coordinates of the LED radiation when the temperature changes from –60 to +55 °C can change considerably, by almost 10 % [13]. Therefore, before starting any measurement of characteristics LED should be warmed for some time, achieving an established thermal equilibrium. All of the approaches for improving the accuracy of measurements described above have been applied in this study.

In 1931 the CIE recommended the data of colors summing for the standard colorimetric observer, embodied in two different, but equivalent color coordinate systems RGB and XYZ [14, 15]. To describe the color characteristics of an optical radiation source the XYZ system is commonly used. There are several characteristics of LEDs that present a practical interest, such as chromaticity coordinates, dominant wavelength, correlated color temperature, color rendering indexes (general and partial) and efficacy of a light source. At this step of the study we will focus on the first three of them.

3. Results and discussion

As objects for study white LEDs SvL-03 manufactured by JSC "Svetlana-Optoelectronics" [16] that are used to create ceiling lamps were selected. Figure 2 shows the emission spectrum of such a LED.

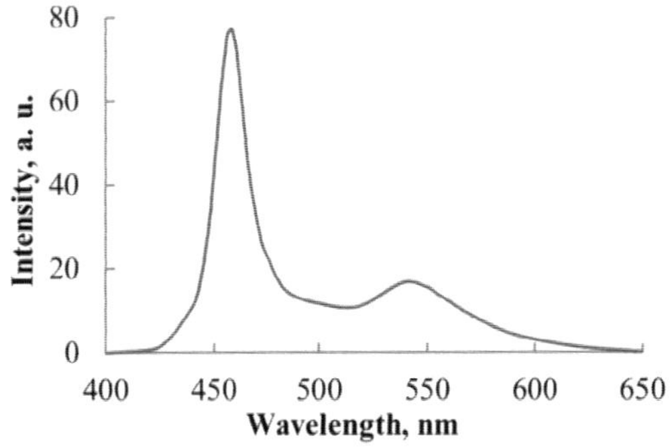

Figure 2. Radiation spectrum of SvL-03 LED.

There are several methods of obtaining white LEDs. In this case, an application on a LED with a blue emission spectrum of a yellow phosphor coating is used. This method is simple in implementation and inexpensive, but has some disadvantages. One of them is the uneven application of the phosphor layer on the blue LED crystal, which may lead to a different perception of the shades of color not only for LEDs of one batch, but even for a single LED at different angles of examination [12, 17].

Figure 3 shows the results of determining the chromaticity coordinates x and y (chromaticity coordinate z is redundant, since it can be determined from the other two) for a sample of 42 LEDs forming a ceiling lamp TIS-15M1. The study was conducted using a described above technique. Line in figure 3 shows the boundary of the chromaticity group B2 [18] stated by the manufacturer as the characteristic of the lamp. The measurements show that all LEDs belong to this chromaticity group. Sorting of the LEDs used in this lamp is made with sufficiently high quality – points corresponding to the chromaticity coordinates of LEDs do not fill all the shown area, but a little more than a half, which leads to the rise of device consumer properties, in particular the homogeneity of the radiation perception.

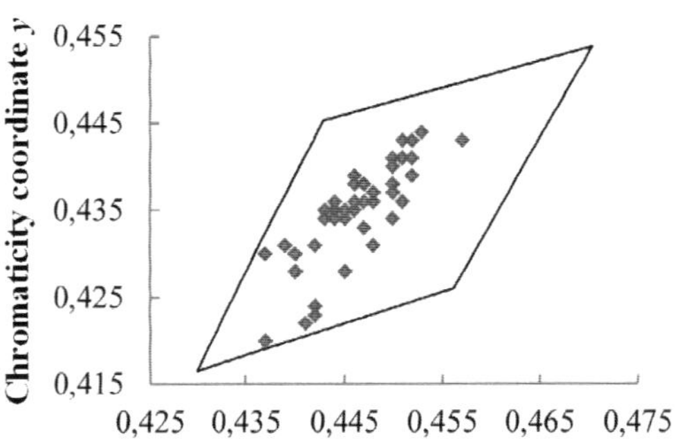

Figure 3. Chromaticity coordinates of 42 LEDs of a TIS-15M1 lamp.

Another important factor characterizing the color of light [19] is a dominant wavelength, which means the wavelength of monochromatic light, located on the edge of the color chart on the shortest distance from the point with the chromaticity coordinates of the source. To determine the dominant wavelength a straight line to the edge of the chart passing through the point of equal energy of a white light source (x = 0,3333, y = 0,3333) and the point with chromaticity coordinates (x, y) of the test LED is drawn. The point of intersection of this line with the outer border of the color chart will determine the dominant wavelength of the source.

For white LEDs the most important colorimetric parameter is a correlated color temperature equal to the temperature of a black body at which it emits radiation with the same chromaticity as considered. This measure is an objective impression of the color of a light source expressed in kelvins. LEDs produced today have a fairly wide range of shades of white color with a correlated color temperature of 2600...10000 K. The most comfortable radiation for a human is sunlight, with a correlated color temperature of about 2800…3500 K. Noonday sunlight has a correlated color temperature of about 5500 K. It follows that the source of illumination shall have approximately the same correlated color temperature value. Knowing the chromaticity coordinates of the LED radiation, its correlated color temperature T_c can be found using the following formula [20]:

$$T_c = 5520{,}33 - 6823{,}3\,P + 3525\,P^2 - 449\,P^3,$$

where $P = (x - 0{,}332)/(y - 0{,}1858)$. Figure 4 shows the values of dominant wavelength and correlated color temperature for 42 LEDs of a ceiling lamp.

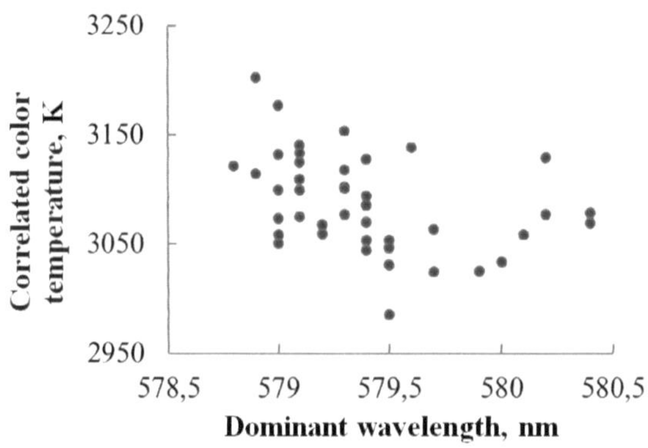

Figure 4. Dominant wavelength and correlated color temperature of 42 LEDs of a TIS-15M1 lamp.

The range of values of correlated color temperature is 2985...3200 K, which corresponds to a “warm” white light with a yellowish tint. The variation in dominant wavelength is 1,6 nm (range – 578,8...580,4 nm). Given that the human eye can distinguish colors that differ by 1 nm, there is a need to use in this lamp of a diffuser with a frosted glass to ensure a greater uniformity of color perception.

As the last part of the research let us show the experimental data demonstrating the angle variation of the dominant wavelength (figure 5), where α – angle between the observation point and the vertical axis of symmetry of the LED, φ – angle of observation in the horizontal plane.

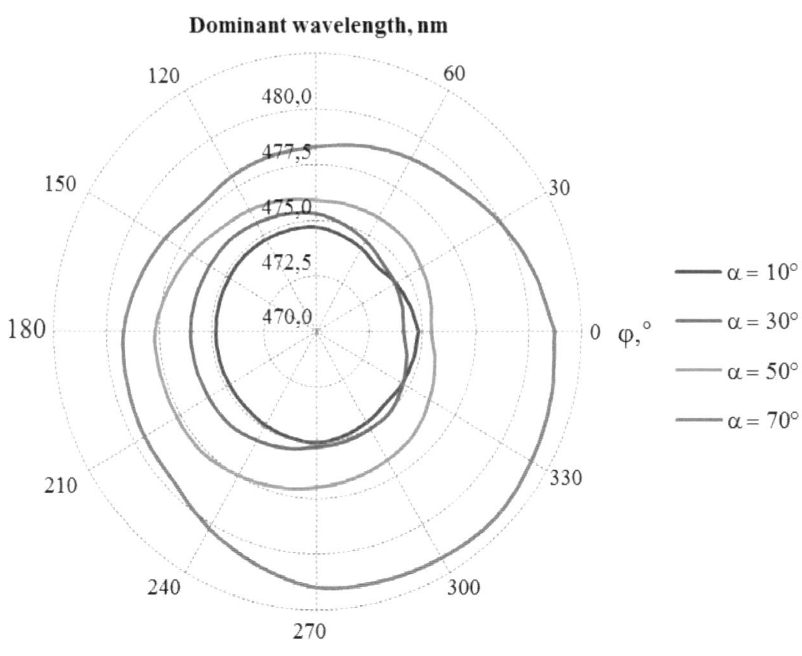

Figure 5. Spatial distribution of the dominant wavelength for a white LED.

The emission color of LED when observed from different angles is almost the same, except for large angles relative to the axis of symmetry. The differences in dominant wavelength at angles $\alpha = 10...50°$ is not more than 2,5 nm.

4. Conclusion

In this paper the general approaches to the measurement of certain color parameters of LEDs with the use of the simplified techniques are presented. The obtained results confirm that the use of the developed spectrometric equipment and such methods allow obtaining sufficiently accurate results of measurements for a rapid assessment of LEDs parameters.

Acknowledgments

This work was performed within the framework of agreement on scientific-technical partnership between Saint Petersburg Electrotechnical University "LETI" and JSC "Svetlana-Optoelectronics".

References

[1] Judd D B and Wyszecki G 1975 *Color in Business, Science, and Industry* (New York: Wiley)
[2] Žukauskas A, Shur M S and Gaska R 2002 *Introduction to Solid-State Lighting* (New York: Wiley)
[3] Ohno Y 2006 *Proceedings of SPIE* **6046** 25
[4] Kostrin D K and Uhov A A 2013 *Biotechnosphere* **3** 21–5
[5] Kostrin D K and Uhov A A 2013 *Testing. Diagnostics* **7** 47–50
[6] MacAdam D L 1942 *Journal of the Optical Society of America* **32** 247–75
[7] Vogels I, Seuntiens P and Sekulovski D 2008 *LED professional Review* **5** 12–6
[8] Ohno Y, Miller C and Zong Y 2008 *LED Professional Review* **5** 17–20
[9] Uhov A A and Kostrin D K 2013 *Proceedings of Saint Petersburg Electrotechnical University "LETI"* **4** 8–12
[10] Kostrin D K and Uhov A A 2013 *Sensors and systems* **5** 13–5
[11] Yudin R V, Kostrin D K, Shishov D I and Uhov A A 2013 *Proceedings of Saint Petersburg Electrotechnical University "LETI"* **3** 8–13
[12] Kostrin D K and Uhov A A 2014 *Testing. Diagnostics* **2** 65–8
[13] Nikiforov S 2006 *Components and technologies* **1** 18–23
[14] Smith T and Guild J 1931 *Transactions of the Optical Society* **3** 73–134
[15] Fairman H S, Brill M H and Hemmendinger H 1998 *Color Research and Application* **4** 259
[16] Molodtsov V 2010 *Solid State Lighting* **6** 16–8
[17] Fairman H S, Brill M H and Hemmendinger H 1997 *Color Research and Application* **1** 11–23
[18] Mazzochette J 2008 *LED professional Review* **5** 30–1
[19] Kostrin D K and Uhov A A 2013 *Proceedings of Saint Petersburg Electrotechnical University "LETI"* **1** 9–12
[20] Baklanov S A 2012 *Proceeding of the Conference "Innovations – design and technology – TUSUR 2012" (Tomsk)*

IOP Conf. Series: Journal of Physics: Conf. Series **1038** (2018) 012070 doi:10.1088/1742-6596/1038/1/012070

Comparative study of impact of random environment on individual and combined Laguerre-Gauss modes

A A Artyukova[1], **M S Kirilenko**[1,2] **and S N Khonina**[1,2]
[1]Samara National Research University, 34, Moskovskoye Shosse, Samara, Russia, 443086
[2]Image Processing Systems Institute – Branch of the Federal Scientific Research Centre "Crystallography and Photonics" of Russian Academy of Sciences, 151, Molodogvardeyskaya, Samara, Russia, 443001

Abstract. In this paper, the passage of individual and combined Laguerre-Gauss modes through the random medium with the given correlation function in the form of the Gaussian function is numerically investigated. The passage of the beams is described with the help of the extended Huygens-Fresnel principle, and the numerical calculation of the propagation operator is realized, using the fast Fourier transform. It is established, that the stability of the state of the orbital angular momentum of single modes, shown earlier is also observed for their superposition, while an increase in the degrees of freedom is provided, which can be used to encode the transmitted information.

1. Introduction

Communication systems, based on the work of space optics are becoming very competitive in the field of broadband access networks [1, 2]. As a consequence, it becomes necessary to study the influence of turbulent perturbations in the atmosphere on the intensity and other properties of laser beams [3].

The propagation of an optical signal in free space can be subject to distortions, associated with the turbulence of the medium [4]. Due to the properties of turbulence, random changes in the atmospheric refractive index may cause a distortion in the intensity of the laser radiation. Classical methods for describing the propagation of the wave through the turbulent atmosphere are based on applications of the Rytov method [5] and the method of parabolic equations [6].

With the help of these methods, the propagation of optical signals from a partially coherent source was investigated [7, 8], and also the features of the propagation of various laser beams in the turbulent medium, including high-order Gaussian beams [9], hollow beams [10], nondiffraction beams, beams of Bessel-Gauss and Airy type [11, 12], as well as cosine-shaped beams [13].
By this it was found, that higher-order Gaussian beams, including vortex beams [14-16], as well as various spatially structured beams, are broadened to a lesser degree by the influence of turbulence, than the fundamental Gaussian beam.

The Laguerre-Gauss modes are example of the beams with so-called screw phase dislocations or phase vortexes. As early as 1992 it was established, that the Laguerre-Gauss laser modes possess an orbital angular momentum [17]. Later it was shown, that the orbital angular momentum is a natural characteristic of all light beams with a vortex phase and can be produced in a conventional optical laboratory [18-21].

The beams, bearing an orbital angular momentum (vortex singular beams [22]) have an infinite number of possible states, which excites a huge theoretical and applied interest in these beams. In particular, they are used to seal the channels of information transfer in free space and in optical fibers

[23, 24]. As a rule, it should be noted, that the individual modes of laser radiation or individual beams of special type are considered. However, different superpositions of multi-mode beams can have a given orbital angular momentum [25, 26], including those, demonstrating other properties, for example, self-reproduction [27]. The stability of such beams to the inhomogeneities of the optical medium has not yet been investigated.

In this paper, the comparative simulation of the passage of individual and combined Laguerre-Gauss modes through a random medium with a given correlation function in the form of Gaussian function is considered. The passage of the beams is described with the help of the extended Huygens-Fresnel principle [28], and the numerical calculation of the propagation operator is realized using, the fast Fourier transform. The effect of random perturbations in the medium on the characteristics of the Laguerre-Gaussian modes, as well as on the superposition of modes with opposite indices of the vortex orders is investigated.

2. Theoretical foundations

The Laguerre-Gauss modes are described by the following expression [29]:

$$GL_{nm}(r,\varphi)=\frac{1}{\sigma_0}\left[\frac{2n!}{\pi(n+|m|)!}\right]^{\frac{1}{2}}\exp\left(-\frac{r^2}{\sigma_0^2}\right)\cdot\left(\frac{\sqrt{2}r}{\sigma_0}\right)^{|m|}L_n^{|m|}\left(\frac{2r^2}{\sigma_0^2}\right)\exp(im\varphi), \tag{1}$$

where $L_n^m(x)$ is the generalized Laguerre polynomial; n and m are the orders of the modes, where m is whole number, and n is whole positive.

To describe the propagation of the laser beam in the medium with random distortions, the extended Huygens-Fresnel principle is used [28]:

$$E(u,v,z,t)=-\frac{ik}{2\pi z}\exp(ikz)\int_{-\infty}^{\infty}\int_{-\infty}^{\infty}E_0(x,y)\exp\left\{\frac{ik}{2z}\left[(x-u)^2+(y-v)^2\right]+\Psi(x,y,u,v,z)-i\omega t\right\}dx\,dy, \tag{2}$$

where $E_0(x,y)$ is the field in the input plane (at $z=0$); $E(u,v,z,t)$ is the field at distance z from the input plane; $\Psi(x,y,u,v,z)$ is random complex phase, associated with the turbulence of the atmosphere. Expression (2) corresponds to the Rytov method [6, 30, 31], and the function describes random deviations of the phase function of the spherical wave, propagating from the initial plane to the output plane.

Due to the fact, that the theoretical and experimental parameters of turbulent media are considered in statistical aspect, the extended Huygens-Fresnel principle is used only to analyze the average characteristics of light beams, such as the average intensity [7-13, 32].

However, in order to investigate the conservation of the state of the orbital angular momentum, in this paper we simulate the propagation of vortex beams and their superpositions through the random environment on the basis of individual statistical realizations.

To illustrate the approach used, it is needed to write down a special case of the extended Huygens-Fresnel principle (2) for the one-dimensional input function, as well as discarding the time dependence:

$$E(u,z)=-\frac{ik}{2\pi z}\exp(ikz)\int_{-\infty}^{\infty}E_0(x)\exp\left\{\frac{ik}{2z}(x-u)^2+\psi(x,u,z)\right\}dx. \tag{3}$$

Let's assume, that the correlation function of the random field does not depend on the specific coordinates x_1 and x_2 but only on their difference. Moreover, it does not depend on coordinates in the output plane, but depends only on the distance z to it [30, 31]:

$$R_U\left\{\exp[\psi(x_1,u_1,0)],\exp[\psi(x_2,u_2,z)]\right\}=\exp\left[-\frac{(x_1-x_2)^2}{\rho_0^2}\right], \tag{4}$$

where $\rho_0(z) = (0{,}545 C_n^2 k^2 z)^{-3/5}$ is the coherence length of the propagation of the spherical wave through the turbulent atmosphere, C_n^2 is the structural constant of the medium, U is the random field.

To determine the state of the orbital angular momentum of the beam, the scalar product of the beam and optical vortexes are used:

$$c_p = \int_0^{2\pi}\int_0^{R} E(r,\varphi,z)\exp(-\mathrm{i}p\varphi) r\,\mathrm{d}\varphi\,\mathrm{d}r, \tag{5}$$

where $E(r,\varphi,z)$ is analyzed beam; $\exp(\mathrm{i}p\varphi)$ is the optical vortex of the p-th order (angular harmonic); c_p is value of the weighting factor of the p–th order.

Perform operation (5) optically, and simultaneously for several orders of optical vortexes it is possible with the help of the multichannel (multi-order) diffractive optical element [34, 35]. The values of the coefficients of the expansion in the angular harmonics (5) make it possible to determine the state of the orbital angular momentum of the laser beam [36, 37]:

$$\mu = \left(\sum_p p\left|c_p\right|^2\right)\left(\sum_p \left|c_p\right|^2\right)^{-1}. \tag{6}$$

3. Modeling results

The simulation of propagation of the individual beams and superpositions of the Laguerre-Gauss modes was performed using a fast Fourier transform. The width of the input area is 2x2 mm, the wavelength is 633 nm, $\sigma_0 = 0.2$ mm.

Figure 1 shows the propagation of the Laguerre-Gauss beam in the random medium at distances of 310 mm, 610 mm and 910 mm. Table 1 shows the weight content of angular momenta, depending on the range of propagation.

As can be seen from Figure 1 and Table 1, even when propagating in the random environment, the content of the corresponding optical vortex in the beam is remained. The greatest values of the weight coefficients are attained at $p = m$.

Table 1. The weight content of the optical vortexes in the single Laguerre-Gauss beams after passing through the random medium.

	z	p = 0	p = 1	p = 2	p = 3	p = 4	p = 5	p = 6	p = 7	p = 8	p = 9	p = 10
$n = 2$, $m = 3$	310	0.0004	0.0394	0.0227	0.7585	0.0706	0.0436	0.0465	0.0023	0.0068	0.0003	0
	610	0.0017	0.0232	0.1258	0.7254	0.1098	0.0015	0.0033	0.0031	0.0041	0.0002	0.0011
	910	0.0004	0.0014	0.0284	0.9295	0.0282	0.003	0.0019	0.004	0.0015	0.0004	0.0008
$n = 3$, $m = 5$	310	0.011	0.0005	0.0117	0.0053	0.0146	0.8909	0.0035	0.0084	0.0087	0.0215	0.008
	610	0.0004	0.0005	0.0046	0.0108	0.0477	0.8377	0.0409	0.0392	0.0158	0.0008	0.0009
	910	0	0	0.002	0.0258	0.0219	0.8055	0.0484	0.0282	0.0531	0.0028	0.0105

Figure 2 shows the propagation of superposition of the Laguerre-Gauss beams with opposite orders of the optical vortexes in opposite directions. Table 2, by analogy with Table 1, shows the content of the optical vortexes after passing through the superposition in the random environment.

IOP Conf. Series: Journal of Physics: Conf. Series **1038** (2018) 012070 doi:10.1088/1742-6596/1038/1/012070

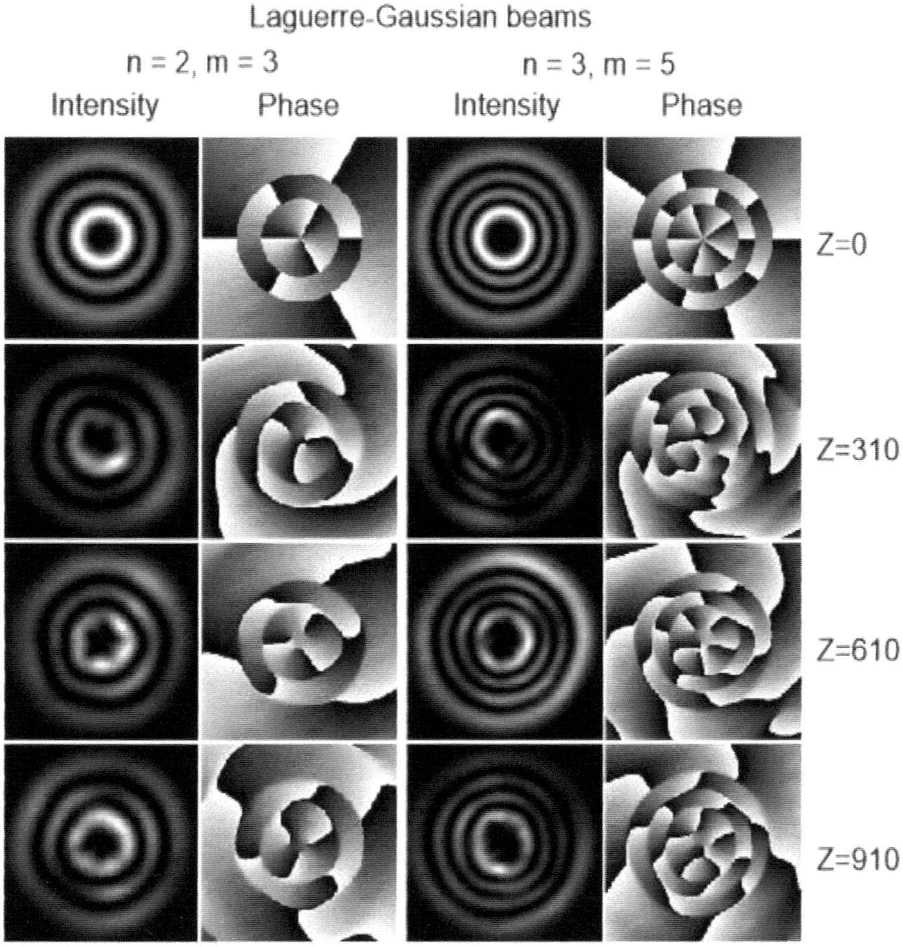

Figure 1. Propagation of the single Laguerre-Gauss beams in the random environment.

Table 2. The weight content of the optical vortexes of superposition of the Laguerre-Gauss beams after passing through the random medium (n = 3).

	z	p = -5	p = -4	p = -3	p = -2	p = -1	p = 0	p = 1	p = 2	p = 3	p = 4	p = 5
$m_1 = 5$, $m_2 = -5$	310	0.4542	0.0187	0.003	0.0008	0.0006	0.0022	0.0015	0.0011	0.0013	0.0172	0.4507
	610	0.3599	0.0339	0.0028	0.0002	0.0003	0.0008	0.0009	0.0027	0.0081	0.0473	0.3627
	910	0.4673	0.0077	0.0069	0	0	0	0	0.0013	0.0011	0.0071	0.4682
$m_1 = 3$, $m_2 = -3$	310	0.0014	0.0113	0.4592	0.0032	0.0021	0.0023	0.0075	0.0255	0.4596	0.0007	0.0012
	610	0.0161	0.0512	0.3963	0.0047	0.0035	0.0083	0.0129	0.0387	0.4013	0.0084	0.0228
	910	0.0003	0.0165	0.4195	0.0453	0.0066	0	0.0002	0.0166	0.4193	0.0387	0.0184

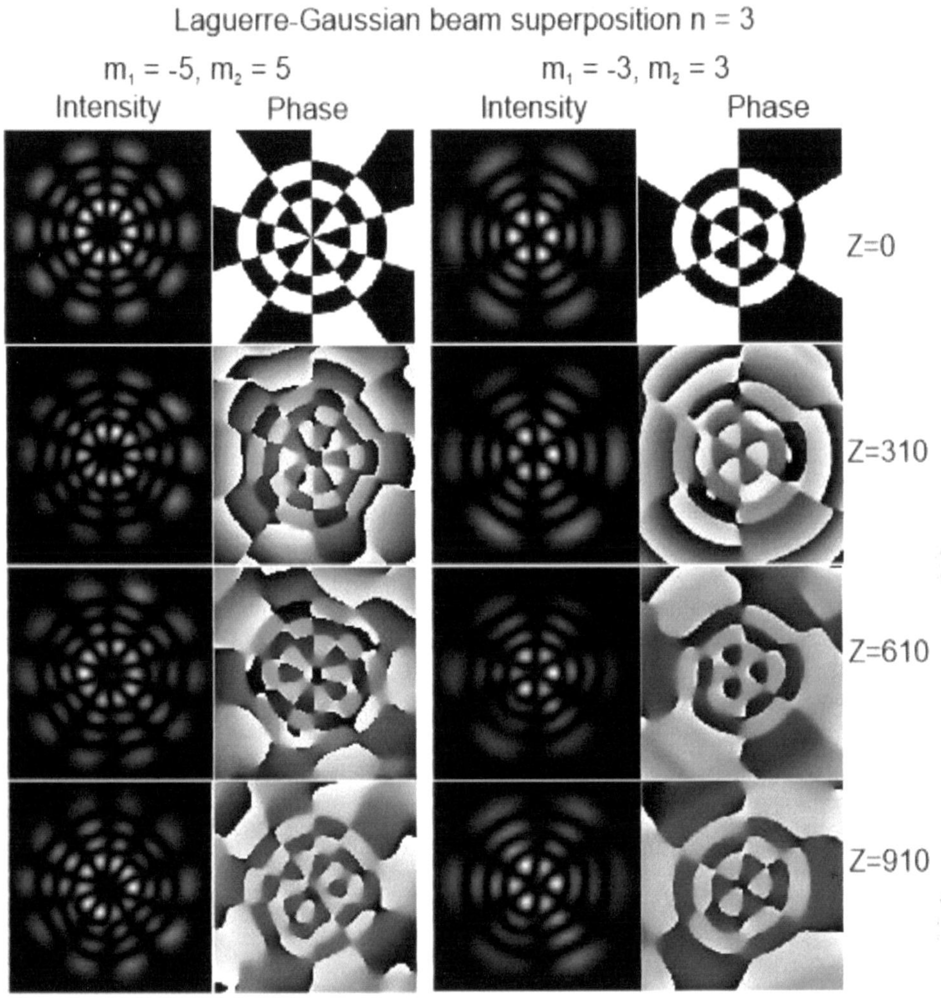

Figure 2. Propagation of the superposition of the Laguerre-Gauss beams in the random environment.

4. Conclusion

In this paper, the propagation of both single Laguerre-Gaussian modes and their combinations were simulated. A comparative analysis of the stability of the characteristics of the propagated laser beams to random perturbations in the medium was made. It is established, that the stability of the state of the orbital angular momentum of single modes, shown earlier is also observed for their superposition, while an increase in the degrees of freedom is provided, which can be used to encode the transmitted information.

Acknowledgments

This work was financially supported by the Ministry of Education of the Russian Federation in the framework of the government-assigned task No. 3.3025.2017/4.6.

References

[1] Majumdar A K and Ricklin J C 2008 *Free-space laser communications: principles and advances, Vol. 2* (New York: Springer Science & Business Media)
[2] Wang F, Liu X and Cai Y 2015 *Progress in Electromagnetics Research* **150** 123-43
[3] Korotkova O 2013 *Random light beams: theory and applications* (Boca Raton, FL: CRC Press, Taylor & Francis Group)

[4] Grabner M and Kvicera V 2012 *Radioengineering* **21** 455-8
[5] Tatarskii V I 1971 *The Effects of the Turbulent Atmosphere on Wave Propagation* (Jerusalem: Israel Program for Scientific Translations)
[6] Manning R M 2008 *Waves in Random and Complex Media* **18** 615-21
[7] Ricklin J C and Davidson F M 2002 *Journal of the Optical Society of America A* **19** 1794-802
[8] Wang S C H and Plonus M A 1979 *Journal of the Optical Society of America A* **69** 1297-304
[9] Young C Y, Gilchrest Y V and Macon B R 2002 *Opt. Eng.* **41** 1097–103
[10] Cai Y and He S 2006 *Opt. Express* **14** 1353–67
[11] Eyyuboglu H T 2007 *Appl. Phys. B* **88** 259–65
[12] Chu X 2011 *Opt. Lett.* **36** 2701–3
[13] Zhou G 2011 *Optics express* **19** 3945-51
[14] Gbur G and Tyson R K 2008 *J. Opt. Soc. Am.* **25** 225
[15] Soifer V A, Korotkova O, Khonina S N and Shchepakina E A 2016 *Computer Optics* **40** 605-24
[16] Porfirev A P, Kirilenko M S, Khonina S N, Skidanov R V and Soifer V A 2017 *Applied Optics* **56** E8-15
[17] Heckenberg N R , McDuff R, Smith C P and White A G 1992 *Opt. Lett.* **17** 221-3
[18] Kotlyar V V, Soifer V A and Khonina S N 1998 *Optics and Lasers in Engineering* **29** 343-50
[19] Khonina S N, Kotlyar V V and Soifer V A 1998 *Optics and Spectroscopy* **85** 636-44
[20] Khonina S N, Kotlyar V V, Soifer V A, Lautanen J, Honkanen M and Turunen J 1999 *Optik* **110** 137-44
[21] Khonina S N, Kotlyar V V, Soifer V A, Honkanen M, Lautanen J and Turunen J 1999 *Journal of Modern Optics* **46** 227-38
[22] Soskin M S and Vasnetsov M V 2001 Singular optics *Progress in Optics* (Amsterdam, North Holland: Elsevier Science) chapter 4 pp 219-76
[23] Bozinovic N, Yue Y, Ren Y, Tur M, Kristensen P, Huang H, Willner A E and Ramachandran S 2013 *Science* **340** 1545-8
[24] Yan Y, Xie G, Lavery M P J, Huang H, Ahmed N, Bao C, Ren Y, Cao Y, Li L, Zhao Z, Molisch A F, Tur M, Padgett M J and Willner A E 2014 *Nature Communications* **5876** 4876
[25] Khonina S N, Skidanov R V, Kotlyar V V, Soifer V A and Turunen J 2005 *Proceedings of SPIE* **5962** 59622W
[26] Kotlyar V V, Khonina S N, Skidanov R V and Soifer V A 2007 *Optics Communications* **274** 8-14
[27] Khonina S N and Volotovsky S G 2007 *Optical Memory & Neural* **16** 167-77
[28] Lutomirski R F and H T Yura 1971 *Appl. Opt.* **10** 1652-8
[29] Yariv A 1985 *Optical Electronics* (New York: Holt, Rinehart and Winston)
[30] Rytov S M 1976 *Introduction to statistical radiophysics. Part I. Random processes* (Moscow: Nauka)
[31] Feizulin Z I and Y A Kravtsov 1967 *Radiophys. Quantum Electron.* **10** 33–5
[32] Young C Y, Gilchrest Y V and Macon B R 2002 *Opt. Eng.* **41** 1097-103
[33] Shirai T, Dogariu A and Wolf E 2003 *J. Opt. Soc. Am. A* **20** 1094-102
[34] Kotlyar V V, Khonina S N, Soifer V A 1998 *Journal of Modern Optics* **45** 1495-506
[35] Khonina S N, Kotlyar V V, Soifer V A, Jefimovs K and Turunen J 2004 *Journal of Modern Optics* **51** 761–73
[36] Khonina S N, Kotlyar V V, Soifer V A, Pääkkönen P, Simonen J and Turunen J 2001 *Journal of Modern Optics* **48** 1543-57
[37] Khonina S N, Kotlyar V V, Soifer V A, Paakkonen P and Turunen J 2001 *Optical Memory and Neural Networks* **10** 241-55

IOP Conf. Series: Journal of Physics: Conf. Series **1038** (2018) 012071 doi:10.1088/1742-6596/1038/1/012071

Propagation of the phase-modulated femtosecond pulses through the optically-dense quasi-resonant medium

A A Preobrazhenskaia[1], A A Pastor[1], P Yu Serdobintsev[1], I A Chekhonin[1], V S Egorov[1]

[1]St. Petersburg State University, St. Petersburg, Russia

E-mail: nakozina_alena@inbox.ru

Abstract. It is known that ultrashort pulse coherent propagation through the resonant absorbing medium leads to the pulse splitting and pulse train generation. First of these pulses is the input pulse and the following pulses are resonant medium response. It is also known that such a propagation mode leads to the «spectrum condensation» effect, at which broad-band input femtosecond laser emission, as it leaves opically-dense medium, centralized in the vicinity of the resonance absorption lines. It's interesting to clear up the dependence of a condensation effect on an input femtosecond pulse phase modulation value (chirp).

1. Introduction

Coherent propagation of an ultrashort pulse through the optically-dense rubidium vapors under the quasi-resonant conditions is studied. Hence the early published articles [1,2] concerned the data of these processes without chirping of the incident laser pulse it is interesting to elucidate the chirping influence on the transformation of an outcoming pulse. The brief results of the experiments combined with the computer simulation are presented below.

In our investigation we used the femtosecond laser «Pulsar-10» with the following parameters: pulse duration – 50 fs, pulse energy – up to 10 mJ, carrier wavelength – 790 nm, pulse spectrum half-width – about 20 nm, light intensity – 0.1 TW/cm^2, beam diameter – 5 mm. To generate phase modulation the compressor, consisted of 2 of diffraction grating, with changeable distances between them, was used.

As a resonant medium the cell (length – 10 cm, diameter – 3 cm) with Rb vapor and buffer gas Ar was taken. To get required Rb vapor pressure we heated the cell to 200-300 °C. Concentration of vapor is about 10^{16} cm^{-3}.

The femtosecond laser «Pulsar-10» was consisted of the oscillator, stretcher, amplifier and compressor, comprised 2 diffraction gratings. Output pulse chirping was attained due to the changing of a distance between these 2 diffraction gratings. Group Delay Dispersion (GDD) was given by following dependence:

GDD(L) = 5250 * L(mm) [fs^2], where L – distance between the gratings, measured with relation to the position, when chirp is 0.

Dependence of the pulse duration and chirp on the compressor detuning is shown in figures 1 and 2 correspondingly.

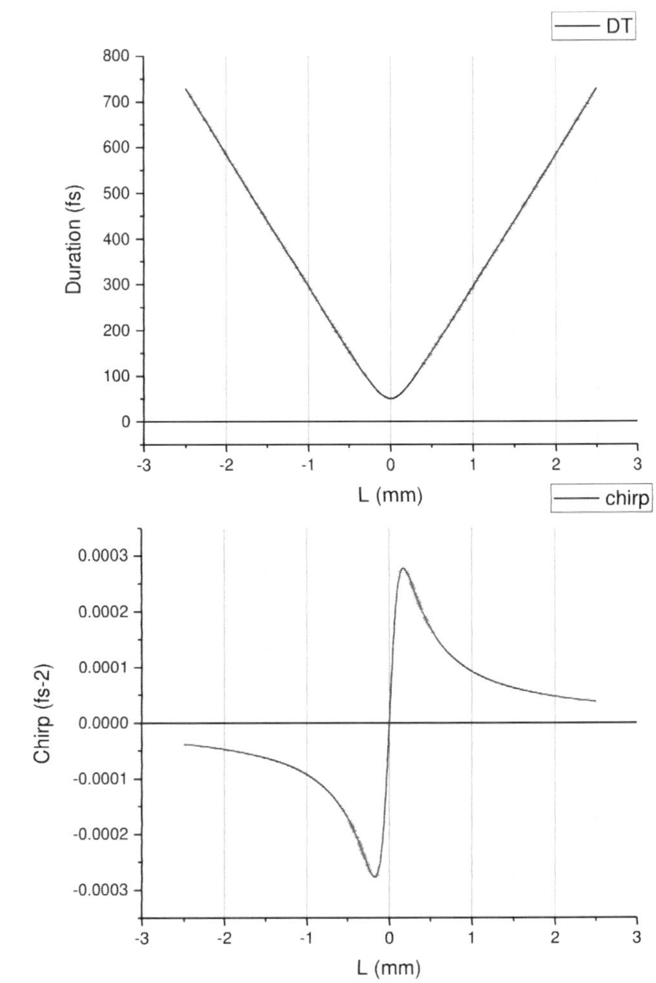

Figures 1 and 2. Dependence of the pulse duration and chirp on the compressor detuning.

Figure 1 shows the pulse half-width changing, depending on the detuning, giving by the following expression:

$\Delta t^2 = (\Delta t_0^4 + 16(\ln 2)^2 \cdot GDD^2)/\Delta t_0^2$

where Δt_0 – pulse half-width in case of compressor zero detuning.

Figure 2 shows the dependence of the chirp on the detuning, giving by the following dependence:

$Chirp = 8\,(\ln 2)^2 \cdot GDD/(\Delta t_0^4 + 16(\ln 2)^2 \cdot GDD^2)$

The results of our experiment show us that a response time of a resonant medium to an input pulse is highly sensitive to a phase modulation sign (chirp). It leads whether to an input signal and a response medium signal constructive interference or, as the result of the change of a chirp sign, to a destructive interference. Examples of the «condensation» and «anti-condensation» of the femtosecond pulse, passed through the Rb vapor cell, were registered on the Rb resonance line (wavelength is 780 nm) in the setting of different chirp signs and are shown in figures 3 and 4.

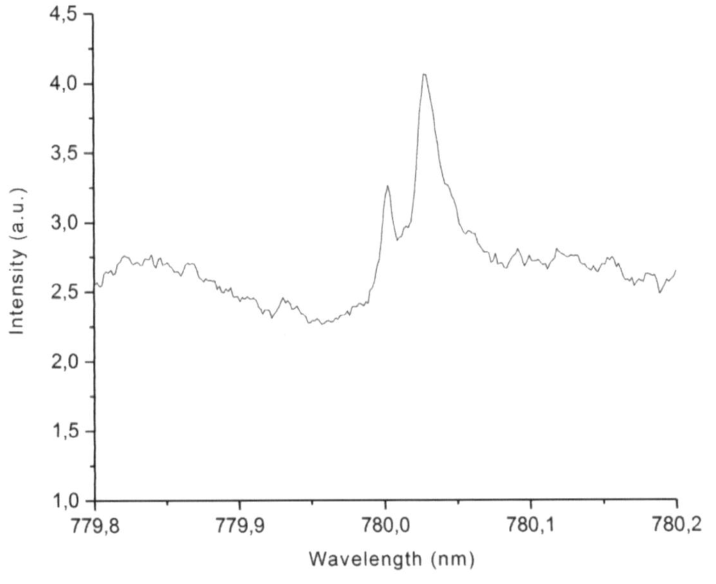

Figure 3. «Spectrum condensation» effect in the vicinity of the Rb resonance line (wavelength is 780 nm). GDD= -2610 fs^2.

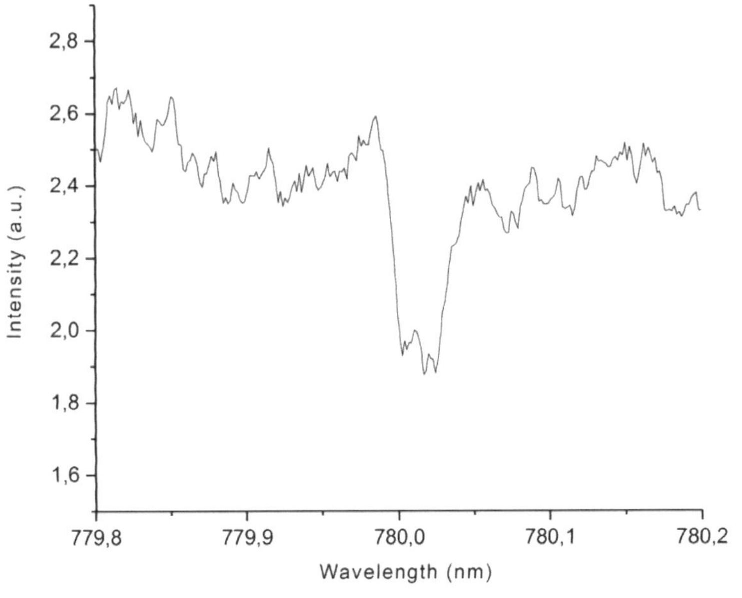

Figure 4. «Spectrum anti-condensation» effect in the vicinity of the Rb resonance line (wavelength is 780 nm). GDD= 2610 fs^2.

It is apparent that under negative chirp value the «spectrum anti-condensation» (spectrum notches in the vicinity of the absorption line, caused by destructive train pulses interference) is observed.

Spectra of laser emission, got through the rubidium cell in «spectrum condensation» regime, was registered with the high resolution spectrograph DFS-8 with a diffraction grating with following parameters: 1200 grooves per mm, dispersion – 0.4 nm/mm. Laser emission spectrum half-width was

20 nm. We used CCD camera with the 7 mm matrix (1000 pixels) for the registration, so the spectral region with a width of 2 nm was registered and it wasn't possible to register the whole spectrum. Panoramic spectrum was registered with the Ocean Optics spectrometer (instrumental width – 0.5 nm). This spectrum is shown in figure 5.

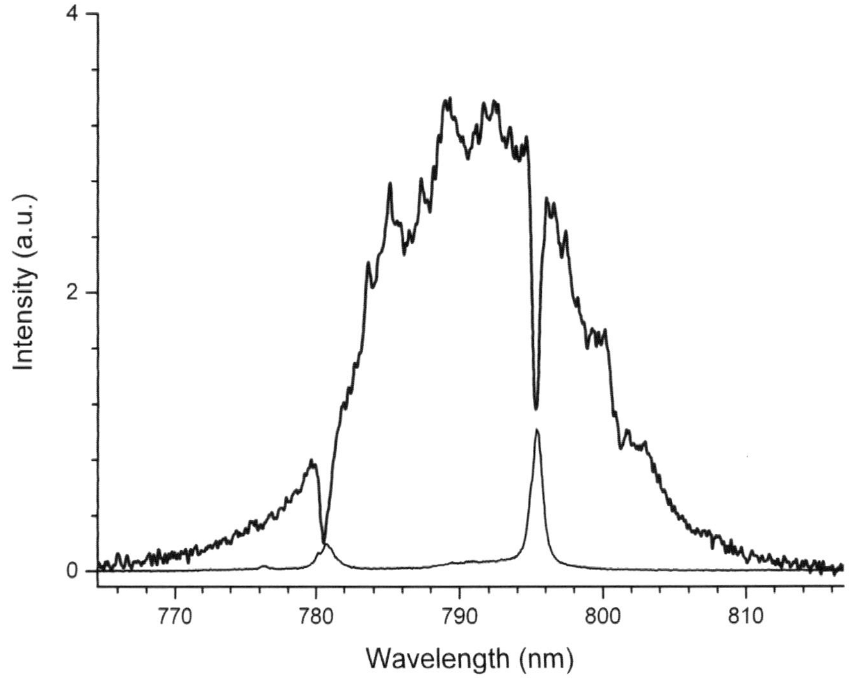

Figure 5. Panoramic spectrum, registered with the Ocean Optics spectrometer.

Upper spectrogram shows the laser spectrum, obtained through the rubidium cell in incoherent propagation conditions. To get these conditions laser beam passed through the teflon diffuser before the cell. You can see 2 rubidium resonance doublet absorption lines. Lower spectrogram shows the laser spectrum, got through the rubidium cell in coherent propagation conditions (without the diffuser). You can see the spectrum condensation effect in the vicinity of the rubidium lines. Condensation effect, as it seen from the panoramic spectrum, become apparent on the both lines (D_1 and D_2) of the rubidium resonance doublet.

Rubidium vapor, consisted of the natural isotopic mixture with the atomic weight 85 and 87, was used. But isotopic splitting of the D_1 and D_2 rubidium lines is much less than the observed line width of the spectra condensation (and anti-condensation).

«Spectrum condensation» effect can be observed with 2 types of emission spectrum on a frequency of the resonant atomic transition. In the case of low density of the rubidium atoms N_0 the narrow superradiance line on a frequency of an atomic resonance ω_{12} is observed. In case of atomic concentration increasing, the intensity of resonance emission increases.

But from some threshold value N_0, the emission spectrum of the atomic ensemble splits into 2 satellites [2]. Their frequency slitting value is defined by the value of the polariton mode frequency splitting value in the resonant medium and increases as N_0 increases.

2. Results of the computer simulation.

We give here some of the model calculations of the resonant absorbing medium polarization response to the femtosecond excitation pulse by taking into account resonance transition frequency detuning in response to the different sign of the input pulse phase modulation.

Model used for computer simulation and obtained results can be described in the following way.

Bloch equations – equations, described precession of the vector ρ around vector Ω.

$$\frac{d\rho}{dt} = [\Omega \times \rho]$$

$\Omega = [-\Omega_{R_1}, -\Omega_{R_2}, \Delta]$ – pump field control vector.

$\rho = [u, v, w]$ – Bloch vector.

Without relaxation ($T_1 = T_2 = \infty$) Bloch vector's squared length is $|\rho|^2 = u^2 + v^2 + w^2 = 1$.

Bloch equations for the projections ρ by taking into account phenomenological decay time T_1 and T_2:

$$du/dt = -\Delta v - u/T_2 - \Omega_{R_2} w$$
$$dv/dt = \Delta u - v/T_2 + \Omega_{R_1} w$$
$$dw/dt = -(w - w_{eq})/T_1 - \Omega_{R_2} u - \Omega_{R_1} v$$

Slow envelope curve of the medium polarization $P = d_{12} N_0 (u + iv)$.

d_{12} – matrix element of the dipole moment operator.

$w = N/N_0$ – normalized population difference.

w_{eq} – equilibrium normalized population difference.

In this case $w_{eq} = -1$.

$\Omega_R = \Omega_{R_1} + i\Omega_{R_2}$ – complex Rabi frequency.

$\Omega_R = d_{12} E/\hbar$, $E = E_1 + iE_2$

$\Delta(t) = \omega_{12} - \omega(t)$, $\Delta(t) = \Delta_0 + (d\Delta(t)/dt)\,\mathrm{t}$ – current laser frequency detuning.

$d\Delta(t)/dt$ – chirp.

Δ_0 – laser pulse frequency detuning without chirp.

All the calculations were made in arbitrary units. But in fact only magnitude relation of $|\Omega_R|$, pulse duration 2σ, Δ_0, T_1, T_2 and chirp $d\Delta(t)/dt$ are important.

Knowing pulse duration and detuning values, we can estimate effects in real conditions.

$|\Omega_R|$ = 3, pulse duration 2σ = 3, frequency laser detuning from exact position of an atomic resonance ω_{12} without chirp Δ_0 = 20, $T_1 = T_2 = 10$.

In figures 6-11 the field amplitude $|E(t)|$ during the pulse is shown. Medium polarization $|u(t) + iv(t)|$ is given in arbitrary units.

In this thin medium layer model the polarization superradiance effect is absent, that's why it decays only by virtue of relaxation (T_1 and T_2 are finite).

IOP Conf. Series: Journal of Physics: Conf. Series **1038** (2018) 012071 doi:10.1088/1742-6596/1038/1/012071

In accordance with the above-described model some analytical data for different sign and values of the chirp is presented below.

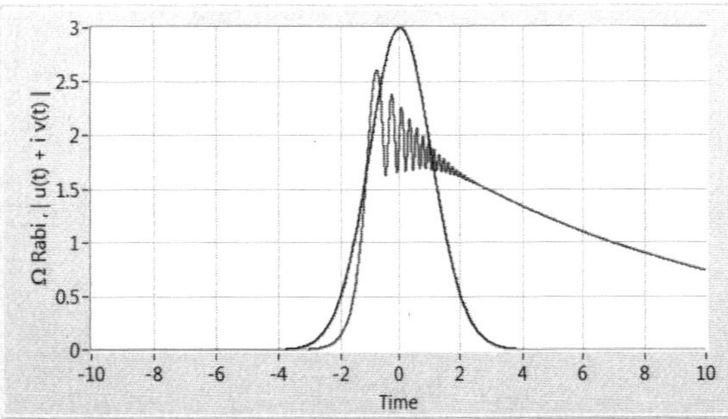

Figure 6. Polarization time dependence in a thin layer on input pulse. The chirp is +15.

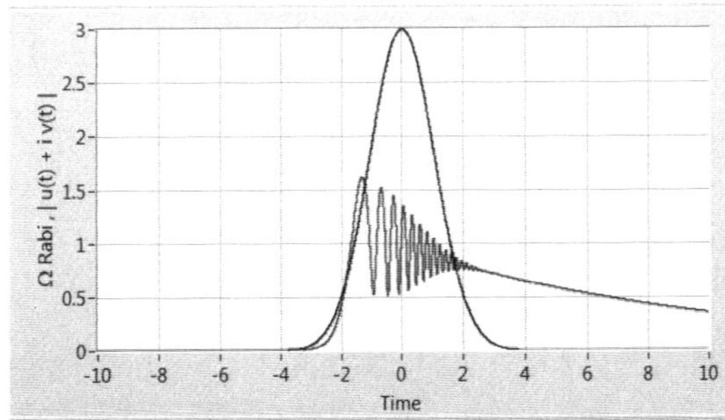

Figure 7. Polarization time dependence in a thin layer on input pulse. The chirp is +10.

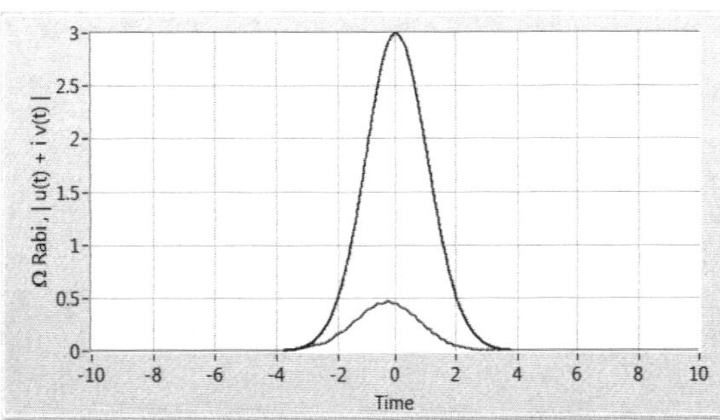

Figure 8. Polarization time dependence in a thin layer on input pulse. The chirp is +5.

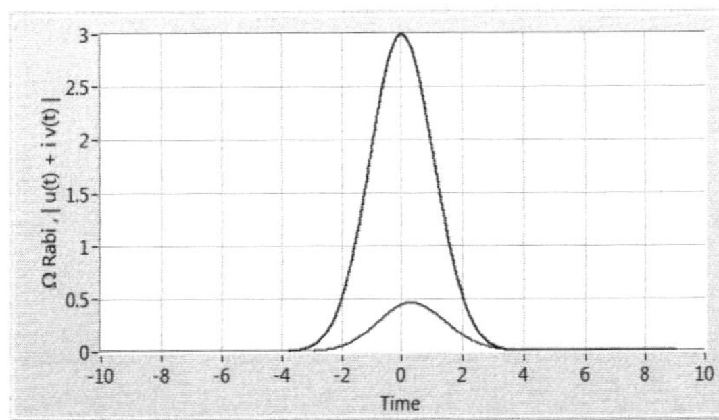

Figure 9. Polarization time dependence in a thin layer on input pulse. The chirp is -5.

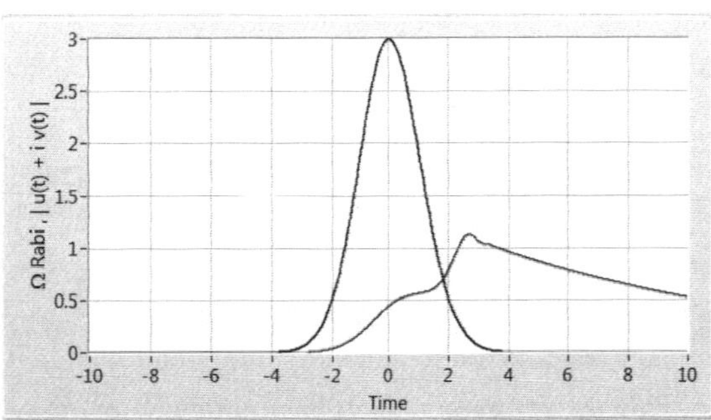

Figure 10. Polarization time dependence in a thin layer on input pulse. The chirp is -10.

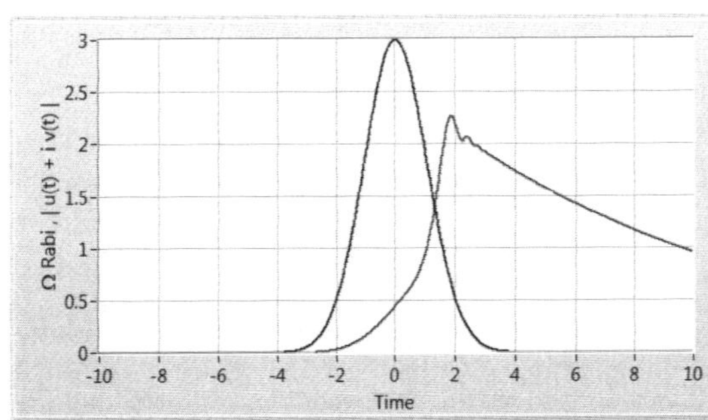

Figure 11. Polarization time dependence in a thin layer on input pulse. The chirp is -15.

In the case of the non-adiabatic excitation of the medium polarization, the superradiant pulse of the rubidium atom ensemble occurs. It's important to say that this superradiant radiation type can occur without achievement of the population inversion w of the resonant medium.

The result of the excitation pulse and superradiant pulse interference is defined by the phase difference of the field oscillation of the excitation pulse and superradiant pulse on the resonance frequency ω_{12} : $\Delta\varphi_1 = \varphi_1$ - atan[$v(t_1)/u(t_1)$] and $\Delta\varphi_2=\varphi_2$-atan[$v(t_2)/u(t_2)$], where t_1 and t_2 - time moments, when medium induced polarization amplitude $|u(t) + iv(t)|$ achieves its' maximum for the negative and positive chirp sign, accordingly; φ_1 and φ_2 - excitation pulse oscillation phases on the frequency ω_{12} in the time moment t_1 and t_2. Constructive interference condition is $\Delta\varphi_1 = 0$, destructive interference condition is $\Delta\varphi_2 = \pm\,\pi$.

3. Conclusion

As can be seen from the computer modeling in figures 6-11, the chirp sign change leads to the radical change of polarization response. In the case of a positive chirp the polarization response appears on the pulse leading edge, and in the case of a negative chirp the substantial delay of the polarization response is observed. We are of the opinion that this delay is the major cause of the transfer from constructive to destructive interference of the pulse train, passing through the optically-dense medium.

Acknowledgements

V.S.E. and I.A.C. thanks Russian Science Foundation (project 17-19-01097) for the partial support of numerical simulations.
Part of the experiments were performed out in the Research Resource Center "Physical methods of surface investigation" of Saint Petersburg State University.

References

[1] Rothenberg J E, Grischkovsky D, Balant A C 1984 Phys.Rev.Lett. **53** (6) 552-555;
[2] Bagaev S N, Egorov V S, Pastor A A, Preobrazhenskii D Yu, Preobrazhenskaya A A, Serdobintsev P Yu, Chekhonin I A, Chekhonin M A 2016 Opt. Spectr. **121** (3) 391-394

International Conference PhysicA.SPb/2017 IOP Publishing
IOP Conf. Series: Journal of Physics: Conf. Series **1038** (2018) 012072 doi:10.1088/1742-6596/1038/1/012072

Holographic formation of phase diffractive elements for light beams conversion with photo-induced absorption coefficient changing in PDLCs

A O Semkin[1], **S N Sharangovich**[1]

[1]Tomsk state university of control systems and radioelectronics, 40 Lenina av., Tomsk, Russia

E-mail: a.o.semkin@gmail.com

Abstract. A theoretical model of the holographic formation of diffractive optical elements for the Gaussian light beams conversion into Bessel-like ones in polymer-dispersed liquid crystals (PDLCs) is developed. The model is based on the solution of the kinetic equations of photopolymerization-diffusion processes of holograms recording in PDLCs by light beams with inhomogeneous amplitude and phase spatial distributions. The model additionally takes into account the photo-induced changings in the absorption coefficient of the PDLC. It is shown that the absorption changings decreases the rate of the refractive index's spatial profile decline over the thickness of the material. In this case, when the characteristic time of the absorption changing is much less than the characteristic polymerization time (the absorption changes more rapidly than the photopolymerization takes place), the diffractive element can be characterized by a homogeneous distribution of the refractive index over the thickness.

1. Introduction

As it's known, any monochromatic collimated beam will travel in free space without significant spreading over the Rayleigh distance' [1]. Looking for types of waves, which propagate at the speed of light and remain focused, has been held by scientists since the early 1980s. Some solutions were found [2, 3], but they had infinite energy, remained theoretically interesting but physically unreal.

Then Ziolkowski [4] found new, exact solutions to the free space wave equation for waves, that could be launched with finite energy from a complex antenna and propagates non-diffracting. Later [5] he proved it experimentally with acoustic waves in water. In 1987, Durnin [6, 7] developed a non-diffracting zero-order Bessel function solution to the free space scalar wave equation and then published experimental evidence for the nearly-non-diffracting nature of these waves, claiming propagation distances far exceeding the normal Rayleigh range before significant beam spreading was observed. Beams of such waves were called Bessel beams, and they still find a lot of applications.

Non-diffracting Bessel-like beams could be generated in several ways: using optical systems [8], holographic methods [9, 10] fixed-length Fabry-Perot cavity [11, 12], etc. The most common method is the transformation of the Gaussian wave's wavefront by the devices of axicon optics [13, 14], as well as by diffraction elements [15, 16].

Polymer-dispersed liquid crystals (PDLCs) with a dye sensitizer, used as a photosensitive medium for holographic diffraction elements recording [17-20, etc.], can potentially be used to form the light beam conversion elements. The aim of this work is to study the holographic formation of diffraction elements based on the PDLCs for the conversion of light beams.

IOP Conf. Series: Journal of Physics: Conf. Series **1038** (2018) 012072 doi:10.1088/1742-6596/1038/1/012072

2. Main equations

Holographic formation of a diffractive optical element (DOE) is performed by two beams of quasi-monochromatic linearly polarized laser radiation with a wavelength lying in the region of the maximum of the dye's photosensitivity. In the general case, in view of the photo-induced changings in the absorption coefficient, the beams can be described by the following expressions:

$$\mathbf{E}_0(\mathbf{r},t)=\mathbf{e}_0\cdot E_0(\mathbf{r})\cdot e^{i\cdot(\omega t-\mathbf{k}_0\cdot\mathbf{r}-\varphi_0(\mathbf{r}))-\alpha(\mathbf{r},t)\cdot(\mathbf{N}_0\cdot\mathbf{r})},\ \mathbf{E}_1(\mathbf{r},t)=\mathbf{e}_1\cdot E_1(\mathbf{r})\cdot e^{i\cdot(\omega t-\mathbf{k}_1\cdot\mathbf{r}-\varphi_1(\mathbf{r}))-\alpha(\mathbf{r},t)\cdot(\mathbf{N}_1\cdot\mathbf{r})}, \quad (1)$$

where $\mathbf{e}_0,\mathbf{e}_1$ - unit polarizations vectors of the beams; $E_0(\mathbf{r})$, $E_1(\mathbf{r})$ - spatial amplitude distributions; $\varphi_0(\mathbf{r})$, $\varphi_1(\mathbf{r})$ - spatial phase distributions; $\mathbf{k}_0$, $\mathbf{k}_1$ - wave vectors; $\mathbf{N}_0,\mathbf{N}_1$ – wave normals; $\alpha(\mathbf{r},t)$ – absorption coefficient; $\omega=2\pi/\lambda$, λ – wavelength of the recording radiation.

In this case, the spatial distribution of the forming field's intencity is determined by the following expression:

$$I(\mathbf{r},t)=I^0(\mathbf{r},t)\cdot[1+m(\mathbf{r},t)\cos(\mathbf{K}\cdot\mathbf{r}+\varphi_0(\mathbf{r})-\varphi_1(\mathbf{r}))], \quad (2)$$

where $m(\mathbf{r},t)=2\sqrt{I_0(\mathbf{r},t)\cdot I_1(\mathbf{r},t)}\cdot(\mathbf{e}_0\cdot\mathbf{e}_1)/(I_0(\mathbf{r},t)+I_1(\mathbf{r},t))$ – local contrast of interference pattern; $I_{0,1}(\mathbf{r},t)=\left|\mathbf{E}_{0,1}(\mathbf{r},t)\right|^2$; $I^0(\mathbf{r},t)=I_0(\mathbf{r},t)+I_1(\mathbf{r},t)$; $\mathbf{K}=\mathbf{k}_0-\mathbf{k}_1$.

In the framework of this paper, the process of holographic formation of DOEs for the Gaussian beams into Bessel-like conversion processes is investigated. The wave vectors of the forming beams lie in the *xOz* plane of the laboratory coordinate system. The recording scheme is assumed to be symmetric, i.e. the angles of the beams incidence are related by expression $\theta_0=-\theta_1$. The PDLC sample is perpendicular to the *xOz* plane, with the *xOy* plane coinciding with the lateral face of the sample. The polarization of the recording beams corresponds to the own extraordinary waves in the PDLC sample. Vectors $\mathbf{e}_0,\mathbf{e}_1$ are parallel to each other and *Oy*-axis. Amplitude distributions of the beams are assumed to be uniform $E_0(\mathbf{r})=E_1(\mathbf{r})=1$. The reference beam has a plane wave front $\varphi_0(\mathbf{r})=0$, while the wavefront of the signal beam is formed by a successively located phase screw and an axicon and is described by the following expression:

$$\varphi_1(x,y)=arctg\left(\frac{|y|}{|x|}\right)+s\cdot\sqrt{x^2+y^2}, \quad (3)$$

where s – the steepness parameter of the forming axicon's cone.

When the light falls on the PDLC, the absorbed quantum leads to the excitation of dye molecules, which interact with the initiator to form the primary radical. After the completion of this stage, the dye molecules do not participate in further chemical reactions, passing into a colorless leucoform. Thus, the concentration of the photosensitive dye during the formation of the DOE decreases, which causes a photo-induced decrease in the absorption coefficient [19]:

$$\alpha(\mathbf{r},t)=\alpha_{sub}+\alpha_0 K_{d0}\cdot\exp(-\beta_{\kappa\text{в}}\alpha_0\times(I_0(x)/\cos\theta_0+I_1(x)/\cos\theta_1)z\cdot t), \quad (4)$$

where α_{sub} – substrate's absorption coefficient; α_0 – absorption of one dye molecule; K_{d0} – initial concentration of the dye; $\beta_{\kappa\text{в}}$ – quantum yield of the dye.

Below on Fig. 1 are the images of the wavefront of the signal beam (Fig. 1a) in the case under consideration, as well as the corresponding spatial distribution of the intensity of the recording field (Fig. 1b). The shown images are received for the described recording geometry and at following values of parameters: $\lambda=633$ *nm*; $\theta_0=-\theta_1=2°$; $s=\sqrt{2}\pi/0{,}001$.

IOP Conf. Series: Journal of Physics: Conf. Series **1038** (2018) 012072 doi:10.1088/1742-6596/1038/1/012072

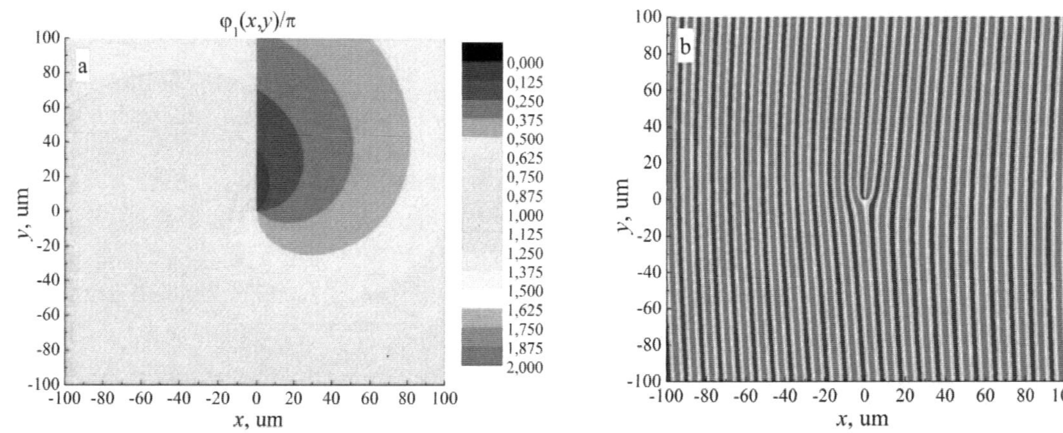

Figure 1. Wavefront of the signal beam (a), spatial distribution of the recording field intensity (b).

Holographic formation of diffractive elements in the PDLC is described by the kinetic equations of the monomer concentration and the refractive index of the material changings [21, 22]:

$$\frac{\partial M(\mathbf{r},t)}{\partial t} = \mathrm{div}\left[D_M \,\mathrm{grad}\, M(\mathbf{r},t)\right] - K_g \cdot \left[\frac{\alpha_0 \beta K_d \tau_0 I(\mathbf{r},t)}{K_b}\right]^{0,5} M(\mathbf{r},t)\,, \tag{5}$$

$$\frac{\partial n(\mathbf{r},t)}{\partial t} = \delta n_p \cdot K_g \cdot \left[\frac{\alpha_0 \beta K_d \tau_0 I(\mathbf{r},t)}{K_b}\right]^{0,5} \frac{M(\mathbf{r},t)}{M_n} + \delta n_{lc}\,\mathrm{div}\left[D_{LC}\,\mathrm{grad}\frac{M(\mathbf{r},t)}{M_n}\right], \tag{6}$$

where M_n – initial concentration of the monomer; K_g, K_b – parameters of the rate of growth and breakage of the polymer chain, respectively; α_0 – absorption of one dye molecule; β – parameter of initiation reaction; K_d – concentration of the dye; τ_0 – the lifetime of the excited state of the dye molecule; D_M, D_{LC} – the diffusion coefficients of the monomer and liquid crystal, respectively; δn_p, δn_{lc} – weight coefficients of the contribution of photopolymerization and diffusion processes; $I(\mathbf{r},t)$ – intencity distribution (2).

Weight coefficients δn_p and δn_{lc} from (6) are found from the Lorentz-Lorentz formula [23]:

$$\delta n_p = \frac{4\pi}{3} \cdot \frac{\left(n_{st}^2 + 2\right)}{6 n_{st}^2} \cdot \left(\alpha_M + \frac{\alpha_P}{l}\right) \cdot \frac{\rho_M}{W_M}, \tag{7}$$

$$\delta n_{lc} = \frac{4\pi}{3} \cdot \frac{\left(n_{st}^2 + 2\right)}{6 n_{st}^2} \cdot \left(\alpha_M \frac{\rho_M}{W_M} + \alpha_{LC} \frac{\rho_{LC}}{W_{LC}}\right), \tag{8}$$

where α_M, α_P, α_{LC} – polarizability of monomer, polymer, and liquid crystal molecules, respectively; ρ_M, ρ_{LC} – density of the monomer and liquid crystal, respectively; W_M, W_{LC} – molecular weights; l – average length of polymeric chains; n_{st} – the refractive index of the composition prior to the start of the recording process, determined by the Lorentz-Lorentz formula from the refractive indices of the monomer and the LC [23]:

$$n_{st} = M_n \cdot \frac{n_M^2 - 1}{n_M^2 + 2} + L_n \cdot \frac{n_{LC}^2 - 1}{n_{LC}^2 + 2}, \tag{9}$$

where n_M, n_{LC} – monomer and liquid crystal refractive indices; M_n, L_n – their initial concentrations.

Equations (5), (6) are differential equations of parabolic type with respect to two spatial coordinates. In the general case, their analytical solution is difficult. Within the framework of the problem under consideration, by dividing the entire *Oy*-axis into N segments Δy, such that $\Delta y << \Lambda$, where Λ - the intensity distribution period (see Fig. 1b), the diffusion of the material components along the *Oy*-axis can be neglected and the problem can be considered as quasi-two-dimensional. So, its solution can be found in the form of a set of all distributions of the refractive index along the *Ox*-axis, defined for each value $y = y_0 + j \cdot \Delta y$, where y_0 - the initial value, $j = 0...N$ - the segment number.

Then the solution of equations (5), (6) for each y_j in accordance with the procedure [21, 22] can be found as a sum of spatial harmonics of the monomer concentration and the refractive index:

$$\begin{gathered} M^j(\mathbf{r},\tau) = \sum_{i=0}^{H} M_i^j(\mathbf{r},\tau)\cos\left[i\cdot\left(K_x\cdot x + \varphi_1^j(x)\right)\right] \\ n^j(\mathbf{r},\tau) = n_{st} + \sum_{i=0}^{H} n_i^j(\mathbf{r},\tau)\cos\left[i\cdot\left(K_x\cdot x + \varphi_1^j(x)\right)\right] \end{gathered}, \tag{10}$$

where $M^j(\mathbf{r},\tau) = M(x, y_j, z, \tau)$; $n^j(\mathbf{r},\tau) = n(x, y_j, z, \tau)$; $\tau = t/T_m$ – relative time; $T_m = \dfrac{1}{D_M \cdot K_x^2}$ – characteristic diffusion time; $K_x = 2k_0 \cdot \sin\theta_0$; the phase shift $\varphi_1^j(x)$ of each harmonic for each y_j is determined by the following expression:

$$\varphi_1^j(x) = \varphi_1(x, y_j). \tag{11}$$

Consequently, the solution of the problem of holographic formation of DOE is reduced to finding the space-time distributions of the harmonic amplitudes $M_i^j(\mathbf{r},\tau)$ and $n_i^j(\mathbf{r},\tau)$.

3. Analytical solution

Amplitudes of spatial harmonics in accordance with the technique [21, 22] can be found from the solution of the system of H+1 kinetic equations for the monomer concentration and the refractive index:

$$\begin{gathered} \frac{\partial M_i^j(\mathbf{r},\tau)}{\partial\tau} = -i^2 M_i^j(\mathbf{r},\tau) + \sum_{l=0}^{H} a_{i,l}^j(\mathbf{r},\tau) M_l^j(\mathbf{r},\tau) \\ \frac{\partial n_i^j(\mathbf{r},\tau)}{\partial\tau} M_n = \delta n_{lc}\cdot i^2 M_i^j(\mathbf{r},\tau) - \delta n_p \cdot \sum_{l=0}^{H} a_{i,l}^j(\mathbf{r},\tau) M_l^j(\mathbf{r},\tau) \end{gathered}, \tag{12}$$

where the coefficient matrix $a_{i,l}^j(\mathbf{r},\tau)$:

$$a_{i,l}^j(\mathbf{r},\tau)=-\begin{Bmatrix} e_1^j & e_2^j & e_3^j & 0 & 0 & . & 0 & 0 \\ 2e_2^j & e_{11}^j & e_2^j & e_3^j & 0 & . & 0 & 0 \\ 2e_3^j & e_2^j & e_1^j & e_2^j & e_3^j & . & 0 & 0 \\ 0 & e_3^j & e_2^j & e_1^j & e_2^j & . & 0 & 0 \\ 0 & 0 & e_3^j & e_2^j & e_1^j & . & 0 & 0 \\ . & . & . & . & . & . & . & . \\ 0 & 0 & 0 & 0 & 0 & . & e_1^j & e_2^j \\ 0 & 0 & 0 & 0 & 0 & . & e_2^j & e_1^j \end{Bmatrix}, \quad \begin{aligned} e_1^j &= \frac{\sqrt{2}}{b_s^j}\left(1-m_s^{j^2}/16\right) \\ e_{11}^j &= \frac{\sqrt{2}}{b_s^j}\left(1-3m_s^{j^2}/16\right) \\ e_2^j &= \frac{\sqrt{2}}{b_s^j}\frac{m_s^j}{4} \\ e_3^j &= -\frac{\sqrt{2}}{b_s^j}m_s^{j^2}/32 \\ m_s^j &= m(\mathbf{r},\tau)=m(x,y_j,z,\tau) \end{aligned} \tag{13}$$

where $b_s^j = b^j(\mathbf{r},\tau) = \dfrac{T_p(\mathbf{r},\tau)}{T_m^j(\mathbf{r})}$ - the ratio of the characteristic polymerization and diffusion times;

$T_p(\mathbf{r},\tau) = \dfrac{e^{\alpha(\mathbf{r},\tau)\cdot\mathbf{r}}}{K_g}\cdot\left[\dfrac{2K_b}{\alpha_0\beta K_d\tau_0}\right]^{0,5}$ – characteristic polymerization time.

In the framework of the problem under consideration, due to the heterogeneity of the phase distribution of the recording field, the characteristic diffusion time, firstly, depends on the coordinate x and, secondly, is determined for each y_j:

$$T_m^j(\mathbf{r}) = T_m\cdot\left(1+\frac{d\varphi_1^j(x)}{dx}\cdot\frac{1}{K_x}\right)^{-2} \tag{14}$$

For the analysis it is convenient to represent the parameter $b^j(\mathbf{r},\tau)$ as:

$$b^j(\mathbf{r},\tau)=b\cdot\left(1+\frac{d\varphi_1^j(x)}{dx}\cdot\frac{1}{K_x}\right)^2 e^{\alpha(\mathbf{r},\tau)\cdot\mathbf{r}},\ b=\frac{T_p}{T_m},\ T_p=\frac{1}{K_g}\cdot\left[\frac{2K_b}{\alpha_0\beta K_d\tau_0}\right]^{0,5}. \tag{15}$$

The ratio of the rates of photopolymerization and diffusion (expressed through a parameter b) determines the character of the formation process and the distribution of the refractive index of the material [21, 22]. In particular, the predominance of the photopolymerization process can lead to the formation of DOE, the spatial profile of which differs significantly from the interference pattern [24]. The study of the kinetics of the amplitude variation of the spatial harmonics of the refractive index will make it possible to trace the transformation of the DOE profile throughout the entire formation process.

For the analysis of the system of equations (12), it is convenient to introduce the coupling coefficients $c_{i,l}^j(\mathbf{r},\tau)=a_{i,l}^j(\mathbf{r},\tau)-i^2\delta_{i,l}$ ($\delta_{i,l}$ – the Kronecker symbol), which characterize the connection of the i-th harmonic with the l-th harmonic. The difference between the coupling coefficients $c_{i,l}^j(\mathbf{r},\tau)$ and $a_{i,l}^j(\mathbf{r},\tau)$ characterizes the contribution of diffusion of the monomer to the recording process and is proportional to the square of the harmonic number. The increase in the diffusion contribution with the number of the harmonic reflected by the term $i^2\delta_{i,l}$ is due to a decrease in the period of the structure and, accordingly, the diffusion time for a given harmonic with respect to the characteristic diffusion time adopted for the first harmonic.

To solve the system of coupled differential equations (12) initial conditions are introduced:

$$M_0^j(\mathbf{r},\tau=0)=M_n,\ M_1^j(\mathbf{r},\tau=0)=0\ldots M_H^j(\mathbf{r},\tau=0)=0, \tag{16}$$

$$n_0^j(\mathbf{r},\tau=0)=0\,,\; n_1^j(\mathbf{r},\tau=0)=0\ldots n_H^j(\mathbf{r},\tau=0)=0 \tag{17}$$

and the operator method is applied similarly to [21, 22]. Then the general solution for the spatial and temporal amplitude distributions of the harmonics of the monomer concentration will be written as:

$$M_i^j(\mathbf{r},\tau)=M_n\sum_{l=0}^{H}A_{i,l}^j(\mathbf{r},\tau)\exp\left[\lambda_l^j(\mathbf{r})\right], \tag{18}$$

where the functional dependences of the coefficients $\lambda_l^j(\mathbf{r})$ are defined as the roots of the characteristic equation $\left|c_{i,l}^j(\mathbf{r},\tau)-\lambda_l^j(\mathbf{r})\right|=0$. Analysis shows that $\lambda_l^j(\mathbf{r})$ are real, different and negative. The coefficients $A_{i,l}^j(\mathbf{r},\tau)$ are found from (12), taking into account the initial conditions (14) as solutions of (H+1) systems of linear algebraic equations

$$\begin{Bmatrix} 1 & 1 & 1 & . & 1 \\ \lambda_0^j & \lambda_1^j & \lambda_2^j & . & \lambda_H^j \\ {\lambda_0^j}^2 & {\lambda_1^j}^2 & {\lambda_2^j}^2 & . & {\lambda_H^j}^2 \\ {\lambda_0^j}^3 & {\lambda_1^j}^3 & {\lambda_2^j}^3 & . & {\lambda_H^j}^3 \\ {\lambda_0^j}^4 & {\lambda_1^j}^4 & {\lambda_2^j}^4 & . & {\lambda_H^j}^4 \\ . & . & . & . & . \\ {\lambda_0^j}^H & {\lambda_1^j}^H & {\lambda_2^j}^H & . & {\lambda_H^j}^H \end{Bmatrix} \times \begin{Bmatrix} A_{i,0}^j \\ A_{i,1}^j \\ A_{i,2}^j \\ A_{i,3}^j \\ A_{i,4}^j \\ \ldots \\ A_{i,H}^j \end{Bmatrix} = M_n \begin{Bmatrix} \delta_{i,0} \\ c_{i,0}^j \\ \sum\limits_{i_0=0}^{H} c_{i,i_0}^j c_{i_0,0}^j \\ \sum\limits_{i_1=0}^{H} c_{i,i_1}^j \sum\limits_{i_0=0}^{H} c_{i_1,i_0}^j c_{i_0,0}^j \\ \ldots\ldots\ldots\ldots \\ \sum\limits_{i_{(N-2)}=0}^{H} c_{i,i_{(N-2)}}^j \ldots \sum\limits_{i_2=0}^{H} c_{i_3,i_2}^j \sum\limits_{i_1=0}^{H} c_{i_2,i_1}^j \sum\limits_{i_0=0}^{H} c_{i_1,i_0}^j c_{i_0,0}^j \end{Bmatrix},$$

where $\lambda_l^j=\lambda_l^j(\mathbf{r})$, $A_{i,l}^j=A_{i,l}^j(\mathbf{r},\tau)$, $c_{i,l}^j=c_{i,l}^j(\mathbf{r},\tau)$.

The general solution for the amplitude of the i-th harmonic of the refractive index for each y_j has the form [23]:

$$n_i^j(\mathbf{r},\tau)=n_i^j(\mathbf{r},\tau)\Big|_p+n_i^j(\mathbf{r},\tau)\Big|_{lc}, \tag{19}$$

where $n_i^j(\mathbf{r},\tau)\Big|_p=\delta n_p\cdot\sum\limits_{l=0}^{H}a_{i,l}^j(\mathbf{r},\tau)\sum\limits_{g=0}^{H}A_{i,g}^j(\mathbf{r},\tau)\dfrac{\exp\left(\lambda_g^j(\mathbf{r})\cdot\tau\right)-1}{\lambda_g^j(\mathbf{r})}$,

$$n_i^j(\mathbf{r},\tau)\Big|_{lc}=\delta n_{lc}\cdot i^2\sum_{g=0}^{H}A_{i,g}^j(\mathbf{r},\tau)\frac{\exp\left(\lambda_g^j(\mathbf{r})\cdot\tau\right)-1}{\lambda_g^j(\mathbf{r})}.$$

Thus, the expressions (10), (18) and (19) represent a general solution of the equations of holographic formation of DOE in a PDLC with a dye-sensitizer and determine the kinetics of the spatial distribution of the monomer concentration and the refractive index.

4. Numerical simulation

To determine the kinetics of the amplitudes of the spatial harmonics of the refractive index of the PDLC $n_i^{N/2}(x_0,\tau)=n_i(x_0,y_0,z=0,\tau)$, $x_0=y_0=0$ (the center of the sample, absorption is not taken into account), the numerical simulations were made by expressions (10), (18) and (19) for $\delta n_{lc}/\delta n_p=0{,}5$ and for two values of the parameter $b=0{,}25$ and $b=5$.

Figure 2 shows the results of numerical simulation of the amplitudes of the refractive index harmonics kinetics obtained from expression (19) for the intensity distribution from Fig. 1b.

IOP Conf. Series: Journal of Physics: Conf. Series **1038** (2018) 012072 doi:10.1088/1742-6596/1038/1/012072

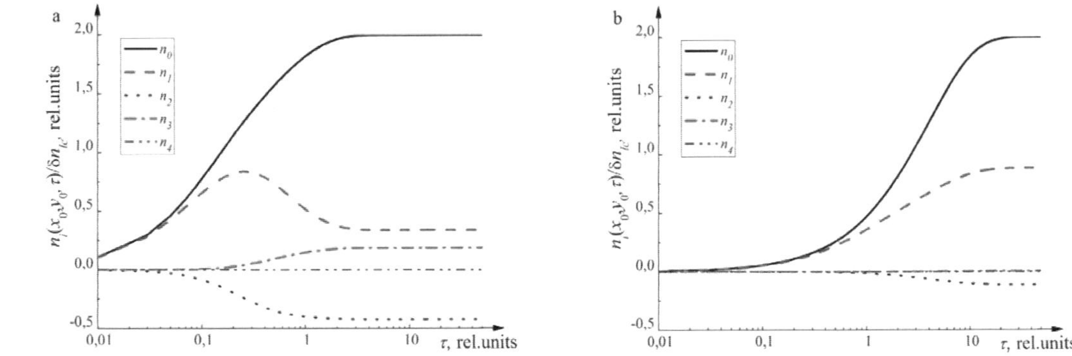

Figure 2. Kinetics of refractive index harmonics for $b=0{,}25$ (a) and $b=5$ (b)

Fig. 2 shows that in the case of predominance of the photopolymerization formation process over the diffusion (Fig. 2a), a significant contribution to the change in the refractive index is made by higher spatial harmonics. In particular, the second one in absolute magnitude exceeds the first harmonic by 0.08 rel. units, starting from $\tau=1$ and up to the completion of the recording process. On the other hand, if the diffusion process predominates (Fig. 2b), the formation process is close to linear and the main (first) harmonic contributes the most to the change in the distribution of the refractive index.

Fig. 3 shows the results of numerical simulations of the spatial distributions of the refractive index of the PDLC after the completion of all the formation processes obtained from expression (10) for the interference pattern (2), (Fig. 1b).

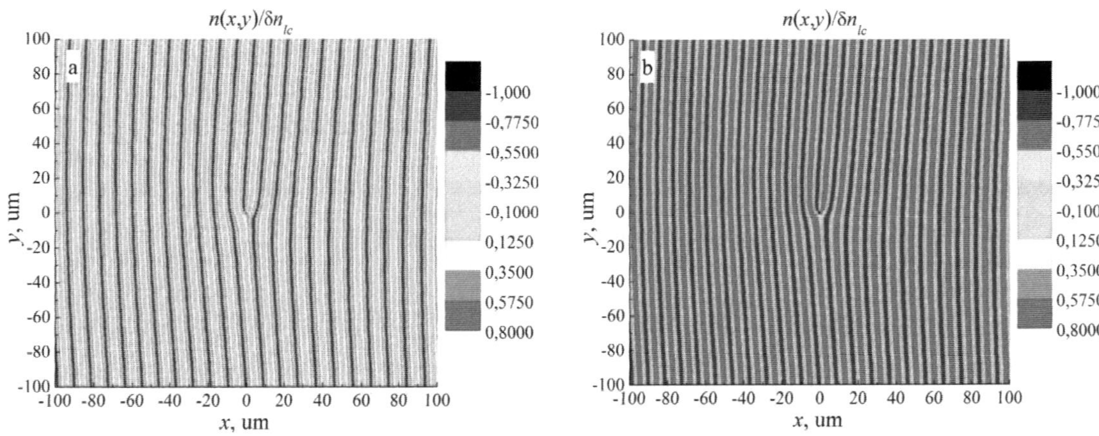

Figure 3. Spatial distributions of refractive index for $b=0{,}25$ (a) and $b=5$ (b)

Fig. 3 shows that if the diffusion process predominates (Fig. 3b), the spatial distribution of the refractive index is close to the interference pattern (see Fig. 1b). This allows us to conclude that the holographic DOE has a minimum of distortion (which is extremely important for the efficient conversion of light beams), if the composition of the PDLC and the formation conditions are selected in such a way that the ratio of the photopolymerization and diffusion rates meets the requirement $b>>1$.

Taking into account the absorption in describing the process of DOE's holographic formation leads to the dependence of the amplitudes of the spatial harmonics on the z coordinate (along the thickness of the sample). In this case, the photo-induced changing of the absorption coefficient causes additional time dependence. Fig. 4 shows the results of numerical simulations of the refractive index harmonics amplitudes as a function of the z coordinate for $b=5$, after the completion of the DOE formation processes for two cases: steady-state attenuation (Fig. 4a) and its photo-induced changing (Fig. 4b). Parameters of photo-induced absorption changing were taken from [19].

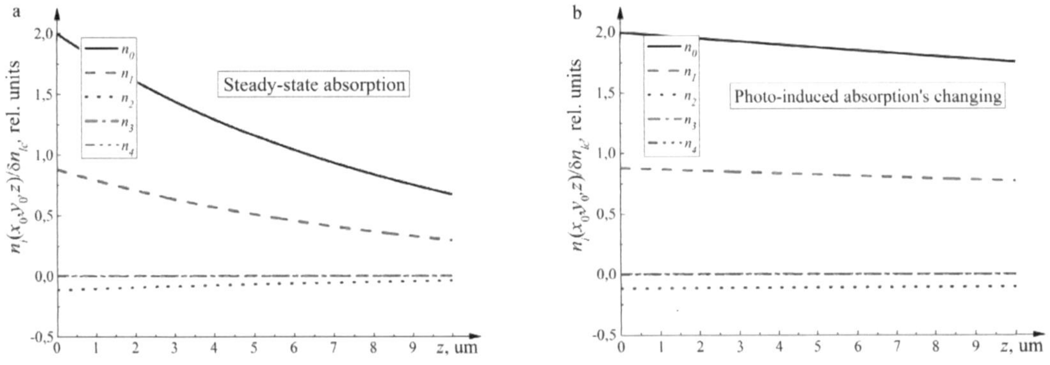

Figure 4. Dependence of harmonics' amplitudes along the samples' thickness for steady-state absorption (a) and for its photo-induced changings (b).

Fig. 4 shows that the changing absorption decreases the rate of decline of the spatial profile of the refractive index over the thickness of the sample. In this case, when the characteristic time of the change in absorption is much less than the characteristic polymerization time (the absorption changes more rapidly than the photopolymerization takes place), the diffraction element can be characterized by a homogeneous distribution of the refractive index over the thickness.

5. Conclusion

Thus, the paper presents a theoretical model of holographic formation of diffractive optical elements (DOE) for the conversion of light beam fields into Bessel-like in PDLCs with a sensitizer dye. The model allows numerical simulation of the course of the formation of DOE depending on the experimental conditions and parameters of the prepolymer composition. Based on the simulation results, it is shown that the ratio of the diffusion and photopolymerization rates during the DOE formation process should be ensured to the level $b >> 1$. In addition, it was shown that the effect of the photo-induced changing of the absorption coefficient causes a nearly uniform distribution of the amplitude of the change in the refractive index along the thickness of the sample.

Acknowledgments

The work is performed as a project part of Government Task of Russian Ministry of Education (project № 3.1110.2017/4.6).

References

[1] Siegman A E 1971 *McGraw-Hill Book Co.* 313
[2] Brittingham J N 1983 *J. Appl. Phys.* **54** 1179
[3] Belanger P A 1984 *J. Opt. Soc. Am.* **1** 723
[4] Ziolkowski R W 1985 *J. Math. Phys.* **26** 861
[5] Ziolkowski R W, Lewis D K, Cook B D 1989 *Phys. Rev. Lett.* **62** 147
[6] Durnin J 1987 *J. Opt. Soc. Am. A* **4** 651
[7] Durnin J, Miceli J J Jr., Eberly J H 1987 *Phys. Rev. Lett.* **58** 1499
[8] Durnin J, Miceli J J Jr., Eberly J H 1988 *Opt. Lett.* **13** 79
[9] Vasara A, Turunen J, Friberg A T 1989 *J. Opt. Soc. Am. A* **6** 1748
[10] Hecht E 1987 *Optics* (Addison-Wesley, Reading, MA)
[11] Cox A J, Dibble D C 1992 *J. Opt. Soc. Am. A* **9** 282
[12] Cox A J, D'Anna J 1992 *Opt. Lett.* **17** 232
[13] Grunwald R, Kebbel V, Griebner U, Neumann U, Kummrow A, Rini M, Nibbering E T J, Piche M, Rousseau G, Fortin M 2003 *Phys. Rev. A* **67** 063820
[14] Zheng C, Petersen T C, Kirmse H, Neumann W, Morgan M J, Etheridge J 2017 *Phys. Rev. Lett.* **119** 174801
[15] Ruan H, Wang L, Wu S, Liu L, Zhou B 2017 *IEEE Photonics Journal* **9** 6500910

[16] Huang Z, Zhu L, Yu J 2017 *Optical Engineering* **56** 103101
[17] Kovalenko E S, Sharangovich S N, Sysuev V V, Dovolnov E A 2000 *Proc. of SPIE* **4513** 75
[18] Sharangovich S N, Dovolnov E A 2002 *Proc. of SPIE* **5129** 236
[19] Dovolnov E A, Ustyuzhanin S V, Sharangovich S N 2006 *Russian Physics Journal* **48** №10 1129
[20] Liu Y J, Sun X W 2008 *Advances in OptoElectronics* **684349** 1
[21] Dovolnov E A, Sharangovich S N 2005 *Russian Physics Journal* **48** №5 501
[22] Dovolnov E A, Sharangovich S N 2005 *Russian Physics Journal* **48** №7 766
[23] Dovolnov E A, Sharangovich S N 2005 *Proc. of SPIE* **6023** 602301-1
[24] Dovolnov E A, Mirgorod V G, Pen E F, Sharangovich S N, Shelkovnikov V V 2007 *Russian Physics Journal* **50** №4 342

IOP Conf. Series: Journal of Physics: Conf. Series **1038** (2018) 012073 doi:10.1088/1742-6596/1038/1/012073

Xenon clusters fragmentation in a supersonic beam under ionization by electrons and photons

A N Arseniev[1], P Yu Serdobintsev[1,2], A S Melnikov[1], L P Rakcheeva[1], A A Pastor[2], M A Khodorkovskii[1]

[1] Peter the Great St.Petersburg Polytechnic University, Saint-Petersburg 195251, Russia
[2] St.Petersburg University, Saint-Petersburg, Saint-Petersburg 199034, Russia

E-mail: arsenievanatoly@gmail.com

Abstract. Measurements of the mass distribution of xenon clusters in a pulsed supersonic beam during ionization by electrons and photons are carried out. For ionization by electron impact, an electron gun was used. The energy of the electron beam varied in the range 15 – 70 eV. Third harmonic of a titanium-sapphire femtosecond laser with a wavelength of 263 nm was used for multiphoton optical ionization. Clusters of xenon Xe_N with $N < 1000$ were observed in a supersonic beam,. It is shown that the shape of the mass spectrum in multiphoton optical and electronic ionizations coincide at the electron energy near the ionization threshold (15 – 20 eV). At the electron energy of 30 – 70 eV the shape of the mass spectrum is substantially distorted in the region N ~ 2 – 150. The change in the shape of the mass spectrum is mainly due to the large fragmentation of multiply ionized xenon clusters due to Coulomb decay.

Introduction

In the supersonic beam, which is created when the gas mixture expands into vacuum through the nozzle, a significant decrease in the gas temperature to several kelvins and the formation of atomic (molecular) clusters occurs. Traditionally, the analysis of the composition of a supersonic beam is carried out by mass-spectrometry with ionization by electron impact.

Ionization potentials of xenon atoms into $^2P^0_{3/2}$ and $^2P^0_{1/2}$ ion states are 12.13 and 13.44 eV, respectively. The ionization potentials of xenon clusters are lower than the ionization potentials of xenon atoms [1]. Lowest adiabatic ionization potentials (E_A) of xenon clusters with N = 2 – 20 were measured in [2] and are 10 - 11eV. Usually, mass spectrometry uses electron energies of $\varepsilon = 70$ eV, which is significantly higher than the ionization potential for xenon clusters.

Ionization of clusters of inert gases was considered in [3]. In the first approximation, it is assumed that the energy of the ionizing electron is not essential, since it only knocks out one electron from the outer shell of the atom and does not lead to the destruction of the cluster. R_2^+ dimer inside the cluster is formed in the case of inert gas cluster ionization by electrons or photons. Energy ~ 1 eV is released during this process and a partial evaporation of the cluster occurs. About 6 atoms are evaporated in the case of xenon clusters [4]. Thus, the shape of the mass spectrum should not depend on the energy of the electron beam.

It is known that there is a significant fragmentation of noble gas clusters during ionization by electron impact [5] and the recorded ion spectrum differs from the original beam composition. We note that the mass composition of the beam can not be obtained from the widely used Hagena formula [6] for a pulsed supersonic beam of a gas mixture with a light carrier gas [7,8].

The purpose of this paper is to study the mass spectra of the supersonic beam of xenon clusters obtained at different energies of ionizing electrons and compare them with the mass spectra obtained in multiphoton ionization of clusters in the near-threshold ionization region.

Experimental setup.

Setup is shown in Fig.1. A pulsed cluster beam was created by an adiabatically expanding gas mixture through a sonic nozzle of 0.2 mm in diameter at room temperature. The gas pulse duration was about 200 μs, the repetition rate was 10 Hz. Stagnation pressure of the mixture (92% He and 8% Xe) ahead of the nozzle ranged from 1 to 9 bar. A skimmer with a diameter of 1.5 mm was located at a distance of 20 mm from the nozzle. The cluster beam speed is about 1 km/s. At a distance of 200 mm from the skimmer, the cluster beam enters the ion source of the time-of-flight reflectron mass spectrometer RFT10 (Kaesdorf, Munich). In the ion source the electron impact ionization of clusters occurs. Electron pulse duration was 5 μs. Electron energy varied in the range 15 – 70 eV. Third harmonic of a titanium-sapphire femtosecond laser with a wavelength of 263 nm (hν = 4.71 eV) was used for optical ionization. Laser pulse energy was 50 μJ, duration was 85 fs, repetition rate was 10 Hz. The laser radiation was focused into the same region of the cluster beam as the electron beam by a lens with a focal length of 160 mm.

The generated ions are extracted into the flight tube by means of a high-voltage pulse delayed by 0.2 μs after the ionization pulse. The MCP and the time analyzer recorded ions with a resolution of 250 ps. The resolution of the mass spectrometer is 10^4 at 130 Da. Xenon of natural isotopic composition was used in the work. Xenon has 9 isotopes: from ^{124}Xe to ^{136}Xe, standard atomic weigth is 131.293 Da. The mass spectrum of the Xe_N cluster is a set of peaks corresponding to Xe_N clusters of different isotopic composition. As an example, Fig. 1b shows the mass spectrum for the Xe_3 cluster. It can be seen that the isotopic distribution contains more than 20 peaks, in the range 386.7-407.7 Da. The width of the isotope distribution increases with increasing N, and at N > 11 the distributions begin to overlap. To ignore a significant number of individual peaks, mass spectrum was smoothed with a resolution of m/z ~ 20, which is sufficient to separate the peaks of the clusters of different sizes. The mass calibration was performed over the first 40 peaks of the clusters.

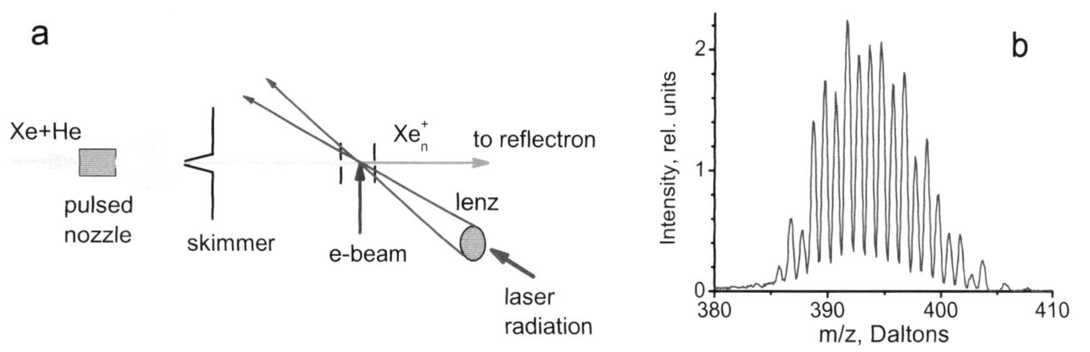

Figure 1. The scheme of the experimental setup (a), mass spectrum of a supersonic beam ionized by electron impact in range 380-410 Da (b).

Experimental results and discussion

Ionization by electron impact

An investigation of the mass spectra at different electron energies was carried out at the maximum stagnation pressure of the gas mixture (9 bar). Mass spectra at various electron energies from the near-threshold energy ε = 15 to 70 eV are presented in Fig.2. Such mass spectra are typical for cluster beams of inert gases [4,9]. The ions of xenon clusters Xe_N^+ with N = 2 – 1000 are observed in the mass spectrum at all electron energies. The shapes of the mass spectrum at ε = 20 eV and ε = 15 eV differs slightly. The relative amount of Xe_N^+ ions with N = 2 – 150 in the mass spectrum increases with increasing electron energy. The strongest changes are observed in the range N = 6 – 60, where the number of ions increases by two orders of magnitude at ε = 70 eV.

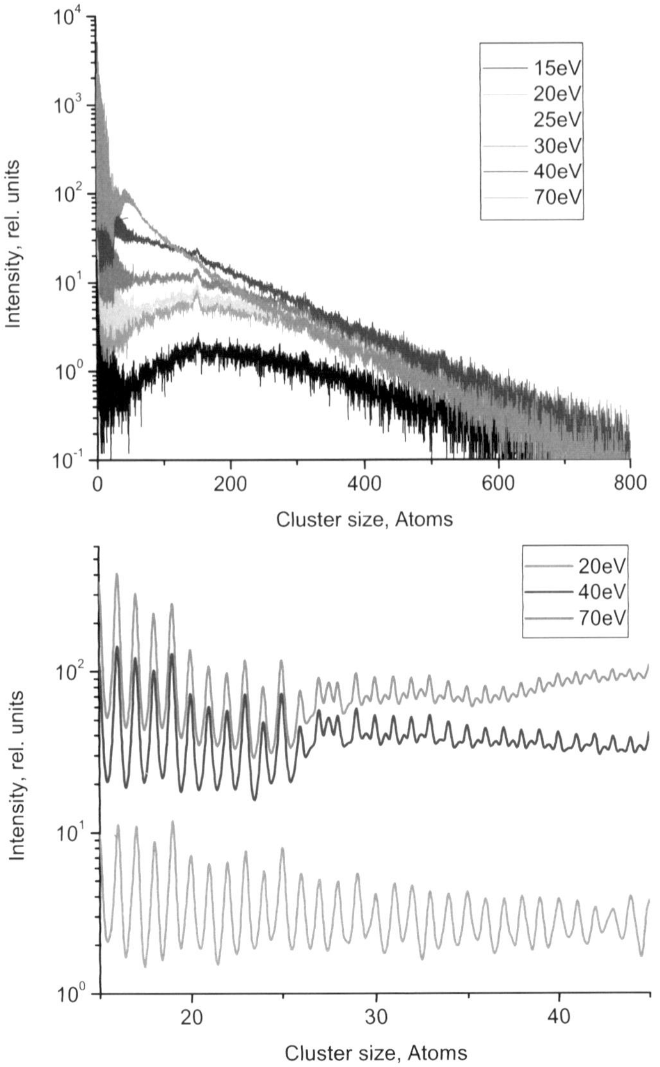

Figure 2. Mass spectra of a supersonic beam ionized by electron impact. Upper: overview spectrum, bottom: part of the spectrum with 15 < N < 45.

The Xe_N^{2+} ions appear in the mass spectrum at electron energies above 30 eV. As is known in the case of small xenon clusters, the Xe_N^{2+} ions decay into singly charged ions with a characteristic time less then 100 ps [10]. The appearance of the ions in the mass spectrum is clearly visible in the region of N = 25 – 40, see Fig.2. In this region of the spectrum there are peaks with a "non-integer" number of atoms in the cluster related to the doubly charged cluster ions with an odd number of atoms. Peaks from doubly ionized clusters with even N coincide with peaks from singly ionized clusters. The first peak from the doubly ionized cluster is at N = 26.5 and corresponds to the Xe_{53}^{2+} cluster. The peak with n = 27.5 has a higher intensity and corresponds to a doubly ionized cluster with the number of atoms equal to the magic number N = 55. The threshold energy of the appearance of Xe_N^{2+} ions is 25 eV, which agrees with the results of [11]. The contribution of Xe_N^{2+} ions to the total intensity of the mass spectrum increases with increasing electron energy. At $\varepsilon = 70$ eV, the contribution of the second ions in the spectral region with N = 26 – 60 is of the order of 50%.

Let us consider the possible variants of the formation of doubly charged clusters.
1. Sequential ionization. This process requires two collisions: in the first one, the cluster ionizes $Xe_N + e \rightarrow Xe_N^+ + 2e$, then the formation of the second ion occurs: $Xe_N^+ + e \rightarrow Xe_N^{2+} + 2e$. The ionization of a neutral atom in a cluster in the second collision requires an energy $E = E_i + E_Q \sim 14$ eV, here E_i is the ionization potential of the xenon atom, E_Q is the energy of the Coulomb interaction. The value of E_Q depends on the arrangement of ions in the cluster and has a value of ~ 0.5 – 3 eV. Thus, the threshold of this reaction is ~ 14 eV.
2. The electron in one collision impacts out two external electrons from one atom of the cluster. The threshold of this process coincides with the potential of the second ion Xe^{2+} (34 eV).
3. The electron in one collision impacts out two external electrons from different atoms of the cluster. The threshold energy of this reaction is $E = 2E_i + E_Q \sim$ 25 eV, this value is close to that observed in the experiment.
4. The electron is knocked out from the inner shell. Next, the intraatomic transition of the electron from the outer shell to the newly formed vacancy occurs. In this case, an energy is released that can ionize another cluster atom via interatomic coulomb decay [12]. For the xenon atom, the energy required to knock out 5s and 4f electrons is 23.4 and 67.5 eV, respectively. Taking into account the ionization potential of the xenon atom the energy release is 11.3 and 57.4 eV. In the case of knocking out 4f electrons of the released energy is sufficient to ionize two more atoms and create Xe_N^{3+} ion. For 5s electron, an energy approximately equal to $E_A + E_Q$ is released, which may be sufficient for ionization of the second atom in clusters.

In the experiment, the doubly charged ions appear at the electron energy of ~ 26 eV, consequently, processes 1 and 2 are unlikely in conditions considered. The ionization thresholds of processes 3 and 4 are close to the observed in the experiment.

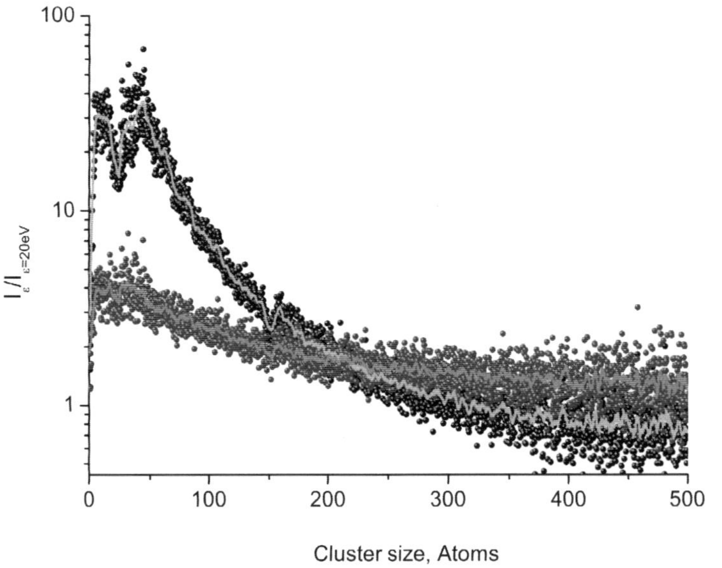

Figure 3. Changes in the shape of the mass spectrum with increasing electron energy. The ratio of the intensities in the mass spectra at $\varepsilon = 70$ (black circles) and 30 eV (blue circles) to the intensity of mass spectrum at $\varepsilon = 20$ eV. Lines are smoothed data.

As mentioned above, the doubly charged ions of clusters with N < 47 are unstable and decay rapidly due to Coulomb repulsion. Note that the intensity of the peaks corresponding to Xe_N^{2+} with N = 47 – 51 observed in [11] was low, in the present work, these ions were not recorded. In [10] it was shown that upon the decay of the Xe_N^{2+} ions with N <51, two singly-charged clusters of size N >> 1 are formed. Stable three-fold ionized clusters of Xe_N^{3+} exist for N > 108 [11].

Figure 3 shows the change in the shape of the mass spectrum at an electron beam energy of 70 eV with respect to the shape of the spectrum at ε = 20 eV. It can be seen that in the region N < 150 a significant increase in the intensity of ions is observed. Two peaks corresponding to N = 6 – 16 and N ~ 40 are also observed. Apparently, the first peak corresponds to fragments formed during the Coulomb decay of doubly charged clusters with N < 47 and triply charged clusters with N < 108. The second peak in the mass spectrum corresponds to doubly and triply charged ions.

Photoionization

For the ionization of xenon clusters with 263 nm radiation, three quanta are required, the total energy of which is ε = 14.1 eV. This energy corresponds to the near-threshold ionization region. The intensity of multiphoton ionization depends nonlinearly on the radiation power density. To reduce the influence of nonlinear effects of high orders, photoionization was carried out at radiation power densities slightly exceeding the threshold for the appearance of ions. In this case, the shape of the mass spectrum in optical ionization (see Fig. 4) is close to the shape of the mass spectrum in electron ionization at ε = 15 – 20 eV.

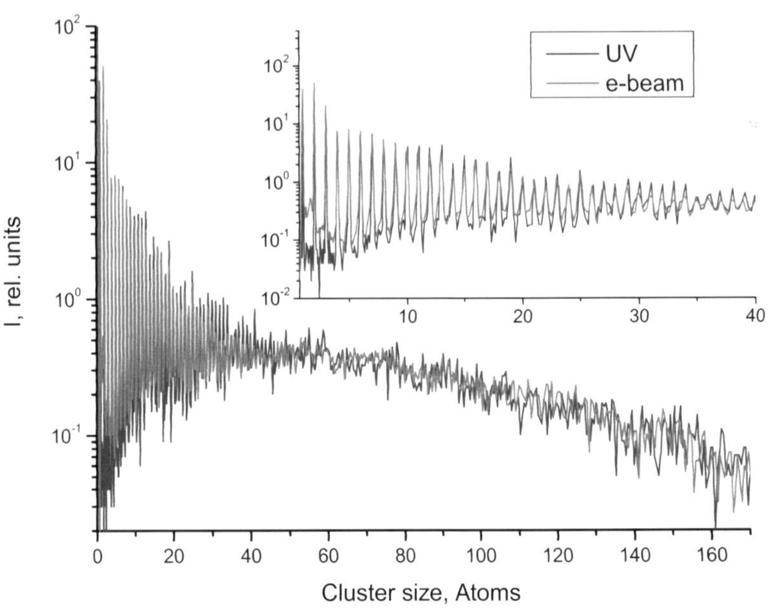

Figure 4. Mass spectra of a supersonic beam with electron (ε = 20 eV) and multiphoton optical ionization (λ = 263 nm).

Conclusion

In optical multiphoton ionization with UV radiation of 263 nm, the mass spectrum coincides with the mass spectrum upon ionization by electron impact in the region of near-threshold energies of 15–20 eV. In the case of ionization of a supersonic beam of clusters by electrons, a significant change in the shape of the mass spectrum is observed with a change in the electron energy from the near-threshold (ε = 15 – 20 eV) to 70 eV. At ε = 70 eV, the intensity of the peaks for clusters with n > 150 decreases, and the number of clusters with n < 150 increases. The intensity of small clusters n = 6–60 increases most strongly. The change in the shape of the mass spectrum is mainly due to the considerable fragmentation of multiply ionized xenon clusters due to Coulomb decay.

At electron beam energies of less than 25 eV, there is no distortion of the shape of the mass spectrum due to the decay of multiply charged clusters.

Acknowledgements
This research has been supported by Russian Federation grant No. 3.8742.201 7/8.9. The experiments were performed out in the Research Resource Center "Physical methods of surface investigation" of Saint Petersburg State University.

References
[1] Carnovale F, Peel J B, Rothwell R G 1991 *J. Chem. Phys.* **95** 1473
[2] Ganteför G, Bröker G, Holub-Krappe E, Ding A 1989 *J. Chem Phys.* **91** 7972
[3] Haberland H 1985 *Surf. Science* **156** 305
[4] Karnbach R, Joppien M, Stapelfeldt J, Wormer J, Moller T 1993 *Rev. Sci. Instrum.* **64** 2838
[5] Poterya V, Fárník M, Buck U, Bonhommeau D, Halberstadt N. 2009 *Int. J. Mass Spectrom.* **280** 78
[6] Hagena O F 1981 *Surf. Science* **106** 101
[7] Serdobintsev P Yu et al. 2015 *J. Chem. Phys.* **143** 114302
[8] Khodorkovskii M, Pastor A, Serdobintsev P 2017 *Progress in Photon Science* Springer International Publishing **115** 143
[9] Miehle W, Kandler O, Leisner T, Echt O. 1989 *J. of Chem. Phys.* **91** 5940
[10] Gay J G, Berne B J 1982 *Phys. Rev. Lett.* **49** 194
[11] Scheier P, Walder G, Stamatovic A, Märk T D 1989 *J. of Chem. Phys.* **90** 4091
[12] Cederbaum L S, Zobeley J., Tarantelli F 1997 *Phys.Rev. Lett.* **79** 4778

IOP Conf. Series: Journal of Physics: Conf. Series **1038** (2018) 012074 doi:10.1088/1742-6596/1038/1/012074

Waveguide and Γ-factor optimization for low-divergence ridge lasers

I K Boikov[1], A V Savelyev[2], A E Zhukov[2]

[1] Department of Applied Mathematics and Physics, St. Petersburg Academic University of the Russian Academy of Sciences, St. Petersburg 194021, Russia

[2] Nanophotonics Lab, St. Petersburg Academic University of the Russian Academy of Sciences, St. Petersburg 194021, Russia

E-mail: ik.boikov@gmail.com

Abstract. The work is aimed for analytical approach to maximize ridge laser modal gain at a given beam divergence in a vertical direction. When the gain/loss profile of planar waveguide is known the equation for optimum near-field profile was obtained. Surprisingly, it is appeared to be Schrödinger type equation where normalized gain/loss profile plays role of potential energy. The developed approach was applied to a few examples widely met in practice. Also it was checked whether usage of leaky waveguides is justified or not.

1. Introduction

Semiconductor ridge emitting lasers have a broad range of applications from medical applications to material processing, therefore, lasers parameters optimization is a matter of particular interest. One of goals is maximizing Γ-factor, which means increased field profile confinement. But it's shown that it results in rising far-field beam divergence, so a trade-off takes place. In this paper we look for analytical approach to this problem.

2. Analytical approach

Consider planar waveguide with TE-polarized wave propagating along x-direction and let $F(z)$ be its near-field profile of y-component of electric field, where z-axis points growth direction in planar waveguide. As it was shown in [1], measure of beam divergence can be introduced as squared average angle Θ:

$$\Theta^2 = -k_0^2 \int\limits_{-\infty}^{\infty} F(z)\frac{d^2F}{dz^2}dz \text{ , while } \int\limits_{-\infty}^{\infty} |F(z)|^2 dz = 1 \text{ ,} \qquad (1)$$

where k_0 is the light wavenumber in vacuum for the lasing wavelength $\lambda_0 = 2\pi/k_0$. Let $g(z)$ be gain profile. Modal gain G can be found via:

$$G = \int\limits_{-\infty}^{\infty} |F(z)|^2 g(z) dz \text{ .} \qquad (2)$$

In order to find best possible modal gain keeping the value of Θ fixed we use variational approach with the Lagrange multipliers, which is described in [2]. To do so, we introduce functional $S[F(z)]$ as well as Lagrange multipliers ρ and λ:

$$S[F(z)] = \int_{-\infty}^{\infty} |F(z)|^2 g(z) dz + \rho \left(\Theta^2 k_0^2 + \int_{-\infty}^{\infty} F(z) \frac{d^2 F}{dz^2}(z) dz \right) + \lambda \left(\int_{-\infty}^{\infty} |F(z)|^2 dz - 1 \right) . \quad (3)$$

Hence, we convert initial task of maximum value of G to the problem of finding extremum of $S[F(z)]$. The conventional variational procedure lead us to the following equation on $F(z)$:

$$-\frac{d^2 F}{dz^2} - \frac{1}{\rho} F(z) g(z) = \frac{\lambda}{\rho} F(z) . \quad (4)$$

One can clearly see here, that this is Shrödinger eigenvalue problem where $-\rho^{-1} g(z)$ plays role of potential energy. Hence we can use well developed methods to solve this problem for different $g(z)$ and to interpret physical meaning of its solution.

We can introduce $\alpha = \rho^{-1}$ as well as $\xi = \lambda \rho^{-1}$ as only their quotient plays role. From this point onwards α is considered a given constant.

$$\frac{d^2 F}{dz^2} + F(z) \left(\alpha g(z) + \xi \right) = 0 \quad (5)$$

This is a Schrödinger's equation for particle in potential $-\alpha g(z)$ with energy ξ. The variable ξ can be found during solution of this eigenproblem. Hence, only α could be varied during optimization procedure. From point of view of Shrödinger equation varying α would lead to changing of depth of "potential well" that corresponds to the gain region. Higher values of the α corresponds to more tight localization of the field profile $F(z)$ around gain region, higher Γ-factor and higher beam divergence. Still the solution at any fixed α is optimal in the sense that Γ could not be better for a given $g(z)$ profile and beam divergence. While $g(z)$ is arbitrary, we cannot proceed any further analytically and bring here few examples.

3. Examples

3.1. Example 1: single gain layer

In that case gain be introduced as $g(z) = A\delta(z)$. Solution of (5) is well known to be

$$F(z) \sim \exp\left(-\sqrt{\alpha A} |z| \right) , \quad (6)$$

where α could be any positive value. Properties of that solution have been studied in details in [1]. We note here that actual value of A is of no importance, since α is arbitrary and should be "scanned" in order to find $F(z)$ with required divergence and, therefore, we ommit it in the following.

3.2. Example 2: two gain layers

Two gain layers in an active media can represent a couple of quantum wells as an example. With the gain profile written as

$$g(z) = g_0(\delta(z - a) + \delta(z + a)) , \quad (7)$$

the solution of (5) is following:

$$F(z) = \begin{cases} N \exp\left(\sqrt{\xi}(z + a)\right) , & z < -a \\ N \operatorname{sech}\left(\sqrt{\xi} a\right) \cosh\left(\sqrt{\xi} z\right) , & z \in [-a, a] \\ N \exp\left(-\sqrt{\xi}(z - a)\right) , & z > a \end{cases} , \quad (8)$$

where N is normalization constant and ξ can be found as the solution of following transcendent equation:

$$\tanh\left(a\sqrt{\xi}\right) = 2\alpha\xi^{-1/2} - 1 \tag{9}$$

The profiles of $F(z)$ are shown in Figure 1 for different values of a. At the Figure 2 best achievable values of Θ are shown as a dependence of modal gain (g_0 is unity for the choice $A = 1$, see above).

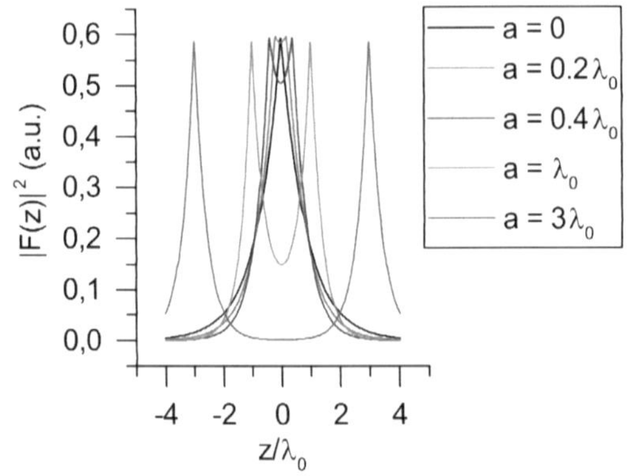

Figure 1: Optimal near-field intensity profiles for different values of a that are shown at the right side of graph.

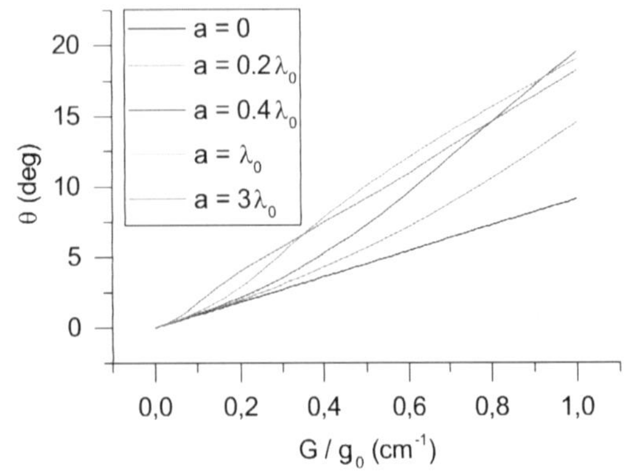

Figure 2: $\Theta(G)$ dependency on the gain for various a.

As we can see from the Figure 2, at low gain and $a < \lambda_0$ small increase in gain results in the proportional increase in divergence for all relevant a due to field distribution being "blob" of width much more than λ_0 so field will not respond to small a change as much as at high gain. Also it is worth noting that best divergence angle is reached at $a = 0$, thus gain profile confinement is preferred.

3.3. Example 3: symmetric slab

There are cases when gain of single layer is not sufficient for effective laser operation, namely, in lasers with quantum dot active media. This problem is typically solved by stacking of gain layers.

As an example, in InGaAs/GaAs quantum dot lasers up to 10 layers may be stacked. The total width of the gain area in that case is up to few hundreds nanometres and is compared with the wavelength in the material. In that case gain profile could be well approximated as a "gain slab" of thickness $2a$ (see Figure 3):

$$g(z) = g_0\theta(a+z)\theta(a-z)\ , \tag{10}$$

where $\theta(z)$ is Heaviside step function.

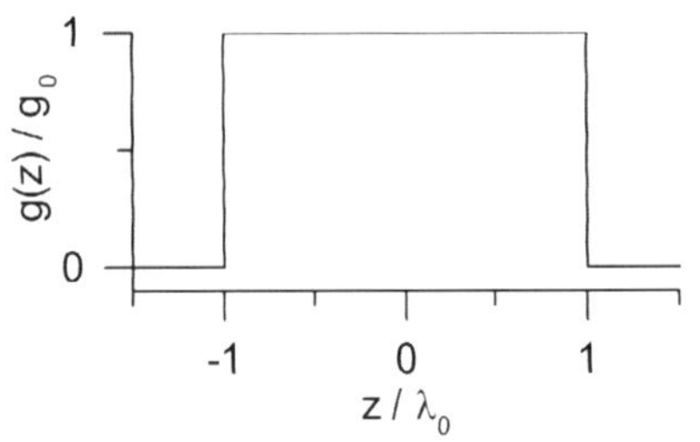

Figure 3: Gain profile with $a = \lambda_0$.

The optimum profile is a symmetric one:

$$F(z) = \begin{cases} N\exp\left(k_1(z+a)\right), & z < -a \\ N\,\mathrm{sech}\left(k_2 a\right)\cosh\left(k_2 z\right), & z \in [-a, a] \\ N\exp\left(-k_1(z-a)\right), & z > a \end{cases} \quad (11)$$

Here $k_1 = \sqrt{-\xi}$ and $k_2 = \sqrt{-\xi - \alpha}$. ξ can be found as solution of the following equation:

$$\tanh(k_2 a) = -\frac{k_1}{k_2} \quad (12)$$

Note that there is modal gain limit g_0, which means field is being contained in $g(z)$ completely. Also, better result is reached for greater slab width.

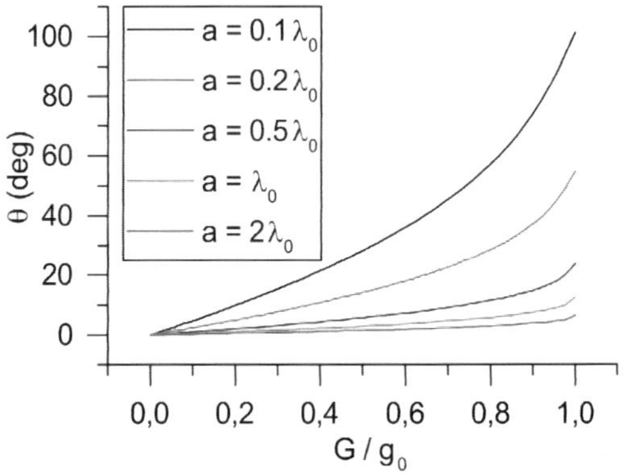

Figure 4: $\Theta(G, a)$ characteristics.

3.4. *Example 4: single gain layer with losses on the sides*

Let us consider laser from Example 1. In emitters electrons and holes absorb light.

$$g(z) = g_0\lambda_0\delta(z) - g_1\theta(z-a) - g_2\theta(-z-a) \quad (13)$$

However, electrons' and holes' absorption differs explicitly, hence (e.g. g_2) is considered negligibly small and omitted. Substituting

$$\begin{cases} k_1 = \sqrt{-\xi} \\ k_2 = \sqrt{-\xi - \alpha g_1} \end{cases}$$

one of possible solutions (symmetric one) of (5) can be written as:

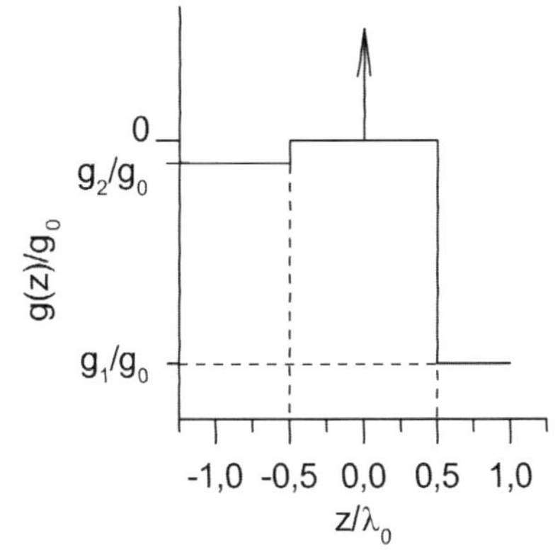

Figure 5: Gain profile with $a = \lambda_0/2$.

$$F(z) = \begin{cases} N\exp\left(k_1(z+a)\right), & z < -a \\ N\left(1 - \frac{\alpha}{k_1}\right)\sinh\left(k_1 z\right) + A\cosh\left(k_1 z\right), & z \in [-a, a] \\ N\left(\left(1 - \frac{\alpha}{k_1}\right)\sinh\left(k_1 a\right) + \cosh\left(k_1 a\right)\right)\exp\left(-k_2(z-a)\right), & z > a \end{cases} \quad (14)$$

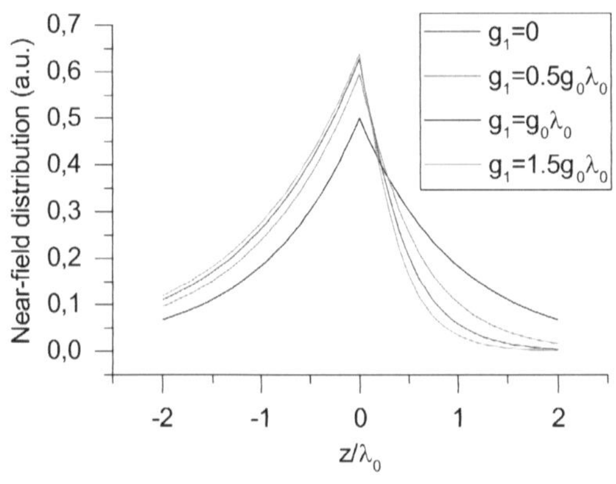

Figure 6: Optimal near-field intensity distributions for fixed modal gain.

where ξ is a solution of the following equation:

$$-\frac{k_2}{k_1} = \frac{\left(1 - \frac{\alpha}{k_1}\right) + \tanh(k_1 a)}{\left(1 - \frac{\alpha}{k_1}\right)\tanh(k_1 a) + 1} \tag{15}$$

The field distribution (Figure 6) visibly shifts to the left as the absorption on the right increases.

4. Optimal refractive index profile equation

Consider n_0 as a substrate's refractive index and $\Delta n(z)$ as a waveguide's refractive index profile.

$$n(z) = n_0 + \Delta n(z) \tag{16}$$

Then, looking at optimal near-field and wave equations, certain symmetry can be seen

$$\begin{cases} \dfrac{d^2F}{dz^2} + k_0^2\left((\Delta n(z))^2 + 2n_0\Delta n(z)\right)F(z) = k_0^2\left(n_{\text{eff}}^2 - n_0^2\right)F(z) \\ \dfrac{d^2F}{dz^2} + \alpha g(z)F(z) = -\xi F(z) \end{cases} \tag{17}$$

$$\begin{cases} n^2(z) = n_0^2 + \alpha k_0^{-2}g(z) = n_0^2 + Ag(z) \\ n_{\text{eff}}^2 = n_0^2 - \xi k_0^{-2} \end{cases} \tag{18}$$

5. Leaky waveguides

Leaky modes appear in lasers when the barrier between waveguide and substrate is thin enough and refractive indexes are very close, allowing electric field to tunnel through barrier. What it allows is to achieve less far-field divergence through wide near-field distribution. In this section we want to compare efficiencies of laser designs with guided and leaky modes in terms of modal gain vs beam divergence.

IOP Conf. Series: Journal of Physics: Conf. Series **1038** (2018) 012074 doi:10.1088/1742-6596/1038/1/012074

Unlike in guided modes, $|F(z)|$ grows exponentially in substrate towards $z = \infty$, so $F(z)$ can't be normalized, rendering (2) useless, thus modal gain should be found by other means. The method we use is based on physical meaning of effective refractive index, n_{losses}.

With leaky modes without pumping $\text{Im}(n_{\text{losses}}) > 0$, which means energy decreases as wave travels through waveguide it flows into substrate. If active layer is being inserted with refractive index $n = \text{Re}(n) - i\gamma$, where $\gamma > 0$, one can write an equation:

$$\text{Im}(n_{\text{losses}}) = \Delta n_{\text{losses}} - A\gamma \tag{19}$$

Here $\Delta n_{\text{losses}} > 0$ stands for energy leak and $A > 0$ is a dimensionless constant which can be found through calculating $\text{Im}(n_{\text{eff},1})$ for $\gamma = 0$ and $\text{Im}(n_{\text{eff},2})$ for fixed known $\gamma \neq 0$. γ won't change field profile and Δn_{losses}, so:

$$\begin{cases} \text{Im}(n_{\text{eff},1}) = \Delta n_{\text{losses}} \\ A = \text{Im}(n_{\text{eff},1} - n_{\text{eff},2})/\gamma \end{cases} \tag{20}$$

Consider Φ as energy flux and α_i as number representing energy losses in a waveguide. Then, for modal gain we can write

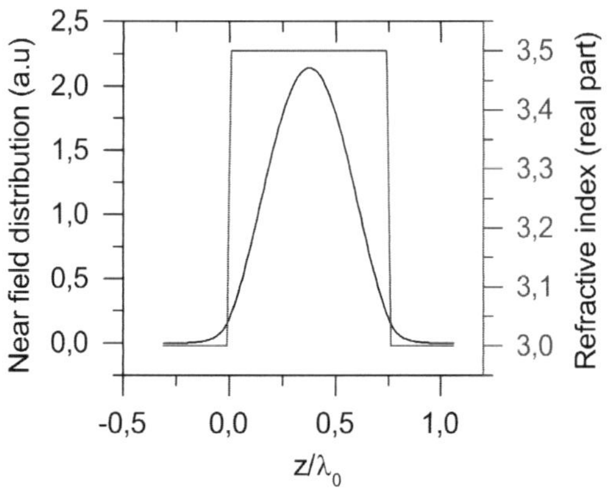

Figure 7: Profile of the guided mode (black) and profile of the refractive index (blue).

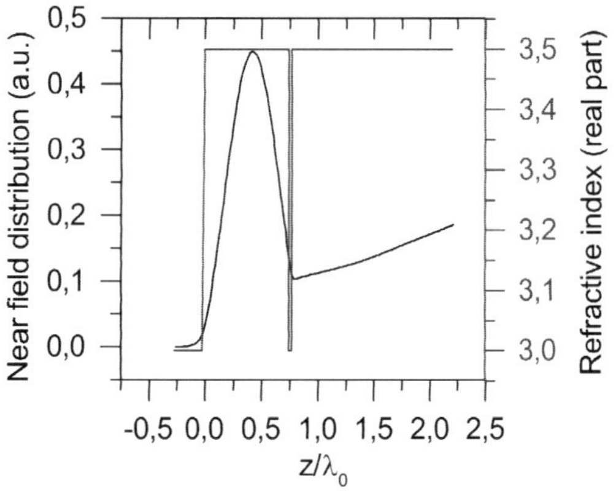

Figure 8: Profile of the leaky mode (black) and profile of the refractive index (blue).

$$\begin{cases} \Phi = \Phi_0 \exp((G - \alpha_i)x) \\ \Phi \sim |E|^2 = \exp((-2\text{Im}(n_{\text{losses}})k_0 x))|F(z)|^2 \end{cases} \tag{21}$$

From here, G can be derived from known variables:

$$G = 2k_0 \text{Im}(n_{\text{eff},1} - n_{\text{eff},2}) \tag{22}$$

To demonstrate, we'll look into simple laser, with varied distance between waveguide and substrate, which is infinity for guided mode and ~ 0.1 micron for leaky.

Consider placing a 0.02 micron quantum well in the middle of a waveguide with $\gamma = 0.0191$ (or 1200 cm^{-1} for 1 μm wavelength). The result as follows:

Table 1: Leakage impact.

Mode	Gap, μm	Gain, 1/cm	Θ, deg
Guided	∞	51.97	24.28
Leaky	0.1	51.44	23.08
Leaky	0.04	49.28	14.23

As seen from the table, with small gain penalty beam divergence can be decreased dramatically. On the contrary, this laser will radiate sideways.

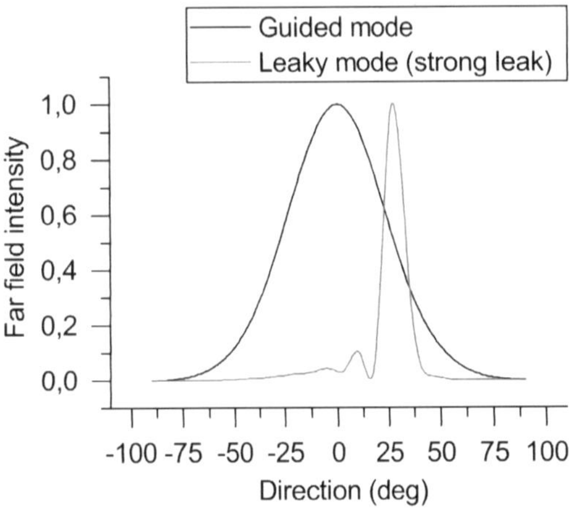

Figure 9: Comparison of far field profiles for leaky and guided modes.

6. Conclusion

We have developed variation approach that in analytical terms describes the near-field profiles that corresponds to best possible modal gain for a given divergence angle or vice versa. Though it is obvious that this approach could not be always used in real material systems due to limited refractive index variations, the approach still is useful to determine how far from the "ideal" any given structure is.

Also it was shown that leaky waveguides are superior to guided ones in terms of modal gain and beam divergence.

7. Acknowledgements

This work is supported by the Russian Science Foundation (Project No. 14-42-00006).

References

[1] Korenev V V, Savelyev A V, Zhukov A E, Maximov M V and Omelchenko A V 2015 Waveguide and active region structure optimization for low-divergence InAs/InGaAs quantum dot comb lasers *Proc. SPIE* **9503** 950305

[2] Dimitri P Bertsekas 1982 *Constrained Optimization and Lagrange Multiplier Methods.* (New-York: Academic Press) p 72

[3] Bruce van Brunt 2003 *The Calculus of Variations.* (Berlin: Springer Science & Business Media) p 28

[4] Casey H C Jr. 1978 *Heterostructure lasers.* (New-York: Academic Press) p 94

[5] Shveikin V I, Bogatov A P, Drakin A E and Kurnyavko Yu V 1999 Directional radiation pattern of quantum dimensional InGaAs/GaAs leaky-mode lasers

IOP Conf. Series: Journal of Physics: Conf. Series **1038** (2018) 012075 doi:10.1088/1742-6596/1038/1/012075

The generation of evanescent beams by means of binary diffraction axicons with high numerical aperture

D A Savelyev[1,2]

[1]Department of Engineering Cybernetics, Samara National Research University, Samara 443086, Russia
[2]Laboratory of Laser Measurements, Image Processing Systems Institute - Branch of the Federal Scientific Research Centre "Crystallography and Photonics" of Russian Academy of Sciences, Samara 443001, Russia

dmitrey.savelyev@yandex.ru

Abstract. The investigations of the generation of evanescent beams by means of binary diffraction axicon with a numerical aperture (NA) were more than one. It is shown that it is possible to generate evanescent beams using axicons. Numeric simulation of radiation passing through the axicon was investigated by the finite difference time domain method (FDTD).

1. Introduction

Interest in evanescent fields arose long ago [1] due to their applications in such fields as near-field optical microscopy, high-density recording, sub-wave lithography. In particular, one of the approaches to evanescent field's production is to illuminate the desired target with radiation passing through an aperture with dimensions smaller than the wavelength [2] and requires very close distances between the target and the aperture. Two fundamental effects cause this circumstance: the damped nature of the vanishing field components generated by the aperture, and the rapid expansion of the remaining field [1].

Near-field optics [3-5] is one of the directions effectively used to achieve super-resolution. In this case, there is no restriction on the size of the light spot, since the localization of the laser study can be arbitrarily small. However, as was shown in [6-9], the compactness of the localization of light essentially depends on the size of the details of the surface relief [6, 8] or the size of the tip of the focusing element [7, 9].

Bessel beams have attracted constant interest since their first mention [10]. Of particular interest at that time were the so-called Bessel-Gaussian beams, which were mathematically interpreted analytically in the paraxial approximation [11]. Later it was also accepted that the definition of Bessel beams could be extended to include additional functions of the same family. It was found that modified Bessel beams do not oscillate in the transverse direction and oblate shape, when it is appropriately apodized by a Gaussian function [12]. These types of beams were subsequently generalized for complex arguments [13]. The generation of evanescent Bessel beams for a radially polarized beam at the interface between dielectric-metal media was considered in [14]. The realization of a evanescent Bessel beam through the surface plasmon interference excited by a radially polarized beam was proposed in [15].

It is also known that the axicon forms a zero-order Bessel beam whose central spot diameter by the half-wave intensity (FWHM) is 0.36 of the wavelength divided by the numerical aperture [16], which is 37% smaller than the size of the Airy disk formed by the lens with the same numerical aperture (NA). Due to the presence of evanescent surface waves in the near zone, it is possible to form a focal spot much smaller than the wavelength, which is of interest in the use of axicon for tight focusing [17-22]. In particular, in [21] considered focusing of laser beams with axicons with different NA (maximum NA = 1.02). In [22], diffraction axioms with a numerical aperture up to NA = 1.2 were also investigated.

However, the model of a thin optical element was used in [21, 22]. In this paper, we investigate the formation of evanescent beams using a binary diffraction axicon with a period shorter than the wavelength in the 3D model. The numerical aperture of the axicon varied from 0.95 to 2.0. To numerically simulate the diffraction of the laser radiation under consideration, the finite time difference (FDTD) method using high-performance computations is used [23], implemented in the Meep software package [24].

2. Generation of evanescent beams

Simulation parameters: the wavelength $\lambda = 0.532$ microns, the size of the computational domain x, y, z $\in [-3.76\lambda; 3.76\lambda]$. The thickness of the absorbing layer PML ~ 1.32λ (0.7 micron). The considered laser beam is the fundamental Gaussian mode with circular polarization (figure 1). The refractive index of the axicon and the substrate is n = 1.46 (corresponding to the material Fused Silica).

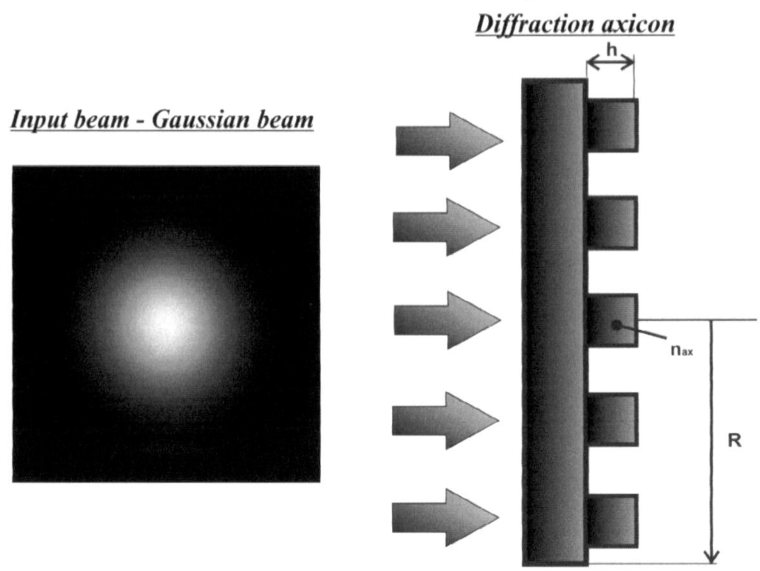

Figure 1. Modelling parameters: (a) Gaussian beam; (b) the optical element.

The results of numerical simulation in the xz plane are shown in figure 2. The results of the calculations showed that when the NA of axicon increases, the formation length of the evanescent beam decreases. It should be noted that when NA is more than 1.3, a ring shadow spot is formed on the optical axis, which means an increase in the longitudinal component of the electric field. Its appearance is shown in the bottom line of figure 2.

Consider the cross-section for the axiocons considered earlier. In figure 3 shows the simulation results at a distance of 0.1 λ from the axicon, in figure 4 - simulation results at a distance of 1.5 λ from the axicon.

It should be noted that for NA > 1, even at a distance of 0.1 λ from the axicon, the formation of a ring shadow spot begins. For NA = 1.5 at a distance of 1.5λ from the axicon, and further, a ring focal spot is formed.

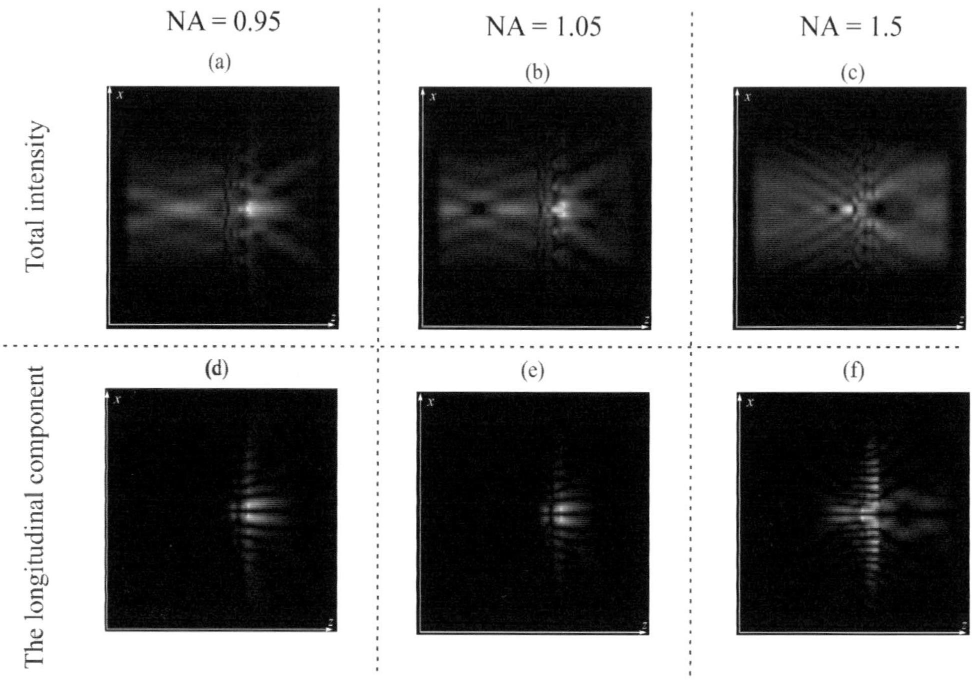

Figure 2. The longitudinal cross section (xz) of propagation of a Gaussian beam through axicons with a different NA: total intensity – (a) NA = 0.95; (b) NA = 1.05; (c) NA = 1.5; the longitudinal component – (d) NA = 0.95; (e) NA = 1.05; (f) NA = 1.5.

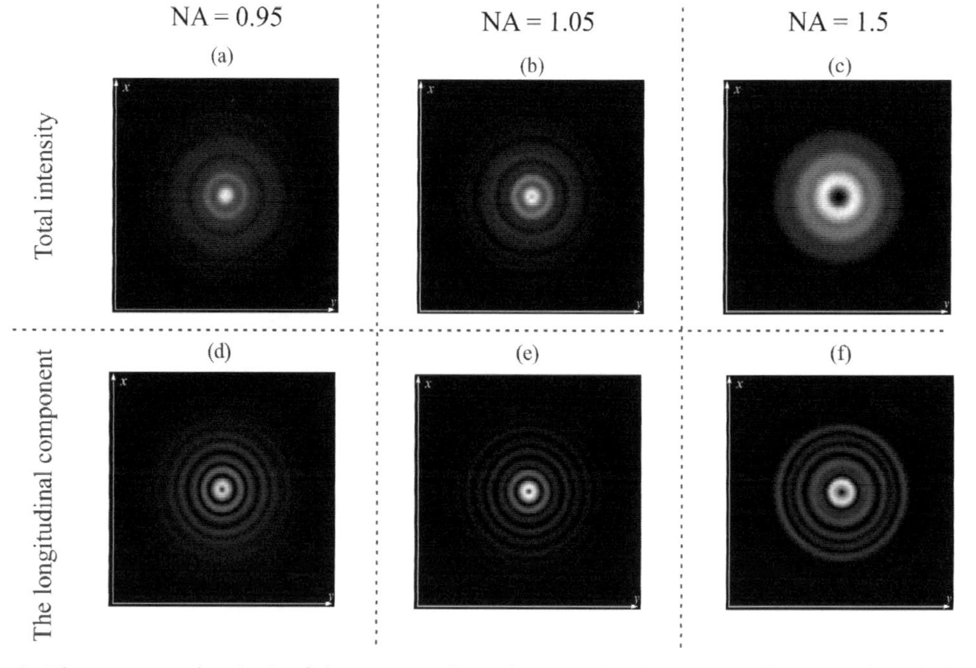

Figure 3. The cross section (xy) of the propagation of a Gaussian beam at a distance of 0.1 λ from the axicon: total intensity – (a) NA = 0.95; (b) NA = 1.05; (c) NA = 1.5; the longitudinal component – (d) NA = 0.95; (e) NA = 1.05; (f) NA = 1.5.

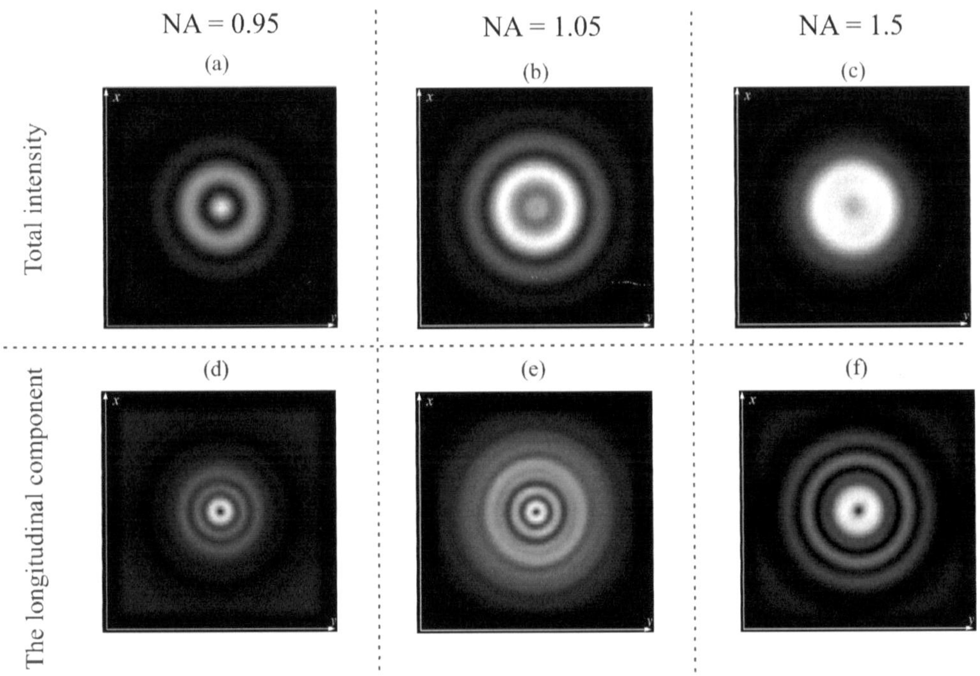

Figure 4. The cross section (xy) of the propagation of a Gaussian beam at a distance of 1.5 λ from the axicon: total intensity – (a) NA = 0.95; (b) NA = 1.05; (c) NA = 1.5; the longitudinal component – (d) NA = 0.95; (e) NA = 1.05; (f) NA = 1.5.

Thus, numerically, using the FDTD method, a comparative study of the diffraction of a Gaussian beam with circular polarization on diffraction axioms with a numerical aperture greater than 1 was performed, and it was shown that it is possible to generate evanescent beams.

3. Conclusion

In this paper, the effects of an increase in the numerical aperture of the optical element (axicon) on the diffraction of a Gaussian beam with circular polarization are considered. Simulation of radiation passing through the axicon was numerically investigated by the finite difference time domain method. It is shown that an increase in the numerical aperture of the element makes it possible to obtain an amplification of the longitudinal component of the electric field, which is expressed in the formation of a ring (shadow) spot.

Acknowledgments

This work was financially supported by the Russian Foundation of Basic Research (RFBR) (16-07-00825, 16-29-11698), by Federal Agency of Scientific Organizations (No. 0026-2018-0104), by grant of the President of the Russian Federation for the scientific school NSh-6307.2018.8 and by the Ministry of Education and Science of Russian Federation.

References

[1] Ruschin S, Leizer A 1998 *JOSA A* **15(5)** 1139
[2] Lewis A M A A, Isaacson M, Harootunian A, Muray A 1984 *Ultramicroscopy*. **13(3)** 227
[3] Paesler M A, Moyer P J 1996 *Near Field Optics: Theory, Instrumentation and Applications* (New York: Wiley Interscience)
[4] Pohl D W 2004 *Phil. Trans. R. Soc. Lond. A* **362** 701

[5] Novotny L 2007 *Progress in Optics* **50(**5) 137
[6] Khonina SN, Ustinov A V, Kovalyov A A, Volotovsky S G 2014 *Optical Memory and Neural Networks* **23(2)** 50
[7] Degtyarev S A, Porfirev A P, Ustinov A V, Khonina S N 2016 *Journal of the Optical Society of America B* **33(12)** 2480
[8] Khonina S N, Ustinov A V 2017 *Optics communications* **391** 24
[9] Khonina S N, Degtyarev S A, Savelyev D A, Ustinov A V 2017 *Optics Express* **25(16)** 19052
[10] Durnin J 1987 *JOSA A* **4(4)** 651
[11] Gori F, Guattari G, Padovani C 1987 *Optics communications*. **64(6)** 491
[12] Ruschin S 1994 *JOSA A*. **11(12)** 3224
[13] Bagini V, Frezza F, Santarsiero M, Schettini G, Spagnolo G S 1996 *Journal of Modern Optics* **43(6)** 1155
[14] Zhan Q. 2006 *Optics letters* **31(11)** 1726
[15] Chen W, Zhan Q 2009 *Optics letters* **34(6)** 722
[16] Kalosha V P, Golub I 2007 *Optics letters* **32(24)** 3540
[17] Khonina S N, Nesterenko D V, Morozov A A, Skidanov R V, Soifer V A 2012 *Optical Memory and Neural Networks (Information Optics)* **21(1)** 17
[18] Khonina S N, Serafimovich P G, Savelyev D A, Pustovoi I A 2012 *Journal of Optical Technology* **79(10)** 626
[19] Khonina S N, Karpeev S V, Alferov S V, Savelyev D A, Laukkanen J, Turunen J 2013 *J. Opt.* **15** 085704
[20] Khonina S N, Savelyev D A 2013 *Journal of Experimental and Theoretical Physics* **117(4)** 623
[21] Kotlyar V V, Stafeev S S, Kovalev A A, Nalimov A G 2012 *Computer Optics* **36(2)** 183
[22] Ustinov A V, Khonina S N 2014 *Computer Optics* **38(2)** 213
[23] Savelyev D A, Khonina S N 2014 *Journal of Physics: Conference Series* **490(1)** 012213.
[24] Oskooi A F, Roundy D, Ibanescu M, Bermel P, Joannopoulos J D, Johnson S G 2010 *Computer Physics Communications* **181(3)** 687

Double magnetic resonance in the hyperfine structure of optically oriented alkali atoms with laser pumping

A A Baranov[1], S V Ermak[1], E A Sagitov[1], and V V Semenov[1]

[1]Dept. of Quantum electronics, Peter the Great St. Petersburg Polytechnic University, Saint Petersburg, 195251, Russia

E-mail: e-sagitov@mail.ru

Abstract. The paper is devoted to the investigation of mutual influence of low-frequency and microwave resonances in ^{87}Rb vapor cell magnetometers system with laser pumping. One of the magnetometers was based on a low-frequency spin generator principle, while the second one was built as passive microwave spectrometer with a resonance frequency lock loop. The paper analyzes the frequency shifts in the low-frequency and microwave channels of the tandem of magnetometers associated with the simultaneous action of the resonant radio fields on the alkali atoms. Such effect is manifested in the frequency shifts of the spin generator for fixed changes in the amplitude of the microwave field.

1. Introduction

The phenomenon of radio-optical resonance is actively used in the development of the precision quantum devices, including the nuclear gyroscopes, as well as the various types of the atomic frequency standards [1-6] and the optically pumped quantum magnetometers [7-9]. The growing demands for stability and measuring accuracy of these devices have determined the search for new circuit solutions in which the semiconductor lasers are used as the pumping sources. Such sources have a higher efficiency, small dimensions and, in contrast to the ordinary spectral lamps, allow effective control of an intensity, a polarization, and a spectrum of the pump radiation. These advantages are of the fundamental importance in the development of the miniature quantum sensors installed in on-board equipment of the mobile placing. In this case, in the variant of the quantum devices based on the effect of coherent population trapping, their circuit solution makes it possible to eliminate the traditional component (microwave cavity) and an isotopic filter, which are necessary when lamp source is used for pumping [10].

In the development of the small-sized atomic clocks with laser pumping, the problem of minimizing the line width of the observed resonance arises, since the miniaturization of the construction requires an increase in the density of the working substance to obtain a sufficiently reliable signal. To achieve this goal, the so-called end resonance observed at transitions between the magnetic sublevels of the hyperfine structure with the extreme values of the atom total angular momentum projection can be implemented [11]. Unlike the radio-optical resonance at the 0-0 transition, used in quantum frequency standards, it is observed with a higher quality factor of the detected signal (the signal-to-noise ratio divided by the resonance line width). This makes it possible to minimize the size of the absorption cell, but requires the use of increased pressures of the inert buffer gas, which is necessary to reduce the rate of the atom diffusion to the walls of the cell. The authors consider an alternative version of the end resonance observed in ^{87}Rb pairs placed in a small-sized absorption cell with an antirelaxation walls coating without a buffer gas [12], [13]. At the same time, the problem of identification the possibilities for the mutual compensation of the various components of the light shift with the aim of increasing the long-term stability of the quantum

IOP Conf. Series: Journal of Physics: Conf. Series **1038** (2018) 012076 doi:10.1088/1742-6596/1038/1/012076

magnetometers was solved. One of that magnetometers was performed on the principle of a low-frequency spin generator; the other, in the form of a passive radio spectrometer with of the resonance frequency self-tuning to the microwave radio-optical resonance. However, the authors did not consider the inevitable mutual influence of LF-microwave channels associated with the effect of a coherence circulation between the states of the atom hyperfine structure [14, 15]. To some extent, this gap is filled by the present work, which analyzes the frequency shifts in the LF and microwave channels of the tandem of magnetometers associated with the simultaneous action of the resonance radio fields on the alkali metal atoms and manifested in the frequency shifts of the spin generator for fixed changes in the amplitude of the microwave field.

2. "Dressed" atom theory implementation for quantum magnetometers system

The physical nature of such frequency shifts is associated with the coherence circulation between the magnetic sublevels of the alkali atoms "dressed" by the spin oscillator LF field. Absolute value of this shift strongly depends on the magnetic-dipole transition number and reaches its maximum at 0-0 microwave transition, which is commonly used in the atomic frequency standards.

In accordance with the quantum theory of the "dressed" atom, simultaneous action of the LF and microwave fields on the alkali atoms yields in multiphoton absorption and reemission of the LF field quants and the probing microwave field absorption [16]. The microwave transitions spectrum between the sublevels of the hyperfine structure takes the form of central and side groups of the satellites lines, separated at intervals equal to the frequency of the LF fields. Probability amplitudes of microwave transitions in the energy structure of the "dressed" atom determine the specific contribution of various magnetic-dipole transitions in the radio-optical resonance signal. In accordance with [10], the probability amplitude for such transitions is expressed through the matrix element of the operator of magnetic dipole moment as follows

$$\langle F^*, m^*|M|F, m\rangle = \sum_{m_1, m_1^*} D^{F^*}_{m_1^*, m^*} D^{F}_{m_1, m} J_0\left[(m + m_1^*)\frac{\omega_1}{\omega}\right] \langle F^*, m_1^*|M|F, m_1\rangle \delta_{m_1, m_1^*+q} \quad (1)$$

where $D^{F^*}_{m_1^*, m^*}$ и $D^{F}_{m_1, m}$ – of the Wigner function, $J_0\left[(m + m_1^*)\frac{\omega_1}{\omega}\right]$ – Bessel functions $\langle F^*, m_1^*|M|F, m_1\rangle$ –the matrix element of the magnetic-dipole transition, δ_{m_1, m_1^*+q} – the Kronecker symbol, where the index q takes the value 0 or 1 depending on the direction of the vector of the RF field relative to the direction of the constant magnetic field, ω_1 characterizes an amplitude of the magnetic field H_1 according to the equation $\omega_1 = \gamma H_1$ and ω is the frequency the low-frequency RF field.

As an example, the possible microwave transitions (arrows) between the hyperfine structure sublevels of alkali "dressed" atom with a nuclear spin of 3/2 is shown in figure 1.

For example, at 0-0 and end resonances frequency transitions, indicated in table 1, are induced in the central series of the absorption spectrum for the magnetic microwave resonances of ^{87}Rb "dressed" atoms. In the second column of the table the probabilities of the corresponding transitions are presented.

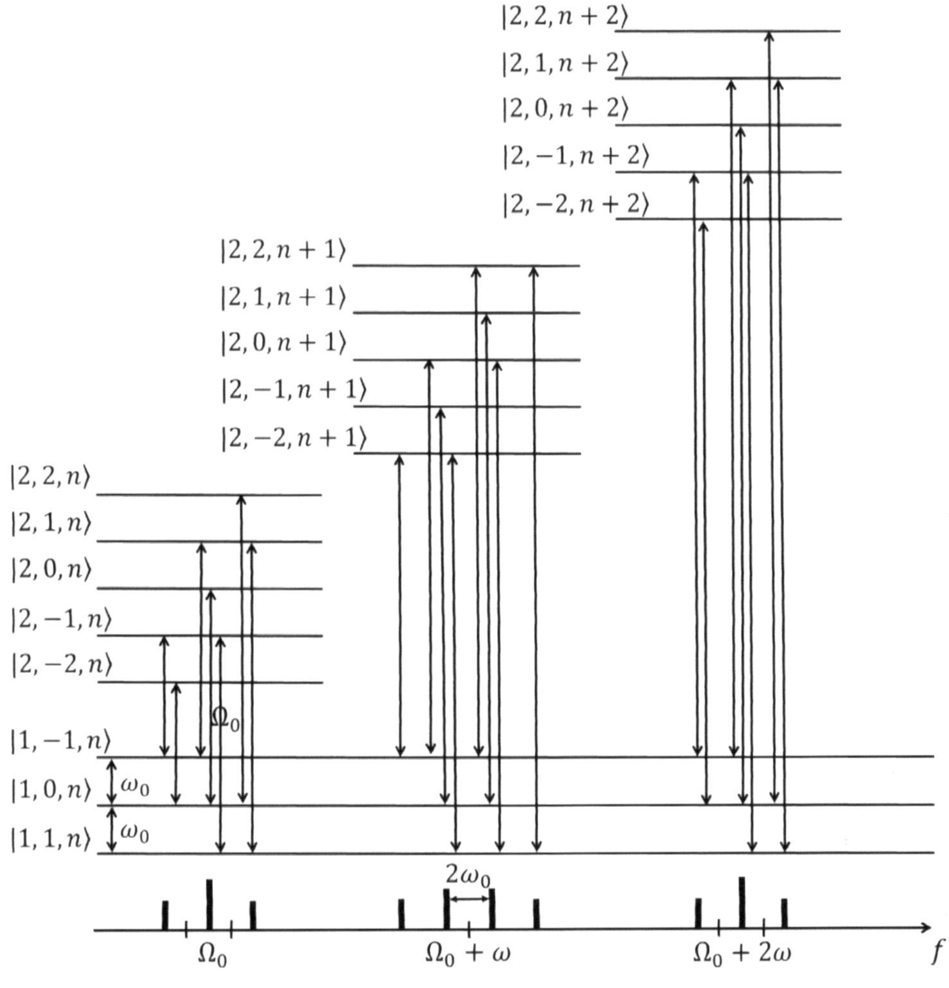

Figure 1. The spectrum of microwave transitions "dressed" alkali atom with a nuclear spin equals 3/2

Table 1. Microwave transitions probabilities atoms of alkali metals

Rubidium 87

F = 1, m_F ↔ F = 2, -m_F

F = 1, $m_F = \pm 1$ ↔ F = 2, $m_F = \mp 1$	$\frac{3}{64}\left[J_0\left(\frac{\omega_1}{\omega}\right) - J_0\left(3\frac{\omega_1}{\omega}\right)\right]^2$
F = 1, $m_F = 0$ ↔ F = 2, $m_F = 0$	$\frac{1}{64}\left[J_0\left(\frac{\omega_1}{\omega}\right) + 3J_0\left(3\frac{\omega_1}{\omega}\right)\right]^2$

F=1, m_F ↔ F=2, m_F ±1

F = 1, m_F = ± 1 ↔ F = 2, m_F = ± 2	$\frac{3}{64}\left[J_0\left(\frac{\omega_1}{\omega}\right)\right]^2$

Rubidium 85

F = 1, m_F ↔ F = 2, -m_F

F = 2, m_F = ±2 ↔ F = 3, m_F = ∓2	$\frac{5}{4}\left[5J_0\left(\frac{\omega_1}{\omega}\right) - 8J_0\left(3\frac{\omega_1}{\omega}\right) + 3J_0\left(5\frac{\omega_1}{\omega}\right)\right]^2$
F = 2, m_F = ±1 ↔ F = 3, m_F = ∓1	$\frac{1}{2}\left[6J_0\left(\frac{\omega_1}{\omega}\right) + 9J_0\left(3\frac{\omega_1}{\omega}\right) - 15J_0\left(5\frac{\omega_1}{\omega}\right)\right]^2$
F = 2, m_F = 0 ↔ F = 3, m_F = 0	$9\left[2J_0\left(\frac{\omega_1}{\omega}\right) + J_0\left(3\frac{\omega_1}{\omega}\right) + 5J_0\left(5\frac{\omega_1}{\omega}\right)\right]^2$

F=2, m_F ↔ F=3, m_F ±1

F = 2, m_F = ± 2 ↔ F = 3, m_F = ± 3	$\frac{15}{128}\left[3 + 4J_0\left(2\frac{\omega_1}{\omega}\right) + J_0\left(4\frac{\omega_1}{\omega}\right)\right]^2$

Cesium 133

F = 3, m_F ↔ F = 4, -m_F

F = 3, m_F = ±3 ↔ F = 4, m_F = ∓3	$7\left[5J_0\left(\frac{\omega_1}{\omega}\right) - 9J_0\left(3\frac{\omega_1}{\omega}\right) + 5J_0\left(5\frac{\omega_1}{\omega}\right) - J_0\left(7\frac{\omega_1}{\omega}\right)\right]^2$
F = 3, m_F = ±2 ↔ F = 4, m_F = ∓2	$3\left[5J_0\left(\frac{\omega_1}{\omega}\right) + 3J_0\left(3\frac{\omega_1}{\omega}\right) - 15J_0\left(5\frac{\omega_1}{\omega}\right) + 7J_0\left(7\frac{\omega_1}{\omega}\right)\right]^2$
F = 3, m_F = ±1 ↔ F = 4, m_F = ∓1	$15\left[3J_0\left(\frac{\omega_1}{\omega}\right) + J_0\left(3\frac{\omega_1}{\omega}\right) + 3J_0\left(5\frac{\omega_1}{\omega}\right) - 7J_0\left(7\frac{\omega_1}{\omega}\right)\right]^2$
F = 3, m_F = 0 ↔ F = 4, m_F = 0	$\left[9J_0\left(\frac{\omega_1}{\omega}\right) + 15J_0\left(3\frac{\omega_1}{\omega}\right) + 5J_0\left(5\frac{\omega_1}{\omega}\right) + 35J_0\left(7\frac{\omega_1}{\omega}\right)\right]^2$

F=3, m_F ↔ F=4, m_F ±1

F = 3, m_F = ± 3 ↔ F = 4, m_F = ± 4	$14\left[10 + 15J_0\left(2\frac{\omega_1}{\omega}\right) + 6J_0\left(4\frac{\omega_1}{\omega}\right) + J_0\left(6\frac{\omega_1}{\omega}\right)\right]^2$

The absorption spectrum of the "dressed" atoms, is shown in figure 1, meets the condition of the significant excess of low frequency field ω over Larmor frequency ω_0. In the tandem magnetometers, the equality of these frequencies was observed and an overlap of the absorption spectra of the neighboring line groups of the satellites was registered. Since the amplitude of the self-generating magnetometer RF field ω_1 was significantly less than its resonant frequency ω in the experiments, the dynamics of the observed frequency shifts was determined only by the center band of the absorption lines. In practice it is necessary to take into account the weight contribution of various magnetic-dipole transitions that depend on the difference of populations in the magnetic sublevels of the hyperfine structure. This parameter, in turn, is determined from the solution of the equations of motion for the diagonal elements of the density matrix.

In the case of using antirelaxation material coated ^{87}Rb vapor cells with circularly polarized laser pumping source tuned to high-frequency transition of the D_1-line, these equations have the following form:

$$\begin{aligned}
\sigma_{11} &= 0.125 - 0.96\Gamma^*\sigma_{11} + 0.144\Gamma^*\sigma_{22} + 0.04\Gamma^*\sigma_{33} \\
\sigma_{22} &= 0.125 - 0.576\Gamma^*\sigma_{22} + 0.04\Gamma^*\sigma_{33} \\
\sigma_{33} &= 0.125 - 0.2\Gamma^*\sigma_{33} \\
\sigma_{44} &= 0.125 \\
\sigma_{55} &= 0.125 + 0.04\Gamma^*\sigma_{33} \\
\sigma_{66} &= 0.125 + 0.144\Gamma^*\sigma_{22} + 0.04\Gamma^*\sigma_{33} \\
\sigma_{77} &= 0.125 + 0.48\Gamma^*\sigma_{11} + 0.144\Gamma^*\sigma_{22} + 0.04\Gamma^*\sigma_{33} \\
\sigma_{88} &= 0.125 + 0.48\Gamma^*\sigma_{11} + 0.144\Gamma^*\sigma_{22}
\end{aligned} \qquad (2)$$

where the index of the diagonal element of the density matrix corresponds to the number of magnetic sublevels of the rubidium atoms in the ground state, Γ^* is expressed in relative units and is coupled with the speed of the pump (the ratio of the rate of pumping to the rate of dark relaxation). Figure 2 shows examples of dependencies of the probability of microwave transitions (F=1, $m_F \leftrightarrow$ F=2, m_F) from the argument obtained based on the data of table 1 by solving the system of equations (2) for the corresponding difference between diagonal elements of the density matrix under various parameters Γ^*

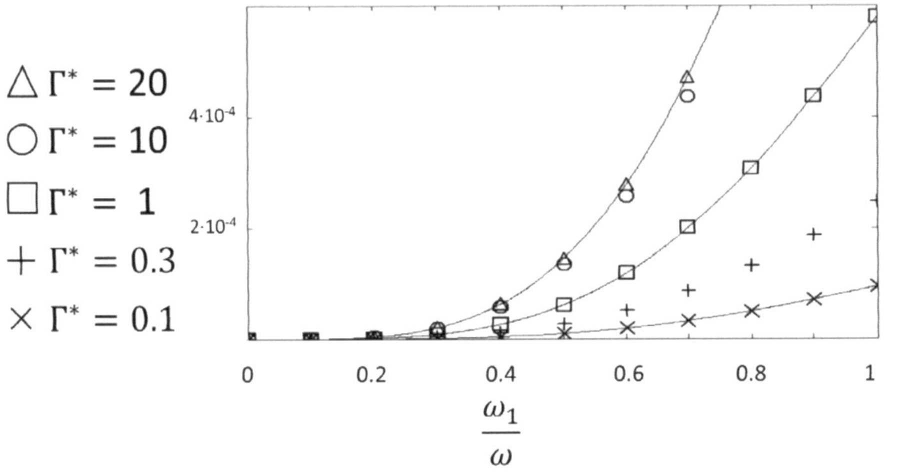

Figure 2. The transition probabilities for the double magnetic microwave resonance of atoms of rubidium-87 in a function of the relative amplitude of RF field spin generator

3. Experiment

The experiments were provided in the conditions of a laser pumping on LF D_2-line component of $S_{1/2}$-$P_{3/2}$ transition from hyperfine structure ground state with $F_g = 2$ where the highest polarization of alkali atoms is reached. When the laser is tuned to a high-frequency component of HFS with $F_g = 1$ of the ground state doesn't give the positive yield because of feedback omission owing to low concentration of atoms on the observed level. A different situation takes place in the case of the D_1-line laser pumping. In this case, as the elementary calculation of the density matrix diagonal elements difference demonstrates, it is possible to reach the higher level of the atoms polarization both on the microwave transition, and in the structure of the magnetic levels with $F_g = 2$ and $F_g = 1$. The experiment with the D_1-line laser pumping the tandem of magnetometers worked both at LF components of electric-dipole transition of this line (transitions of $F_g = 2 \leftrightarrow F_e = 1$ and $F_g = 2 \leftrightarrow F_e = 2$) and at a short-wave component of $F_g = 1 \leftrightarrow F_e = 2$, though with considerably smaller signal-to-noise ratio. The last fact has a basic role in a choice of the optimal pumping mode which would allow to realize the lowest level of a flicker noise of Allan deviation [2]. The equality of the detected signals intensity in the microwave and LF channels of a magnetometers tandem is satisfied to such conditions as both of these channels have equivalent value. The mutual influence of the microwave and LF signals is shown as the changing in dynamics of their intensity and spectrum. As an example, the experimental recordings of the self-generating magnetometer signal and a microwave absorption spectrum of Rb^{87} atoms are shown in figure 2, The record was carried out when the frequency of the microwave field was scanning.

The relative influence of LF and the microwave channels in magnetometers tandem was checked during the experiment in two modes: a) measurements of the frequency shift of the spin generator $\delta\nu_{SG}$ at the fixed detuning of the microwave frequency $\delta\nu_{UHF}$; and b) measurements of the frequency shift for the microwave resonance $\Delta\nu_{UHF}$ at the fixed detuning of the spin generator frequency $\Delta\nu_{SG}$ due to the change of the phase shift in the feedback circuit.

The experiment showed a significant difference in the magnetometers tandem LF and microwave channels interaction for operation modes “a” and “b”. For example, when the same artificial deviations $\delta\nu_{UHF}$ and $\Delta\nu_{SG}$ in magnetic field of 0.01 Oe $\delta\nu_{SG}$ with respect to $\Delta\nu_{UHF}$ was the order of magnitude of 10^2.

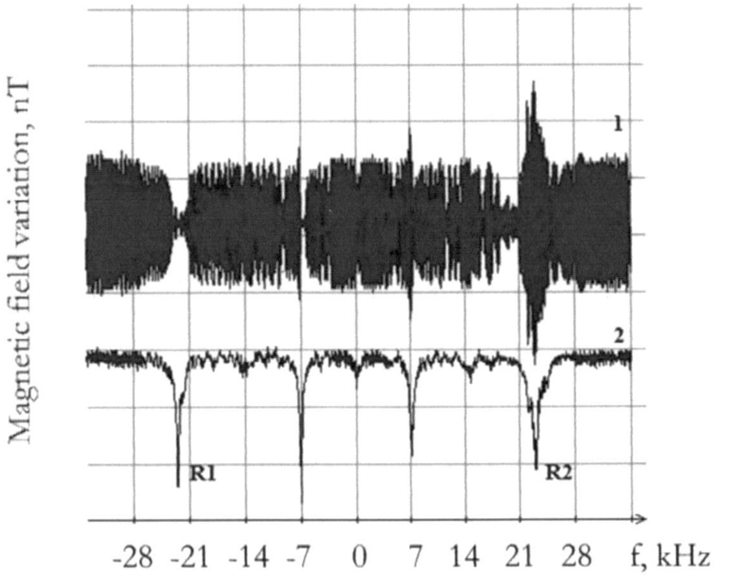

Figure 3. The recordings of the self-generating magnetometer signal and a microwave absorption spectrum of the optical-oriented ^{87}Rb atoms: 1 – signal of the self-generating magnetometer, 2 – absorption signal of RF resonance in a case of the scanning frequency field of microwave range. The absorption signals R1 and R2 correspond to end resonances of transitions between UHS sublevels

4. Conclusions

The presented results allow us to conclude that there is a significant mutual influence of the signals in quantum magnetometers tandem, one of which is constructed as a traditional self-generating magnetic resonance device. Such effect can be noticeably weakened by using of the rotating radio frequency field. In this case, it is possible to eliminate the non-resonance component of the field, and, hence, to eliminate the Bloch-Siegert frequency shift [17]. However, from our point of view, a good solution for this problem is the use of the modulated optical pumping in a self-generating circuit of the magnetometers tandem [18]. In this case, the errors induced by the radio-frequency field are eliminated, as well as the errors due to the light shifts of the resonance frequency, which are coupled with circularly polarized radiation of the lamp [19]. As a preliminary calculation shows, the distinctive feature of a self-generating magnetometer with modulated pumping is the absence of a shift in the resonance frequency associated with the action of the feedback loop, which allows one to predict its better long-term stability in comparison with the schemes using the radio-frequency field.

References

[1] J.C. Fang and J. Qin Advances in Atomic Gyroscopes: A View from Inertial Navigation Applications 2012 *Sensors* **12** 6331-6346
[2] Riehle F. Frequency Standards: Basics and Applications 2004 *Wiley-VCH*
[3] E.B.Alexandrov and A.K.Vershovski 2009 *Phys.Usp.* **52 (5)** 573-601
[4] Ermak S V, Petrenko M V and Semenov V V 2016 *Tech. Phys. Lett.* **42** 127
[5] Fedorov M I, Ermak S V, Petrenko M V, Pyatyshev E N and Semenov V V 2016 *J. Phys.: Conf. Ser.* **769** 012046
[6] Ermak S V, Semenov V V, Petrenko M V and Pyatyshev E N 2016 *Int. J. Modern Phys.: Conf. Ser.* **41** 1660138
[7] Fedorov M I, Ermak S V, Petrenko M V and Semenov V V 2016 *J. Phys.: Conf. Ser.* **741** 012162
[8] Sagitov E A, Ermak S V, Petrenko M V and Semenov V V 2016 *J. Phys.: Conf. Ser.* **769** 012044
[9] Baranov A A, Ermak S V, Sagitov E A, Smolin R V and Semenov V V 2016 *Tech. Phys. Lett.* **42** 186
[10] G.Alzetta, A.Gozzini, L.Moi, G.Orriols 1976 *Nuovo Cimento: B.* **36** 5-20
[11] Y.-Y Jau, A.B.Post, N.N.Kuzma, A.M.Braun, M.V.Romalis and W.Happer 2004 *Phys. Rev. Letters* **92 (11)** 110801-1 – 110801-4
[12] P.V.Zimnitskii, R.V.Smolin, S.V. Ermak, V.V.Semenov 2014 *Technical Physics Letters* **40 (3)** 271-273
[13] Baranov A. A., Ermak S. V., Sagitov E. A., Smolin R. V. and Semenov V. V. 2015 *JETP*, **121(3)**, 393-403
[14] Mathur B. S., Tang H., Happer W. 1968 *Phys. Rev.* **171(1)** 11-19
[15] Kornienko L. S., Mayorshin V. V., Kotkin L. A. and Umarkhodgaev R. M. 1982 *Optics and Spectroscopy* **53(6)** 370-372
[16] Haroche S. 1972 *Ann. De Phys*.- **6 (4-5)** 189-387
[17] Novikov L. N., Skrotskii G. V. 1978 *Sov. Phys. Usp.* **21** 589-610
[18] Bell A., Bloom A.L. 1961 *Phys. Rev.Lett.* **6 (6)** 11-19 280-281
[19] W.Happer 1972 *Review of Modern Phys.* **44(2)** 170-249

IOP Conf. Series: Journal of Physics: Conf. Series **1038** (2018) 012077 doi:10.1088/1742-6596/1038/1/012077

Inscription and visualization of tilted fiber Bragg gratings

Frolov E.A., Konnov K.A., Gribaev A.I., Zakharov V.V., Mikhneva A.A., Novikova V.A., Varzhel S.V.
ITMO University, 197101, St. Petersburg, Russia

Email: fralik95@list.ru

Abstract. In this work the inscription and investigation of tilted fiber Bragg gratings induced in a single-mode isotropic optical fiber with increased photorefractivity are demonstrated. Using the Talbot interferometer and the KrF excimer laser system, tilted fiber Bragg gratings with different angles were fabricated, the spectra of the formed diffraction structures were analyzed, and their photographic images were taken with a confocal laser scanning fluorescence microscope in transmitted light. The calculated dependence of the tilt angle of tilted fiber Bragg gratings fabricated in the Talbot interferometer on the angle between the interference fringes created in the interferometer and the cross section of the optical fiber is experimentally verified.

1. Introduction

Tilted fiber Bragg gratings (TFBGs) are a special kind of fiber-optic gratings. They have a periodic modulation of the refractive index along the fiber axis, but they differ from the standard fiber Bragg gratings (FBG) by having a certain angle between the grating planes and the fiber cross-section, which causes a more complex mode coupling of radiation propagating along the optical fiber (OF). In addition to the coupling of the core modes propagating in opposite directions, the core modes propagating in the forward direction and the cladding modes propagating in the forward and reverse directions interact. Such structures are characterized by an all-fiber design, small size, protection from electromagnetic fields, durability, stability, low insertion loss and flexible spectral transmission characteristics [1-4].

TFBGs find their practical application in the creation of optical sensors for measuring physical parameters such as temperature, deformation, vibration, refractive index of the external environment, and also they can be used in telecommunications as part of an optical erbium-doped fiber amplifier, an optical add-drop multiplexer, optical fiber polarizer [5-7].

For optimization the fabrication of TFBGs in the domestic optical fiber, it is required to know exactly the dependence of the tilt angle of the grating fabricated in the Talbot interferometer on the angle between the interference fringes produced in the interferometer and the cross section of the optical fiber. To achieve this goal, the following tasks were solved: fabrication of TFBGs with different angles, visualization of TFBG samples on an optical microscope, and analyzing photographic images; a comparison of the experimental data and the calculated dependence of the tilt angle of TFBG on the angle between the interference fringes and the cross section of the optical fiber.

IOP Conf. Series: Journal of Physics: Conf. Series **1038** (2018) 012077 doi:10.1088/1742-6596/1038/1/012077

2. Tilted fiber Bragg gratings inscription scheme

For this work, an interferometric fabrication method was used, which is described in [8]. This method makes it possible to change the period of the interference pattern over a wide range, which allows fabrication of TFBG with different periods, using only one phase mask for the amplitude separation of the laser beam. The functional scheme of the Talbot interferometer, optimized for TFBG inscription, is shown in Figure 1.

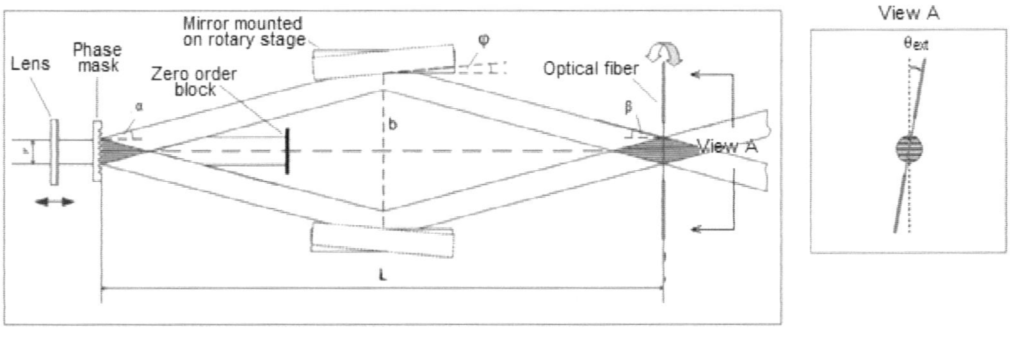

Figure 1. Functional scheme of Talbot interferometer

The laboratory setup was modified by installing the rotary holder of OF. By varying the angle of rotation of the OF θ_{ext} (the external tilt angle) relative to the interference fringes, it is possible to obtain TFBG with different angles.

3. Experimental results of tilted fiber Bragg gratings inscription

For experiments a single-mode isotropic OF with increased photorefractivity (quartz cladding diameter is 125 μm) was used. The photorefractivity of the optical fiber was enhanced by increasing the concentration of germanium dioxide to 12 mol.% at the stage of the preform formation, and then by hydrogen treatment. The conditions for the hydrogen loading of OF are the same as in [9].

After removing the protective acrylate coating, the optical fiber is installed in the fiber holder. For establishing the position of the OF, in which its axis would be perpendicular to the lines of the interference fringes, it is necessary to fabricate the gratings by rotating the OF with a rather small step (in the present work this step was 0.2°), while monitoring in the reflection spectrum the change of the position of the peak corresponding to the Bragg wavelength. When λ_{Bragg} reaches its minimum value, this position of the OF is taken as 0°.

To fabricate a single TFBG, the OF was irradiated for 1 minute at a frequency of 10 Hz, while the energy density on the surface of the fiber was 220 mJ/cm^2. In order to track not only the change in the peak corresponding to the Bragg wavelength but also the set of peaks corresponding to the coupling of the core mode and the cladding modes, the transmission spectrum was recorded using the scheme shown in Fig. 2. The spectrum was observed on the Yokogawa AQ6370C optical spectrum analyzer with a range measurements of 600-1700 nm and a resolution of 20 pm.

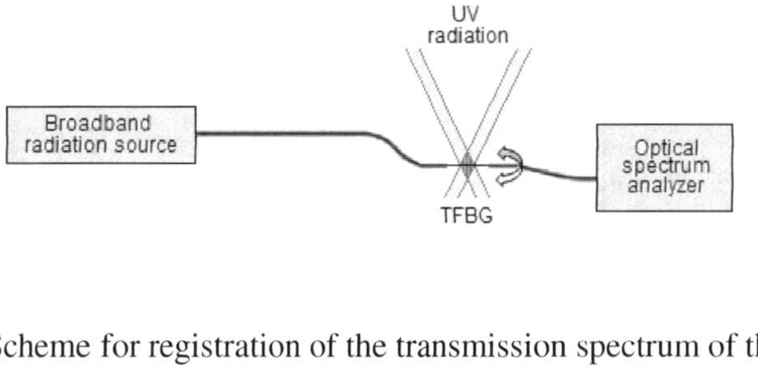

Figure 2. Scheme for registration of the transmission spectrum of the grating

In the work TFBG samples with various external angles θ_{ext} were inscribed. With increasing of θ_{ext}, the value of the Bragg wavelength λ_{Bragg} also increases, but the diffraction efficiency of the Bragg resonance of the grating decreases, as shown in the graphics presented in Fig. 3. These dependences are obtained for the isotropic optical fiber without hydrogen treatment. The dependence shown in Fig. 3(a) can be used to inscribe FBG by the phase mask method, the main disadvantage of which is the fixed value of the period of the interference pattern formed behind the phase mask. Thus, by introducing a tilt angle θ_{ext}, it is possible to adjust the wavelength of the Bragg resonance of the induced grating. Taking into account the fact that the larger the angle, the smaller the coupling of the core modes propagating in the forward and backward directions (Fig. 3 (b)), such adjustment can be effective only in the region of small angles.

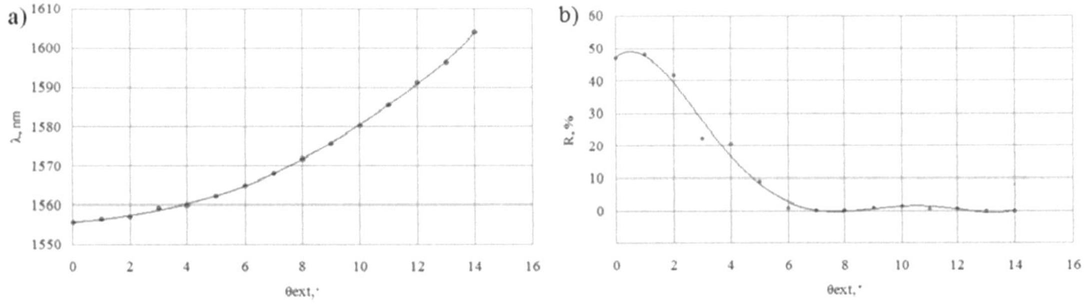

Figure 3. Dependence of Bragg wavelength λ_{Bragg} (a) and reflection coefficient R (b) of TFBG on external tilt angle θ_{ext}

Figure 4 shows the transmission and reflection spectra of TFBG with an external tilt angle of 4°, the Bragg wavelength in this case is λ_{Bragg} = 1562.74 nm. TFBG is induced in an hydrogen-loaded isotropic optical fiber.

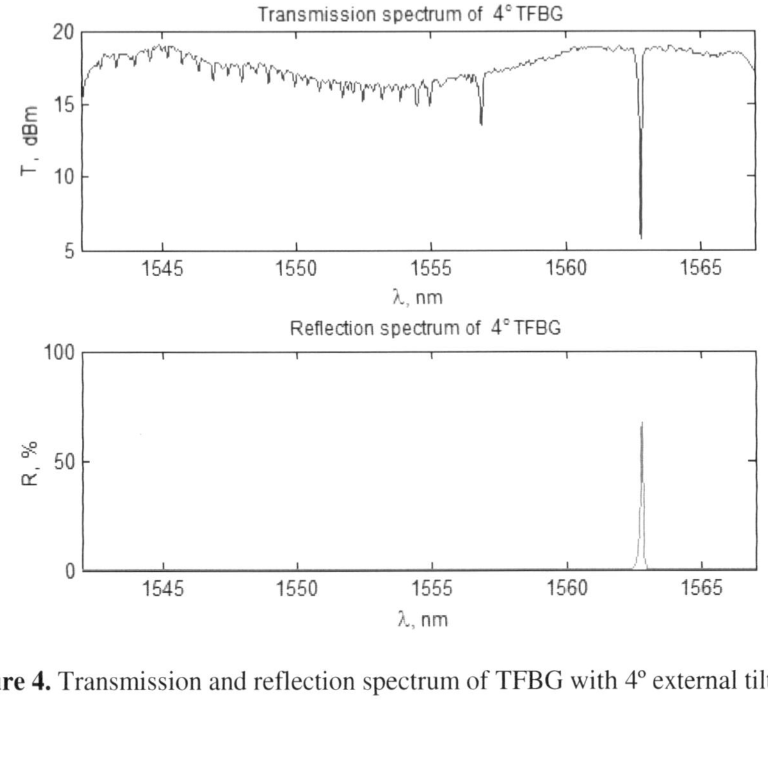

Figure 4. Transmission and reflection spectrum of TFBG with 4° external tilt angle

4. Visualization of tilted fiber Bragg gratings

Using the Zeiss LSM 710 confocal laser scanning fluorescence microscope in transmitted light, when a radiation source with a wavelength of 405 nm (diode laser) was used, experiments were performed on visualization of TFBG. Figure 5 presents photographic images of TFBGs with external tilt angles of 4°, 10° and 20°. TFBG is induced in a hydrogen-loaded isotropic optical fiber. By rotation of the OF relative to its axis the position in which the TFBG strokes were perpendicular to the OF axis was attained, and then OF with the grating was rotated by the rotary holder by 90º, which corresponded to the maximum tilt angle of the TFBG. It can be seen from Fig. 5 that the internal tilt angle of the grating θ_{int} differs from the external θ_{ext} (the angle between the interference fringes and the cross section of the optical fiber) because of the appearance of a prismatic effect when the radiation passes through the OF because of its cylindrical shape. The measured values of internal tilt angles θ_{int} of the gratings were 5.4º, 12.86º and 26.57º corresponding to external tilt angles θ_{ext} of 4°, 10° and 20°.

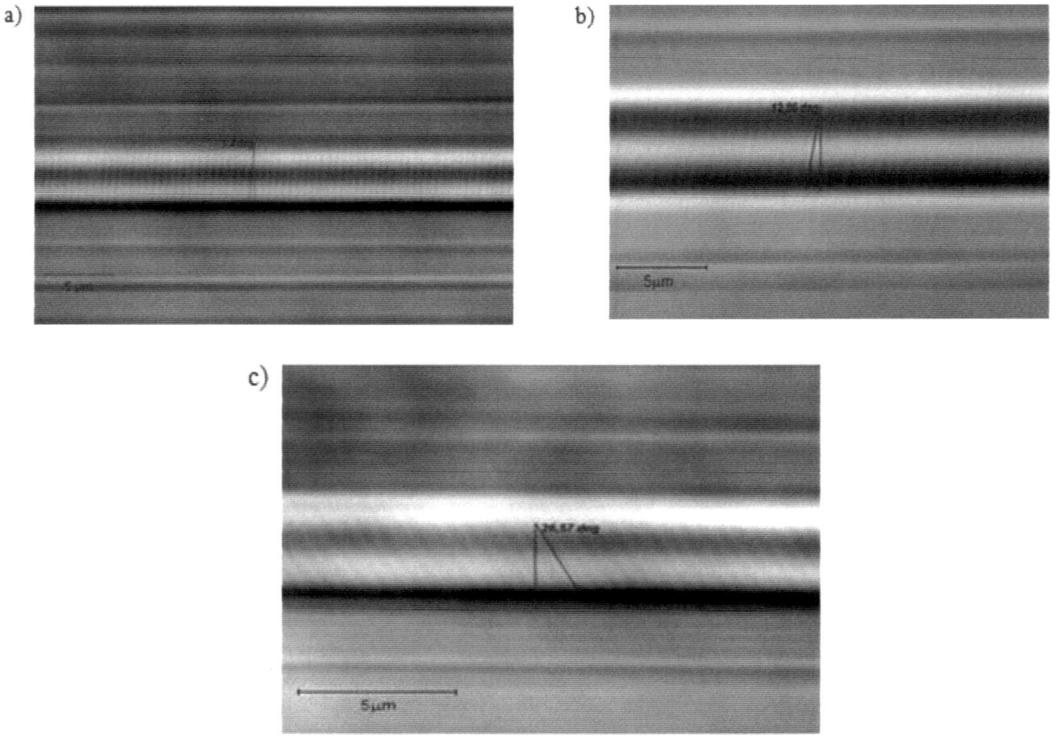

Figure 5. Photographic image of TFBGs with external tilt angles (a) 4°, (b) 10° and (c) 20°

The experimental values obtained are close to the theoretical values calculated from the formula (1) [6], as evidenced by the data presented in Figure 6:

$$\theta_{int} = \frac{\pi}{2} - tan^{-1}\frac{1}{n\,tan\,\theta_{ext}}, \qquad (1)$$

where n – refractive index of the OF cladding.

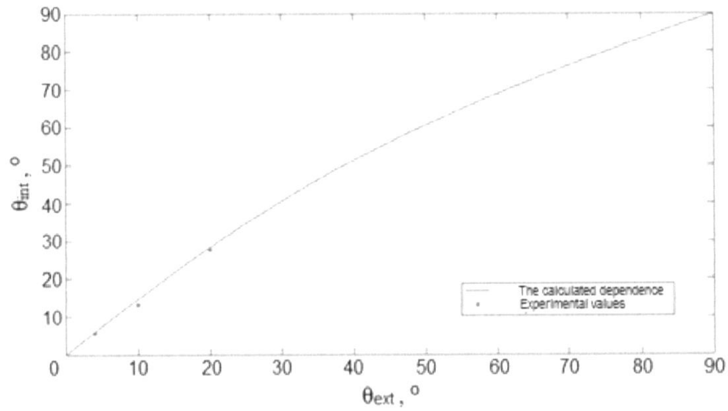

Figure 6. Calculated and experimental dependences of internal angle θ_{int} on external angle θ_{ext} of TFBG

5. Conclusion

TFBG is a universal device for selectively exciting of certain cladding modes, whereby it is possible to record a change in the state of the environment by monitoring the spectral response of the grating, which can be realized in optical sensors. Since the response to the external change in cladding modes with different effective refractive indices and core mode is different, a compact sensor can be created on the basis of TFBG to measure the impact of several physical quantities at once. Also, such structures provide the flexibility to adjust the wavelength of the Bragg resonance when the TFBGs are fabricated by the phase mask method.

In this work the Talbot interferometer was upgraded to fabricate TFBGs. TFBGs with different external tilt angles were fabricated, and also photographic images of TFBGs with external angles of inclination of 4°, 10° and 20° were obtained and analyzed. On the basis of the obtained data, the calculated dependence of the internal angle of TFBG on the external angle between the interference fringes and the cross section of the OF is experimentally verified.

Acknowledgements

This work has been done at the ITMO University and supported by the Ministry of Education and Science of the Russian Federation (Unique identifier of the Project: RFMEFI57816X0202, Contract No. 14.578.21.0202).

References

[1] Hill K.O., Fujii Y., Johnson D.C., Kawasaki B.S. 1978 *Appl. Phys. Lett.* V. **32**. № 10. P. 647-649
[2] Hill K.O., Meltz G. 1997 *J. Lightwave Technol.* V. **15**. № 8. P. 1263-1276
[3] Kashyap R. Fiber Bragg gratings, 2nd Ed. *Academic Press*. 2011
[4] Othonos A., Kalli K. Fiber Bragg Gratings: fundamentals and applications in telecommunications and sensing. *Artech House*. 1999
[5] Albert J., Shao L.Y., Caucheteur C. 2013 *Laser & Photonics Reviews*. V. **7**. № 1. P. 83-108
[6] Chen X. Optical Fibre Gratings for Chemical and Bio – Sensing. *Current Developments in Optical Fiber Technology*. InTech. 2013
[7] Park H.S. 2001 *IEEE Photonics Technology Letters*. V. **13**. № 5. P. 460-462
[8] Gribaev A.I., Pavlishin I.V., Stam A.M., Idrisov R.F., Varzhel S.V., Konnov K.A. 2016 *Opt Quant Electron*. 2016. V. **48**. Article 540. P. 1-7
[9] Varzhel' S.V., Mun'ko A.S., Konnov K.A., Gribaev A.I., Kulikov A.V. 2016 *J. Opt. Technol.* V. **83**. № 10. P. 638-641

IOP Conf. Series: Journal of Physics: Conf. Series **1038** (2018) 012078 doi:10.1088/1742-6596/1038/1/012078

Simulation of light focusing by two-layer microcylinder

E S Kozlova[1,2]

[1]Image Processing Systems Institute - Branch of the Federal Scientific Research Centre "Crystallography and Photonics" of Russian Academy of Sciences, 151 Molodogvardeyskaya st., Samara 443001, Russia
[2]Samara National Research University, 34 Moskovskoe Shosse, Samara 443086, Russia

kozlova.elena.s@gmail.com

Abstract. The formation of a narrow nanojet by polyester microcylinder with metall shell was presented by using the finite element method implemented in COMSOL Myltiphysics. Linear polarized light at a wavelength of 532 nm was used as incident light. The presence of a metal layer allows to increase the focal depth.

1. Introduction

The resolution of optical devices is restricted by diffraction limit. A lot of papers are devoted to the diffraction limit overcoming[1-5]. Tight focusing is used in such a field as optical memory systems [6,7], Raman spectroscopy [8], nano-structuring [9], optical manipulation [10-11], and nanolithography [12].

Subwavelength focal spots created by microoptical elements are called photonic nanojets[13-15]. Dielectric cylinders are widely used for producing nanojets. In [16,17] the method was shown which allows to achieve focusing of TE-polarized radiation in a region that is more than two times smaller than the diffraction limit in the two-dimensional case (the focus full width at the half maximum of the intensity was FWHM = 0.44λ). In [18] author has simulated using high-resolution finite-difference time-domain (FDTD) numerical modeling localized nanoscale photonic jets with waists 160-230 nm generated at the shadow-side surfaces of micron-scale, circular dielectric cylinders illuminated by a plane wave. The detailed analysis of localized photonic nanojets generated at the shadow side surfaces of dielectric (n=1.5) elliptical microcylinder illuminated by a plane wave a wavelength of 500 nm is reported [19]. The focusing of light at by the cylinder with the major axis of 2.5 μm and a minor axis of 1.25 μm is simulated by using FDTD method. The photonic nanojet has smallest FWHM (230 nm) for the rotation angle equal to 0°. Special attention is paid to multilayer cylinders and microspheres [20-22]. In some works, only dielectrics are used as the material of devices [20,21], while in others works metal parts are additionally considered [22]. In [20], using the exact Mie theory, it is shown that ultralong nanojets can be generated using a glass-based two-layer (BaF and LaSF) microsphere with total radius of 5λ. Propagation of laser light at a wavelength of 632.8 nm through this microsphere results in the formation of ultra-long nanojet with FWHM=0.89λ (563 nm) and the length equal to 0.22λ. The whispering gallery modes and their influence on the subwavelength light focusing was investigated in [21] by using the existing analytical solution of the diffraction problem on a cylinder. The FWHM of focal spot is equal to $(0.155 \pm 0.001)\lambda$. Simulations based on FDTD method shows the presence of a nanojet while propogation of laser light with a wavelength of 532 nm through the dielectric microcylinder (n=1.5) with a 10 nm gold shell [22]. The focal spot is formed at a distance equal to the wavelength and has FWHM=250 nm.

In this paper we also consider the focusing of laser light by dielectric cylinders with a metal shell. All simulation are carried out by finite difference method (FEM) implemented in COMSOL Multiphysics software package. The wavelength of the radiation is 532 nm. As materials of the shell, silver and gold were considered. In this paper, the dependence of the focal spot characteristics, such as the maximum intensity and FWHM, on the thickness of the metal layers is investigated.

2. Simulation of light focusing by microcylinder with one metal shell

In the early works we considered the resonant focusing of radiation by a polyester [23]. In this study resonant cylinder radii equal to $r_P = 2.1749\lambda$. However, this peak is formed near of the cylinder. Thin metal films can increase the length of the focus, forming the so-called nanojet [22].

We consider a dielectric cylinder from polyester with a radius r_P=2.1749λ, covered by a thin metal film with the width of Δr. Figure 1 shows the optical scheme. Silver and gold were chosen as materials of the film.

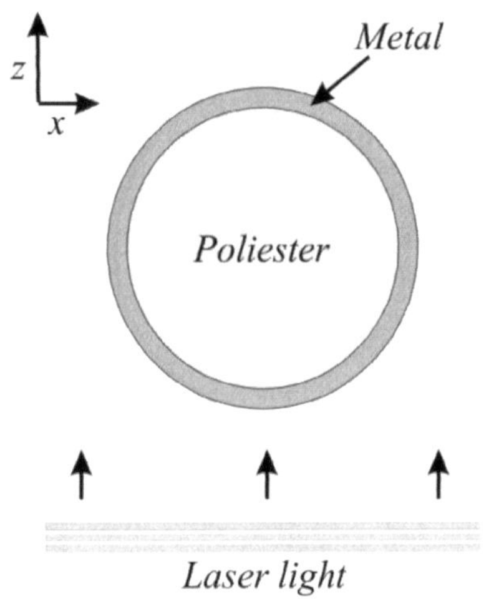

Figure 1. Simulation schem for one shell.

Gold and silver which dielectric permittivity is described by the Drude-Lorentz model [24] were used as materials of shell.For the considered wavelength, we obtain the following values of the permittivity (refractive index) for silver and gold, respectively: ε_{Ag}=-9,1375+0,8025i (n_{Ag}=0,1326+3,0257i) and ε_{Au}=-4,4602+2,5355i (n_{Au}=0,5789+2,1815i).

We simulate the propagation of TM-polarized light with a wavelength of 532 nm through the microcylinders by using the COMSOL Multiphysics package. We vary the thickness of the metal layer from 10 nm to 30 nm in steps of 1 nm while simulations. Figure 2 shows the dependence of the maximum intensity I_{max} and the full width at half maximum of intensity (FWHM) d_{FWHM} of the focal spot on the thickness of the metal layer. It should be noticed that focal spot near the surface of the cylinder has two peaks located close to each other. As a consequence, the maximum intensity was measured at a distance of 150 nm from the cylinder.

After that, we simulate focusing of TE-polarized light by the proposed cylinders and do the similar investigation of the dependence of the maximum intensity and FWHM of the focal spot on the thickness of the metal layer. It should be noticed that in this case the measurement was done near the cylinder surface since in the case of using TE-polarization the focus has one peak.

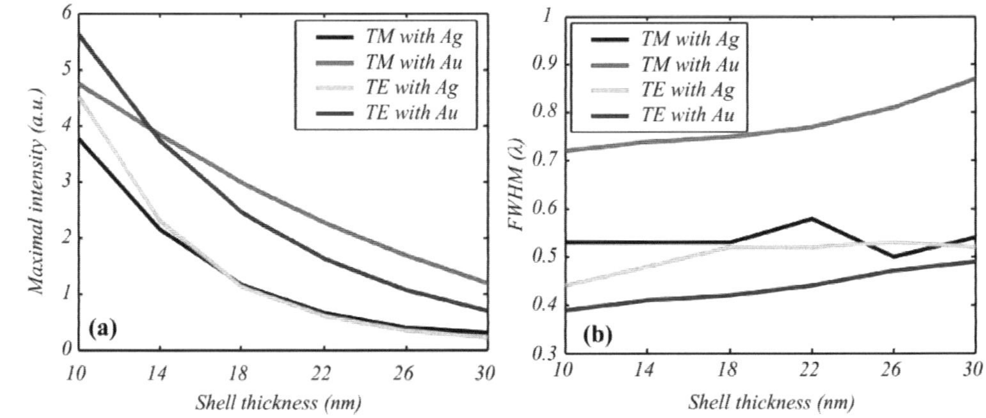

Figure 2. Dependence of the maximal intensity (a) and the FWHM (b) of the focal spot on the thickness of the metal layer for TE polarization and TM polarization.

It can be seen from Figure 2 that the maximum intensity is obtained in focal spots formed by a multilayer cylinder with a gold shell. It is because gold has a lower absorption coefficient than silver at this wavelength. However, the focal spot with minimal FWHM is formed by the cylinder with a silver shell. In both cases, increasing of the metal layer thickness leads to a decreasing of the maximum intensity and increasing of the FWHM. The optimal thickness of the metal shell was 10 nm in both cases. Figure 2 also shows that focal spots formed by a multilayered cylinder with a gold shell has the maximal intensity. However, unlike the previous case, the focal spot with the minimal FWHM is also formed by the cylinder with a gold shell. In both cases, increasing of the metal layer thickness leads to a decreasing of the maximum intensity and increasing of the FWHM. The optimal thickness of the metal shell was 10 nm in both cases. Figure 3 presents the simulation results for a multilayer cylinder with a gold shell which thickness is 10 nm.

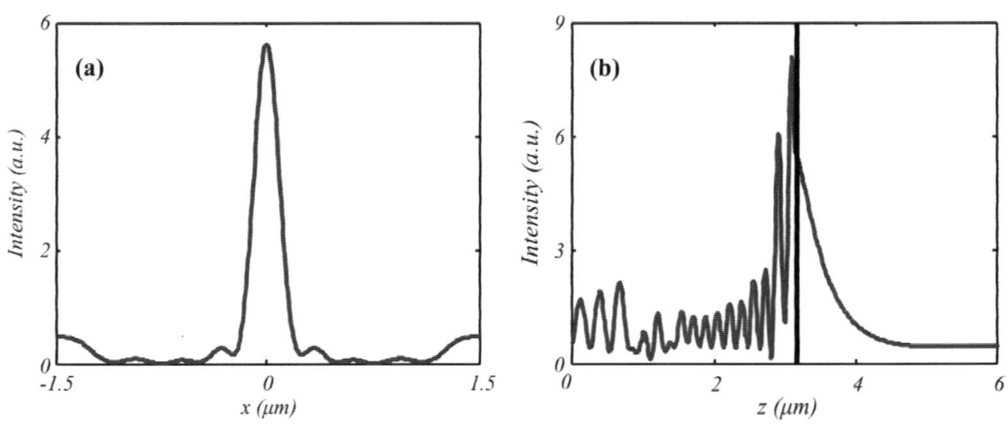

Figure 3. The intensity distribution i along the X axis immediately behind the cylinder (a) and along the Z axis at x = 0 (b) after the TE-polarized beam propagating through a dielectric cylinder with a 10-nm gold metal shell.

Figure 4 shows that focus overcomes the diffraction limit. The nanojet length at the half maximum of intensity is 0.72λ.

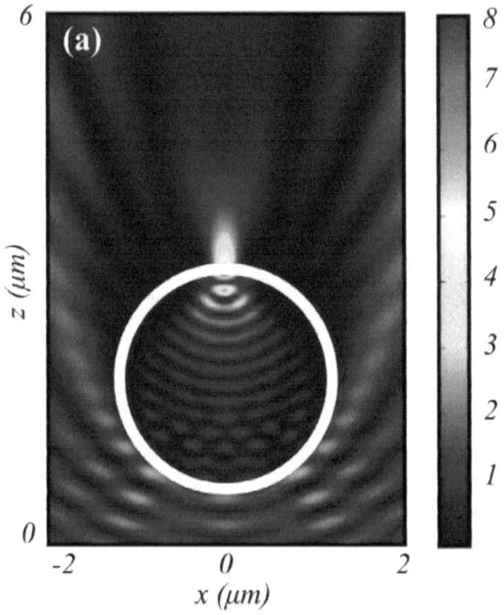

Figure 4. The intensity distribution in the XZ-plane after the TE-polarized beam propagating through a dielectric cylinder with a 10-nm gold metal shell.

3. Conclusion

In this paper, we consider the focusing of laser light by polyester cylinder with a silver and gold shell. All simulation are carried out by using COMSOL Multiphysics software package. The radius of the cores was equal to $r_P = 2.1749\lambda$. The wavelength of the incident light was 532 nm. The analysis of shell thickness influence on the light focusing shows that the design with the thinner metal shell is optimal. Comparative analysis also showed that the gold shell provides less absorption of light and the greater intensity in the focus. Moreover, a narrower focus can be achieved while propagating of TE-polarized light. As a result, a microcylinder with a gold shell of 10 nm was chosen. The simulation shows the presence of the nanojet with FWHM=0.39λ (207 nm). The presence of a metal layer allows to increase the focal depth till 0.72λ (383 nm). The obtained results can find their application in the systems of processing and storage of information, spectroscopy, and nanolithography.

Acknowledgments

The work was supported by the Ministry of Education and Science of the Russian Federation (# SP-4375.2016.5), a Russian Federation President's grant for support of leading scientific schools (# NSh-9498.2016.9) and the Russian Foundation for Basic Research (##17-47-630420, 17-47-630417, 16-47-630483, 15-07-01174).

References

[1] Monticone D G, Katamadze K, Traina P, Moreva E, Forneris J, Ruo-Berchera I, Olivero P, Degiovanni I P, Brida G, and Genovese M 2014 *Phys. Rev. Lett.* **113** 143602
[2] Savelyev D A, Khonina S N 2015 *Computer Optics* **39** 654-62
[3] Geints Yu E, Zemlyanov A A, Panina EK 2015 *Quantum Electronics* **45** 743-7
[4] Zhao Z, Pu M, Gao H, Jin J, Li X, Ma X, Wang Y, Gao P, and Luo X 2015 *Scientific Reports* **5** 15781
[5] Kozlova E S, Kotlyar V V 2013 *Computer Optics* **37** 31-8.
[6] Li X, Cao Y and Gu M 2011 *Opt. Lett.* **36** 2510-2

[7] Li X, Cao Y, Tian N, Fu L, and Gu M 2015 *Optica* **2** 567-70.
[8] Yi K J, Wang H, Lu Y F, Yang Z Y 2007 *J. Appl. Phys.* **101** 063528.
[9] Bhuyan M K, Velpula P K, Colombier J P, Olivier T, Faure N, and Stoian R 2014 *Appl. Phys. Lett.* **104** 021107
[10] Taylor M A, Janousek J, Daria V, Knittel J, Hage B, Bachor H-A, and Bowen W P 2013 *Nature Photonics* **7** 229–33
[11] Porfirev A P, Kovalev A A, Kotlyar V V 2016 *Computer Optics* **40** 152-7
[12] McLeod E, Arnold C B 2008 *Nature Nano* **3** 413-7.
[13] Mahariq I, Astratov V N, Kurt H 2016 *J. Opt. Soc. Am. B.* **33** 535-42
[14] Geints Yu E, Zemlyanov A A, 2016 *J. Appl. Phys.* **119** 153101
[15] Kotlyar V V, Stafeev S S, Feldman A 2014 *Appl. Optics* **53** 5322-9.
[16] Kozlov D A, Kotlyar V V 2014 *Computer Optics* **38** 393-6.
[17] Kotlyar V V, Kovalev A A, Kozlov D A 2016 *Optik* **127** 3803-8.
[18] Chen Z, Taflove A, Backman V 2004 *Opt. Express.* **12** 1214-20.
[19] Liu C-Y, Chang L-J 2014 *Optik* **125** 4043-6
[20] Shen Y, Wang L V, and Shen J-T 2014 *Opt. Lett.* **39** 4120-3.
[21] Kozlov D A, Kozlova E S, Kotlyar V V 2017 *Optical Memory and Neural Networks* **26** 280-8.
[22] Liu C-Y 2012 *Phys. Lett. A.* **376** 1856-60
[23] Kozlova E S, Kotlyar V V, Degtyarev S A 2015 *J. Opt. Soc. Am. B.* **32** 2352-7.
[24] Rakic A D, Djurisic A B, Elazar J M, and Majewski M L 1998 *App. Optics* **37** 5271-83.

IOP Conf. Series: Journal of Physics: Conf. Series **1038** (2018) 012079 doi:10.1088/1742-6596/1038/1/012079

Comparative spectral analysis of the extra-cell matrixes surface of heart valves before and during the process of their decullularization

E V Timchenko[1], P E Timchenko[1], L T Volova[2], D A Dolgushkin[2], P Y Shalkovskaya[1], D S Trapeznikov[1]

[1]Samara National Research University, 34, Moskovskoye shosse, Samara, 443086, Russia
[2]Samara State Medical University, 89, Chapayevskaya St., Samara, 443099, Russia

*E-mail: laser-optics.timchenko@mail.ru

Abstract. There are presented the results of the application of Raman-scattering spectroscopy method (RS) for the qualitative analysis of rams' heart valves surfaces before and during their decellularization. While analyzing RS spectra, it was found that basic differences appear at wave numbers 812 cm^{-1}, 1062 cm^{-1}, 1340 cm^{-1} and 1440 cm^{-1}, corresponding to phosphodiester linkage of RNA; OSO-3 corresponds to symmetrical stretching of glycosaminoglycans and chondroitin-6-sulfate; corresponds to deformation mode of proteins and nucleic acids (DNA); proteins, lipids. Optical analysis has shown that while analyzing decellularization on the valves surfaces, the content of glycosaminoglycans, proteins and lipids decreases; retains a high DNA content. It was found that with the aid of entered optical numbers it is possible to control the process efficiency of decellularization of the heart valves.

1. Introduction

The issue of treating people's heart valve diseases is one of the priorities of modern medicine. One of the most radical methods of treatment is the replacement of valves [1, 2]. However, the quality, design and properties of the prosthetic cardiac valve are constantly improving; they can not be compared with native valves in their properties. Thuswise, clinical cardiosurgery is in need of creating new types of implants and improving the technology of their production. [3, 4].

Because of generous amount of complications while using valve bioimplants there is a need of high-quality processing of biomaterials. Decellularization is one of the supplementary methods of tissue engineering of heart valves. This process is aimed at removing cells from the tissue with preservation of extra-cellular matrix and three-dimensional material structure [5, 6]. A number of authors consider that in order to reduce the tissue antigenicity during the process of decellularization there is a necessity for absolute elimination of cellular components, in particular: membranes and connected membrane proteins with it, bioplasts, nuclei and nucleic acids that contained in them [7, 8].

Currently there is no commonly used methodology for decellularization of heart valves. Furthermore, there is no universally received ways to control its effectiveness. With this object in mind, histological, histochemical, biochemical and immunological methods are currently used. Their main disadvantage is the destruction of the analyzed samples along with the labor-consuming nature and high cost [5-8]. Consequently, the search for appropriate way of analyzing the qualitative composition of heart valves in the process of decellularization is the priority.

The Raman-scattering spectroscopy method can be effective in assessing the decellularization effectiveness of the surface of heart valve samples, since it is able to determine the content of the main matrix components and does not require destruction of the biomaterial during the study [9-11].

The goal of this research is to analyze the qualitative composition of the heart valves surface by using the Raman-scattering spectroscopy method before and after their decellularization.

2. Materials and methods of research

Aortic valves of sexually mature rams are used as a research material. Valves decellularization was carried out in the modification according to the protocols [11, 12, 13] in Samara State Medical University (SamGMU). There is a phase 1 of decellularization before fermentative treatment and a phase 2 after fermentative treatment. Biomaterials samples were stored in phosphatic-salt solution with antibiotics at a temperature of 4 ° C before the study.

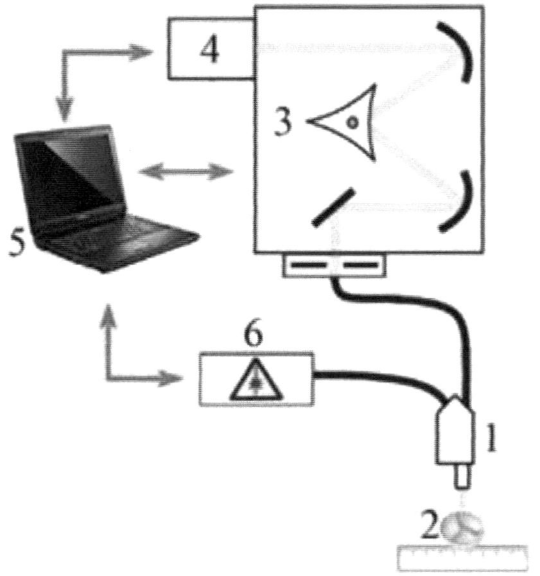

Figure 1 - Experimental stand: 1 - RP probe RPB785, 2 - object, 3 - Shamrock sr-303i spectrometer, 4 - ANDOR DV-420A-OE digital camera, 5 - electronic computing machine (ECM), 6 - LuxxMaster LML-785.0RB-04 laser module

Spectral-response characteristics of the samples were studied using an experimental stand including a high-resolution digital spectrometer Shamrock sr-303i with built-in cooling chamber DV420A-OE, a fiber-optics probe RPB-785 for Raman spectroscopy, combined with laser module LuxxMaster LML-785.0RB-04 (with controlled output up to 500 mW, and a wavelength of 785 nm). The radiation power of 500 mW of the 6 laser within the exposure time up to 300 seconds does not cause destructive changes in the samples.

RP probe focused laser radiation on the object at a distance of 7.5 mm from the output window with a focal spot diameter of less than 0.2 mm and collected radiated emission.

3. Results and discussion

Figure 2 shows the average RS spectra for the surfaces of aortic valve samples before decellularization (control), and RS spectra of valve surfaces in the process of performing decellularization before (step 1) and after (step 2) fermentative treatment of the samples.

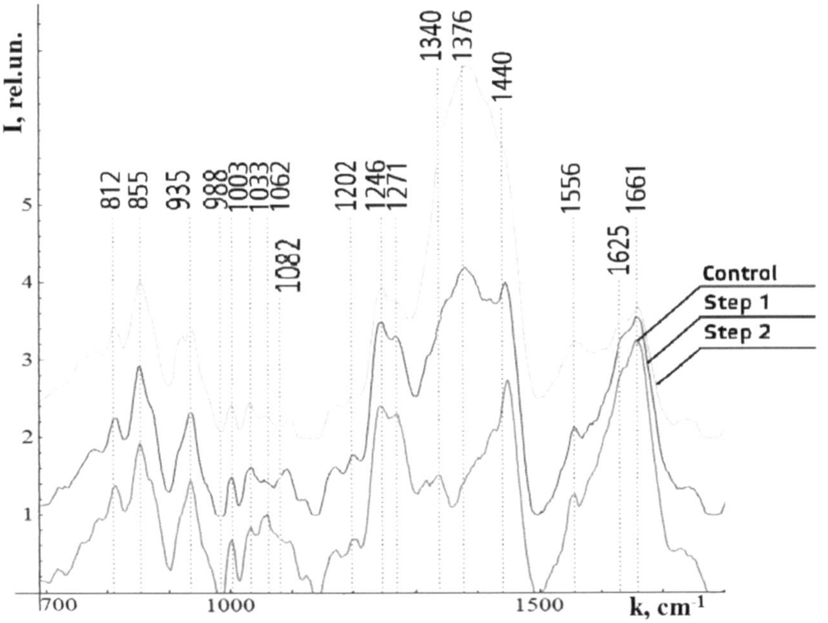

Figure 2 - Average spectra of surfaces of aortic heart valves before and during their decellularization

While carrying out RS spectroscopy of the valve surfaces before and during their decellularization, we obtained qualitatively identical Raman bands corresponding to certain oscillation modes (Table 1).

Table 1. The results interpretation of RS spectroscopy of aortic valves surfaces

Wave number, cm^{-1}	Substance, oscillation
812	Phosphodiester linkage RNA[15]
855	ν(C–C) Proline, oxyproline, tyrosine [16, 17]
935	ν(C–C) Proline, oxyproline [18]
988	C-H Lipide bend [19]
1003	Phenylalanine [18]
1033	CH oscillation of phenylalanine [20], CN stretching in protein [20,21]
1062	OSO-3 symmetrical stretching of glycosaminoglycans, chondroitin-6-sulfate [18]
1082	C-H in a mode of a stretching of fibers (and a mode of lipids to a lesser degree) [21]
1202	Hydroxyproline, tyrosine [23]
1246, 1271	Amide III [21]
1340	Deformation of proteins and nucleic acids (DNA) [20]
1376	CH oscillations [12]
1440	δ (CH2) proteins, lipids [18]
1556	Elastin (protein) [20]
1625,1661	Amide I (C = O) [22]

From Fig. 2 it can be seen that there was a reduction of intensity during the first phase of decellularization at wave numbers 812 cm^{-1}, 1062 cm^{-1} and 1440 cm^{-1}, corresponding to the phosphodiester linkage RNA; OSO-3 symmetrical stretching of glycosaminoglycans of chondroitin-6-sulfate; proteins, lipids. As the second phase of decellularization completed there was noted a little decrease in the intensity at a wave number of 1340 cm^{-1}, corresponding to the deformation mode of proteins and nucleic acids (DNA).

The relatively constant surface samples component of the study before and during decellularization was amide III, corresponding to the line intensity at a wave number of 1246 cm^{-1} [14]. Therefore, the number magnitude was used as a denominator at calculating the entered optical number (coefficient) f:

$$f = \frac{I_i}{I_{1246}}$$

I_i - is the intensity value at the wave number of the analyzed component, I_{1246} - is the intensity value at the wave number of amide III.

Figure 3 presents two-dimensional diagrams of optical numbers (coefficients) reflecting a change in the composition of the main surface components of aortic heart valves before and during the process of decellularization at different phases.

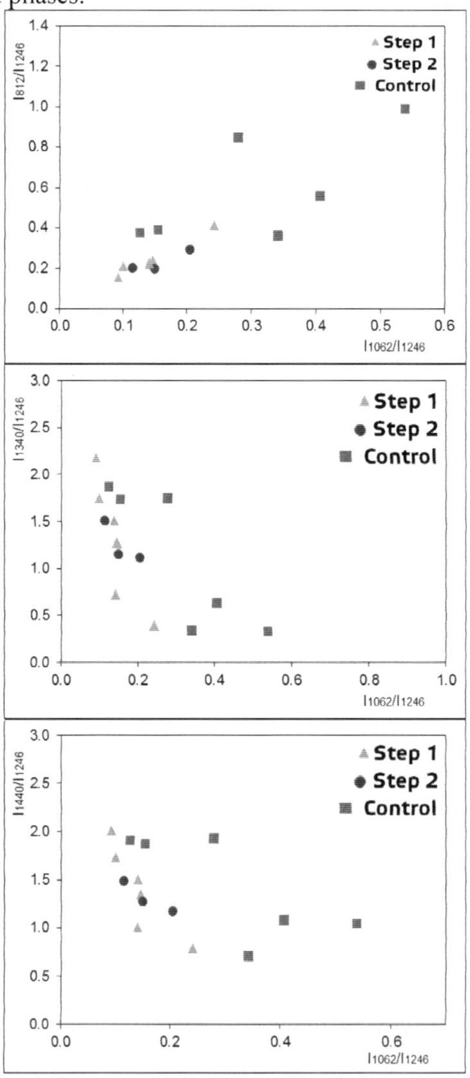

Figure 3 - Two-dimensional diagrams of the entered optical numbers (coefficients) (Control - samples before decellularization, step 1 - samples before fermentative treatment, step 2 - samples after fermentative treatment)

The two-dimensional dependence analysis showed that in the phases of performing decellularization there was observed a gradual decrease of the optical numbers (coefficients) I_{812}/I_{1246}, I_{1062}/I_{1246} and I_{1440}/I_{1246} in comparison with the values obtained during the study of the surfaces of intact aortic valves. The optical number (coefficient) I_{1340}/I_{1246} slightly decreases even after the second phase of decellularization.

4. Conclusions

During the study of surface of aortic valves before and during their decellularization by using RS spectroscopy, it was found that even after the first phase of decellularization there occurred intensity at the wave numbers 812 cm^{-1}, 1062 cm^{-1} and 1440 cm^{-1} corresponding to the phosphodiester linkage RNA ; OSO-3 symmetrical stretching of glycosaminoglycans of chondroitin-6-sulfate; proteins, lipids. After the second phase of decellularization was completed there was noted a little decrease in the intensity at a wave number 1340 cm^{-1} corresponding to the deformation mode of proteins and nucleic acids (DNA).

Entering optical numbers (coefficients) and two-dimensional analysis, we elucidated the effectiveness of the decellularization process of aortic valves. This effectiveness was indirectly manifested as a decrease in the content of lipids, proteins and glycosaminoglycans on the surface. Nevertheless, the DNA content in the valve samples insignificantly decreased even before the completion of the second phase of decellularization.

With the help of the entered optical numbers (coefficients), it is possible to control the effectiveness of decellularization process of the heart valves.

Acknowledgments
The work was supported by the of the Russian Science Foundation project N18-75-10003.

References

[1] Astapov D A, Nazarov V M, Zheltovsky Y V, Isayan M V, Demidov D P, Kaganskaya N A 2013 *A xenobiological prosthesis in the left heart* (Siberian Medical Journal - No. 1) pp. 13-17

[2] Bokeria L A, Kagramanov I I, Kokshenev I V, Britikov D V 2009 *Biomaterials in Cardiovascular Surgery* (Moscow) p. 350

[3] Gendlin G E, Storozhakov G I, Vavilov P A et al. 2008 *Indications for surgical treatment of patients with heart valve diseases* (Heart) Vol. 7, № 2, pp. 113-117

[4] Dohmen P M, Konertz W 2009 *Tissue-engineered heart valve scaffolds* (Ann. Thorac. Cardiovasc. Surg.) 15(6), pp. 362–367

[5] He M, Callanan A 2013 *Comparison of methods for whole-organ decellularization in tissue engineering of bioartificial organs* (Tissue Eng Part B Rev) Vol. 3, № 19, pp. 194-208

[6] Zia S, Mozafari M, Natasha G, Tan A, Cui Z, Seifalian A M 2015 *Hearts beating through decellularized scaffolds: whole-organ engineering for cardiac regeneration and transplantation* (Crit Rev Biotechnol)

[7] Lavresin A V, Nasredinov A S, Kurapeev D I, Anisimov S V, Mitrofanova L B, Beshchuk O V 2014 *Decellularization of aortic allografts and their morphological evaluation* (Genes and cells) Vol. 9, № 1

[8] Assmann A, Delfs C, Munakata H, Schiffer F, Horstkötter K, Huynh K, Barth M, Stoldt V R, Kamiya H, Boeken U, Lichtenberg A, Akhyari P 2013 *Acceleration of autologous in vivo recellularization of decellularized aortic conduits by fibronectin surface coating* (Biomaterials) Vol. 25, № 34, pp. 6015-6026

[9] Krafft C, Dietzek B, Popp J 2009 *Raman and CARS microspectroscopy of cells and tissues* (Analyst Vol. 6, № 134, pp. 1046-1057

[10] Enrique Uceda Otero, Sokki Sathaiah, Landulfo Silveira Jr., Pablo Maria Alberto Pomerantzeff, Carlos Augusto Gonçalves Pasqualucci 2004 *Raman spectroscopy for diagnosis of calcification in human heart valves* (Spectroscopy IOS Press) Vol. 18, pp. 75–84

[11] Sara Mangialardo, Valentina Cottignoli, Elena Cavarretta, Loris Salvador, Paolo Postorino, Adriana Maras 2012 *Pathological Biominerals: Raman and Infrared Studies of Bioapatite Deposits in Human Heart Valves* (Applied Spectroscopy) Vol. 66, pp. 1121-1127

[12] Lichtenberg A et al. 2007 *Biological scaffolds for heart valve tissue engineering* (Methods Mol. Med.) Vol. 140, № 2, pp. 309-317

[13] Chan, J.W., Taylor, D.S., Zwerdling, T., Lane, S.T., Ihara, K., and Huser, T. 2006 *Micro-Raman spectroscopy detects individual neoplastic and normal hematopoietic cells* (Biophysical Journal) Vol. 90, pp. 648–656

[14] Lichtenberg A, Tudorache I, Cebotari S et al. 2006 *A Preclinical testing of tissue-engineered heart valves re-endothelialized under simulated physiological conditions* (Circulation) Vol. 114, pp. 1559-1565

[15] Chan J W, Taylor D S, Zwerdling T, Lane S T, Ihara K, Huser T 2006 Micro-*Raman spectroscopy detects individual neoplastic and normal hematopoietic cells* (Biophysical Journal) Vol. 90, pp. 648–656

[16] E Brauchle, S Noor, E Holtorf, C Garbe, K Schenke-Layland, and C Busch 2014 *Raman spectroscopy as an analytical tool for melanoma research* (Clinical and Experimental Dermatology) Vol. 39, pp. 636–645

[17] Nguyen T T, Gobinet C, Feru J, Brassart–Pasco S, Manfait M 2012 *Piot Characterization of Type I and IV Collagens by Raman Microspectroscopy: Identification of Spectral Markers of the Dermo-Epidermal Junction* (Spectroscopy) Vol. 27, pp. 421–427

[18] Aliz Kunstar, Anne M Leferink, Paul I Okagbare, Michael D Morris, Blake J Roessler, Cees Otto, Marcel Karperien, Clemens A van Blitterswijk, Lorenzo Moroni and Aart A van Apeldoorn 2013 *Label-free Raman monitoring of extracellular matrix formation in three-dimensional polymeric scaffolds* (Journal royal society)

[19] Xu H, Xu B, Yang Q, Li X, Ma X, Xia Q, Zhang Y, Zhang C, Wu Y, Zhang Y 2014 *Comparison of Decellularization Protocols for Preparing a Decellularized Porcine Annulus Fibrosus Scaffold* (PLoS ONE) Vol. 1, № 9

[20] Sana Tfaili, Cyril Gobinet, Gwendal Josse, Jean-Franc¸ois Angiboust, Michel Manfaita and Olivier Piot 2012 *Confocal Raman microspectroscopy for skin characterization: a comparative study between human skin and pig skin* (The Royal Society of Chemistry Analyst) Vol. 137, pp. 3673–3682

[21] Kristy L Cloyd, Ismail El-Hamamsy 2012 *Suwimon Boonrungsiman Characterization of Porcine Aortic Valvular Interstitial Cell 'Calcified' Nodules* (PLoS ONE) Vol. 7, pp. 1-9

[22] Enrique Uceda Otero, Sokki Sathaiah, Landulfo Silveira Jr, Pablo Maria Alberto Pomerantzeff, Carlos Augusto Gonçalves Pasqualucci 2004 *Raman spectroscopy for diagnosis of calcification in human heart valves* (Spectroscopy IOS Press) Vol. 18, pp. 75–84

[23] E Brauchle, S Noor, E Holtorf, C Garbe, K Schenke-Layland1, and C Busch 2014 *Raman spectroscopy as an analytical tool for melanoma research* (Clinical and Experimental Dermatology) – Vol. 39, pp. 636–645

Effect of Ga^+ focused ion beam etching on photoluminescence of AlGaAs/GaAs heterostructure.

G V Voznyuk[1], I V Levitskii[1,2], M I Mitrofanov[1,2] D N Nikolaev[2], V P Evtikhiev[2]

[1]ITMO University, St. Petersburg, Russia
[2]Ioffe Institute, St. Petersburg, Russia

Corresponding author: glebufa0@gmail.com

Abstract. The effect of focused ion beam (FIB) etching by 30 keV Ga^+ on photoluminescence of AlGaAs/GaAs heterostructure is studied. During etching process, high-energy ions induce radiation defects that lead to a decrease of heterostructure internal quantum efficiency of luminescence. We used the SRIM software to simulate the radiation defects penetration depths in AlGaAs/GaAs heterostructure, then carried out FIB etching guided by received information. Annealing of the structure at 300 °C showed partial recovery of the internal quantum efficiency. Subsequent annealing at 620 °C showed almost full recovery of quantum efficiency depending on the etching depth. Experimental findings allowed us to affirm that FIB etching with subsequent annealing is a potential tool for making photonic nanodevices.

1. Introduction

Nowadays there is an increased interest in photonic integrated circuits (PICs). PICs contain a plurality of optically interconnected elements that mutually perform optical signal processing on a single chip. In comparison with electronic integrated circuits, optical signal use provides significant advantages - less heat generation, higher data transfer rates, greater bandwidth due to the variety of the multiplexing techniques (WDM, DWDM and etc.) and low crosstalk. The most common material for the manufacture of photonic circuits is silicon due to the high silicon technology development level in microelectronics [1]. Hybrid coupling technique allows integration of silicon photonic circuits with microlasers based on direct-band A^3B^5 semiconductors [2,3]. It is possible to realize monolithic PIC design completely based on A^3B^5 heterostructures, in this case the reliability of devices is increased and the cost of production is reduced. There are different microfabrication methods for prototyping parts of integrated circuits. Focused ion beam lithography technique is one of the most promising for prototyping PICs due to direct resist-free and maskless etching that on the one hand allows etching three-dimensional (3D) structures (where is hard or impossible to use mask) and on the other hand significantly decreases circuits fabrication cost. The factor that strongly limits the FIB application is the damages of crystalline structure that accompany the etching. This problem is rather good investigated for the ion implantation process where ions had high energy (100-300 keV) [4]. The influence of ions with energy typical for FIB (10-30 keV) on A^3B^5 semiconductors has been investigated insufficiently.

2. SRIM calculations

When accelerated ion strikes the solid, it loses kinetic energy due to an interaction with the atoms of the sample. Transferred energy from the ion to the solid leads to a number of different processes such as sample heating, atomic sputtering, ion emission, etc. The most widespread concept of ion-nucleus interactions is a cascade collision model [5,6]. We used one of the most popular software package SRIM (Stopping and Range of Ions in Matter) to calculate the range of ions in solid. SRIM gives opportunity to calculate important parameters such as track length, distribution of vacancies in the target, recoil atoms, etc. TRIM (transport of ions in matter) is a part of SRIM software that uses the Monte Carlo method to calculate the interaction of ions with a target. In our work we used gallium ion source with energy of 30 keV and the target was $Al_{0.18}Ga_{0.82}As$ layer. The distribution profiles of Ga+ ions were calculated with the following parameters: accelerating voltage 30 keV and ion beam incidence angle 0°. Based on the simulation results, the distribution profiles of gallium ions are plotted. Figure 1 shows the distribution of implanted Ga s and Al, Ga, As recoiled atoms along Y-axis and vacancies distribution profile. According to the SRIM calculation the point defect penetration depth is about 70 nm.

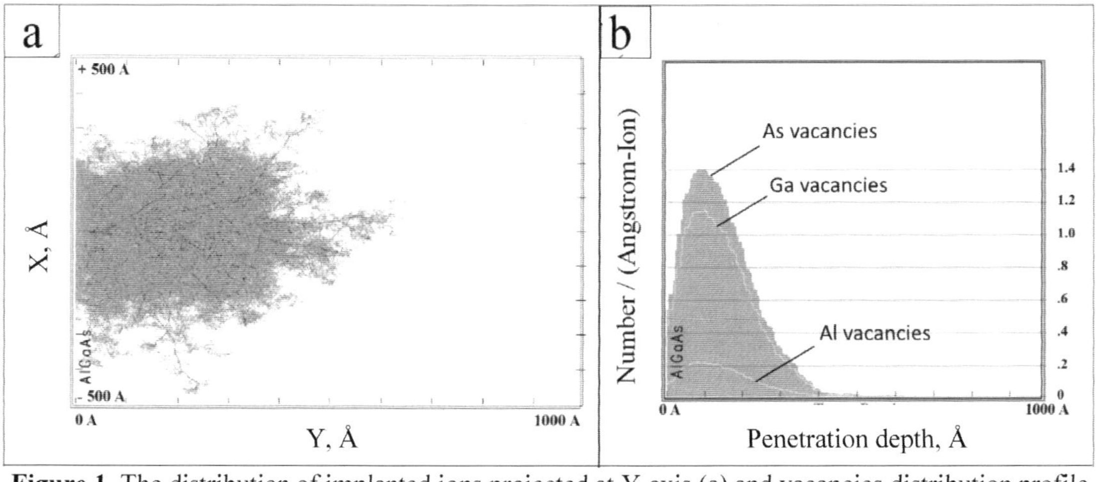

Figure 1. The distribution of implanted ions projected at Y-axis (a) and vacancies distribution profile (b).

3. Experiment

To study the effect of Ga^+ focused ion beam etching on the luminescence of a AlGaAs/GaAs heterostructure, the test double heterostructure consisting of a 1-μm thick GaAs layer enclosed between two 1 μm thick $Al_{18}Ga_{82}As$ barriers was fabricated (figure 2). On the heterostructure surface 7 squares 50 x 50 μm with depths of 1, 150, 180, 330, 850, 1200 and 1500 nm were etched using Ga^+ ions at 30 keV. Figure 3 shows the optical micrograph of etched region and depth profiles of etched squares, obtained by profilometer (Ambios XP1). We study the influence of radiation defects on the quantum efficiency using photoluminescence measurement at room temperature. The pump source was a solid-state laser with 671 nm wavelength, which was focused into 10 μm spot with 0.85 NA microscope objective.

Squares 1-7 were measured. In all cases, a strong luminescence quenching was observed. At the 5, 6, 7 squares, at 0.34 kW/cm^2 pump level, photoluminescence was not observed. To improve the quantum yield we carried out low-temperature annealing of the sample in vacuum conditions at 300 °C. After this stage, photoluminescence from the first three squares (etching depths 1, 150 and 180 nm) was restored almost completely to the initial level. At the next stage, the structure was annealed in the arsenic atmosphere at 620° C over a period of 20 minutes.

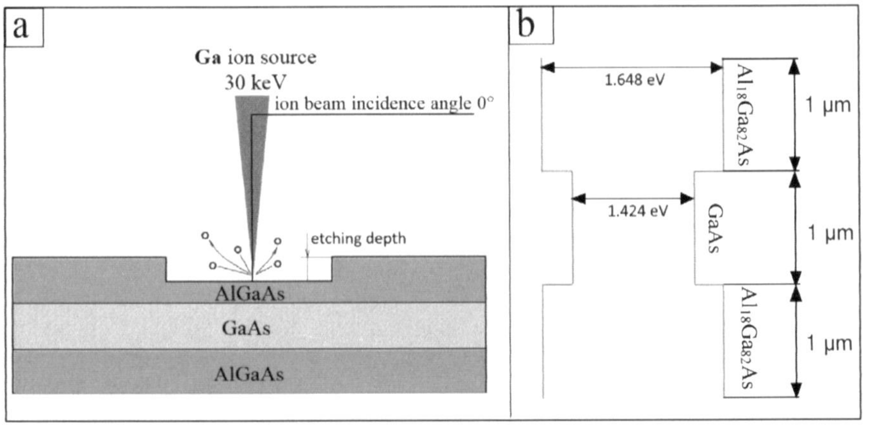

Figure 2. Principal scheme of the sample (a) and bandgap diagram of the etched sample (b).

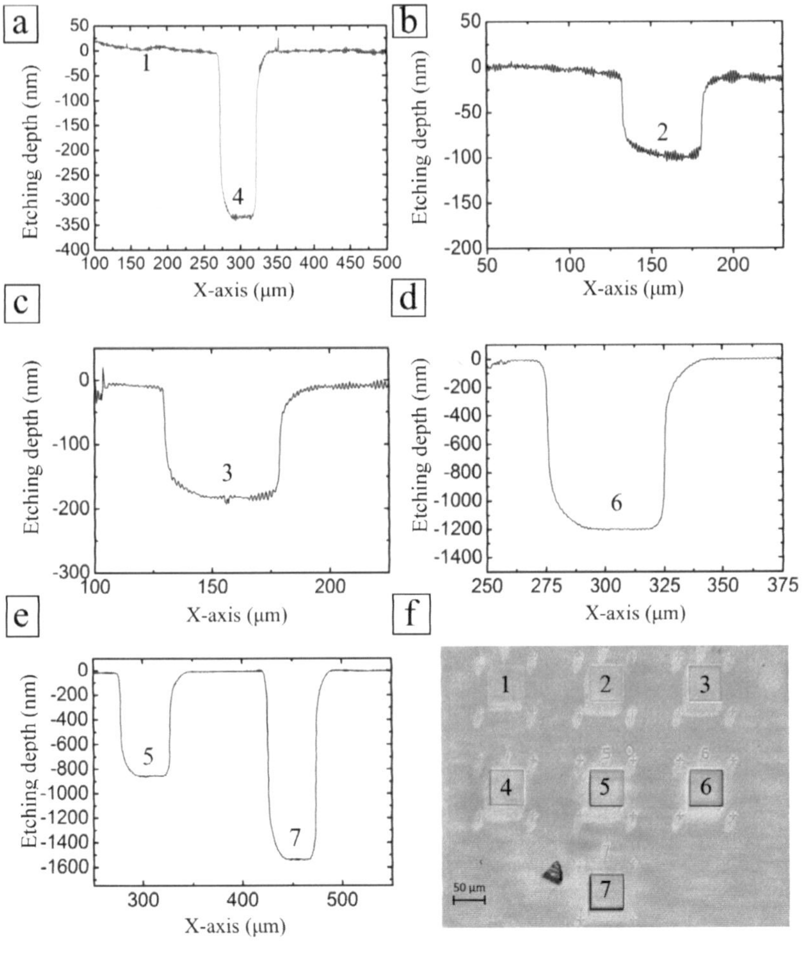

Figure 3. Depth profiles of the etched squares obtained by Ambios XP 1 profilometer (a-e) and the optical micrograph of the surface (f) (numbers 1-7 link squares with corresponding depth profiles).

Annealing at 620 °C gave practically full intensity recovery up to the level of non-etched surface in the first five squares. Figure 4 demonstrates comparison in photoluminescence (PL) spectrums between etched (5th square) and non-etched region together with the results of measurements of quantum efficiency before and after the annealing. Squares, triangles and circles denote different temperature of the annealing. The sixth and the seventh squares luminescence was not observed even after the second annealing. We suppose that increasing radiation defects penetration depth (compared with the results of SRIM calculations) is related to the high local temperature and to the diffusion and accumulation of point defects closer to the heterointerface in depth of the sample. Nevertheless, our results show the FIB etching at the depth of 850 nm (distance to the heterointerface 150 nm) could be used for photonics applications.

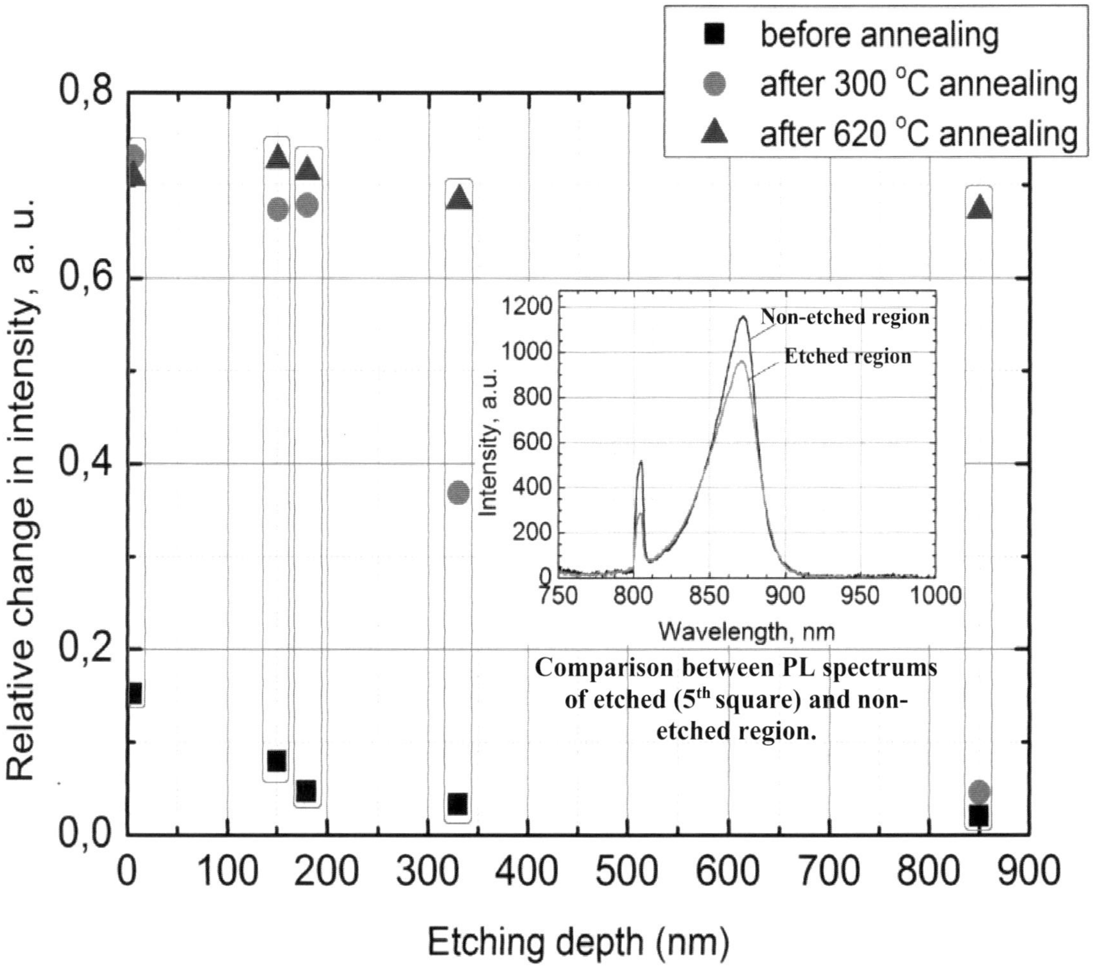

Figure 4. The evolution of relative change in luminescence intensity for five different etching depths before and after two subsequent annealing.

4. Conclusion

In spite of the fact that FIB etching generate defects in a structure, subsequent annealing helps to heal radiation defects. Obtained results demonstrate the possibility of recovering luminescent properties of the structure after FIB etching. Our experimental findings confirm prospects of using FIB etching technique for the formation of PICs elements.

IOP Conf. Series: Journal of Physics: Conf. Series **1038** (2018) 012080 doi:10.1088/1742-6596/1038/1/012080

References

[1] R. G. Beausoleil 2011 Large-Scale Integrated Interconnects *ACM Journal on Emerging Technologies in Computing Systems* Vol. 7. No. 2 Article 6
[2] G. Roelkens, L. Liu, D. Liang et al. 2010 III-V/Silicon photonics for on-chip and inter-chip optical interconnects *Laser & Photonics reviews* Vol. 4 pp 751–779
[3] H. D. Wanzenboeck, S. Waid 2011 Focused Ion Beam Lithography Recent Advances in Nanofabrication Techniques and Applications pp 27-50
[4] A.J. Steckl, P. Chen, A.G. Choo et al. 1992 Enhanced photoluminescence from AIGaAs/GaAs superlattice gratings fabricated by Si FIB implantation, *MRS Online Proceedings Library (OPL)* Vol. 281 pp 319-324
[5] P. K. Kashkarov 1999 The formation of point defects in semiconductor crystals *Soros Educational Journal* Vol. 1 pp 105-112
[6] J. Gierak 2014 Focused Ion Beam nano-patterning from traditional applications to single ion implantation perspectives *Nanofabrication 2014* Vol. 1 pp 35–52

IOP Conf. Series: Journal of Physics: Conf. Series **1038** (2018) 012081 doi:10.1088/1742-6596/1038/1/012081

Multichannel IR Fourier transform spectrometer

I Sh Khasanov [1], V A Vagin[1] and Il S Golyak[2]

[1] Scientific and Technological Center of Unique Instrumentation of the Russian Academy of Sciences, Moscow, 117342, Russia
[2] Bauman Moscow State Technical University, Moscow, 105005, Russia

Abstract. The authors propose to implement a multi-channel continuous scan Fourier spectrometer for simultaneous recording and analysis of the spectral characteristics of several objects. To implement such a scheme is used multiprobe fiber in which several optical fibers, at one end, are connected in one optical connector and fixed in the output of interferometer. In this scheme, the Fourier spectrometer is used as a modulator. On the other hand, each fiber individually mated with the test sample and its own radiation detector. The spectral resolution of the proposed system is 1 cm-1. For the system the aperture of the spectrometer from the condition of minimum spectral resolution and optical fibers parameters is calculated. Using the proposed scheme emission spectra of helium neon lamp have been registered for each single fiber with diameter of 1 mm and a numerical aperture NA= 0,22.

1. Introduction

To solve some of modern scientific, scientific-technical and production tasks, sometimes it becomes necessary to simultaneously measure the spectral characteristics of an object of research by various spectral methods and in different spectral regions. In this paper we propose and examine the physical possibility of organizing such measurements using a single spectrometer.

The effectiveness of modern production is determined not only by the emergence of new production technologies and the introduction of new high-tech tools, but also by the quality of continuous (rapid) monitoring of the various stages of the production process, with the ability to react quickly to its change (i.e., active management of these processes). To ensure this, it is necessary to effectively measure the parameters (characteristics) of the process at its key stages. Since the measurement of the chemical composition of the input, intermediate and final product is the most important condition in many technological processes (in the food, chemical, oil refining, pharmaceutical and other industries), development of a system for diagnostics and control over complex chemical processes providing such a measurement is extremely topical. One of the most effective means of measuring various characteristics of the objects under study is spectral instruments.

Spectral methods make it possible to perform a spectral analysis of the sample under study and determine the presence of a substance from the optical absorption lines and then, by measuring the optical density at these lines, to measure the concentration of the desired substance in the sample. In this case, the measurement can be made both in the sample taken and with the help of special devices (for example, optical fiber probes) directly inside the process.

The use of fiber-optic probes matched with spectral instruments allows carrying out various spectral measurements in remote, hard-to-reach and dangerous places without the need for sample preparation. Fiber-optic probes were used in the first place in conventional spectrometers [1], as one of the

IOP Conf. Series: Journal of Physics: Conf. Series **1038** (2018) 012081 doi:10.1088/1742-6596/1038/1/012081

improvement. But in the future specialized spectrometers were created, oriented only to fiber-optic probes [2, 3, 4].

An important task was the creation of spectrometers, which could work directly with several probes and carry out measurements on each channel. In this case, one device allows to control several phases of the technological, physical or chemical process. For such measurements, the MATRIX-F Fourier spectrometer (Bruker, Germany) was developed. This spectrometer uses a 6-channel multiplexer for the possible simultaneous connection of 6 probes. It allows the use of one instrument for sequential process control in 6 reactors. The original optical-mechanical system provides the necessary switching of the spectrometer from one probe to another and allows sequential spectral measurements at 6 points of the process under study.

Thus, the actual task is the development of a continuous scanning spectrometer, which would allow simultaneous measurements at several points in the technological process and obtain measurements in real time.

2. Scheme for simultaneous registration of a signal from several samples

We propose a scheme of an IR Fourier spectrometer for continuous scanning, which will allow simultaneous measurements of the investigated object with several probes at different points [5]. To implement such a scheme, several optical fibers on one side are combined into one optical bundle (figure 1) and fixed in the corresponding fiber optical connector at the output of the interferometer.

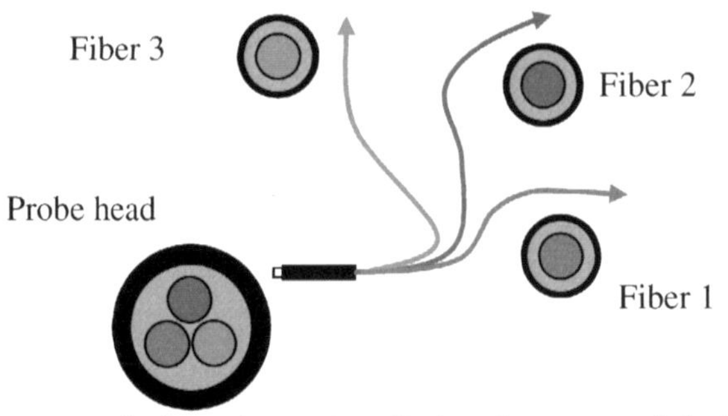

Figure 1. Structure of the optical fiber probe.

Using an optical condenser installed at the output of the Fourier spectrometer, the radiation emitted by the interferometer is applied to each optical fiber. Thus, the proposed Fourier spectrometer becomes a multichannel instrument. Each individual probe, through the appropriate fiber, monitors its object through its receiver (figure 2). As a result, it is possible to simultaneously measure several objects or different stages of a certain process, for example, in a production line.

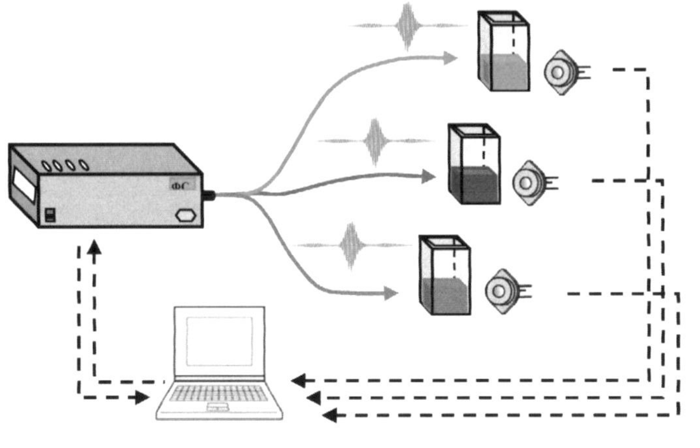

Figure 2. A scheme for simultaneous registration of a signal from several samples

In such a scheme, the radiation source is placed at the input of the interferometer. With the aid of a special optical system (for example, an off-axis mirror paraboloid) in the focus of which the radiation source is located, a collimated flux of light is directed to the input of the interferometer. An interferometer with continuously variable optical path difference acts as an optical modulator of radiation, creating an initial interferogram. Then the interfering radiation is directed to the optical connector, to which is connected an optical fiber bundle consisting of several separate optical probes. Each probe ends with its IR receiver. Thus, it is possible to perform sequential or simultaneous recording of interferograms from each individual receiver.

3. Estimating the luminosity of the proposed scheme of the Fourier transform spectrometer

With such a system of operation, it is necessary to calculate the Fourier transform spectrometer's luminosity, depending on the characteristics and number of probes. The Fourier transform spectrometer is a high-speed instrument, its luminosity is determined by the value of ΩS, where S is the effective cross section of the light flux incident on the beamsplitter, calculated by the formula:

$$S = \pi D^2 / 4 , \qquad (1)$$

where D is the effective beam splitter diameter, and Ω is the solid angle, which determines the beam divergence of this light flux. For small α, the solid angle is given by:

$$\Omega = \pi\alpha^2 , \qquad (2)$$

where α is the maximum deviation angle from the radiation axis. It is known [6] that the requirement for the beam divergence of the above-mentioned light flux is determined by the maximum spectral resolving power (R), which one wants to obtain on the Fourier transform spectrometer, namely

$$\Omega R \leq 2\pi . \qquad (3)$$

In the case under consideration, the element limiting the luminosity of the spectral system is optical fiber. The diameter of the optical fiber (d) usually does not exceed 1 mm. All the radiation emitted from the optical fiber is concentrated near its axis in the angular range (φ) of approximately ± 15 °. Thus, the Lagrange-Helmholtz invariant leads to the relation

$$\varphi d = D\alpha . \qquad (4)$$

From (2) and (3) it follows

$$\alpha^2 \pi R \leq 2\pi \qquad (5)$$

or

$$\alpha \leq (2 / R)^{1/2} . \qquad (6)$$

Thus, a Fourier transform spectrometer with an effective beamsplitter diameter is sufficient to work with such an optical fiber

$$D = \varphi d / \alpha = \varphi d / (2 / R)^{1/2} . \qquad (7)$$

With a spectral resolution of the order of 1 cm^{-1}, realized in the proposed Fourier spectrometer with a short-wave boundary of the spectral range of 5000 cm^{-1}, we will have R = 5000.
Substituting the values of the resolving force R, the diameter d, and the fiber aperture into the resulting equation (7), the effective beam splitter diameter will be D = 12.5 mm.
With a spectral resolution of 4 cm-1, the diameter of the beam splitter is D = 6 mm.
It can be seen from the calculations that the interferometer required for fiber-optic measurements can be small, with a beam splitter diameter (and corresponding dimensions of other optical elements of the interferometer and matching optics) not exceeding 10-20 mm. On its basis, a portable, portable Fourier spectrometer or a spectral system of input, intermediate and output monitoring of technological lines can be created.
In the case of working with several probes, it is not necessary to increase the size of the beam splitter. It will be necessary to increase the light size of the radiation source in order to illuminate the entire input area of the fiber optic harness at the output of the interferometer. Accordingly, each fiber will have a part of the radiation satisfying the relation (4).

The multi-probe IR Fourier transform spectrometer will differ significantly from the multi-probe NIR Fourier transform spectrometer due to the materials of the optical fibers used and to the spectral sensitivity of the photodetectors.
In the near IR range are used quartz optical fibers [7]. Optical absorption in them is much smaller than in polycrystalline and chalcogenide optical fibers used in the middle IR range. Accordingly, the multi-probe Fourier spectrometer of the near-IR range can use significantly longer probes (up to 10 meters or more), which is especially important for working with the objects that are remote from the spectrometer.

4. Experimental proof of work of a breadboard model

To study the possibilities of working in multi-channel mode with several probes, preliminary measurements were made on the model of the Fourier spectrometer of the visible and near IR bands (figure 3). The spectral resolution of the spectrometer is 1 cm^{-1}, which is ensured by using corner reflectors (tetrahedra) instead of flat mirrors. Such a design allows to reduce the requirements for the stability of their spatial position, since the incident beam is reflected from them strictly parallel to the original one [8].

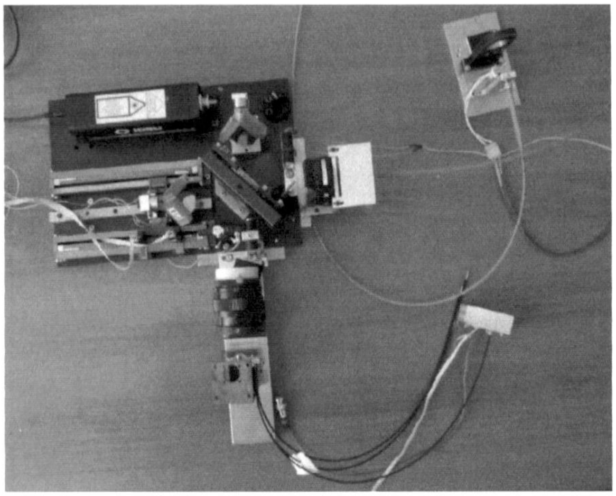

Figure 3. Fourier transform spectrometer breadboard model.

The input of the Fourier spectrometer was illuminated by the emission of a gas-discharge neon lamp. The radiation emitted from the interferometer was focused at the input of a shortened version of the fiber probe-a bundle (figure 4) consisting of three optical fibers of a larger cross-section with a core diameter of the order of 0.6-1.0 mm (fiber type OKK800 / 880 / 1200T with a numerical aperture NA = 0.22).

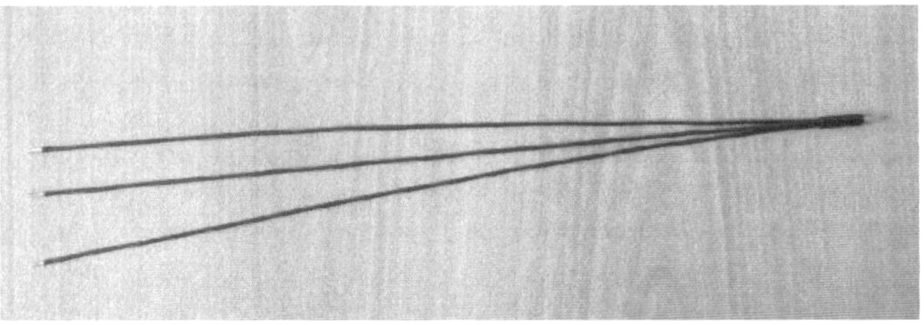

Figure 4. Fiber harness for adjusting and calibrating a multi-probe Fourier transform spectrometer for the visible and near-IR bands.

Then, successively, each optical fiber was connected to a photodetector and the spectrum of radiation transmitted through this optical fiber was recorded. Thus, the variant of operation in the multichannel mode was simulated. In figure 5 shows the spectra obtained on the optical fiber 1 (a) and optical fiber 2 (b).

IOP Conf. Series: Journal of Physics: Conf. Series **1038** (2018) 012081 doi:10.1088/1742-6596/1038/1/012081

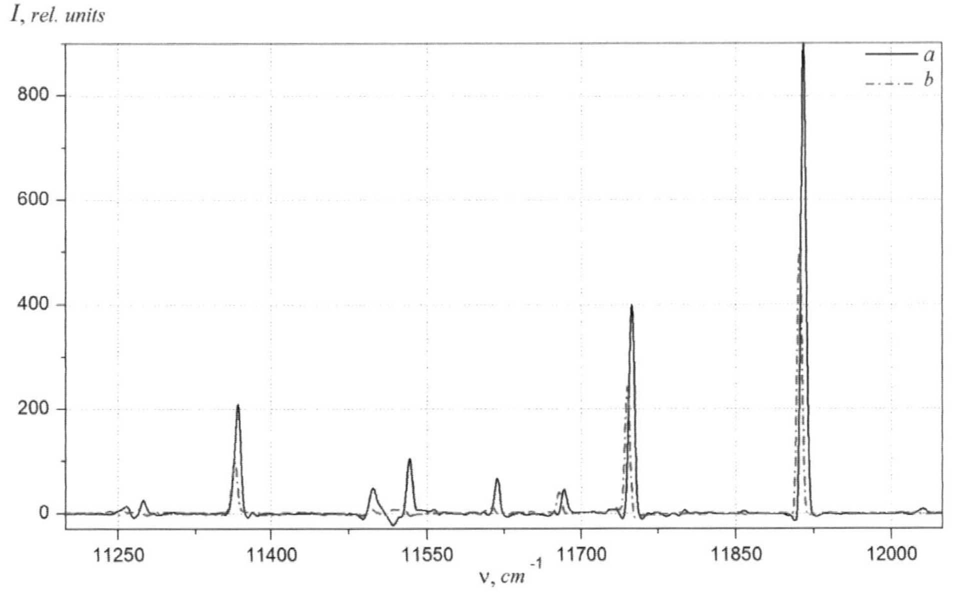

Figure 5. A fragment of the emission spectrum of a gas-discharge neon lamp (fiber 1 (*a*), fiber 2 (*b*)).

As you can see, this is one and the same spectrum, differing in intensity and in reference to wave numbers. The first is due to the fact that the inputs of fiber optic cables are not equally illuminated, and also due to the difference in their optical characteristics. The relative shift in wave numbers is due to the fact that the optical axes of light beams coming from the interferometer to the inputs of fiber-optic probes (cables) are not parallel to each other, but turn out to be at a small angle. Accordingly, these axes and with respect to the optical axis of the laser radiation in the reference channel turn out to be at various random (very small) angles. This leads, when processing interferograms, to a relative frequency shift. In figure 6 shows the emission spectrum of an individual line of a gas-discharge neon lamp for two different fibers.

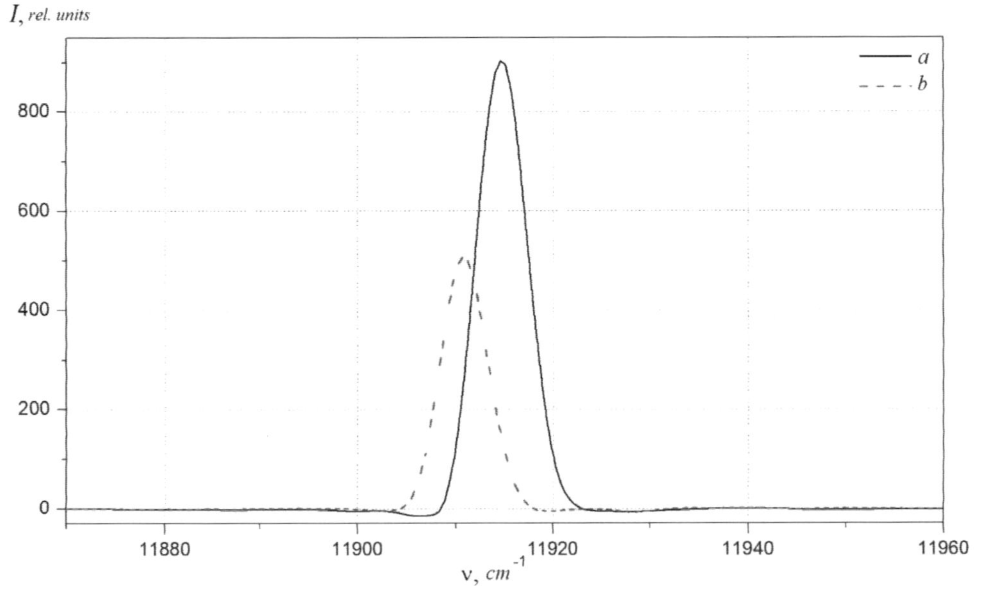

Figure 6. The emission spectrum of a separate line of a gas-discharge neon lamp (fiber 1(*a*), 2 (*b*)).

Indeed, if the axis of the light beam is inclined at an angle α to the optical axis of the interferometer, then when the moving mirror moves at a distance d, the optical path difference (δ) for the above ray will change according to the law $\delta = 2d\cos\,\alpha$, which explains the above frequency shifts in obtained spectra [9].
Thus, when working in a multi-probe variant, each fiber channel must be pre-calibrated in intensity, if necessary and, most importantly, in frequency (wave numbers).

Conclusion
A wide variety of commercially available optical-mechanical and electronic components makes it possible to create a portable IR Fourier spectrometer operating simultaneously with several fiber-optic probes. Due to the use of probes of various designs, effective radiation and reception of the light flux are performed for a variety of spectroscopic applications. Further research should be aimed at creating a multi-probe Fourier spectrometer and on its basis an integrated, multifunctional and highly optimized system that solves the problems of spectral analytics of technological processes.

Acknowledgments
The proposed spectrometer was developed and manufactured at the "Scientific and Technological Center of Unique Instrumentation" of the Russian Academy of Sciences (STC UI RAS).
The work was carried out within the framework of the state budgetary work 40.4 (Fundamental research program of the State Academies of Sciences "Investigation of the possibilities of creating a specialized Fourier transform IR spectrometer with a fiber-optic probe system for on-line molecular spectroscopy").

References

[1] Balashov A A, Vagin V A, Viskovatykh A V, Zhizhin G N, Pustovoit V I and Khorokhorin A I An AF-1 analytical Fourier-transform spectrometer for a wide field of applications 2003 *Instruments and Experimental Techniques* **46** (2) 219-221
[2] Balashov A A, Vagin V A, Moshkin B E, Khitrov O V and Khorokhorin A I Fiber-optic Fourier-Spectrometer 2009 *Instruments and Experimental Techniques* (6) 143
[3] Balashov A A, Vagin V A and Khorokhorin A 2016 *Instruments and Experimental Techniques* № 1 158. DOI: 10.7868/S0032816216010304
[4] Danielyan G L, Bazhanov Yu V, Savosin S V and Markov S N Development of a wide-range sensor - a mini-spectrometer with fiber input for spectral analysis of biological structures and liquids 2007 *Optical J* **74***(12)* 55-61
[5] Balashov A A, Vagin V A, Viskovatykh A V, Zhizhin G N, Pustovoit V I and Khorokhorin A I Multichannel dynamic IR Fourier-spectrometer 2017 *J. of Appl. Spectroscopy* **84** (4) DOI: 10.1007/s10812-017-0526-z
[6] Morozov A N and Svetlichnyi S I *Fundamentals of Fourier Spectroradiometry* 2014 (Moscow: Nauka) 456 p. [in Russian]
[7] Burkov V D and Ivanov G A *Physico-Technological Fundamentals of Fiber-Optic Technology* 2007 (Moscow: MGUL) 222 p. [in Russian]
[8] Griffiths P R, Haseth J A De and Winefordner J D *Fourier Transform Infrared Spectrometry*. 2007 2nd ed. (Wiley) 560 p.
[9] Morozov A N, Svetlichnyi S I and Fufurin I S 2015 Reports of the academy of sciences 464 (2) 156-159 DOI: 10.7868/S0869565215260114

IOP Conf. Series: Journal of Physics: Conf. Series **1038** (2018) 012082 doi:10.1088/1742-6596/1038/1/012082

Investigation of multimodality effect in quantum dots InGaAs/GaAs grown by MOVPE

I S Kosarev[1], A M Nadtochiy[1,2], R A Salii[2], N A Kalyuzhnyy[2]

[1]St. Petersburg Academic University, 8/3 Khlopina, St Petersburg 194021, Russia

[2]Ioffe Institute, 26 Polytechnicheskaya, St Petersburg 194021, Russia

e-mail: vnksarev@rambler.ru

Abstract. Self-assembled InGaAs quantum dots with In content 80% grown on vicinal GaAs substrates by MOVPE method have been investigated by photoluminescence technique in a wide range of temperatures and excitation densities. Multimodality of distribution of size and (or) shape in array of vertically stacked quantum dots synthesized under different growth conditions as well as in a single layer structure was found on the ground of appearance of a set of peaks in photoluminescence spectra at lowered temperatures. Influence of growth conditions on density and radiation wavelength of quantum dot modes has been analyzed. Investigation of photoluminescence of samples as a function of temperatures has allowed determining activation energies of three channels of the losses that were attributed to different modes of quantum dots.

1. Introduction

A lot of modern semiconductor optoelectronic devices, such a lasers or solar cells often contain quantum dots (QDs) in the active region. The unique properties of QDs (delta-like energy spectrum, low surface density) determine their leading position as the active media in low-threshold lasers and single photon sources. Besides, the effect of self-organization and consequent relaxation of elastic stress during pseudomorphic growth allows one to obtain multilayered structure with QDs (vertically stacked QDs) having high volume density, which is crucial for realization of quantum dots based solar cells [1]. Due to selforganization there is a probability of forming QDs of different size and/or shape. Each type of QDs generates photons with specific wavelength and, as a result, array of QDs has broadened radiation spectrum. On the other hand this feature may be used in positive way in comb-lasers [2], microdisk resonators [3,4] and tunable lasers, which require broad gain spectrum. In this paper we address to optical properties of the QDs grown by metal-organic vapor-phase epitaxy (MOVPE) and having multimodality in distribution of size and (or) shape [5].

2. Experimental

Experimental structures have been grown by using the MOVPE technique on vicinal GaAs (100) substrates with 6° of misorientation. The structures contained single layer or vertical array (stack) of five layers of self-assembled $In_{0.8}Ga_{0.2}As$/GaAs QDs. The growth of QDs was carried out at the temperature of 520 °C while the other layers were grown at 700 °C. To avoid the degradation of deposited QDs during substrate heating we used GaAs cap layer grown at 520 °C and covering QDs (Fig. 1). QD layers in stacked structures were separated by 40-nm GaAs spacer. $Al_{0.3}Ga_{0.7}As$ barriers were used to prevent of escape of carriers from QD area. Structures studied varied in the thickness of deposited InGaAs layer forming QDs and the thickness of GaAs cap layer. The structures with stacked QDs differed by the quantity of deposited $In_{0.8}Ga_{0.2}As$ material and thickness of cap-layer, which were following: 3 monolayers of $In_{0.8}Ga_{0.2}As$ and 8.5 nm of cap layer (C8I3), 2 monolayers of $In_{0.8}Ga_{0.2}As$ and 8.5 nm of cap layer (C8I2), and 2 monolayers of $In_{0.8}Ga_{0.2}As$ and 5 nm of cap layer (C5I2). Forth structure contained one layer of QDs similar to C8I3 and was called CS8.

Optical properties of structures grown were investigated by photoluminescence (PL). PL spectra were excited by second harmonic (532 nm) of Nd:YAG laser with maximum excitation density about $5kW/cm^2$. Variation of excitation density was achieved by using several neutral optical filters. PL spectra were registered by cooled Ge diode using conventional lock-in technique. Closed-cycle helium cryostat was used to study PL spectra at temperature range of 10-320 K.

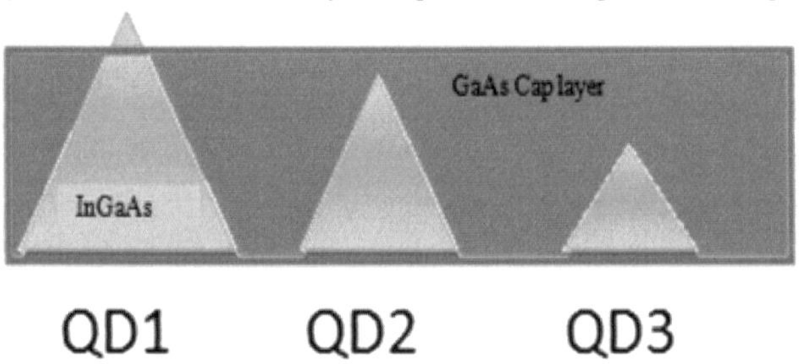

Figure 1. Schematic representation of investigated structures with QDs. Different QD modes are shown.

3. Results and discussion

3.1. Multimodality in InGaAs/GaAs quantum dots grown by MOVPE

Evolution of PL spectrum of CS8 structure with excitation density at 300 K (a) and 20 K (b) is shown on Fig. 2.

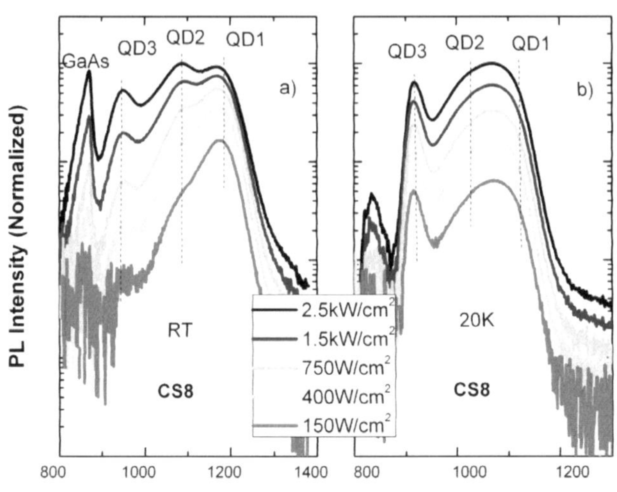

Figure 2. PL spectra obtained from CS8 structure at 300K (a) and 20K (b) under different excitation density.

At room temperature conditions increase of excitation modifies spectrum from single-peak shape (QD1) to the shape with a number of pronounced peaks (QD2 and QD3) demonstrating saturation-like behaviour (Fig. 2, a). This observation may remind behaviour of conventional MBE-grown quantum dots (see, for instance [6,7]), where peaks QD2 and QD3 are attributed to the first and the second excited states.

But this idea is refuted while analysing low temperature PL spectrum (Fig. 2, b). At low temperature conditions transport of carriers from one quantum dot to another is suppressed. Carriers captured by a certain quantum dot, cannot escape from it, and, as a result, they take part in optical recombination in this QD. This results in random population of QDs by carriers. So, appearance of a set of the peaks at low temperature and low excitation density points to a multimodal distribution of QD while intensity of PL peak being proportional to modal density of QD. One can see on Fig. 2, b that the multipeak shape of the PL spectrum doesn't change through the wide range of excitation density. So, it can be stated, that observed PL peaks resulted from radiative transitions from ground states of different QD modes (QD1, QD2, QD3), and excited states has no influence on PL spectrum, unlike classic quantum dots [8]. Similar behaviour was observed for all investigated samples, and figure 2 shows PL spectra for one-layer structure because of the fact that optical effects due to QD multimodality are most clearly seen for this structure.

In this letter we investigate mostly optical properties of QD arrays caused by multimodality while structural properties and growth aspects of self-assemble of such structures are the subjects of further research.

Evolution of PL spectrum of structure CS8 with temperature is shown on Fig. 3, a.

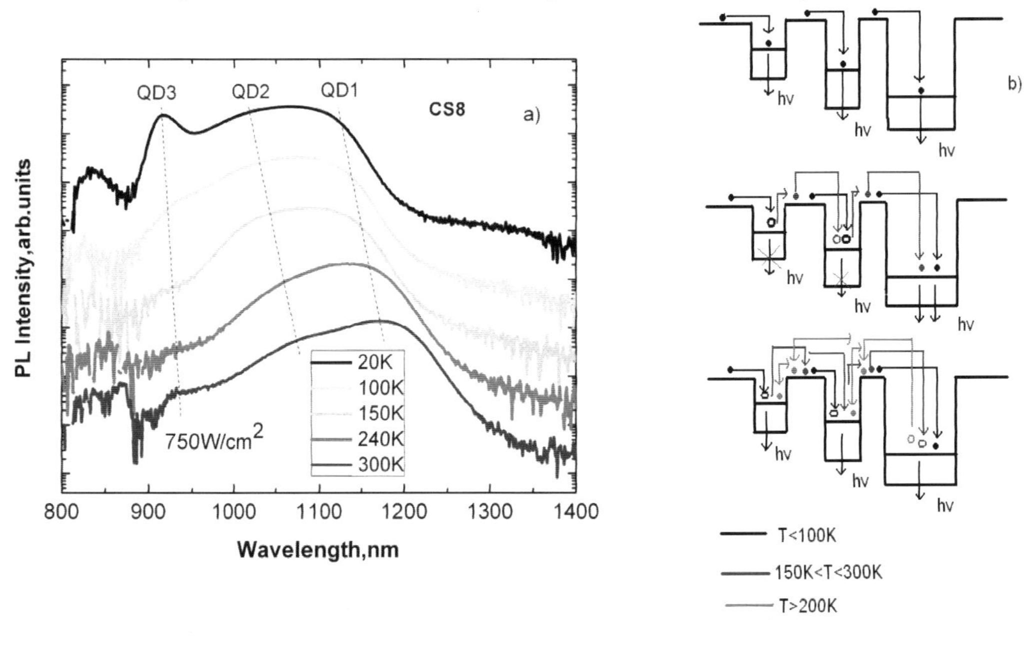

Figure 3. Evolution of the PL spectrum of CS8 structure with temperature (a), spectra are shifted for clarity. Schematic model of filling of multimodal QD array by carriers (b).

Redistribution of intensity between PL peaks with increase of temperature is clearly seen. This observation is in agreement with model [9], schematically presented on Fig. 3, b, which describes carrier transport between QDs in terms of thermal escape of carriers into matrix and consequent recapture. At temperatures below 100K transport of carriers is suppressed. Photoinduced carriers are populating QDs randomly (black colour process in Fig. 3, b), realizing so-called nonequlibrium carrier distribution. PL spectrum at this stage demonstrates only ground states for each of the modes of QD array. This situation is continued till the temperatures about 100K. Increasing of temperature up to 150K activates thermal escape of carriers from predominantly small QDs (type QD3 or QD2) into matrix followed by recapture into big QDs (QD1). Due to such unidirectional transport of carriers between QDs, distribution deviates from nonequilibrium one (blue colour process in Fig. 3, b). Intensity of QD3 peak degrades dramatically relative to QD2 and QD1, and the PL spectrum is narrowed. Further temperature increase (above 240K) opens possibility of carrier escape into matrix from big QDs - QD1 (red colour process in Fig. 3, b). Carrier transport is becoming random and filling of QDs is occurring according to Fermi-Dirac statistics, i.e. equilibrium distribution. At this stage weak QD3 peak is emerging in PL spectrum. So, we can conclude that temperature modification of PL spectrum of multimodal QDs has a complex character.

3.2. Influence of growth parameters on optical properties of multimodal QD array

Fig. 4 shows PL spectra for structures C8I3, C8I2, and C5I2. At temperatures about 20 K PL peaks of each type of quantum dots are clearly seen. At temperature of about 320 K one can see only peaks of long wavelength types of QDs (QD1 and QD2) due to mechanism of filling QDs described above. Note, that the width at half maximum of PL spectrum, which is composed from ground states of different types of QDs, reaches 140 nm for structure C5I2 at 20K.

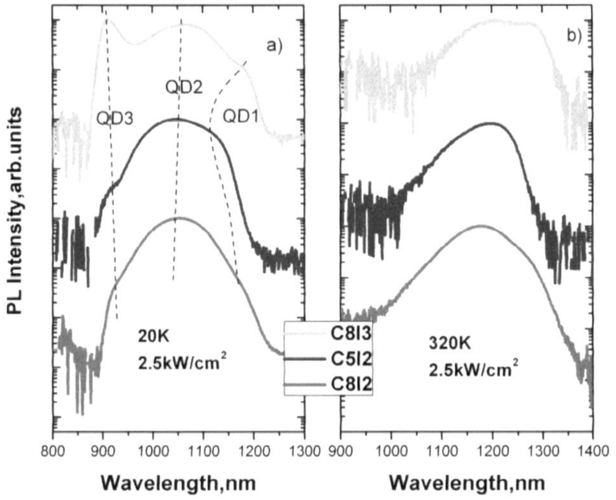

Figure 4. PL spectra (shifted for clarity) of structures C8I3, C8I2, and C5I2 taken at 20 K (a) and 320 K (b).

The results of analysis of PL spectra on Fig.4 are shown on Figure 5. Accurate position of PL peaks for each type of QDs was determined via analysing of differentiated PL spectrum.

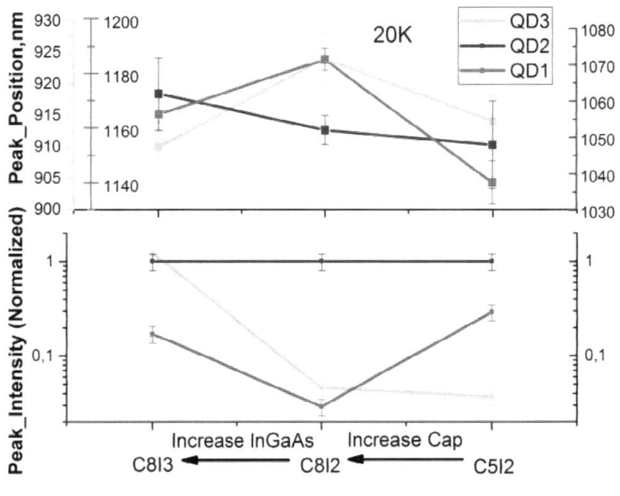

Figure 5. Position and intensity of each type of QDs for different structures. Peak intensities of structures are normalized with QD2 peak. 95% confidence intervals are shown.

Analysis of Fig. 4 and 5 is giving a number of observations. Firstly, increasing of the thickness of GaAs cap layer (Fig.1) leads to slight increase of the size of big QDs (QD1) and decrease of their density, as following from redshift of QD1 peak and reduce its intensity. So, one can conclude that size of big QDs (QD1) is limited to the thickness of cap layer. At the same time increasing of the thickness of cap layer has a little effect on mode QD3. Secondly, increasing of the effective thickness of deposited InGaAs material leads to significant increase of density of small QDs (QD3) while remaining their size (peak position has small variations only). Besides, deposition of thicker InGaAs layer results in remarkable increase of density of big QDs (QD1) with insignificant decrease of the size of the dots.

3.3. Temperature dependence of integrated photoluminescence

Temperature dependences of integrated PL intensity for investigated structures are shown on Fig. 6.

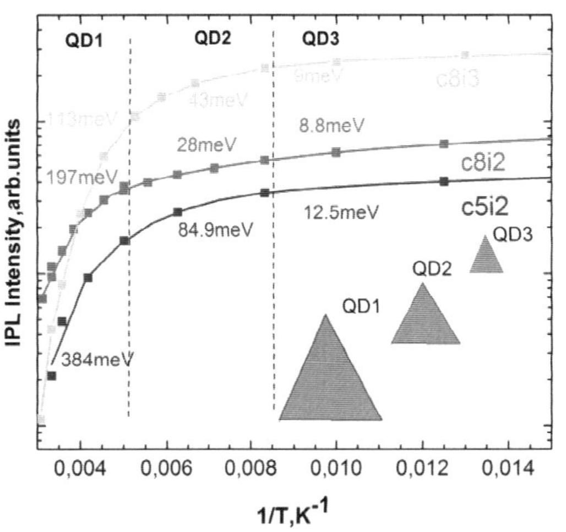

Figure 6. Temperature dependences of Integrated PL intensity for investigated structures. Temperature ranges corresponding to channels of losses of QD1, QD2, and QD3 are shown schematically. Activation energies are also shown.

Experimental data (symbols) were approximated (solid line) with the model [10] which includes three channels of losses. Activation energies have been determined for each of the channels. Fitting of the experimental data with models with less than three of channels of losses has given unsatisfactory results. These channels of losses of carriers were associated with thermally induced carrier escape from QDs of three types (QD1, QD2, QD3) into matrix and consequent non-radiative recombination. For all structures investigated determined activation energies have fall into three groups with characteristic energies about 10 meV, 30-80 meV, and 100-400 meV, which were attributed to QD3, QD2, and QD1, correspondingly. However, there is no strong correlation between activation and localization energies (determined on the ground of position of the peaks of ground states for each types of QDs), which is the subject of further research.

4. Conclusions

Heterostructures containing self-assembled $In_{0.8}Ga_{0.2}As/GaAs$ quantum dots grown by metal-organic chemical vapor epitaxy were investigated by means of photoluminescence (PL). All investigated structures have demonstrated multimodality of distribution of size and(or) shape of QD array. Multimodality of QDs resulted in complex evolution of the shape of PL spectra with temperature increase. To protect QDs from degradation during growth process GaAs cap layer deposited at lowered temperature was used. It was found that cap layer thickness has significant influence on the size of QDs of long wavelength type. Density of QDs of short wavelength type was found to be strongly sensitive to effective thickness of deposited InGaAs layer. Fitting of temperature dependencies of integrated PL intensity has allowed determining activation energies of three channels of the losses that were attributed to different modes of QDs.

Acknowledgments

The work was supported by the Russian Science Foundation (project 16-12-10269).

References

[1] Blokhin S A, Sakharov A V, Nadtochy A M, Pauysov A S, Maximov M V, Ledentsov N N, Kovsh A R, Mikhrin S S, Lantratov V M, Mintairov S A, Kaluzhniy N A, and Shvarts M Z 2009 *Semiconductors.* **43**(4) 514

[2] Zhukov A E and Kovsh A R 2008 *Quantum Electronics.* **38**(5) 409

[3] Kryzhanovskaya N V, Maximov M V and Zhukov A E 2014 *Quantum Electronics.* **44**(3) 189

[4] Zhang L and Hu E 2003 *Appl. Phys. Lett.* **82** 319

[5] Salii R A, Mintairov S A, Brunkov P N, Nadtochiy A M, Payusov A S and Kalyushnyy N A 2015 *Semiconductors.* **49**(8) 1136

[6] Grundman M, Ledentsov N N, Stier O, Böhrer J, Bimberg D, Ustinov V M, Kop'ev P S and Alferov Zh I 1996 *Phys. Rev. B.* **53**(16) R10 509

[7] Heitz R, Born H, Guffarth F, Stier O, Schliwa A, Hoffmann A and Bimberg D 2001 *Phys. Rev. B* **64** 241305

[8] Buckle P D, Dawson P, Hall S A, Chen X, Steer M J, Mowbray D J, Skolnick M S and Hopkinson M 1999 *J. Appl. Phys.* **86** 2555

[9] Nadtochiy A M, Mintairov S A, Kalyuzhnyy N A, Rouvimov S S, Shernyakov Yu M, Payusov A S, Maximov M V and Zhukov A E 2015 *Semiconductors.* **49** (8) 1090

[10] Heitz R, Mukhametzhanov I, Madhukar A, Hoffmann A and Bimberg D 1999 *Journal of Electronic Materials.* **28**(5) 520

IOP Conf. Series: Journal of Physics: Conf. Series **1038** (2018) 012083 doi:10.1088/1742-6596/1038/1/012083

Self-focusing of the light in transparent nanosuspension

K V Platonov, V I Ivanov and A V Myagotin

Department of Physics and Theoretical Mechanics, Far Eastern State Transport University, Khabarovsk, Serysheva street, 47, Russia

tmeh@festu.khv.ru

Abstract. It was analyzed the self-focusing regime of the Gaussian beam in transparent nanosuspension with electrostrictive nonlinearity. The theoretical analysis of the light-induced mass transfer in the nanosuspension was carried out for large intensities of the Gaussian beam radiation, when the concentration change is comparable to the primary value. The nonlinear lens in this mode is the nonlinear function of the incident light intensity. It is shown that the critical power value decreases significantly for high intensity beam. The results are relevant to the study of the radiation self-action in the nanosuspension and optical diagnostics of such materials.

1. Introduction

The self-focusing (defocusing) of the light beam is well known nonlinear effect [1-3]. The light-induced thermal lens in a homogeneous fluid is formed as a result of thermal expansion of a medium. In two-component liquid the heat flow also can cause concentration stream arising from occurrence of thermodiffusion (Soret effect [2]). Another mechanism of optical nonlinearity of the medium is due to the forces operating on the particles of the dispersed phase in gradient light field [3-5]. Gradient electrostrictive forces may cause concentration streams in nanosuspension (artificial Kerr media) [5]. Depending on the sign of polarizability nanoparticles drift from areas with higher intensity of the electromagnetic wave (if their refractive index is more than one for liquid). This optical nonlinearity was studied experimentally in nanosuspensions and microemulsions [6-7]. The theoretical analysis is restricted usually by the case of weak intensities of radiation.

The aim of this work is to analyze the self-focusing regime in transparent nanosuspension at high intensities of radiation.

2. Electrostrictive mechanisms of cubic nonlinearity in nanosuspension

Usually in the case of non-resonant mechanism (for weakly absorbing media) is used the parameter of cubic nonlinearity - $n_2^{eff}[m^2/W]$, which characterizes the change of the refractive index n of the medium under the influence of incident radiation:

$$n = n_0 + n_2^{eff} I, \tag{1}$$

where n_0 - the refraction index of the medium in the absence of radiation, I - radiation intensity, $n_2^{eff} = (dn/dI)$ - the coefficient of effective cubic nonlinearity.

The coefficient of cubic nonlinearity for the electrostrictive mechanism is determined by the concentration nonlinearity:

$$n_2^{eff} = (dn/dC)(dC/dI) \tag{2}$$

where C is the concentration of nanoparticles.

In nanosuspension the particle radius is much smaller than the radiation wavelength λ, therefore the refractive index of the medium is proportional to the concentration of particles [8]:

$$n = n_1(1 + f\delta), \tag{3}$$

where $\delta = (n_2 - n_1)/n_1$; n_1 *u* n_2 are the refractive indices of the liquid and the dispersed phase, respectively; $f = v_0 C$ is the volume fraction of the dispersed phase, r is the radius of the nanoparticle, $v_0 = (4/3)\pi r^3$ is the volume of the nanoparticle.

Balanced equation describing the dynamics of concentration of nanoparticles considering diffusion and electrostrictive flows can be written as [4]:

$$\frac{\partial C}{\partial t} = D\Delta C - div(\gamma C \nabla I), \tag{4}$$

where D is the diffusion coefficient, $\gamma = 4\pi\beta D(\bar{c} n k_B T)^{-1}$, β is the particle polarizability, k_B is the Boltzmann constant, $\bar{c}$ is the velocity of light in vacuum, $I(r)$ is the intensity of radiation.

In the stationary mode equation (4) takes the next form (diffusion and electrostrictive flows are equal):

$$0 = -D\nabla C + \gamma C \nabla I. \tag{5}$$

This equation can be linearized in the case of low concentrations ($C << 1$) and small it changes:

$$(C(r,t) = C_0(1 + C'(r,t)),\ C'(r,t) << 1). \tag{6}$$

$$0 = -D\nabla C' + \gamma \nabla I. \tag{7}$$

The exact solution of the equation (7) is:

$$C' = I(r) I_s^{-1}. \tag{8}$$

where $I_s = \gamma^{-1} D$ is the “saturation” intensity.

Thus the changing of the particle concentration (and effective refractive index) is directly proportional to radiation intensity. It is a case of the classic cubic nonlinear response of the material. The coefficient of effective cubic nonlinearity of nanosuspension can be calculated using expression (2-3) and (8) [7]:

$$n_2^{eff} = n_1 \delta f_0 I_s^{-1}, \tag{9}$$

where $f_0 = v_0 C$ is the initial volume fraction of the dispersed phase.

The self-focusing regime in the cubic nonlinearity medium is characterized by the critical beam power [5]:

$$P_s = \frac{\lambda^2}{64\pi^2 n_2^{eff}}. \tag{10}$$

where λ is the radiation wavelength.

Using (3-4) one can found the critical beam power for nanosuspension with electrostrictive response:

$$P_{s0}^{el} = \frac{\lambda^2 I_s}{64\pi^2 n_1 \delta f_0}. \tag{11}$$

For ordinary parameters of nanosuspension formula (5) gives the values $P_{s0}^{el} \approx 1 \div 10 \quad W$.

The above analysis is valid only for small intensities of radiation ($I << I_s$), when the coefficient of effective cubic nonlinearity is a constant parameter of the medium.

3. Self-focusing of the power light beam

For great intensity beam ($I >> I_s$) the nonlinear response of the nanosuspension does not match to the cubic nonlinearity, because the concentration is not a linear function of the light intensity. The task of mass transfer in the light field requires the solving of the equation (5), because the concentration change is comparable to the primary value in this case.

We will consider the transparent nanosuspension under the influence of laser radiation with Gaussian intensity profile [9]:

$$I = I_0 e^{-r^2/r_0^2}, \tag{12}$$

where I_0 is the radiation intensity on the beam axis, r is the radial distance from the beam axis, r_0 is the beam radius.

The general solution of the equation (5) is looking for in the form of:

$$C_{st} = B\, exp\{I_0 I_s^{-1} e^{-r^2/r_0^2}\}, \tag{13}$$

where B is the normalization constant.

The constant of normalization B is given from conservation of particle number:

$$\int_0^R C_{st} 2\pi r dr = \pi R^2 C_0, \tag{14}$$

where R is the radius of the cylindrical cell, C_0 is the initial concentration of dispersed particles.

Let's introduce dimensionless parameter of light intensity $\alpha = I_0 I_S^{-1}$. Figure 1 shows the concentration of the nanoparticles versus the normalized distance from the axis of the beam $\rho = (r/r_0)$ for the different intensity values α (it was calculated for $R/r_0 = 5$).

The change of the concentration ΔC_{st} on the axis of the beam:

$$\Delta C_{st} = B(\exp\alpha - 1). \tag{15}$$

where $\Delta C_{st} = (C_{st} - 1)$.

Figure 2 plots the change of the concentration ΔC_{st} on the axis of the beam ($\rho = 0$) versus normalized intensity α. The change of the particle concentration is not proportional to the radiation intensity (as opposed to the usual cubic nonlinearity).

The self-focusing regime is analyzed usually in paraxial approximation [5]. In our case the refractive index profile in the nanosuspension near the axis beam is expressed from (3) and (13):

$$n = n_1 + n_1 \delta v_0 B e^{\alpha}[1 - \alpha(r^2/r_0^2)], \tag{16}$$

The analysis of the critical power for this case gives the next expression:

$$P_s^{'} = P_s / P_{s0}^{el} = \alpha(e^{\alpha} - 1)^{-1}. \quad (17)$$

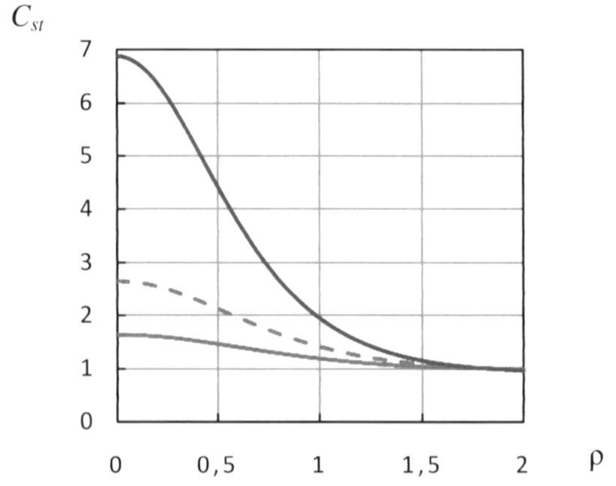

Figure 1. The concentration of nanoparticles $C_{st} = C / C_0$ (a.u.) versus distance ρ (a.u.) from the axis of the beam for different values of the intensity of the light: $\alpha_1 = 0,5$ (——); $\alpha_2 = 1$ (·······); $\alpha_3 = 2$ (– –).

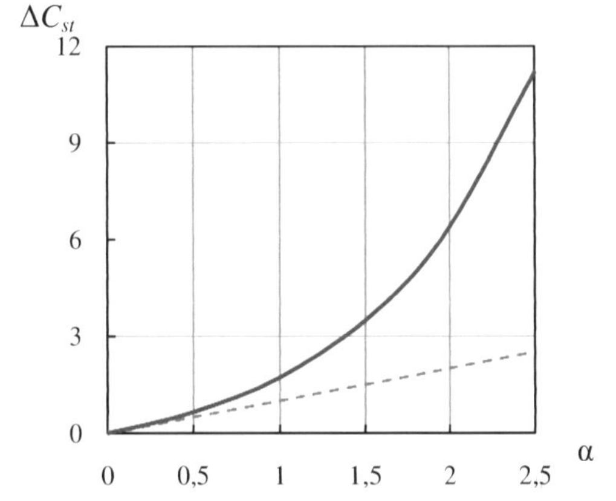

Figure 2. The change of the concentration ΔC_{st} on the axis of the beam versus normalized intensity α (——) . The linear dependence $\Delta C_{st} = \alpha$ is shown for comparison (- -).

Figure 3 plots the critical power of the beam versus normalized intensity α (according to the formula (17)).

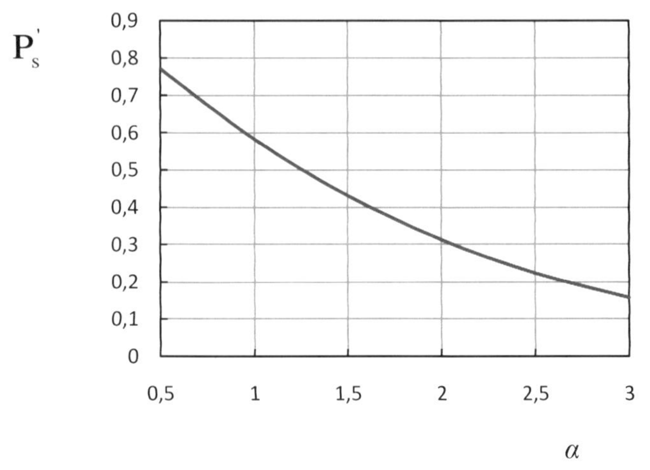

Figure 3. The critical power of the beam (arb. un.) versus normalized intensity α .

We can see the sharply decreasing of the critical power value when α>>1. Thus this region is more convenient for the experimental realization.

4. Conclusions

We have analyzed the two-dimensional diffusion in the nanosuspension with electrostrictive nonlinearity in a Gaussian beam of radiation. In case of high intensity beam the nonlinear response of the nanosuspension does not match to the cubic nonlinearity. The expression was achieved for light induced stationary concentration response in nanosuspension. The resulting expression shows that the critical power value decreases significantly for high value α.

The results are relevant to nonlinear optics of nanosuspension [10-12], including the optical diagnostics of such nano-materials [13-14].

References

[1] Freysz E, Afifi M, Ducasse A, Pouligny B and Lalanne J R 1984 *J. Opt. Soc. Amer. B* **1** 433
[2] Ivanov V I and Ivanova G D 2016 *Journal of Physics: Conference Series* **735** 012037
[3] Vicary L 2002 *Philos. Mag.* B. **82** 447-452
[4] Ivanov V I, Ivanova G D and Khe V K 2017 *Proc. SPIE* **10176** 1017607
[5] Smith P W, Maloney P J and Ashkin A 1982 *Opt. Lett.* **7**(8) 347.
[6] Ivanov V I , Ivanova G D, Okishev K N and Khe V K 2016 *Proc. SPIE* **10035** 100354Y
[7] Lee W, El-Ganainy R, Christodoulides D, Dholakia K and Wright E 2009 *Optics Express* **17** 10277
[8] Ivanov V, Ivanova G, Okishev K and Khe V 2017 *IOP Conf. Ser.: Mater. Sci. Eng.* **168** 012045
[9] Leppla C and Wiegand S 2003 *Philosoph. Mag.* **83** 1989
[10] Giglio M and Vendramini A 1974 *Appl. Phys. Lett.* **25** 555
[11] El-Ganainy R, Christodoulides D, Rotschild C and Segev M 2007 *Opt. Express* **15** 10207
[12] Smith P W, Ashkin A and Tomlinson W J 1981 *Opt. Lett.* **6** 284
[13] Chintamani P, Shalini M, Agnel P, Meera V, Tejas I H and Radha S 2014 *International Journal of Chemical and Physical Sciences* **3**(5) 44
[14] Sheldon S J, Knight L V and Thorne J M 1982 *Appl. Opt.* **12** 1663

The synovial fluid analysis by using Raman Scattering spectroscopy in order to educe the synovial joint pathology

E Timchenko[1], P Timchenko[1], L Volova[2], D Dolgushkin[2], M Markova[1] and E Yagofarova[1]

1 - Samara National Research University, Moskovskoye highway 34, Samara, 443086, Russia

2 - Samara State Medical University, Chapayevskaya St. 89, Samara, 443099, Russia

E-mail: laser-optics.timchenko@mail.ru

Abstract. In this paper, we present the results of experimental studies by using Raman-scattering spectroscopy method of synovial fluid that was obtained from the joint cavity during the surgery. Analyzing the composition of synovial fluid, it was identified that with the progress of degenerative-dystrophic process in synovial fluid of the affected joint, the gross amount of components increases at wave numbers: 1155 cm^{-1} (hyaluronic acid (C-O, C-C)) and 1250cm^{-1} (Amide III). The entered optical numbers allow us to characterize synovial fluid (SF) in osteoarthrosis (OA). This specific Raman-scattering spectroscopy method may become a new diagnostic screening in order to educe the synovial joint pathology.

1. Introduction

The contained synovial fluid (SF) in the joint cavity being a biological medium is unique in biophysical and physical-chemical content and properties. It provides with the sliding of the bones' articular surfaces playing a damper function. Its damage causes a change in the cell composition SF [1]. SF or synovia is a sort of life activity indicator of the joint, which is formed and replenished by means of substances originating from the blood plasma and secreted covering layer of articular membrane - is typified as a boundary layer between stratum synoviale, cartilage and subchondral bone [2]. In light of this, for early and accurate diagnostics of joint disease and, thereby, for the more accurate treatment order of a patient, there is a necessity to carry out an analysis of synovial fluid.

Unfortunately, people conduct a study on synovial fluid very rare, because analysis' interpretation of results causes considerable difficulties. Standard diagnostics of synovial fluid is comprised of macro panel whereby we can define the volume, colour, fluid viscosity, turbidity and other external feature; cell counting; as well as microscopic analysis of native preparation; cytologic screening of the stained preparation [3].

Raman scattering spectroscopy method has certain advantages in comparison with the analytical methods that mentioned above. The most important of them are the simplicity of sample preparation, minimal invasiveness and a big amount of received information. Spreading this method in medicine can improve the results of surgical treatment of patients as well as to choose rational therapeutic approach for patients, and put in perspective the functional outcomes of surgical interferences that attempted to correct synovial joint pathology.

The aim of this work is to study the joint synovial fluid in the treatment of osteoarthrosis by using Raman scattering spectroscopy method.

2. Research materials

Patients' samples of SF were obtained by knee joint puncture with a disposable syringe. This procedure was performed according to the standard technic under aseptic conditions. 5-10 ml of normal saline solution were preliminarily injected in the joint in order to obtain a right amount of SF. The sampling was taken before diagnostic arthroscopy from patients without any obvious clinical signs of joint disease, the absence of which was confirmed (group 1). The patients with chronical damages of infra-articular structures before doing arthroscopy (group 2). Before doing knee arthroplasty in connection with the destructively-dystrophic osteoarthrosis disease at the IV stage (group 3).

The samples of synovial fluid were carried out using a stand that implements the SRS method. This stand includes a high-resolution digital spectrometer Shamrock sr-303i with a wavelength range of 200-1200 nm, with built-in cooling chamber DV420A-OE, a fiber-optics probe RPB-785 for Raman spectroscopy, combined with laser module LuxxMaster LML-785.0RB -04 with laser emission wavelength of 785 nm and with a line width of 0.2 nm. [4]

The processing of Raman spectra was carried out in the Wolfram Mathematica 9 program. The unknown spectrum was removed from noise during the processing by a smoothing median filter (5 points). We had been determining approximating straight line of an auto-fluorescent component in the picked interval 700-2200 cm^{-1} by using iterative algorithm (a fifth-degree polynom), then we subtracted this component, and as a result, we obtained the allocated Raman spectrum. The separation of the overlapped lines in the spectrum was carried out in the software environment of MagicPlotPro.

3. Results and discussion

The averaged Raman spectra of the SF samples of three groups are shown in the figure 1. The curves for better visibility are shifted vertically relative to each other for 2 relative units. The value of the intensities for the case of the 1155 cm^{-1} in figure 1 is difficult to compare, but it is used below in figure 2.

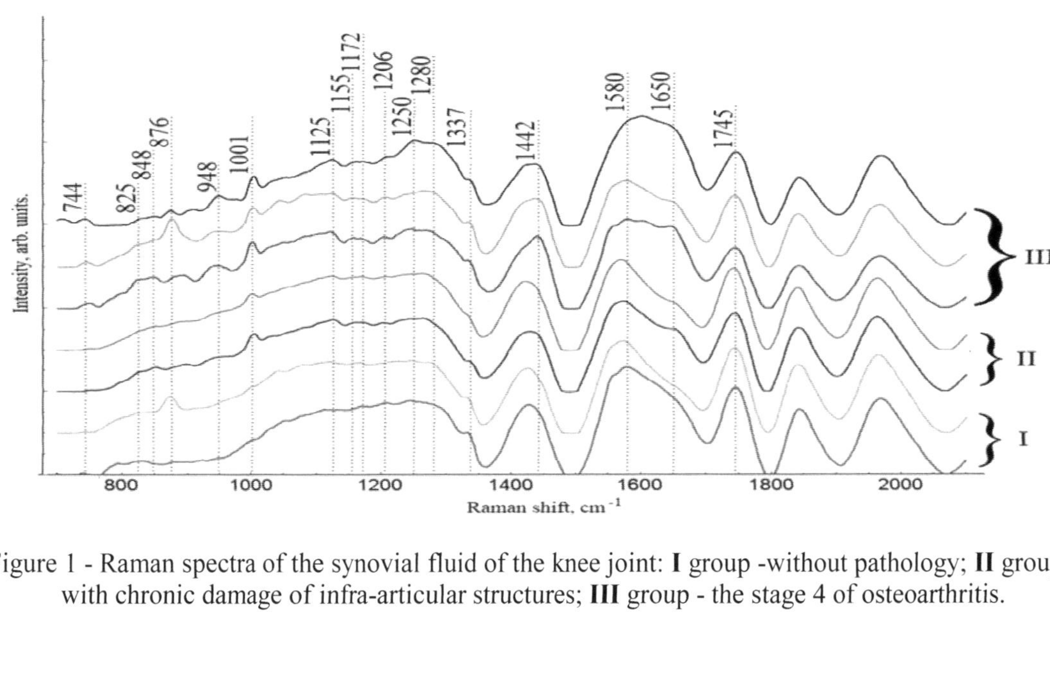

Figure 1 - Raman spectra of the synovial fluid of the knee joint: **I** group -without pathology; **II** group - with chronic damage of infra-articular structures; **III** group - the stage 4 of osteoarthritis.

In a comparative analysis of the spectra of SF samples, it turned out that in the case of chronic damage to the structures of the knee joint and destructively-dystrophic diseases in SF, components at 744 cm-1 (Protein (C-C-O), deformation twisting) [5], 948 cm-1 (secondary structure of protein (N-Ca-C stretching, α-helix))[6], 1001 cm-1 (phenyl alanine E-ring) [7]. Whereas during the SF analysis of joints without any pathology such components were not determined. Also in the SF samples of the second and third groups, the content of the components progressively increased at wave numbers: 1155 cm-1 (Hyaluronic acid (C-O, C-C)) [5], 1206 cm-1 (Hyaluronic acid, phenylalanine) [5] and 1337 cm-1 (Hyaluronic acid, nucleic acids) [5], 1125 cm-1 (C-C, C-O stretching of glycosidic linkage) [6], 1250 cm-1 (Amide III)[6], 1442 cm-1 (CH2 / CH3 deformational twisting) [7].

Relatively constant components of SF in all three groups of samples were (C = O) lipids corresponding to a wave number of 1745 cm-1 [8]. This number used as a denominator in the input optical numbers (A1-6):

$$A_i = \frac{I_i}{I_{1745}}$$

Where Ii – the intensity values at the corresponding wave numbers:

i= 1 – 1001 cm^{-1} (phenyl alanine E-ring);

i= 2 - 1125 cm^{-1} (C-C, C-O stretching of glycosidic linkage),

i= 3 - 1155 cm^{-1} (hyaluronic acid (C-O, C-C)),

i= 4 - 1206 cm^{-1} (hyaluronic acid, phenyl alanine),

i= 5 - 1250 cm^{-1} (Amide III),

i= 6 - 1442 cm^{-1} (CH2/CH3 deformational twisting).

Based on the obtained values of the optical numbers A1-6 there were constructed two-dimensional diagrams (Figure 2).

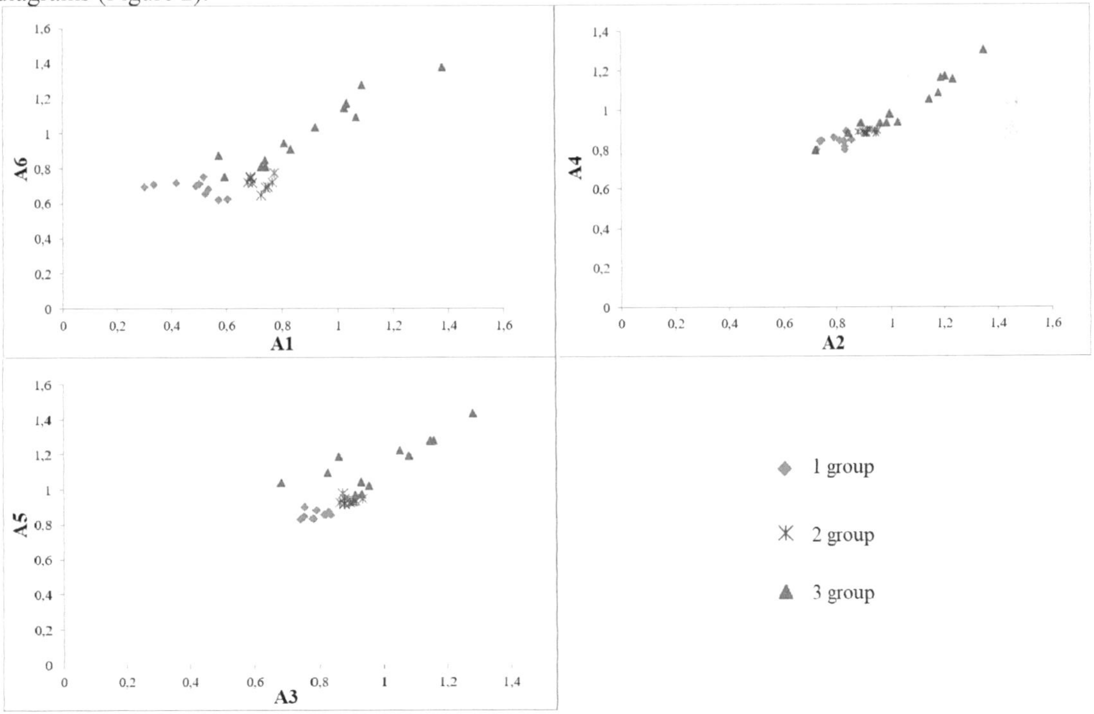

Figure 2 - Two-dimensional groups diagrams of samples' comparison of synovial fluid.

The two-dimensional diagrams show that all the optical numbers (A1-6) progressively increase due to intensity enhancement of the destructively dystrophic processes in joint from the first to the second and third groups of the study samples. In such a way the number A1 changes its magnitude from 0.3-0.65 to 0.6-1.4, the number A2 changes its magnitude from 0.7-0.8 to 0.7-1.4, the numbers A5 and A6 change their magnitude from 0,8-0.9 to 0.9-1.5 and from 0.6-0.8 to 0.75-1.4. This makes perfect sense because SF from a transudation becomes an exudation during the involvement in the pathological process, which have an effect on its component composition - increase in the protein content. This phenomenon can be explained not only by the synovial membrane hyperpermeability during inflammatory state but also by the increase of g-globulins output by synovial cells.

We attribute this increase of the numbers A3 (from 0.7-0.8 to 0.7-1.3) and A4 (from 0.8-0.9 to 0.8-1.35) to increase of the synovial fluid viscosity along with a progressive increase of disrupture product of an articular hyaline cartilage in the second and third groups of SF samples. The two-dimensional diagrams analysis showed that with the aid of the entered optical numbers it is possible to verify the degree of development of destructively dystrophic diseases of the knee joint for sure.

4. Conclusion

The component composition of synovial fluid can be identified in the setting of various pathologies of the joint by using Raman-scattering spectroscopy method.

In the case of destructively dystrophic diseases in the synovial fluid of the affected joint the amount of protein components and disrupture product of the cartilaginous tissue increases - that can be judged by the growth of the wave numbers of 1001 cm^{-1} (phenyl alanine E-ring), 1155 cm^{-1} (hyaluronic acid (C-O, C-C)), 1206 cm^{-1} (hyaluronic acid, phenyl alanine), 1125 cm^{-1} (C-C, C-O stretching of glycosidic linkage), 1250 cm^{-1} (Amide III), 1442 cm^{-1} (CH2 / CH3 deformational twisting), as well as 744 cm^{-1} (protein (C-C-O)), 948 cm^{-1} (secondary structure of protein (N-Ca-C stretching, α- spiral)).

By using entered optical numbers, it is possible to verify the degree of destructively dystrophic diseases development of the knee joint, which may become a new diagnostic screening method in order to educe the synovial joint pathology.

The Ministry of Education and Science of The Russian Federation.

5. References

[1] Chernyakov Yu and Sementovsky E A 2005. Synovial fluid: structure, properties, laboratory methods for research Medical news 2 p 9

[2] Nesterenko S 2000 Orthopedics, traumatology and prosthetics 3. Syndrome of disequilibrium dysplastic patella Genesis in terms of the concept of multifactorially **17-19**

[3] Kamyshnikov V 2009 Representors Methods of clinical laboratory studies: the textbook p 752

[4] Timchenko E, Timchenko P, Volova L, Ponomareva Y and Taskina L 2014 The Study of organicmineral structure of bone implantation Raman scattering Quantum electronics 44 pp 696-700

[5] Коваленко А, Елисеев А 2011 Методическая разработка Спектроскопия комбинационного рассеяния с 37

[6] Esmonde-White K, Esmonde-White F, Morris M and Roessler B 2014 Characterization of Biofluids Prepared by Sessile Drop Formation Analyst p 139

[7] Hou W, Li W, Keyszer G, Weber E, Klein L, quantification Gravallese E and Goldring S Dieter Brömme 2002 Comparison of cathepsins K and S expression within the rheumatoid and osteoarthritic synovium Arthritis Rheum 46 pp 663-674

[8] Yoshida M and Marumo K 2004 Expression analysis of three isoforms of hyaluronan synthase, hyaluronidase in synovium in knees in osteoarthritis, rheumatoid arthritis by quantitative realtime reverse transcriptase polymerase chain reaction Arthr Res Ther 6 pp 1-14.

Simulation of the propagation of the vortex eigenfunctions of the two-lens system in the parabolic fiber

M S Kirilenko[1,2]
[1]Samara National Research University, 34, Moskovskoye Shosse, Samara, Russia, 443086
[2]Image Processing Systems Institute – Branch of the Federal Scientific Research Centre "Crystallography and Photonics" of Russian Academy of Sciences, 151, Molodogvardeyskaya, Samara, Russia, 443001

Abstract. In this paper, the limited optical two-lens system, which describes the transmission of light beams in the parabolic profile fiber is considered. Vortex eigenfunctions, calculated for the given system form the set of the orthogonal functions, through which another optical field can be represented. Numerical simulation of the passage of the eigenfunctions through the system of the multiple lenses is carried out. It is shown, that eigenfunctions, whose eigenvalues are close to unity by absolute value, can pass through a large number of consecutive two-lens systems with practically no distortion.

1. Introduction

The current level of use of the optical fiber in terms of the time and frequency domain characteristics tends to the bandwidth limit [1]. However, the requirements for increasing the volume of global traffic are constantly growing. To ensure the correspondence of communication networks to ever-growing bandwidth requirements, additional approaches are considered for the capacity increasing of the optical fiber channels. One of such approaches is the mode division multiplexing (MDM) [2,3]. A special advantage for increasing the capacity of the information channel is achieved with the help of optical beams with orbital angular momentum and having an infinite number of possible quantum states [4, 5]. The significant success of this method has already been demonstrated in optical fibers [6] and in free space [7,8]. For generation and analysis of the vortex beams, the diffractive optical elements are used [9-11], and the lens systems are used to introduce them into the optical fiber [12-15].

The propagation of the vortex laser beam of the m-th order through the spherical lens can be described using the Hankel transform of the m-th order. In real lens systems, there is a spatial limitation, and finite (spatially-restricted) propagation operators are used to describe the propagation of the optical signal [16, 17]. Due to the spatial limitation, both in the object and spectral regions, it is impossible to obtain an ideal image in the two-lens system. In order to understand how the optical signal is distorted, it is necessary to expand them according to the eigenmodes of the lens system. In connection with this, the concept of communication modes is widely used [18, 19]. The communication modes for square apertures and Fresnel transformations are an elongated angular spheroidal functions [20,21], which are widely studied and applied in optics [22-25]. The communication modes for round apertures and finite Hankel transform are the circular [26] and generalized [27] spheroidal functions.

In this paper, the set of vortex eigenfunctions of the bounded circular two-lens system is calculated. The set allows analyzing the distortion of the optical signal during transmission on the basis of approximation by eigenfunctions.

2. Theoretical foundations

The propagation of the optical signal in the parabolic profile fiber is analogous to its propagation through the successive system of lenses. The passage of the vortex beam through the bounded by radius two-lens system can be described as follows [28]:

$$H_{R,P}\left[f(r)\exp(im\varphi)\right](r',\varphi') = \frac{kP}{2\pi f}\int_0^R L(r,r';m,P) f(r) r\,\mathrm{d}r \cdot \exp(im\varphi'),$$

$$L(r,r';m,P) = \int_0^{2\pi} \frac{J_1\left(\frac{kP}{f}\sqrt{r^2+r'^2+2rr'\cos\theta}\right)}{\sqrt{r^2+r'^2+2rr'\cos\theta}}\exp(im\theta)\,\mathrm{d}\theta, \qquad (1)$$

where J_1 is the Bessel functions, k is the wavenumber, f is the focal length of both lenses, R is the radius of the first aperture, P is the radius of the second aperture, m is the order of the optical vortex. The eigenfunctions of such system have modal properties and are orthogonal.

As the radius of the aperture increases, an increase in the number of significant eigenvalues (eigenvalues close to unity by absolute value) is observed, which means an increase in the number of orthogonal states supported by the system in the spatial degree of freedom. In other words, when the aperture radius is increasing, the more number of eigenfunctions can be transmitted without losses. So, more complex optical signals propagate through the system without distortions and with less energy losses.

3. Numerical simulation

Simulation was carried out with the given parameters: R = 1 μm, P = 5 μm, m = 2, and $k/2\pi f = 1$. It should be noted, that with increasing parameters R, P and m, the calculations can also be performed, but they become more complex and may require the use of high-performance supercomputers. The simulation of the propagation of the vortex eigenfunctions of the operator (1) in the bounded parabolic fiber is carried out with the aid of the finite Hankel transform:

$$H_R\left[f(r)\exp(im\varphi)\right](r',\varphi') = \frac{k}{f}\int_0^R f(r) J_m\left(\frac{k}{f}rr'\right) r\,\mathrm{d}r \cdot \exp(im\varphi'). \qquad (2)$$

In this case, each odd transformation H_R occurs in the region, bounded by the radius R, and is equivalent to passing the beam through the first half of the two-lens system. Every even transformation H_P takes place in the region, bounded by the radius P, and corresponds to the passage of the optical beam through the second half of the two-lens system.

Figure 1 shows the graph of the moduli of eigenvalues of the operator (1) with the parameters described above. The trajectory of the graph has step behavior: starting from a certain moment all eigenvalues are close to zero, whereas before that their moduli are approximately equal to 1. The graph shows only the first 15 values, since the remaining ones are approximately equal to zero.

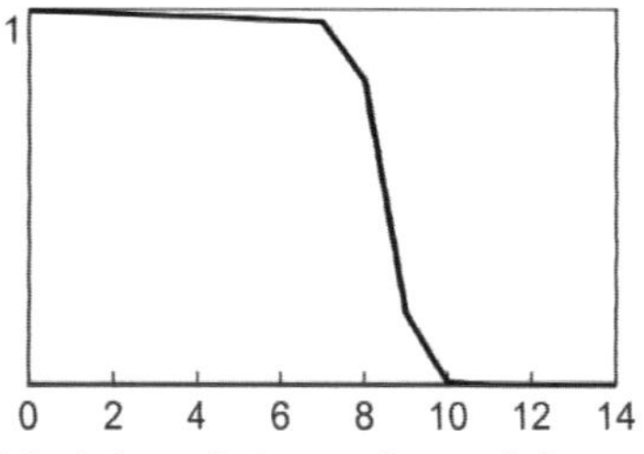

Figure 1. Modules of eigenvalues of the operator (1).

Figure 2 shows one-dimensional graphs of the original eigenfunction $\psi_{nm}(r,\varphi) = \psi_n(r)\exp(im\varphi)$ with the index n = 2, as well as its passage through the system of 10 lenses (the left graph). It can be

IOP Conf. Series: Journal of Physics: Conf. Series **1038** (2018) 012085 doi:10.1088/1742-6596/1038/1/012085

seen from the Figure 2, that the shape of the function remains practically unchanged. This is because the absolute value of its eigenvalue is close in value to 1. On the right, the spectrum of this eigenfunction, i.e. its passage through only 1 lens is shown. It can be seen, that the spectrum differs greatly from the original function.

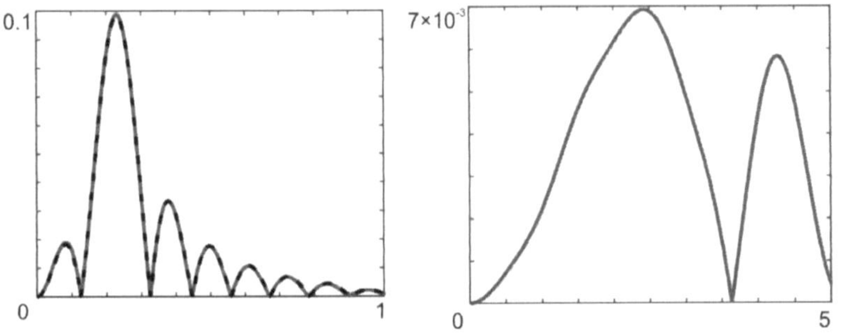

Figure 2. Amplitude of the eigenfunction $\psi_2(r)$ (blue colour, left) and its passage through 10 lenses (black colour, left): the graphs coincide; the amplitude of the spectrum of the original function (right).

As the index of eigenfunctions increases, the moduli of their eigenvalues decrease, therefore, they will be less resistant to the diffraction effects, associated with domain limitation. Figure 3 shows analogous to the previous graphs, but for the function with the index $n = 7$. Unlike the previous case, the graphs do not coincide already, but differ slightly from each other.

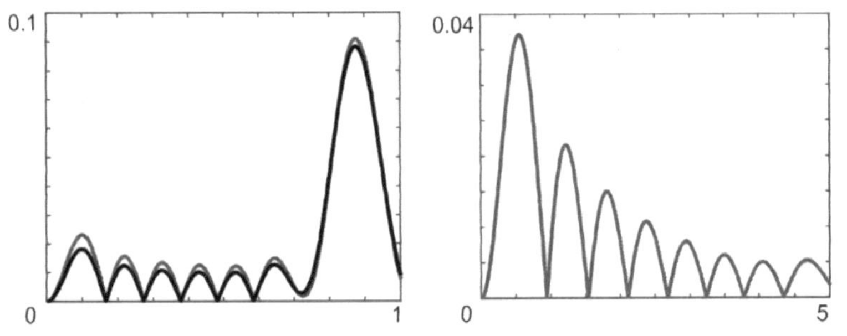

Figure 3. Amplitude of the eigenfunction $\psi_7(r)$ (blue colour, left) and its passage through 10 lenses (black colour, left): the graphs slightly diverge; the amplitude of the spectrum of the original function (right).

As can be seen from Figure 1, the modulus of the eigenvalue of the function with index $n = 8$ is less than one, so the corresponding eigenfunction will be distorted even more, than the previous one, as shown in Figure 4. Part of the energy is lost and goes beyond the system when passing each lens. In this case, the spectrum pattern becomes similar to the original function.

IOP Conf. Series: Journal of Physics: Conf. Series **1038** (2018) 012085 doi:10.1088/1742-6596/1038/1/012085

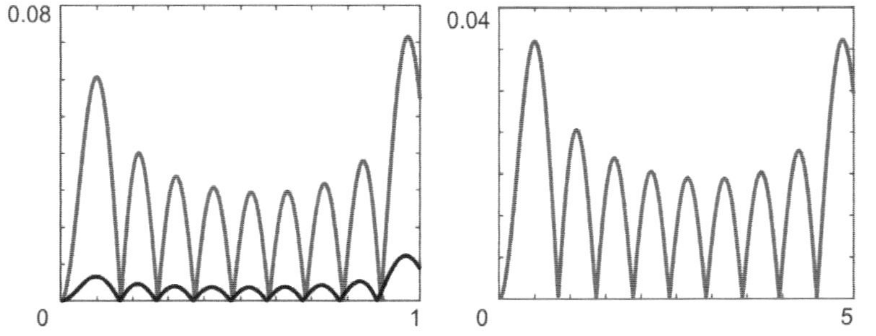

Figure 4. Amplitude of the eigenfunction $\psi_8(r)$ (blue colour, left) and its passage through 10 lenses (black colour, left): the graphs vary greatly; the amplitude of the spectrum of the original function (right).

For clarity, Table 1 lists the 2D-images of the regarded eigenfunctions.

Table 1. Two-dimensional form of eigenfunctions and their spectra.

n	Eigenfunction		Spectrum	
	Amplitude	Phase	Amplitude	Phase
2	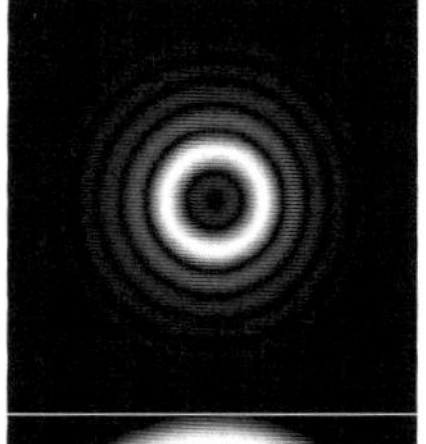	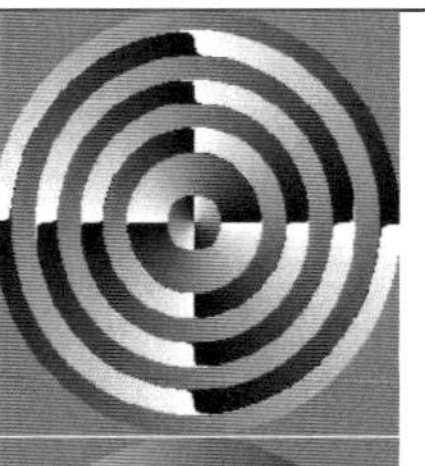	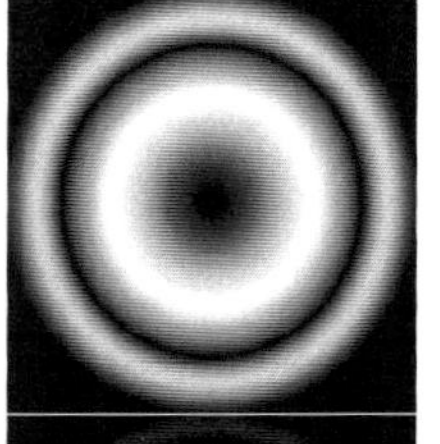	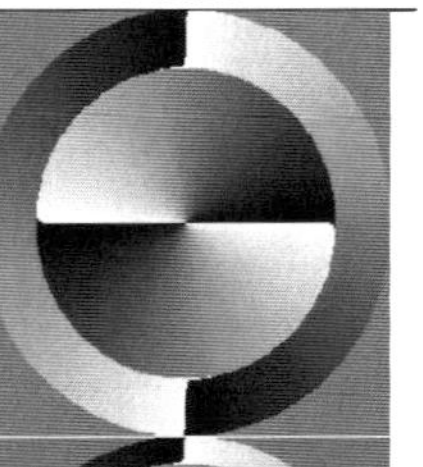
7	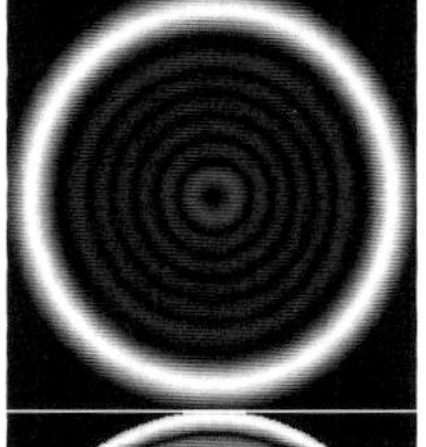	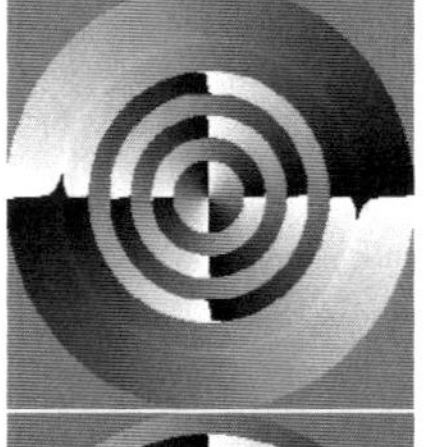	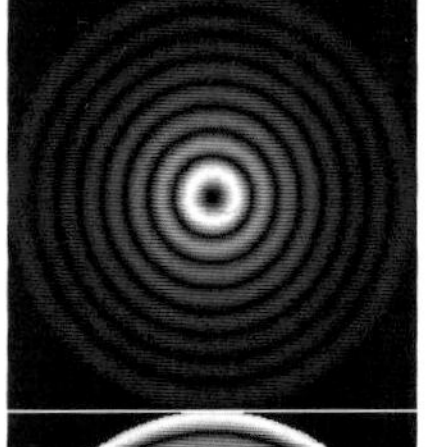	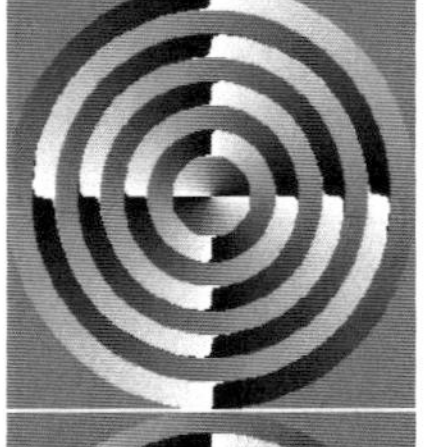
8	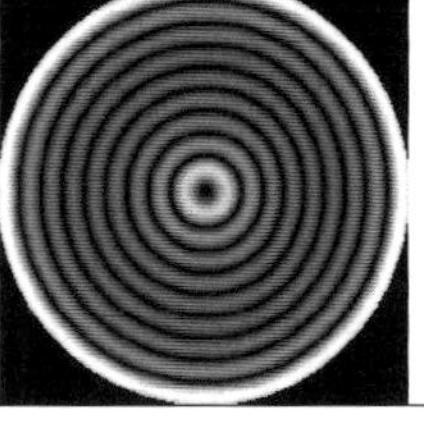	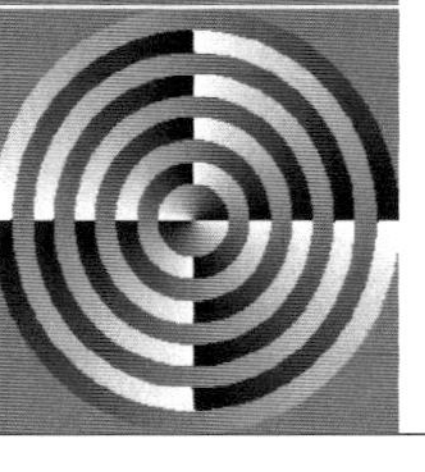	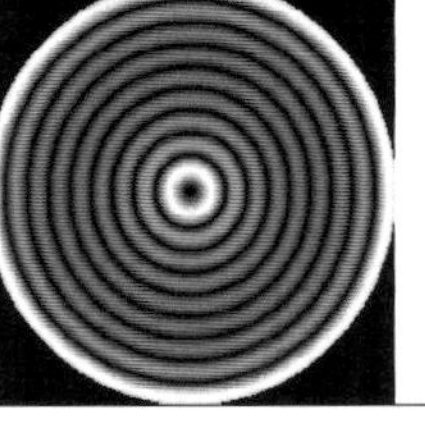	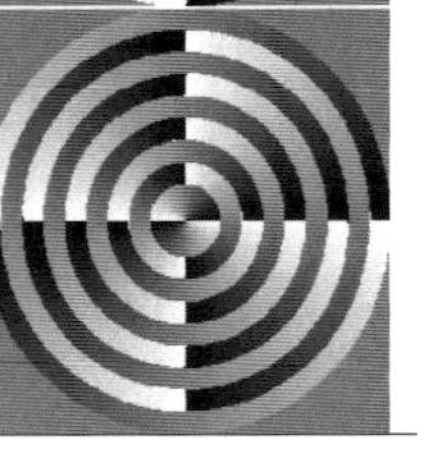

Thus, proposed propagation operator (1) allows us to simulate the propagation of vortex beams through the bounded by radius two-lens system or the optical fiber with parabolic refractive index. Vortex eigenfunctions of the propagation operator (1) with eigenvalues close to 1 propagate through these systems without deformation, energy loss or information distortion. So, the proposed calculation algorithm makes it possible to estimate the number of orthogonal states supported by the system in the spatial degree of freedom or the number of optical vortices that can be transmitted by the systems under consideration. The number of the system states defines the information capacity (each vortex in this case is a bit of an information word). Such estimations can be useful for communication systems based on vortex MDM.

4. Conclusion

The calculation of the set of vortex eigenfunctions of the bounded circular two-lens system is performed. A simulation of the propagation of the functions, obtained in the bounded optical fiber with parabolic refractive index, was conducted. The results showed, that the beams, corresponding to the eigenfunctions with small indices, remain in the fiber while passing and retain the properties of orthogonality. This means, that any signals, that can be represented as a superposition of such eigenfunctions, can be transmitted without losses.

Acknowledgement

This work was financially supported by the Russian Foundation for Basic Research (grant No. 16-47-630546).

References

[1] Essiambre R, Kramer G, Winzer P J, Foschini G J and Goebel B 2010 *J. Lightw. Technol.* **28** 662-701
[2] Berdague S and Facq P 1982 *Appl. Optics* **21** 1950-5
[3] Karpeev S V, Pavelyev V S, Soifer V A, Khonina S N, Duparre M, Luedge B and Turunen J 2005 *Proc. SPIE* **5854** 1-13
[4] Soskin M S and Vasnetsov M V 2001 *Progress in Optics* **4** 219-76
[5] Khonina S N, Kazanskiy N L and Soifer V A 2012 *Recent Progress in Optical Fiber Research* 327-52
[6] Bozinovic N, Yue Y, Ren Y, Tur M, Kristensen P, Huang H, Willner A E and Ramachandran S 2013 *Science* **340** 1545-8
[7] Yan Y, Xie G, Lavery M P J, Huang H, Ahmed N, Bao C, Ren Y, Cao Y, Li L, Zhao Z, Molisch A F, Tur M, Padgett M J and Willner A E 2014 *Nature Communications* **5** 4876
[8] Soifer V A, Korotkova O, Khonina S N and Shchepakina E A 2016 *Computer Optics* **40** 605-24
[9] Khonina S N, Kotlyar V V, Soifer V A, Honkanen M, Lautanen J and Turunen J 1999 *Journal of Modern Optics* **46** 227-38
[10] Khonina S N, Kotlyar V V, Soifer V A, Jefimovs K and Turunen J 2004 *Journal of Modern Optics* **51** 761–73
[11] Kirilenko M S and Khonina S N 2013 *Optical Memory and Neural Networks* **22** 81–9
[12] Thornburg W Q, Corrado B J and Zhu X D 1994 *Opt. Lett.* **19** 454-6
[13] Dubois F, Emplit Ph and Hugon O 1994 *Opt. Lett.* **19** 433-5
[14] Khonina S N and Karpeev S V 2004 *Computer Optics* **26** 16-26
[15] Karpeev S V and Khonina S N 2007 *Optical Memory & Neural Networks* **16** 295-300
[16] Sneddon I N 1972 *The Use of Integral Transforms* (New York & Boston: McGraw-Hill)
[17] Debnath L and Bhatta D 2007 *Integral Transforms and their Applications* (Boca Raton, FL: Goo Chapman and Hall/CRC Press)
[18] Miller A R 2000 *Applied Optics* **39** 1681-99
[19] Martinsson P, Ma P, Burval A and Friberg A T 2008 *Optik-International Journal for Light and Electron Optics* **199** 103-11
[20] Slepian D and Pollak H O 1961 *Bell System Technical Journal* **40** 43-63
[21] Landau H J and Pollak H O 1961 *Bell System Technical Journal* **40** 65-84
[22] Khonina S N, Volotovskij S G and Sojfer V A 2001 *Doklady Akademii Nauk* **63** 30-3
[23] Khonina S N and Kotlyar V V 2001 *Pattern Recognition and Image Analysis* **11** 521-8
[24] Khonina S N and Kotlyar V V 2001 *Optical Memory and Neural Networks (Allerton Press)* **10** 267-76
[25] Kirilenko M S and Khonina S N 2014 *Proc. SPIE* **9156** 91560J
[26] Karoui A and Moumni T 2009 *Journal of Computational and Applied Mathematics* **233** 315-33
[27] Yoshinobu I 1970 *Journal of the Optical Society of America* **60** 10-4
[28] Kirilenko M S and Khonina S N 2018 *Optics Communications* **410** 153-9

Theoretical calculations of resonant signals in the atomic-beam quantum frequency standard with laser pumping and detection

A Yu Rumyantsev[1] and M V Petrenko[1,2]

[1] Russian Institute of Radionavigation and Time, 120 Obukhovskoy Oborony prospect, St. Petersburg, 192012, Russia
[2] Ioffe Institute, 26 Politekhnicheskaya Street, St. Petersburg, 194021, Russia

E-mail: m.petrenko@mail.ioffe.ru

Abstract. We consider the theoretical model of the resonant signals formation in cesium atomic-beam QFS depending on input parameters, such as the number of atoms, the power and the linewidth of the laser radiation in detection area. On the basis of the model calculations were made and the values of fluorescence signal, which are specific for compact cesium QFS, were obtained. The influence of fundamentally irremovable noise sources on the characteristics of atomic-beam frequency standards was considered. On the basis of theoretical expressions, the values of the spectral power density of the noises were calculated and the signal-to-noise ratio was obtained. It was shown that the signal-to-noise ratio for compact cesium atomic-beam QFS does not exceed the value 5×10^4, which leads to an estimation of the best achievable value of the short-term frequency instability by the quantity 7×10^{-13} for 1 second.

1. Introduction

At the present time the optical pumping of atoms is one of the most promising ways for increasing the characteristics of quantum devices, such as miniature quantum frequency standards (QFSs) based on a vapor cell [1, 2], quantum magnetometers [3, 4], and atomic-beam QFSs. In particular, one of the most used sources of reference frequency is the atomic-beam QFS with laser pumping and detection [5, 6]. The frequency instability of such a standard directly depends on both the quality of the resonant fluorescence signal and the amount of noise introduced by various sources into the output signal of the standard. In this paper we present the theoretical model that allows one to determine the characteristics of a resonant signal of atomic-beam QFS based on ^{133}Cs atoms depending on input parameters such as the atomic flux, the linewidth and the power of laser radiation. In the presented model the processes occurring during the interaction of cesium atoms with pumping and detection radiation and with the microwave field in the resonator are taken into account. We also consider fundamentally irremovable noise sources that affect the fluorescence signal of atoms when they are exposed to laser radiation and, as a consequence, the final frequency instability of atomic-beam QFS.

In this paper we investigate atomic-beam QFS on ^{133}Cs atoms, which scheme is shown in figure 1. This QFS works according to the following principle. A beam of ^{133}Cs atoms *(a)* is emitted from the source *(b)* and propagates along the axis of the atomic-beam tube. In the pumping area *(c)* atoms in the presence of a static magnetic field interact with a field of laser radiation *(d)*, which frequency is tuned to the frequency of the $F = 4 \rightarrow F' = 4$ transition of the D_2 line of ^{133}Cs atoms, the radiation in this case is linearly polarized in the direction perpendicular to the magnetic field (σ-polarization). Part of the atoms goes to the level with $F = 3$, $m_f = 0$ of the ground state depleting the level with $F = 4$. Then, while passing through the Ramsey cavity *(e)* and interaction with a microwave field tuned to the hyperfine transition frequency $F = 3$, $m_f = 0 \rightarrow F = 4$, $m_f = 0$ (9.19263 GHz for ^{133}Cs), these atoms transfer to the state $F = 4$, $m_f = 0$ with probability that depends on the frequency detuning of the microwave field from the frequency of the hyperfine transition. Further, in the detection area *(f)* atoms in the state $F = 4$, $m_f = 0$ interact with laser radiation *(g)*, tuned to the frequency of the cyclic transition $F = 4 \rightarrow F' = 5$ of D_2 line, and emit fluorescence photons that are detected by the photodiode *(h)*.

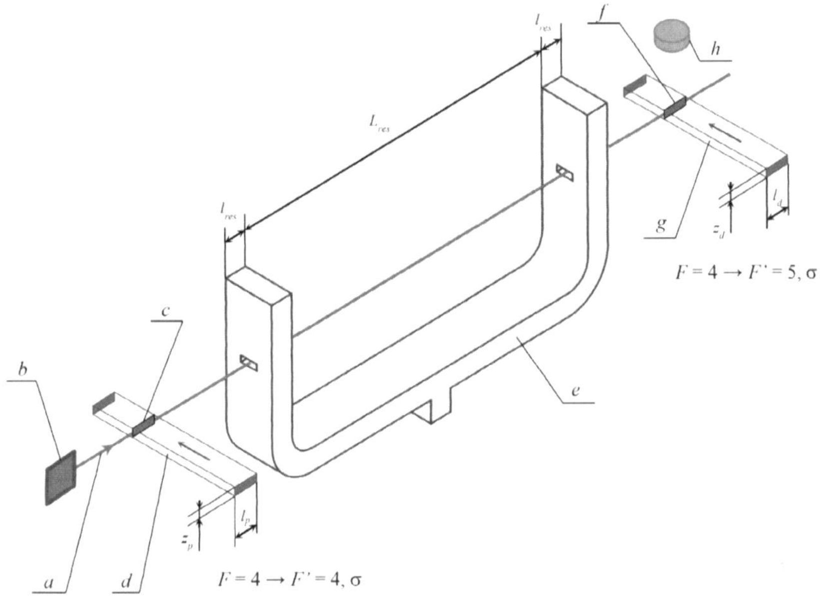

Figure 1. The scheme of atomic-beam QFS with laser pumping and detection.

2. Resonant signals calculation

Let us consider the method of formation of resonant signals in compact cesium atomic-beam QFS with laser pumping and detection. The values of the parameters used in this method, which are specific for compact cesium QFS, are presented in Table 1.

The source of the atomic beam in the atomic-beam QFS is a volume filled with cesium vapor and heated to a temperature of the order of 120 °C. The cesium atomic beam is emitted from this volume, passing through a collimator formed of microchannels, and propagates along the atomic-beam tube. Such a source [6, 7] can provide a flow of ^{133}Cs atoms in the detection region at level $I_{\text{aт}} = 10^{10}$ – 10^{11} atoms/s.

In the optical pumping area, according to [8], in the considered model of a compact atomic-beam QFS the maximum pumping rate of ^{133}Cs atoms to the $F = 3$ level of the ground state is realized and is equal to $\rho_p = 0{,}155$ of the total number of atoms in the beam.

In the Ramsey cavity a transition between the hyperfine sublevels $F = 3$, $m_f = 0 \rightarrow F = 4$, $m_f = 0$ of the ^{133}Cs ground state occurs under the influence of the microwave field. The transition probability in the absence of phase shift of the microwave field between the arms of the resonator is given by the expression [5, 6]

$$P(v,\Delta,b) = \frac{4b^2}{\alpha^2}\sin^2\frac{\alpha\tau_{res}(v)}{2}\left[\cos\frac{\alpha\tau_{res}(v)}{2}\cos\frac{\Omega T_{res}(v)}{2} - \frac{\Omega}{\alpha}\sin\frac{\alpha\tau_{res}(v)}{2}\sin\frac{\Omega T_{res}(v)}{2}\right]^2, \quad (1)$$

where v – velocity of the atom, Ω – frequency detuning of the microwave field from the frequency of the hyperfine transition, b – the parameter of the power of the microwave field (Rabi frequency of the magnetic dipole interaction), $\Delta = \Omega/2\pi$ – frequency detuning in Hz, $\alpha = (\Omega^2+b^2)^{1/2}$, $t_{res}(v) = l_{res}/v$, $T_{res}(v) = L_{res}/v$. When the frequency of the microwave field is equal to the frequency of the hyperfine transition, the maximum transition probability for the most probable velocity of atoms in the beam is realized for $b_{\max} = \pi/t_{res}(v_T) = 34690{,}2\ s^{-1}$.

Table 1. Parameters of the compact atomic-beam cesium QFS.

Parameter	Symbol	Value
Atomic beam source		
Source temperature, °C	T_{Cs}	120
Pumping area		
Power of the pumping laser, mW	P_p	1 – 10
Length of the pumping area, mm	l_p	5
Height of the pumping area, mm	z_p	2
Ramsey cavity		
Interaction length of an atom with microwave field, cm	l_{res}	1
Flight length of the cavity, cm	L_{res}	15
Detection area		
Atomic flow in the detection area, atoms/s	I_{at}	$10^{10} – 10^{11}$
Power of the detection laser, mW	P_{las}	1 – 10
Detection laser linewidth, Hz	ν_{las}	$10^5 – 10^7$
Length of the detection area, mm	l_d	5
Height of the detection area, mm	z_d	2
Silicon photodiode		
Efficiency of the fluorescence photons collection	ρ_{pd}	0,2
Photodiode quantum efficiency	η	0,95
Photodiode shunt resistance, Ω	R_{pd}	10^6
Photodiode temperature, °C	T_{pd}	40
Dark current, A	I_{dark}	10^{-8}
Rejection factor of the direct laser light	r_{pd}	10^{-6}

Thus, atoms in the state $F = 4$, $m_f = 0$, the number of which is proportional to the pumping rate and is determined by the expression (1), enter the detecting area. In the detection area the fluorescence of these atoms occurs when interacting with the laser radiation tuned to the frequency of $F = 4 \rightarrow F' = 5$ transition. The number of photons emitted by an atom during fluorescence is described by [9]

$$\beta(\gamma_{las}, I_{las}, v) = \Gamma\Pi(\gamma_{las}, I_{las})\tau_d(v), \quad (2)$$

where $I_{las} = P_{las}/(l_d \cdot z_d)$ is the power density of the laser radiation in the detection area, $\gamma_{las} = 2\pi\nu_{las}$, $\Gamma = 3.2815\times10^7\ s^{-1}$ is the rate of free radiative transition $F' = 5 \rightarrow F = 4$ [10], $\tau_д(v) = l_д/v$ is the spent time of an atom in the laser field, $\Pi(\gamma_{las}, I_{las})$ is the average population of the exited state $F' = 5$, which is calculated according to [9, 10, 11].

The collection of fluorescence photons in the detection area is performed by a silicon photodiode whose parameters are given in Table 1. Combining expressions (1) and (2) and using the data from Table 1, we can write the expression for the microwave resonance signal registered by the photodiode in volts:

$$S_{res}(\gamma_{las}, I_{las}, I_{at}, \Delta, b) = eR_{pd}\eta\rho_{pd}\rho_p I_{at}\int_0^\infty \beta(\gamma_{las}, I_{las}, v)P(v, \Delta, b)f_M(v, T_{Cs})\mathrm{d}v, \quad (3)$$

where e is the electron charge, $f_M(v,T_{Cs})$ is Maxwell velocity distribution of the atoms in the beam for the source temperature T_{Cs}.

Figure 2 shows the result of the calculations of the microwave resonance signals. Calculations have shown that the values of the recorded signals are tens of millivolts. With a decrease in the microwave power parameter, the width of the resonance reduces along with the predicted decrease in the signal amplitude. This is apparently due to the shift of the maximum of the probability of a hyperfine transition under the action of a microwave field to the region of low atomic velocities in a beam. When integrating according to formula (5), this leads to a decrease in the width of the resonant signal.

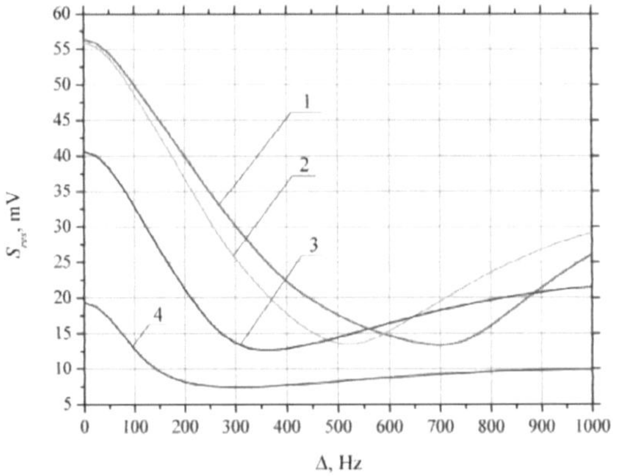

Figure 2. Resonance signals in dependence on microwave frequency detuning for the case $I_{at} = 5\times10^{10}$ atoms/s, $P_{las} = 10$ mW, $\nu_{las} = 1$ MHz. 1 – $b = b_{max}$, 2 – $b = 0{,}7\times b_{max}$, 3 – $b = 0{,}4\times b_{max}$, 4 – $b = 0{,}2\times b_{max}$.

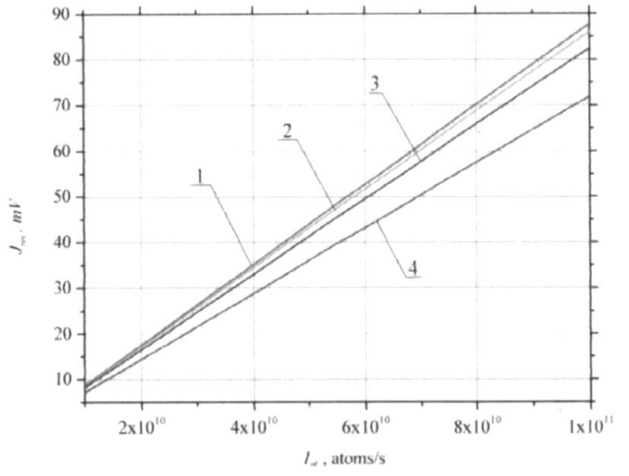

Figure 3. Atomic-beam QFS signals in dependence on atomic flow for the case $P_{las} = 10$ mW. 1 – $\nu_{las} = 100$ kHz, 2 – $\nu_{las} = 1$ MHz, 3 – $\nu_{las} = 3$ MHz, 4 – $\nu_{las} = 10$ MHz.

In practice the difference in voltage values at the photodiode output at the maximum and minimum of the resonant curve (at $\Delta = 0$ Hz and $\Delta = \Delta_{min}$ respectively) is considered as a value of the output signal of the atomic-beam QFS. The expression for the signal can be written as

$$J_{res}(\gamma_{las}, I_{las}, I_{at}) = S_{res}(\gamma_{las}, I_{las}, I_{at}, 0\text{ Hz}, b) - S_{res}(\gamma_{las}, I_{las}, I_{at}, \Delta_{min}, b)\,. \quad (4)$$

Here and below, the value of the microwave power parameter $b = b_{max}$ will be considered. In this case, the minimum signal corresponds to the frequency detuning value $\Delta = \Delta_{min} = 700$ Hz. Signals calculated by formula (4) are shown in figure 3.

3. Signal-to-noise ratio calculation

The main parameter, that characterizes the quality of the QFS signal and its relative frequency instability, is the signal-to-noise ratio. The operation of the QFS is influenced by noise from various sources, including instrumental. In this paper we consider the influence of fundamentally irremovable noises. It is assumed, as in [9, 11, 12], that the contribution of noise is considered only in the low-frequency region. The spectral power density of noise is assumed to be constant in the considered frequency range and is described by expressions for its asymptotic values at frequencies close to zero. All expressions are written for the noise spectral power density in the 1 Hz band.

The expression for the shot noise contribution to the noise power spectral density of the output signal can be written in the form [12]

$$S_{shot}(\gamma_{las}, I_{las}, I_{at}) = 2(eR_{pd})^2 \eta\rho_{pd}\rho_p I_{at}\left[\int_0^\infty \beta(\gamma_{las}, I_{las}, v)P(v, 0\text{ Hz}, b_{max})f_M(v, T_{Cs})\mathrm{d}v + \eta\rho_{pd}\int_0^\infty [\beta(\gamma_{las}, I_{las}, v)]^2 P(v, 0\text{ Hz}, b_{max})f_M(v, T_{Cs})\mathrm{d}v\right]. \quad (5)$$

Here we take into account both the contribution of the photon noise associated with fluctuations in the number of fluorescence photons emitted by one atom and the shot atomic noise associated with fluctuations in the number of atoms in the beam.

One of the main parameters of the interaction of laser radiation with atoms that affect the noise of the output signal is also the spectral purity of the radiation. The laser frequency noise leads to additional fluctuations in the number of photons emitted by the atoms, which in the case of a wide laser emission line can significantly decrease the signal-to-noise ratio. According to [9, 11], the laser-frequency fluctuations induce coupled fluctuations of the fluorescence signals emitted by two different atoms of the interaction area. The spectral power density of the output signal noise in this case is expressed by the formula [9]

$$S_{las}(\gamma_{las}, I_{las}, I_{at}) = [S_{res}(\gamma_{las}, I_{las}, I_{at}, 0\ \text{Hz}, b_{\max})]^2 \frac{4(3\Gamma + \gamma_{las})}{[\Omega_R^{\,2}(I_{las}) + \Gamma(\Gamma + \gamma_{las})]} \times \frac{\gamma_{las}^2 \Gamma}{\left[\Omega_R^{\,2}(I_{las})(3\Gamma + 4\gamma_{las}) + \frac{\Gamma}{2}(3\Gamma + \gamma_{las})(\Gamma + 2\gamma_{las})\right]}, \tag{6}$$

where $\Omega_R^2(I_{las})$ is the square of the Rabi frequency of the electric dipole interaction.

Additional noise in the output signal is made by a photodiode that collects fluorescence photons. The spectral power density of this noise can be written in the form

$$S_{pd}(I_{las}) = (eR_{pd})^2 (2\eta r_{pd} \frac{P_{las}}{h\nu} + 2\frac{I_{dark}}{\text{e}} + 4\frac{kT_{pd}}{e^2 R_{pd}}), \tag{7}$$

The first term describes the noise caused by the direct illumination of the photodiode by laser radiation with the r_{pd} rejection factor (shielding), the second term is the noise caused by the dark current of the photodiode and the third term is the thermal noise of the photodiode current on the shunt resistance.

The signal-to-noise ratio for the output signal of the atomic-beam QFS in the 1 Hz band is written taking into account formulas (4), (5) – (7) in the form [9, 12]

$$SNR(\gamma_{las}, I_{las}, I_{at}) = \frac{J_{las}(\gamma_{las}, I_{las}, I_{at})}{\sqrt{S_{shot}(\gamma_{las}, I_{las}, I_{at}) + S_{las}(\gamma_{las}, I_{las}, I_{at}) + S_{pd}(I_{las})}}. \tag{8}$$

The signal-to-noise ratios calculated from formula (8) are shown in Figure 4 and Figure 5. In Figure 4 it is shown that the signal-to-noise ratio increases substantially with increasing number of atoms in the beam and with a decrease in the linewidth of the laser radiation in the detection area. Calculations have shown that the use of lasers with a linewidth more than 3 MHz in atomic-beam QFSs can be considered inexpedient, since it leads to a noticeable decrease of the signal-to-noise ratio. According to the obtained data, it is necessary to investigate the possibilities of increasing the number of atoms in the beam to improve the signal-to-noise ratio. However, the method proposed in [6] for increasing the density of atoms by raising the source temperature to 130 – 150 °C leads to a significant increase in the consumption of the working material and a degradation of the angular distribution of the atomic flux [7, 13].

Figure 5 shows the behavior of the signal-to-noise ratio as a function of the power of the detection laser. Calculations have shown that with laser power increasing the signal-to-noise ratio tends to its limit value. It is due to the influence of the noise caused by the laser frequency noise, which is at high power of the laser radiation becomes dominant over the other noises. It should be noted that if the photodiode is not shielded from direct laser light with a suppression factor exceeding 10^{-4}, the contribution of the photodiode noise will increase with laser power increasing, which will lead to a decrease in the signal-to-noise ratio. It also follows from the results of calculations that when using detection lasers with a power of less than 5 mW, the effect of the laser linewidth on the signal-to-noise ratio becomes more critical. The signal-to-noise ratio significantly decreases when the linewidth exceeds 500 kHz.

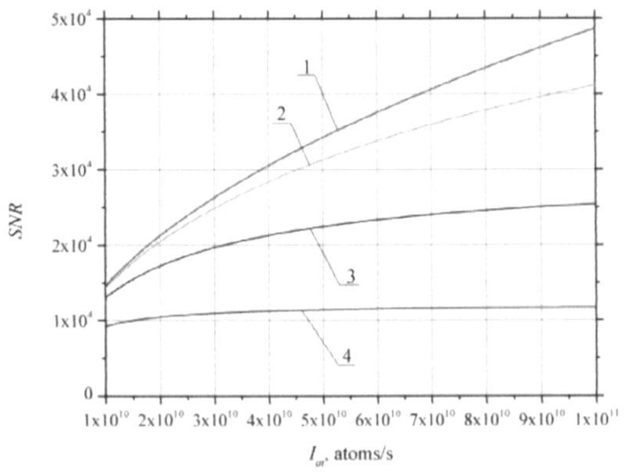

Figure 4. Signal-to-noise ratio in dependence on atomic flow for the case P_{las} = 10 mW. 1 – ν_{las} = 100 kHz, 2 – ν_{las} = 1 MHz, 3 – ν_{las} = 3 MHz, 4 – ν_{las} = 10 MHz.

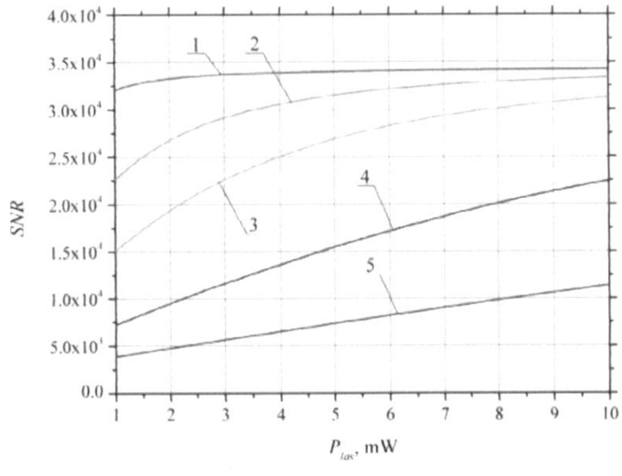

Figure 5. Signal-to-noise ratio in dependence on power of detection laser for the case $I_{at} = 5\times10^{10}$ atoms/s. 1 – ν_{las} = 100 kHz, 2 – ν_{las} = 500 kHz, 3 – ν_{las} = 1 MHz, 4 – ν_{las} = 3 MHz, 5 – ν_{las} = 10 MHz.

4. Conclusion

In conclusion, let us consider the instability of the atomic-beam QFS with laser pumping and detection. The relative short-term frequency instability of the atomic-beam QFS with rectangular modulation of the frequency of the microwave field can be calculated by the formula

$$\sigma = \frac{\sqrt{2}\Delta f}{\pi \cdot SNR(\gamma_{las}, I_{las}, I_{at}) \cdot f_0 \sqrt{\tau}}, \tag{9}$$

where f_0 = 9.19263 GHz is the frequency of the hyperfine transition, Δf is the halfwidth of the microwave resonance (Δf = 700 Hz for b_{max}) and τ is the measuring time. Calculations of the signal-to-noise ratio showed that its value does not exceed 5×10^4 in the range of the parameters considered, which are specific for compact cesium atomic-beam QFSs. This leads to an estimation of the best possible value of the relative short-term frequency instability with a value of 7×10^{-13} for 1 second.

References

[1] Ermak S V, Petrenko M V and Semenov V V 2016 *Tech. Phys. Lett.* **42** 127
[2] Fedorov M I, Ermak S V, Petrenko M V, Pyatyshev E N and Semenov V V 2016 *J. Phys.: Conf. Ser.* **769** 012046
[3] Baranov A A, Ermak S V, Sagitov E A, Smolin R V and Semenov V V 2016 *Tech. Phys. Lett.* **42** 186
[4] Sagitov E A, Ermak S V, Petrenko M V and Semenov V 2016 *J. Phys.: Conf. Ser.* **769** 012044
[5] Vanier J and Audoine C 1989 *The quantum physics of atomic frequency standards* (Philadelphia: IOP Publishung Ltd) p 453
[6] Pimenov A V and Pleshanov S A 2010 *Electronnaya Tekhnika. SVCH Tekhnika* **507** 16
[7] Rumyantsev A Yu, Petrenko M V, Poniaev S A, Shustrov Yu A, Kochegarov V P and Schennikov D L 2017 *J. Phys.: Conf. Ser.* **929** 012095
[8] Avila G et al 1987 *Phys. Rev.* A **36** 3719
[9] Dimarcq N, Giordano V, Cerez P and Theobald G 1994 *Appl. Phys.* B **59** 135
[10] Steck D A *Cesium D Line Data* (http://steck.us/alkalidata)
[11] Halswanter Th, Ritsch H, Cooper J and Zoller P 1988 *Phys. Rev.* A **38** 5652
[12] de Clercq E, Clairon A, Dahmani B, Gerard A and Aynie P 1989 *Frequency standards and metrology* ed. A De Marchi (Berlin, Heidelberg: Springer-Verlag) p 120
[13] Olander D R and Kruger V 1970 *J. Appl. Phys.* **41** 2769

IOP Conf. Series: Journal of Physics: Conf. Series **1038** (2018) 012087 doi:10.1088/1742-6596/1038/1/012087

Diagnostic of semiconductor device structures by spin-labeled electrons

R.I. Dzhioev, M. Kotur, N.K. Poletaev

Ioffe Physico-Technical Institute of the RAS, 194021 St. Petersburg, Russia

E-mail: mladenikotur@gmail.com

Abstract. The influence of non-uniform distribution of spin density in the temperature range from 5 to 80 K under the conditions of optical orientation of electrons in semiconductors is analyzed. Possibility to determine the kinetic parameters of carriers by depolarization of recombination radiation in a magnetic field is shown. Electron diffusion length and mobility is defined for materials with a hole conductivity. An estimated value of the rate of recombination on a free surface and its influence on the distribution of the concentration of non-equilibrium electrons from the excitation surface into the bulk is presented.

1. Introduction

The first experiments on optical orientation in direct band gap semiconductors [1, 2] allowed for determination of the lifetime of minority carriers preferential spin orientation. Most of the findings were obtained under steady-state conditions for bulk III-V crystals and solid solutions based on gallium arsenide.

Results of theoretical studies of spin relaxation mechanisms, propped by experimental work, are collected in the review book [3]. To better demonstrate the role of mechanisms of spin relaxation in the actual experimental conditions were not taken into account, which were cast aside as an insignificant correction.

The edge luminescence of the active layer of GaAs photocathodes with hole conductivity [4, 5, 6] and active layer of semiconductor laser diode structures [7] was studied in detail.

In this paper we investigated polarized luminescence from layers of AlGaAs stuctures with different manufacturing technologies: GaAs photocatodes buffer layer and the AlGaAs laser diode heterostructure.

2. Samples and experimental procedure

The experiments were carried out in the temperature range from 5 to 80 K, where, at low-temperature, we deal with localized carriers. As the temperature increases, the carrier mobility reaches its maximum value [8] and the diffusion transport of carriers should manifest itself. Investigated layers had thickness d of the order of magnitude of the diffusion length of the nonequilibrium carriers L_e (thin) and substantially exceeding it ($d >> L_e$) (thick). Layers of p-type $Al_{0.3}Ga_{0.7}As$ with doping concentrations $p \sim 10^{16}$ cm^{-3} and thickness $d \approx 2$ μm and $d \approx 20$ μm were grown on GaAs substrates (Fig. 1(a) and (b)). The effect of optical orientation of electrons was observed in reflection geometry. Sample was excited by a circularly polarized light from a He-Ne laser. The degree of circular polarization ρ was measured using the

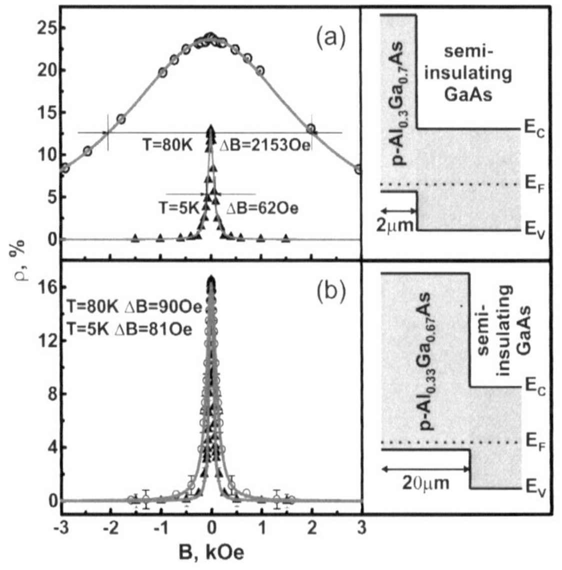

Figure 1. The PL depolarization curves in a magnetic field and the space-energy structure of (a) thin and (b) thick sample. Circles correspond to measurements at $T = 80$ K and triangles at $T = 5$ K.

photoelastic modulator and a linear polarizer with synchronous counting of left and right-hand polarized photons.

3. Results

The lifetimes of electrons τ and of the preferential orientation of the spin of minority carriers T_s are related to the photoluminescence (PL) polarization degree and the half-width of PL depolarization curve in a magnetic field given by:

$$\tau = \frac{0.25}{\rho} \cdot T_s \qquad T_s = \frac{h}{g\mu} \cdot \frac{1}{\Delta B} \tag{1}$$

where ρ is the degree of circular polarization, g is the electron g-factor in the conduction band, μ is the Bohr magneton, h is the Planck constant and ΔB is a magnetic field at which the degree of PL polarization decreases twice from its original value [3].

Noticeable difference is observed in how the PL polarization degree and spin lifetime T_S changes with increasing temperature. At low temperature, $T = 5$ K, the half-width of the PL depolarization curves for both p-AlGaAs are approximately the same (Fig. 1). Nonetheless, broadening of the PL depolarization curve for a thin layer is evident with the temperature rise. When increasing the temperature to 80 K, the half-width ratio is $\Delta B_{T=80\mathrm{K}}/\Delta B_{T=5\mathrm{K}} \approx 30$ for thin (Fig. 1(a)) and $\Delta B_{T=80\mathrm{K}}/\Delta B_{T=5\mathrm{K}} \approx 2.7$ (Fig. 1(b)) for thick p-AlGaAs sample.

Values for τ and τ_s determined from the experiment for a thin ($d \approx 2\mu$m) and thick ($d \approx 20\mu$m) sample at two temperatures $T = 5$ K and $T = 80$ K are given in Table 1. The value of the electron g-factor, $g = 0.41$, for $Al_{0.3}Ga_{0.7}As$ is taken from Ref. [9].

In thin structure (Fig. 1(a)), the entire thickness of the AlGaAs layer is subjected to the photoexcitation. Difference in the electron energy in the AlGaAs layer and the GaAs substrate leads to the energy relaxation of electrons at the heterojunction to substrate. A diffusion flux is created in the direction of the heterojunction. The increase in the polarization degree to the limiting values and the broadening of the depolarization curves in a magnetic field with increasing temperature is explained by a decrease of the electron lifetime. Under the condition

Table 1. τ and τ_s for thin and thick sample.

d (μm)	τ (ns)		τ_S (ns)	
2	7.0 (5K)	0.2 (80K)	9.0 (5K)	-
20	4.9 (5K)	4.4 (80K)	8.8 (5K)	7.9 (80K)

$d < \alpha^{-1}$, where α is the absorption coefficient of substance in the spectral region of PL (in our case for $d = 2\mu$m, $\alpha = 3 \cdot 10^4\, cm^{-1}$ [10]) it is valid to assume that the excitation intensity is the same over the entire thickness of the layer: $g_0 \approx \alpha \cdot I_0$, where g_0 is the generation rate of carriers. Then, the diffusion equation is:

$$D \cdot \frac{d^2 n(z)}{dz^2} - \frac{n(z)}{\tau} + g_0 = 0. \tag{2}$$

Here D is the carrier diffusion coefficient, and z is the distance from the surface. The rate of surface recombination can be assumed to be negligibly small [6] while the rate of disappearance of electrons at the heterojunction infinitely large. The solution of equation (2) will then be:

$$n_z = g_0 \cdot \tau + A \cdot exp(\lambda \cdot d) + exp(-\lambda \cdot d), \tag{3}$$

where $A = -\dfrac{g_0 \cdot \tau}{exp(\lambda \cdot d) + exp(-\lambda \cdot d)}$ and $\lambda = \dfrac{1}{\sqrt{D \cdot \tau}} = \dfrac{1}{L_e}$, L_e is the electron diffusion length.

This makes it possible to find the effective lifetime of nonequilibrium carriers in the layer, as the average lifetime:

$$\bar{\tau} = \frac{1}{d} \cdot \int_0^d \frac{n_z}{g_0} dz = \tau \cdot \left(1 - \frac{L_e}{d} \cdot \frac{exp(2 \cdot d/L_e) - 1}{exp(2 \cdot d/L_e) + 1}\right). \tag{4}$$

The carrier lifetimes in the bulk material τ were taken from experiments on a thick ($d >> L_e$) sample.

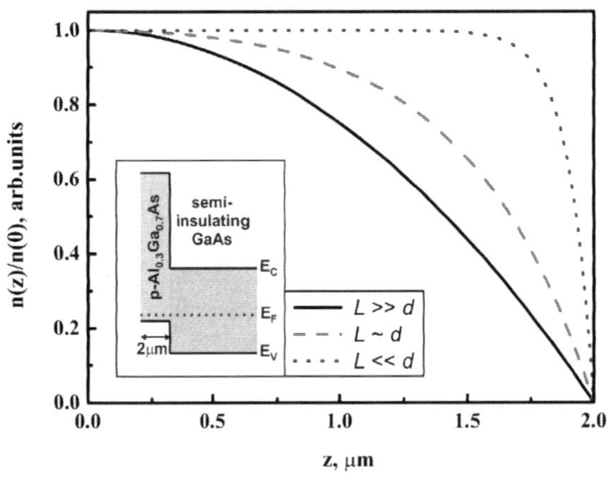

Figure 2. Dependence of the concentration distribution of photoexcited carriers in the layer on distance from the free surface for different values of the diffusion length.

Comparing the calculated lifetime τ with the experimentally determined one, we can determine the diffusion length of the electrons and diffusion coefficient. For temperature $T = 80$ K, the best agreement with experiment is achieved when $L_e = 13.8$ μm, which corresponds to $D = 240$ cm^2/s. Using the Einstein relation, one can estimate the mobility

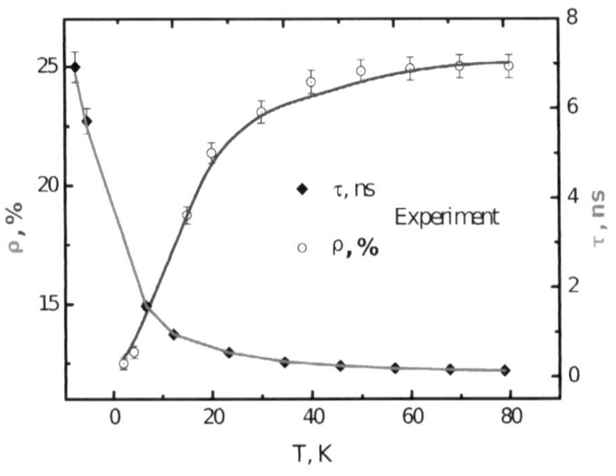

Figure 3. Dependence of the PL polarization degree for a thin layer, $d = 2$ μm, and the lifetime of electrons on temperature. Solid lines are calculation.

of electrons $\mu_e = 21500$ cm^2/V $\cdot$ s, i.e. the mobility of the minority carriers is experimentally determined.

Taking into account the dependence of the mobility μ on the temperature reported in Ref. [8] and exploiting the Einstein relation, we calculated $\bar{\tau}$, the degree of PL polarization, for the whole temperature range of interest (Fig. 3).

We used the approach realized in Ref. [5, 6] to estimate the rate of surface recombination. Analysis of diffusion in a semi-infinite sample makes it possible to calculate the increase in the half-width of the depolarization curve as a function of the rate of surface recombination. The results of such calculations for several values of the diffusion coefficient D are given in Fig. 4.

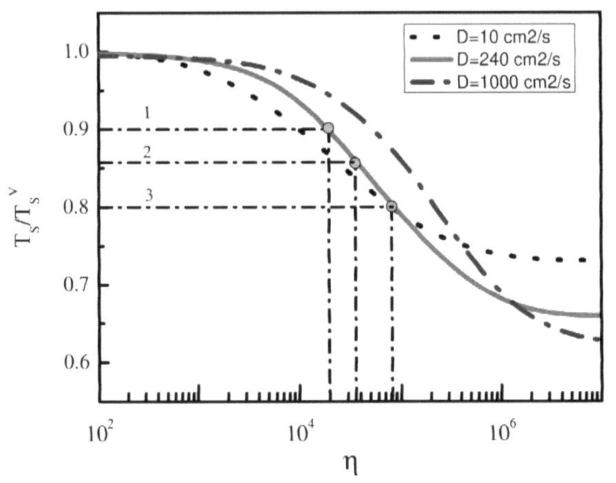

Figure 4. Calculation of the effect of surface recombination on the lifetime of spin T_S in semi-infinite sample with different diffusion coefficients. Points on the figure corresponds to the experimental data given in Table 2.

The diffusion coefficient $D \approx 240$ cm^2/s is taken from our experiments. Dependencies for other values are given for comparison and illustration of the scale of possible changes of the spin lifetime, influenced by the surface recombination. With an increase in the surface recombination rate T_S decreases as a result of the loss of carriers on the surface, however, at high surface recombination rates η, the spin lifetime tends to a constant value, which is determined by the diffusion coefficient of the electrons.

Rate of surface recombination and electron diffusion coefficient increase with temperature. From the comparison of T_S at the interface in considered temperature range, the rate of surface recombination at T = 80 K can be predicted. Table 2 gives estimates of the rate of surface recombination performed on the basis of our data and published in the literature.

Table 2. Estimates of the rate of surface recombination

Point number on the Fig.4	$T_S(77)/T_S(5)$	η cm/s	Ref.
-	0.97	$10^2 <$	[11]
-	0.978	$10^2 <$	[12] Table 1, sample 4
1	0.86	$4 \cdot 10^4$	[12] Table 1, sample 1
2	0.8	$8 \cdot 10^4$	[13]
1	0.88	$2 \cdot 10^4$	our data, Fig.4

Measuring the rate of depolarization in a transverse magnetic field in the buffer layer of the GaAs photocathode (Fig. 1(a)), we practically repeated the first study of diffusion transport of optically oriented (spin labeled) electrons at 77 K [14]. We have proved that the decrease in the spin lifetime T_S is related, not with recombination on a free (etched) surface, as claimed by the authors, but with the loss of electrons at the interface.

4. Conclusions

Introduction of experimental conditions and an adequate description of the processes of generation, relaxation, and transport of spin in the conditions of its non-equilibrium distribution can give us a tool for studying spin dynamics and diagnostics of the parameters of semiconductor device structures.

5. Acknowledgment

This work was supported by the Ministry of Education and Science of the Russian Federation (Contract No. 14.Z50.31.0021 with the Ioffe Institute, Russian Academy of Sciences, and leading researcher M. Bayer).
We are grateful to Yu.G. Kusrayev for stimulating interest and to K.V. Kavokin for useful discussion.

References

[1] Parsons RR 1969 *Phys. Rev. Lett.* **23** 1152
[2] Ekimov AI and Safarov VI 1970 *JETP Lett.* **12** 1
[3] Meier F and Zakharchenya BP (eds) 1984 *Optical orientation* (North-Holland, Amsterdam)
[4] Dzhioev R and Kavokin K 1991 *Sov. Phys. Solid State* **33** 1654
[5] Dzhioev RI, Zakharchenya BP, Ichkitidze RR, Kavokin KV and Pak PE 1993 *Phys. Solid State* **35** 1396
[6] Dzhioev RI, Zakharchenya BP, Kavokin KV and Pak PE 1994 *Phys. Solid State* **36** 1501
[7] Dzhioev RI, Kavokin KV, Kusrayev YuG and Poletaev NK 2015 *Semiconductors* **49** 1531–1535
[8] Blakemore JS 1982 *J. Appl. Phys.* **53** R123
[9] Chadi DJ, Clark AH and Burnham RD 1976 *Phys. Rev. B* **13** 4466
[10] Monemar B, Shih KK and Pettit GD 1976 *J. Appl. Phys.* **47** 2604
[11] Fishman G and Lampel G 1977 *Phys. Rev. B* **16** 820
[12] Clark AH, Burnham RD, Chadi DJ and White RM 1975 *Phys. Rev. B* **12** 5758
[13] Zerrouati K, Fabre F, Bacquet G, Bandet J, Frandon J, Lampel G and Paget D 1988 *Phys. Rev. B* **37** 1334
[14] Garbuzov DZ, Merkulov IA, Novikov VA and Fleisher VG 1976 *Sov. Phys. Semiconductors* **10** 552

On the possibility of recording absorption spectra in weak magnetic fields by the method of nuclear magnetic resonance

N S Myazin[1], **V V Davydov**[1, 2, 3], **V V Yushkova**[4], **V Yu Rud'**[3]

[1]Higher School of applied physics and space technologies, Peter the Great Saint Petersburg Polytechnic University, Saint Petersburg 195251, Russia
[2]Department of photonics and communication lines, The Bonch-Bruevich Saint Petersburg State University of Telecommunications, Saint Petersburg 193232, Russia
[3]All-Russian Research Institute of Phytopathology, Moscow Region 143050, Russia
[4]Saint Petersburg University of Management Technologies and Economics, Saint Petersburg 190103, Russia

myazin.n@list.ru

Abstract. The new magnetic system construction and the registration circuit of the nuclear magnetic resonance signal in the weak magnetic field are considered. The new elaborated construction of the small nuclear magnetic spectrometer gives a possibility to register signals from different nuclei of investigated medium, having magnetic moment. That expands greatly the functional possibilities of the nuclear magnetic resonance in the investigation of media in the express-mode. The experiment investigation results for different media are represented.

1. Introduction

One of the urgent problems of applied physics today is the elaboration of high speed and reliable methods of express control of condensed media condition [1–5]. They are especially important for the medium condition control before carrying out different physical and chemical experiments (in particular after long their storage, container changing or transportation). Also this is suitable during ecological monitoring of difficult to access places of water objects and their coastal zones, during quality control of production, etc. [4–9]. A great number of devices using different physical phenomena are constructed for carrying out the medium investigation in the express mode [1–9].

Their technical characteristics analysis carried out by us, as well as analysis of data that we get while studying various media [4, 5, 7–10], showed that most of them can be used only for control of a small number of media or for the solution of narrow range of problems. For example, express analyzer "IT-1" for determination of the water percentage in engine or transformer oil and diesel fuel, the portable photometer pHotoFlex (for determination of the water condition and pH), the visible light portable spectrophotometer DR 190 manufactured by HACH-LANGE for the liquid media condition control with the turbidity not more than 10 FNU, etc.

Such devices are unsuitable for work in which it is proposed to study a wide range of substances with different physical characteristics (turbidity, viscosity), as well as containing various chemical elements (e.g. in environmental monitoring). Previous studies of condensed media using several models of these devices have confirmed this [3, 6, 10–12]. Therefore, as one of the solutions to this problem, it has been proposed to use the method of nuclear magnetic resonance to conduct studies of media in the express mode. In the proposed method, it was suggested to measure the relaxation constants T_1 and T_2 of the sample at the sampling site and, comparing them with the basic ones, corresponding to the standard state of the medium under study, instantly determine the degree of its deviation from the standard state [7, 9-12].

However, experience with the operation of a small-sized NMR relaxometer has shown that during environmental monitoring this may not be enough to make a decision at the measurement site for the further use of the investigated medium. The conducted experiments showed that the aqueous media, depending on the natural conditions, might contain chemical elements (for example, fluorine, potassium, etc.) that are harmless in small quantities for living organisms. Their effect on the shape of the NMR signal is difficult to predict: these elements cause significant changes in T_1 and T_2. For this reason, after measuring T_1 and T_2, it may be concluded that the researched medium represents a potential hazard. Thus, it is possible to make wrong further decisions, which leads to additional spending on researching samples of this medium.

In this situation, an important role will be played by a method that will allow us to determine the presence of the chemical elements considered above in the investigated medium. One of them can be nuclear magnetic resonance. The nuclei of these chemical elements (fluorine, potassium, etc.) have a magnetic moment, which makes it possible to record the NMR signal at their resonance frequencies in a magnetic field. In a compact magnetic system, it is extremely difficult to change magnetic field induction during media express control. Therefore, it became necessary to develop a small-size NMR spectrometer, in which the frequency of the NMR signal detection f_{nmr} varies widely. This makes it possible to detect NMR signals at different resonant frequencies f_{nmr} from nuclei with magnetic moments contained in the medium under study (like a spectrum in stationary high-resolution spectrometers). The implementation of this device in most cases will solve the problems discussed above, which arise when investigating media with a compact NMR relaxometer in the express mode.

2. Design of a compact NMR spectrometer

Figure 1 shows the structural diagram of a compact NMR spectrometer with the new construction elements developed by us.

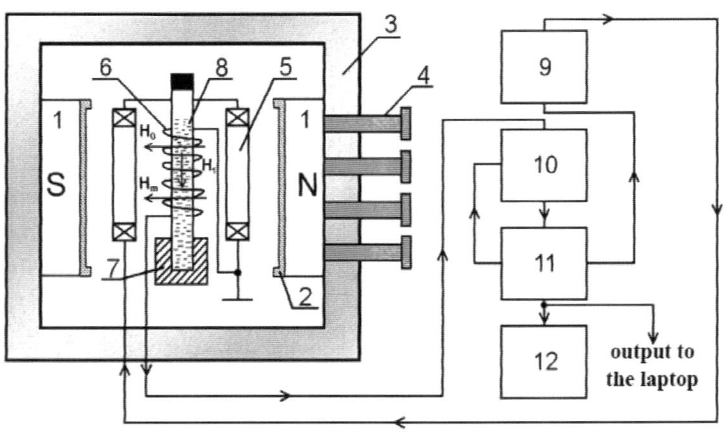

Figure 1 Structural scheme of compact NMR spectrometer: 1 — permanent magnet; 2 — inserts; 3 — neutral for the placement and alignment of the magnets; 4 — adjusting screws; 5 — modulation coil; 6 — NMR signal registration coil; 7 — locking device for the container with the researched medium; 8 — container with the researched medium; 9 — magnetic field modulation generator; 10 — registration scheme including RF autodyne generator; 11 — processing and control unit; 12 — oscilloscope

The small-sized magnetic system was made of the material NdFeB in the form of a disk 1 with a large residual induction. This allowed at the diameter of the magnet 1 poles $d_m = 92$ mm and distance between them $d_z = 16$ mm to provide in the area of the registration coil 6 the inhomogeneity $0.5 \cdot 10^{-3}$ cm^{-1} at the induction equal to $B_0 = 0.132$ T. The magnetic field modulation frequency can vary from 1 to 200 Hz. We have experimentally established that the optimal value is $f_m = 50$, since it provides a high value of the signal noise ratio and at the same time allows to investigate a large number of media [12]. The weight of the new developed magnetic system together with the coils of modulation 5 and registration 6, and the locking device of the container 7 appeared to be equal to about 3.2 kg.

To register NMR signals in weak magnetic field from nuclei with low sensitivity to the NMR method, a new scheme of a weak oscillation generator (an autodyne detector) was developed [1, 4, 6, 12]. This scheme is assembled on the basis of an amplifying cascade with drain detection of the NMR signal and its subsequent amplification using modern planar field-effect transistors. Such construction of the circuit allows to create a minimum level of oscillations in the receiver-transmission loop of the autodyne detector, in which the sample with the investigated medium is placed, to obtain the greatest sensitivity (the ratio of the generation amplitude in the loop to the change in its Q) for NMR signal registering.

In the new design of the compact NMR spectrometer, the scheme of AFC on a resonance is based on the STM32 microcontroller (ARM Cortex M3, STM32F100RBT6B). In addition, on its basis, we have created an accumulation scheme and the new auto-tuning circuits to the maximum S/N ratio for the generation level of the autodyne detector (field H_1), the modulation frequency f_m and amplitude H_m of the B_0 field.

Figure 2 shows, as an example, the registered NMR signals (the change in the voltage U_S at the output of the autodyne detector 10 from the time t) from the aqueous solution of sodium hydroxide NaOH at the resonance frequency of the sodium nuclei f_{Na} = 1488591 Hz.

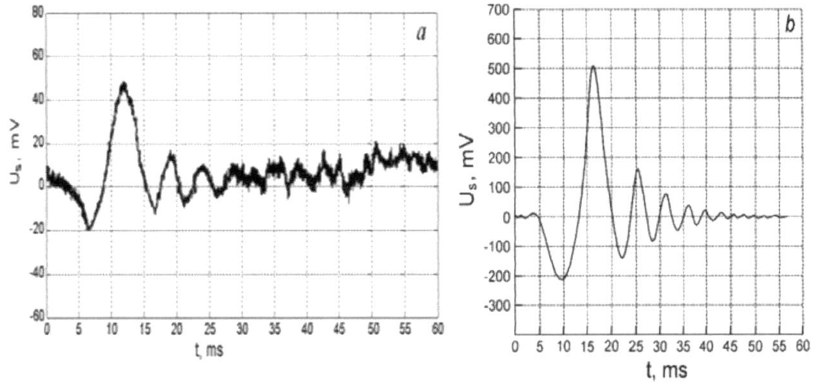

Figure 2 The registered NMR signal from sodium hydroxide at T = 291.3 K: a) without accumulation; b) output of accumulation scheme

In the earlier developed design of a compact NMR relaxometer [4, 7, 9, 10] researchment of this media was possible only by using the NMR signal from protons registered at the resonant frequency f_p. Analysis of the presented on figure 2 NMR signals shows that ratio S/N > 1.3 while registering signals on the sodium nuclei resonant frequency f_{Na}. This makes it possible to perform AFC f_{nmr} on the resonance of sodium nuclei. However, measuring of the relaxation constants T_1 and T_2 with an error less than 1.0% (which allows us to uniquely determine the medium state [1–4, 6, 8]) impossible without using accumulation scheme since the ratio S/N < 3.0 (figure 2.a). In addition, there are noises at the peaks of the NMR signal. The subsequent accumulation of the NMR signal allows us to obtain a ratio S/N> 10.0 (figure 2.b), ensuring the measurement of T_1 and T_2 with the required accuracy.

Nevertheless, when investigating media for the presence of nuclei with low sensitivity to the NMR method, a situation may occur when the S/N of the detected signal is less than 1.3. In this case, the operation of the auto-tuning circuit will be impossible, and the signal cannot be registered. Therefore, in order to increase the S/N ratio in the new developed compact magnetic system, we proposed a new technical solution to the problem posed to increase the volume of researched medium V_R and reduce the value of ΔB. At the same time, an optimum between the values of B_0, ΔB and V_R was achieved, which makes it possible to obtain S/N > 1.3 of the detected signal from different nuclei of the investigated medium with low sensitivity to the NMR method. In addition, this allows measuring the number of peaks ("wiggles") in the NMR signal of at least 5 (figure 2.b). This makes it possible to measure T_2 with an error not greater than 1.0% [1, 4–6, 9]. Reduction in the degree of inhomogeneity of the magnetic field was achieved by placing inserts 2 (figure 1) at the poles of the magnets in the form of steps (shims) of soft magnetic material (ARMCO iron). The inserts that we have developed are disks with a diameter of 92 mm, a thickness of 8 mm. From the edge of the disc, along its diameter, two strips were made in the form of steps with a recess to the center (the width of each step is 3 mm and a height is 2 mm). As a result, inside of each of the inserts 2 (figure 1) was formed a free space along the axis of action of the field B_0 in the form of hollows (of cylindrical shape with a depth of 4 mm and a diameter of 80 mm), where we have put carcasses with modulation coils. Therefore, to accommodate the manufactured inserts, the distance d_z between the poles of the magnets 1 was increased by only 10 mm compared to the original (d_z = 16 mm), the new value d_z = 26 mm. In the newly developed design of the magnetic system, the possibility of mechanical alignment of the mutual arrangement of the magnet poles, previously used in [4, 7, 9, 10], was also retained.

In addition, in the developed new design of a compact NMR spectrometer we can measure amplitude change U_S of the detected NMR signal by changing the frequency of NMR signal detection (autodyne detector 10) at a constant value of the magnetic field induction B_0. This allows us to register the absorption spectrum of the investigated medium using nuclei with magnetic moments. Figure 3–5 shows, as an example, the dependence of the amplitude U_S of the detected NMR signal on the frequency f_{nmr} at B_0 = 0.132 T and at temperature T = 293.3 K.

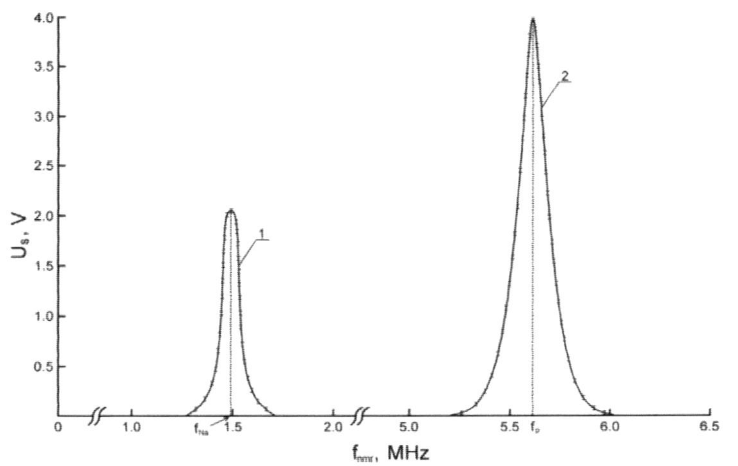

Figure 3 The absorption spectrum of an aqueous solution of sodium hydroxide. Graph 1 corresponds to the detected NMR signal from sodium nuclei, graph 2 — from protons

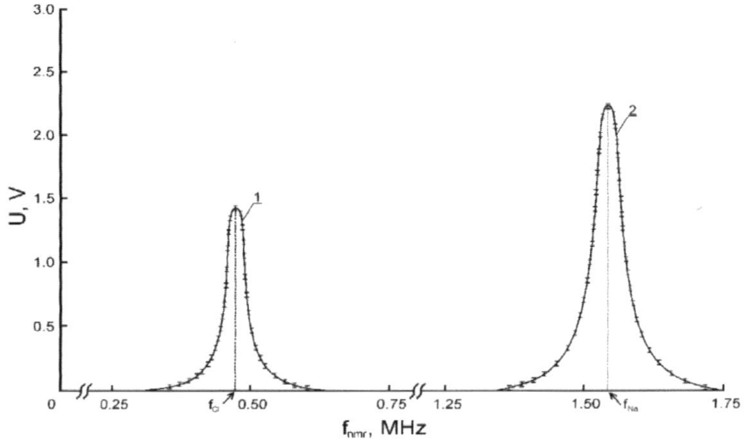

Figure 4 The absorption spectrum of sodium chloride. Graph 1 corresponds to the NMR absorption signal on chlorine nuclei, graph 2 — from sodium

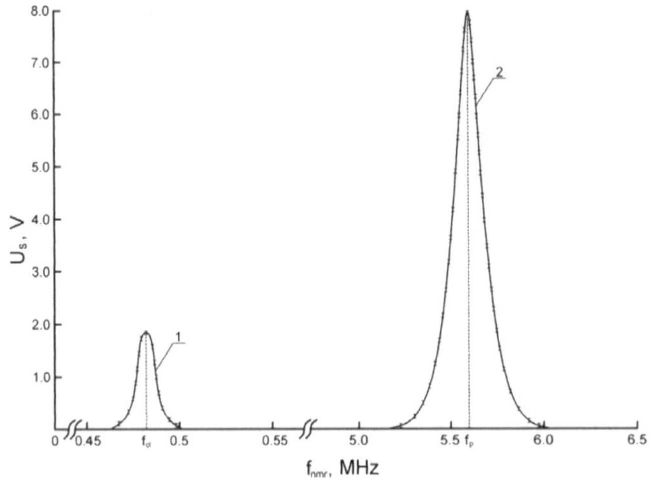

Figure 5 The absorption spectrum of sodium chloride. Graph 1 corresponds to the NMR absorption signal on chlorine nuclei, graph 2 — from protons

Using the dependence of U_S on f_{nmr} (absorption spectrum) obtained in figures 3–5 it is possible to determine which nuclei with magnetic moments are present in the investigated medium, their relative concentrations, acidity of the mediums (pH), etc. Similar dependences are obtained in stationary high-resolution NMR spectrometers while studying various mediums.

3. Conclusion

The obtained experimental results have shown that the small-size NMR spectrometer developed by us makes it possible to carry out express control over the state of a much larger number of media than the previously considered instruments in [4, 7, 9, 10].

For the first time, the possibility of registering NMR signals in a weak field from a large number of nuclei with magnetic moments in the investigated medium (except for carbon, nitrogen and sulfur nuclei) was realized for a frequency tuning f_{nmr} with its automatic tuning to resonance. This allows us to consider this device as a small-sized NMR spectrometer, since it differs from the NMR relaxometers considered in [4, 7, 9].

In the future, our researches will be aimed at increasing the sensitivity of the autodyne detector circuit. This is necessary to solve the problem of registering the NMR signal from the nuclei of chemical elements that are not very sensitive to the NMR method (for example, calcium, etc.) but cause significant changes in medium's T_1 and T_2 even at their small concentrations.

4. References

[1] Arkhipov V V 2012 *Instruments and Experimental Techniques* **55** 692–695
[2] Kashaev R S, Gazizov E G 2010 *Journal of Applied Spectroscopy* **77** 321–328
[3] Mussil V V, Pilipenko V V, Lemeshevskaya E T, Keremzhanov K D 2011 *Instruments and Experimental Techniques* **54** 397–399
[4] Karseev A Yu, Cheremiskina A V, Davydov V V, Velichko E N 2014 *J. of Physics: Conference Series* **541** 012006
[5] Alexandrov A S, Archipov R V, Ivanov A A, Gnezdilov O I, Gafurov M R, Skirda V D 2014 *Applied Magnetic Resonance* **45** 1275–1287
[6] Filippov A V, Artamonova M R, Rudakova M F, Gimatdinov R G, Skirda V D 2012 *Magnetic Resonance in Chemistry* **50** 114–119
[7] Karseev A Yu, Vologdin V A, Davydov V V 2015 *Journal of Physics: Conference Series* **643** 012108

IOP Conf. Series: Journal of Physics: Conf. Series **1038** (2018) 012088 doi:10.1088/1742-6596/1038/1/012088

[8] Zaporozhets O V, Shkurdoda V F, Peregudov O N, Zaporozhets V K 2010 *Instruments and Experimental Techniques* **53** 718–722
[9] Davydov V V, Velichko E N, Dudkin V I, Karseev A Yu 2015 *Instruments and Experimental Techniques* **58** 234–238
[10] Davydov V V, Dudkin V I, Karseev A Yu 2014 *Measurement Techniques* **57** 912
[11] Davydov V V, Myazin N S 2017 *Measurement Techniques* **60** 183–189
[12] Davydov V V, Myazin N S, Velichko E N 2017 *Technical Physics Letters* **43** 607–610

IOP Conf. Series: Journal of Physics: Conf. Series **1038** (2018) 012089 doi:10.1088/1742-6596/1038/1/012089

The universal optical method for condition control of flowing medium

N M Grebenikova[1], K J Smirnov[2, 4], V V Artemiev[3], V V Davydov[1] and S V Kruzhalov[1]

[1]Higher School of applied physics and space technologies, Peter the Great Saint Petersburg Polytechnic University, Saint Petersburg, 195251, Russia
[2] OJSC "NRI "Electron", Saint Petersburg, 194223, Russia
[3]Saint Petersburg National Research University of Information Technologies, Mechanics and Optics, Saint Petersburg, 197101, Russia
[4]The Bonch-Bruevich Saint Petersburg State University of Telecommunications, Saint Petersburg 193232, Russia

Abstract. In the article peculiarities of determination of dissolved substances concentration change in flowing liquid medium by measuring its refractive index with refractometer are considered. The measuring results of different media refractive indices are presented. The investigations of their refractive index dependencies on different factors (temperature, concentration, etc.) are carried out. The capabilities of refractometer applications for flowing medium state control at light-shadow borders displacement from refracted laser radiation are showed.

1. Introduction

Experimental and theoretical studies of liquid medium streams are one of the typical problems of physics [1-3]. The research results are important for improving the design of various meters of physical quantities of the liquid medium flow, developing new methods for measuring them and monitoring the state of the flowing medium [3-7].

The most difficult is to study the flow of aggressive and dangerous media (for example, benzene, heptane, concentrated sulfuric acid, etc.), as well as in cases when sterility is required (for example, biological solutions, medical suspensions, part of food products, etc.) [1, 3, 6]. Measurements in most of the considered flows of liquid media must be carried out by methods in which there is no direct contact between the measuring elements of the instrument and the medium under study. In some cases, small contact of the measuring elements with the flow of the medium is possible (for example, optical materials, etc.) [7-9]. But the following condition must be satisfied. With long contact of such elements with the medium under study, they should not alter its structure and chemical composition (sterility is ensured by installing such instruments on the pipeline for measurement).

Today there is a constant increase in competition in the food industry, which requires the manufacturer to produce high quality products. On the other hand, it is necessary to optimize the costs of its production. Reducing the cost of manufacturing products in large food production industries is provided mainly through the automation of the technological process, ensuring its continuity and control at all stages of production. Therefore, in contrast to other cases, additional requirements are imposed on instruments for monitoring the state of the flowing medium (production products, for example, juice, oil, wine, etc.) that are a part of the overall system for controlling the technological process of production

[7, 8]. In addition to reliable operation for a long period of time, real-time monitoring of the environment, simplicity of construction and relatively low cost (they are quite numerous - more than 10, the control of the medium should not change its taste qualities, etc. The latter is especially important, as experience shows in the production of liquid food products, control of its quality must be carried out at various stages of production for the rapid elimination of failures and violations of technology.

Most of the measuring instruments successfully used to solve problems of contactless monitoring of the flowing medium in various areas of industrial production and services could not satisfy the additional requirements. The exception was only nuclear magnetic spectrometers, but they are expensive, even with a wholesale purchase (more than 10 devices) [3, 10, 11]. Therefore, the search for solutions to this problem is quite urgent and in demand.

One of the possible solution is to use a refractometer. Using signals based on measuring the refractive index n_m of the flowing medium (for example, juice, etc.), a refractometer mounted directly on the pipeline can monitor and control the production process.

2. Refractometer for controlling liquid stream condition

The method of refractometry makes it possible to study stationary and flowing stream of liquid media by measuring their refractive indices n_m. Using refraction ratios, according to the results of measurements n_m, chemical compounds can be qualified, physicochemical parameters of various media can be determined, and structural and quantitative analysis can be carried out [8, 12, 13].

Today a large number of refractometers of various designs have been developed for measuring n_m in steady-state and in the flowing state of liquid media with high accuracy (the measurement error in most cases does not exceed 0.3%). But they are designed, especially flowing, to solve local problems. For example, one of the best is the CM-800 alpha-SW refractometer-salt meter designed to measure the NaCl salt concentration in water with an error of 0.1% in continuous mode. Refractometer models that allow measuring n_m in the range from 1.33000 to 1.55700 with an error of 0.05% (for example, PRM-100 alpha (ATAGO, Japan)) are quite expensive.

There are various chemical compounds (acids, alcohols, etc.), in which the values of n_m are less than 1.33 (at T = 293 K). For example, hydrochloric acid HCl has the values of n_m = 1.2540, methyl alcohol CH3OH — 1.3296, etc. With increasing T, the refractive index of the medium n_m decreases insignificantly (in general, less than 10% with increasing T up to 373 K). These media are used in technological processes in various industries (food and chemical, etc). Their condition must be controlled by contactless methods because, for example, HCl is an aggressive medium.

Using the described above refractometers does not allow solving this problem. These devices are mainly designed to control the production of various fertilizers (for example, aqueous solutions of urea, etc.), technical liquids (for example, aqueous solutions of ethylene glycol and propylene glycol that are used at airports to remove icing from the hull of the aircraft, etc.) etc.

Our research has shown that the expansion of the measurement range of n_m by a refractometer is primarily related to the parameters of its optical part. Therefore, we made an optical system based on the newly developed sapphire prism design. Its structural scheme is shown in figure 1.

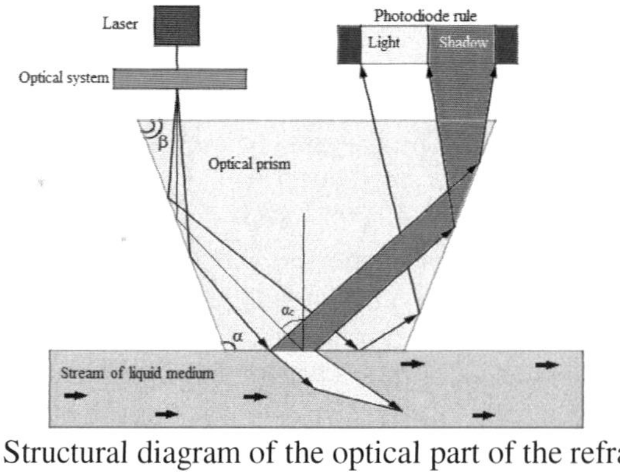

Figure 1. Structural diagram of the optical part of the refractometer

IOP Conf. Series: Journal of Physics: Conf. Series **1038** (2018) 012089 doi:10.1088/1742-6596/1038/1/012089

In the developed design of the optical part of the refractometer, the source of laser radiation is set up in such a way that its rays after passing through the optical system reach the boundary of the prism, which contacts the flowing medium at different angles. Part of the rays with angle of incidence greater than the critical angle α_c (figure 1) completely reflects from the inner surface of the prism and, emerging from it, forms the light part of the image on the photodiode ruler. The critical angle α_c is determined by the following relation:

$$\alpha_c = \arcsin (n_m/n_p) \tag{1}$$

where np is the refractive index of the material from which the prism is made; nm is the refractive index of the flowing medium.

The remaining rays, the angle of incidence of which is less than the critical α_c, are partially refracted and pass into the solution, partially reflected and form the dark part of the image on the photodiode ruler (figure 1). In a small range of angles from α_c, an interface between light and shadow is formed. The contrast of this interface between light and shadow depends on the relation between the refractive indices of the material (1) from which the prism and the investigated medium are made, as well as the parameters of the laser radiation (wavelength λ, divergence angle θ and spatial coherence length L_k).

On the basis of our studies, the following has been established. The coefficient of reflection of laser radiation from the prism-medium boundary to the value of the critical angle changes monotonically (the Fresnel formula). In a small range of incidence angles, before reaching the value of α_c, the reflection coefficient increases sharply (with an increase of one order of magnitude higher than before). For laser radiation rays incident on the prism-medium interface at angles $\alpha \geq \alpha_c$, the reflection coefficient has a value of 1. If the center of the beam pattern of a laser falls on the interface between two media at an angle α_c, then the light-shadow boundary contrast ratio is maximal. The presence of a peak in the intensity of the detected laser radiation along the length of the photodiode line (photodetector) is due to two factors. The distribution of intensity in the rays of laser radiation forms a cone (with a maximum in the center). For a coefficient of 1, this distribution does not change. A sharp change in the reflection coefficient before reaching the critical angle of incidence appears. This change makes the intensity maximum distinct.

In case of a drop in the center of the radiation pattern at a different angle, the light-shadow boundary widens (the contrast decreases). This can possibly lead to an error in measurements. That is why the refractometer needs a preliminary calibration before measurements.

When the composition of the medium changes, its refractive index n_m changes and the position of the light-shadow boundary recorded by the photodiode ruler is shifted. We propose to monitor the state of the flowing liquid using this method, in contrast to previously used in refractometers based on the measured value of the refractive index n_m.

Another distinguishing feature of the developed refractometer from other models is the use of a new design of a prism made of leucosapphire in its optical system which geometric configuration and dimensions are shown in figure 2.

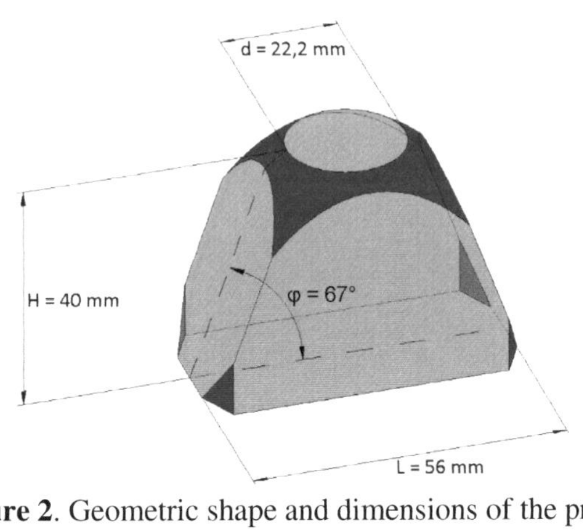

Figure 2. Geometric shape and dimensions of the prism

The research showed that the calculated and implemented geometry of the prism provides the possibility of the refractometer to monitor the state of the flowing medium in the range of n_m from 1.2246 to 1.6120. This result was achieved, because this design of the prism allowed us to use laser radiation with a plane angle of the radiation pattern of about 22.6° (after the optical system) and to move the position of the semiconductor laser along the base of the prism within 8 mm. To cover this working range, we used a photodiode line TSL1406R with a photo-sensitive layer length of 38.77 mm. This ruler is placed at a small distance from the base of the prism (4-5 mm) and can move freely along it. In the design of the model TSL1406R, a plate with an enlightenment is placed in front of the photosensitive layer. On the inner side of this plate a special layer is applied, which makes unimportant the influence of radiation reflection effects inside the photodiode ruler on the measurement result. Between the prism and the photodiode ruler there are no focusing optical elements. Parasitic reflections can occur only between the outer edges of the protective plastic of the photodiode ruler and the prism.

Let I_0 be the maximum intensity of the radiation incident on the plate. The reflection coefficient of the plate is less than 3%. The reflection coefficient of sapphire for laser radiation incident from air is not more than 7%. The value of the parasitic intensity received secondarily on the plate is less than $0.0021 \cdot I_0$. The experiments have shown that with a sharp change in the reflection coefficient at the prism-current interface, the phenomenon considered does not significantly affect the contrast value of the light-shadow boundary.

Besides, the expansion of the n_m measurement range was ensured by using a cone shaped gasket in the measuring probe of the refractometer with the prism (figure 2). This gasket makes the influence of the vignetting effect of the laser beam on the faces not significant as compared to the previously used in the refractometer prism of trapezoidal geometry with an annular gasket. The conical gasket also provided greater sealing reliability, which is very important in a pipeline with fast fluid flow of high pressure. To process information from the photodiode line, we have developed a program that allows us to determine the refractive index values n_m, the relative concentration N_m, as well as the temperature coefficient dn_m/dT after the preliminary calibration of the instrument.

The calibration of the refractive index scale n_m was carried out using the normalized output signal A from the photodiode array:

$$A = \frac{\sum_{i=1}^{256} A_i}{A_{max}} \tag{2}$$

where A_{max} is the maximum value of the illumination signal from the photodiode ruler, A_i is the amplitude of the signal from the photodiode cell element, and i is the element number.

To test the performance of the refractometer we used reference aqueous solutions of ethylene glycol and 1,2 propylene glycol with different relative concentrations of N from 0 to 100%.

3. The results of experimental studies of the flowing liquids

Figure 3 shows the results of measuring different refractive indices $n_m(T)$ for liquid aqueous solutions of ethylene glycol and 1.2 propylene glycol with different relative (mass) concentration N. The solid line in figure 3 shows a linear interpolation of these dependencies.

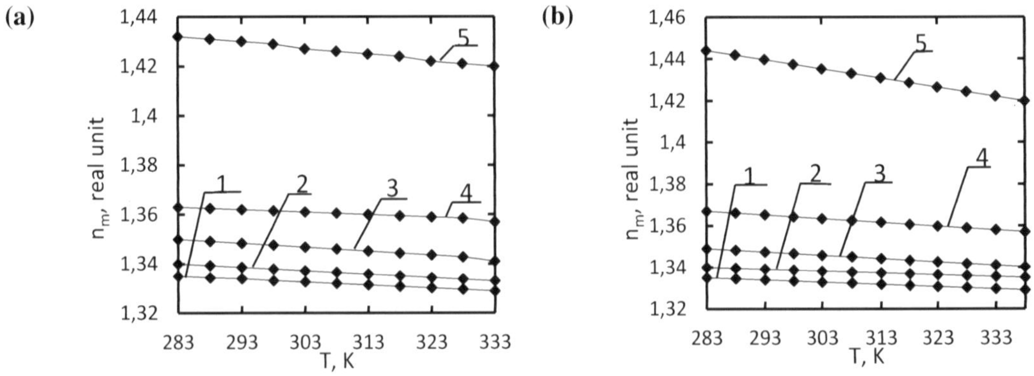

Figure 3 (a,b). Temperature dependences of the refractive index n_m of aqueous solutions of ethylene glycol **(a)** and 1,2 propylene glycol **(b).** For graphics 1, 2, 3, 4 and 5, the following concentration of N in % corresponds to **(a)**: 0.00; 5.22; 14.85; 31.40; 100.00, for **(b)**: 0.00; 4.59; 13.41; 27.59; 100.00.

Analysis of the figure 3 dependencies shows that, at T of about 333 K, the dependency $n_m(T)$ is linear and coincides with the linear interpolation, which was calculated for each graph. Also, for each graph (figure 3), we calculated the mean square convergence Δn_m of interpolation dependence calculations and experiments. Table 1 presents the results of our calculations.

Table 1. Linear interpolation formulas for the variation of n_m from the temperature T for different concentrations of N aqueous solutions of ethylene glycol and 1.2 propylene glycol and the values of Δn_m.

Aqueous solution of ethylene glycol			An aqueous solution of 1.2% propylene glycol		
N, %	n_m, rel.un.	Δn_m, rel.un	N, %	n_m, rel.un.	Δn_m, rel.un
0.00	1.3357 – 0.00010339·T	0.00155	0.00	1.3357 – 0.00010339·T	0.00155
5.22	1.3407 – 0.00013007·T	0.00087	4.59	1.3415 – 0.00012569·T	0.00088
14.85	1.3506 – 0.00015162·T	0.00085	13.40	1.3509 – 0.00014340·T	0.00083
31.40	1.3683 – 0.00019586·T	0.00081	27.59	1.3682 – 0.00018155·T	0.00078
100.00	1.4370 – 0.00027520·T	0.00061	100.00	1.4385 – 0.00030048·T	0.00065

We have established that the error of linear interpolation for the concentration $N \leq 5\%$ is about 0.00155. For values $N > 5\%$, the values of Δnm do not exceed 0.00088. The obtained results coincided within the measurement error of previously obtained results in [14-16] using the "Expert pro" refractometer (the medium under study is in a stationary state).

Experimental studies have shown that in the developed refractometer with the use of the new prism design, an exceptional feature that is inherent in some models of optical measuring instruments has been implemented. It is connected with the fact that the optical scheme of the refractometer is based on the use of reflection and light transmission only within the working prism. Therefore, factors such as the transparency of the flowing medium, the presence of light scattering insoluble elements in it (for example, pulp in juice, etc.), gas bubbles, as well as the flow rate do not affect the measurement results, unlike other methods (except NMR), used to monitor the stream of liquid medium. The light-shadow boundary shifts only when n_m changes. This allows us to accurately determine the presence of changes in the composition of the flowing medium, for example, on the concentration of N or T. Figure 4, as an example of the refractometer operation, studies of flowing aqueous solutions of various media at a temperature T = 293.1 K.

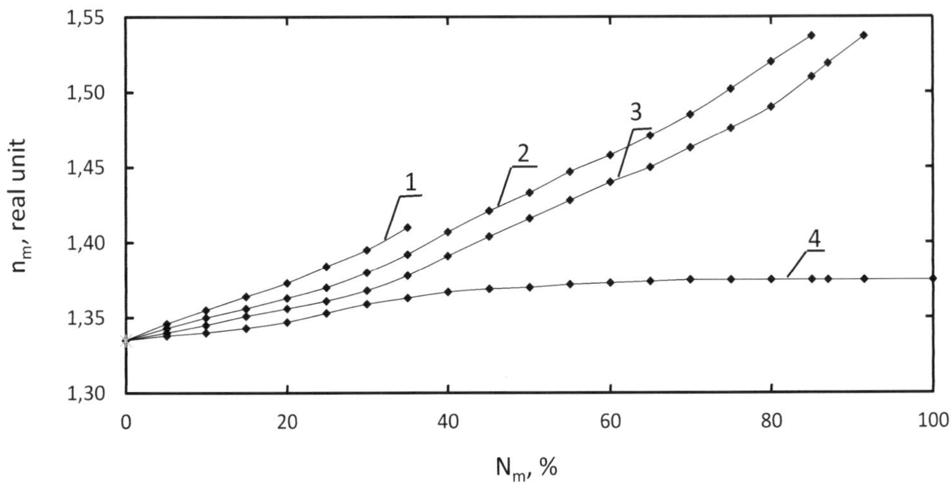

Figure 4. Dependence of the refractive index of aqueous solutions of some substances on concentration: 1-gelatin, 2 - sodium glutamate, 3 - sucrose, 4 - ethyl alcohol

An analysis of the obtained experimental results showed that the variation of n_m from N for most media is nonlinear, in contrast to the dependence of n_m (T). At low concentrations of N (not more than 10%) and T no more than 293 K for some media, the device allows measuring small changes in n_m on N. This allows using it to monitor the state of a large number of liquid media during their flow in the pipeline.

Figure 5 shows, for example, the experimental dependencies of the change in the relative density of the aqueous solution in Brix units (Brix is the most common calibration scale for refractometers, expresses the concentration of a solution of chemically pure sucrose in distilled water in mass percent) with the addition of sugar in various concentrations from temperature T. As the reference point for the change in the relative concentration, the value of $\Delta N = 0$ at $T = 293$.

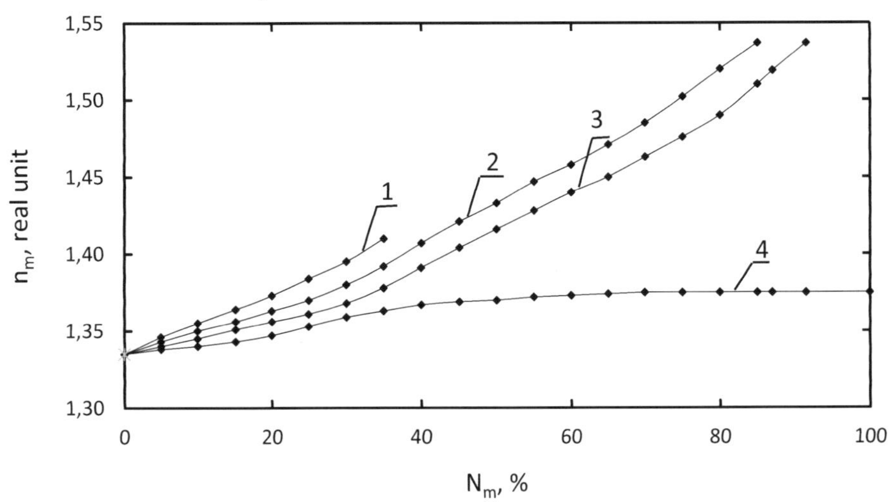

Figure 5. Temperature corrections for the concentration of aqueous solutions of sucrose: 1-10%, 2 -30%, 3 -60%

Experimental studies have established that the most important factor that can make errors in determining the state of the medium is T. To account the effect of change in T of the investigated flowing medium on the results of monitoring its state in the flowing liquid, a temperature sensor is installed in the pipeline in front of the prism. Based on the readings of this sensor, a correction is taken into account in the displacement of the light-shadow boundary by means of the gradient tables. The discrepancy between the amount of boundary displacement and the data of the table means that there are additional changes in the state of the medium that are not related to changes in the temperature T. Similarly, the temperature factor must be taken into account when controlling the state of the medium in terms of nm [8, 14-16].

4. Conclusion

The research showed that the use of the optical system developed by us with a new design of a sapphire prism makes it possible to monitor the state of the flowing liquid medium with n_m from 1.2246 to 1.6120 with an error of 0.1%. We have established that in order to improve the accuracy of monitoring the state of the flowing medium, it is necessary for each variety to select the appropriate wavelength and laser radiation power at which the light-shadow border contrast will be maximum. The obtained result allows using the developed refractometer to control the state of the flowing medium of bromobenzene n_m = 1.5601, which is used to prepare reagents for various chemical reactions. Also this refractometer can be used in the medical industry, e.g. to control the state of the phenylhydrazine (n_m = 1.6105) in the manufacture of medicines (e.g. antipyrine, amidopyrine). Today other types of instruments are used for their control.

Moreover, it has been discovered that to ensure a stable operation of the refractometer to monitor the state of the flowing of liquid medium, regardless of its flow rate, it must be installed on the vertical section of the main line. Sometimes the pipeline is filled with an incompletely flowing liquid in the technological process of production, especially food liquid products. (In the technological process of production, especially food liquid products, there are quite often situations where the pipeline is filled with an incompletely flowing liquid.) The contact of the upper face of the prism with the measured medium will be absent. This circumstance will not allow the measurement of various parameters of the flowing medium and the control of its state with the required accuracy.

The restriction in the application of the refractometer is also the diameter of the pipeline through which the liquid medium flows. Because the flow at the borders of the pipeline is much less than in the center, and if a refractometer is placed on the vertical section of the pipeline, the prism should be the upper edge partially flooded into the steam of liquid mediun. The smaller the diameter of the pipeline, the smaller the geometric dimensions of the prism, the smaller the detectable dynamic range of the light-shadow boundary and the measuring range n_m.

5. References

[1] Rozanov N N 2012 *Optics and Spectroscopy* **113** 613
[2] Popovac M and Hanjalic K 2007 *Flow, Turbulence and Combustion* 78 177-186
[3] Davydov V V, Dudkin V I and Vologdin V A 2015 *Journal of Applied Spectroscopy* **82** 936
[4] Davydov V V, Dudkin V I and Karseev A Yu 2014 *Optical Memory & Neural Networks (Information Optics)* **23** 170
[5] Davydov V V, Dudkin V I and Karseev A Yu 2014 *Optical Memory & Neural Networks (Information Optics)* **23** 259
[6] Pavlov I N, Rinkevicius B S and Tolkachev A V 2011 *Measurement Techniques* **53** 1130
[7] Vologdin V A, Davydov V V and Velichko E N 2016 *Journal of Physics: Conference Series* **741** 012095
[8] Shur V L, Naidenov A S and Lukin A Ya 2006 *Measurement Techniques* **49** 815
[9] Belov N P, Lapshov S N, Maiorov E.E, Sherstobitova A S and Yaskov A D 2012 *Journal of Applied Spectroscopy* **79** 514
[10] Karseev A Yu, Vologdin V A and Davydov VV 2015 *Journal of Physics: Conference Series* **643** 012108
[11] Davydov V V, Dudkin V I and Karseev A Yu 2014 *Technical Physics Letters* **40** 895
[12] Akmarov K A, Belov N P, Smirnov Yu Yu, Sherstobitova A S, Scherbakova Yu E and Yaskov A D 2013 *Scientific and Technical Journal of Information Technologies, Mechanics and Optics* **87** 39
[13] Fekete Z A, Hoffmannz E A, Kortvelyesi T and Penke B 2007 *Molecular Physics* **105** 19

[14] Tsierkezos N G and Molinou I E 1998 *Journal Chem. Eng. Data* **43** 989
[15] Zhou Y, Li S, Zhai Q, Jiang Y and Hu M 2010 *Journal Chem. Eng. Data* **55** 1289
[16] Belov N P, Prokopenko V T, Sherstobitova A S and Yaskov A D 2012 *Scientific and Technical Journal of Information Technologies, Mechanics and Optics* **78** 1

IOP Conf. Series: Journal of Physics: Conf. Series **1038** (2018) 012090 doi:10.1088/1742-6596/1038/1/012090

Raman spectroscopy method for the evaluation of bone bioimplants made using the "Lyoplast" technology from cadaveric and in vivo resected bone tissue

P E Timchenko[1], E V Timchenko[1], L T Volova[2], D A Dolgushkin[2], V V Boltovskaya[1], O O Frolov[1]

[1]Samara National Research University, 34, Moskovskoye shosse, Samara, 443086, Russia
[2]Samara State Medical University, 89, Chapayevskaya St., Samara, 443099, Russia

E-mail: laser-optics.timchenko@mail.ru

Abstract. The results of application of the Raman spectroscopy method for estimating alternative sources for the production of bone spongy implants using the "Lioplast" technology, namely, the femoral head heads resected in the operation of hip replacement surgery, are presented. It is shown that Raman spectroscopy can be used to assess the component composition of the bone implants surface during their processing. Comparing different sources of sponge bone production before and after demineralization, no significant differences were found, but there are differences in the ratio of the Raman peaks intensities at wave numbers 1555 cm^{-1} and 1665 cm^{-1} corresponding to amide II and amide I, and also in the intensity of Raman peaks at the wave numbers 429 cm^{-1} (PO_4^{3-} (ν_2) (PO symmetric vibrational)), 1068 cm^{-1} (CO_3^{2-} (ν_1) B-type substitution (C-O planar valence)), 850 cm^{-1} (benzene ring proline), 1000 cm^{-1}, (aromatic phenylalanine ring).

1. Introduction

Recently, reconstructive and regenerative medicine is actively developing. In traumatology, orthopedics, dentistry, oncology and purulent surgery the search for approaches to the treatment of the supporting and integumentary tissues pathology of the human body assumes not only the means of its elimination, but also the preservation of the original form, structure and functions destroyed by traumas and bone diseases [1, 2]. Providing full regeneration of bone tissue in the defective bone sites area, in spite of the accumulated knowledge in this matter – is one of the most acute problems of modern medicine. It can be solved by creating optimal conditions for regenerative processes in the zones of its resorption. One of the ways is the use of osteoplastic materials [3, 4]. Among them, allogeneic implants from human tissues are the optimal materials for the reconstruction of musculoskeletal system injuries. When applied, unlike auto– and xenoplasty and the use of synthetic drugs, the homeostasis and metabolism of connective tissues and the functions of the life–support systems of the recipient are not impaired [5]. Allogenic materials after their special treatment almost completely lose their antigenicity and when placed in the body do not have a negative impact on it. They play the role of a matrix, conductor, gradually dissolve completely, and in their place a new bone tissue is formed [6].

With the increase in the number of operations performed in the region of hip arthroplasty, it became possible to use resectable intraoperatively femoral heads to produce new bone spongy bioimplants using the "Lioplast" technology.

To compare the quality of bioimplants obtained from cadavers and resected intraoperatively spongy bones treated with the Lioplast technology, it is advisable to use physical methods of investigation.

Raman spectroscopy has certain advantages, allows real-time non-invasive quantitative and qualitative analysis of the composition of biological objects, and provides information on the molecular structure with high spatial resolution. For example, in [7, 8, 9], the Raman spectroscopy method was used to evaluate the content of key components of human bone tissue – hydroxyapatite and amides of the I–III type. Comparing the amplitudes of the Raman peaks at wavenumbers in the range from 400 to 1800 cm^{-1}, the authors revealed significant signal changes for samples with different degrees of demineralization.

The aim of this study: a comparative assessment of the component composition of mineralized and demineralized bone bioimplants surfaces manufactured using the "Lioplast" technology from cadaveric and in vivo resected spongy bone.

2. Materials and methods of research

Materials of the study were samples of a spongy bone bioimplants in the form of a cube 5 * 5 * 5 mm in size, manufactured using the "Lioplast" ® technology (TU–9398–001–01963143–2004). The samples were divided into four groups. The first group consisted of demineralized samples obtained from cadaveric tissue, the second group – mineralized samples from the same source. The third group formed demineralized samples made from the femoral head heads obtained intraoperatively in hip replacement, the fourth group – mineralized samples from the same source. In the process of bone bioimplants production, at the first stage, they performed their low-frequency ultrasound treatment (24–40 kHz for 2-3 minutes), as a result of which the samples were degreased, all the elements of the bone marrow were removed from the interballal spaces of the bone. To obtain demineralized samples, bone tissue was placed in a hydrochloric acid solution of 2.4 normality. At the final stage of the treatment, lyophilization of the bones was carried out by means of the ALPHA 2–4 LSC, which consisted in freeze drying of the samples. As a basic method for studying bioimplants, the Raman spectroscopy method was used, realized using an experimental stand (Figure 1), described in detail in the work [10]. The spectra were taken for each sample on each side of the cube at three different points.

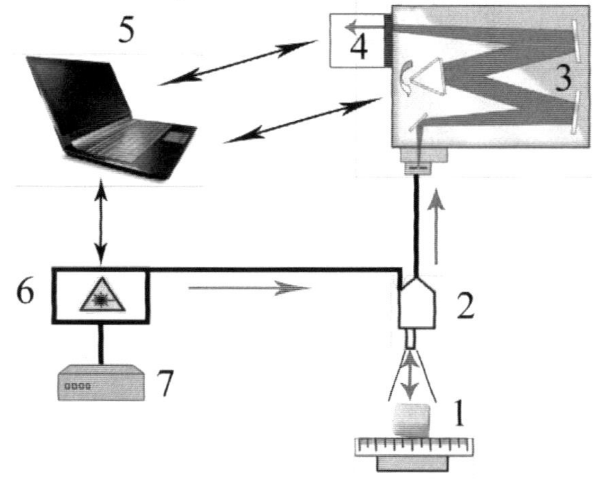

Figure 1 – Experimental stand:
1 – object of the study;
2 – Raman probe RPB785;
3 – Andor Shamrock sr–303i spectrometer;
4 – built-in cooled camera Andor DV420A–OE;
5 – computer;
6 – laser module LuxxMaster LML–785.0RB–04;
7 – power supply of the laser module.

The Raman probe 2 focused laser radiation at object 1 (spot diameter 0.2 mm at a distance of 7.5 mm from the exit window of the tube), collected the radiation scattered by the object and focused the radiation filtered out in the spectral range 790–1200 nm to the input end of the receiving optical fiber, through which it was transported to input slit of the Shamrock sr-303i 3 spectrometer with integrated cooling chamber 4. To reduce the noise level, the camera array 4 was frozen to a temperature of -60°C, providing a spectral resolution of 0.15 nm (~ 1 cm^{-1}). The error of the method did not exceed 4%. Spectrum processing was carried out in the software environment Wolfram Mathematica 10 and consisted in removing the noise by a smoothing median filter of 7 points. Then, on the selected interval 400-2200 cm^{-1}, using the iterative algorithm [11], an approximating line (a polynomial of the fifth degree) of the auto-fluorescent component was determined, and then this component was subtracted, obtaining a separate Raman spectrum.

3. Results and discussion

To study the Raman spectra of bones the intensities at the wavenumbers of 959 cm^{-1} (vibrations of the phosphate ion PO_4^{3-} (ν_1) (P–O symmetric valence)), 1068 cm^{-1} (carbonate ion CO_3^{2-} (ν_1) B-type substitution (C–O in-plane stretch)), 1555 cm^{-1} (amide II (C–N–H valence) (deformation vibration N–H)), 1665 cm^{-1} (amide I (C=O valence) (peptide bond)), 850 cm^{-1} (a benzene ring of proline), 870 cm^{-1} (a benzene ring of hydroxyproline), 1000 cm^{-1} (aromatic ring breathing of phenylalanine (protein assignment)), 1030 cm^{-1} (phenylalanine (CH_2CH_3 bending modes) (collagen assignment)), 917 cm^{-1} (ribose (RNA)) and 1298 cm^{-1} (fatty acids (CH_2, CH_3 twisting and wagging) (lipids)) were chosen [7, 9, 12].

Figure 2 shows the averaged Raman spectra for mineralized and demineralized samples obtained from different sources.

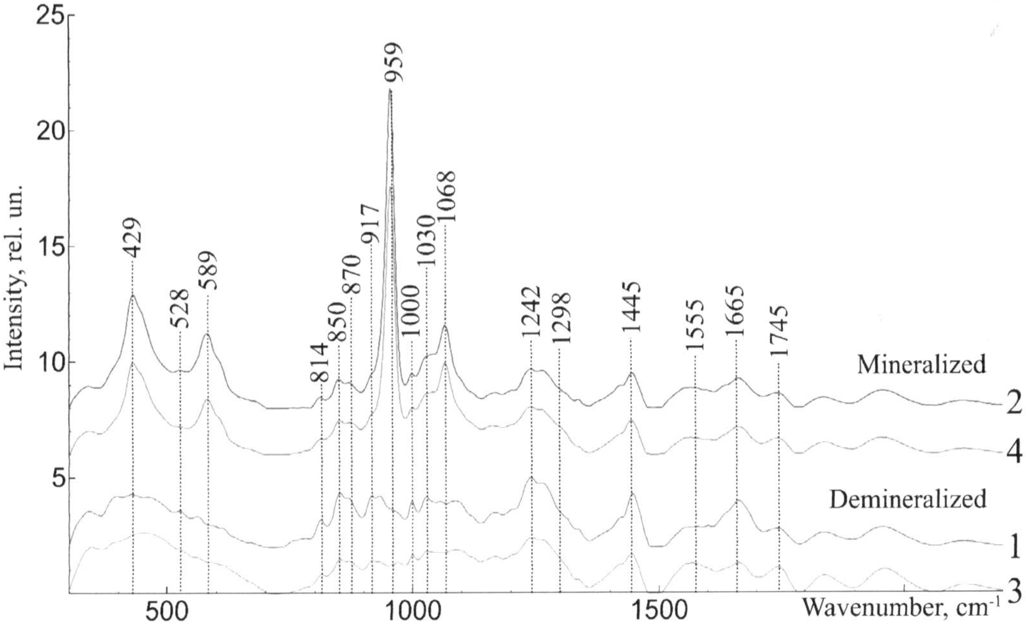

Figure 2 - Average Raman spectra for mineralized and demineralized samples of bioimplants obtained from cadaveric (1,2) and intravital (3,4) bone tissue

The interpretation of Raman spectra for samples of bone bioimplants is presented in Table 1.

Table 1. Interpretation of Raman spectra for samples of bone bioimplants

Raman shift, (cm^{-1})	Assignments
429	PO_4^{3-} (ν_2) (P–O stretch) [13]
528	Lysozyme (oscillation of the disulfide S–S in protein) [13]
589	PO_4^{3-} (ν_4) (P–O bending) [13]
814	Phosphodiesterase (DNA/RNA) (C'$_5$–O–P–O–C'$_3$) [12]

850	Benzene ring proline [14]
870	Benzene ring of hydroxyproline [14]
917	Ribose (RNA) [12]
959	PO_4^{3-} (ν_1) (P-O symmetric stretch) [13]
1000	Aromatic ring breathing of phenylalanine (protein assignment) [12]
1030	Phenylalanine (CH_2CH_3 bending modes (collagen assignment)) [12]
1044	PO_4^{3-} (ν_3) (P–O asymmetric stretch) [14]
1062	GAG's (O–SO_3^- symmetric stretch)
1068	CO_3^{2-} (ν_1) B-type substitution (C–O in-plane stretch) [13]
1177	Tyrosine (DNA/RNA) [12]
1230–1289	Amide III (C–N–H stretch) [13]
1298	Fatty acids (CH_2, CH_3 twisting and wagging (lipids))
1416	Deoxyribose (DNA/RNA) [12]
1445–1465	Proteine, CH_2 twisting [13]
1555–1565	Amide II (C–N–H stretch) [12]
1655-–1675	Amide I (C=O stretch) [13]
1745	Phospholipids (C=O stretch) [12]

As can be seen from Figure 2, significant differences in the intensities of the wavenumber peaks when comparing bioimplants, for the production of which bone tissue from different sources was used, was not found. For mineralized bioimplants of both groups, was found an increase in the intensities at the wavenumbers of 959 cm^{-1}, 1068 cm^{-1}, 429 and 589 cm^{-1}, corresponding to PO_4^{3-} (ν_1) (P–O symmetric valence), CO_3^{2-} (ν_1) substitution B-type, PO_4^{3-} (ν_2) (PO symmetric vibrational), PO_4^{3-} (ν_4) (PO deformation.) After demineralization a decrease in the intensity was observed at these wave numbers both in samples obtained from cadaveric and in vivo. For mineralized specimens the correlation coefficient between both groups was in the range of $R_m = \{0.924; 0.987\}$. The calculated correlation coefficient for samples of demineralized bioimplants of both groups was $R_d = \{0.834, 0.963\}$.

It should be noted that in all samples of bone bioimplants manufactured using the Lioplast technology, no high intensity was observed at wave numbers – 1298 cm^{-1}, 1745 cm^{-1}, corresponding to fatty acids (CH_2, CH_3 twisting and wagging (lipids)) and C=O valence) phospholipids. The use of low-frequency ultrasound at the stages of spongy bone treatment made it possible to maximize the removal of lipids and bone marrow elements, which was confirmed in the spectral characteristics of the bioimplant samples surfaces under study. Thus, according to the results of the Raman spectra analysis, the source of bone tissue production (cadaveric or intravital spongy bone) did not affect the component composition of the implants surface created using the Lioplast technology.

However, for a more complete quantitative assessment of the component composition of the bone bioimplants surface, we introduced optical coefficients. A relatively constant component in the samples of bone tissue studied was amide I [15], corresponding to a wave number of 1665 cm^{-1}, so it was used as a denominator in the introduced optical coefficients (k):

$$k = \frac{I_i}{I_{1665}},$$

where I_i – the intensity values at the wave numbers of the analyzed components.

To analyze the organomineral composition of bone tissue and biomaterials based on hydroxyapatite, the following ratios can be used:

The ratio I_{959}/I_{1665} determines the degree of leaching of mineral components in the process of demineralization [16].

The optical coefficient I_{1068}/I_{1665} (Carbonate / Organic Matrix ratio) is used to estimate the degree of carbonate mineralization of the biomaterial [16].

Both coefficients show the dependence of the mineral content in relation to the organic component.

I_{1555}/I_{1665} coefficient (ratio Amide II / Amide I) characterizes the change in Amide I with respect to Amide II [17].

I_{1068}/I_{1665} and I_{1068}/I_{959} coefficients (ratio of Carbonate / Amide I and Carbonate / Phosphate) indicate the degree of carbonate inclusion in the crystal lattice of hydroxyapatite [17].

The ratio I_{850}/I_{870} (proline / hydroxyproline) is an indicator of damage to the structure of the collagen helix with a decrease in the relative concentration of proline [18].

A two-dimensional analysis of the introduced optical coefficients was carried out. Figure 3 shows two-dimensional diagrams representing the quantitative content of the main components of mineralized and demineralized bone implants made from cadaveric and in vivo resected spongy bones.

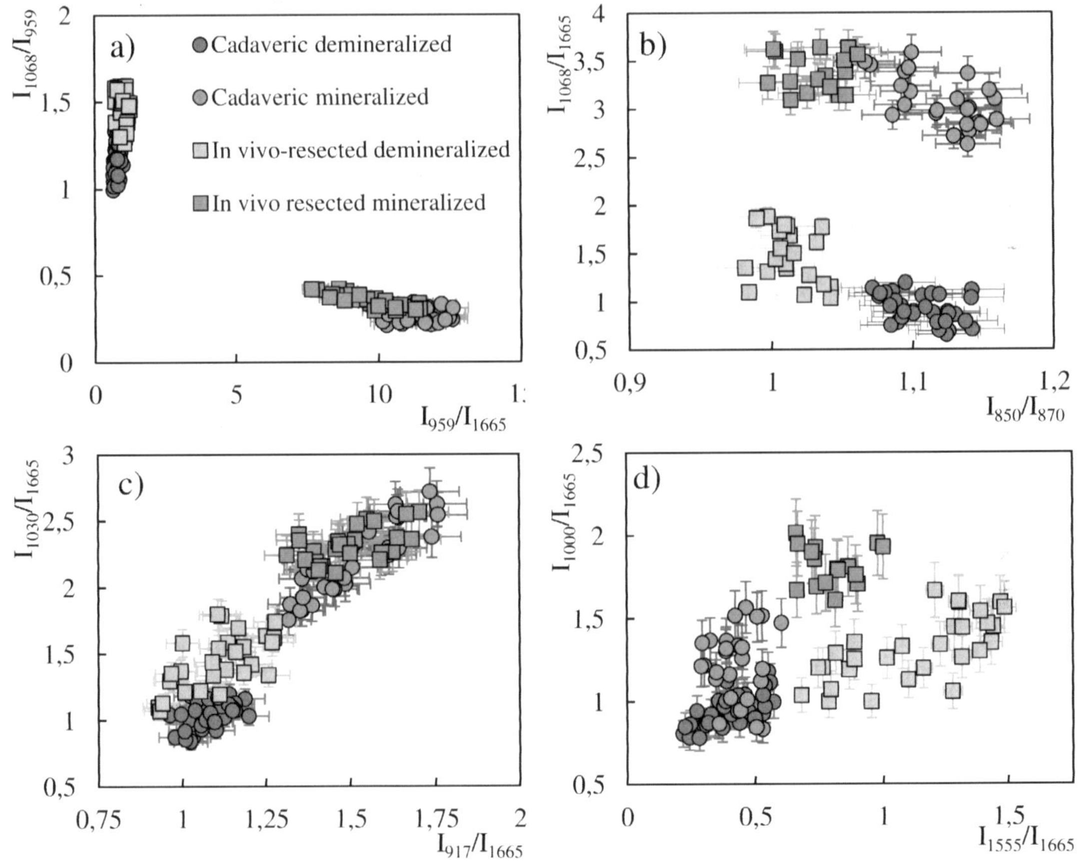

Figure 3 - Two-dimensional diagrams of the introduced optical coefficients

It can be seen from the analysis of figure 3a that for demineralized samples a smaller value of the coefficient I_{959}/I_{1665} – "mineral / organic matrix" ($0.8 < I_{959}/I_{1665} < 1.3$) was characteristic, compared to mineralized samples ($7 < I_{959}/I_{1665} < 13$). In the process of demineralization using the "Lioplast" technology, the relative concentration of mineral components was sharply reduced. At the same time, the values of the I_{959}/I_{1665} coefficient for demineralized tissue samples were close, despite the difference in the sources of spongy bone sampling for the manufacture of implants.

The I_{1068}/I_{1665} ratio (Carbonate / Organic Matrix ratio) was used to assess the degree of mineralization of biomaterials. It can be seen from figure 3b that for the demineralized samples the coefficient I_{1068}/I_{1665} varies within the limits of $0.5 < I_{1068}/I_{1665} < 2$) and for mineralized samples in the range $2.5 < I_{1068}/I_{1665} < 4$.

Also it can be seen from Figures 3a, 3b that higher values of I_{1068}/I_{959} and I_{1068}/I_{1665} coefficients are characteristic for samples obtained from intravital resected bone tissue than for cadaveric bone, and, consequently, a higher content of B–type carbonate carbonate CO_3^{2-} (ν_1), which is more resistant to the effect of acid during demineralization [17], which manifests itself in an increase in the difference between materials in the process of demineralization for the coefficients I_{1068}/I_{959} and I_{1068}/I_{1665}. The ratio I_{850}/I_{870} shows the difference in the ratio of the peaks of proline and hydroxyproline benzene rings due to deformations in the structure of collagen [18], and allows differentiating cadavers ($1.07 < I_{850}/I_{870} <1.16$) and samples from intravital resected bone tissue ($0.98 < I_{850}/I_{870} < 1.05$). For demineralized and mineralized samples, the values of the ratio I_{850}/I_{870} were in the same intervals.

The process of bone implants demineralization using the "Lioplast" technology made it possible to lower the level of the content of such a bone component as ribose (RNA), which reflected the change in the optical coefficient I_{917}/I_{1665} from $1.3 < I_{917}/I_{1665} < 1.75$ to $0.9 < I_{917}/I_{1665} < 1.25$ after demineralization. The values of the coefficient I_{917}/I_{1665} did not depend on the source of bone tissue production. Reducing the amount of RNA during processing of biomaterials using the "Lioplast" technology reduced their immunogenicity for the recipient organism.

The values of the optical coefficient I_{1030}/I_{1665} vary from the interval $0.8 < I_{1030}/I_{1665} <1.6$ for demineralized and to the interval $1.7 < I_{1030}/I_{1665} < 2.6$ for mineralized samples (figure 3c), which reflects a decrease in the relative concentration of phenylalanine with respect to secondary protein structures (amide I) in the process of demineralization, and for the samples obtained from intravital resected bone tissue this process proceeds more slowly.

Analysis of diagram 3d shows that the value of the optical coefficient I_{1555}/I_{1665} does not significantly depend on the degree of demineralization of the samples, but there is a separation of the regions for samples obtained from the femoral head in the intraoperative ($0.66 < I_{1555}/I_{1665} < 1.5$) and cadaveric $0.2 < I_{1555}/I_{1665} < 0.6$) of bone implants, which may be due to the different reaction of groups of samples to heat treatment.

4. Conclusions

A comparative spectral evaluation of the surfaces component composition of mineralized and demineralized bone bioimplants manufactured using the "Lioplast" technology from cadaveric tissue and femoral bone heads resected during the operation of hip arthroplasty was performed.

Comparing different sources of sponge bone production before and after demineralization, no significant differences were found, but there are differences in the ratio of the Raman peaks intensities at wave numbers 1555 cm^{-1} and 1665 cm^{-1} corresponding to amide II and amide I, and also in the intensity of Raman peaks at the wave numbers 429 cm^{-1} (PO_4^{3-} (ν_2) (P–O symmetrical vibrational)), 1068 cm^{-1} (CO_3^{2-} (ν_1) B-type substitution (C–O planar valence), 850 cm^{-1} (benzene ring of proline), 1000 cm^{-1} (aromatic ring of phenylalanine).

Optical coefficients were introduced and a two-dimensional analysis was carried out, which showed that the values of I_{1068}/I_{959} and I_{1068}/I_{1665} are higher for intravital resected bone tissue than for cadaveric bone, and therefore a higher content of B-type carbonate carbonate CO_3^{2-} (ν_1), which is more resistant to the effects of acid during demineralization.

The introduced optical coefficients made it possible to confirm that the content of the main components of bioimplants necessary for the realization of their osteoinductive and osteoconductive properties turned out to be similar, both in mineralized and demineralized samples obtained from the femoral head intraoperatively, and from samples made from cadaveric material.

5. Acknowledgments

The reported study was funded by RFBR according to the research project № 18-32-00004.

References

[1] Popkov A V 2014 *Biocompatible implants in traumatology and orthopedics (review of the literature)* (Geniy ortopedii) №3

[2] Berchenko G N 2008 *Bone grafts in traumatology and orthopedics* (Biomaterials) №9 p. 4–5

[3] Muslimov S A 2000 *Morphological Aspects of Regenerative Surgery* (Ufa: Bashkortostan) p. 168

[4] Lekishvili M V 2005 *Technologies for manufacturing bone plastic material for use in reconstructive surgery* (dissertation of Doctor of Medical Sciences: 14.00.41, 14.00.22 Moscow) p. 47

[5] Saveliev V I, Kornilov N V, Ivankin D E, Linnik S A 2001 *Allotransplantation of formalinized bone tissue in traumatology and orthopedics* (Saint Petersburg, MORSARAB) p. 208

[6] Ladonin S V, Belozertseva E A 2007 *Application of allogeneic demineralized bone implants in the treatment of chronic osteomyelitis in an experiment* (Topical issues of tissue and cell transplantology: Moscow, CITO) p. 27

[7] Ager III J W, Nalla R K, Balooch G, Kim G, Pugach M, Habelitz S, Marshall G W, Kinney J H, O'Ritchie R 2006 *On the Increasing Fragility of Human Teeth With Age:A Deep-UV Resonance Raman Study* (Journal of Bone and Mineral Research vol 21) pp. 1879-1887

[8] Draper E R C, Morris M D, Camacho N P, Matousek P, Towrie M, Parker A W, Goodship A E 2005 *Novel Assessment of Bone Using Time-Resolved Transcutaneous Raman Spectroscopy* (Journal of Bone and Mineral Research) 20 (11), p. 1968

[9] Tarnowski C P, Ignelzi Jr M A, Morris M D 2002 *Mineralization of Developing Mouse Calvaria as Revealed by Raman Microspectroscopy* // (Journal of Bone and Mineral Research) 17, № 6. pp 1118–1126

[10] Timchenko E V, Timchenko P E, Volova L T, Ponomareva Yu V, Taskina L A 2014 *Investigation of the organomineral structure of bone implants by Raman scattering* (Quantum Electronics) 44 (7), pp. 696 - 700

[11] Zhao J, Lui H, Mclean D I, Zeng H 2007 *Automated autofluorescence background subtraction algorithm for biomedical Raman spectroscopy* (Society for applied spectroscopy) 61(11), pp. 1225–1232

[12] Moaghi Z, Rehman S, Rehman I 2007 *Raman Spectroscopy of Biological Tissues* (Applied Spectroscopy Reviews) 42, pp. 493–541

[13] Olsztyńska-Janus S, Gasior-Glogowska M, Szymborska-Malek K, Komorowska M, Witkiewicz W, Pezowicz C, Szotek S, Kobielarz M 2012 *Spectroscopic techniques in the study of human tissues and their components. Part II* (Raman spectroscopy Acta of Bioengineering and Biomechanics) Vol. 14, № 4

[14] Polomska M, Kubisz L, Kalawski R, Oszkinis G, Filipiak R, Mazurek A 2010 *Fourier Transform Near Infrared Raman Spectroscopy in Studies on Connective Tissue* (Acoustic and Biomedical Engineering) Vol.118 №1

[15] Maher J R, 2013 *Transcutaneous Raman Spectroscopy of Bone* (University of rochester) 173 p.

[16] Timchenko P E, Timchenko E V, Pisareva E V, Vlasov M Yu, Volova L T, Frolov O O 2017 *Experimental studies of hydroxyapatite* (Journal of Optical Technology)

[17] Lim N S, Hamed Z, Yeow C H, Chan C, Huang Z 2011 *Early detection of biomolecular changes in disrupted porcine cartilage using polarized Raman spectroscopy* (J. Biomed. Opt.) 16 (1): 017003

Thermal lens spectroscopy in two-component liquid

O O Ovseychook, V I Ivanov and G D Ivanova

Department of Physics and Theoretical Mechanics, Far Eastern State Transport University, Khabarovsk, Serysheva street, 47, Russia

tmeh@festu.khv.ru

Abstract. In two-component fluid (binary mixture) the heat flow can cause concentration stream arising from occurrence of thermodiffusion phenomenon (Soret effect). As a result these phenomenon changes the magnitude of the transport coefficients of the mixture. In this paper the theoretical analysis of the light-induced thermodiffusion mass transfer in two-components liquid in a field of Gaussian beam was carried out. It was calculated a concentration contribution to the thermal lens response. The results are relevant to the study of the radiation self-action in the nanosuspension and optical diagnostics of such materials.

1. Introduction

Nonlinear optical diagnostics of materials involves methods based on different mechanisms of light induced modulation of optical constants of medium [1-3]. For example, nonlinear optical method is used in thermal lens spectrometry and thermo-optical diagnostics of materials [4,5]. A thermal lens response has its own peculiarities in two-component liquid, because in addition to normal thermal response associated with thermal expansion of the liquid, concentration flows can occur here arising from the phenomenon of thermal diffusion (Soret effect) and electrostriction.

The purpose of this work is the theoretical analysis of the contribution of concentration mechanisms nonlinearity in the formation of light induced response in the two component liquid.

2. Thermal lens scheme

Thermal lens technique is widely used for the optical diagnostics of materials [6-8]. The light-induced thermal lens in a homogeneous fluid is formed as a result of thermal expansion of a medium. In two-component fluid the heat flow also can cause concentration stream arising from occurrence of thermodiffusion (Soret effect). Another mechanism of optical nonlinearity of the medium is due to the forces operating on the particles of the dispersed phase in gradient light field. A change in the concentration of dispersed components in the liquid as a result of these phenomena changes the magnitude of medium thermal lens response.

Let's take a two beam thermal lens scheme (see Figure 1). The reference (Gaussian) beam generates the thermal field in the optical cell with nanosuspension. The second (Gaussian) beam tests the thermal lens.

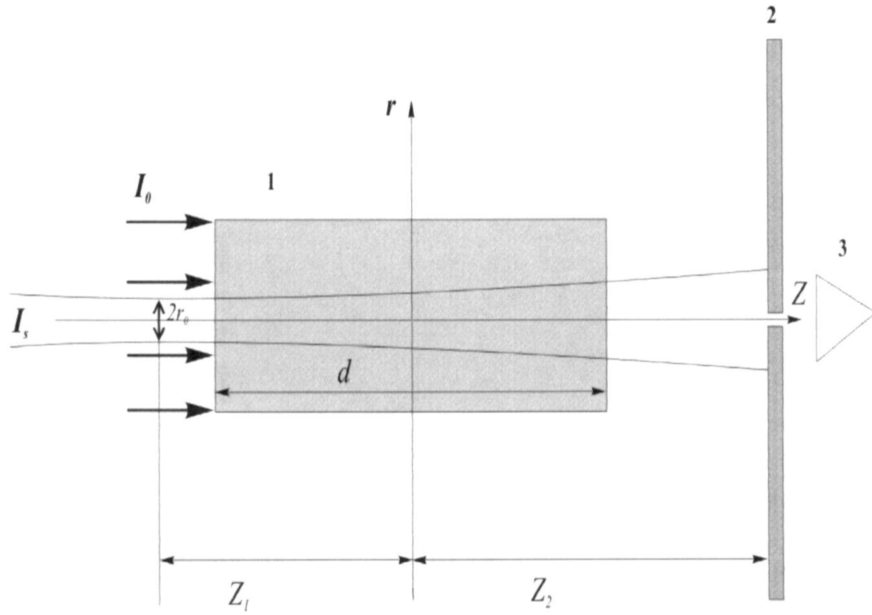

Figure 1. A thermal lens scheme.

Thermal lens signal shows the change of the beam intensity on the optical axis behind the screen:

$$\vartheta(t) = \frac{I(t) - I(0)}{I(0)} \quad (1)$$

To calculate the thermal lens signal we use the term for the lens transparency of the cell [5]:

$$\vartheta_{st} = -\frac{2(z_1/l_0)\Phi_{nl}(0)}{(1+z_1^2/l_0^2)(1+3z_1^2/l_0^2)} \quad (2)$$

Where z_1, z_2 are the distance from the cell center to the Gaussian beam waist and to screen, respectively (fig. 1), $l_0 = \pi r_0^2/\lambda$ is the confocal parameter , r_0 is the radius of the Gaussian beam waist, $\Phi_{nl}(0)$ is the nonlinear phases in optical cell on the beam axis.

Nonlinear phase consists of two deposits: the first appears due to thermal expansion of the liquid phase and the second is associated with a change in the concentration of dispersed particles:

$$\Phi_{nl}(0) = \Phi_T(0) + \Phi_C(0) \quad (3)$$

$$\Phi_T(0) = k\int_0^d \left(\frac{\partial n}{\partial T}\right)\Delta T(z, r=0)dz \quad (4)$$

$$\Phi_C(0) = k\int_0^d \left(\frac{\partial n}{\partial C}\right)\Delta C(z, r=0)dz \quad (5)$$

Where $k = 2\pi/\lambda$ is the wave vector of the probing beam, d is the thickness of a layer of liquid.

3. Light induced lens response in nanosuspension

To determine the value of thermal lens response we should consider the thermal and mass transport processes. The next system of balanced equations describes the processes that occur when a binary mixture is exposed by light field [8]:

$$c_p \rho \frac{\partial T}{\partial t} = \chi \nabla^2 T + \alpha I_0 \tag{6}$$

$$\frac{\partial C}{\partial t} = -\mathrm{div}(J_d + J_{el} + J_{thd}) \tag{7}$$

Here T is the liquid temperature, C is concentration of the nanoparticles, χ is coefficient of thermal conductivity; c_p, ρ are the specific heat capacity and density of the fluid, respectively; α – radiation absorption coefficient, I_0 is the beam intensity.

We have three concentration flows:

$$J_d = -D\nabla C, \tag{8}$$

$$J_{el} = \gamma C \nabla I, \tag{9}$$

$$J_{thd} = D_T C(1-C)\nabla T, \tag{10}$$

here J_d is diffusion flow, J_{el} is electrostrictive flow, J_{thd} is thermal diffusion flow. D и D_T are the diffusion and thermal diffusion coefficients, respectively, $\gamma = (2\beta b / cn_1)$, (β and b are polarizability and mobility of nanoparticle, respectively), c is the light velocity [9].

We will consider the case of low concentrations ($C << 1$) and small it changes. This approximation allows the presenting of the desired concentration in the form of sum of unperturbed parts C_0 and perturbed C_N:

$$C(r,t) = C_0 + C_N(r,t) = C_0(1 + C'(r,t)), \tag{11}$$

$$C'(r,t) = \frac{C_N(r,t)}{C_0} << 1. \tag{12}$$

We have next boundary (13) and initial (14) conditions:

$$C(R,t) = C_0,\ T(R,t) = T_0, \tag{13}$$

$$C(r,0) = C_0, T(r,0) = T_0, \tag{14}$$

where R is the radius of the cylindrical cell, T_0 is the temperature at the border cell.

After the necessary integration we obtain an exact solution:

$$T(r,t) = T_0 + \alpha I_0 \lambda^{-1} r_1^2 \sum_{n=1}^{\infty} \frac{J_0(\mu_n r / R) \exp\{-\mu_n^2 r_1^2 / 4R^2\}[1 - \exp\{-\mu_n^2 t / \tau_{th}\}]}{\mu_n^2 J_1^2(\mu_n)} \tag{15}$$

$$\tau_{th} = 4R^2 c_P \rho (\mu_n \chi)^{-1}, \tag{16}$$

where μ_n is the positive roots of the equation $J_0(\mu_n)=0$; τ_{th} is relaxation time of the thermal response.

Further it is assumed that the temperature field is set in the system faster than the particle concentration distribution. It allows solving the task in terms of stationary temperature [10].

Decision of the task (6-7) can be obtained by using the Green's function:

$$C'(r,t)=\xi D^{-1}r_1^2\sum_{n=1}^{\infty}\frac{J_0(\mu_n r/R)\exp\{-\mu_n^2 r_1^2/4R^2\}[1-\exp\{-\mu_n^2 t/\tau_{thd}\}]}{\mu_n^2 J_1^2(\mu_n)}+\gamma I_0 D^{-1}C_0\exp\{-r^2/r_1^2\}[1-\exp\{-t/\tau_d\}] \quad (17)$$

$$\tau_d = r_1^2 D^{-1} \quad (18)$$

$$\tau_{thd} = R^2 D^{-1} \quad (19)$$

where τ_{thd} is relaxation time of the thermodiffusion response, τ_d. is relaxation time of the diffusion response.

First and second terms describe the thermal diffusion and electrostrictive effects, respectively.

The obtained results allow us to calculate the summarized light induced lens response (for example, in stationary case):

$$\vartheta_{st}^{tot}=\frac{z_1}{l_0}\frac{kI_0 d}{4}\frac{\left(\alpha\chi^{-1}S_T C_0 r_1^2\left(\frac{\partial n}{\partial C}\right)-\alpha\chi^{-1}r_1^2\left(\frac{\partial n}{\partial T}\right)+\gamma C_0 D^{-1}\left(\frac{\partial n}{\partial C}\right)\right)}{\left(1+z_1^2/l_0^2\right)\left(1+3z_1^2/l_0^2\right)} \quad (20)$$

Thus, the expression was achieved for the light induced lens response in the binary mixture.

The most interesting case realises for nanosuspension. In nanodispersive medium particle radius is much smaller than the radiation wavelength λ and the refractive index of the medium is proportional to the concentration of particles (for dilute systems):

$$n=n_1(1+f\delta), \quad (21)$$

where $\delta=(n_2-n_1)/n_1$; n_1 and n_2 are the refractive indices of the liquid and the medium of the dispersed phase, respectively; $f=(4/3)\pi r^3 C$ is the volume fraction of the dispersed phase, r - the radius of the microparticle, C - the concentration of nanoparticles.

The experimental results show that for nanosuspension all three components in (20) can be compared by size [11].

4. Conclusions

We have analyzed the two-dimensional diffusion in the nanosuspension with two concentration nonlinearities (thermodiffusion and electrostrictive) in a Gaussian beam radiation field. As a result the exact expressions were achieved for the light induced lens response in the binary mixture. We get the exactly expression describing the kinetics of the thermal expansion, thermal diffusion and electrostrictive effect [6,7].

The results are relevant to nonlinear optics of nanosuspension [10-12], including the optical diagnostics of such nano-materials [13-14].

References

[1] Chintamani P, Shalini M, Agnel P, Meera V, Tejas I H and Radha S 2014 *International Journal of Chemical and Physical Sciences* **3**(5) 44
[2] Freysz E, Afifi M, Ducasse A, Pouligny B and Lalanne J R 1984 *J. Opt. Soc. Amer. B* **1** 433
[3] Ivanov V I and Ivanova G D 2016 *Journal of Physics: Conference Series* **735** 012037
[4] Vicary L 2002 *Philos. Mag.* B. **82** 447-452
[5] Ivanov V I, Ivanova G D and Khe V K 2017 *Proc. SPIE* **10176** 1017607
[6] Smith P W, Maloney P J and Ashkin A 1982 *Opt. Lett.* **7**(8) 347.
[7] Ivanov V I , Ivanova G D, Okishev K N and Khe V K 2016 *Proc. SPIE* **10035** 100354Y
[8] Lee W, El-Ganainy R, Christodoulides D, Dholakia K and Wright E 2009 *Optics Express* **17** 10277
[9] Ivanov V, Ivanova G, Okishev K and Khe V 2017 *IOP Conf. Ser.: Mater. Sci. Eng.* **168** 012045
[10] Leppla C and Wiegand S 2003 *Philosoph. Mag.* **83** 1989
[11] Giglio M and Vendramini A 1974 *Appl. Phys. Lett.* **25** 555
[12] El-Ganainy R, Christodoulides D, Rotschild C and Segev M 2007 *Opt. Express* **15** 10207
[13] Smith P W, Ashkin A and Tomlinson W J 1981 *Opt. Lett.* **6** 284
[14] Sheldon S J, Knight L V and Thorne J M 1982 *Appl. Opt.* **12** 1663

IOP Conf. Series: Journal of Physics: Conf. Series **1038** (2018) 012092 doi:10.1088/1742-6596/1038/1/012092

Diffraction on random fractal structures

O A Mossoulina[1], S G Volotovsky[2]

[1]Samara National Research University, 34, Moskovskoye Shosse, Samara, Russia, 443086
[2]Image Processing Systems Institute – Branch of the Federal Scientific Research Centre "Crystallography and Photonics" of Russian Academy of Sciences, 151, Molodogvardeyskaya, Samara, Russia, 443001

Abstract. In this paper, the formation of diffraction patterns for regular and random fractals of Cantor is numerically studied. The calculation of diffraction on one-dimensional, two-dimensional and radial fractal structures is performed using a fractional Fourier transform. This propagation operator makes it possible to obtain a field distribution (with the accuracy of scale) in any paraxial region, both in the Fresnel diffraction zone and in the far field.

1. Introduction

Laser beams with increased resistance to the effects of random fluctuations of the optical medium, including atmospheric turbulence, are an important factor in improving the quality of optical communication in free space, as well as remote sensing systems and laser radar systems [1, 2]. Among such beams, the laser modes of high order [3, 4], including those with a vortex phase singularity [5-9]. Coherence and polarization are also important characteristics of beam stability [10, 11]. When using the Bessel type communication beams, the possibility of coding the polarization state [12] provides an additional degree of information protection, which previously assumed only amplitude and phase coding.

There are various ways of modeling random environments [2, 13-16], including the turbulence of the atmosphere and reservoirs. One way is using of the random fractal structures [17, 18]. The regular fractals are characterized by self-similarity and fractional dimension [19]. The spatial spectrum of such structures also has self-similarity properties, and therefore the diffraction pattern of optical radiation on the fractal structure [20] can be used to determine the characteristics of the structure itself [21, 22]. Moreover, the property of self-similarity makes it possible to use even a small part of the spatial spectrum to obtain an image of the original object [23].

The investigation of diffraction by fractal gratings [20, 24, 25] is promising for solving other important problems – the formation of periodically self-reproducing fields [26-29], the creation of multifocus [30-33] or specified longitudinal distributions [34, 35], and in achromatic depicting systems [36-39].

One of the most important characteristics of fractals is the spatial spectrum [21, 22, 40, 41]. For optical fractal structures, it is also important to have a longitudinal distribution, namely, the formation of a number of foci, which is characteristic of binary zone plates [42].

Random fractal structures preserve some properties of regular fractals, in particular, dimension. Moreover, in the diffraction pattern it is possible to observe the speckle structures, that differ from the speckle patterns of the ordinary random fields [41, 43]. Thus, the spatial spectrum contains important information even about random structures. Even more information can be obtained about the object of research, if to consider not only the focal plane, but some region of diffraction, although not exclusively in the far zone (the Fraunhofer zone).

In this paper, the simulation of formation of the diffraction patterns for regular and random fractals of Cantor on the basis of the fractional Fourier transform is performed [44]. This operator makes it possible to obtain the field distribution (with the accuracy of scale) in any paraxial region, both in the Fresnel diffraction zone and in the far zone [45].

2. Theoretical part and simulation results

The spatial spectrum for the one-dimensional triadic Cantor fractal [46] can be estimated from the formula:

$$F_S(u) = 2^{-S}\left[\prod_{s=0}^{S-1}\cos\left(2\pi\cdot 3^s u\right)\right]\frac{\sin(\pi u)}{\pi u}, \qquad (1)$$

where S – level of the fractal.

As follows from formula (1) in the Fraunhofer diffraction zone (or in the focal plane of the lens), a superposition of the set of cosines with increasing frequency will increase as the level of the fractal increases. The envelope of this superposition depends on the size of the fractal structure.

Figure 1 shows the distributions of the spatial spectrum for fractals of various levels, from which a self-similar structure is clearly visible.

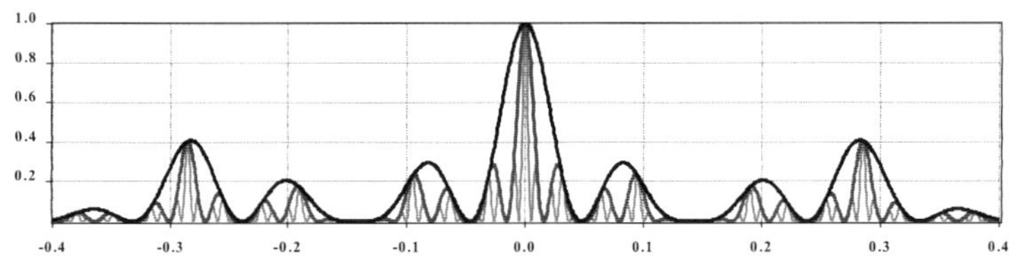

Figure 1. Spatial spectrum (intensity) for S = 2 (black colour), S = 3 (blue colour), S = 4 (red colour).

The spatial spectrum shown in figure 2 corresponds to the Fraunhofer diffraction zone or the focal plane of the lens. To obtain the field distribution in any paraxial region, including in the Fresnel diffraction zone, as well as during the passage of the lens system, a fractional Fourier transform (FFT) can be used [44, 45, 47, 48]:

$$E(u,z) = \left(-\frac{ik}{2\pi f\sin(\alpha z)}\right)^{\frac{1}{2}}\exp\left\{\frac{iku^2\cos(\alpha z)}{2f\sin(\alpha z)}\right\}\int_{-R}^{R} g(x)\exp\left[\frac{ikx^2\cos(\alpha z)}{2f\sin(\alpha z)} - \frac{ikxu}{f\sin(\alpha z)}\right]\mathrm{d}x \qquad (2)$$

where $k = \frac{2\pi}{\lambda}$, λ – the wavelength of the radiation, f – the focal length of the lens, $\alpha = \frac{\pi}{2f}$, z – the distance from the input plane, $2R$ – the size of the field within input plane.

Figure 2 shows the longitudinal diffraction patterns, formed at the distance of two foci (f = 100 mm) from the input plane, when the fractal structure of 1 mm in size is illuminated by radiation with the wavelength of 633 nm.

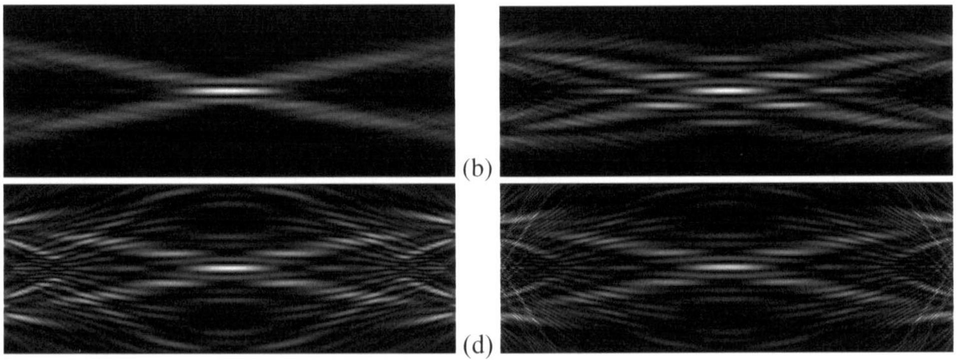

Figure 2. Longitudinal diffraction patterns (intensity, $z \in [0.1f, 1.9f]$) for different fractal levels: a) S=1; b) S=2, c) S=3, d) S=4.

The fields in figure 2 in the central part with the accuracy of scale, correspond to the Fraunhofer diffraction zone (focal region), the vertical cross section at the center gives a picture of the spatial spectrum (as in figure 2). The rest part allows to see the distribution in the Fresnel diffraction zone and get more information about the fractal structure. In particular, at the edges of the pictures, traces of diffraction on fine details, characteristic of high-level fractals are clearly visible.

Note, that the diffraction properties of real fractals are different from theoretical consideration [49], especially in the near-field diffraction.

Let's consider further examples of formation of the random one-dimensional fractals.

Fractal modeling takes place by generating the random Cantor set. Each random Cantor set corresponds to the same random process, that is, the removal of the part of the segment, and the length of the segment itself and the parts separated by it are randomly assigned within the allowed value.

At the same time, it retains a certain degree of regularity, since at each step of the process of constructing the fractal, the deletion is repeated at random on both sides of the remote segment.

Figure 3 shows examples of the random fractals, derived from the Cantor set and the Cantor ladder.

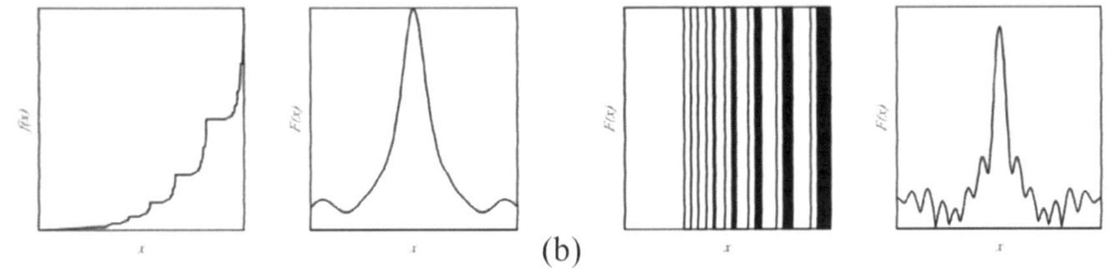

Figure 3. The form of the random one-dimensional fractal structure (on the left) and its spatial spectrum (on the right) for different fractal levels: a) Cantor's ladder, b) Cantor's set.

It is possible to construct the two-dimensional fractal, simply as the product of the one-dimensional fractures, and the scale transformation can be introduced along different axes. Moreover, in this case, it is possible to use fractals of different levels. Then the spatial spectrum will have the following form:

$$F_{S,P}(u,v) = 2^{-S-P}\left[\prod_{s=0}^{S-1}\cos\left(2\pi\cdot 3^{s}\,\alpha u\right)\right]\left[\prod_{p=0}^{P-1}\cos\left(2\pi\cdot 3^{p}\,\beta v\right)\right]\operatorname{sinc}(\alpha u)\operatorname{sinc}(\beta v), \tag{3}$$

where $\operatorname{sinc}(x) = \dfrac{\sin(\pi x)}{\pi x}$, α, β – the scale factors.

The propagation operator (2) can be generalized in the same way:

$$E(u,v,z)=-\frac{ik}{2\pi f\sin(\alpha z)}\exp\left\{\frac{ik\left(u^2+v^2\right)\cos(\alpha z)}{2f\sin(\alpha z)}\right\}\times$$
$$\times\int_{-R}^{R}\int_{-R}^{R}g(x,y)\exp\left[\frac{ik\left(x^2+y^2\right)\cos(\alpha z)}{2f\sin(\alpha z)}-\frac{ik(xu+yv)}{f\sin(\alpha z)}\right]\mathrm{d}x\,\mathrm{d}y \quad (4)$$

To generate the random two-dimensional fractal, a unit square on the first step are used $E_0=[0,1]\times[0,1]$. At the next step (level), the fractal is specified as $E_1=([0,a_1]\cup[b_1,1])\times([0,a_2]\cup[b_2,1])$, where a_1, a_2, b_1 and b_2 – the fractal parameters, specified in the range $(0,1)$, with $a_1<b_1$and $a_2<b_2$. Generation of the two-dimensional fractal, as for the one-dimensional case, occurs according to the principle, shown in figure 1, with some corrections, so that the length at each step is given randomly. If the parameters will be set as equal, then the fractals, which are shown in figure 4. will be received.

The spatial spectrum was calculated using the direct Fourier transform:

$$F(X)=\Im\left[f(X)\right](U)=\int_{R^n}f(X)\exp(-2\pi iXU)\,\mathrm{d}^nX, \quad (5)$$

where $f(X)$ – the input function, given in the form of the vector, representing the given structure; $X=(x_1,x_2,...,x_n)$ – the input values of the function; $F(U)$ – the output function received; $U=(u_1,u_2,...,u_n)$ – the output values of the function; $\Im[\cdot]$ – Fourier transform operator.

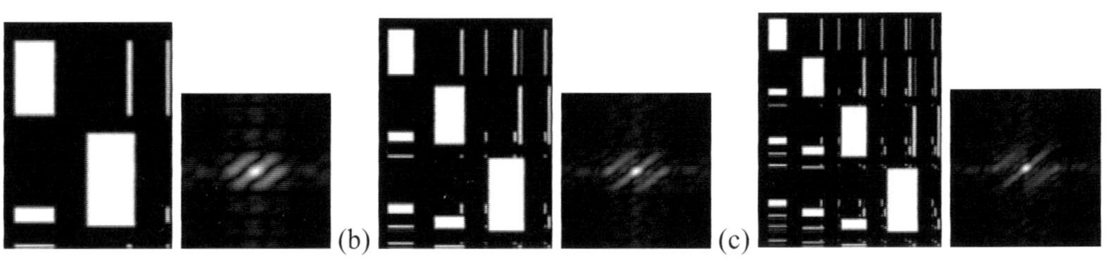

(a) (b) (c)

Figure 4. The form of the random fractal structure (on the left) and its spatial spectrum (on the right) for different fractal levels: a) *S*=3; b) *S*=4; c) *S*=5.

An interesting type of fractals are radial fractals [50]. On their basis, the fractal lenses and axicons were considered [30, 51-52].

The propagation of the radially symmetric field through the paraxial lens system can be described with using of the FFT in polar coordinates [49, 50]:

$$E(\rho,\theta,z)=-\frac{ik}{2\pi f\sin(\alpha z)}\exp\{ikz\}\exp\left\{\frac{ik\rho^2}{2f\tan(\alpha z)}\right\}\times$$
$$\times\int_0^{2\pi}\int_0^{\infty}E_0(r,\varphi)\exp\left\{\frac{ikr^2}{2f\tan(\alpha z)}\right\}\exp\left\{-\frac{ik}{f\sin(\alpha z)}\rho r\cos(\theta-\varphi)\right\}r\,\mathrm{d}r\,\mathrm{d}\varphi \quad (6)$$

Figure 5 shows the longitudinal diffraction patterns formed at the distance of two focal points (*f* = 300 mm) from the input plane, when the fractal structure with the radius of 1 mm is illuminated by radiation with the wavelength of 633 nm.

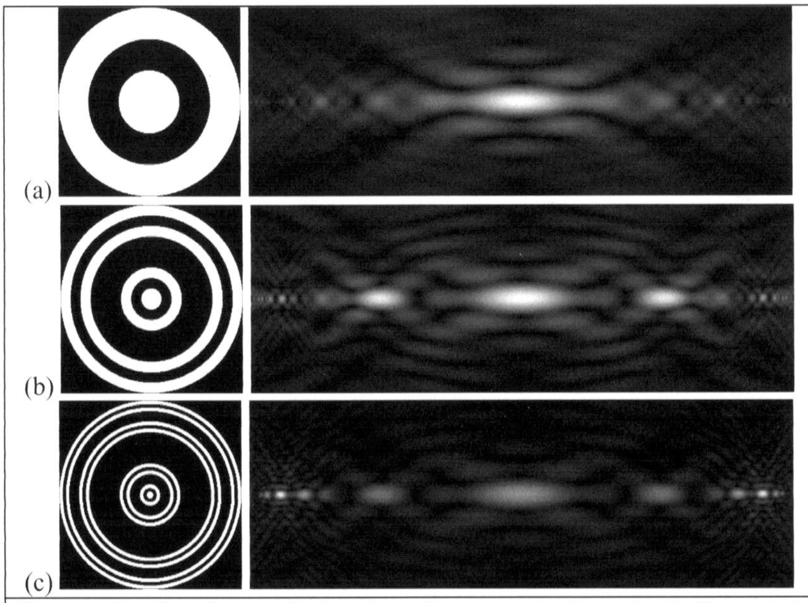

Figure 5. The form of the fractal structures (on the left) and the longitudinal diffraction patterns (amplitude, $z \in [0.1f, 1.9f]$) for different fractal levels: a) S=1; b) S=2; c) S=3.

In Figure 5, the appearance and amplification of the local foci in the near-field diffraction zone is clearly visible, due to the increase in the level of the fractal and the narrowing of the width of the ring lines.

3. Conclusions

In this paper, the one-dimensional, two-dimensional, and radial case of obtaining the fractal structure of the regular and random type are presented. The spatial spectra of the formed fractals are obtained, as well as diffraction patterns in the Fresnel zone. It is shown, that the diffraction properties of the regular fractals in the near diffraction zone differ substantially from the diffraction patterns in the Fraunhofer zone. The difference is particularly noticeable for fractals with high levels, which is due to appearance of many local foci in the near zone due to diffraction on small details.

References

[1] Majumdar A K and Ricklin J C 2008 *NewYork: Springer Science & Business Media* **2** 418
[2] Korotkova O 2013 *CRC Press* Boca Raton FL 362
[3] Young C Y, Gilchrest Y V and Macon B R 2002 *Opt. Eng.* **41** 1097-103
[4] Eyyuboglu H T 2007 *Appl. Phys. B* **88** 259-65
[5] Gbur G and Tyson R K 2008 *J. Opt. Soc. Am. A* **25** 225
[6] Wang T, Pu J and Chen Z 2009 *Optics Communications* **282** 1255-59
[7] Soifer V A, Korotkova O, Khonina S N and Shchepakina E A 2016 *Computer Optics* **40** 605-4
[8] Kirilenko M S, Porfirev A P and Khonina S N 2017 *Proc. SPIE* **10342** 103420B
[9] Porfirev A P, Kirilenko M S, Khonina S N, Skidanov R V and Soifer V A 2017 *Applied Optics* **56** E8-15
[10] Wolf E 2007 *Cambridge U. Press* 59
[11] Wang F, Liu X and Cai Y 2015 *A Review, Progress In Electromagnetics Research* **150** 123-43
[12] Khonina S N and Golub I 2015 *Optics Letters* **40** 4070-3
[13] Coles Wm A, Filice J P, Frehlich R G and Yadlowsky M 1995 *Applied Optics* **34** 2089-101
[14] Frehlich R 2000 *Applied Optics* **39** 393-7

[15] Hyde IV M W, Basu S and Schmidt J D 2015 *Optics Letters* **40** 233-6
[16] Kirilenko M S and Khonina S N 2016 *CEUR Workshop Proceedings of International Conference Information Technology and Nanotechnology* **1638** 55-65
[17] Sreenivasan K R and Meneveau C 1986 *J. Fluid Mech* **173** 357-86
[18] Mazzi B and Vassilicos J C 2004 *J. Fluid Mech* **502** 65-87
[19] Mandelbrot B B 1982 The Fractal *Geometry of Nature, Freeman, New York* 468
[20] Berry M V 1979 *J. Phys. A: Math. Gen* **12** 781-97
[21] Uozumi J, Kimura H and Asakura T 1990 *J. Mod. Opt.* **37** 1011-31
[22] Zunino L and Garavaglia M 2003 *J. Mod.* Opt. **50** 717-27
[23] Moocarme M and Vuong L T 2015 *OpticsExpress* **1509** 04761
[24] Berry M V and Klein S 1996 *J. Mod. Opt.* **43** 2139-64
[25] Jaggard A D and Jaggard D L 1998 *Opt. Commun.* **158** 141-8
[26] Hou B, Xu G, Wen W and Wong G K L 2004 *Appl. Phys. Lett.* **85** 6125-7
[27] Mendez D C and Lehman M 2004 *Optik.* **115** 439-42
[28] Khonina S N, Kotlyar V V and Soifer V A 1999 *Technical Physics Letters* **25** 489-91
[29] Khonina S N and Volotovsky S G 2007 *Optical Memory & Neural Networks (Information Optics) Allerton Press* **16** 167-77
[30] Saavedra G, Furlan W D and Monsoriu J A 2003 *Optics letters.* **28** 971-3
[31] Mihailescu M, Preda A M, Sobetkii A and Petcu A C 2009 *Optoelectronics review.* **17** 330-7
[32] Kotlyar V V, Khonina S N and Soifer V A 1993 *Proccedings of SPIE.* **1780** 263-72
[33] Soifer V A, Doskolovich L L and Kazanskiy N L 1994 *Optical Engineering.* **33** 3610-15
[34] Khonina S N, Kotlyar V V and Soifer V A 1993 *Journal of Modern Optics.* **40** 761-9
[35] Khonina S N and Ustinov A V 2015 *Optics Communications.* **341** 114-21
[36] Furlan W D, Saavedra G and Monsoriu J A 2007 *Opt. Lett.* **32** 2109-11
[37] Andersen G and Tullson D 2007 *Phys. Rev. A.* **46** 3706-8
[38] Karpeev S V, Ustinov A V and Khonina S N 2016 *Computer Optics* **40** 173-8
[39] Kharitonov S I, Volotovsky S G and Khonina S N 2016 *Computer Optics* **40** 331-7
[40] Allain C and Cloitre M 1987 *Phys. Rev.* **36** 5751-7
[41] Horvath P, Smid P, Vaskova I and Hrabovsky M 2010 *Optik.* **121** 206-13
[42] Khonina S N, Ustinov A V, Skidanov R V and Porfirev A P 2015 *Applied Optics* **54** 5680-5
[43] Uozumi J, Ibrahim M and Asakura T 1998 *Opt. Commun* **156** 350-8
[44] Mendlovic D and Ozaktas H M 1993 *I, J. Opt. Soc. Am. A* **10** 1875-81
[45] Alieva T, Bastiaans M J and Calvo M L 2005 *J. Appl. Signal Processing* **10** 1-22
[46] Allain C and Cloitre M 1986 *Phys. Rev. B* **33** 3566-9
[47] Alieva T 1996 *J. Opt. Soc. Am. A* **13** 1189-92
[48] Kirilenko M S, Zubtsov R O and Khonina S N 2015 *Computer Optics* **39** 332-8
[49] Rodriguez Merlo D, Rodrigo Martín-Romo J A, Alieva T and Calvo M L 2003 *An Experimental Study 1, Optics and Spectroscopy* **95** 13103
[50] Jaggard A D and Jaggard D L 1998 *Opt. Commun.* **158** 141-8
[51] Monsoriu J A, Furlan W D, Andres P and Lancis J 2006 *Opt. Express* **14** 9077-82
[52] Mossoulina O A, Kirilenko M S and Khonina S N 2016 *Journal of Physics: Conference Series* **741** 012142

IOP Conf. Series: Journal of Physics: Conf. Series **1038** (2018) 012093 doi:10.1088/1742-6596/1038/1/012093

New method of researches of the magnetic fields force lines structure

S E Logunov[1], V V Davydov[1, 2, 3], M G Vysoczky[1], M S Mazing[1]
[1]Peter the Great Saint-Petersburg Polytechnic University, Saint Petersburg 195251, Russia
[2]The Bonch-Bruevich Saint - Petersburg State University of Telecommunications, Saint Petersburg 193232, Russia
[3]Department of Ecology, All-Russian Research Institute of Phytopathology, 143050, Moscow Region, Odintsovo district, B.Vyazyomy, Russia

Abstract. A new method for constructing an optical image of the configuration of magnetic field lines using a ferrofluid cell with a magnetic fluid is considered. The carried out experimental researches have shown that the proposed method allows to determine the volumetric inhomogeneity of the field in the magnetic system. It is shown that the optical image can be used in real time to adjust the magnetic system to the minimum induction values ΔB in three planes.

1. Introduction.

One of relevant problems of fundamental physics is the research of structure of force lines of magnetic fields [1-5]. The greatest difficulties arise at the research of structure of force lines of the magnetic field in the interpolar space of magnetic systems of various designs [3-6]. When studying various media using the magnetic systems (spectrometers, tomographs, relaxometrs, etc.) it is desirable to know the picture of force lines of this magnetic system in three planes [6-13]. The use of Hall sensor in most cases allows to obtain information on value of induction B in a point of its placement location [6-11]. For definition of degree of inhomogeneity of magnetic field only in one plane it is necessary to move Hall sensor or place two, three and more sensors along one straight line between poles. It allows to estimate the change of heterogeneity of the magnetic field from the center of the magnetic system to its edges, but it is impossible to construct the picture of force lines of the magnetic field (the distance between force lines is considerably less than between sensors). When moving the Hall sensor an error of determination of the coordinates between poles is sufficiently high [6, 8-11]. The use of sensors on the basis of the phenomenon of the nuclear magnetic resonance (NMR) allows to determine heterogeneity by width of the line ΔB in the volume from which the nuclear magnetic resonance signal is registered [6, 8, 12]. But at its movement there appear the same problems, as with Hall sensor.

One of the possible solutions of this problem can be the method of investigation of force lines of the magnetic field developed by us with the using of a ferrofluid cell.

2. The experimental setup and measurement technique.

As a result of experiments [3, 7, 14] that were conducted by us earlier it was established that nanoparticles in magnetic liquid are located along force lines of a magnetic field. They form the transparent and opaque zones in the ferrofluid cell for a laser radiation with the definite period. This period depends on the inhomogeneity of a magnetic field (i.e. density and configuration of force lines).

Besides, the carried out experiments showed that as magnetic liquid better to use an aqueous solution of one-domain nanoparticles of hematite with a volume concentration 0.054 together with the superficial

IOP Conf. Series: Journal of Physics: Conf. Series **1038** (2018) 012093 doi:10.1088/1742-6596/1038/1/012093

active material (SAM) oxyhydroxide a tetraethylammonium [3, 7, 14]. Now data on structure of force lines in pole space of magnetic systems with "neutral", which are used in spectrometers, flowmeters, magnetometers and etc., are most demanded [5-7, 12, 14]. Information on the degree of inhomogeneity of a magnetic field in three planes of magnetic system as well as the force lines density is necessary for carrying out researches and measurements in various media with use of these devices.

Considering features of magnetic systems with "neutral" (restriction on the directions of impact of a laser radiation on a ferrofluid cell and conditions of registration the scattered radiation passed through it) [5-8, 12, 14] for a research of structure of force lines of a magnetic field in three planes. We constructed two variants of the experimental setup. In one case the diffraction image from the laser radiation which passed through a ferrofluid cell was registered. In two others – the laser radiation reflected from a cell was registered. In the first case used the ferrofluid cell with two transparent sides and the semiconductor laser with λ = 613 nm. In two other cases a cell with one transparent face and a laser with λ = 515 nm was used.

In Figure 1 the block diagram of experimental setup for registration of the diffraction pattern from the laser radiation which passed through a ferrofluid cell is represented. Thickness of a layer of magnetic liquid in it in the direction of action of a laser radiation is equal to d_k = 1 mm.

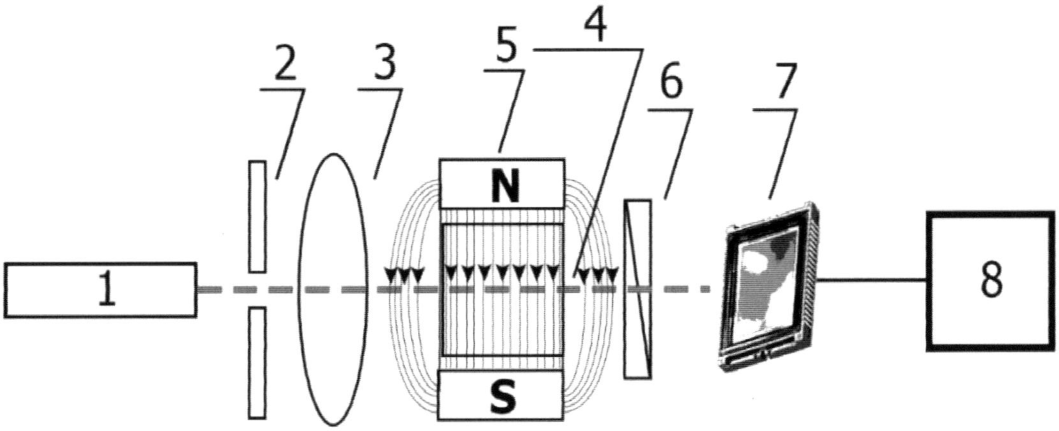

Figure 1. Structural scheme of experimental setup 1 – semiconductor laser; 2 – diaphragm; 3 – lens; 4 – ferrofluid cell; 5 – magnetic system; 6 – polarizer; 7 – photosensitive element; 8 – processing device.

As a result of the conducted researches we found peculiarities of measurement of a grating constant of d_r which corresponds to distance between force lines of a magnetic field. Unlike classical consideration of provision of the diffraction peaks and minima location in case of ferrofluid cell using it is necessary to consider at their definition thickness of d1 of the transparent walls of a ferrofluid cell and thickness of a layer of magnetic d_k liquid in it.

Taking into account the established feature the measuring technique of the diffraction peaks location was developed and the corresponding calculation were carried out executed for them. In Figure 2 the scheme of distribution of a laser radiation through a ferrofluid cell in a magnetic field of century is shown the refraction of laser radiation. On borders of media air-glass, glass - magnetic liquid, magnetic liquid-glass and glass-air is taking account refraction of a laser radiation. It is found that the intensity of twice reflected laser radiation from limits of sections of media (glass - magnetic liquid and glass-air) in passed radiation has not significant effect on formation of the diffraction image. Therefore, the derivation of the ratio for the diffraction maxima this part of radiation is not considered. In an experiment the photoreceiving device (screen 3) is placed at the distance L from a lateral face of a ferrofluid cell. In its plane the diffraction image is registered. The location of each maximum on the screen with respect

IOP Conf. Series: Journal of Physics: Conf. Series **1038** (2018) 012093 doi:10.1088/1742-6596/1038/1/012093

its center (a point 0) is defined by order of diffraction k and depends on the period of the formed diffraction grating d_r.

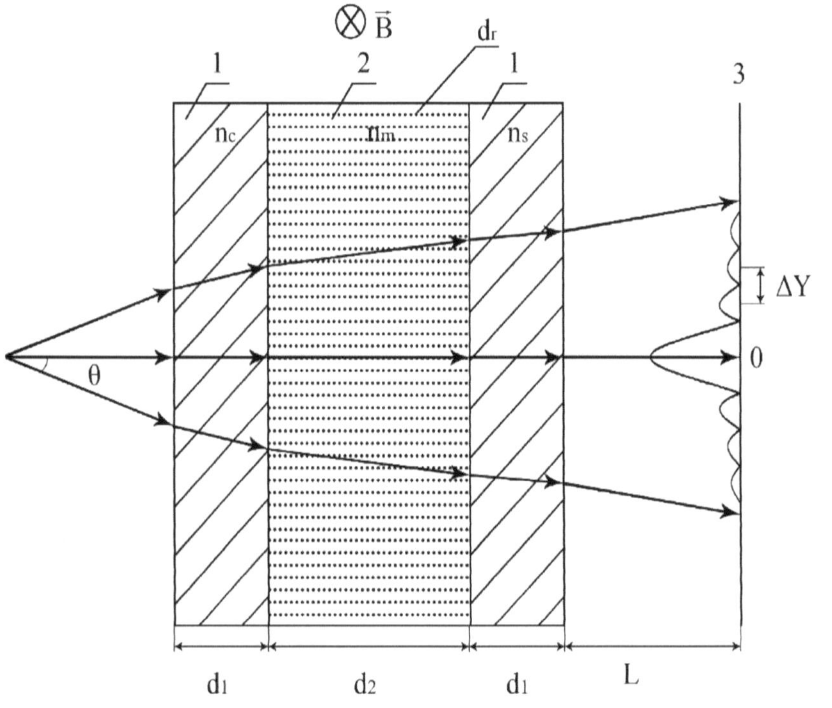

Figure 2. Scheme of propagation of laser radiation rays in a ferrofluidic cell when a diffraction pattern is recorded in transmitted light: 1 - walls of a ferrofluidic cell; 2 - magnetic fluid layer in the direction perpendicular to the magnetic field with the thickness equal to d_r; 3 - the screen.

As a result of the made experiments by us it was determined that on distance L = 20 cm value of the additional padding shift along the axis of OY which is formed due to the for refraction of radiation on borders of media is of infinitesimal size in comparison with the distance between two adjacent maxima long the ΔY. This shift in calculating the position of the maxima in the radiation passing through the ferrofluidic cell, in contrast to the consideration of the case with the registration of the radiation reflected from it, is inadvisable. Therefore, to determine the distance ΔY between the peaks k_1 and k_2 in the diffraction pattern on screen 3 (Figure 2) we obtained the following relation:

$$\frac{d_r^2-\lambda^2 k_2^2}{\lambda^2 k_2^2}\Delta Y^2 + 2L\frac{k_1^2}{d_r^2-k_1^2\lambda^2}\frac{d_r^2-k_2^2\lambda^2}{k_2^2}\Delta Y + 4L^2\frac{k_1^4\lambda^2}{\left(d_r^2-k_1^2\lambda^2\right)^2}\frac{d_r^2-k_2^2\lambda^2}{k_2^2} - L^2 = 0 \quad (1)$$

where L – distance between the ferrofluid cell and camera; k_1 and k_2 - orders of diffraction maxima ($k_2 > k_1$).

The recorded diffraction pattern measures ΔY (the distance between peaks) for the corresponding values of k_1 and k_2. From the measured value of ΔY, using (1), the value of d_r is determined as the order of the diffraction grating (the distance between force field lines in investigation plane of the magnetic system space between poles).

In Figure 3 the block diagram of the experimental setup for registration of the diffraction pattern in the reflected laser radiation is represented. In the conducted researches the ferrofluid cell 4 with one transparent side is used. Thickness of a layer of magnetic liquid in a ferrofluid cell is equal to d_k = 3 mm. It is enough that the reflected intensity from an opaque side of a ferrofluid cell did not influence formation of the diffraction pattern in the reflected radiation.

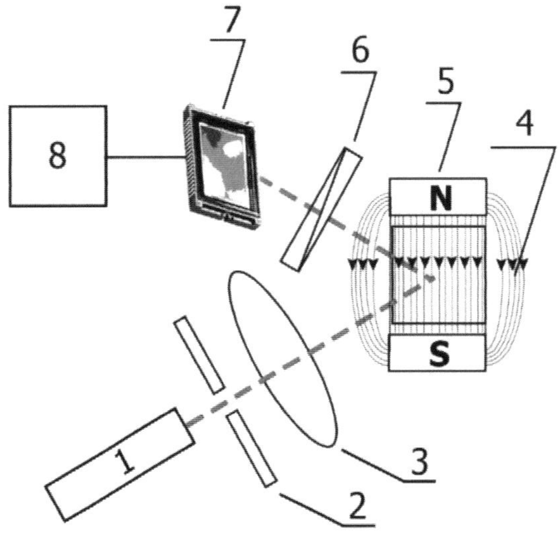

Figure 3. Structural scheme of experimental setup 1 – semiconductor laser; 2 – diaphragm; 3 – lens; 4 – ferrofluid cell; 5 – magnetic system; 6 – polarizer; 7 – photosensitive element; 8 – processing device.

In Figure 4 the scheme of distribution of beams of a laser radiation in a ferrofluid cell is represented.

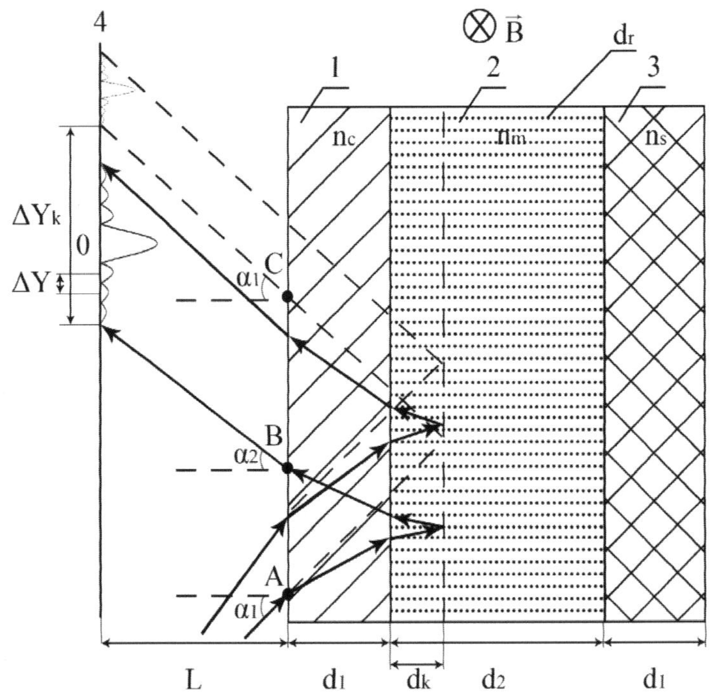

Figure 4. Scheme of propagation of laser radiation rays in a ferrofluid cell when a diffraction pattern is recorded in reflected light: 1,3 - walls of the ferrofluidic cell; 2 – the layer of the magnetic fluid with the thickness equal to d_r in the direction perpendicular to the magnetic field; 4 - the screen.

For the recorded diffraction pattern in a reflected light positions of the diffraction peaks along the axis OY were determined as well as the value of their shift ΔY_k, arising because of refraction of a laser radiation on borders of four media: air – glass, glass – magnetic liquid, magnetic liquid – glass, glass –

air. The screen 7 (a photosensitive element) when carrying out calculations was placed at the distance L from a ferrofluid cell. The laser radiation illuminated the ferrofluid cell in point A at an angle $\alpha_1 = 45^o$. Then radiation after refraction on cell walls once again feel on the diffraction grating formed by nanoparticles with the period equal to d_r. The structure of the diffraction grating from which the "reflection" takes place at the distance d_k from the internal wall of the ferrofluid cell. In this case for calculation the value ΔY we derived the following ratio:

$$\Delta Y = \frac{d_k}{d_r}\cdot\left[\frac{t_1}{\sqrt{1-\frac{t_1^2}{d_r^2}}} - \frac{t_2}{\sqrt{1-\frac{t_2^2}{d_r^2}}}\right] + d_1\cdot n_m\cdot\left[\frac{\sqrt{1-\frac{t_1^2}{d_r^2}}}{\sqrt{n_c^2 - n_m^2\left(1-\frac{t_1^2}{d_r^2}\right)}} - \frac{\sqrt{1-\frac{t_2^2}{d_r^2}}}{\sqrt{n_c^2 - n_m^2\left(1-\frac{t_2^2}{d_r^2}\right)}}\right] +$$

$$+L\cdot\left[\frac{\sqrt{n_c^2 - n_m^2\left(1-\frac{t_1^2}{d_r^2}\right)}}{\sqrt{1-\left(n_c^2 - n_m^2\left(1-\frac{t_1^2}{d_r^2}\right)\right)}} - \frac{\sqrt{n_c^2 - n_m^2\left(1-\frac{t_2^2}{d_r^2}\right)}}{\sqrt{1-\left(n_c^2 - n_m^2\left(1-\frac{t_2^2}{d_r^2}\right)\right)}}\right] \quad (2)$$

$$t_1 = (2k_1 + 1)\cdot\lambda/2;\ t_2 = (2k_2 + 1)\cdot\lambda/2 \quad (3)$$

where n_m – the refractive index of a magnetic fluid, n_c – refractive index of the transparent glass wall of the cuvette, d_k – distance from the cell wall to the first layer of nanoparticles in a magnetic fluid that forms a diffraction grating, k_1 и k_2 – orders of diffraction maxima ($k_2 > k_1$).

Our calculations showed that the difference (α_1 - α_2) between the angles of the laser radiation leaving the ferrofluidic cell at point B (after all 4 refractions) from the direction of the beam at point C (the classical diffraction case in reflected radiation) is less than 0°30', the value of ΔY_k - less than 10^{-3} mm. Therefore when determining the position of the diffraction peaks this shift on an axis of OY was taken into account in (2) unlike the case of the diffraction pattern in registration in the radiation passed through a ferrofluid cell.

Measuring distance between maxima of the diffraction pictures ΔY which is registered by means of photosensitive element 7 (Figure 3), it is possible to calculate by means of (2) of the diffraction grating d_r in a zone of its placement in pole space of magnetic system. And it is also possible define structure of force lines of a magnetic field in this plane.

The experiments carried out showed that it is much more difficult to take into account to shift value ΔY_k in the reaction (2) when in a case angle of incidences of a laser radiation are considerably differing from 45°. Therefore, at a research of magnetic systems of the complex configuration there can be difficulties when determining size ΔY in various points of pole space. It is the single possible restriction in application of this method.

3. Results and its discussion.

In Figure 5 is represented as an example the diffraction pattern (after computer processing) from the laser radiation which passed through a cell in the direction perpendicular to an induction B of a magnetic field. Transparent faces of the cell are located perpendicular to the laser radiation incident on them and the side faces of the "neutral" of the magnetic system. In addition, the transparent faces of the cell are parallel to the bases of the "neutral" magnetic system. For the magnetic field lines that are located in the interpolate space of the magnetic system the direction must be perpendicular to the pole tips for both the electromagnets and the permanent magnets. At present use cylindrical tips in magnetic systems. The side surfaces of cylindrical tips must be strictly parallel to each other and the lateral faces of the "neutral". This ensures the maximum value of B and the minimum value of ΔB in the interpolar space.

The angle between the surfaces of the side faces and the base should be 90^0. Numerous experiments have shown the inexpediency of using magnetic systems in which the pole tips are parallel to one another, but located at different angles to the side faces and the base of the "neutral".

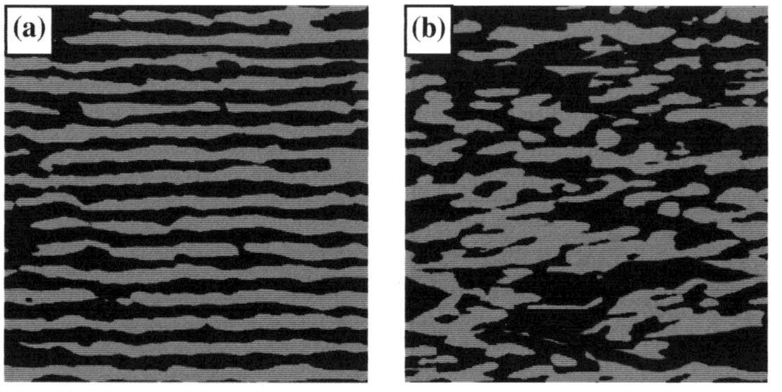

Figure 5 (a, b). The diffraction pattern of the laser radiation in the case of the magnetic fluid placing: **(a)** in a uniform magnetic field; **(b)** in an inhomogeneous magnetic field.

In Figure 6 the dependence of distribution of intensity of the recorded laser radiation on the vertical section of a ferrofluid cell at the choice of a line on the center of the image (Figure 5) recorded by a photoreceiving device is presented as an example.

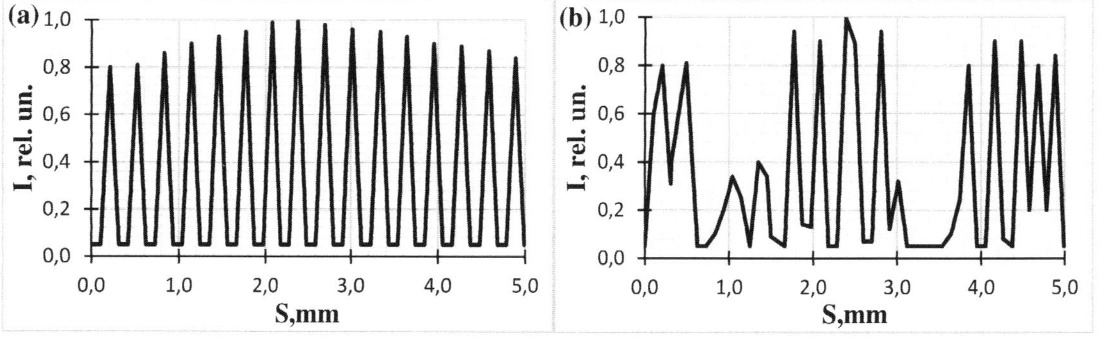

Figure 6 (a, b). The graph of dependence of intensity as function of force lines of a magnetic field **(a)** in a uniform magnetic field **(b)** in an inhomogeneous magnetic field.

The analysis of the received results allows on distribution of intensity on the section of a ferrofluid cell to make a conclusion on degree of uniformity of a magnetic field and about the distribution of its force lines in this zone of pole space. In case of the homogeneous magnetic field the intensity of the recorded peaks and distance between them with respect to the central maximum are equal. Small deviations in the form of peaks (Figure 6a) can be explained by the irregularity of magnetite particles distribution over the ferrofluid cell volume for various reasons. In case of the inhomogeneity field (Figure 6b) the symmetrical distribution between peaks is absent. The pronounced maximum of intensity in Figure 6b is also absent. It is means that the pole tips of the magnetic system, for example, of the electromagnet, are not parallel to one another or their centers are not located on a straight line that is perpendicular to the side face of the "neutral".

The experiments carried out that it is more convenient to use the dependences presented in Figure 6 for tuning of magnetic system at the minimum of inhomogeneity of a magnetic field in real time for substance placement in the subsequent in this zone the samples of with investigated medium etc. [5, 7, 12, 14].

During the experiments, additional problems arise in the case of placing transparent faces of the ferrofluid cell not parallel to the lateral bases of the "neutral". Selecting the position of the photoreceiver behind the ferrofluid cell, decoding and subsequent spatial interpretation of the recorded diffraction images into force lines and etc. For example, to obtain information about magnetic field lines in the z-plane, the cell will need to be moved (its coordinates in the x and y plane). On the z axis, it is more efficient to move the laser and the photodetector. The conducted studies are shown inadvisability to have transparent faces of the ferrofluidic cell at angles to the lateral faces and the base of the "neutral", which are not equal to 90^0 and 180^0. During the experiments it was established that the study of the structure of magnetic fields using ferrofluid cells to obtain results with an error of not more than 1% should be carried out in the temperature range T from 283 to 313 K. The temperature of the ferromagnetic liquid in the ferrofluid cell in the interpolar space of the magnetic system was controlled by an infrared remote temperature meter Testo 845 (Testo AG-Germany) along the end of the ferrofluid cell. The sensitivity to the temperature change of this device is 0.1 K. The temperature of the poles of the magnetic system was controlled by Testo 845. In experiments we used an electromagnet to create a magnetic field. That made it possible to create a magnetic field with induction B from 0.0005 to 0.86T. The value of B was controlled by a magnetic induction meter SH1-1 (NMR sensor - magnetic field is measured at the resonance frequency of protons) and measure AC and DC magnetic field DX-160 (Hall sensor). The inhomogeneity of the magnetic field was measured by SH1-1 along the line width Δf_{nmr} of the NMR signal at the resonance frequency. In modern stationary NMR spectrometers and relaxometers (for example, ER-100M or Minispec - Bruker), an electromagnet is used in vibrating sample magnetometer (for example, VSM model 7410 - Lake Shore Cryotronics Inc.) to generate magnetic fields with induction up to 2.6 T [14 -16]. In desktop NMR spectrometers (for example, Pulsar - from Oxford Instruments) to create magnetic fields B up to 2 T the permanent magnets are used. In portable (small-size) NMR spectrometers and relaxometers, systems with the permanent magnets are used too [12, 16, 17]. At present, in all the devices considered for the study of condensed matter in the temperature range T from 283 to 313 K the magnetic systems are stabilized.

4. Conclusion

The received results of experimental investigations showed that on the measured values of a grating constant in three planes of pole space of magnetic system with use of two types the ferrofluid of cells it is possible to construct a volume picture of force lines of a magnetic field. Such image of a configuration of force lines cannot be received when carrying out even approximate calculations of fields of magnetic systems.

As a result of researches we also found that in the ferrofluid cells for a research of structure of force lines of magnetic fields it is desirable to make its transparent sides of a quartz glass. It considerably reduces influence of various re-reflected radiation on intensity and position of maxima in the recorded diffraction image.

5. References

[1] Suharnikov K V, Rychkov M M 2015 *Radiophysics and Quantum Electronics* **58** 132-138
[2] Cherepov C V, Moroz O H, Derecha D A 2011 *Radiophysics and Quantum Electronics* **54** 137-145
[3] Logunov S E, Koskin A Yu, Davydov V V, Petrov A A 2016 *Journal of Physics: Conference Series* **741(1)** 012092
[4] Zaitsev N I, Ilyakov E V, Kulagin I S, Shevchenko A S 2006 *Radiophysics and Quantum Electronics* **49** 120-127
[5] Karseev A Yu, Vologdin V A, Davydov V V 2015 *Journal of Physics: Conference Series* **643(1)** 012108
[6] Davydov V V, Dudkin V I, Karseev A Yu 2015 *Technical Physics* **60** 456-460
[7] Davydov V V, Dudkin V I 2016 *Technical Physics* **61** 1115-1119
[8] Davydov V V, Dudkin V I, Karseev A Yu 2015 *Instruments and Experimental Techniques* **58** 787-793

[9] Kudasov Yu B, Makarov I V, Maslov D A, Platonov V V, Popov E Ya, Surdin O M, Voronov S L, Malyshev A Yu, Korotkov S V, Vodovozov V M 2015 *Instruments and Experimental Techniques* **58** 781-787

[10] Stolypko A L 2013 *Instruments and Experimental Techniques* **56** 649-654

[11] Cherepov S V, Moroz O Kh, Derecha D A, Hesse O P 2011 *Instruments and Experimental Techniques* **54** 273-279

[12] Davydov V V, Myazin N S 2017 *Measurement Techniques* **60** 183-189

[13] Petrov A A, Davydov V V 2017 *Journal of Communications Technology and Electronics* **62** 289 – 293

[14] Davydov V V, Myazin N S 2017 *Measurement Techniques* **60** 491-496

[15] Davydov V V, Cheremiskina A V, Velichko E N, Karseev A Yu 2014 *Journal of Physics: Conference Series* **541(1)** 012006

[16] Davydov V V, Velichko E N, Dudkin V I, Karseev A Yu 2015 *Instruments and Experimental Techniques* **58** 234-238

[17] Davydov V V, Dudkin V I, Karseev A Yu 2015 *Russian Physics Journal* **58** 146 – 152.

Model of ultra-wideband signal transmission and reception using the pseudorandom carrier

S V Valin[1], V A Glukhov[1], A V Siasko[1] and Yu A Tolmachev[1]

[1] St. Petersburg State University, Universitetskaya St. 7/9, 199034 St. Petersburg, Russia

E-mail: valin.sergey@gmail.com

Abstract. We show the possibility of transmission and subsequent decoding of an ultra-wideband signal in an optical communication channel using a noise-like carrier. For an example, the simulation of the transmission of word encoded with Baudot telegraph code was realized. The necessary sequence of symbols was introduced into the noise-like carrier represented by a pseudo-random sequence of ultra short pulses. The possibility of the transmitted word decoding to a high degree of reliability was shown by calculation of auto- and cross-correlation functions. To some extent, the method can be treated as an analog of pulse-position modulation.

1. Introduction

One of the crucial points in the development of information transmission technical means using electromagnetic radiation of visible and even shorter wavelength optical range is the development of technology of all-optical methods allowing to implement the theoretical limit of the speed of information processing. This process is associated with the development and upgrading of femto- and attosecond generators of high power laser pulses containing only a small number of the electromagnetic field oscillations, as well as with improving of pseudo-random "white" light sources, as the emitters of so-called supercontinuum radiation characterized with very broad spectrum that leads to very small temporal coherence, but characterized with high spatial coherence [1, 2].

On the way to solution of general problem formulated above, are the problems associated with

- the introduction of digital information into the carrier wave using various methods of modulation of the light flow,
- the decoding of information at the end point of transmission line, and
- the reliability and security of the communication channel.

For the first two points, most of existing methods are based on the pulse-position modulation of ultrashort laser pulses [3]; in this case, the signal is formed by time-pulse modulation. The only information parameter is the position in time of the rising edge of the pulse. With this type of modulation, the position of each pulse depends on the instantaneous value of the modulating signal and varies with respect to the position of periodic reference pulses generated by the receiver itself. In order to transmit a logical zero, the working pulse has to be sent, for example, slightly earlier than its predetermined position in the pulse sequence, and vice versa, for the transmission of the logical unit.

One of the most promising approaches to the solution of optical communication problems is in the development of optical technologies based on the use of so-called "complex" signals and noise-like data carriers that are of the greatest information capacity [4]. As a rule, the signal generated by the source of white light is transmitted by the amplitude modulation of the light flow. At the same time, other methods of an optical signal formation are possible, which are close to phase and pulse-code modulation by their implication. The detection of a signal in this case is possible by applying correlation methods of analysis of the light flow, which outwardly represents a random process. It is well known that the measurement of auto- and cross-correlation functions is the optimum method of a signal of known form detection, against the background of random additive noise [5].

In [6], we experimentally demonstrated the possibility of using one of the simplest optical systems for introducing the information into a random (noise) light flow of a thermal source with a "white" spectrum, and consequent identification of the corresponding signal, i.e. decoding. In fact, we suppose that it is more promising to use the impulses of supercontinuum as the data carrier in a similar communication system [1, 2].

The easiest technique to introduce one bit of information into a random or pseudorandom light flow is in the delay of the corresponding wave for a fixed time at the condition of the exact reservation of wave amplitude dependence on time. The presence or absence of a pulse at a certain "location" due to delay of the copy of the part of wave with respect to the carrier is transferred to the binary sequence of 0 and 1, which allows to decode the transmitted signal. This procedure is equivalent to pulse-code modulation and can be applied to the radiation of both a thermal source and a pulsed supercontinuum source.

In this paper, different methods of introducing information into the light flow, based on the use of bipolar sequences and allowing to encode alphanumeric symbols are studied. The obtained results form the basis for the subsequent experimental implementation of coding, transmission and decoding of information by the optical method.

2. Model

2.1. Carrier

For the transmission of one bit into an ultra-wideband pseudo-random continuous or pulsed light flow, an additional implementation of the same process is made, but time-shifted by a certain interval. The spectrum of such coded signal by its external characteristics (amplitude and phase) can only slightly differ from the spectrum of the original signal, however, the correlation properties of the generated light flow vary significantly. In the cross-correlation function, additional maxima appear, corresponding to the presence of shifted copies of the original signal in general sequence of pulses. Based on the analysis of the position of maxima of the cross-correlation function, information is decoded.

For the formation of optical signal, the Legendre sequence is used as a pseudo-random sequence in our study. The binary Legendre or quadratic residue sequences exist for all lengths L which are prime. They can be constructed using the Legendre symbol (i/L), a Legendre sequence $a_0, a_1 \dots a_{L-1}$ is then formed by writing $a_i = (i/L)$, $0 < i < L$, and the value of a_0 can be taken either as 1 or -1 [7].

The main characteristics of pseudo-random sequences are an autocorrelation function, defined for a discrete real-valued sequence y as

$$R_{yy}(l) = \sum_{n \in Z} y(n)y(n-l), \tag{1}$$

and the cross-correlation function of two real-valued sequences f and g:

$$CC(n) = \sum_{m=-\infty}^{\infty} f(m)g(m+n). \tag{2}$$

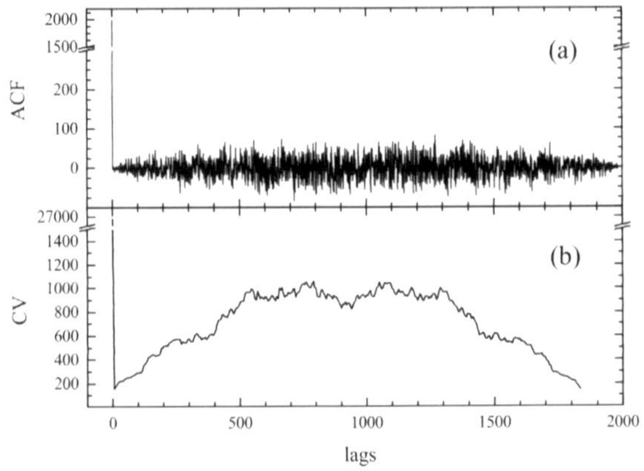

Figure 1. Autocorrelation function of the Legendre sequence for $p = 1987$ (a). Current dispersion of the autocorrelation function (b).

The "current" dispersion as a function of x_i is calculated than on the intervals $[i_k, i_k + N]$, $k = 1, 2, 3 \ldots$:

$$CV(x_i) = \frac{1}{N} \sum_{i_k}^{i_k+N} (x_{i_k} - \mu_k)^2, \quad \mu_k = \frac{1}{N} \sum_{i_k}^{i_k+N} x_{i_k}. \tag{3}$$

The example of autocorrelation function of such a sequence is shown in Fig. 1 together with its "current" dispersion.

As can be seen from Fig. 1, the Legendre binary sequence composed of -1 and 1 has distinctive correlation properties. In particular, the dispersion of the autocorrelation function is small at small shifts and increases with increasing delay.

2.2. Message

Telegraph codes like the 5–bit Baudot code [8] are the most easily implemented in the optical range.

We give an example of encoding and decoding of the short word "SPB". According to the Baudot code table, such a message can be written as a sequence [10100 01101 10011] (control characters are not included).

3. Results and Discussion

All calculations were performed for bipolar sequences generated using the Legendre symbol. The Legendre sequence for the prime number $p = 3571$ was used for the carrier sequence. Additional implementations of the pseudo-random process are introduced with an initial delay of 500 code elements. The delay between two logical units in a message is 20 elements. Logical "1" in the message is the introduction of an additional implementation of the original carrier with an appropriate delay of $500 + 20 \times (i - 1)$, where i is the running number of the unit in the message.

The results of transmission simulation are shown in Figs. 2 - 4.

In Fig. 2 the autocorrelation function is depicted of the initial sequence containing additional copies corresponding to the message "SPB". The "zero" peak has an amplitude of $A = 33143$. With a short message, already at this stage, the autocorrelation function allows one to confidently determine the message itself ("code" in Fig. 2) with the signal-to-noise ratio ≈ 4. But it should be noted that, besides the message, the autocorrelation function contains some beats ("beat" in Fig. 2), which make it difficult to detect the signal, especially if the message length is large

IOP Conf. Series: Journal of Physics: Conf. Series **1038** (2018) 012094 doi:10.1088/1742-6596/1038/1/012094

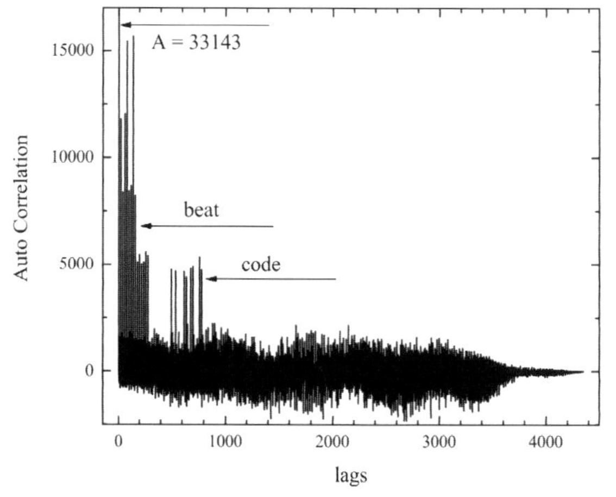

Figure 2. The autocorrelation function of the sequence containing the encoded message.

enough: $x_2 - x_1 > D$, where x_1, x_2 are the positions of the first and last characters in the message, D is the delay between the carrier start and x_1.

At the same time, the cross-correlation of the signal with the added copies and the original carrier yields only the message itself with no beats at the output (Fig. 3).

As it can be seen from Fig. 3, the cross-correlation gives the set of maxima corresponding only to logical units in the encoded message. In addition, the signal-to-noise ratio increases by two orders of magnitude, as is clearly seen in Fig. 4, where the neighborhood of maxima of the cross-correlation function is represented.

4. Conclusion

We show the principal possibility of the digital information introduction into an ultra-wideband pseudorandom carrier for its transmission and subsequent decoding using the simplest Baudot code. The high signal-to-noise ratio and rather simple engineering feasibility makes it possible

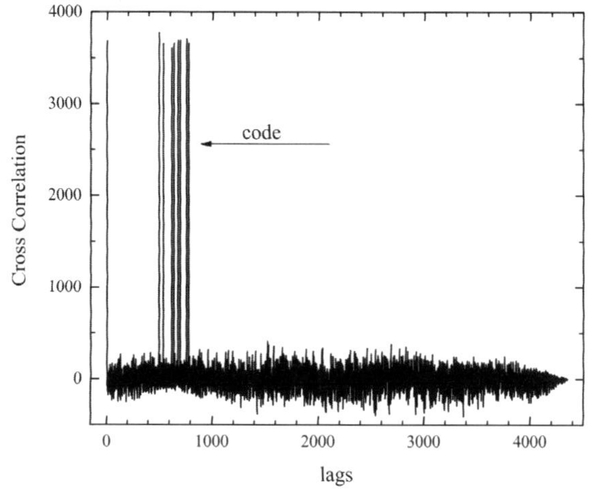

Figure 3. Cross-correlation of the original carrier and the sequence containing additional copies.

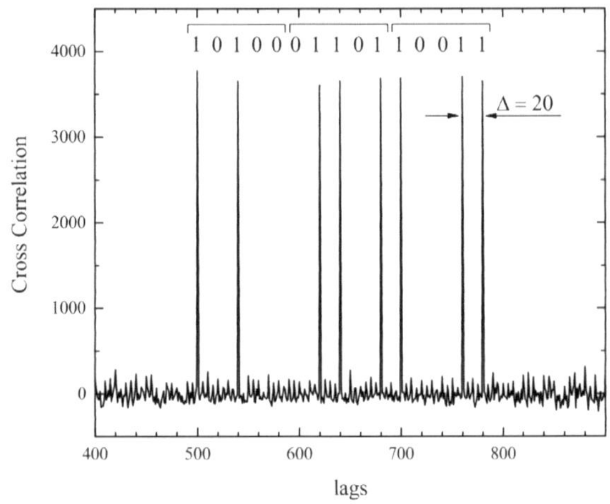

Figure 4. The neighborhood of maxima of the cross-correlation function given in Fig. 3.

to apply such a code for the experimental realization of the transmission and reception of signal using the optical correlator described in [7].

Acknowledgments
The authors thanks Russian Science Foundation (17 – 19 – 01097) for the support.

References

[1] Dianov E M and Kryukov P G 2001 *Quantum Electronics* **31** 877–882
[2] Zheltikov A M 2006 *Uspekhi Fizicheskikh Nauk* **176** 623–649
[3] Wong K T 2007 *European Transactions on Telecommunications* **18** 193–197 ISSN 1541-8251
[4] Schuster H G 1995 *Deterministic Chaos: An Introduction* 3rd ed (Wiley-VCH) ISBN 3527293159
[5] Kharkevich A 2009 *Fighting Noises* (Communications Technology)
[6] Glukhov V, Koulikov I and Tolmachev Y 2016 *Journal of Physics Conference Series* **735**
[7] Bai E and Zhang B 2004 *Progress on Cryptography: 25 Years of Cryptography in China* (Boston, MA: Springer US) chap Legendre Sequences and Modified Jacobi Sequences, pp 9–16 ISBN 978-1-4020-7987-0
[8] Ralston A, Reilly E and Caryl A 1993 *Encyclopedia of Computer Science & Engineering* (New York: Van Nostrand Reinhold) chap Baudot Code 3rd ed ISBN 0-442-27679-6

Phase-shifted fiber Bragg gratings fabrication method

Novikova V.A., Varzhel S.V., Konnov K.A., Gribaev A.I., Mikhneva A.A., Frolov E.A.
ITMO University, 197101, St. Petersburg, Russia

E-mail: novivial@mail.ru

In this paper, we showed a technique for the formation of phase-shifted fiber Bragg gratings. The analysis of the efficiency of the obtained structures was carried out with the help of an optical spectrum analyzer with the subsequent calculation of the full width at half maximum of a bandpass in reflection spectrum. Fiber Bragg gratings inscription was performed on the Talbot interferometer, the KrF excimer laser system was used as a UV radiation source, and a phase shift was introduced by the electrical discharge of an arc fusion splicer. The experimental dependence between the length of the obtained structure and the full width at half maximum of a bandpass in reflection spectrum of a phase-shifted grating was presented, the value of the last parameter varies in the range from 24 pm to 96 pm.

1. Introduction

The formation of Bragg gratings representing the modulation of the refractive index along the direction of wave propagation in the optical fiber core was first described in a 1978 [1]. Later, in 1989 [2], an interferometric inscription of the refractive index gratings through a side surface into a germanium-silicate optical fiber (with an increased value of photorefractivity) by exposure to UV radiation was presented. This publication [2] initiated the propagation of fiber Bragg gratings (FBGs), which make it possible to obtain spectral characteristics depending on the tasks assigned. Currently, fiber Bragg gratings are used in fiber lasers, in fiber optic communication systems, in sensors of various physical quantities (for example, pressure, temperature, mechanical stresses, etc.).

With the development of fiber optics, Bragg grating production technologies were also improved, and special structures were developed that made it possible to obtain spectral characteristics different from the standard refractive index gratings. Some of these structures are fiber Bragg gratings with a phase shift, the study of which was first described in 1994 [3, 4]. The distinctive feature of such gratings is the presence of a narrow transmission region in the reflection spectrum, which full width at half maximum could be on the order of a few tens, sometimes units of picometers. This type of refractive index gratings is used in the manufacture of spectral filters [5-7], lasers with distributed feedback [8], sensors of various physical quantities with increased resolution (in comparison with standard FBGs) [9, 10].

Many approaches have been proposed to the formation of Bragg gratings with a phase shift. One of the simplest techniques for the formation of such structures is the use of phase masks with a pre-formed phase shift [3]. The disadvantage of this method is the absence of the possibility of changing the grating period, as well as in a high cost of such masks manufacturing. Different techniques of phase-shifted FBGs formation are presented: with overlapping of one grating period by a region with an increased refractive index [11]; with the introduction of an air gap between the two ends of the fibers on which the FBG is formed [12]; with ultraviolet post-processing of a narrow region in the center of the FBG formed [4]. In general, the disadvantages of the proposed techniques are associated with the use of high-precision optical and/or mechanical devices and, correspondingly, their high cost, as well as with a relatively complex adjustment of the specific position of the fibers.

The technique using an arc fusion splicer for introducing a phase shift was described in [12] and was used as a prototype. But, unlike our method, in work [12] the electric arc was effected in the center of a single FBG. Thus, firstly, there were increased requirements for an accurate determination of the center of the Bragg grating, and secondly, the diffraction efficiency of such a structure decreased after the influence of the electric arc in comparison with the originally induced standard FBG.

In this paper, we propose a technique for the formation of phase-shifted FBGs, which are inscribed on a Talbot interferometer, while a phase shift is introduced by means of the electrical discharge of an arc fusion splicer between two formed gratings. The advantages of this method consist in the possibility of adjusting the parameters (period and wavelength) of the induced Bragg grating, the absence of high-precision devices at the stage of phase shift formation and high diffraction efficiency of the formed structures.

2. Experimental setup

In this work the KrF excimer laser system MOPA CL-7550 (Optosystems Ltd, Russia) was used as a radiation source. The advantage of using such Master Oscillator–Power Amplifier laser system is a high degree of spatial (> 5 mm) and temporal (> 10 mm) coherence, which reduces the requirements for the alignment of the interferometer, and also allows obtaining an interference pattern of high contrast on the surface of the optical fiber. The radiation wavelength of the laser system is 248 nm.

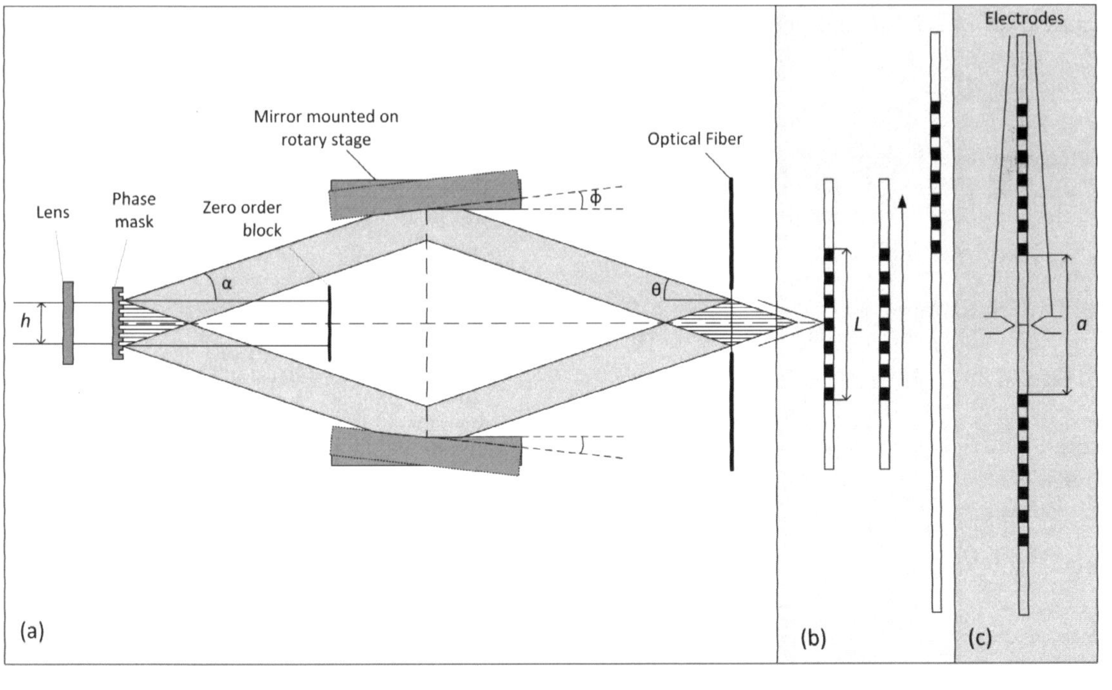

Figure 1. FBG inscription scheme by means of a Talbot interferometer and a phase shift introducing stage by means of an electrical discharge between electrodes

The FBG inscription was performed on the Talbot interferometer [13], which functional scheme is shown in Fig. 1(a). The laser beam passing through the lens is focused on the phase mask, which is used for amplitude separation of the beam. To block the zero order of diffraction, an opaque screen is installed in the scheme. +1 and -1 diffraction orders, reflected from the dielectric mirrors mounted on rotary stages, interfere with each other and as a result of the interference a fiber Bragg grating is formed in the core of the optical fiber. The rotary stages on which the mirrors are fixed allow one to change the angle of incidence of the rays onto the optical fiber θ, thereby adjusting the grating period (and hence the wavelength of the Bragg resonance) by changing the angle φ of the mirrors by equation (1):

$$\Lambda = \frac{\lambda_{laser}}{2\sin(\alpha - 2\varphi)}, \quad (1)$$

where Λ is the grating period, α is the 1st order diffraction angle, φ is the angle of rotation of the mirrors, λ_{laser} is the radiation wavelength of the excimer laser system.

After inscription of the first FBG, the fiber is shifted along its axis to a certain distance and the second FBG is induced (Fig. 1(b)); the formation of both Bragg gratings takes place under identical conditions while preserving the main parameters, so the period and length of the successively inscribed gratings, as well as the value of the refractive index modulation and the Bragg resonance wavelength, are the same. Then, a π-shift is induced between two FBGs, leading to destructive interference and the appearance of a narrow bandpass in a reflection spectrum of the diffraction structure.

Figure 1(c) shows a schematic representation of the phase shift formation stage. The phase difference is induced using the electrical discharge of an arc fusion splicer: the influence of the electric arc produced between the electrodes slightly changes the geometry of the fiber. The result of the described manipulations is being monitored in real time on the optical spectrum analyzer, and the number of effects of the electric arc can be different until the most appropriate spectrum is achieved. The exposure time of the electrical discharge of an arc fusion splicer was 0.3 seconds, the current value was 8 mA.

In our experiments the gratings were spaced from each other by a distance equal to 1 mm (*a* in Fig. 1(c)). The distance control was carried out with the help of a linear micrometric stage, the error of which determines the case when the inscription of the second grating allows to induce the π-phase-shift, and there is no need for further exposure to the electric arc, but such a random hit has a small probability of occurrence.

3. Experimental results

To analyze the effectiveness of the developed technique, it is necessary to consider the spectral characteristics of obtained phase-shifted FBGs. The reflection spectra were recorded according to the scheme shown in Figure 2 by means of the Yokogawa AQ6370C optical spectrum analyzer.

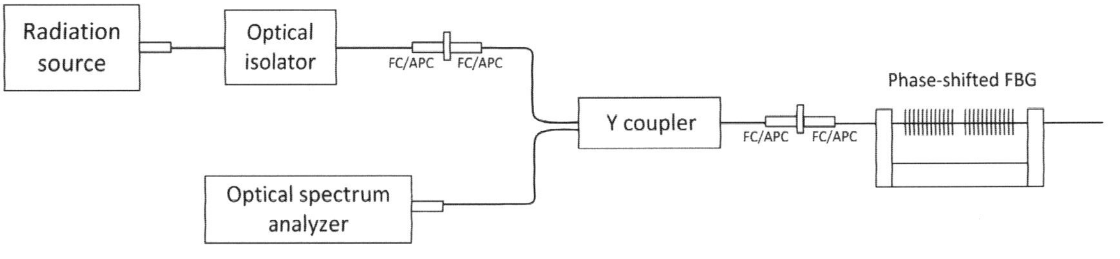

Figure 2. Principal scheme for monitoring the phase-shifted FBG reflection spectrum

Optical radiation from a broadband source is coupled through the Y-coupler to an optical fiber into which a π-shifted Bragg grating is inscribed. Reaching the FBG, part of the radiation is reflected and goes to the optical spectrum analyzer through the Y-coupler. The optical isolator transmits light from the radiation source to the induced FBG, but prevents its passage backwards in order to prevent the influence of back reflections on the spectrum of the radiation source.

According to this scheme, the reflection spectra shown in Figure 3 were obtained. As was pointed out earlier, the parameters of two gratings consecutively inscribed in an optical fiber and spaced apart by a distance of 1 mm are the same, hence, the reflection spectra of these gratings coincide. In Figure 3, the green dashed line represents the reflection spectrum of one of the two FBGs, and the blue solid line represents the phase-shifted FBG reflection spectrum.

For FBG inscription the optical fiber was irradiated for 1 minute at a frequency of 10 Hz, while the energy density on the surface of the fiber was 130 mJ/cm^2.

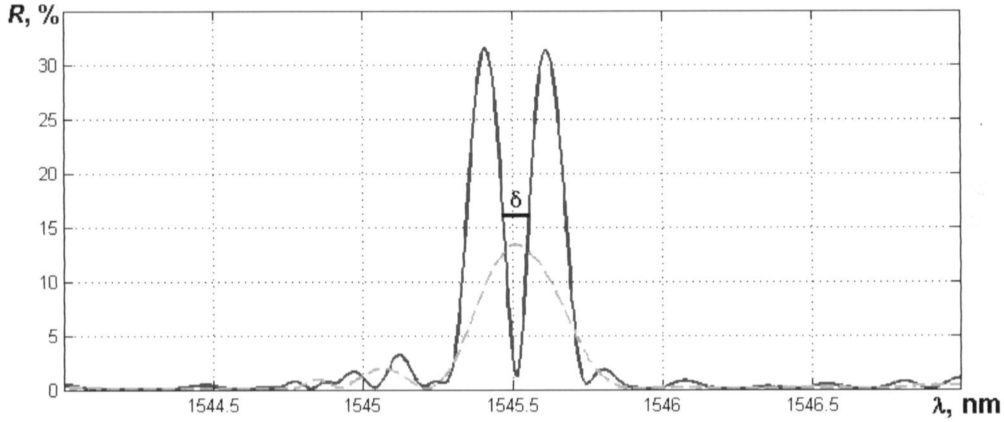

Figure 3. Reflection spectrum of a phase-shifted fiber Bragg grating: the green dashed line - the reflection spectrum of one of the two FBGs, the blue solid line - the phase-shifted FBG reflection spectrum

The shift of the Bragg wavelength for the phase-shifted FBG relative to the standard FBG, that could be seen on Figure 3, is explained by the fact that the inscription of Bragg gratings is carried out in a fiber under small tension. After inscription of the grating, the optical fiber is removed from the fiber holders and placed in the arc fusion splicer in order to induce a phase shift by means of an electrical discharge.

In Figure 3, the variable δ is the value of the full width at half maximum of a bandpass in reflection spectrum. This parameter makes it possible to compare Bragg gratings with π-shifts among themselves. When analyzing gratings of different lengths, it is noted that as the grating length increases, the full width at half maximum of a bandpass in reflection spectrum decreases. This is due to the fact that the number of spaced interfering waves increases with the length of the grating. In turn, lateral values of the reflection band make less impact and, being out of phase with each other, attenuate, leading to a decrease in the full width at half maximum of the reflection spectrum, hence the width of a bandpass in reflection spectrum also decreases. Figure 4 shows the experimental dependence of the full width at half maximum of a bandpass in reflection spectrum on the length of the induced FBG with a π-shift.

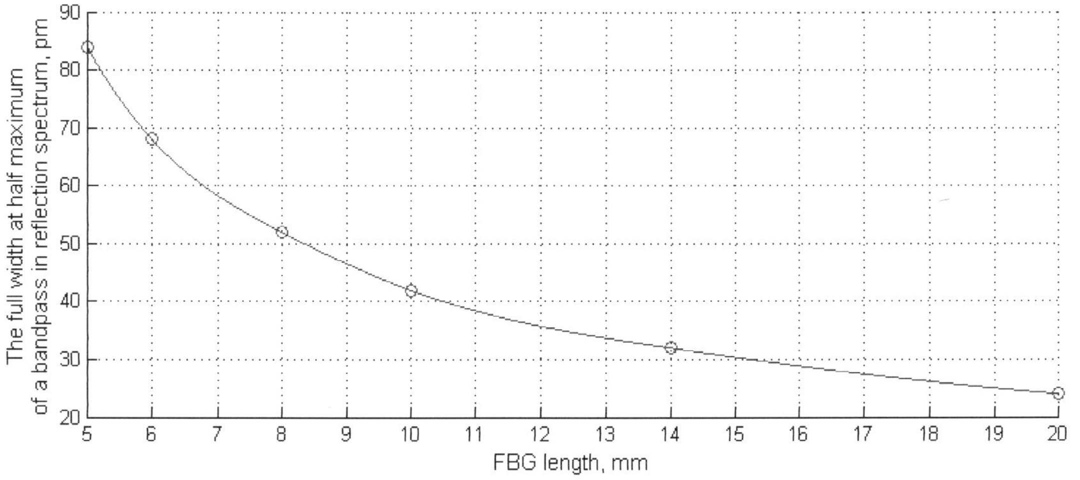

Figure 4. The blue open points - the experimental values of the full width at half maximum of a bandpass in reflection spectrum for the different grating lengths, the blue solid line - first degree spline, which was built for these experimental values

The π-shifted FBG length in mm is calculated as the total value of the two lengths of Bragg gratings without taking into account the distance to which they are spaced (a). The minimum value of the full width at half maximum of a bandpass in reflection spectrum was achieved with a grating length of 20 mm (two FBG of 10 mm) and equaled 24 pm.

4. Conclusion

In this paper we present an inscription of fiber Bragg gratings on the Talbot interferometer and the subsequent introduction of a phase shift by means of the electrical discharge of an arc fusion splicer between two FBGs. This method of forming such structures makes it possible to exclude the use of high-precision and expensive devices at the stage of phase-shift formation. In comparison with the method presented in [12] taken as a prototype, in the method described by us, the electrical discharge of an arc fusion splicer is produced not in the center of the generated FBG, but between two inscribed FBGs, which avoids mashing the diffraction structure induced in the optical fiber. This circumstance has a positive effect on the quality of the obtained FBG with a phase shift.

The main value for comparing the gratings was the full width at half maximum of a bandpass in reflection spectrum - this parameter depends mainly on the length of the grating. As a result of the experiments, this value varies in the range from 24 to 96 pm, which allows us to conclude that it is possible to obtain a narrowband passing radiation according to the procedure described in this paper.

Acknowledgements

This work has been done at the ITMO University and supported by the Ministry of Education and Science of the Russian Federation (Unique identifier of the Project: RFMEFI57816X0202, Contract No. 14.578.21.0202).

References

[1] Hill K.O., Fujii Y., Johnson D.C., Kawasaki B.S. 1978 *Appl. Phys. Lett.* V. 32. N 10. P. 647-649
[2] Meltz G., Morey W.W., Glenn W.H. 1989 *Opt. Lett.* V. **14**. N 15. P. 823-825
[3] Kashyap R., McKee P.F., Armes D. 1994 *Electron. Lett.* V. **30**. N 23. P. 1977-1978
[4] Canning J., Sceats M.G. 1994 *Electron. Lett.* V. **30**. N 16. P. 1344-1345
[5] Bakhti F., Sansonetti P. 1996 *Electron. Lett.* V. **32**. P. 581-582

[6] Zengerle R., Leminger O. 1995 *J. Lightwave Technol.* V. **13**. P. 2354-2358
[7] Wei L., Lit J.W.Y. 1997 *J. Lightwave Technol.* V. **15**. P. 1405-1410
[8] Kashyap R. Fiber Bragg Gratings. 1999 San Diego, CA: *Academic Press*. 478 p.
[9] Hao Wang. A Phase-shifted Fiber Bragg Grating Based Humidity Sensor. 2013 *Thesis for MS degree.* 55 p.
[10] LeBlanc M., Vohra S. T., Tsai T. E., Friebele E. J. 1999 *Opt. Lett.* V. **24**. N 16. P. 1091-1093
[11] Edmon Chehura, Stephen W. James, Ralph P. Tatam. 2009 *Proc. of SPIE*. V. 7503. 750379. P. 1-4
[12] Yuanhong Yang, Xuejing Liu, Wei Jin. 2013 *ACP/IPOC*, paper ATh3D.5
[13] Gribaev A.I., Pavlishin I.V., Stam A.M., Idrisov R.F., Varzhel S.V., Konnov K.A. 2016 *Opt Quant Electron*. V. **48**. Article 540. P. 1-7

IOP Conf. Series: Journal of Physics: Conf. Series **1038** (2018) 012096 doi:10.1088/1742-6596/1038/1/012096

Optical method for assessing the effectiveness of treatment of staphylococcal infection of tonsils

P E Timchenko[1], E V Timchenko[1*], A A Asadova [1], Yu D Ityaksov[1]

[1]Samara National Research University, 34, Moskovskoye shosse, Samara, 443086, Russia.

*E-mail: laser-optics.timchenko@mail.ru

Abstract. The results of studies of the efficacy of treatment of staphylococcal infection in palatine tonsils using «Amoxiclav» by a method of Raman spectroscopy. Spectral changes in the treatment of palatine tonsils with the antibiotic «Amoxiclav» are revealed. The coefficients allowing to estimate the effectiveness of treatment of staphylococcal infection with antibiotics «Amoxiclav» are introduced. It is established that the antibiotic «Amoxiclav» is more effective when exposed to a strain of staphylococcus culture (I).

1.Introduction

According to the report on the state of sanitary and epidemiological welfare of the population in the Russian Federation, the leading place in the structure of infectious and parasitic diseases in 2016, as in previous years, is acute infections of the upper respiratory tract of multiple and unspecified localization (ARVI). The proportion of which is 84%, for acute tonsillitis - 3.7% [1].

The most significant bacterial pathogen of acute tonsillitis (angina) is the hemolytic streptococcus of group A. Less commonly, acute tonsillitis is caused by viruses and other streptococci, extremely rarely by mycoplasma and chlamydia. The causative agent is transmitted by airborne droplets. Sources of infection are patients, and in some cases carriers of the disease, who do not have obvious symptoms. Chronic tonsillitis is a consequence of repeated angina or insufficient treatment [2].

Treatment of acute tonsillitis is carried out through the use of various antibacterial drugs, the most common among them was the antibiotic «Amoxiclav», consisting of Amoxicillin and clavulanic acid [3].

The reasons for the development of resistance of microorganisms to antibiotics are: unreasonable prescription of antibacterial agents, mistakes in the choice of an antibacterial drug, errors in the choice of the dosing regimen of an antibacterial drug, errors in the combined use of antibiotics, errors associated with the duration of antibiotic therapy [4].

Therefore, monitoring the effectiveness of antibiotic treatment is an urgent task.

To monitor the occurrence of complications and the effectiveness of antibiotic therapy, a number of laboratory studies are necessarily conducted at the beginning and at the end of treatment of acute tonsillitis. This is a general blood test and a general urine test, an electrocardiogram and a biochemical blood test. In most cases, the treating specialists neglect to conduct a laboratory analysis of pharyngeal tonsillar flora. However, in the treatment of tonsillitis it is important to observe the correct concentration of antibiotic for the course of treatment. Usually it is checked by chemical and microbiological methods [5].

At present, optical methods of investigation have been widely used in medical problems [6-11], such as backscattering [6,7], Raman spectroscopy [8], IR spectroscopy [9,10], and confocal microscopy [11].

Raman spectroscopy can also be used to study saliva, with larynx diseases. Thus, it was established in [8] that a decrease in the intensity at wave numbers 635 cm-1 (tyrosine) and 812 cm-1 (L-serine) and an increase of 1619 cm-1 (tryptophan) indicates a change in the amino acids that make up the saliva , and corresponds to the growth of tumor cells in the nasopharynx.

The article [9] considers the conduction of infrared spectroscopy of saliva in patients with clinically diagnosed chronic tonsillitis by determining the average values of the transmittance of infrared radiation in the ranges: 3085-2832 cm^{-1}, 2120-1880 cm^{-1}, 1600-1535 cm^{-1}, 1543-1425 cm^{-1}, 1430-1210 cm^{-1}, 1127-1057 cm^{-1}. Depending on the mean values obtained over several ranges, either compensated or decompensated chronic tonsillitis is diagnosed. This method is painless fast, and can be performed on an outpatient basis.

The aim of the study is to analyze the effectiveness of treatment with «Amoxiclav» of staphylococcal infection in palatine tonsils using Raman spectroscopy.

2. Materials and methods of research

During experiments investigated the 12 samples containing Staphylococcus strain ATCC№29923 (culture I) and ATCC№35591 (culture II), are in saliva of patients and in saline. Half of the samples were treated with an antibiotic, such as «Amoxiclav», the other half was a control.

Samples of the studies were provided by the professor, MD. Zarubina E.G. (Samara Medical University "Reaviz"). The classification of the test samples is shown in Table 1.

Table 1. Classification of test samples

Label	Description
1.1	1 ml. saliva + culture (I)
1.2	1 ml. saline solution + culture (I)
1.3	1 ml. saliva + culture (I) + antibiotic 3 mg.
1.4	1 ml. saline solution + culture (I) + antibiotic 3 mg.
2.1	1 ml. saliva + culture (II)
2.2	1 ml. saline solution + culture (II)
2.3	1 ml. saliva + culture (II) + antibiotic 3 mg.
2.4	1 ml. saline solution + culture (II) + antibiotic 3 mg.

Spectral characteristics were studied using an experimental stand including the high-resolution digital spectrometer Andor Shamrock sr-303i with the built-in cooled DV420A-OE camera, the fiber-optic probe for Raman spectroscopy RPB785, combined with the LuxxMaster LML-785.0RB-04 laser module adjustable power up to 500 mW, wavelength 785 nm [12, 13]. Processing of Raman spectra was carried out in the program Wolfram Mathematica 8. The investigated spectrum during processing was cleaned of noise by a smoothing median filter (5 points). At the chosen interval of 400-2200 cm^{-1}, using an iterative algorithm by the method of polynomial approximation of the fluorescent component, an approximating line (a polynomial of the fifth degree) of the autofluorescent component was determined, and then this component was subtracted, obtaining a selected Raman spectrum [14].

3. Results and discussion

Figure 1 shows the averaged Raman spectra of the saliva samples (a) and saline (b).

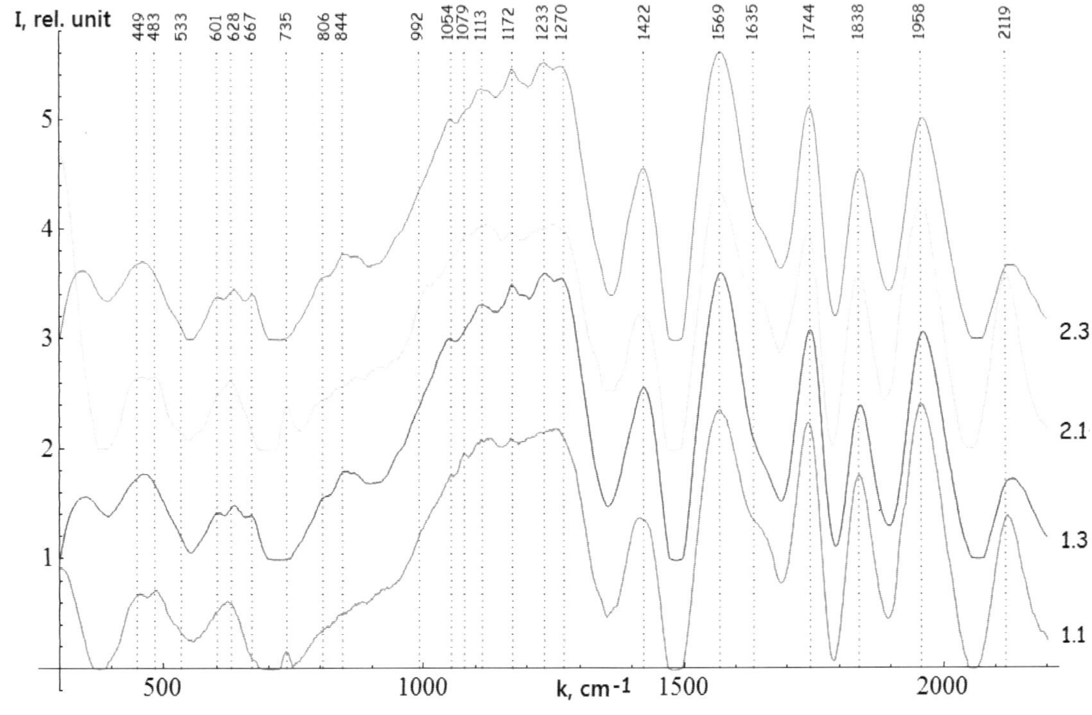

a)

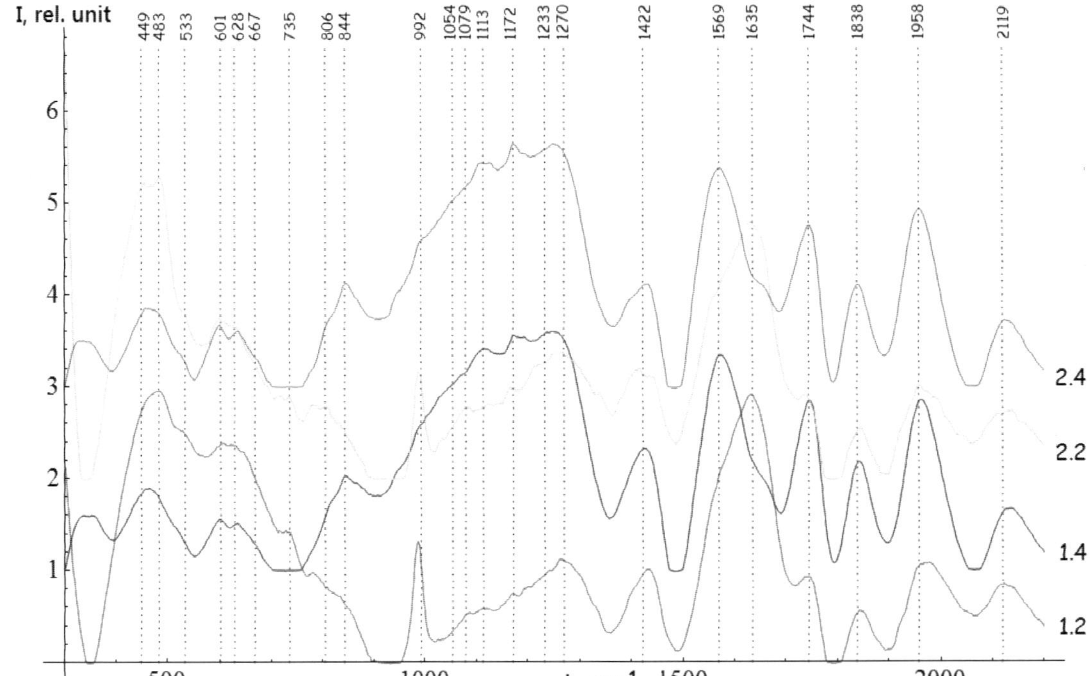

b)

Figure 1- Averaged Raman spectra of the investigated samples of saliva (a) and saline (b)

Figure 1 shows the spectral changes at wave numbers: 628 cm^{-1}, 667 cm^{-1}, 735 cm^{-1}, 844 cm^{-1}. Changes in the intensity of the lines at wave numbers of 628 cm^{-1} and 844 cm^{-1} correspond to the lines of the introduced antibiotic in the samples under study.

The line intensities at 667 cm^{-1}, 735 cm^{-1} wave numbers correspond to changes in the nitrogenous bases of DNA, such as adenine and guanine, contained in staphylococcal strains. In Fig. 1, spectral features are also noticeable for samples with cultures without antibiotic, which manifest themselves in a change in line intensities at wave numbers of 992 cm^{-1} and 1635 cm^{-1}, which is associated with a change in glycine and proteins.

Changes in the intensity of these lines can be used to assess the effectiveness of treatment of staphylococcal infection with the antibiotic «Amoxiclav».

To evaluate the effectiveness of treatment of staphylococcus in the palatine tonsils with antibiotic «Amoxiclav», coefficients (1-4) were introduced. The denominator in the introduced coefficients was the line intensity at the wave number of 1744 cm^{-1}, corresponding to the acetates.

$$A = \frac{I_{667}}{I_{1744}} \quad (1)$$

$$B = \frac{I_{735}}{I_{1744}} \quad (2)$$

$$E = \frac{I_{992}}{I_{1744}} \quad (3)$$

$$F = \frac{I_{1635}}{I_{1744}} \quad (4)$$

I_{667} – the Raman intensity at a wave number of 667 cm^{-1}, proportional to the content of guanine;
I_{735} – the Raman intensity at a wave number of 735 см$^{-1}$, proportional to the content of adenine;
I_{992}– the Raman intensity at a wave number of 992 см$^{-1}$, corresponds to glycines.
I_{1635}– the Raman intensity at a wave number of 1635 см$^{-1}$, corresponds to proteins.
I_{1744} – the Raman intensity at a wave number of 1744 см$^{-1}$, proportional to the content of acetates.
Based on the introduced coefficients, two-dimensional dependences were constructed (Figure 2).

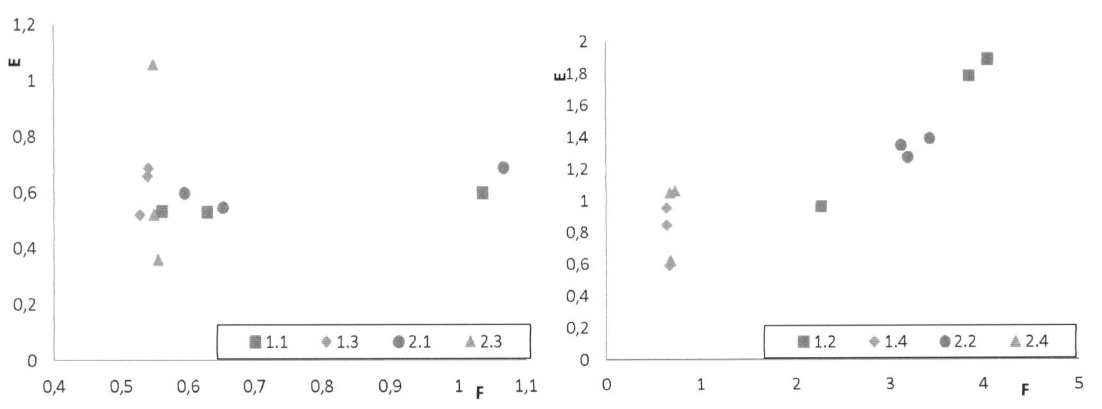

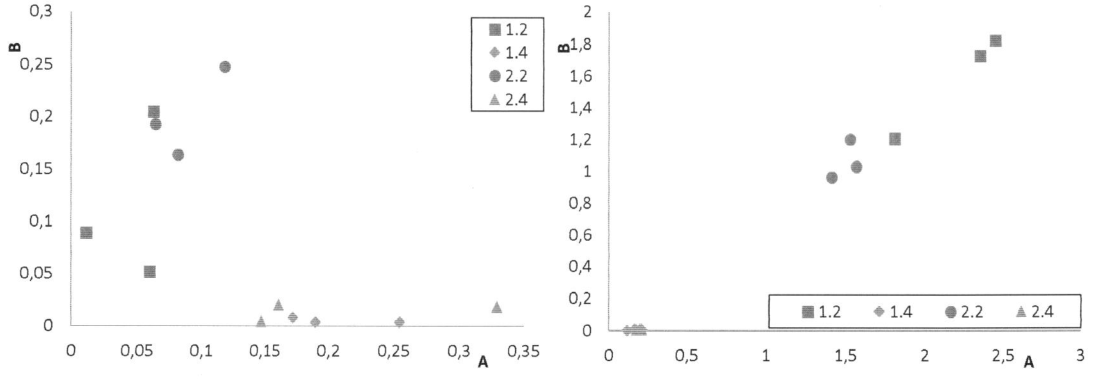

Figure 2 - E dependence coefficient of the coefficient F (top), the coefficient B from the coefficient A (below) for saliva samples (left) and samples based on saline (right)

As can be seen from the figure, the addition of an antibiotic causes a decrease in the coefficients F and B for both saliva samples and saline samples. These coefficients show a decrease in the relative concentration of proteins (1635 cm^{-1}) and adenine (735 cm^{-1}), indicating a decrease in the concentration of bacteria in the samples and can be used to monitor the effectiveness of treatment. Furthermore, the antibiotic in saliva samples and spectra, and the figure shows a decrease in the coefficient A, indicating the relative concentration of guanine (667 cm^{-1}). In the spectra of saline solution this is not observed, which is possibly connected with overlapping of this line by other stronger lines. A similar situation is observed for the coefficient E showing the relative concentration of glycine: for saline solution it is well resolved with respect to other lines and its decrease is clearly visible, and for saliva it is lost against the background of other peaks.

Table 2 shows the results of the calculation of the change in the relative concentration of staphylococcal infection of different strains under the influence of the antibiotic «Amoxiclav» with the help of the introduced coefficient B.

Table 2 - Coefficient B for evaluation of the effectiveness of treatment of staphylococcus with antibiotic «Amoxiclav» with a dosage of 500 mg / 10 ml

Samples	Coefficient $B=I_{735}/I_{1774}$
1.1 – 1 ml. saliva + culture (I)	100,0%
1.3 - 1 ml. saliva + culture (I) + antibiotic 3 mg.	3,5%
2.1 – 1 ml. saliva + culture (II)	100,0%
2.3 – 1 ml. saliva + culture (II) + antibiotic 3 mg.	6,9%
1.2 – 1 ml. saline solution + culture (I)	100,0%
1.4 – 1 ml. saline solution + culture (I)+ antibiotic 3 mg.	0,02%
2.2 – 1 ml. saline solution + culture (II)	100,0%
2.4 – 1 ml. saline solution + culture (II) + antibiotic 3 mg.	0,4%

It can be seen from Table 2 that when the antibiotic is exposed, the values of the optical coefficient B are reduced to 3.5% -6.9% (depending on the staphylococcus strain) for saliva and up to 0.02% - 0.4% for saline. Antibiotic «Amoxiclav» is more effective when exposed to the strain of staphylococcus culture (I), from 100% to 3.5% and from 100% to 0.02% compared to exposure to the strain of Staphylococcus culture (II), from 100% to 6 , 9% and from 100% to 0.4%.

4. Conclusions

Result of the study:

-Spectral changes were detected in the treatment of palatine tonsils with antibiotic "Amoxiclav", which appear to change lines at wave numbers 667 cm^{-1}, 735 cm^{-1}, 992 cm^{-1}, 1635 cm^{-1}, corresponding to guanine, adenine, glycine and proteins.

- The coefficients allowing to estimate the effectiveness of treatment of staphylococcal infection with antibiotic «Amoxiclav» are introduced. It is established that the antibiotic «Amoxiclav» is more effective when exposed to a strain of staphylococcus culture (I)

Acknowledgments
The work was supported by the Ministry of Education and Science of the Rassian Federation.

References

[1] On the state of sanitary and epidemiological well-being in the Russian Federation in 2016: State report .- Moscow: Federal Service for Supervision of Consumer Rights Protection and Human Welfare, 2017.- p. 220
[2] Kryuchkov T.A., Tkachenko O. Ya., Shpekht T.V.. The problem of tonsillitis in pediatric practice // Child's health - 2010. - №. 1. - p. 22
[3] Belov B.S., Grishaeva T.P. A-streptococcal tonsillitis: modern aspects of antibiotic therapy // Pediatric Pharmacology. - 2007. - T. 4. - №. 3
[4] Volosovets AP, Krivopustov SP, Yulish EI Contemporary views on the problem of antibiotic resistance and its overcoming in clinical pediatrics // Child's Health. - 2007. - № 6. - p. 9
[5] Kulapina EG, Modern Methods for Determining Antibiotics in Biological and Medicinal Media. Kulapina, O.V. Barinova, O.I. Kulapin // Antibiotics and chemotherapeutics. 2009, p. 9-10, 53-60
[6] Zaharov V.P., Timchenko E.V., Timchenko P.E., Zolotuhina A.D., Alembekov S.V. Alteration of hydrosphere optical properties by synthetic active compounds // Computer Optics, 2011 - 35 (2), pp. 238-242
[7] V.P. Zakharov, O.N. Makurina, E.V. Timchenko, P.E. Timchenko, S.P. Kotova, Valliylov Ecological monitoring of megapolis on the basis of differential backscattering control of the wood culture // Laser Physics, 2009, 19 (6), pp. 1366-1372
[8] Li XZ, Yang TY and Ding JH: Surface enhanced Raman spectroscopy (SERS) of saliva for the diagnosis of lung cancer. Guang Pu Xue Yu Guang Pu Fen Xi. 32:391–393. 2012
[9] Nauman D, Keller S, Helm D, Schultz C, Schrader B. FT-IR Spectroscopy and FT-Raman Spectroscopy are powerful analytical tools for the non-invasive characterization of intact microbial cells. J. Mol. Struct. 1995; 347:399-406
[10] P E Timchenko, E V Timchenko, E V Pisareva, M Yu Vlasov, N A Red'kin and O O Frolov Spectral analysis of allogeneic hydroxyapatite powders // IOP Conf. Series: Journal of Physics: Conf. Series 784 (2017) 012060 doi:10.1088/1742-6596/784/1/012060
[11] Timchenko P.E., Zakharov V.P., Volova L.T., Boltovskay V.V., Timchenko E.V. Diagnostics of bone implantat and control of their process osteointegration with of a method confocal microscopy // Computer Optics, 2011 - 35 (2), pp. 183-187
[12] E.V. Timchenko, P.E. Timchenko, N.V. Tregub, L.A. Taskina, E.A. Selezneva Mapping of the Samara city by definition of areas with hydrogen degassing using Raman spectroscopy // Proc. of SPIE, 2014. - Vol. 9448. - № 94480K. doi: 10.1117/12.2179510
[13] E.V. Timchenko, P.E. Timchenko, L.T. Volova, S.V. Pershutkina, P.Y. Shalkovsky OPTICAL ANALYSIS OF AORTIC IMPLANTS, Optical Memory and Neural Networks, 2016, V.25, №3, pp.192-197
[14] E.V. Timchenko, P.E. Timchenko, L.A. Taskina, L.T. Volova, M.N. Miljakova; N.A. Maksimenko Using Raman spectroscopy to estimate the demineralization of bone transplants

IOP Conf. Series: Journal of Physics: Conf. Series **1038** (2018) 012096 doi:10.1088/1742-6596/1038/1/012096

during preparation // JOURNAL OF OPTICAL TECHNOLOGY,2015 – V. 82 – №3 Pp.153-157

Carrier lifetime in InAs(Ga,Sb,P) heterostructures

A A Semakova[1,2], N L Bazhenov[2] and K D Mynbaev[1,2]

[1] ITMO University, Saint-Petersburg 197101, Russia
[2] Ioffe Institute, Saint-Petersburg 194021, Russia

antonina.semakova@gmail.com

Abstract. Carrier lifetime limited by radiative and most probable non-radiative recombination processes was calculated for InAs(Ga,Sb,P) heterostructures under injection. At low temperatures (T<130 K), the carrier lifetime was mostly determined by radiative recombination. With temperature increasing, influence of Auger processes became crucial. Among these processes, the dominating role of CHHS process, in which the energy of a recombining electron-hole pair is transferred to a heavy hole transitioning to the spin-orbit-splitted band, was established at low temperatures.

1. Introduction

Devices based on infrared light sources are useful for medical and industry applications as well as for environmental monitoring. The most common infrared sources are thermal ones, but their efficiency is limited by the maximum temperature that materials can withstand without thermal degradation. In this regard, devices based on light-emitting diodes (LEDs) are very promising. One of the main disadvantages of middle-wavelength infrared (MWIR) LEDs is their low power. Therefore, study of recombination processes in such devices is very important, as the mechanisms of recombination determine the lifetime of charge carriers, and consequently, the efficiency of the devices. Recently, there has been much interest in theoretical calculations of carrier lifetime in materials for MWIR devices, including InAs-based materials. Currently, there exist different approaches to calculation of the carrier lifetime limited by non-radiative processes in such materials [1-3]. For the most rigorous calculations, it is important to consider the band structure of the semiconductor, in particular, the non-parabolicity of the electron and light hole bands, characteristic of narrow-bandgap materials. In this paper, we report on the results of calculations of carrier lifetime for InAsSb active layer of InAs(Ga,Sb,P) heterostructures limited by radiative and most probable non-radiative processes, which were performed within the frames of the Kane model. The calculations were based on the results of the recent study of electroluminescence of InAs(Ga,Sb,P) LED heterostructures as presented in Ref. [4], so the carrier concentrations considered varied from 1×10^{16} to 5×10^{16} cm^{-3}.

2. Calculation of the carrier lifetime

The heterostructures were grown with the use of metal-organic chemical vapour deposition on InAs substrates at Microsensor Technology, LLC. An active layer of the heterostructures was made of $InAs_{1-x}Sb_x$. A typical thickness of the active layer was 2–3 μm. On top of the active layer, an InAsSb(Ga,P) barrier layer was grown. Parameters of the heterostructures are given in Table 1.

Electroluminescence of the heterostructures was studied in the temperature range T=4.2–300 K. At low temperatures (4.2–50 K), stimulated emission was observed. With temperature increasing, the radiation became spontaneous. This transition from stimulated to spontaneous emission occurred at different temperatures for structures with different content of InSb in the active layer [4]. To analyze these results in more detail, the necessity to calculate carrier lifetimes limited by various recombination mechanisms became apparent, and this was the task of the current work.

Table 1. Parameters of the studied heterostructures.

Structure type	Active layer	Barrier layer
A	InAs	$InGa_{0.15}As_{0.57}Sb_{0.28}$
B	$InAs_{0.94}Sb_{0.06}$	$InAs_{0.40}Sb_{0.20}P_{0.40}$
C	$InAs_{0.93}Sb_{0.07}$	$InAs_{0.70}Sb_{0.10}P_{0.20}$
D	$InAs_{0.91}Sb_{0.09}$	$InAs_{0.48}Sb_{0.18}P_{0.34}$

It is known that for semiconductors, in which the energy of spin-orbit-splitting is smaller or comparable to the width of the band gap, in general the most probable processes are the Auger process, in which the energy of a recombining electron-hole pair is transferred to a heavy hole transitioning to the spin-orbit-splitted band (CHHS), and the process with the participation of two electrons and a heavy hole with excitation of an electron to a more energetic state (CHCC) [5]. As mentioned above, there were several approaches to the calculation of the carrier lifetime limited by Auger processes. For example, in Ref. [1] in the calculation of Auger processes the matrix element was taken to be constant. However, expressions for the matrix element include the overlap integrals which should be calculated at the threshold values of momentum of the particles participating in the recombination. In Ref. [6], it was shown that it was necessary to take into account the behavior of the overlap integral near the Auger recombination threshold. As a result, the overlap integrals are not constant, but depend on temperature. Indeed, in Ref. [2] it was shown that for InAsSb materials the values of overlap integrals vary with temperature. Still, in lifetime calculations the overlap integrals in narrow-bandgap III–V semiconductors for simplicity are often assumed to be temperature-independent [3].

In our study, we performed calculation of the carrier lifetime limited by radiative and Auger processes within the framework of the Kane model [7,8]. This calculation appears to be the most accurate since the model considers both electron-heavy hole and electron-light hole transitions with an account for non-parabolic dependence of the energy of the conduction band and the valence band of light holes on the wave vector.

The detailed calculation of the carrier lifetime with an account for the CHCC and CHHS processes was performed in Ref. [7]. The lifetime defined by these processes was:

$$\tau_{CHCC} = [R_{CHCC}(n+p)n]^{-1},\ \tau_{CHHS} = [R_{CHHS}(n+p)p]^{-1}.$$

where n and p are concentrations of electrons and holes, respectively, R_{CHCC} and R_{CHHS} are corresponding recombination coefficients.

An important role in recombination processes is played by the radiative recombination, which dominates at low temperatures. In this work, calculation of the carrier lifetime limited by radiative recombination was based on Ref. [8]. The lifetime of radiative recombination is related to the corresponding recombination coefficient γ_{ph} as follows:

$$\tau_R = [\gamma_{ph}(n+p)]^{-1}.$$

Parameters of the materials were taken from Ref. [9]. Values of carrier concentrations, which were used in the calculations, were determined in Ref. [4] by fitting calculated electroluminescence spectra to experimental ones with the Fermi level (carrier concentration) being the parameter. The carrier concentration under injection for all types of heterostructures varied from 3×10^{16} cm^{-3} up to 5×10^{16} cm^{-3}.

3. Results and Discussion

Figure 1(a) shows the results of calculations of the inverse lifetime limited by CHCC and CHHS Auger processes, and by radiative recombination for heterostructure of type A at carrier concentrations $np=4\times10^{16}$ cm^{-3}. As can be seen, with the temperature increasing, the inverse lifetime limited by radiative recombination decreases and that limited by Auger recombination increases. At the considered injection level at $T<210$ K, the carrier lifetime is mostly determined by radiative recombination, and at $T<170$ K the most intensive Auger process is CHHS. However, at $T>170$ K, the value of τ^{-1} for the CHCC process becomes greater than that for CHHS recombination and increases with temperature. At $T>80$ K, the value of τ^{-1} for CHHS recombination is practically constant. Figure 1(b) shows τ^{-1} limited by the sum of non-radiative processes and by radiative recombination. As can be seen in Fig. 1(b), the "equi-rate temperature" T_{eq}, i.e., the temperature where the value of τ^{-1} of radiative recombination equals that of Auger recombination, is $T_{eq}\approx210$ K.

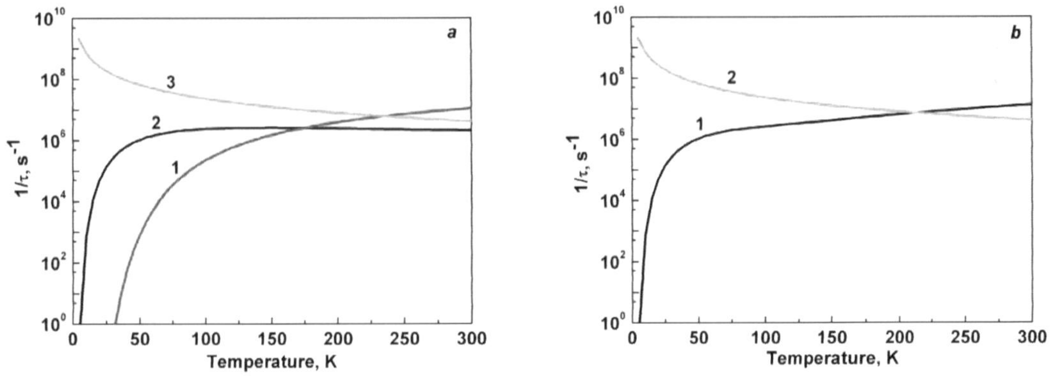

Figure 1. Calculated temperature dependences of inverse lifetime limited by CHCC (1), CHHS (2) and radiative (3) processes (*a*), and the sum of the two non-radiative (1) and the radiative (2) processes (*b*) for the heterostructure of type *A*.

The results of calculations of the inverse lifetime for heterostructure of type D at carrier concentrations $np=3\times10^{16}$ cm^{-3} showed that the nature of studied processes was similar to that for the heterostructure of type A. At low temperatures ($T<115$ K), the CHHS process still played the dominant role. For this heterostructure, T_{eq} equaled 150 K. Similar results were obtained for heterostructures of other types: CHHS process held the dominant role among the non-radiative processes at low temperatures. With temperature increasing, the CHCC process became more effective than CHHS.

Figure 2 shows the results of calculation of T_{eq} for various content of InSb x in the alloy at various carrier concentrations. We calculated τ^{-1} for the set of x from 0 up to $x=0.12$, InAsSb alloy with this chemical composition is also widely used in fabrication of photodetectors and LEDs [10,11]. As can be seen, with x increasing, which corresponds to decrease of the bandgap of the active area, T_{eq} decreases. For example, increasing x from $x=0$ to $x=0.12$ leads to decrease of T_{eq} from 233 to 135 K at $np=3\times10^{16}$ cm^{-3}. Also, our results show that with the carrier concentration increasing, non-radiative recombination becomes stronger than radiative one at lower temperatures: for $x=0$, changing carrier concentration from 1×10^{16} to 5×10^{16} cm^{-3} decreases T_{eq} by 127 K.

The temperature of 'quenching' of stimulated radiation observed in the experiment in Ref. [4] appeared to be always lower than T_{eq} (for example, for heterostructure of type A at $np=4\times10^{16}$ cm^{-3}, 80 K and 210 K, respectively, and for heterostructure of type D at $np=3\times10^{16}$ cm^{-3}, 50 K and 150 K, respectively). Thus, contribution of other effects to the 'quenching' could be expected, which was natural as the studied LED heterostructures were not designed to be lasers. Still, the calculated lifetime relations obviously contributed to the 'quenching' of the stimulated emission in the experiment.

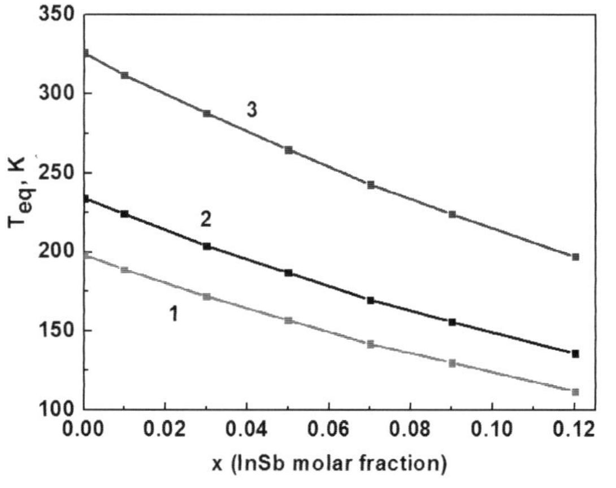

Figure 2. Calculated dependence of T_{eq} on InSb content in InAsSb alloy at carrier concentrations: np=5×10^{16} cm^{-3} (1), np=3×10^{16} cm^{-3} (2) and np=1×10^{16} cm^{-3} (3).

4. Conclusion

Carrier lifetime limited by radiative and most probable non-radiative processes for InAsSb active layer (with InSb content varying from 0 up to 0.012) in InAs(Ga,Sb,P) LED heterostructures under injection (carrier concentrations varying from 1×10^{16} up to 5×10^{16} cm^{-3}) was calculated. At low temperatures the lifetime was mostly determined by radiative recombination. As expected, with temperature increasing, influence of Auger processes increased. The dominating role of different Auger processes was established, with the CHHS process being crucial at low temperatures, and the CHCC process at higher temperatures. A strong effect of both InSb content and carrier concentration on the temperature where the value of lifetime of radiative recombination equals that of Auger recombination was found for the studied materials.

References

[1] Kabanau D M, Lebiadok Y V and Yakovlev Yu P 2017 *J. Appl. Spectrosc.* **84** 786
[2] Wen H, Pinkie B and Bellotti E 2015 *J. Appl. Phys.* **118** 015702
[3] Olson B V, Shaner E A, Kim J K, Klem J F, Hawkins S D, Flatte M E and Boggess T F 2013 *Appl. Phys. Lett.* **103** 052106
[4] Mynbaev K D, Bazhenov N L, Semakova A A, Stoyanov N D, Kizhaev S S, Chernyaev A V, Lipsanen H, Bougrov V E and Salikhov Kh M 2017 *Infrared Phys. Techn.* **85** 246
[5] Abakumov V N, Perel V I and Yassievich I N 1991 Nonradiative Recombination in Semiconductors *Modern Problems in Condensed Matter Sciences* vol 33 (Amsterdam: Horth-Holland) chapter 11
[6] Gelmont B L 1978 *J. Exp. Theor. Phys.* **75** 536
[7] Aydaraliev M, Zotova N V, Karandashov S A, Matveev B A, Remennyi M A, Stus' N M and Talalakin G N 1992 *Semiconductors* **26** 246
[8] Gelmont B L and Zegrya G G 1991 *Semiconductors* **25** 246
[9] Vurgaftman I, Meyer J R and Ram-Mohan L R 2001 *J. Appl. Phys.* **89** 5815
[10] Mynbaev K D, Shilyaev A V, Semakova A A, Bykhanova E V and Bazhenov N L 2017 *Opto–Electron. Rev.* **25** 209
[11] Il'inskaya N D, Karandashev S A, Lavrov A A, Matveev B A, Remennyi M A, Stus' N M and Usikova A A 2018 *Infrared Phys. Techn.* **88** 223

IOP Conf. Series: Journal of Physics: Conf. Series **1038** (2018) 012098 doi:10.1088/1742-6596/1038/1/012098

Modification of the CCD photodetectors for the suppression of interference in their internal structure

A N Ramazanov, V A Simon, A A Uhov, D K Kostrin, V A Gerasimov and L M Selivanov

Department of electronic instruments and devices, Saint Petersburg Electrotechnical University "LETI", 197376, Saint Petersburg, Russia

E-mail: ramazanow.alexander@yandex.ru

Abstract. In this work the reasons leading to the occurrence of interference in the internal structure of the common CCD photodetectors are discussed, that result in degradation of the metrological characteristics of the spectrometers based on them. Methods of removing a protective glass of the photodetector are shown. Possibility of modification of the CCD photodetectors for suppressing the interference by applying the surface phosphor layer is analyzed. It is experimentally confirmed that the proposed method allows achieving almost complete suppression of interference and also increasing the sensitivity of the detector to the radiation in the UV part of the spectral range.

1. Introduction

Photodetectors, based on the charge coupled device (CCD) technology, use the properties of metal–oxide–semiconductor structure (MOS) to accumulate minority charge carriers in the potential wells at the boundary of the semiconductor and oxide [1]. During the generation process under the influence of the incident radiation in the individual cells of the accumulation section charge occurs, depending on the area of the element, the registered radiation intensity and the accumulation time [2].

The most common "standard" (not back-thinned) CCD design is schematically shown in figure 1. On the surface of the CCD photodetector there is a thin layer of a natural oxide. Further, depending on the type of photodetector, the polysilicon electrodes and dielectric system silicon oxide–silicon nitride, situated between the electrodes and the substrate, are located.

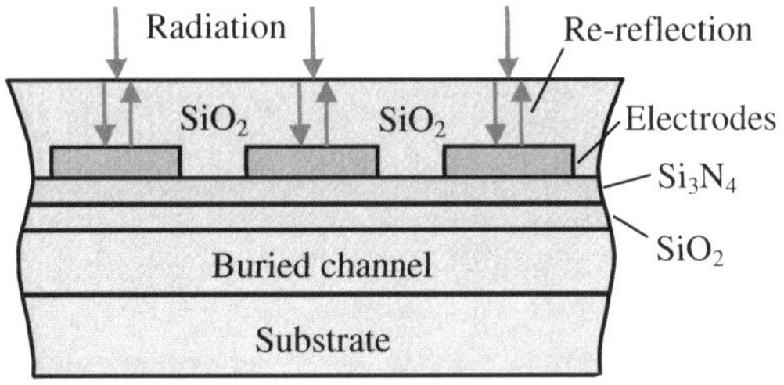

Figure 1. Scheme of the CCD internal structure.

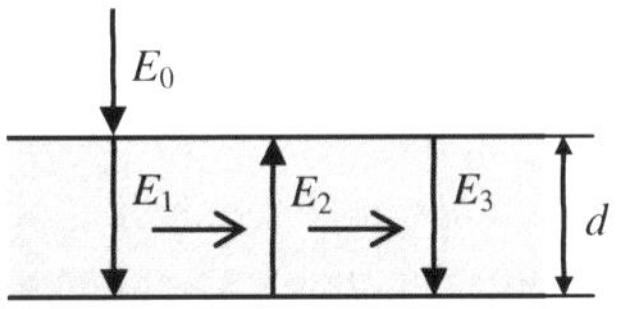

Figure 2. Path of a light beam.

In the shown structure each layer is characterized by its own thickness and refractive index. The refractive index, in turn, has a real part n and the extinction coefficient k (imaginary part), which is associated with the absorption coefficient α by the following relationship [3]: $\alpha = 4\pi k/\lambda$.

2. Reasons leading to the occurrence of interference in the CCD structure

Let us assume that onto the surface layer with thickness d a light beam E_0 falls (figure 2) and consider the interference from the primary beam E_0 and the two-times reflection from the boundaries of a film E_3. From the theory of interference it is known that in the presence of two radiation sources with equal amplitudes, the amplitude at the point of observation is equal to:

$$S = 2a^2 \cos\left(2\pi\left(L_1 - L_2\right)/\lambda + \varphi/2\right),$$

where a – amplitude of the primary beam E_0; L_1 – path traversed by the first beam E_1; L_2 – distance travelled by the second beam ($E_1 + E_2 + E_3$); λ – wavelength; φ – initial phase difference.

For the case of coherent waves the initial phase difference is constant, the difference ($L_1 - L_2$) is also constant and, therefore, the difference in intensity of radiation depends only on the wavelength.

The phase difference resulting from the path difference of the waves is defined as

$$\Psi = 2\pi(L_1 - L_2)/\lambda.$$

The path difference can be expressed through the wavelength $\Delta = L_1 - L_2 = m\lambda$, where m – any integer or a fractional number. Then the phase difference will be equal to $\Psi = 2\pi m$. The intensity of the two interfering waves (without consideration of the amplitudes and under the condition of equality of the initial phases) can be written in the form

$$I \sim A^2 = 2a^2 \cos\left(2\pi\left(L_1 - L_2\right)/\lambda\right) = a^2 \cos^2\left(\pi m\right).$$

Integer values m correspond to the difference in the phase by $2\pi m$, and the intensity is proportional to a^2. When m is half-integer the phase of the emerging waves is opposite and the intensity is zero (for equal amplitudes). In general case assuming the inequality of the amplitudes a_1 and a_2 we have:

$$I \sim A^2 = a_1^2 + a_2^2 + 2a_1a_2 \cos^2\left(2\pi m\right).$$

For integer m we will have maxima $A^2 = (a_1 + a_2)^2$, and for half-integer – minima $A^2 = (a_1 - a_2)^2$.

Knowing the parameters of the photodetector layers it is possible to calculate the transmittance T, defined as the ratio of radiation power passed into the substrate to the radiation power incident on the first boundary. At a fixed layer thickness dependence of the transmittance from the wavelength has a periodical character for the presented system air–oxide–polysilicon–nitride–oxide–silicon.

The surface oxide film does not affect the waveform of a signal in the case of a non-monochromatic radiation. However, in the optical scheme of the spectrometer [4] each pixel of the sensor is exposed to an almost monochromatic light. As a result, at the interface of two media with different refractive indices the interference occurs. In each pixel the level of interference will be unique, because the wavelength of the incident radiation is continuously changing over the photodetector length. This interference causes a redistribution of the signal amplitude, which leads to a significant alteration of the measurement results [5]. For example, during registration of signals from the sources of radiation with a continuous spectrum the signal at the output of a standard CCD will have an obvious modulation (figure 3).

The location of this oscillation extrema is determined by thickness and refractive index of the oxide film on the photodetector surface and the detected spectral range [6]. The depth of the signal modulation depends on the spectral resolution of the device in which the photodetector is installed.

In the spectrometer with TCD1304 CCD photodetector the modulation depth of the spectral signal shown in figure 3 is 38.35 % at amplitude of 99.04 % in the selected local maximum of the signal with the maximum modulation depth. As a result the normalized modulation depth is 39 % at a wavelength of 603 nm. The photodetector has a signal-to-noise ratio of 300, which gives the amplitude of the noise approximately 0.3 % of the maximum signal. The amplitude of the noise can be considered the same for both the maximum and the minimum of the interference oscillations. Thus, for the minimum of the interference oscillations comparing to the neighboring maximum the signal-to-noise ratio worsens by 39 % to 180. The best solution to this problem is the use of a back-thinned CCD. The absorption in the silicon layer prevents interference at short wavelengths, however, in NIR range the interference is still possible, particularly for substrates having a thickness of several micrometers. It should be noted that the cost of a linear CCD of this type is quite high, that prevents their use in the low-cost mass

IOP Conf. Series: Journal of Physics: Conf. Series **1038** (2018) 012098 doi:10.1088/1742-6596/1038/1/012098

production spectral instruments. In order to get rid of the wavy modulation of the signal in the most common CCD photodetectors a correction of their spectral sensitivity is required [7–9].

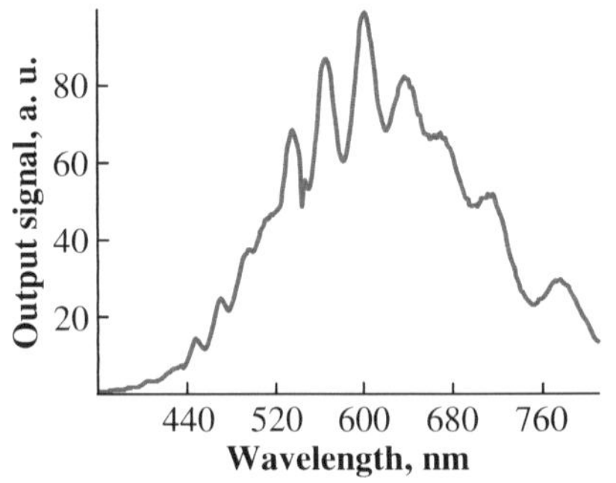

Figure 3. Spectrum with modulation.

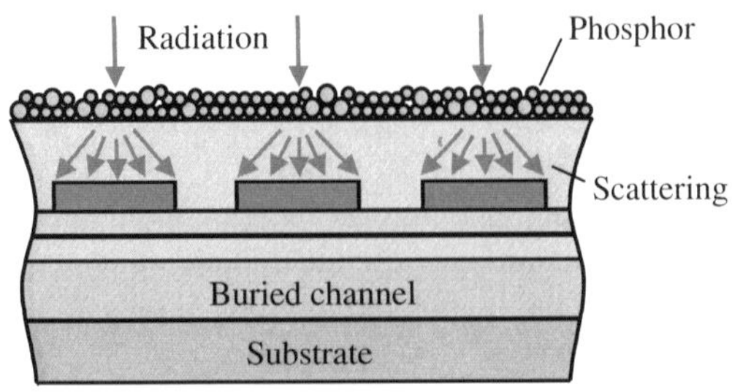

Figure 4. Layer of phosphor on a CCD surface.

3. Suppression of interference by applying the surface phosphor layer

One of the possible ways of suppressing interference is deposition on the irradiated surface of MOS structure of a thin layer of material with strong scattering [10]. As a result, the surface of the accumulation section will be irradiated not by coherent radiation, but by diffuse radiation, resulting in occurrence of interference only for a small fraction of the input luminous flux. When passing through the diffusing surface light has two components: diffuse and directed, whose influence must be reduced to a minimum. As such a surface we will consider a film containing scattering centers on the surface, that are spherical particles which dimensions are small compared with wavelength (figure 4). The most suitable material for such modification of the detector is phosphor. A phosphor layer coated on the surface of the accumulation section can also increase the UV sensitivity of the photodetector, although this will reduce its overall sensitivity due to the partial absorption of the radiation in the phosphor layer. Therefore, the transparency of the phosphor over the whole operating spectral range should be high enough.

Phosphor in condition that it is properly applied on the CCD surface is a parallel layer of light-scattering medium. We can assume that each elementary area, equal to an area of one pixel, is exposed to a monochromatic radiation. Therefore, to solve the given problem, let us consider a flat layer, which is uniformly and diffusely irradiated from one side with monochromatic light (detailed consideration of this problem is given in [10]). Characteristics of the phosphor layer very strongly dependent on the distribution of particle size. The grain size of the phosphor affects the quality of its fixing on the surface of the CCD. Moreover, the strength of fixing of the phosphor increases with decreasing grain size; however, it simultaneously decreases the efficiency of excitation of the phosphor by increasing the reflection coefficient of the exciting radiation. The size of the particles affects the scattering index s. Let us view a phosphor as a set of particles with diameter d and the number of particles per unit volume z. Then the cross sectional area of all particles S per unit volume has the form

$$S = 0.25\pi d^2 z. \tag{1}$$

For the dense packing z is inversely proportional to the particle volume:

$$z \sim 1/V = 6/\pi d^3. \tag{2}$$

The scattering index s is directly proportional to the cross sectional area of particles per unit volume, then taking into account (1) and (2) it can be written as

$$s \sim S \sim 1/d. \tag{3}$$

From (3) follows, that with increasing in the phosphor of fine particles quantity s also increases, which in turn leads to an increase of the reflection coefficient of the layer ρ, and as a result to a decrease in the efficiency of converting radiation into luminescence. Proceeding from the above, it is possible to formulate main requirements to phosphor for the modification of the CCD photodetectors:

- high efficiency of excitation (high quantum yield);

IOP Conf. Series: Journal of Physics: Conf. Series **1038** (2018) 012098 doi:10.1088/1742-6596/1038/1/012098

- high transmittance of visible radiation;
- line of emission of the phosphor should match the maximum quantum efficiency of a CCD;
- high stability in the process of operation of the device;
- afterglow time of the phosphor must be coordinated with a clock frequency of CCD.

Due to the fact that the greatest interest from the point of view of increasing the sensitivity of the device causes the UV region, special attention must be paid to the efficiency of excitation at wavelengths from 200 to 400 nm when choosing a type of phosphor. It is desirable that the wavelength of phosphor luminescence coincides with the maximum quantum efficiency of radiation absorption in the CCD. At a present time sufficiently efficient phosphors are produced; for example, Phosphor Technology offers phosphors, which have an excitation wavelength of 254 or 365 nm. In addition, the line of radiation is in the red region, which corresponds to the maximum quantum efficiency of radiation absorption of most spectral instruments built using CCD photodetectors.

4. Results and discussion

Surface modification of the CCD photodetector is impossible without direct access to the crystal (MOS structure) which, as a rule, is sealed by a protective glass. Without such glass a crystal of the photodetector can be easily damaged even with a slight mechanical action. There are companies, such as Framos and Eureca that offer services to remove the glass from almost any CCD photodetectors, however, without revealing the technological features of the process. The cost of such services is quite high, and the probability of failure of the sensor is about 10 %. Meanwhile it is stated that after years of use of the CCD photodetectors with removed glass their parameters practically do not change.

For the removal of this glass the heating of the device can be used, followed by a long exposure at sufficiently high temperature [11]. As a result of the temperature exposure adhesive, with which the protective glass is attached to the casing of the sensor, is destroyed and the glass can be removed. This method, however, cannot be applied to all multi-element photodetectors. Experiments carried out with a popular CCD sensor of TCD1304 type showed that its glue does not undergo significant changes in mechanical properties even after prolonged exposure to a temperature of 300 °C. Further increase in temperature allowed to remove the glass, but led to the CCD photodetector damage. For the removal or at least softening of the adhesive, as which epoxy compounds are usually used, it is possible to use a special solvent – dichloromethane (methylene chloride) or dimethylformamide. The process of removing glue lasts up to a month, and solvents themselves are quite toxic. In addition, manufacturers of CCD photodetectors use various types of glues and one universal solvent for all of them simply does not exist.

Another method of the glass removal, which is practically independent of the type of the photodetector casing and used glue is the mechanical "cut" of the glass piece that do not have direct adhesive contact with the shell. This method also does not guarantee the safety of the detector, since during the mechanical destructive effect on the glass, the small fragments fall into the casing and at high speed bombard the surface of the photodetector, causing its damage (figure 5(a)).

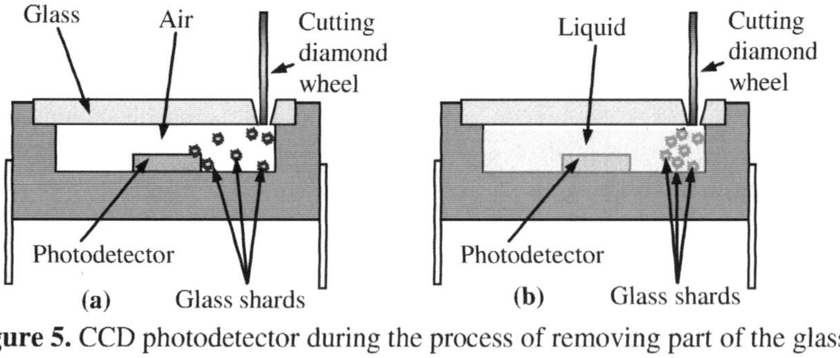

Figure 5. CCD photodetector during the process of removing part of the glass.

To prevent the damage of the sensor the following technology of glass removal was designed. At the first phase in the glass at the farthest point from the photodetector a minimized cut is made to gain access to the internal volume of the casing. Then under the glass a fluid, not harmful to the photodetector, for

IOP Conf. Series: Journal of Physics: Conf. Series **1038** (2018) 012098 doi:10.1088/1742-6596/1038/1/012098

example, alcohol, is pumped. The next step is a basic procedure for removal of glass using a diamond cutting wheel (figure 5(b)). Liquid, due to a sufficiently high density, prevents glass shards to reach the surface of the CCD sensor at speeds enough to its damaging. After removal of the glass the sensor is released from the liquid and small pieces, after which the surface can be modified. The results of the glass removal from a CCD photodetector TCD1304 using two different ways are shown in figure 6.

(a) **(b)** **(c)** **(d)**

Figure 6. CCD photodetector TCD1304 with removed glass.

A photodetector with glass, for comparison, is shown in figure 6(c). Minimal destruction of the casing was obtained using the thermal method (figure 6(a)), however, during the process the sensor itself was damaged. Results of the glass removal by the mechanical means (figure 6(b), a fragment with notches is shown in figure 6(d)) are acceptable from the point of view of yield of the devices because the number of faulty sensors does not exceed 10 %.

The most common method of applying the phosphor coating is deposition from the aqueous suspensions [12]. Before the beginning of the process, the phosphor powder is sifted: only particles with diameter less than 12 μm are left. Then a suspension of phosphor in the distilled water is made, sometimes with addition of potassium silicate. After that with the help of ultrasonic treatment, the resulting suspension is mixed. Depending on at what point in time to put the substrate under the deposited stream particles of a certain faction would be precipitated. To control the thickness of the applied layer the phosphor is also simultaneously deposited on a control sample, which is weighed before and after the operation. Later the resulting coating is dried.

In this way samples of phosphors were deposited on the surface of the CCD. The experiment gave a positive result, and the technology of deposition phosphors from the aqueous suspensions does not worsen the properties of the photodetector. For the improvement of CCD parameters two phosphors were investigated: K-77 with the specific gravity of 0.77, 0.82, 1.12 and 2.06 mg/cm^2 and FK-6 with the specific gravity 1.21 and 2.27 mg/cm^2, each of them been deposited on a separate photodetector. The phosphor K-77 is a cathode phosphor with composition Y_2O_3:Eu, maximum diameter of selected fractions was 4 μm. The luminescence spectra of the phosphor K-77 has a red color and is excited by radiation in the range 210...280 nm. The maximum brightness of the phosphor was observed at the excitation wavelength of 250 nm. The phosphor FK-6 is a phosphorus-containing photoluminescent phosphor. The maximum diameter of the selected fractions does not exceed 4 μm. The phosphor has a red-orange color of luminescence. The exciting radiation lies in a wide interval of wavelengths from UV to visible: 250...512 nm. The maximum brightness was observed at excitation wavelength equal to 400 nm. The luminescence intensity of the phosphor with a specific gravity 2.27 mg/cm^2 was higher than for a thinner phosphor with a specific gravity 1.21 mg/cm^2.

The luminescence spectrum of the K-77 phosphor has a maximum coincidence with the maximum spectral sensitivity of the CCD photodetector. The luminescence spectrum of the FK-6 phosphor lies in a rather wide interval of wavelengths from 550 to 750 nm and does not have a pronounced maximum. The transmittance of phosphor K-77 is 5–10 %, and weakly depends from the specific gravity. A maximum transmittance has a K-77 phosphor with a specific gravity 0.82 mg/cm^2. The transmittance of phosphor

FK-6 varies in the range of 8–50 %, intensively depending on the specific gravity of the phosphor and the wavelength. The highest transmittance has FK-6 phosphor with a specific gravity of 1.21 mg/cm^2.

The K-77 phosphor with a specific gravity 1.12 mg/cm^2 was deposited on the surface of the accumulation section of the TCD1304AP photodetector manufactured by Toshiba. The application of the phosphor layer does not affect the resolution of the spectrometer – in both cases (with and without phosphor) it was ~3 nm. The application of the phosphor layer completely eliminated the wavy modulation of the spectral characteristic of the spectrometer (figure 7).

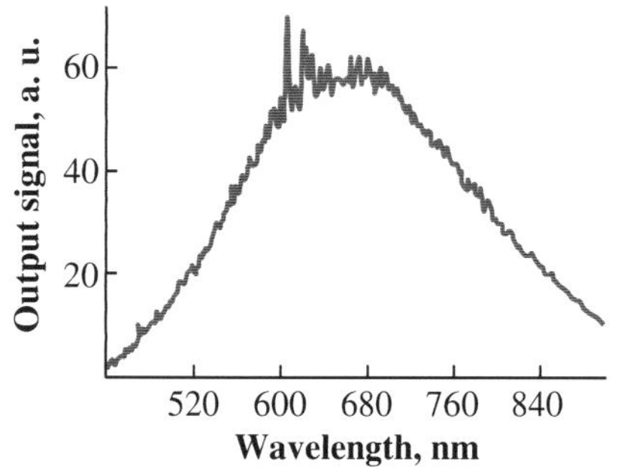

Figure 7. Spectrum after modification of a CDD with a phosphor layer.

Comparing the spectra of the incandescent lamp, obtained using a spectrometer having a CCD without the phosphor coating and with it, it can be concluded that wavy modulation has been completely suppressed. The noise arising on the spectrum is due to the presence of foreign particles in the deposited phosphor layer. These deviations are stationary in time and can be further eliminated by improving the technology of the phosphor deposition.

5. Conclusions

The method of modifying a CCD photodetector considered in this paper, that includes the stages of removing the protective glass and applying a layer of phosphor on MOS structure allows to obtain a sensor for spectrometric applications with metrological characteristics close enough to much more expensive models. However, it should be noted that the photodetector with a removed glass is almost not protected from mechanical damage and requires very careful handling.

References

[1] Kostrin D K, Lisenkov A A and Uhov A A 2016 *Electronic means of controlling technological processes* (Saint Petersburg: Saint Petersburg Electrotechnical University “LETI”)
[2] Kostrin D K and Uhov A A 2013 *Testing. Diagnostics* **5** 40–2
[3] Krylova T N 1973 *Interference coatings. Optical properties and methods of research* (Leningrad: Mashinostroenie)
[4] Kostrin D K and Uhov A A 2013 *Biotechnosphere* **3** 21–5
[5] Kostrin D K and Uhov A A 2013 *Sensors and systems* **5** 13–5
[6] Pikhtin A N, Kolguine E A and Uhov A A 2003 *Proceedings of Saint Petersburg Electrotechnical University “LETI”* **2** 7–10
[7] Uhov A A 2016 *Application of multi-element photodetectors in optical spectrometers* (Saint Petersburg: Saint Petersburg Electrotechnical University “LETI”)
[8] Kostrin D K and Uhov A A 2014 *Vacuum technique and technology* **1** 34–8
[9] Simon V A, Kostrin D K, Uhov A A, Ramazanov A N and Lisenkov A A 2017 *Proceedings of the 2017 IEEE Russia Section Young Researchers in Electrical and Electronic Engineering Conference* pp 342–5
[10] Uhov A A, Kostrin D K, Gerasimov V A, Selivanov L M, Simon V A 2017 *Proceedings of Saint Petersburg Electrotechnical University “LETI”* **7** 5–11
[11] Ryć L, Jablonski S, Czarnecka A, Parys P and Wołowski J 2007 *34th EPS Conference on Plasma Physics* **31** 1266–9
[12] Gugel B M 1967 *Phosphors for electrovaccum industry* (Moscow: Energiya)

IOP Conf. Series: Journal of Physics: Conf. Series **1038** (2018) 012099 doi:10.1088/1742-6596/1038/1/012099

Eigen-frequencies of whispering gallery modes of disk dielectric resonators: a dimensional quantization method

G.A. Zaretskaya[1], A. V. Drozdovskii[1], A. B. Ustinov[1] and B. A. Kalinikos[1]

[1]Department of physical electronics and technology, Saint Petersburg Electrotechnical University "LETI", Saint Petersburg, 197376, Russia

E-mail: zaretskajaga@gmail.com

Abstract. A method for calculation of the resonant frequencies as well as free spectral range was developed for the high-Q whispering gallery mode disk dielectric resonators. The method is based on application of the size quantization to the dispersion equation derived for an infinite dielectric cylinder. The size quantization is caused by the requirement of the phase balance for standing waves in the resonator.

1. Introduction

During the past decade or so a high research interest in the field of microwave photonics is evident. The microwave photonics is an interdisciplinary field comprising microwave electronics, wave guiding and integrated optics, and other research areas [1-3]. One of the key elements of the microwave photonics is an optical resonator with whispering gallery modes (WGMs). We assume that the WGMs in such resonators are modes with a high azimuthal index that are "pressed" to the resonator side surface and have a Q-factor much greater than the lowest types of oscillations [4, 5]. Such optical resonators may have a Q-factor greater than 10^{11} [6] and also a high stability with respect to the external microwave influences. This makes them very attractive for various practical applications [7-9].

Resonators in the form of balls [10-12], toroidal and disk resonators [13, 14] and resonators of "bottle" form [15] are typical examples of the axisymmetric resonators with WGMs. At present, the planar disk resonators of micrometer thickness compatible with silicon technology [13, 14, 16] are developed. It opens up wide possibilities for their fabrication.

Several analytical methods for estimation of the resonant frequencies of the disk dielectric WGM resonators are already known. Among them are the method of partial regions [4], the effective dielectric permittivity method [17], and the magnetic wall method [18]. These methods are based on the simultaneous solution of two dispersion equations having an implicit form. Therefore, finding WGM resonant spectrum is quite laborious and require considerable computational powers.

The purpose of the present work is to develop a theory for calculation of the resonant frequency spectra of the disk WGM resonators.

2. Theoretical investigation

The theory was developed in two stages. In stage 1, a dispersion equation for electromagnetic waves propagating in dielectric cylinder waveguide of an infinite length was obtained. We began from Maxwell's equations and constitutive equation taking into account the appropriate boundary conditions for the electromagnetic field on the side surface. As a result, we obtained the following components of the electric and magnetic fields

$$
\begin{aligned}
E_z &= \begin{matrix} A \\ B \end{matrix} Z_m(k_\perp \rho) e^{-i\beta z} \begin{matrix} \cos(m\phi) \\ \sin(m\phi) \end{matrix}, \\
H_z &= \begin{matrix} D \\ C \end{matrix} Z_m(k_\perp \rho) e^{-i\beta z} \begin{matrix} \sin(m\phi) \\ \cos(m\phi) \end{matrix}, \\
H_\phi &= -\frac{i}{k_\perp^2}\left(\begin{matrix} D \\ -C \end{matrix} \frac{\beta m}{\rho} Z_m(k_\perp \rho) + \begin{matrix} A \\ B \end{matrix} \varepsilon_0 \varepsilon \omega Z_m'(k_\perp \rho) \right) e^{-i\beta z} \begin{matrix} \cos(m\phi) \\ \sin(m\phi) \end{matrix}, \\
H_\rho &= \frac{i}{k_\perp^2}\left(\begin{matrix} -A \\ B \end{matrix} \frac{m\varepsilon_0 \varepsilon \omega}{\rho} Z_m(k_\perp \rho) - \begin{matrix} D \\ C \end{matrix} \beta Z_m'(k_\perp \rho) \right) e^{-i\beta z} \begin{matrix} \sin(m\phi) \\ \cos(m\phi) \end{matrix}, \\
E_\phi &= -\frac{i}{k_\perp^2}\left(\begin{matrix} -A \\ B \end{matrix} \frac{\beta m}{\rho} Z_m(k_\perp \rho) - \begin{matrix} D \\ C \end{matrix} \mu_0 \mu \omega Z_m'(k_\perp \rho) \right) e^{-i\beta z} \begin{matrix} \sin(m\phi) \\ \cos(m\phi) \end{matrix}, \\
E_\rho &= -\frac{i}{k_\perp^2}\left(\begin{matrix} D \\ -C \end{matrix} \frac{m\mu_0 \mu \omega}{\rho} Z_m(k_\perp \rho) + \begin{matrix} A \\ B \end{matrix} \beta Z_m'(k_\perp \rho) \right) e^{-i\beta z} \begin{matrix} \cos(m\phi) \\ \sin(m\phi) \end{matrix},
\end{aligned}
\tag{1}
$$

where A, B, C, and D are the amplitude constants determined from the boundary conditions, Z_m is the cylindrical function of a particular form depending on a range of field definition and the stated problem, m is an order of the cylindrical function showing quantity of the azimuthal variation of the field, β is a longitudinal wave number, $k_\perp$ is a transversal wave number, ρ is a radial coordinate, ϕ is an azimuthal angle, ε is a dielectric permittivity, which equals to a dielectric permittivity ε_{res} of the resonator materials for $\rho \le R$ and to a dielectric permittivity of the space surrounding the resonator (ε_{air}) for $\rho \ge R$, where R is the radius of the resonator.

The transverse wave number is determined by the following expression:

$$
k_\perp = \begin{matrix} \sqrt{k_{\text{res}}^2 - \beta^2} \;\; \text{for } \rho \le R \\ \sqrt{k_{\text{air}}^2 - \beta^2} \;\; \text{for } \rho \ge R \end{matrix},
\tag{2}
$$

where $k_{\text{res}} = \frac{\omega}{c} \cdot \sqrt{\varepsilon_{\text{res}}}$ is a wave number in the resonator material and $k_{\text{air}} = \frac{\omega}{c} \cdot \sqrt{\varepsilon_{\text{air}}}$ is a wave number of the space surrounding the resonator.

In solving the spectral problem, two cases can be distinguished: the case of a "thick" and a "thin" disk. When $\beta > k_{\text{air}}$ we will consider the case of "thick" disk. For this condition $k_\perp$ is real inside and outside the resonator and standing waves inside the resonator are described by the Bessel functions, i.e. $Z_m(k_\perp \rho) = J_m(k_\perp \rho)$. Outside the resonator, the waves are described by the Hankel functions of the first and second kinds, i.e. $Z_m(k_\perp \rho) = H_m^{(1)}(k_\perp \rho) + H_m^{(2)}(k_\perp \rho)$, which corresponds to divergent and incoming spherical waves. For the chosen harmonic time dependence ($e^{i\omega t}$), it is necessary to take the second Hankel function, which satisfies the Sommerfeld radiation condition [5].

Applying the boundary conditions of electrodynamics, in the case of a "thick" disk we obtain the following dispersion equation:

$$\frac{m^2}{R^2}\left(\frac{1}{k_{\perp 1}{}^2}-\frac{1}{k_{\perp 2}{}^2}\right)\left(\frac{\varepsilon_{\rm res}}{k_{\perp 1}{}^2}-\frac{\varepsilon_{\rm air}}{k_{\perp 2}{}^2}\right)= \\ =\left(\frac{\varepsilon_{\rm res}}{k_{\perp 1}}\frac{J_m'(k_{\perp 1}R)}{J_m(k_{\perp 1}R)}+\frac{\varepsilon_{\rm air}}{k_{\perp 2}}\frac{H_m'(k_{\perp 2}R)}{H_m(k_{\perp 2}R)}\right)\times\left(\frac{1}{k_{\perp 1}}\frac{J_m'(k_{\perp 1}R)}{J_m(k_{\perp 1}R)}+\frac{1}{k_{\perp 2}}\frac{H_m'(k_{\perp 2}R)}{H_m(k_{\perp 2}R)}\right), \tag{3}$$

where $k_{\perp 1}=\sqrt{k_{\rm res}{}^2-\beta^2}$ and $k_{\perp 2}=\sqrt{k_{\rm air}{}^2-\beta^2}$. It is seen from this equation that the transverse oscillations in the "thick" disk has a hybrid nature, namely, it is possible to distinguish HE_{ml} и EH_{ml} modes in the solution.

The condition $k_{air}<\beta<k_{res}$ is satisfied for the "thin" disk approximation. Similar to the case of a "thick" disk the standing waves inside the resonator are described by the Bessel functions. The transverse wave number $k_\perp$ has imaginary value outside the resonator. Therefore, the waves outside the resonator are described by the modified Bessel functions of the first and second kind. Using the condition of the vanishing field at infinity, we take the modified Bessel function of the second kind as the solution, i. e. the MacDonald function. Below, this function is denoted by K_m. Applying the boundary conditions of the electrodynamics, we obtain the following dispersion equation for the case of a "thin" disk:

$$\frac{m^2}{R^2}\left(\frac{1}{k_{\perp 1}{}^2}+\frac{1}{(ik_{\perp 2})^2}\right)\left(\frac{\varepsilon_{\rm res}}{k_{\perp 1}{}^2}+\frac{\varepsilon_{\rm air}}{(ik_{\perp 2})^2}\right)= \\ =\left(\frac{\varepsilon_{\rm res}}{k_{\perp 1}}\frac{J_m'(k_{\perp 1}R)}{J_m(k_{\perp 1}R)}+\frac{\varepsilon_{\rm air}}{ik_{\perp 2}}\frac{K_m'(ik_{\perp 2}R)}{K_m(ik_{\perp 2}R)}\right)\left(\frac{1}{k_{\perp 1}}\frac{J_m'(k_{\perp 1}R)}{J_m(k_{\perp 1}R)}+\frac{1}{ik_{\perp 2}}\frac{K_m'(ik_{\perp 2}R)}{K_m(ik_{\perp 2}R)}\right). \tag{4}$$

As is seen from this equation, the modes in the "thin" disk have a hybrid nature. They are named as HE_{ml} and EH_{ml}. Note that for the case $m=0$ the modes are not hybridized. In this case, Eq. (4) splits into two simpler equations describing the transverse TE_{0l} and TM_{0l} modes.

In stage 2, we derive the equation for the resonant frequencies of electromagnetic oscillations for the disk of the finite thickness through quantization of the longitudinal wave number of the infinite cylinder. Applying the phase balance condition, we determine the unknown values of the longitudinal wave number as

$$\beta_n=\frac{\pi\cdot n-\varphi}{d}, \tag{5}$$

where n is any integer determining the number of the field maxima along the axial coordinate, d is the thickness of the resonator, φ is a fitting parameter having sense of a phase shift appearing from the wave reflection at the media interfaces. We will assume further that φ equals zero. Substitution the β_n to the dispersion equations (3) and (4) allows one to calculate resonant frequencies and free spectrum range.

It should be noted that the resonant types of oscillations of the dielectric disk cavity and the frequencies corresponding to them are characterized by three indices m, l and n. The index m appears in Eqs. (3) and (4) and shows the number of wavelengths that fit along the angular coordinate. In other words, this is an azimuthal mode index. The index l characterizes the number of field variations along the radial coordinate. It is determined by the number of solutions of Eqs. (3) and (4), for which there is an infinite set of roots. The index n shows the number of variations of the field along the axis of the disk.

Figure 1 shows the results of the resonant frequency calculation for EH_{mln} modes of a disk with a dielectric permittivity of 2.074 and a diameter of 3 μm for m from 2 to 12 and for l = 1. The square, triangular and round symbols in the figure depict the resonant frequencies corresponding to the cases of the disks with thicknesses of 1.2 μm, 3 um, and 5 μm, respectively. As is clear from the figure, with decreasing the thickness of the resonator its resonance frequencies shift to the region of larger longitudinal wave numbers, which leads to decreasing the free spectral range and increasing the resonance frequency. With an increase in the index m, an increase in the free spectral range occurs, and for $m >> 1$ the resonance frequency spectrum becomes practically equidistant, and the distance between neighboring harmonics tends to the value of the free spectral range of an infinite length rod.

3. Conclusions

Thus, the paper presents a simpler and less laborious method for calculating the resonant frequencies of the high-Q WGM disk dielectric resonators in comparison with the known ones [4, 17, 18]. The influence of the thickness of the disk resonator on its own resonant frequencies is considered.

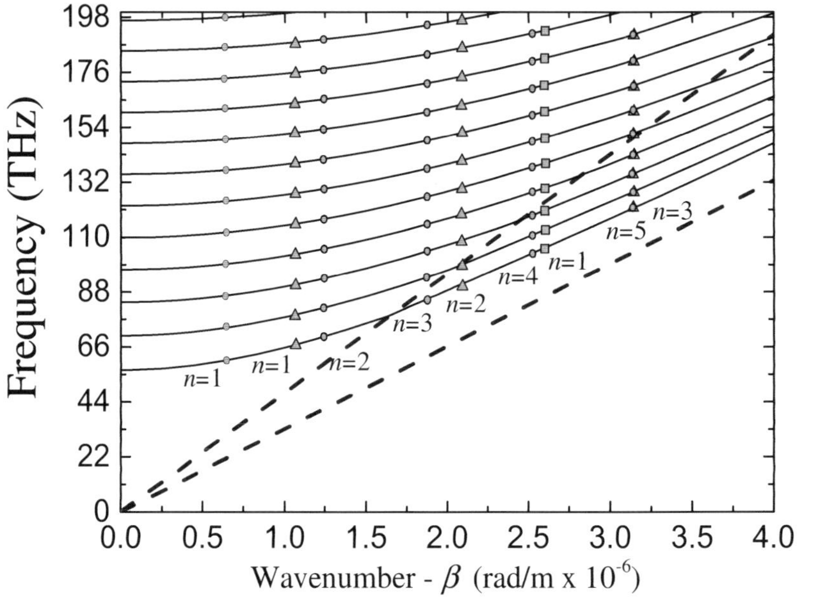

Figure 1. Results of calculation of resonant frequencies EH_{mln} of a disk dielectric resonator with azimuthal oscillations

Acknowledgments
The experimental investigation was supported in part by Ministry of Education and Science of Russian Federation (Project "Goszadanie"). The development of a theoretical model fabrication and characterization of the magnetic samples was supported in part by Russian Science Foundation (Grant# 16-12-10440).

References

[1] Capmany, J., & Novak, D. 2007 *Nature photonics* **1(6)** 319-330.
[2] Yao, J. 2009 *Journal of Lightwave Technology* **27(3)** 314-335
[3] Marpaung D et al. 2013 *Laser & Photonics Reviews* **7(4)** 506-538
[4] Ilchenko M 1989 *Dielectric resonators (in russian)* (RiS)
[5] Gorodetsky M 2011 *Optical microresonators with a giant Q-factor (in russian)* (Fizmatlit)
[6] Savchenkov A A et al. 2007 *Optics Express* **15(11)** 6768-6773.
[7] Ayazi A et al. 2008 *Optics express* **16(3)** 1742-1747

[8] Matsko A B et al. 2003 *Journal of modern optics* **50(15-17)** 2523-2542
[9] Merrer P H et al. 2012 *Applied optics* **51(20)** 4742-4748
[10] Ilchenko V S et al. 1998 *Optics communications* **145(1)** 86-90
[11] Vernooy D W et al. 1998 *Optics letters* **23(4)** 247-249
[12] Vassiliev V V et al. 1998 *Optics Communications* **158(1)** 305-312
[13] Armani D K et al. 2003 *Nature* **421(6926)** 925-928
[14] Lee H et al. 2012 *Nature Photonics* **6(6)** 369-373
[15] Pöllinger M et al. 2009 *Physical review letters* **103(5)** 053901
[16] Kippenberg T J et al. 2003 *Applied Physics Letters* **83(4)** 797-799
[17] Knox, R. M. 1976 *IEEE Transactions on Microwave Theory and Techniques* **24(11)** 806-814.
[18] Tsibizov K N 1981 *Zarubezh. Radioelektron* **11** 21
[19] Braginsky V B, Gorodetsky M L, Ilchenko V S 1989 *Physics Letters A* **137(7)** 393-397

IOP Conf. Series: Journal of Physics: Conf. Series **1038** (2018) 012100 doi:10.1088/1742-6596/1038/1/012100

Photosensitivity of structures based on $A^{II}B^{III}{}_2C^{VI}{}_4$ monocrystals

I.A. Zharikov [1], V.Yu. Rud ' [1,2], Yu.V. Rud ' [3], E.I. Terukov [3], V.V. Davydov [1], N.N. Bykova [4]

[1]Peter the Great Saint Petersburg Polytechnic University, Saint Petersburg 195251, Russia

[2]Department of Ecology, All-Russian Research Institute of Phytopathology, 143050, Moscow Region, Odintsovo district, B.Vyazyomy, Russia

[3]Department of Solid State Physics, Ioffe Institute, Russian Academy of Sciences, St Petersburg 194021, Russia

[4]Saint Petersburg State University of Economics, Saint Petersburg, 191023, Russia

Abstract. Monocrystals of ternary compounds of $A^{II}B^{III}{}_2C^{VI}{}_4$ semiconductors ($HgGa_2S_4$ and $CdGa_2S_4$) and various photosensitive structures based on them were investigated. The photoelectric properties of the structure obtained were studied using natural and linearly polarized light at 300 K. The results show that the structures created can find application as no polarized radiation photodetectors, and with allowance for high-tetragonal compression of the crystal lattice and linearly polarized radiation.

1. Introduction

The importance of the results of studying the fundamental properties of multicomponent semiconductors consists in the discovery of new nonlinear media with record values of nonlinear polarizability among them. For nonlinear optics, it is very important that when the atomic composition becomes more complicated with respect to elementary and binary semiconductor compounds, the efficiency of radiation conversion increases. In particular, the triple compound $HgGa_2S_4$ is of interest for the conversion of CO_2 laser radiation with an efficiency of up to 60%, as well as for recording radiation in the 10-12 μm transparency band of the atmosphere and for visualizing the spectra of fast processes in the 10-12 μm range. Since these substances have a large band gap (up to 4 eV), the development of technologies and instrument structures based on them can solve to some extent the problem of expanding the range of semiconductor electronics materials of the short-wave spectral range. It is important to note that the level of optical absorption in the region of their transparency slows down a wide application of these materials in nonlinear optics and photonics.

2. Results and discussion

Single crystals of ternary compounds of $A^{II}B^{III}{}_2C^{VI}{}_4$ semiconductors ($HgGa_2S_4$ and $CdGa_2S_4$) and various photosensitive structures based on them were investigated. The monocrystals were produced by directional crystallization of the triple melt compound close to the stoichiometry in a horizontally located boat made of glassy carbon ($HgGa_2S_4$), and also by the Bridgman-Stockbarger method from pre-synthesized single-phase polycrystalline ingots of this compound ($CdGa_2S_4$).

2.1 Structure on the basis of $HgGa_2S_4$ monocrystals

Monocrystals of $HgGa_2S_4$ were grown by directional crystallization of the triple melt compound close to the stoichiometry in a horizontally located boat made of glassy carbon. In the process of slow directional cooling of the melt at rates of 0.1-0.3°C / h, the axial temperature gradient is 1-3°C / cm, which provides effectively suppressed crack formation, mass transfer and disturbances in the composition of the solid phase. The obtained bars with average dimensions of 10 × 15 × 200 mm were uniformly colored in a light orange color. The unit cell parameters of these crystals coincided with those known for $HgGa_2S_4$. Electrical measurements made on specially non-oriented single-crystal plates show that such crystals have a concentration of free electrons of 10^8-10^9 cm^{-3} and a resistivity of 10^9-10^{10} Ω cm at T = 300 K. It should be noted that attempts to obtain crystals of mercury thiogallate have been carried out for a long time. However, the small size of crystals and their heterogeneity determines the need for further research to study the optical and photoelectric properties of such crystals [1].

The contact of vacuum-deposited layers of pure indium with a chipped or polished surface of $HgGa_2S_4$ reveals a clear straightening, with the throughput direction in Schottky barriers In / $HgGa_2S_4$ always being realized with positive polarity of the external displacement on the semiconductor.

The rectification coefficient in the best barriers reached 15, and the reverse current increased with the power law and did not exceed (2-5)· 10^{-9} A for U ≤ 5B and T = 300 K. When illuminated Schottky barriers, natural radiation generates a photovoltage and a semiconductor charges positively, which agrees with the direction of rectification. The voltage photosensitivity of the best Schottky barriers reaches 200 V/W at T = 300 K and dominates when illuminated by the barrier contact.

For all the Schottky barriers In / $HgGa_2S_4$ obtained at the place of the short-wavelength decay typical for the case of illumination from the substrate side, η, a sharp increase in the photosensitivity occurs at hω> 2.3 eV during the transition to illumination from the barrier contact. In the range from 2.3 to 2.8 eV this growth is close to the root law, as evidenced by the rectification of the photosensitivity spectra in the coordinates $(\eta h\omega)^{1/2}$-hω. Extrapolation of this dependence $(\eta h\omega)^{1/2} \rightarrow 0$ gives the value of the energy B_1 = 2.32 eV, which can be compared to indirect optical transitions in $HgGa_2S_4$. At hω> 2.85 eV, the growth of the photosensitivity becomes close to the exponential one and can be characterized by the steepness $s = \partial (\ln \eta) / \partial (h\omega) \cong 12$ eV^{-1}, which gives grounds to compare it with the onset of direct interband transitions with an energy $E_2 \cong 2.85$ eV, which is close to the estimate of the bandgap width $HgGa_2S_4$ given in the literature [2]. The increase of η in the region hω> E_2 up to 3.5 eV allows us to conclude that the obtained Schottky barriers are sufficiently effective in collecting pairs which photogenerated on the semiconductor surface.

2.2 Structure on the basis of $CdGa_2S_4$ monocrystals

$CdGa_2S_4$ monocrystals were grown by the Bridgman-Stockbarger method from pre-synthesized single-phase polycrystalline ingots of this compound. For the synthesis, binary components CdS and Ga_2S_3 were used, which were loaded into cone-shaped crucibles of pyrolytic boron nitride with an internal diameter of ~ 16 mm and a length of 90 mm, and then crucibles filled with substance were placed in quartz ampoules. Ampoules were supplied with a quartz rod, with which they were attached to a mechanical vibrator. The pumped and sealed ampoule was placed in a single-zone vertical furnace, the temperature of which increased at a rate of ~ 100°C / h to 1020-1050°C, at which it was kept for 3 hours with continuous vibrational stirring, after which the melt was cooled at a rate of ~ 100°C / h to room temperature. After synthesis, the ampoule was transferred to a vertical two-zone furnace in which the growth of monocrystals was carried out. The temperature of the melt zone was set ~ 1030°C, the annealing zone ~ 830°C. The melt under these conditions was held for 24 hours, after which the ampoule with the melt dropped through the crystallization zone at a rate of ~ 0.26 mm / h at a temperature gradient of ~ 40°C / cm. After the crystallization of the melt was completed, homogenizing annealing of the resulting crystals was usually carried out for 72 hours. The monocrystalline ingots (16 mm in diameter, 40 mm in length) obtained in this method were transparent enough and homogeneously colored in a light yellow color.

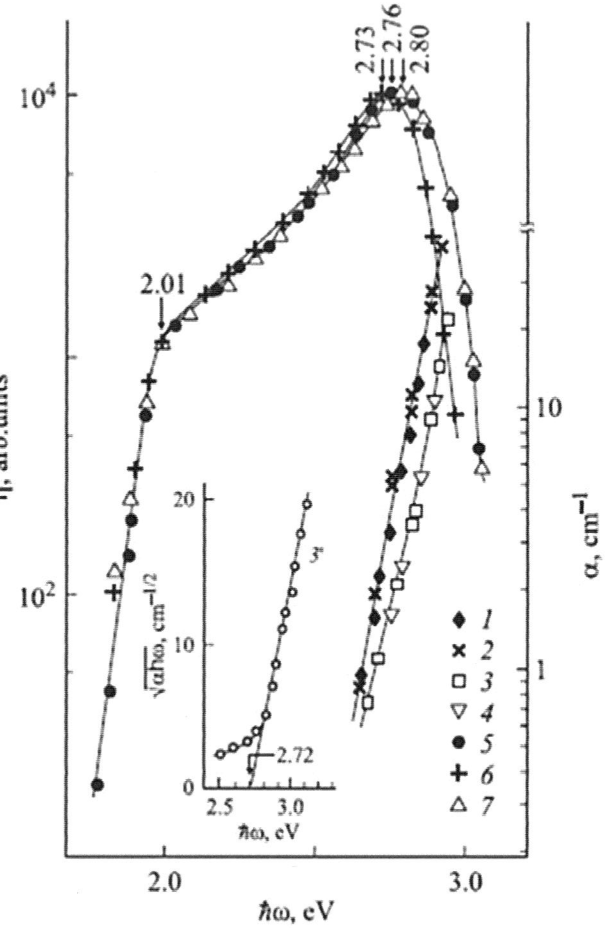

In the absence of intentional doping, $CdGa_2S_4$ monocrystals grown by both methods indicated an electronic conductivity type and resistivity $\rho \approx 10^8$-10^{10} Ω·cm at 300 K, with the highest-resistivity crystals obtained by the gas-phase method. An estimate of the concentration and Hall mobility of

Figure. 1. Spectral dependences of optical absorption for $CdGa_2Se_4$ monocrystal (sample K10- curves 1,3,3', sample B7- with oriented in (100) plate with thickness d, mm: 1,3 – 0.33; 2,4 – 0.15; with polarization of radiation: 1,2 – **E**∥**c**, 3,4 – **E**⊥**c**) and relative quantum efficiency of p-GaSe/n-CdGa2Se4 (sample 8, nonpolarized radiation – curve 5, **E**∥**c** – curve 6, **E**⊥**c** – curve 7) at room temperature. Spectral resolution setup is 1meV.

electrons on the basis of measurements of the Hall coefficient in $CdGa_2S_4$ monocrystals yields $n \approx 10^8$-10^9 cm^{-3} and $u_n \approx 1$-10 cm^2 /V·s at 300 K.

Since $CdGa_2S_4$ monocrystals can only grow n-type conductivity, the method of planting various semiconductors for optical contact with each other was used to obtain heterostructures.

Mechanically, and then the chemically polished plane of the single crystal $CdGa_2S_4$ was brought into contact with the surfaces of the newly obtained natural cleavages of the layered GaSe semiconductor compound with a hole concentration of $\approx 10^{14}$ cm^{-3} and a Hall mobility of ≈ 40 cm^2 / Vs at T = 300 K. In Fig. 1 shows typical spectral dependences of the α (hω) crystal of the $CdGa_2S_4$ crystal and η (hω) heterostructure from this crystal in linearly polarized radiation (LPI). It can be seen that in the LPI the spectrum of the optical absorption coefficient splits in such a way that $\alpha^{\|} > \alpha^{\perp}$, as a result of which the spectrum is shifted in parallel to the short-wave region by ~ 80 meV with a transition from the polarization **E** ‖ **c** to **E** ⊥ **c**. This means that indirect interband transitions are predominantly resolved in polarization **E** ‖ **c**, as for uniaxial $A^{II}B^{IV}C^{V}{}_2$ crystals [3,4]. Polarization photosensitivity occurs only in heterostructures produced on $CdGa_2S_4$ (100) plates. The long-wavelength edge of the photosensitivity of heterostructures, determined by the interband absorption

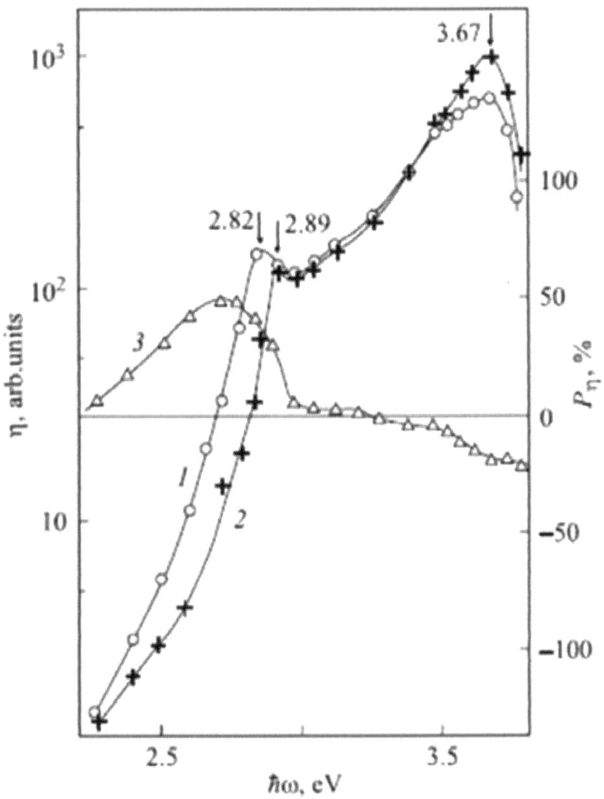

Figure 2. Spectral dependences of related quantum efficiency of photoconversion (curve 1- **E**‖**c**, curve 2 – **E**⊥**c**) and coefficient of natural photopleochroism (curve 3) of H_2O/n-$CdGa_2S_4$ photoelectrochemical cell , T=300K, radiation from the side of H_2O. Spectral resolution setup is 1 meV.

of GaSe when radiation enters it along the isotropic direction, as expected [], is insensitive to the polarization of the radiation. Polarization photosensitivity occurs only at hω> E_g of gallium selenide and can be related to the anisotropy of photoactive absorption in $CdGa_2S_4$ crystals. This anisotropy turns out to be very peculiar. In polarized radiation, the spectrum η (hω) splits (Fig. 1, curves 6 and 7) by ~ 80 meV, as does the α(hω) spectrum of the plate from which the heterostructure is obtained. In this case, a photoisotropic point at which $\eta^{\|} = \eta$ is observed in the spectrum of η(hω) at $h\omega_i \approx 2.76$ eV, and below $h\omega_i$ when $\eta^{\|} > \eta^{\perp}$, while above this energy, $\eta^{\|} < \eta^{\perp}$. Consequently, only in the long-wave spectral region (hω <$h\omega_i$) the photosensitivity anisotropy agrees with the anisotropy of the optical absorption coefficient (Fig. 2), whereas in the region hω> $h\omega_i$ the sign of the photosensitivity anisotropy $\eta^{\perp} > \eta^{\|}$ becomes opposite to the dichroism of the optical absorption ($\alpha^{\|} > \alpha^{\perp}$).

This difference may be due to the fact that the photosensitivity of GaSe/$CdGa_2S_4$ heterostructures in the region hω> $h\omega_i$ begins to be determined by the optical transmission of the $CdGa_2S_4$ plate because the radiation E ⊥ c slightly absorbed in $CdGa_2S_4$ still continues to contribute to the photosensitivity of the heterostructure, whereas the absorption depth in $CdGa_2S_4$ radiation with polarization E ‖ c is already shorter than the diffusion displacement length of the photoholes in the $CdGa_2S_4$ plate.

Direct photoelectrochemical contact of $CdGa_2S_4$ with H_2O, like solid structures, reveals rectification and photovoltaic effect. Straightening in the structures of $H_2O/CdGa_2S_4$ at U~10 V reaches ~10^2, and under their illumination a photocurrent arises. Among the photoelectrochemical (PHCN) cells we obtained, the $H_2O/CdGa_2S_4$ cells showed the highest photosensitivity, with the photoelectrode always having a negative potential, which agrees with the direction of rectification. The main results of this cycle are shown in Fig. 2 and are as follows. When the photocell is illuminated from the electrolyte side along the normal to the photoelectrode (100) plane, the long-wavelength photosensitivity edge in the region of indirect optical transitions exhibits the same polarization dependence as the optical absorption coefficient of $CdGa_2S_4$. This dependence consists in shifting the long-wave part of the spectra η(h ω) to the short-wavelength region with a change in the polarization **E ‖ c** by **E ⊥ c** because the inequality $\eta^{|} > \eta^{\perp}$ is satisfied.

In the case of $GaSe/CdGa_2S_4$ heterostructures, the short-wavelength photodispertivity drop due to the effect of optical absorption in $CdGa_2S_4$ turned out to be opposite to the inequality $\alpha^{|} > \alpha^{\perp}$. And only in the PHCN, where it was possible to exclude the influence of absorption by the $CdGa_2S_4$ crystal on the photoconversion process, the anisotropy of α and η was found to be consistent. A correspondence between the absorption and photoconversion process has been achieved for the PHCN, as a result of which the inequality $\eta^{|} > \eta$ is satisfied.

The shift of the spectra of η (h ω) with a change in the polarization of the radiation from **E ‖c** to **E ⊥ c** by ≈ 70-80 meV (Fig. 2, curves 1, 2) can be a consequence of the removal of degeneracy in the free band and its splitting due to the natural tetragonal compression in $CdGa_2S_4$ crystals. In the region of indirect interband transitions in $CdGa_2S_4$, the natural photopleochrism coefficient P_η turned out to be positive and reaches a maximum value of ~ 50% near the photon energy ~ 2.7 eV (Fig. 2, curve 3), which is close to the energy of indirect interband transitions in $CdGa_2S_4$. With increasing hω> 2.9 eV (Fig. 2, curve 3), the natural photopleochroism coefficient remains low (Pη≅0) up to 3.4 eV. This may be a consequence of competition between optical transitions with opposite selection rules and only with an approach to the energy of direct optical transitions does the amplitude of the negative coefficient of photopleochroism increase to ~ 20%. This means that direct interband transitions in $CdGa_2S_4$ are predominantly resolved in the polarization E ⊥ c. Unfortunately, detailed theoretical calculations of the band structure of $CdGa_2S_4$ crystals are currently not available [5,6,7,8], which makes it impossible to compare now our results with the theory.

3. Conclusions

Presented here results open up new opportunities for structures based on anisotropic semiconductors $HgGa_2S_4$ and $HgGa_2S_4$ as the photodetectors of natural and linearly polarized radiation. Also we show that the Shottky barriers can be used for monitoring the optical quality of $HgGa_2S_4$ monocrystals as wide-band photodetectors.

References

[1] Levine B F, Bethea C G, Kasper H M, Thiel F A, 1976 *IEEE Journal Quantum Electron* QE 12 367

[2] Shay J L, Tell B, Kasper H M, and Schiavone L M. 1972 *Physics Review* B 3 5003.

[3] Bairamov B Ch, Rud' V Yu, Rud' Yu V 1998 MRS Bull 23 91.

[4] Shpunt V Ch, Rud' Yu V, Tsendin K D 1993 *Journal Technical Physics* **19** 13 41

[5] Kesamanly F P, Rud' V Yu, Rud' Yu V 1999 *Semiconductors* **5** 483

[6] Rud`V Yu, Rud` Yu V, Shpunt V Ch, Terukov E I 2016 *Optical Memory & Neural Networks (Information Optics)* **25** 1 40
[7] Myazin N S, Smirnov K J, Davydov V V, Logunov S E 2017 Journal of Physics: Conference Series. **929**-1 012064
[8] Rud V Yu, Rud Yu V, Terukov E I 2016, *J. Opt. Technol.* **83**-5-275

IOP Conf. Series: Journal of Physics: Conf. Series **1038** (2018) 012101 doi:10.1088/1742-6596/1038/1/012101

Drift-diffusion numerical simulation of UTC photodiodes for on-chip optical interconnections

I V Pisarenko and E A Ryndin

Department of Electronic Apparatus Design, Institute of Nanotechnology, Electronics, and Electronic Equipment Engineering, Southern Federal University, 2 Shevchenko St., Taganrog, 347922, Russia

E-mail: ivan123tgn@yandex.ru, earyndin@sfedu.ru

Abstract. $A^{III}B^{V}$ laser-modulator is an advanced optoelectronic device offering new possibilities in the field of on-chip optical interconnecting. To realize the laser-modulator-based interconnections, high-performance integrated photodetectors are required. In this paper, we researched uni-travelling-carrier photodiodes (UTC-PDs). We proposed the extended drift-diffusion model taking into account the effects of carrier drift velocity saturation and electron inter-valley transition. For the implementation of the model, we developed the finite difference numerical simulation technique and dedicated software. These aids were applied for the simulation of *InP*/*InGaAs* UTC-PD. According to the simulation results, the device response time is about 3 ps. Thus, it is reasonable to consider the methods of UTC-PD performance improvement.

1. Introduction

Nowadays, the improvement of on-chip interconnection performance seems to be an urgent problem in the fields of micro- and nanoelectronics. The minimum feature size of mass-produced integrated circuits (IC) has been scaled down to 10 nm [1]. Fabrication technology is getting closer to the predicted fundamental limit of traditional interconnecting techniques [2]. Therefore, the dramatic reduction in the characteristics of traditional metal on-chip conductors is forthcoming. The problem comes into even sharper focus in devices based on the beyond-CMOS (complementary metal-oxide-semiconductor) technologies. To solve it, different methods are proposed [3]–[5]. The optoelectronic one is a promising technique contemplating the optical interconnecting of on-chip elements [6], [7]. For example, hybrid or monolithically integrated optoelectronic systems can replace critical metal inter-core connections in multi-core ultra-large-scale ICs [8].

The development of dedicated optoelectronic devices suitable for on-chip optical interconnecting is a relevant research line [9]–[11]. In [12], [13], we developed the $A^{III}B^{V}$ injection laser with the functionally integrated modulator of optical radiation. The device applies the principle of the controlled spatial relocation of charge carrier density peaks in quantum regions of valence and conduction bands. It provides the generation of optical signals with terahertz modulation frequency and subpicosecond laser pulses [14]. According to our theoretical estimations, the laser-modulator offers new possibilities for IC optical interconnecting [15]. However, the laser-modulator-based interconnections require specialized photodetectors characterized by fabrication compatibility, subpicosecond response time, optimized bandwidth, high reliability, energy performance, sensitivity, and signal-to-noise ratio.

Uni-travelling-carrier photodiode (UTC-PD) uses electrons as the only type of active carriers in contrast with conventional p-i-n device [16]. The capability of low-voltage operation, high output saturation current, and fast photoresponse are the main advantages of UTC-PD [17]. The superior high-speed performance of the device is caused by the small thickness of space-charge region and high electron mobility. According to [18], the response time of the *InP/InGaAs* UTC-PD with 350-nm collection layer is less than 1 ps. Thus, the device being considered seems to be the promising version of photodetector for high-speed optical interconnections.

This paper is aimed at the development of the model, simulation technique, and software designed for the research of charge carrier transport and accumulation in UTC-PDs. To implement the device model, we chose the drift-diffusion (DD) approximation of the semiclassical approach [19]. We extended the base DD model by the simplified energy-balance equation in order to take into account the inter-valley transition of electrons in $A^{III}B^{V}$ materials.

2. Numerical model and simulation technique

In this paper, we consider the DD transport model in terms of $\{\varphi, F_n, F_p\}$, where φ is the electrostatic potential; F_n, F_p are the exponents of electron and hole imrefs [19]. The transformation formulas are given by

$$F_n = \frac{n}{n_i} \cdot \exp\left(\frac{-\varphi - V_n}{\varphi_T}\right); \tag{1}$$

$$F_p = \frac{p}{n_i} \cdot \exp\left(\frac{\varphi - V_p}{\varphi_T}\right), \tag{2}$$

where n, p are the electron and hole densities; n_i is the intrinsic carrier concentration; φ_T is the temperature potential; V_n, V_p are the offsets of valence and conduction bands. After the scaling by standard factors, the basic DD equation system becomes

$$\frac{\partial}{\partial t}\left[F_n \cdot \exp(\varphi + V_n)\right] = \nabla\left[\mu_n \cdot \exp(\varphi + V_n) \cdot \nabla F_n\right] + G - R; \tag{3}$$

$$\frac{\partial}{\partial t}\left[F_p \cdot \exp(-\varphi + V_p)\right] = \nabla\left[\mu_p \cdot \exp(-\varphi + V_p) \cdot \nabla F_p\right] + G - R; \tag{4}$$

$$\nabla(\varepsilon \cdot \nabla\varphi) = F_n \cdot \exp(\varphi + V_n) - F_p \cdot \exp(-\varphi + V_p) - N_D + N_A, \tag{5}$$

where ε is the dielectric permittivity of semiconductor; t is time; μ_n, μ_p are the electron and hole mobilities; G, R are the generation and recombination rates of electron-hole pairs; N_D, N_A are the densities of ionized donors and acceptors. At ohmic contacts, boundary conditions for equations (3)–(5) has the following scaled form:

$$F_n = \exp[-U(t)]; \tag{6}$$

$$F_p = \exp[U(t)]; \tag{7}$$

$$\varphi = \ln\left[\frac{N_D - N_A}{2} + \sqrt{\left(\frac{N_D - N_A}{2}\right)^2 + 1}\right] + U(t) = -\ln\left[-\frac{N_D - N_A}{2} + \sqrt{\left(\frac{N_D - N_A}{2}\right)^2 + 1}\right] + U(t), \tag{8}$$

where $U(t)$ is the bias voltage applied to the contact at the moment t.

Field strengthening leads to the significant mobility decline in $A^{III}B^{V}$ semiconductors. This process has non-linear character. It is caused by the saturation of carrier drift velocity and electron inter-valley transition. Due to intensive electric fields in UTC-PDs, high-field effects are addressed in this paper. However, the aforementioned mechanisms can not be directly described by the DD formulation. We took them into account by the comprehensive mobility model involving the following components [20]:

- the Caughey-Thomas analytical model which considers the phonon and impurity scattering;
- the model of carrier drift velocity saturation:

$$\mu_{c,\text{sat}} = \mu_{c,\text{LF}} \cdot \left(1 + \left(\frac{\mu_{c,\text{LF}} \cdot E_c}{\upsilon_{c,\text{sat}}}\right)^{\chi}\right)^{-\frac{1}{\chi}}, \tag{9}$$

where $c = \text{n},\text{p}$; $\mu_{c,\text{LF}}$ is the Caughey-Thomas mobility; $\upsilon_{c,\text{sat}}$ is the carrier saturation velocity; χ is the adjustable parameter; E_c is the effective electric field equal to imref gradient;

- the two-valley mobility model for electrons:

$$\mu_{\text{n}} = \frac{\mu_{\text{n,sat,low}} + \mu_{\text{n,up}} \cdot P_{\text{val}}}{1 + P_{\text{val}}}; \tag{10}$$

$$P_{\text{val}} = Q \cdot \exp\left(-\frac{\Delta E_{\text{val}}}{kT_{\text{n}}}\right), \tag{11}$$

where $\mu_{\text{n,sat,low}}$, $\mu_{\text{n,up}}$ are the lower- and upper-valley mobilities calculated with saturation formula (9); Q is the the ratio between the state densities in valleys; k is the Boltzmann constant; T_{n} is the electron temperature; ΔE_{val} is the energy gap between the corresponding minimums of conduction band.

To calculate the electron temperature distribution, we applied the simplified energy-balance equation for electrons [21]. Under the assumption that the average electron energy is defined by $3kT_{\text{n}}/2$, it can be rewritten as follows:

$$\frac{dT_{\text{n}}}{dt} = \frac{2e}{3k}\mu_{\text{n}}E_{\text{n}}^2 - \frac{T_{\text{n}} - T}{\tau_{\text{E}}}, \tag{12}$$

where e is the elementary charge; T is the lattice temperature; τ_{E} is the energy relaxation time. The last parameter was estimated using the analytical T_{n}-dependent model represented in [20]. We solved equation (12) consistently with DD model (3)–(5) and mobility equations (9)–(11).

We developed the numerical simulation technique designed for the implementation of extended DD model (3)–(12). It is based on the finite difference method and has the following key features:

- The discretization scheme of the DD equations is similar to the one proposed in [22].
- To discretize the energy balance equation, we use the explicit difference scheme. The appropriate finite difference equation is given by

$$\frac{T_{\text{n},i}^{j+1} - T_{\text{n},i}^{j}}{\Delta t^{j}} = \frac{2e}{3k}\mu_{\text{n},i}^{j}\left(E_{\text{n},i}^{j}\right)^2 - \frac{T_{\text{n},i}^{j} - T}{\tau_{\text{E},i}^{j}}, \tag{13}$$

where i is the index of one-dimensional coordinate grid points; j is the index of time grid points; Δt^j is the j-th time grid step.

- We carry out the stationary simulation in order to calculate the initial condition for the working value of supply voltage and illumination absence. The Newton numerical method is applied.
- DD equation system (3)–(5) is solved by the Newton numerical method at each time step. Between the method iterations, we refine carrier mobilities using the energy-balance equation and formulas (9)–(11).
- Linear equation systems are solved by the direct method.

The above technique of extended DD simulation demonstrates numerical stability and quick convergence over a wide range of time and grid steps.

We implemented the specialized software package intended for the numerical simulation of semiconductor devices within the approach being considered. The program is written in the Octave programming language. We executed it with the use of the GNU Octave 4.2.1 software. The applied modelling aids allow to simulate various heterostructure devices at different control actions.

3. Simulation results and discussion

We performed one-dimensional DD simulation of the waveguide *InP/InGaAs* UTC-PD. The device structure is shown in figure 1(a). Lateral faces of the photodiode are covered by half and totally reflecting mirrors 2 and 4. They form the resonant cavity which enhances the quantum efficiency of the device. The peak value of this quantity equals to 0.7 at the resonant cavity length of 1 μm.

Figure 1(b) shows the energy band diagram of UTC-PD with *InP*/$In_{0.53}Ga_{0.47}As$ heterojunctions at the reverse bias voltage of 0.5 V. The thickness of p-$In_{0.53}Ga_{0.47}As$ absorption layer is 150 nm. The collector one is 350 nm. The plot demonstrates the influence of illumination on electron imref. The light action leads to the emergence of imref gradient in the collector. As a result, we observe the effective field growth and mobility reduction.

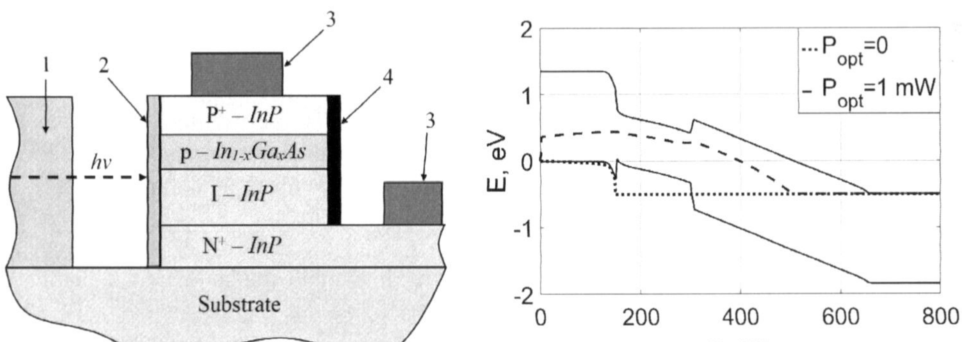

(a) (b)

Figure 1. (a) The cross-section of *InP/InGaAs* UTC-PD: 1 – integrated optical waveguide; 2 – half reflecting optical mirror; 3 – ohmic contacts to device regions; 4 – totally reflecting optical mirror. (b) The band diagram of P^+-*InP*/p-$In_{0.53}Ga_{0.47}As$/I-*InP*/N-*InP* UTC-PD. Dashed lines refer to electron imref. Solid ones are valence band top and conduction band bottom.

Figure 2 shows the results of UTC-PD numerical simulation. To calculate these curves, we applied the extended DD model (3)–(12). Carrier transport in UTC-PD is dominated by the electron drift through the collection layer (see figure 2(a)). It means that the low mobility of holes does not significantly affects the device performance in contrast with conventional p-i-n photodiode. However, the electron mobility in collector is comparable to the hole one due to the effects of drift velocity saturation and inter-valley transition (see figure 2(b)). Nevertheless, the UTC-PD photoresponse is a bit faster than the p-i-n one at the detection of 1-ps rectangular optical pulse (see figure 2(c)). The curve shown in figure 2(d) indicates that the *InP/InGaAs* UTC-PD has the rise time of 3.3 ps (90 % of amplitude) and fall time of 3.2 ps (10 % of amplitude).

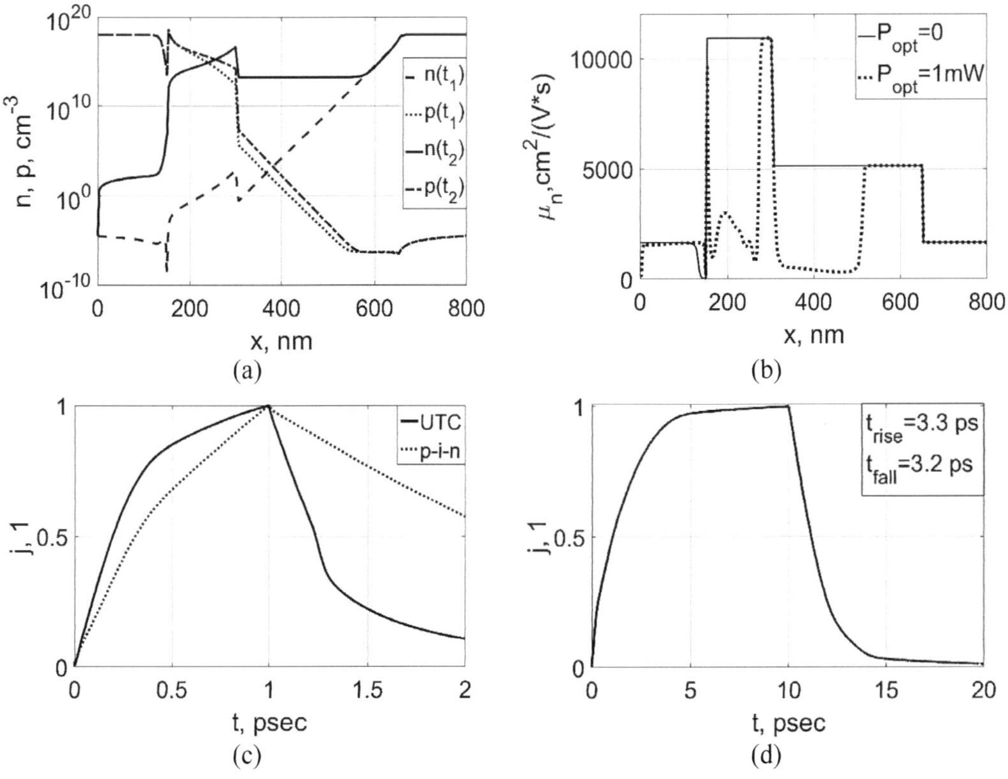

Figure 2. (a) The distributions of electron (n) and hole (p) density in *InP/InGaAs* UTC-PD before (t_1) and after (t_2) illumination. (b) The impact of UTC-PD illumination on electron mobility. (c) The comparison of UTC and p-i-n high-speed performance. (d) UTC-PD photocurrent versus time characteristic at the detection of 10-ps rectangular laser pulse.

4. Conclusion

In this paper, we proposed the DD model extended over high-field transport by the comprehensive carrier mobility. To calculate electron temperature included in the mobility model, we solved the simplified energy-balance equation. The aforementioned approach allows to address the carrier drift velocity saturation and electron inter-valley transition in $A^{III}B^{V}$ materials within the DD approximation. In UTC-PD collector, these effects determine the carrier mobility decline.

We designed the finite difference simulation technique in order to implement the model. The technique is based on the Newton method and provides the optimized balance between the consumption of computational and time resources and simulation results' adequacy. To realize the technique, we developed the specialized software package. The tools being considered allows the simulation of UTC-PDs and other photosensitive devices with various physical and technical parameters.

We performed the simulation of the *InP/InGaAs* UTC-PD. According to the results, the photocurrent pulse edges are longer than 3 ps. Thus, it is reasonable to develop the improved UTC structure providing the desired subpicosecond response time.

Acknowledgements
The research is supported by the funds of "The Development Program of Southern Federal University up to 2021" (Project VnGr-07/2017-10).

References
[1] http://www.samsung.com/semiconductor/minisite/Exynos/Solution/MobileProcessor/Exynos_9_Series_8895.html
[2] Ceyhan A and Naeemi A 2013 *IEEE Trans. Electron Devices* **60** 374–82
[3] Das D and Rahaman H 2014 *Carbon Nanotube and Graphene Nanoribbon Interconnects* (CRC Press)
[4] Rakheja S and Naeemi A 2011 *IEEE Trans. Electron Devices* **58** 1319–28
[5] Tam S-W, Socher E, Chang M-C F, Cong J and Reinman G D 2010 RF-Interconnect for Future Network-On-Chip *Low Power Networks-on-Chip* ed C Silvano, M Lajolo and G Palermo (Boston: Springer) pp 255–80
[6] Miller D A B 2010 *Appl. Opt.* **49** F59–70
[7] Belkin M and Sigov A 2012 *Nanoindustry* **31** 8–14
[8] Moiseev K, Kolodny A and Wimer S 2014 An Overview of the VLSI Interconnect Problem *Multi-Net Optimization of VLSI Interconnect* (New York: Springer-Verlag) pp 1–9
[9] Thubthimthong B, Sasaki T and Hane K 2015 *IEEE Photonics Journal*, **7**, 1–11
[10] Chen L, Huang X, Li M, Huang Y, Wang Y, Yan G and Li C 2015 *Optoelectronics Lett.* **11** 195–8
[11] Hung S-C, Lin S-J, Chao J-J, Chang C-Y, Lin M-J and Lin C-F 2017 *Journal of Lightwave Technology* **35** 2266–72
[12] Pisarenko I, Ryndin E and Denisenko M 2015 *J. Phys.: Conf. Ser.* **586** 012015
[13] Konoplev B G, Ryndin E A and Denisenko M A 2013 *Tech. Phys. Lett.* **39** 986–9
[14] Konoplev B G, Ryndin E A and Denisenko M A 2015 *Russian Microelectronics* **44** 190–6
[15] Konoplev B G, Ryndin E A and Denisenko M A 2011 *Izvestiya SFedU. Engineering Sciences* **117** 21–7
[16] Li C, Xue C-L, Li C-B, Liu Z, Cheng B-W and Wang Q-M 2013 *Chin. Phys. B* **22** 118503
[17] Nagatsuma T and Ito H 2011 High-Power RF Uni-Traveling-Carrier Photodiodes (UTC-PDs) and Their Applications *Advances in Photodiodes* ed G-F Dalla Betta (InTech) chapter 14 pp 291–314
[18] Filachev A M, Taubkin I I and Trishenkov M A 2011 *Solid-state Photoelectronics. Photodiodes* (Moscow: Fizmatkniga)
[19] Vasileska D, Goodnick S M and Klimeck G 2010 *Computational Electronics: Semiclassical and Quantum Device Modeling and Simulation* (CRC Press)
[20] Palankovski V and Quay R 2004 *Analysis and Simulation of Heterostructure Devices (Computational Microelectronics)* (Wien: Springer-Verlag)
[21] Dragunov V P, Neizvestnyj I G and Gridchin V A 2004 *Fundamentals of Nanoelectronics* (Novosibirsk: NGTU Publishing)
[22] Pisarenko I V and Ryndin E A 2016 *Proc. of SPIE* **10224** 102240Z

IOP Conf. Series: Journal of Physics: Conf. Series **1038** (2018) 012102 doi:10.1088/1742-6596/1038/1/012102

Photocathodes for near infrared range devices based on InP/InGaAs heterostructures

K J Smirnov[1,2], V V Davydov[3], S F Glagolev[2], N S Rodygina[2], N V Ivanova[2]
[1] OJSC "NRI "Electron", Saint Petersburg, Russia
[2] Bonch-Bruevich Saint – Petersburg State University of Telecommunications, Saint Petersburg 193232, Russia
[3] Peter the Great Saint – Petersburg Polytechnic University, Saint Petersburg 195251, Russia

Abstract. Results of indium phosphide structures research, showing the possibility of using in the near infrared range photocathodes of InP/InGaAs, are represented. An optimal method of obtaining the atomically clean indium phosphide surface was suggested. The spectral characteristics of InP/InGaAs heterostructure were given. The results of an experimental study of pin-diode, which was used as the receiver of photoelectrons, are presented.

1. Introduction

At present time photodetection devices widely use in various fields of science and technology (night-vision devices, control sensors, fiber optic systems, etc.) [1-4]. One of the most perspective way of development such photodetectors is creation of devices, which work in near infrared range (NIR). [5]. The main element for these devices could be photocathode based on heterostructures of InP/InGaAs. Photosensitive modules for NIR with photocathodes could be used for special purposes, where high processing speed of detection reflected radiation of special laser systems (for example quantum-dot lasers) from the object is needed [6].

The creation of the photocathode, which operates in the NIR, also known as the transferred electron (TE) photocathode, was proposed by Bell [7]. Such photocathodes can move the long-wavelength threshold of photoemission by increasing the internal energy of the photoelectrons (with the help of the metal electrode, which is applied to the effective surface of the photocathode). This structure has been realized on the basis of heteropair indium phosphide – solid solution of indium – gallium arsenide. The InGaAs forbidden band is narrow enough to detect near–infrared range. This narrow–band material ("absorber") cannot provide by itself the negative electron affinity (NEA) on the surface, which is a determinant factor for obtaining a high quantum efficiency for such structures. However, InGaAs can be perfectly matched to the InP crystal lattice, which is a conductor (emitter) of electrons generated in the absorber, allows to get on its surface the effective state of the NEA.

Schottky barrier is needed for creation of the TE photocathode. It is a thin metallic layer activated by Cs and O which allows to apply a pulling field between two layers of the photocathode structure. This barrier is high enough to have a sufficiently low hole current in case of applying of the locking hole voltage. The bias voltage expands the pulling field to the boundary (absorber - emitter), carrying out the transfer of electrons through a heterojunction (fig. 1). Electrons from the conduction band of the emitter layer due to the pulling field can go over to the upper levels of the conduction band. Electrons from the upper valleys of the conduction band can emit into a vacuum.

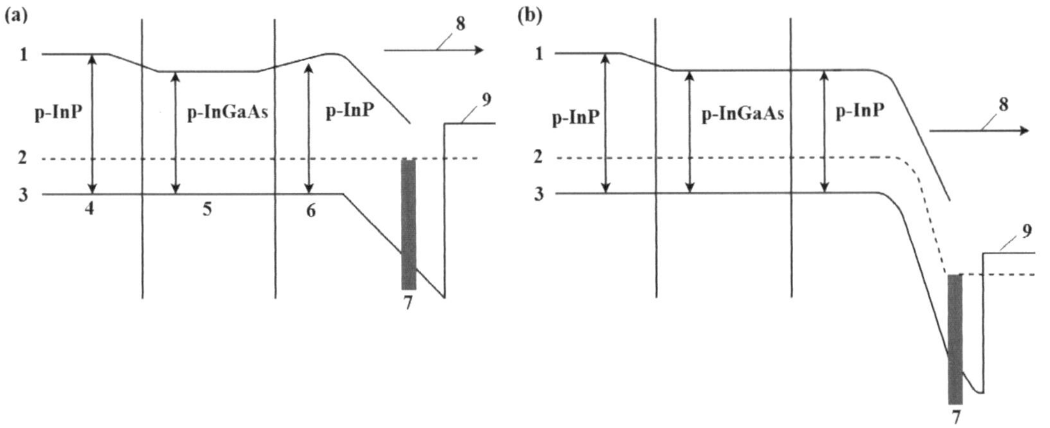

Figure 1. Energy band diagram of transferred electron photocathode **(a)** without applying bias voltage **(b)** with applied bias voltage, where 1 – conduction band, 2 – Fermi level, 3 – valence band,4 – substrate, 5 – absorber, 6 – electron emitter, 7 – metallic layer (Schottky barrier), 8 – photoemission direction, 9 – vacuum level

Semiconductor photocathodes activated to negative electron affinity (NEA) serve as electron beam sources for many applications. InP photocathode is an emitter of InP/InGaAs heterostructure. These heterostructures as compared to other semiconductor materials allows getting greater quantum efficiency in case of detection reflective laser beam in near infrared range. The purposes of the work were to develop the optimal method of obtaining atomic clean photocathode surface, to obtain high quantum efficiency of the InP/InGaAs samples, to estimate the parameters of processing speed and amplification of pin-diode as the photoelectron receiver and to apply the result of the researches for obtaining real photodetector for special purposes, which works in near infrared range.

2. Purifying of InP surface

For effective InP/InGaAs photocathodes, it is necessary to achieve atomic clean surface of structure emitter – InP. The structures based on the indium phosphide are characterized by low thermal resistance and heating over with the temperature of 593 K lead to the degradation of these structures, so that the typical cleaning methods for metals and thermally stable materials cannot be used. Sulfur passivation leads to stable surface termination, but the chemisorbed sulfur atoms cannot be removed completely by thermal annealing at temperatures below the decomposition temperature of InP. Process of obtaining a clean InP surface is rather difficult, especially pre–vacuum chemical cleaning step. The treatment with HCl–C_3H_8O and subsequent vacuum annealing is useful method for obtaining the atomically clean GaAs surfaces, another semiconductor of the A_3B_5 group [8]. However, etching with these chemical compounds is ineffective for InP surfaces. Therefore, we used the efficient three-step method of chemical etching combined with a thermal cleaning in vacuum to obtain the atomic clean surface of InP [9, 10]. Etching time was increased, as compared to [9], and the mode was adjusted to increase in rate of the removing of a thick natural oxide layer by the use of ultrasound in the etching process. Besides that the intermediate etching step was added: a weak solution of sulfuric acid was used. Strong acid solutions with HCl or H_2SO_4, which are used in third step of chemical cleaning, are able to remove the surface oxide and leave the InP surface passivated with elemental P which is, in turn, terminated with H [9]. This yields a hydrophobic surface and allows lower temperatures to be used during annealing. All steps of chemical cleaning are shown in table 1:

Table 1. Optimized three-step method of chemical cleaning

	Solution	Concentration
1	H_2SO_4:H_2O_2:H_2O	4:1:100
2	H_2SO_4:H_2O	1:10
3	H_2SO_4:H_2O	1:1

Protective layer formed on InP surface are also allows to carry etched structures from chemical laboratory bench to vacuum installation in atmosphere. This is an important achievement in terms of surface oxidation and uncontrolled contamination of surfaces. We studied the effect of etching on the titanium grid and the possibility of etching agents use without degradation of the grid structure. As a result, we found that etching is equally effective for two different acid solutions HCl and H_2SO_4, however, in case of HCl titanium grid is exposed to chemical degradation. In view of the fact that indium phosphide is emitter of the InP/InGaAs heterostructures, accordingly chemical cleaning of this structures identical to the InP.

3. Activation of InP and InP/InGaAs structures

As a result of the etching process the hydrophobic surface of InP and InP/InGaAs structures was obtained [9]. The ultra-high vacuum installation was used to conduct the research in a vacuum, which could reach the level of $1\cdot10^{-9}$ Pa [11]. After heating in a vacuum chamber at a temperature of about 573 K, the net emitter atomic surface was received, that is ready for the activation with Cs and O_2, which allows to form effective state of NEA on photocathode surface. The method of the activation structure called «yo-yo» [12], based on a continuous simultaneous supply of cesium and oxygen was used. The process for obtaining the maximum photoemission current was about 90 minutes.

Research was conducted on InP structures which etched by solutions of C_3H_8O+Br, C_3H_8O+HCl and our new three stage method of chemical cleaning with solutions of H_2SO_4. After graphical analysis, we found that the results of photoemission current using the new method of chemical cleaning increase 6 times as compared to previous experiments.

Measurements were carried out in range of 500 to 950 nm for InP samples. We managed to get quantum efficiency greater than 10% at a wavelength of 630 nm, and more than 1% for 900 nm on Zn-doped InP(111) structures with concentration $p=1\cdot10^{18}cm^{-3}$[13]. In [10] the QE at 632 nm was in the range 9-12% and spectral characteristics were not present. The photolithographic grid stamped on of InP, has demonstrated not only the overall photoemission gain increase, but also provided the greater sensitivity at longer wavelengths when a voltage was applied to the grid with the help of the pressure contact.

Results of the investigations of indium phosphide were applied on InP/InGaAS heterostructures, which were grown by MOCVD methods. Heteropair indium phosphide – solid solution of indium – gallium arsenide was realized on substrate InP with thickness 350 mkm. Etching and vacuum annealing were conducted identical to InP structures Spectral characteristics of the InP/InGaAs heterostructure with developed surface grid titanium electrode are shown in the figure 2. The electrode has a grid-lattice structure in form of parallel conductive titanium bands. It is manufactured by photolithographic methods and provides the uniform distribution of bias voltage on the operating surface of the photocathode structure. The thickness of the electrode is about 100 angstroms.

Analyzing obtained dependencies (fig. 2), we found that quantum efficiency of the proposed heterostructure, which operates in reflection mod, reaches 4% at the wavelengths from 900 to 1600 nm. Our studies showed that results of QE obtained on the investigated InP/InGaAs structures are enough to create reliable sensor with such types of photocathodes.

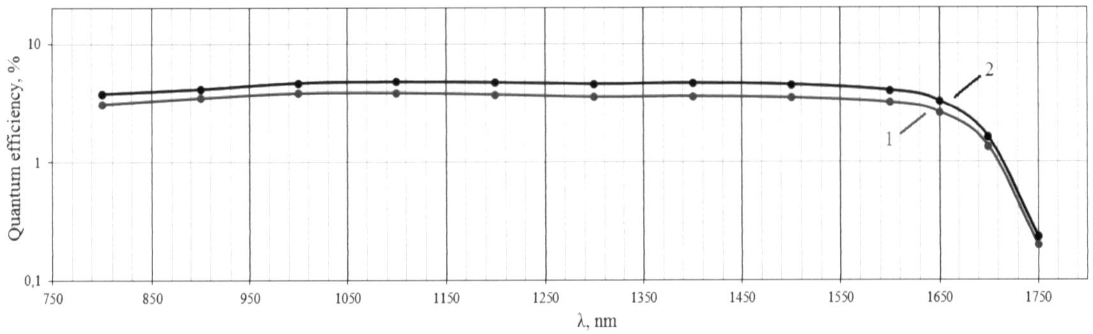

Figure 2. Spectral characteristics of InP/InGaAs, **1** – with grid at U_s= 2 V; **2** – with grid at U_s= 3V

4. Receiver of photoelectrons

In device layout with InP/InGaAs heterostructure the silicon pin-diode was selected as the receiver of the photoelectrons and amplifying element. It is a vertical structure based on pure high-resistance silicon obtained by the zone melting method. To investigate the processing speed, pin-diode was irradiated with a laser light emitting diode with a wavelength of 650 nm and pulse duration of no more than 50 ps (pulse repetition rate 1 kHz). The results of the impulse response are shown in figure 3(a,b). As the supply voltage increased from 190 V to 280 V, the impulse response front decreased from 2.1 ns to 1.1 ns, and the impulse response duration decreased from 5.5 ns to 1.9 ns.

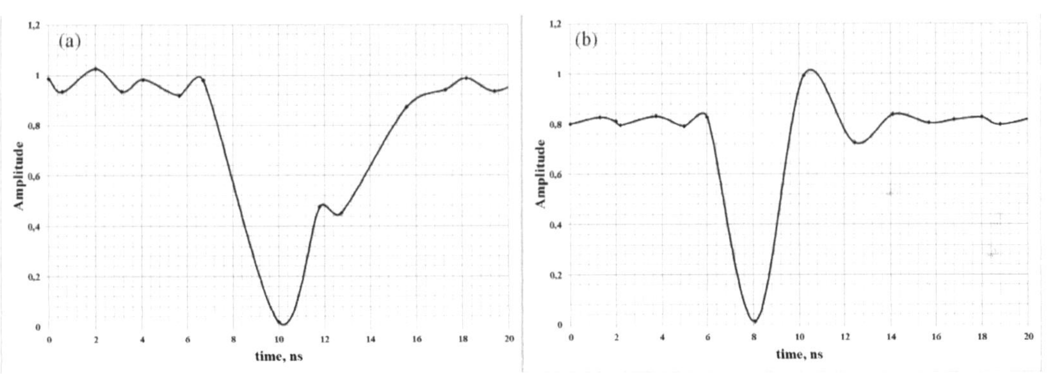

Figure 3. Impulse characteristic of the diode with supply voltage
(a) U_d=190V; **(b)** U_d=280V

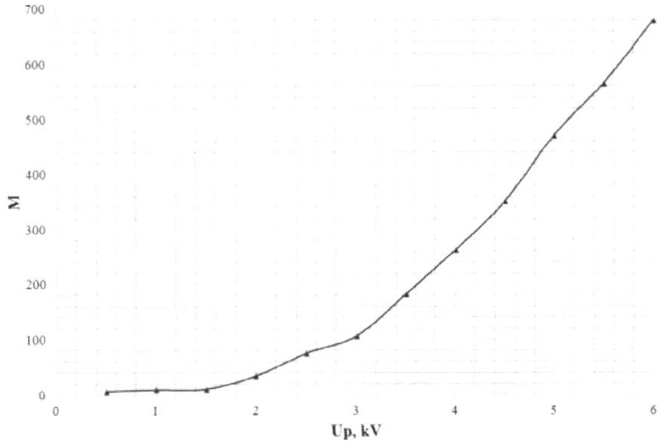

Figure 4. Amplification factor (M) depending on the energy of the photoelectrons Up (the voltage between the photocathode and the pin-diode).

Amplification is achieved due to the occurrence in the solid-state element (diode structure) of a large number of charge carriers by the influence of high energy (several keV) electrons emitting from the photocathode. Dependence of the amplification factor (M) on the energy of photoelectrons (on the voltage between photocathode and pin-diode) are shown in figure. 4. M is determined by the ratio of output to input current of pin-diode. Value of amplification factor shows the possibility of achieving amplification several hundred times at an accelerating voltage up to 6 kV. It is mainly determined by the electric properties of silicon, it does not depend on the type of conductivity or resistance of the silicon wafer.

5. Conclusion

Photoelectron vacuum devices, where emitter and receiver of photoelectrons are consisted in one vacuum value, have a low level of internal noise as compared to devices with non-vacuum parts. Our results of processing speed and amplification factor, obtained on one pin-diode structure, show the possibility of creation photosensitive devices for near infrared range with line or matrix of pin-diodes. It would help to realize fast, high-sensitive multichannel scanning systems. In addition, amplification factor of photoelectron receiver could be increased up to 10^5 with the help of avalanche photodiode. Preliminary experiments showed that obtained quantum efficiency on the InP/InGaAs samples is enough for reliable operation of real device because of internal amplification. Our studies showed that increase of QE could be achieved by optimizing thickness and the basic carrier concentration of heterostructure layers. As such parameters influence the minimum energy that is required for generation and emission of photoelectrons.

References

[1] Petrov A A, Davydov V V 2017 *Journal of Communications Technology and Electronics* **62** 289 – 293

[2] Petrov A A, Vologdin V A, Davydov V V, Zalyotov D V 2015 *Journal of Physics: Conference Series* **643(1)** 012087

[3] Tarasenko M Yu, Davydov V V, Sharova N V, Lenets V A, Yalunina T R 2017 *Lecture Notes in Computer Science (including subseries Lecture Notes in Artificial Intelligence and Lecture Notes in Bioinformatics)* **10531 LNCS** 694-702

[4] Davydov V V, Sharova N V, Fedorova E V, Vologdin V A, Karseev A Yu 2015 *Lecture Notes in Computer Science (including subseries Lecture Notes in Artificial Intelligence and Lecture Notes in Bioinformatics)* **9247** 712-721

[5] Ermolaev A N, Krishpents G P, Davydov V V, Vysoczkiy M G *2016 J. of Phys: Conference Series* **741** 012171.

[6] Kovsh A, Zhukov A, Livshits D, Krestnikov I 2009 US Patent N7561607

[7] Bell R L 1974 US Patent N23958143

[8] Tereshchenko O E, Chikichev S I, Terekhov A S 1999 *Appl. Surface Science* 142 75–80

[9] Sun Y, Liu Z, Machuca F, Pianetta P, Spicer W 2005 *SLAC-PUB*-11018

[10] Sun Y, Liu Z, Pianetta P 2007 *SLAC-PUB*-12710

[11] Myazin N S, Smirnov K J, Davydov V V, Logunov S E 2017 *Journal of Physics: Conference Series* **929** 012080

[12] Chanlek N, Herbert J, Jones R, Jones L, Middleman K, Militsyn B 2014 *J. Phys. D: Appl. Phys.* **47** 055110

[13] Smirnov K J, Medzakovskiy V I, Davydov V V, Vysoczky M G, Glagolev S F 2017 *Journal of Physics: Conference Series* **917(6)** 062019

IOP Conf. Series: Journal of Physics: Conf. Series **1038** (2018) 012103 doi:10.1088/1742-6596/1038/1/012103

Creation and investigation of OLED-structures with inclusion of colloidal quantum dots

Eremeev Mark Anatolyevich, Mikhailov I.I., Tarasov S.A., Lamkin I.A., Tadtaev P.O., Degterev A.E.

Saint Petersburg Electrotechnical University «LETI» (ETU), Saint-Petersburg, Russia

markeremeev22@gmail.com

Abstract. During the work organic light-emitting structures FTO/PEDOT:PSS//TPD/TPD+CQD CdSe(650)/Alq3/Al with an efficiency value of about 2% were created. The technique for introducing CdSe CQD (650) into the structure of OLED was developed. The change in the electroluminescence spectra in structures with colloidal quantum dots is demonstrated. The decrease in OLED switching on voltage in structures with the PEDOT: PSS layer is demonstrated.

1. Introduction

The creation and investigation of colloidal quantum dots (CQD) and devices based on them are some of the most perspective trends in modern electronics. At the same time, in recent years organic semiconductor materials and devices based on them (OLED) have become increasingly important, since their use makes it possible to create screens with low power consumption, high speed of operation, and intensive and rich color rendering [1]. The addition of CQD even to the simplest organic light-emitting structures should significantly improve their performance [2].

The typical structure of OLED consists of a transparent substrate, a transparent anode, two thin layers of organic compounds acting as transport layers for holes and electrons, respectively, and also a cathode [3]. More complex devices, characterized by improved characteristics, contain more layers [4]. For example, the application of the PEDOT:PSS hole-injecting layer allows to significantly reduce the switching on voltage of the structure, and the introduction into the structure of the TAZ material, which blocks holes and excitons, helps to highlight the emission peak of the active layer [5]. In recent years, studies have begun of organic light-emitting diodes that contain CQD acting as the active layer. Colloidal quantum dots encapsulated in the OLED structure allow to increase the efficiency of the device due to the high quantum yield of luminescence and make it possible to control the emittion properties [6,7].

2. Experiment

2.1. Structures and methods of their creation

Various types of organic light-emitting diodes were created (Figure 1), some of which were encapsulated by CQD arrays. The structures consisted of a hole transport layer, which was TPD (N, N'-bis (3-methylphenyl) -N, N'-diphenylbenzidine) and PEDOT: PSS (poly (3,4-

ethylenedioxythiophene) -poly (styrene sulfonate)), as well as an electronic transport layer of Alq3 (tris (8-hydroxyquinoline) aluminum) located between the aluminum cathode and the transparent anode of indium-tin oxide (ITO). The creation of OLED-structures was carried out by the methods of vacuum thermal evaporation and spin-coating.

The decision to use an additional of PEDOT:PSS layer despite a large number of studies of structures containing only powder materials [2, 5] is due to the more technologically simple application of PEDOT:PSS, and also to the improvement of the electrical parameters of the entire structure being created [1, 8] . As substrates were used glass 3x3 cm.

The main optical and electrical parameters of the structures were investigated: current-voltage and watt-ampere characteristics, efficiency characteristics of structures, as well as their luminescence spectra were obtained. The analysis of the current-voltage characteristics confirms the presence of the rectification effect in the created structures. The obtained luminescence spectra clearly demonstrate the possibility of changing the emission spectrum of OLED by introducing into the structure a layer of QCDs of different composition and sizes. This will allow without significant changes in the process technology to create devices with the necessary spectral characteristics and increased emission power. Figure 1 shows that adding to the structure of the PEDOT:PSS layer allowed to reduce the OLED switching on voltage.

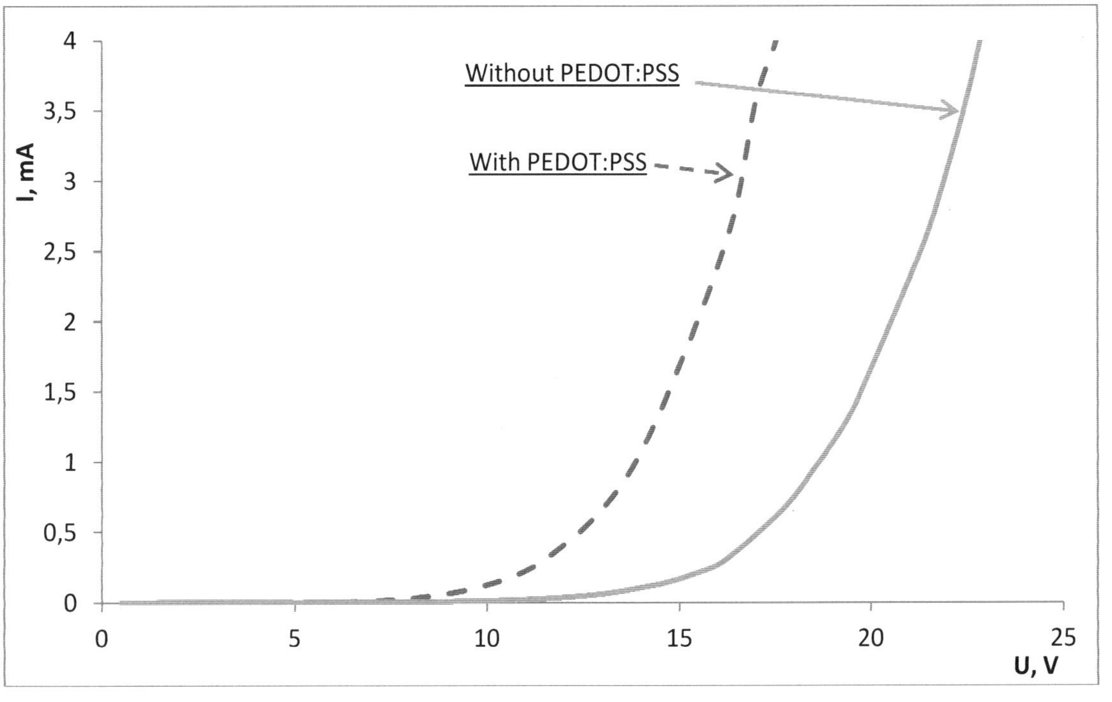

Figure 1. Current-voltage characteristic of the ITO/TPD/Alq3/Al and ITO/PEDOT:PSS/TPD/Alq3/Al.

2.2. Samples characterization

Emitters containing CQD based on cadmium selenide coated with a wide-zone zinc sulphide shell were created in this work (Fig. 2). Figure 3 shows the photoluminescence spectrum of colloidal quantum dots CdSe/ZnS (650) and their image obtained using the method of scanning electron spectroscopy. This image was used to estimate the size of CQD. The possibility of changing the emission wavelength in the range of 500-650 nm due to the introduction of CdSe-based nanoparticles into the structure is shown (Fig. 4). An increase of the efficiency of OLED radiation was revealed when CQDs were introduced into them. The obtained efficiencies exceeded the value of 2%, which is a rather high index for low-layer structures that do not contain dopants based on iridium. Further enhancement of efficiency of the created OLEDs and extension of the range of their radiation is discussed.

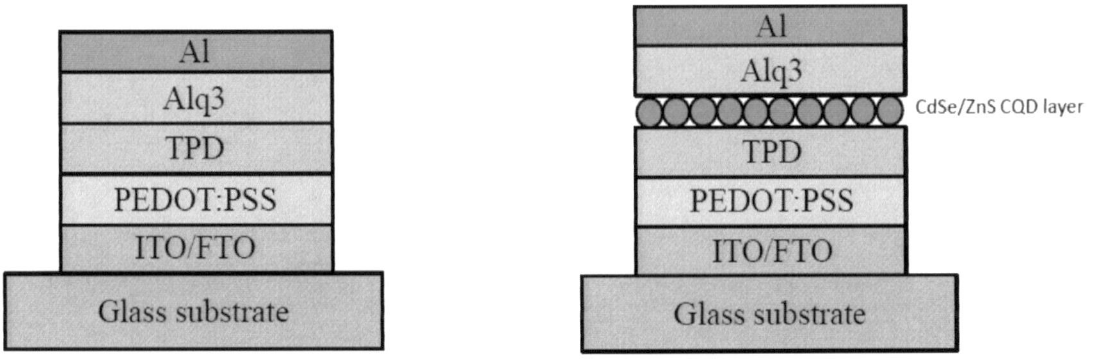

Figure 2. Schematic representation of the organic light-emitting structures PEDOT:PSS/TPD/Alq3/Al and PEDOT:PSS/TPD/TPD+CQD_CdSe_650/Alq3/Al.

To create these structures, methods of spin-coating and vacuum thermal deposition were used. The polymer compound PEDOT:PSS was applied to the ITO substrate by spin-coating. The time of spin-coating was 30 seconds, the speed was 3000 r/s, the number of layers was 2. The TPD + CQD_CdSe_650 layer was also applied by spin-coating. TPD and CQD were mixed in toluene. The spin-coating time was 30 seconds, the speed was 3000 r/s, the number of layers was 3. The layers TPD, Alq3 and aluminum contacts were made by the method of vacuum thermal evaporation.

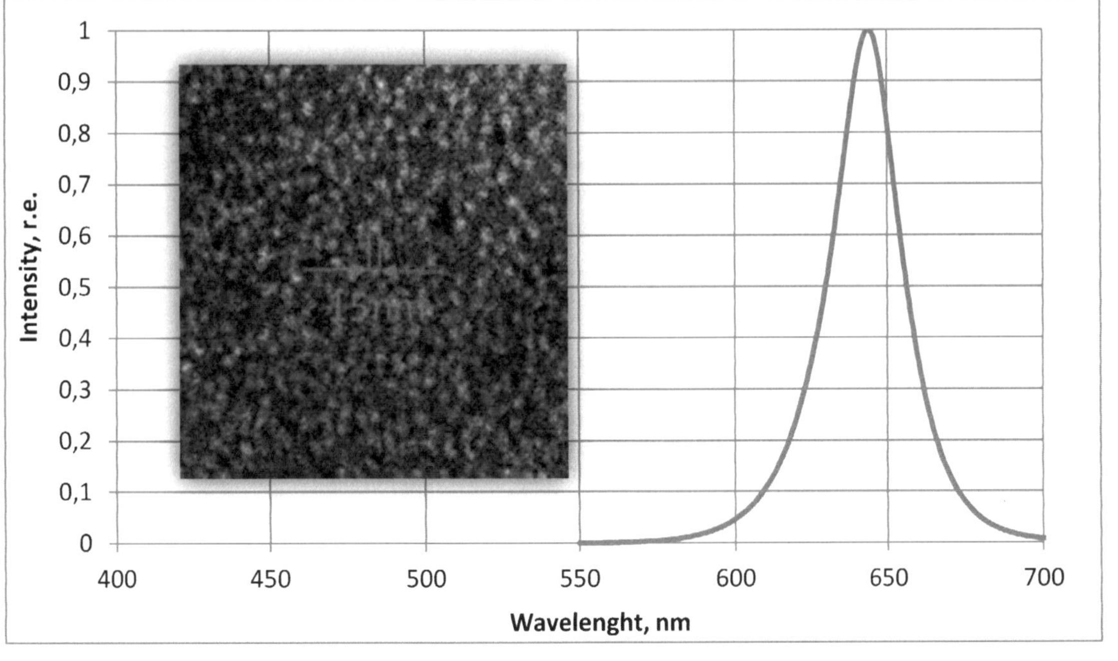

Figure 3. The photoluminescence spectrum of colloidal quantum dots CdSe/ZnS (650).

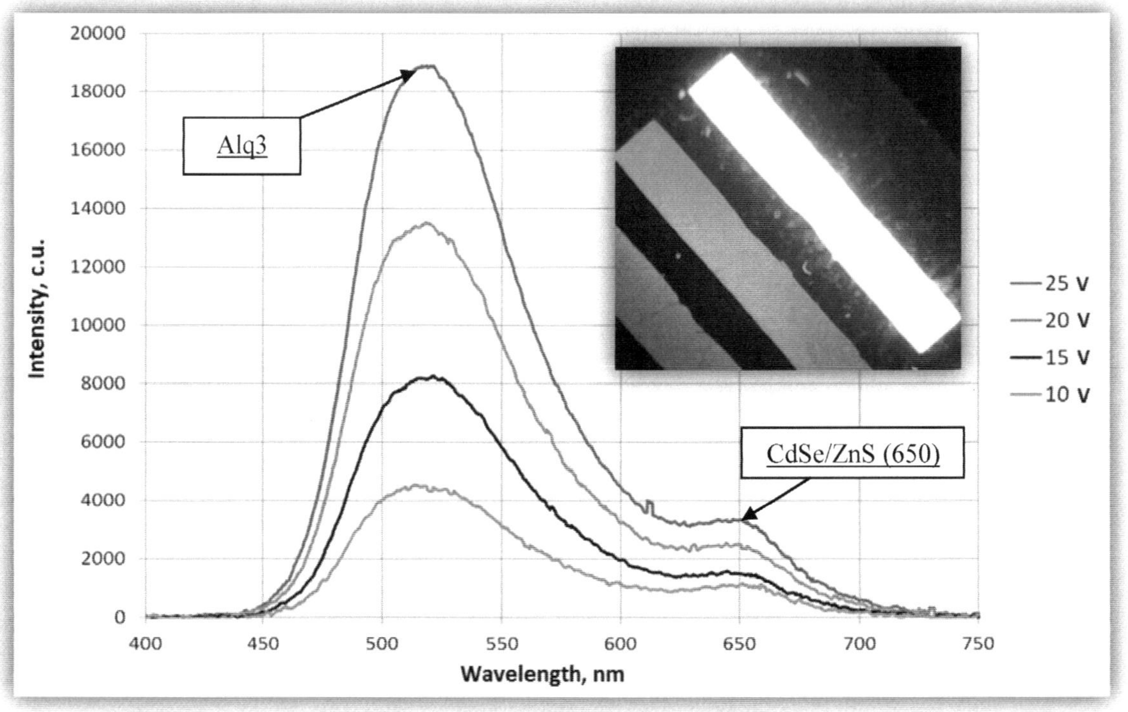

Figure 4. Electroluminescence spectra of the organic structure containing colloidal quantum dots in the active layer.

The switching on voltage was 6 volts. This voltage is comparable with the voltage of inclusion of similar organic structures [8]. The presence of a "bulge" on the current-voltage characteristic in the vicinity of 6 volts is explained by the formation of current-conducting shunt channels (Fig. 5). These channels are formed due to the presence of unevenness on the surface of the layers. As a result, regions with small thicknesses of layers appear which, when the structure is first switched on, "break through" and become conductive channels that are not involved in the electroluminescence process. To eliminate irregularities on the surface of the layers, work is underway to improve the technology of creating organic structures.

Figure 6 shows the watt-ampere characteristics of created structures. It shows that in a sample with a CQD, the optical power for the same current is lower, this is due to the additional expenditure of energy to excite radiation in the CQD layer. The shapes of both characteristics are similar to those of the watt-ampere characteristics of inorganic light-emitting diodes.

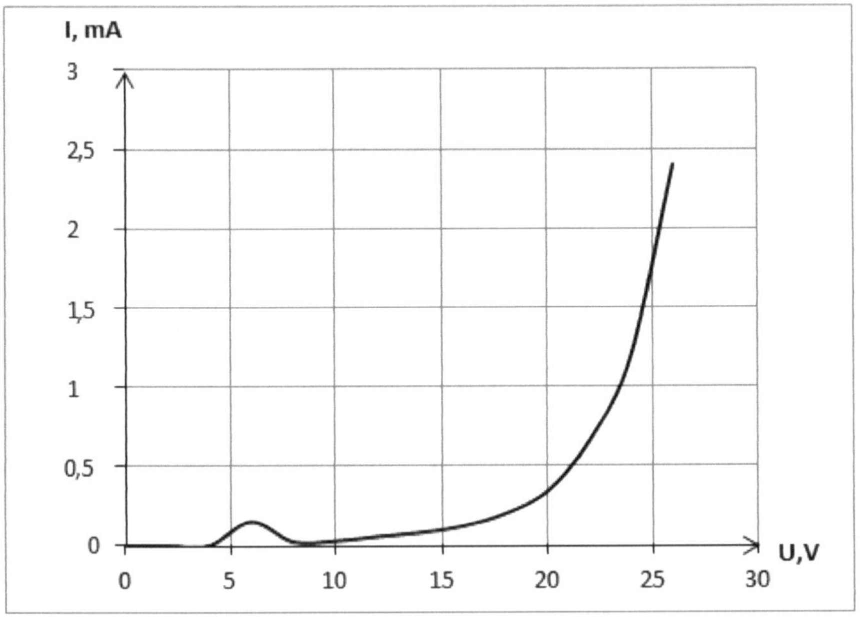

Figure 5. The current-voltage characteristic of the ITO/PEDOT:PSS/TPD/TPD+CQD_CdSe_650/Alq3/Al structure.

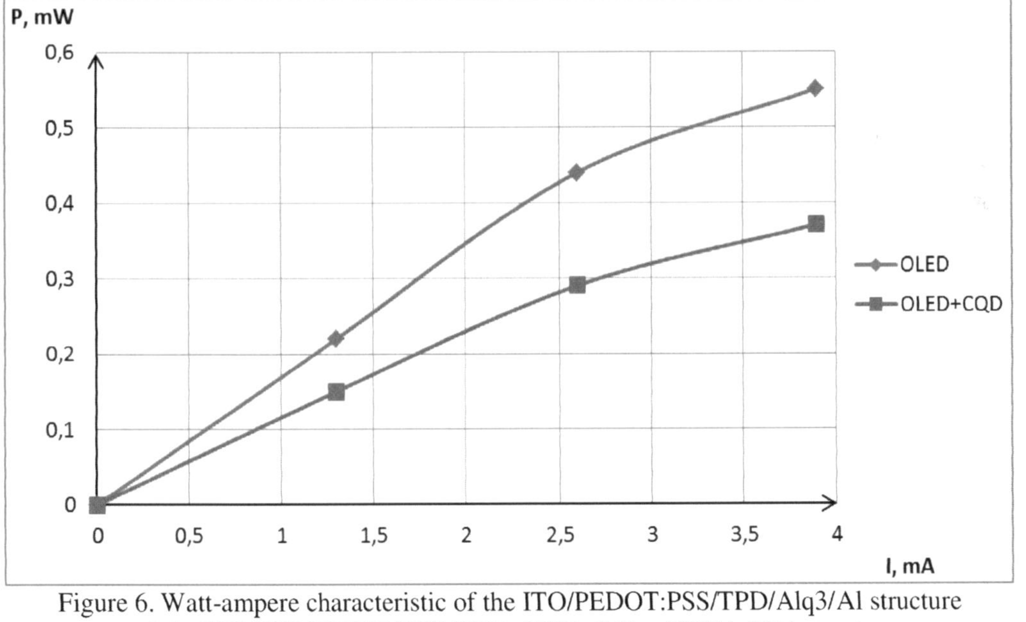

Figure 6. Watt-ampere characteristic of the ITO/PEDOT:PSS/TPD/Alq3/Al structure and the ITO/PEDOT:PSS/TPD/TPD+CQD_CdSe_650/Alq3/Al structure.

3. Conclusions

The technology of creating OLED structure FTO/PEDOT:PSS/TPD/Alq3/Al was developed by methods of spin-coating and vacuum thermal evaporation. It is shown that adding to the structure of the PEDOT:PSS layer allowed to reduce the OLED switching on voltage. A technique for introducing CdSe CQD (650) into the OLED structure was developed. An increase of the efficiency of OLED radiation is shown when introducing CQD into them. The obtained efficiencies exceeded the value of

2%, which is a rather high index for structures based on FTO/PEDOT:PSS//TPD/TPD+CQD CdSe (650)/Alq3/Al, which do not contain dopants based on iridium.

At present, work is under way to increase the brightness of the radiation from the CQD array due to the optimization of the procedure for the distribution of quantum dots in the structure. Preliminary work is being done to introduce into the composition of the structure a multicomponent CQD array of different composition and size, having a predetermined PL spectrum [9, 10]. The use of such arrays will make it possible to create organic structures with an EL spectrum that is close to the solar emission spectrum with a color rendering index of CRI above 95 or with any other form of the emission spectrum.

The work was supported by the Ministry of Education of the Russian under project of the state task in the field of scientific activity, project number 16.1750.2017/4.6.

References

[1] K. S. Cho, E. K. Lee, W. J. Joo, E. Jang, T. H. Kim, S. J. Lee, S. J. Kwon, J. Y. Han, B. K. Kim, B. L. Choi, and J. M. Kim, Nature Photon. 3, 341 (2009).https://doi.org/10.1038/nphoton.2009.92

[2] V. L. Colvin, M. C. Schlamp, and A. P. Alivisatos, Nature 370, 354 (1994).https://doi.org/10.1038/370354a0

[3] Y. Karzazi // Organic Light Emitting Diodes: Devices and applications // J. Mater. Environ. Sci. 5 (1) (2014) 1-12.

[4] E. M. Stepanov, I. I. Mikhailov, S. A. Tarasov, A. V. Solomonov, Proceedings of the 2016 IEEE North West Russia Section Young Researchers in Electrical and Electronic Engineering Conference, ElConRusNW, 82 – 84, (2016);

[5] S. Coe, K. W. Woo, M. G. Bawendi, and V. Bulovic, Nature 420, 800 (2002).https://doi.org/10.1038/nature01217

[6] Fengjuan Zhang, Shujie Wang, Lei Wang // Nanoscale. 2016. V. 8. P. 12182−12188. doi: 10.1039/C6NR02922A

[7] Mikhailov I.I., Tarasov S.A., Lamkin I.A., Tadtaev P.O., Kozlovich L.I., Solomonov A.V., Stepanov E.M. // J. Phys. Conf. Ser. 2016. V. 741. P. 012103.

[8] N. S. Kurochkin, A. A. Vashchenko, A. G. Vitukhnovsky, P. N. Tananaev // Effect of the Length of Ligands Passivating Quantum Dots on the Electrooptical Characteristics of Organic LightEmitting Diodes // ISSN 10637826, Semiconductors, 2015, Vol. 49, No. 7, pp. 953–958.

[9] P. O. Tadtaev, M. O. Gurevich ; L. I. Kozlovich ; I. I. Mikhailov The study of CdSe/ZnS and CdSeZnS/ZnS colloidal quantum dots structures as flexible electronics components // IEEE. – 2016. 10.1109/EIConRusNW.2016.7448126, 85-87.

[10] E. M. Stepanov; I. I. Mikhailov; S. A. Tarasov; A. V. Solomonov Light-emitting structures based on colloidal quantum dots of cadmium sulphide having a high color rendering index // IEEE. – 2016. 10.1109/EIConRusNW.2016.7448125, 82-84.

IOP Conf. Series: Journal of Physics: Conf. Series **1038** (2018) 012104 doi:10.1088/1742-6596/1038/1/012104

Organic photodetective device based on metal phthalocyanine

M D Pavlova, I A Lamkin, S A Tarasov, A V Solomonov

Saint Petersburg Electrotechnical University «LETI» (ETU), Russia

E-mail: minaepanae@mail.ru

Abstract. During the work, organic photosensitive structures were created. The photosensitivity spectra of the samples were analyzed. According to the spectrum of photosensitivity, were made assumptions about possible transitions that take place in structures.

1. Introduction

In the 21st century, humanity faced with a problem of depletion of organic fuel sources. In the future, the resource of such minerals as oil, coal, gas will be exhausted, so the issue of obtaining energy from alternative sources is actual. In the future, out of all available sources, one of the most capacious is the energy of the Sun, coming to the surface of the Earth in huge numbers, many times exceeding the humanity's global needs. For converting it, we use solar panels. Each year their efficiency increases. To replace silicon batteries, new materials and types of structures come. The most interesting alternative to silicon batteries are organic materials, because they allow you to create flexible structures that are easy to create, as well as cheaper and more environmentally friendly. In addition, the level of development of modern chemistry will create materials with the necessary properties [1].
Organic photosensitive structures can be divided into 2 groups: the elements in which the photoactive elements are applied by separate layers - the structures of the layered type.

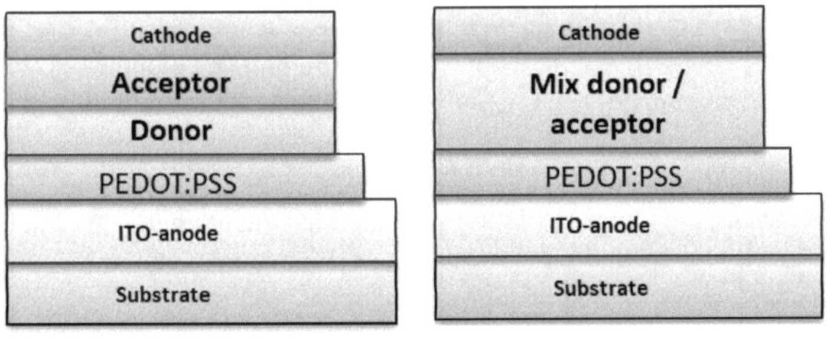

Fig. 1 - Diagram of the layered type structure (on the left) and a diagram of a structure with a bulk heterojunction (on the right).

The second group includes elements in which only one photoactive layer, which is a mixture of n- and p-types of electrical conductivity, called structures with a bulk heterojunction. Under the influence of electromagnetic radiation, a photoinduced transfer of electrons from the donor to the acceptor occurs, this phenomenon does not depend on the type of the photosensitive structure. But in elements of a layered type this process occurs at the boundary of the donor and acceptor material. In elements with a bulk heterojunction, charge redistribution occurs throughout the entire volume of the active layer. After the separation of charges, it is necessary to prevent the recombination of the charge carriers and to place them on the electrodes. In structures of layered type, electrons and holes can freely move to the electrode in the layer of acceptor and donor materials, respectively. The probability that charge carriers will fall into a layer with opposite electrical conductivity is extremely small, and as a consequence, the probability of recombination is small also. The suppression of recombination and charges transport in structures with a bulk heterojunction is realized only with a clear donors ordering and acceptor phases. [2]

2. Experiment

Device design and fabrication

The structures were based on FTO (Fluorine-doped Tin Oxide)/CuPc (Copper(II) phthalocyanine)/PTCDA (Perylenetetracarboxylic dianhydride)/Al and FTO/ZnPc (Zinc phthalocyanine):C60 (Fullerene)/C60/Al, deposited on a glass substrate. Layers of organic materials and aluminum contact have deposited by vacuum thermal evaporation at a residual gas pressure $2 \cdot 10^{-5}$ torr. During the sputtering process of organic layers, the substrate was heated to 60 °C, and when the metal contact was applied, the substrate temperature was maintained at 70 °C.

Measurements and results

For created structures, were explored the spectrum of absorption and transmission, the I-V characteristic, and the photosensitivity spectrum, which received the highest attention. Investigation of the photosensitivity spectrum is extremely important, because of for maximal range of solar radiation, the sensitivity spectrum of the solar cell should be similar to the spectrum of the solar radiation

The photosensitivity spectra were studied using a diffraction grating monochromator and Keithley 6483 picoammeter, an incandescent lamp was used as a radiation source.

The dependence of the photosensitivity on the wavelength for FTO / CuPc / PTCDA / Al type structures has 4 peaks, the first peak corresponds to a wavelength of 535 nm, the peak width at half-height is in the wavelength range from 481 nm to 591 nm, it corresponds to the absorption in the PTCDA layer. The second peak lies at a wavelength of 648 nm, whose half-width lies in the wavelength range from 632 nm to 688 nm, it is probably associated with indirect interband transitions. The maximum of the third peak is at wavelength of 754 nm, the width of the peak is at half-height 92 nm (716 – 808 nm), is related to absorption in the CuPc layer. The fourth peak corresponds to a wavelength of 1106 nm, the width of the peak is at half-height from 1078 nm to 1128 nm. The short-wave edge of photosensitivity (0.05) is at a wavelength of 446 nm, and the long-wavelength edge is at a wavelength of 1135 nm. But the lack of this structure is low sensitivity in the yellow-red region of the spectrum. Therefore, in the future study, materials with high sensitivity are used in this wavelength range.

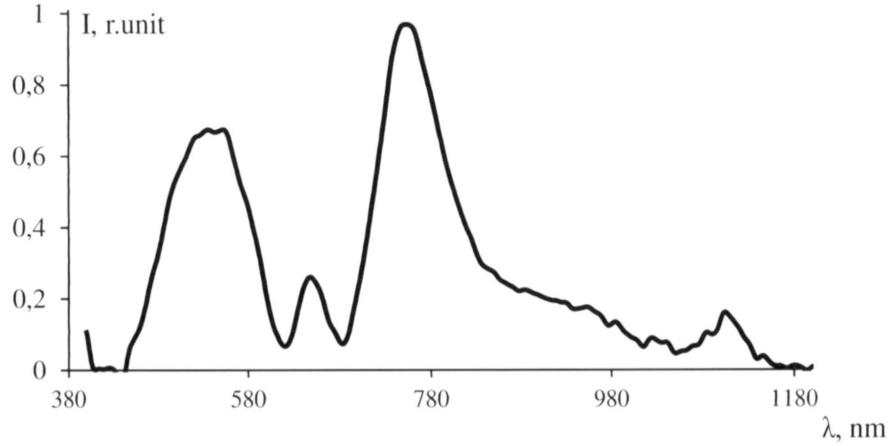

Fig. 2 - Spectrum of photosensitivity of structures of the type FTO/ CuPc/ PTCDA/ Al

The photosensitivity spectrum of FTO / ZnPc: C60 / C60 / Al structures has 3 peaks. The first corresponds to a wavelength of 627 nm, its width at half-height lies in the wavelength range from 623 nm to 649 nm. The second peak occurs at a wavelength of 693.5 nm, the half-height of this peak lies in the range of 683 - 705 nm, it corresponds to the absorption in the layer of a mixture of fullerene and zinc phthalocyanine. The last maximum is at a wavelength of 944 nm, its width at half-height 833-1008 nm corresponds to the absorption of light by zinc phthalocion. However, the structure under consideration has a dip in the near-IR region, a thin layer of Bphen (batofenantroline) was added to its structure for eliminating of a dip.

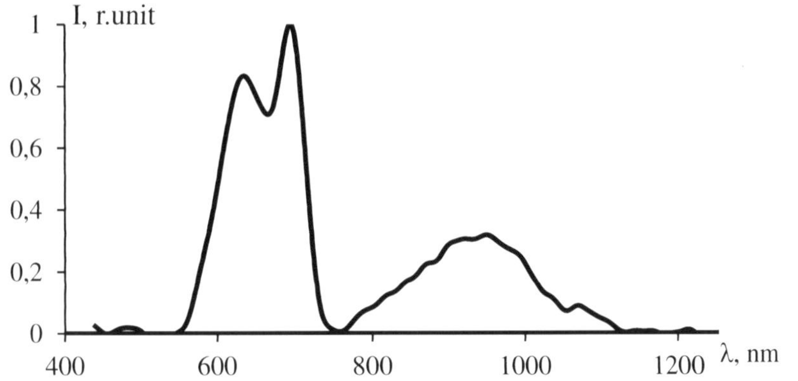

Fig. 3 - Spectrum of photosensitivity of structures of the type FTO/ ZnPc:C_{60}/C_{60}/Al.

Introduction to the FTO/ ZnPc: C_{60}/ C_{60}/ Bphen/ Al structure of the additional Bphen transport layer has gave a possibility to lower the values of the barriers that arise at the interface of the layers and improve the collection of photocarriers. As a result, a structure with a high sensitivity in the range of 450-1 1150 nm was obtained. Two maximum are observed on the photosensitivity spectrum of the FTO/ ZnPc: C_{60}/ C_{60}/ Bphen/ Al structures. The first peak corresponds to a wavelength of 597 nm, its width at half-height is in the range from 573 nm to 635 nm. The maximum of the second peak is located at 698 nm, the peak half-width covers a significant wavelength range, in the red and infrared region, at 677 nm - at 948 nm. The maximum of the second peak is located at 698 nm, the peak half-width covers a significant wavelength range, in the red and infrared region, at 677 nm - at 948 nm. The short-wave edge of photosensitivity (0.05) is at a wavelength of 455 nm, and the long-wave edge is at a wavelength of 1160 nm.

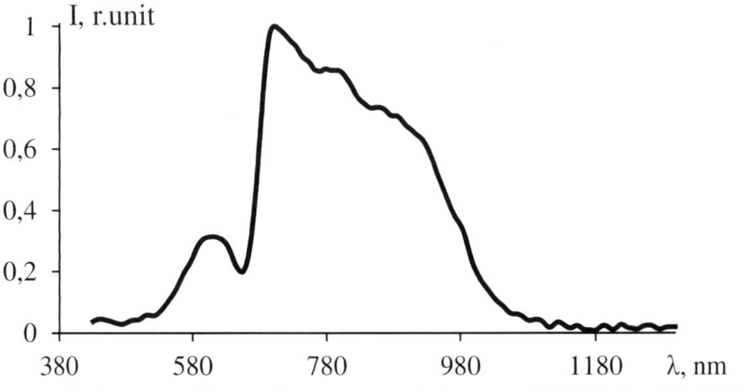

Fig. 4 - Spectrum of photosensitivity of structures of the type FTO/ZnPc:C60/C60/Bphen/Al.

The features of the obtained dependences from the spectral position of the sensitivity extrema are determined by the band diagram of researched materials and qualitatively consistent with the optical transmission spectra of the deposited layers.

3. Conclusion

Prototypes of organic solar cells based on FTO / CuPc / PTCDA / Al and FTO / ZnPc: C60 / C60 / Al structures have been created and investigated in the work. They have achieved a significant increase in the efficiency of solar radiation conversion using to optimization of transport layers and introduction into the organic matrix inclusions of fullerene. The creation of an energy barrier for holes by the introduction of the Bphen layer prior to the deposition of the Al layer substantially increased the width at half-height of the photosensitivity spectrum, increasing the coefficient of solar energy's conversion into electric energy. The resulted structures can be used as an alternative energy source.

Acknowledgments

The work was supported by the Ministry of Education of the Russian under project of the state task in the field of scientific activity, project number 16.1750.2017/4.6.

References

[1] Egidius R R 2017 *International Journal of Photoenergy* vol 2017
[2] Toroshin P A, Lyubovskaya R N, Razumov V F 208 *Reviews. Russian nanotechnology* vol 3 pp 56 - 77

IOP Conf. Series: Journal of Physics: Conf. Series **1038** (2018) 012105 doi:10.1088/1742-6596/1038/1/012105

Series spreading resistance in single- and multi-junction concentrator solar cells

M A Mintairov[1], V V Evstropov[1], S A Mintairov[1], M Z Shvarts[1] and N A Kalyuzhnyy[1]

[1] Photovoltaics Lab, Ioffe Institute, Saint Petersburg 194021, Russia

Abstract. The series resistance of concentrator solar cells is determined by the processes of current spreading under the contact grid. The influence of the number of subcells in multi-junction solar cells on these processes is considered in the paper. Using the tube model of current spreading, it is shown that with the increase in the number of subcells, the current distribution of tubes becomes more uniform. This affects the shape of the IV-curve of the spreading resistance in such a way that the magnitude of the resistive losses at the maximum power point decreases.

1. Introduction

The properties of series resistance significantly affect the maximum efficiency of the concentrator solar cell (SC). These properties are determined by the processes of current spreading under the contact grid. Previously, to describe the regularities of current spreading, a tube model was developed [1]. The model described current spreading with the help of a large number of (50 or more) parallel-connected branches consisting of series-connected unequal resistances, and identical p-n junctions. The resistive part of the model is specified by two parameters: R_L – lateral resistance and R_V - vertical resistance. In [1] it is shown that the model allows calculating the distribution of currents in tubes and obtaining an IV curve of resistive losses (spreading resistance IV-curve). In this paper, the model was used to study the changes in the process of current spreading in the solar cell with an increase of the subcells number.

In the paper, the existing tube model was extended to multi-junction (MJ) SCs case and the influence of the number of subcells on the spread resistance IV-curve was determined. The extended tube model showed that the number of subcells affects the shape of the spreading resistance IV-curve. At the same time, in the current region near the maximum power point, the resistive losses decrease with increasing of the subcells number. This, as shown, leads to decrease in the lumped linear equivalent of the series resistance R_s.

2. Tube model of current spreading

To study the current spreading in the MJ SC, we used the tube model developed previously [1]. The approach, based on the representation of current lines (correspondingly, on current tubes [2]) is used to describe the resistance. If one presents the current flow through the structure by current lines, the latter will follow along the model paths (Fig.1). In this case, the path of charge carriers in the longitudinal direction is much longer than that in the transverse one (the ratio of the light-sensitive surface area to the shaded by the contact grid one is >>1). The total resistance on this portion will be mainly determined by the longitudinal propagation (spreading) of charge carriers.

The SC is represented as consisted of two regions: spreading region which has h_s height and vertical one with h_v height (Fig. 1). Each current tube consists of 4 portions: 1 - vertical portion of height h_1 in the current spreading region under the face contact, 2 - horizontal portion of length l_2 in the current spreading region, 3 - vertical portion of height h_3 in the current spreading region and 4 - vertical portion of height h_4. Both current spreading resistivity and vertical resistivity we denote to be ρ_S and ρ_V correspondingly. In considering the current tubes to be thin, their thickness on the vertical portions can be written in a form of a coordinate increment along the X axis: $dx_1 = \frac{W_c}{h_s} dz$ and $dx_4 = dx_3 = \frac{W_{0.5}}{h_s} dz$. Correspondingly, the horizontal portion length is given by $l_2 = dx_4 - dx_1 = \frac{W_{0.5} - W_c}{h_s} z$. Thickness of any transverse tube dz can be expressed through the amount of current tubes n: $dz = \frac{h_s}{n}$. The number of a current tube (index $i = 1 \ldots n$) can be described in the following way: $i = \frac{z}{dz}$. According to [1], applying above mentioned inputs and the scheme of the SC (Fig. 2) results in the following analytical formula of the i-tube resistance:.

$$R_i = R_V \cdot n + R_L \cdot i \quad (1)$$

where: $R_V \approx \rho_V h_V$ and $R_L \approx \frac{\rho_S W_{0.5}^2}{h_S}$

Note that R_V is mostly determined by the resistance of the substrate due to the large substrate thickness $h_{sub} \approx h_V$. Therefore, $R_V \approx \rho_{sub} h_{sub}$, where ρ_{sub} is the resistivity of the substrate. So R_V can be found from the substrate specifications. The formula for R_L has a form of sheet resistance and can be measured by the multimeter.

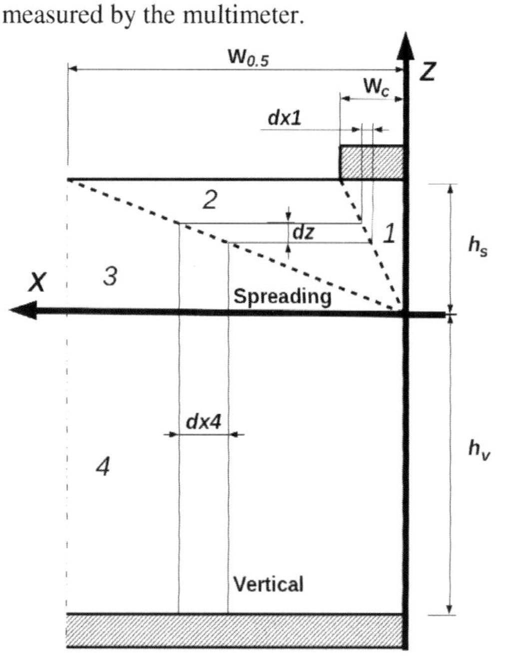

Figure 1. The tube model for current spreading in a SC. The SC is presented in a form of several regions: 1, 2, 3 are regions, in which current spreading takes place. Shaded area represents output metallic contacts.

According to the tube model, the equivalent circuit of SC is constructed from parallel branches (Fig. 2) with different resistances. The resistances of i-branch can be calculated by Eq. (1). All branches have a number of similar diodes. Note that series-connected diodes can be replaced, as shown in [1], by one diode with the resulting ideality factor (A) and saturation current (J_o) obtained by following formulas:

$$A = \sum_{i=1}^{n} A_i \qquad J_0 = \sqrt[A]{\prod_{i=1}^{n} J_{oi}^{Ai}} \tag{2}$$

The IV curves of the branches can be obtained by solving simple mono-exponential equation:

$$J_i = \frac{J_g - J_0\left[\exp\left(\frac{q(V - J_i \cdot R_i)}{AkT}\right) - 1\right]}{n} \tag{3}$$

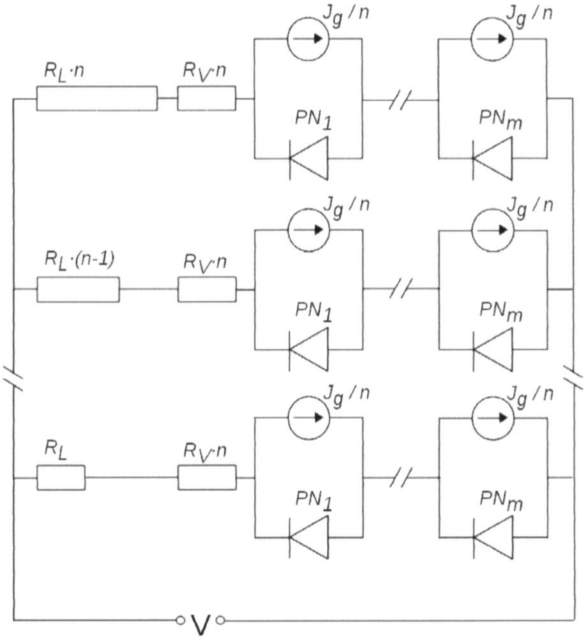

Figure 2. The distributed equivalent circuit of solar cells used in tube model.

Thus, the application of Eq. (3) together with Eq. (2) and Eq. (1) allows calculating the distribution of currents in the branches of an equivalent circuit (Fig. 2) and obtaining its IV-curve. Repeating the calculation for the case when all the resistances in the equivalent circuit are equal - zero ($R_i=0$) one can get a resistive-less IV-curve. The difference between such characteristics gives the required for the study IV-curve of the spreading resistance.

3. Calculation of the current spreading processes in MJ SC

The purpose of the calculation was to consider the influence of the number of p-n junctions (N_{pn}) in the MJ SC on the resistive losses. For this purpose, according to the tube model described in Section 2, a calculation of the IV-curves for MJ SC with different (from 1 to 8) number of p-n junctions was carried out. The number of current tubes in calculations was 50. In the calculations, the photogenerated current J_g was constant and equal to 7 A/cm^2. Basically, the results of the calculation do not depend on the choice of this parameter, so we chose the value of Jg at which the maximum efficiency of triple-junction GaInP/GaAs/Ge SC was achieved [2]. Note that J_0 was also constant in all calculations and equal to $J_0 = 1 \cdot 10^{-20}$ A/cm^2. This was made due to the fact that J_0 is obtained as a geometric average (see eq. 2) of the saturation currents of all subcells. It is approximately constant because with an increase in the number of subcells, the average value of the forbidden bandgaps is approximately a constant [3, 4].

In the calculations of resistive losses, the values of R_L and R_V were fixed. The values $R_L = 2.6 \cdot 10^{-2}$ Ohm·cm^2 and $R_V = 0.4 \cdot 10^{-4}$ Ohm·cm^2 were obtained from the analysis of experimental data for GaInP/GaAs/Ge SC [5]. Basically, the results of calculations do not change with other parameters R_L and R_V. Partially this is demonstrated in Figure 6.

4. Result and discussion

Figure 3 shows the distribution of current through tubes for SCs with a different number of subcells. It is seen that with the N_{pn} increase, the distribution becomes more uniform. This may be due to the fact that as the number of subcells increases, the role of the series resistance decreases in comparison with the resistance of the diodes. The change in the distribution of currents across the tubes affects the resistive losses. The IV-curves of the spreading resistance is shown in Fig. 4. It is seen that the shape of the curve changes, and there are regions in which the spreading resistance increases with an increase in the number of pn-junctions (approximately from 0 to 5 A/cm2 in Fig. 4) and vice versa decrease (from 5 to 7 A / cm2). The second region is of practical interest since it contains the maximum power point (MPP).

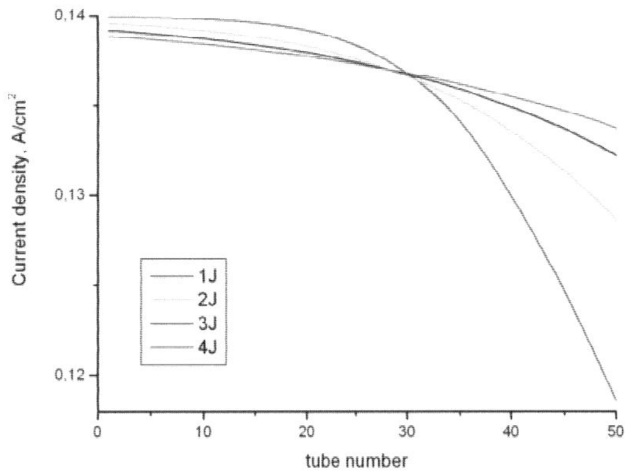

Figure 3. Current distribution in tubes for SCs with a different number of subcells.

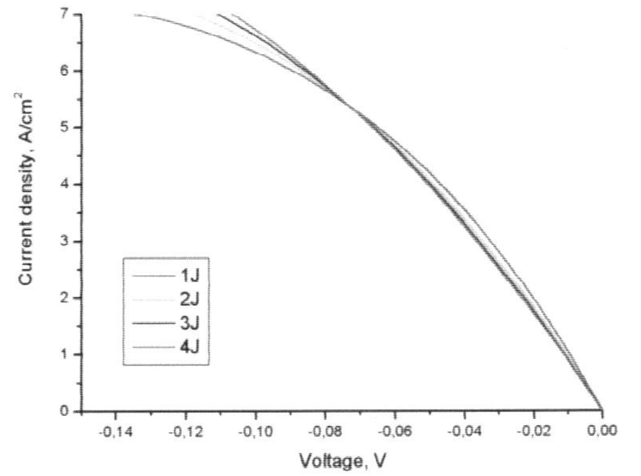

Figure 4. IV-curves of the spreading resistance for SCs with a different number of subcells.

To study the reduction of resistive losses at the MPP, the lumped resistance at the point was used. This resistance was calculated as the ratio of the spreading resistance value at the point to the its current. Note that due to the fact that such resistance connected with MPP it is value considered to be a lumped linear resistance equivalent (R_s). Such equivalent is widely used in photovoltaics because it allows describing the SC IV-curve in practically important cases (up to the maximum of efficiency) without taking into account the distributed character of the spreading resistance [2, 6]. Dependence of Rs on the number of subcells is shown in Figure 5. It is evident that this resistance decreases with the N_{pn} increasing. Thus, a change in the current distribution in the tubes of the SC (Fig. 3) reduces the resistive losses at the operating point, which effectively reduces the linear lumped resistive equivalent R_s.

In addition, the calculation of the influence of the spreading resistance parameters (R_L and R_V) on the value of the R_s was carried out. The parameters R_L and R_V were varied by deviation from the standard ones. The result of the calculation is shown in Figure 6. It can be seen that R_s is much more dependent on R_L than on R_V. At the same time, the general form of the regularity (the decrease of R_s with N_{pn} increasing) is preserved. Also the tendency toward saturation is observed with an increase in the number of subcells.

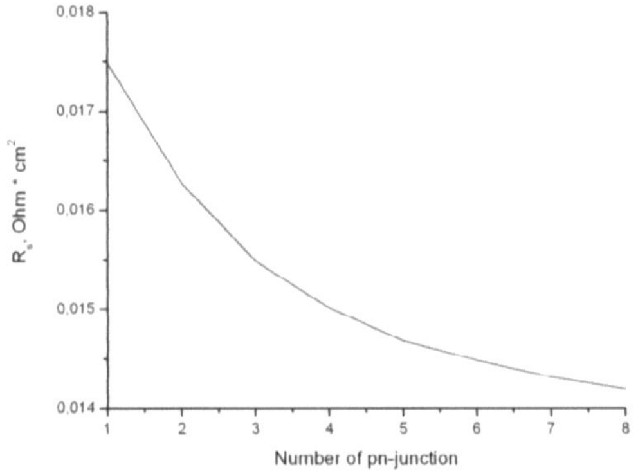

Figure 5. Lumped series resistance of SCs with a different number of subcells.

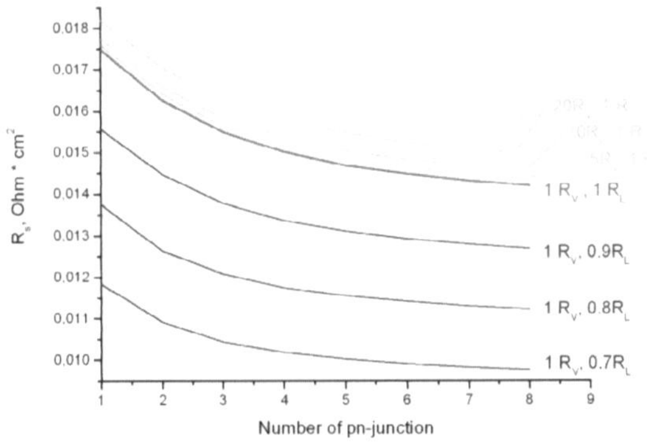

Figure 6. The influence of spreading resistance parameters (R_L and R_V) on the dependence of lumped series resistance of SCs with a different number of subcells.

5. Conclusion

In the paper, a tube model is used to describe the process of current spreading under the contact grid of solar cells with any structure including MJ SCs. It has been calculated that the increase of the subcells number in MJ SC affects the process of current spreading. The distribution of currents in tubes with increasing number of subcells becomes more uniform. It is assumed that this this may be occurs due to the fact that the role of the series resistance decreases in comparison with the resistance of diodes.

As a consequence of the change in the tube currents distribution, the IV-curve of the spreading resistance changes. In the region around the maximum power point current, the spreading resistance decreases with increasing number of subcells. Calculation of the linear lumped equivalent (of the series resistance) R_s has shown that R_s decreases with an increasing a number of subcells. In was also shown that R_s critically depends on the lateral resistance R_L.

Thus a change in the current distribution in the tubes of the SC with increasing of its subcells number reduces the resistive losses at the maximum power point, which effectively reduces the lumped resistive equivalent R_s.

Acknowledgments

The work has been supported by the Russian Foundation for Basic Research (grant No. 16-08-00742).

References

[1] M.A. Mintairov, V.V. Evstropov, S.A. Mintairov, N.Kh.Timoshina, M.Z.Shvarts, N.A. Kalyuzhnyy 2016 *Semiconductors* **60** pp 970-975

[2] M.A. Mintairov V.V. Evstropov, N.A. Kalyuzhnyy, S.A. Mintairov, N.K. Timoshina, M.Z. Shvartz, V.M. Lantratov 2012 *Semiconductors* **46** pp 1051-1058

[3] B. Minnaert, P. Veelaert, 2012 *Materials*, **5** pp 1933-1953

[4] L. Zhu, T. Mochizuki,M. Yoshita, S. Chen, C. Kim, H. Akiyama, and Y. Kanemitsu 2016 *Optics Express* **24** pp A740-A751

[5] M.A. Mintairov, V.V. Evstropov, S.A. Mintairov, R.A. Salii, M.Z. Shvarts, N.K. Timoshina, N.A. Kalyuzhnyy 2015 *Proc. of the 31th EU PVSEC* 1470-1473

[6] M.A. Mintairov, N.A. Kalyuzhnyy, V.V. Evstropov, V.M. Lantratov, S.A. Mintairov, M.Z. Shvarts, V.M. Andreev, A. Luque 2015 *IEEE Journal of Photovoltaics* **5** pp 1229 – 1236

A Theoretical Model of Dual Tunable Optoelectronic Oscillator

V.V. Vitko, A.A. Nikitin, A.B. Ustinov, B.A. Kalinikos

St. Petersburg Electrotechnical University, Department of Physical Electronics and Technology, St. Petersburg, 197376 Russia

E-mail: vitaliy.vitko@gmail.com

Abstract. A theoretical model of an electronically tunable multi-loop optoelectronic oscillator (OEO) based on photonic and microwave multiferroic delay lines has been developed. The theory takes into account a dispersion of electromagnetic waves in optical fibers and a dispersion of spin-electromagnetic waves in ferrite-ferroelectric structures. An expression for transfer function of the OEO circuitry was derived. The theory enables one to calculate transmission characteristics and suppression of spurious harmonics of the multi-loop OEOs. The features of both magnetic and electric tunabilities are investigated.

Electronically tunable oscillators are one of the most important components of various radioelectronic systems. Optoelectronic oscillators (OEOs) based on high-Q whispering gallery mode optical resonators or long optical fibers offer the best spectral purity and the short-term frequency stability as well as the low phase noise. Due to these unique features OEOs have attracted a great attention of researchers in the last two decades [1].

Oscillation frequency in a classical configuration of the oscillator based on the single fiber loop is determined by the phase shift in the fiber and microwave (MW) power regeneration in a feedback containing an amplifier and a bandpass filter [2], [3]. An increase in the optical fiber length provides increasing the delay time and rising the Q-factor, hence decreasing the phase noise level. Nevertheless, it causes simultaneously a decrease in a free spectral range (FSR) that leads to appearance of unwanted spurious modes. Earlier, in the work [4]–[6] the multi-loop configuration was suggested to solve this problem.

The tunable OEOs are based on tunable filters in a microwave frequency domain [7] or an optical domain [8]. Spin-wave delay lines based on epitaxial yttrium iron garnet films were recently suggested for the magnetic tuning of an oscillation frequency of the OEO [9]. In order to increase functionality and realize a dual electric and magnetic tunable OEO one is required to use a multiferroic delay lines, which are usually fabricated with a combination of ferrite and ferroelectric layers [10].

The purpose of this work is to develop a theoretical model of the multi-loop optoelectronic oscillators based on photonic and microwave multiferroic delay lines. In contrast to earlier works, the developed theory takes into account the specific features of wave processes in both delay lines.

A theoretical model was developed for a common multi-loop OEO circuitry shown in Fig. 1. The OEO has a ring circuitry consisting of the microwave and optical paths. The optical path includes a monochromatic optical signal source (LD), an electro-optical modulator (EOM) of Mach–Zehnder type, an optical fiber splitter with N channels and splitting coefficients η_j for each channel, N single-

IOP Conf. Series: Journal of Physics: Conf. Series **1038** (2018) 012106 doi:10.1088/1742-6596/1038/1/012106

mode optical fibers of lengths l_j, an optical fiber combiner, and a photodetector (PD). The main elements of the MW path are a MW amplifier with gain coefficient G and a multiferroic delay line.

Let us consider a circulation of microwave signal in the oscillator ring circuitry. The MW signal applied to the driving port of the electro-optical modulator modulates amplitude of the monochromatic optic radiation. The modulated radiation is divided by the splitter between the single-mode fibers. The optical signals propagating in different fibers accumulate different phase shifts and then interfere in the combiner. The optical radiation reaches the photodetector. After detection the electric signal of MW frequency propagates in the microwave path reaching the multiferroic delay line. The MW signal excites hybrid spin-electromagnetic wave that propagates in multiferroic structure and accumulates a delay time in accordance with its dispersion law. After the multiferroic delay line, the MW signal is introduced to an amplifier. The MW amplifier together with a variable attenuator controls amplification in the ring. A small part of the circulating MW signal is taken out of the ring by the output directional coupler. The other part of the MW signal is supplied to the driving input of the electro-optical modulator. As a result, the ring becomes closed. When the amplification compensates the total loss of the signal in the ring and an open-loop gain reaches unity, the microwave generation begins.

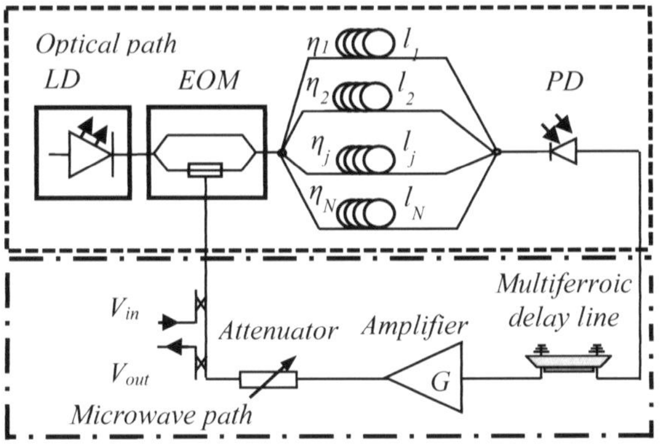

Figure 1. Block-diagram of the multi-loop ferrite-ferroelectric OEO

Let us describe the proposed theoretical model of a multi-loop resonator. The complex transmission coefficient $\mathbf{H}(\omega)$ is calculated as the ratio of the complex amplitudes of the output and input signals:

$$\mathbf{H}(\omega) = A_{out}(\omega)/A_{in}(\omega). \tag{1}$$

We assume that the input signal is monochromatic. For definiteness, let the phase of the input signal at the initial instant of time be equal to zero:

$$A_{in}(\omega) = A_0 \exp(i\omega t). \tag{2}$$

As was discussed above modulated optical radiation is divided by the splitter with splitting coefficient η_j between the single-mode fibers of length l_j. Note, that $\sum_j \eta_j = 1$. During the signal transmission in the optical waveguide it accumulates a phase shift proportional to the fiber length and the wavenumber of the carrier optical signal. The optical signals are combined and interfere before the photodetector. Then the signal propagates through the remaining elements of the ring. The transfer coefficient $\mathbf{T}(\omega)$ of M optical fibers connected in parallel is described by the following expression:

$$\mathbf{T}(\omega) = \sum_{j=1}^{M} \eta_j \exp\left(-ik_{opt_j}(\omega) l_j - \alpha_{opt_j}(\omega) l_j\right), \tag{3}$$

where $k_j(\omega)$ is the wavenumber of the carrier optical signal propagating in the optical fiber having a number j, $\alpha_j(\omega)$ is a spatial damping decrement in the fiber.

Dispersion equation for optical waves in the fiber [11]:

$$\left[\frac{\varepsilon_1}{\varepsilon_2}\frac{a\gamma_j^2}{\kappa_j}\frac{J_\nu^{(1)'}(\kappa_j a)}{J_\nu^{(1)}(\kappa_j a)}+i\gamma_j a\frac{H_\nu^{(1)'}(i\gamma_j a)}{H_\nu^{(1)}(\kappa_j a)}\right]\left[\frac{a\gamma_j^2}{\kappa_j}\frac{J_\nu'(\kappa_j a)}{J_\nu^{(1)}(\kappa_j a)}+i\gamma_j a\frac{H_\nu^{(1)'}(i\gamma_j a)}{H_\nu^{(1)}(i\gamma_j a)}\right]=\left[\nu\left(\frac{\varepsilon_1}{\varepsilon_2}-1\right)\frac{k_j k_2}{\kappa_j^2}\right]^2, \quad (4)$$

where ε_1 and ε_2 are dielectric permittivities of the fiber core and cladding, respectively; a is a core radius; $J_\nu^{(1)}(\kappa_j a)$ is the Bessel function of the 1-st kind in the core; $J_\nu^{(1)'}(\kappa_j a)$ is the derivative of the Bessel function of the 1-st kind; $H_\nu^{(1)}(i\gamma_j a)$ is the Hankel function of the 1-st kind in the cladding; $H_\nu^{(1)'}(i\gamma_j a)$ is the derivative of the Hankel function of the 1-st kind; ν is the order of Bessel and Hankel functions; $k_2^2=\omega^2\mu_0\varepsilon_0\varepsilon_2$; $\kappa_j^2=\omega^2\mu_0\varepsilon_0\varepsilon_1-k_j^2$; $\gamma_j^2=k_j^2-\omega^2\mu_0\varepsilon_0\varepsilon_2$; $\mu_0=4\pi\cdot10^{-7}$ H·m^{-1} is a magnetic permeability and $\varepsilon_0=8.85\cdot10^{-12}$ F·m^{-1} is an electric constant.

Propagation of the hybrid spin-electromagnetic wave through the multiferroic delay line utilizing ferrite-ferroelectric layered structures is described by the term $\exp\left[-i\left(k_{SEW}(\omega)-i\alpha_{SEW}(\omega)\right)l_{SEW}\right]$, where $k_{SEW}(\omega)$ is a wavenumber of spin-electromagnetic wave, $\alpha_{SEW}(\omega)$ is a damping decrement, l_{SEW} is a length of a multiferroic delay line. For the quasi-surface spin-electromagnetic wave in a two-layered ferrite-ferroelectric structure the wavenumber $k_{SEW}(\omega)$ is calculated from an appropriate dispersion equation [12]

$$\frac{\tan(\kappa_d a_d)}{\kappa_d}\left(\left(\kappa_0^2-\kappa_d^2\right)\mu_\perp-P(\omega,k)\right)+\left(\kappa_0^2\mu_\perp^2-\kappa^2{}_f-\frac{k_{SEW}^2\mu_a^2}{\mu^2}\right)\frac{\tan(\kappa_f a_f)}{\kappa_f}+2\kappa_0\mu_\perp=0, \quad (5)$$

where a_d and a_f are thicknesses of ferroelectric and ferrite layers, respectively; $\kappa_d=\sqrt{\omega^2\mu_0\varepsilon_0\varepsilon_d-k_{SEW}^2}$, $\kappa_0=\sqrt{k_{SEW}^2-\omega^2\mu_0\varepsilon_0}$, $\kappa_f=\sqrt{\omega^2\mu_0\varepsilon_0\varepsilon_f\mu_\perp-k_{SEW}^2}$, ε_d and ε_f are dielectric permittivities of ferroelectric and ferrite layers, respectively; $P(\omega,k)=\tan(\kappa_f a_f)\left[\left(\frac{k_{SEW}\mu_a}{\mu}+\kappa_0\mu_\perp\right)\left(\frac{\kappa_0 k_{SEW}\mu_a}{\kappa_f\mu}+\frac{\kappa_0^2\mu_\perp}{\kappa_f}\right)+\kappa_0\kappa_f\right]$; $\mu=\frac{\omega_H(\omega_H+\omega_M)-\omega^2}{\omega_H^2-\omega^2}$, $\mu_a=\frac{\omega\omega_M}{\omega_H^2-\omega^2}$, $\mu_\perp=\frac{\mu^2-\mu_a^2}{\mu}$, $\omega_H=|\gamma|H$, $\omega_M=4\pi|\gamma|M_0$, $|\gamma|$ = 2.8 MHz/Oe is the gyromagnetic ratio for electrons, H is an external magnetic field, M_0 is a saturation magnetization. The spatial damping decrement for the quasi-surface spin-electromagnetic wave is calculated as [13]

$$\alpha_{SEW}(\omega)=\frac{\partial k_{SEW}(\omega)}{\partial H}\Delta H+\left|\frac{\partial k_{SEW}(\omega)}{\partial\varepsilon}\right|\varepsilon_d\tan\delta, \quad (6)$$

where ΔH is a half of the ferromagnetic resonance line-width and $\tan\delta$ is a dielectric loss tangent.

The microwave signal passed the multiferroic delay line is fed to the microwave amplifier used to compensate the propagation loss in the elements of the scheme. The signal transmission through the microwave amplifier is described by a factor $\exp(g(\omega))$ assuming that the amplifier does not introduce an additional phase shift. Note, that transfer functions of electro-optical modulator and photodetector could be taken into account with using of the small signal approximation as described in [2].

$$A_{out}(\omega) = A_0 \left[\sum_{n=1}^{\infty} \left(\mathbf{T}(\omega) \exp\left(g(\omega) - \alpha_{SEW}(\omega) l_{SEW} - i k_{SEW}(\omega) l_{SEW} \right) \right)^n \right], \quad (7)$$

where $\mathbf{T}(\omega)$ is the transmission coefficient of parallel connection described by expression (3).

Using the mathematical approach described above, the complex transmission coefficient (1) can be represented as:

$$\mathbf{H}(\omega) = \sqrt{\left| H_p(\omega) \right|} \exp\left(i\varphi(\omega) \right), \quad (8)$$

where $H_p(\omega)$ is the power transfer function and $\varphi(\omega)$ is the phase shift between the input and output signals.

The numerical analysis of the single-, dual-, and triple-loop OEO resonant spectra were carried out. In order to do it, we calculate the transmission characteristics in the cases of equal optical power division, i.e. $\eta = 0.5$ for the dual-loop, and $\eta = 1/3$ for the triple-loop OEO. For the calculations, we utilized the typical parameters for the single-mode optical fiber: $\varepsilon_1 = 2.1543$, $\varepsilon_2 = 2.0875$, $a = 4.1$ μm, and $\alpha_j(\omega) = 0.2$ dB/km; and for the ferrite-ferroelectric structure composed of yttrium iron garnet film and barium strontium titanate (BST) ceramic layer: $M_0 = 1750$ G, $H = 1500$ Oe, $\Delta H = 0.5$ Oe, $\varepsilon_f = 14$, $a_f = 5.2$ μm, $a_d = 500$ μm, $\varepsilon_d = 1500$, $\tan\delta = 0.01$, and $l_{SEW} = 3$ mm.

The results of the numerical calculations of the power transmission coefficient H_p for the single-, dual- and triple-loop OEOs are shown in Fig. 2. The simulations were carried out for the ring gain just below the self-generation threshold. The characteristics of the single-loop OEO with fiber length of 1000 m, 500 m, and 250 m are shown in Fig. 2 (a), (b), and (c), respectively. The characteristic of the dual-loop OEO based on the two fibers with lengths of 500 m and 1 km is shown in Fig. 2 (d). The characteristic of the triple-loop OEO contained the three fibers with lengths of 250 m, 500 m, and 1000 m and the same OEO with multiferroic delay line is shown in Fig. 2 (e).

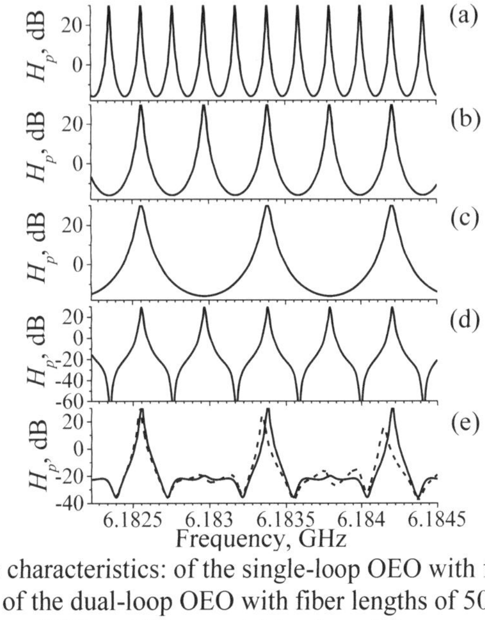

Figure 2. Transmission characteristics: of the single-loop OEO with fiber length of 1000 m (a), 500 m (b), and 250 m (c); of the dual-loop OEO with fiber lengths of 500 m and 1 km (d); the triple-loop OEO with fiber lengths of 250 m, 500 m, and 1 km by solid curves and the same triple-loop OEO with multiferroic delay line by dashed curves (e)

One can see that the transmission characteristic of single-loop OEO consists of series of resonant peaks. The increase of fiber length from 250 m to 1000 m provides rising the Q-factor from $1.9\cdot10^5$ up to $7.9\cdot10^5$, but simultaneous decrease in the FSR from 800 kHz down to 200 kHz (see Fig. 2 (a), (b), and (c)). Transmission characteristic of dual-loop configuration consists of the series of stopbands and passbands (see Fig. 2 (d)). The passbands correspond to the resonant frequencies that are resulted by constructive interference of the optical signals in the combiner. Positions of the stopbands are defined by destructive interference of the optical signals. It gives possibility to suppress odd spurious harmonics in the resonant spectrum, therefore the FSR of dual-loop configuration is equal to the FSR of single-loop with fiber length of 500 m (400 kHz). The triple-loop circuity provides increase of the FSR up to 800 kHz (see solid line in Fig. 2 (e)). Therefore, increasing of short loop number allows to suppress the spurious at higher offsets from the oscillation frequency, otherwise, to achieve higher FSR. Number of short loops is limited by decreasing of the total Q-factor. The Q-factor of multi-loop configuration is described by following expression: $Q_{\Sigma} = \sum_{j=1}^{M} \eta_j Q_j$. It is seen form the Fig. 2 (e) that Q-factor of the triple-loop configuration (solid line) $Q = 5.2\cdot10^5$ less than Q-factor of single-loop $Q = 7.9\cdot10^5$ (see Fig. 2 (a)). Multiferroic delay line implements frequency selection as well as introduces an additional time delay (see dash line in Fig. 2 (e)).

The magnetic and electric tuning of oscillation frequency of the triple-loop OEO with fiber lengths of 250 m, 500 m, and 1000 m connected with the multiferroic delay line are shown in Fig. 3 (a) and (b), respectively.

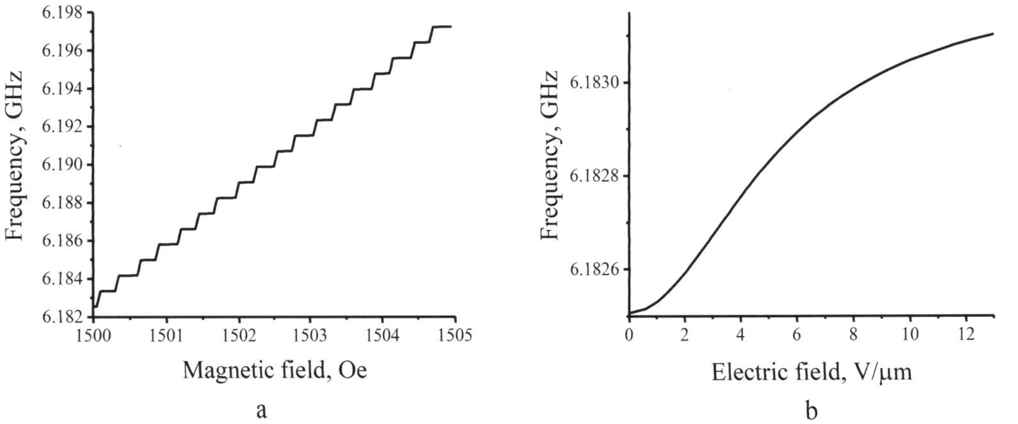

Figure 3. Magnetic (a) and electric (b) tunability of the triple-loop OEO contained the three fibers with lengths of 250 m, 500 m, and 1000 m connected with the multiferroic delay line

The electric tunability of the multiferroic delay line is realized due to the dependence of dielectric permittivity of the ferroelectric layer on external electric field. In particular, for the BST layer in paraelectric state the dependence of the dielectric permittivity on the electric field can be approximated from the experimental characteristic as

$$\varepsilon_d(E) = \frac{\varepsilon_d(0) - \varepsilon_d(E_{\max})}{1 + E^2/E_0^2} + \varepsilon_d(E_{\max}), \tag{9}$$

where $\varepsilon_d(0)$ and $\varepsilon_d(E_{\max})$ are the dielectric permittivities in the absence of electric field and in case of maximum electric field, respectively, and E_0 is an electric field on the half of experimental characteristic of the dielectric permittivity versus electric field. The magnetic tuning is provided by a dependence of magnetic permeability of ferrites on the bias magnetic field. Therefore, the resulting spectrum of spin-electromagnetic waves is dually controllable by the both electric and magnetic fields.

As is seen in Fig. 3 (a), an increase of the magnetic field shifts the oscillation frequency towards to higher frequencies in accordance with the shift of the wave spectrum in the multiferroic delay line. The range of magnetic tuning is typical for spin-wave device based on yttrium iron garnet film and belongs to the range from 3 GHz up to 9 GHz for the magnetic field variation in the range from 500 Oe up to 2500 Oe. One can see, that magnetic tuning has step-like behavior. The value of steps equal to the distance between the resonant frequencies. For the triple-loop configuration described above it is equal to 800 kHz. An increase of the electric field shifts the wave spectrum in the multiferroic delay line towards low wavenumbers that gives possibility to smooth tuning the oscillation frequency in the free spectral range (see Fig. 3 (b)).

In conclusion, the theory has been developed for electronically tunable multi-loop optoelectronic oscillators that are based on optical and microwave multiferroic delay lines. An expression for transfer function of the OEO circuitry consisting of an arbitrary number of optical delay lines in optical path and a dispersive multiferroic delay line in the microwave path was derived. Numerical simulations of the resonant frequency spectra showing its transformation with increasing a number of optical loops were carried out. The theory enables one to investigate transmission characteristics and suppression of spurious frequency harmonics of the multi-loop OEOs. It was shown that oscillation frequency of multi-loop OEO with multiferroic delay line are dual tunable by both electric and magnetic fields.

The investigation was supported by Ministry of Education and Science of Russian Federation (agreement 14.575.21.0157, unique identifier RFMEFI57517X0157).

References

[1] Maleki L. Optoelectronic oscillators for microwave and mm-wave generation 2017 *18th International Radar Symposium (IRS) IEEE* 1-5
[2] Yao X S and Maleki L Optoelectronic microwave oscillator 1996 *J. Opt. Soc. Am. B* **13** 1725-1735
[3] Belkin M E, Loparev A V, Semenova Y et al Tunable RF-band optoelectronic oscillator and optoelectronic computer-added design model for its simulation 2011 *Microwave and Optical Technology Letters* **53** 2474-2477
[4] Yao X S and Maleki L Multiloop optoelectronic oscillator 2000 *IEEE Journal of Quantum Electronics* **36** 79-84
[5] Bánky T, Horváth B and Berceli T Optimum configuration of multiloop optoelectronic oscillators 2006 *J. Opt. Soc. Am. B* **23** 1371-1380
[6] Hong J, Zhanga S, Yaoa S et al Comparison of both type injection locked and parallel dual-loop OEO 2015 *Opt.-Int. J. for Light and Elect. Opt.* **126** 4410-4413
[7] Eliyahu D and Maleki L Low phase noise and spurious level in multi-loop opto-electronic oscillators 2003 *Proc. of the 2003 IEEE Int. Frequency Control Symp. and PDA Exhibition Jointly with the 17th European Frequency and Time Forum* 405-410
[8] Cen Q, Dai Y, Yin F et al Rapidly and continuously frequency-scanning opto-electronic oscillator 2017 *Optics Express* **25** 635-641
[9] Ustinov A B, Nikitin A A and Kalinikos B A Magnetically Tunable Microwave Spin-Wave Photonic Oscillator 2015 *IEEE Magnetics Letters* **6** 1-4
[10] Ustinov A B, Srinivasan G and Kalinikos B A Ferrite-ferroelectric hybrid wave phase shifters *Applied physics letters* 2007 **90** No 3 031913
[11] Agrawal G P Nonlinear fiber optics 2007 *Academic press*
[12] Nikitin A A, Ustinov A B, Vitko V V et al Spin-electromagnetic waves in planar multiferroic multilayers 2017 *Journal of Applied Physics* **122** 014102
[13] Ustinov A B, Kalinikos B A, Tiberkevich V S et al Q factor of dual-tunable microwave resonators based on yttrium iron garnet and barium strontium titanate layered structures 2008 *Journal of Applied Physics* **103** No 6 063908

Development of the technology of manufacturing connecting elements in cascade photodetectors

A E Marichev[1,2], R V Levin[1,2], N D Prasolov[1], E V Kontrosh[1] and B V Pushnyi[1,2]

[1] Ioffe Institute, St. Petersburg, Russia

[2] Submicron Heterostructures for Microelectronics Research and Engineering Center of the RAS, St. Petersburg, Russia

E-mail: segregate1@yandex.ru

Abstract. In this paper, we report on the initial studies of cascade photodetectors. The heterostructures used in this work are based on InP. InP is the most suitable material for converting the solar spectrum in the range from 0.95 to 1.2 μm. It is proposed to use nanocrystalline inclusions as a connecting element. For this, nanocrystals GaP are best suitable. Because this material (GaP) does not create an absorption of the incident radiation.

1.Introduction

In recent years, interest in wireless transmission of energy through the laser beam has increased. So experts from Laser Motive and Lockheed Martin (USA) charged the batteries of an unmanned aerial vehicle, which is in flight by laser radiation directed at the photodetector. The residence time in the air has increased by 20 times.

The most promising are photoelectric converters for laser radiation with a wavelength of 1.06 μm. This radiation is "located" at the local maximum transmittance of the Earth's atmosphere, and lasers have a high power of up to 10 kW [1].

To convert high-power radiation (more than 100 W/cm^2), it is preferable to use many cascade photodetectors (analogous to cascade solar cells). Tunnel *p* ++-*n* ++ junctions are now used to join FPP cascades. But in a tunnel junction at high powers of the incident radiation, the generated photocurrent can exceed the peak current of the tunnel junctions(peak current is not more than 14 A/cm^2). This leads to a sharp increase in the resistance of the entire structure and a decrease in the efficiency of photoconversion [2]. To solve this problem, it has been proposed to create conduction channels by introducing an array of nanocrystalline inclusions into the space charge region (SCR) of the connecting *p-n* junctions [3]. In this case, it becomes possible to substantially reduce the extremely high levels of doping of *p*++ and *n*++ layers in tunnel junctions and the effect of smearing of the tunneling doping profile caused by the diffusion of electrically-active impurities. During the subsequent growth of more wide-gap materials. The *p-n* junction is no longer rectifying and provides current transmission through the conduction channels in the layer of nanocrystalline inclusions. In [3], the properties of tunnel and photovoltaic p-n junctions in GaSb were fabricated and investigated. Crystalline inclusions were introduced in the SCR of these p-n transitions. Based on the results of this work, a patent of the Russian Federation was obtained [4].

The material of crystalline inclusions for creating conduction channels in the space charge region must satisfy the following conditions:

1. Weak absorption of the radiation being converted.

2. It should not form a continuous layer on the boundary with the photoactive material, but only individual crystals with dimensions larger than the thickness of the space charge region.

For InP, this condition is satisfied by GaP. It has a bandgap width of 2.261 eV greater than that of InP (1.351 eV) and, consequently, has a weak absorption of light in the photosensitivity region InP. The lattice constant of GaP 5.45117 Å (in InP -5.86875 Å), which guarantees the production of crystallites, rather than a continuous layer (the Stranski-Krastanov model).

In the present work, data are presented on the technology of growing junction transitions and the results of studies of the photoelectric parameters of the InP-based structures obtained. These structures are used in solar cells and photodetectors of high-power laser radiation with 1.06 μm.

2. Experimental details

All structures of InP were grown by metalorganic chemical vapor deposition (MOCVD) using an AIX-200 setup operating at a reduced pressure. As the In, Ga and P sources, Trimethylindium(TMIn), Triethylgallium (TEGa) and phosphine(PH3) were used. The growth temperature was 600°C, the reactor pressure was 100 mbar. As the substrates, *n*- InP(Te) (100) misorientation at 4 degrees in the plane (111) wafers with an electron concentration of (1–5) $\times 10^{18}$cm^{-3} were used.

To analyze the GaP - inclusion sizes and shape, technological experiments on GaP crystal growth on an *n*- InP substrate surface were performed. It became possible to vary the GaP - inclusion size by varying the triethylgallium flow, providing low ohmic loss in the connecting *p–n* junctions. The following conditions of crystal growth were chosen as optimal for the study: the temperature Tg=600°C and time t_g= 1 min.

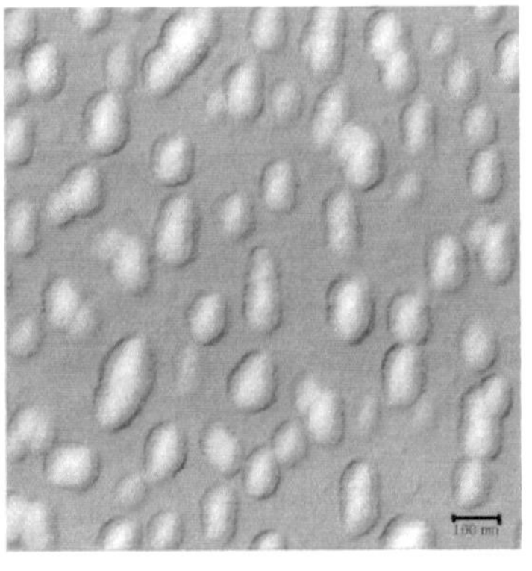

Figure1. AFM image of GaP nanocrystals grown on the surface of the *n*-InP substrate – the ratio of PH3/TEGa flows is 300, t_g = 60 seconds

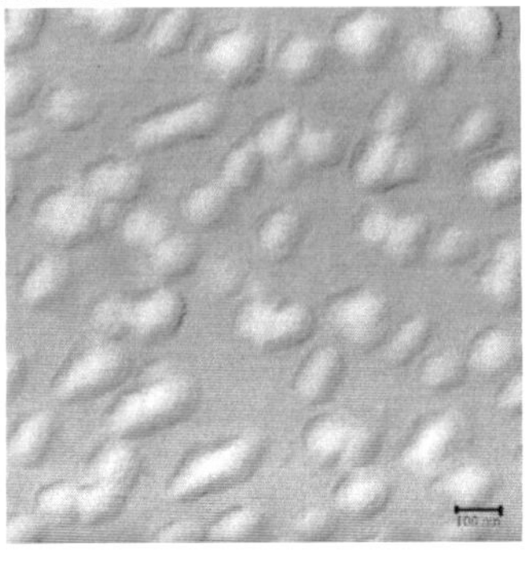

Figure 2. AFM image of GaP nanocrystals grown on the surface of the *n*-InP substrate – the ratio of PH3/TEGa flows is 300, t_g = 30 seconds

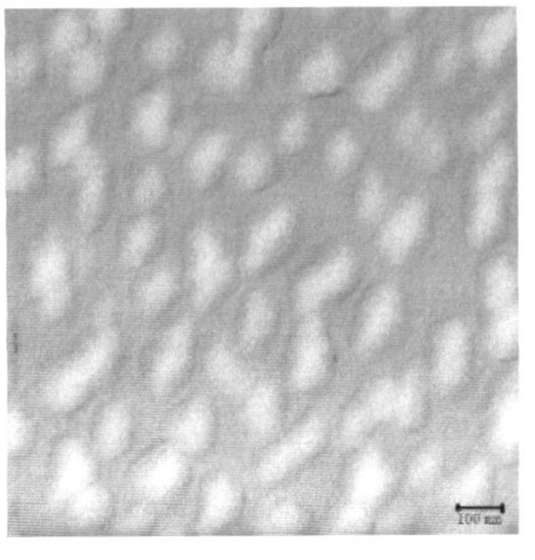

Figure 3. AFM image of GaP nanocrystals grown on the surface of the *n*-InP substrate – the ratio of PH3/TEGa flows is 590, t_g = 30 seconds

Figure 1–3 shows a micrograph for the GaP crystals in the SCR, obtained using an atomic force microscope. As a result of the analysis of the surface images of the analogous ones shown in the figure, graphs of the dependence of the effect of growth conditions on geometric dimensions (figure 4) and the density of GaP nanocrystals (figure 5) were plotted. Thus, it was found that an increase in the growth time at a constant PH3/TEGa ratio leads to an increase in the average dimensions of GaP nanocrystals (height by an average of 40%, and width by an average of 15%) (figure 4). From the dependencies shown in figure 4-5, it can also be noted that an increase in the PH3/TEGa ratio from 300 to 590 leads to a decrease in geometric dimensions.

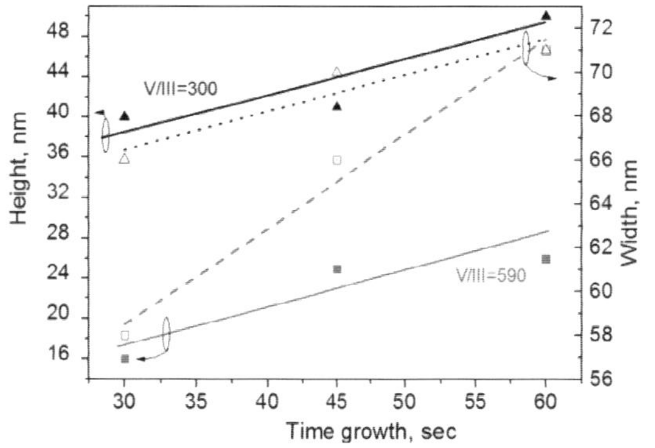

Figure 4. Dependence of the size of nanocrystals on the growth time.

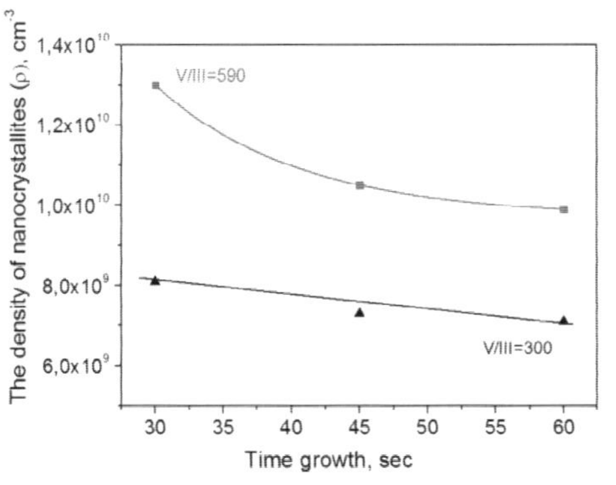

Figure 5. Dependences of crystal densities on growth time.

The ratio of PH3/TEGa was changed due to a decrease in TEGa in the gas phase, as a result of which less material of the element of the third group entered the growth zone, which is equivalent to a decrease in the growth rate. The density of GaP nanocrystals decreased with increasing growth time, due to an increase in the geometric dimensions of the crystallites. It was also noted that an increase in the PH3/TEGa ratio led to an increase in the density of GaP nanocrystals, for example, figure 1-3.

As noted earlier, preservation of the qualitative characteristics of photoactive transitions grown on top of the layers containing GaP nanocrystals is extremely important. So after determining the growth regimes of GaP nanocrystals, optical characteristics (from the photoluminescence spectra) of epitaxial layers grown over layers containing GaP nanocrystals were studied.
For these studies it was grown structure presented in figure 6.

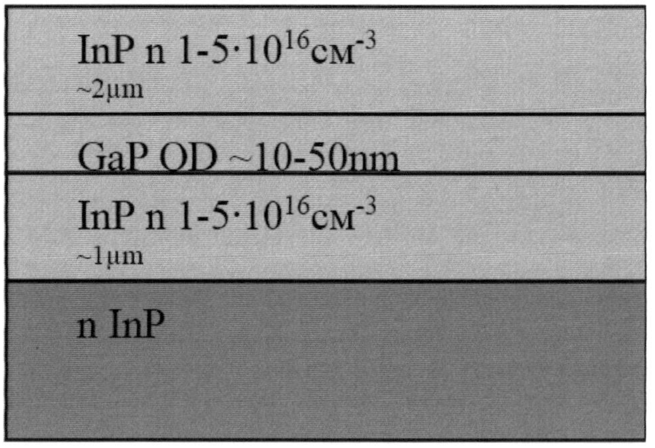

Figure 6. A sample of the *n*-n transition with GaP crystallites.

The studies were carried out by photoluminescence at 77 K in the wavelength range 600-1500 nm. Figure 7 shows the photoluminescence spectra of the samples under study. Thus, for InP layers without QDs and with QDs (2 and 3 in the graph in figure 7), the intensities of peaks with an energy of 1.41 eV (interband transitions in InP) differ by less than 20%, which may indicate that the defects arising from the growth epitaxial layers on a layer of GaP nanocrystals, is practically absent.

When the layer of GaP nanocrystals is introduced into the InP layers, which is equivalent, in fact, to the introduction of defects, the quality of the obtained layers does not deteriorate, as evidenced by a change in the photoluminescence half-width by 1 meV laying within the measurement error.

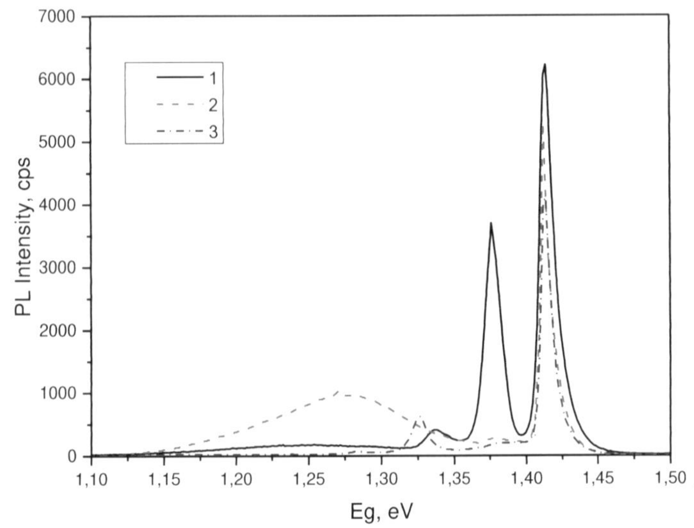

Figure 7. The photoluminescence spectrum of the samples under study. 1 - InP substrate spectra, 2 - spectrum of the InP layer grown on the InP substrate, 3 - the QD spectrum grown between two layers of InP

Conclusion

The paper presents the initial stages of work on the creation of new connecting elements for use in monolithic multi-transient solar cells and photoconverters of powerful optical radiation based on InP. GaP nanocrystals were grown as conductivity channels in the IPF, with dimensions 60-85 nm in height and 78-92 nm in width. At the same time, it is shown that the quality of epitaxial, intentionally non-doped InP layers grown on GaP nanocrystals is not deteriorated.

Acknowledgements

The authors wish to thank P.N. Brunkow, V.S. Kalinovsky and A.S. Vlasov for their valuable contribution to the research and discussion of the results.

This work was supported by Russian Science Foundation (Grant Agreement №17-79-30035).

References

[1] Jian Dong, Xue-Sheng Liu, Chao Peng, You-Qiang Liu and Zhi-Yong Wang, 2015 *Appl. Sci.* **5** 1837

[2] D. Masson, F. Proulx and S. Fafard 2015 *Progress in photovoltaics: Research and appications* **23** 1687

[3] V.S. Kalinovsky, R.V. Levin, B.V. Pushniy, M.N. Mizerov, V.D. Rumyntsev, V.M. Andreev 2013 *Semiconductors* **12** 1677

[4] V.M. Andreev, M.N. Mizerov, V.D. Rumyntsev, V.S. Kalinovsky, R.V. Levin, B.V. Pushniy, *Invention* **RU 2 461 093 C1** Date of publication: 10.09.2012 Bull. 25

IOP Conf. Series: Journal of Physics: Conf. Series **1038** (2018) 012108 doi:10.1088/1742-6596/1038/1/012108

Optical emission spectroscopy of gallium phosphide plasma-enhanced atomic layer deposition

A V Uvarov[1], A S Gudovskikh[1,2]

[1]St. Petersburg Academic University, St. Petersburg 194021, Russia
[2]St. Petersburg Electrotechnical University "LETI", St. Petersburg 197376, Russia

Abstract. The capability of optical emission spectroscopy for *in situ* study and control of plasma-enhanced atomic-layer deposition (PE-ALD) of gallium phosphide from phosphine and trimethylgallium carried by hydrogen was explored. The gas composition changing during the PE-ALD process was monitored by *in situ* measurements of optical emission intensity for phosphine and hydrogen lines. For PE-ALD process where phosphorus and gallium deposition steps are separated in time a negative influence of excess phosphorus accumulation on the chamber walls was observed. Indeed, the phosphorus deposited on the walls during PH_3 decomposition step is etched by hydrogen plasma during the next trimethylgallium decomposition step leading to uncontrollable and unwanted conventional plasma-enhanced chemical vapor deposition. To reduce this effect, it has been proposed to introduce a step of hydrogen plasma etching, which allows one to etch excess phosphorus before the beginning of gallium deposition step and achieve atomic-layer deposition growth mode.

1. Introduction

Gallium phosphide grown on silicon substrates is of interest for optoelectronics, in particular for photovoltaics. Silicon based heterojunction solar cells with GaP wide-gap emitter have an advantage compared to a-Si:H emitter due to the wider band gap of GaP and therefore lower optical losses [1]. Plasma-chemical atomic layer deposition (PE-ALD) makes it possible to obtain thin layers of high-quality binary material such as TiO_2[2] and ZnS[3] at deposition temperatures of less than 400 °C. This is possible due to cyclic deposition and the self-limiting mechanism of the surface reaction [4]. In this case, one monolayer of obtained material is deposited by one cycle. Thus, it is possible to obtain an exact thickness value by specifying a certain number of cycles. Growth process of gallium phosphide by PE-ALD should consist of cyclic sequential deposition of gallium and phosphorus in the presence of radio frequency (RF) plasma discharge. It is important to separate in time the steps of gallium and phosphorus deposition in order to avoid a simultaneous precursor reaction in the chamber, because this will lead to conventional plasma enhanced chemical vapour deposition (PECVD). Therefore a purge step is introduced between the deposition steps to remove the remains of unreacted components. However, an effect of phosphorus deposition on the chamber walls in the red allotropic modification during the plasma assisted decomposition of phosphine as well as phosphorous etching by atomic hydrogen was demonstrated by Bruno et al [5]. The combination of the both effects could affect the growth mechanism of GaP in case of time modulated process when plasma stimulation is involved, for instance PE-ALD. To study the process of cyclic deposition, it is necessary to investigate *in situ* dynamics of gas composition changes in the chamber. One of the most attractive methods for controlling plasma deposition processes is optical emission spectroscopy (OES), which on the one hand allows one to determine the composition of the precursors from the optical spectrum of gas discharge [6] and on the other hand does not require a significant modification of the deposition setup. The intensity of the spectral lines of gas components is proportional to their concentration. Using a high-speed charge-coupled device (CCD) spectrometer, it is possible to carry out *in situ* independent measurements of several spectral lines during the deposition process. This technique can be successfully used as for monitoring the technological process as well as for study the growth mechanism of PE-ALD.

IOP Conf. Series: Journal of Physics: Conf. Series **1038** (2018) 012108 doi:10.1088/1742-6596/1038/1/012108

2. Experimental study

PE-ALD of GaP was realized using Oxford Plasmalab 100 with 13.56 MHz RF generator. Trimethylgallium ($Ga(CH_3)_3$) and phosphine (PH_3) were used as precursors of gallium and phosphorus, respectively. Hydrogen was used as carrier gas for trimethylgallium. The plasma spectrum was studied using a CCD spectrometer Solar laser systems S100 in the wavelength range 400-1000 nm. Optimal integration time values of 250 ms were used to provide maximal sensitivity and time resolution being equal to 300 ms. Control of the spectrometer for study of time dependencies was implemented by using LabView software package. The schematic presentation of the measurement setup is shown in figure 1.

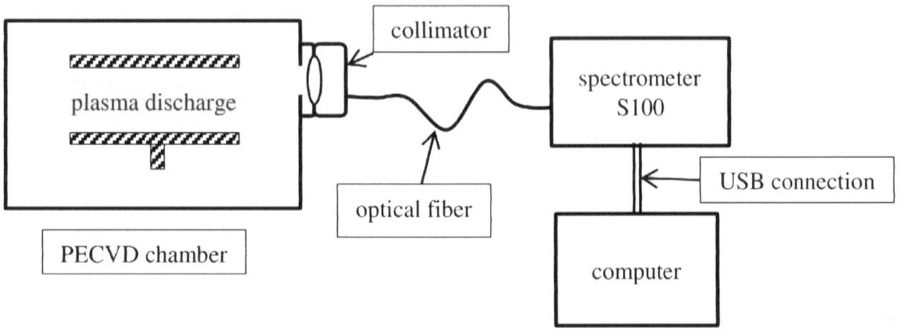

Figure 1. The schematic presentation of OES measurement with CCD spectrometer

The pure hydrogen plasma has the most intense line in the visible spectrum of H_α from the Balmer series at a wavelength of 656 nm due to the emission of atomic hydrogen. The phosphine plasma spectrum has maxima at wavelengths of 341, 666, 717 nm, which corresponds to the emission lines of phosphorus radicals PH and PH_2 [5]. Among them the most intense is the emission line of the PH_2 radicals at a wavelength of 666 nm (figure 2).

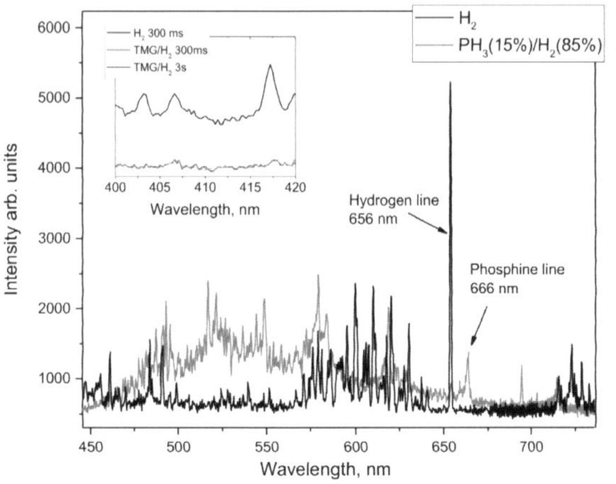

Figure 2. The emission spectrum of hydrogen plasma and a mixture of hydrogen and phosphine. The inset shows the spectrum of hydrogen emission and a mixture of hydrogen and TMG at integration times of 300 ms and 3 s.

This line is of the greatest interest due to the fact that at this wavelength there is no emission of molecular and atomic hydrogen, which allows one to obtain information on the concentration of phosphorus radicals PH_2 in the gas plasma of hydrogen and phosphine. It was noted that when low concentrations of phosphine are added to the chamber, the intensity of the H_α line significantly decreases. This is due to the fact that ionization energy of phosphine is much lower compared to that of atomic hydrogen (1,2). Such property of a mixture of these gases can be used as an indirect method to measure the concentration of phosphine in the chamber.

$$H_2 + e \rightarrow H^* + H + e \ (E = 18.0 \text{ eV}) \tag{1}$$

$$PH_3 + e \rightarrow PH_2^* + H + e \ (E = 13.4 \text{ eV}) \tag{2}$$

The most intense lines of gallium emission in the visible region correspond to wavelengths of 403.3 and 417.2 nm[7]. But because of the low concentration of gallium, the intensity of these lines is insufficient to measure at short integration times. With an integration time of 3 s, one can observe clear maxima corresponding to the emission lines of gallium (figure 2). However, for large values of the integration time, the time resolution will not be enough to study the cyclic deposition process. Using a CCD spectrometer, we study the change of gas composition during the plasma enhanced atomic-layer deposition of gallium phosphide. The data are presented in the form of the time dependence of the RF power, gas flows and lines intensity of the hydrogen and phosphine (figure 3).

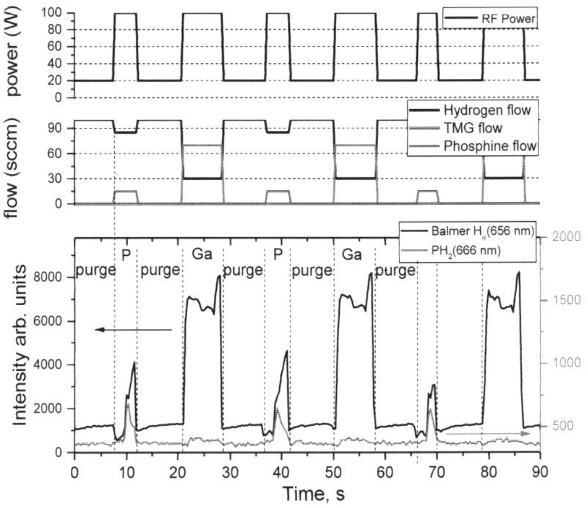

Figure 3. Timing diagram of the RF power, gas flows and emission lines intensity of hydrogen and phosphine during the PE-ALD process

During the deposition step, the plasma power was set to 100 W and was reduced to 20 W during the purge step. The plasma power during the purge stage was not set to zero in order to keep the plasma discharge throughout the entire deposition process. As can be noted at the beginning of the step of phosphorus deposition, a sharp drop in the intensity of the hydrogen line is observed. This occurs because the dissociation energy of the phosphine molecules is sufficiently small and most of the plasma power is converted to internal energy. Further at this step, the intensity of the phosphine line increases indicating that a considerable part of the molecules are in the ionized state. At the next purge step, the gradual increase in the intensity of hydrogen line can be noted; hence the concentration of PH_3 in the chamber decreases due to change in the gas composition. However, at the step of gallium deposition, response of the phosphine line can also be noted, which is an important issue. It indicates a presence of phosphine in the chamber and therefore a "parasitic" conventional PECVD process occurs. This can be either because of insufficient purging, or because of deposition and subsequent evaporation of phosphorus from the walls of the chamber. In any case, the presence of traces of PH_3 in the gallium deposition step can adversely affect to the deposition process. At the step of gallium deposition phosphine reacts with trimethylgallium in the chamber. This leads to suppression of the atomic-layer deposition growth mode and significant increase of growth rate, which exceeds one monolayer by cycle as well as decrease of the thickness uniformity. To determine the source of phosphine in the chamber, the following experiment was carried out. First, phosphorus was deposited from the phosphine plasma for 30 seconds in continuous mode. Then, the chamber was purged with hydrogen for 1 min to exclude the influence of insufficient purge time. After purging RF power of 100 W was applied and emission lines of hydrogen and phosphine were observed in the process of continuous hydrogen plasma etching (figure 4). It was shown that within 20 s the intensity of the PH_3 line gradually decreased and the hydrogen lines increased accordingly. This indicates that the solid phosphorus deposited on the walls of the chamber, under the ambient of the hydrogen plasma transforms back to the phosphine as was previously described in [5].

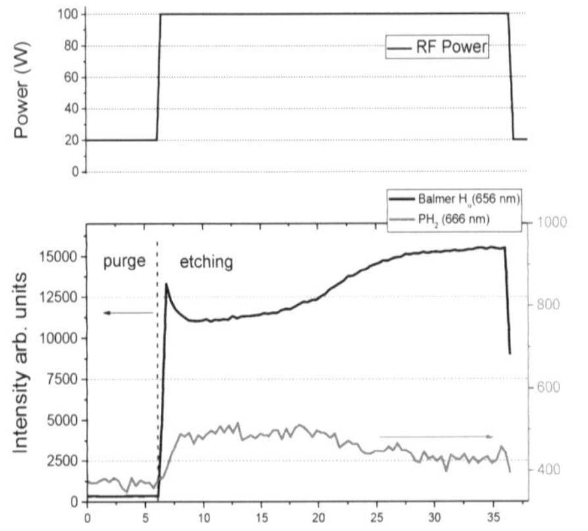

Figure 4. Timing diagram of the RF power and emission lines intensity on etching phosphorus from the walls of the chamber

To reduce this effect during PE-ALD, it has been proposed to introduce the etching step of hydrogen plasma, which allows one to hydrogenate the excess phosphorus before beginning the gallium deposition step. After a step of phosphorus deposition, a step of hydrogen plasma with an increased power of 100 W and duration of 9 s was added. This allowed one to remove phosphorus from the chamber walls before the gallium deposition step (figure 5).

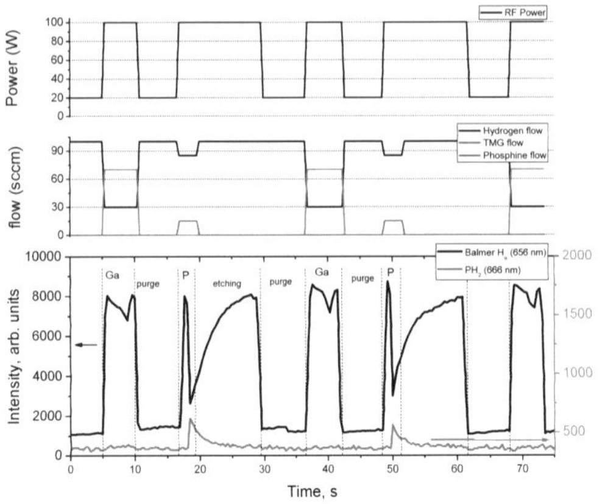

Figure 5. Timing diagram of the RF power, gas flows and emission lines intensity of modified deposition process with additional hydrogen etching step

No response of the phosphine line was observed at the step of gallium deposition by OES study in the modified process. At the step of hydrogen etching with increased power we can note a gradual increase of the hydrogen line intensity up to a maximum value, which corresponds to the complete removal of phosphorus traces. As a result, the deposited gallium phosphide layers became more homogeneous and the deposition rate decreased to a value of one monolayer by cycle indicating atomic layer deposition growth mode [8].

3. Conclusions

It was found that the increased deposition rate of gallium phosphide by the PE-ALD is associated with the accumulation of phosphorus on the walls of the chamber. At a subsequent step of gallium deposition in the presence of hydrogen, phosphorus is hydrogenated and passes into the gas phase. This results in a simultaneous reaction of PH_3 and $Ga(CH_3)_3$ and destroys atomic-layer deposition growth mechanism. To reduce this effect, it has been proposed to introduce a step of hydrogen plasma etching, which allows one to etch excess phosphorus before the beginning of the gallium deposition

IOP Conf. Series: Journal of Physics: Conf. Series **1038** (2018) 012108 doi:10.1088/1742-6596/1038/1/012108

step. The proposed approach allows one to achieve the atomic layer deposition growth mode by PE-ALD of GaP.

Acknowledgements
This work was supported by the Russian Science Foundation (№ 17-19-01482)

References
[1] Feifel M *et al.* 2016 *IEEE Journal of Photovoltaics* **6** 1
[2] Das C *et al.* 2015 *J. Vac. Sci. Technol.* A **33** 01A144
[3] Kuhs J and Dobbelaere T 2017 *J. Vac. Sci. Technol.* A **35** 01B111
[4] Puurunen R L 2005 *J. Appl. Phys.* **97** 121301
[5] Bruno G, Losurdo M and Capezzuto P 1995 *J. Vac. Sci. Technol.* A **13** 349
[6] Gabriel O, Kirner S, Klick M, Stannowski B and Schlatmann R 2014 *EPJ Photovoltaics* **5** 55202
[7] Shirai T, Reader J, Kramida A E and Sugar J 2007 *J. Phys. Chem. Ref. Data* **36** 509
[8] Gudovskikh A S, Morozov I A, Uvarov A V, Kudryashov D A, Nikitina E V, Bukatin A S, Nevedomskiy V N and Kleider J-P 2018 *J. Vac. Sci. Technol.* A **36** 021302

IOP Conf. Series: Journal of Physics: Conf. Series **1038** (2018) 012109 doi:10.1088/1742-6596/1038/1/012109

Sensitivity and directivity measurement of ultrasonic transducer with polymer-powder matching layer

Marsel Fazlyyyakhmatov
Kazan (Volga Region) Federal University, 18 Kremlyovskaya str., Kazan 420008, Russian Federation

E-mail: mfazlyjy@kpfu.ru

Abstract. The results of an experimental study of the sensitivity and directivity pattern of the ultrasonic transducers are presented. Data was obtained for transducers with central frequency at 2.5 and 5 MHz with protective matching layers, in comparison with transducers without matching layers. Increase of transducer sensitivity is shown. In addition, it is shown that the directivity patterns for transducers with the matching layer and without it are practically identical. Also was observed that the directivity pattern gets sharper as the frequency of the transducer increases.

1. Introduction

Ultrasound is the most popular method of non-destructive testing and diagnostics in various fields including medicine. In [1,2] it is shown results of new methods and apparatus development for non-destructive testing multiphase and biological objects.

Ultrasonic transducer is a critical component that used for the transformation of acoustic energy and electric energy in ultrasonic nondestructive detection system. Despite significant advances in the development of new materials that have piezoelectric properties in recent years, such as polymers with a piezoelectric effect [3], relaxor ferroelectrics PMN-PT, PZN-PT, etc. [4], the most common are piezoceramic materials based on lead zirconate titanate (PZT). The specific acoustic impedance of the PZT ceramics ranges from 22×10^6 to 35×10^6 Rayl. This is a limiting factor for immersion and medical sensors. The specific acoustic impedance of water and biological tissues about 1.5×10^6 Rayl [5]. To improve the acoustic impedance matching of medium with piezoceramics matching layers (ML) are used. According to the theory of transmission lines, to ensure the transmission of 100% acoustical energy, the thickness of the matching layer should be equal to a quarter of the wavelength in the layer. The specific acoustic impedance of the matching layer (Z_m) is defined by the equation [6]:

$$Z_m = (Z_1 Z_2)^{1/2}, \qquad (1)$$

where Z_1 – the specific acoustic impedance of the piezoceramics, Z_2 – the specific acoustic impedance of the media.

For most tasks of ultrasound diagnostics and probing the frequency range 0.5-15 MHz is used. Ultrasonic vibrations in this frequency range provide sufficient resolution in the low-level vibration damping. In previous paper [7] it was shown that the thickness of the matching layer the piezoceramic at these frequencies is 10-300 um.

Various polymers are used as the material for the protective matching layer: polycarbonates, polyester films, polytetrafluoroethylene, acrylonitrile-butadiene-styrene, poly-para-xylylene, polypropylene, polystyrene, etc. In Russian medical devices, a copolymer of vinyl chloride and ethyl acrylate was widely used. Despite a large number of developed materials, there is currently no long-term matching protective coating capable of ensuring a long service life of the transducer.

In [8], a method for applying as a matching layer of polymer powder coatings in an electrostatic field of corona discharge is proposed. Polyester, epoxy-polyester and polyurethane powder coatings are considered.

In this paper, we report the results of the sensitivity and directivity pattern studies of ultrasonic transducers with protective matching layers, in comparison with transducers without matching layers.

2. Methods

2.1. Sensitivity measurement

Sensitivity is an important parameter to describe the electro-acoustic transducer energy conversion efficiency, and is a key indicator of transducer performance [9].

Sensitivity is simply defined as the ratio of an output quantity to an input quantity. For an ultrasonic transducer characterized as a two-port network, there are the mechanical quantities of force and velocity (F, v) at the acoustic port and quantities of voltage and current (V, I) at the electrical port. Since an ultrasonic transducer can be used as ether a transmitter or a receiver, there are a variety of sensitivities that can be defined from these quantities [10].

A complete theory of the ultrasonic waves propagation and the description of the transducers characteristics is described in detail in [9,11].

There are several common methods for measuring the ultrasonic transducer sensitivity. In 1940s was developed a first method of measuring electro-acoustic transducer sensitivity based on the reciprocity principle. In the process, a reciprocal transducer is used. In [12] it was proposed an improved reciprocal method. During the measurement, three transducers were arranged in a straight line along a calibrated support. The measured transducer is placed in the middle of the transmitter and the reciprocity transducer. In [13] it was proposed a pulse-echo technique. The transducer was excited by electrical pulses, and the ultrasonic waves were reflected by the reflector and received by the measured transducer. In another method a laser interferometer was used to measure the particle vibration displacement in the acoustic field [14]. Method of measuring the transducer sensitivity by using a calibrated hydrophone is described in [15]. In laboratory tanks, the pressure is usually measured with a calibrated hydrophone by reference to its sensitivity curve. The direct field of the sound source is obtained by time-gating the signal sent to the transducer so that the hydrophone receives the direct signal in the absence of reflections.

In this paper, the author used a simplified method of measuring sensitivity. Method based on reciprocity and pulse-echo methods. One transducer always used as receiving. Two another transducers were used sequentially as transmitters. The radiating sensitivities of these transducers were compared with each other.

The sensitivity in logarithmic units can be written as [16]:

$$S = 20\log_{10}\left(\frac{V_{out}}{V_{in}}\right), \tag{2}$$

where V_{in} – voltage applied to transmitting transducer, V_{out} – amplitude of voltage recorded at receiving transducer.

2.2. Directivity pattern measurement

Beam directivity is the most fundamental aspect used in characterizing the transducer. Each ultrasound transducer has its own specific directivity pattern. The directivity pattern, also known as beam pattern or radiation pattern, is an important far-field characteristic of a transducer. Directivity pattern consists of a main lobe and side lobes. Radiation intensity is dominant mainly in the front region of the transducer source, so the main lobe is directly in front of the ultrasound transmitter, followed by side lobes sidewise with null region in between these lobes. In general, the directivity patterns are the same whether the transducer is used as a transmitter or as a receiver [17].

The sound pressure of the plane rectangular transducer is no longer symmetrical, but elliptical in the far field. The mathematical expression for the normalized directivity pattern in far-field of the plane rectangular transducer is:

$$I = \frac{S}{\lambda r}\frac{\sin(a_1 k \sin\theta_1)}{a_1 k \sin\theta_1}\frac{\sin(a_2 k \sin\theta_2)}{a_2 k \sin\theta_2}, \quad (3)$$

where k – is the wave number, r – distance from the transmitting transducer, λ – wavelength, a_1, a_2 – length and width of the transducer, θ_1 – beam divergence on length direction, θ_2 – beam divergence on width direction.

Transmitter and the receiver should be separated by the minimum acceptable distance, called far-field distance, to minimize interference from reflections. The far-field distances were determined to be approximately beyond 20 cm and 40 cm for the 2.5 MHz and 5 MHz transducers, respectively. These distances were obtained using following equation, which are the standard criteria for uniform rectangular transducer:

$$X > \frac{\pi S}{\lambda}, \quad (4)$$

where X – the far-field distance, S – the square of the transducer, λ – the wavelength of the ultrasound waves in the medium.

The boundary of the field is conventionally considered to be the locus of points where the amplitude falls by a factor of 10 compared to its value on the axis, i.e., by 20 dB. The width of the directivity pattern in the far zone is in practice characterized by a 2-fold decrease in amplitude, i.e., by 6 dB.

3. Experimental setup

To measure the sensitivity and the directivity pattern, two sets of transducers were made. Central frequency of them 2.5 and 5 MHz. Each set consists of one transducer with the matching layer and two without the matching layer. The transducers are made of piezoceramics PZT-19 (the closest analogue of PZT-5A) (Aurora-ELMA, Volgograd, Russia).

Piezoelements with a central frequency of 2.5 MHz (±5%) are rectangular shape with geometric dimensions of 10×5×0.8 mm. Piezoelements with a central frequency of 5 MHz (±5%) are rectangular shape with geometric dimensions of 10×5×0.4 mm.

The specific acoustic impedance of such a piezoceramic is 22.4×10^6 Rayl. To match with water or biological tissue, the specific acoustic impedance of the coating should be about 5.8×10^6 Rayl [6]. The damper material is an epoxy resin with the addition of red lead and tungsten powder.

All sensitivity and directivity pattern measurements here were made at a temperature of 25 °C.

3.1. Emitter circuit

The electrical circuit are shown in figure 1 emits electrical pulses to excite the emitting transducer. Power to the circuit is supplied from the line power source AKIP-1137 (PriST, Russia).

Transducer X1 emit ultrasonic pulses with a filling frequency of 2.5 to 5 MHz. The operation of the circuit is based on the effect of avalanche breakdown of transistor Q1.

The capacitor C_1 starts charging through the resistor R_2 when the supply voltage is turned on 150 – 200 V. The time constant of the chain R_2C_1 determines the charge time of the capacitor C_1. The transistor Q_1 is closed at the initial moment. Voltage at the collector Q_1 increases, approaching the supply voltage. However, when the voltage on the collector reaches the avalanche breakdown value of the p-n junction, the resistance of the transistor between the collector and the emitter drops sharply and the capacitor C_1 is connected to the ground by the left terminal. The voltage at the capacitor C_1, equal to the breakdown voltage of the transistor (about 130 – 140 V), at this moment is applied to the transducer X_1 and resistor R_3. X_1 – is a resonant circuit with a center frequency of 2.5 or 5 MHz. X_1 emits a short pulse with a filling frequency equal to the resonance frequency.

After C_1 is quickly discharged through X_1 and R_3, the voltage on it falls and the transistor Q_1 closes. Further, C_1 is recharged through R_2 and the process is periodically repeated.

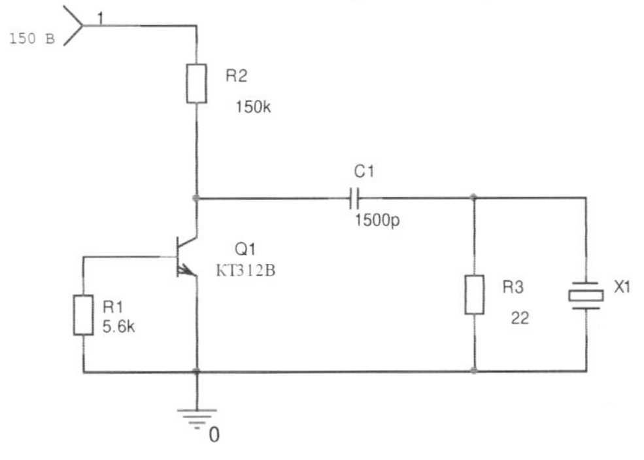

Figure 1. The electrical scheme of the emitter circuit

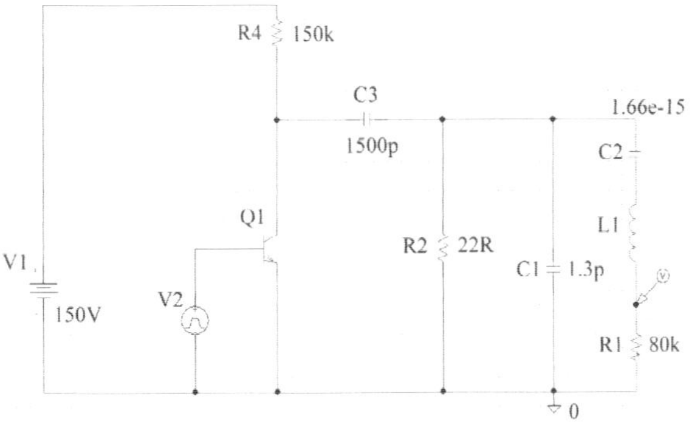

Figure 2. The equivalent electrical scheme of the emitter circuit

The simulation of the described emitter circuit was carried out. To take into account the process of avalanche breakdown of transistor, an equivalent circuit is developed (figure 2).

The KT312B transistor in the simulation circuit has been replaced by an analog.

The following results of the simulation were obtained. The shape of the signal on the equivalent resistor of the replacement circuit of the resonant emitter is shown in figure 3. Damped oscillations at the resonant frequency are seen.

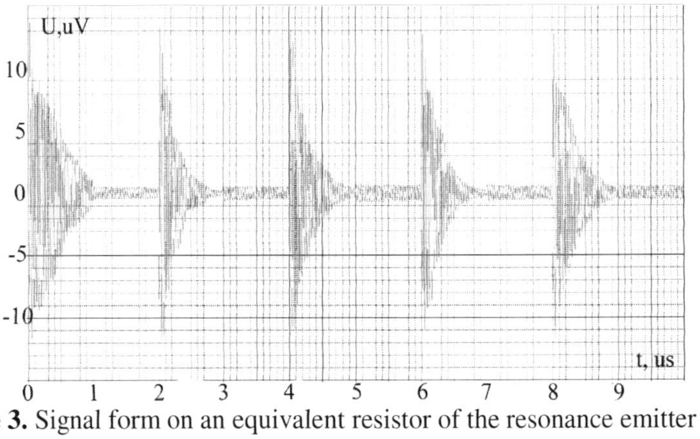

Figure 3. Signal form on an equivalent resistor of the resonance emitter circuit

3.2. Sensitivity measurement

The measurements were carried out in a tank with transformer oil. To measure the sensitivity, the reciprocity method (figure 4a) and pulse-echo method (figure 4b) were used. Other of the described methods can not be implemented because of the lack of a calibrated hydrophone and a laser vibrometer. Reciprocity method uses three sensors. One transducer is constantly used as a receiver. The transducer with matching layer and then the transducer without matching layer is alternately used as a transmitter. The transmitting and receiving transducer were placed opposite each other at a distance of 20 to 100 cm. For emitting transducer, short electrical pulses with an amplitude of up to 70 V were applied. As the emitter the circuit is based on the effect of avalanche breakdown of a KT312B series transistor was used.

To receive electrical signals the Agilent DSO 9104A digital oscilloscope was used (Agilent Technologies, US).

The second method uses a reflector instead of a receiving transducer. The transmitting transducer is also receiving a reflected signal. A metal plate with dimensions 20×20 cm was used as a reflector. The distance from the transducer to the reflector also varied from 20 to 100 cm.

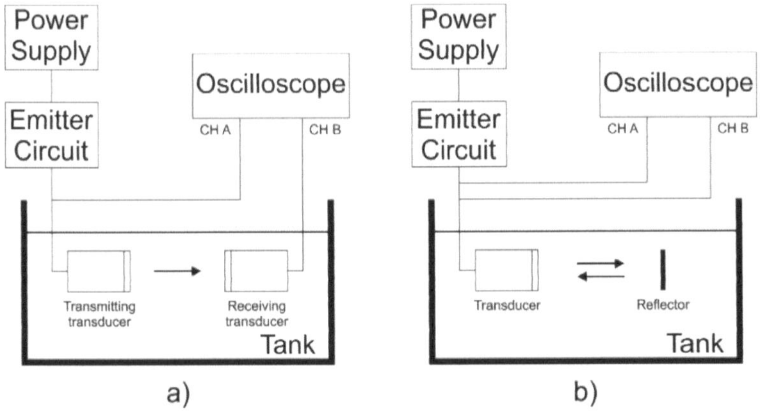

Figure 4. Sensitivity measuring system: a) reciprocity method, b) pulse-echo method

3.3. Directivity pattern measurement
In this work, the beam directivity was analyzed by assuming values of the sound pressure radiated from transmitting transducer and redistricted by receiving transducer.

Testing was done by mounting the receiving transducer in a motorized fixture driven by a step motor. The scheme of the measuring system is shown in figure 5.

The pair of transducers with an identical resonance frequency were used in each measurement. One transducer was used as a transmitter and the other as a receiver. Then transmitting transducer was changed to another from the set. After the one set of transducers having the same resonance frequency, the second set was placed in the measurement system.

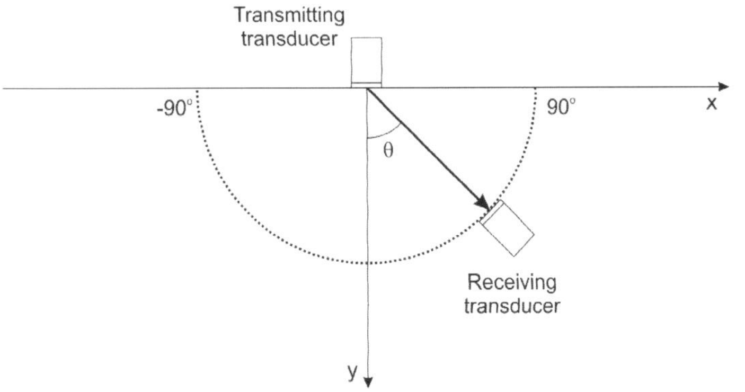

Figure 5. Directivity pattern measuring principal scheme

Directivities were obtained from –45 degrees to 45 degrees by plotting the peak amplitude on the y-axis. The alignment was based on the recording of the maximum amplitude of the received signal on

IOP Conf. Series: Journal of Physics: Conf. Series **1038** (2018) 012109 doi:10.1088/1742-6596/1038/1/012109

the oscilloscope. Next, the directivity pattern measurement was carried out with the fixed receiver. The transmitter was rotated using program based on Arduino from –45 to 45 degrees in 1 degrees increments (a total of 91 readings). For each angle, the peak-to-peak voltage of the received signal was recorded and the data were saved. The obtained data were normalized and then compared to the theoretical results using the directivity pattern model provided in equation 3. This procedure was repeated for two sets of transducers, having resonance frequencies of 2.5 MHz and 5.0 MHz.

4. Experimental results and discussion

The results of sensitivity measurements are shown below. Figure 6 and 7 shows the results for 2.5 MHz transducers without matching layer and with matching layer respectively. Signal amplitude for the transducer with the matching layer is about 1.8 V, while for the transducer with the matching layer is about 1.1 V.

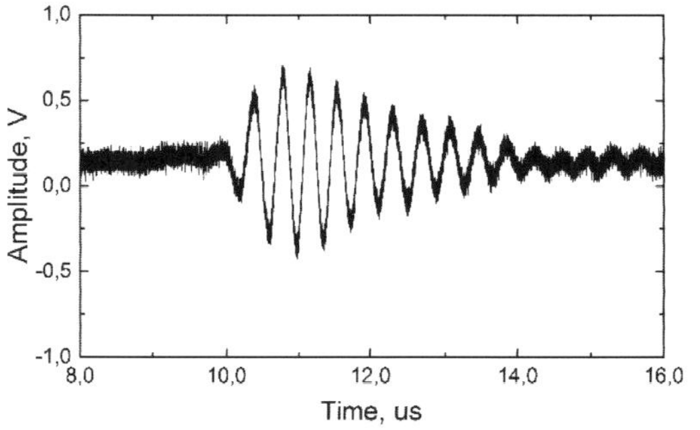

Figure 6. 2.5 MHz without matching layer

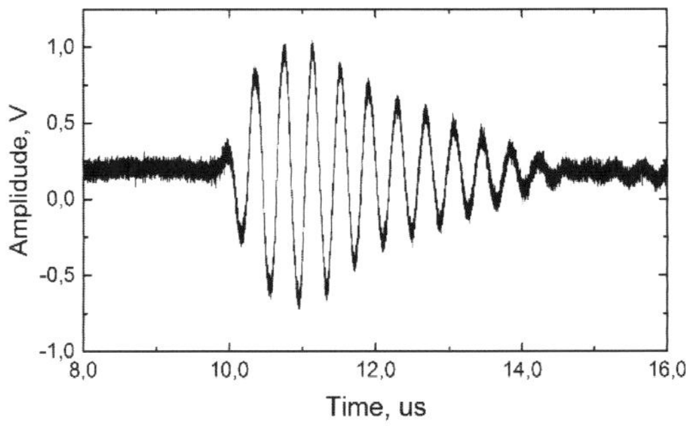

Figure 7. 2.5 MHz with matching layer

Figure 8 and 9 shows the results for 5 MHz transducers without matching layer and with matching layer respectively. Signal amplitude for the transducer with the matching layer is about 3.5 V, while for the transducer with the matching layer is about 1.7 V.

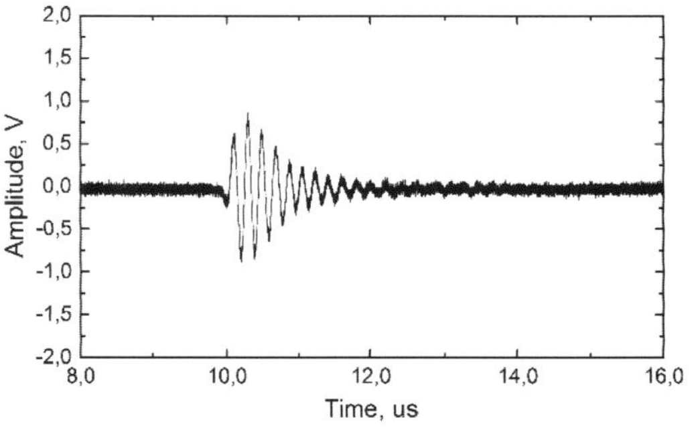

Figure 8. 5 MHz without matching layer

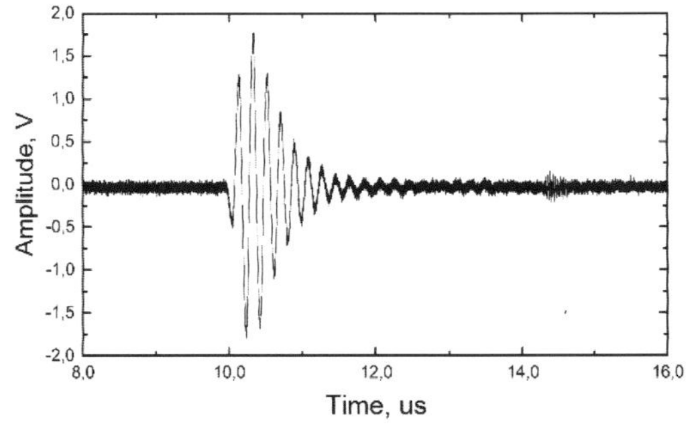

Figure 9. 5 MHz with matching layer

The obtained results shows increasing of transducers with a matching layer sensitivity in 1.6 ... 2 times in comparison with transducers without matching layer. The next step will be an experimental comparison with transducers coated with lavsan and a two-component epoxy polymer EPOTEK 301.

Then directivity pattern measurements of transducers were made. Figure 10 shows the signal along the sensor axis, i.e. for $\theta = 0^{\circ}$. Figure 11 shows the signal for $\theta = 30^{\circ}$.

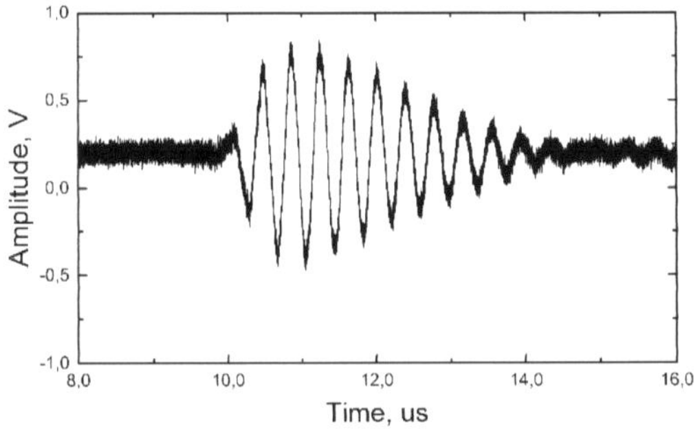

Figure 10. 2.5 MHz with matching layer
$\theta = 0^o$

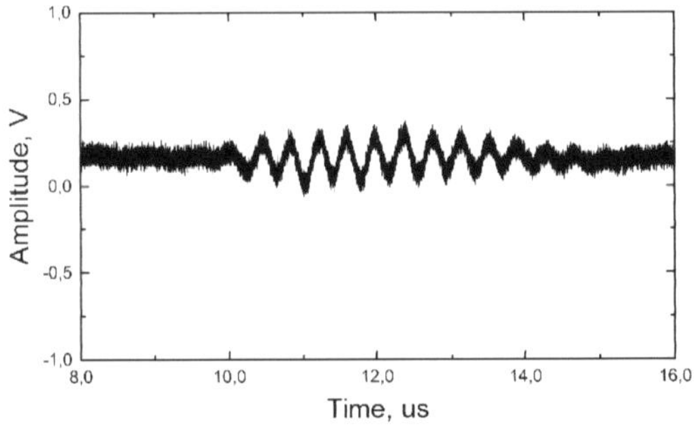

Figure 11. 2.5 MHz with matching layer
$\theta = 30^o$

The assuming results of the directivity pattern are presented for transducers with matching layer and without matching layer in figures 12-13.

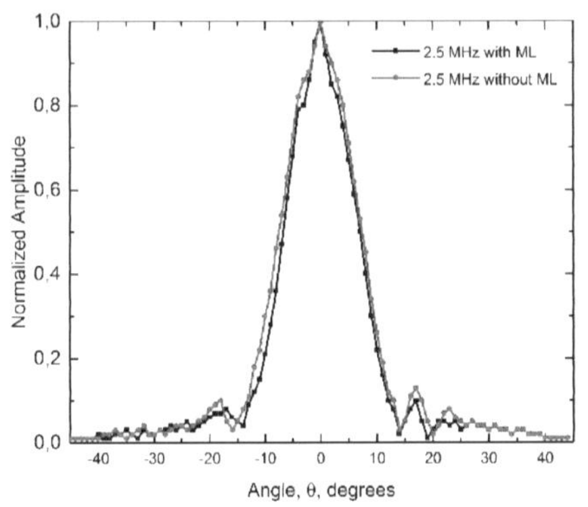

Figure 12. Directivity pattern of transducers with central frequency at 2.5 MHz

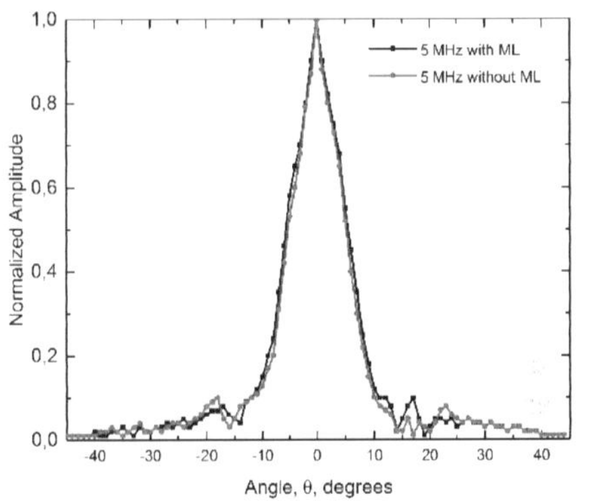

Figure 13. Directivity pattern of transducers with central frequency at 5 MHz

The results of the measurements show that the directivity patterns for transducers with a matching layer and without it are practically identical. This indicates that the proposed coating does not spoil the radiation pattern.

Figure 12 and 13 shows that the directivity pattern gets sharper as the frequency of the transducer increases. Half power beam width of 2.5 MHz transducer is about 12 degrees, 5 MHz transducer is about 8 degrees.

5. Conclusions

In conclusion, the sensitivity and directivity patterns were measured experimentally. The key aspect of this study was described by the directivity, which shows amplitude as a function of radiation angle at a specific radius. The analytical expressions for beam directivity were derived from the far field pressure distribution.

The results of the experiments showed increase of the transducers sensitivity by 1.5 ... 2 times in the case of using a polymer powder coating as a matching layer. It was also shown that using the matching layer does not change the radiation pattern of the transducers.

IOP Conf. Series: Journal of Physics: Conf. Series **1038** (2018) 012109 doi:10.1088/1742-6596/1038/1/012109

In addition, the results presented above confirmed that the beam pattern of ultrasound waves emitted from transducers of varying resonance frequencies is dependent on the frequency of the transducers. Also was observed that the directivity pattern gets sharper as the frequency of the transducer increases.

References

[1] Denisov E S, Temyanov B K, Sagdiev R K and Fazlyyyakhmatov M G 2014 *IOP Conf. Series: Mater. Sci. Eng.* **69** 012014

[2] Sagdiev R K, Denisov E S, Evdokimov Yu K, Fazlyyyakhmatov M G and Kashapov N F 2014 *IOP: Conf. Ser. Mater. Sci. Eng.* **69** 012012

[3] Khan A, Abas Z, Soo Kim H and Oh I K 2016 *Smart Mater. Struct.* **25(5)** 053002

[4] Zhang S, Li F, Jiang X, Kim J, Luo J and Geng X 2015 *Prog. Mater Sci.* **68** 1–66

[5] Zhou Q, Lau S, Wu D and Shung K K 2011 *Prog. Mater Sci.* **56(2)** 139–174

[6] Gavrilova V A, Fazlyyyakhmatov M G and Kashapov N F 2014 *J. Phys.: Conf. Series* **567** 012023

[7] Fazlyyyakhmatov M and Kashapov N 2014 *High Temperature Material Processes* **18(4)** 273–279

[8] Gavrilova V A, Fazlyyyakhmatov M G and Kashapov N F 2013 *J. Phys.: Conf. Series* **479** 012010

[9] Xiao D, Fan Q, Xu C and Zhang X 2016 *Ultrasonics* **68** 150–154

[10] Dang C, Schmerr L W and Sedov A 2002 Res. Nondestr. Eval. **14** 203–228

[11] Van Neer P L M J, Vos H J and De Jong N 2011 *Ultrasonics* **51** 1–6

[12] Luker L D and Buren V 1981 *J. Acoust. Soc. Am.* **70(2)** 516–519

[13] Lopez-Sanchez A L and Schmerr L W 2006 *IEEE Trans. Ultrason. Ferroelectr. Freq. Contr.* **53 (11)** 2101–2112

[14] Bacon D R 1988 *IEEE Trans. Ultrason. Ferroelectr. Freq. Contr.* **35 (2)** 157–160

[15] Everitt S J and Humphrey V F 2000 *Ultrasonics* **38** 118-121

[16] Moles D C 2004 *Introduction to Phased Array Ultrasonic Technology Applications* (Waltham, MA: Olympus NDT)

[17] Umchid S 2009 *International Journal of Applied Biomedical Engineering* **2(1)** 39–43

IOP Conf. Series: Journal of Physics: Conf. Series **1038** (2018) 012110 doi:10.1088/1742-6596/1038/1/012110

The investigation of InGaAs quantum dot growth peculiarities for GaAs intermediate band solar cells

R A Salii[1], S A Mintairov[1], M A Mintairov[1], A M Nadtochiy[1,2], M Z Shvarts[1] and N A Kalyuzhnyy[1]
[1] Ioffe Institute, Russia, Polytechnicheskaya str. 26, 194021 St.-Petersburg
[2] St Petersburg Academic University, Russia, Hlopina str. 8/3, 194021 St.-Petersburg

E-mail: r.saliy@mail.ioffe.ru

Abstract. In this work the growth peculiarities of InGaAs quantum dots (QDs) on GaAs surface have been investigated by metal-organic vapor-phase epitaxy (MOVPE). The influence of the main structural parameters, such as the amount of deposited InGaAs material and the thickness of the GaAs cap layer, on the optical properties of QDs has been considered. For these parameters the optimal values at which achieved the best photoluminescence intensity of QDs embedded into the matrix of the GaAs-based light-emitting structure have been established. The possibility of using several layers of QDs with the preservation of the optical properties of the investigated structures has been demonstrated. The design of solar cell (SC) based on GaAs with QD arrays in the active area has been proposed. It was shown that 20 layers of QDs embedded into the proposed SC structure contribute to the photogenerated current up to 0.97 mA/cm^2 for AM0 spectra and up to 0.77 mA/cm^2 for AM1.5D spectra, while maintaining the high quality of the p-n junction.

1. Introduction

Solar cells (SC) based on A3B5 semiconductor heterostructures are the most effective solutions among renewable energy sources. However in this field of photovoltaics there are still a lot of unresolved issues, such as fundamental losses, which impedes approaching the efficiency to the theoretical limit [1]. Currently, one of the most successful approaches which allow reducing the fundamental losses are multi-junction (cascade) solar cells (MJSC) [1, 2], in particular based on the lattice-matched InGaP/GaAs/Ge material system.

Today there is an alternative approach to reduce the fundamental losses aimed, in particular, at incomplete absorption in SC structure based on a single p-n junction. In the past two decades theoretical models which describe SC with an intermediate band (IBSC) have been actively developing. In these models the spectral sensitivity (and thus the efficiency) increases due to multiphoton absorption processes in semiconductor structures based on a single p-n junction [3, 4]. IBSC is proposed as an approach that allows transforming an additional part of the solar radiation spectrum by capturing photons with energy below the band gap of the matrix, to generate current while maintaining a high output voltage. As a candidate to practical realization of such approach the quantum dot arrays (QDs) are proposing [5]. The bulk layers containing QDs embedded in the SC matrix creates additional energy levels for the absorption of low-energy photons (so-called subbandgap photons), thus extend the spectral sensitivity of the device and increasing the photogenerated current.

Currently, QDs widely used in active area of lasers and allow increasing the radiation wavelength, improving thermal stability and reducing the threshold current, compared with traditional devices. However, first successful results of the application QDs in photovoltaic began publishing only recently and demonstrate the expansion of spectral sensitivity of the SC via photoelectric effect in InAs/GaAs QD [6-8]. In these works a noticeable but small contribution of QDs to the internal quantum efficiency of single-junction SC has been obtained. Wherein, one of the main problems in the practical realization of the approach are the structural stresses caused by a large number of QD layers in a bulk SC material that lead to the quality degradation of the p-n junction and decrease in the quantum yield in the short-wave part of the spectrum.
In this paper we propose introducing a small amount of Ga (~ 20%) into the composition of InAs QDs in order to reduce the difference between the lattice parameters of the QDs and the GaAs SC matrix and, thus, reduce the stresses in the structure of the device. This will preserve the structural quality of the p-n junction in SC without introducing technologically complex layers that compensate the stresses in the structure. In the paper, technological features of $In_{0,8}Ga_{0,2}As$ QD grown in the GaAs matrix are discussed in detail. SCs with QDs in active area which demonstrate the high quality of the p-n junction in the GaAs absorption region have been obtained.

2. Experimental procedure

All structures have been grown using the MOVPE technique by means of R&D installation with a horizontal reactor at low pressure (100mbar). Metal-organic compounds were used as sources of Group III elements (trimethylgallium (TMGa), trimethylaluminum (TMAl) and trimethylindium (TMIn)). Arsine (AsH_3) and phosphine (PH_3) were used as a sources of the Group V elements (As and P). Experiments were carried out on (100) GaAs substrates misoriented to the (111) direction by 6°. InAs QDs with a density of up to $1 \cdot 10^{10}$ cm^{-2} were obtained for misoriented substrates [9].
In the study two series of light-emitting structures based on GaAs with InGaAs QD in the active area were grown to estimate their optical parameters by the photoluminescence (PL) method. In the first structure the active area included a single $In_{0.8}Ga_{0.2}As$ QD layer placed in the center of a 500 nm GaAs waveguide surrounded by wide-gap $Al_{0.3}Ga_{0.7}As$ barriers to prevent carrier leakage to the surface or substrate (Fig. 1a).

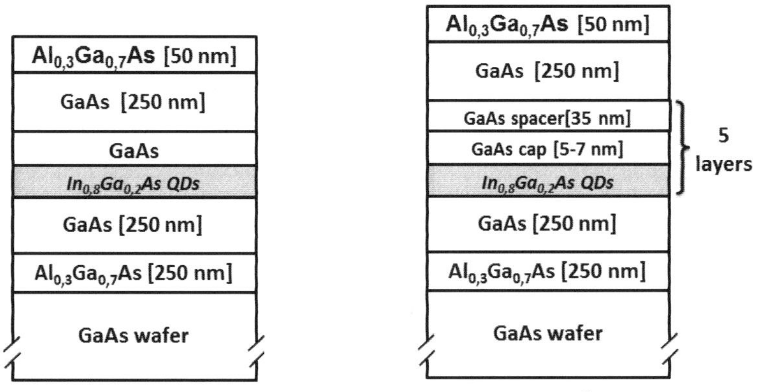

Figure 1. Experimental structures for studying the photoluminescence spectra of single layers of QDs (a), and array of QDs (b).

The growth was carried out according to the following technological process: after growing 250 nm $Al_{0.3}Ga_{0.7}As$ and 250 nm GaAs at 700°C the reactor cooled down to the growth temperature of QDs (520°C). Then the $In_{0.8}Ga_{0.2}As$ material was deposited for the nucleation and formation of QDs by Stranski-Krastanov mode [10]. After that the QDs were covered with a layer of GaAs at the same temperature in order to protect them from degradation during the subsequent reactor heating. Then the rest of the GaAs waveguide and wide-band $Al_{0.3}Ga_{0.7}As$ were grown at 600 °C.

In the second type of experimental structures (Fig. 1b) the absorption was increased due to 1, 5, 10 and 20 layers of QDs grown through the GaAs spacers with a thickness 35 nm. The spacers grew at elevated temperature 600°C which was also necessary for preparing a planar surface for each subsequent layer of QD [8]. These structures were designed to investigate the possibility of embedding several layers of QDs into the active region of GaAs SC.
In both types of structures the main technological parameter that varied depending on the experiment was the amount of $In_{0.8}Ga_{0.2}As$ deposited material. The amount of InGaAs material was varied from 1 to 4 monolayers (ML). The thickness of GaAs cap layer was chosen based on the known lateral sizes of QDs [7, 8] and was 5 nm for all experimental structures. Nevertheless, an additional experiment was performed to evaluate the effect of this parameter on the PL spectrum in which the thickness of GaAs cap layer was 7 nm with the amount of InGaAs material of 2 ML. The bulk of the light-emitting structure was grown at growth rates of 1.5 µm/h while the GaAs cap layers grew at a reduced growth rate 0.75 µm/h since in [7] the authors showed a positive effect of decrease the GaAs cap layer growth rate on the optical properties of the obtained QDs. The deposition of InGaAs in all structures occurred at the constant growth rate 0.167 ML/s and temperature of 520°C.
To obtain the PL spectra a 532nm YAG:Nd laser of 350mW power was used. All measurements were carried out by means of a cooled Ge photodetector using a standard lock-in amplifier. The multifunctional installation has been used for measuring the external quantum efficiency (EQE) of the SCs. The measuring hardware included an unblocked ultra-violet halogen light source, a grating monochromator with 2 nm/mm dispersion within 300–1200 nm wavelength scanning range, and an optical chopper of 90 Hz and high sensitivity lock-in electronics. The EQE measurement was controlled by a laptop. The lock-in technique allows weak output signals at the QD sensitivity range to be detected confidently.

3. Results and discussion

3.1. A growth peculiarities of InGaAs QDs on GaAs surface

Optical properties of single-layer structures (Fig. 1 a) were studied by PL method to determine optimal growth parameters for QDs. This method allows estimating the physical parameters of QDs in terms of position, intensity and number of PL peaks. In this case, the criterion for optimality of growth parameters was the highest PL intensity of QDs. Fig. 2 shows PL spectra at the different amount of InGaAs material. The PL intensity in the absorption region of QDs (950-1300 nm) was estimated and it was shown that with the deposition 2 ML InGaAs and 5 nm GaAs cap layer QDs with the highest PL intensity are formed compared with samples grown with a larger amount of InGaAs material.
It can be seen that the deposition of 1 ML InGaAs is not enough to start the formation of coherent islands since the obtained spectrum has a shape close to the quantum well one with two apparent peaks: 870 nm from the GaAs matrix and 900 nm from the wetting layer of the QD. The resulting optimum thickness of 2 ML exceeds the critical thickness for the formation of InAs QDs on the GaAs surface, which is 1.7 ML like it has been shown in [8]. In fact, $In_{0.8}Ga_{0.2}As$ has a smaller difference in the parameters of the crystal lattices with GaAs so when the QD growth in the Stranski-Krastanov mode the more material is required to begin the formation of the islands.
The spectra obtained have the same shape and consist of two apparent peaks that are red-shifting after deposition of more than 3 ML of InGaAs. This indicates the formation of QDs of two different types with differing lateral sizes. In Fig. 2 they are highlighted as QD1 and QD2 for small and large QDs respectively. With increasing the quantity of MLs QD sizes also increase proportionally which leads to the red-shift of the PL peaks (in Fig. 2 it is shown by arrows). The PL intensity significantly drops with increasing the quantity of MLs which is the result of an increase the population of defective QDs that were formed after reaching the critical thickness of InGaAs due to coalescence processes (after QDs formation the migration of In atoms along the uncapped surface proceeds towards energetically favorable QDs with large lateral sizes). Also, QDs of high altitude after capping with 5nm of GaAs layer became not completely covered and dissolve during further growth.

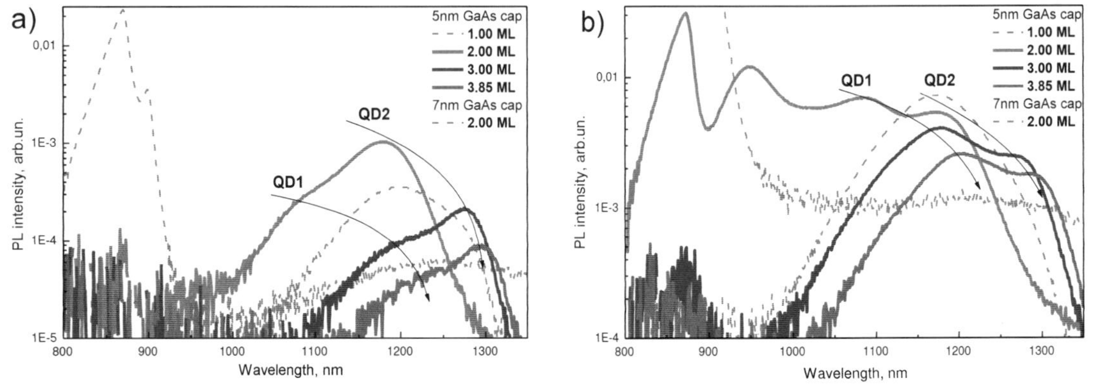

Figure 2. The PL spectra at room temperature of single QD layers at the different amount of $In_{0.8}Ga_{0.2}As$ material with 5 nm GaAs cap layer at laser pump of $30W/cm^2$(a) and $1kW/cm^2$ (b).

In Fig. 2 the red dotted line shows the PL spectrum for QDs covered with 7 nm of GaAs cap layer. It can be seen that it is red-shifted compared with QDs which are covered with 5-nm GaAs cap. PL from QD1 is suppressed in this case which is apparently due to the poor quality of thicker low-temperature GaAs cap. Wherein, the appearance of an apparent long-wave shoulder indicates an increase the contribution of large QDs which before were partially dissolved after thin capping.
Based on the single-layer structure a series of experiments with 5 layers of QDs was carried out in order to increase the absorption in the further SC structure. A wide range quantity of MLs from 1.75 to 3 ML was investigated. Figure 3 shows how the shape of the PL spectrum changes with increasing the amount of InGaAs material at 5 nm GaAs cap layer.

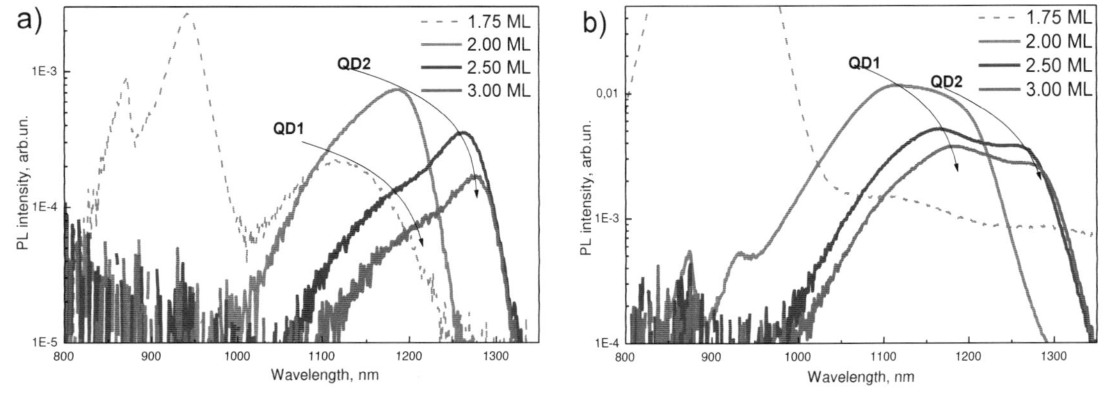

Figure 3. PL spectra at room temperature for 5 layers of QDs depending on the amount of $In_{0.8}Ga_{0.2}As$ material with 5 nm GaAs cap layer at laser pump of $30W/cm^2$ (a) and $1kW/cm^2$ (b).

After growing several layers of QDs the evolution of the spectra still unchanged which is indicated by the position of the PL peaks and by the half-width of the spectral lines. With increasing the number of QD layers a red-shift of the entire spectrum is also observed, accompanied by a decreasing of the PL intensity comparable with single-layer QD-structures. This shows that the optimum amount of InGaAs material still the same after transition from single-layer technology to the technology of QDs arrays and for the solid solution $In_{0.8}Ga_{0.2}As$ this value is 2 ML. Wherein, the optimum thickness for GaAs cap layer is 5 nm for QD arrays. These growth parameters were used for embedding QD arrays into the active area of GaAs SC.

3.2. Embedding InGaAs QDs in active area of GaAs SC

The heterostructure of the GaAs SC (Fig. 4 a) was grown with an increased i-region for embedding the QD arrays. This structure was used as a reference in the study of the spectral characteristics of SC with QD. The arrays containing 1, 5, 10 and 20 $In_{0.8}Ga_{0.2}As$ QD layers were embedded into four SC heterostructures in the middle of the i-region (Fig. 4 b).

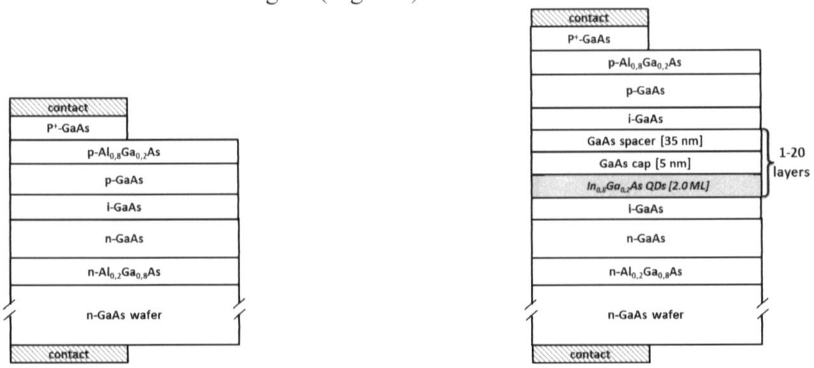

a b

Figure 4. Experimental structures of reference GaAs SC(a) and SC with QDs in active area (b).

The total thicknesses of the base, emitter and i-region layers for all SC structure were the same to ensure comparable charge carrier collection. Experimental SC samples were created by forming the front and rear solid Ni contacts and etching out the contact layer of p+-GaAs from the surface free from the front contact (without using antireflection coating).

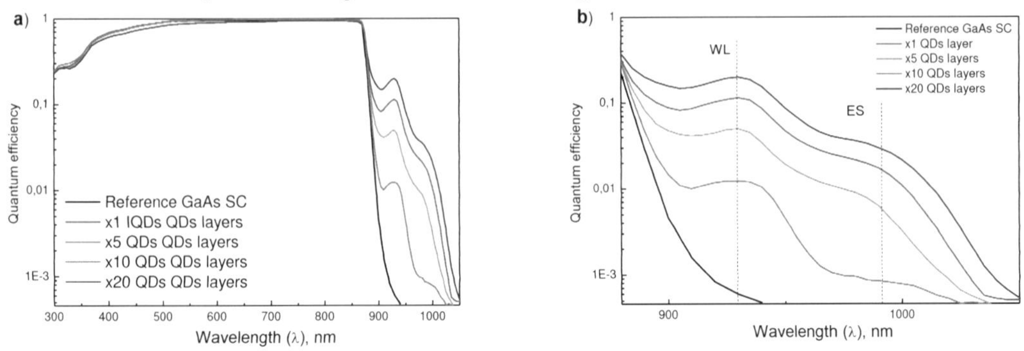

Figure 5. Spectral characteristics of the internal quantum yield of GaAs SC with embedded InGaAs QD arrays in comparison with the reference GaAs SC without QD: a - the whole absorption spectrum, b - the absorption range of QD.

Figure 5 a shows the spectral characteristics of the internal quantum yield of the obtained SC in comparison with the reference GaAs SC without QD arrays. In the long-wave part of the spectrum (880-1100 nm) beyond the GaAs absorption limit the extended spectral sensitivity was observed caused by photocurrent generated in the QD by absorbing subbandgap photons. The greatest contribution to the spectral sensitivity have been demonstrated by wetting layer at wavelength of ~ 940 nm (marked as WL) and an excited state of QDs with small lateral sizes at wavelength of 980 nm (marked as ES).

For QD array of 20 layers the values of 0.97 mA/cm^2 for AM0 and 0.77 mA/cm^2 photogenerated current are achieved. This corresponds to 0.05 mA/cm^2 and 0.04 mA/cm^2 after recalculated per 1 QD layer. Such increase in the photogenerated current does not definitely insure an increase of SC efficiency, however the technology of QD arrays can be used for current balance in MJSC. It is important to note that as a result of the embedding QD arrays (up to 10 layers) the quality of the p-n junction is still comparable with the reference GaAs SC as evidenced by the absence of a critical drop in the spectral sensitivity in the GaAs absorption range. A small drop in the short-wavelength spectral sensitivity is observed for a SC with 20 layers of QDs.

4. Conclusions

Thus, in this work an experimental analysis of InGaAs QD arrays formation in a GaAs matrix by the MOVPE method in the Stranski-Krastanov mode was carried out. In the framework of single QD layer technology the optimal growth parameters at which the high PL intensity of test light-emitting structures was observed have been found. With the deposition 2 ML of $In_{0.8}Ga_{0.2}As$ covered with 5 nm thick GaAs cap layer QDs with maximum PL intensity were formed. As in the case of the InAs/GaAs system the bimodal distribution of QDs by size after $In_{0.8}Ga_{0.2}As$ deposition on the GaAs surface by the MOVPE method was observed. The position of the peaks and the bimodal nature of the PL spectra still the same for QD arrays.

The developed technology was used for embedding QD arrays in the active area of GaAs SC. The maximum calculated contribution of QDs in the range from 880 to 1100 nm for a sample with 20 layers was 0.97 mA/cm^2 for space spectrum (AM0), and 0.77 mA/cm^2 for terrestrial spectrum (AM1.5D). Wherein, critical drop in the spectral sensitivity was not observed in the absorption region of GaAs which indicates the quality of the p-n junction comparable to the reference GaAs SC.

Acknowledgements

The work has been supported by the Russian Foundation for Basic Research (grant No. 16-08-01004). The authors wish to thank N. Kh. Timoshina for the SC spectral dependence measurements.

References

[1] Andreev V M, Grilikhes V A and Rumyantsev V D 1997 *Photovoltaic Conversion of Concentrated Sunlight* (John Willey & Sons Ltd.)

[2] Cavicchi B T, Krut D D, Lillington D R, Kurtz S R and Olson J M 1991 *Photovoltaic Specialists Conference* (Las Vegas) pp 63–67

[3] Luque A and Marti A 1997 *Phys. Rev. Lett.* **78** pp 5014–5017

[4] Luque A and Mellor A V 2015 *Photon Absorption Models in Nanostructured Semiconductor Solar Cells and Devices* (Springer Briefs in Applied Sciences and Technology)

[5] Martí A, Cuadra L and Luque A 2002 Quasi-Drift Diffusion Model for the Quantum Dot Intermediate Band Solar Cell *Transactions on electron devices* **49** pp 1632–1639

[6] Blokhin S A, Sakharov A V, Nadtochy A M, Pauysov A S, Maximov M V, Ledentsov N N, Kovsh A R, Mikhrin S S, Lantratov V M, Mintairov S A, Kaluzhniy N A and Shvarts M Z 2009 Semiconductors **43** pp 514–518

[7] Hubbard S M, Cress C D, Bailey C G, Raffaelle R P, Bailey S G and Wilt D M 2008 *Appl. Phys. Lett.* **92** 123512

[8] Kalyuzhnyy N A, Mintairov S A, Salii R A, Nadtochiy A M, Payusov A S, Brunkov P N, Nevedomsky V N, Shvarts M Z, Martí A, Andreev V M and Luque A 2016 *Progress in Photovoltaics* **24** pp 1261–1271

[9] Salii R A, Mintairov S A, Brunkov P N, Nadtochiy A N, Payusov A S, Kalyuzhnyy N A 2015 Semiconductors **49** pp 1111–1118

[10] Stranski I N and von Krastanobv L 1939 (Akad. Wiss. Lit. Mainz Abh. Math. Naturwiss. Kl.)

IOP Conf. Series: Journal of Physics: Conf. Series **1038** (2018) 012111 doi:10.1088/1742-6596/1038/1/012111

The thermovoltaic effect in rare-earth semiconductors based on *SmS* and the conversion of thermal energy into electrical energy on its basis

M A Grevtsev, G D Havrov, S A Kazakov and V V Kaminskii

194021, Russian Federation, St Petersburg, 26 Politekhnicheskaya, Ioffe Institute

E-mail: vladimir.kaminski@mail.ioffe.ru

Abstract. The article emphasizes the mechanism of the thermovoltaic effect in the rare-earth semiconductor material of samarium sulfide (*SmS*). The maximum achieved parameters of the *SmS* based thermal energy converters are given. Comparison of them with the parameters of classical thermoelements proves the practicability of further developments in this field.

1. Introduction

The aim of the work was to analyze the possibilities of applying the thermovoltaic effect in rare-earth semiconductor materials based on samarium sulfide (*SmS*) for the conversion of thermal energy into electrical energy.

The thermovoltaic effect was first discovered at the Ioffe Institute of the Russian Academy of Sciences in 2000 [1, 2] in semiconductor samarium sulphide (*SmS*). It was subsequently discovered in other semiconductors: *ZnO*, $Pr_{0.6}\,Ca_{0.4}MnO_3$, *Si*-based structures, etc. It consists in the appearance of the electrical voltage between opposite sides of the sample while heating up to *400-500K* without external temperature gradients. The value of the generated voltage on *SmS* single crystals reached *2.5V*, and in thin films it reached *4.5V*. The mechanism of the effect is described in [3]. The effect is based on the fact that defective Sm^{2+}-ions, which create impurity donor levels, are distributed in the volume of the sample irregularly and make a certain concentration gradient. Gradient direction in this case approximately coincides with the direction of the sample's electrodes.

2. Appearance of electrons local concentration

When the area of the sample, adjacent to one of the electrodes, is heated, a higher local concentration of conduction electrons and a gradient of n appear throughout the entire sample ('figure 1'). Due to the diffusion of current carriers, an electric current with a density

$$j = eDgrad\,(n),$$

appears in the sample. Here, *D* is the electron diffusion coefficient in *SmS*.

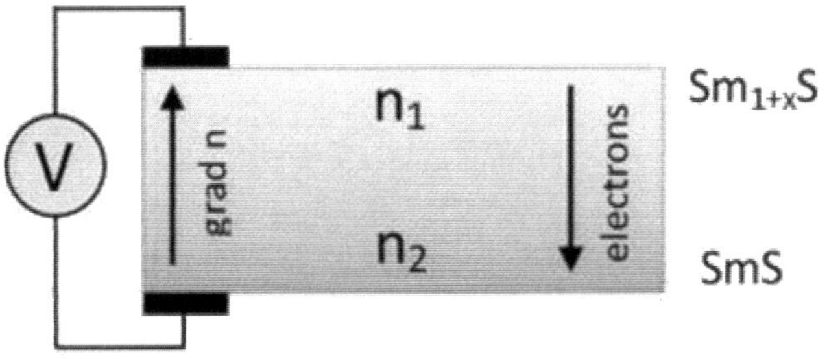

Figure 1. General scheme for measurement of the thermovoltaic effect.

Thus, only the mechanism of the formation of a single voltage pulse can be explained. However, the generation of pulses can continue indefinitely ('figure 2').

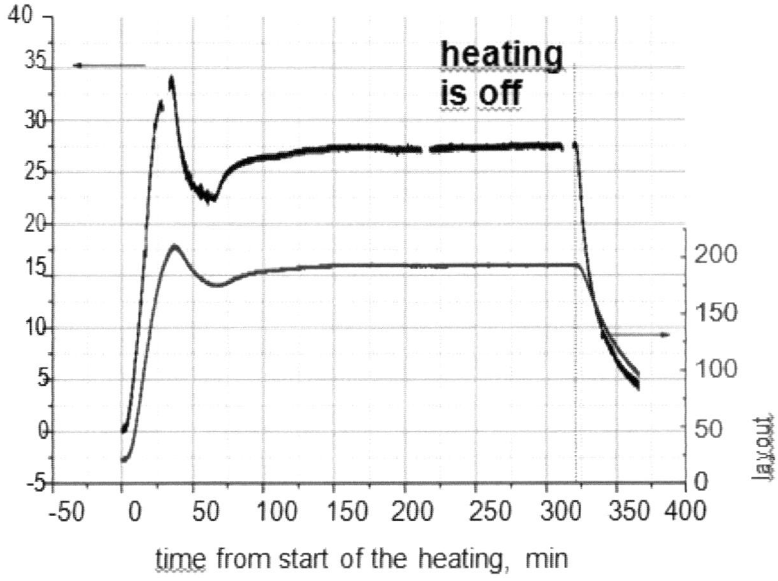

Figure 2. Generation during stationary heating of the thermoelement on the basis of thermovoltaic effect in *SmS*.

To explain this fact [3], it is necessary to consider the specificity of *SmS* band structure ('figure 3').

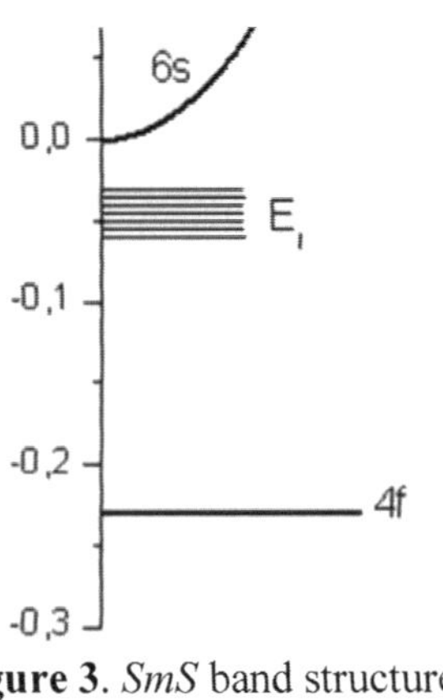

Figure 3. *SmS* band structure.

IOP Conf. Series: Journal of Physics: Conf. Series **1038** (2018) 012111 doi:10.1088/1742-6596/1038/1/012111

Semiconducting *SmS* has a *NaCl* type crystal structure. *SmS* can only have an *n*-type of conductivity. In the forbidden band at a distance of *0.23 eV* there are localized *4f*-levels of *Sm* ions. In addition, there are impurity donor levels with the positon $Ei = (0.045 \pm 0.015)eV$ and concentration $N_i = 10^{20} - 10^{21}\ cm^{-3}$. The impurity donor levels are connected with Sm^{2+}-ions, which are located not in the regular sites of samarium sublattice but, perhaps, in vacancies of sulfur one. Such defects are caused by the peculiarities of the samples creation technology.

Upon heating of the sample to a certain temperature T_b , a critical electron concentration n_b is accumulated in the conduction band. At this moment, the Coulomb potential of the impurity Sm^{2+}-ions becomes completely screened and the electrons delocalize from the E_i-levels to the conduction band. This collective process is spasmodic and occurs primarily in the area of the sample with the maximum value of N_i , leading to the appearance of a high local concentration of conduction electrons $n = 10^{20} - 10^{21}\ cm^{-3}$ in this area. This is the reason for the gradual increase in *grad (n)*.

3. Electric voltage pulse generation

The continious process of electric voltage pulse generation can be presented as follows. Upon heating the sample to temperature Tb, the electron concentration in the conduction band grows stepwise to $n \approx N_i$ in the sample's region with maximum N_i ; consequently, the value of grad *(n)* increases. As a result, a voltage pulse occurs. The heat absorption accompanying this pulse lowers the temperature of the sample to $T_0 < T_b$. Thus, the sample is prepared to generate the next pulse. The heater continues operating, the temperature of the sample rises to the value T_b again, and the process is repeated. Figure 4 schematically shows the dynamics of temperature *(T)* and local electron concentration *(n)* in the conduction band in the sample's area with the maximum value of N_i , as well as the generated voltage *(U)*.

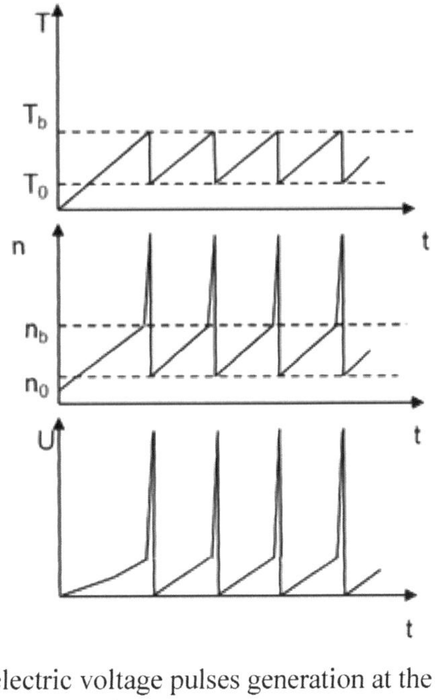

Figure 4. Scheme of the electric voltage pulses generation at the thermovoltaic effect in *SmS*.

Thus, we used the thermovoltaic effect in *SmS* to convert thermal energy into electric energy in the absence of temperature gradient. Thermoelements have been developed based on both thin film structures [4] and bulk samples. The principle of these thermoelements construction is based on the fact that bulk or film ("sandwich" type) samples consist of two layers with different concentrations of impurity donor levels. This was achieved in various ways: by diffusion, by sintering of layers with different impurity concentrations, by successive deposition of thin films of two or more materials, by creating different values of *X*-ray coherent scattering regions in different parts of the sample. The conversion efficiency up to *36%* was experimentally obtained [4].

4. Conclusion

A thermoelectric generator was constructed, containing a bulk sample of *SmS* semiconducting material, located between the metal current contacts ('figure 5'). The method for manufacturing of thermoelectric generator comprises soldering of *SmS* sample to a plate of refractory metal with a metal samarium.

Figure 5. A sample of *SmS/Sm/Mo* multilayer bulk structure.

Table 1. Maximum achieved results.

Characteristic	*SmS*	Classical thermoelements
Specific generating power, W/g	under 1.8	<0.2
Voltage generated by a single thermoelement, V	under 5	<0.1
Internal resistance, Ohm	0.2-2	0.2-2
Radiation resistance of electrical parameters (exposure up to the 1% change of the parameter, Roentgen)	10^{10}	10^{9}
Maximum power obtained from a single thermoelement, mW	5	unites of W
Transformation efficiency, %	36	under 10

In conclusion, the use of the thermovoltaic effect in *SmS* looks promising for converting thermal energy into electric energy. This transformation principle has the following advantages:
1) allows conversion in the absence of an external temperature gradient;
2) allows to increase by an order of magnitude the specific power of the generated electrical signal in comparison with the classical transformation (*Seebeck*-effect);
3) allows to *3* times increase the conversion efficiency;
4) it allows to increase by *1-2* orders the magnitude of the electrical voltage generated by a single thermoelement;

5) can be used at low temperatures.

The disadvantage of this principle of transformation is the low power of the signal generated by a single element. This is due to the fact that only electrons that are delocalized from impurity levels participate in the generation process. To eliminate this drawback, when creating a generator, a large number of thermoelements must be plugged.

References

[1] Kazanin M M, Kaminskii V V and Solov`ev S M 2000 *Tech. Phys.* **70** 136
[2] Kaminskii V V and Solov`ev S M 2001 *Physics of the Solid State* **43** 423
[3] Kaminskii V V 2013 *Tech. Phys. Lett.* **39** 12
[4] Egorov V M, Kaminskii V V, Kazanin M M, Solov`ev S M and Golubkov A V 2015 *Tech. Phys. Lett.* **41** 50

IOP Conf. Series: Journal of Physics: Conf. Series **1038** (2018) 012112 doi:10.1088/1742-6596/1038/1/012112

Photon-coupled characteristic of a multijunction solar cell

S A Levina, E D Filimonov and M Z Shvarts

Ioffe Institute, St. Petersburg, 194021, Russia

E-mail: kozhukhovskaya@mail.ioffe.ru

Abstract In the work, optical coupling arising in a multijunction solar cell under powerful laser radiation is investigated. To describe the cascade of interconnected luminescent processes in a structure with optical coupling, the photon-coupled characteristic is used. Investigation for two specimens with different effeciency of luminescent interaction was carried out.

1. Introduction

Most AIIIBV semiconductor materials applied in modern photovoltaics are direct-bandgap structures with high crystallographic perfection, for which domination of radiative recombination over nonradiative one is typical. Radiative recombination mechanisms can and largely have an impact on design and optimization solutions for solar cells (SCs) and on procedure approaches in studying their photo-electrical characteristics.

In recent years, raised interest is observed to investigating such a phenomenon as optical (photon) coupling in multijunction (MJ) SCs. It is an effect, at which photons arising in a wide bandgap p-n junction, as a result of radiative recombination of electron-hole pairs, are absorbed in the adjacent narrowband subcell generating in the latter an additional photocurrent. Importance of this process was considered in detail in [1-3]. However, up to now similar investigations are based only on considering the interconnection within pair of neighboring subcells being in a direct optoelectronic contact, whereas processes inside a MJ SC are more complicated.

2. Photon-coupled characteristic

In the present work, optical coupling phenomenon in GaInP-Ga(In)As-Ge SCs under powerful laser radiation was investigated. In the experiment, laser sources (corresponding by radiation wavelength to the sensitivity range of GaInP and Ga(In)As subcells) were in a pulse mode with different radiation modulation frequencies for possibility to separate and to register the Ge subcell response arising apart from each wide bandgap junction. It was established [4] that optical interaction between subcells occurs even if they are not in the immediate electrical contact (in a multilayer semiconductor structure, it is a contact through a tunnel diode). Hence, any variation of the current state in one of the subcells will immediately reflect on the current imbalance of the whole SC. The problem becomes more complicated in applying bias voltage (required for isolating a photoresponce from the tested subcell), what leads to abrupt change of currents flowing through emitting p-n junctions [4-6].

To describe processes occurring inside MJ structures, with allowing for photonic interaction between subcells, it was proposed to use the photon-coupled characteristic [4], i.e. the dependence of current induced by luminescent emission in the absorbing subcell on photocurrent resulting from action of external radiation in the top wideband p-n junction.

Let us consider in detail the principle of experimental formation of photon-coupled characteristic of a triple junction GaInP-Ga(In)As-Ge SC. At the first stage, the light bias is provided by a laser radiation source corresponding to the sensitivity range of the top wide bandgap subcell (blue laser with wavelength of 450nm), what results in proportional rise of its photocurrent. In narrow bandgap subcells, due to optical coupling in the GaInP-Ga(In)As and Ga(In)As-Ge pairs, photogenerated charge carriers also arise. Registration of photocurrents induced in the narrow bandgap junctions is performed in the voltage bias modes ($V_{BIAS1}\approx$ 1V for Ga(In)As and $V_{BIAS2}\approx$ 2.2V for Ge), what corresponds to the short circuit states for subcells. To determine current of GaInP, additional radiation sources are required (red and IR lasers), the application of which allows transferring the whole MJ SC into the state of current limitation from the top subcell side. With increasing light power of the main blue laser, the procedure for registering photocurrents of the subcells repeats in many times. Thus, a massive of data is formed, where three values of photocurrents of separate subcells ($J^{TOP}, J_{LC}^{MID}, J_{LC}^{BOT}$) correspond to each level of the initial light bias. On the basis of the obtained massive, two dependencies $J_{LC}^{MID}(J^{pn})$ and $J_{LC}^{BOT}(J^{pn})$ are formed (Fig.1a), where J^{pn} – internal current through the luminescent GaInP p-n junction equal to ($J^{TOP} - J_{LC}^{MID}$) in the short circuit mode of Ga(In)As and equal to ($J^{TOP} - J_{LC}^{BOT}$) in a similar mode for Ge.

Combining photocurrents of narrow bandgap subcells on a plane by means of J^{pn} allows determining graphically "equivalent" current states, that is, MJ SC states at different bias voltages with fulfilling the equality $J^{TOP} - J_{LC}^{MID} = J^{TOP} - J_{LC}^{BOT}$ (points x, y in Fig.1a). Set of "equivalent" current states forms the main photon-coupled characteristic of a tested SC – the dependence $J_{LC}^{BOT}(J_{LC}^{MID})$ (Fig.1b).

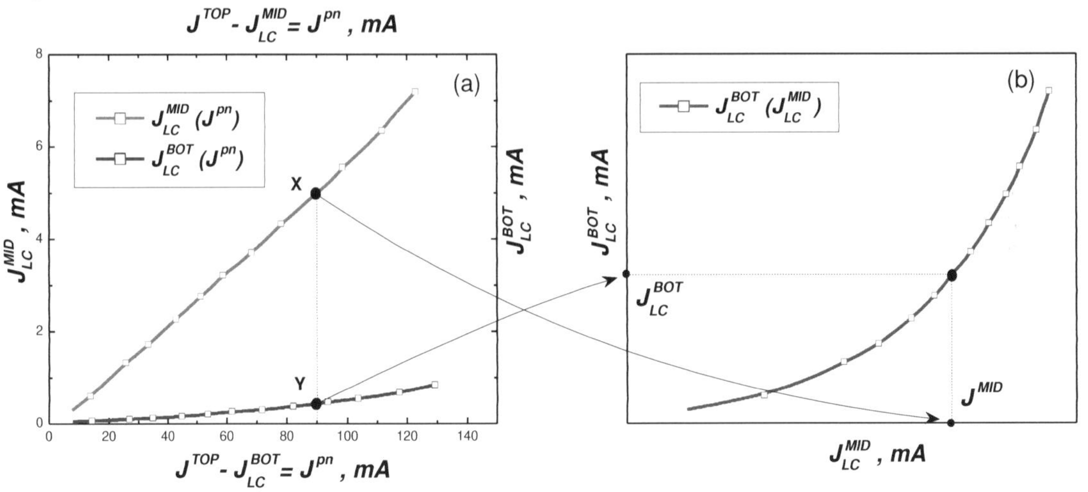

Figure 1. a – dependences of current induced by luminescence in Ga(In)As (red) and Ge (blue) on current through emitting GaInP subcell obtained in the short circuit mode at voltages V_{BIAS1} (for Ga(In)As) and V_{BIAS2} (for Ge). Points x, y – "equivalent" current states of Ga(In)As (J^{MID}) and Ge (J_{LC}^{BOT}), respectively. b – photon-coupled characteristic formed of "equivalent" current states.

At the second stage, a light bias from the source oriented on the middle cell spectral range (red laser with wavelength of 792nm) is added. This allows widening the photon-coupled characteristic possibility to describe processes inside a MJ SC. Forming the main photon-coupled characteristic in conditions of multi-wave illumination is similar to the case with one laser considered above. Then, an arbitrary level of light bias from a blue laser is selected, and the initial measurement condition corresponding to this level is fixed (points 1, 2, 3 or 4 in Fig.2). In varying the red laser intensity, photocurrents of narrow bandgap subcells in corresponding short circuit modes are registered. The GaInP subcell photocurrent at the first and second stages is determined identically. Depending on the chosen initial conditions, the shape of the photon-coupled characteristic changes, and the position of the additional branches (b, c, d, e) will be different (Fig.2).

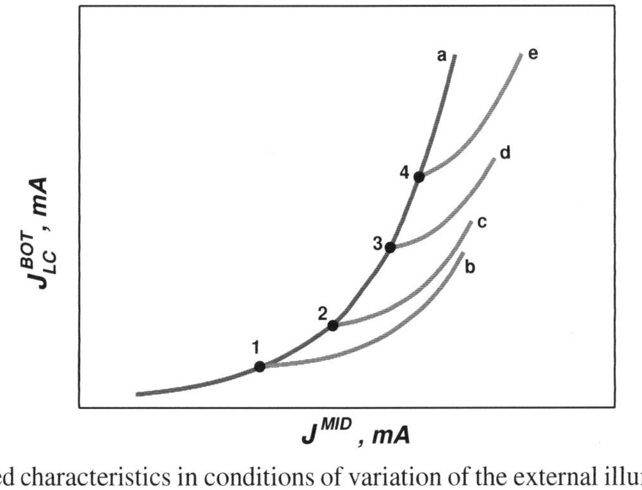

Figure 2. Photon-coupled characteristics in conditions of variation of the external illumination of GaInP only (branch a) and at addition of a light bias by red laser radiation (branches b, c, d, e) corresponding to different initial conditions of illumination (1, 2, 3, 4).

3. Results and discussion

In the work, photon-coupled characteristics for two GaInP-Ga(In)As-Ge SCs have been investigated: #1 – lattice-matched (LM) and #2 – lattice-mismatched metamorphic (MM). It has been found that photon-coupled characteristics of such cells are significantly different (Fig. 3). LM SC is characterized by a rapid increase in the induced current in Ge as the light bias of GaAs increases, and, consequently, the additional branch of the photon-coupled characteristic goes sharply upward. On the contrary, in MM SC the growth rate of induced current J_{LC}^{BOT} is lower, so the additional branch passes under the main dependence. The difference in velocities can be estimated from the angles of the tangents α, β (Fig. 3). It is evident, that α > β for the metamorphic SC.

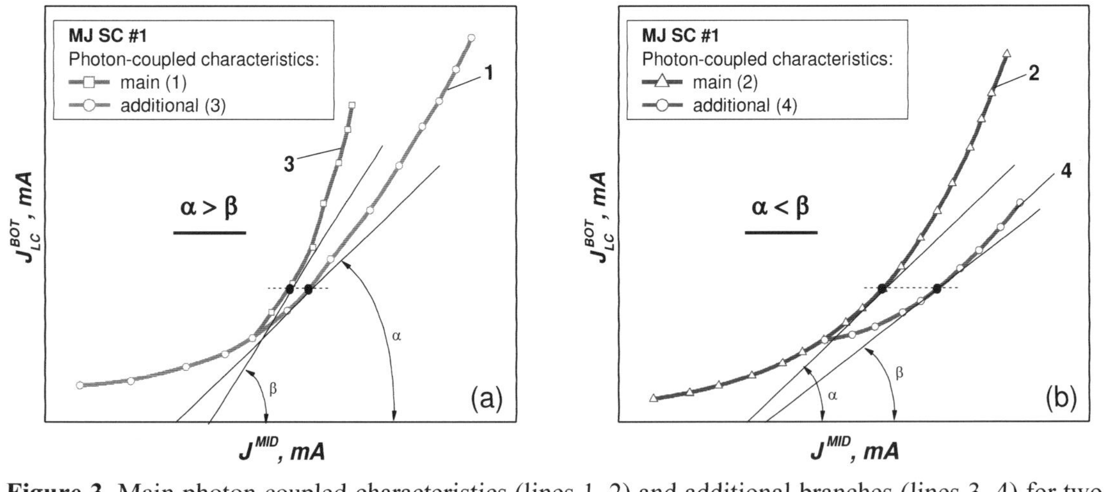

Figure 3. Main photon-coupled characteristics (lines 1, 2) and additional branches (lines 3, 4) for two MJ SCs: (a) – #1 LM; (b) – #2 MM. Angles α, β – tangents to the main and additional branches, respectively.

Such behavior can be explained by the quality of the optical interaction in the pairs of middle-bottom subcells, which is defined by the efficiency of the luminescent coupling process.

In general case luminescent coupling efficiency γ is determined by the ratio of the current induced by reemission to that flowing through the emitting p-n junction. Physically, parameter γ helps to understand what the portion of charge carriers in one of the subcells will pass to the external circuit, and what part will recombine radiatively in order for luminescent photons to be absorbed in the layers of the adjacent narrow bandgap subcell giving contribution to the photocurrent. In a triple-junction structure, the photon-coupling efficiency (Fig.4) is characterized by equations (1) and (2) for GaInP-Ga(In)As and Ga(In)As-Ge pairs, respectively:

$$\gamma_{TOP-MID} = \frac{J_{LC}^{MID}}{J^{TOP} - J_{LC}^{MID}} \qquad (1)$$

$$\gamma_{MID-BOT} = \frac{J_{LC}^{BOT}}{J_{LC}^{MID} - J_{LC}^{BOT}} \qquad (2)$$

Depending on a MJ SC design and properties of semiconductor layers, $\gamma_{TOP-MID}$ and $\gamma_{MID-BOT}$ will be different. Thus, for LM structures the efficiency of luminescent coupling is greater (Fig. 4a) than for MM SC (Fig. 4b), especially for the middle-bottom pair of subcells.

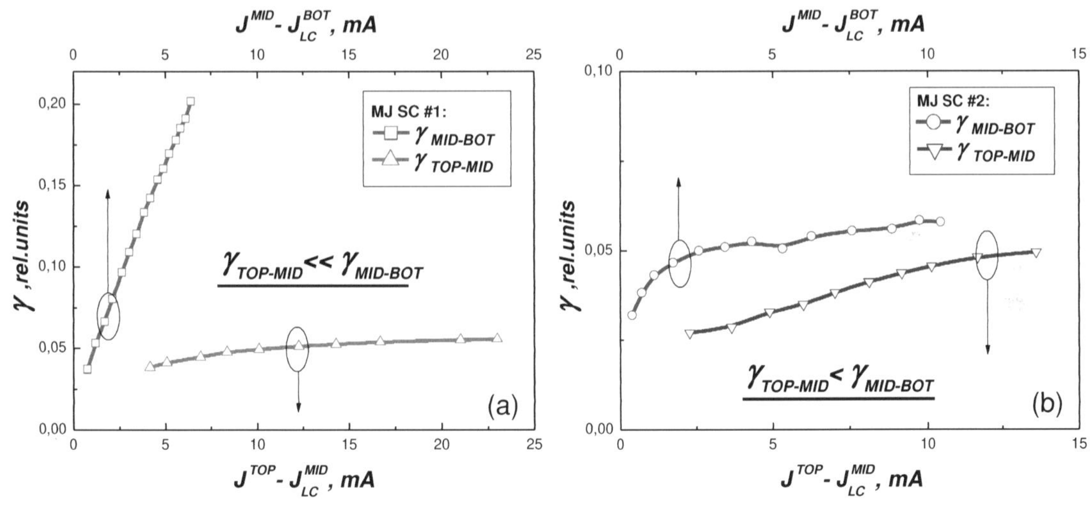

Figure 4. Dependencies of the luminescent coupling efficiency on the internal current through the emitting subcell for two MJ SCs: (a) – #1 LM: (b) – #2 MM. The $\gamma_{TOP-MID}$ and $\gamma_{MID-BOT}$ characteristics describe the optical coupling within the pairs of top-middle and of middle-bottom subcells, correspondingly.

4. Conclusion

In the work, the optical coupling in a MJ SC was investigated. A principle for forming the main photon-coupled characteristic and its additional branches is described and results for two specimens of MJ SC (based on lattice-matched GaInP-GaAs-Ge and metamorphic GaInP-GaInAs-Ge structures) are presented. It has been shown, that the luminescent coupling for the pair "middle-bottom" subcells is more pronounced in LM structure than in MM one, which is reflected in the faster growth of the values of J_{LC}^{BOT} for the corresponding additional branches of photon-coupled characteristics with increasing J^{MID}.

Acknowledgments
The work has been supported by the Russian Science Foundation (grant №17-79-30035).

References
[1] Meusel M., Baur C., Let'tay G., Bett A.W., Warta W., Fernandez E. 2003 Spectral response measurements of monolithic GaInP/Ga(In)As/Ge triple-junction solar cells: measurement artifacts and their explanation *Progress in Photovoltaics: Research and Applications* 11 499–514
[2] Lim S H, Li J J, Steenbergen E H and Zhang Y H 2011 Luminescence coupling effects on multi-junction solar cell external quantum efficiency measurements *Progress in PV: Research and Applications* 21 344
[3] Li J J, Lim S H, Allen C R, Ding D and Zhang Y H 2011 Combined effects of shunt and luminescence coupling on external quantum efficiency measurements of multi-junction solar cells *IEEE J. Photovoltaics* 1 225–230
[4] Shvarts M Z, Mintairov M A, Emelyanov V M, Evstropov V V, Lantratov V M, and Timoshina N Kh 2013 Method for direct measurements of luminescent coupling efficiency in concentrator MJ SCs *AIP Conference Proceedings* 1556 147
[5] Shvarts M Z, Emelyanov V M, Mintairov M A, Evstropov V V, and Timoshina N Kh 2015 Temperature influence on luminescent coupling efficiency in concentrator MJ SCs *AIP Conference Proceedings* 1679 120003
[6] Shvarts M Z, Emelyanov V M, Evstropov V V, Mintairov M A, Filimonov E D, and Kozhukhovskaia S A 2016 Overcoming the luminescent coupling effect in experimental search for the actual quantum efficiency values in multi-junction solar cells *AIP Conference Proceedings* 1766 060005

Control of ferroelectrics polarization for increasing of alternative energy device's efficiency coefficient

V I Zubtsov[1], **E V Zubtsova**[2], **V V Senterova**[3]

[1] Polotsk State University, 29, Blochin st., Novopolotsk, 211440, Belarus
[2] Republican Unitary Enterprise "Belaeronavigatsia", 19, Korotkevich st., Minsk, 220039, Belarus
[3] Belarusian State Academy of Aviation, 77, Uborevich st., Minsk, 220096, Belarus

E-mail: subcv@rambler.ru

Abstract: A plant, developed with the use of ferro-piezoelectric ceramics for generation of electric power, is studied in the article. The use of electrochemical generator in the plant makes it possible to increase the efficiency of electric power generation by means of polarization of ferro-piezoelectric ceramics control. Polarization control is proved by a mathematical model. Consumption of 1 joule electric power, using mechanical power, generates 3,5...5 joule of electric power output. Efficiency factor of the power plant is about 50...55 percent and depends on ceramics modification and electric circuit.

1. Introduction
The importance of finding balance between scientific and technical achievements and nature are felt badly today. Thus, manufacturing of e-mobiles, as a means of transport, is a possible solution to current serious ecological problems. The most practical of all electric drive means are e-motorcycles, e-scooters, e-bicycles and minivans, which belong to lightweight vehicles. A traditional electric drive comprises a power source (one or several storage batteries), a controller for motor control, an electric motor itself and a mechanism (absent in certain cases) for translation of motion to the movable unit. Storage batteries have a low energy content and provide a low run of an e-vehicle during one charging, and a sufficient weight and size, making the vehicle hard to place.
Problems are seemed to be easily solved by so-called primary electric power sources, generating energy directly from the fuel, e.g. oxygen or hydrogen. Despite a high efficiency factor (about 50 percent), fuel cells have very specific drawbacks: disability for energy recovery during braking, low specific output under great specific energy, etc. Thus the mass of energy cells for lightweight vehicles' capacities (and e-vehicles in general) is great. As a result, power units with driving motors for them are absolutely unacceptable. Today condensing energy storage systems, having a high specific output, are tried to be used together with fuel cells. But it is not efficient, as capacitor-based energy storage systems are characterized with a big size and mass, and they are highly explosive.

2. Construction and Mathematical Model of an Effective Electric Drive. Dependence of Mass Variation on Electric Drive Power
Taking into account the foregoing, an environment friendly unit of alternative technology with the use of electro-chemical generator (ECG), based on ferro-piezoelectric ceramics, is proposed. Such a power plant is characterized with both a high specific output and energy.
Theoretical and experimental research [1-4] prove, that a unit based on ferroelectrics, artificial ferro-piezoelectric ceramics is applied, can be used as the most efficient alternative to devices and technologies, used today for electric drive by foreign companies.
Experimental dependence [3], gained on small-sized full-cycle element (⌀ 2,3×1 mm) made of PZT (lead zirconatetitanate), proves the possibility of current I generation, equal to 4,5 mA, caused by a

fluctuating mechanical force F, equal to 15 H, Figure 1. The PZT film composition is solid solutions of $PbZrO_3$ and $PbTiO_3$. The electromechanical parameters (piezoelectric modulus, electromechanical coupling coefficient, dielectric permittivity, etc.) reach their maximum near certain composition of 52% $PbZrO_3$ and 48% $PbTiO_3$. Modification of this composition with oxides of various metals has led to the creation of various PZT mode materials (domestically produced ceramics PZT - 19, PZT - 21, PZT - 23, etc.). Figure 1 shows the dependence for PZT – 23. In order to increase the power density is necessary to consider mainly the modification of the solid solution, electrical circuit, mechanical strength and ferroelectric element configuration when the ratio of the warmth radiation surface to the element volume will be the maximum possible. Furthermore, select of electric-field vector direction and mechanical stress components are taken into account. Dimensions of an element's certain form depend on its mass, the mass depends on electric power. Recalculation for a PZT unit of a bigger mass goes well with calculations made, proving that the specific output and specific energy are enough for the run of a lightweight vehicle.

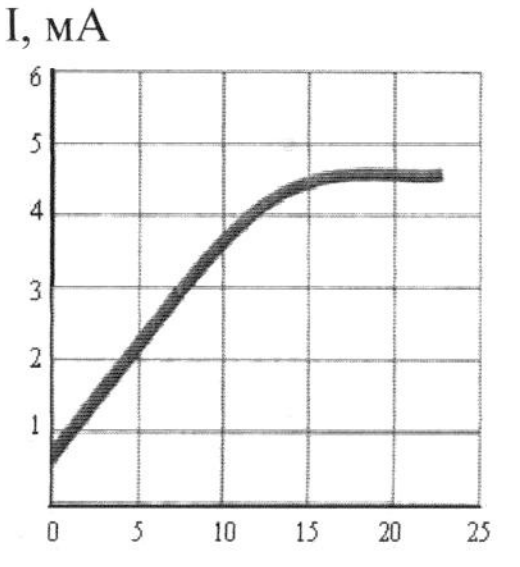

Figure 1. Dependence of the current on mechanical load of a PZT system element

Moreover, today there are multicomponent ferro-piezoelectric ceramics of a considerably higher electromechanical coupling factor K (a kind of efficiency factor of material's role in transformation of mechanical power into electric and vice versa) than a PZT system, expression (9). It also means that there are ways of generating bigger currents (bias currents in dielectrics) [4] describes experimental dependence of output voltage on mechanical load of ferro-piezoelectric ceramic elements of various modifications and closed circuits.

A system, allowing use less storage battery energy and improve the efficiency of transformation of mechanical power into electric, about by an order, due to electric characteristics of ferro-piezoelectric ceramics and physical-and-technical solutions (technologies), is required for efficient use of ferroelectrics. A created system, based on ferroelectrics, can transform up to 90 percent of gained energy into net energy [5-7]. In brief, the idea of ECG operation, the main joint of the plant, includes a release of "frozen" energy of an oxidant chemical reaction and a ferro-piezoelectric ceramic element, a multicomponent system of a solid solution. This energy is generated by mechanical energy, ferroelectric second-kind transition, and interlayer and dipole polarization [5].

Structurally, ECG is a ferro-piezoelectric ceramic element of a certain size and form with metal contacts and attached leads for its engagement into electric circuit. Efficiency rate of the plant, subject to the use of mechanical energy, makes 50…55 percent.

High efficiency of generation electric power in the plant (efficiency of the polarization of ferroelectric increase) is possible due to modification of ferro-piezoelectric ceramics, electric circuit, conditions of mechanic load, and, it was aforesaid, interlayer and dipole polarization. The plant comprises also a power supply (a storage battery for e-vehicle), a device for mechanical energy generation and electromechanical transducer Figure 2. Energy consumption of 1 joule, using mechanical energy, makes it possible to get the output of 3,5…5 joule of electric energy. Used mechanical energy is generated by a device of a simple construction.

Possibility to increase the efficiency of polarization of ferroelectrics' control is explained further on Figure 3 is a simplified mathematical model (electromechanical analog) of the proposed plant [4]. Direction of the electric field vector *E*, exiting a ferro-piezoelectric ceramic element, and that of the mechanical stress component *T* (mechanical power *F*) are the same. Two ferro-piezoelectric ceramic elements are required: the lower is excited by electric means and forces the upper one to fluctuate; the

IOP Conf. Series: Journal of Physics: Conf. Series **1038** (2018) 012113 doi:10.1088/1742-6596/1038/1/012113

upper one is excited by a running component of the mechanical stress. These processes are described by the equation of piezoelectric effect.

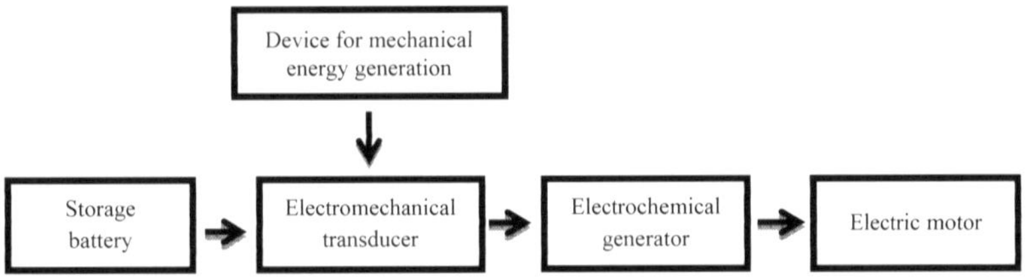

Figure 2. Functional diagram of a power plant.

$$P^E = d \cdot T + \varepsilon^T \cdot E,$$
$$S^T = d_t \cdot E + s^E \cdot T \quad (1)$$

Equations (1) for this case are:

$$P_i^E = d_{ij} T_j,$$
$$S_j^T = d_{ij} \cdot E_i, \quad (2)$$

P_i and S_j are tensor components of polarization and deformation, d–ferro-piezoelectric module. Index t goes for transposition. Indexes ij have following values: i = 1÷ 3; j = 1÷ 6

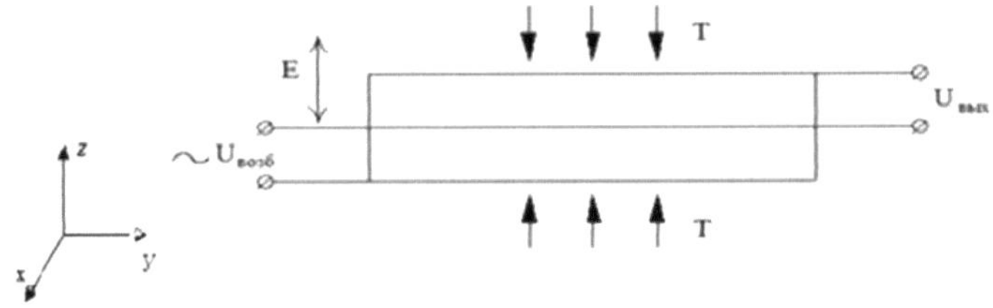

Figure 3. Flow diagram of the plant's electromechanical analog

Upper indexes E and T go for constancy of electric field voltage and mechanical load, respectively.
First equation (2) goes for the run of ferroelectric piezoelectric ceramic element, excited by the strain. The second equation is true to the run of ferroelectric piezoelectric ceramic element, excited by the electric field.
Subject to the known matrix of piezoelectric modules, e.g. for a PZT material [8]

$$d_{ij} = \begin{Vmatrix} 0 & 0 & d_{13} \\ 0 & 0 & d_{13} \\ 0 & 0 & d_{33} \\ 0 & d_{15} & 0 \\ d_{15} & 0 & 0 \\ 0 & 0 & 0 \end{Vmatrix} \quad (3)$$

the equation for ferro-piezoelectric ceramic element, excited by the electric field, has a tensor form after transformation:

$$S_1 = d_{31} \cdot E_1$$
$$S_2 = d_{32} \cdot E_3$$
$$S_3 = d_{33} \cdot E_3$$
$$S_4 = d_{24} \cdot E_2 \quad (4)$$
$$S_5 = d_{15} \cdot E_1$$
$$S_6 = 0$$

Equation for ferro-piezoelectric ceramic element, excited by the mechanical stress, after transformation is following:

$$P_1 = d_{15} \cdot T_5$$
$$P_2 = d_{15} \cdot T_4 \quad (5)$$
$$P_3 = d_{31} \cdot T_1 + d_{31} \cdot T_2 + d_{33} \cdot T_3$$

Analysis of the equations (3), (4) and (5) makes it clear, that excitement of deformation of one ferro-piezoelectric ceramic element along the axes Z is required, i.e. S_3, excited only by the electric field vector E_3 (4) for S_3, as far a correspondent mechanical stress component is concerned, e.g. T_3true to the axes Z. The other ferro-piezoelectric ceramic element, fluctuating in the same direction under the influence of the strain component T_3 and moving in the same direction as the electric field E_3, is characterized with the existence of a polarization vector P_3, (5) for P_3.
Active mechanical stress components T1 and T2 also cause polarization in ferro-piezoelectric ceramic element and, consequently, charges. However, these charges are not effective because of the static mode of this element's work in T_1andT_2directions. Dynamic mode of ferro-piezoelectric ceramic element's work is achieved in the direction of a mechanical stress component T_3.
Control for polarization of ferroelectrics is achieved due to the proper choice of vector of electric field's mechanical stress E, which coincides with the strain component T direction.
Moreover, the equations of transformation for the plant of this construction and ferro-piezoelectric ceramics under active mechanical stress and closed output electrodes and$E$=0 is following:

$$S^E = s^E T \quad (6)$$
$$D^E = P^E = dT$$

S^E goes for deformation under field strength of, E=0; s^E is compliance under E=0;D goes for displacement when E=0; P means polarization when E=0.
The field with mechanical stress equal to E is used in case of open electrodes and the lack of strain (T=0), when the element can be easily deformed. The equation is following:

$$S^T = dE \quad (7)$$
$$D^T = \varepsilon_a^T \mathrm{E}$$

S^T means deformation under T=0, different from S^E, D^T also differs from D^E.
If then decided that the electrodes are open and the field strain E and mechanical stress T influence simultaneously, and displacement D=0, then an interdependence between E and T can be found:

$$T = -\frac{\varepsilon_a^T}{d} E, \quad or$$

$$S^D = \left(S^E - \frac{d^2}{\varepsilon_a^T}\right)T = s^D T$$

Where *D* is less than S^E. It is clear that electric field E can be applied for changing elastic compliance s of ferroelectrics, i.e. for control of its rigidity (compressibility).
If *E*and *T*influences imultaneously, and deformation *S*=0, i.e. the element is pressed and does not deform, then an interdependence between mechanical stress *T* and electric field strain *E* is observed:

$$T = -\frac{d}{s^E}E$$

displacement *D* is found by the expression [8]:

$$D^T = \left(\varepsilon_a^T - \frac{d^2}{s^E}\right)E = \varepsilon_a^T \quad (8)$$

Absolute dielectric constants of the pressed element $\varepsilon_a^S$is obviously less than ε_a^T(7) and they are described in equation (9), *K* is an electromechanical coupling factor:

$$\varepsilon_a^S = \varepsilon_a^T\left(1 - \frac{d^2}{\varepsilon_a^T s^E}\right) = \varepsilon_a^T(1 - K^2) \quad (9)$$

s^E goes for electric field strain compliance *E*=0
$\varepsilon_a^T$is an absolute dielectric constant under mechanical stress *T*=0.
Thus, control of compressibility and electric elasticity of ferro-piezoelectric ceramic elements in the plant allows to increase sensitivity of the output electric voltage of these elements to the acting mechanical load [4,9]. The function of ECG transformation in the plant is following:

$$U_{вых}=K_H \frac{d_{ij}\cdot F}{C_{эхг}+C_H} \quad (10)$$

Uвых means an output electric voltage of ECG, *F* is acting mechanical force, *Сэхг* goes electro capacity of ECG, *Сн* is electrical load capacity (load electrical devices), *Кн*– an electrical voltage increase rate due to control of polarization of ferroelectrics, *dij*–piezoelectric module, induced polarization per unit of mechanical stress.
Equation (9) proves that electromechanical coupling factor K depends on relations of dielectric constants of pressed and free ferroelectrics, i.e. change of dielectric capacitivity under mechanical load. In case of *K* = 0,5 (an average value) the ratio is equal to 0,75. Which, in its turn, is highly important for output electrical voltage (power density) of the plant (10), moreover, K of modern ferro-piezoelectric ceramics is 0,6…0,7 (ε_a^Tdiffers greatly from ε_a^T, as dielectric constant and electric capacity are in direct proportion). Tests proved that 2 times increase (decrease) of a load electric power (Pн) causes 2 $\sqrt{2}$ times, 3 times - $3\sqrt{3}$ etc., increase (decrease) of ECG mass (Mecg). All the changes follow the principles of geometric progression. The mass of mechanical transducer together with the mass of the mechanical energy generating device (Fig.2) is 10 – 15 percent more than Mecg and depends on Pн as well.
Figure 4 describes a Mecg - Pн growth-increase diagram. The foresaid and the diagram make it is possible to calculate the plant mass (without a storage battery and actuating device) taking into account Pн. In the diagram one may find a point where Pн = 0,182 kilowatt and Mecg = 0,826 kg, around which Pн and Mec are changing almost in direct proportion. Mecg grows slower below this point and faster above it.

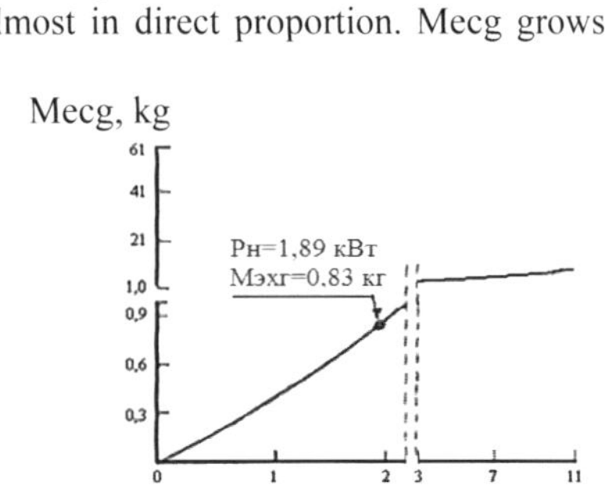

Figure 4 Mecg - Pн growth-increase diagram

As foresaid, 1 joule energy consumption from the storage battery gives an electric power output of 3,5 .... 5 joule under mechanical stress.

As 4 joule of energy is generated per 1 joule of storage battery energy, required for the run of the plant, 4 times energy increase in comparison with the energy, lost by the storage battery, is generated, not taking into account the loss of energy for heat and partial charge of the storage battery. It means that an e-vehicle run charged from one battery is also 4 times longer (an average value) in comparison with the traditional electric drive. The increase of the plant's mass is insignificant.

Thus, an alternative electric drive has certain advantages to a traditional one. But the use of the plant with P_H more than 10 kilowatt is problematic, and more than 50 kilowatt is unreasonable at this stage of experiment, as it cannot compete with a traditional electric drive. The reason is that the mass of the proposed plant increases faster than its power and requires specially developed ferro-piezoelectric ceramic materials of advanced features. Providing a certain level of ferroelectric devices leakage current is required for working in statics and with step electrical voltage change. For example, when objects are precisely moved with the help of actuators having a static leakage current not more than 5 μA. Ferroelectric high electrical resistance reduces the leakage current up to μA units or less. The leakage current can slightly increase, due to improper operation of devices (usually - it's unacceptable temperatures, humidity and electrical voltage), as it causes the activation of electrode material migration into ceramics. In practice, maintaining dynamic conditions (as in the proposed setup) and proper operation it can be neglected. Especially at low electrical voltage and low mechanical stress.

Certain obstacles are also caused by the importance of uniform distribution of great mechanical loads on fragile ferro-piezoelectric ceramic elements [10] and the initial polarization of bigger elements for generating of greater powers. The use of modern methods helps to overcome the obstacles easily.

The plant proposed can also be used as an environment friendly source of electric power, reserved and autonomous (charging from a battery or a power unit). It is important to highlight, that the plant mass is of no importance, as in case of an electric drive, as emergency sources of energy are permanent. Moreover, the plant can be used for charging of small-sized UAVs, wheelchairs, office and country buildings, etc.

3. Conclusion

The advantages of the proposed way of generating electric power to those of solar- or wind energy are: it does not depend on climate and time of the day. Alas, despite all the advantages, the technology has not yet been properly developed.

So, the control of ferroelectrics' polarization is found by means of:

- modification of ferroelectrics and electric closed circuit;
- mechanic load (design features of the plant);
- interlayer and ion polarization of ferroelectrics in the band of about 1 - 1,5 (10^3 - 10^5) Hz.

References

[1] Zubtsov V I, Bosylev V V, Zubtsova E V 2001 Design and Tests of Piezotransducers for Control of Stress in Deformable Medium *Devices and Systems. Management, Control, Diagnostics* No 2 pp 45-50

[2] Zubtsov V I, Zubtsova E V. 2012 Piezotransducers for Control of Stress in Deformable Medium *Instruments and Systems. Management, Control, Diagnostics* No 6 pp 51-57

[3] Trofimov A I 1979 *Piezoelectric Transducers of Static Loads* (Moscow: Mechanical Engineering) p 128

[4] Zubtsov V I 2013 *Control of the Physical Properties of Materials Using Piezoelectrics* (Saarbrücken: LAP LAMBERT Academic Publishing) p 262

[5] Zubtsov V I 2003 Increase in the Range of Measurements of Static Mechanical Stresses in Controlled Environments *Instruments and systems. Management, Control, Diagnostics* No 11 pp 32-35

[6] Zubtsov V I, Zubtsova E V 2013 Increase of the Specific Power and Specific Energy of a Device Based on Ferro-piezoelectric Ceramics for an Electric drive *Materials of International Conference on Physics and Astronomy* (Ioffe PTI Saint-Petersburg) pp 243-244

[7] Zubtsov V I, Zubtsova E V 2015 Physical-and-Technical Aspects of Ferroelectrics Application for the Increase of Specific Capacity and Specific Energy of Alternative Energy Devices *Materials of International Conference on Physics and Astronomy* (Ioffe PTI Saint-Petersburg) pp 267-268

[8] Okadzaki K 1979 Handbook on Electric Materials (Moscow Mechanical Engineering) p 432

[9] Zubtsov V, Baranov V, Emelyanov V. 2002 Piezocrystal Transformers for Measuring Static Mechanical Stresses in Situ *Proceedings of SPIE International Symposium* (Cannes (France) pp 815-820
[10] Zubtsov V I 2005 Nondestructive Evaluation of Elastic Materials' Durability *Devices and Systems. Management, Control, Diagnostics* No 1 pp 28-32

Thermoelectric element on the basis of the sandwich metal-ferroelectric-metal structure

Yu O Perkov and V I Ivanov

Department of Physics and Theoretical Mechanics, Far Eastern State Transport University, Khabarovsk, Russia

E-mail: tmeh@festu.khv.ru

Abstract. It was described the thermoelectric effect in thin sandwich metal-ferroelectric-metal system. The effect was observed in doped lithium niobate crystals with two electrodes of different metals. The effect is observed only in doped lithium niobate crystals and has a maximum for concentrations of impurities of iron around 0.3 % weight. It was proposed the electrets model of the investigated phenomena resulting from field contact potential difference on the borders of section of metal-ferroelectric material.

1. Introduction

There are many photo- and thermoelectric phenomena, which can be used to generate and transform energy [1,2]. However the interest of researchers aims to find new and more effective mechanisms and ways to generate electrical energy. There are a series of nonequilibrium phenomena in ferroelectrics, which do not have a definite theoretical description. In paper [3] the thermovoltaic effect consisting in initiation of the electric voltage between the opposite faces of the semiconductive sample of samarium sulphide (SmS) on heating to 400–500K is described. The effect has appeared in the presence of the donor impurity concentration gradient directed towards these faces.

In our work [4] the thermally-stimulated quasi-stationary current arising only in the highly-alloyed with iron crystal of lithium niobate with the evaporated electrodes from a pair of different metals was described. A sign of the thermo-EMF is determined by location of electrodes applied by vacuum evaporation on the opposite faces of crystal and does not depend on orientation of the crystallographic axes of the sample in reference to electrodes.

In this paper the results obtained in experimental studies of the thermoinduced current in the thin-layer metal-lithium niobate-metal (MLM) structure are presented. This effect is similar to photovoltaic effect, but it is characterized by thermal nature. It was proposed the electrets model of the phenomenon.

2. Experimental resultes

In contrast with the classic pyroelectric effect, the value of observed thermal response is proportional to the crystal temperature increase and does not depend on the rate of its change. It was experimentallymeased the coefficient characterizing the thermoinduced electro-moming force (EMF) value in the mode of voltage U_n measurement:

$$P_{el} = (R_n S)^{-1} \Delta U_n / \Delta T, \text{(A/K·cm}^2\text{)}, \quad (1)$$

where S is the area of crystal surface covered with sputtered (evaporated) electrode; ΔU_n is the change of voltage across load impedance R_n when the sample temperature changes by ΔT; ΔT is the difference between the initial and final temperatures of sample [5].

The study of the effect was performed using the slow modulation of the crystal temperature under the conditions of thermostat and the thermo-EMF was recorded under the stationary conditions (relaxation was not observed for observation time $\tau \sim 10^5$ s). The measurements showed that the thermo-voltaic effect has approximately equivalent values in the Y- and Z-cut crystals. Because the sufficiently high rates of heating [about 0.1 K/s] were used in the installation, the pyroelectric current for the polar Z-cut crystals reached the considerable value. Therefore, the principal measurements were performed for the Y-cut lithium niobate crystals where the thermal response could be observed in pure form.

The metal electrodes were applied by vacuum evaporation with thickness varying from 0.1 to 1 μm. The samples with different pairs of electrodes (aluminum (Al) – chromium (Cr), indium (In) – chromium (Cr), aluminum (Al) – copper (Cu), silver (Ag) – aluminum (Al)) were studied. For studying the dependence of the thermo-EMF value on the concentration of impurity, the lithium niobate crystals with the same areas and thicknesses [S = 0.5 sm^2, d = 1 mm] were used. The iron concentration varied in the range of $1.3 \cdot 10^{-2}$ wt% to 0.6 wt%. For the impurity concentration of less than 0.25 wt%, the thermo-voltaic effect is not observed. Beginning with the impurity concentration of 0.25 wt%, the coefficient P_{el} rises sharply and reaches maximum (P_{el} = $5 \cdot 10^{-12}$ A/cm^{2}K) at concentration of 0.3 ÷ 0.4 wt% of Fe.

3. Thermoelectric model of the thermovoltaic effect in the MLM structure

The known contact phenomena in the sandwiched structures with ferroelectrics do not explain the observed quasi-stationary currents [6]. The presence of initial voltage U_n^0 across the sample allows us to suggest the non-equilibrium nature of the thermally stimulated current.

Consider the electrets model of the MLM structure. In this model the thermoinduced changing of the conductivity of the crystal leads to change in the stationary current on the load impedance:

$$P_{el} = E_0 L \rho' (R_n S + \rho L)^{-2}, \quad (2)$$

where E_0 is the electrets EMF; L is the thickness of crystal; ρ is the specific resistivity of the crystal; $\rho' = (\partial\rho / \partial T)$.

The temperature dependence of resistivity of lithium niobate crystals (with a concentration of more than 0.3 wt.% Fe) is describe by Mott law [7-9]:

$$\rho_k = \rho_0 \exp(T_0^{0,25} T^{-0,25}), \quad (3)$$

Where ρ_0 and T_0 are empirical constants that depend on the concentration of Fe .
Using (2-3) one can get:

$$P_{el} = E_0 L \rho_k T_0^{0,25} T^{-1,25} \left[R_n S + \rho_0 L \exp(T_0^{0,25} T^{-0,25})\right]^{-2}. \quad (4)$$

The comparison of analytical dependencies with experimental ones allows check the adequacy of the proposed model.

When increasing the heating range, it was found that P_{el} depends nonlinearly on the sample temperature (Fig. 1a).

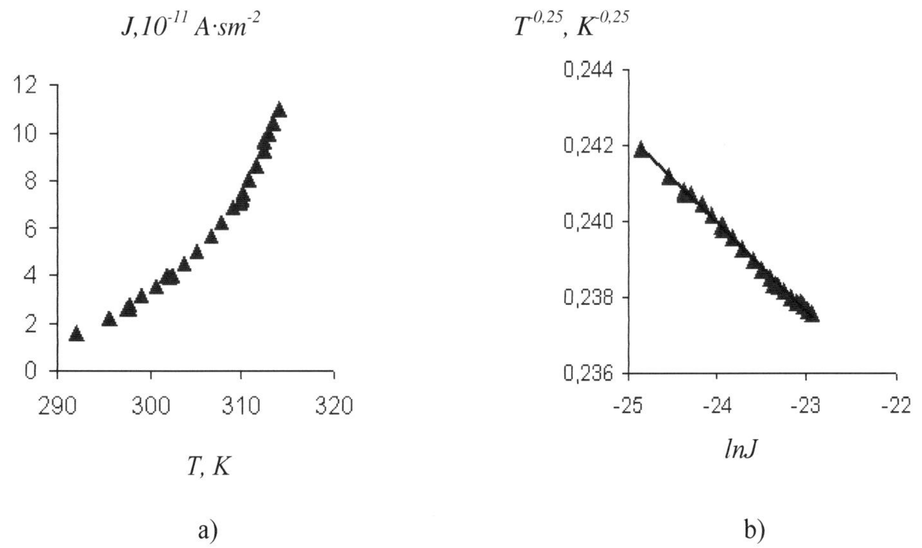

Figure 1. The thermally induced current (normalized to the electrode surface area) versus the crystal temperature ($LiNbO_3$:Fe – 0.43 wt%, 10x5x1mm^3, Y – cut; Al–Cr electrodes) – (a) and its linearized approximation according to formula (5) – (b)

This dependence can be approximated by linearized function is obtained from the formula (4):

$$\ln J = \ln\left(E_0 L_e \rho_k (L_0 \rho_0)^{-1}\right) - 0{,}25 T_0^{\,0{,}25} T^{-0{,}25}. \quad (5)$$

This linearized function is shown on the figure 1b.

When investigating the dependence of the coefficient P_{el} value on the sample geometry, it was identified that P_{el} value increases sharply when the crystal thickness decreases (Fig.2a).

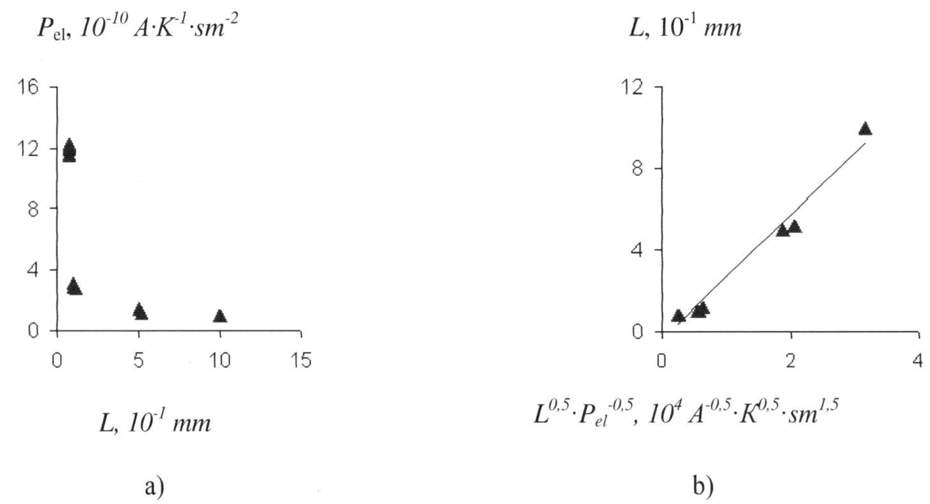

Figure 2. The coefficient P_{el} versus the crystal thickness (S=5 mm^2; 0.3 wt% of Fe; Al–Cr electrodes) – (a) and its linearized approximation according to formula (6) – (b)

By the same way last dependence can be approximated by linearized function is obtained from the formula (4):

$$L = L^{0,5} P_{el}^{-0,5} (E_0 \rho_k ')^{0,5} \rho_k^{-1} - R_n S \rho_k^{-1}, . \qquad (6)$$

This linearized function is shown on the figure 2b.
Figure 3a shows the coefficient P_{el} versus the area of crystal electrode surface. The linearized function is obtained from the formula (4):

$$S = P_{el}^{-0,5} (E_0 \rho ' L)^{0,5} R_n^{-1} - \rho_k L R_n^{-1}, . \qquad (7)$$

This linearized function is shown on the figure 3b.

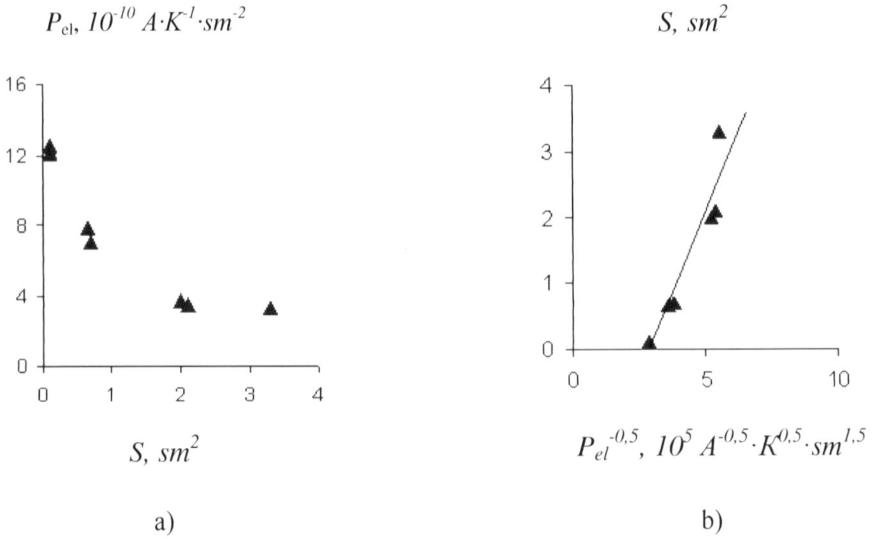

Figure 3. The coefficient P_{el} versus the crystal crystal electrode surfaces (L=1 *mm*; 0.3 wt% of *Fe*; Al–Cr electrodes) – (a) and its linearized approximation according to formula (7) – (b)

We can see that all experimental dependences are in accordance with the proposed model. Using (2-5) and the figures 1-3 one can define the value of specific resistivity of the crystal ($10^9 \div 10^{10}$ Ωsm) and electrets EMF ($E_0 \approx 10^{-2}$ V) [10].

4. Conclusions

In the thin-layer metal-ferroelectric-metal system the thermo-voltaic response exists which increases essentially with decreasing the crystal thickness. A sharp increase in the thermally induced current for small thicknesses of crystal (less than 100 μm) points to the crucial role of the near-contact zones with thicknesses of less than 100 μm.

In this paper the electrets model of the studied phenomenon caused by the field of contact potential difference at the metal-ferroelectric interface was proposed. In case of the same electrode materials, the contact potential differences are equal and oppositely directed. Therefore, the resulting EMF is equal to zero. In the electrets model the thermoinduced changing of the conductivity of the crystal leads to change in the stationary current on the load impedance. All experimental dates are in accordance with the proposed model [11].

The results obtained can be used for investigating the properties of the sandwiched ferroelectric structures as well as for developing new types of the thermo-electrical energy converters.

IOP Conf. Series: Journal of Physics: Conf. Series **1038** (2018) 012114 doi:10.1088/1742-6596/1038/1/012114

References

[1] Serway R. A. 1990 *Physics for Scientists & Engineers* (3rd ed.) Saunders p 1150
[2] Glass A, Von der Linde D and Nergran T 1974 *Appl. Phys.* **25** p 233
[3] Kaminskii V 2013 *Technical Physics Letters* **39** p 673
[4] Itskovskii M, Shchedrina L and Kladkevich M 1980 *Ferroelectrics* **29** p 167
[5] Ivanov V, Karpets Yu, Marchenkov N and Zdorovtsev G 2005 *Proc. SPIE* **5851** p 415
[6] Zdorovtsev G, Ivanov V, Karpets Yu and Klimentyev S 2007 *Proceedings of Tomsk Polytechnic University* **311** p 102
[7] Pichugin V F, Bulycheva A A, Yakovlev V Yu and Kim I W 2005 *Physica status solidi (c)* **1** pp 208-211.
[8] Barkan I.B., Baskin E.M., Entin M.V. Mechanism of Conductivity of Fe-Doped $LiNbO_3$ Crystal // phys. stat. sol. (a) 59, K97 (1980).- P. 12-25.
[9] Barkan, I.B. Mechanism of Conductivity of a Fe-Doped $LiNbO_3$ crystal / I.B. Barkan, E.M. Baskin, M.V. Entin // Phys. Stat. Sol. 1980.- P.97-102.
[10] Kim I W, Park B C, Jin B M, et al. 1995 *Material Letters* **24** p 157-160.
[11] Ivanov V I, Karpets Y M and Perkov Y O 2017 *Proc. SPIE* **10176** p 1017625

IOP Conf. Series: Journal of Physics: Conf. Series **1038** (2018) 012115 doi:10.1088/1742-6596/1038/1/012115

Investigation of the electrocaloric effect in strontium barium niobate (SBN) ceramics with rare-earth dopants

A V Es'kov[1], A S Anokhin[1], M T Bui[1], O V Pakhomov[1], A A Semenov[2], P Yu Belyavskiy[2], A B Ustinov[2]

[1]International Laboratory: "Materials and Structures for Electro- and Magnetocaloric Energy Conversion", ITMO University, St. Petersburg 197101, Russia
[2]Department of Physical Electronics and Technology, St. Petersburg Electrotechnical University, St. Petersburg 197376, Russia

E-mail: asanokhin@corp.ifmo.ru

Abstract. In ceramics, based on a solid solution of strontium barium niobate (0.7SBN) doped with cerium and gadolinium in various concentrations, an investigation was made of dielectric constant, electrical polarization, and the electrocaloric effect. For 0.7SBN ceramics, it was found that the addition of cerium (0.2%) leads to a decrease in the electrocaloric effect compared to pure 0.7SBN, and the addition of gadolinium (<1%) to an increase in the electrocaloric effect.

In the recent decades, due to the decreasing size and increasing power of electronic components the problem of cooling and temperature control is one of the hottest in the modern electronics and electrical engineering. The effective heat transformation could be implemented in the solid state structures using electrocaloric effect [1, 2]. Many laboratories in the world are engaged in developing new cooling principles based on the use of the electrocaloric effect. One of the most important tasks is still the search and development of new highly effective electrocaloric materials. Now researchers are very interested in lead-free materials based on solid solutions of barium strontium niobate and titanate with various dopants. According to previously published information, the addition of cerium (Ce) in monocrystals of strontium barium niobate leads to an increase in the electrocaloric effect [3], and the addition of gadolinium (Gd) to an increase in the pyroelectric effect [4]. Since the pyroelectric effect is proportional to the electrocaloric effect, it is logical to assume that materials possessing a large pyroelectric effect will have a greater electrocaloric effect. But in previous works, comparative data on the magnitude of the electrocaloric effect in pure and doped strontium barium niobate ceramics and single crystals were not given. In this paper, we investigate and compare the dielectric and electrocaloric properties of ceramics based on strontium barium niobate (SBN), pure and doped with various rare-earth metals (Ce, Gd) at different concentrations.

For research, we chose a solid solution SBN with 70 percent of strontium. The temperature of the phase transition of this composition is close to room temperature, which is convenient for measurements and further use in solid-state cooling elements.Sr0.7Ba0.3Nb2O6 ceramics were fabricated by the solid-state reaction method. The raw materials of BaCO3 and Nb2O5, SrCO3 and Nb205 were grinded by planetary micro mill Fritsch "PULVERISETTE 7 premium line" for 20 min in distillated water using zirconia balls of 3mm diameter as milling medium. The resultant mixtures were calcined at 1200 °C for 2 hours in air and the resultant BaNb2O6 and SrNb2O6 were milled again.

Then, the resulting materials were mixed in the required stoichiometric proportions and the necessary amount of rare earth oxides (CeO2 and Gd2O3) was added therein. After grinding granulation using 4 wt% aqueous solution of methylcellulose binder-lubricant, the powders were dry-pressed into a disc with a diameter of 10 mm and a thickness of 1 mm under a pressure of 6 MPa. The green samples were sintered at 1350 °C in air atmosphere and then cooled in furnace. The sintered samples were double-side polished to thicknesses 0.5 mm gradually. The obtained ceramic samples are shown in table 1.

Table 1. Designation of studied materials.

Designation	Composition
0.7SBN	$Sr_{0.7}Ba_{0.3}Nb_2O_6$
0.7SBN – 0.2%Ce	~ $Sr_{0.698}Ce_{0.002}Ba_{0.3}Nb_2O_6$
0.7SBN – 0.2%Gd	~ $Sr_{0.698}Gd_{0.002}Ba_{0.3}Nb_2O_6$
0.7SBN – 1%Gd	~ $Sr_{0.69}Gd_{0.01}Ba_{0.3}Nb_2O_6$
0.7SBN – 5%Gd	~ $Sr_{0.65}Gd_{0.05}Ba_{0.3}Nb_2O_6$

First, we investigated the temperature dependence of the dielectric constant in our materials. To study the dielectric constant, we used Agilent E4980A Precision LCR Meter. Conductivity measurements of the samples showed a resistance value greater than 0.5 teraohm. Since the SBN samples are aging, in order to avoid the effects of aging when measuring the dielectric constant in the small-signal mode, the measurements were carried out in a mode of temperature decrease after heating the sample above the Curie point [5]. In measurements of electrical polarization in the mode with the using full voltage amplitude, the aging effects of the samples do not affect the values of the electric polarization. Figure 1 presents the temperature dependence of dielectric constant for strontium barium niobate ceramics with rare-earth dopants.

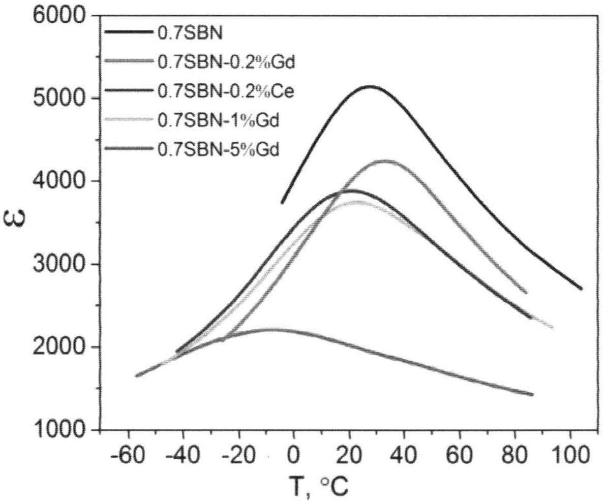

Figure 1. The temperature dependence of dielectric constant for strontium barium niobate ceramics with rare-earth dopants.

It can be seen from the curves that additions of rare earth dopants lead to a decrease in the permittivity as compared to the pure composition 0.7SBN ceramics.

The ferroelectric hysteresis loops were measured at 10 Hz within -50 ~ 80 °C using a Sawyer–Tower circuit based on the thermostatic bath Julabo FP40-MA Refrigerated/Heating Circulator, arbitrary waveform generator Agilent 33522A, a high-voltage amplifier Trek 609E-6, nanovoltmeter Agilent 34420A for temperature control and Agilent MSOX2004A Mixed Signal Oscilloscope for observing ferroelectric hysteresis loops. The temperature step at which each hysteresis loop was taken was 0.5 degrees. Figure 2 presents ferroelectric hysteresis loops at some different temperatures for strontium barium niobate ceramics with rare-earth dopants.

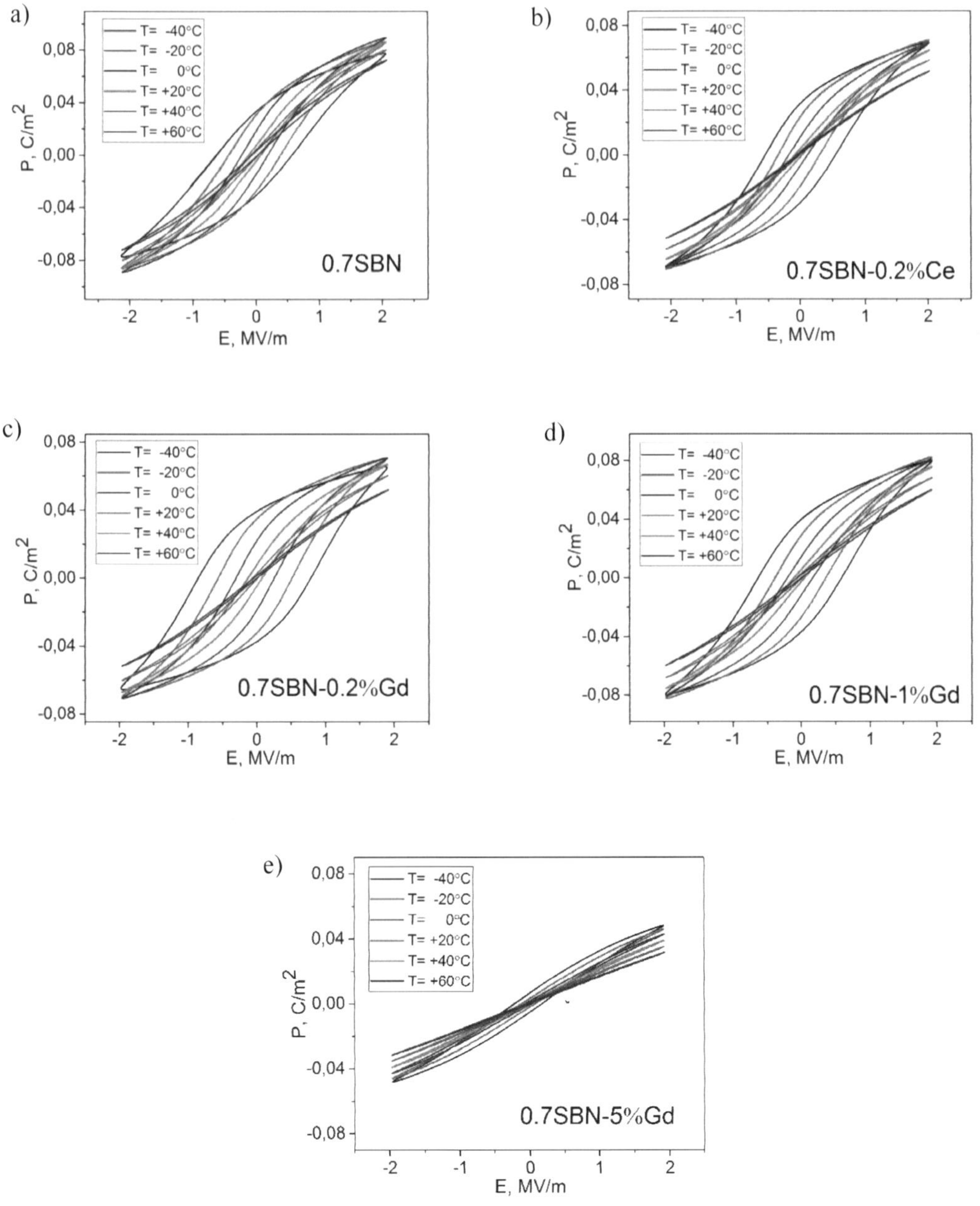

Figure 2. Ferroelectric hysteresis loops at some different temperatures for ceramic samples based on 0.7SBN (a), 0.7SBN-0.2%Ce (b), 0.7SBN-0.2%Gd (c), 0.7SBN-1%Gd (d) and 0.7SBN-5%Gd (e).

From these dependences, well noticeable decrease in the area of the ferroelectric hysteresis loop at the transition from the ferroelectric phase to the paraelectric phase. It should be noted that there is a large difference in the hysteresis loops in the ferroelectric phase and in the paraphrase for 0.7SBN – 0.2%Gd ceramic. Figure 3 presents the dependence of electric polarization versus temperature at different electric field for strontium barium niobate ceramics with rare-earth dopants.

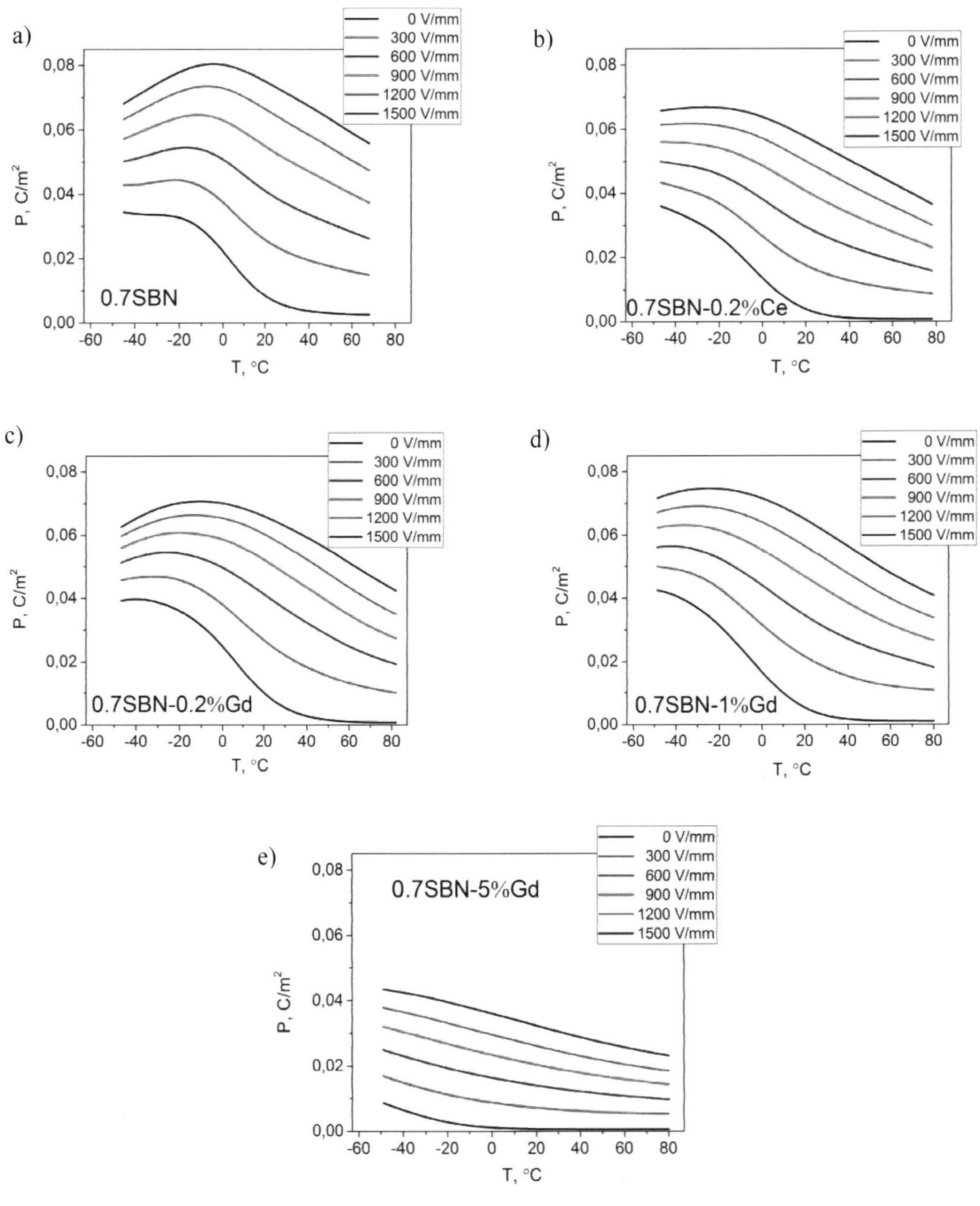

Figure 3. The dependence of electric polarization versus temperature at different electric field for ceramic samples based on 0.7SBN (a), 0.7SBN-0.2%Ce (b), 0.7SBN-0.2%Gd (c), 0.7SBN-1%Gd (d) and 0.7SBN-5%Gd (e).

IOP Conf. Series: Journal of Physics: Conf. Series **1038** (2018) 012115 doi:10.1088/1742-6596/1038/1/012115

Based on the thermodynamic Maxwell relation and the obtained temperature dependences of the electric polarization, the electrocaloric adiabatic ΔT was calculated using the following formula:

$$\Delta T = -\int_0^{E_1} \frac{T}{\rho C_E} \left(\frac{\partial P(T,E)}{\partial T}\right)_E dE \qquad (1)$$

where ρ is the measured density and CE is the specific heat under constant electrical fields. The calculation used the density value of 5328 kg/m3 and the value of specific heat 440 J/kg*K. The maximum value of the electric field E1 was 1500 V/mm. Figure 4 presents the temperature dependence of the electrocaloric effect for strontium barium niobate ceramics with rare-earth dopants.

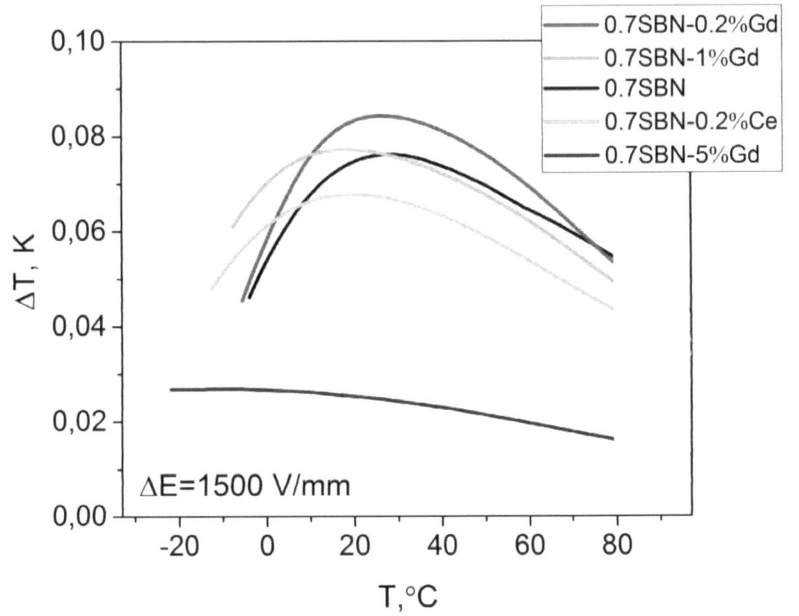

Figure 4. The temperature dependence of the electrocaloric effect.

From the analysis of the obtained dependences, it can be concluded that the addition of cerium in our ceramics leads to a decrease in the maximum magnitude of the electrocaloric effect as compared to pure 0.7SBN. The addition of gadolinium (<1%) leads to an increase in the maximum the magnitude of electrocaloric effect, which agrees well with the assumptions advanced in article [4]. The maximum magnitude of the electrocaloric effect among all our ceramics was obtained in the composition with 0.2% gadolinium supplement.

Acknowledgments
The research was supported by FASIE. The research was partly supported by Ministry of Education and Science of Russian Federation (Project "Goszadanie").

References
[1] Valant M 2012 *Prog.in Mater. Sci.* 57(6) 980-1009
[2] Ozbolt M, Kitanovski A, Tusek J, Poredos A 2014 *Int. J. of Refrigeration* 40174-188
[3] Le Goupil F, Axelsson A K, Valant M, Lukasiewicz T, Berenov A, Alford N M 2014 *Applied Physics Letters* 104(22) 222911
[4] Yao Y, Mak C L, Wong K H, Lu S, Xu Z 2009 *International Journal of Applied Ceramic Technology* 6(6) 671-678
[5] Van Damme N S, Sutherland A E, Jones L, Bridger K, Winzer S R 1991 *Journal of the American Ceramic Society* 74(8) 1785-1792

IOP Conf. Series: Journal of Physics: Conf. Series **1038** (2018) 012116 doi:10.1088/1742-6596/1038/1/012116

Features of spin-wave envelope solitons of the terahertz frequency range in thin hexaferrite films

M A Cherkasskii[1], A V Drozdovskii[1], A V Es'kov[2]

[1]Department of Physical Electronics and Technology, St. Petersburg Electrotechnical University, St. Petersburg, 197376 Russia
[2]International Laboratory: "Materials and Structures for Electro- and Magnetocaloric Energy Conversion", ITMO University, St. Petersburg 197101, Russia

E-mail: macherkasskii@hotmail.com

Abstract. The paper is devoted to the study of the formation and propagation features of terahertz bright envelope solitons of dipole-exchange spin waves in thin ferrite and hexaferrite films. The influence of the exchange interaction and the uniaxial anisotropy field on the soliton formation threshold is studied.

1. Introduction

In recent years, the increase of development terahertz and sub-terahertz devices is observed in the electronics. For that reason, the investigation of the terahertz physics effects that can serve as a basis for the new devices is needed. The ferrites with the hexagonal crystal structure, so called hexaferrites, demonstrate unique waveguide properties in the 0.1-1 THz frequency range. The device operation frequency can be increased due to usage of the hexaferrites in the relatively low magnetic bias field. This is possible because the hexaferrites have the high effective fields of uniaxial magnetic crystallographic anisotropy [1].

From the radiophysics point of view, the ferrites and hexaferrites are nonlinear dispersive waveguide media. In this type of media stability wavepackets, namely envelope solitons can exist. The solitons are studied in the ferromagnetic films [2, 3], optical fibers [3], Bose-Einstein condensate [4], etc. In spite of that, the solitons are well investigated in the mentioned media, the features of bright spin-wave envelope solitons in thin hexaferrite films are not revealed.

It is well known, that the spin wave (SW) in ferrites can exist due to dipole-dipole and exchange interactions. The relative contribution of the each type of the interactions are different in the different wavenumber ranges. It is worth noting a paper [5] that is devoted to a soliton investigation in hexaferrite films, but there the solitons were studied in such wavenumber range where only dipole-dipole interaction contribute to the wave process. Thus, the influence of the exchange interaction on the formation and propagation soliton processes is not well studied.

The aim of this paper is to investigate theoretically the impact of the anisotropic properties and exchange interaction of the hexaferrite thin films on terahertz bright envelope solitons. Note that the investigation was carried out in the terahertz and sub-terahertz frequency ranges that can be significant for the applications.

2. Theoretical Investigation

The investigation in this paper was carried out in several steps. First, a dispersion characteristic of SW propagating in the hexaferrite thin film was evaluated. Second, coefficients of the nonlinear Schrodinger equation (NLSE) that described the soliton process were calculated. In the end, influence of the exchange interaction and the anisotropy on the threshold of soliton formation was analyzed.

In the investigation the $BaFe_{12}O_{19}$ M-type (BaM) hexaferrite thin film that was magnetized to saturation with the magnetization M_0=4800 Gs and the uniaxial anisotropy field H_a=17000 Oe was studied [1]. The film thickness was L=5.6 μm. The bias field had value H_0=2000 Oe and was orientated in the plane. The wave propagating along the bias field was investigated.

The dispersion characteristic based on the dispersion equation for dipole-exchange SW, which was obtained in [6], was evaluated numerically. The evaluation was fulfilled for two sets of parameters. The first set was mentioned above; it is the hexaferrite thin film. The second corresponds to an yttrium iron garnet (YIG) film that had the same parameters exclude the saturation magnetization and the anisotropy. The YIG film has the saturation magnetization M_0=1750 Gs and the cubic anisotropy field H_a=70 Oe. The obtained dispersion characteristic is plotted in Figure 1.

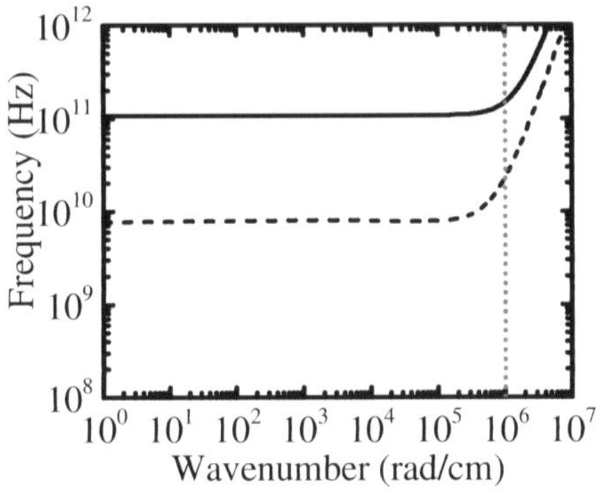

Figure 1. The dispersion characteristic of SW. The solid line is BaM, the dash line is YIG, the gray dotted line shows wavenumber

It is obvious from the figure that the frequency range of the propagating SW in the hexaferrite BaM film lies above in compare with the waves in the ferrite YIG film due to the anisotropy. The exchange interaction differently impacts on the dispersion curve of SW in different films, the slope of the BaM curve starts at greater wavenumber than the slope of the YIG curve.

Hereafter the wavenumber of the propagating SW was fixed at $k_0=10^6$ rad/cm. This value is showed with the gray dotted line in the Figure 1. At the selected wavenumber, the frequency of the propagating SW in the hexaferrite equals 0.15 THz; the frequency of the SW in the ferrite equals 23 GHz.

It is well known that the propagating wave processes is described with the NLSE in the nonlinear, dispersion and dissipative media [2]:

$$i\left(\frac{\partial u}{\partial t}+V_g\frac{\partial u}{\partial z}\right)+\frac{D}{2}\frac{\partial^2 u}{\partial z^2}-N|u|^2u=-i\Gamma u, \qquad (1)$$

where $V_g=\partial\omega/\partial k$ is the group velocity, $D=\partial^2\omega/\partial k^2$ is the dispersion coefficient, $N=\partial\omega/\partial|u|^2$ is the nonlinear coefficient, Γ is the damping decrement. For the mentioned sets of the parameters these coefficients are following for the BaM film: $V_g=8.3\cdot10^4$ cm/s, $D=8.3\cdot10^{-2}$ cm²/(rad·s), $N=-1.4\cdot10^{11}$ rad/s, for the YIG film $V_g=3\cdot10^4$ cm/s, $D=3\cdot10^{-2}$ cm²/(rad·s), $N=-1.7\cdot10^{10}$ rad/s. The damping decrement was evaluated based on the results of the paper [1].

The soliton processes was analyzed with a numerical modeling of NLSE. The modeling results for the both films concluded that the exchange interaction lead to a reducing of the influence of dispersion spreading effect on soliton processes. The compare of the modeling results for the films showed that the increasing of the anisotropy field results in an enhancement of the self-phase modulation effect.

Let us turn to description of the threshold of the bright spin-wave envelope solitons in the both films. The study was carried out with the formulae from the paper [7]. The formulae were modified taking into account the anisotropy and the exchange interaction. The investigation results are showed in the Figure 2. The plotted curves are described the relation between the threshold $|u_{th}|^2$ and an input pulse duration τ, which is needed for the formation of the solitons.

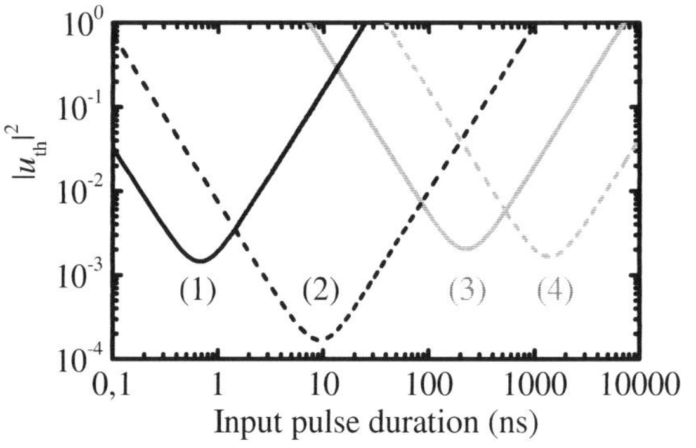

Figure 2. The bright soliton thresholds for several cases:
(1) is for the BaM film, (2) is for the YIG film, (3) is for the BaM film without the exchange interaction, (4) is for the YIG film without the exchange interaction.

One can infer two conclusions from the Figure 2. First, the increase of the anisotropy field leads to the increase of the threshold and the decrease the input pulse duration. Second, the exchange interaction significantly reduces the input pulse duration for which minimum threshold is observed both in the ferrite and hexaferrite films.

Let us give a comparison of the obtained results and results from work [5]. Note that the material parameters of the investigated hexaferrite and a hexaferrite from the work [5] are close. At the same time, carrier frequencies are distinguished drastically. It provides different physical background for soliton excitation because of in our work the carrier SW wavenumber is much higher than in work [5]. Such a high value corresponds to exchange spectrum of spin waves. Namely, in the work [5] the upper boundary of the soliton carrier frequency is 37.4 GHz and the duration of the single soliton at the generation threshold to be 2 ns. This frequency is much less than the soliton carrier frequency in our investigation that is 0.15 THz. As a result, we obtained twice-smaller duration of the single soliton. This was possible by taking into account the exchange interaction.

Acknowledgments
This work was supported in part by Russian Science Foundation (Grant# 14-12-01296-P).

References

[1] Nicholson D B 1990 Hewlett-Packard Journal 41 59
[2] Remoissenet M, Waves called solitons: Concepts and Experiments (Springer-Verlag, Berlin, 1999)
[3] Kivshar Y S, Agrawal G P, Optical solitons. From Fibers to Photonic Crystals (Academic Press, 2003)
[4] Kevrekidis P G, Frantzeskakis D J, Carretero-González R, Emergent Nonlinear Phenomena in Bose-Einstein Condensates (Springer-Verlag, Berlin, 2008)
[5] Kalinikos B, Slavin A 2001 Appl. Phys. Lett. 79 1576
[6] Belyakov S V, Kalinikos B A, Kozhus N V 1989 Electronic Technology 415 22

[7] Slavin A, Benner H 2003 Phys. Rev. B 67 174421

IOP Conf. Series: Journal of Physics: Conf. Series **1038** (2018) 012117 doi:10.1088/1742-6596/1038/1/012117

Polarization switching in single crystals and films of 2-methylbenzimidazole

F B Svinarev[1], E V Balashova[1], G A Pankova[2] and B B Krichevtsov[1]

[1] Ioffe Institute, 26 Politekhnicheskaya, St Petersburg 194021, Russia
[2] Institute of macromolecular compounds, 31 Bolshoy pr., St Petersburg 199004, Russia

svinarev@mail.ioffe.ru

Abstract. The polarization switching was studied in single crystals and thin films of 2-methylbenzimidazole (MBI) obtained by evaporation method from an MBI ethanol solution. Dielectric hysteresis loops were measured in the temperature interval 290-390 K and frequency range 20-1000 Hz for different amplitudes of the electric field. The Kolmogorov β-model with account of Mertz law is used for simulation of hysteresis loops. The temperature dependence of domain wall motion activation field E_a has been obtained. Mechanisms of polarization switching in single crystals and films of MBI are discussed. In MBI films, a comparison of polarization switching for in-plane and out-of-plane electric field orientation is presented.

1. Introduction

The existing interest to new organic ferroelectrics and film structures based on them is related to various potential applications of these materials for development of memory devices, sensors of different fields, bio-harmonized and print production technologies for electronic devices. Particular attention is paid to cheap materials with low values of coercive fields E_c and high values of spontaneous polarization P_s at room temperature. One such perspective material is 2-methylbenzimidazole (MBI) in which the ferroelectric properties were discovered in 2012 [1].

The molecule of MBI includes a benzene ring, an imidazole ring, and a methyl group. MBI is a molecular crystal with hydrogen bonds between molecules. The crystal structure belong to pseudo tetragonal symmetry $P4_2/n$ (Pn) with lattice parameters a = 13.95 Å, b = 7.192 Å [2]. MBI has a layered crystal structure with orientation of the layers perpendicular to the pseudo tetragonal $[001]_{tetra}$ axis (see figure 1). Each MBI molecule is linked with two neighboring molecules by hydrogen bonds N-H ··· N, which form chains along the $[110]_{tetra}$ axis in one layer and along the $[1\text{-}10]_{tetra}$ axis in the neighboring one. The ordering of protons in the hydrogen bonds leads to appearance of spontaneous polarization P_s ~ 5 μC/cm^2, which orientation can be controlled by an external electric field. The polarization consists of two components oriented along the hydrogen bond chains. Thus, polarization switching in MBI is possible in any direction normal to pseudo tetragonal axis. The coercive field in MBI is E_c ~ 3 V/μ at room temperature and decreases to E_c ~ 1 V/μ at T ~ 400 K [1]. The melting point of MBI is T_m ~ 450 K.

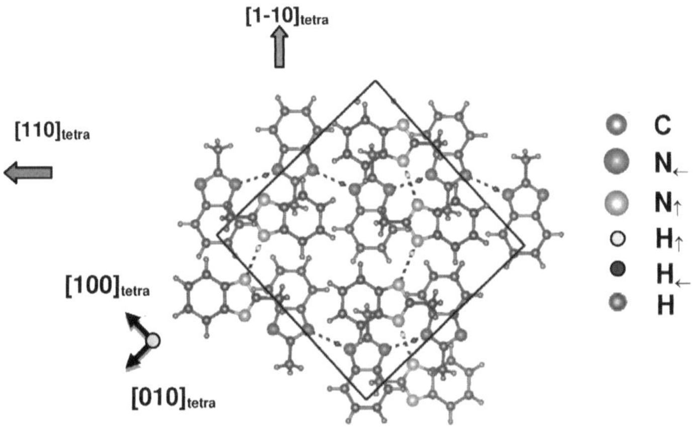

Figure 1. Projection of the MBI unit cell onto $(001)_{tetra}$ plane. The nitrogen and hydrogen atoms participating in the formation of hydrogen bonds oriented along the $[110]_{tetra}$ and $[1\text{-}10]_{tetra}$ axes, as well as carbon atoms and hydrogen atoms that do not participate in the formation of hydrogen bonds, are indicated by different colors.

The MBI films were obtained in 2015 by solution shearing method on substrates covered by structure of hydrophilic and hydrophobic bands. A few Volt polarization switching has been demonstrated in [3]. In [4] the continuous MBI films were obtained by evaporation method from a solution of MBI crystals in ethanol on various substrates without the use of special coatings. The technology is cheap, simple, and the ferroelectric properties of the obtained films are close to those known from the literature [1,3].

It is worth noting that despite of studies of MBI crystals and films carried out up to now the investigation of mechanisms of polarization switching is absent. The purpose of this work was a detailed study of the polarization switching in MBI single crystals and films as a function of the temperature, amplitude and frequency of the external electric field and a description of the results obtained within the Kolmogorov-Avrami-Ishibashi (KAI) model.

2. Experimental

MBI crystals and films were grown by evaporation and slow cooling method from a solution of MBI powder in ethanol. Polarization microscopy and X-ray diffraction analysis were used to determine the orientation of the crystallographic axes and the block structure. MBI single crystals used in our experiments have the form of a parallelepiped (~ 4x1x1 mm^3) elongated along the $[001]_{tetra}$ axis.

MBI films were grown on dielectric substrates (α-Al_2O_3, crystalline and fused SiO_2, $Bi_{12}GeO_{20}$, $NdGaO_3$) with interdigital system of gold electrodes (figure 2a) and could subsequently be covered by an upper metallic electrode with an area of the order of 20 mm^2 (figure 2b). In this case, between the ferroelectric film and the upper electrode, there was thin dielectric layer of oil.

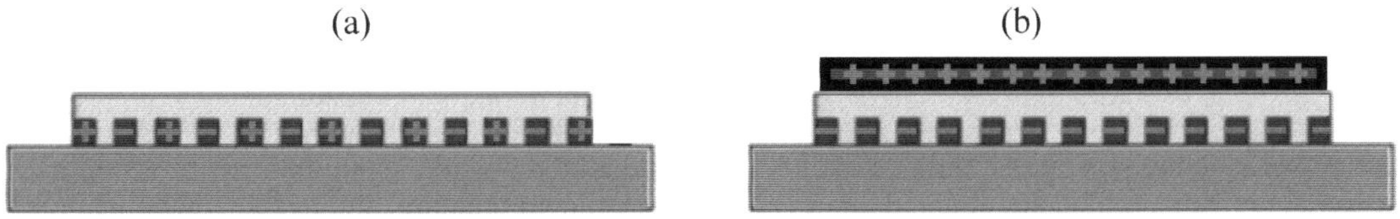

Figure 2. Schematic images of the film structure grown on the substrate with IDS of electrodes without (a) and with (b) upper electrode. + and – shows a distribution of potential after applying an electric voltage.

The strong-signal dielectric response in MBI single crystals and film structures was investigated in Sawyer - Tower scheme. Sinusoidal electric field with a frequency from 30 to 200 Hz for single crystals and 20-1000 Hz for films, and amplitude up to E_{max} = 4.5 V/μ was applied to a sample. The voltage U_P, proportional to the polarization P was detected from the reference capacitor C_{ref} = 0.1 μF. The dielectric hysteresis loops were recorded by digital oscilloscope GDS-71062A. The measurements were carried out in the temperature interval 290 – 390 K. The processing of the results of measurements of the strong-signal dielectric response has been performed in the "MatLab" software package. The small-signal dielectric response was measured with LCR-meters MIT 9216A and E7-20 (MNIPI Belarus) in frequency range 10^2-10^6 Hz. The driving voltage applied to the sample was U_E ~ 1V that corresponds to electric field several orders of magnitude smaller than in strong-signal measurements.

As shown by XRD analysis MBI films are textured with out-of-plane orientation of $[110]_{tetra}$ axis [4]. The films consist of blocks, which are formed by a number of needle-shaped crystallites diverging radially from the crystallization center. The crystallite growth axis coincides with the pseudo tetragonal $[001]_{tetra}$ axis and lies in the film plane. When the blocks are observed in a polarizing microscope with crossed polarizers, a cross-shaped dimming region appears in each block (figure 3a). The dark rays of the crosses are oriented along and perpendicular to the polarization of the incident light. Orientation of cross does not depend on azimuth of the sample. This can be if the block is a split crystal such as spherulite. For any sample azimuth, the block contains crystallites oriented along and perpendicular to the polarization of the incident light. Because main directions of crystallites optical indicatrix are along and perpendicular to $[001]_{tetra}$ axis, they darken, and the center of the cross coincides with the crystallization center of the split crystal. In rare cases, the films were a set of split crystals of the early stages of the formation of spherulite - sheaves, double-leafed and spherical crystals (see figure 3b).

(a) (b)

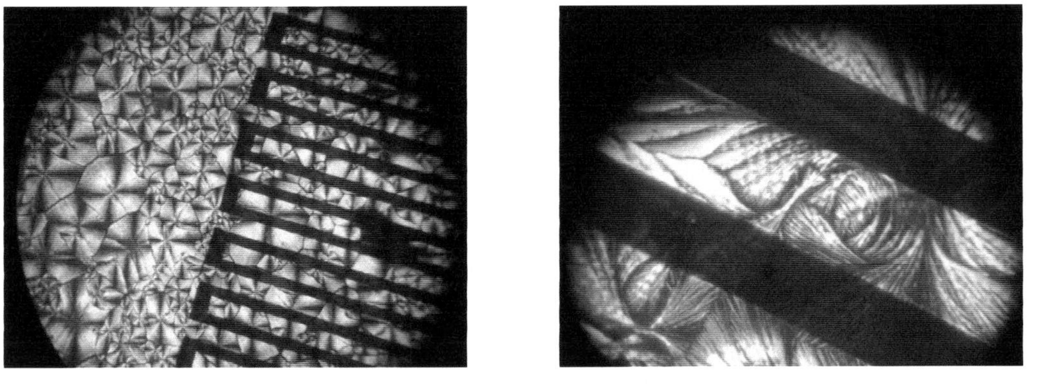

Figure 3. Images of MBI films grown on fused silica (a) and Al_2O_3 (b) substrates observed in the polarization microscope with crossed polarizers. The alternating light and dark bands represent the structure of the electrodes, the width of one strip is 50 μ. (a) Film consisting of spherulites, (b) Film consisting of split crystals corresponding to the early stages of spherulite formation.

It was found that the cross-shaped dimming region rotates as the polarizers are rotated relative to each other. The rotation of the cross is caused by the birefringence in crystallites which form spherulite.

Thus, MBI films are a textured set of split crystals, the sizes of which vary from one to hundreds of micrometers, the thickness of the films being ~ 0.2 - 5 μ. The measurements of the small-signal dielectric response and the dielectric hysteresis loops were carried out in the same MBI film for both in-plane and out-of-plane direction of electric field by applying a voltage between the interdigital electrodes (in the planar capacitor scheme) and in the direction perpendicular to the film plane (in the scheme of a plane-parallel capacitor). The distance between the electrodes of the IDS 50 μ was much larger than the thickness of the film. The height of the IDS electrodes 100 nm was much smaller than the thickness of the film. This allowed us to consider the field to be almost homogeneous in both cases.

3. Dielectric hysteresis in MBI single crystals

Measurements of small- and strong-signal dielectric response has been performed for orientation of electric field along the crystallographic $[110]_{tetra}$ axis, i.e. along the chains of hydrogen bonds. The capacity of the samples was several tenths of picofarad. This corresponds to a dielectric constant $\varepsilon \sim 20$ in agreement with the data known from the literature [1]. Dielectric hysteresis loops observed in crystals depend on the amplitude, frequency of applied field E, and temperature T. The conductivity of single crystals varied from sample to sample and is increased with heating. The influence of conductivity on strong-signal dielectric response is negligible up to T = 370 K. For higher temperatures the contribution of conductivity was compensated during the processing of the results.

Figure 4 shows the dielectric hysteresis loops measured for different amplitudes of the electric field at T = 350 and 381 K. The linear $P(E)$ dependence is subtracted. The saturation field E_{sat} decreases with temperature increase to T = 381 K. In electric fields less than the saturation field, incomplete hysteresis loops are observed in the form of a rounded rectangle and the coercive field E_c is close to the applied field E_{max}. Increase of the electric field is accompanied by formation of saturated hysteresis loop typical for ferroelectrics.

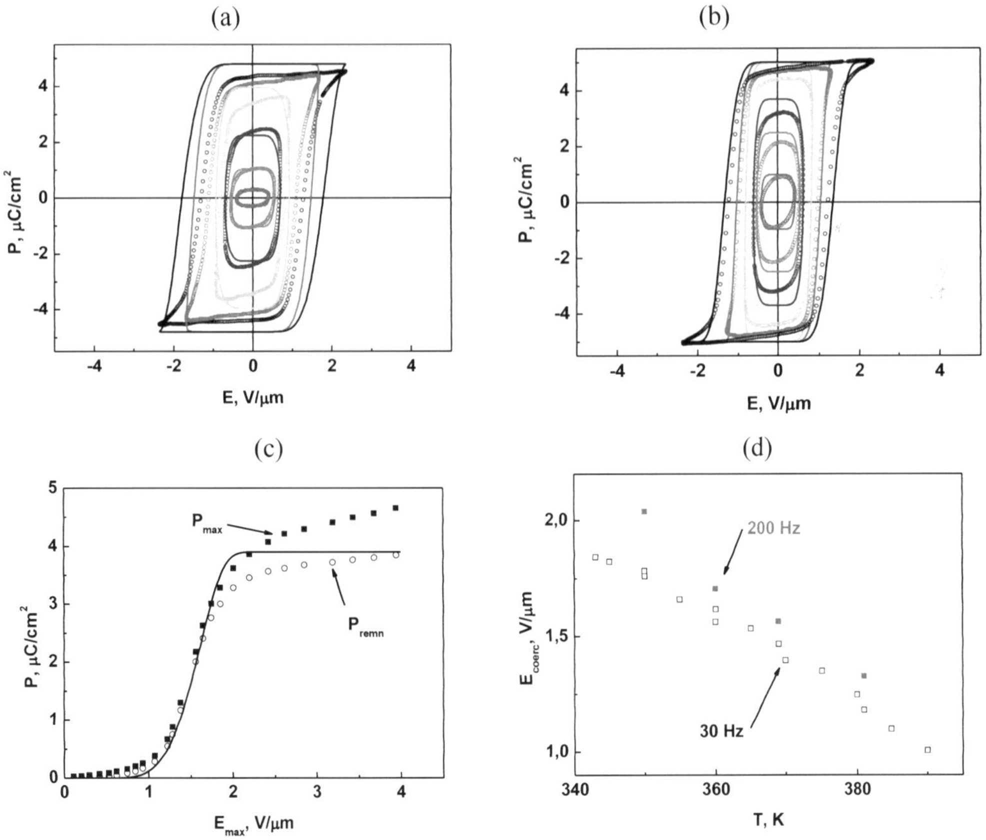

Figure 4. Dielectric hysteresis loops in an MBI single crystal for f = 50 Hz and different values of the amplitude E_{max} of the applied field for T = 350 K (a) and T = 381 K (b). The lines show the simulations. Dependence of the maximum polarization P_{max} and remnant polarization P_{rem} on the amplitude of the applied field for f = 50 Hz, T = 334 K (c). Temperature dependence of coercive field E_c for f = 30 Hz and 200 Hz, E_{max} = 2.4 V/μ (d).

The maximum value of the spontaneous polarization P_{max} of loops is close to the remnant polarization P_{rem} when $E_{max} < E_{sat}$. For $E_{max} > E_{sat}$, a considerable difference between P_{rem} and P_{max} is observed (figure 4c). The difference can not be explained by the manifestation of dielectric constant and may arise due to backswitching [5]. Coercive field E_c decreases with temperature (figure 4d) in agreement with data known from literature [1]. Frequency increase from 30 Hz to 200 Hz does not result in considerable change of E_c.

Simulation of the strong signal dielectric response in MBI crystals was carried out in frames of Janta model [6] which is a particular case of Kolmogorov-Avrami-Ishibashi (KAI) model [7]. Janta two-dimensional model suppose the existence in crystal of non-switching thin cylindrical domains. Polarization switching is caused predominantly by the motion of 180-degree domain walls in a direction perpendicular to the external field E. The velocity of domain walls v is determined by the Mertz law [8]:

$$v = v_\infty \exp(-\frac{E_a}{E}), \qquad (1)$$

where v_∞ is a domain wall velocity in extremely strong field, E_a the activation field. The application of a sinusoidal electric field with the amplitude E_{max} leads to appearance of time dependent polarization $P(t)$:

$$P(t) = P_s - 2P_s \exp\left[-\left(\theta + \frac{\tilde{\omega}}{\omega}\int_0^{\omega t} \exp\left[-\frac{E_a}{E_{\max}\sin\varphi'}\right]d\varphi'\right)^2\right], \qquad (2)$$

where P_s is a spontaneous polarization, θ is found from the loop symmetry condition and determines the residual polarization [5], $\tilde{\omega} = 2\pi\tilde{f}$, where $\tilde{f}$ is the critical frequency that limits the frequency range in which saturated hysteresis loops are observed.

The lines in figure 4 correspond to the calculation in Janta model. The model adequately describes the evolution of the hysteresis loop as function of the amplitude of external field for above room temperatures. The parameters of the model E_a and $\tilde{\omega}$ were obtained by best fitting procedure of array of experimental hysteresis loops for different amplitude E_{max} and frequency f. Figure 5 shows the temperature dependence of the activation field E_a. The magnitude of E_a decreases from $E_a \sim 8$ V/μ at room temperature to $E_a \sim 2$ V/μ at $T = 381$ K.

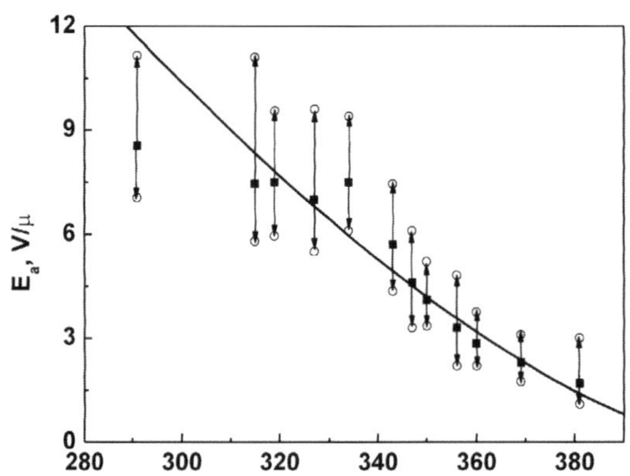

Figure 5. Temperature dependence of the activation field E_a in MBI single crystal, obtained by fitting the strong-signal dielectric response.

The fitting gives the value $\tilde{f} \sim 1$ kHz that is in agreement with experiment. At the same time the calculations of coercive field E_c gives some discrepancy with experimental data (figure 6). According to the Janta model the coercive field E_c goes to zero with frequency in a power law with an exponent of about $\approx 1/3$. In contrast to that the experiment shows approximately linear decrease, and E_c does not tend to zero. This discrepancy may be removed if we suppose that for beginning of the switching process we should apply an electric field $-E < -E_0$, where E_0 is start field. The field E_0 may be found from following condition:

$$\sqrt{\ln 2}\,\frac{\omega}{\tilde{\omega}} = \int\limits_{\arcsin(E_0/E_{\max})}^{\arcsin(E_c/E_{\max})} \exp\left[-\frac{E_a}{E_{\max}\sin\varphi'}\right] d\varphi', \qquad (3)$$

Calculated frequency dependence of E_c for $\tilde{f}$ ~ 5 kHz is shown in figure 6. Start field $E_0 = \lim\limits_{\omega\to 0} E_c$ depends on temperature and is varied from E_0 = 1.1 V/μ to 1.7 V/μ.

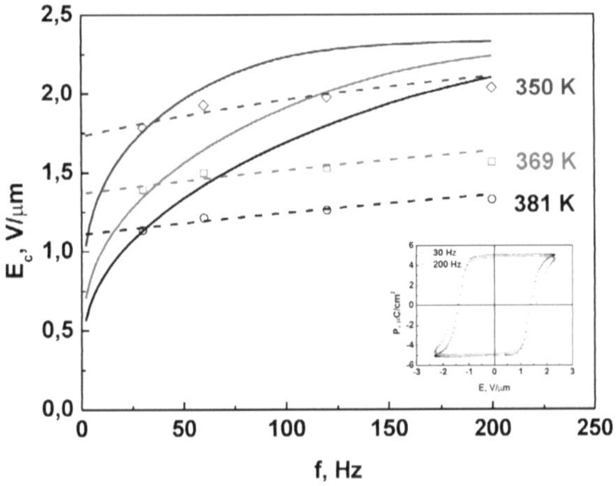

Figure 6. Frequency dependence of the coercive field E_c in the MBI crystal at T = 381, 369, 350 K for the amplitude of the applied field E_{max} = 2.4 V/μ. Lines correspond to calculation in frames of Janta model with (dash lines) and without (solid lines) accounting the start field E_0. Inset shows hysteresis loops for f = 30 Hz and 200 Hz at T = 369 K.

4. Dielectric hysteresis in MBI films

Ferroelecrtic properties of MBI films grown on different substrates are approximately the same. Figure 7a shows dielectric hysteresis loops in MBI/SiO_2 structure measured in geometry of planar capacitor circuit for different values of electric field amplitude E_{max}. Dependence of remnant polarization P_{rem} on E_{max} is presented in figure 7b. At room temperature, the magnitude of P_{rem} and P_{max} in film is less than in the MBI crystals. This is due to the fact that in the single crystals the chains of hydrogen bonds are oriented along the applied field. In contrast to that they are arbitrarily oriented in the film plane because of spherulite blocks structure.

The line in figure 7b presents $P_{rem}(E_{max})$ dependence calculated in frames of Janta model for film formed by spherulite blocks. Every block was modeled as a circle consisted of 40 sectors possessing the properties of a bulk MBI single crystal, but oriented at different angles to the external field. The polarization response was obtained by summation of P_{rem} from all sectors. We see (figure 7b) that simulations are close to experimental points. The difference may be attributed to existence of a bound charges at boundaries between crystallites forming split crystal, and, also, at the boundaries between blocks. Note, that in crystallites oriented parallel to the external field the polarization switching does not occur. These crystallites can be considered as the dielectric passive layers, which can lead to a decrease in the remnant polarization [5]. It has been experimentally found that in MBI films with large blocks, the magnitude of the remnant polarization is greater.

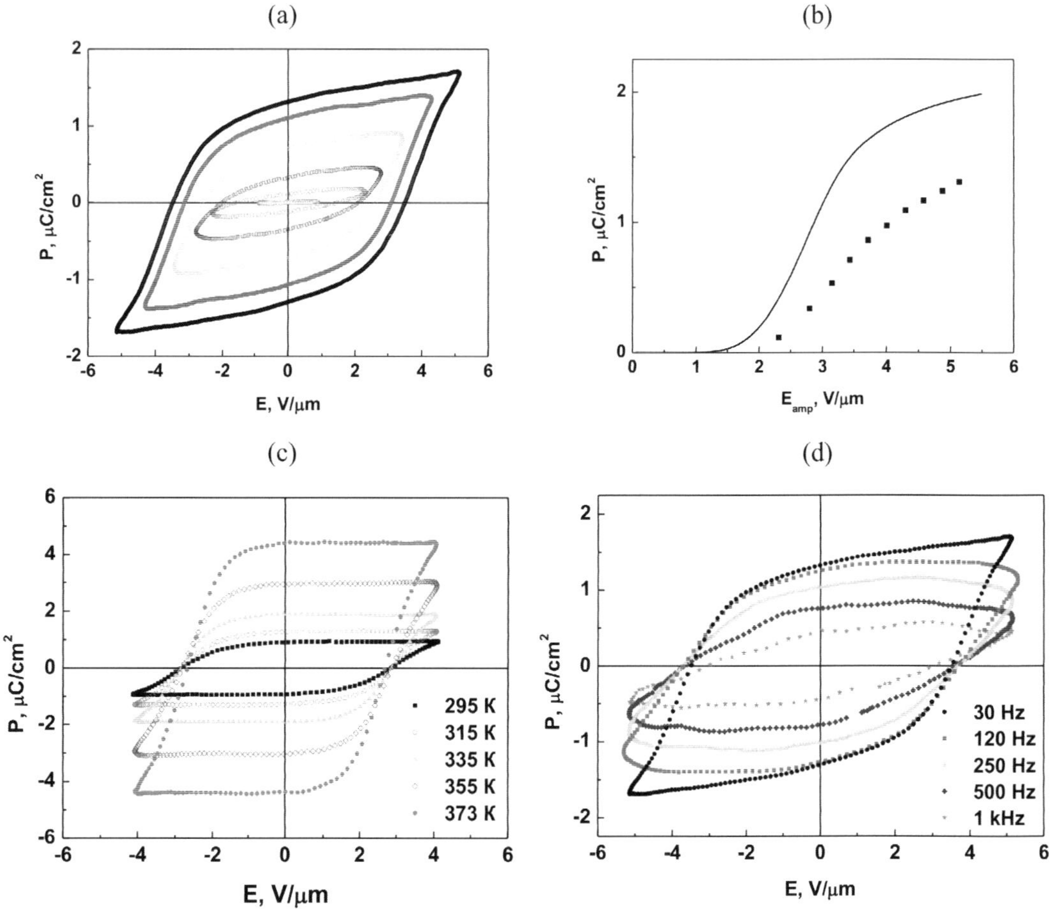

Figure 7. Dielectric response in MBI films grown on fused silica (a,b,d) and Al_2O_3 (c) substrates. (a) – Hysteresis loops at room temperature for different values of E_{max}. Linear contribution from capacity of the substrate is removed. (b) Remnant polarization as a function of E_{max} for f = 30 Hz. Points – experiment, lines – calculation (see text). (c) Hysteresis loops at f = 60 Hz for different temperatures. Linear contribution is removed from every loop. (d) Hysteresis loops at room temperature for different frequencies.

The remnant polarization measured in MBI/Al_2O_3 structure increases sharply with heating and reaches $P_{rem} \approx 4.5$ μC/cm^2 at 373 K, while the coercive field remains almost unchanged (figure 7c). This behavior can be explained by taking into account the temperature dependence of the coercive field in MBI single crystals and the disorientation of crystallites in the film. Heating leads to a considerable decrease in the coercive field (see figure 4d) in each individual crystallite, increasing the range of angles in which the polar axes of switching crystallites can lie. The more crystallites are switched, the greater is the remnant polarization.

The loop shape and P_{rem} in MBI films strongly depend on the frequency f (figure 7d). Frequency increase from 30 Hz to 1 kHz results in a decrease in the polarization by a factor of three. This allows us to assume that the estimates for the critical frequency $\tilde{f} \sim 1 \div 5$ kHz in the single crystals are correct in order of magnitude.

Figure 8 presents dielectric hysteresis loops measured in MBI/Al_2O_3 structure for in-plane (a) and out-of-plane (b) orientation of the electric field at room temperature. Thin dielectric layer between upper electrode and film results in appearance of depolarizing field and polarization switching is observed at higher voltage applied to the film.

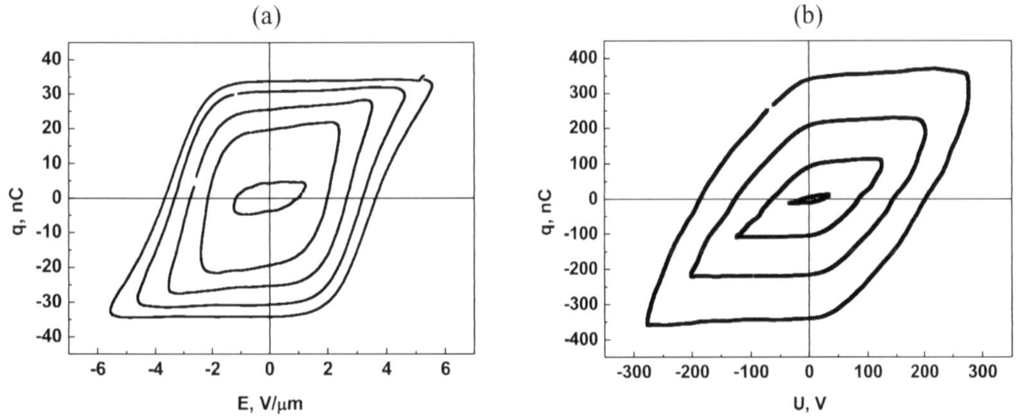

Figure 8. Hysteresis dielectric loops measured in the same MBI film grown on Al_2O_3 substrate for in-plane (a) and out-of-plane (b) orientation of electric field at T = 338 K. Vertical axis shows the charge on electrodes.

5. Conclusions

This work shows that MBI crystals and films grown from ethanol solutions clearly manifest ferroelectric properties at and above room temperature. Polarization switching is allowed in directions which are perpendicular to pseudo tetragonal axis. Dielectric hysteresis loops in MBI crystals are adequately described in frames of the Kolmogorov β-model with accounting for the Mertz law. Domain wall motion activation field E_a in MBI is lower in comparison with PZT crystals. Small values of coercive field E_c, relatively large values of polarization P_s above room temperature, and possibility to switch polarization in various directions makes MBI attractive for applications. Textured MBI films obtained in this work consist of blocks that are split crystals in which the pseudo tetragonal axis makes a full turn around the center of the block. Polarization switching takes place for both in-plane and out-of-plane orientation of electric field.

References

[1] Horiuchi S et. al. 2012 *Nature Communications*, 3:1308 doi: 10.1038/ncomms2322
[2] Obodovskaya A E, Starikova Z A, Belous S N, Pokrovskaya I E 1991 *J. Struct. Chem.* **32** 421
[3] Noda Y et al. 2015 *Adv. Mater.* **27** 6475
[4] Balashova E V, Krichevtsov B B, Svinarev F B, Zaitseva N V 2016 *Scientific and Technical Journal of Information Technologies, Mechanics and Optics* **16** 801
[5] Tagantsev A K, Cross L E, Fousek J 2010 *Domains in Ferroic Crystals and Thin Films* (Springer Science+Business Media LLC)
[6] Janta J 1971 *Ferroelectrics* **2** 299
[7] Ishibashi Y and Takagi Y 1971 *J. Phys. Soc. Jap.* **31** 506
[8] Merz W J 1954 *Phys. Rev.* **9** 690

Acknowledgements

The work was supported in part by RFBR project N 16-02-00399 and (for FBS) by grant of Committee on Science and Higher School of St.-Petersburg Government.

IOP Conf. Series: Journal of Physics: Conf. Series **1038** (2018) 012118 doi:10.1088/1742-6596/1038/1/012118

Investigation of conductivity mechanisms in ferroelectics based on the doped barium titanite

I L Mylnikov[1,2], A I Dedyk[1,2], Yu V Pavlova[1,2], A P Burovihin[1], P Yu Belyavskiy[1], A A Semenov[1] and O V Pakhomov[2]

[1]Saint Petersburg Electrotechnical University «LETI», ul. Professora Popova 5, 197376, St. Petersburg, Russia
[2]ITMO University, 49 Kronverksky Pr., 197101 St. Petersburg, Russia

mylnikov.il@gmail.com

Abstract. The results of studies of the current-voltage characteristics of ceramic structures based on manganese doped barium titanate were presented. It was found that the electrical conductivity of ceramic samples based on barium titanate at high field strengths is determined by the Schottky emission. Moreover, the addition of manganese increases the resistance of the samples.

Ferroelectrics with a perovskite structure are widely used both in developing of microwave microelectronic devices [1] and switching devices for accelerating technology [2] due to the nonlinear field strength and temperature dependences of the permittivity, and possibility of varying properties with change the stoichiometric composition. Thin ferroelectric films are not compatible to use in high power tunable devices. Therefore, it requires the use of bulk ceramics. Ceramics based on barium titanate (BTO) have despite the high tunability but have high dielectric permeability, high dielectric losses and significant dielectric hysteresis in the microwave region [3]. Synthesis of heterophase materials with BTO doped with manganese can reduce dielectric permeability, integral losses and hysteresis of the material while maintaining controllability of the dielectric constant. The doping of Mn leads to the influence of the magnetic field on the composite structure, i.e. the properties of multiferroics appear [4].

Leakage currents and charge carriers transport mechanisms are an important aspect of the use of manganese (Mn) doped ferroelectrics ceramic for high-power microwave devices and for the fabrication of multiferroic materials. Therefore, the aim of this work was to study the electrical conductivity of BTO ceramics with different manganese content.

This paper presents the results of studies of the current-voltage characteristics (CVC) of the plane-parallel Ag/ BTO/Ag structures with a Mn impurity concentration variation from 0 to 10 mol%, namely: 1) pure $BaTiO_3$ ceramics (BTO); 2) $BaTiO_3$ with 5 mol. % Mn (BTO5M), 3) $BaTiO_3$ with 10 mol. % Mn (BTO10M). Samples were prepared using ceramic technology [5] at a sintering temperature about 1350 °C. The composition and content of the manganese-containing impurity were chosen in such a way that at the sintering temperature it is necessary to ensure a weak interaction of the components with each other and to preserve the ferromagnetic properties of the Mn impurity. Silver-electrodes was fabricated on the plane-parallel polished surfaces of the ceramic discs, the disks thickness varied within h = 0.25 - 1.35 mm, the discs area was S = 25-120 mm^2. The CVC measurements were made at room temperature (T = 283 K).

The current-voltage dependences shows linear and nonlinear regions. The regions of ohmic conductivity corresponded to weak electric fields (up to $E \sim 5 \cdot 10^4$ V/m), for which the sample resistance was estimated. The resistance was decreased with an increase the Mn impurity concentration from ~ 10^{12} Ω for pure BTO to ~ 10^{11} Ω for BTO10M samples (at sample thickness 1mm).

Figure 1 shows the current-voltage characteristics for BTO10M samples at different thicknesses. Nonlinear regions on the CV characteristic was started at field intensities ~ $5\cdot10^6$ V/m. The nonlinearity of the CVC in dielectrics and ferroelectrics is usually associated with the following mechanisms: Schottky emission, field emission and the Frenkel-Pool effect.

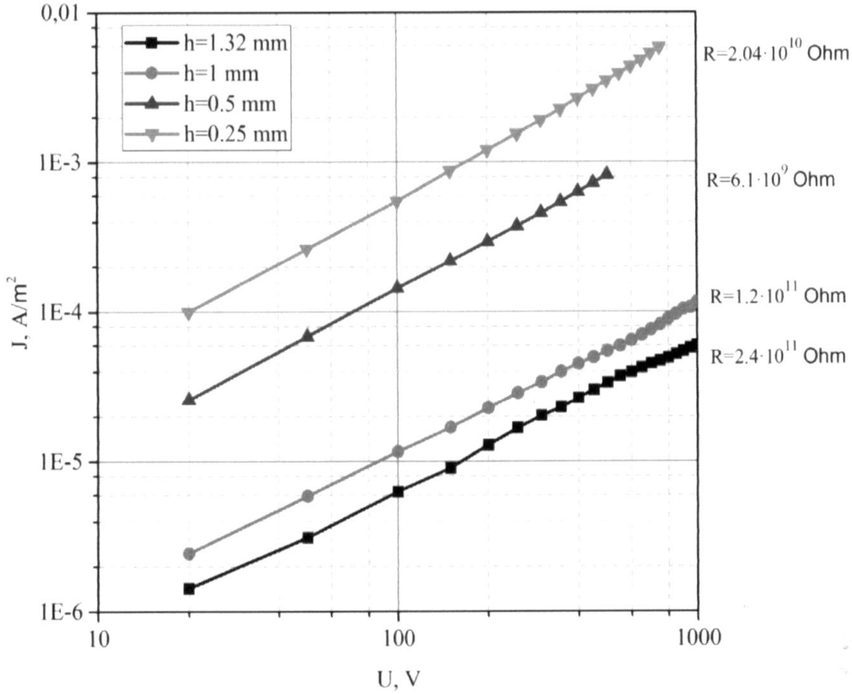

Figure 1. The current-voltage characteristics for different thickness samples BTO10M.

The conduction mechanism due to Schottky emission is described by the following expression [6]:

$$J = AT^2 \exp\left(\frac{1}{kT}\sqrt{\frac{q^3E}{4\pi\varepsilon_0\varepsilon_r}}\right), \quad (1)$$

where A is the Richardson constant, T is the absolute temperature, k is the Boltzmann constant, q is the electron charge, E is the field strength, ε_0 is the electric constant ε_r is the relative permittivity.

Figure 2 shows the BTO10M samples CVC of different thickness in $\ln(J/T^2)$ from $E^{1/2}$ coordinates. All the curves are well approximated by linear dependences at field intensities $\geq 5\cdot10^6$ V/m. That presupposes the presence of a conduction mechanism associated with Schottky emission, while the values of the Pearson agreement (χ) were sufficiently high (in Figure 2, values of the Pearson test).

Thermal ionization of traps and its ionization by an electric field depending on concentration of the traps and their depth affects on current throw thin dielectric layers. The thermal ejection of trapped electrons into the conduction band, known as the Frenkel-Pool effect, gives the following expression for the current-voltage characteristic [6]:

$$J \approx \exp\frac{1}{kT}\left[\left(57.7\frac{eV}{\varepsilon S}\right) - E_l\right], \quad (2)$$

where E_t is the depth of traps, and S is the gap between the electrodes. The presence of a linear section on CVC at ln $(J/E^2) = f(1/E)$ coordinates indicates the presence of Frenkel-Pool effect. All the experimental CVCs shown in Figure 1 were rearranged in the ln $(J/E^2) = f(1/E)$ coordinates and showed

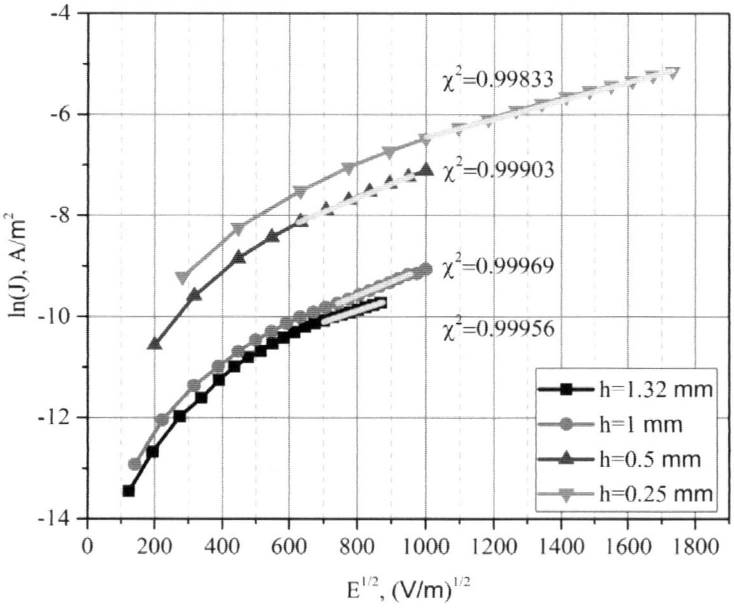

Figure 2. The current-voltage characteristics for different thickness samples BTO10M in Schottky coordinates.

the dependencies close to linear. However, the values of the Pearson criterion for the approximation corresponding to the Frenkel-Pool effect were lower than for the Schottky emission.

Field emission from the traps begins to be observed with a large electric field strength or with heating during the CVC recording. It consists of trapping through a potential barrier. This mechanism is described by the following expression:

$$J = C_t E^2 \exp\left(-\frac{\alpha\varphi_t^{3/2}}{E}\right),\ \alpha = \frac{8\pi\sqrt{2m_{ox}q}}{3h}, \tag{3}$$

where m_{ox} is the effective mass of the electron in the oxide, C_t is the concentration density of the traps, and φ_t is the depth of the traps. In fact, the expression describing the process of field emission from traps, like the well-known Fowler-Nordheim tunneling [7]. The linear approximations of the CVC of the samples in the coordinates $\ln(J/E^2) = f(1/E)$ were also characterized by Pearson's criterion.

In figure 3 shows the values of Pearson's agreement criterion for all the approximations for different thicknesses of the samples. At large sample thicknesses, the values of χ for the Schottky emission and for the field emission practically coincide, and the values of χ for the Frenkel-Pool effect are much smaller. Consequently, for large sample thicknesses, electrical conductivity is mainly determined by Schottky emission and field emission. At sample thicknesses ≤ 0.5 mm, the mechanism associated with Schottky emission is most likely to predominate. At a thickness of 0.25 mm, the Frenkel-Pool effect can also make a significant contribution to the electrical conductivity.

A plane-parallel structure is the series connection between the volume sample resistance and the resistance of the contact region. Therefore, for resistances of unit area we can write the expression:

$$R \cdot S = \frac{h}{\sigma} + R_k , \tag{4}$$

where R is the measured resistance of the sample, R_k is the unit contact area resistance, S is the sample area, h is the sample thickness, and σ is the electrical conductivity of the ceramic.

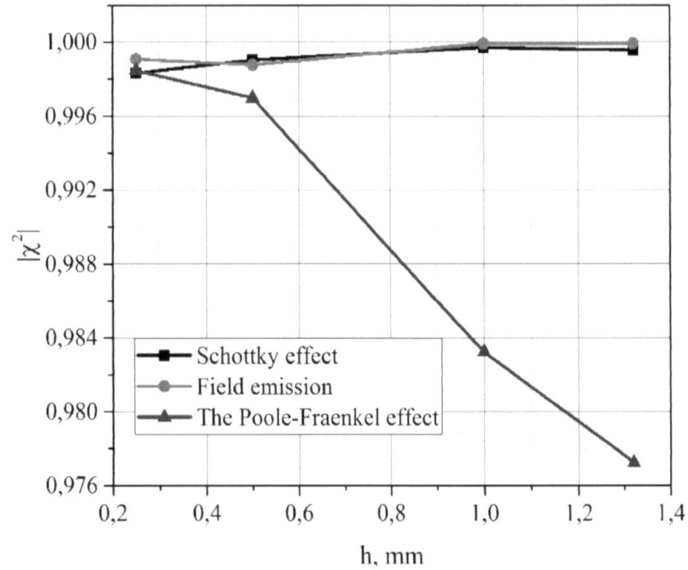

Figure 3. Dependence of the Pearson's agreement criterion for various conduction mechanisms on the thickness of

Figure 4 shows the experimental dependence $R{\cdot}S(h)$. After its approximation by the method of least squares, the extrapolation resulting straight line to the ordinate axis gives us the resistance value of the unit area of the contact area R_k and the angle value of the resulting line gives us the volume sample resistance. The obtained value of the electrical conductivity of the ceramic BTO10M is $\sigma = (2\pm2)\cdot10^{-10}$ $\Omega^{-1}\cdot\text{m}^{-1}$, the value of the contact area resistance is $R_k = (9\pm9)\cdot10^{5}$ $\Omega\cdot\text{m}^2$.

IOP Conf. Series: Journal of Physics: Conf. Series **1038** (2018) 012118 doi:10.1088/1742-6596/1038/1/012118

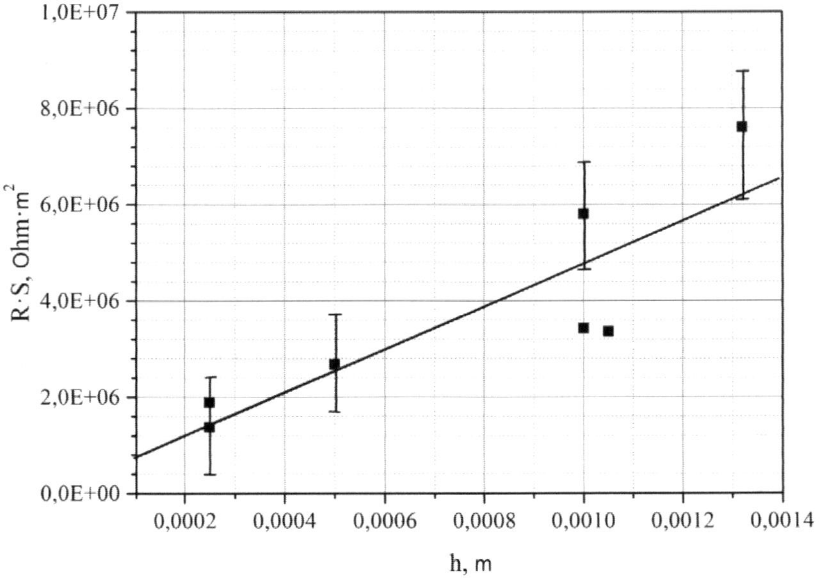

Figure 4. Resistance per unit area for structures Ag/BTO10M/Ag different thicknesses.

Thus, the CV characteristics of the all studied samples has the initial sections corresponding to the fulfillment of Ohm's law, and for field intensities ~ $5\cdot10^6$ V/m, Schottky emission and field emission from traps contribute significantly to the through conductivity currents. When the thickness of the samples is lower than 0.5 mm, the main contribution to the conductivity is made by contact phenomena due to Schottky emission.

The estimation of the electrical conductivity of the ceramics BTO10M is $\sigma = (2\pm2)\cdot10^{-10}\ \Omega^{-1}\cdot m^{-1}$, and the resistance of the unit of the contact area Ag/BTO10M/Ag is $R_k = (9\pm9)\cdot10^5\ \Omega\cdot m^2$.

The investigation was supported by Russian Science Foundation (Grant# 14-12-01296-P).

References

[1] Vendik O G 1979 *Ferroelectrics in Microwave Devices* (Moscow: Soviet Radio) p. 272
[2] Nenasheva E A, Kartenko N F, Gaidamaka I M, Trubitsyna O M, Redozubov S S, Dedyk A I and Kanareykin A D 2010 *J. of Eur Cer. Soc.* **30** 395–400
[3] Dedyk A I, Karmanenko S F, Melkov A A, Pavlovskaya M V, Sakharov V I and Serenkov I T 2003 *Ferroelectrics* **286** 267–278
[4] Semenov A A, Dedyk A I and Nikitin A A 2016 *J. of Materials Science* **51** 7803–13
[5] Nenasheva E A, Kanareikin A D, Dedyk A I and Pavlova Yu V 2009 *Phys.Stat.Sol* **51** 1468–71
[6] Shao T C and Joseph Y L 2002 *Applied Physics Letters* **80** 655-657
[7] Collins J H 1968 *Applied Physics Letters* **13**

IOP Conf. Series: Journal of Physics: Conf. Series **1038** (2018) 012119 doi:10.1088/1742-6596/1038/1/012119

Resonant Bragg structures with GaN/AlGaN Quantum Wells

D S Arteev[1], A V Sakharov[1], W V Lundin[1,2], E E Zavarin[1], S O Usov[3], V V Chaldyshev[1], A S Bolshakov[1], M A Yagovkina[1], A F Tsatsulnikov[2,3]

[1] Ioffe Institute, 26 Politekhnicheskaya, 194021 St. Petersburg, Russia
[2] National Research University of Information Technologies, Mechanics and Optics, Kronverskyi 49, 197101 St. Petersburg, Russia
[3] Submicron Heterostructures for Microelectronics, Research & Engineering Center, RAS, 26 Politekhnicheskaya, 194021 St. Petersburg, Russia

E-mail: dima0724@gmail.com

Abstract. Optical properties of the resonant Bragg heterostructures with 10 and 30 GaN/AlGaN quantum wells were studied. The increasing of reflectivity at the resonance wavelength under condition of the Bragg wavelength and optical transition wavelength matching was observed experimentally at room temperature. The computer simulation of the optical transition wavelength in quantum wells and the optical reflectivity spectra at different reverse bias was implemented to evaluate radiative and non-radiative broadening parameters of the exciton in GaN/AlGaN quantum wells.

1. Introduction

Bragg reflector is a periodic structure typically formed from two alternating layers with different refractive index. If one of the layers is thin enough to form a quantum well and the wavelength of the electron-hole transition is equal to the Bragg wavelength, the effect of exciton-polariton resonance is observed. When the multiple quantum well system of N quantum wells is used, a superradiant exciton-polariton mode is formed by electromagnetic coupling of excitons in all quantum wells. Such structures, called Resonant Bragg Structures (RBS), were first theoretically proposed by E L Ivchenko *et al.* [1].

Formation of the superradiant mode was observed in RBS based on different material systems such as GaAs/AlGaAs [2–4], InGaAs/GaAs [5–6], CdTe/CdZnTe [7] and CdTe/CdMgTe [8]. It is worth mentioning that exciton-polariton resonance effect in these material systems occurs only at cryogenic temperatures. However, using InAlGaN-based periodic structure enables obtaining the effect at room temperature due to large exciton binding energy in GaN (~20 meV in bulk GaN compared to 4 meV in GaAs [9]). Resonant exciton enhancement in RBS based on InGaN quantum wells separated by GaN barriers at room temperature was demonstrated recently [10]. Furthermore, applying the external voltage to the structure makes it possible to change the optical transition energy of the exciton due to strong quantum-confined Stark effect (QCSE) in III-nitride quantum well. This effect was demonstrated in InGaN/GaN and GaN/AlGaN electroabsorption modulators in lateral geometry [11], but vertical geometry seems to be preferable in terms of usability. The totality of these factors makes InAlGaN RBS attractive for using as electro-optic modulator operating at room temperature.

In this paper for the first time we present efficient reflectivity modulation in vertical geometry resonant Bragg structure based on GaN/AlGaN quantum well system at room temperature.

2. Samples and experimental details.

Two samples were grown by metalorganic chemical vapor deposition (MOCVD) on sapphire (0001) substrates. Trimethylgallium and trimethylaluminum were used as the group-III source materials; ammonia was used as the nitrogen source. A set of several GaN quantum wells (10 and 30 QWs in the first and the second sample, respectively), separated by $Al_{0.12}Ga_{0.88}N$ barriers, was grown on double $Al_{0.18}Ga_{0.82}N/Al_{0.12}Ga_{0.88}N$ buffer layer. In contrast to the InGaN/GaN-based heterostructures, conventionally grown on well-established GaN buffer layer, GaN/AlGaN-based optoelectronic structures require AlGaN buffer layer to avoid optical absorption. The buffer layer was Si-doped and the others were unintentionally doped. The samples were capped *in-situ* with dielectric Si_3N_4 layer. The hybrid top contact, consisting of transparent ITO, encircled by NiAu (figure 1), was formed by optical lithography, sputtering and etching in order to enable applying voltage to active region of RBS.

X-Ray diffraction measurements were performed using D8 Discover Bruker diffractometer. Room temperature photoluminescence studies were carried out using excitation by a He-Cd laser (325 nm) focused in a 0.3 mm^2 spot. AvaSpec-2048 Fiber Optic Spectrometer was used to obtain PL spectra. Normal-incidence optical reflectivity measurements were performed using a deuterium lamp as a white light source and CCD-spectrometer. Focusing of the incident and reflected light was carried out with UV transmitting lenses.

According to the XRD-analysis, thicknesses of GaN QWs and AlGaN barriers in 10 QW structure are 2.7 and 70 nm, respectively. It should be noted, that the sample with 30 QWs was intentionally grown having barrier thickness gradient along the diameter of the structure, while the sample with 10 QWs has constant barrier thickness throughout the structure.

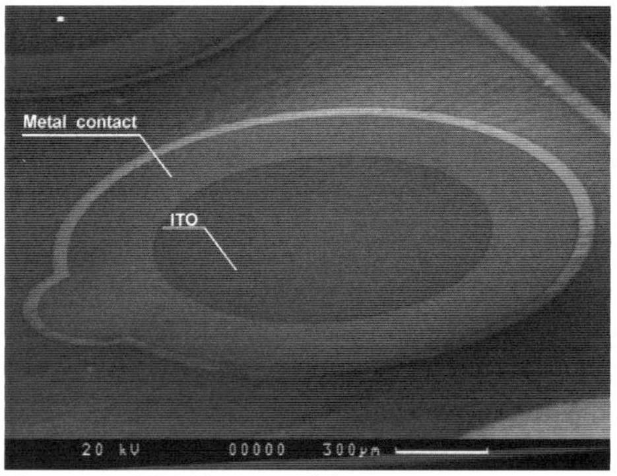

Figure 1. SEM image of the top contact

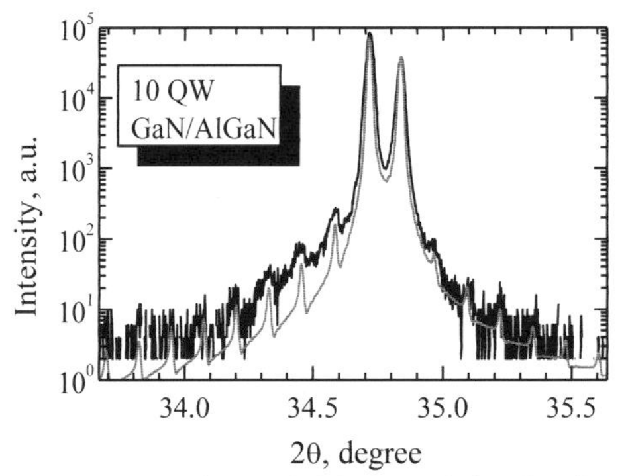

Figure 2. Experimental (black) and simulated (red) XRD diffraction curve of 10 QW sample.

A numerical simulation was implemented to evaluate exciton parameters. Optical reflectance spectra of the structures were calculated by Transfer Matrix Method taking into account existing of exciton in QW, developed by E L Ivchenko [1]. The approach is described in detail in ref.12. Parameters for calculation of $Al_xGa_{1-x}N$ refractive index were taken from [13], the values of the thickness of each layer and aluminum content in barrier and buffer layers were taken from XRD analysis. The band diagram, electron and hole energy levels in quantum wells were calculated by solving self-consistent Schrödinger-Poisson equations in Silvaco TCAD. Note that $Al_{0.12}Ga_{0.88}N$ lattice constant was used as the "substrate" lattice constant in the calculation of the piezoelectric polarization as the structures were grown on the thick relaxed $Al_{0.18}Ga_{0.82}N/Al_{0.12}Ga_{0.88}N$ buffer layer.

3. Results.

The experimental and simulated normalized zero-voltage PL spectra at room temperature are plotted in figure 3. As one can see, they are in good agreement with each other. Unfortunately, micro-PL

IOP Conf. Series: Journal of Physics: Conf. Series **1038** (2018) 012119 doi:10.1088/1742-6596/1038/1/012119

measurements at different voltage were made impossible by low photoluminescence efficiency resulted from the absence of localization centers in GaN/AlGaN QW as well as from higher dislocation density compared to InGaN/GaN heterostructures. Therefore, computer simulation of the dependence of the reverse bias on the optical transition energy in QWs was implemented. Figure 4 shows that the optical transition wavelength has minimum at ~45 V, corresponding to the flat bands in QW. It should be noted that the transition wavelength on figure 4 was derived from $\lambda = E_{eh}/hc$, where E_{eh} is difference between calculated electron and hole levels in quantum well, without any broadening taken into account, while the PL spectrum on figure 3 was simulated using Lorentzian broadening function, leading to the redshift, and therefore a small discrepancy between the PL peak wavelength and optical transition wavelength exists. Conduction and valence bands in wurtzite GaN-based polar QW are tilted due to the build-in internal electric field caused by spontaneous and piezoelectric polarization. Applying the reversed bias decreases the electric field in QW, resulting in the blueshift of the interband transition wavelength, until the internal field is fully compensated by the external one and the bands in QW become flat. With further reverse voltage increasing, after the external field has become greater than the internal one, bands tilt in the opposite direction, causing redshift (figure 5).

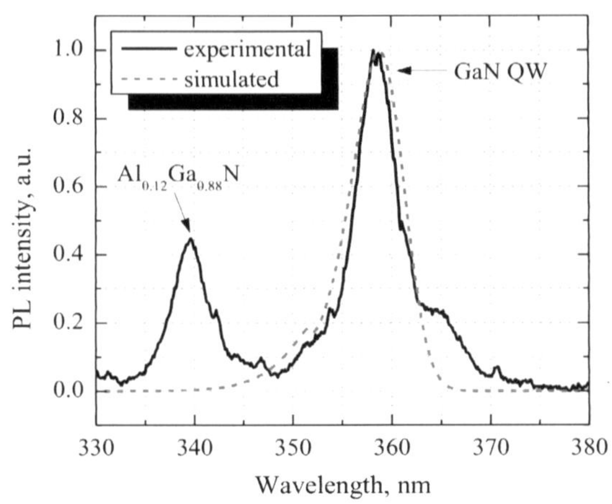

Figure 3. Experimental and simulated photoluminescence spectra of 10 QW structure.

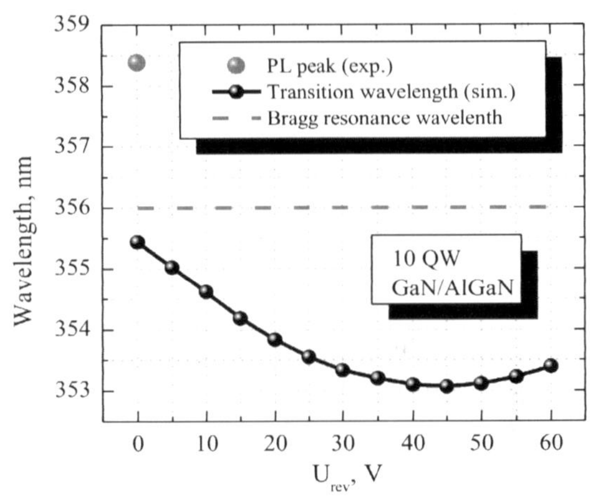

Figure 4. Simulated electron-hole transition energy.

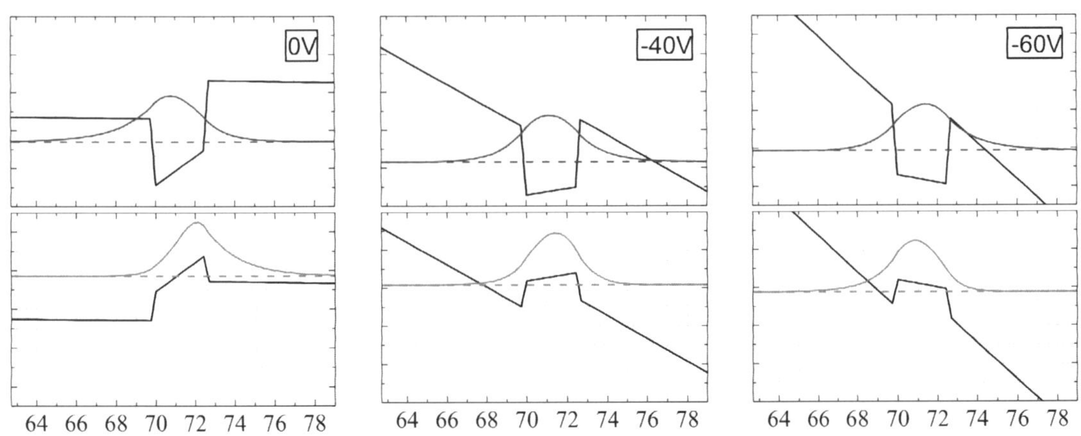

Figure 5. Conduction and valence bands (black lines), electron (blue) and hole (red) wave functions (color solid lines) and energy levels (color dashed lines) in quantum well at the voltage of 0V, -40V and -60V.

The obtained values of the transition wavelength were used to simulate optical reflectance spectra and compare them with the experimental ones. The relative change in optical reflection $\Delta R/R(0)=(R(U)-R(0))/R(0)$ is plotted in figure 6. Good agreement is found between experimental and simulated curves; the maximum value of the relative optical reflection (OR) change is ~5.5% at the Bragg resonance wavelength.

In order to enhance exciton-polariton resonance effect, the structure with 30 quantum wells has been grown. A special growth conditions with non-uniform precursor gases distribution were used to obtain gradient of growth rates along the diameter of the structure. As a result the structure has different thicknesses of the layers and therefore different values of the transition wavelength and the Bragg resonance wavelength. The PL measurements show that the peak of the QW emission is nearly constant, while OR peak varies significantly (figure 7). Therefore, the constant quantum wells thickness of 2.7 nm and varying $Al_{0.12}Ga_{0.88}N$ barrier thickness can be considered. Thus, we can get a match between optical transition and Bragg resonance wavelengths at normal incidence via choosing the appropriate point on the surface of the structure. In case of that match (at 10 mm) a dramatic increase in reflectance (more than 50%) is observed, indicating exciton-polariton resonance. Unfortunately, high reverse leakage current through the Schottky contact, causing heating of the structure and changing of the refractive index, arising from insufficient crystal quality, prevented OR measurements of the biased structure.

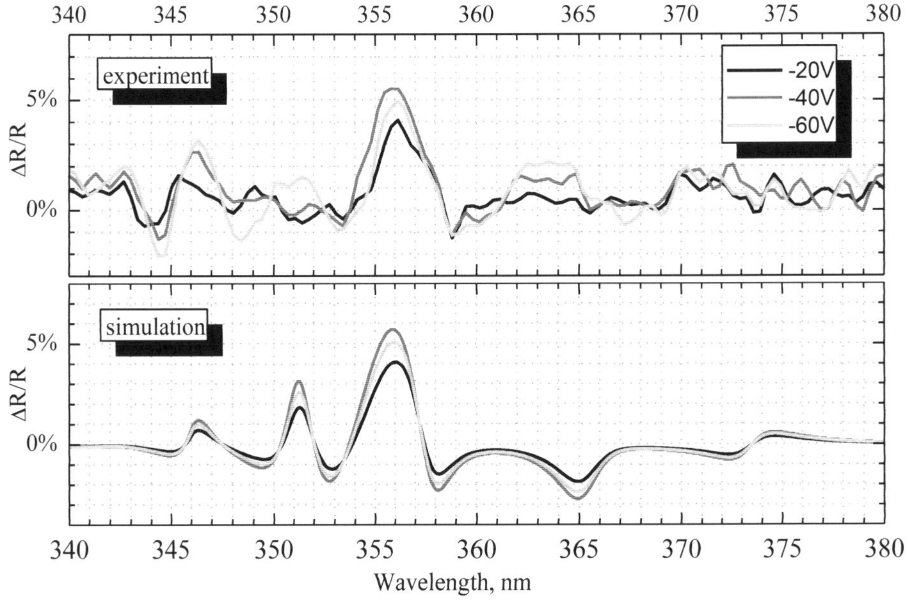

Figure 6. Relative change in optical reflection spectra with voltage.

Simulation of OR spectra in different points was performed as follows. The Bragg resonance wavelength position was fitted by changing AlGaN barrier thickness; slight adjustments of the buffer layer thickness were done to fit the Fabry-Pérot oscillations. Simulation at each point was performed with and without taking into account exciton in quantum well. When the Bragg wavelength and the exciton wavelength are detuned (at -10 mm point), the difference between calculated spectra with and without exciton contribution is almost negligible (figure 8, top). But if the double resonance condition is met, the values of the reflectance, obtained with and without excition contribution, differ greatly in the area near the resonance wavelength (figure 8, bottom). As can be seen, exciton accounting gives the value of reflectance 52.7% at the resonance wavelength (which is in a good agreement with the experimental value of ~53%). Disregarding of exciton contribution gives underestimated value of the reflectance (~43%); even the shape of the spectrum curve is qualitatively different from the experimental one in the area near the resonance wavelength. The totality of the experimental and simulated results provides strong evidence that the effect of exciton-polariton resonance takes place in the structure. The exciton parameters in GaN QW estimated from the simulation are as follows: non-radiative broadening parameter Γ is 40 meV at RT; radiative broadening parameter $N\Gamma_0$ for systems of 10 QWs and 30 QWs is 4 meV and 11 meV, respectively, giving Γ_0=0.37–0.4 meV.

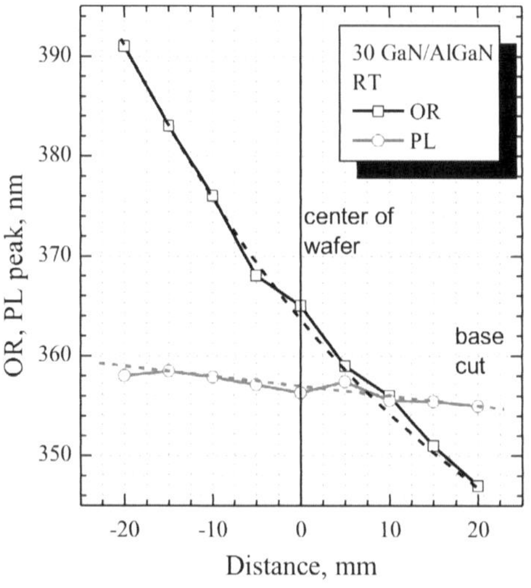

Figure 7. Photoluminescence (PL) and optical reflection (OR) peak wavelengths depending on the coordinate.

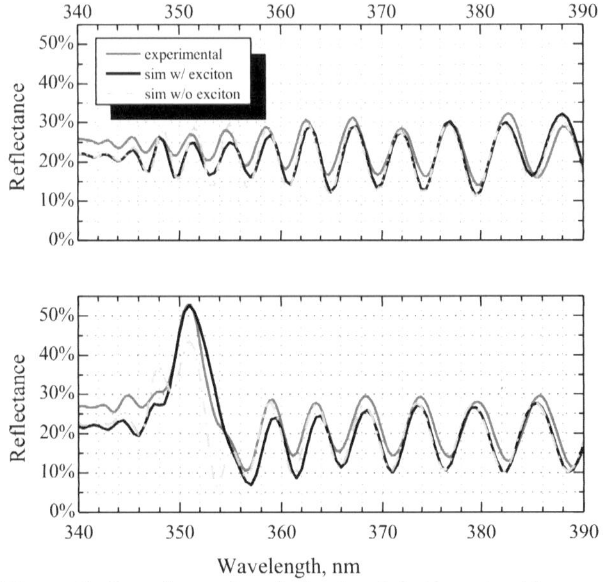

Figure 8. Experimental and simulated (with and without taking into account exciton in QWs) spectra at -10 mm and 10 mm.

4. Conclusions.

The possibility of fabrication of the electro-optic modulator based on periodical resonant Bragg structure with GaN/AlGaN quantum wells, operating at room temperature was demonstrated. The obtained maximum value of the relative modulation of the reflectivity is ~5.5% for the structure with 10 quantum wells. Computer simulation results provide strong evidence of exciton-polariton resonance existence in the structure. The calculated values of the exciton non-radiative Γ and radiative Γ_0 parameters are 40 meV and 0.37–0.4 meV, respectively.

Acknowledgements

The XRD characterizations were performed using equipment owned by the Joint Research Center "Material science and characterization in advanced technology" (Ioffe Institute, St.-Petersburg, Russia).

References

[1] Ivchenko E L, Nesvizhskii A I, Jorda S 1994 *Phys. Sol. State* Bragg reflection of light from quantum-well structures **36** 7 1156–61
[2] Chaldyshev V V, Shkol'nik A S, Evtikhiev V P and Holden T 2006 *Semiconductors* Optical reflection and contactless electroreflection from GaAlAs layers with periodically arranged GaAs quantum wells **40** 1432–5
[3] Goldberg D, Deych L I, Lisyansky A A, Shi Z, Menon V M, Tokranov V, Yakimov M and Oktyabrsky S 2009 *Nat. Photonics* Exciton-lattice polaritons in multiple-quantum-well-based photonic crystals **3** 662–6
[4] Chaldyshev V V, Chen Y, Poddubny A N, Vasil'ev A P and Liu Z 2011 *Appl. Phys. Lett.* Resonant optical reflection by a periodic system of the quantum well excitons at the second quantum state **98** 073112

[5] Hübner M, Prineas J P, Ell C, Brick P, Lee E S, Khitrova G, Gibbs H M and Koch S W 1999 *Phys. Rev. Lett.* Optical Lattices Achieved by Excitons in Periodic Quantum Well Structures **83** 2841–4

[6] Hayes G 1999 *Physica B* Suppression of exciton–polariton light absorption in multiple quantum well Bragg structures **272** 488–90

[7] D'Aubigné Y M, Wasiela A, Mariette H and Dietl T 1996 *Phys. Rev. B* Polariton effects in multiple-quantum-well structures of $CdTe/Cd_{1-x}Zn_xTe$ **54** 14003–11

[8] Ivchenko E L, Kochereshko V P, Platonov A V, Yakovlev D R, Waag A, Ossau W and Landwehr G 1997 *Phys. Sol. State* Resonance optical spectroscopy of long-period quantum-well structures **39** 1852–8

[9] Muth J F, Lee J H, Shmagin I K, Kolbas R M, Casey H C, Keller B P, Mishra U K and Denbaars S P 1997 *Appl. Phys. Lett.* Absorption coefficient, energy gap, exciton binding energy, and recombination lifetime of GaN obtained from transmission measurements **71** 2572–4

[10] Chaldyshev V V, Bolshakov A S, Zavarin E E, Sakharov A V, Lundin W V, Tsatsulnikov A F, Yagovkina M A, Kim T and Park Y 2011 *Appl. Phys. Lett.* Optical lattices of InGaN quantum well excitons **99** 251103

[11] Ozel T, Sari E, Nizamoglu S and Demir H V 2007 *J. Appl. Phys.*Violet to deep-ultraviolet InGaN/GaN and GaN/AlGaN quantum structures for UV electroabsorption modulators. **102**(11), 113101. doi:10.1063/1.2817954

[12] Bolshakov A S, Chaldyshev V V, Zavarin E E, Sakharov A V, Lundin W V, Tsatsulnikov A F and Yagovkina M A 2017 *J. Appl. Phys.* Room temperature exciton-polariton resonant reflection and suppressed absorption in periodic systems of InGaN quantum wells **121** 133101

[13] Tisch U, Meyler B, Katz O, Finkman E and Salzman J 2001 *J. Appl. Phys* Dependence of the refractive index of $Al_xGa_{1-x}N$ on temperature and composition at elevated temperatures **89** 2676–85

Single photon emitters based on hybrid microcavities with $InAs/Al_xGa_{1-x}As$ quantum dots

K G Belyaev, M V Rakhlin, G V Klimko, U M Zadiranov, M M Kulagina, I V Sedova, S V Ivanov, and A A Toropov

Ioffe Institute, 26 Politekhnicheskaya str., 194021, St. Petersburg, Russia

E-mail: belyaev.kirill@mail.ioffe.ru

Abstract. We report on fabrication by molecular beam epitaxy and optical studies of hybrid semiconductor-dielectric micropillar structures with distributed Brag reflectors and a microcavity containing InAs/AlGaAs quantum dots. The single photon emission in the visible spectral range with the autocorrelation function $g^{(2)}(0)<0.2$ is detected in such structures with a photon count rate above 1 MHz.

1. Introduction

Recently, the sources of single photon emission and entangled pairs of photons have been intensively developed for applications in quantum cryptography and quantum computing industry. The fabrication of single-photon emitters based on heterostructures with either InAs/InGaAs [1-3] or InAs/InAlGaAsP [4-5] quantum dots (QDs) have been reported for telecommunication spectral bands (1.3 and 1.55 μm). Less developed are bright single-photon emitters intended for red part of the visible spectral range, which is suitable for the realization of protected lines of atmospheric optical communication and also covers the region of high sensitivity of most efficient single-photon avalanche photodiodes based on silicon. Until recently, single-photon emission in this range has been achieved either using InP based QDs [6-7] or GaAs QDs grown by droplet epitaxy [8]. The structures of first type suffer from a lack of a lattice matched binary distributed Bragg reflector (DBR) system with high contrast of refractive index that complicates growth of cavity structures, while the droplet-induced QD formation requires precise tuning of growth parameters.

One more promising approach to development of the efficient single-photon emitters for the red part of the visible spectral range can be fabrication of cavity heterostructures based on InAs/AlGaAs QDs, whose emission wavelength can be tuned in a widest spectral range between 650 nm and 1000 nm by only varying Al content in the AlGaN barrier layers [9-10]. The potential advantage of this approach is also availability of well-developed epitaxial technique for the Stranski-Krastanow QD formation in the InAs/AlGaAs material system. On the other hand, the available lattice-matched DBR system for such heterostructures is AlGaAs/AlAs, whose refractive index contrast progressively decreases with an increase of Al content in the AlGaAs layers. This circumstance set a limit to the shortest wavelength of bright single-photon emission achievable in this material system.

In this paper, we investigate emission properties of a microcavity structure with InAs/AlGaAs QDs, focusing on elucidation of the single-photon emission characteristics of individual InAs/AlGaAs QDs and exploration of the possibility to use dielectric DBR instead of the AlGaAs/AlAs one.

2. Experiment and structures

A semiconductor part of the hybrid structure, grown by molecular beam epitaxy, contains 19 periods of GaAs/$Al_{0.3}Ga_{0.7}As$ quarter wavelength pairs, which form bottom DBR and an $Al_{0.3}Ga_{0.7}As$ cavity with a layer of InAs/AlGaAs QDs in the middle (figure 1). A height of the 1λ cavity is 224 nm. Thicknesses of the GaAs and $Al_{0.3}Ga_{0.7}As$ DBR constituting layers are 56 and 62 nm correspondingly. The top DBR is formed after growth by magnetronic sputtering and consists of 3 periods of 125 nm SiO_2 and 91 nm Ta_2O_5. At the last technological stage, submicron size micropillar structures were formed by photolithography and reactive ion dry etching (figure 2).

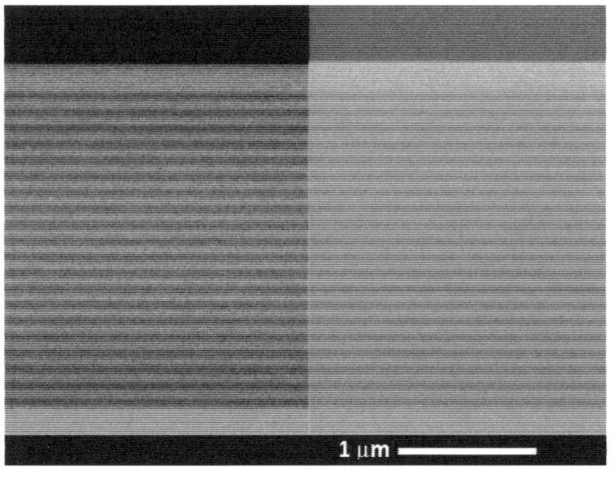

Figure 1. Cross-section transmission electron microscopy image of the as-grown heterostructure with the 19 periods of the GaAs/$Al_{0.3}Ga_{0.7}As$ bottom DBR and the 224 nm $Al_{0.3}Ga_{0.7}As$ cavity with the QDs insertion in the middle.

Figure 2. Scanning electron microscopy image of the micropillar structure with a top dielectric mirror containing 3 periods of SiO_2/Ta_2O_5 pairs.

Micro-photoluminescence measurements were performed in a standard confocal optical scheme with a 50x micro-objective and a grating monochromator equipped with a liquid-nitrogen cooled CCD camera (spectral resolution is estimated as ~90 μeV). Both cw and picosecond pulse lasers with the emission wavelength 404 nm were used for the PL excitation. Photon correlation measurements were performed with a Hanbury Brown – Twiss intensity interferometer exploiting two single-photon avalanched Si photodiodes with a time resolution better than 40 ps. The measurements were performed at the 8 K temperature in a continuous flow He cryostat with a nanopositioning piezo-driver.

3. Results

As compared to emission of a semiconductor heterostructure with only bottom DBR the hybrid structure emission has significantly narrower spectrum with full width at half maximum about 1 meV and a narrow reflection dip, which is characteristic of a DBR microcavity (figure 3).

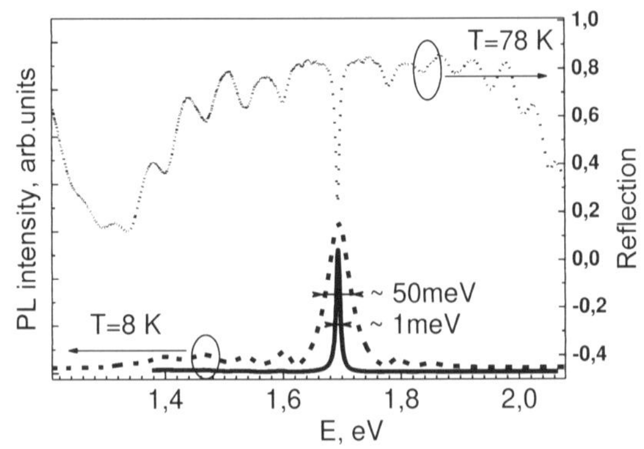

Figure 3. Bottom curves represent photoluminescence spectra of the as-grown structure (dashed curve) and of the hybrid structure with top dielectric DBR (solid curve). An upper dotted curve represents reflectivity spectrum of the hybrid structure.

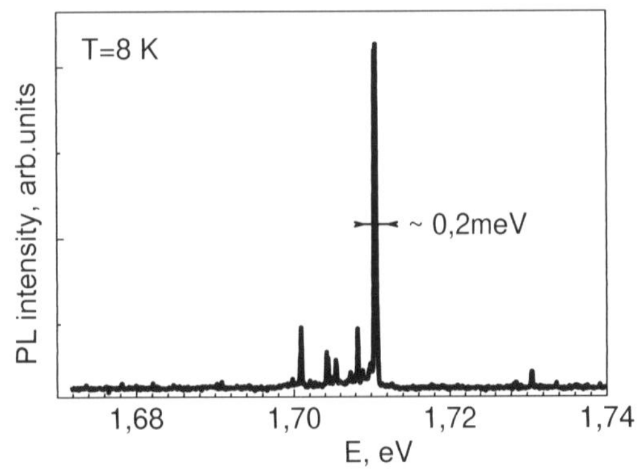

Figure 4. Micro-photoluminescence spectrum of a micropillar structure with a submicron lateral size.

The surface QDs density estimated from atomic force microscopy measurements of uncapped QD structures is as low as $\sim 10^{10}$ cm^{-2} and has fluctuations at a submicron size scale. As a result, the fabricated array of several hundreds of micropillars includes cavities with essentially different amounts of QDs. For measurements, we chose several micropillars with spectrally resolved single-QD emission lines located within the microcavity resonance, where one spectral line was strongly dominating (figure 4). We assume that these bright lines may be attributed to QDs situated close to the center of the cavity mode. The brightest spectral lines of this kind were identified as the emission of charged excitonic states (trions) in single InAs/AlGaAs QDs, rather than neutral excitons, as they do not possess any visible linear polarization splitting. These lines possess full with at half maximum about 0.2 meV that is typical for the emission of single-QDs (figure 4). Autocorrelation measurements performed at the excitation power 2-4 times lower than that corresponding to saturation of the QD emission demonstrate the $g^{(2)}$ function value at zero delay less than 0.2 that clearly confirms the single-photon quantum nature of the emission (figure 5). The estimated count rate of the single-photon emission "at the first lens" exceeds 1 MHz.

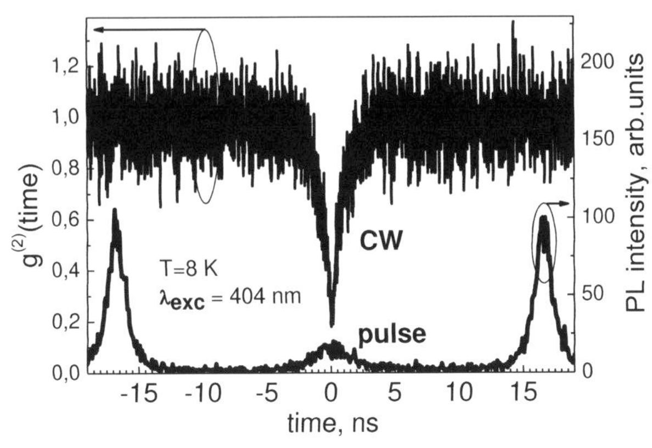

Figure 5. The second-order correlation function for the brightest line in the μ-PL spectrum shown in figure 4 under both cw and pulse laser excitations.

As compared to the single-photon emission in etched mesa-structures without DBRs, fabrication of the hybrid microcavity results in the intensity enhancement approximately of one order of magnitude. As-grown flat InAs/AlGaAs QDs structures possess sufficiently lower intensity (2 - 5 times weaker than in mesa-structures).

4. Conclusions

We have fabricated micropillar structures with a hybrid semiconductor-dielectric DBR microcavity with low enough InAs/AlGaAs QDs surface density, which are suitable for single-photon applications in the red part of the visible spectral range. The obtained count rate value of the single-photon emission is above 1 MHz with the value of $g^{(2)}(0)$ below 0.2 at non-resonant above-barrier excitation.

Acknowledgments

The authors acknowledge the financial support of the Russian Science Foundation (project #14-22-00107).

References

[1] Chen Z -S, Ma B, Shang X -J, Ni H -Q, Wang J –L, and Niu Z -C 2017 *Nanoscale Research Lett.* **12** 378
[2] Paul M, Olbrich F, Höschele J, Schreier S, Kettler J, Portalupi S L, Jetter M, and Michler P 2017 *Appl. Phys. Lett.* **111** 033102
[3] Al-Khuzheyri R, Dada A C, Huwer J, Santana T S, Skiba-Szymanska J, Felle M, Ward M B, Stevenson R M, Farrer I, Tanner M G, Hadfield R H, Ritchie D A, Shields A J, and Gerardot B D 2016 *Appl. Phys. Lett.* **109** 163104
[4] Benyoucef M, Yacob M, Reithmaier J P, Kettler J, Michler P 2013 *Appl. Phys. Lett.* **103** 162101
[5] Birowosuto D, Sumikura H, Matsuo S, Taniyama H, van Veldhoven P J, Nötzel R, Notomi M 2012 *Sci. Rep.* **2** 321
[6] Rossbach R, Reischle M, Beirne G J, Jetter M, and Michler P 2008 *Appl. Phys. Lett.* **92** 071105
[7] Ugor A, Kremling S, Hatami F, Höfling S, Worschech L, Forchel A, and Ted Masselink W 2012 *Appl. Phys. Lett.* **100** 023116
[8] Mano T, Abbarchi M, Kuroda T, McSkimming B, Ohtake A, Mitsuishi K, and Sakoda K 2010 *Appl. Phys. Express* **3** 065203
[9] Finley J J, Mowbray D J, Skolnick M S, Ashmore A D, Baker C, Monte A F G, and Hopkinson M 2002 *Phys Rev B* **66** 153316
[10] Grijseels S C M, J van Bree, Koenraad P M, Toropov A A, Klimko G V, Ivanov S V, Pryor C E, Silov A Yu 2016 *J. Luminescence* **176** 95

Estimation of the area of field emission of a carbon nanotube using modelling in COMSOL Multiphysics

M A Chumak[1], S V Filippov[2], A G Kolosko[2], E O Popov[2*]

[1]Peter the Great St.-Petersburg Polytechnical University, ul. Polytechnitscheskaya 29, St.-Petersburg, 195251, Russia

[2]Ioffe Institute, ul. Polytechnitscheskaya 26, St.-Petersburg, 194021, Russia
*e.popov@mail.ioffe.ru

Abstract. In this paper, the computer simulation methods were used to study the effect of the mutual shielding in the multi-point system of a nanoscale field cathode on its current-voltage characteristics and the characteristic parameter of the field emission area. Modeling is carried out in the COMSOL Multiphysics software package. Geometric parameters of the model correspond to the parameters of carbon nanotubes. As a result of the simulation, the dependence on the distance between the emission centers of the shape of the current-voltage characteristic of the central emitter, the effective current density of the entire emission system, and the size of the field emission area were constructed.

1. Introduction

Field emitters are complex systems for analyzing their properties and characteristics. The process of field emission depends on many external and internal factors. Computer modeling is one of the effective methods for studying such complex systems, while analytical studies are difficult. The emission characteristics of field emitters are affected by the distribution of the field enhancement factor β, which in turn depends on the geometry of the emission system, namely, the shape, aspect ratio and radius of the rounding of the emitter vertices, and the magnitude of the interelectrode gap.

The decrease in the distance between nanoscale emitters increases their number, which can be located at a given area of the cathode, and accordingly the total emission current, but at the same time, the mutual screening effect reduces the fields on their surface and accordingly the total current falls. Thus, it is obvious that there is an optimal distance between the emitters, which makes it possible to achieve the maximum emission current from the cathode of a given area.

The paper [1] presented a three-dimensional model for estimating the electrostatic properties of a single-point field emitter characterized by the geometry of the hemisphere on the cylinder. Numerical simulation was performed by the finite element method using the Ansys-Maxwell software. As a result, the distribution of the electric field on the surface and near the emitter was obtained. The modeling and analytical calculation correlated well with each other. The dependences of the β on the aspect ratio of the emitter and on the ratio of the interelectrode distance to the emitter height were obtained. In work [2] a simulation of electrostatic fields in COMSOL Multiphysics has been created to predict the field enhancement factor β of real emitters with different tip radii. In article [3], the effect of mutual screening of emitters consisting of an array of carbon nanotubes (CNTs) was investigated using 3D modeling in SIMION 8.0 software and its effect on the efficiency of electron emission was demonstrated. The simulated case of nine CNTs arranged in three rows of three columns showed that the local electric field of the central CNT is lower than one of its neighbors, but less than one of an isolated nanotube. The analytical dependence of the field enhancement factor β on the distance between the emitters in triangular and quadrangular arrays was calculated in [4] using the LCM (Line Charge Model) method. With the help of the dependence obtained, the dependences of the current density on the value of the external electric field were calculated. The obtained dependences have peaks on which the optimum distances between the emitters corresponds to the maximum currents was determined. Similar calculations were conducted in article [5]. The effect of screening at the edges of arrays of tips with the help of analytical modeling by the LCM method was investigated in article [6]. The authors showed that a decrease in the distance between the external and adjacent emitter of the array increases the screening and, consequently, decreases the field enhancement factor of the emitters near the edge of the array.

In this paper we modeled the distribution of the electrostatic field in a symmetric system of multi-point field cathode consisting of several CNTs and carried out a study of the influence of the mutual screening effect on the current-voltage characteristic (IVC) and the field emission area.

2. Modeling of the emission system

The software package COMSOL Multiphysics solves the Laplace equation by the finite element method and extracts values of the electric field at any point of the surface and volume of the investigated model. The system of a multi-point field cathode consists of the nine tips (CNTs) which have a form of cylinders with hemispheres on the vertices. The parameters of the tips in our model are the same: radius $r = 0.5$ nm and height $h = 500$ nm. Tips are located on the "cathode" plane below the "anode" plane (distance between the planes $D_{A\text{-}C}$ was set 2 μm). The Dirichlet boundary condition was specified at the electrodes surfaces: a ground state for "anode" and a positive potential (applied voltage U) for "cathode". To avoid the edge effect, we applied the Neumann boundary condition, so the potential distribution at the edges of the model region (total cylinder where all electrodes were located) has linear changes only in the direction of the Z axis. The distance between central and peripheral emitters d was varied from $0.05h$ to $5h$. The selected voltage range produces an acceptable level of the electrostatic field at the tips vertices (up to $1.5\cdot10^{10}$ V/m) and level of the current which is appropriate for nanotubes (registered and non-destructive).

After the settings of the basic parameters of model in COMSOL the module "electrostatics" automatically selected the morphology and density of the mesh (Fig. 1a). The regions around tips are critical and require the high density of the mesh, therefore we set a triangular grid with side dimensions equal to $5\cdot10^{-10}$ m at the surface of tips vertices (Fig. 1b). For the anode and cathode planes the triangular grid was constructed with the minimum and maximum dimensions of the triangles sides $7.2\cdot10^{-8}$ and $4\cdot10^{-7}$ m, respectively.

As a result of the modeling the potential distribution in the space and on the surface of the system (Fig. 1c) and the contour lines of the electric field distribution (Fig. 1d) were obtained. The maximum value of the electric field was $E_{max} = 8.6\cdot10^{9}$ V/m at a radial distance between the central and peripheral emitters $d = 0.05h$. It is clear that distortion of the electric field contour lines caused by the screening effect of the tips. The field distributions on the cathode surface was modeled for the applied voltage range 17.5 ÷ 55 V at different radial distances d.

3. Calculation of the current density and emission area

To calculate an emission current density j, which depends on the local electric field E, we used the Elinson [7] analytical formula:

$$j = A_{\varphi}(E)^2 \cdot 10^{\frac{B\varphi}{E}}, \tag{1}$$

where A_φ and B_φ are functions that depend on the work function of the material. The work function value we took from [8, 9] for CNT: $\varphi = 4.6$ eV.

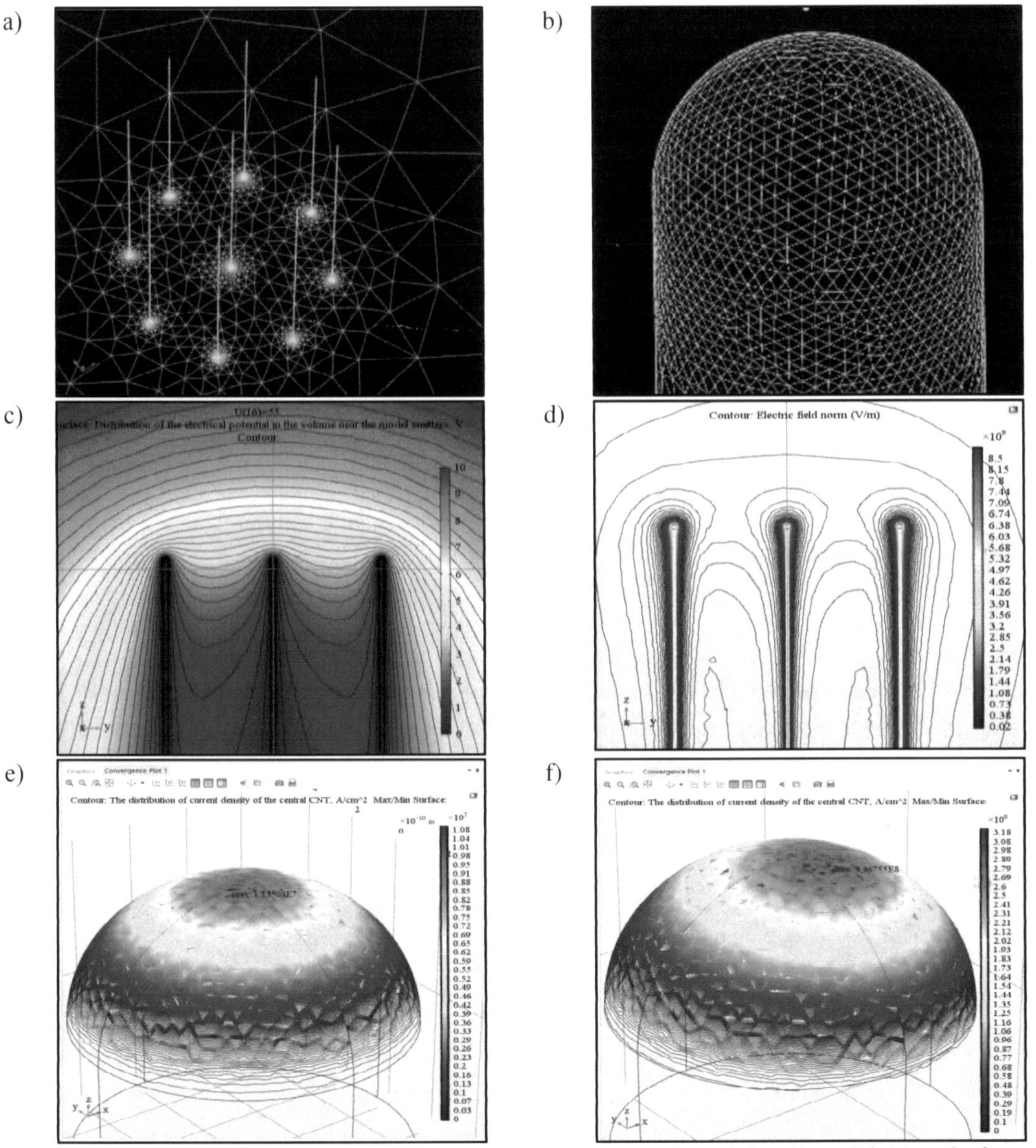

Figure 1. The COMSOL model at $U = 55$ V: calculating meshes of the all model (a) and of one of the tips (b), distribution of the electric potential (c) and of the electric field (d), the distribution of the current density on the surface of central (e) and peripheral (f) tips.

The calculated current density distribution on the surface of the central and outside tips are presented at Fig. 1e,f. The distribution of the outside tip is displaced from the center of the vertex due to influence of the neighboring tips.

To investigate the mutual screening effect of the tips we varied the distance d in the range from $5h$ to $0.05h$ and calculated total current dependence on voltage of the central tip. The corresponding IVCs are presented at Fig. 2a. It is obvious that the IVCs, starting with a distance equal to $5h$ weakly depends on the distance since it has become too large and the tip fields affect each other in a less degree. When the distance d is increasing the compression of the IVC to the abscissa axis is observed as a consequence of the increase of mutual screening on the tips. At $d = 0.05h$ the emission current is of the order of 1 pA.

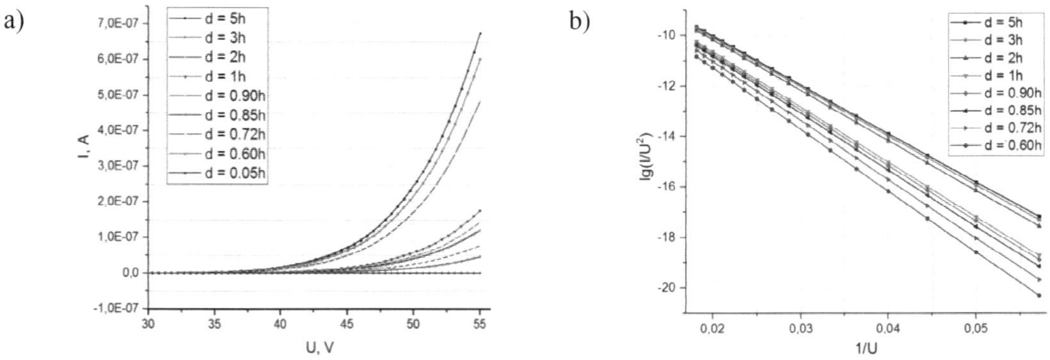

Figure 2. Calculated IVCs (a) and IVC-FNs (b) of the central tip for different inter-tip distances.

Current-voltage characteristics were calculated by integrating the current density of an array of emission objects by COMSOL. As noted earlier, on the one hand, the magnitude of the emission current depends on the number of emitters per unit of surface, but on the other hand the increasing effect of their mutual screening reduces the fields on their surface that entails a total current drop.

The effective current density J_{eff} of the all system:

$$J_{eff}(U,d) = \frac{I_\Sigma}{\pi d^2}, \tag{2}$$

where I_Σ is the total current of the whole array of emitters; *d* is the radial distance between the central and peripheral tips.

To estimate an optimal ratio of radial distance to height of the tips *d/h* we calculated an effective current density J_{eff} at different voltages (Fig. 3a). From the constructed dependences it follows that each curve in the family has a maximum at the corresponding ratio *d/h*. These maxima were shifted to the left with increasing in the applied voltage: from 1.1 to 0.7. Fig. 3b shows this shifting of the maximum for the minimum and maximum voltages (37.5 V and 55 V).

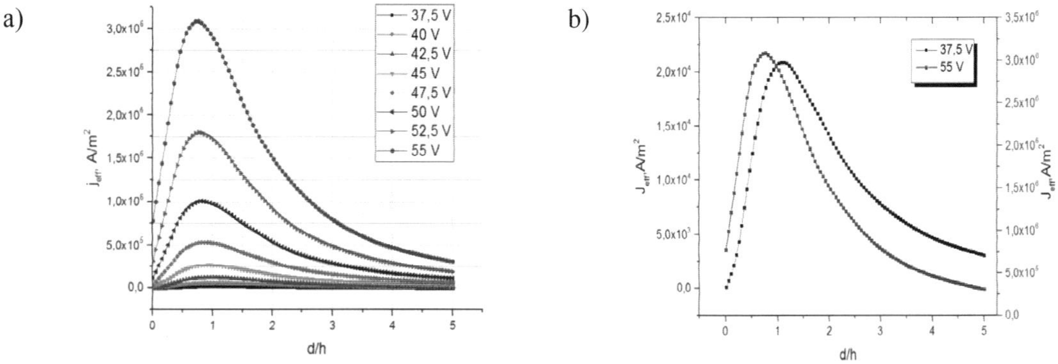

Figure 3. Dependence of the calculated effective current density J_{eff} of all emitter array on the ratio *d/h* for different applied voltages (a) and for minimum and maximum applied voltages (b).

The calculation of the field emission area was made in two ways. The first way is analytical. It requires calculation of the current density at the top of the emitter, an experimental measurement of which cannot be performed [10, 11]. Corresponding area S_c (so called "notional area") is:

$$S_c = \frac{I}{j_{max}} \tag{3}$$

where j_{max} – the maximum density of the emission current calculated by formula (1), I – the current computed by integrating j over the surface of the CNT (rounded end face of the model emitter) by means of COMSOL.

The obtained family of dependencies S_c on the applied voltage U at the different distances d are presented at Fig. 4a. As can be seen, the value of S_c nonlinear decreases with the decreasing in inter tip distance d at a fixed voltage.

The second way is experimental, since it uses the values of the experimental IVCs. It requires calculation of the effective slope b and intercept a of the IVC in the Fowler-Nordheim coordinates (IVC-FN), which can be obtained using approximation of the IVC-FN by a straight line. Corresponding effective area S_{eff} [9] is:

$$S_{eff} = \frac{10^a b^2}{A_\varphi B_\varphi^2} \tag{4}$$

Note that S_{eff} is an integral characteristic, averaging the IVC over the entire voltage range.

To estimate the effective field emission area S_{eff} and its dependence on the voltage, we translated the IVCs of the central tip into the Fowler-Nordheim coordinates (Fig. 2b) and estimated cutoffs of the corresponding trend lines.

The effective field emission area dependence on the inter tip distance d is presented at Fig. 4b. The emission area increased to 1 nm^2. It is 63% of the total area of the hemisphere vertex which is equal to 1.57 nm^2. Comparison of the emission areas S_c and S_{eff} has shown that the effective area exceeds the conditional area over the entire voltage range and for all values of d/h. The ratio S_{eff}/S_c for d/h = 0.6 with the increase in voltage varies from 4.9 to 1.9, and for d/h = 5 from 3.7 to 1.5.

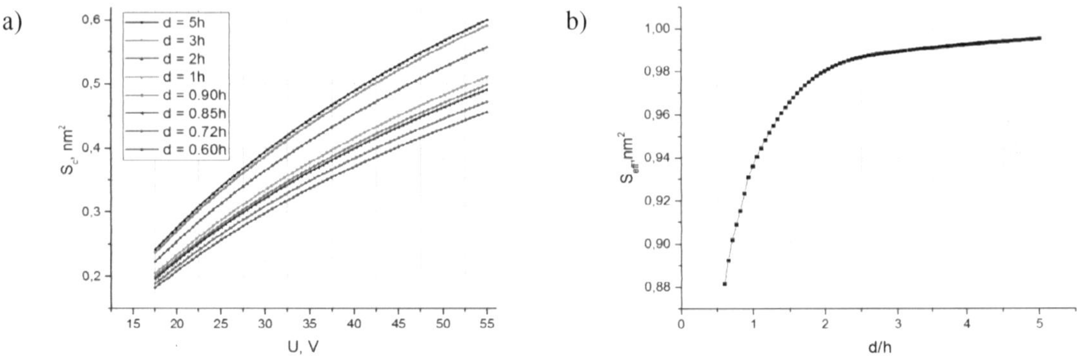

Figure 4. Calculated field emission areas of the central tip for variance of the ratio d/h from 0.6 to 5: the notional area S_c depending on the applied voltage (a) and the effective area S_{eff} averaging the IVC over the entire voltage range (b).

4. Conclusion

The distribution of the electric fields on the surface of a nanostructured multi-point field cathode consisting of the nine parallel conducting tips was modeled in COMSOL Multiphysics. By post-processing of the simulation results the distribution of the emission current density over the emitter surface was obtained and corresponding IVCs were constructed.

To estimate an optimal inter-tip distances d when effective current density j_{eff} is maximal we calculated J_{eff} at different voltages in range 37.5 to 55 V. It is established that these maxima are subject to displacement to the left with increasing applied voltage. The obtained results confirm the similar calculations made in the works of leading foreign scientists in the branch of field emission [4-5].

The change in the value of the optimum distance between the emitters with increasing voltage level is associated with a change in the relative contribution of each CNT to the total I-V characteristic. This change is reflected in the IVC-FN form, making it different from the straight line. This curvature can be seen on the plot of the effective area versus the voltage Fig. 4b, which is determined by cutting off the trend line.

To estimate the effective emission area S_{eff}, the IVCs of the central tip were translated into the FN coordinates. It's clearly seen, that emission area increased to 1 nm^2, what is 63% of the total area of the vertex of the tip, equal to 1.57 nm^2.

The dependence of the notional field emission area S_c of the central tip on the applied voltage was obtained. From the obtained family of dependences, it is shown there is a nonlinear decrease with the decreasing in the inter tip distance d at a fixed voltage.

It is worth noting that the presented solution for field distribution is valid for any system with appropriate scaling, especially for carbon nanotubes. But the shielding effect most important for multi-tip nanoscaled field emission systems with distributed current load. Our results are of interest for investigations of field emission systems with a dense arrangement of CNTs [12]. Such systems are widely used in field emitters, which consist of ordered CNT and tips arrays [13-14].

References

[1] Roveri D S, Sant'Anna G M, Bertan H H, Mologni J F, Alves M A R and Braga E S 2016 Simulation of the enhancement factor from a nindividual 3D hemisphere-on-post field emitter by using finite elements method *Ultramicroscopy* **160** 247–251

[2] Dusberg F 2014 *Electron Field Emission from Silicon Emitter Arrays* (Munchen: Hochschule Munchen) 41–43

[3] Smith R C and Silva S R P 2009 Maximizing the electron field emission performance of carbon nanotube arrays *Appl. Phys. Lett.* **94** 133104-1–7

[4] Harris J R, Jensen K L and Shiffler D A 2015 Dependence of optimal spacing on applied field in ungated field emitter arrays *AIP Advances* **5** 087182-1–7

[5] Bocharov G S, Eletskii A V, Sommerer T J 2011 Optimization of the parameters of a carbon nanotube-based field emission cathode *Tech. Phys.* **56** 540–545

[6] Harris J R, Jensen K L and Shiffler D A 2016 Edge enhancement control in linear arrays of ungated field emitters *J. Appl. Phys.* **119** 043301-1–8

[7] Elinson M I 1974 *Cold Cathodes* (Moscow: Sovetskoe Radio) 336

[8] Gao R, Pan Z and Wang Z 2001 *Appl. Phys. Lett.* **78** (12) 1757–1759

[9] Liu H, Kato S and Saito Y 2009 Empirical expression for the emission site density of nanotube film emitters *Nanotechnology* **20** 275206-1–6

[10] Podenok S, Sveningsson M, Hansen K and Campbell E E B 2006 Electric field enhancement factors around a metallic, end-capped cylinder *NANO: Brief Reports and Reviews* **1** (1) 87–93

[11] Tang W W, Shiffler D A, Harris J R, Jensen K L, Golby K, LaCour M and Knowles T 2016 Field emission characteristics of a small number of carbon fiber emitters *AIP Advances* **6** 095007-1–9

[12] Fujii S, Honda S-i, Machida H, Kawai H, Ishida K, Katayama M, Hiroshi F, Takashi H, Kenjiro O 2007 Efficient field emission from an individual aligned carbon nanotube bundle enhanced by edge effect *Appl. Phys. Lett.* **90** 153108-1–3

[13] Li Z, Yang X, He F, Bai B, Zhou H, Li C, Dai Q 2015 High current field emission from individual non-linear resistor ballasted carbon nanotube cluster array *Carbon* **89** 1–7

[14] Pandey A, Prasad A, Moscatello J P, Engelhard M, Wang C and Yap Y K 2013 Very stable electron field emission from strontium titanate coated carbon nanotube matrices with low emission thresholds *ACSNANO* **7** (1) 117–125

IOP Conf. Series: Journal of Physics: Conf. Series **1038** (2018) 012122 doi:10.1088/1742-6596/1038/1/012122

Intraband light absorption by holes in InGaAsP/InP quantum wells

N V Pavlov[1], **G G Zegrya**[1]

1 Ioffe Institute, 194021, Russian Federation, St Petersburg, Politekhnicheskaya st., 26

E-mail: pavlovnv@mail.ru

Abstract. A microscopic analysis of the radiation intraband absorption mechanism by holes with their transition to a spin-split band for quantum wells based on InGaAsP/InP solid solutions is performed within the framework of the four-band Kane model. The calculation is made for two polarizations of the incident radiation: along the crystal growth axis and in the plane of the quantum well. It is shown that this process can be the main mechanism of internal radiation losses for quantum well lasers. It is also shown that the dependence of the absorption coefficient on the width of the quantum well has a maximum at a well width from 40 to 60 A.

1. Introduction

The mechanisms radiation intraband absorption mechanisms in QW lasers are studied both theoretically and experimentally for many years [6]. Experimental results [7,8] show that the laser radiation intraband absorption coefficient is substantially higher than predicted by the theory [9]. One of the candidates for explaining these results is the process radiation intraband absorption process by holes with the transition to the spin-split (so) zone. In this paper, we will use the modification of the four-band Kane model [10] proposed by Polkovnikov and Zegrya [11,12], which is based on the use of the 8×8 ***kP*** Hamiltonian, and allows us to obtain explicit analytical expressions for energy spectra and wave functions of charge carriers, as well as matrix elements of transitions. The authors proposed a modification of this method, allowing to take into account the elastic stresses arising in mismatched heterostructures.

The aim of this work is to calculate the radiation intraband absorption coefficient by holes with their transition to the so-zone for QWs based on A_3B_5 semiconductors, and also to study the dependence of the results obtained on the electromagnetic wave polarization, temperature, hole concentration, and QW width. The calculations are based on the example of the InGaAsP/InP heterostructure, which is widely used in the construction of semiconductor lasers with a radiation wavelength of 1.55 μm, and the structure parameters are taken from [13].

2. Basic Equations

We used 8×8 ***kP*** Hamiltonian [11] that takes interaction with the higher bands and terms arising from elastic stresses into account up to quadratic terms by the wave vector but neglects the relativistic linear terms and the term with heavy electron mass.

For our calculations, we choose the following representation of the basis wave functions:

$$|s\uparrow\rangle, |s\downarrow\rangle, |x\uparrow\rangle, |x\downarrow\rangle, |y\uparrow\rangle, |y\downarrow\rangle, |z\uparrow\rangle, |z\downarrow\rangle$$

where the spinors s and x, y, z are Bloch functions of s- and p-type with angular momentum 0 and 1, respectively. The s-functions describe the conduction band states and the p-functions describe the valence band states at the Γ-point. The arrows indicate the spin direction. The carrier wave function ψ can be represented in the form:

$$\psi = \Psi|\boldsymbol{u}\rangle = \Psi_s|s\rangle + \Psi_p|\boldsymbol{p}\rangle$$

Where $\boldsymbol{\Psi}$, Ψ_s, $\boldsymbol{\Psi}_p$ are spinors.

The Kane equations in the spherical approximation near the Γ-point are given by following expressions:

$$(E_C - E)\Psi_s - i\hbar\gamma\nabla\Psi_p = 0,$$

$$(E_V-\delta-E)\Psi_p-i\hbar\gamma\nabla\Psi_s+\frac{\hbar^2}{2m_0}(\gamma_1+4\gamma_2)\nabla(\nabla\Psi_p)-$$
$$\frac{\hbar^2}{2m_0}(\gamma_1-2\gamma_2)[\nabla\times[\nabla\times\Psi_p]]+i\delta[\sigma\times\Psi_p]=0 \quad (1)$$

Here E_C and E_V are the energies of conduction and valence band edges, γ is the Kane matrix element with dimension of velocity, $\delta=\Delta_{so}/3$ is the spin-orbit splitting constant, m_0 is the free electron mass, γ_1 and $\gamma_2=\gamma_3$ are the generalized Luttinger parameters, $\boldsymbol{\sigma}$ is the Pauli matrices, $\boldsymbol{k}$ is the wave vector and E is the energy.
The absorption coefficient per one QW in the framework of the Kane model can be calculated from the formula:

$$\alpha_p^d=\frac{2\gamma^2e^2}{\hbar^2c}\frac{1}{n\omega a}\sum_{k_h}\sum_{k(0)}m_r^d\sigma_p^d f(E_h), \quad (2)$$

for the transitions into so-holes discrete spectrum and

$$\alpha_p^c=\frac{\gamma^2e^2}{2\pi\hbar^2c}\frac{m_h}{n\omega a}\sum_{k_h}\int_{E_h^{\min}}^{E_h^{\max}}\frac{\sigma_p^c f(E_h)}{k\left|\frac{\partial E(k^2)}{\partial k^2}\right|}dE_h, \quad (3)$$

for the transitions into so-holes continuous spectrum.
Here, the index p denotes the polarization of light and takes on a value $\perp$ for a wave polarized along the crystal growth axis (i. e. when the amplitude of the vector potential A_0 is parallel to the x axis) and the value $\|$ for a wave polarized in the QW plane, n is the refractive index, ω is the optical transition frequency, a is the QW width, k and k_h are the wave vector x components of the so and heavy holes, respectively, E and E_h are the energy of the so and heavy holes, respectively, $f(E)$ is the Fermi-Dirac distribution function, $m_h=\frac{m_0}{\gamma_1-2\gamma_2}$ is the heavy hole effective mass, $m_r=\frac{m_hm_{so}}{|m_h-m_{so}|}$, m_{so} is the so-hole effective mass. The value of σ is determined by the overlap integrals of the wave functions:

$$\sigma_\perp=2\pi\left|I_x^{j,a}\right|^2$$
$$\sigma_\|=\pi(\left|I_y^{j,s}\right|^2+\left|I_z^{j,s}\right|^2), \quad (4)$$

Where $I_i=\int_{-a/2}^{a/2}\Psi_{h,i}^*\Psi_s dx$.

3. Results

In Fig. 1a and 1b the frequency dependences $\alpha_\perp^d$ and $\alpha_\|^d$ at different temperatures in the InGaAsP/InP heterostructure at a=80A and the hole concentration $p=10^{12}$ cm^{-2} are shown. It can be seen that at T=150K more sharp peaks with a larger amplitude are observed than at T=300K. It is also clear that as the temperature is raised, the transition energy corresponding to the absorption peaks increases. It is also noticeable that the radiation polarized in the plane of the QW is absorbed more efficiently than the radiation of the transverse polarization. In the calculations, we assumed that the carrier concentration is constant.

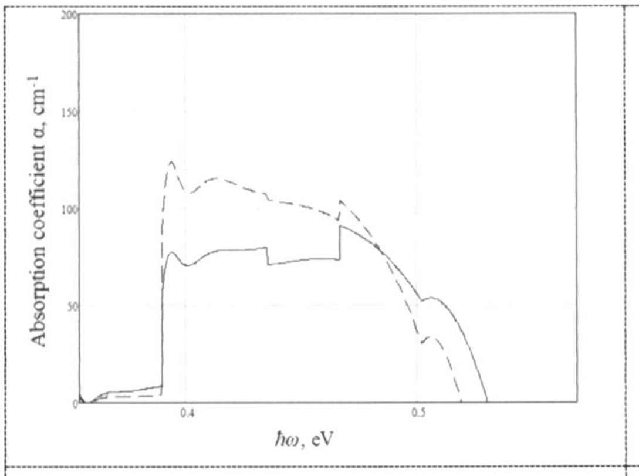

Figure 1a. Dependence of the absorption coefficient $\alpha_{\parallel}^{d}$ on the discrete levels of so-holes on the incident radiation frequency for different temperatures in the InGaAsP/InP heterostructure with the QW width a=80A and the hole concentration $p=10^{12}$ cm^{-2}. The solid curve shows the calculation result for the temperature T=300K, dotted for the temperature T=150K.

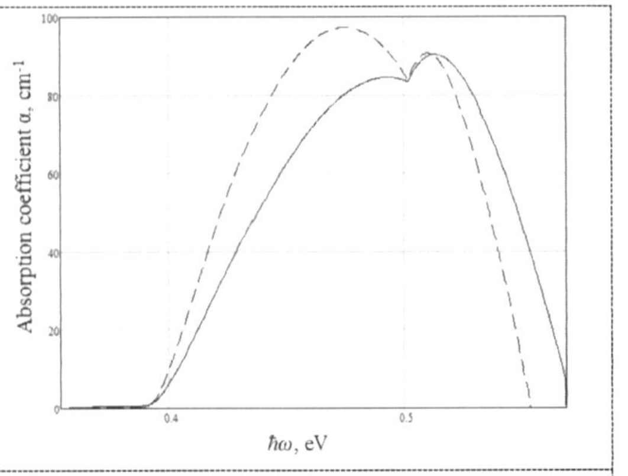

Figure 1b. Dependence of the absorption coefficient $\alpha_{\perp}^{d}$ on the discrete levels of so-holes on the incident radiation frequency for different temperatures in the InGaAsP/InP heterostructure with the QW width a=80A and the hole concentration $p=10^{12}$ cm^{-2}. The solid curve shows the calculation result for the temperature T=300K, dotted for the temperature T=150K.

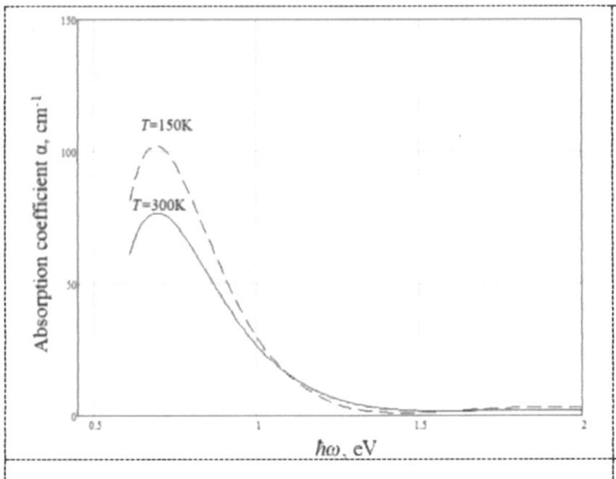

Figure 2a. Dependence of the absorption coefficient $\alpha_{\parallel}^{c}$ on the continuous spestrum of so-holes on the incident radiation frequency for different temperatures in the InGaAsP/InP heterostructure with the QW width a=80A and the hole concentration $p=10^{12}$ cm^{-2}. The solid curve shows the calculation result for the temperature T=300K, dotted for the temperature T=150K.

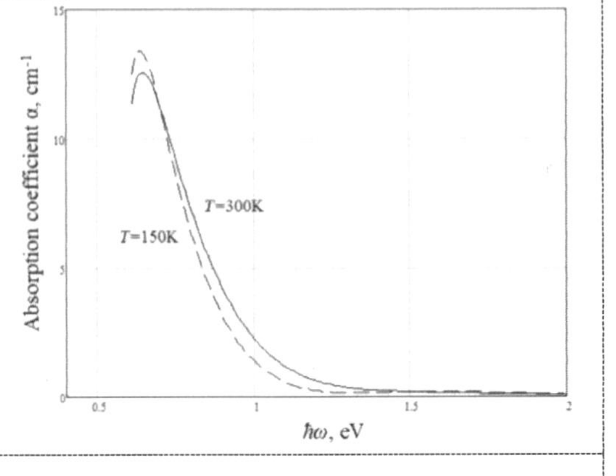

Figure 2b. Dependence of the absorption coefficient $\alpha_{\perp}^{c}$ on the continuous spestrum of so-holes on the incident radiation frequency for different temperatures in the InGaAsP/InP heterostructure with the QW width a=80A and the hole concentration $p=10^{12}$ cm^{-2}. The solid curve shows the calculation result for the temperature T=300K, dotted for the temperature T=150K.

In Fig. 2a and 2b the frequency dependences $\alpha_{\perp}^{c}$ and $\alpha_{\parallel}^{c}$ at different temperatures in the InGaAsP/InP heterostructure at a=80A and the hole concentration $p=10^{12}$ cm^{-2} are shown. First, it should be noted that the absorption maximum with transition to the continuous spectrum is observed at a transition energy of about 0.7 eV, which is close to the value of the optical transition energy at a wavelength of 1.55 μm. Hence it can be concluded that the light absorption by heavy holes with the transition to the continuous spectrum of so-holes can make a large contribution to the value of internal radiation losses and explain the large value of the internal loss coefficient observed in the experiment [7,8]. Another result is that for transitions to the continuous spectrum, the predominance of absorption of radiation polarized in the QW plane is much stronger than for transitions to the discrete spectrum.

In Fig. 3 the dependence of the absorption coefficient at the generation wavelength on the QW width at room temperature and hole concentration $p=10^{12}$ cm^{-2} is shown. It can be seen that the maximum absorption will be observed at values of the QW width from 40 to 60 A, and for the values less than 30 A and greater than 80 A, the absorption will be much weaker.

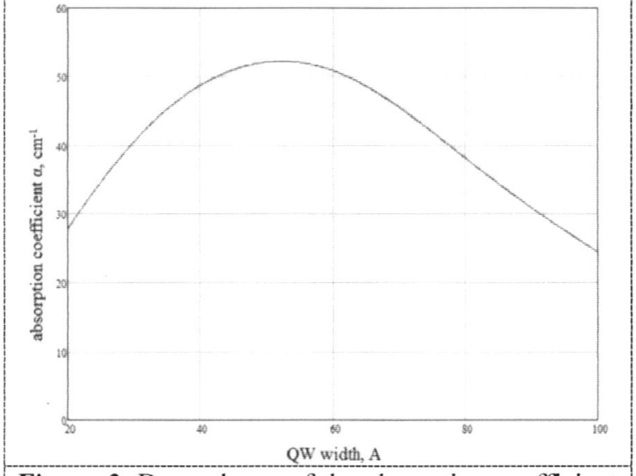

Figure 3. Dependence of the absorption coefficient to the continuous spectrum of spin-split holes on the width of a quantum well in a heterostructure InGaAsP/InP at room temperature and hole concentration 10^{12} cm^{-2}.

4. Conclusion

A microscopic analysis of the radiation intraband absorption mechanism by holes with a transition to the so-band for QWs based on A_3B_5 semiconductors is performed. The cases of transverse and longitudinal polarization of the incident radiation are considered separately. The analysis is performed for the cases of transition to a discrete and a continuous spectrum of so-holes. It is shown that the intraband absorption can be the main mechanism of internal optical losses for semiconductor lasers on QWs.

In quantum wells, the absorption that occurs with the transition to the discrete spectrum of the so-holes is stronger than the absorption that occurs with the transition to the continuous spectrum. The frequency dependence of the absorption coefficient due to transitions to the discrete spectrum of so-holes has pronounced maxima, the position and magnitude of which depends on the temperature. With increasing temperature, the peak shifts towards higher energies, and the peak value of the absorption coefficient decreases. Also, the absorption of radiation polarized in the plane of the quantum well predominates over the absorption of the radiation of transverse polarization.

The value of the absorption coefficient with transition to the discrete spectrum decreases with increasing value of the energy of the forbidden band of the material. For transitions to the continuous spectrum, a pronounced dependence on the width of the forbidden band is not observed. In this case the inverse proportion of the absorption coefficient to the transition energy plays a more significant role.

5. Aknowledgements

This work is supported in part by RFBR grant № 16-08-01130-A.

References

[1] Thijs P J A, van Dongen T, Tiemeijer L F, Binsma J J M 1994 *J. Lightwave Technol.* **12** 28
[2] Piprek J, Abraham P, Bowers J E 2000 *IEEE J. Quantum Electron.* **36** 366
[3] Silver M, O'Reilly E P 1995 *IEEE J. Quantum Electron.* **31** 1193
[4] Agrawal G P, Dutta N K 1993 *Long-Wavelength Semiconductor Lasers* Van Nostrand Reinhold, New York
[5] Zory P S 1993 *Quantum Well Lasers* Academic Press
[6] Rosencher E, Levine B 1992 *Intersubband Transitions in Quantum Wells* Plenum, New York
[7] Ackerman D A, Shtengel G E, Hybertsen M S, Morton P A, Kazarinov R F, Tanbun-Ek T, Logan R A 1995 *IEEE Select. Topics Quantum Electron*. **1** 250
[8] Phillips A F, Sweeney S J, Adams A R, Thijs P J A 1998 IEEE *Select. Topics Quantum Electron.* **5** 401
[9] Asryan L V, Gun'ko N A, Polkovnikov A S, Zegrya G G, Suris R A, Lau P.-K, Makino T 2000 *Semicond. Sci. Technol.* **15** 1132
[10] Kane E O 1957 *J. Phys. Chem. Sol.* **1** 249
[11] Zegrya G G and Polkovnikov A S 1998 *J. Exp. Theor. Phys.* **86** 815
[12] Polkovnikov A S, Zegrya G G 1998 *Phys. Rev. B.* **58** 4039
[13] Zegrya G G, Pikhtin N A, Skrynnikov G V, Slipchenko S O, Tarasov I S 2001 *Semiconductors* **35** 962

IOP Conf. Series: Journal of Physics: Conf. Series **1038** (2018) 012123 doi:10.1088/1742-6596/1038/1/012123

Interplay between angular and quantum magnetoresistance oscillations

Grigoriev P.D.[1,2,3], **Mogilyuk T.I.**[4]
[1] L. D. Landau Institute for Theoretical Physics, 142432 Chernogolovka, Russia
[2]National University of Science and Technology MISiS, Moscow 119049, Russia
[3]P.N. Lebedev Physical Institute, RAS, 119991, Moscow, Russia
[4]National Research Centre "Kurchatov Institute", Moscow, Russia
E-mail: grigorev@itp.ac.ru

Abstract. We investigate the mutual influence of angular (AMRO) and quantum oscillations (MQO) of interlayer magnetoresistance $R_{zz}(\theta)$ in quasi-two-dimensional (quasi-2D) layered metals. The MQO and the shape of Landau levels (LL) affects the AMRO amplitude. The influence of AMRO on MQO leads to the new qualitative effect: the angular oscillations of the amplitude of MQO, called "false spin zeros", that could lead to a wrong determination of electron g-factor from experiment.

1. Introduction

Determination of the electronic structure of strongly-correlated electronic conductors is a fundamental problem of modern condensed state physics. Most actively studied conductive materials, such as high-temperature cuprate and iron-based superconductors, organic metals and other strongly correlated electron systems, are usually characterized by high quasi-2D anisotropy. Angular and field dependences of magnetoresistance (MR) and magnetization, in particular quantum magnetic oscillations, make it possible to reveal even subtle details of the electronic dispersion and the geometry of the Fermi surface.

The Fermi surface in quasi-2D system under study is a slightly corrugated cylinder with strongly anisotropic spectrum, approximately given by:

$$\epsilon_{3D}(\mathbf{k}) \approx \epsilon_{2D}(k_x,\, k_y) - 2t_z \cos(k_z d), \tag{1}$$

where $\epsilon_{2D}(k_x,\, k_y) = \hbar^2(k_x^2 + k_y^2)/(2m^*)$ represents an isotropic in-plane dispersion with effective mass m^*, t_z is the interlayer transfer integral, which is much less than the Fermi energy E_F and is assumed to be independent on in-plane momentum, and d is the interlayer distance.

The angular magnetoresistance oscillations in quasi-2D metals were discovered in [1], and explained in [2] and [3]. The first expression for interlayer conductivity as a function of tilt angle θ of magnetic field was obtained in tau-aproximation from the semiclassical equations of electron motion [3]:

$$\frac{\sigma_{zz}(\theta)}{\sigma_{zz}^0} = J_0^2(\kappa) + 2\sum_{p=1}^{+\infty} \frac{J_p^2(\kappa)}{1+(p\omega_c\tau)^2}, \tag{2}$$

where $\sigma_{zz}^0 = 2e^2 m^* t_z^2 \tau d/(\pi\hbar^4)$ is the Drude conductivity along the z-direction in zero magnetic field, $J_p(\kappa)$ is the Bessel's function, τ is the electron mean free time, the cyclotron frequency ω_c in quasi-2D metal depends on the tilt angle θ of magnetic field as $\omega_c \equiv eB_z/(m^*c) = \omega_{c0}\cos(\theta)$, where B_z is component of magnetic field perpendicular to the conducting layers, c is light velocity, $\kappa \equiv k_F d \tan\theta$, and k_F is the in-plane Fermi momentum.

Eq. (2) predicts sharp minima of interlayer magnetoconductivity $\sigma_{zz}(\theta)$ at the so-called Yamaji angles, corresponding to the zeros of the Bessel function $J_0(\kappa)$. These Yamaji angles have been observed in many experiments [4] and used to extract the value of $k_F d$. At the Yamaji angles the first term of (2) is zero, and the second one predicts the quadratic field dependence of interlayer conductivity. Nevertheless, there are experimental data in organic metalls, that can not be described with quadratic field dependence.

Beforementioned Eq. (2) was confirmed in Ref. [5] with the help of Kubo formula in the two-layer model in the limit $\Gamma_0 \gg t_z$ ($\Gamma_0 = \hbar/(2\tau_0)$ is the electron-level width due to scattering on impurities in zero-magnetic field), where electrons on average scatter on impurities many times before tunneling on the adjacent conducting layer. Therefore, this limit was called "weakly incoherent" in Ref. [5]. However, since the obtained formulas for conductivity turned out to be the same as in the coherent limit, in Ref. [6] this limit was named as "weakly coherent". Also, in Ref. [6, 7] it was suggested that in the limit $t_z \ll \hbar\omega_c$ a crossover between two qualitatively different regimes is driven by another parameter $\hbar\omega_c/\Gamma_0$. At $\hbar\omega_c/\Gamma_0 > \pi$ the LLs become isolated, because the growth of the LL broadening [8, 9] $\Gamma(B_z) \propto \sqrt{B_z}$ is slower than that of the LL separation $\hbar\omega_c \propto B_z$. This leads, e.g., to longitudinal interlayer magnetoresistance [13]–[17] contradicting the prediction of tau-approximation. Within the two-layer model it was shown that at $\hbar\omega_c/\Gamma_0 \gg 1$ MQO generate this monotonic growth [13]–[16] of the interlayer magnetoresistance $R_{zz}(B_z) = 1/\sigma_{zz}$. This result was confirmed within the anisotropic 3D model [17]. In Ref. [17] it was also shown that the two-layer model is applicable in the limit $t_z \ll \sqrt{\hbar\omega_c\Gamma_0}$.

The effect of MQOs on AMROs was explored in Ref. [18]. It was shown that the shape of LLs influences AMRO. The amplitude is stronger for the Gaussian or dome-like LL shapes, suitable for the microscopic models at $t_z \ll \Gamma_0 \ll \hbar\omega_c$, than for commonly used Lorentzian LL shape. The effect of AMROs on MQOs was investigated in Ref. [19], where it was shown that factorization of magnetoconductivity by purely angle $\Phi_{AMRO}(\theta)$ and field factors $\sigma_{zz}^{MQO}(B_z)$ is not generally valid: $\sigma_{zz}(\mathbf{B}) \neq \Phi_{AMRO}(\theta)\sigma_{zz}^{MQO}(B_z)$. Escpecially it fails at large $\omega_c\tau$ and near Yamaji angles. This leads to the appearance of "false spin zeros" on the angular dependence of MR, which may be confused with the true spin zeros used to extract the electron g-factor.

In spite of this extensive investigation, a number of important problems remained. In particular, a detailed theoretical study of the behavior at the Yamaji angles is required to describe the experimental data in organic metals and in other quasi-two-dimensional structures.

In this work, we derive a more convenient expression for the angular dependence of interlayer conductivity and with its help we calculate the field dependence of MR at the Yamaji angles. In doing so, we consider various cases. In particular, the presence of a reservoir is necessary to describe the experimental data in some layered compounds, where the reservoir may come from quasi-one-dimensional parts of the Fermi surface. It is interesting to see how the reservoir states affect conductivity. Since Eq. (2) was obtained within the framework of the approximate model, namely, in the τ-approximation [3] or neglecting the MQO [5], in our calculations we use a more rigorous two-layer model that takes into account MQO in the limit $t_z \ll \sqrt{\hbar\omega_c\Gamma_0}$.

2. General formulas

In Ref. [19] withing the two-layer model, applicable at $t_z \ll \sqrt{\hbar\omega_c\Gamma_0}$, for the case of coherent (i.e., conserving the in-plane momentum) interlayer tunneling and neglecting the electron-electron interaction (approximately valid when many LL are filled) the following formula for interlayer conductivity in tilted magnetic field was derived:

$$\frac{\sigma_{zz}(\epsilon)}{\sigma_{zz}^{0}} = \frac{\Gamma_0}{\tilde{\Gamma}(\epsilon)} \sum_{p=-\infty}^{+\infty} S_p \left[J_p(\kappa)\right]^2 , \tag{3}$$

where $\varepsilon^* = \varepsilon - Re\Sigma^R(\varepsilon)$, $\gamma \equiv -2\pi Im\Sigma^R(\varepsilon)/\hbar\omega_c$, $\alpha \equiv 2\pi\varepsilon^*/\hbar\omega_c$, the function

$$S_0 \equiv \sum_{n\in Z} \frac{(2/\pi)\hbar\omega_c\tilde{\Gamma}^3}{\left[\left(\epsilon^* - {}^-h\omega\left(n+\frac{1}{2}\right)\right)^2 + \tilde{\Gamma}^2\right]^2} = \frac{\sinh(\gamma)}{(\cos(\alpha)+\cosh(\gamma))} - \gamma\frac{1+\cos(\alpha)\cosh(\gamma)}{(\cos(\alpha)+\cosh(\gamma))^2} \tag{4}$$

in agreement with Eq. (23) of Ref. [20], and for $p \neq 0$

$$S_p \equiv \sum_{n\in Z} \frac{(2/\pi)\,\hbar\omega_c\tilde{\Gamma}^3}{\left[(\epsilon^* - {}^-h\omega(n+1/2))^2 + \tilde{\Gamma}^2\right]} \frac{1}{\left[(\epsilon^* - {}^-h\omega(n+p+1/2))^2 + \tilde{\Gamma}^2\right]} = \frac{\sinh(\gamma)}{(\cos(\alpha)+\cosh(\gamma))\left[1+(p\pi/\gamma)^2\right]}. \tag{5}$$

In [21] one can find the following relation:

$$\sum_{k=1}^{+\infty} \frac{J_k^2(z)}{(k^2-a^2)} = \frac{J_0^2(z)}{2a^2} - \frac{\pi J_a(z)J_{-a}(z)}{2a\sin(\pi a)}. \tag{6}$$

Replacement $a \to ia$ gives us the following formula:

$$\sum_{k=1}^{+\infty} \frac{J_k^2(z)}{(k^2+a^2)} = -\frac{J_0^2(z)}{2a^2} + \frac{\pi J_{ai}(z)J_{-ai}(z)}{2a\sinh(\pi a)}. \tag{7}$$

Using the property $J_{ai}(z) = J_{-ai}^*(z)$ and relation (7) one can deduce from (3):

$$\frac{\sigma_{zz}}{\sigma_{zz}^0} = \gamma_0 \left(\frac{|J_{\gamma i/\pi}(\kappa)|^2}{(\cos(\alpha)+\cosh(\gamma))} - J_0^2(\kappa)\frac{1+\cos(\alpha)\cosh(\gamma)}{(\cos(\alpha)+\cosh(\gamma))^2} \right), \tag{8}$$

where $\gamma_0 \equiv 2\pi\Gamma_0/(\hbar\omega_c)$. In Eq. (8) $\cos(\alpha) = 1$ corresponds to the minimum of interlayer conductivity or maximum of resistivity, and $\cos(\alpha) = -1$ vice versa.

This formula is quite general and convenient for numerical calculations. It differs from Eq. (2) in two aspects. First, it contains field-dependent self-energy part γ, which leads to qualitatively different behavior in the high-field limit $\omega_c\tau \gg 1$. Second, it does not contain infinite series, which is especially helpful in the low-field limit $\omega_c\tau \ll 1$.

3. Influence of quantum on angular oscillations of magnetoresistance

The LL quantization leads to the MQO of the electron DoS on the Fermi level. While the period of the DoS oscillations is determined by the Fermi-surface geometry, the shape of DoS oscillations depend on the compound, on type of disorder, on the strength of magnetic field, etc Below we analyze how the shape of LLs influences AMRO.

The case of Lorentzian LL shape is simple: the electron self-energy $Im\Sigma = \Gamma$ is independent of energy and $|ImG(\epsilon, n)| = \Gamma/[(\epsilon-\epsilon_n)^2 + \Gamma^2]$. The Gaussian LL shape is given by

$$|ImG(\epsilon, n)| = \sqrt{2\pi}/\Gamma \exp[-2(\epsilon-\epsilon_n)^2/\Gamma^2]. \tag{9}$$

Now, to find $\gamma(\epsilon)$ we assume that the following identity is approximately valid even in the case of Gaussian LL shape, as it takes place in the self-consistent Born approximation (SCBA):

$$Im\Sigma(\epsilon) \approx n_i u^2(0) \sum_n ImG(\epsilon, n), \tag{10}$$

where n_i is volume concentration of short-range and randomly distributed impurities. Unfortunately, we do not have the strict proof of (10), but the inclusion of diagrams with the intersection of impurity lines in addition to SCBA gives $|ImG(\epsilon, n)|$ close to Gaussian [22]. In the case of SCBA the imaginary part of self-energy function is [7]:

$$\frac{\gamma}{\gamma_0} = \frac{\sinh(\gamma)}{\cosh(\gamma) + \cos(\alpha)}, \tag{11}$$

while the real part of self-energy function is (influence of magnetostriction is neglected):

$$\delta \equiv \alpha - \frac{2\pi\varepsilon}{\hbar\omega_c} = \frac{\gamma_0 \sin(\alpha)}{\cosh(\gamma) + \cos(\alpha)}, \tag{12}$$

Also we took into account semi-phenomenological amendment to γ in magnetic field: $\Gamma \to \Gamma(B_z) \approx \Gamma_0[1 + (2\hbar\omega_c/(\sqrt{\pi}\Gamma_0))^4]^{1/8}$. This approximate formula describes the width of LLs in both limits and during the crossover [7] from the weak to strong-field behavior.

In Figs. 1 and 2 one can see that the maximum of conductivity is not at $\theta = 0$, which clearly demonstrates how quantum oscillations affect on AMRO. This happens because for taken parameters Fermi-level stays in the middle between two closest LLs at $\theta = 0$, where density of electron states is minimal. Suprizingly, on Fig. 2 oscillations for all three shapes of LLs are close to each other.

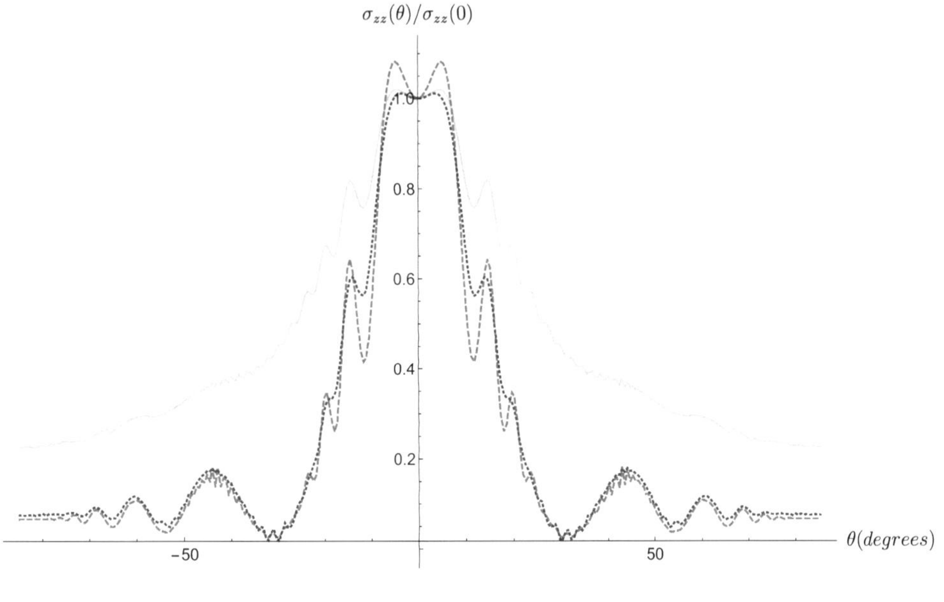

Figure 1. The angular dependence of normalized interlayer conductivity for the Lorentzian LL shape (thin solid green curve), Gaussian LL shape (dashed red curve), SCBA shape (dotted blue curve) with $\Gamma_0 = 4\,K$. The other parameters are $k_F d = 4$, $\mu = 500\,K$, $T = 2\,K$, and $B_0 = 17.4\,T$, which for cyclotron mass $m^* = m_e$ and for $\theta = 0$ corresponds to $\hbar\omega_c = 15\,K$.

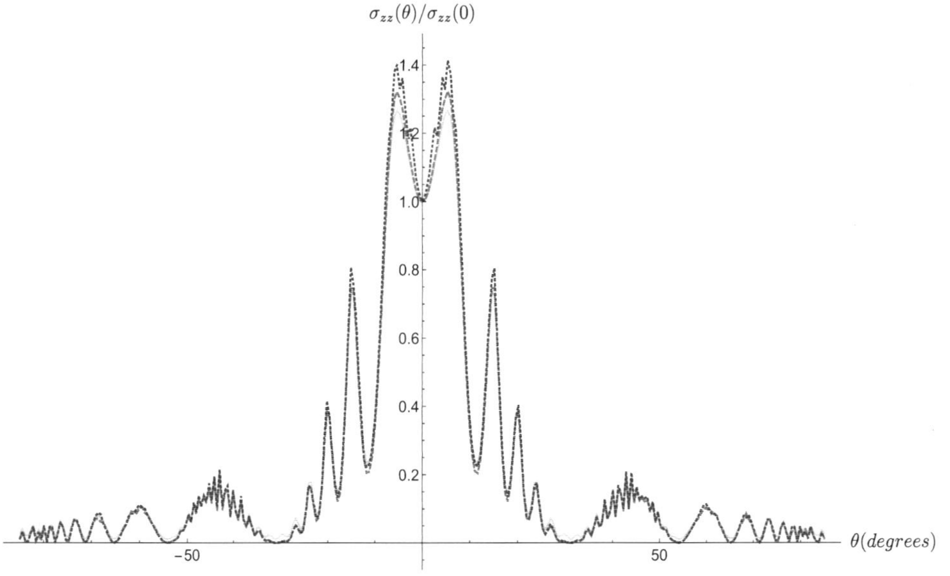

Figure 2. The angular dependence of normalized interlayer conductivity for the Lorentzian LL shape (thin solid green curve), Gaussian LL shape (dashed red curve), SCBA shape (dotted blue curve) with $\Gamma_0 = 0.1\,K$. The other parameters are the same as in Fig. 1.

4. Effect of angular on quantum magnetoresistance oscillations. False spin zeros

If in SCBA we take into account reservoir states that do not give addition to quantum oscillations, we find the imaginary part of self-energy function [20]:

$$\frac{\gamma}{\gamma_0} = \frac{R}{R+1} + \frac{1}{R+1}\frac{\sinh(\gamma)}{\cosh(\gamma) + \cos(\alpha)}, \tag{13}$$

where R is relation of density of states of the reservoir to the density of states of the rest of the system in zero magnetic field. If we neglect the effect of magnetostriction, we find the following expression for the real part of self-energy function (new result):

$$\delta \equiv \alpha - \frac{2\pi\varepsilon}{\hbar\omega} = Const + \frac{1}{R+1}\frac{\gamma_0 \sin(\alpha)}{\cosh(\gamma) + \cos(\alpha)}, \tag{14}$$

where $Const$ depends on the model of reservoir. We consider the case in which the conductivity of reservoir states σ^i_{zz} is much smaller than σ^0_{zz}. This is true if the reservoir states are localized or conductivity is suppressed by magnetic field. The total conductivity is a sum of the coherent and incoherent conductivity channels: $\sigma^{tot}_{zz} = \sigma_{zz} + \sigma^i_{zz}$. Fig. 3 shows the angular dependence of the amplitude of resistivity quantum oscillations for different strengths of reservoir R. One can see strong angular oscillations of the MQO amplitude, which are somewhat similar to the effect of spin-zeros, originating from Zeeman splitting. However, the proposed "false spin zeros" have completely different origin. They come from the interplay of AMRO and MQO, and the minima of MQO amplitude approximately correspond to the Yamaji angles.

Acknowledgments
The work was partially supported by the RFBR grants # 16-02-00522, 18-32-00205, 18-02-00280, 18-02-01022. Sec. 4 was supported by the Russian Science Foundation grant # 16-42-01100.

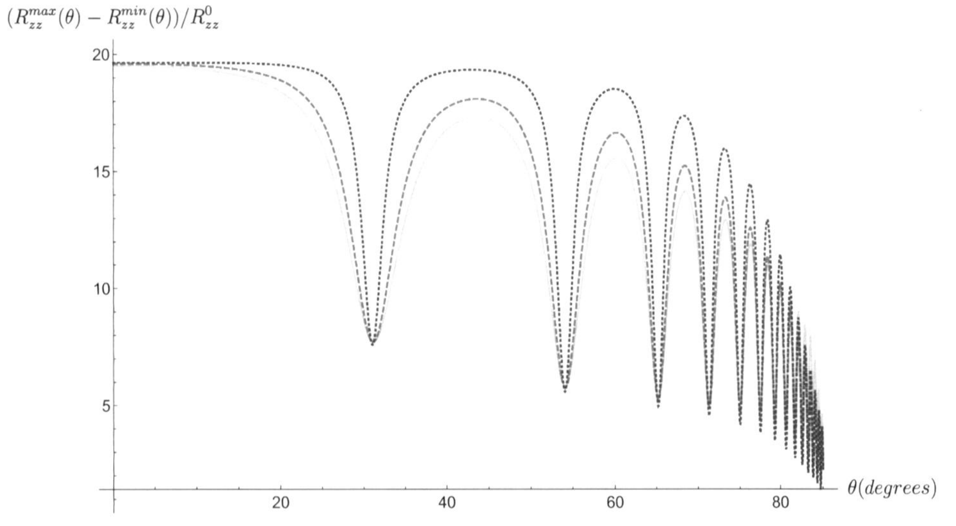

Figure 3. The angular dependence of amplitude of oscillations normalized interlayer resistivity $1/\sigma_{zz}^{tot}$, calculated in SCBA for $R = 0.01$ (green curve), $R = 1$ (dashed red curve) and $R = 100$ (dashed blue curve) with $\Gamma_0 = 0.5\,K$, $\sigma_{zz}^{i} = 0.05\sigma_{zz}^{0}$, $R_{zz}^{0} = 1/\sigma_{zz}^{0}$. The other parameters are the same as in Fig. 1.

References

[1] Kartsovnik M V, Kononovich P A, Laukhin V N, Shchegolev I F 1988 *JETP Lett.* **48** 541
[2] Yamaji K 1989 *J. Phys. Soc. Jpn.* **58** 1520
[3] Yagi R, Iye Y, Osada T, Kagoshima S 1990 *J. Phys. Soc. Jpn.* **59** 3069
[4] Kartsovnik M V 2004 *Chem. Rev.* **104** 5737
[5] Moses P, McKenzie R H 1999 *Phys. Rev.* B **60** 7998
[6] Grigoriev P D, Kartsovnik M V, Biberacher W 2012 *Phys. Rev.* B **86** 165125
[7] Grigoriev A D, Grigoriev P D 2014 *Low Temp. Phys.* **40** 367
[8] Ando T 1974 *J. Phys. Soc. Jpn.* **36** 959
[9] Ando T 1974 *J. Phys. Soc. Jpn.* **36** 1521
[10] Grigoriev P D 2011 *JETP Lett.* **94** 47
[11] Grigoriev P D 2011 *Low Temp. Phys.* **37** 738 [2011 *Fiz. Nizk. Temp.* **37** 930]
[12] Grigoriev P D 2013 *Phys. Rev.* B **88** 054415
[13] Grigoriev P D 2011 *Phys. Rev.* B **83** 245129
[14] Grigoriev P D 2011 *JETP Lett.* **94** 47
[15] Grigoriev P D 2011 *Low Temp. Phys.* **37** 738 [2011 *Fiz. Nizk. Temp.* **37** 930]
[16] Grigoriev P D, Kartsovnik M V, Biberacher W 2012 *Phys. Rev.* B **86** 165125
[17] Grigoriev P D 2013 *Phys. Rev.* B **88** 054415
[18] Grigoriev P D, Mogilyuk T I 2014 *Phys. Rev.* B **90** 115138
[19] Grigoriev P D, Mogilyuk T I 2017 *Phys. Rev.* B **95** 195130
[20] Champel T, Mineev V P 2002 *Phys. Rev.* B **66** 195111
[21] Prudnikov A P, Brychkov Yu A, Marichev O. I. *Integrals and Series* vol 2 (New York: Gordon and Breach)
[22] Ando T 1974 *J. Phys. Soc. Jpn.* **37** 622
[23] McKenzie R H, Moses P 1998 *Phys. Rev. Lett.* **81** 4492
[24] Grigoriev P D 2011 *JETP Lett.* **94** 47
[25] Grigoriev P D 2011 *Low Temp. Phys.* **37** 738 [2011 *Fiz. Nizk. Temp.* **37** 930]
[26] Shoenberg D 1984 *Magnetic oscillations in metals* (Cambridge: Cambridge University Press)
[27] Abrikosov A A 1988 *Fundamentals of the theory of metals* (Amsterdam: North-Holland)
[28] Ziman J M 1972 *Principles of the Theory of Solids* (Cambridge: Cambridge University Press)
[29] Gradshteyn I S, Ryzhik I M 1980 *Table of Integrals, Series and Products* (New York: Academic Press) 1980
[30] Abramowitz M, Stegun I A (ed) 1970 *Handbook of Mathematical Functions* (New York: Dover Publications)
[31] Kartsovnik M V, Peschansky V G 2005 *Low Temp. Phys.* **31** 185 [2005 Fiz. Nizk. Temp. **31** 249]

IOP Conf. Series: Journal of Physics: Conf. Series **1038** (2018) 012124 doi:10.1088/1742-6596/1038/1/012124

Investigation of GaAs/AlGaAs superlattice by photoreflectance method

V D Goryacheva*, M S Mironova, O S Komkov*

Saint Petersburg Electrotechnical University "LETI", Micro- and Nanoelectronics Department, Professora Popova st. 5, 197376, St. Petersburg, Russia

*Corresponding authors: gor.ler177@gmail.com, okomkov@yahoo.com

Abstract. The GaAs/AlGaAs superlattice grown by metal-organic chemical vapour deposition (MOCVD) was investigated by photoreflectance (PR) spectroscopy. Optical transitions over the whole band structure were measured at room temperature. Different laser pump beams (632 and 532 nm) were used to clarify the nature of the PR signal formation. Charge carriers transport over minibands was revealed by using long-wave modulation. An acceptable correlation of theoretical and experimental results was achieved for low-energy transitions. Optical transitions relating to the edges of the first and the second minibands for electrons, heavy and light holes were clearly observed experimentally.

1. Introduction

Superlattices have been an important part of micro- and nanoelectronics ever since their unique properties were discovered in [1]. Due to the strong quantum-mechanical coupling between the quantum wells, it is possible to create structures with the desired band structure parameters. Superlattices have proved to be suitable for the realization of the infrared lasers based on the quantum cascade concept [2]. GaAs/AlGaAs superlattices are also used in semiconductor lasers for mode selection. In addition, the presence of negative differential conductivity in superlattices enables creation of amplifiers, generators and detectors of the GHz and THz ranges.

Photoreflectance (PR) spectroscopy has a number of advantages over conventional diagnostic methods because of derivative spectra character, high sensitivity, as well as its contactless and non-destructive nature. The PR method can reveal optical transitions both at various critical points of the Brillouin zone, and between discrete (or quasidiscrete) energy levels. A high sensitivity is another advantage of PR: typical modulation amplitudes are about 10^{-4}-10^{-5} of the source light intensity, while the noise level can be kept below 10^{-6} [3]. Energy transitions from higher levels in low-dimensional structures can also be observed in room-temperature photoreflectance measurements. Additionally, such structure properties as free carrier concentration [4, 5], built-in electric fields, the Fermi level pinning at the surface [6] or interface, the degree of strain relaxation [7] can be determined by PR.

There are a number of papers describing results of investigating different semiconductor superlattices using the photoreflectance method. Authors of article [10] present the investigation the GaAs/AlGaAs superlattice at room and low temperature (10K). The influence of the growth parameters (QW and barrier widths) was obtained on GaAs/AlGaAs superlattices with Fourier transform photoreflectance by authors of paper [8]. The work [12] contains results of the room-temperature measurements on GaAs superlattices that demonstrate the influence of the build-in potential. Papers [9, 13] present the investigating CdTe/ZnTe and GaAs/InGaAs superlattices using photoreflectance method respectively. Miniband and carrier transport properties in the GaAs superlattice structure were studied with photoreflectance, photoluminescence, and photothermal spectroscopies in work [11]. But there are no works where the nature of the PR signal formation in superlattices would be shown clearly. In this work, we used photoreflectance spectroscopy with different wavelength of modulating laser beam to investigate optical transitions in GaAs/AlGaAs superlattice.

2. Experimental setup and sample parameters

The setup employed for our photoreflectance measurements was described in [14]. The light of a 100 W tungsten-halogen lamp went through a 0.6 m monochromator and was focused on the sample as a probe beam. The modulating laser pump beam was directed into the same place on the sample. During the measurements, different pump beams were used – one was provided by a He-Ne laser (632 nm line) and the other by a Nd:YAG laser (532 nm line). The pump beam was mechanically chopped at a frequency of 36 Hz.

The reflected probe beam, which carried information about sample parameters, was focused by a collimator on the OP913WSL silicon photodetector input window. Phase sensitive detection of the modulation-induced change in reflectance was performed using a SR830 lock-in amplifier. An optical filter was employed to prevent diffused laser light from entering the detector window.

The investigated sample contains a $GaAs/Al_{0.45}Ga_{0.55}As$ superlattice grown by metal-organic chemical vapour deposition (MOCVD). It contains 50 periods with 42 Å barrier thickness and 65 Å quantum well width and has an impurity concentration (Si) of $n = 7\cdot10^{15}$ cm^{-3}.

3. The possible origin of the superlattice photoreflectance signal

When the superlattice sample is irradiated with a modulating pump beam of the red laser ($h\nu_2$), electron-hole pairs are formed in the minibands (Fig. 1 *a*). They move through the minibands due to the influence of the surface built-in electric field. Approaching the surface, the holes compensate the surface states, the schematic image of which is shown in the figure. The field F_S decreases to F'_S, and, consequently, the reflection of the probe beam changes. In the absence of modulating radiation, everything returns to its initial state (Fig. 1 *b*).

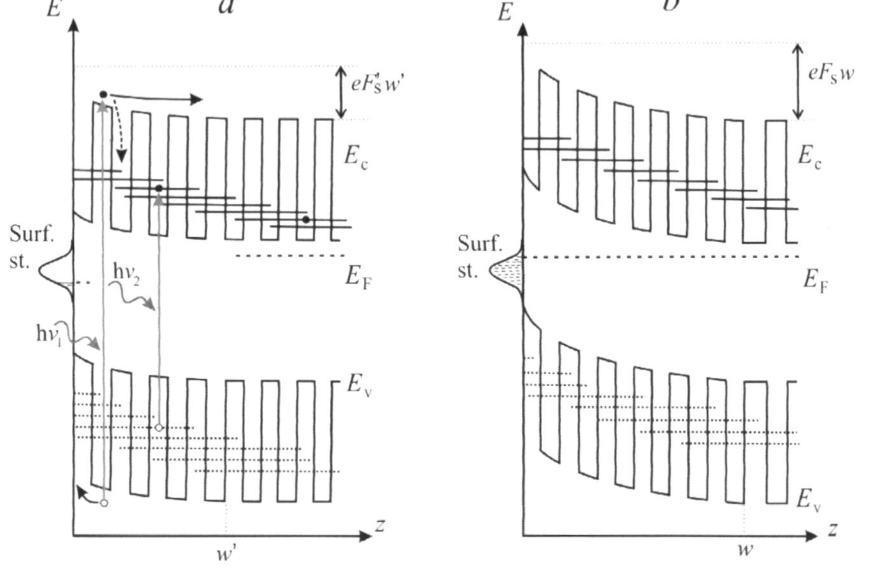

Figure 1. The effect of laser radiation on the superlattice band structure.
a - the superlattice is pumped by a laser beam ; *b* – the initial state of the superlattice;

Surf. st. – surface states;
$F'_S < F_S$ - the built-in electric field with laser irradiation on and off, respectively;
$w' < w$ - width of the depletion region with and without laser irradiation

The change in the reflectance ΔR of the superlattice is based on the quantum-dimensional Stark effect. The edges of the minibands shift in energy as the field is varied, thus the dielectric function and, consequently, the reflectance R changes. Therefore, the registered PR is a derivative signal.
When the modulation with the green laser takes place (the laser photon energy $h\nu_1$ is greater than the $Al_{0.45}Ga_{0.55}As$ band gap at room temperature), the mechanism of surface electric field change can be the same as for a bulk semiconductor material. The charge carriers are separated by the field, and the holes compensate the surface states. However, there is also a probability of relaxation of photoinduced charge carriers to quantum wells of the superlattice. Then the situation becomes identical to the case described above with carrier transport in the miniband.

4. Results and discussion

Figure 2 shows the resulting ΔR/R spectra measured with the different modulating laser radiation. The red line is related to He-Ne radiation, the green one shows the signal modulated by Nd:YAG laser.
Since PR is a derivative signal, the spectra were transformed with the method described in [15] to determine the energies of the interband optical transitions. The resulting transformed spectra are shown in Figure 3. Here, the energy of each peak corresponds to the energy of the optical transition.

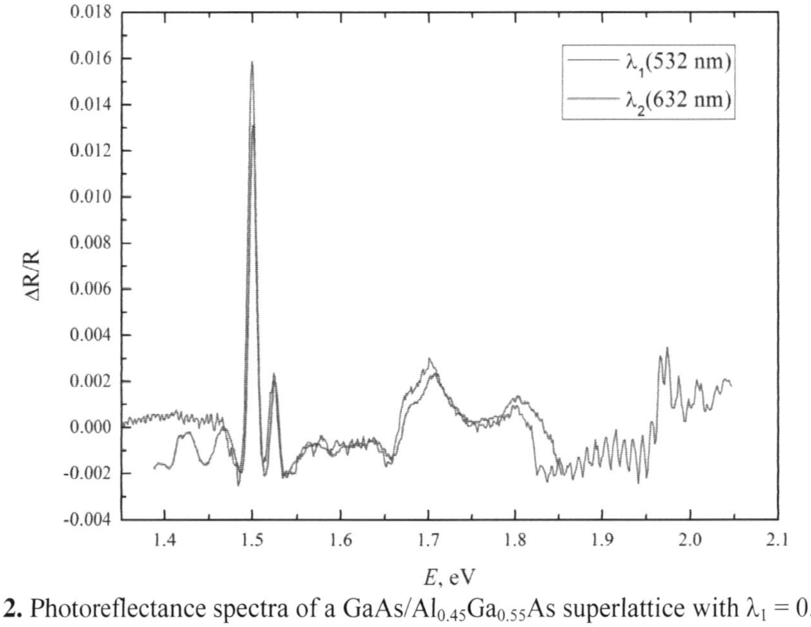

Figure 2. Photoreflectance spectra of a $GaAs/Al_{0.45}Ga_{0.55}As$ superlattice with $\lambda_1 = 0.53$ μm and $\lambda_2 = 0.63$ μm pump beams, measured at room temperature

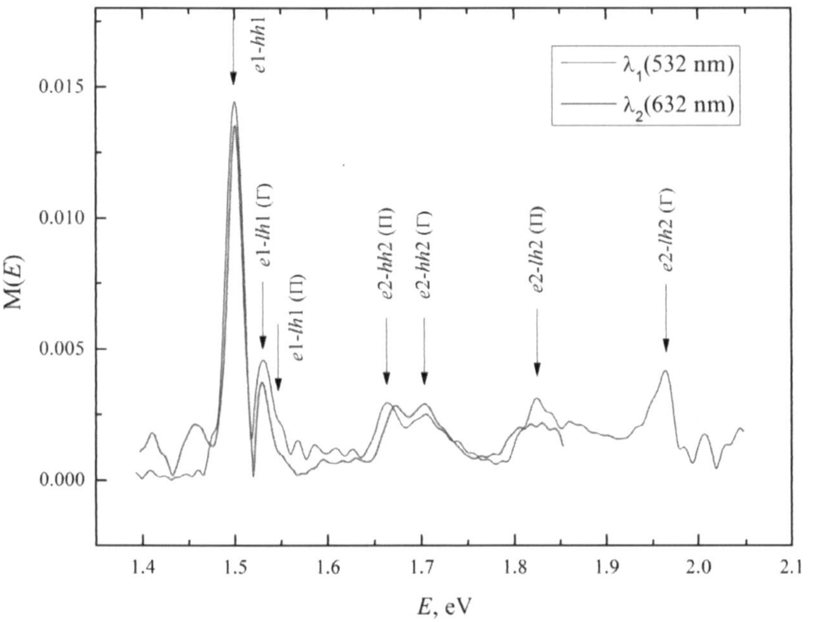

Figure 3. The photoreflectance spectra from Fig. 2 transformed by the method described in [15]

The following interband transitions were determined (they are pointed with arrows in Fig. 3 and shown on the band diagram in Fig. 4):

- the ground state transition *e*1-*hh*1 between the first electron miniband and the first miniband of heavy holes;
- transitions involving light holes of the first miniband (*e*1-*lh*1) at the Γ and Π critical points, corresponding to the lower and upper edges of the miniband, respectively. These transitions are located very close on the energy scale and practically merge;
- transitions of heavy holes and electrons from the second minibands: *e*2-*hh*2 (Π) and *e*2-*hh*2 (Γ);
- *e*2-*lh*2 (Π) and *e*2-*lh*2 (Γ) transitions, related to the second miniband of electrons and the second miniband of light holes.

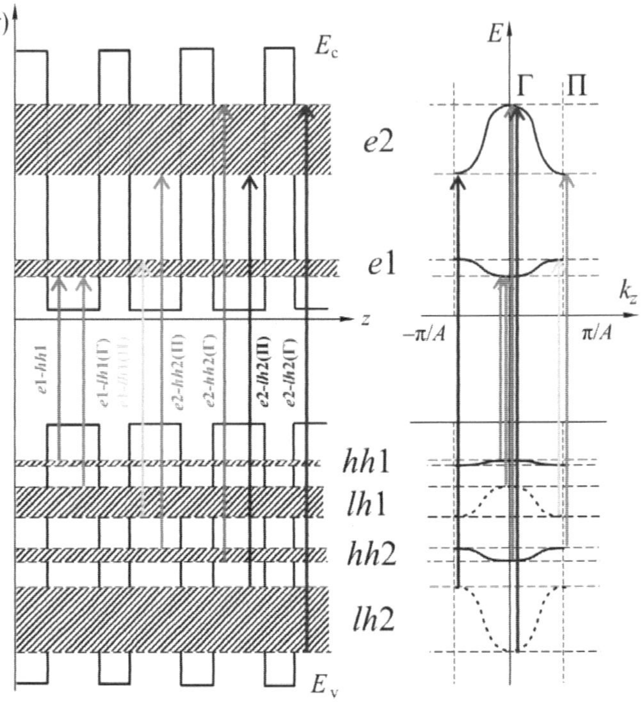

Figure 4. A schematic image of the observed energy transitions within the superlattice band structure

The obtained experimental results were compared with the theoretical energy values of the band structure transitions calculated within the effective mass approach assuming a parabolic dispersion law for both electrons and holes. The experimental and theoretical transition energy values are summarized in Table 1.

Table 1. Experimental and theoretical values of the transition energies

Energy transition	-	-	*e*1-*hh*1	*e*1-*lh*1(Γ)	*e*1-*lh*1(Π)	*e*1-*hh*3	*e*2-*hh*2(Π)	*e*2-*hh*2(Γ)	*e*2-*lh*2(Π)	*e*2-*lh*2 (Γ)
E_{theor}, **eV**	-	-	1.500	1.532	1.5375	1.6395	1.741	1.7595	1.811	1.856
$E_{at\ \lambda 1}$, **eV**	-	-	1.500	1.531		-	1.664	1.705	1.824	1.964
$E_{at\ \lambda 2}$, **eV**	1.411	1.456	1.501	1.529		-	1.672	1.703	1.824	-

As can be seen from Table 1, both experimental spectra (at λ_1 and λ_2) are in good agreement in the 1.48-1.8 eV energy range. The neodymium laser enables measurement of the spectrum over a wider wavelength range, which allows for detection of higher-energy transitions. The experimental result for the ground state transition is in good correlation with theoretical calculations. It is revealed that the first energy level in the valence band and in the conduction band is practically discrete. According to experimental data, the light hole minibands and the second electron miniband are wider due to the smaller degree of charge carriers localization in quantum wells. The difference between experiment and theory for high-energy minibands could be related to the chosen calculation model based on the parabolic assumption.

5. Conclusion

In conclusion, the GaAs/$Al_{0.45}Ga_{0.55}As$ superlattice band structure has been optically investigated using photoreflectance spectroscopy. Different pump beams (632 and 532 nm) were used. As a result the energy transitions were identical for both wavelengths, so the mechanisms of PR signal formation are very close. Charge carrier transport over minibands is revealed by using long-wave modulation. When the modulation photon energy exceeds the barrier band gap, an additional mechanism of surface electric field change, analogous to a bulk material, could also take place. An acceptable agreement of theoretical and experimental results is achieved for the measurements at room temperature.

Acknowledgments

The results were obtained within the project of the Ministry of Education and Science of the Russian Federation (No. 16.1750.2017/4.6).
The authors are grateful to the technologists of Lobachevsky University for the grown sample.

References

[1] Keldysh L V 1962 *Fiz. Tv. Tela* **4** 2265
[2] Faist J, Capasso F, Sivco D L, Sirtori C, Hutchinson A L, Cho A Y 1994 *Science* **264** 553
[3] Yu P, Cardona M 2002 Fundamentals of Semiconductors. Physics and Materials Properties 3rd edition, Springer
[4] Komkov O S, Pikhtin A N, Zhilyaev Yu V, Fedorov L M 2008 *Technical Physics Letters* **34** (1) 37
[5] Komkov O S, Firsov D D, Lvova T V, Sedova I V, Semenov A N, Solov'ev A N, Ivanov S V 2016 *Physics of the Solid State* **58** (12) 2394
[6] Yin X, Chen H M, Pollak F H, Chan Y, Montano P A, Kirchner P D, Pettit G D, Woodall J M 1991 *Appl. Phys. Lett.* **58** (3) 260
[7] Gelczuk L, Motyka M, Misiewicz J, Dabrowska-Szata M 2008 *Materials Science and Engineering (B)* **147** (2–3) 166
[8] Motyka M, Janiak F, Misiewicz J, Wasiak M, Kosiel K, Bugajski M 2011 *Opto-electronics Review* **19** (2) 151
[9] Ferizovic D, Peng L, Sultana H, Mukherjee P, Witanachchi S, Tamargo M C, Munoz M 2011 *Journal of Applied Physics* **110** 093703
[10] Motyka M, Sek G, Janiak F, Ryczko K, Misiewicz J, Koziel K, Bugajski M 2009 *Optica Applicata* **39** (4) 897
[11] Fukuyama A, Matsuochi K, Nakamura T, Takeda H, Suzuki H, Toprasertpong K, Sugiyama M, Nakano Y, Ikari T 2016 IEEE Nanotechnology Materials and Devices Conference (NMDC)
[12] Shen X C, Dhen H, Parayanthal P, Pollak F H G1986 *Superlattices and Microstructures* **2** (6) 513
[13] Tang Y S 1989 *Semicond. Sci. Technol.* **4** 871
[14] Komkov O S, Pikhtin A N, Zhilyaev Yu V 2012 *Russian Microelectronics* **41** (8) 508
[15] Pikhtin A N, Komkov O S, Bazarov K V 2006 *Semiconductors* **40** (5) 592

IOP Conf. Series: Journal of Physics: Conf. Series **1038** (2018) 012125 doi:10.1088/1742-6596/1038/1/012125

Effect of the toroidal magnetic field the on energy and fast particle confinement in the Globus-M spherical tokamak

A Yu Telnova, V K Gusev, N N Bakharev, G S Kurskiev, M I Patrov, Yu V Petrov, E O Kiselev, V B Minaev, N V Sakharov, P B Shchegolev

Ioffe Institute, 194021, St. Petersburg, Politehnicheskaya st, 26.

e-mail: anna.telnova@mail.ioffe.ru

Abstract. The experiments with auxiliary heating by neutral beam injection (NBI) were carried out before the spherical tokamak Globus-M was upgraded to Globus-M2. The aim of the experiments was to determine the dependence of the confinement time on toroidal magnetic field. The total stored energy was obtained by diamagnetic loops, and verified with ASTRA modelling based on kinetic measurements. The absorbed heating power was estimated using 3D fast ion tracking modeling. The obtained dependence of the energy confinement time is similar to the MAST and NSTX results but contradicts the conventional IPB98(y,2) scaling. The toroidal Alfvén eigenmodes study was performed in a set of discharges with NBI at the early stage of the discharge. The experiments have shown that fast particle losses decrease with the increase of the toroidal magnetic field and plasma current.

1. Introduction

One of the main reasons for the energy and fast particle losses in spherical tokamaks is a low toroidal magnetic field. Therefore the toroidal magnetic field will be increased by 2.5 times (from 0.4 T to 1 T) in the upcoming Globus-M2 machine [1]. Improvement in fast particle confinement is therefore expected. Toroidal magnetic field in the Globus-M [2] tokamak was increased up to 0.5 T during the last experimental campaign and a series of experiments were performed with auxiliary heating by means of neutral beam injection (NBI). The total plasma stored energy and the energy confinement time were estimated using kinetic and magnetic measurements. A significant increase of the plasma total stored energy was observed with the increasing magnetic field, while heating power wasn't change significantly. This phenomenon is associated mainly with the confinement time rise. The fast particle confinement was also improved due to decrease of the orbital losses and losses due to MHD instabilities, including toroidal Alfvén eigenmodes (TAE).

2. Thermal energy confinement time dependence on the toroidal magnetic field.

The experiments were carried out for the values of the toroidal magnetic field B_T = 0.4 T and B_T =0.5 T, while the plasma current I_p = 200 kA remains at the same level. The ohmic heating power was about ~ 0.23 MW. The auxiliary heating was performed using the deuterium beam (beam energy E_b=26 keV, beam power P_b=0.7 MW), that was injected into deuterium plasma at the steady state stage of the discharge during the current flat-top phase. The beam duration (τ_{beam} = 40 ms) significantly exceeded the typical value of the energy confinement time τ_E = 3÷5 ms. The line average electron density was changed for different discharges in the range of $1.5\text{-}5\cdot10^{19}$ m^{-3}.

Electron temperature and density profiles were measured by the Thomson scattering (TS) diagnostics [3], the line average density was monitored by the microwave interferometer. Investigation of the ion component of the plasma was performed with the help of the two neutral particle analyzers that measured charge exchange atomic fluxes outgoing from the plasma column in tangential and transversal to toroidal magnetic field directions. This data was used to estimate the fast particle losses level, ion temperature, ion composition and the quantity of the absorbed beam power.

To determine the energy confinement time plasma total stored energy was estimated using two different approaches. The first approach is based on the diamagnetic measurements, the second one is the integration of the kinetic profiles using ASTRA modelling [4] and Zero-dimensional code [5]. Figure 1 shows the energy confinement time τ_E as a function of the average electron density $<n_e>$. The significant difference in energy confinement time for different magnetic fields appears for the densities higher than $3\cdot10^{19}m^{-3}$. The dependence of the confinement time on toroidal magnetic field for the Globus-M tokamak was calculated for this range as $\tau_E \sim B^{0.9\pm0.2}$. This result corresponds well to preliminary analysis presented in [6].

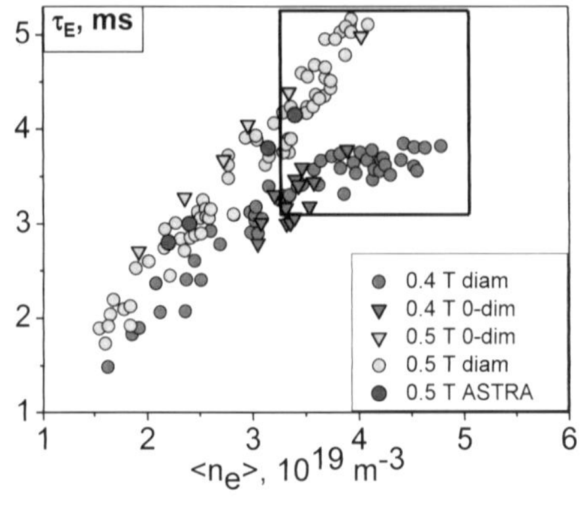

Figure 1. The energy confinement time dependence on the line average electron density. Comparison of the different energy confinement time estimation approaches (ASTRA code, zero-dimension code and diamagnetic measurements) for the different experimental conditions: the same plasma current (200 kA) and various toroidal magnetic fields (0.4 T and 0.5 T).

ASTRA simulations were used to determine thermal diffusivity coefficients for ions and electrons, for plasma with different densities. The modeling was performed for the quasi-stationary stage of the discharges using the following assumptions. The ion temperature profile was calculated assuming neoclassical transport coefficients that were calculated with the help of NCLASS transport code [7]. This approach is valid, since the recent experiments carried out on Globus-M demonstrated neoclassical behavior of the ion transport [8,9] like in other spherical tokamaks [10,11]. The magnetic equilibrium was reconstructed using the EFIT code [12]. The effective plasma charge (Z_{eff}) was set as a constant along the minor radius. The Z_{eff} value was estimated by matching the calculated loop voltage with the experimental one. Carbon was assumed as the main impurity. The profiles of the neutral beam absorbed power were determined by a standard block of the ASTRA code. Additional correction for the first orbit losses of the fast particles was made using the three-dimensional modelling algorithm [13].

The electron heat diffusivity profiles were calculated using TS profiles. The neoclassical ion heat diffusivity varies slightly with density, while the electron heat diffusivity decreases dramatically.

The dependence of the confinement time on toroidal magnetic field was studied for the same plasma current and for the different densities $<n_e> \sim 1.5$ - $4.2\cdot10^{19}m^{-3}$. It was found that the influence of the magnetic field on energy confinement is considerable in Globus-M. However, the ITER scaling predicts [14] weak dependence of the confinement time on toroidal magnetic field ($\tau_E \sim B^{0.15}$). It is opposite to well-known spherical tokamak experiments [15,16] (where are $\tau_E \sim B^{1.4}$ in MAST and $\tau_E \sim B^{1.08}$ in NSTX). The energy confinement time that was calculated by IPB98(y,2) scaling shows that is rather H-factor weak 0.7-0.9 for the range of densities $<n_e> \sim 2$ - $3.5\cdot10^{19}m^{-3}$.

3. Toroidal Alfvén Eigenmodes

Instabilities, identified as toroidal Alfvén eigenmodes, were found in the experiments with NBI in the early stage of the discharges [17]. To study dependence of the fast particle losses induced by TAE on the toroidal magnetic field value, the experiments with the increased magnetic field (0.5 T) were carried out.

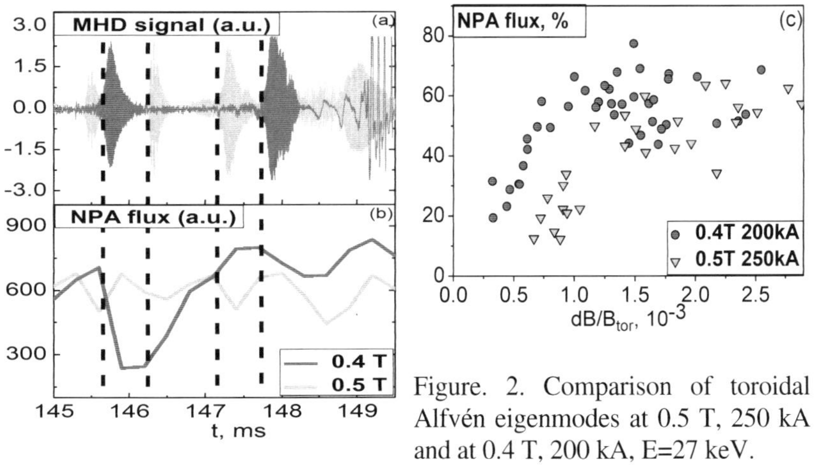

Figure. 2. Comparison of toroidal Alfvén eigenmodes at 0.5 T, 250 kA and at 0.4 T, 200 kA, E=27 keV. (a) – MHD-signals, (b) –NPA flux, (c) - the dependence of the NPA flux in the case of the TAE amplitude.

The results of the experiments with the increased toroidal magnetic field and plasma current were compared with the previous experiments (0.4 T, 200 kA). Figure 2 demonstrates MHD signal (fig. 2a) and the NPA flux (fig. 2b) for two cases: 0.4 T, 200 kA – red lines and 0.5 T, 250 kA – green lines. Correlation between the TAE burst and the NPA flux drops is marked by the dashed lines. From these figures one can see that at the increased magnetic field and plasma current the bursts of the Alfvén modes become more frequent, and the levels of fast particle losses during the burst decrease. The increase in the frequency of the flushes is probably associated with the better accumulation of the fast particles due to the lower classical and TAE induced losses.

In figure 2 (c) the dependence of the drop in the NPA flux (energy of particles is 28.5 keV) on the amplitude of the Alfvén mode is presented. The figure demonstrates that increase of the magnetic field and plasma current leads to decrease of the fast particle losses induced by TAE.

4. Conclusions

The energy confinement time increases with toroidal magnetic field growth, especially for high densities. Results of the diamagnetic measurements, zero-dimension code and ASTRA code calculations are in a good agreement. With the rise of electron density the H-factor increases approaching IPB98(y,2) scaling value. The strong dependence of the energy confinement time on the magnetic field ($\tau_E \sim B^{0.9\pm0.2}$) found on Globus-M, is in a good agreement with other spherical tokamaks.

The experiments demonstrate that increase of the magnetic field and plasma current leads to decrease of the fast particle losses induced by TAE while the modes became more frequent. The increase in the mode frequency is probably associated with the better accumulation of fast particles due to lower classical and TAE losses.

Acknowledgments
The study of the energy confinement time (part 2) was supported by RFBR, research project No. 16-32-60114 mol_a_dk. tokamaks.

The results of TAE influence on fast particle confinement (part 3) were obtained with support of RSF research project № 17-12-01177.

The work is performed on the Unique Scientific Facility "Spherical tokamak Globus-M", which is incorporated in the Federal Joint Research Center "Material science and characterization in advanced technology", with financial support by the Ministry of Education and Science of the Russian Federation (id RFMEFI62117X0018).

References

[1] Minaev V B et al 2017 *Nucl. Fusion* **57** 066047
[2] Gusev V K et al 1999 *Tech. Phys.* **44** 1054
[3] Kurskiev G S et al 2012 *Problems of Atomic Science and Technology Ser. Thermonucl. Fusion* **2** 81-88 (in Russian)
[4] Pereverzev G and Yushmanov P N 2002 Max-Plank IPP Report 5/98
[5] Kurskiev G S et al 2016 *Problems of Atomic Science and Technology Ser. Thermonucl. Fusion* **39** 02023822 (in Russian)
[6] Telnova A Yu et al 2017 *J. Phys.: Conf. Ser.* **907** 012014
[7] Houlberg W A et al 1997 *Phys. Plasmas* **4** 3230
[8] Avdeeva G.F. et al 2016 *J. Phys.: Conf. Ser.* **666** 012002
[9] Kurskiev G S et al 2017 *Plasma Phys. Control. Fusion* **59** 045010
[10] Akers R J et al 2003 *Plasma Phys. Control. Fusion* **45** A175
[11] LeBlanc B P et al 2004 *Nucl. Fusion* **44** 513
[12] Lao L L et al 1985 *Nucl. Fusion* **25** 1611
[13] Bakharev N N et al 2010 *Nucl. Fusion* **55** 043023
[14] ITER Physics Basis 1999 *Nucl. Fusion* **39** 2175, 2007 *Nucl Fusion* **47** S18
[15] Kaye S M et al 2006 *Nucl. Fusion* **46** 848–57
[16] Valovic M et al 2009 *Nucl. Fusion* **49** 075016
[17] Petrov Yu V et al 2015 *J. Plasma Phys.* **81** 515810601

IOP Conf. Series: Journal of Physics: Conf. Series **1038** (2018) 012126 doi:10.1088/1742-6596/1038/1/012126

Role of resonance radiation trapping in the mechanisms of constriction of the glow discharge. Theory and experiment.

Yu B Golubovskii[1], A V Siasko[1], D V Kalanov[1] and V O Nekuchaev[2]
[1]Faculty of Physics, St Petersburg State University, Ulianovskaia ul. 3, St Petersburg 198504, Russian Federation
[2]Ukhta State Technical University, Pervomaiskaia ul. 13, Ukhta 169300, Russian Federation

E-mail: aleksei.siasko@gmail.com, y.golubovskiy@spbu.ru

Abstract. The paper presents latest results of constricted glow discharge simulations. Detailed self-consistent plasma model includes a large set of plasma chemical reactions, considering inhomogeneous gas heating and accurate solution of radiation transport equation. Results of the computer simulations were compared with an experimental data of positive column constriction in neon and argon. Experimental plasma diagnostics was performed using classic methods of emission and absorption spectroscopy and a line ratios method. Comparison of the obtained results allows to reveal the influence of resonance radiation trapping on the plasma parameters of a constricted discharge.

1. Introduction

Constriction is an interesting phenomenon in gas discharges which is observed as an abrupt compression of plasma to a thin bright cord on the discharge axis when pressure and current reach the critical values. This phenomenon has been studied in detail both theoretically and experimentally [1–6]. However, the problem of radiation trapping was ignored in these works so far as the Holstein effective lifetime approximation (escape-factor approximation) was used, which assumes that excitation and destruction of resonant atoms are local. Radiation transport leads to a spatial redistribution not only of resonant atoms but also metastable atoms and other plasma components related to the resonant atoms through the collisional - radiative processes. It is of great interest to reveal the influence of radiation trapping in the mechanism of constriction. Constricted discharge is quite a convenient object for the analysis since in diffuse state at low currents spatial distributions of plasma components are close to the fundamental modes and traditional description in terms of effective diffusion and radiation lifetimes is applicable. But when the plasma contracts to a thin cord higher radiation and diffusion modes start playing a crucial role [7]. First attempt of correct account of radiation trapping in the mechanism of constriction was made in paper the [8] using a simple discharge model.

Current work presents a detailed model of a constricted glow discharge with a large set of plasma chemical reactions, including accurate solution of kinetic equation and account of inhomogeneous gas heating. The model allows one to compare traditional approach of the Holstein escape factor with a precise treatment of the resonance radiation. The role of radiation trapping is demonstrated on a formation of radial profiles of various plasma components and current-voltage

characteristics.
Along with simulations, experimental study of positive column constriction in neon and argon was performed. In particular, field-current characteristics in neon and argon, densities and radial profiles of the excited states $2p^53s$ and $2p^53p$ in neon and $3p^54s$ and $3p^54p$ in argon (1s and 2p in Pashen's notation) were measured. Densities and radial profiles were obtained using classic methods of emission and absorption spectroscopy [9] as well as the line ratios method [10]. Similar features and differences in the mechanisms of discharge constriction in neon and argon are discussed. Obtained results show good agreement with simulations.

2. Self-consistent glow discharge model

A detailed model of the glow discharge presented in this paper is similar to a model of argon discharge described in [11]. The model allows to treat the resonance radiation transport accurately and makes it possible to obtain a continuous hysteresis in transition between diffuse and constricted regimes.
All elementary processes, included in the model of neon, are listed in table I with the references.

Table 1. Plasma-chemical reactions included in the collisional-radiative model.

Process	Notation	Ref. (Neon)
Elastic scattering	$Ne + e \rightarrow Ne + e$	[12]
Excitation	$Ne + e \rightarrow Ne^*(1s, 2p) + e$	[13]
Intermixing	$Ne^* + e \rightarrow Ne^* + e$	[14]
Direct ionization	$Ne + e \rightarrow Ne^+ + 2e$	[12]
Stepwise ionization	$Ne^* + e \rightarrow Ne^+ + 2e$	[12]
Chemoionization	$Ne^* + Ne^* \rightarrow Ne_2^+ + e$	[15]
Ion conversion	$Ne^+ + 2Ne \rightarrow Ne_2^+ + Ne$	[16]
Dissociative recombination	$Ne_2^+ + e \rightarrow Ne^* + Ne$	[17]
Dissociation	$Ne_2^+ + e \rightarrow Ne^+ + Ne + e$	[18]
Excimer formation	$Ne^* + 2Ne \rightarrow Ne_2^* + Ne$	[19]
Resonance radiation	$Ne^*(1s_2, 1s_4) \rightarrow Ne + h\nu$	[20]
Excimer radiation	$Ne_2^*(^3\Sigma_u) \rightarrow 2Ne + h\nu$	[21]
Visible radiation	$Ne^{**} \rightarrow Ne^* + h\nu$	[22]
Metastable diffusion	$Ne^* \rightarrow$ wall	[23]

Most of the rate constants of processes with electrons were calculated by an averaging of the cross sections over the electron energy distribution function (EEDF). The EEDF is obtained by solving the Boltzmann equation. Densities and radial profiles of plasma components were found from the balance equations. Resonant atoms are described by an integral equation of resonance radiation transport while all other plasma components are described by the diffusion equation. Papers [8, 11] give a detailed description of applied balance equations and a thermal balance equation describing gas heating which affects the filament radius. The applicability limits of the model and solution algorithm are also discussed in detailsin mentioned papers.

3. Experimental plasma diagnostics

3.1. Experimental set-up

The experiment was conducted with the argon and neon cylindrical discharge tubes. The neon-filled tube with horizontal orientation had a radius of 1.5 cm, pressure p =60 Torr and a length

between the electrodes $d = 70$ cm. The argon tube was oriented vertically and spectrally pure argon was fed through the vacuum system with liquid-nitrogen cooled traps. The radius of the tube was 2.3 cm, pressure p =42 Torr and a length between the electrodes $d = 50$ cm. Electrical characteristics of DC glow discharges were studied in a range of currents 5 - 70 mA/cm.
To perform spatial absorption measurements, a second tube with ICP discharge in 2 Torr argon or neon (depending on the main source) was placed behind the main source. A back window of the source tube was uniformly illuminated with a lens. A registration system consisted of a monochromator Acton SpectraPro 2300i with a high-speed camera pco.1200hs. Radial scanning was performed with the high-speed camera, so an instrumental function of the experimental set-up was determined by a pixel size of the camera. Scheme of the experimental set-up with fragments of constricted neon and argon discharges are presented in figure 1.
Both constricted discharges represented a thin plasma cord located at the axis of a discharge.

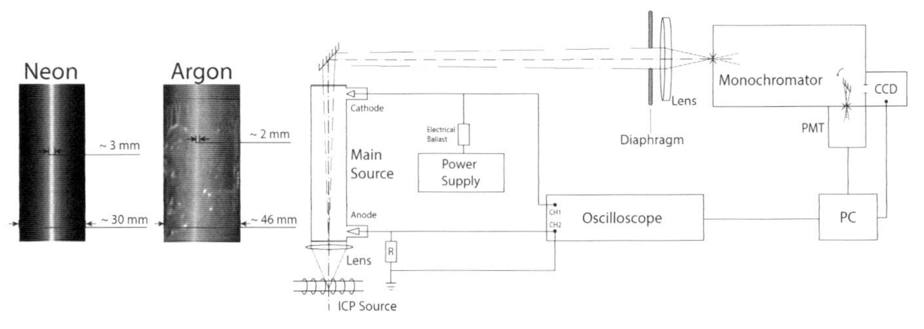

Figure 1. Scheme of the experimental set-up with the fragments of constricted neon and argon discharges.

Reduced diameter of a filament in neon $(d/R)_{Ne} \approx 0.1$ is much bigger than in argon $(d/R)_{Ar} \approx 0.04$. Spatial distribution of a luminous cord determines the excitation and ionization zones in the constricted discharges. The discharges are surrounded with a weak glow which is continuum spectrum of electrons deceleration by neutral species. The intensity of the glow is related to the electron density and electron temperature.
Densities and radial profiles of the $1s$ and $2p$ states in argon and neon were obtained using conventional emission and absorption spectroscopy technique [9] and a relatively new line ratios method [10]. Application of the classic emission and absorption spectroscopy method is described in details in [11] so only a theory of line ratios method will be discussed below.

3.2. Line ratios method

One of the method that was implemented to measure radial profiles and densities of lower excited $1s$ states was a line ratios method which is based on obtaining a set of algebraic equations from photon flux ratios for lines radiating from same upper state to different lower states. For a specific transition $i \to j$ a photon flux that is projected onto the entrance slit of the spectral device will be written as:

$$\Phi_{ij} \propto N_i A_{ij} h\nu_{ij} L S(k_0 L), \tag{1}$$

where N_i is a density of radiating atoms, A_{ij}- transition probability, $h\nu_{ij}$- energy of quant and $S(k_0L)$ is a Ladenburg function which takes into account the reabsorption of the photons along the plasma source. Thus, a ratio of the photon fluxes from the same upper level i to different lower levels j and k will be

$$\frac{\Phi_{ij}}{\Phi_{ik}} = \frac{A_{ij}\nu_{ij} S\left(K_{0j}L\right)}{A_{ik}\nu_{ik} S\left(K_{0k}L\right)} \tag{2}$$

The Ladenburg function can be represented as a function of the lower state density

$$S(k_0L) = S(a_{ij}N_j),$$

$$a_{ij} = \frac{\lambda_{ij}^2 g_i}{8\pi g_j} A_{ij}\varepsilon_0 L, \qquad \varepsilon_0 = \frac{2\sqrt{ln2}}{\pi^{3/2}} \frac{a}{\Delta\nu_D} \int_{-\infty}^{\infty} \frac{exp(-y^2)dy}{a^2+y^2}.$$

ε_0 is an emission coefficient in the center of a spectral line assuming Voigt lineshape, g_i and g_j are statistical weights, λ_{ij}- central wavelength of the transition and parameter $a = \sqrt{ln2}\Delta\nu_L/\Delta\nu_D$ characterizes the ratio of a Lorentz half-width to a Doppler half-width. Using equation (12) the densities of lower-lying states can be determined. A system of homogeneous non-linear equations will have a form

$$S(a_{ij}N_j) - \frac{\Phi_{ij}A_{ik}\nu_{ik}}{\Phi_{ik}A_{ij}\nu_{ij}} S(a_{ik}N_k) = 0.$$

Ratio of the photon fluxes Φ_{ij}/Φ_{ik} is a measured value. Representing the Ladenburg function in the form of series expansion one can solve the equation system using the least-squares method. An increase of the number of equations will enhance the accuracy of the method.

4. Results

Figure 2 contains volt-ampere characteristics measured in the neon and argon discharges under described conditions compared to the simulations carried out in case of traditional effective lifetime approximation and precise treatment of resonance radiation transport. Electric field in

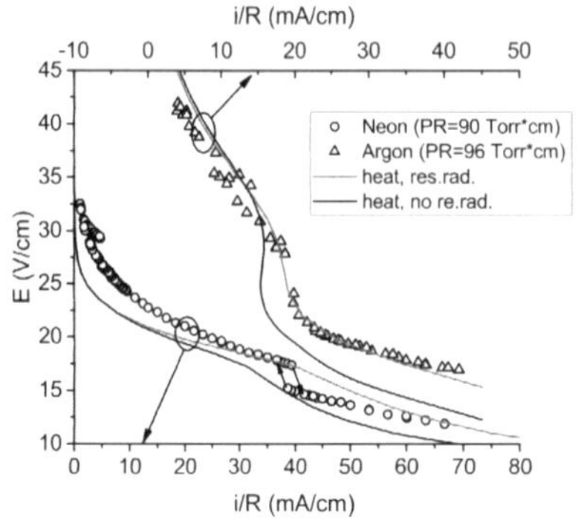

Figure 2. Measured electric field in argon (△) and neon (○) with corresponding numerical results: ignoring radiation transport (black lines), account of radiation transport (red lines).

argon is much larger than in neon. Even though the ionization potential in argon is lower than in neon, difference in cross-sections of electron - atom elastic collisions, which is much larger in argon, leads to a depletion of fast electrons in the EEDF in case of argon. Lack of fast electrons causes a decrease of ionization frequency. To maintain the balance of ionization with diffusion electric field in argon increases.

In measured electric fields diffuse states switch to constricted ones with abrupt changes of discharge voltage when the current exceeds a critical value - about 40 mA/cm for neon and 10 mA/cm for argon. In neon transition from the diffuse to the constricted state and in the opposite direction is accompanied by a perceptible hysteresis.

In the simulations, precise account of radiation transport resulted in a shift of critical currents towards higher values so the electric field increases in the constricted states. Gas heating which was taken into account led to a smoothing of the characteristics near the critical currents so no hysteresis is observed.

Figure 3 shows absolute densities of $1s$ states in argon (a) at $i = 40$ mA and in neon (b) at $i = 70$ mA. For the neon discharge, absolute densities were obtained using classic absorption method

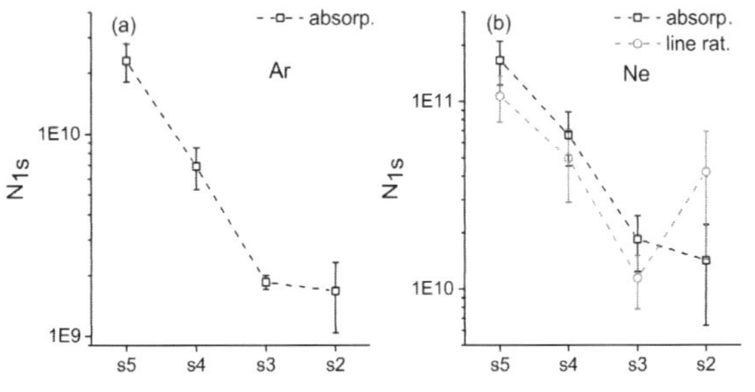

Figure 3. Absolute densities in argon (a) and neon (b) measured using classic method of emission and absorption spectroscopy (black lines) and line ratios method (red line).

and the line ratios method. As it can be seen both methods give identical densities within the measuring and statistical error except for the higher $1s2$ level, where the intensity of the lines is very weak that cause big errors. Densities in neon are higher on the order of magnitude than in argon.

Radial profiles of higher $2p$ states, which are presented in figure 4 for constricted discharges at

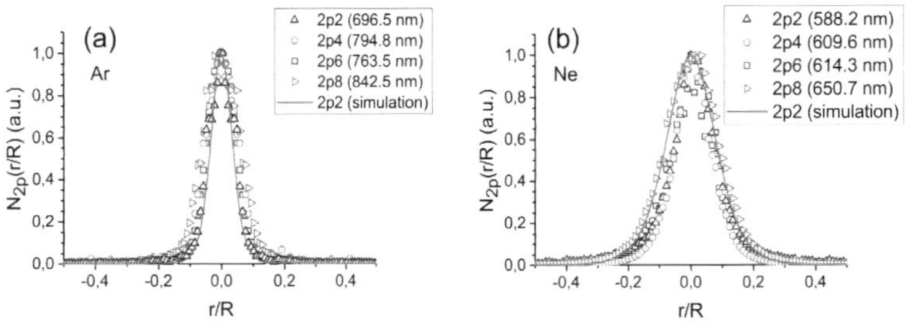

Figure 4. Normalized radial profiles of $2p-$states in argon (a) and neon (b) over different wavelengths with simulations results.

40 mA in argon and 70 mA in neon over different wavelengths are more narrow in argon. As in volt-current characteristics, observed difference is caused by a distinction of elastic electron - atom scattering cross sections. Higher cross section in argon leads to a bigger ionization rate which means that the radial profile in argon should be more constricted than in neon to maintain the balance of ionization, diffusion and recombination at high current.

Red lines in figure 4 represent the result of simulations for $2p_2$ state radial profile, which are

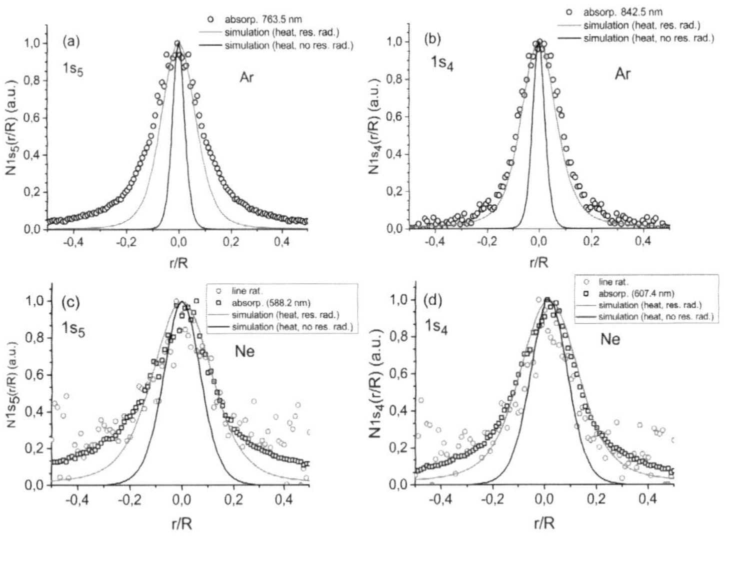

Figure 5. Normalized radial profiles of $1s_5$ and $1s_4$ states in argon (a, b) and neon (c, d) obtained from classic absorption method and line ratios method with theoretical simulations.

in good correlation with the experimental results.

Figure 5 shows radial profiles of metastable 1*s*5 and resonant 1*s*4 states in argon (a, b) and neon (c, d) discharges. As before, radial profiles in argon are more narrow than in neon. For both discharges black lines correspond to the simulations in the approximation of the effective lifetime ignoring redistribution of resonant atoms due to resonance radiation trapping. When the resonance radiation transport is taken into account precisely (red lines), radial profiles become broadened. It is caused by the emergence of resonant atoms outside the excitation zone due to a reabsorption of the photons. The correct account of radiation transport gives better correlation to the experimental results.

Radial profiles in neon were measured using classic absorption method and the line ratios method. As it can be seen in the central part of the discharge both methods give identical results. However, moving to the outer region, the error of the line ratios method grows rapidly. It can be explained by a higher sensitivity of the line ratios method to the light intensity which is very low far from the discharge center in the constricted discharge.

5. Conclusion

Detailed self-consistent model which allows simulating constricted glow discharge using traditional approximation of Holstein effective lifetime and accurate treatment of radiation transport was applied to argon and neon noble gases. In constricted discharges, precise consideration of radiation transport leads to a broadening of radial profiles of plasma components, decrease of absolute densities at the discharge axis as a result of radial redistribution and a shift of a constriction current on the volt-ampere characteristics towards higher values.

Demonstrated radial profiles in argon discharge were obtained to be more narrow than in neon and the reasons of this phenomena were discussed as well as the difference in volt-ampere characteristics.

Experimental diagnostics performed with classic absorption method and line ration method show good correlation between the results. But the error given by the line ratios method becomes very high when the light intensity is insufficient.

6. References

[1] Petrov G M and Ferrerira C M 1999 *Phys. Rev.* **59** 3571
[2] Ionikh Y Z, Meshchanov A V, Petrov F B, Dyatko N A and Napartovich A P 2008 *Plasma Phys. Rep.* **34** 867
[3] Dyatko N A, Ionikh Y Z, Kochetov I V, Marinov D L *et al.* 2008 *J. Phys. D: Appl. Phys.* **41** 055204
[4] Shkurenkov I A, Mankelevich Y A and Rakhimova T V 2008 *Plasma Phys. Rep.* **34** 780
[5] Shkurenkov I A, Mankelevich Y A and Rakhimova T V 2009 *Phys. Rev. E* **79** 046406
[6] Gnybyda M, Loffhagen D and Uhrlandt D 2009 *IEEE Trans. Plasma Sci.* **37** 1208
[7] Golubovskii Y B, Siasko A V and Nekuchaev V O 2017 *Plasma Sources Sci. Technol.* **26** 015012
[8] Golubovskii Y B and Mayorov V A 2015 *Plasma Sources Sci. Technol.* **24** 025027
[9] Loureiro J and Amorim J 2016 *Kinetics and spectroscopy of low temperature plasmas* (Springer Int. Publ.)
[10] Schulze M, Yanguas-Gil A, Keudell A and Awakowicz P 2008 *J. Phys. D: Appl. Phys.* **41** 06520
[11] Golubovskii Y, Kalanov D and Maiorov V 2017 *Phys. Rev. E* **96** 023206
[12] Zatsarinny O and Bartschat K 2012 *Phys. Rev. A* **85** 062710
[13] Zatsarinny O and Bartschat K 2012 *Phys. Rev. A* **86** 022717
[14] Dyatko N A, Latyshev F E, Mel'nikov A S and Napartovich A P 2006 *Plasma Physics Reports* **32** 158
[15] Sheverev N A, Stepaniuk V P and Lister G G 2002 *Journal of Applied Physics* **92** 3453
[16] Hackman R 1966 *British Journal of Applied Physics* **17** 197
[17] Frommhold L, Biondi M A and Mehr F J 1968 *Phys. Rev.* **165** 44
[18] Marchenko V S 1983 *Zh. Eksp. Teor. Fiz.* **85** 500
[19] Virin L I, Dzhatspanyan R V, Karachevtsev G V, Potapov V K and Tal'rose V L 1979 *Ion-molecule reactions in gases* (Moscow: Nauka)
[20] Fuhr J A and Wiese W L 1998 *CRC Handbook of Chemistry and Physics* (Boca Raton, Fl: CRC Press)
[21] McCusker M 1984 *The rare gas excimers* (Berlin, Heidelberg: Springer Berlin Heidelberg) p 47
[22] Krebs K 1936 *Z. Phys.* **101** 604
[23] Phelps A V and Molnar J P 1953 *Phys. Rev.* **89** 1202

The external circuit effect on the steady states of a vacuum diode with a decelerating electron beam

V.I.Kuznetsov, A.B.Gerasimenko
Ioffe Institute, 26 Politekhnicheskaya, St Petersburg, 194021, Russian Federation
E-mail: gerasimenko.alexander@mail.ioffe.ru

Abstract. It was found out that a reactive load presence in the external circuit of a diode with an electron beam can crucial change its operation modes. Using a non-relativistic Bursian diode with a decelerating electron beam as the example, it was demonstrated that the presence of inductance in the external circuit leads to the development of instability in the diode–external circuit system. Values of the parameters for which the diode can generate oscillations are determined. The instability growth rates and frequencies are calculated.

1. Introduction

Microwave oscillations in electron tubes were discovered about 100 years ago (i.e., the Barkhausen-Kurtz generator). Study of non-linear oscillations in diodes with an electron beam remains to be up today. This is due to various applications of such diodes [1], [2]. As a rule, large amplitude oscillations occur when the particles move collisionless in the diode interelectrode gap. The existence of such oscillations is due to the Bursian-Pierce instability [3] – [5].

The study of the dispersion properties of plasma diodes with an electron beam has a long history. In [6], for example, authors made an attempt to explain the results of their experiments in the Bursian diode (i.e., a diode in which a monoenergetic electron flow moves through the inter-electrode gap) with an electric field that slows down the electrons. They analysed the solutions of the dispersion equation with a purely active external circuit. In this paper, we'll show that instability can not develop in such a system. We'll show that in the diode studied in [6], oscillations can develop due to the presence of an parasitic inductance in the external circuit.

When studying non-linear oscillations in vacuum diodes, as a rule, numerical calculations of the processes in the inter-electrode gap of the diode are carried out taking into account the change in the voltage at the collector caused by the presence of an reactance in the external circuit. There is no analyse at what parameters of the system such oscillations can arise. In the current paper we analyse the solutions of the dispersion equations for the Bursian diode, taking into account the presence of inductance in the external circuit. As a result, we could opportunity construct the diode generation regions and determine the growth rates and frequencies of increasing perturbation.

2. Problem statement

We consider the planar electrodes placed at a distance d from each other and infinitely extent in the transverse direction. Across them, a potential difference U is applied. The z-axis is directed

IOP Conf. Series: Journal of Physics: Conf. Series **1038** (2018) 012127 doi:10.1088/1742-6596/1038/1/012127

perpendicular to the emitter surface ($z = 0$). A non-relativistic monoenergetic electron flow is supplied by the emitter with density n_0 and injection velocity v_0 perpendicular to the emitter surface with no proper magnetic field (Fig. 1). The electrons move in a self-consistent electric

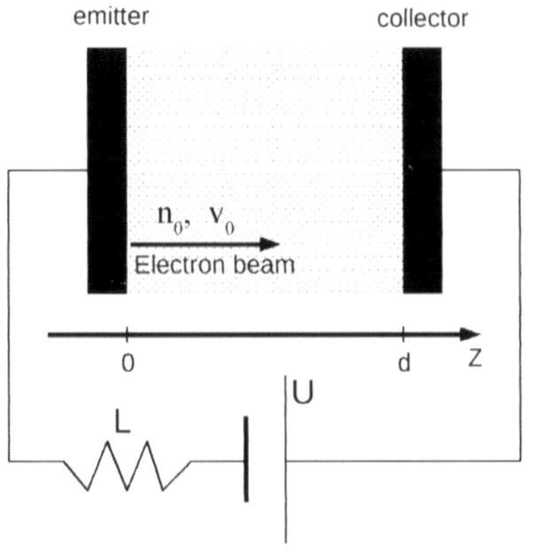

Figure 1. Diode scheme.

field with no collisions and are absorbed at the electrode if they reach it. Such a device was called the Bursian diode in the literature. We use λ_D and W_0 as units of length and energy, respectively:

$$\lambda_D = \left[\frac{2\epsilon_0 W_0}{e^2 n_0}\right]^{1/2} \approx 0.3240 \cdot 10^{-2} \frac{V_0^{3/4}}{J_0^{1/2}}[cm], \quad W_0 = \frac{1}{2}m_e v_0^2 = eV_0. \tag{1}$$

Here $J_0 = en_0v_0$ is the current density of electrons entering in the inter-electrode gap, V_0 is the accelerating voltage, e and m are the charge and mass of the electron, and $\epsilon_0 \approx 8.854 \cdot 10^{-12}C^2/Nm^2$. The dimensionless coordinate $\zeta = z/\lambda_D$, the time $\tau = t\omega_0$, the velocity $u = v/v_0$, the potential $\eta = e\Phi/(2W_0)$, the electric field strength $\varepsilon = eE\lambda_D/(2W_0)$ and the current density $j = J/J_0$. The characteristic frequency ω_0 is determined by the formula

$$\omega_0 = \frac{v_0}{\lambda_D} \approx 0.3754 \cdot 10^8 \frac{J_0^{1/2}}{V_0^{1/4}}[c^{-1}]. \tag{2}$$

We denote the inductance of the external circuit in the dimensionless form by $L = \tilde{L}/L_0$

$$L_0 = 5.3277 \cdot 10^{-8} \frac{V_0^{5/4}}{SJ_0^{3/2}}[H]. \tag{3}$$

Here S is the surface area of the electrode. Examples of typical values of the parameters are given in the table 1.

3. Results

3.1. Stationary states

The steady states of the Bursian diode are corresponded to monotonically decreasing distributions of the potential or one minimum ones. The corresponding solutions are determined by only two dimensionless parameters: the inter-electrode distance $\delta = d/\lambda_D$ and the external voltage $V = eU/(2W_0)$. The solutions are written in terms of the Lagrangian coordinates as

$$\zeta(\tau) = \frac{1}{6}\tau^3 - \frac{\varepsilon_0}{2}\tau^2 + \tau, \quad \eta(\tau) = \frac{1}{2}(u^2 - 1), \quad u(\tau) = \frac{1}{2}\tau^2 - \varepsilon_0\tau + 1. \tag{4}$$

Table 1. Typical values of the parameters.

V_0,V	J_0,A/cm^2	λ_D,cm	v_0,cm/c	ω_0,c^{-1}	$L_0\, S, \mu H$
10	1	0.0182	$0.384\, 10^6$	$21.11\, 10^6$	0.95
100	1	0.1025	$1.22\, 10^6$	$11.87\, 10^6$	16.85

The relation between the time-of-flight of an electron between electrodes T and the values δ and V takes the following form [7]:

$$\delta = -\frac{1}{12}T^3 + \frac{1}{2}\left(1 + \sqrt{1+2V}\right)T. \tag{5}$$

$$T = 2\sqrt{P}\cos\left[\frac{1}{3}\left(\pi + \arccos\frac{6\delta}{P^{3/2}}\right)\right], \quad P = 2\left(1 + \sqrt{1+2V}\right). \tag{6}$$

Steady state solutions is convenient to display by continuous curves in the plane $\{\varepsilon, \delta\}$, corresponding to fixed value V (Fig. 2) [5]. All these solutions are stable ones.

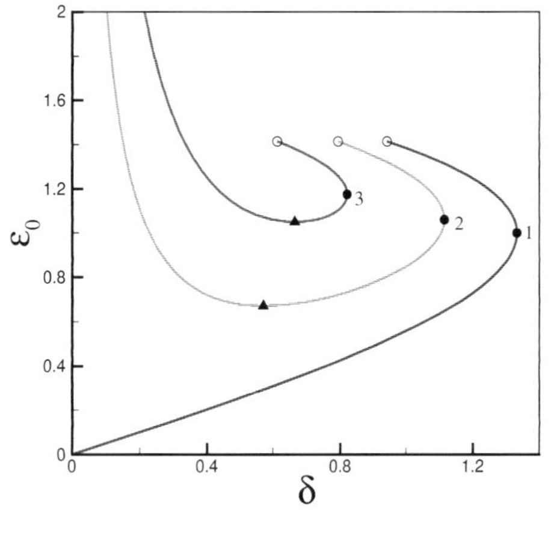

Figure 2. Branches of solutions for various values of an external voltage V: (1) – V = 0, 2 – -0.2, 3 – -0.4. Closed circles mark position of the SCL point, closed triangles – the minimum ε_0 point, circles – the boundaries of solutions with no electron reflection.

3.2. The dispersion equation

The condition for matching the diode with the external circuit: $Z + Z_{ext} = 0$ with diode impedance $Z = -\tilde{U}/\tilde{I}$ and the external circuit impedance Z_{ext}, takes the form

$$-\frac{1}{\Omega^4}\left[(2 - i\Omega T)\exp(i\Omega T) - i\Omega^3\delta - i\Omega T - 2\right] - i\Omega L = 0, \tag{7}$$

Here T is related to δ and V by the formula (6).

We have studied the solutions of the dispersion equation (6). It is known that there are a countable number of dispersion branches, i.e. dependences of growth rate Γ and frequency Ω of the eigenmode on δ, for each fixed value of the V. Three of the branches, that have the largest growth rates, for the diode with a purely active external circuit ($L = 0$) are shown in Fig. 3.

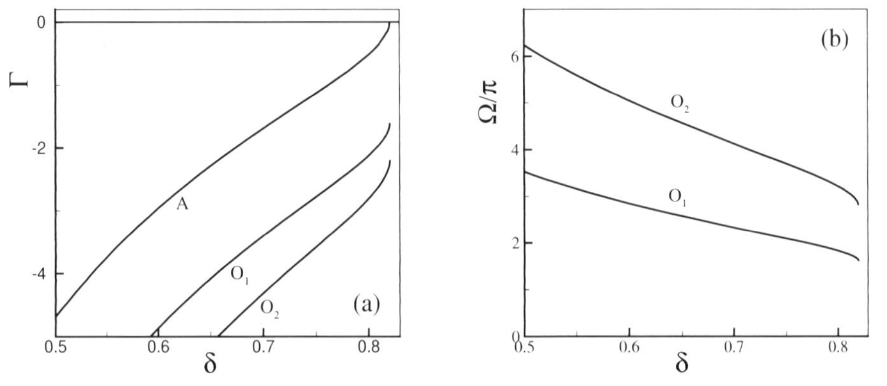

Figure 3. Dispersion curves of the Bursian diode with a purely active external circuit ($L = 0$) corresponding to the branch 3 ($V = -0.4$) in Fig. 2. A and O are aperiodic and oscillation branches, respectively; (a) relates to the growth rates, (b) – to the frequencies.

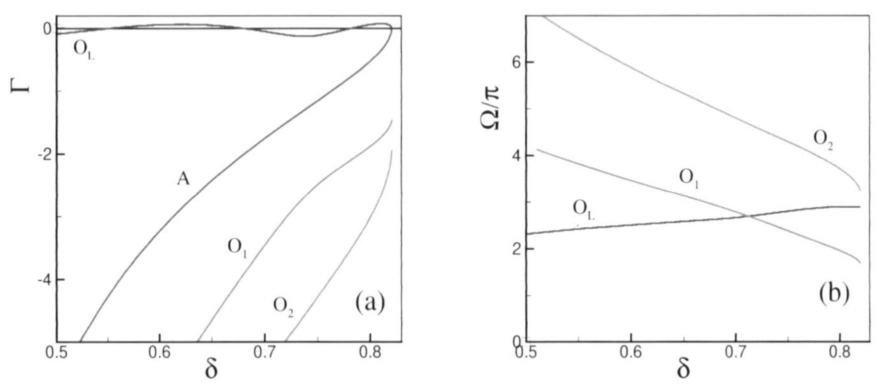

Figure 4. Dispersion curves of the Bursian diode with an inductive external circuit ($L = 0.01$) corresponding to branch 3 ($V = -0.4$)in Fig. 2. The lower digit of the symbol O indicate an oscillation branch corresponding to a similar branch in the diode with $L = 0$; the lower symbol L indicates a new branch appearing in a diode with $L > 0$.

Occurrence of the inductance L in the external circuit leads to the appearance of a new branch (it is lettered by the symbol L in Fig. 4.) Its growth rate runs into the region of positive values, i.e. the solutions turn out to be unstable at relevant values of δ. In this case, nonlinear oscillations have to develop in the diode, which, apparently, were observed authors [6]. We have also revealed that for each value of the external voltage V, the instability develops only for inductance values being fallen in the interval $(0, 0.0311 \cdot (1 + \sqrt{1 + 2V})^{5/2})$. Dependencies of new dispersion branch on the inductance value are shown in Fig. 5.

4. Conclusion

We studied the problem of an external circuit effect on the stability of steady state solutions of the Bursian diode. It is shown that an inductive load in the external circuit can initiate the

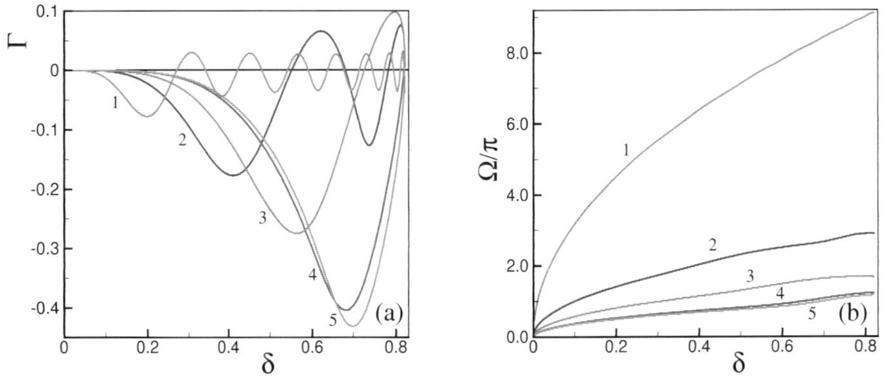

Figure 5. Dispersion curves O_L of the Bursian diode with an inductive external circuit corresponding to branch 3 ($V = -0.4$) in Fig. 2 for various values of an inductance L: (1) – $L = 0.001$, (2) – 0.01, (3) – 0.03, (4) – 0.07 and (5) – 0.08.

oscillation instability in the diode regime with a decelerating electron flow. The value of the inductance at which the instability develops is estimated. The regions in which the diode can generate are found. The growth rates and frequencies being specific for the instability in the generation region are calculated. In order to see to what state the system diode – the external circuit will fall as the perturbation develops, the non-linear stage of the process must be studied. For this purpose, it is necessary to carry out calculations of the process at the non-linear stage.

References

[1] Luginsland J W and Arman M J and Lau Y Y 1997 *Phys. Plasmas* **4**, 4404-8
[2] Zhang P and Valfells A and Ang L K and Luginsland J W and Lau Y Y 2017 *Appl. Phys. Rev.* **4**, 011304
[3] Bursian V R and Pavlov V I 1923 *J. Russ. Phys. Chem. Soc.*, **55**, 71
[4] Pierce J R 1944 *J. Appl. Phys.* **15**, 721-6
[5] Ender A Ya and Kuznetsov V I and Schamel H and Akimov P V 2004 *Phys. Plasmas* **11**, 3212-23
[6] Paschenko A V and Zalkind V M and Lonin Yu V 2008 *Tech. Phys. Letters* **34**, 1079-84
[7] Kuznetsov V I and Ender A Ya 2010 *Plasma Phys. Rep.* **36**, 226-35

IOP Conf. Series: Journal of Physics: Conf. Series **1038** (2018) 012128 doi:10.1088/1742-6596/1038/1/012128

Numerical Investigation of Cavitating Flows with Liquid Degassing

U Iben[1], A Makhnov[2] and A Schmidt[2, 3]
[1] Robert Bosch GmbH, Germany
[2] Peter the Great Saint-Petersburg Polytechnic University, Russia
[3] Ioffe Physical and Technical Institute, Russia

E-mail: a_makhnov@mail.ru

Abstract. Cavitation is a phenomenon of formation of bubbles (cavities) in liquid as a result of pressure drop. Cavitation plays an important role in a wide range of applications. For example, cavitation is one of the key problems of design and manufacturing of pumps, hydraulic turbines, ship's propellers, etc. Special attention is paid to cavitation erosion and to performance degradation of hydraulic devices (noise, fluctuation of the mass flow rate, etc.) caused by formation of a two phase system with an increased compressibility. One more important problem accompanying cavitation is liquid degassing due to diffusion of the dissolved gas into the cavities. Determination of the degassed air content is important for the performance forecast of devices. Mathematical models of the degassing in cavitating flows are not yet developed and presented in the literature satisfactorily. Therefore, development of a model for simultaneous cavitation and liquid degassing is an important fundamental and applied task. To validate the algorithm simulations of unsteady flow of a cavitating liquid in a channel have been conducted. The obtained results are in satisfactory agreement with the experimental data and demonstrate the efficiency and robustness of the formulated model and the algorithm.

Introduction

Cavitation means formation of bubbles (cavities) as a result of pressure drop. This phenomenon has been a subject of many experimental and theoretical investigations for more than one hundred years. The first fundamental studies on cavitation date back to the beginning of the 20th century and belong to Lord Rayleigh. Cavitation plays a very important role in power engineering (pumps, hydraulic turbines), in various liquid injection systems, in shipbuilding and in many other applications. Formation of a two phase system with an increased compressibility can lead to performance degradation of hydraulic devices (noise, fluctuation of the mass flow rate, etc.). Collapse of cavitation bubbles and clouds often leads to structure erosion. Therefore, the cavitation phenomenon greatly affects performance and lifetime of hydraulic devices.

All liquids contain some amount of a dissolved gas. It is known that the concentration of dissolved air in water at normal conditions varies from 8 vol % to 14 vol % [1]. Typical mineral-oil-based hydraulic liquids can also dissolve high amounts of air (7-12 vol %) [2,3]. The dissolved gas transported from the liquid into growing cavitation bubbles due to diffusion through the interfaces is a reason of the higher bubble volume fraction [4].

Therefore, physically cavitation is a superposition of two mechanisms: phase transition and vapor bubbles formation and release of the dissolved gas into the bubbles. Thus, the bubbles formed during cavitation contain both vapor and gas. Nowadays, quite a few is known about the interaction of these two processes. Mathematical description of these processes is not completed yet. The goal of the present study is development and validation of a combined mathematical model for numerical (CFD) simulation of cavitating flows with liquid degassing.

The present paper is divided into four sections. In the first section, a description of the developed mathematical model is given. The second section is dedicated to the numerical approach that is used in simulations. In the third section, the most important results of the study are presented and discussed. The last section summarizes the main conclusions drown from the obtained results.

1. Mathematical model

The model developed and validated in the present study includes a model of vapor cavitation as well as a model of liquid degassing due to gas diffusion into the vapor cavities. The total system of governing equations (1) is a combination of the Navier-Stokes equations, the developed cavitation model, and the developed degassing model:

$$\begin{cases} \dfrac{\partial \rho}{\partial t} + \nabla \cdot (\rho \boldsymbol{u}) = 0 \\ \dfrac{\partial}{\partial t}(\rho \boldsymbol{u}) + \nabla \cdot (\rho \boldsymbol{u}\boldsymbol{u}) - \nabla \cdot (\mu \nabla \boldsymbol{u}) = -\nabla p \\ \textbf{model of cavitation} \\ \textbf{model of liquid degassing} \end{cases} \tag{1}$$

Here ρ, $\boldsymbol{u}$ and p are for the density, for the velocity and for the pressure fields, respectively.

Viscosity of both liquid and vapor phases is taken into account (μ is the dynamic viscosity coefficient of the mixture). Heat transfer is not considered, therefore the energy equation is not solved. The two-phase mixture of liquid and vapor is assumed to have constant and uniform temperature.

1.1. Model of vapor cavitation

A vapor cavitation model formulated in the present study is based on the homogeneous equilibrium approach and barotropic equation of state. Liquid phase is described using a weakly compressible fluid linearization (2), vapor phase is assumed to follow the ideal gas law.

$$\rho(p) = \rho_0 + \frac{1}{c_L^2}(p - p_v) \tag{2}$$

Here ρ_0 is for the liquid density under saturation conditions and p_v is the saturation pressure. The speed of sound in the liquid phase, c_L, is assumed to be constant.

When the pressure of the two-phase mixture of liquid and vapor becomes less than the saturation pressure, the vapor mass fraction is calculated using the barotropic model [5]. This model provides a coupling between the vapor mass fraction and the pressure:

$$m(p) = -\frac{1}{r_{evap}}\left(\frac{dh'}{dp} - \frac{1}{\rho_0}\right)(p - p_v) = -K(p - p_v) \tag{3}$$

Here m is for the vapor mass fraction, p is the pressure, r_{evap} is for the evaporation enthalpy, h' is the enthalpy of the liquid phase. The coefficient K is assumed to be constant in the present study.

Formula (3) is derived from a simplified version of The First Law of Thermodynamics for the considered two-phase mixture. Simplifications are based on the assumption that, firstly, the considered system is adiabatic, and, secondly, entropy change due to dissipation can be neglected.

1.2. Model of degassing

A degassing model formulated in the present study provides a coupling between the dissolved gas diffusion process and the local characteristics of the cavitating flow (pressure and vapor volume fraction). The model is based on an analytical solution of the diffusion equation in a spherical cell corresponding to a single bubble:

$$D_s \frac{1}{r^2}\frac{\partial}{\partial r}\left(r^2 \frac{\partial c_s}{\partial r}\right) = 0 \tag{4}$$

Here D_s is for the dissolved gas diffusivity, c_s is for the dissolved gas concentration, r is the radial coordinate (distance from the bubble center).

The Henry's law is used to formulate a boundary condition for gas concentration at the bubble surface.

Equation (4) describes stationary diffusion of the dissolved gas. Transient term is neglected because validation cases considered in the present study are characterized by much smaller time scales (by 2-3 orders of magnitude) in comparison with the diffusion time scale.

2. Numerical method

The system of governing equations is solved numerically based on a finite volume approach. Open-source CFD toolbox OpenFOAM [6] is used in the present study for simulations. The main advantage of the OpenFOAM toolbox is open access to codes which provides possibility to implement original models and algorithms.

The following settings of the algorithm were specified:

- PISO scheme for pressure-velocity coupling;
- Approximation of convective terms using *Gauss limitedLinear 1* scheme;
- Approximation of diffusive terms using central scheme *(Gauss linear).*

3. Results and discussion

The developed algorithm has been validated on two test cases. The first test case is simulation of a single cylindrical bubble collapse. The second test case is the cavitating flow in the planar throttle for which the experimental data is available.

3.1. Single cylindrical bubble collapse

The main goal of this test case is to study the capability of the solver to capture a complex structure of interacting compression and rarefaction waves. Initial conditions correspond to cylindrical vapor domain located in a square liquid domain (see Fig. 1). Boundaries of the square liquid domain are solid walls with no-slip condition for the velocity and symmetry boundary condition for the pressure.

Size of the square domain is 100 mm x 100 mm. Radius of the cylindrical vapor domain is 0.0015 mm. Fluid properties correspond to water at the temperature of 20^{0}C: ρ_0 = 998 kg/m^3, p_v = 2330 Pa, c_L = 1485 m/s. Initial pressure inside the vapor and liquid regions is specified to 2000 Pa and $2{\cdot}10^7$ Pa, respectively.

The computational mesh consists of 1000 cells at both directions. The mesh is uniform, and all the cells have a shape of square. The time step is 10 ns.

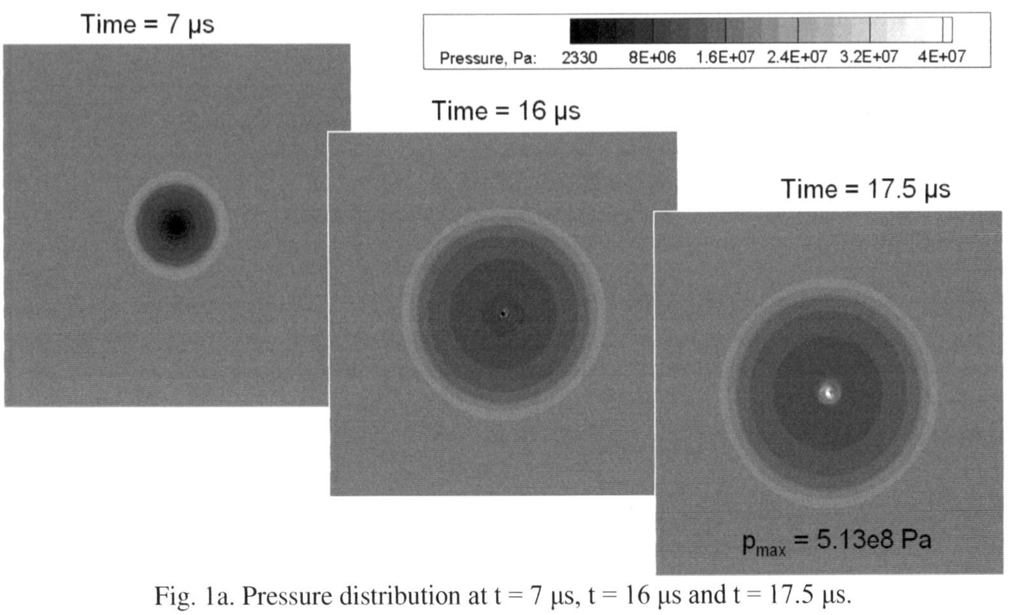

Fig. 1a. Pressure distribution at t = 7 μs, t = 16 μs and t = 17.5 μs.

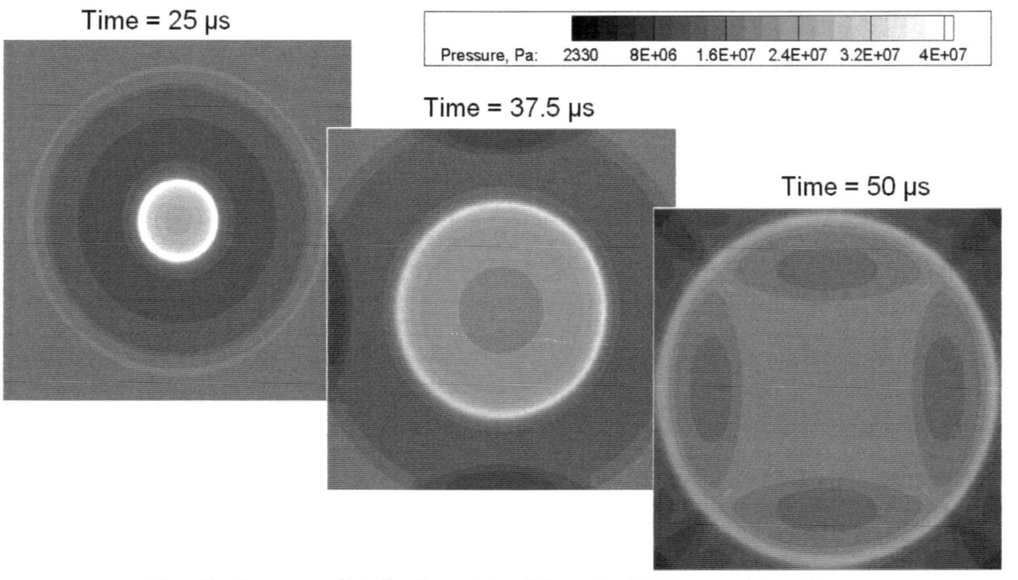

Fig. 1b. Pressure distribution at t = 25 μs, t = 37.5 μs and t = 50 μs.

Pressure distributions at different time moments are presented in Fig. 1a and 1b. The first set of pictures (Fig. 1a) illustrates two processes. On one hand, a rarefaction wave is propagating from the vapor region into the liquid. On the other hand, the vapor region in the center is being compressed, liquid is accelerated towards the center and finally this leads to the collapse. The second set of pictures (Fig. 1b) visualizes the propagation of a pressure wave generated by the collapse. Complex wave structure of the flow formed by interaction of compression and rarefaction waves is seen in Fig. 1b at time t = 50 μs.

3.2. Cavitating flow in a planar throttle

The main goal of the simulations presented in this subsection is to validate the implemented solver on the experimental data. Geometry of the channel and fluid properties were selected in accordance with the experiment [7]. Fig. 2 shows a scheme of the considered channel. Inlet total pressure is 300 bar, outlet static pressure is 114 bar.

The most important problems caused by cavitation in hydraulic devices are mass flow rate fluctuations and cavitation erosion. That is why special attention in the present study is paid to the mass flow rate and to the wall loading. Left plot in Fig. 3 demonstrates oscillations of the mass flow rate with high amplitude. Right plot shows that despite the decrease of the average wall pressure due to cavitation, formation and collapse of cavities lead to local pressure peaks up to 1000 bar. Such pressure peaks can result in erosion.

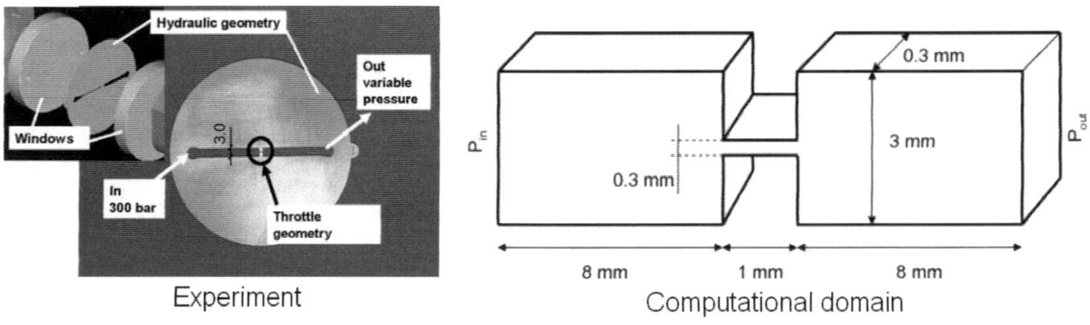

Fig. 2. Experimental throttle geometry [7] and the computational domain.

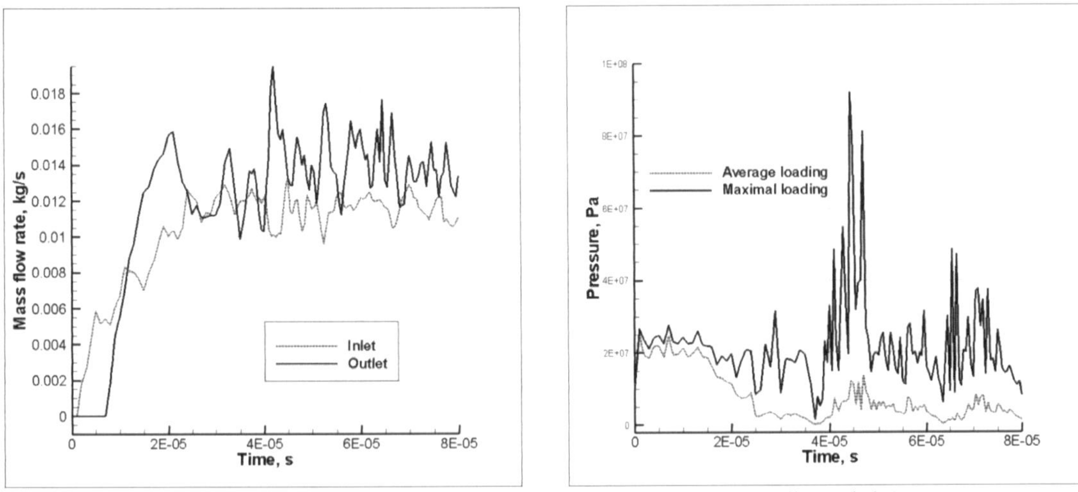

Fig. 3. The mass flow rate (left) and the throttle wall loading (right).

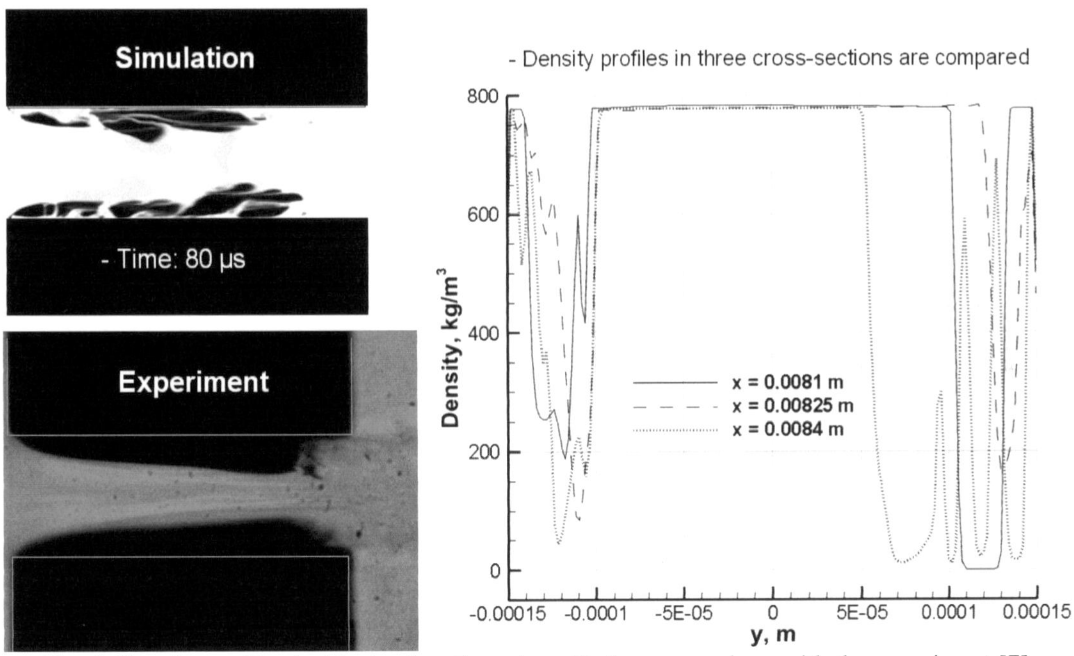

Fig. 4. Density profiles inside the throttle and qualitative comparison with the experiment [7].

Simulations enabled to observe inception and development of the cavitation zones. Comparison shows that the predicted cavities are in qualitative agreement with experimental observations [7].

4. Conclusions

A model for cavitating flows with stationary degassing is formulated based on the Euler-Euler description and barotropic equation of state for equilibrium two-phase mixture. Results of the validation show the capability of the model to predict inception of cavitation, cavity development and collapse. Comparison with the experimental data demonstrates the efficiency and robustness of the formulated model and algorithm.

Acknowledgements
Andrei Makhnov thanks the German-Russian Interdisciplinary Science Center (G-RISC) funded by the German Federal Foreign Office via the German Academic Exchange Service (DAAD) for the support of these investigations.

References
[1] Brennen C E 1995 *Cavitation and Bubble Dynamics* (Oxford University Press)
[2] Bock W, Braun J, Puhl N, and Heinemann H 2010 *Air Release Properties of Hydraulic Fluids* (Internationals Fluidtechnisches Kolloquium (IFK), Vol. 7)
[3] Iben U, Wolf F, Freudigmann H-A, Froehlich J, and Heller W 2015 *Optical Measurements of Gas Bubbles in Oil Behind a Cavitating Micro-Orifice Flow* (Exp. Fluids, 56(6))
[4] Crowe C T 2006 *Multiphase Flow Handbook* (Taylor & Francis Group, LLC)
[5] Iben U 2015 *Vorlesungsskript Ein- und Mehrphasenströmungen - Modellierung und Anwendun- gen in der Industrie. Institut für Aerodynamik und Gasdynamik* (Universität Stuttgart)
[6] OpenFOAM: *the open source CFD toolbox* // URL: https://www.openfoam.com/
[7] Skoda R, Iben U, Morozov A, Mihatsch M, Schmidt S J, and Adams N A 2011 *Numerical Simulation of Collapse Induced Shock Dynamics for the Prediction of the Geometry, Pressure and Temperature Impact on the Cavitation Erosion in Micro Channels* (Warwick: WIMRC 3rd International Cavitation Forum 2011)

IOP Conf. Series: Journal of Physics: Conf. Series **1038** (2018) 012129 doi:10.1088/1742-6596/1038/1/012129

Modification of the surface layers with plasma of a vacuum-arc discharge by controlling the energy of precipitating particles

A A Rikov, M I Yurchenkov, M I Pikus, D K Kostrin and A A Lisenkov

Department of electronic instruments and devices, Saint Petersburg Electrotechnical University "LETI", 197376, Saint Petersburg, Russia

E-mail: rikov247@gmail.com

Abstract. In this paper we consider one of the most effective ways to improve the performance properties of metals – their treatment by intense flows of metal plasma generated by a vacuum-arc discharge. The change in the thermal regime and the duration of the plasma treatment allows controlling both the thickness of a modified layer and the distribution of implanted ions into its depth. At the first stage of part processing the preliminary cleaning of the surface is carried out using a glow discharge in argon. In the second stage, vacuum-arc plasma sources are used, that provide surface treatment with ions of the sprayed material. To weaken and purify the plasma flow special separation systems were used to ensure the creation of an impassable barrier for the droplet fraction. Modification of the treated surface layer is achieved either due to the high energy of the deposited particles, or due to the implementation of the diffusion processes deep into the substrate. The presence of a modified surface layer allows improving the quality of the working coating that is deposited further.

1. Introduction

One of the most effective ways to improve the performance properties of metals is their treatment with intense streams of charged particles and, in particular, flows of metal plasma generated by a vacuum-arc discharge. The advantage of this method is that not only the high-energy ions, but also a high flux density has an impact on the conditions of interaction. The latter contributes to an increase of the target temperature to ~1000 K, which allows forming in the surface layer of compounds, the synthesis of which is impossible under ordinary conditions, and as well obtaining intermetallic compounds and solid solutions with a high concentration of nanosized phases of implementation [1].

The change in the thermal regime and the duration of the plasma treatment allows controlling both the thickness of a modified layer and the distribution of implanted ions into its depth, while the observed modification of the target volume is achieved to a sufficient depth with the absence of a sharp boundary of separation. For example, during the deposition of carbon the following sequence of generated layers is observed under certain conditions [2]: $(Me_{base} - Me_2C - MeC)_{sub} - C_{coat}$, where "Me" denotes a metal, particularly molybdenum or tungsten; indexes "sub" and "coat" mean substrate and coating, respectively.

It should be noted that a significant part of the carbide phases of transition metals has a wide area of homogeneity, within which the change in the carbon content occurs without the restructurization of the crystal lattice ($MeC_{1-x} - MeC$) [3]. Carbides, having a close to stoichiometric composition, possess the least resistance. With the increase of the imperfection degree for carbon of the carbide lattice the resistance increase, that is caused by a significant decrease of the mobility of conductivity electrons due to a strong scattering by vacancies in the carbon sublattice, which are in this case the centers of scattering. The microhardness of the resulting carbides is defined by the energy sustainability of the d^n-configurations that leads to the disruption of sp^3-configurations of carbon and strong delocalization of electrons.

2. Results and discussion

Technology of the processing of metal parts with a flow of plasmachemical synthesis of compounds includes the following technological operations (figure 1).

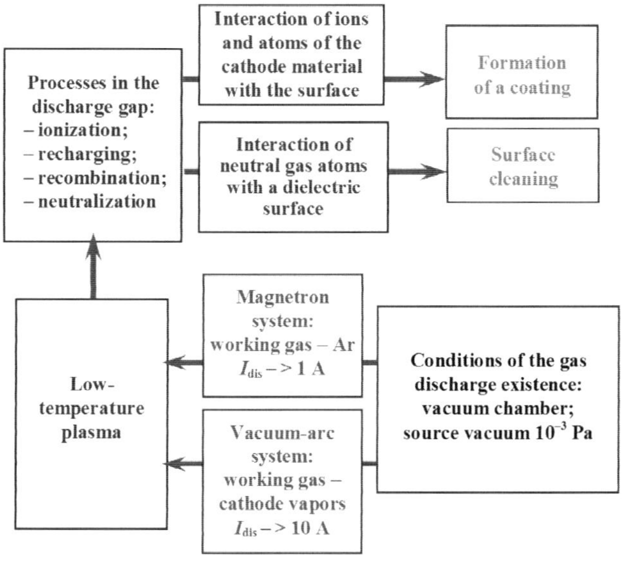

Figure 1. Technological ion-plasma processes (I_{dis} – discharge current).

The first stage is the pre-cleaning of a treated surface using a glow discharge in argon at a pressure of 10^{-1} Pa. At this stage, along with cleaning of the surface through its sputtering the heterogeneities (microscopic protrusions) are removed, capable of causing breakdowns in the system "plasma – device" during the high-energy treatment of the substrate by metal ions at the stage of formation of a pseudodivision layer. Smoothing of the microscopic irregularities at an average current density of ions from the gaseous medium of about 1 mA/cm^2 and their energy of 600...700 eV takes few minutes.

At the second stage vacuum-arc plasma sources (figure 2), providing a surface treatment using ions of the sprayed material, are used [4, 5]. High energy of ions (several keV) is achieved under condition that on the treated surface a negative potential $-U_{bias}$ = 1…2 kV is set.

For the weakening and cleaning of a plasma flow a special separation system, ensuring the creation of an impenetrable barrier to droplet fraction, is used [6]. This unit allows to allocate the stream of charged particles and to control the density of the charged particles flow arriving at the treated part. In addition, it is possible to provide a high electric strength of the system, i.e. to exclude the possibility of breakouts.

The output of the charged particles (ions and electrons) from the plasma flow is defined as by the above-mentioned conditions, and also by the design of the vacuum installation and parameters of the deposition process: bias voltage; value of the discharge current; distance between the substrate and the cathode; distribution of the magnetic field; degree of rarefaction (pressure is below 10^{-3} Pa) and the use of additional gases contributing to an increase in the degree of ionization of the plasma flow (Ar, He) and ensuring the flow of chemical reactions with the formation of chemical compounds (hydrocarbon compounds – C_xH_y).

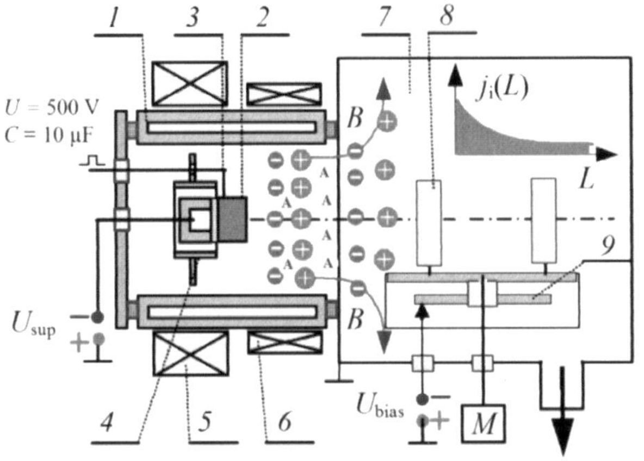

Figure 2. Design of a coaxial vacuum-arc plasma source (U_{sup} – supply voltage; j_i – ion current density): *1* – anode; *2* – cathode; *3* – igniting electrode; *4* – screen; *5*, *6* – stabilizing and focusing solenoids; *7* – working volume; *8* – workpiece; *9* – planetary gear.

The formation of the sublayer or, otherwise, the modification of the treated surface layer is achieved either due to a high energy of the deposited particles (directional velocity of the plasma flow is 10^4 m/s), or due to the implementation of the diffusion processes into the depth of a substrate (figure 3).

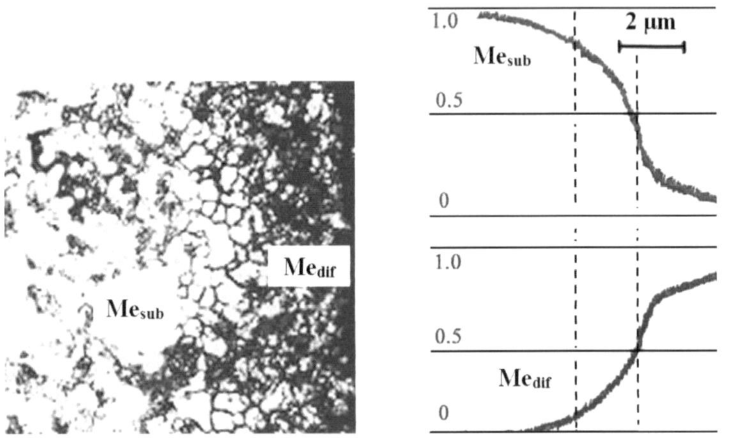

Figure 3. Distribution of the concentration of the deposited material into the substrate material (Me_{sub} – metal of the substrate; Me_{dif} – metal diffusing into the substrate).

The conditions for the occurrence of diffusion processes between the deposited particles and atoms of the surface are determined by the energy and properties of the particles themselves, as well as the process conditions: substrate temperature and the composition of its surface.

Changing the nature of the flow of diffusion processes the quality and composition of the coating, and also its adhesion and structure can be varied [7].

The occurrence of the diffusion processes leads to a distribution of deposited atoms through the thickness of the substrate $n_{dif}(x, t)$. The diffusion coefficient $D(x) = D_\tau + D^*(x)$ is represented as the sum of the thermal (D_τ) and radiation-enhanced diffusion $D^*(x)$. In accordance with the theory [8] $D^*(x)$ is chosen proportional to the concentration of radiation defects $C(x)$:

$$D^*(x) = [D_\tau D_d / (D_s N_0)]C(x),$$

where D_s, D_d – self-diffusion and defects diffusion coefficients, respectively. The coefficient of the thermal diffusion of implanted atoms is estimated by the formula of Arrhenius [9]:

$$D_\tau = D_0 \exp(-E_a/(kT)),$$

where D_0 – pre-exponential factor; E_a – activation energy; k – Boltzmann constant; T – temperature of the target. Due to the high diffusion mobility of the interacting components, provided by the thermal conditions, the diffusion zone in the substrate reaches a significant size [10]. The diffusion coefficient along grain boundaries is much higher than the coefficient of volume diffusion, due to a large value of the energy of volume diffusion [11, 12].

3. Conclusions

Diffusion of carbon atoms into the substrate leads to the chemical reactions with the formation of chemical compounds of Me_xC type, which in turn leads to the emergence of structural inhomogeneities and internal interface surfaces, complicating the modeling of further diffusion process. In addition, the saturation of the surface layer reduces the number of carbon atoms involved in diffusion processes and contributes to the subsequent growth of the coating formed from the plasma of the vacuum-arc discharge. It should be noted that the presence of a modified surface layer allows improving the quality of the working coating that is deposited further.

Parallel use of the hydrocarbon compounds introduced into the working volume with the simultaneous spraying of the graphite cathode leads to the fact that in the area of the treated part exists a zone with a temperature equal to or higher than the decomposition temperature for the hydrocarbon compounds. This condition changes the kinetics and mechanism of formation of the deposited coating based on carbon.

Thus, it is shown that the treatment of the substrate by an intense flow of charged particles allows controlling the temperature of the target and form compounds in the surface layer, the synthesis of which is impossible under normal conditions. The obtained results can be applied in the production technology of electrovacuum devices, where refractory metals molybdenum and tungsten are used as a structural material, and their processing is carried out by carbon-containing particles that change their emission and mechanical properties.

References

[1] Kostrin D K and Lisenkov A A 2017 *Vakuum in Forschung und Praxis* **3** 34–7
[2] Lisenkov A A, Vetrov N Z and Kostrin D K 2017 *Technical Physics Letters* **8** 55–62
[3] Kostrin D K and Lisenkov A A 2016 *Materials Science Forum* **870** 371–6
[4] Bystrov Yu A, Kostrin D K, Lisenkov A A and Vetrov N Z 2015 *Vakuum in Forschung und Praxis* **2** 22–5
[5] Kostrin D K and Lisenkov A A 2016 *Materials Science Forum* **843** 278–83
[6] Bystrov Yu A, Vetrov N Z, Lisenkov A A and Kostrin D K 2014 *Vakuum in Forschung und Praxis* **5** 19–23
[7] Pout J, Tu K, Meyer J 1982 *Thin films. Interdiffusion and reactions* (Moscow: Mir)
[8] Makarets N V, Falco G L and Fedorchenko A M 1984 *Physics of Metals and Metallography* **5** 29–34
[9] Ryssel H and Ruge I 1983 *Ion implantation* (Moscow: Nauka)
[10] Krasil'nikov V V and Savotchenko S E 2015 *Technical Physics* **7** 1031–8
[11] Knyazeva A G 2003 *Journal of Applied Mechanics and Technical Physics* **3** 373–84
[12] Vakhnii T V, Vershinin G A, Sharkeev Yu P, Kurzina I A, Eroshenko A Yu, Grekova T S and Gritsenko B P 2010 *Journal of Surface Investigation. X-ray, Synchrotron and Neutron Techniques* **2** 353–8

IOP Conf. Series: Journal of Physics: Conf. Series **1038** (2018) 012130 doi:10.1088/1742-6596/1038/1/012130

Comparative analysis of transition models in prediction of flow over NACA-0012 airfoils in tandem

A A Matyushenko, A S Stabnikov, A V Garbaruk
Peter the Great St.Petersburg Polytechnic University
Russia, Saint-Petersburg, 195251, Politechnicheskaya, 29

E-mail: alexey.matyushenko@gmail.com

Abstract. Comparative analysis of several semi-empirical turbulence models taking into account laminar-turbulent transition (γ-$Re_{\theta t}$ SST, γ-SST, γ-SA and e^N-SA) was carried out for flow around an airfoil in the wake generated by another upstream airfoil for different Reynolds numbers and distances between airfoils. Two dimensional Reynolds-averaged Navier Stokes equations were solved in different CFD codes on unstructured and chimera-type structured grids. Results of the considered transition models are virtually independent of the computational code. Results of the SST-based transition models are in reasonable agreement with experimental data while SA-based models predict too early transition.

1. Introduction

The flow over the blades of turbomachinery are usually highly three dimensional, turbulent, and unsteady due to their complex geometry and interaction between the rotating and nonrotating blades. The flows are transitional over a considerable part of the blade, especially in machines of compact design and small size. Therefore correct prediction of wall characteristics in the transitional boundary layer are important for developing highly efficient compressor blades.

In engineering practice these characteristics are mostly predicted with semi-empirical turbulence models which take into account laminar-turbulent transition (transition models). Generally, transition models are calibrated on simple flat plate boundary layer with zero and non-zero pressure gradient, or flows around airfoils with uniform freestream while the incoming flow on the blades of turbomachinery is no uniform. Thus, to determine the applicability of the transition models for prediction flow around turbine blades it is necessary to study the accuracy of these models when the blade is mounted in the wake of another blade. In this paper we considered the relatively simple flow over NACA-0012 airfoils in tandem which demonstrates the accuracy of different transition models and gives the recommendation of the most appropriate transition model for flow over blade cascade.

2. Considered transition models and CFD codes.

Incompressible numerical simulations of this flow were carried out using the commercial ANSYS Fluent and the academic NTS (Numerical Turbulence Simulation) codes using Reynolds-averaged Navier Stokes (RANS) equations in combination with four transition models solving additional equations together with k-ω SST [1] or SA [2] turbulence models. The considered transition models use different approaches for prediction of the locations of transition. The first method is based on solving additional transport equations for the turbulence intermittency (γ-$Re_{\theta t}$ SST [3], γ-SST [4], and γ-SA [5]). The last model based on the e^N method is e^N-SA [6]. SST based transition models (γ-$Re_{\theta t}$ SST and γ-SST) are

IOP Conf. Series: Journal of Physics: Conf. Series **1038** (2018) 012130 doi:10.1088/1742-6596/1038/1/012130

implemented in both considered CFD codes, while SA based models (γ-SA and e^N-SA) are available only in the NTS code.
The double precision version of the solvers was used in all the computations. For all transport equations, including the equations for turbulent variables, a second order discretization scheme for the convective terms was selected. For each test case, all simulations were started from the same uniform initial guess and the simulations were performed until the normalized maximum residuals of all the equations were below a value of 10^{-5}.

3. Problem definition

Flow around NACA-0012 airfoils in tandem mounted in the centre of the wind tunnel was experimentally carried out for different regimes [7]. The Reynolds number based on incoming uniform flow and airfoil chord *C* was varied from 2×10^5 to 6×10^5 together with the distance between the airfoils. The parameters of flow configurations are shown in table 1. In the experiment strips of sandpaper were pasted on both sides of the upstream airfoil between *X/C*=0.58 and 0. 65 to generate a stable wake flow.

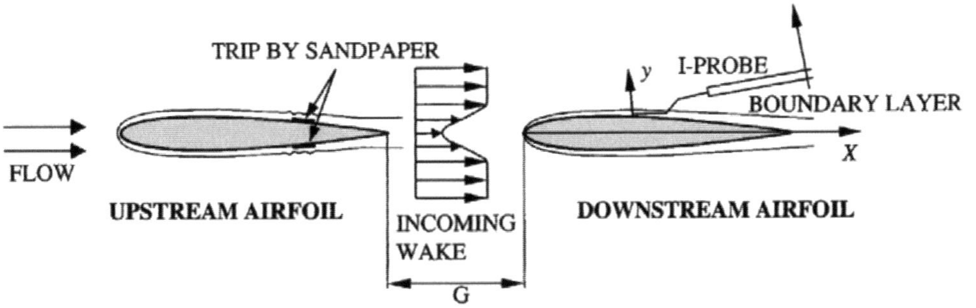

Figure 1. Schematic of the arrangement of airfoils

Table 1. Flow regimes of flow around the airfoils

Case	Distance between airfoils, G	Reynolds Number ($\times10^5$)
Case-1	1.00	6.00
Case-2	1.00	2.00
Case-3	0.50	2.00
Case-4	0.25	2.00

The height of the computational wind tunnel H=2C corresponds to the experimental one. Inlet and outlet boundaries were located 10C upstream of the leading edge of the first airfoil and downstream of the trailing edge of the second airfoil correspondingly.
A constant velocity is specified at the inlet section of the computational domain. No-slip conditions are used on the airfoil boundary and constant pressure is specified on the outlet. Symmetry boundary condition was specified on the wind tunnel upper and lower walls for imitation of the slip-walls. For k-ω based models, the inlet turbulent kinetic energy provides experimental turbulence intensity $Tu = 0.3\%$ near the leading edge of the upstream airfoil and the specific dissipation rate is specified as $\omega = 10\cdot U_\infty/C$ [2]. For the SA-based model only the analogous value of turbulent viscosity was specified.
In order to reproduce the sound paper on the upstream airfoil the onset of transition in the simulations was tripped at $X/C = 0.58$. The efficiency of this method is shown for distribution of skin friction coefficient over the upstream airfoil in Figure 2.

IOP Conf. Series: Journal of Physics: Conf. Series **1038** (2018) 012130 doi:10.1088/1742-6596/1038/1/012130

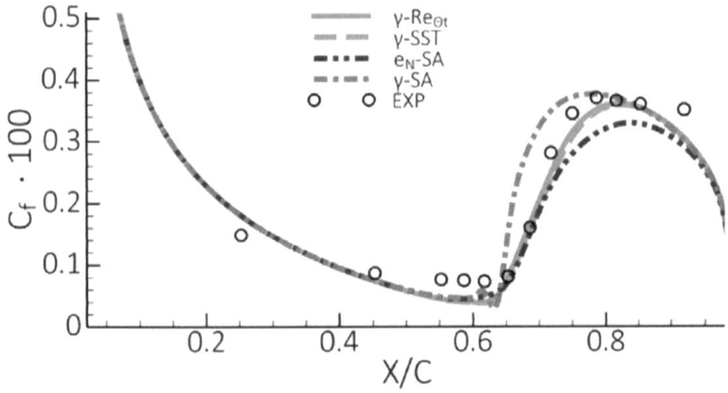

Figure 2. Distribution of the computational and experimental skin friction coefficient on the upstream airfoil for the Case-1

4. Computational grids

Considered CFD codes used different types of grids. For ANSYS Fluent a hybrid unstructured grid was generated using ANSYS Meshing. A hex grid was generated in the thin area around the airfoil and downstream in the wake while triangle cells were generated in the rest of the computational domain. A chimera type grid was used was used in the NTS code with structured grids generated in three overlaping areas: the domain around upstream and downstream airfoil and wind tunnel domains. In both codes Y^+ in the first near wall cell was less than 1, about 400 points were specified on the airfoil and the total number of cells was about 10^5. The grid structure and parameters are presented in the Table.2.

Table 2. Grid structure and parameters for ANSYS Fluent and NTS code.

Fluent	**NTS code**
Grid structure	
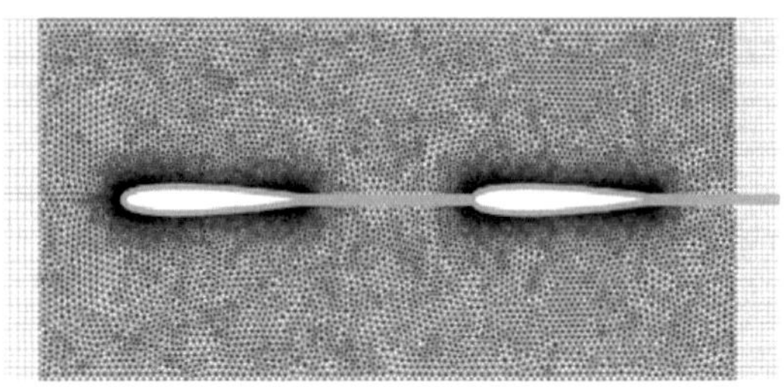	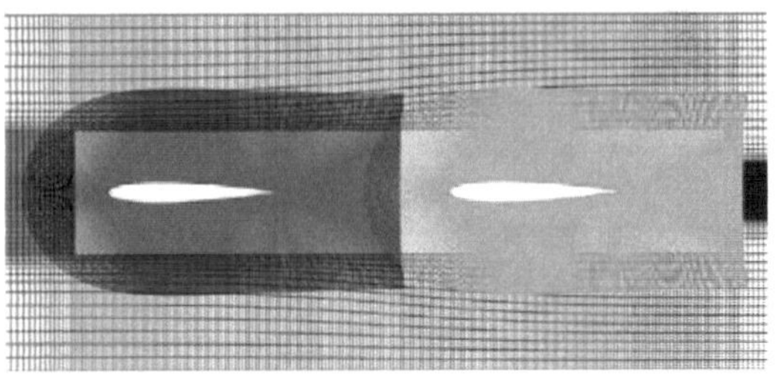
Grid in the vicinity of the airfoil	
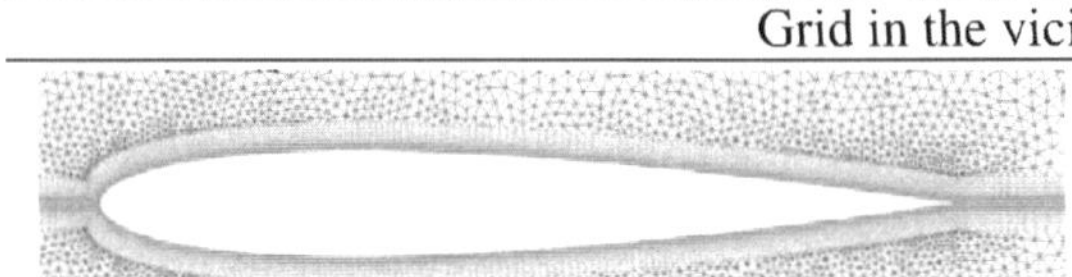	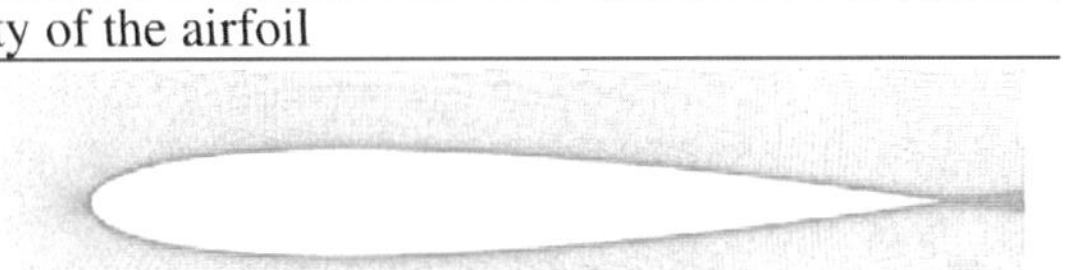
400 points on the airfoil	440 points on the airfoil
$\Delta Y^+_{1,max} \sim 1.0$	$\Delta Y^+_{1,max} \sim 0.1$
Aspect ratio in normal-wall direction	
1.1	1.2
Totlan number of cells	
10^5	1.2×10^5

5. Results and discussions

5.1 Cross-code comparison

Cross-code comparison of the skin friction on the downstream airfoil (Figure 3) shows good or even perfect agreement between transition models implemented in both ANSYS Fluent and NTS codes. The results of the γ-$Re_{\theta t}$-SST model are virtually identical for all the cases, only small differences are observed in the transitional area. The more visible difference for the γ-SST model in Case-1 (Figure 2 Left) contains higher sensitivity of the model to different numerical implementation.

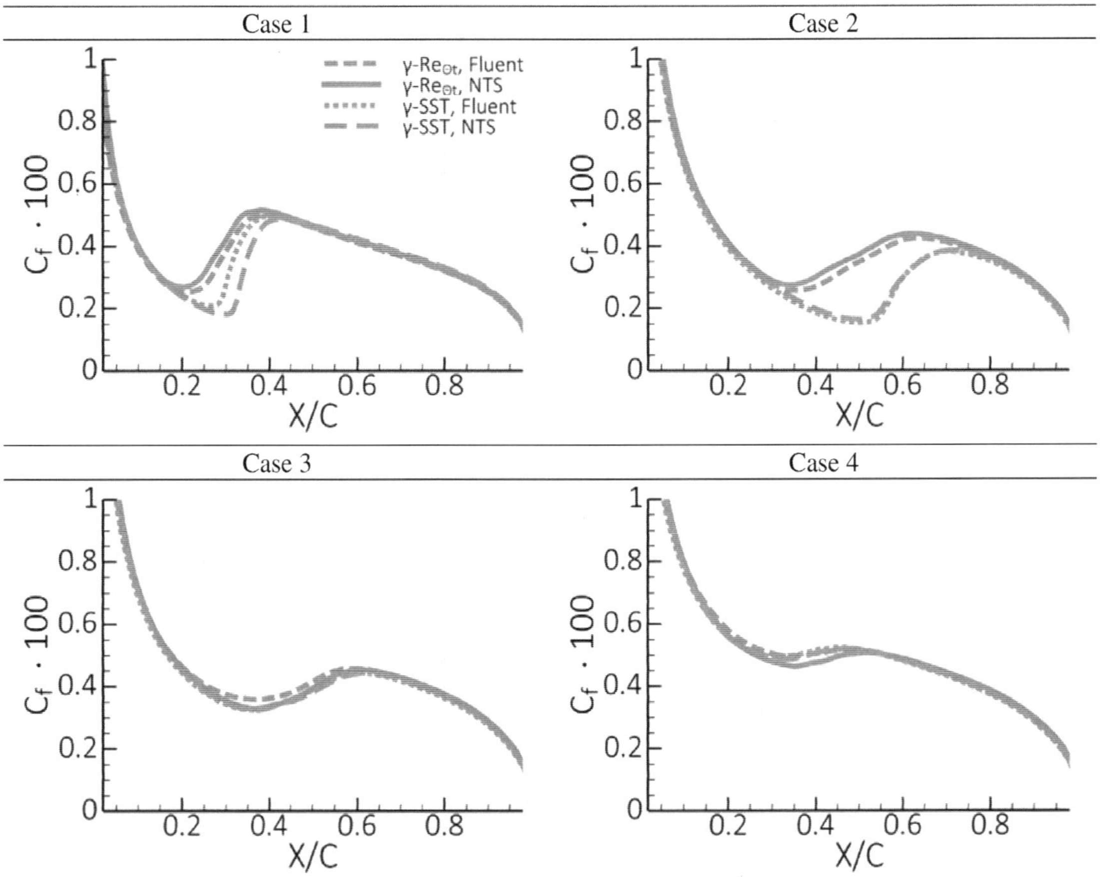

Figure 3. Distribution of skin friction coefficient on the downstream airfoil for Case-1 (Left) and Case-4 (Right) predicted by the different transition models implemented in considered CFD codes.

5.2 Comparison of CFD results with experimental data

The comparison of computational and experimental distributions of skin friction coefficient over the downstream airfoil is shown in Figure 4. The SA-based transition models (γ-SA and e^N-SA) tend to predict too early laminar-turbulent transition. Meanwhile e^N-SA developed for the aerodynamic flows with low turbulence incoming flow predicts transition on the leading edge. The results of the SST-based transition models are in reasonable agreement with the experiment. The onset of transition, predicted by these models is close to the experimental value. However accuracy of the models also leaves much to be desired. These models as well as the γ-SA model predict shorter transition length defined as the distance between minimum and maximum values of skin friction in comparison to the experiment for all the cases (Table 3, Length of transition region).

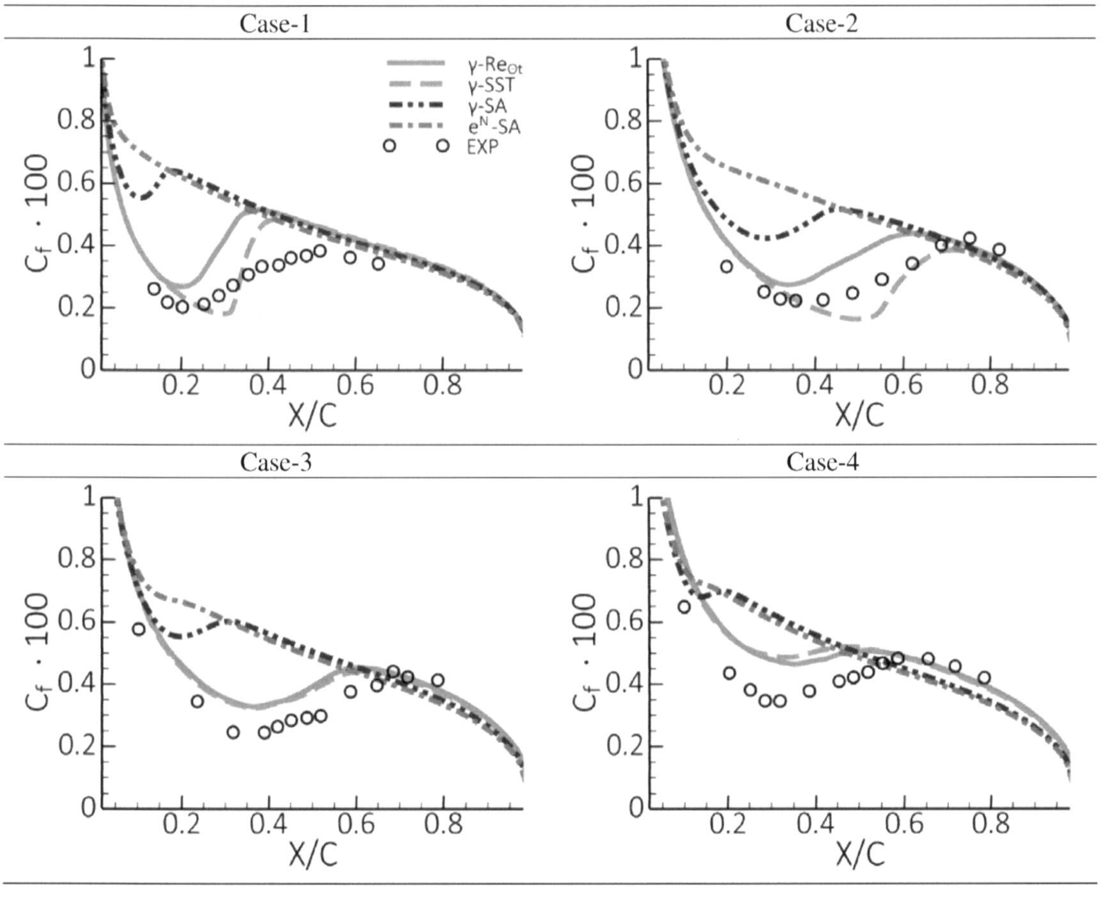

Figure 4. Distribution of the computational and experimental skin friction coefficient on the downstream airfoil for different cases.

Table 3. Transition extents in cases tested

Case	EXP	γ-Re$_{\theta t}$-SST	γ-SST	γ-SA
		Onset of transition, X_S/C		
Case 1	0.20	0.20	0.30	0.11
Case 2	0.35	0.35	0.50	0.30
Case 3	0.32	0.37	0.37	0.20
Case 4	0.28	0.35	0.32	0.14
		Length of transition region, L_T/C		
Case 1	0.32	0.18	0.14	0.07
Case 2	0.40	0.27	0.21	0.15
Case 3	0.33	0.24	0.24	0.10
Case 4	0.30	0.17	0.16	0.05

6. Conclusions

The applicability of four different transition models for prediction of flow around turbine blades was carried out using the example of numerical simulations of flow over NACA-0012 airfoils in tandem in various regimes. The results of the considered transition models are virtually independent of the CFD code and type of the computational grid (hybrid unstructured vs. chimera-type structured). It is shown that the most accurate results correspond transition models based on the SST model (γ-$Re_{\theta t}$-SST or γ-SST). The SA-based transition models tend to predict too early laminar-turbulent transition (γ-SA) or even fully turbulent flow (e^N-SA) on the downstream airfoil. Therefore, the use of the γ-$Re_{\theta t}$-SST or γ-SST transition models can be recommended for prediction flow over blade cascade.

Acknowledgements

The results of the work were obtained using computational resources of Peter the Great Saint-Petersburg Polytechnic University Supercomputing Center (http://www.spbstu.ru).

References

[1] Menter F. R., Zonal Two Equation k-ω Turbulence Models for Aerodynamic Flows, AIAA Paper 93-2906, 1993

[2] Spalart P.R., Allmaras, S.R., A one-equation turbulence model for aerodynamic flows, AIAA Paper 92-043, 1992

[3] Menter F.R., Langtry R.B., Correlation-Based Transition Modeling for Unstructured Parallelized Computational Fluid Dynamics Codes, AIAA J. 47(12), pp2984-2906, 2009

[4] Menter F.R., Smirnov P. E., Liu T., et al., A One-Equation Local Correlation-Based Transition Model, Flow Turbulence Combust, 95, pp583–619, 2015

[5] Coder J.M., Maughmer M. D., One-Equation Transition Closure for Eddy-Viscosity Turbulence Models in CFD. AIAA Paper 2012-0672, 2012

[6] Coder J.M., Maughmer M. D., Computational Fluid Dynamics Compatible Transition Modelling using an Amplification Factor Transport Equation, AIAA J Vol. 52, No. 11, pp2506-2512, 2014

[7] Lee H., Kang S-H., Flow Characteristics of Transitional Boundary Layers on an Airfoil in Wakes, ASME, J. Fluids Engineering 9 122 pp522-532, 2000

IOP Conf. Series: Journal of Physics: Conf. Series **1038** (2018) 012131 doi:10.1088/1742-6596/1038/1/012131

Erosion of rod electrodes of the air AC plasma torch

Subbotin D I[1,2,3], Surov A V[1], Kuznetsov V.E.[1], Popov V.E.[1], Dudnik J D[1], Kuchina J D[1], Obraztsov N V[4]

[1]Institute for Electrophysics and Electric Power of Russian Academy of Sciences (IEE RAS), Dvortsovaya emb. 18, 191186, St. Petersburg, Russia
[2]Saint-Petersburg State Institute of Technology, Moskovsky pr. 26, 190013, St. Petersburg, Russia
[3]Saint-Petersburg State University, Universitetskaya emb. 7/9, 199034, St. Petersburg, Russia
[4]Peter the Great St. Petersburg Polytechnical University Polytechnicheskaya 29, 195251, St.Petersburg, Russia
subbotin1987@mail.ru
Abstract. Depending on the plasma-forming mixture, various types of electrode materials are used. If air, carbon dioxide and steam are used, it is optimal to use copper electrodes, since copper has a high thermal conductivity. Copper rod electrodes of an air AC plasma torch are considered. In the binding zone of the electric arc, the electrode material melts and boils, which leads to physical erosion of the electrode. Scanning electron microscopy with elemental analysis, X-ray phase analysis and IR Fourier spectroscopy were used to analyze the products of electrode erosion. The erosion products content copper, iron, nitrogen, oxygen, and other metals from the plasma torch case.

1. Introduction

Air plasma can be used as an oxidizer in the process of plasma gasification, a heat carrier for heating gases or liquids. The greatest load during operation of the plasma torch is tested by electrodes, therefore the electrode material must have good electrical conductivity and thermal conductivity, as well as a high melting point. Tungsten, molybdenum, zirconium, hafnium, yttrium, aluminum, copper, iron and their alloys are used as electrode materials. The choice of material depends on the nature of the working gas [1]. Tungsten is used only in the case of working on non-oxidizing gases, since in the presence of oxygen at high temperatures it is oxidized to oxides. Electrodes are made of zirconium or hafnium, the most stable materials when plasma torches operate in oxidizing environments. On the surface of these materials an oxide film is formed, on the one hand, it conducts electric current well at high temperatures, and on the other hand, protects the metal from further rapid oxidation. For gases containing oxygen, it is more rational to use copper-based alloys. Copper provides a high level of thermal conductivity, which provides low temperatures on the surface of the cooled electrode, and also has a low cost compared to hafnium and zirconium. To increase the melting and boiling point, copper alloys with components that reduce the intensity of evaporation of electrode material during the creation of plasma are often used. For example, iron, chromium, and nickel can serve as such substances [2].

When using in the experimental plasma chemical plants in the IEE RAS, high-voltage AC plasma torches with electric arcs that burn in individual cylindrical channels have proved to be well established. In the power range up to 50 kW three-phase plasma torch with rod water-cooled electrodes are designed. The duration of continuous operation of these devices reaches 200 hours and is limited by the resource of the electrodes. The flow rate of plasma-forming air is up to 35 g / s, the thermal efficiency is up to 95% [3]. Further increase in power due to an increase in current on rod electrodes leads to their more intensive wear. As a result, a high-voltage AC plasma torch with power up to 500 kW with cylindrical hollow electrodes was created (Fig. 1) [4].

Figure 1. 500kW AC plasma torch

The principle of operation of such a plasma torch is similar to the above described high-voltage plasma torches stabilized by a swirling gas flow of an arc also burn in separate cylindrical channels, closing one another at the exit from the body. The difference lies in the fact that arc bindings are not installed on a small-sized end face of the rod electrodes, but continuously rotate along the inner surface of the hollow electrodes. The thermal efficiency of such a plasma torch is about 92%. Resource tests have shown that such a plasma torch can work continuously for 2,000 hours [5].

2. Experimental section

Sampling is carried out on a water-cooled stainless steel surface (Tab. 1). After cooling the erosion products were mechanically separated from the cooled surface.

Table 1 Sampling conditions

Sample	1	2
Electrode	Copper, hollow	
Plasma torch power, кW	200	450
The mean temperature, 10^{-3} K	2.8	3.2-3.3

The following physicochemical methods were used to study the compositions and structures of the erosion products of plasma torch electrodes: elemental analysis (EDAX); scanning electron microscopy (SEM); IR spectral analysis (FTIR); X-ray phase analysis (XRD); differential thermal analysis (DTA).

2.1 Elemental analysis and scanning electron microscopy

Element analysis (energy dispersive X-ray spectral analysis) was carried out using the EDAX FP 2012/12 microprobe attachment on a Quanta 200 FEG Environmental Scanning Electron Microscope (ESEM) device to determine the percentage of chemical elements included in the samples.

The error in determining the content of elements by this method varies depending on the atomic number of the element and on the average is ± 0.3 mass. %. Table 2 shows the conditions for obtaining microphotographs of the structure of the samples with an increase of 2500.

Table 2 – Conditions for obtaining photomicrographs of samples

Sample	1	2
Voltage, kV	20	
Working distance, mm	9,6	10,3
Scanning speed, sec	91,19	36,48
Temperature	room indoor temperature	
Pressure	vacuum	

IOP Conf. Series: Journal of Physics: Conf. Series **1038** (2018) 012131 doi:10.1088/1742-6596/1038/1/012131

2.2 IR spectral analysis
To determine the groups of atoms and the structure of the samples, IR spectral analysis was performed on a SHIMADZU IRTracer-100 IR-Fourier spectrometer under vacuum at room temperature.
IR transmission spectra were measured in the range from 350 to 4000 cm^{-1} with a resolution of 2 cm^{-1}, the moving speed of the moving mirror of the interferometer was 2.8 mm / s. To conduct IR spectral measurements, the solid samples and powder of dry potassium bromide in the ratio 1: 200 were triturated together.
2.3 X-ray phase analysis
To determine the composition of the crystalline phase, an X-ray diffraction analysis was performed on a powder X-ray diffractometer SHIMADZU XRD-7000 on Cu Kα radiation (λ = 1.5046 Å). The parameters for the analysis are shown in Table 3.

Table 3 - Analysis parameters

Sample	1	2
Voltage, kV	35	
Current, A	30	
Temperature	room indoor temperature	
Measurement step, deg.	0,02	
Range of Bragg angle, 2Θ, deg.	10 – 80	
Holding time at point, s	3	8

The data obtained were processed in the PDWin 4.0 software package with a PDF2 powder diffraction database. The Rietveld method was used to determine the crystal phase ratios in the samples.
2.4 Differential-thermal analysis
To determine endothermal and exothermic transitions as a function of temperature, a differential thermal analysis was carried out using the SHIMADZU DTG-60A. The analysis was carried out under an air atmosphere. As a reference sample, aluminum oxide was used. The samples were heated from room temperature. Table 4 shows the measurement conditions and the end-of-analysis temperature.

Table 4 - Data of differential thermal analysis

Sample	1	2
Weight of sample, mg	16,097	19,944
Temperature of the end of heating, °C	1100	

Heating of samples 1 and 2 at 200-300°C and 300-1100°C was carried out at a rate of 3 deg/min. and 10 deg/min, respectively.

3. Results and discussion
Samples 1 and 2 are light green powders with particles of different sizes. Table 5 shows the results of elemental analysis of the surface of these samples. The presence of iron in the samples indicates that the electric arc interacts not only with the material of the electrode, but also with the case of the plasma torch made of stainless steel. Especially probable is the entrainment of the material of the plasma torch housing at the time of its launch, when the electric arc of one of its bindings moves along the inner surface of the arc channel. The presence of carbon in the sample is explained by the presence of carbon dioxide in the air used to generate the plasma.

From the analysis of the samples, it can be concluded that with an increase in the power of the AC plasma torch, the content of iron and copper in the products of erosion of the copper electrode increases, and the remaining elements, respectively, decreases.

Table 5 - Content of elements in the products of erosion of copper electrodes of AC plasma torches of various powers

Element	Content, mol. %	
	Sample 1 (200 kW)	Sample 2 (450 kW)
Oxygen	50,29	47,04
Copper	24,97	32,80
Nitrogen	13,34	10,47
Carbon	10,69	8,11
Iron	0,71	1,58

X-ray diffraction patterns of the samples are shown in Fig. 2. X-ray diffraction patterns a and b have a similar character, the difference is in the intensities of the peaks. Comparison of the diffraction patterns with a file of powder diffractograms PDF 2 shows the presence of several crystalline phases in the investigated samples: orthorhombic copper (II) orthogranite, monoclinic copper (II) oxide and cubic copper oxide (I), the percentage of which is given in Table 6.

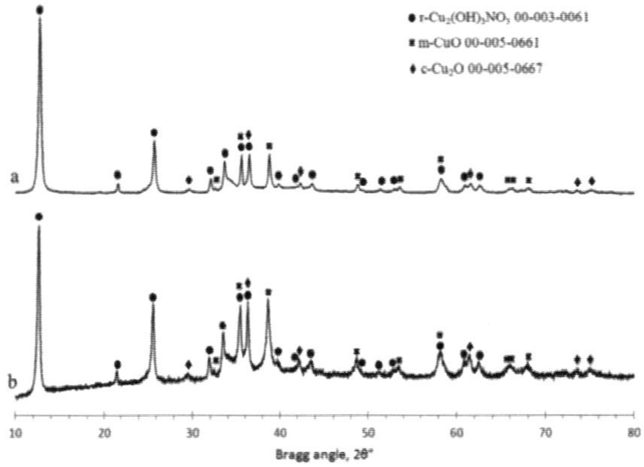

Figure 2. X-ray diffraction patterns of erosion products of copper electrodes of AC plasma torches of different capacities: a - sample 1, b - sample 2

Table 6 - Phase fraction in samples

Phase	Content, mass. %	
	Sample 1 (R_{wp}=5,38 %)	Sample 2 (R_{wp}=6,67 %)
Copper (II) trihydroxonitrate	50,23	76,33
Copper oxide (II)	6,86	18,34
Copper oxide (I)	42,91	5,33

These assumptions are confirmed by the results of IR spectral analysis. Fig. 3 shows IR spectra of samples of erosion products of copper electrodes of AC plasma torch working in air and various capacities. They represent a similar complex picture with whole series of absorption spectra. The infrared spectra of sample 1 (Fig. 3a) obtained by operating an AC plasma torch with a power of 200 kW and sample 2 (figure 3b) of 450 kW are compared. Both spectra are characterized by the presence of bands of OH-groups: 3546 (υs), 3433 (υs) cm^{-1} and NO_3^- groups: 1384 (υas), 1344 (υas), 1047 (υs), 840 (δs), 787 (δas) cm^{-1} [6]. Also on the IR spectra of the samples are peaks with wave numbers 1631 (δs), 1562 (δs) cm^{-1}, related to the deformation vibrations of the H-O-H bond in adsorbed water. Obviously, these groups are part of the compound $Cu_2(OH)_3NO_3$ [7].

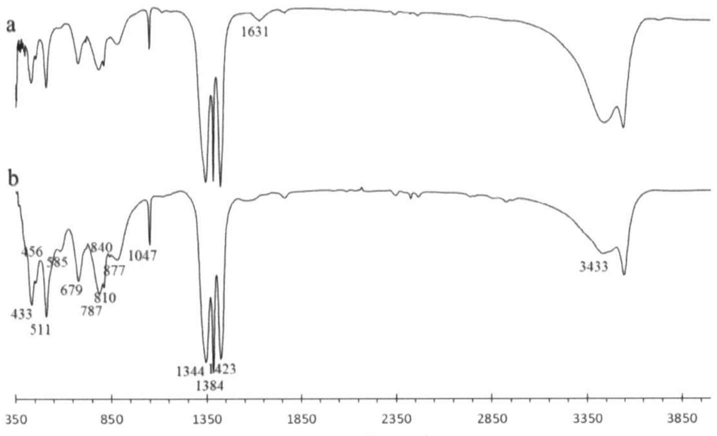

Figure 3. IR absorption spectra of the samples under study: a - sample 1; b - sample 2

Also on these spectra there are peaks with wave numbers 1423 (υas), 1047 (υs), 877 (δs), 679 (δas) cm^{-1}, which can be attributed to the vibrations of atoms in the CO_3^{3-}- [6] group. However, during the X-ray phase analysis of the test samples, the presence of carbonates is not revealed, therefore, it can be assumed that the carbonates are in the amorphous phase.

Peaks with wave numbers 787 (υs), 585 (υs), 511 (υas), 456 (δ), 433 (δ) cm^{-1} can be attributed to vibrations of Cu-O-Cu bond atoms in the copper (I) oxide molecule [8]. Peaks with wave numbers of 585, 456 cm^{-1} - to the stretching vibrations of Cu-O bond atoms in the molecule of copper (II) oxide [8, 9]. A peak with a wave number of 585 cm^{-1} can be related to the stretching vibrations of the Fe-O bond in the Fe_3O_4 molecule [9, 10].

Fig. 4 shows the thermograms of the samples under study. Analyzing the change in mass (Fig. 4a curve 1) and the change in enthalpy (Fig. 4a curve 2), when heating sample 1, we see that at the beginning of heating, moisture begins to evaporate from the sample under investigation, which, possibly, the sample was enriched when exposed to humid air. About 208 ° C there is a peak on curve 2, which corresponds to the endothermic effect, i.e. decomposition occurs, presumably copper (II) hydroxycarbonate (reaction 2) ($\Delta m1 = 0.282$ mg) [11]. A temperature of 220 ° C was followed by the release of structural water from copper (II) trihydroxonitrate and its further decomposition to form copper (II) oxide (reaction 3) [12, 13] ($\Delta m2 = 1.514$ mg). In the temperature range 300 - 1025 ° C, curve 2 has the form of a plateau, which indicates the stability of chemical compounds in this temperature range and the absence of chemical transformations. Near 1000 ° C, curve 1 shows an increase in mass, indicating the conversion of copper (I) oxide to copper (II) oxide. The peak on curve 2, observed at 1050 ° C, presumably corresponds to the decomposition of copper (II) oxide on copper (I) oxide and oxygen (reaction 4) ($\Delta m3 = 0.545$ mg).

$$(CuOH)_2CO_3 \rightarrow 2CuO + CO_2 + H_2O \quad (1)$$
$$4Cu_2(OH)_3NO_3 \rightarrow 8CuO + 6H_2O + 4NO_2 + O_2 \quad (2)$$
$$4CuO \rightarrow 2Cu_2O + O_2 \quad (3)$$

At a temperature of 1089 ° C, an endothermic effect is observed, accompanied by an increase in the mass of sample 1 ($\Delta m4 = 0.09$ mg). This fact could not be explained by chemical transformations.

In the case of sample 2 (Fig. 4b), a similar situation is observed, but the temperatures of the thermal effects are shifted by 4-7 °C, which is permissible. At the same time, $\Delta m1 = 0.45$ mg, $\Delta m2 = 4.669$ mg, $\Delta m3 = 1.327$ mg.

In the literature [14] it is pointed out that in the air, magnetite Fe_3O_4 already oxidizes to γ-Fe_2O_3 at low temperatures and with slight heating, α-Fe_2O_3 is formed with further heating above 300 ° C. Perhaps because of so small a content of magnetite in the sample, the thermogram does not notice the course of these processes.

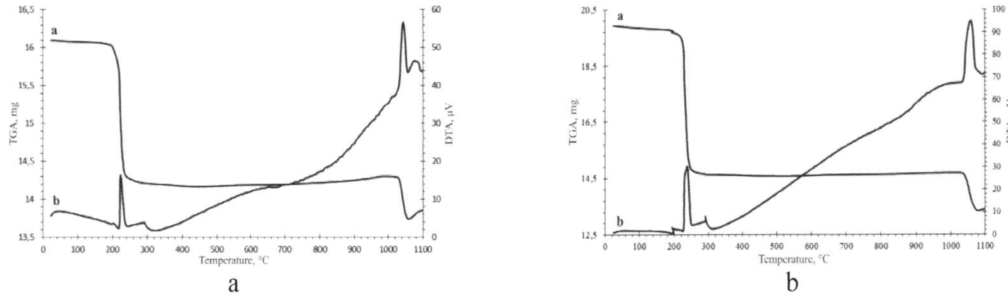

a b

Figure 4. Thermograms of the erosion products of copper electrodes of plasma torches operating in air and at various powers: a - sample 1, b - sample 2: 1 - change in mass upon heating, 2 - change in enthalpy upon heating

Thus, the use of a set of physicochemical methods of investigation makes it possible to determine the main composition of the products of erosion of a copper electrode taking into account the amorphous phase of copper (II) hydroxycarbonate.

The obtained data indicate that during operation of copper electrodes in an AC plasma torch operating in air, the products of interaction between the material of the electrodes and air are formed on them. In the course of the study, physicochemical methods of analysis found that the products of erosion of the copper electrode of an AC plasma torch operating in air consist of copper (II) oxide trihydrocoxonitrate, copper (II) oxide, copper (I) oxide, copper (II) hydroxycarbonate, as well as a small amount of magnetite. The percentage of these substances in the products of erosion is determined by the thickness of the plasma torch. At relatively low powers (120-200 kW), the major portion is copper (II) trihydroxonitrate and copper (I) oxide. Increasing the power of the plasma torch of alternating current, the content of trihydroxonitrate in the products of erosion of the copper electrode substantially increases, the content of copper (II) oxide and magnetite also increases. Based on the composition of the products of erosion, one can put forward a hypothesis about their formation. In the binding zone of the electric arc, copper is heated, accompanied by melting, evaporation, spraying, etc. The ionized air stream interacts with copper particles: copper is oxidized to copper (I) oxide (reaction 4), oxidation of the formed copper (I) oxide to copper (II) oxide (reaction 5) also occurs, copper reacts with oxygen, carbon dioxide and water (reaction 6), nitrogen reacts with oxygen (reaction 7), copper (I) oxide reacts with nitric acid, hydroxyl and steam (reaction 8).

$$4Cu + O_2 \rightarrow 2Cu2O \quad (4)$$

$$2Cu_2O + O_2 \rightarrow 4CuO \quad (5)$$

$$2Cu + O_2 + CO_2 + H_2O \rightarrow (CuOH)_2CO_3 \quad (6)$$

$$N_2 + 2O_2 \rightarrow 2NO_2 \quad (7)$$

$$Cu_2O + 2OH^- + HNO_3 \rightarrow Cu_2(OH)_3NO_3 \quad (8)$$

At the moment of plasma torch launching, the electric arc interacts with the plasmatron body, while the material is carried away, proceeding from the elemental composition of the erosion products-iron. Iron interacts with the plasma flow of air, forming an iron scale (reaction 9).

$$3Fe + 2O_2 \rightarrow Fe_3O_4 \quad (9)$$

4. Conclusion

When copper electrodes operate in an AC plasma torch operating on various mixtures, the interaction products of the electrode material with the plasma-forming medium are formed on them. The erosion of the electrode material can affect the operating mode of the plasma torch and the physico-chemical processes taking place in it, and erosion products can contaminate the products of plasma synthesis, and also participate in the chemical process as a raw material or a catalyst.

In the course of the study, physico-chemical analysis methods found that the products of erosion of the copper electrode of an AC plasma torch operating in air are copper (II) trihydroxonitrite, copper (II) oxide, copper (I) oxide, copper (II) hydroxycarbonate, and also a small amount of magnetite. The percentage of these substances in the products of erosion is determined by the thickness of the plasma torch. Increasing the power of the plasma torch, the content of trihydroxonitrate, copper (II) oxide and magnetite in the products of erosion of the copper electrode substantially increases.

Acknowledgments
The work is supported by RFBR grant 15-08-05909-a.

References

[1] Rutberg Ph. G. Physics and Technology of High Current Discharges in Dense Gas Media and Flows / Ph. G. Rutberg. - New York: Nova Science Publishers, 2009. – 226 pp
[2] Szente R N, Munz R J, Drouet M G 1992 *Plasma Chemistry and Plasma Processing* **12** 327-343
[3] Surov A V, Popov S D, Popov V E, Subbotin D I, Serba E O, Spodobin V A, Nakonechny Gh V, Pavlov A V 2017 *Fuel* **203** 1007-1014
[4] Surov A V, Popov S D, Serba E O, Nakonechny G V, Spodobin V A, Ovchinnikov R V, Kumkova I I, Shabalin S A 2012 *Journal of Physics: Conference Series* **402 1** 1-7
[5] Rutberg F G, Kuznetsov V A, Serba E O., Nakonechnyi G V, Nikonov A V, Popov S D., Surov A V 2013 *High Temperature*. **51 5** 608-614
[6] Nakamoto K Infrared and Raman Spectra of Inorganic and Coordination Compounds, Part A. Theory and Applications in Inorganic Chemistry, 6th ed. (Wiley, New York, 2009)
[7] Zhan Y, Zhou X, Fu B, Chen Y 2011 *Journal of Hazardous Materials* **187** 348-354
[8] Korshunov A V, Il'in A P 2009 *Journal of Applied Chemistry* **82 7** 1164-1171
[9] Ma C, Li C, He N, Wang F, Ma N, Zhang L, Lu Z, Ali Z, Xi Z, Li X, Liang G, Liu H, Deng Y, Xu L, Wang Z 2012 *Journal of Biomedical Nanotechnlogy* **8** 1000-1005
[10] Du G H, Liu Z L, Xia X, Chu Q, Zhang S M 2006 *Journal Of Sol-Gel Science And Technology* **39** 285-291
[11] Brown I W M, Mackenzie K J D , Gainsford G J 1984 *Thermochimica Acta* **75 1–2** 23-32
[12] Ryu S K, Lee W K, Park S J 2004 *Carbon Science* **5 4** 180-185
[13] Biswick T, Jones W, Pacuła A, Serwicka E 2006 *Journal of Solid State Chemistry* **179** 49-55
[14] Cornell R M The Iron Oxides: Structure, Properties, Reactions, Occurrences, and Uses. Weinheim: Wiley, 2003. – 703 p.

Unsteady processes in a natural convective plume

E F Khrapunov, Y S Chumakov

Peter the Great St.Petersburg Polytechnic University
Russia, 195251 St.Petersburg, Polytechnic st., 29

Email: hrapunov.evgenii@yandex.ru

Abstract. The paper presents the results of an experimental and numerical study of the structure of a plume formed over a heated disk under conditions of conjugate heat transfer in a range of small and average Grashof numbers. Numerical simulation was performed using the Ansys Fluent code, where the problem of the flow of a compressible gas in conditions of conjugate heat transfer in an unsteady formulation was solved under the assumption of laminar flow regime. One of the main goal of this paper is to compare results of modelling of the plume in assumption of axisymmetric and three-dimensional flow regime. As a result of a comparison of the results of physical and numerical experiments, there is a conclusion about a good qualitative concurrence of the flow structure in a range of a small Grashof numbers.

1. Introduction

Natural convection is the determining type of motion in many applied problems, for example, meteorological, cooling and ventilating tasks, and the problems of creating a microclimate. Modern approaches to the study of free convection flows are, as a rule, either in physical modeling using modern contactless methods of diagnostics or in numerical modeling within the framework of vortex-prone approaches.

The analysis of known works shows that special difficulties arise in the study of flows formed over horizontal heated surfaces in unbounded space - free convection plumes. One of the main difficulties in physical modeling is the instability of the flow, sensitivity to small perturbations. Even the use of actively developing PIV techniques in such problems is associated with a large number of difficulties, although, in the opinion of the authors, it is the most promising direction [1], [2]. In the numerical modeling, the main difficulty lies in the need to resolve a large number of different-scale turbulent structures and their interaction. At the same time, the use of necessary computational grids and approximation algorithms is mostly impossible due to the lack of adequate computing power.

The need to calculate correctly both the free-running motion of air masses and the heat transfer characteristics of heated horizontal surfaces led to the emergence of a large number of simplified approaches to modeling. Thus, for example, there are known works with the use of the boundary-layer approximation [3] to calculate flow characteristics. However, in the case of an upward flow over a heated surface, this approach is justified over a small portion of this surface. Obviously, the use of this approximation is no longer able to detect and describe the flow at the proper level. However, it should be noted that in a number of cases it is still possible to achieve a significant simplification of the formulation of the problem. So in the previous works of the authors it was shown that the flow in a freely convective plume at small Grashof numbers can be considered close to axisymmetric.

The purpose of this study is to compare the results of the numerical simulation of a free convection plume in a three-dimensional formulation and on the assumption of axisymmetry of flow for small and moderate Grashof numbers (the defining parameter in the problem), as well as the comparison of axisymmetric and three-dimensional formulations with experimental data in the case of moderate Grashof numbers.

2. Experimental stand and computational model

The scheme of the experimental stand is shown in Figure 1. The main element of the stand is a heated brass disk with a diameter of 190 mm and a thickness of 8 mm. The lower surface of the disk contacts the heater, and to improve the thermal contact, a paste with a high coefficient of thermal conductivity is used. Into the layer of this paste a thermocouple is placed to control the temperature of the lower surface of the disk. The temperature mode of the disk is controlled by the controller, which switches on / off the heating when the set temperature is reached. So, on the lower surface of the disk there are realized conditions close to the conditions of constant temperature, the temperature of the upper surface is determined by the conditions of the conjugate heat exchange between brass and air. The thermal localization of the disk is carried out by using the water heat exchanger that maintains the temperature of the horizontal near-disk surface at room temperature. To eliminate the influence of external disturbances around the disk, a grid is installed at a distance of two diameters, and the whole system is placed in a leaky chamber whose walls are made of Plexiglas, which makes it possible to conduct the visualization of the flow. Measurements of the temperature field are carried out using a resistance thermometer. To visualize the flow, the entire measuring volume is seeded with particles that are illuminated with a laser plane.

One of the purposes of this work is to determine the characteristics of heat exchange between a heated disk and air. For this purpose, has been developed a technique based on the assumption of a thin, wall-mounted heat-conducting layer, distribution of the mean temperature in which is linear. On the basis of this assumption, it is possible to calculate the derivative of the temperature along the normal to the disk, and then the heat flux, the heat transfer coefficient, and local Nusselt numbers.

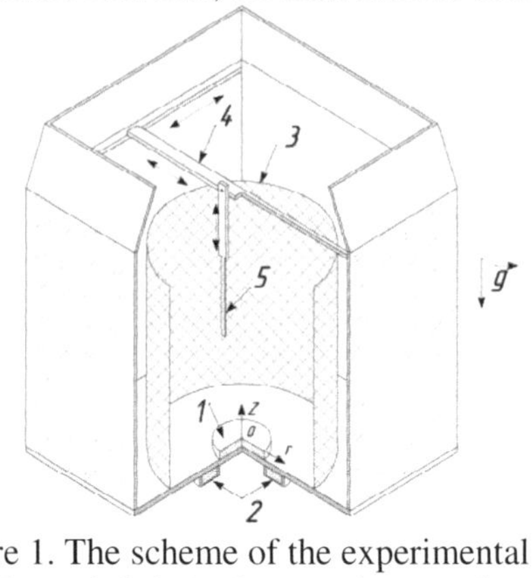

Figure 1. The scheme of the experimental stand
1 - heated disk, 2 - heat exchanger element,
3 - protective grid, 4 - coordinate device,
5 - resistance thermometer

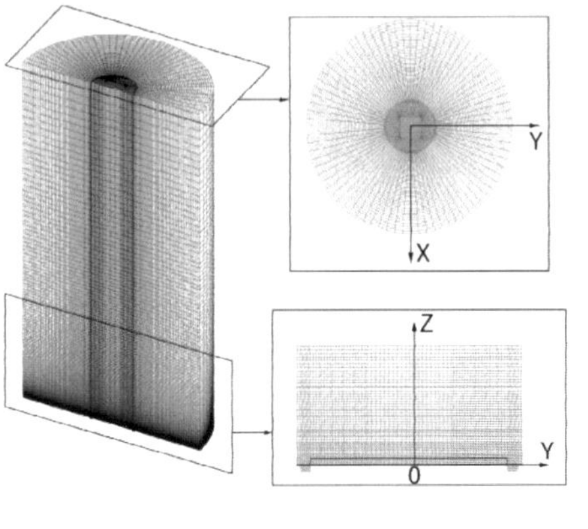

Figure 2. Calculation mesh for numerical simulation

Numerical simulation was performed using the ANSYS Fluent code and consists in solving the Navier-Stokes equations in the Boussinesq approximation and assuming laminar flow in a three-dimensional region whose size correspond to the size of the experimental stand and is close to the size of the computational domains described in the other authors' works. The problem is solved both in steady and unsteady formulations. The air parameters are set as follows: specific heat

$c_p = 1006.43 \left[\dfrac{J}{kgK}\right]$, thermal conductivity $\lambda = 0.0242 \left[\dfrac{W}{mK}\right]$, thermal expansion coefficient $\beta = 0.00366 \left[\dfrac{1}{K}\right]$.

The constructed quasistructured mesh (Figure 2) contains $N_{cell} \approx 10^6$ elements and has a thickening to the expected boundaries of the plume and to wall horizontal surfaces. On the lateral and upper surfaces, the outlet boundary conditions are set, as well as the conditions of the existence of only a normal component of the velocity for the backflow. The lower horizontal surface is a solid wall, maintained at room temperature (20 °C). On the lower surface of the disk, temperature condition of the first type is set, corresponding to the temperature in the experiment. The temperature of the upper surface is determined by the conditions of the conjugate heat transfer.

3. The results analysis

As it was noted, the Grashof number based on the characteristic temperature difference between the temperature of the lower surface of the disk T_w (controlled temperature in the experiment and given as a boundary condition for numerical simulation) and room temperature T_a is the defining parameter in the problem. The radius of the disk R is selected as the linear scale.

$$Gr = \frac{g\beta(T_w - T_a)R^3}{\nu^2} \tag{1}$$

The paper presents the results for two characteristic Grashof numbers: Gr = 4.3×10^6 and Gr = 1.4×10^7. In the first case, there is a stationary solution for the flow, in the second, the flow characteristics are determined mainly by nonstationary processes-the destruction of the wall layer. Temperature profiles will present as dimensionless temperature Θ, determined by the formula

$$\Theta = \frac{T - T_\infty}{T_w - T_\infty} \tag{2}$$

where T_∞ - is the ambient air temperature, T_w - is the wall temperature in the section under investigation. In the case of a stationary flow (for small Grashof numbers) it is convenient to single out a wall region in the flow whose temperature profiles are represented in Figure 3. The lines in the figure represent the profiles obtained as a result of solving the axisymmetric problem, and the symbols represent the solution of the three-dimensional problem. The profiles of the dimensionless temperature coincide throughout the wall region. Figure 4 shows the same profiles in self-similar variables presented in [4]. The exponent in this case is $n = 0.35$, which is in satisfactory agreement with [3] and [4].

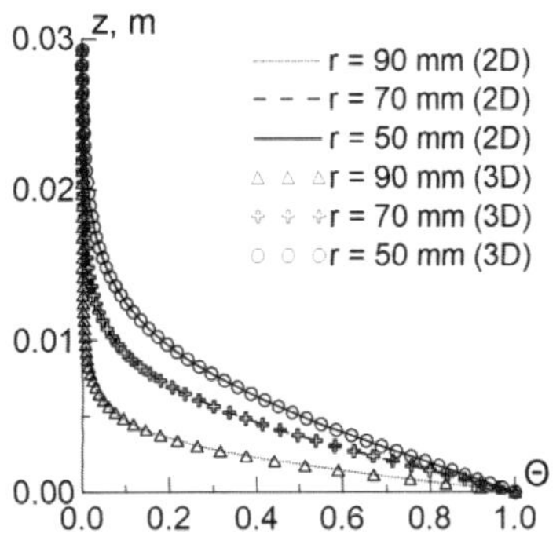

Figure 3. Dimensionless profiles of temperature in the wall region at Gr = 4.3×10^6

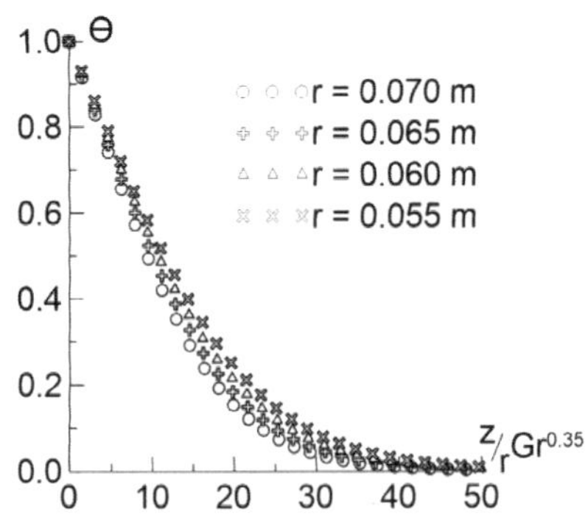

Figure 4. Dimensionless profiles of temperature in the wall region at Gr = 4.3×10^6 in self-similar variables

In one of the main papers devoted to natural convection [5], there was considered free convection plume formed over a point source of buoyancy. As one of the flow invariants, it is suggested to consider the total energy transferred by the flow

$$Q = 2\pi \int \rho c_p U \Delta T r dr \qquad (3)$$

where U is the axial velocity, ΔT is the air overheating at a given point. In our case, the source of the flow is a disk of finite radius. The total amount of heat that participates in the formation of the flow above it can be determined as

$$Q^{Theory} = 2\pi \int q_W r dr \qquad (4)$$

However, the question about the applicability of expression 3 in this case for the analysis of the flow is arise. Figure 5 shows the change in the value of the energy Q of the flow in height in axisymmetric and three-dimensional cases. It is clearly seen that in the case of a heated horizontal surface, the expression for the invariant is valid only at some distance from the surface. Near the source, convective transfer does not play a significant role, which leads to not saving the selected expression. It can be seen that the calculated values correspond well to each other.

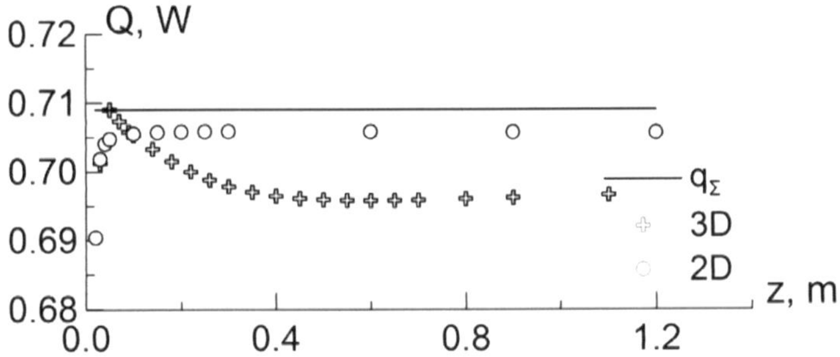

Figure 5. Change in the value of the energy Q of the flow

Now let's consider a change in the flow characteristics in the presence of unsteady processes in a plume at Gr = 1.4×10^7. The left side of Figure 6 shows the mean temperature field for the axisymmetric case (averaging over 20s), and on the right is the instantaneous field of the actual temperature. It is clearly seen that because of the process of destruction of the wall layer, the isotherms of the mean temperature field are deformed. The change of the temperature in time on the axis of symmetry at a distance of 1mm from the surface is periodic (Figure 7). Also we note that the temperature change at the monitoring point in the three-dimensional and two-dimensional case is very different both in amplitude and frequency. The reason for this may be the presence of a symmetry axis as a strictly defined boundary condition

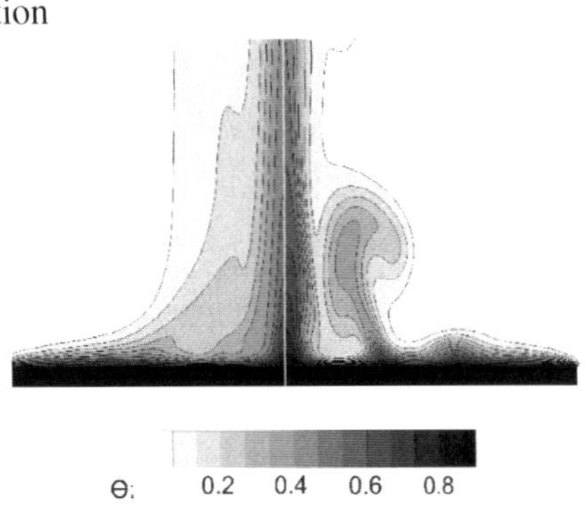

Figure 6. Average temperature field (left) and the actual temperature (right) for axisymmetric formulation for Gr = 1.4×10^7

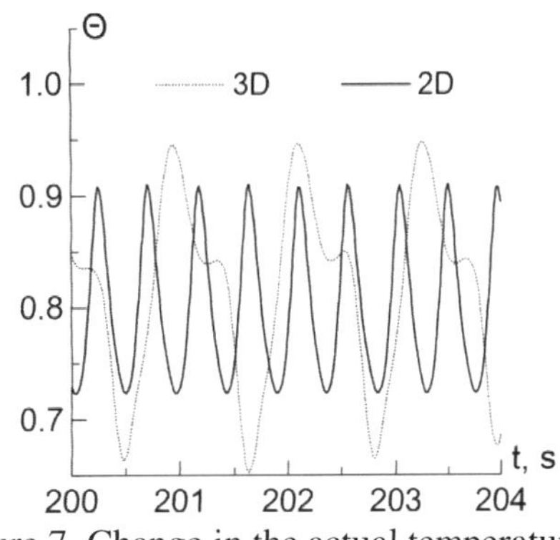

Figure 7. Change in the actual temperature on the symmetry axis at a height of 1 mm from the surface at Gr = 1.4×10^7

Figure 8 shows the thickness of the wall layer at both Grashof numbers. As the layer thickness, a coordinate z was chosen in which the value of the average temperature composes 5% of the surface temperature, i.e. $\delta = z|_{T=0.05T_w}$. To determine the thickness of the layer, temperature profiles averaged over 20 s were used. It can be seen that with a small Grashof number, the thicknesses obtained as a result of both sets coincide. Note the region $r < 0.04m$, near which the wall layer is destroyed and the rising plume is formed. With a larger Grashof number, the wall layer on the periphery is thinner, but its destruction occurs at $r < 0.05m$.

Figure 9 shows the change in the temperature radius (b_T) of the jet in height for both Grashof numbers. The temperature radius was defined as a radial coordinate in which the average temperature is half of the maximum in the selected plane, i.e. $b_T = r|_{T=0.5T_m}$. It is seen that for both regimes a narrowing of the flow near the surface of the disk and a further linear expansion are characteristic. However, in the case of a larger Grashof number, the radius of the plume increases more intensively, taking in this the smaller values at the surface of the disk.

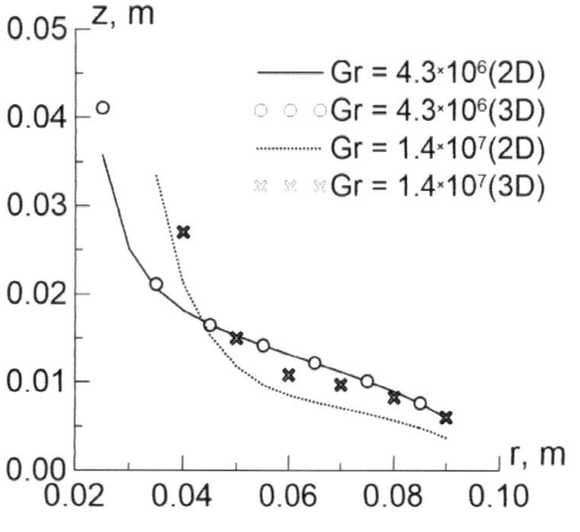

Figure 8. Thickness of wall layer (δ)

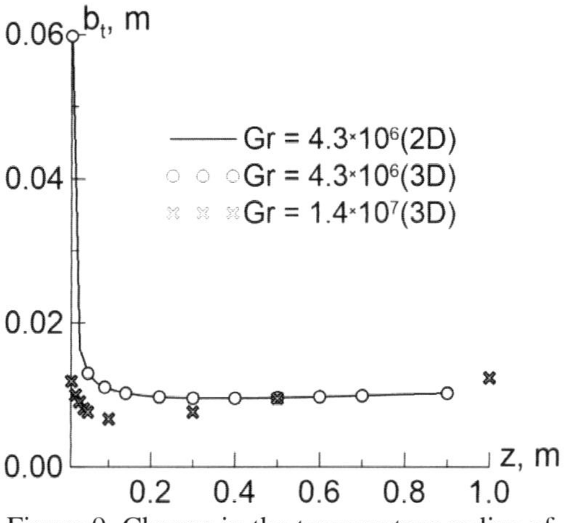

Figure 9. Change in the temperature radius of the plume (b_T) for both Grashof numbers

As a heat transfer characteristic, let us consider the change in the heat transfer coefficient in both sets. Figures 10 and 11 show the variation of α in two-dimensional, three-dimensional formulations and a certain experimental one for both Grashof numbers. It can be seen that the change in the coefficient is monotonic, and all the data are in good agreement with each other. In the case of nonstationary effects, the change in the averaged coefficient is nonmonotonic, local maxima are observed in the range $r \in [0.01;0.03]m$. We also pay attention to the large mismatch of coefficients on the flow axis, which is also explained by the presence of the symmetry axis as a boundary condition in one of the productions. In addition, the maximum heat transfer coefficient in the experiment is shifted and can be observed at coordinates $r = 0.05m$.

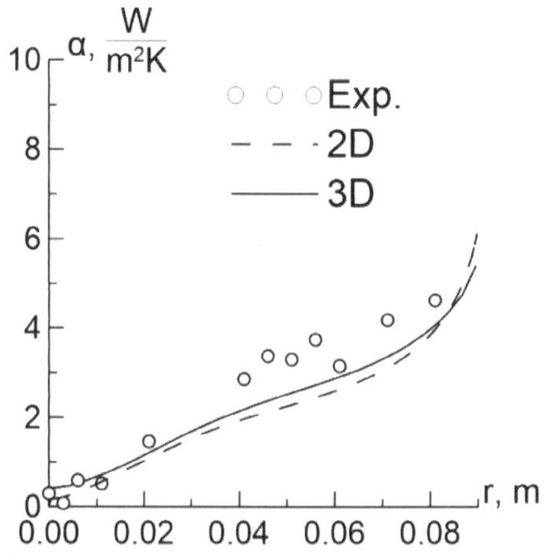

Figure 10. The coefficient of heat transfer, determined numerically and by the experimental method at Gr = 4.3×10^6

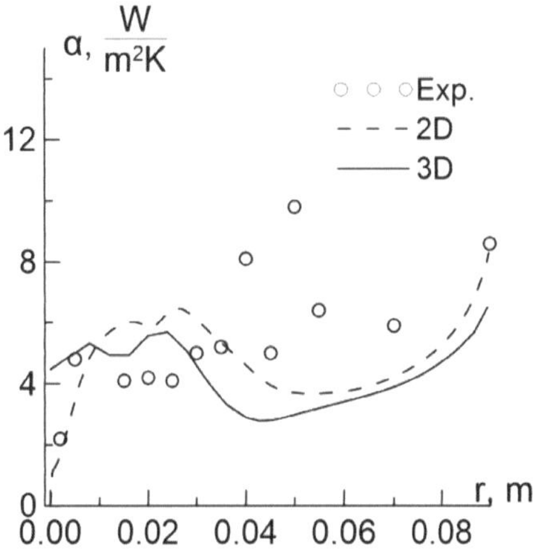

Figure 11. The coefficient of heat transfer, determined numerically and by the experimental method at Gr = 1.4×10^7

4. Conclusion

As a result of experimental and numerical simulation of a natural convective plume, two flow regimes corresponding to the small and moderate Grashof numbers were investigated. It is found that the presence of unsteady effects of fracture of the near-wall flow strongly affects the flow characteristics. It is shown that for small Grashof numbers it is possible to correctly perform numerical simulation within the framework of an axisymmetric approximation. In the case of moderate Grashof numbers, the presence of a symmetry axis as a boundary condition does not allow obtaining reliable results, although the results remain qualitatively similar to a more general formulation.

References

[1] Pham M V, Plourde F and Doan K S 2005 Three dimensional characterization of a pure thermal plume *J. Heat Transfer* **127** pp 624-636

[2] Plourde F, Pham M V, Doan K S and Balachandar S 2008 Direct numerical simulations of a rapidly expanding thermal plume: structure and entrainment interaction *J. of Fluid Mechanics* **604** pp 99-103

[3] Gebhart B, Pera L 1973 Natural Convection Boundary Layer Flow over Horizontal and Slightly Inclined Surfaces *Int. J. of Heat and Mass Transfer* **16** pp 1131-1146

[4] Yousef W W, Tarasuk J D and McKeen W J 1982 Free convection heat transfer from upward-facing isothermal horizontal surfaces *J. Heat Transfer* **104** pp 493-500

[5] Morton B R, Taylor G and Turner J S 1956 Turbulent gravitational convection from maintained and instantaneous sources *Proc. R. Soc. Lond.* **234** pp 1-23

IOP Conf. Series: Journal of Physics: Conf. Series **1038** (2018) 012133 doi:10.1088/1742-6596/1038/1/012133

Some aspects of numerical modeling of inviscid supersonic flow in a duct with a central wedge

E V Kolesnik and E M Smirnov

Department of Fluid Dynamics, Combustion and Heat Transfer,
Peter the Great St.Petersburg Polytechnic University,
Polytechnicheskaya str. 29, 195251, St.Petersburg, Russia

E-mail: kolesnik_ev@mail.ru

Abstract. On the example of the supersonic inviscid flow in a duct with a central wedge, the problem of increased errors of the solution arising in the vicinity of the fractures of a streamlined surface is investigated. The effectiveness of various ways to improve the solution quality is evaluated. Calculations based on the MUSCL approach for increasing the approximation order are performed using the finite-volume unstructured SINF/Flag-S code developed at the Peter the Great St. Petersburg Polytechnic University. The data obtained with this code are compared with the results of calculations performed using the ANSYS Fluent.

1. Introduction

It is well-known that transonic and supersonic flows may contain gasdynamic discontinuities, therefore special approaches in their numerical simulation are required. The main issue is how to approximate convective fluxes. Numerical scheme for convective flux evaluation should provide sufficiently accurate resolution of discontinuities in the absence of the spurious oscillations caused by solution discontinuities. Godunov-type schemes based on solving the local Riemann problem for a computational cell interface satisfy these requirements.

Usually, Godunov-type schemes are described in the literature with respect to calculation of fluxes at inner faces of computational cells. Because the evaluation of numerical fluxes at wall boundary faces requires pressure value at each boundary face centre, the method of wall pressure reconstruction has a significant effect on the solution, especially in the case when the streamlined surface contains a fracture.

The present paper covers description and application of different methods of wall pressure reconstruction in the framework of the first-order and second-order finite-volume method as applied to inviscid flow computations. Comparative analysis of quality of the solutions obtained with different approaches of wall pressure reconstruction are performed considering a representative test case that is supersonic flow in a duct with a central wedge. Since well-known analytical solution exists for this problem, basic characteristic of the flow are well defined, in particular, the position of the oblique shock wave and pressure distribution along the wedge. This test case is widely used in literature to assess the efficiency of various numerical schemes [1-3]. However, there is a rather limited amount of published data on the effect of numerical treatment of wall boundary condition on the solution quality.

2. Computational Method

2.1. Finite volume method and flux computations

Balance equations for the dynamics of an inviscid compressible gas in a finite-volume formulation can be written as:

$$\int_{\Omega} \frac{\partial \vec{w}}{\partial \tau} d\Omega + \sum_{M} \int_{S_m} \vec{F} dS = 0, \tag{1}$$

where Ω is the control volume, N – number of finite-volume faces, S_m – area of the current face, m=1,N, $\vec{w} = \left[\rho,\rho u,\rho v,\rho w,\rho H \right]$ – vector of conservative variables, $\vec{F} = \left[\rho V_n, \rho u V_n + p n_x, \rho v V_n + p n_y, \rho w V_n + p n_z, \rho H V_n \right]$ – flux vector, n_x, n_y, n_z – components of the normal to the face.

Convective flux at a finite volume face can be written as a sum of the central-scheme flux and an artificial dissipative flux: $\vec{F}_f = \langle \vec{F} \rangle - \vec{D}$. In the case of the first-order scheme, the central-scheme flux is calculated as an average of values for the neighbouring ("left" and "right") computational cells (control volumes), and the dissipative flux is evaluated by using the Roe's approximate Riemann solver [4]:

$$\langle \vec{F} \rangle = \frac{1}{2}\left(\vec{F}(\vec{w}_L) + \vec{F}(\vec{w}_R) \right), \tag{2}$$

$$\vec{D} = \frac{1}{2}\left| \tilde{A}(\vec{w}_R, \vec{w}_L) \right| (\vec{w}_R - \vec{w}_L), \tag{3}$$

where Jacobian $\tilde{A} = (\partial \vec{F} / \partial \vec{w})$ on the face is calculated with the Roe's average values of variables [4]. The absolute value of the Jacobian is computed as the product $|\tilde{A}| = R|\tilde{\Lambda}|R^{-1}$, where $|\tilde{\Lambda}|$ is the diagonal matrix composed of the absolute values of the matrix $\tilde{A}$ eigenvalues; R is the matrix composed of right eigenvectors as columns.

The first-order accuracy is constructed under assumption of the piecewise-constant approximation in each control volume. To achieve the second-order spatial discretization, the MUSCL approach [5] is used. The fluxes are computed using reconstructed values of the variables at the cell faces. To prevent non-physical numerical oscillations arising in the vicinity of shock waves special limiter functions should be introduced while reconstructing of linear distributions. These limiters modify the slopes of variables distributions in the cells.

With this approach second-order Roe's scheme can be written as:

$$\langle \vec{F} \rangle = \frac{1}{2}\left(\vec{F}(\tilde{\vec{w}}_L) + \vec{F}(\tilde{\vec{w}}_R) \right), \tag{4}$$

$$\vec{D} = \frac{1}{2}\left| \tilde{A}(\tilde{\vec{w}}_R, \tilde{\vec{w}}_L) \right| (\tilde{\vec{w}}_R - \tilde{\vec{w}}_L), \tag{5}$$

where $\tilde{\vec{w}}_{L/R}$ denotes results of piecewise-linear reconstruction of the solution in the left/right cells.

Piecewise-linear reconstruction of the solution in each cell is performed using gradients of variables at the cell centre. Variable on the face is evaluated as

$$\tilde{u}_{L/R} = u_{L/R} + \varphi_{L/R}(\nabla u)_{L/R} \cdot \vec{r}_{L/R}, \tag{6}$$

where u stands for any of the variables, $\varphi_{L/R}$ – value of the scalar limiter in the left/right cell, $\vec{r}_{L/R}$ – vector from the left/right cell center to the face center. For the calculations presented below, we used the scalar limiter proposed by Venkatakrishnan [6] and modified in [7].

2.2. Boundary conditions for pressure on the wall

In order to evaluate numerical fluxes at boundary faces belonging to the wall, it is necessary to reconstruct pressure at each boundary face centre. In this section, possible variants of wall pressure reconstruction are given.

Figure 1 illustrates a variable arrangement for the near wall computations. Here, p stands for value of pressure in the center of the first inner cell adjacent to the boundary, P is a restored value of pressure at boundary face center, V_n is the normal component of velocity.

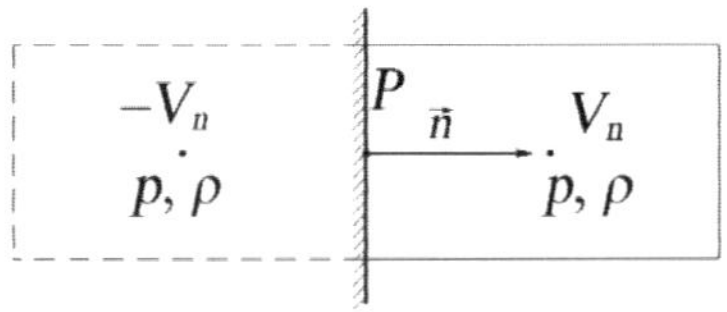

Figure 1. Boundary cell.

The simplest approach to get evaluation of wall pressure is to use the first-order extrapolation from the near wall cell center:

$$P = p \tag{7}$$

However, this approach does not take into account the effect of normal-to-wall momentum that can be of great importance [8]. Indeed, if the value of V_n is negative (opposite to the wall normal vector), there is a compression wave towards the surface. Therefore, the pressure at the face should be greater than in the cell center. If V_n is positive then there is a rarefaction region near the wall and pressure value should decrease in comparison with the value at the cell center.

The approach based on the Riemann problem of discontinuity breakdown allows to get a more physical value of reconstructed wall pressure. With this approach, the values of the variables at the dummy outer cell (marked by dash line in figure 1) are defined first. In the case of zero boundary value of normal velocity component, the dummy cell velocity value of V_n has the opposite sign to the value of the inner near-wall cell, whereas the dummy cell pressure and density values can be taken same as in the inner cell. The boundary value of pressure is determined from solution of the discontinuity breakdown problem as the value on the contact surface, which is assumed to be located on the boundary face. That gives the following expressions [8]:

$$P = p\left(1 - \frac{k-1}{2}\frac{V_n}{c}\right)^{\frac{2k}{k-1}}, \qquad \text{if } V_n > 0 \tag{8}$$

$$\frac{P}{\sqrt{\rho\left(\frac{k+1}{2}P + \frac{k-1}{2}p\right)}} = -V_n, \qquad \text{if } V_n < 0 \tag{9}$$

Both the above given approaches are of the first-order accuracy. A second-order boundary condition can be obtained by replacing value of V_n in (9), (10) by a reconstructed value, denoted as $\tilde{V}_n$. To calculate this value, reconstruction procedure (7) is used, either without introducing the limiter (the third variant of wall pressure reconstruction) or with the limiter (the forth variant).

The fifth and the sixth variant consists in using a weighted scheme for evaluation of V_n:

$$\tilde{V}_n^{weighted} = d \cdot V_n + (1-d) \cdot \tilde{V}_n \ , d \in [0,1], \tag{10}$$

where d – weight.

IOP Conf. Series: Journal of Physics: Conf. Series **1038** (2018) 012133 doi:10.1088/1742-6596/1038/1/012133

Denotation of the above described variants of wall pressure reconstruction are given in Table 1.

Table. 1 Variants of wall pressure reconstruction

№	*denotation*	calculation of *P*
1	*P1*	formula (7)
2	*R1*	formulas (8), (9)
3	*R2*	formulas (8), (9) with reconstructed value of $\tilde{V}_n$ (φ=1.0)
4	*R3*	formulas (8), (9) with reconstructed and limited value of $\tilde{V}_n$
5,6	*R2(d), R3(d)*	formulas (8), (9) with weighted value $\tilde{V}_n^{weighted}$

2.3. CFD codes used

All the variants have been implemented in the in-house unstructured-grid CFD-code SINF/Flag-S that is under permanent development at the St.-Petersburg Polytechnic University. Using this code, calculations of inviscid supersonic flow in a duct with a central wedge were carried on the base of an implicit scheme of the stepping method (in increments) with the stabilizing operator discretized by the first-order scheme with scalar dissipation.

For the sake of comparisons, numerical solutions of the problem were obtained also by the ANSYS Fluent 16.2 choosing the second-order Roe scheme and the modified Venkatakrishnan limiter, same as implemented in SINF/Flag-S.

3. Results and discussion

Inviscid supersonic flow in a duct with a central wedge was calculated for different wedge angle tangent: tg(α)=0.3(3); 0.5; 0.6 (figure 2(a)). It is assumed that supersonic flow enter the duct with the inlet Mach number Min = 3.0. The outflow is also supersonic. For all cases, the quasi-structured computational mesh used consists of 4800 quadrangle elements (figure 2(b)).

Figure 3 illustrates Mach number maps calculated for tg(α)=0.3(3) and 0.6. As seen, the flow consists of interacting shocks and expansion waves. In the case of tg(α)=0.6 the shock front begins to bend near the upper wall due to interaction with rarefaction waves(figure 3(b)).

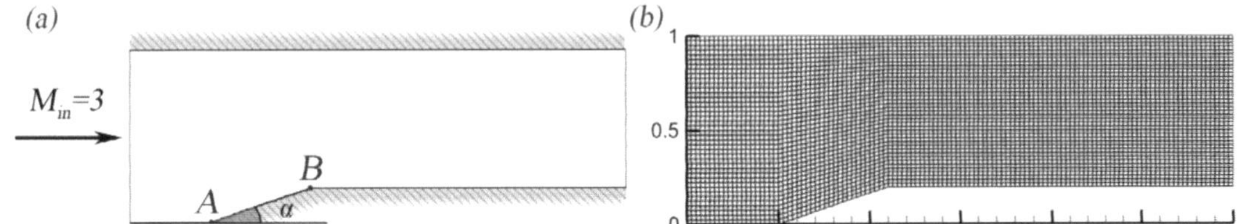

Figure 2. Geometry of the computational domain (a) and mesh (b).

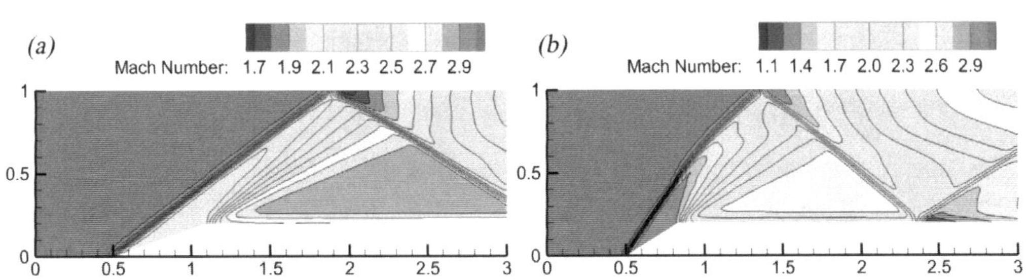

Figure 3. Fields of Mach number for the case tg(α)=0.3(3) (a) and tg(α)=0.6 (b).

IOP Conf. Series: Journal of Physics: Conf. Series **1038** (2018) 012133 doi:10.1088/1742-6596/1038/1/012133

In order to investigate effect of approaches for wall pressure reconstruction on regions near fractures of boundary surface (points A and B in figure 2), pressure distributions on the wedge of surface are analyzed. Figure 4 presents the pressure distributions computed for the case of tg(α)=0.3(3) with different variants of wall pressure calculations.

As seen, even in the case of the first-order scheme, figure 4 (a), the simplest approach (variant *P1*) results in nonmonotonic solution in the vicinity of the fractures. In the case of the *R1* approach, there are no nonmonotonicities near the surface fractures, and the SINF/Flag-S solution is closer to the analytical results than the solution obtained with the ANSYS Fluent.

Results obtained with different wall pressure reconstruction approaches based on the second-order scheme for flux computations are compared in figure 4 (b). Same as in the case of the first-order flux scheme, the *P1* approach leads to appearance of regions with strong nonmonotonicity in the pressure distribution, while the *R1* technique provides monotone solution in the vicinity of both point A and point B.

Figure 4 (b) shows also that the attempt to increase the order of accuracy in the wall pressure reconstruction approach based on the Riemann problem of discontinuity breakdown (variant *R2*) leads to a nonmonotonicity in the pressure distribution, although it is much less expressed than in the case of the simplest approach (variant *P1*). Appearance of these non-physical oscillations is prevented via introducing the limiter together with employment of the weighted scheme setting a small weight of the first order scheme (variant *R3(d)*). For instance, completely monotone solution in the vicinity of both point A and point B is obtained in the case of *R3(d=0.3)*. Using the "pure" weighted scheme leads to worse result (see variant *R2(d=0.5)*). Figure 5 illustrates the scalar limiter field from computations with the *R2* approach. As seen, the limiter is «turned on» not only along the shock, but also in the region between the shock and the wedge, and it is «turned off» near other walls.

Figure 6 presents an additional comparison of the ANSYS Fluent data with results obtained for tg(α)=0.3(3) and 0.6 with the best choice from the wall pressure reconstruction variants implemented in the in-house code. Choosing variant *R3(d=0.3)* for tg(α)=0.3(3) and *R3* for tg(α)=0.6, it is possible to get the best quality of the solutions.

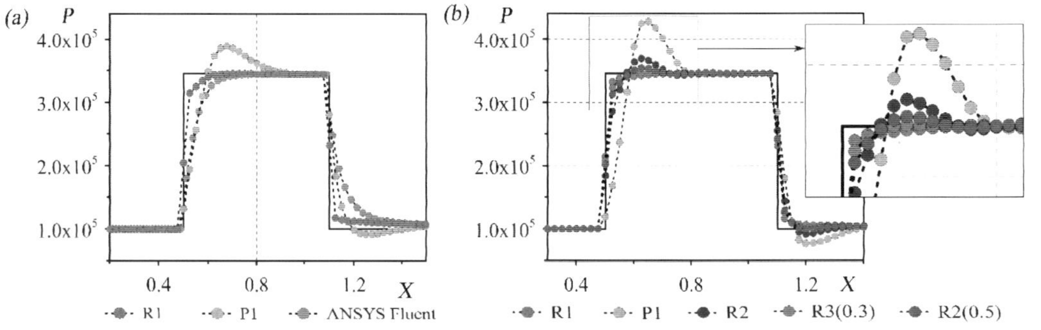

Figure 4. Pressure distribution on lower wall for first-order scheme for fluxes (a), second-order scheme for fluxes (b).

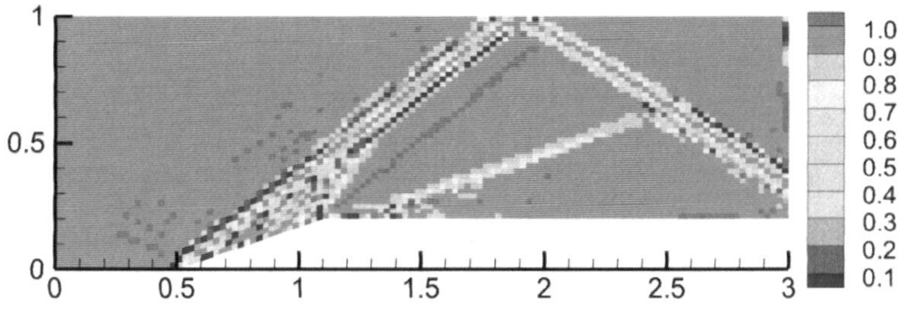

Figure 5. Field of scalar limiter.

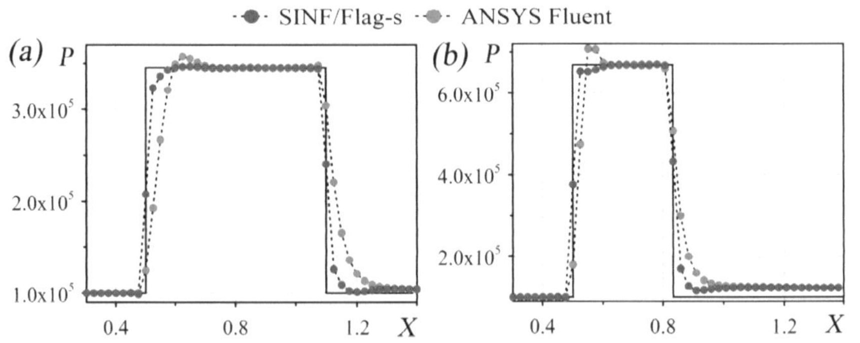

Figure 6. Pressure distribution on lower wall for the case tg(α)=0.3(3) (a) and tg(α)=0.6 (b).

4. Conclusions

Numerical solutions of inviscid supersonic flows can contain areas with non-monotone solution in the vicinity of the boundary surface fractures. The quality of the solution strongly depends on wall pressure reconstruction.

Several variants of wall pressure reconstruction, implemented in the in-house finite volume code, have been described. Comparative analysis of quality of the solutions obtained with different approaches of wall pressure reconstruction has been performed considering a representative test case being supersonic flow in a duct with a central wedge.

First-order wall pressure reconstruction based on the solution of the Riemann problem of discontinuity breakdown results in monotone solution in the vicinity of the surface fractures, both for the first-order and the second-order scheme for convective fluxes at inner faces of computational cells.

An attempt to increase the order of accuracy in the Riemann-problem-solution wall pressure reconstruction, without introducing any limitation procedure, has resulted in appearance of pressure distribution nonmonotonicity in the vicinity of the surface fractures. This undesirable effect can be fully suppressed by introducing a limiter in combination with the weighted scheme. Finally, all this allows getting high quality results for the whole flow region.

References

[1] Bruner C. and Walters R. 1997 Parallelization of the Euler equations on unstructured grids, *AIAA Paper No. 97-1894*

[2] Michalak K. and Ollivier-Gooch C. 2008 Limiters for unstructured higher-order accurate solutions of the Euler equations *in: Proceedings of the AIAA 46th Aerospace Sci. Meeting*

[3] Park J.S., Yoon S.H. and Kim C. 2010 Multi-dimensional limiting process for hyperbolic conservation laws on unstructured grids *J. Comput. Phys* **229** 3 pp 788–812

[4] Roe P.L. 1981 Approximate Riemann Solvers, parameter vectors, and difference schemes *J. Comput. Phys.***43** 2 pp 357–372

[5] Van Leer B. 1979 Towards the ultimate conservative difference scheme. V. A second order sequel to Godunov's method *J. Comput. Phys.* **32** 1 pp 101–136

[6] Venkatakrishnan V. 1995 Convergence to steady state solutions of the Euler equations on unstructured grids with limiters *J. Comput. Phys.* **118** pp 120-130

[7] Wang Z. J. 2000 A Fast Nested Multi-grid Viscous Flow Solver for Adaptive Cartesian/Quad Grids *International Journal for Numerical Methods in Fluids* **33** pp 657–680

[8] Godunov S. K., Zabrodin A. V.; Ivanov M. Ia., Kraiko A. N. and Prokopov G. P. 1976 *Numerical solution of multidimensional problems of gas dynamics*, (Moscow: Nauka) p 400

IOP Conf. Series: Journal of Physics: Conf. Series **1038** (2018) 012134 doi:10.1088/1742-6596/1038/1/012134

Assessment of two approaches to accelerate RANS to LES transition in shear layers in the framework of ANSYS-FLUENT

E K Guseva, M S Gritskevich, A V Garbaruk

Peter the Great St. Petersburg Polytechnic University, 29 Polytechnicheskaya str., 195251, St. Petersburg, Russia

E-mail: katia.guseva@inbox.ru

Abstract. An assessment of two approaches aimed at acceleration of the RANS-LES transition in separated flows has been performed in the framework of Delayed Detached Eddy Simulation (DDES). The former approach is based on the use of the shear-layer adapted subgrid scale (DDES Δ_{SLA}) and the latter involves an alternative subgrid model (σ-DDES). Simulations of two separated flows, namely the backward-facing step flow and flow over a wall-mounted hump have been performed with the use of ANSYS-FLUENT. The results were compared with those obtained with the use of the in-house NTS code and with experimental data. It is shown that both RANS-LES acceleration techniques implemented in ANSYS-FLUENT allow more accurate and less code-sensitive results to be obtained in comparison to the original DDES formulation.

1. Introduction

Nowadays, hybrid RANS-LES approaches for turbulence modelling are considered as a compromise between the high accuracy of Large Eddy Simulation (LES) in separated flow regions and the low computational costs of methods based on the Reynolds Averaged Navier-Stokes equations (RANS). One of the most widely used approaches of this type is Detached Eddy Simulation (DES) [1] and its enhanced modifications. These methods are acknowledged as powerful tools in computational fluid dynamics and can be used in a wide range of applications.

Several modifications of DES aimed at acceleration of three-dimensional turbulent structure formation due to development of the Kelvin-Helmholtz instability in separated shear layers have been recently proposed [2],[3]. It has been shown [2],[3],[4] that such techniques lead to significant improvement of the results for separated flows. These modifications, however, have been implemented and tested only with the use of in-house codes, including NTS code [5], in which high-order low dissipation schemes are used. The results of eddy-resolving simulations and the development of the Kelvin-Helmholtz instability in shear layers depends on the discretization scheme [6], and thus the high dissipation of numerical schemes typically employed in commercial general purpose CFD codes can influence the efficiency of these enhanced approaches.

Therefore, the principal goal of the current paper is to assess the accuracy of the above mentioned techniques for the RANS-LES transition acceleration in the general purpose CFD code ANSYS-FLUENT [7]. For that purpose, these two approaches, namely an approach based on the substitution of the standard sub-grid length scale with that adapted to shear-layers (Δ_{SLA}) [2] and an approach using an alternative sub-grid model [3] (σ-model) coupled with Delayed DES (DDES) [8] based on the Shear

Stress Transport (SST) model [9] are implemented in ANSYS-FLUENT. The obtained results are compared with experimental data and with those of the NTS [5] code.

2. Considered approaches to accelerate RANS-LES transition in shear layers

2.1. SST DDES combined with shear layer adapted sub-grid scale (DDES Δ_{SLA})

A detailed outline of the physical background of this model can be found in [2]. The key ingredient of the model is the sub-grid length-scale, which reads as:

$$\Delta_{SLA} = \tilde{\Delta}_{\omega} \cdot F_{KH}(<\mathrm{VTM}>)$$

Here $\tilde{\Delta}_{\omega}$ is the sub-grid length scale accounting for high anisotropy of the grids in initial regions of shear layers by excluding the size of the grid edge aligned with the vorticity vector [3] and F_{KH} is anempirical function aimed at additional reduction of the subgrid length-scale defined as follows:

$$F_{KH} = \begin{cases} 1.0 & f_d < 0.99 \\ \max(0.1,\ \min(1.0,\ 0.1 + 6 \cdot (<\mathrm{VTM}> - 0.15))) & f_d \geq 0.99 \end{cases}$$

$$\tilde{\Delta}_{\omega} = \frac{1}{\sqrt{3}} \max_{n,m=1,8} |(\mathbf{I}_n - \mathbf{I}_m)|$$

Here, $\mathbf{I}_n = \mathbf{n}_{\omega} \times \mathbf{r}_n$, $\mathbf{n}_{\omega}$ is the unit vector aligned with the vorticity vector, $\mathbf{r}_n$ is the radius-vector of the *n*-th vertex of the considered grid cell, f_d is the shielding function of DDES [8]. The argument of the F_{KH} function is the Vortex Tilting Measure (VTM), which is a kinematic parameter used to identify quasi-2D regions of the flow:

$$\mathrm{VTM} \equiv \frac{\sqrt{6}\,|(\hat{\mathbf{S}} \cdot \boldsymbol{\omega}) \times \boldsymbol{\omega}|}{\omega^2 \sqrt{3tr(\hat{\mathbf{S}}^2) - [tr(\hat{\mathbf{S}})]^2}}$$

$\hat{\mathbf{S}}$ is the strain rate tensor, *tr*( ) denotes the trace operation, and $\boldsymbol{\omega}$ is the vorticity vector averaged over the current and closest neighbouring cells. It should be mentioned, that averaged over the current and neighbouring cells VTM quantity, denoted by <VTM>, is close to zero in quasi two-dimensional regions of the flow, whereas in regions with fully developed turbulence it is about 0.3 and higher.

The shear layer adapted length-scale Δ_{SLA} simply replaces the original length scale Δ_{max} in the definition of the hybrid length scale of DDES. This substitution ensures reduction of the length scale in the initial part of separated shear layers leading to a strong drop of the eddy viscosity, which then unlocks the Kelvin-Helmholtz instability.

2.2. SST σ-DDES approach

Modification of the sub-grid model aimed at acceleration of the RANS-LES transition in separated shear layers for the DDES approach based on the *k*-ω SST turbulence model has been proposed in [10] based on the ideas of Mockett et. al. [3]. The subgrid model of SST σ-DDES performs as the algebraic σ-model [11] rather than as the Smagorinsky model.

The main difference between the SST σ-DDES and the original model is the substitution of the strain tensor invariant $S^2 = 2S_{ij}S_{ij}$ in the production terms of the turbulence kinetic energy and specific dissipation rate equations by the differential operator $S^2_{\sigma-DDES}$, which reads as follows:

$$S^2_{\sigma-DDES} = S^2 - pos(f_d - 0.99) \cdot pos(L_{RANS} - L_{LES}) \cdot (S^2 - B_{\sigma}^2 \cdot S_{\sigma}^2)$$

Here, L_{RANS} and L_{LES} are respectively the RANS and LES length scales of DDES, f_d is the shielding function of DDES [8], C_{DES} is the DES constant [12] , $B_{\sigma} = 57$ is the empirical constant calibrated based on simulations of the decay of isotropic homogeneous turbulence and operators *pos*(a) and S_{σ} are defined as follows:

$$pos(a) = \begin{cases} 0 & a \le 0 \\ 1 & a > 0 \end{cases}$$

$$S_\sigma = \sigma_3(\sigma_1 - \sigma_2)(\sigma_2 - \sigma_3)/\sigma_1^2$$

In the equation above $\sigma_1 \ge \sigma_2 \ge \sigma_3 \ge 0$ are singular values of the velocity gradient tensor. In quasi two-dimensional flow regions, S_σ is close to zero, which ensures the decrease of the eddy viscosity in shear layers by rapidly reducing the production terms of DDES.

Finally, similarly to the DDES Δ_{SLA} approach described in the previous section, the sub-grid length scale of DDES approach is replaced by one accounting for high anisotropy of the grids ($\tilde{\Delta}_\omega$).

3. Flows under consideration, computational setups, and numerics

3.1. Flow over a wall-mounted hump

This flow (the so-called 2D NASA wall-mounted hump) is experimentally studied in [13]. The Reynolds number based on the hump length c and reference velocity U_0 (maximum free stream velocity at the inlet of the domain) is $Re_c = 9.36 \cdot 10^5$, whereas the height of the hump is $h = 0.128c$.

The computational domain and grid are shown in Figure 1. The grid has 510 and 126 cells in x- and y-directions respectively. The size of the domain in the spanwise direction is $0.4c$ and the grid-step is $\Delta z = 5 \cdot 10^{-3}c$ resulting in a total grid size of about $5.0 \cdot 10^6$ cells. The time step in the simulations is $0.001c/U_0$, which ensures the CFL number is less than one in the entire domain.

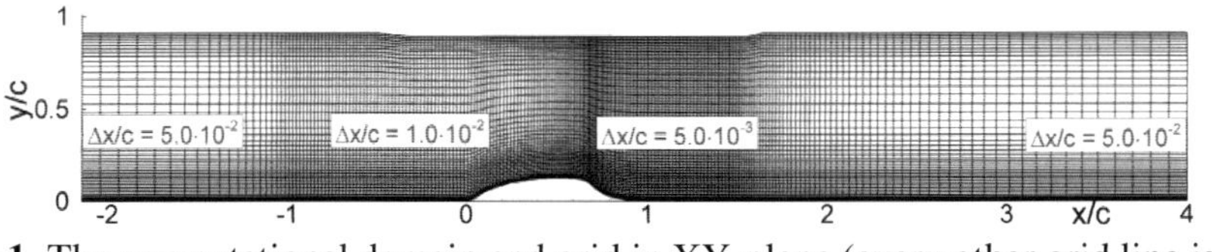

Figure 1. The computational domain and grid in XY-plane (every other grid line is shown)

3.2. Backward-facing step

The Backward-Facing Step (BFS) flow [14] is a flow in the plane channel with a straight upper wall and a step on the lower wall with the channel expansion ratio of 5/4. The experimental Reynolds number based on the step height H is $Re_H = 28000$. The grid used consists of about 2.5 million cells. The XY-plane of the grid is shown in Figure 2. The grid clusters towards the walls and near the step. The size of the domain in the spanwise direction is $4H$ with a spanwise grid step of $\Delta z = 0.05H$. The time step in the simulation is equal to $0.02H/U_0$ ensuring the CFL number is less than one.

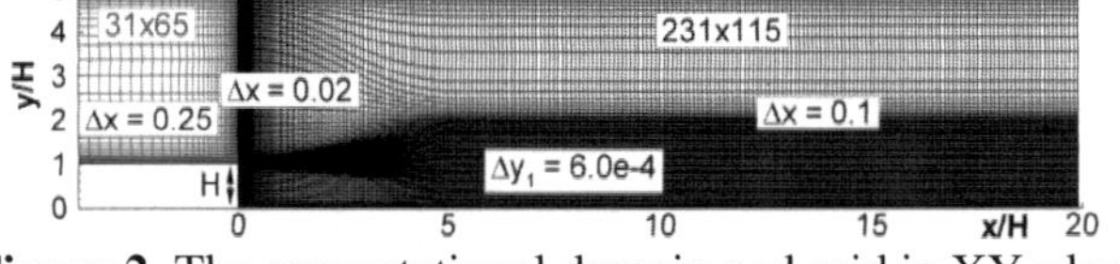

Figure 2. The computational domain and grid in XY-plane

3.3. Numerics used in the simulations

The considered versions of the DDES approach, described in Section 2, are implemented in ANSYS-FLUENT [7] codes using User Defined Functions (UDF).

In ANSYS-FLUENT the governing equations are written in the transient formulation for the incompressible fluid and solved with the use of the SIMPLEC method [15]. The pressure is interpolated with the use of "Standard" scheme [16] and for velocity the second-order bounded central difference scheme [17] is used. Gradients are calculated at cell centers with the use of the Green-Gauss theorem [18]. Time derivatives are calculated with the use of the three-layer second-order backward difference scheme [18]. Finally, the discrete equations are iteratively solved with the use of

IOP Conf. Series: Journal of Physics: Conf. Series **1038** (2018) 012134 doi:10.1088/1742-6596/1038/1/012134

the implicit point Gauss-Seidel method in conjunction with the algebraic multi-grid approach (10 sub-iterations are employed for each time step).

The most pronounced difference between ANSYS-FLUENT and the in-house code NTS is in approximation of the inviscid fluxes in the LES zones (fourth-order central-difference in NTS and second-order bounded central-difference in ANSYS-FLUENT), or in other words, in the dissipative properties of the schemes, which have significant influence on the development of the Kelvin-Helmholtz instability in shear layers and, therefore, could result in noticeably different performance of the considered DDES versions.

4. Results of the simulations

Figures 3 and 4 present visualizations of the considered flows obtained with different DDES versions. It could be seen, that similarly to NTS code, usage of the methods for RANS-LES transition acceleration in the framework of ANSYS-FLUENT leads to more rapid development of turbulent structures in the separated shear layer and to resolution of finer eddies in the recirculation zone, which could be seen in the results of both computational codes. Note however, that the fourth-order centered scheme of NTS code allows resolving slightly smaller structures than the second-order bounded central difference scheme of ANSYS-FLUENT.

Another important observation is the relatively high sensitivity of the original DDES approach to the code numerics, which is observed for both flows considered (the difference is slightly more pronounced for the backward facing step). In contrast, the modified DDES versions seem to be much less sensitive to peculiarities of the employed numerical schemes and yield almost identical results in both codes considered.

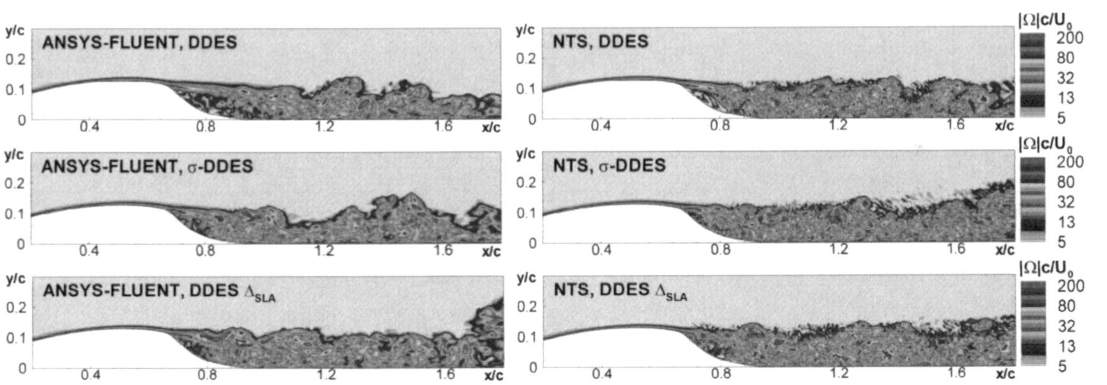

Figure 3. Contours of the vorticity magnitude for the wall-mounted hump flow

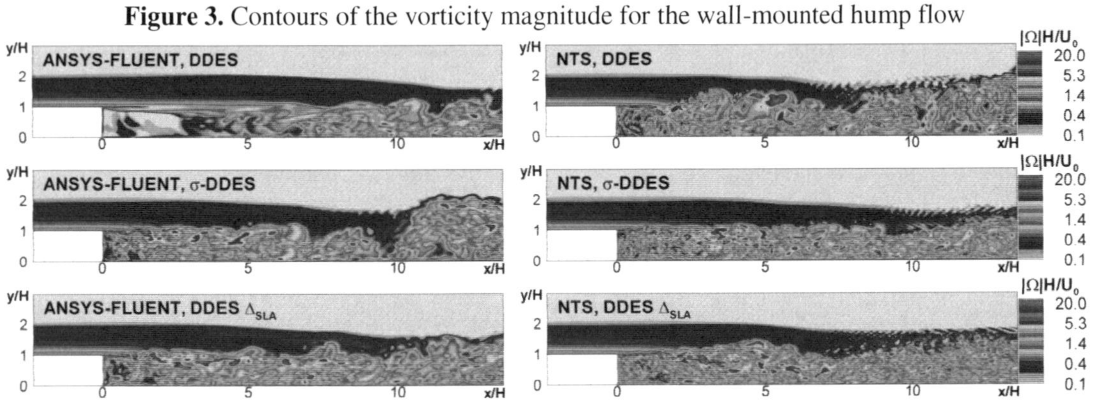

Figure 4. Contours of the vorticity magnitude for the backward facing step flow

A significantly better agreement between the computed and experimental data is observed in both codes due to the using of the methods for acceleration of the turbulent structure formation. Indeed, as could be seen from distributions of the mean skin friction coefficient (Figures 5, 6), the length of the

IOP Conf. Series: Journal of Physics: Conf. Series **1038** (2018) 012134 doi:10.1088/1742-6596/1038/1/012134

recirculation zone predicted by the enhanced DDES versions is in substantially better agreement with the experimental data compared to the standard one.

Moreover, mean results of the enhanced DDES versions are found to be less dependent on the numerics employed in the CFD code compared to the results of the original one, which agrees with the previous conclusions made from the instantaneous vorticity contours. Particularly, the longer RANS-LES transitional region observed in the standard DDES results is caused by the high dissipation of the ANSYS-FLUENT second-order bounded central difference scheme leading to accuracy deterioration in comparison to the NTS code.

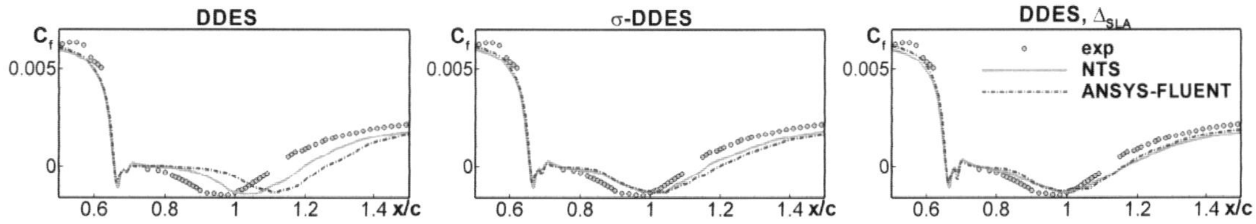

Figure 5. Distributions of the skin-friction coefficient for the wall-mounted hump flow

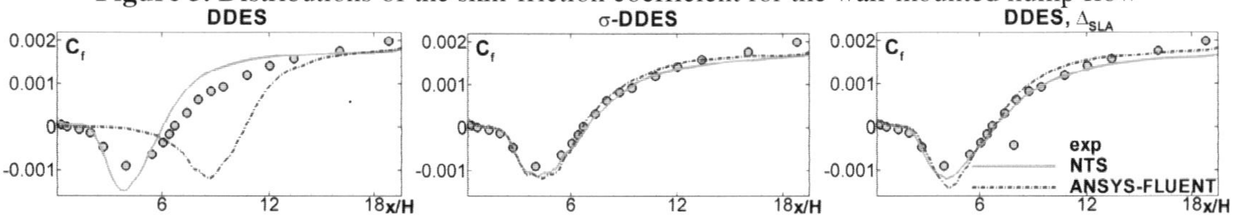

Figure 6. Distributions of the skin-friction coefficient for the backward facing step flow

The above mentioned trends could be also seen from figures 7, 8 where the resolved velocity fluctuations are depicted. Particularly, a delay in the formation of the three-dimensional structures is observed for DDES, which, in addition, highly depends on the employed numerical schemes, whereas similar results are obtained for enhanced DDES versions in both codes considered.

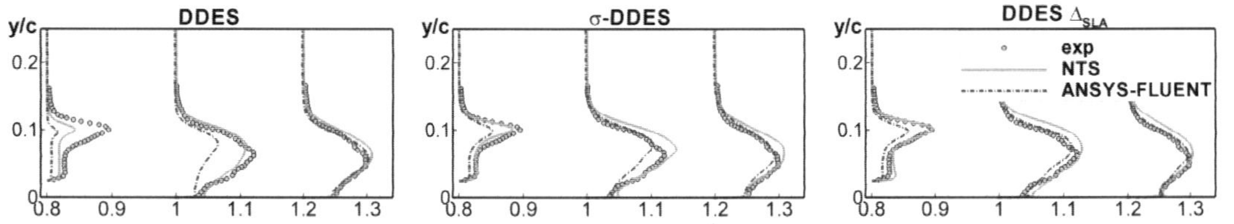

Figure 7. Profiles of x-velocity fluctuations ($2\langle u'u'\rangle/U_0^2 + x/c$) for the wall-mounted hump flow

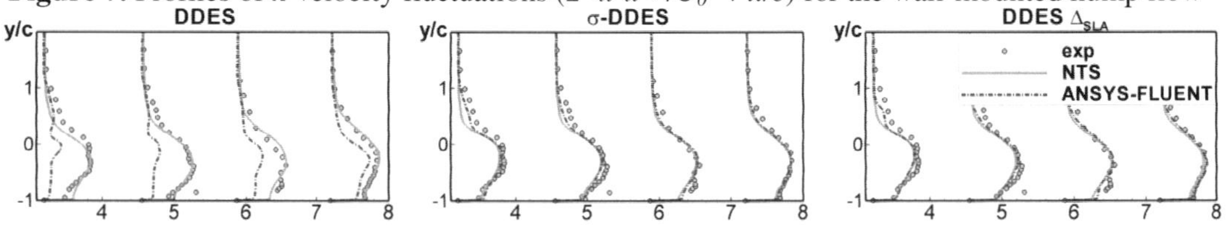

Figure 8. Profiles of x-velocity fluctuations ($4U_{RMS}/U_0 + x/H$) for backward facing step flow

Conclusions

Simulations of two separated flows have been performed with the use of DDES combined with approaches aimed at acceleration of RANS-LES transition in separated shear layers. The results indicate that usage of such enhanced DDES versions within the ANSYS-FLUENT code allows significantly better agreement with both the experimental data and the results of NTS code to be obtained. Therefore, these approaches can be recommended for use within general purpose CFD codes.

Acknowledgments
The results of the work are obtained with the use of the computational resources of Peter the Great Saint-Petersburg Polytechnic University Supercomputing Center (http://www.spbstu.ru).

References

[1] Spalart P R, Jou W-H, Strelets M and Allmaras S R 1997 Comments on the feasibility of LES for wings, and on a hybrid RANS/LES approach *Proceedings of first AFOSR international conference on DNS/LES* (Ruston, Louisiana, USA) pp 137–47
[2] Shur M L, Spalart P R, Strelets M K and Travin A K 2015 An enhanced version of DES with rapid transition from RANS to LES in separated flows *Flow Turbul. Combust.* **95** 709–37
[3] Mockett C, Fuchs M, Garbaruk A, Shur M, Spalart P, Strelets M, Thiele F and Travin A 2015 Two non-zonal approaches to accelerate RANS to LES transition of free shear layers in DES *Notes Numer. Fluid Mech. Multidiscip. Des.* **130** 187–201
[4] Guseva E K, Garbaruk A V and Strelets M K 2017 Assessment of Delayed DES and Improved Delayed DES Combined with a Shear-Layer-Adapted Subgrid Length-Scale in Separated Flows *Flow, Turbul. Combust.* **98** 481–502
[5] Shur M L, Strelets M K and Travin A K 2004 High-Order Implicit Multi-Block Navier-Stokes Code: Ten-Year Experience of Application to RANS/DES/LES/DNS of Turbulence *https://cfd.spbstu.ru//agarbaruk/doc/NTS_code.pdf*
[6] Kok J C and Ven H Van Der 2012 *Capturing free shear layers in hybrid RANS – LES simulations of separated flow*
[7] Mathur S R and Murthy J Y 1997 A pressure-based method for unstructured meshes *Numer. Heat Transf.* **32** 195–215
[8] Spalart P R, Deck S, Shur M L, Squires K D, Strelets M K and Travin A 2006 A New Version of Detached-eddy Simulation, Resistant to Ambiguous Grid Densities *Theor. Comput. Fluid Dyn.* **20** 181–95
[9] Menter F R 1994 Two-equation eddy-viscosity turbulence models for engineering applications *AIAA J.* **32** 1598–605
[10] Guseva E K, Garbaruk A V and Strelets M K 2017 Development and testing of σ-DDES approach based on k-ω SST turbulence model *Therm. Process. Eng.* **9** 434–40 [in russian]
[11] Nicoud F, Toda H B, Cabrit O, Bose S and Lee J 2011 Using singular values to build a subgrid-scale model for large eddy simulations *Phys. Fluids* **23** 1–12
[12] Gritskevich M S, Garbaruk A V., Schutze J and Menter F R 2012 Development of DDES and IDDES formulations for the k-w shear stress transport model *Flow, Turbul. Combust.* **88** 431–49
[13] Greenblatt D, Paschal K B, Chung-Sheng Y and Harris J 2005 A Separation Control CFD Validation Test Case Part 2. Zero Efflux Oscillatory Blowing *AIAA Pap.* **2005-0485** 1–24
[14] Vogel J C and Eaton J K 1985 Combined Heat Transfer and Fluid Dynamic Measurements Downstream of a Backward-Facing Step *J. Heat Transfer* **107** 922–9
[15] Patankar S V. 1980 *Numerical Heat Transfer and Fluid Flow*
[16] Rhie C M and Chow W L 1983 Numerical Study of the Turbulent Flow Past an Airfoil with Trailing Edge Separation *AIAA J.* **21** 1525–32
[17] Jasak H, Weller H G and Gosman A D 1999 High resolution NVD differencing scheme for arbitrarily unstructured meshes *Int. J. Numer. Methods Fluids* **31** 431–49
[18] Kim S E, Mathur S R, Murthy J Y and Choudhury D 1998 A reynolds averaged Navier-Stokes solver using unstructured mesh-based finite-volume scheme *AIAA Pap.* **98–0231**

IOP Conf. Series: Journal of Physics: Conf. Series **1038** (2018) 012135 doi:10.1088/1742-6596/1038/1/012135

Numerical simulation of aeroacoustical noise from a wing-flap configuration

K. Nikiforova[1] and A. Garbaruk[2]

Peter the Great Saint-Petersburg Polytechnic University, Russia, 195251, Saint-Petersburg, Polytechnicheskaya, 29

bkv.nikiforova@gmail.com[1], agarbaruk@mail.ru[2]

Abstract. A noise-free volumetric synthetic turbulence generator has been applied for Embedded LES of the flow over the wing-flap configuration experimentally studied by Lemonie et al. Good agreement of CFD results with experimental data provides some evidence for the correctness of the chosen computational strategy, which includes a noise-free volumetric synthetic turbulence generator to create turbulent content, "sponge layers" to avoid the reflection of pressure waves from the external boundaries and the Ffowcs Williams, Hawkings acoustic model for calculating far-field noise.

1. Introduction

Prediction of noise generated by flow over the airframe elements, in particular, noise from a wing-flap configuration, is one of the most important aeroacoustic tasks. Since in this case the noise is caused by turbulent vortices reacting with each other and with solid surfaces, it is necessary to use methods which resolve at least a part of the turbulence structures. However, the use of scale resolving simulation (even LES with near-wall RANS modelling) for attached boundary layers at high Reynolds numbers is too costly. Therefore a zonal method, combining RANS and LES, such as Embedded LES (ELES), is a more preferable approach. ELES assumes using more accurate WMLES only in a restricted flow region where resolved turbulence is essential, whereas the rest of the flow is treated with economical RANS.

ELES of attached flow requires generation of turbulent content at the RANS-LES interface to ensure transition from the RANS zone where the whole turbulence is modeled to the LES zone where it is partly resolved. For aeroacoustic tasks the quality of artificial turbulent content has a crucial effect since the creation of artificial vortices can lead to the generation of spurious noise. A noise-free Volumetric Synthetic Turbulence Generator has been developed by NTS [1] specifically to use in aeroacoustic applications. NTS VSTG has been modified to implement it in ANSYS Fluent and tested by the example of compressible boundary layer with zero pressure gradient [2], however its efficiency for aeroacoustic tasks has not been estimated yet.

The goal of this work is application of the modified VSTG in the framework of ELES of flow over a wing-flap configuration in Fluent. The results of the ELES solution are compared with experimental data to assess the efficiency of the considered generator.

IOP Conf. Series: Journal of Physics: Conf. Series **1038** (2018) 012135 doi:10.1088/1742-6596/1038/1/012135

2. Brief description of Volumetric Synthetic Turbulence Generator

In the frame of VSTG, synthetic velocity fluctuations are introduced within a 3D forcing zone (VSTG on Figure 1) by means of an additional volume source term into the momentum equations:

$$f_i = \frac{C_{BF} \cdot U_0 \cdot \rho \cdot u_i^{'}(\mathbf{r},t)}{L_{VSTG}}$$

Here U_0 is the maximum of velocity over the interface updated every time step, L_{VSTG} - the length of the forcing zone in the streamwise direction, $u_i^{'}(\mathbf{r},t)$ - synthetic fluctuations of the velocity field and C_{BF} - empirical constant. The field of velocity fluctuations $u_i^{'}(\mathbf{r},t)$ is defined by means of NTS STG [3] and based on the RANS solution at the interface. Due to smooth introducing turbulent fluctuations inside the forcing zone the appearance of spurious noise is avoided and VSTG can be applied for simulation of an aeroacoustic task.

Figure 1. ELES with VSTG.

3. Problem definition

The wing-flap configuration consists of two aligned airfoils: the wing is a long plate with a sharp trailing edge and the flap is a NACA0012 airfoil. Based on the free-stream velocity ($U_0 = 50$ m/s) and the flap chord ($c = 0.1$ m) the Reynolds number of a compressible flow over the wing-flap configuration is $Re_{flap} = 3.3 \cdot 10^5$, and the Mach number is $M = 0.15$.

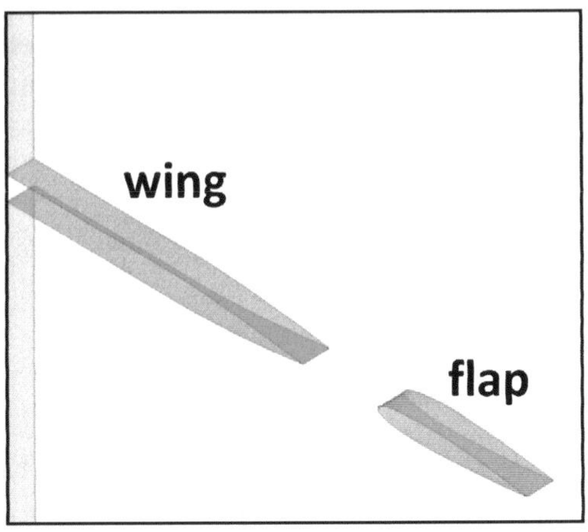

Figure 2. Wing-flap configuration.

4. Computational strategy

Aeroacoustical noise simulation was carried out in two stages: in the first stage turbulence simulation was performed by means of ELES and in the second stage far-field noise was calculated with the use of Ffowcs Williams, Hawkings acoustic model [4].

4.1. Embedded LES

In the frame of ELES of the flow over the wing-flap configuration, turbulence in the RANS sub-domain is modelled by means of *k-ω* SST model and algebraic WMLES is applied in LES zone. The

RANS-WMLES interface is located 0.2 m upstream of the wing trailing edge. The boundary layer thickness at the RANS-WMLES interface is $\delta_0 = 6 \cdot 10^{-3}$m. Simulation of the considered flow is performed in a computational domain with the size $L_x = 11$c, $L_y = 12$c, and $L_z = 0.18$c in the streamwise, wall-normal and spanwise directions respectively.

The grid in the WMLES sub-domain is designed in accordance with the guidelines of wall-modeled LES. Particularly, the near-wall step is $\Delta y_1 = 8 \cdot 10^{-6}$ m, which ensures $\Delta y_1^+ = 1.2$ in wall coordinates everywhere in the domain. The computational grid is uniform in the X and Z directions with grid steps $\Delta x = 0.1\delta_0$, $\Delta z = 0.05\delta_0$, which translate into $\Delta x^+ = 90$ and $\Delta z^+ = 45$ in wall coordinates. The total grid count is around 12 million cells.

Developed velocity and turbulent characteristics profiles are specified at the inlet boundary. The top and bottom boundaries are slip walls. Constant static pressure is set at the outlet. On the wall, a no-slip condition is assumed. Periodic boundary conditions are used in the Z direction. VSTG is employed for creation of the turbulent content. The volume source is activated at the RANS-WMLES interface ($x = -0.2$ m) and the length of the volume forcing zone is equal $L_{VSTG} = 3\delta_0$. In order to avoid the reflection of pressure waves from the external boundaries a procedure using "sponge layers" [3] near the boundaries is implemented in Fluent via UDF. Inside these layers at each time step a current WMLES density field is defined by "weighting" it with the constant RANS density.

The transient pressure based solver of Fluent using the SIMPLEC algorithm is employed. Time integration is performed using a second order accurate scheme. The dimensional time step is $\Delta t = 6 \cdot 10^{-3}$ s, ensuring that the CFL number is less than unity in the entire computational domain. For the approximation of convective terms in the momentum equations, the weak bounded central differencing scheme (WBCD scheme) combining central differencing and bounded central differencing scheme is used for both sub-zones (RANS and WMLES). For the turbulence equations the second order upwind scheme (SOU scheme) is utilized. The interpolation of variables on faces is performed by the Green-Gauss cell based method and the Standard scheme is used for the pressure interpolation scheme.

4.2. Ffowcs Williams, Hawkings acoustic model

The far-field noise in Fluent is calculated by means of Ffowcs Williams, Hawkings acoustic model. In the frame of the FWH method the far-field noise is calculated as a superposition of point sources enclosed by the control FWH surfaces. Two FWH control surfaces (Figure **3**) for the noise calculations are considered:

- Solid wall surface s_0, including the whole flap surface and the part of the wing surface starting at $x = x_{inflow}$.
- Permeable surface s_1 starting at $x = x_{inflow}$ and ending at $x = x_{outflow}$.

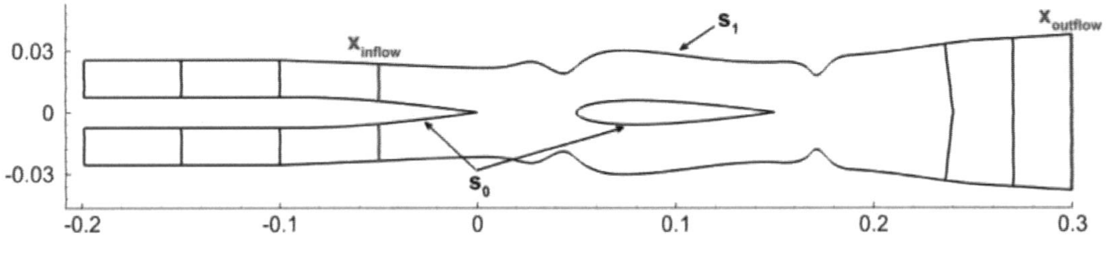

Figure 3. FWH surfaces.

Since the simulations are carried out using a computational domain which is narrow compared to the domain studied in the experiment, a "long span body" correction [5] is employed in order to take into account the contribution of noise caused by the whole span. This correction assumes that the inputs of separate spanwise patches of length L_z into the total noise are non-correlated.

5. Results

5.1. Flow structure
Shown is a typical flow pattern obtained by means of an eddy-resolving simulation in Figure **4**. The use of the WBCD scheme with a fine mesh permits to resolution of rather small turbulent structures inside the boundary layer. Turbulent structures generated by VSTG are similar to those which are observed downstream of the forcing zone. This means that the turbulent content generated by VSTG is similar to the turbulence which occurs in reality.

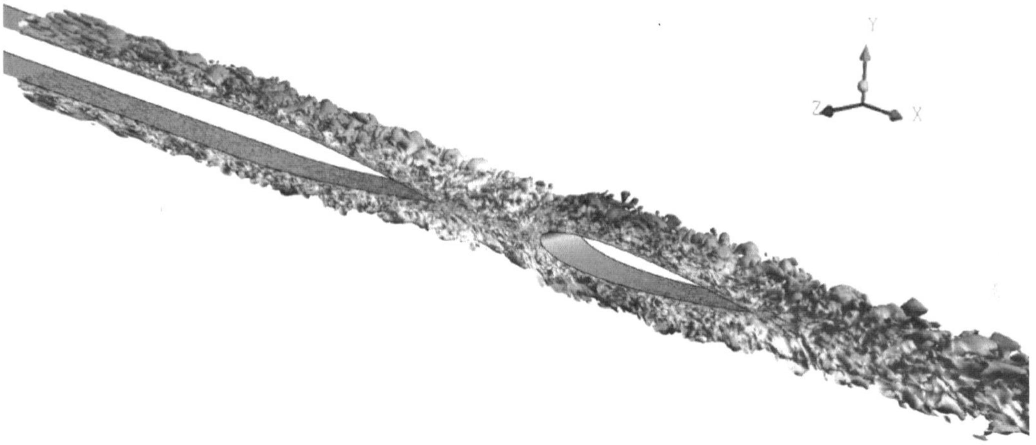

Figure 4. Q-criterion colored by streamwise velocity.

Figure **5** shows that there is no spurious noise caused by the synthetic turbulence generation thus confirming the efficiency of the VSTG method for aeroacoustic applications. As seen from the acoustic pressure fields, the noise is radiated from the both elements of the airframe evenly and the source of the noise is the vortices populating the boundary layer.

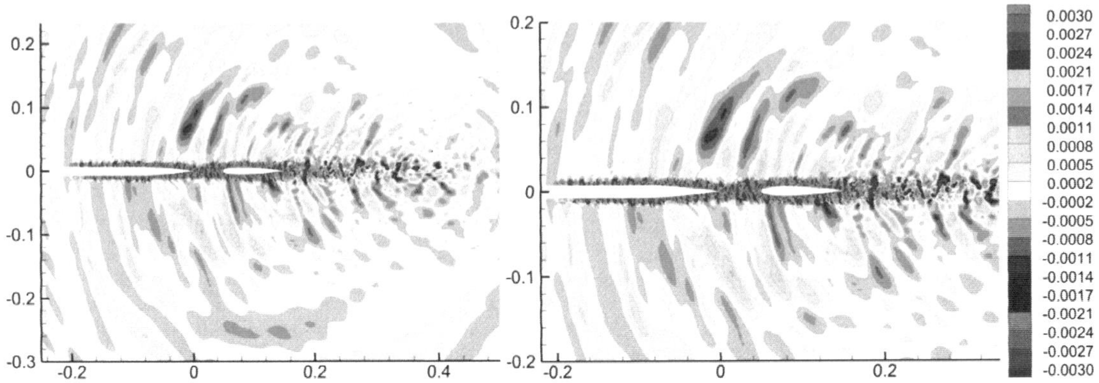

Figure 5. Snapshots of pressure time-derivative in the acoustic range.

5.2. Comparison of CFD results with the experimental data
The wing-flap configuration has been experimentally studied by Lemonie et al [6]. A comparison of computational spectra with the experimental data has been carried out for 15 control points distributed over the wing and flap surfaces. For all of them fairly good agreement of Fluent results with the experiment is observed. Figure **6** shows the comparison of spectra for some of probes where it can be seen that the simulation correctly reproduces the variation of both the spectral shapes and levels over the main wing and flap surfaces.

IOP Conf. Series: Journal of Physics: Conf. Series **1038** (2018) 012135 doi:10.1088/1742-6596/1038/1/012135

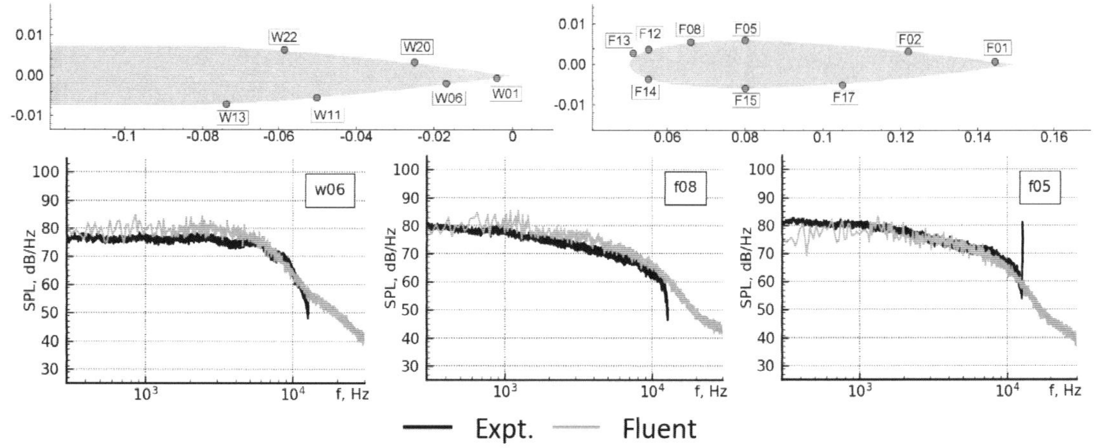

Figure 6. Wall-pressure spectra at the control points.

Far-field noise in Fluent is calculated by means of FWH model with the use of two types of control FWH surfaces. Considering solid surface s_0 as a control FWH surface, the noise sources populating the boundary layer are taken into account partially: monopole and dipole noise are accounted for while quadrupole noise is not considered. Due to accounting for quadrupole noise, the use of permeable surface s_1 enables estimation of whole noise spectrum and significantly improves the agreement of Fluent results with measurements in the high-frequency range (Figure 7). The remaining part of the spectrum is predicted quite accurately using as the control surface ether the solid or the porous surface.

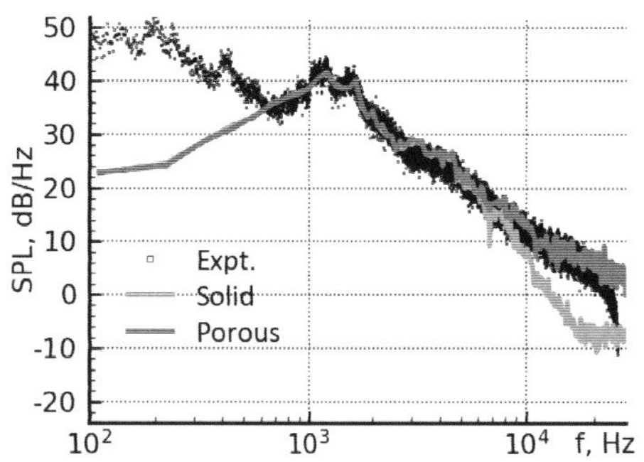

Figure 7. A comparison of far-field noise obtained by integration over solid and porous FWH surfaces.

Figure 7 shows that two closely-located low frequency peaks observed on the experimental distribution are reproduced in the computations. The difference between CFD and experimental data in the low frequency range is caused by background wind-tunnel noise existing in the experiment [3].

6. Conclusions

The VSTG has been successfully applied for simulation of an aeroacoustic task concerning the flow over the wing-flap configuration. The good agreement of CFD results with the experimental results confirms the efficiency of the computational tools used, including a volumetric noise-free synthetic turbulence generator, “sponge layers” and the Ffowcs Williams, Hawkings acoustic model.

Acknowledgements

The results of the work were obtained using computational resources of Peter the Great Saint-Petersburg Polytechnic University Supercomputing Center (http://www.spbstu.ru).

References

[1] Shur M et al 2018 Improved embedded approaches *Go4Hybrid: Grey Area Mitigation for Hybrid RANS-LES Methods* p 51-87

[2] Nikiforova K, Garbaruk A, Menter F and Smirnov P 2017 Volumetric synthetic turbulence generator in ansys fluent *Extended abstracts of the School-Seminar of Young Scientists and Specialists under supervision of Prof. A.I. Leontiev* (Saint-Petersburg)

[3] Shur M, Spalart Ph, Strelets M and Travin A 2014 A synthetic turbulence generator for unsteady content at RANS-LES interfaces in zonal simulation of aerodynamic and aeroacoustic problems *Flow, Turbulence and Combustion*

[4] Ffowcs-Williams J and Hawkings D 1969 Sound Generation by Turbulence and Surfaces in Arbitrary Motion *Philosophical Transactions of the Royal Society of London* vol 264 p 321–342.

[5] Seo J H, Chang K W and Moon Y J 2006 Aerodynamic noise prediction for long-span-bodies *AIAA Paper*

[6] Lemoine B, Roger M and Legriffon I 2011 Aeroacoustics of a model non-lifting wing-flap system in a parallel flow *AIAA Paper*

IOP Conf. Series: Journal of Physics: Conf. Series **1038** (2018) 012136 doi:10.1088/1742-6596/1038/1/012136

Large Eddy Simulation of airflow in a test ventilated room

N G Ivanov and M A Zasimova

Department of Fluid Dynamics, Combustion and Heat Transfer,
Peter the Great St.Petersburg Polytechnic University,
Polytechnicheskaya str. 29, 195251, St.Petersburg, Russia

E-mail: zasimova_ma@mail.ru

Abstract. The goal of the study is the validation of Wall-Modeled Large Eddy Simulation (WMLES) of ventilation in an isothermal room at conditions of experiments by Nielsen et al. Two values of the inlet slot width were considered. Calculations were carried out using the ANSYS Fluent 16.2 software with an algebraic WMLES subgrid-scale model. The uniform computational meshes up to 58 million cells were used. The focus in the results discussion is on the accuracy of mean and rms velocity fields prediction both in the jet zone and the low-velocity occupied zone. It is concluded that to predict the jet spread the meshes used are sufficient. However, there is visible difference between the computed and measured data in the occupied zone especially in the essentially 3D case with narrow inlet slot.

Design and optimization of indoor heating, ventilation and air conditioning (HVAC) systems with respect to space air diffusion aspects are often carried out using simplified analytical and empirical models (see, e.g., [1]). Along with the methods assuming complete air mixing, during last decades Computational Fluid Dynamics (CFD) simulation became popular in HVAC industry, as CFD models are able to produce detailed field information concerning air quality, e.g. location of stagnation and high-velocity zones, temperature distributions and contaminant transport information.

In engineering practice, the CFD models based on the Reynolds-Averaged Navier-Stokes (RANS) equations solution are widely used due to relatively moderate computational resources required. RANS approach has been applied to solve complex ventilation problems such as design of HVAC systems in an airplane cabin [2] and in an ice rink arena [3], as well as development of life support systems on board of the International Space Station (ISS) [4, 5]. However, RANS approach solves a set of transport equations obtained from the Navier-Stokes equations by means of the averaging procedure that results in the unknown Reynolds stress tensor emergence. To close the equations, the Reynolds stresses must be modeled, and the commonly used method is to apply the Boussinesq turbulent viscosity approximation [6]. To define the turbulent viscosity, a semi-empirical turbulence model should be involved, and the RANS results depend strongly on the particular turbulence model, especially when a fully developed turbulent flow is coupled with a moderate Reynolds number flow that is typical for ventilation problem [7].

Opposite to RANS, the Large Eddy Simulation (LES) approach solves the filtered Navier-Stokes equations that provides resolving large scales of motion, while small scales must be modeled with an appropriate subgrid-scale (SGS) model. The overview and outlook of LES could be found in [8]. LES, being an eddy-resolving technique, requires very significant computational resources. However, as accurate prediction of turbulent flows becomes more and more important, LES starts to be used in wide range of industrial application. With respect to ventilation performance prediction, LES has been applied for the first time in [9] to simulate the flow in a 3D test ventilated room for which accurate velocity measurements was performed by Nielsen et al. [10]. Later LES method was used several times to solve model indoor airflows, see [11] and references therein. LES approach was applied also in [12] to the ISS Columbus module ventilation modeling and the simulation results were compared with the measurement data available. Though satisfactory agreement with experimental data was reported in the papers cited, relatively coarse computational meshes of 0.5-2 million cells were used, and a revision of early LES data is anticipated. An example of recent LES simulation is a cross-ventilation flow study [13], but even in this paper it was concluded that due to computational limitations it was possible to run LES using a relatively coarse mesh.

The growing need for CFD verification and validation in the indoor air applications is emphasized by many authors, see, e.g., [6, 13]. The key factor required for CFD results validation is an appropriate experimental data. Experiment by Nielsen et al. [10] is one of the well-documented isothermal tests where the buoyancy effects were excluded. Accurate mean velocity data and corresponding rms values were obtained with the Laser-Doppler Anemometry (LDA) technique, with the uncertainty less than 0.5% for mean velocities [10] (see also [14] and references therein for detailed description of the test). The goal of the present study is the validation of the LES of ventilation airflow. Numerical solutions were obtained with the CFD package ANSYS Fluent 16.2.

1. Problem formulation and computational settings

Figure 1a shows the geometry model adopted for the computations of the indoor airflow that corresponds to the test facility [10]. The rectangular room with the height of $H = 3$ m was considered, the longitudinal and transverse dimensions of the room were $L/H = 3.0$ and $W/H = 1.0$. The rectangular inlet slot of height $h_{in}/H = 0.056$ was placed under the ceiling (in the center of the wall with respect to z-coordinate). The origin of the coordinate system is in the corner of the room (see figure 1a). Two cases with various values of the slot width, w_{in}, were considered: Case 1 with $w_{in}/H = 1.0$ and Case 2 with $w_{in}/H = 0.5$. The dimensions of the outlet slot located on the opposite wall near the floor were $w_{out}/H = 1.0$ and $h_{out}/H = 0.16$. An outlet channel with the length of $L_{out}/H = 0.5H$ was included into the computational domain.

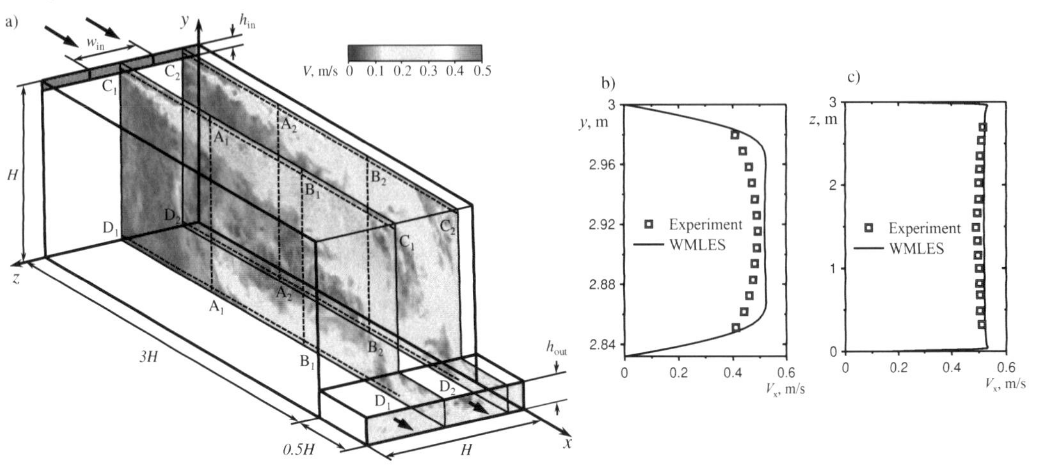

Figure 1. (a) Computational domain; instantaneous velocity magnitude distributions at cross-sections $z/H = 0.1$, 0.5 are shown; (b, c) inlet velocity profiles for Case 1 at (b) $z/H = 0.5$ and (c) $y/H = 0.972$

The measurement data [10, 14] are available at eight dashed lines shown in figure 1a: vertical lines A-A located at $x/H = 1.0$ and B-B at $x/H = 2.0$ and horizontal lines C-C located at $y/H = 0.972$ (starting at the mid-height of the inlet slot) and D-D located at $y/H = 0.028$ (that is equal to $h_{in}/2$). The subscript «1» in the line notation corresponds to $z/H = 0.5$, while subscript «2» – to $z/H = 0.1$.

Besides the main 3D problem with side walls, an additional 3D problem with the periodicity boundary condition in the transversal z-direction was set to perform mesh sensitivity study at lower computational cost. The transverse size of the computational domain for the parametric computations was $W_{per}/H = 1/6$ (note, that an increase of W_{per}/H by two times did not change the flow pattern).

Air was assumed as an incompressible fluid with constant physical properties ($\rho = 1.225$ kg/m^3, $\mu = 1.8\times10^{-5}$ kg/m·s). The boundary conditions approximately reproduced the experimental conditions [10]. To provide this, LES computations of turbulent flow in a straight rectangular duct with dimensions $L_{duct}/H = 2.0$, $W_{duct}/H = 1.0$, $H_{duct}/H = 0.056$ were performed prior to the main room airflow simulation. The time-averaged velocity profile extracted from the duct flow solution was set as the boundary condition at the inlet slot section. The computed inlet profiles are compared with the experimental data in figure 1b,c. The bulk velocity at the inlet slot was $V_{in} = 0.455$ m/s that corresponded to the Reynolds number $Re = \rho h_{in} V_{in}/\mu = 5233$. For the periodic problem, the 1D profile corresponding periodic duct flow was set. To model the fluctuating velocity at the inlet slot, the vortex method available in ANSYS Fluent was used.

For the incompressible fluid, the filtered Navier-Stokes equations are written as follows:

$$\begin{cases} \nabla \cdot \vec{V} = 0 \\ \dfrac{\partial \vec{V}}{\partial t} + \nabla \cdot (\vec{V}\vec{V}) = -\dfrac{1}{\rho}\nabla \cdot p + 2\nu(\nabla \cdot \underline{S}) - \nabla \cdot \underline{\tau}^{SGS} \end{cases} \quad (1)$$

where $\vec{V}$ is the velocity vector with components (u, v, w), $\underline{S}$ is the strain rate tensor for the resolved motion, and $\underline{\tau}^{SGS}$ is the SGS stress term arising from the spatial filtering procedure. The filtering operation [8] for a variable f determines the filtered (resolved) and small-scale (non-resolved) components $\tilde{f}$ and f' as follows:

$$\tilde{f}(x,t) = \int_{Vol} G(x - x', \Delta)\, f(x',t)dx'^3, \quad f' = f - \tilde{f}, \quad (2)$$

where $G(x - x', \Delta)$ is a filter function that determines the size and structure of the small scales, x is a coordinate of the point under consideration, and Δ is the filter width.

To determine the SGS stress term, the generalized Boussinesq hypothesis is used:

$$\tau_{ij}^{SGS} - \frac{1}{3}\tau_{kk}\delta_{ij} = -2\nu_{SGS} S_{ij} \quad (3)$$

where ν_{SGS} is the SGS viscosity, which must be determined by a SGS model:

The algebraic Wall-Modeled LES (WMLES) S-Omega SGS model available in ANSYS Fluent was applied in the current study. The model realization in the code is based on [15], and the SGS viscosity is calculated with the use of a hybrid length scale and the wall-damping function:

$$\nu_{SGS} = min\left\{(\kappa d_w)^2, (C_S\Delta)^2\right\} | S - \Omega | \left(1 - exp\left\{\left(-y^+/25\right)^3\right\}\right) \quad (4)$$

where S and Ω are the strain rate and vorticity magnitude, $C_S = 0.2$ is the Smagorinsky constant, $\kappa = 0.41$ is the von Karman constant, d_w is the distance to the nearest wall, y^+ is the normal to the wall inner scaling. The grid scale is defined as follows:

$$\Delta = \min\left\{\max\left(C_w d_w, C_w \Delta_{max}, \Delta_{wn}\right), \Delta_{max}\right\} \quad (5)$$

where Δ_{max} and Δ_{wn} are the maximum local grid spacing and the grid step in the wall-normal direction, and $C_w = 0.15$ is the empirical constant.

The non-iterative time advancement scheme (NITA) based on the fractional step method was used. The spatial discretization was performed with the central-differencing scheme for convective terms and the second-order central scheme for viscous terms, the second order pressure interpolation was used. For all the cases computed, the uniform meshes with cubic cells were used. The size of the meshes used is discussed below, in Section 2. For all the meshes the maximum values of the normalized distance from the first computational cell to the wall did not exceed 0.5 in the occupied zone and were up to 20 in the jet zone near the inlet slot.

The second order implicit time integration was used. The value of a time step, Δt, is equal to 0.01 s, and it was chosen to provide the Courant number in the computational domain less than 1 for all the meshes considered. Note that a special time-step sensitivity study was performed with $\Delta t = 0.006$ s, and no changes in the averaged characteristics were fixed. To accumulate representative statistics, it was required to calculate samples of about 150,000 time steps (1500 s). It was checked that the averaging time has been sufficient to obtain statistically steady data. It was possible to perform massively parallelized computations (up to 512 cores) due to computational resources of the Peter the Great St.Petersburg Polytechnic University supercomputer center (scc.spbstu.ru).

2. Mesh-sensitivity analysis for periodic problem

The mesh-sensitivity analysis was performed for the periodic problem, as it was not possible to complete a mesh dependence study for the room problem with side walls due to enormous computational cost. The initial coarse mesh consisted of 536×179×30 uniform cubic cells that corresponds to about 3.0 million cells. The initial mesh was refined three times, and the total cell number for each successively refined mesh was increased by a factor of $8^{0.5}$ (i.e., $2^{0.5}$ times in each direction), so that mesh #2 had 8 million cells, #3 – 23, and #4 – 58 million cells.

Sure, the flow pattern computed with the periodic assumption differs much from the airflow in the 3D room with side walls ($W/H = 1.0$) even at the mid-plane. Time-averaged velocity magnitude fields computed with two problem formulations are compared in figure 2 (here and after the dimensional velocities are presented). The use of the periodic assumption results in more pronounced recirculation flow as there is no side walls that slow down the flow, and the averaged flow pattern was uniform in z-direction. However, in both the cases the flow patterns could be separated into two zones: the relatively high-velocity zone of the near-wall jet with the mixing layer region and the relatively low-velocity occupied zone with the clockwise recirculation flow. Conclusions from the mesh dependence discussion for the periodic problem could be extended to the problem with side walls.

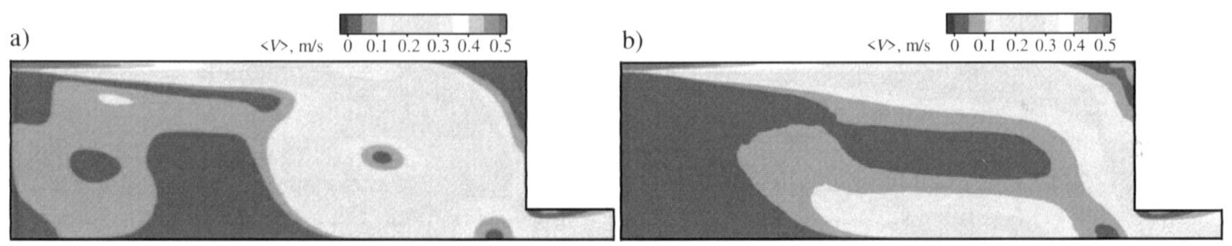

Figure 2. Mean velocity fields at mid-sections: (a) the periodic problem; (b) the problem with $W/H = 1.0$

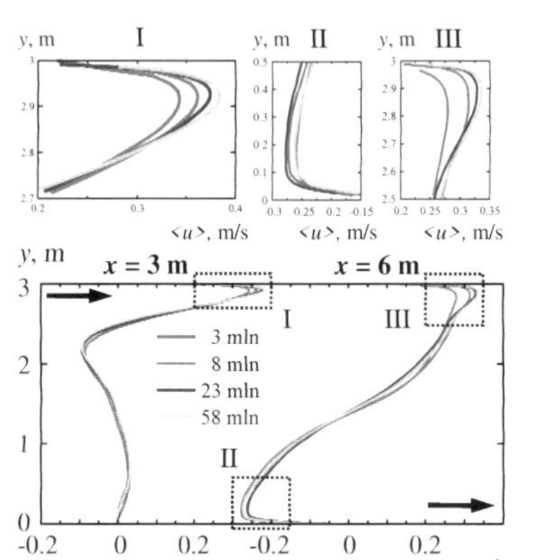

Figure 3. Longitudinal velocity profiles for the periodic problem

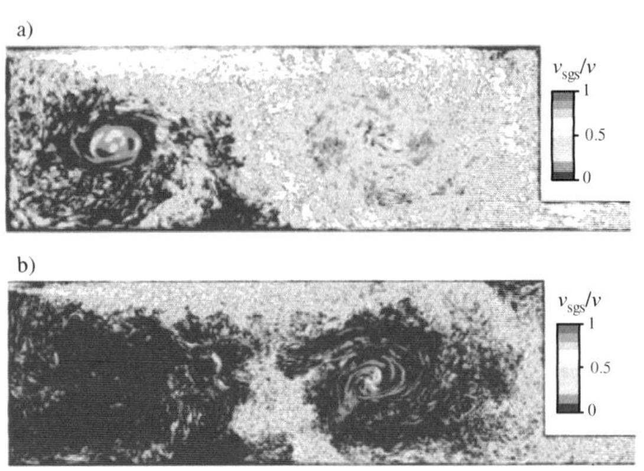

Figure 4. SGS to molecular viscosity ratio for the periodic problem; meshes of (a) 8 and (b) 58 million cells

Figure 3 shows the mean longitudinal velocity profiles along two lines in the vertical mid-plane, with zoomed near-wall regions. The profiles demonstrate monotonous growth of velocity values with mesh refinement in the jet zone. The deviation from results obtained with the finest mesh seems to be

IOP Conf. Series: Journal of Physics: Conf. Series **1038** (2018) 012136 doi:10.1088/1742-6596/1038/1/012136

critical for mesh #1, and could be treated as acceptable for mesh #2. In the occupied zone, the velocity values also depend on mesh size, but not as much as in the jet zone. Thus, mesh #2 provides nearly mesh independent results. Figure 4 gives additional information on the contribution of the non-resolved scales for the meshes used: for the finest mesh the ratio of the SGS to molecular viscosity is less than 3.0 everywhere in the computational domain (figure 4b).

Based on the results obtained for the periodic problem, for the room with side walls the basic mesh of almost 48 million cells was used (751×252×250 cells), and the results obtained with this mesh are discussed below in Section 3. The mesh provides the cell length in each direction of 12 mm. Note that in this case the Kolmogorov scale for the mixing layer region was about 0.7 mm, and in the occupied zone it was about 2.5 mm.

3. Comparison of WMLES results with the experimental data

Figure 5 compares computed and measured [10] profiles of mean longitudinal velocity and its pulsations: the upper row of the plots corresponds to the mid-section of the room, while the lower row demonstrates flow behavior near the side wall. It is visible that the jet spreading is reproduced quite accurate (the only deviation is in the <*u*> values at large *x*, figure 5b that corresponds to slight overprediction of the mixing in the jet, visible also in figure 5d). At the same time there is a pronounced difference between the computational velocity profiles and the experimental data in the recirculation zone (figure 5c,f); the pulsations in this zone are underpredicted.

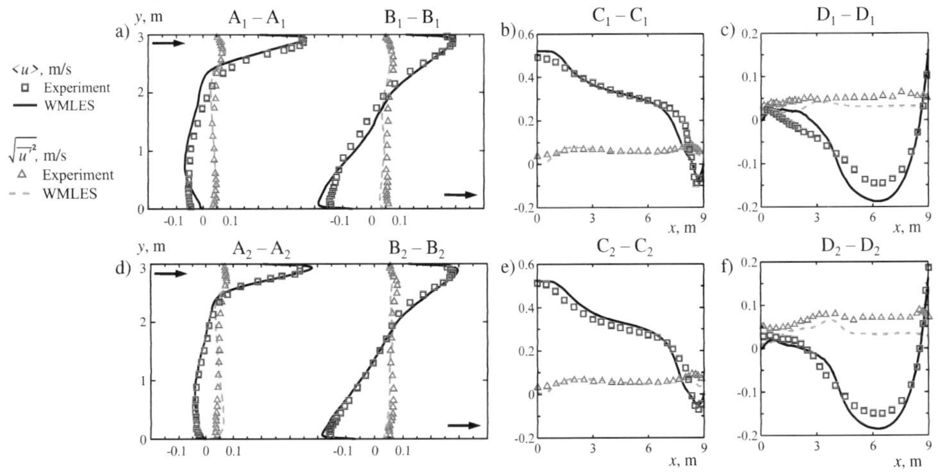

Figure 5. Mean longitudinal and rms velocity, Case 1, $w_{in}/H = 1.0$: (a – c) $z/H = 0.5$ and (d – f) $z/H = 0.1$

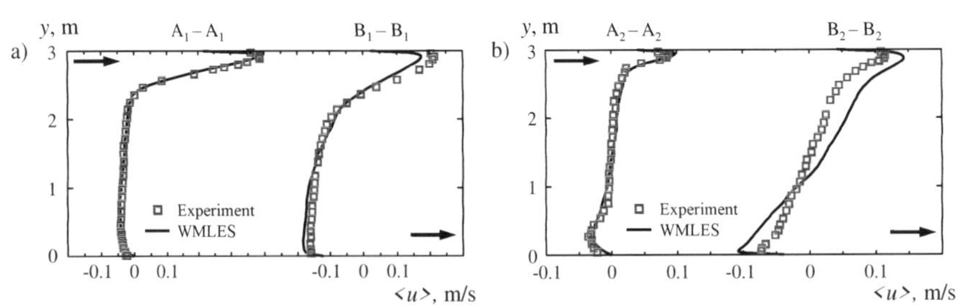

Figure 6. Mean longitudinal velocity, $w_{in}/H = 0.5$: (a) $z/H = 0.5$ and (b) $z/H = 0.1$

The difference between the computed and measured data is even larger if Case 2 with narrow inlet slot is considered. Inlet slot width reduction leads to appearance of additional mixing layers and 3D effects are more pronounced. The computed mean longitudinal velocity profiles are compared with the measurements in figure 6 (unfortunately, experimental pulsation data are not available for this case).

The main difference is visible in figure 6b, at line B_2-B_2 that corresponds to the turn of the stream from the inlet. This difference due to inaccuracy of mixing prediction (see line B_1-B_1); another reason for it may be in non-symmetrical recirculation flow pattern that will be analyzed in the future.

4. Conclusions

Wall-Modeled Large Eddy Simulation of indoor air flow in an isothermal ventilated room at conditions of benchmark test experiments at Re of 5×10^3 have been performed using the ANSYS Fluent 16.2 CFD package. The mesh-dependence study has been performed for a simplified periodic problem formulation using the uniform meshes up to 58 million cells. The conclusion is that for the mesh with 250 cells per room height provides nearly mesh-independent results.

Comparison of WMLES results with the experimental data on mean and rms velocities results in conclusion that the mesh used is sufficient to predict accurately the jet spread, though there is slight overprediction of the jet mixing. At the same time, there is some difference between the computed and measured data in the low-velocity occupied zone, especially for the essential 3D case with narrow inlet slot. The future work will focus on the detection of possible reasons for this difference, as well as on the detailed analysis of the unsteady 3D structure of the recirculation flow.

References

[1] Wang S K 2001 *Handbook of air conditioning and refrigeration* (New York: McGraw-Hill)
[2] Liu W, Wen J, Chao J, Yin W, Shen C, Lai D, Lin C H, Liu J, Sun H and Chen Q 2012 Accurate and high-resolution boundary conditions and flow fields in the first-class cabin of an MD-82 commercial airliner *Atmospheric Environment* **56** pp 33-44
[3] Palmowska A and Lipska B 2016 Experimental study and numerical prediction of thermal and humidity conditions in the ventilated ice rink arena *Building and Environment* **108** pp 171-82
[4] Son C H, Ivanov N G, Telnov D S and Smirnov E M 2015 Integrated computational fluid dynamics ventilation model for the international space station *Proc. of the 35th Int. Conf. on Environmental Systems (Rome, Italy, July 11-14 2015)* SAE Paper 2005-01-2794 pp 1-7
[5] Son C H, Ivanov N G, Telnov D S and Smirnov E M 2012 Numerical Study of Ammonia Leak and Dispersion in the International Space Station *Proc. of the 42nd Int. Conf. on Environmental Systems (San Diego, California, USA, July 15-16 2012)* pp 1-10
[6] Hinze J O 1975 *Turbulence* (New York: McGraw-Hill)
[7] Li Y and Nielsen P V 2011 CFD and Ventilation Research *Indoor Air* **21** pp 442-53
[8] Piomelli U 2014 Large eddy simulations in 2030 and beyond *Phil. Trans. R. Soc. A* **372** pp 1-13
[9] Davidson L and Nielsen P V 1996 Large eddy simulation of the flow in a three-dimensional ventilation room *Proc of the 5th Int. Conf. on Air Distribution in Rooms, ROOMVENT'96 (Yokohama, Japan, July 17-19, 1996)* pp 161-68
[10] Nielsen P V, Restivo A and Whitelaw J H 1978 The velocity characteristics of ventilated room *ASME J. Fluids Engineering* **100** pp 291-98
[11] Jiang Y, Su M and Chen Q 2003 Using Large Eddy Simulation to Study Airflows in and around Buildings *ASHRAE Transactions* **109** (2) pp 517-26
[12] Smirnov E M, Ivanov N G, Telnov D S and Son C H 2006 CFD modeling of cabin air ventilation in the International Space Station: a comparison of RANS and LES data with test measurements for the Columbus Module *Int. J. of Ventilation* **5** (2) pp 219-28
[13] Hooff T, Blocken B and Tominaga Y 2017 On the accuracy of CFD simulations of cross-ventilation flows for a generic isolated building: Comparison of RANS, LES and experiments *Building and Environment* **114** pp 148-65
[14] Nielsen P V 1990 Specification of a two-dimensional test case *Aalborg: Instituttet for Bygningsteknik, Aalborg Universitet, Gul serie* **R9040** (8) pp 1-15
[15] Shur M L, Spalart P R, Strelets M K and Travin A K 2008 A hybrid RANS LES approach with Delayed DES and Wall Modelled LES capabilities *Int. J. of Heat and Fluid Flow* **29** (6) pp 1638-49

IOP Conf. Series: Journal of Physics: Conf. Series **1038** (2018) 012137 doi:10.1088/1742-6596/1038/1/012137

Modelling of heating of plasma-chemical reactor in Comsol Multiphysics.

N V Obraztsov[1,2], D I Subbotin[2,3,4], V E Popov[2], V Y Frolov[1] and A V Surov[2]
[1] Peter the Great Saint-Petersburg Polytechnic University, Polytechnicheskaya 29, Saint-Petersburg 195251, Russia
[2] Institute for Electrophysics and Electric Power of the Russian Academy of Sciences, Dvortsovaya emb. 18, Saint-Petersburg 191186, Russia
[3] Saint-Petersburg State Institute of Technology (Technical University), Moskovsky prospect, 26, Saint-Petersburg, 190013, Russia
[4] Saint-Petersburg University, 11 Lieutenant Shmidt Emb., Saint-Petersburg 199034, Russia

E-mail: nikita.obrazcov@mail.ru

Abstract. For the thermophysical calculation of the reactor, a mathematical model was developed in Comsol Multiphysics, describing physical processes in the plasma-chemical reactor at a stationary mode. To simplify the calculation at this stage, it was carried out for an air plasma torch. The model includes two processes: fluid dynamic and heat transfer ones. As a result of the mathematical modelling the following parameters were obtained: temperature distribution in the reactor at steady state, average mass temperature in the reaction volume, velocity field and others.

1. Introduction
At present, mathematical modeling has become an integral part of any research. It allows assessing correctness of the choice of a particular technical solution at the design stage. At the same time, mathematical modeling enables us to approach the final result with no physical models or prototypes [1,2]. Plasma technologies have been used in various industries for a long time. In recent years, the plasma technologies are becoming more and more popular since they aid to recycle waste [3,4] and plasma coating [5] more efficiently than traditional methods. The main processes of decomposition and destruction of waste by the plasma method take place in a plasma-chemical reactor. For example, to decompose organochlorine compounds it is necessary to keep it in a zone with a temperature of 1500 K for at least 2 seconds [6,7]. For this purpose, the reaction space is preheated with an air plasma torch for several hours. In the mathematical modeling of a plasma chemical reactor the uniformity of heating of the whole reaction space, the velocity field and the thermal losses of the reactor when bringing into the operation mode are of particular interest.

2. Model of plasma-chemical reactor
2.1. Assumptions and simplifications of calculation model

For the thermophysical calculation of the reactor, a mathematical model was developed, which describes the physical processes in a plasma-chemical reactor at a steady state. At this stage, the calculation was carried out for an air plasma torch. The model includes two physical processes: gas flow and heat transfer in gas and solids. In this work the model of the processes under consideration are simplified. Nevertheless, the results of modeling are scientifically and practically valuable.
The following simplifications are assumed:

- Plasma-forming medium is air.
- Chemical interaction between air components is lacking.
- Electric arc processes are excluded.
- Outflow from plasma torch is modeled by the flow rate, the power and the average working gas temperature [7,8].
- Radiation losses are not taken into account

The aims of modeling the plasma-chemical reactor are:

- Design model of the existing plasma chemical reactor;
- Calculation of the velocity, temperature and pressure fields;
- Determination of the thermal losses power of the process, estimation of its share in the total energy input;

2.2. Geometry of computational domain
The scheme of the plasma-chemical reactor used in the Institute of Electrophysics and Electric Power of the Russian Academy of Sciences is presented in Fig.1.

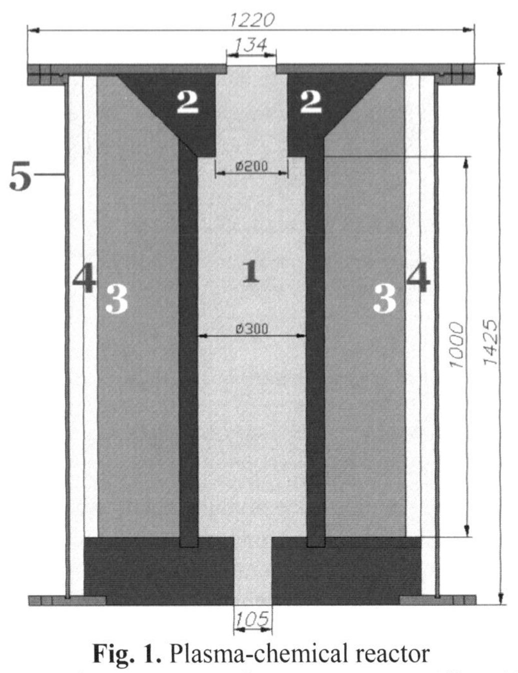

Fig. 1. Plasma-chemical reactor
1 - reaction space; 2- refractory concrete Alkor-90;
3 - insulating concrete Alaks-1.6-1600; 4 – insulating board MKRP-340;
5- vessel of 12X18H10T steel.

IOP Conf. Series: Journal of Physics: Conf. Series **1038** (2018) 012137 doi:10.1088/1742-6596/1038/1/012137

2.3 Set of equations.
Model of the plasma-chemical reactor is described by equations of gas dynamics and heat transfer. The Navier-Stokes equation (1) describing motion of the compressible gas and the continuity equation are to be solved (4.2). Equation (4.3) describes heat transfer in the gas and reactor vessel.

$$\rho(\mathbf{u}\cdot\nabla)\mathbf{u}=\nabla\cdot(-p\mathbf{I}+\mu(\nabla\mathbf{u}+(\nabla\mathbf{u})^T)-2/3\ \mu(\nabla\cdot\mathbf{u})\mathbf{I})+\mathbf{F} \quad (1)$$

$$\nabla\cdot(\rho\mathbf{u})=0 \quad (2)$$

$$\rho\cdot Cp\cdot\mathbf{u}\cdot\nabla T+\nabla(-k\nabla T)=Q \quad (3)$$

Here ρ is the gas density; u is the velocity field; p is the pressure; μ is the viscosity; F are the external forces acting on the gas; Cp is the heat capacity; k is the thermal conductivity; q is the heat flow.

2.4 Initial and Boundary Conditions:
Initial pressure in the reaction volume is P = 101325 Pa. At the inlet to the reaction space, the velocity field at the exit from the plasma torch nozzles (Fig. 2) is set from the simulation model of the air ac plasma torch. Since it is difficult to simulate processes in AC plasma torches, the plasma flow is modeled by velocities and mean mass temperatures.

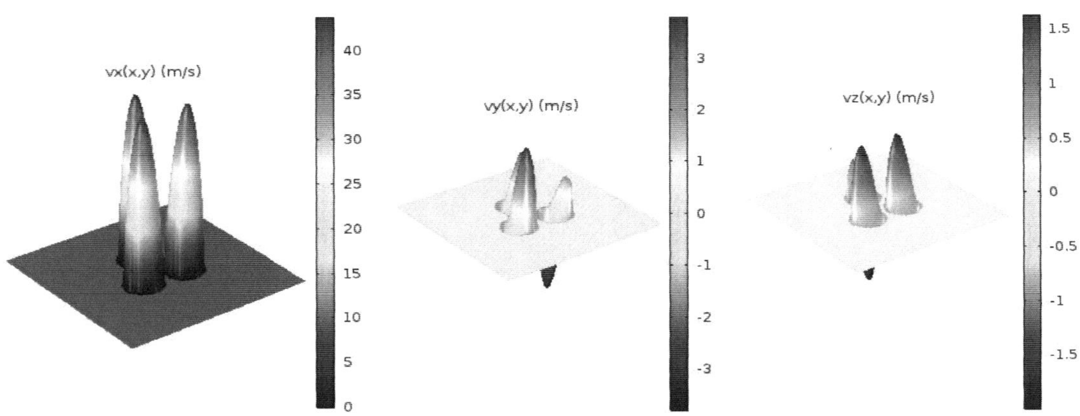

Fig. 2. Field of gas velocities at the exit from the plasma torch in the form of component expansion

Rarefication of 0.2 kPa is maintained at the outlet from the reaction space. Condition of non-percolation (the normal velocity component $V_n = 0$) is set on the walls of the reaction space of the reactor. The influence of gravity is taken into account. At the inlet into the reaction space, the temperature distribution is determined (Figure 3), which is also obtained from the model of the three-phase ac plasma torch.

IOP Conf. Series: Journal of Physics: Conf. Series **1038** (2018) 012137 doi:10.1088/1742-6596/1038/1/012137

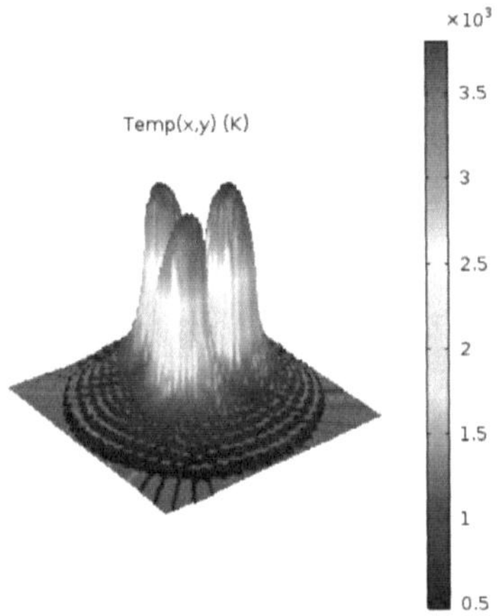

Fig.3 Temperature field of the gas at the inlet into the reaction volume

The initial temperature of the entire design area is set T = 293.15 K. The convective heat flux (4) is specified from the boundaries of the design area.

$$q=h\cdot(T_{ext}-T);h=h\_air\ (L;p\_A;T\_ext) \qquad (4)$$

The main difficulty in using heat transfer coefficients is to calculate or determine the corresponding value of the coefficient h. This coefficient depends on the cooling flow, the properties of the flow material and the surface temperature and for forced convection cooling from the flow velocity. In addition, the coefficient is affected by the geometric configuration. The heat exchange interface provides built-in functions for heat transfer coefficients. For most technical purposes, applying these coefficients is an accurate and numerically efficient approach to modeling. Handbooks on heat transfer usually contain a large set of empirical and theoretical corrections for the coefficients h. The heat transfer module contains such a subset. These expressions are based on the following dimensionless quantities: Nu (Nusselt number), Re (Reynolds number), Pr (Prandtl number), Ra (Rayleigh number), Gr (Grashof number).

In the model under consideration, heat transfer is given by external natural convection, where the geometric arrangement must be specified to calculate the heat transfer coefficient. For calculation, the following parameters are used: h – heat transfer coefficient (W/(m2·K)), L – linear dimension (m), ΔT – The difference between surface temperatures and the volume of the cooling stream (K), g – acceleration of gravity (m/s2), k – flow thermal conductivity (W/(m·K), ρ – flux density (kg/m3), U – volumetric flow rate (m/s), μ – dynamic viscosity (P·s), Cp – flow capacity (J/(kg·K)), β – thermal expansion coefficient (1/K).

Vertical wall:

$$h=\begin{cases}\frac{k}{L}\left(0.68+\frac{0.67Ra^{1/4}}{\left(1+\left(\frac{0.492k}{\mu C_p}\right)^{9/16}\right)^{4/9}}\right), & Ra\le 10^9\\ \frac{k}{L}\left(0.825+\frac{0.387Ra^{1/6}}{\left(1+\left(\frac{0.492k}{\mu C_p}\right)^{9/16}\right)^{8/27}}\right)^2, & Ra>10^9\end{cases} \qquad (5)$$

The horizontal plate, the upper side of the reactor: If T> Text, then

$$h = \begin{cases} \frac{k}{L}0.54Ra^{0.25}, & \text{Ra} \leq 10^7 \\ \frac{k}{L}0.15Ra^{0.33}, & \text{Ra} > 10^7 \end{cases} \quad (6)$$

While T ≤ Text, then

$$h = \frac{k}{L}0.27Ra^{0.25} \quad (7)$$

Ra is calculated in the same way as for a vertical wall.
The horizontal plate, the lower side of the reactor. Equation (6) is used for T ≤ Text, and equation (7) is used for T> Text. Otherwise it is also used as for a horizontal surface from the upper side. Geometric parameters are determined for the regions, for which the condition of convective heat transfer and ambient temperature Text = 293.15 K are set. The heat exchange interface provides built-in functions for the heat transfer coefficients. For the most technical purposes application of these coefficients is an accurate and numerically efficient approach to modeling.

2.5. Calculation method

Thermophysical calculation of the reactor is a stationary problem. For this purpose, a stationary solver was chosen where the solution was divided into sub-steps (segregated steps). This approach is called a separate solution. To solve a linear system of equations one can use an optimal solver for every process either direct, or iterative one, which is less resource-intensive. Each separate step may be a non-linear problem in itself, which may be solved with the desired error and controlled relaxation depending on a specific physical task. In the first step, we solve the equations for temperature, in the second for velocity and pressure. The total number of iterations was set to 300, the solution was achieved by Newton's method with damping coefficients of 0.5 in the first case and 0.8 in the second one. In each step the PARDISO solver was used.

3. Results

The calculation took 2 hours 50 minutes 20 seconds, it was carried out on a computer with AMD A10-7850K Radeon R7, 4 cores, 16 GB. The following parameters were obtained from the mathematical model:
1. Temperature distribution in the reactor at steady state is presented in Fig. 4. The average mass temperature in the reaction volume exceeds 1500 K.
2. Velocity fields are presented in Figs. 5-7. The data obtained indicate a turbulent flow regime in the reactor that promotes better mixing and improves completeness of transformation of the starting materials.
3. Estimation of the thermal losses power shows it makes up 9.3% (5.8 kW) of the plasma torch power (62.3 kW). Taking into account the radiative heat transfer, this value will be even greater.

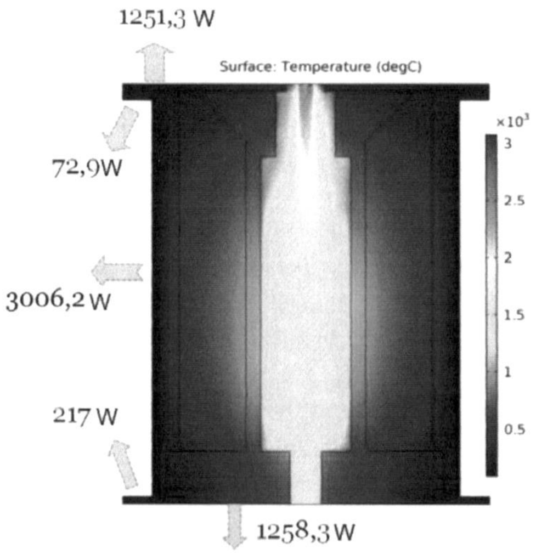

Fig.4. Distribution of temperature and heat loss from outer surface of the reactor.

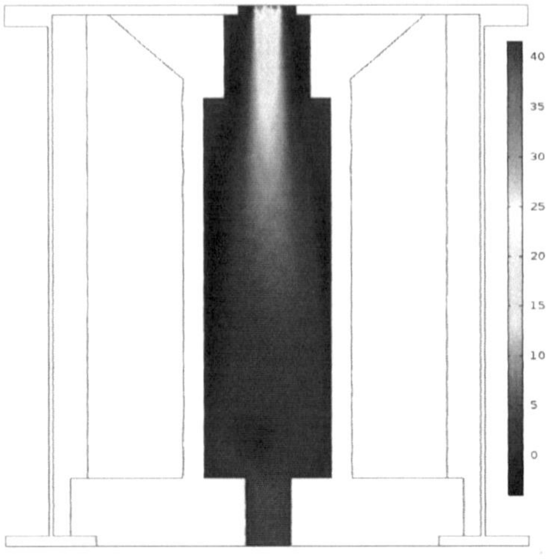

Fig. 5. Distribution of projection of gas velocity vector on longitudinal axis of the reactor (X axis, directed downwards).

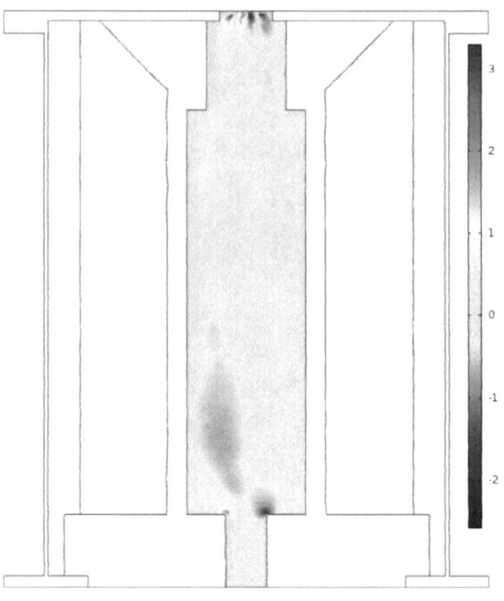

Fig.6. Distribution of projection of gas velocity vector on Y axis.

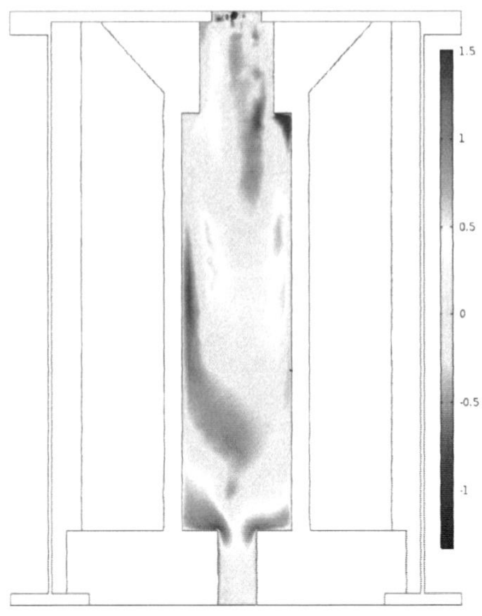

Fig. 7. Distribution of projection of gas velocity vector on Z axis.

4. Conclusion

The average mass temperature in the reaction volume exceeds 1500 K that meets the reaction conditions (decomposition of organochlorine compounds). Thermal losses are 9.3%. The high level of heat loss is due to the experimental type of the reactor and inherent in the small-capacity plants. When designing industrial installations, heat losses will be minimized by introducing additional technological chains for its useful utilization.

References

[1] Alain Gleizes Perspectives on Thermal Plasma Modelling Plasma 2015 *Chem. Plasma Process* **35** 455–69

[2] Dayal A. R., Pfluger D., Kearney T. N., Western R. J. and McAllister T. The Design Optimization of a Plasma Reactor for Chemical Waste Destruction 2004 *Plasma Chemistry and Plasma Processing* **24** 573-84

[3] Rutberg Ph.G., Bratsev A.N., Kuznetsov V.A., Popov V.E., Ufimtsev A.A., Shtengel' S.V., On efficiency of plasma gasification of wood residues 2011 *Biomass and Bioenergy* **35** 495-504

[4] Rutberg Ph.G., Kuznetsov V.A., Serba E.O., Popov S.D., Surov A.V., Nakonechny Gh.V., Nikonov A.V. Novel three-phase steam–air plasma torch for gasification of high-caloric waste 2013 *Applied Energy* **108** 505-14

[5] Murashov I.V., Frolov V. Y., Ivanov D.V. Numerical simulation of DC air plasma torch modes and plasma jet instability for spraying technology 2016 *IEEE NW Russia* 625-28

[6] Surov A.V., Popov S.D., Popov V.E., Subbotin D.I., Obraztsov N.V., Kuchina J.A., Serba E.O., Nakonechny Gh.V., Spodobin V.A., Pavlov A.V., Nikonov A.V. Steam, methane and carbon dioxide thermal plasma interaction with perhalocarbons 2017 *Journal of Physics: Conf. Series* **825** 012016

[7] Surov A.V., Popov S.D., Popov V.E., Subbotin D.I., Serba E.O., Spodobin V.A., Nakonechny Gh.V., Pavlov A.V. Multi-gas AC plasma torches for gasification of organic substances 2017 *Fuel* **203** 1007–14

[8] Rutberg Ph. G., Nakonechny Gh. V., Pavlov A. V., Popov S. D., Serba E. O., Surov A. V. AC plasma torch with a H2O/CO2/CH4 mix as the working gas for methane reforming 2015 *J.Phys. D: Appl. Phys.* **48** 245204 –16

IOP Conf. Series: Journal of Physics: Conf. Series **1038** (2018) 012138 doi:10.1088/1742-6596/1038/1/012138

Numerical investigation of swirling flow in the graft with a spiral ridge

Y F Radchenko[1], Y A Gataulin[1], A D Yukhnev[1], V N Vavilov[2], A A Moiseev[2]

[1]Department of Fluid Dynamics, Combustion and Heat Transfer, Peter the Great St. Petersburg Polytechnic University, 29 Polytechnicheskaya Str., 195251, St. Petersburg, Russia

[2]Department of Faculty Surgery, Pavlov First St. Petersburg State Medical University, 6-8 L. Tolstogo Str., 197022, St. Petersburg, Russia

E-mail: radfn94@mail.ru

Abstract. In the present work, numerical simulation was conducted for a typical connection of a vascular graft (prosthesis) and an artery to assess the effect of inducing swirling flow, which is believed to remove unfavourable flow environment. A parametric study on the spiral ridge geometrical features that was conducted showed that the ridge height and pitch have significant effects on inducing swirling flow, and revealed the potential of improving the efficiency of such design. Simulations were carried out for two models – a standard graft and the one with an optimal spiral ridge. Hemodynamic parameters were compared and the results showed that the graft with a spiral ridge is more effective. The induced swirling flow was generally found to be decreasing particle residence time within the connection of the graft and artery region and the host artery, which may be beneficial to the graft patency rates.

1. Introduction

Currently available synthetic vascular grafts perform well as large-caliber replacements, but their long-term flow capacity is discouraging in small-caliber applications such as lower extremities surgery. This failure is mainly a result of an unfavorable healing process with surface thrombogenicity, due to the damage of endothelial cells and thickening of the inner layer of a blood vessel caused by hemodynamic disturbances [1]. In this case there is a need for repeated surgery.

Numerous experimental studies performed recently have made it possible to register the helical blood flow in different parts of the cardiovascular system using such methods as phase contrast cardiovascular magneticresonance and the ultrasound colour flow imaging [2]. While the physiological importance of secondary motion in circulation has clearly been highlighted in the literature, the benefits of helical/spiral prostheses in vascular conduits have been firmly established [3].According to clinical studies, the swirl number characterized by the ratio of the maximum circumferential to the maximum axial velocity is 0.1-0.3 for the femoral artery[4].

There are two main types of experimental vascular grafts with swirling flow. The first type is a helical graft developed by Caro and colleagues [5]. However, its spatial tortuosity can be destroyed by surrounding tissues. The second is a graft with an internal spiral ridge introduced by Vascular Flow Technologies Ltd [6]. Both types have been clinically tested and introduced into clinical practice. These types of grafts reduce the thickening of the inner layer of a blood vessel but not fully solve this problem.

The aim of the present work is the numerical simulation of blood flow in the graft with a spiral ridge model, its geometric parameters optimization and comparison with the standard linear graft.

2. Methods

2.1. Geometric models

The geometry studied in this work represents a connection of a graft with inner diameter D = 6 mm and an artery (D = 6 mm) at an angle of 40°. This model has a one-pitch-long internal spiral ridge with spiral pitch (S), height (h) and width (w = 2.5 mm). The effects of the ridge have been studied through the variation of spiral pitch (S = 30; 40; 60 mm) and height (h = 1.8; 2; 2.2 mm).

2.2. Mathematical model

Three-dimensional unsteady Navier–Stokes equations were solved in order to simulate the flow in the graft model. The computations were performed with the ANSYS CFX code by the control volume method with second-order accuracy in space and time. Navier-Stokes equations were used as governing equations.

A uniform velocity profile and a variation in the mean flow velocity during the cycle were specified at the inlet boundary. The mean velocity curve was obtained from the flow rate curve reproducing the results of clinical measurements of blood flow in the femoral artery of healthy volunteers by phase-contrast magnetic resonance imaging [7]. The time period of the flow cycle T = 0.9 s. The velocity increase phase makes up 22% of the total cycle time. The maximum mean flow velocity for the period $V_{b\,max}$ = 0.94 m/s. The phase of the mean flow velocity decrease is characterized by the reverse flow associated with the motion of the pulse wave through the vessels. At the outlet boundaries of the graft the flow rates were specified as 1/4. No-slip boundary condition is applied to all walls and a rigid wall model is assumed.

The computations were performed for a liquid with properties similar to blood: the dynamic viscosity coefficient μ= 0.004 Pa·s, the density ρ= 1000 kg/m^3. The maximum Reynolds number percycle for these parameters was $\mathrm{Re} = 2\rho R V_{b\max}/\mu$ ≈ 1500 and the Womersley number was $W_0 = R\sqrt{2\pi\rho/\mu T}$ ≈4.

2.3. *Choice of the computational mesh and time step*

The computational domains used here are based on finite-volume hybrid mesh consisting of prismatic elements for the near-wall and tetrahedral elements for the core regions and were generated using ANSYS-Meshing (Version 16.2). A series of rigorous mesh independency tests were conducted on the graft model with a spiral ridge. The computational domains with 1.5 million elements were considered to be sufficient here. The time-step size was taken to be 0.01s. In order to eliminate the start-up effects of transient flow, the computation was carried out for three periods, and the three period results are presented.

3. Results

3.1. Hemodynamic parameters

There is currently extensive and increasing evidence, correlating the localisation of atherosclerosis, thickening of the inner layer of a blood vessel to different local hemodynamic metrics - relative residence time (RRT), time-averaged wall shear stress (TAWSS) and oscillatory shear index (OSI) [8]. They are calculated according to equations

$$RRT = \frac{1}{(1-2\cdot OSI)TAWSS} \qquad TAWSS = \frac{1}{T}\int_0^T \left|\overrightarrow{\tau_w}\right| dt \qquad OSI = \frac{1}{2}\left(1 - \left|\int_0^T \overrightarrow{\tau_w} dt\right| \Big/ \int_0^T \left|\overrightarrow{\tau_w}\right| dt\right)$$

where $\overrightarrow{\tau_w}$ is wall shear stress vector, T is the time period of the flow cycle.

The swirl parameter $V_{\varphi\,\max} / V_{z\,\max}$, often used to characterize flow swirling, is defined as the ratio of the maximal circumferential velocity $V_{\varphi\,\max}$ to the maximal axial velocity $V_{z\,\max}$. It is within the range of 0.1 – 0.3 for the femoral artery [4].

3.2. Graft optimization

A series of pulsatile flow simulations were conducted for the model of the graft with an internal ridge to evaluate the effects of different geometrical parameters, including spiral pitch (S) and height (h) of the spiral ridge. Fig.1 shows normalized by the standard graft wall-averaged RRT for different S and h. The optimal graft parameters (S/D = 5; h/D = 0.33) correspond to minimal RRT. RRT is almost 20% less than the one for the standard graft.

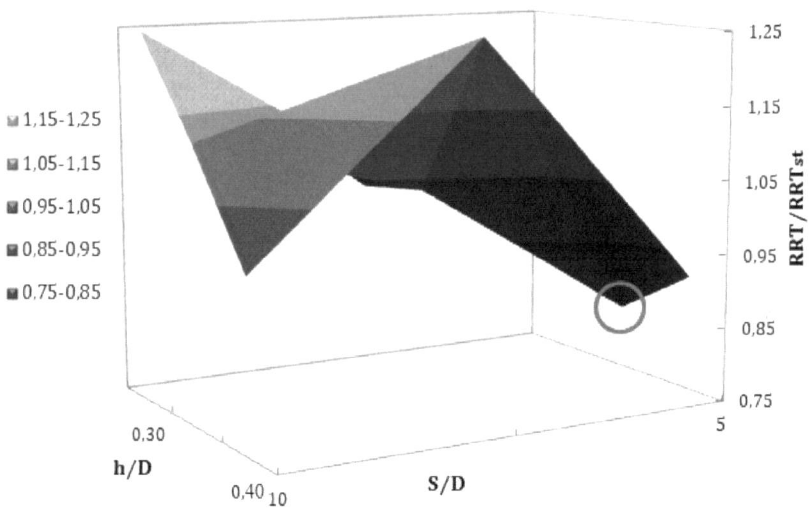

Figure 1. Normalized wall-averaged RRT

Fig. 2 shows the variation of the swirl parameter along the length of the graft and artery connection (coordinate x = 0 showed in Fig. 3). The swirl number is in the physiological range for all examined values of the spiral ridge geometric parameters. The swirl number dependence on the spiral ridge pitch is stronger than on the spiral ridge height.

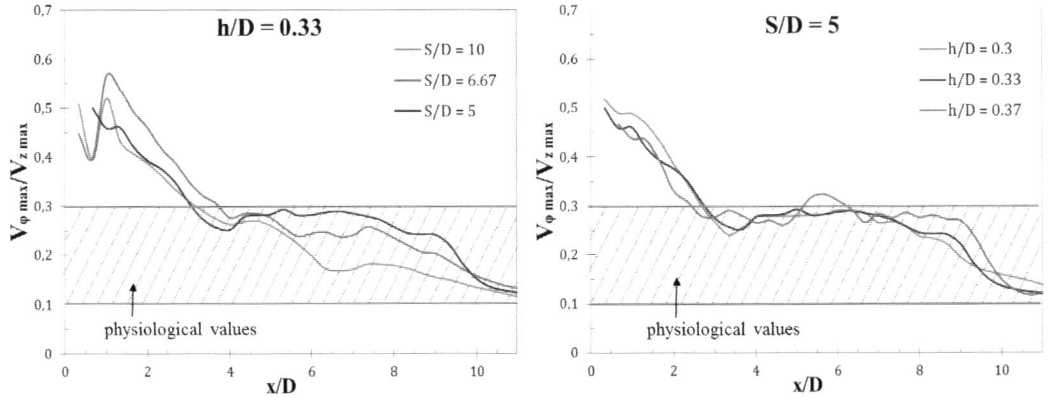

Figure 2. Variation of the swirl parameter over the length of the graft and artery connection for different geometrical parameters

3.3. Comparison with standard graft

The effectiveness of two graft designs –the standard graft and the novel graft with a spiral ridge - has been investigated in this study. While the main geometrical feature of the standard graft has already been investigated numerically by a number of researchers [1], the effects of introducing a spiral ridge inside the graft has not been studied before. The main findings from the present research will be discussed below.

Distribution of time-averaged RRT is shown in Fig.3. Elevated-RRT levels can be seen at the occluded segment of the artery and in the flow separation regions downstream of the point x = 0. The swirling flow induced by the graft with a spiral ridge has reduced the RRT values, by eliminating the flow separation in this region.

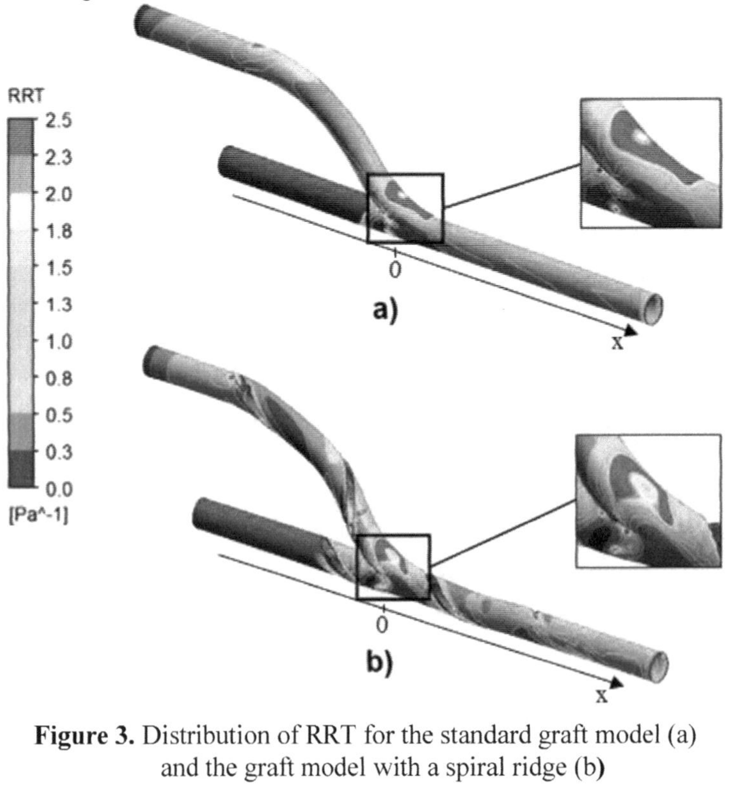

Figure 3. Distribution of RRT for the standard graft model (a) and the graft model with a spiral ridge (b)

The computations revealed that Dean vortex pairs in which the fluid rotated in opposite directions were formed in the standard graft model. The Dean vortices in Fig. 4 are visualized by Q-criterion isosurfaces; they have the form of two structures of similar shape, elongated along region downstream of point x = 0. In the graft model with a spiral ridge the Dean vortices are absent. The spiral ridge generates swirling flow with one single vortex. The most intense swirling was generated close to the branching point. The swirl attenuated downstream of the spiral ridge end.

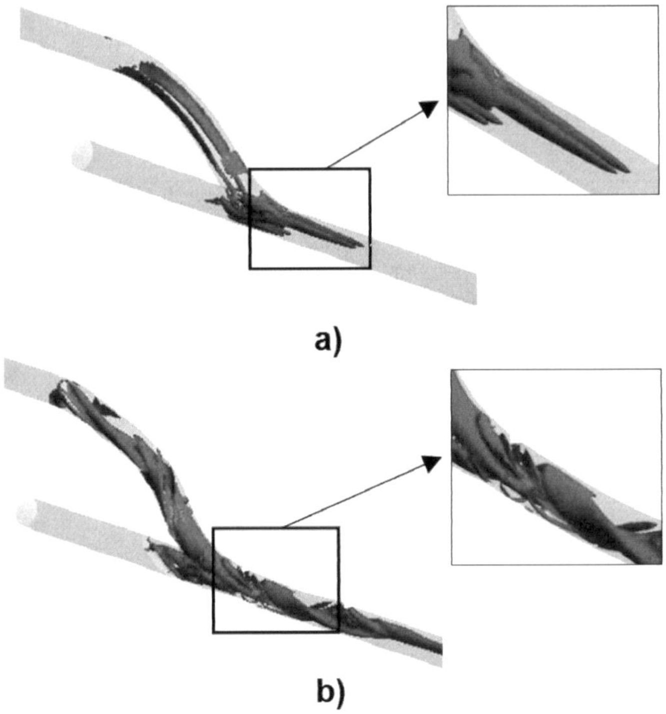

Figure4. Q-criterion isosurface (level 0.08) at the peak flow phase for the standard graft model (a) and the graft model with a spiral ridge (b)

4. Conclusions

The present study shows that inducing swirling flow into the grafts leads to positive flow features and more favorable distribution of hemodynamic parameters in the connection between the graft and artery region and the host artery, which possibly enhances the patency and longevity of the graft.

The graft model with examined values of the spiral ridge geometric parameters forms swirling flow with physiological values of swirl number 0.1-0.3 that corresponds to a healthy femoral artery. The graft with pitch S/D = 5 and height h/D = 0.33 is more effective. For it relative residence time averaged over the wall is over 20% less than the one for the standard graft model.

References

[1] Kuryanov PC, Razuvaev AS, Vavilov VN 2008 Intimal hyperplasia within a vascular anastomosis *Angiology and vascular surgery* **14** pp 146–150

[2] Bockeria LA, Gorodkov AY, Nikolayev DA, Kiknadze GI, Gachechiladze IA 2003 An analysis of the velocity field of a swirled blood flow in the aorta on the basis of three-dimensional mapping by MP velocimetry *Bulletin of Bakoulev CCVS RAMS for Cardiovascular Surgery* **4** pp 70–74

[3] Houston JG, Gandy SJ, Milne W, Dick JB, Belch JJ, Stonebridge PA 2004 Spiral laminar flow

in the abdominal aorta: a predictor of renal impairment deterioration in patients with renal artery stenosis? *Nephrol Dial Transplant* pp 1786–1791

[4] Stonebridge PA, Suttie SA, Ross R, Dick J 2016 Spiral laminar flow: a survey of a three-dimensional arterial flow pattern in a group of volunteers *Eur J Vasc Endovasc Surg* **52** pp 674–680

[5] Huijbregts HJTAM, Blankestijn PJ, Caro CG, Cheshire NJW, Hoedt MTC, Tutein Nolthenius RP, Moll FL 2007 A helical PTFE arteriovenous access graft to swirl flow across the distal anastomosis: results of a preliminary clinical study *Eur J Vasc Endovasc Surg* **33** pp 472–475

[6] Stonebridge PA, Vermassen F, Dick J, Belch JJF, Houston G 2012 Spiral laminar flow prosthetic bypass graft: medium-term results from a first-in-man structured registry study *Ann Vasc Surg. Annals of Vascular Surgery Inc.* **26** pp 1093–1099

[7] Klein WM, Bartels LW, Bax L, van der GY, Mali WP 2003 Magnetic resonance imaging measurement of blood volume flow in peripheral arteries in healthy subjects *Journal of Vascular Surgery* **38** pp 1060–1066

[8] Lee S-W, Antiga L, Steinman DA 2009 Correlations among indicators of disturbed flow at the normal carotid bifurcation *Journal of Biomechanical Engineering* **131**061013

IOP Conf. Series: Journal of Physics: Conf. Series **1038** (2018) 012139 doi:10.1088/1742-6596/1038/1/012139

Phase-structural irregularity of the mechanically activated saponite-containing material surface

M V Morozova[1], M A Frolova[1], T A Makhova[1] and V S Lesovik[2]

[1]Northern (Arctic) Federal University named after M.V. Lomonosov, Severnaya Dvina Emb. 17, Arkhangelsk, 163002 Russia

[2]Belgorod State Technological University named after V.G.Shoukhov, Kostyukov str., 46, Belgorod, 308012 Russia

E-mail: m.morozova@narfu.ru

Abstract. The purpose of the studies was to evaluate the possibility of using mechanically activated saponite-containing material (SCM) as an additional binder in cementitious binders of hydration type of hardening. According to the working hypothesis, the activated SCM is able to show the binding properties. To confirm this fact, the phase-structural heterogeneity investigation of the material was made. As a result, after mechanical dispersion the content of the amorphous phase increases by 2 times. Investigations of the microstructure of the binder samples with and without saponite showed that: in the first case, there are two types of crystals: spongy and needle. The SEM and IR spectroscopy data showed that when the additive is introduced, the formation of hydrolides of the tobermorite group is observed, which play the role of an additional binder in the cement hardening process.

1. Introduction

The papers [1, 2] show that the saponite-containing material (SCM) extracted from recycled water suspension in the diamond industry can exhibit sorbent properties through the extended surface. Moreover, the paper [3] establishes that SCM samples with a specific surface value over 35000 m2/kg demonstrate the optimal water absorption.

The additive based on this material can stabilize system oversaturation with water relative to the solid phase in concrete compositions [4. 5]. This provides for evenly distributed products of clinker mineral hydration reaction throughout the composite.

An imbalance in original sample weights was noted when studying water sorption and desorption by the highly dispersed mineral additive. In our opinion, it is associated with additional calcium silicate hydrates formed through activation of chemical compounds in the saponite-containing material (elemental composition equivalent to oxides: silicon - 56%, magnesium - 22%, aluminum - 7%, iron - 8%, calcium - 5% [3]). As mentioned above, intensive mechanical activation of the SCM promotes its moisture sorption [7 – 8]; besides, this process is accompanied by active amorphous phase formation in the material that consequently improves its binding properties. This is due to structural features of the saponite-containing material. Therefore, according to [6] this mineral has a three-layer structure with two layers of silicon–oxygen tetrahedron and aluminium-oxygen, octahedral, dioctahedral and trioctahedral layers between them. The layered structure is destroyed and silicon dioxide is released during the mechanical activation of raw materials.

IOP Conf. Series: Journal of Physics: Conf. Series **1038** (2018) 012139 doi:10.1088/1742-6596/1038/1/012139

The selected object of research was evaluating the possibility to use the mechanically activated saponite-containing material as an additional binder in hydration-type hardening binding compositions.

2. Materials and methods

2.1. Materials
The saponite-containing material was extracted from the suspension circulating water in the process of kimberlite ore enrichment during industrial diamond mining at the M.V. Lomonosov deposit located in the Archangelsk region by electrolytic coagulation based on transforming superfine particles to the state close to an isoelectric one.

2.2. Methods
The saponite-containing material extracted from recycle water was brought to the constant mass at 105 ^{0}C.

Dry milling at the Retsch PM100 planetary ball mill was used to reduce the material to the required particle size. Optimal dispersion parameters were selected for each test sample to ensure minimum particle sizes in samples and high reproducibility (at least three parallel tests). Rotor speed was 420 rpm. Dispersion was performed using tungsten carbide grinding balls (20 pcs).

Particle sizes were determined at the Delsa Nano zeta potential and submicron particle size analyzer by measuring dynamic and electrophoretic light scattering.

A superfine sample obtained was characterized by nitrogen sorption at the Autosorb-iQ-MP analyzer using the S_{sp} specific surface value.

The Zeiss Sigma VP scanning electronic microscope was used to take photographs of the superfine SCM (at 20,000X magnification with a resolution up to 1.3 nm) at the Arktika Common Use Center (Arktika CUC).

In addition, phase structural heterogeneity of saponite-containing material samples before and after SCM mechanical activation was studied by X-ray diffraction using the XRD-7000S X-ray diffractometer at the Arktika CUC.

Then infrared spectra of samples were recorded using the Vertex 70v FT-IR spectrometer. Spectra were recorded under the following conditions: range 4000-600 cm^{-1}, resolution 4 cm^{-1}

3. Resuits and discussion
Dispersion produced the saponit-containing material samples with a particle size range of 1 μm to 400 nm and specific surface of 18610±10 m^2/kg to 50670±20 m^2/kg respectively.

Correlation between the SCM specific surface and dispersion time have shown that the increase in grinding time over 90 min. does not lead to higher $S_{sp,}$ and the increase in machining time over 100 min. demonstrates conglomeration of particles in the system. Therefore, the fraction with S_{sp} = 50670 m^2/kg and an average particle size of 445±40 obtained within 90 min was chosen for further research.

The material is found to have a flaky nature with numerous voids that can be filled by water (Figure 1).

a b

Figure 1. The microphoto of pre-production SCM sample, obtained on raster electronic microscope Sigma VP: a – the original (porosity – 0,088 cm^3/g); b – crushed (porosity – 0,001 cm^3/g).

The phase-structural heterogeneity of saponite-containing material samples studied before and after SCM mechanical activation demonstrated that the amorphous phase content in comparison with initial raw materials is doubled (Figure 2).

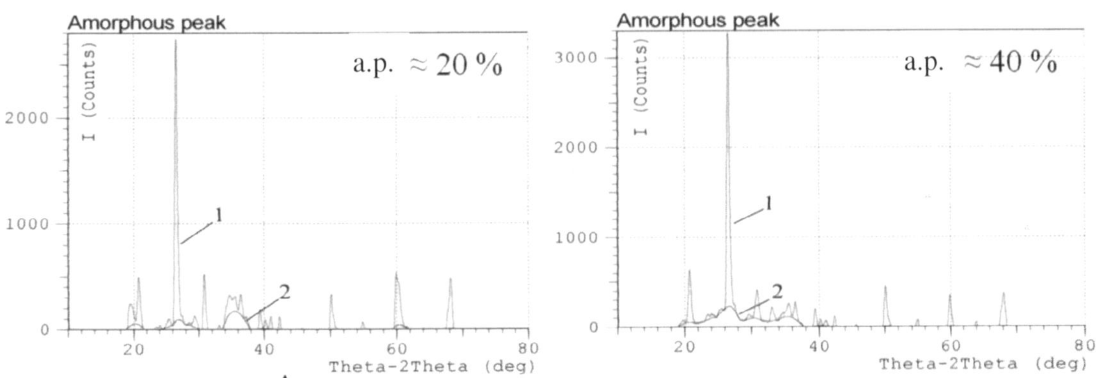

a b

Figure 2. Diffractogram of the sample SCM: a – the original; b – sample after mechanoactivation; 1 – crystal part; 2 – amorphous part (a.p.).

Further studies focused on making samples with the reference (original) and mechanically activated saponite-containing material having the following composition: sand – 55%, SCM – 30%, water – 15%. Electron micrographs of test objects were made after 28-day aging (time of clinker mineral hydration reaction) (Figure 3).

IOP Conf. Series: Journal of Physics: Conf. Series **1038** (2018) 012139 doi:10.1088/1742-6596/1038/1/012139

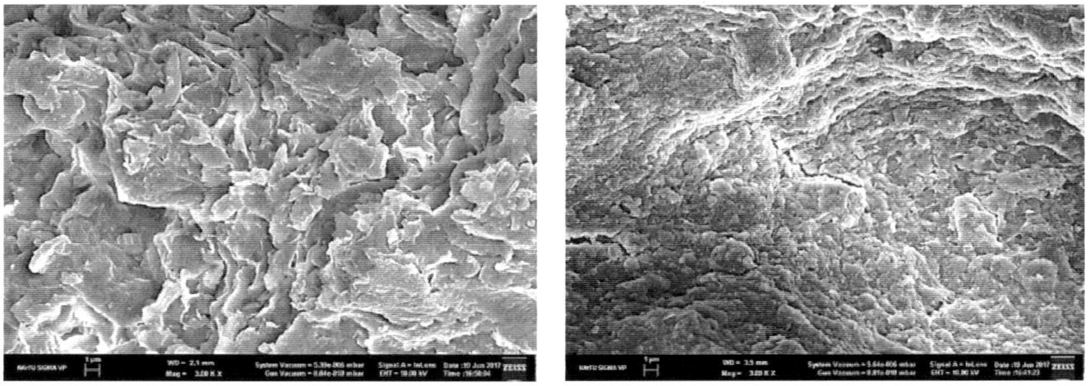

a b

Figure 3. The microphoto of SCM sample: a – control (average size of particles SCM – 1 mm); b – prototype (average size of particles SCM – 445 nm).

The results of SEM have shown that the sample microstructure is presented as spongy crystals. However, numerous voids (Figure 3a) predominate in reference samples, while the surface of test samples (with activated SCM) is denser and more uniform (Figure 3b).

Therefore, based on the data obtained the saponite-containing material can be considered not only as a sorbent that optimizes structure formation by aqueous phase sorption, but also as an active mineral component in hydration-type hardening binding compositions.

The binder samples (portland cement) with and without addition of the saponite-containing material (20%) were produced in order to confirm binding properties of mineral additives. The IR spectra of compositions under test were recorded after mixing with water and aging for 28 days (Figure 4).

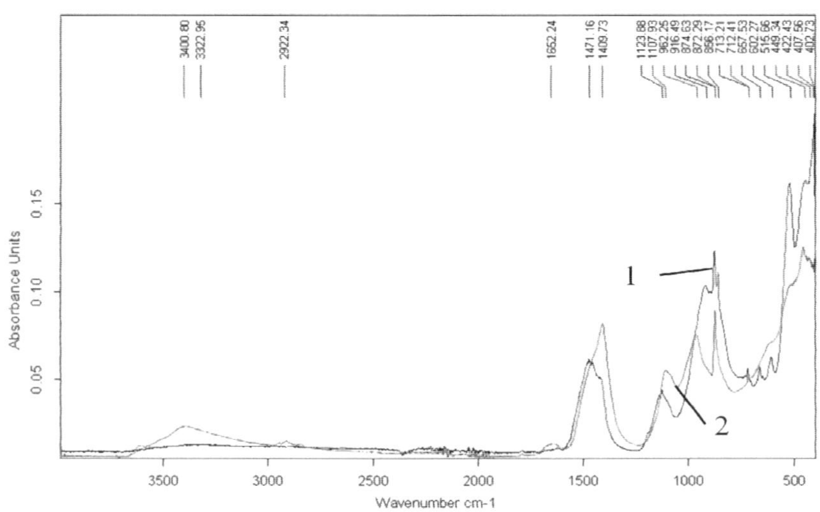

Figure 4. IR spectrums of cement examples of structure: 1 – skilled with SCM additive; 2 – control without SCM additive.

Spectroscopic data show the presence of new formations in the form of silicate groups (oscillations at 975 cm^{-1}). However, the intensity of oscillations for the initial saponite-containing material under the same wave numbers (975 cm^{-1}) is several times higher than for control samples without SCM.

The electron micrographs of samples confirmed changes in the cement structure when introducing a mineral additive. According to SEM data, the formed hydrosilicates are conglomerates of particles with grain sizes ranging from 2 to 20 μm. Particles have different shapes, but basically two types of particles can be distinguished: spongy particles (Figure 5a) with a developed microporous surface sized from 5 to 10 μm; needle-shaped particles (Figure 5b) with a length of 0.5 to 5 μm (needle diameters are about 0.5 μm).

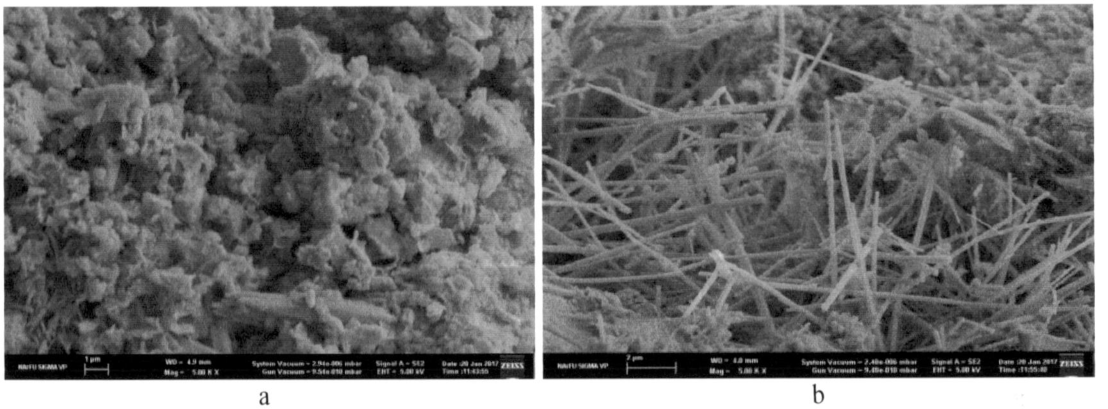

a b

Figure 5. Microstructure of a cement example of structure: a – control; b – prototype.

By comparing the results of scanning electron microscopy and IR spectroscopy of cement samples (Figure 4), the following conclusion can be drawn. The presence of absorption maxima in the IR range at 1400-1600 cm^{-1} and of a broad band at 3300-3500 cm^{-1}, as well as the formation of needle-shaped crystals in the sample with the saponite-containing material added, confirms the presence of tobermorite group hydrosilicate submicrocrystals additional generation.

Synthesis of these crystals (alite and belite) from clinker minerals can occur according to the following pattern:

$$2\underbrace{(3CaO \cdot SiO_2)}_{\text{alite}} + 6H_2O \rightarrow 3CaO \cdot 2\,SiO_2 \cdot 3H_2O + 3Ca(OH)_2 \quad (1)$$

$$2\underbrace{(2CaO \cdot SiO_2)}_{\text{belit}} + 4H_2O \rightarrow 3CaO \cdot 2\,SiO_2 \cdot 3H_2O + Ca(OH)_2 \quad (2)$$

$$5Ca(OH)_2 + 6SiO_2 \rightarrow \underbrace{5CaO \cdot 6\,SiO_2 \cdot 5H_2O}_{\text{tobermorite}} \quad (3)$$

The result of the chemical reaction (3) is the synthesis of tobermorite crystals through the interreaction of cement clinker hydration products and saponite-containing material minerals.

Tobermorite crystals growing in all directions act as an additional binder during cement hardening. The obtained findings about shapes and sizes of particles and spectral features of hydrosilicates are in line with literature sources.

Therefore, based on the data obtained the saponite-containing material can be considered not only as a sorbent that optimizes structure formation by aqueous phase sorption, but also as an active mineral component in hydration-type hardening binding compositions.

4. Conclusion

Summarizing the experimental results we can draw the following conclusions:

1. It has been proved that chemical compounds in saponite-containing material can form calcium hydrosilicates after the mechanical activation of raw materials.

IOP Conf. Series: Journal of Physics: Conf. Series **1038** (2018) 012139 doi:10.1088/1742-6596/1038/1/012139

2. Surface amorphization (up to 40%) after the SCM mechanical dispersion and new growth of silicate groups with different basicity may indicate a potential for using raw materials to improve binding properties in hydration-type hardening compositions.

5. Acknowledgements

The research was carried out on the unique scientific equipment "Physical Chemistry of Surfaces of Nano-Dispersed Systems".

References

[1] Morozova M V, Frolova M A and Mahova T A 2016 *Abstracts of the international youth conference "Physicist. SPb"* p 118
[2] Morozova M V, Ayzenshtadt A M and Tutyigin A S 2013 *J. Industrial and civil construction* **11** 29-31
[3] Morozova M V, Ayzenshtadt A M, Mahova T A and Frolova M A 2015 *15th International Multidisciplinary Scientific GeoConference & EXPO SGEM 2015, Nano, bio and green — technologies for a sustainable future* **1** 135-142
[4] Strokova V V, Cherevatova A V, Zhernovskiy I V and Voytovich E V 2012 *J. Construction Materials* **7** 9-12
[5] Izotov V S and Ibragimov R A 2010 *J. News of the KazSACU* **2** 229-233
[6] Ovcharenko F D 1961 *Hydrophilicity of clay and clay minerals* (Kiev: National Academy of Sciences of Ukraine) p 292
[7] Harhardin A N, Strokova V V and Kozhuhova M I 2012 *J. Higher Education News (Construction)* **10** 109-115
[8] Andrievskiy R A and Glezer A M 2009 *J. Successes of physical sciences* **4** 337-358
[9] Glezer A M 2002 *J. Russian chemical magazine* **5** 57–63

IOP Conf. Series: Journal of Physics: Conf. Series **1038** (2018) 012140 doi:10.1088/1742-6596/1038/1/012140

Protolytic properties influence of the dispersion medium on the process of silicic acid polycondensation

V E Danilov, A M Ayzenshtadt

Department of Composite Materials and Construction Ecology, Higher Engineering School, Northern (Arctic) Federal University named after M.V. Lomonosov, Severnaya Dvina Emb. 17, Arkhangelsk, 163002, Russia

v.danilov@narfu.ru

Abstract. The paper presents a procedure for the preparation of sodium polysilicate from silica hydrosol for use as an additional binding agent in the production of composite materials based on wood matrix. It is proved that 3-stages mechanical activation of low quality sand makes it possible to obtain silica particles with sizes on the upper boundary of the ultradispersed range (100 nm). The effect of dimensional characteristics, the zeta potential of the surface of silica particles, and the protolytic properties of the dispersion medium on the gel formation process is studied. The obtained dependence of the zeta potential of particles from the pH of diluent has shown that in acidic region have equivalence point and a low zeta-potential area in which the colloidal particles become unstable (which may lead to rapid gelification in the system). It is proposed to accelerate the process of depolarization of silica particles by maintain the pH value of the dispersion medium equal to 2-4, this will quickly produce a large yield of gel-like sodium polysilicate.

1. Introduction

It is known the property of colloidal silica to form a gel when interacting with acids, alkalis and the simplest alcohols [1-3]. Moreover, along with an increase in viscosity of the system and binding of particles occur imparting heat resistant properties to the material [4, 5]. In turn, this feature of colloidal silica allowed it to be used both for the production of new heat-shielding, fire-resistant nanocomposites, and for modifying the already known and traditional materials for the construction field [6, 7]. The regulation of the gelling process, in our opinion, in most cases should be limited to controlling the protolytic properties of the solvent (dispersion medium), which largely determine the aggregative stability of the dispersed phase substance. Therefore, the research task, described in this paper, was the study of the dispersion medium pH influence on the process of silicic acid polycondensation.

The appearance of a double electrical layer on the particles of the silica hydrosol is due to the ionization of their surface. The SiO_2 molecules on the surface of the particles interact with the dispersion medium, they hydrate and form a silicic acid which is capable of ionizing:

$$H_2SiO_3 = SiO_3^{2-} + 2H^+$$

In this case, the silicate ions remain on the surface of the particle, which causes its negative charge.

$$\{m[H_2SiO_3]nSiO_3^{2-}2(n-x)H^+\}^{-2x}2xH^+$$

As the acidity of the dispersion medium increases (for example, due to the addition of HCl) on the surface of the nucleus excessive amounts of protons (H^+) are sorbed, because, according to the Fajans-Paneth rule, H^+ ions are part of the crystal lattice of the nucleus. As a result of this adsorption, excess surface energy is compensated at the interface between the phases in the "silicic acid" - "aqueous HCl solution" system. The nucleus together with the adsorbed proton layer forms a positively charged aggregate. A part of the counterions (Cl^-) is adsorbed on the surface of the aggregate. The micelle formula in this case:

$$\{m[H_2SiO_3]nH^+(n-x)Cl^-\}^{+x}\,xCl^-$$

Under certain conditions (heating, mixing, addition of other electrolytes), the adsorption layers of ions on the nucleus are partially or completely destroyed. In this case, the compensation of excess surface energy at the interface between the phases occurs by aggregation (agglomeration of silicic acid particles). This process, leading to the coarsening of the phase particles without precipitation, is characterized by gelling.

With the addition of $NaOH$, it is possible to form a sodium silicate Na_4SiO_4 (soluble in water) as a solid phase. Further, due to hydrolysis, colloidal solutions are formed from the true solutions. Subsequently, a white precipitate is deposited from such systems.

2. Materials and methods

2.1. Materials

Dusty polymineral quartz-feldspar quarry sand from the Kholmogorskoe deposit (Arkhangelsk region) was chosen as an object of study.

The sodium liquid glass with a density of 1.36 g/cm^3 and a silicate module of 2.6 was used for sodium polysilicate synthesis.

2.2. Methods

The mineral composition of sand was determined using X-ray fluorescence spectroscopy using the Shimadzu EDX-800 HS spectrometer.

The sand was mechanically activated in three stages: by mechanical dry grinding for 30 minutes at a rotor speed of 420 rpm by 2 cm large grinding balls in the Retsch PM100 planetary ball mill at the first stage; by mechanical wet grinding for 30 minutes at a rotor speed of 420 rpm by grinding balls of 1 cm in diameter in the Retsch PM100 planetary ball mill at the second stage; by mechanical wet grinding for 10 min at a rotor speed of 16000 rpm in the MK module (a colloid mill) of the Magic Lab portable mini lab equipped with the Micro-Plant attachment (product circulation equipment) at the third stage. The synthesized concentrated silica sol was diluted with water to the 16 Wt% silica hydrosol with a density of 1100 kg / m^3.

The silica particle size and their zeta potential were determined using the Delsa Nano C analyzer by photon-correlation spectroscopy and electrophoretic velocity measurement, respectively.

NaOH and HCl aqueous solutions were added to vary protolytic properties of the dispersion medium in the obtained silica sols. The Expert-001-3 fluid analyzer was used for pH control.

Sodium polysilicate was synthesized as follows: liquid glass was diluted with water to obtain a 20% aqueous solution of sodium silicate with a density of 1250 kg/m^3; then silica hydrosol was added at the ratio of 1:1.5 under continuous stirring at 95°C for 1 hour and aging at this temperature for 30 minutes.

In order to study the microstructure, gel test samples were placed on a specimen slide and left for 24 hours to self-level the surface. In total, 12 specimen samples were prepared for the study using the LOMO MICMED-6 light microscope. Macro and micrographs were made using the Nikon COOLPIX P500 digital camera.

3. Results and discussions

Table 1 shows the mineral composition of sand from the Kholmogorskoe deposit expressed as oxides.

Table 1. Mineral composition of sand expressed as oxides.

The content of oxides, Wt. %							
SiO_2	MgO	Al_2O_3	Na_2O	Fe_2O_3	CaO	K_2O	other oxides
77.48	1.80	12.13	2.76	1.61	1.45	2.09	0.68

As is seen from the table, the SiO_2 content in sand is less than 90%, which indicates a low raw material quality. In addition, a significant amount of Al_2O_3 is observed, which indicates clay (aluminosilicate) inclusions.

Figure 1 specifies the average particle sizes of mechanically activated finely dispersed sand.

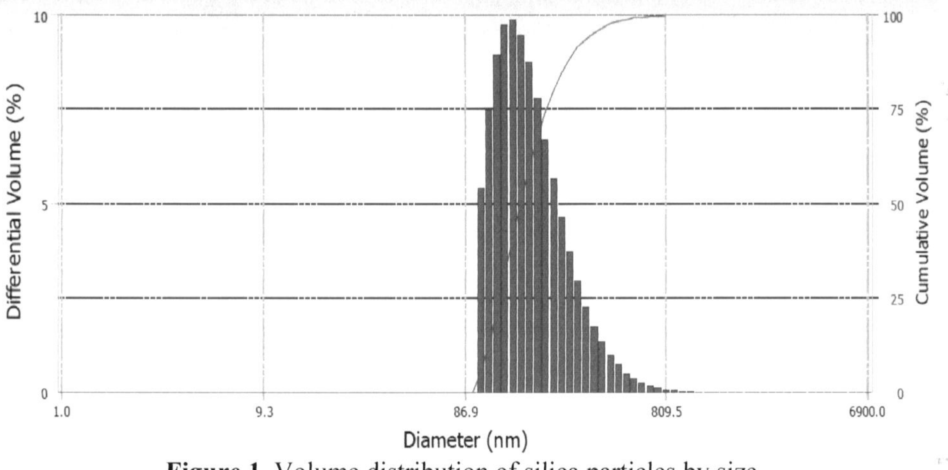

Figure 1. Volume distribution of silica particles by size.

The obtained data show that mechanical activation of sand makes it possible to obtain silica particles with sizes on the upper boundary of the ultradispersed range.

Figure 2 shows the functional relation between zeta potential change and pH of medium.

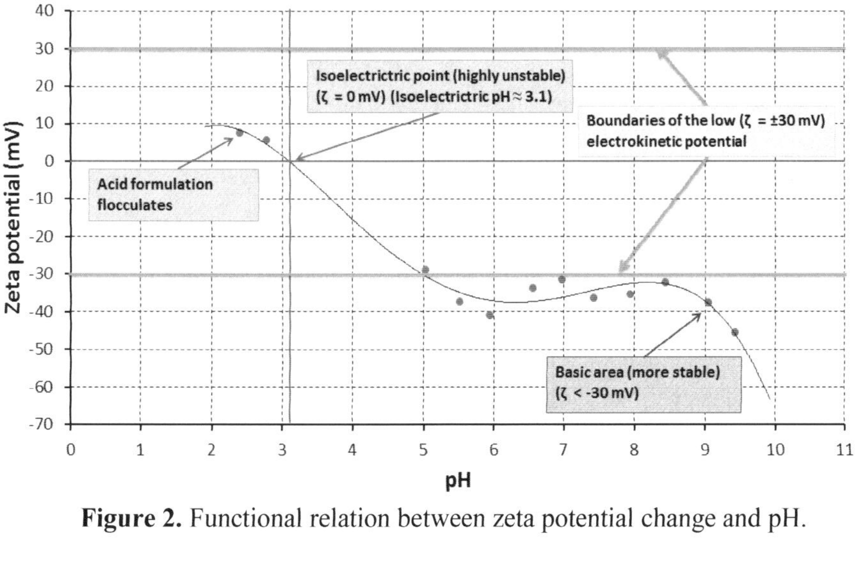

Figure 2. Functional relation between zeta potential change and pH.

IOP Conf. Series: Journal of Physics: Conf. Series **1038** (2018) 012140 doi:10.1088/1742-6596/1038/1/012140

It is known that the aggregative stability of colloidal silica is strongly influenced by the electrostatic barrier which in turn depends on the zeta potential level of particles. Therefore, an accelerated particle growth and subsequent gel forming should be expected in the system under low absolute values of ζ.

According to experimental data, higher pH brings zeta potential of particles beyond boundaries of the low electrokinetic potential zone ($\zeta = \pm$ 30 mV) where colloidal silica particles begin losing their aggregation stability according to the working hypothesis. In order to confirm this assumption, the particle sizes in suspensions were measured repeatedly after 24-hour aging. The experimental results are shown in figures 3, 4.

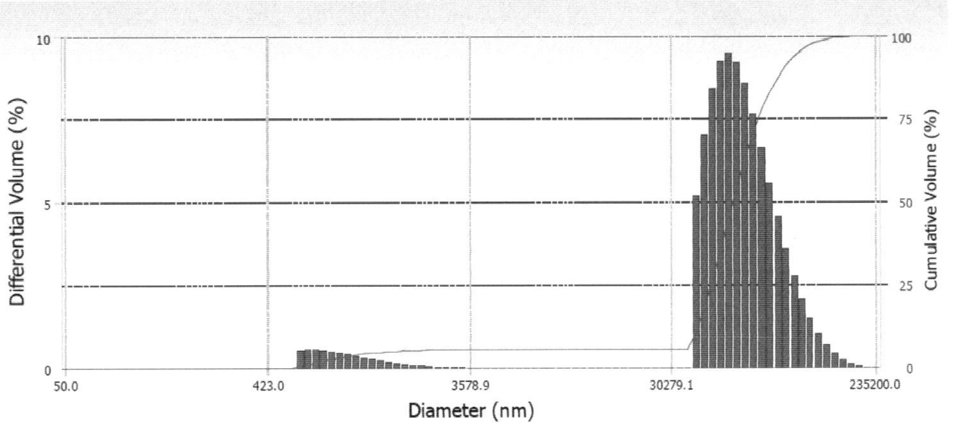

Figure 3. Volume size distribution of silica particles after 24-hour aging (pH 2.4).

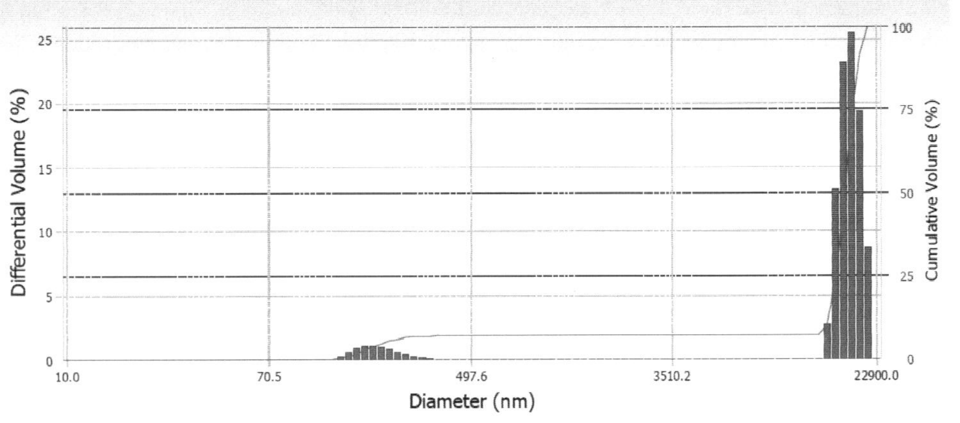

Figure 4. Volume size distribution of silica particles after 24-hour aging (pH 9.0).

Repeated measurement results show that the particle size in the silica sol increases both in acidic and alkaline dispersion medium, however, particle size characteristics at pH = 2.4 are significantly higher than at pH = 9. In addition, particles of the original ultradispersed and finely dispersed range in amounts of 7.1% can be observed in the latter.

In contrast, the large sedimenting white flocks and residue visible to the naked eye were found in silica sol at pH = 2.4. This gel-like residue was extracted from the sol and moved to a specimen slide by a sampler for microstructure study (see figure 5).

The remaining samples with different pH values were used for sodium polysilicate synthesis. Upon synthesis completion, polysilicate identification, and further free water removal, all samples had gels formed in them at different times (see figure 6).

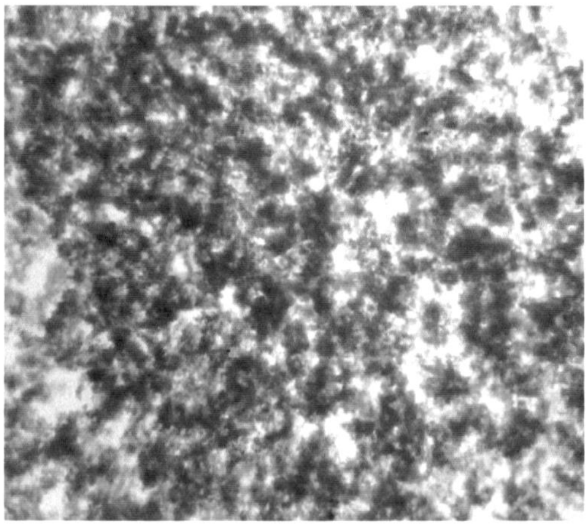

Figure 5. Photo of gel-like residue surface.

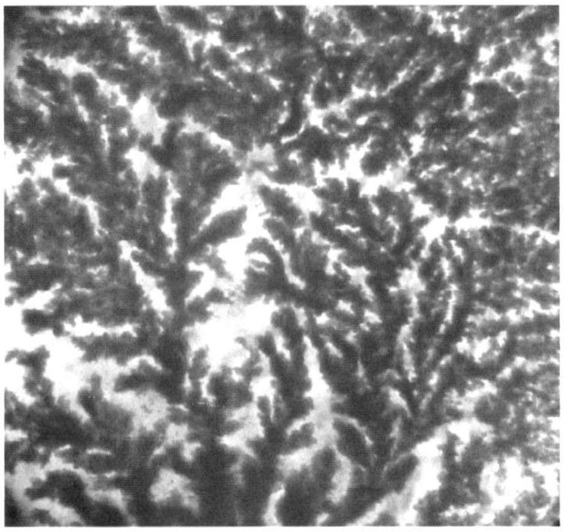

Figure 6. Photo of gel surface.

It was observed that the formation time, particle size, color, transparency, and viscosity of gels in samples with different pH values differ significantly. This fact indicated the need for further study of these regularities.

4. Conclusion

Thus, it can be concluded based on the conducted studies that low grade silica-containing raw materials can be used to produce finely and nanodispersed silica. In order to accelerate the particle depolarization process it is necessary to maintain the pH value of the dispersion medium in the suspension equal to 2-4.

References

[1] Iler R K 1979 *The Chemistry of Silica* (New York: Wiley). Translated under the title *Khimiya kremnezema* (Moscow: Mir) part 1 p 712

[2] Shabanova N A and Sarkisov P D 2004 *Bases of sol-gel technology of nanodispersion silica* (Moscow: IKC Akademkniga) p 208

[3] Khlebnikov V N 2009 *Regularities of gelling in acid sols of aluminosilicates and silicates* (Moscow: Gubkin Russian State University of Oil and Gas) vol 2 pp 25–31

[4] Ke Y C and Stroeve P 2005 *Polymer-Layered Silicate and Silica Nanocomposites* (Amsterdam: Elsevier) p 394

[5] Shabanova N A and Sarkisov P D 2012 *Sol–Gel Technologies. Nanodispersed Silica* (Moscow: BINOM. Laboratoriya Znanii) p 328

[6] Orlov E IU, Belskikh G N and Kuzmenko A P 2015 *Sovremennye materialy, tekhnika i tekhnologii [Modern materials, techniques and technologies]* **1(1)** pp 177-179

[7] Gonchikova E V, Arkhincheva N V and Dorzhieva E V 2009 *Stroitel'nye materialy [Construction materials]* **9** pp 68-69

Surface tension determination in glyoxal-silica dispersed system

Y V Sokolova[1], A M Ayzenshtadt [1], V V Strokova[2] and V S Malkov[3]

[1]Northern (Arctic) Federal University named after M.V. Lomonosov, Severnaya Dvina Emb. 17, Arkhangelsk, 163002 Russia
[2]Belgorod State Technological University named after V.G. Shoukhov, Kostyukov str. 46, Belgorod, 308012 Russia
[3]National Research Tomsk State University, Lenina Avenue 36, Tomsk, 634050 Russia

E-mail: yu.sokolova@narfu.ru

Abstract. Preliminary studies have shown the possibility of using glyoxal-silica dispersed system for soil stabilization in groundwork base and utility systems. However, the problem associated with optimizing the composition can be solved by measuring energy characteristics. Since the interaction in glyoxal-silica system takes place at the interphase the surface tension was chosen as an energy parameter. The aim of research was to compare calculated by Zisman's and Owens-Wendt-Rabel-Kaelble (OWRK) methods surface tension values of glyoxal-silica composition. Polymineral sand was used as the silica model. The surface tension value calculated by Zisman's method is σ_c=27.47 mN/m. According to the OWRK method, the polar component of surface tension for test material is σ_S^P=12.61 mN/m, dispersion component is σ_S^D=13.50 mN/m, and the total value of surface tension is σ_S=26.11 mN/m. Comparison of surface tension values calculated by different methods has shown good repeatability. However, the OWRK method provides a more detailed overview of interaction in the glyoxal-silica system and makes it possible to estimate the contribution of polar and dispersion components to the total value of a surface energy characteristic. Therefore, the OWRK method can be recommended for characterizing the structure formation process when optimizing the organomineral compound composition.

1. Introduction

Interfacial processes in coarse, micro and nanodispersed systems are characterized by surface energy changing (E_S). Surface tension (σ) – a free surface energy measure (numerically equal to work of the surface unit formation) – is a criterion that allows E_S value quantification in the structure formation process of the substance. This statement is true for both liquids and solid bodies. However, σ quantitative measurement for solid body surfaces is a difficult experimental task.

There are different methods of surface tension quantitative measurement for solid body surfaces such as "zero creep", crystal cleaving, powder dissolution, "scratch healing", "neutral drop" [1]. But the most effective one is the "neutral drop" method (G.A. Zisman's [2-4] and OWRK (Owens-Wendt-Rabel-Kaelble methods) [5-8]) – a non-destructive method based on measuring the contact angle of solid surface wetting with service fluids.

Zisman's method based on modified Antonov's equation determines the critical value of surface tension (σ_c) for solid surfaces through plotting functional dependency (1) and extrapolating it to $\cos\theta = 1$.

$$\cos\theta = f(\sigma_L) \tag{1}$$

where σ_L is surface tension of service fluids determined experimentally, N/m;

θ is the contact angle of wetting of test material surface.

However, this method has applicability limitations related to the selection process of solution and test structure types. The contact angle of wetting should be measured using low-polar or non-polar liquids. The homologous series of liquids is the most appropriate way. With the liquid surface tension more than 35 mN/m this method generates a high error as it is necessary to take into account the contribution of polar forces to the adhesive interaction value. G.A. Zisman's method is applicable only when the adhesive interaction between contacting phases is determined by dispersion (Van der Waals) forces only, and surfaces have no chemically active centers.

In comparison with the first one, the OWRK method is more universal and based on the fact that free surface energy is a sum of polar and dispersion components (equation 2) and calculated by linear regression (equation 3) plotted based on the results of measuring of the contact angle of solid surface wetting with service fluids with known polar and dispersion elements. Ethylene glycol, ethanol, glycerin and decane with different molecular chemical structure are commonly used as service fluids.

$$\sigma_S = \sigma_S^P + \sigma_S^D \tag{2}$$

where σ_S is surface tension for the test material, N/m;

σ_S^P, σ_S^D are polar and dispersion components of surface tension for the test material, N/m.

$$\frac{\sigma_L(\cos\theta + 1)}{2\sqrt{\sigma_L^D}} = \frac{\sqrt{\sigma_S^P}\sqrt{\sigma_L^P}}{\sqrt{\sigma_L^D}} + \sqrt{\sigma_S^D} \tag{3}$$

where σ_L^P, σ_L^D are polar and dispersion components of surface tension for service fluids, N/m. The values of these elements are shown in the paper [9].

The polar component of the surface energy is formed by means of electrostatic forces originated from polar groups of molecular structures that lost some atoms, and by means of chemical bonds forming electrically neutral dipoles. The dispersion component is defined by Van der Waals interaction formed by the molecular internal energy in the surface layer of material.

The study presented in this paper focuses on comparing calculated by Zisman's and OWRK methods surface tension values of glyoxal-silica composition which is used for soil injection stabilization in groundwork base and utility systems [10, 11].

2. Method

A sedimentary rock – polymineral river sand from the Krasnoflotsky-Zapad deposit (Arkhangelsk region) containing 91% silicon dioxide was used as the silica model.

Sand was preliminarily washed free from clay intrusions and dried to its constant mass at 105±5°C. Grain-size composition is determined by Focused Beam Reflectance Measurement technology using the Lasentec D600E laser particle size analyzer.

The original 40% aqueous solution of glyoxal was diluted to obtain the 20% solution. The 20% aqueous solution of glyoxal was mixed with a mineral component for producing test samples. To obtain the homogeneity of composition the quantity of organic additive has to be 2.3% of sand weight. The resulting mixture having been compacted by the PLG-20 press with a working force of 40 kN, samples were kept in natural conditions at 25±2°C until the glyoxal polymerization completion which was controlled by measuring the weight of test material. The criterion of polymerization completion was to achieve constant weight.

The DataPhysics OCA 35 automatic unit was used to determine the contact angle of wetting at 20±2°C. Service fluids were aqueous solutions of industrial ethyl alcohol with a water volume ratio from 4 to 50% for Zisman's method and ethylene glycol, ethanol (96%), glycerin and decane for the OWRK method. The contact angle of wetting was determined at during the period of the first contact between the liquid and test material surface which varied within 1.0±0.5 s. This is due to the fact that evaporation, diffusion, capillary phenomena and other processes related to surface irregularity are negligible in this period. The set of experiments consisting of three mandatory parallel measurements with a maximum allowable difference in contact angle of wetting values of 2-3° having been completed, functional dependencies (1) and (3) were plotted and the surface tension of test material was calculated.

3. Results

Table 1 contains grain-size composition of sand.

Table 2 contains the surface tension of service fluids and average contact angles of wetting for the glyoxal/silica (sand) system at 2.3% glyoxal content determined by Zisman's method.

Table 3 contains the surface tension of service fluids and average contact angles of wetting for the glyoxal/silica (sand) system at 2.3% glyoxal content determined by OWRK method.

Figure 1 shows functional dependence (1) necessary to determine the critical value of surface tension for the test material at 2.3% glyoxal content.

Figure 2 shows functional dependence (3) necessary to determine the polar and dispersion components and the total surface tension for the test material at 2.3% glyoxal content.

Table 1. Sand particle size distribution obtained by the Lasentec D600E analyzer.

Particle size, μm Content, %						
300…500	150…300	50…150	10…50	5…10	1…5	0.5…1
0.27	4.23	28.50	39.69	9.79	17.53	0.00

Table 2. The surface tension of service fluids and average contact angles of wetting for the glyoxal/silica system at 2.3% glyoxal content determined by Zisman's method.

Aqueous solutions of ethanol, % vol.	$\sigma_L \cdot 10^3$, N/m	θ,°	$\cos\theta$
96	28.3	8.89	0.988
90	28.7	17.25	0.955
80	29.3	15.42	0.964
70	29.5	21.41	0.931
60	30.6	22.63	0.923
50	30.7	23.79	0.915

Table 3. The surface tension of service fluids and average contact angles of wetting for the glyoxal/silica system at 2.3% glyoxal content determined by OWRK method.

Service fluids	$\sigma_L \cdot 10^3$, N/m	$\sigma_L^P \cdot 10^3$, N/m	$\sigma_L^D \cdot 10^3$, N/m	θ,°	$\cos\theta$
glycerin	63.4	26.4	37.0	75.94	0.243
ethanol	21.4	2.6	18.8	8.73	0.988
ethylene glycol	48.8	16.0	32.8	59.31	0.510
decane	23.8	0	23.8	59.60	0.506

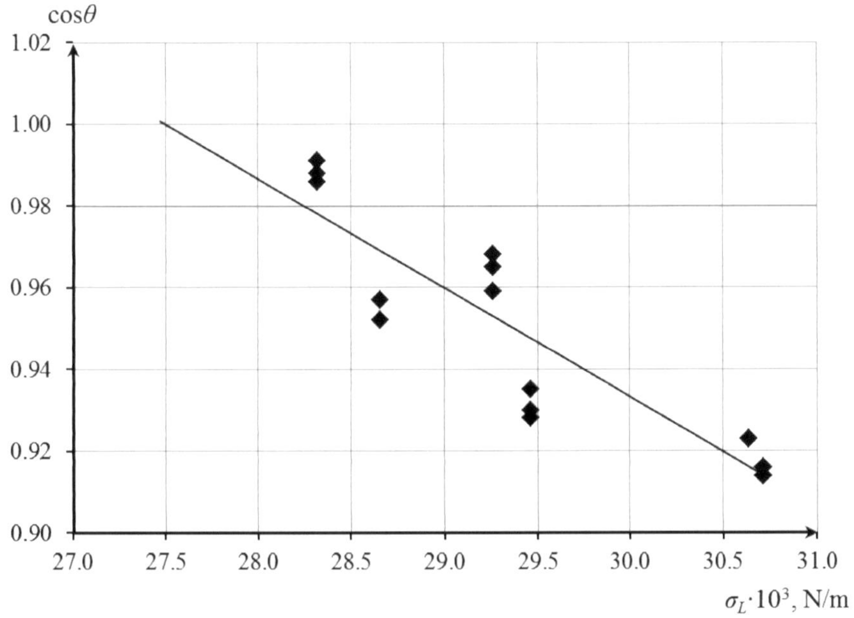

Figure 1. The functional dependence $\cos\theta = f(\sigma_L)$
for the glyoxal/silica system at 2.3% glyoxal content (Zisman's method).

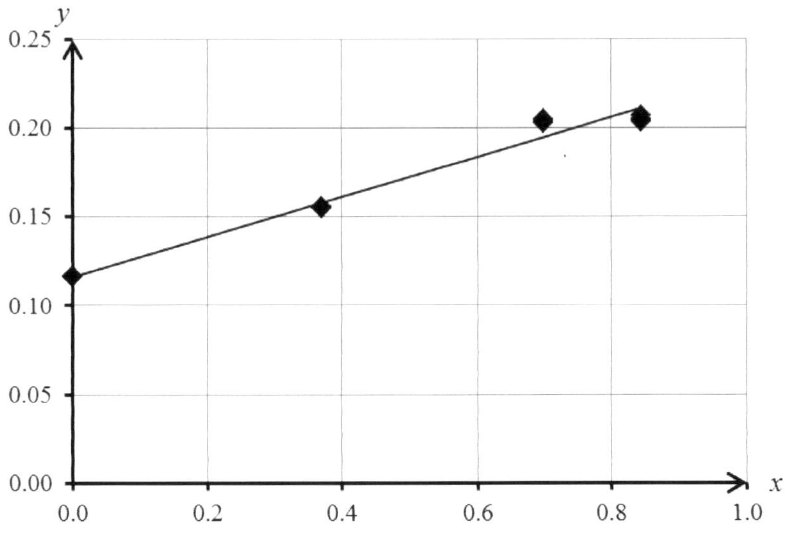

Figure 2. The functional dependence $y = f(x)$
for the glyoxal/silica system at 2.3% glyoxal content (OWRK method).

where $y = \dfrac{\sigma_L(\cos\theta + 1)}{2\sqrt{\sigma_L^D}}$; $x = \sqrt{\dfrac{\sigma_L^P}{\sigma_L^D}}$

4. Discussion

Grain-size analysis of river polymineral sand from Krasnoflotsky-Zapad deposit using the Lasentec D600E particle size analyzer demonstrated that the average size of its grains is 30 μm. Therefore, the investigated silica model is a coarsely dispersed system (particle size exceeds 10^{-5} m). It is found that glyoxal addition does not change the original dispersed structure of sand, but forms a periodic colloidal system by synthesizing the sealing layer of organic polymer that strengthens dispersion interaction between soil particles.

The obtained functional dependencies (Figure 1, 2) are described by linear equations (4, 5) with high values of approximation certainty factor (*R*), which may indicate a potential for Zisman's and OWRK methods to measure surface tension in the glyoxal-silica dispersed system:

$$\cos\theta = -0.0267\sigma_L + 1.7335,\ R^2 = 0.8 \quad (4)$$

$$\frac{\sigma_L(\cos\theta + 1)}{2\sqrt{\sigma_L^D}} = 0.1123\sqrt{\frac{\sigma_L^P}{\sigma_L^D}} + 0.1162,\ R^2 = 0.9 \quad (5)$$

The critical value of surface tension for glyoxal-silica dispersed system calculated by Zisman's method is σ_c=27.47±1.15 mN/m. According to the OWRK method, the polar component of surface tension for test material is σ_S^P=12.61 ±2.02 mN/m, dispersion component is σ_S^D=13.50±0.07 mN/m, and the total value of surface tension is σ_S=26.11±2.67 mN/m.

In our opinion, comparison of surface tension values calculated by different methods has shown good repeatability despite Zisman's method limitations. Relative value difference does not exceed 5%.

Based on current theories, G.A. Zisman's method allows to determine Van der Waals interaction in the dispersed system only. It should be noted, however, that total surface tension (critical surface tension) experimentally calculated by this method includes both dispersion and polar components. But in comparison with the above, the OWRK method provides a more detailed overview of interaction in the glyoxal-silica system and makes it possible to estimate the contribution of polar (chemical) and dispersion (physical) components to the total value of a surface energy characteristic.

Therefore, the OWRK method can be recommended for characterizing the structure formation process when optimizing the organomineral compound composition.

Acknowledgments

The research was carried out on the unique scientific equipment "Physical Chemistry of Surfaces of Nano-Dispersed Systems".

References

[1] Roldugin V I 2011 *Physicochemistry of the surface* (Dolgoprudny: Publishing House "Intellect") p 568
[2] Frolova M, Tutygin A, Ayzenstadt A, Makhova T and Pospelova T 2011 *Nanotechnologies in Construction: A Scientific Internet-Journal* **6** pp 13-25
[3] Volkov V A 2015 *Colloid chemistry. The surface phenomena and disperse systems* (St. Petersburg: Lan) p 659
[4] Özen S, Şenay V, Pat S and Korkmaz Ş 2016 *Journal of Physics* D: *Applied Physics* **49(10)** 105303
[5] Yakavets N, Krut'ko N and Opanasenko O 2012 *Collection of articles: Sviridov readings* **8** pp 253–60
[6] Ariawan D, Mohd Ishak Z A, Salim M S, Mat Taib R and Ahmad Thirmizir M Z 2017 *Polymer Composites* **38(3)** pp 507-15
[7] Encinas N, Pantoja M, Abenojar J and Martínez M A 2010 *Journal of Adhesion Science and Technology* **24(11-12)** pp 1869-83
[8] Karl C W and Klüppel M 2011 *Chemicke Listy* **105**(15 SPEC. ISSUE) pp s275-6
[9] Strokova V V, Aizenshtadt A M, Sival'neva M N, Kobzev V A and Nelubova V V 2015

Construction Materials **2** pp 3-9
[10] Gayda J, Ayzenshtadt A, Tutygin A and Frolova M 2016 *Procedia Engineering* **143** pp 90–7
[11] Sokolova Y V, Ayzenshtadt A M and Strokova V V 2016 *Abstracts of Int. Youth Conf. PhysicA.SPb/2016* November (St. Petersburg) p 193

Institute of Physics
Dirac House, Temple Back
Bristol BS1 6BE UK

ISSN: 1742-6588
ISBN 978-1-5108-6491-7

CURRAN ASSOCIATES INC.
proceedings.com

2nd International Telecommunication Conference on Advanced Micro- and Nanoelectronic Systems and Technologies (AMNST 2017)

IOP Conference Series: Materials Science and Engineering
Volume 498

Moscow, Russia
1 – 2 June 2017

IOP
Institute of Physics